REGULATED CHEMICALS DIRECTORY 1995

Compiled by ChemADVISOR®, Inc.
Pittsburgh, Pennsylvania

VNR SPRINGER SCIENCE+BUSINESS MEDIA, LLC

For more information contact:

Van Nostrand Reinhold
115 Fifth Avenue
New York, NY 10003

International Thomson Publishing GmbH
Königswinterer Str. 418
53227 Bonn
Germany

International Thomson Publishing Europe
Berkshire House,168-173
High Holborn, London WC1V 7AA
England

International Thomson Publishing Asia
221 Henderson Road #05-10
Henderson Building.
Singapore 0315

Thomas Nelson Australia
102 Dodds Street
South Melbourne 3205
Victoria, Australia

International Thomson Publishing Japan
Hirakawacho Kyowa Building, 3F
2-2-1 Hirakawacho
Chiyoda-ku, 102 Tokyo
Japan

Nelson Canada
1120 Birchmount Road
Scarborough, Ontario
Canada M1K 5G4

International Thomson Editores
Campos Eliseos 385, Piso 7
Col. Polanco
11560 Mexico D.F. Mexico

This edition of *Regulated Chemicals Directory* ™ was prepared in collaboration with ChemADVISOR®, Inc., Regulatory Compliance Products & Services, Pittsburgh, PA, from the ChemADVISOR LOLI™ database, using software developed by Automated Publishing/ Pre-press Services, Phoenix, AZ.

1 2 3 4 5 6 7 8 9 10 BBR 01 00 99 98 97 96 95

ISSN 1058 1707
ISBN 978-0-442-02124-5   ISBN 978-94-011-4910-5 (eBook)
DOI 10.1007/978-94-011-4910-5

Regulated Chemicals Directory 1995

Contents

Acknowledgments

The *Regulated Chemicals Directory*™ is prepared from a database of chemical regulatory information maintained by ChemADVISOR®, Inc. Maintenance of the database is currently under the supervision of Scott A. Amoroso. The electronic database is updated on almost a daily basis to keep pace with regulatory changes and additions. The technical content and structure of the database are under the supervision of Patricia Dsida.

The *Regulated Chemicals Directory*™ is the product of a close collaboration between Van Nostrand Reinhold and Chapman and Hall scientific publishers and ChemADVISOR®, Inc. We wish to thank Dr. Barbara Goldman, Editor at Chapman & Hall, who originally approached Patricia Dsida with the idea of developing a reference book based on the ChemADVISOR® database, and who worked closely with ChemADVISOR® to create the *RCD*™.

Values from the ACGIH publication, "Threshold Limit Values and Biological Exposure Indices for 1994-95," were used for the ChemADVISOR® database with the kind permission of the American Conference of Governmental Industrial Hygienists.

Preface

The *Regulated Chemicals Directory*™ is meant to be a convenient source of information for everyone who needs to keep up-to-date regarding the regulations and recommendations that pertain to chemical substances. The *RCD*™ is designed to be the first reference book to consult when beginning compliance efforts. Every regulatory or advisory list used in the *RCD*™ is keyed to its source, to help readers who need more detailed information on regulations, recommendations, or guidelines readily locate source documents.

Some organizations now center their compliance efforts on computerized information stored in cross-referenced databases. A unique feature of the *RCD*™ is the availability of an electronic version suitable for use on IBM-compatible personal computers, download onto mainframes and CD-ROM players. Both the print and electronic versions are updated with the same timeliness. For more information on the electronic versions of the *Regulated Chemicals Directory*™, contact ChemADVISOR®, Inc. directly (750 William Pitt Way, Pittsburgh, PA 15238, phone 1-800-466-3750).

Many companies working on product development need information on what may be regulated in the future. The *RCD*™ provides selected information on pending regulations and in-progress testing lists, which can provide a starting place for tracking future regulatory considerations.

Information for the *RCD*™ is continually gathered and updated. Suggestions from readers for information that should be added to the *RCD*™ or for other ways to improve the book are welcomed by Van Nostrand Reinhold.

—Patricia L. Dsida, Pres.
ChemADVISOR®, Inc.

Part B.
Regulatory Summaries

Each entry in the REGULATORY SUMMARIES section is listed alphabetically under its Reference Name, the most frequently used regulatory synonym for a given chemical substance. The CAS number is included in each Regulatory Summary to confirm the identity of the substance. Reference Names may be found by looking up a particular substance either by name in the ALPHABETICAL INDEX BY CHEMICAL NAME, or by CAS number in the NUMERICAL INDEX BY CAS NUMBER (See Part A, Section 3).

Each Regulatory Summary entry integrates data from health and safety, environmental, state, and international lists, and includes proposed changes to regulatory or advisory status. All the regulations and reports on which that substance appears, along with the applicable data values, are included, along with any required footnotes.

For detailed information on the list sources or an explanation of the data values, see LIST DESCRIPTIONS (Part A, Section 2).

REGULATORY SUMMARIES

ABAMECTIN [AVERMECTIN B1]   71751-41-2
ENVIRONMENTAL LISTS
  CERCLA/SARA - Section 313 - Emission Reporting
  form R reporting required

ABIES ALBA OIL   8021-27-0
INTERNATIONAL LISTS
  Canada - WHMIS: Ingredient Disclosure
  1% item 1 (972)

ABIETIC ACID   514-10-3
INTERNATIONAL LISTS
  Canada - WHMIS: Ingredient Disclosure
  0.1% item 2 (50)

ABIETIC ACID, DIETHYLENE GLYCOL ESTER   10107-99-0
ENVIRONMENTAL LISTS
  List of Pesticide Product Inert Ingredients
  [present]

ABIETIC ACIDS, SODIUM SALT   14351-66-7
ENVIRONMENTAL LISTS
  List of Pesticide Product Inert Ingredients
  [present]

ABIETIC ACID, ZINC SALT   6798-76-1
ENVIRONMENTAL LISTS
  List of Pesticide Product Inert Ingredients
  [present]

ABS RESIN   9003-56-9
HEALTH AND SAFETY LISTS
  IARC - Group 3 (not classifiable)
  [present]

ACENAPHTHENE   83-32-9
SEE ALSO:
  K035-HAZARDOUS WASTES
  F039-HAZARDOUS WASTES
HEALTH AND SAFETY LISTS
  U.S. DOT - Appendix A Table 1 - Hazardous Substances
  final RQ = 100 pounds (45.4 kg)
ENVIRONMENTAL LISTS
  ATSDR Priority List
  Rank (of 275): 140
  CERCLA/SARA - Hazardous Substances and their Reportable Quantities
  final RQ = 100 pounds (45.4 kg)
  CAA - HON Rule - SOCMI Chemicals
  compliance by Oct. 23, 1995
  Clean Water Act - Priority Pollutants
  [present]
  Clean Water Act - Toxic Pollutants
  [present]
  RCRA - Basis for Listing - Appendix VII
  Included in waste streams: F039, K001, K035
  RCRA - TSD Facilities Ground Water Monitoring
  TM 8100 = 200 ug/L PQL;   TM 8270 = 10 ug/L PQL
  RCRA - Universal Treatment Standards (LDR)
  WW: 0.059 mg/l; NWW: 3.4 mg/kg

INTERNATIONAL LISTS
  Canada - WHMIS: Ingredient Disclosure
  1% item 3 (2)
  Mexico - Wastewater - Organic Toxic Pollutants and Heavy Metals
  Listed under [Organic Toxic Pollutants]
  Mexico - Drinking Water - Ecological Criteria
  0.02 mg/l
STATE LISTS
  California - Directors List of Hazardous Substances (8 CCR 339)
  [present]
  Massachusetts Right To Know List
  [present]
  Pennsylvania Right to Know List
  environmental hazard

ACENAPHTHYLENE   208-96-8
SEE ALSO:
  K001-HAZARDOUS WASTES
  F039-HAZARDOUS WASTES
HEALTH AND SAFETY LISTS
  U.S. DOT - Appendix A Table 1 - Hazardous Substances
  final RQ = 5000 pounds (2270 kg)
ENVIRONMENTAL LISTS
  CERCLA/SARA - Hazardous Substances and their Reportable Quantities
  final RQ = 5000 pounds (2270 kg)
  Clean Water Act - Priority Pollutants
  [present]
  RCRA - Basis for Listing - Appendix VII
  Included in waste stream: F039
  RCRA - TSD Facilities Ground Water Monitoring
  TM 8100 = 200 ug/L PQL;   TM 8270 = 10 ug/L PQL
  RCRA - Universal Treatment Standards (LDR)
  WW: 0.059 mg/l; NWW: 3.4 mg/kg
INTERNATIONAL LISTS
  Mexico - Wastewater - Organic Toxic Pollutants and Heavy Metals
  Listed under [Aromatic Hydrocarbons]
STATE LISTS
  Massachusetts Right To Know List
  [present]
  Pennsylvania Right to Know List
  environmental hazard

ACEPHATE   30560-19-1
ENVIRONMENTAL LISTS
  CERCLA/SARA - Section 313 - Emission Reporting
  form R reporting required

ACETAL   105-57-7
HEALTH AND SAFETY LISTS
  U.S. DOT - Substances From 49 CFR 172.101
  regulated by DOT (UN1088)
  U.S. DOT - Hazard Classes
  DOT hazard class = 3
  NFPA - Flash Points
  flash point = -5 degrees F (-21 degrees C)
  NFPA - Hazard Identification Ratings
  health-2; flammability-3; reactivity-0
  NIOSH - Selected LD50s and LC50s
  Oral, rat: LD50 = 4600 mg/kg

ENVIRONMENTAL LISTS

CAA - HON Rule - SOCMI Chemicals
compliance by Oct. 23, 1995

STATE LISTS

Florida Hazardous Substance List
[present]

Massachusetts Right To Know List
[present]

NJ Right to Know List (Total)
sn 2057

Pennsylvania Right to Know List
[present]

ACETALDEHYDE   75-07-0

HEALTH AND SAFETY LISTS

ACGIH 1995 - Ceiling Limits
C 25 ppm; C 45 mg/m3

ACGIH 1995 - Carcinogens
A3-animal carcinogen

AIHA - Odor Threshold Values
geometric mean air odor threshold = 0.067 ppm (detectable)

U.S. DOT - Substances From 49 CFR 172.101
regulated by DOT (UN1089)

U.S. DOT - Hazard Classes
DOT hazard class = 3

U.S. DOT - Appendix A Table 1 - Hazardous Substances
final RQ = 1000 pounds (454 kg)

IARC - Group 2B (sufficient animal data)
[present]

NFPA - Flash Points
flash point = -38 degrees F (-39 degrees C)

NFPA - Hazard Identification Ratings
health-3; flammability-4; reactivity-2

NIOSH - Selected LD50s and LC50s
Inhalation, rat: LC50 = 37 gm/m3 Oral, rat: LD50 = 1930 mg/kg

NIOSH 1990 - Pocket Guide - IDLHs
10,000 ppm IDLH (not considering carcinogenic effects)

NIOSH 1990 - Pocket Guide - Carcinogens
occupational carcinogen

NIOSH 1990 - Pocket Guide - Target organs
respiratory system, skin, kidneys

NTP Seventh Report - Suspect Carcinogens
suspect carcinogen

OSHA - Vacated PELs - Time Weighted Averages
100 ppm TWA; 180 mg/m3 TWA

OSHA - Vacated PELs - Short Term Exposure Limits
150 ppm STEL; 270 mg/m3 STEL

OSHA - Final PELs - Time Weighted Averages
200 ppm TWA; 360 mg/m3 TWA

OSHA - Possible Select Carcinogens
[present]

OSHA - List of Highly Hazardous Chemicals
threshhold quantity = 2500 pounds

ENVIRONMENTAL LISTS

CERCLA/SARA - Section 313 - Emission Reporting
form R reporting required for 0.1% de minimus concentration

CERCLA/SARA - Hazardous Substances and their Reportable Quantities
final RQ = 1000 pounds (454 kg)

Clean Air Act (1990) - List of Hazardous Air Contaminants
[present]

CAA - Flammable Substances for Accidental Release Prevention
threshold quantity = 10,000 lbs

CAA - HON Rule - SOCMI Chemicals
compliance by Jan. 23, 1995

CAA - HON Rule - Organic HAPs
[present]

Clean Water Act - Hazardous Substances
[present]

RCRA - U Series Wastes
waste number U001 (Ignitable waste)

RCRA - Hazardous Constituents-Appendix VIII
waste number U001

RCRA - Substances Banned From Land Disposal
[present]

TSCA - Code of Federal Regulations Citations
40 CFR 712.30(x); 40 CFR 716.120(d)

TSCA - PAIR - Reporting List
Reporting Date: November 27, 1991

TSCA - Health and Safety Reporting List
Effective Date: September 30, 1991

INTERNATIONAL LISTS

Australian Exposure Standards - Time Weighted Averages
100 ppm TWA; 180 mg/m3 TWA

Australian Exposure Standards - Short Term Exposure Limits
150 ppm STEL; 270 mg/m3 STEL

Australian Exposure Standards - Under Review
exposure limits under review

Canada - WHMIS: Ingredient Disclosure
1% item 4 (3)

Canada - NPRI (National Pollutant Release Inventory)
[present]

Canada - Alberta - 8 Hour Occupational Exposure Limit
100 ppm TWA; 180 mg/m3 TWA

Canada - Alberta - 15 Minute Occupational Exposure Limit
150 ppm STEL; 270 mg/m3 STEL

Canada - British Columbia - 8 Hour Exposure Limits
100 ppm TWA; 180 mg/m3 TWA

Canada - British Columbia - 15 Minute Exposure Limits
150 ppm STEL; 270 mg/m3 STEL

Canada - Ontario - OHSA - TWAEVs
100 ppm TWAEV; 180 mg/m3 TWAEV

Canada - Ontario - OHSA - STEVs
150 ppm STEV; 270 mg/m3 STEV

Canada - Quebec - Time-Weighted Average Exposure Values
100 ppm TWAEV; 180 mg/m3 TWAEV

Canada - Quebec - Short-term Exposure Values
150 ppm STEV; 270 mg/m3 STEV

Canada - Quebec - Carcinogens
C3 carcinogen: effect detected in animals

German (DFG) - MAK Values
50 ppm MAK; 90 mg/m3 MAK

German (DFG) - Peak Limitations
2 x normal MAK (5 min momentary value); don't exceed 8 times during shift

German (DFG) - Carcinogens
suspected carcinogen

German (DFG) - Pregnancy
classification not yet possible

Israel - Time Weighted Averages
100 ppm TWA; 180 mg/m3 TWA

Israel - Short Term Exposure Limits
150 ppm STEL; 270 mg/m3 STEL

Israel - Action Levels
50 ppm AL; 90 mg/m3 AL

Mexico - Instruction No. 10 - TWAs
100 ppm TWA; 180 mg/m3 TWA

Mexico - Instruction No. 10 - STELs
150 ppm STEL; 270 mg/m3 STEL

STATE LISTS

California - Air Bill 2588 Appendix A-I
known or potential carcinogen

California - Prop. 65 - Cancer list
carcinogen - initial date 4/1/88

California - Prop. 65 - No Significant Risk Levels
inhalation: no significant risk level = 90 ug/day

California - Exposure Limits - PELs
100 ppm PEL; 180 mg/m3 PEL

California - Exposure Limits - STELs
150 ppm STEL; 270 mg/m3 STEL

California - Directors List of Hazardous Substances (8 CCR 339)
[present]

Florida Hazardous Substance List
[present]

Massachusetts Right To Know List
carcinogen; extraordinarily hazardous

Minnesota Hazardous Substance List
carcinogen

NJ Right to Know List (Total)
sn 0001

NJ Special Hazardous Substances
(flammable - fourth degree; mutagen; reactive - second degree)

Pennsylvania Right to Know List
environmental hazard

PROPOSED REGULATIONS

Canada - Ontario - Proposed Occupational TWAEVs
25 ppm TWAEV; 45 mg/m3 TWAEV

Canada - Ontario - Proposed Occupational STEVs
50 ppm STEV; 90 mg/m3 STEV

ACETALDEHYDE AMMONIA   75-39-8

HEALTH AND SAFETY LISTS

U.S. DOT - Substances From 49 CFR 172.101
regulated by DOT (UN1841)

U.S. DOT - Hazard Classes
DOT hazard class = 9

STATE LISTS

NJ Right to Know List (Total)
sn 0002

ACETALDEHYDE, (1,3-DIHYDRO-1,3,3-TRIMETHYL-   84-83-3
2H-INDOL-2-YLIDENE)

ENVIRONMENTAL LISTS

TSCA - Code of Federal Regulations Citations
40 CFR 712.30(x); 40 CFR 716.120(d)

TSCA - PAIR - Reporting List
Reporting Date: November 27, 1991

TSCA - Health and Safety Reporting List
Effective Date: September 30, 1991

ACETALDEHYDE, HOMOPOLYMER   9002-91-9

STATE LISTS

Pennsylvania Right to Know List
[present]

ACETALDEHYDE OXIME   107-29-9

HEALTH AND SAFETY LISTS

U.S. DOT - Substances From 49 CFR 172.101
regulated by DOT (UN2332)

U.S. DOT - Hazard Classes
DOT hazard class = 3

STATE LISTS

NJ Right to Know List (Total)
sn 0003

ACETAMIDE   60-35-5

HEALTH AND SAFETY LISTS

IARC - Group 2B (sufficient animal data)
[present] (Degree of evidence in animals revised on the basis of data that appeared after the most recent monograph and/or on the basis of present criteria)

NIOSH - Selected LD50s and LC50s
Oral, rat: LD50 = 7000 mg/kg

OSHA - Possible Select Carcinogens
[present]

ENVIRONMENTAL LISTS

CERCLA/SARA - Section 313 - Emission Reporting
form R reporting required for 0.1% de minimus concentration

CERCLA/SARA - Hazardous Substances and their Reportable Quantities
final RQ = 1 pound (.454 kg)

Clean Air Act (1990) - List of Hazardous Air Contaminants
[present]

CAA - HON Rule - SOCMI Chemicals
compliance by Jan. 23, 1995

CAA - HON Rule - Organic HAPs
[present]

TSCA - Code of Federal Regulations Citations
40 CFR 704.225(a)

TSCA - CAIR - Reporting List
reporting required by: manufacturer; importer

INTERNATIONAL LISTS

German (DFG) - Carcinogens
suspected carcinogen

STATE LISTS

California - Air Bill 2588 Appendix A-I
known or potential carcinogen

California - Prop. 65 - Cancer list
carcinogen - initial date 1/1/90

California - Prop. 65 - No Significant Risk Levels
no significant risk level = 10 ug/day

California - Directors List of Hazardous Substances (8 CCR 339)
[present]

Massachusetts Right To Know List
[present]

Minnesota Hazardous Substance List
carcinogen

NJ Right to Know List (Total)
sn 2890

Pennsylvania Right to Know List
environmental hazard

ACETAMIDE, N-[5-[BIS(2-(ACETYLOXY)ETHYL)   3956-55-6
AMINO]-2-[(2-BROMO-4,6-DINITROPHENYL)AZO]-
4-ETHOXYPHENYL]-

ENVIRONMENTAL LISTS

TSCA - Code of Federal Regulations Citations
40 CFR 712.30(t); 40 CFR 716.120(a)

TSCA - PAIR - Reporting List
Reporting Date: February 12, 1987

TSCA - Health and Safety Reporting List
Effective Date: December 15, 1986; Sunset Date: November 9, 1993

ACETAMIDE, N-(5-(BIS(2-(ACETYLOXY)ETHYL) AMINO)-2-((2-CHLORO-4,6-DINITROPHENYL)AZO)-4-METHOXYPHENYL)-   3618-73-3

ENVIRONMENTAL LISTS

TSCA - Code of Federal Regulations Citations
40 CFR 712.30(u); 40 CFR 716.120(a)

TSCA - PAIR - Reporting List
Reporting Date: August 18, 1987

TSCA - Health and Safety Reporting List
Effective Date: June 19, 1987; Sunset Date: November 9, 1993

ACETAMIDE, N-[5-[BIS[2-(ACETYLOXY)ETHYL] AMINO]-2-[(2-CHLORO-4,6-DINITROPHENYL)AZO]-4-ETHOXYPHENYL]-   21429-43-6

ENVIRONMENTAL LISTS

TSCA - Code of Federal Regulations Citations
40 CFR 712.30(u); 40 CFR 716.120(a)

TSCA - PAIR - Reporting List
Reporting Date: August 18, 1987

TSCA - Health and Safety Reporting List
Effective Date: June 19, 1987; Sunset Date: November 9, 1993

ACETAMIDE, 2,2-DICHLORO-N-(1,3-DIOXOLAN-2-YLMETHYL)-N-2-PROPENYL-   79660-25-6

ENVIRONMENTAL LISTS

List of Pesticide Product Inert Ingredients
[present]

ACETAMIDE, N,N-DIETHYL-   685-91-6

ENVIRONMENTAL LISTS

TSCA - Code of Federal Regulations Citations
40 CFR 716.120(a)

TSCA - Health and Safety Reporting List
Effective Date: June 1, 1987; Sunset Date: November 9, 1993

ACETAMIDE, N-9H-FLUOREN-1-YL-   28314-03-6

STATE LISTS

Pennsylvania Right to Know List
[present]

ACETAMIDE, N-[4-(PENTYLOXY)PHENYL]- ACETAMIDE, N-[2-NITRO-4-(PENTYLOXY)PHENYL]-, AND ACETAMIDE, N-[2-AMINO-4-(PENTYLOXY) PHENYL]   RR-01650-2

ENVIRONMENTAL LISTS

TSCA - Chemicals with Significant New Use Rules
PMN numbers: P-92-31, P-92-32, P-92-33

ACETAMIDE, N-PHENYL-   103-84-4

HEALTH AND SAFETY LISTS

NFPA - Flash Points
flash point = 337 degrees F (169 degrees C)

NFPA - Hazard Identification Ratings
health-3; flammability-1; reactivity-0

NIOSH - Selected LD50s and LC50s
Oral, rat: LD50 = 800 mg/kg

ENVIRONMENTAL LISTS

CAA - HON Rule - SOCMI Chemicals
compliance by Jan. 23, 1995

STATE LISTS

Florida Hazardous Substance List
[present]

Massachusetts Right To Know List
[present]

Pennsylvania Right to Know List
[present]

P-ACETAMIDOBENZENESTIBONIC ACID, SODIUM SALT   138-31-8

SEE ALSO:
ANTIMONY

INTERNATIONAL LISTS

Canada - WHMIS: Ingredient Disclosure
1% item 5 (4)

ACETHYDRAZIDE   1068-57-1

HEALTH AND SAFETY LISTS

NIOSH - Selected LD50s and LC50s
Oral, bird: LD50 = 42200 ug/kg

ACETIC ACID   64-19-7

HEALTH AND SAFETY LISTS

ACGIH 1995 - Time Weighted Averages
10 ppm TWA; 25 mg/m3 TWA

ACGIH 1995 - Short Term Exposure Limits
15 ppm STEL; 37 mg/m3 STEL

AIHA - Odor Threshold Values
geometric mean air odor threshold = 0.074 ppm (detectable)

U.S. DOT - Substances From 49 CFR 172.101
regulated by DOT (UN2790, UN2789)

U.S. DOT - Hazard Classes
DOT hazard class = 8

U.S. DOT - Appendix A Table 1 - Hazardous Substances
final RQ = 5000 pounds (2270 kg)

NFPA - Flash Points
flash point = 103 degrees F (39 degrees C)

NFPA - Hazard Identification Ratings
health-3; flammability-2; reactivity-0

NIOSH - Selected LD50s and LC50s
Inhalation, mouse: LC50 = 5620 ppm 1 hr Oral, rat: LD50 = 3530 mg/kg Skin, rabbit: LD50 = 1060 mg/kg

NIOSH 1990 - Pocket Guide - RELs
10 ppm TWA; 25 mg/m3 TWA; 15 ppm STEL; 37 mg/m3 STEL

NIOSH 1990 - Pocket Guide - IDLHs
1000 ppm IDLH

NIOSH 1990 - Pocket Guide - Target organs
respiratory system, skin, eyes, teeth

OSHA - Vacated PELs - Time Weighted Averages
10 ppm TWA; 25 mg/m3 TWA

OSHA - Final PELs - Time Weighted Averages
10 ppm TWA; 25 mg/m3 TWA

ENVIRONMENTAL LISTS

CERCLA/SARA - Hazardous Substances and their Reportable Quantities
final RQ = 5000 pounds (2270 kg)

CAA - HON Rule - SOCMI Chemicals
compliance by Jan. 23, 1995

Clean Water Act - Hazardous Substances
[present]

List of Pesticide Product Inert Ingredients
[present]

INTERNATIONAL LISTS

Australian Exposure Standards - Time Weighted Averages
10 ppm TWA; 25 mg/m3 TWA

Australian Exposure Standards - Short Term Exposure Limits
15 ppm STEL; 37 mg/m3 STEL

Canada - WHMIS: Ingredient Disclosure
1% item 6 (51)

Canada - Alberta - 8 Hour Occupational Exposure Limit
10 ppm TWA; 26 mg/m3 TWA

Canada - Alberta - 15 Minute Occupational Exposure Limit
15 ppm STEL; 39 mg/m3 STEL

Canada - British Columbia - 8 Hour Exposure Limits
10 ppm TWA; 25 mg/m3 TWA

Canada - British Columbia - 15 Minute Exposure Limits
15 ppm STEL; 37 mg/m3 STEL

Canada - Ontario - OHSA - TWAEVs
10 ppm TWAEV; 25 mg/m3 TWAEV

Canada - Ontario - OHSA - STEVs
15 ppm STEV; 37 mg/m3 STEV

Canada - Quebec - Time-Weighted Average Exposure Values
10 ppm TWAEV; 25 mg/m3 TWAEV

Canada - Quebec - Short-term Exposure Values
15 ppm STEV; 37 mg/m3 STEV

United Kingdom - Occupational Exposure Standards - TWAs
10 ppm TWA; 25 mg/m3 TWA

United Kingdom - Occupational Exposure Standards - STELs
15 ppm STEL; 37 mg/m3 STEL

German (DFG) - MAK Values
10 ppm MAK; 25 mg/m3 MAK

German (DFG) - Peak Limitations
2 x normal MAK (5 min momentary value); don't exceed 8 times during shift

Israel - Time Weighted Averages
10 ppm TWA; 25 mg/m3 TWA

Israel - Short Term Exposure Limits
15 ppm STEL; 37 mg/m3 STEL

Israel - Action Levels
5 ppm AL; 12.5 mg/m3 AL

Mexico - Instruction No. 10 - TWAs
10 ppm TWA; 25 mg/m3 TWA

Mexico - Instruction No. 10 - STELs
15 ppm STEL; 37 mg/m3 STEL

STATE LISTS

California - Exposure Limits - PELs
10 ppm PEL; 25 mg/m3 PEL

California - Exposure Limits - STELs
15 ppm STEL; 37 mg/m3 STEL

California - Exposure Limits - Ceilings
C 40 ppm

California - Directors List of Hazardous Substances (8 CCR 339)
[present] (exempt in solutions of less than 10% or when present in food or beverages)

Florida Hazardous Substance List
[present]

Massachusetts Right To Know List
[present]

Minnesota Hazardous Substance List
[present]

NJ Right to Know List (Total)
sn 0004

NJ Special Hazardous Substances
(corrosive)

Pennsylvania Right to Know List
environmental hazard

PROPOSED REGULATIONS

Canada - Ontario - Proposed Occupational TWAEVs
5 ppm TWAEV; 13 mg/m3 TWAEV

Canada - Ontario - Proposed Occupational STEVs
10 ppm STEV; 25 mg/m3 STEV

ACETIC ACID, CYANO- 372-09-8

HEALTH AND SAFETY LISTS

NIOSH - Selected LD50s and LC50s
Oral, rat: LD50 = 1500 mg/kg

ENVIRONMENTAL LISTS

CAA - HON Rule - SOCMI Chemicals
compliance by Jan. 23, 1995

STATE LISTS

Florida Hazardous Substance List
[present]

Massachusetts Right To Know List
[present]

Pennsylvania Right to Know List
[present]

ACETIC ACID, DODECYLAMINE SALT 2016-56-0

ENVIRONMENTAL LISTS

List of Pesticide Product Inert Ingredients
[present]

ACETIC ACID, 2-ETHYLBUTYL ESTER 10031-87-5

HEALTH AND SAFETY LISTS

NFPA - Flash Points
flash point = 130 degrees F (54 degrees C)

NFPA - Hazard Identification Ratings
health-1; flammability-2; reactivity-0

STATE LISTS

Pennsylvania Right to Know List
[present]

ACETIC ACID, ISOCYANATO-, ETHYL ESTER 2949-22-6

ENVIRONMENTAL LISTS

TSCA - Code of Federal Regulations Citations
40 CFR 712.30(x); 40 CFR 716.120(d)

TSCA - PAIR - Reporting List
Reporting Date: December 27, 1990

TSCA - Health and Safety Reporting List
Effective Date: October 29, 1990; Sunset Date: November 9, 1993

ACETIC ACID, OCTYL ESTER 112-14-1

HEALTH AND SAFETY LISTS

NIOSH - Selected LD50s and LC50s
Oral, rat: LD50 = 3000 mg/kg

STATE LISTS

Pennsylvania Right to Know List
[present]

ACETIC ACID, SEC-PENTYL ESTER 53496-15-4

STATE LISTS

Pennsylvania Right to Know List
environmental hazard

ACETIC ACID, 2-PHENYLHYDRAZIDE 114-83-0

HEALTH AND SAFETY LISTS

NIOSH - Selected LD50s and LC50s
Oral, mouse: LD50 = 270 mg/kg

ACETIC ACID TRICHLORO-, COMPD. WITH 3-(P-CHLOROPHENYL)-1, 140-41-0

HEALTH AND SAFETY LISTS

NIOSH - Selected LD50s and LC50s
Oral, rat: LD50 = 2300 mg/kg

STATE LISTS

California - Directors List of Hazardous Substances (8 CCR 339)
[present]

ACETIC ACID, (2,4,5-TRICHLOROPHENOXY)-, 2-BUTOXYETHYL ESTER   2545-59-7

HEALTH AND SAFETY LISTS

U.S. DOT - Appendix A Table 1 - Hazardous Substances
final RQ = 1000 pounds (454 kg) (Listed under '2,4,5-T esters')

ENVIRONMENTAL LISTS

CERCLA/SARA - Hazardous Substances and their Reportable Quantities
final RQ = 1000 pounds (454 kg) (Listed under '2,4,5-T esters')
Clean Water Act - Hazardous Substances
[present]

STATE LISTS

California - Directors List of Hazardous Substances (8 CCR 339)
[present] (Listed under '2,4,5-T esters')
Massachusetts Right To Know List
[present]
NJ Right to Know List (Total)
sn 2937
Pennsylvania Right to Know List
environmental hazard

ACETIC ACID, (2,4,5-TRICHLOROPHENOXY)-, BUTYL ESTER   93-79-8

HEALTH AND SAFETY LISTS

U.S. DOT - Appendix A Table 1 - Hazardous Substances
final RQ = 1000 pounds (454 kg) (Listed under '2,4,5-T esters')
NIOSH - Selected LD50s and LC50s
Oral, rat: LD50 = 619 mg/kg

ENVIRONMENTAL LISTS

CERCLA/SARA - Hazardous Substances and their Reportable Quantities
final RQ = 1000 pounds (454 kg) (Listed under '2,4,5-T esters')
Clean Water Act - Hazardous Substances
[present] (Listed under '2,4,5-T esters')

STATE LISTS

California - Directors List of Hazardous Substances (8 CCR 339)
[present] (Listed under '2,4,5-T esters')
Massachusetts Right To Know List
[present]
NJ Right to Know List (Total)
sn 1802
Pennsylvania Right to Know List
environmental hazard

ACETIC ACID, (2,4,5-TRICHLOROPHENOXY)-, COMPOUND WITH 1-AMINO-2-PROPANOL(1:1)   1319-72-8

HEALTH AND SAFETY LISTS

U.S. DOT - Appendix A Table 1 - Hazardous Substances
final RQ = 5000 pounds (2270 kg) (Listed under '2,4,5-T amines')

ENVIRONMENTAL LISTS

CERCLA/SARA - Hazardous Substances and their Reportable Quantities
final RQ = 5000 pounds (2270 kg) (Listed under '2,4,5-T amines')
Clean Water Act - Hazardous Substances
[present] (Listed under '2,4,5-T amines')

STATE LISTS

California - Directors List of Hazardous Substances (8 CCR 339)
[present] (Listed under '2,4,5-T amines')

Massachusetts Right To Know List
[present]
NJ Right to Know List (Total)
sn 2931
Pennsylvania Right to Know List
environmental hazard

ACETIC ACID, (2,4,5-TRICHLOROPHENOXY)-, COMPOUND WITH N,N-DIETHYLETHANAMINE   2008-46-0

HEALTH AND SAFETY LISTS

U.S. DOT - Appendix A Table 1 - Hazardous Substances
final RQ = 5000 pounds (2270 kg) (Listed under '2,4,5-T amines')

ENVIRONMENTAL LISTS

CERCLA/SARA - Hazardous Substances and their Reportable Quantities
final RQ = 5000 pounds (2270 kg) (Listed under '2,4,5-T amines')

STATE LISTS

Massachusetts Right To Know List
[present]
NJ Right to Know List (Total)
sn 2932
Pennsylvania Right to Know List
environmental hazard

ACETIC ACID, (2,4,5-TRICHLOROPHENOXY)-, COMPOUND WITH 2,2',2"-NITROTRIS (ETHANOL) (1:1)   3813-14-7

HEALTH AND SAFETY LISTS

U.S. DOT - Appendix A Table 1 - Hazardous Substances
final RQ = 5000 pounds (2270 kg) (Listed under '2,4,5-T amines')

ENVIRONMENTAL LISTS

CERCLA/SARA - Hazardous Substances and their Reportable Quantities
final RQ = 5000 pounds (2270 kg) (Listed under '2,4,5-T amines')
Clean Water Act - Hazardous Substances
[present] (Listed under '2,4,5-T amines')

STATE LISTS

California - Directors List of Hazardous Substances (8 CCR 339)
[present] (Listed under '2,4,5-T amines')
Massachusetts Right To Know List
[present]
NJ Right to Know List (Total)
sn 2933
Pennsylvania Right to Know List
environmental hazard

ACETIC ACID, (2,4,5-TRICHLOROPHENOXY)-, COMPOUND WITH N-METHYLMETHANAMINE   6369-97-7

HEALTH AND SAFETY LISTS

U.S. DOT - Appendix A Table 1 - Hazardous Substances
final RQ = 5000 pounds (2270 kg) (Listed under '2,4,5-T amines')

ENVIRONMENTAL LISTS

CERCLA/SARA - Hazardous Substances and their Reportable Quantities
final RQ = 5000 pounds (2270 kg) (Listed under '2,4,5-T amines')
Clean Water Act - Hazardous Substances
[present] (Listed under '2,4,5-T amines')

STATE LISTS

California - Directors List of Hazardous Substances (8 CCR 339)
[present] (Listed under '2,4,5-T amines')
Massachusetts Right To Know List
[present]

NJ Right to Know List (Total)
sn 2935

Pennsylvania Right to Know List
environmental hazard

ACETIC ACID, (2,4,5-TRICHLOROPHENOXY)-, 6369-96-6
COMPOUND WITH TRIMETHYLAMINE

HEALTH AND SAFETY LISTS

U.S. DOT - Appendix A Table 1 - Hazardous Substances
final RQ = 5000 pounds (2270 kg) (Listed under '2,4,5-T amines')

ENVIRONMENTAL LISTS

CERCLA/SARA - Hazardous Substances and their Reportable Quantities
final RQ = 5000 pounds (2270 kg) (Listed under '2,4,5-T amines')

Clean Water Act - Hazardous Substances
[present] (Listed under '2,4,5-T amines')

STATE LISTS

California - Directors List of Hazardous Substances (8 CCR 339)
[present] (Listed under '2,4,5-T amines')

Massachusetts Right To Know List
[present]

NJ Right to Know List (Total)
sn 2934

Pennsylvania Right to Know List
environmental hazard

ACETIC ACID, (2,4,5-TRICHLOROPHENOXY)-, 2- 1928-47-8
ETHYLHEXYL ESTER

HEALTH AND SAFETY LISTS

U.S. DOT - Appendix A Table 1 - Hazardous Substances
final RQ = 1000 pounds (454 kg) (Listed under '2,4,5-T esters')

ENVIRONMENTAL LISTS

CERCLA/SARA - Hazardous Substances and their Reportable Quantities
final RQ = 1000 pounds (454 kg) (Listed under '2,4,5-T esters')

Clean Water Act - Hazardous Substances
[present] (Listed under '2,4,5-T esters')

STATE LISTS

California - Directors List of Hazardous Substances (8 CCR 339)
[present] (Listed under '2,4,5-T esters')

Massachusetts Right To Know List
[present]

NJ Right to Know List (Total)
sn 2936

Pennsylvania Right to Know List
environmental hazard

ACETIC ACID, (2,4,5-TRICHLOROPHENOXY)-, 25168-15-4
ISOOCTYL ESTER

HEALTH AND SAFETY LISTS

U.S. DOT - Appendix A Table 1 - Hazardous Substances
final RQ = 1000 pounds (454 kg) (Listed under '2,4,5-T esters')

ENVIRONMENTAL LISTS

CERCLA/SARA - Hazardous Substances and their Reportable Quantities
final RQ = 1000 pounds (454 kg) (Listed under '2,4,5-T esters')

Clean Water Act - Hazardous Substances
[present] (Listed under '2,4,5-T esters')

STATE LISTS

California - Directors List of Hazardous Substances (8 CCR 339)
[present] (Listed under '2,4,5-T esters')

Massachusetts Right To Know List
[present]

NJ Right to Know List (Total)
sn 1801

Pennsylvania Right to Know List
environmental hazard

ACETIC ACID, (2,4,5-TRICHLOROPHENOXY)-, 1- 61792-07-2
METHYL PROPYL ESTER

HEALTH AND SAFETY LISTS

U.S. DOT - Appendix A Table 1 - Hazardous Substances
final RQ = 1000 pounds (454 kg) (Listed under '2,4,5-T esters')

ENVIRONMENTAL LISTS

CERCLA/SARA - Hazardous Substances and their Reportable Quantities
final RQ = 1000 pounds (454 kg) (Listed under '2,4,5-T esters')

Clean Water Act - Hazardous Substances
[present] (Listed under '2,4,5-T esters')

STATE LISTS

California - Directors List of Hazardous Substances (8 CCR 339)
[present] (Listed under '2,4,5-T esters')

Massachusetts Right To Know List
[present]

NJ Right to Know List (Total)
sn 2938

Pennsylvania Right to Know List
environmental hazard

ACETIC ACID, WATER SOLUTIONS RR-00780-7

STATE LISTS

Pennsylvania Right to Know List
environmental hazard

ACETIC ANHYDRIDE 108-24-7

HEALTH AND SAFETY LISTS

ACGIH 1995 - Time Weighted Averages
5 ppm TWA; 21 mg/m3 TWA

AIHA - Odor Threshold Values
geometric mean air odor threshold < 0.14 ppm (detectable); 0.36 ppm (recognizable)

U.S. DOT - Substances From 49 CFR 172.101
regulated by DOT (UN1715)

U.S. DOT - Hazard Classes
DOT hazard class = 8

U.S. DOT - Appendix A Table 1 - Hazardous Substances
final RQ = 5000 pounds (2270 kg)

FDA - Controlled Substances Act - Essential chemicals
Import/Export threshold volume = 250 gallons; weight = 1,023 kilograms; Domestic threshold volume = 250 gallons; weight = 1,023 kilograms

NFPA - Flash Points
flash point = 120 degrees F (49 degrees C)

NFPA - Hazard Identification Ratings
health-3; flammability-2; reactivity-1 (avoid use of water)

NIOSH - Selected LD50s and LC50s
Inhalation, rat: LC50 = 1000 ppm 4 hr Oral, rat: LD50 = 1780 mg/kg Skin, rabbit: LD50 = 4000 mg/kg

NIOSH 1990 - Pocket Guide - RELs
C 5 ppm; C 20 mg/m3

NIOSH 1990 - Pocket Guide - IDLHs
1000 ppm IDLH

NIOSH 1990 - Pocket Guide - Target organs
respiratory system, eyes, skin

OSHA - Vacated PELs - Ceiling Limits
C 5 ppm; C 20 mg/m3

OSHA - Final PELs - Time Weighted Averages
5 ppm TWA; 20 mg/m3 TWA

ENVIRONMENTAL LISTS

CERCLA/SARA - Hazardous Substances and their Reportable Quantities
final RQ = 5000 pounds (2270 kg)

CAA - HON Rule - SOCMI Chemicals
compliance by Jan. 23, 1995

Clean Water Act - Hazardous Substances
[present]

EPA - Master Testing List
[present]

List of Pesticide Product Inert Ingredients
[present]

INTERNATIONAL LISTS

Australian Exposure Standards - Time Weighted Averages
Peak Limitation: 5 ppm; 21 mg/m3

Canada - WHMIS: Ingredient Disclosure
1% item 7 (227)

Canada - Alberta - Ceiling Occupational Exposure Limit
C 5 ppm; C 21 mg/m3

Canada - British Columbia - Ceiling Exposure Limits
C 5 ppm; C 20 mg/m3

Canada - Ontario - OHSA - CEVs
5 ppm CEV; 21 mg/m3 CEV

Canada - Quebec - Ceiling Limits
P 5 ppm; P 21 mg/m3

United Kingdom - Occupational Exposure Standards - STELs
5 ppm STEL; 20 mg/m3 STEL

German (DFG) - MAK Values
5 ppm MAK; 20 mg/m3 MAK

German (DFG) - Peak Limitations
2 x normal MAK (5 min momentary value); don't exceed 8 times during shift

Israel - Ceiling Exposure Limits
C 5 ppm; C 21 mg/m3

Mexico - Instruction No. 10 - TWAs
5 ppm TWA; 20 mg/m3 TWA

STATE LISTS

California - Exposure Limits - Ceilings
C 5 ppm; C 20 mg/m3

California - Directors List of Hazardous Substances (8 CCR 339)
[present]

Florida Hazardous Substance List
[present]

Massachusetts Right To Know List
[present]

Minnesota Hazardous Substance List
[present]

NJ Right to Know List (Total)
sn 0005

NJ Special Hazardous Substances
(corrosive)

Pennsylvania Right to Know List
environmental hazard

ACETOACETANILIDE   102-01-2

HEALTH AND SAFETY LISTS

NFPA - Flash Points
flash point = 365 degrees F (185 degrees C)

NFPA - Hazard Identification Ratings
health-2; flammability-1; reactivity-0

NIOSH - Selected LD50s and LC50s
Oral, rat: LD50 = 5400 mg/kg

ENVIRONMENTAL LISTS

CAA - HON Rule - SOCMI Chemicals
compliance by April 24, 1995

EPA - Master Testing List
[present]

TSCA - Code of Federal Regulations Citations
40 CFR 712.30(x); 40 CFR 716.120(d)

TSCA - PAIR - Reporting List
Reporting Date: November 27, 1991

TSCA - Health and Safety Reporting List
Effective Date: September 30, 1991

STATE LISTS

Florida Hazardous Substance List
[present]

Massachusetts Right To Know List
[present]

Pennsylvania Right to Know List
[present]

O-ACETOACETANISIDIDE   92-15-9

HEALTH AND SAFETY LISTS

NFPA - Flash Points
flash point = 325 degrees F (168 degrees C)

NFPA - Hazard Identification Ratings
health-2; flammability-1; reactivity-0

INTERNATIONAL LISTS

Canada - WHMIS: Ingredient Disclosure
1% item 8 (38)

STATE LISTS

Florida Hazardous Substance List
[present]

Massachusetts Right To Know List
[present]

Pennsylvania Right to Know List
[present]

ACETOACET-ORTHO-TOLUIDIDE   93-68-5

HEALTH AND SAFETY LISTS

NFPA - Flash Points
flash point = 320 degrees F (160 degrees C)

NFPA - Hazard Identification Ratings
health-2; flammability-1; reactivity-1

NIOSH - Selected LD50s and LC50s
Oral, rat: LD50 = 1600 mg/kg

STATE LISTS

Florida Hazardous Substance List
[present]

Massachusetts Right To Know List
[present]

Pennsylvania Right to Know List
[present]

M-ACETOACET XYLIDIDE   97-36-9

HEALTH AND SAFETY LISTS

NFPA - Flash Points
flash point = 340 degrees F (171 degrees C)

NFPA - Hazard Identification Ratings
health-2; flammability-1; reactivity-0

STATE LISTS

Florida Hazardous Substance List
[present]

Massachusetts Right To Know List
[present]

Pennsylvania Right to Know List
[present]

P-ACETOAMINOANILINE   **122-80-5**

INTERNATIONAL LISTS

Canada - WHMIS: Ingredient Disclosure
1% item 9 (40)

ACETOCHLOR   **34256-82-1**

STATE LISTS

California - Air Bill 2588 Appendix A-II
known or potential carcinogen: 9/89

California - Prop. 65 - Cancer list
carcinogen - initial date 1/1/89

ACETOHEXAMIDE   **968-81-0**

HEALTH AND SAFETY LISTS

NTP Chemical Status Reports - Testing Status and NTIS Number
Technical reports printed (PB284673/AS)

NTP Chemical Status Reports - Evidence of Carcinogenicity
male rat-negative; female rat-negative; male mice-negative; female mice-negative

STATE LISTS

Massachusetts Right To Know List
[present]

ACETOHYDROXAMIC ACID   **546-88-3**

STATE LISTS

California - Air Bill 2588 Appendix A-II
9/90

California - Prop. 65 - Developmental Toxicity
developmental toxicity - initial date 4/1/90

ACETOMINOPHEN (4-HYDROXYACETANILIDE)   **103-90-2**

HEALTH AND SAFETY LISTS

IARC - Group 3 (not classifiable)
[present]

NTP Chemical Status Reports - Testing Status and NTIS Number
Technical reports printed (no NTIS number given)

NTP Chemical Status Reports - Evidence of Carcinogenicity
male rat: no evidence; female rat: equivocal evidence; male mice: no evidence; female mice: no evidence

2-ACETONAPHTHONE   **93-08-3**

HEALTH AND SAFETY LISTS

NIOSH - Selected LD50s and LC50s
Oral, mouse: LD50 = 599 mg/kg

1-ACETONAPHTHONE   **941-98-0**

HEALTH AND SAFETY LISTS

NIOSH - Selected LD50s and LC50s
Oral, rat: LD50 = 1560 mg/kg

ACETONE   **67-64-1**

SEE ALSO:
F003-HAZARDOUS WASTES
F039-HAZARDOUS WASTES
KETONES, LIQUID, N.O.S.

HEALTH AND SAFETY LISTS

ACGIH 1995 - Time Weighted Averages
750 ppm TWA; 1780 mg/m3 TWA

ACGIH 1995 - Short Term Exposure Limits
1000 ppm STEL; 2380 mg/m3 STEL

ACGIH 1995 - Biological Exposure Indices
Acetone in urine: 100 mg/L creatinine, end of shift (B,Ns)

AIHA - Odor Threshold Values
geometric mean air odor threshold = 62 ppm (detectable); 130 ppm (recognizable)

U.S. DOT - Substances From 49 CFR 172.101
regulated by DOT (UN1090)

U.S. DOT - Hazard Classes
DOT hazard class = 3

U.S. DOT - Appendix A Table 1 - Hazardous Substances
final RQ = 5000 pounds (2270 kg)

FDA - Controlled Substances Act - Essential chemicals
Import/Export threshold volume = 500 gallons; weight = 1,500 kilograms; Domestic threshold volume = 50 gallons; weight = 150 kilograms

NFPA - Flash Points
flash point = -4 degrees F (-20 degrees C)

NFPA - Hazard Identification Ratings
health-1; flammability-3; reactivity-0

NIOSH - Selected LD50s and LC50s
Oral, rat: LD50 = 5800 mg/kg Skin, rabbit: LD50 = 20 gm/kg

NIOSH 1990 - Pocket Guide - RELs
250 ppm TWA; 590 mg/m3 TWA

NIOSH 1990 - Pocket Guide - IDLHs
20,000 ppm IDLH

NIOSH 1990 - Pocket Guide - Target organs
respiratory system, skin

NIOSH - Health Standards - Exposure Limits
250 ppm TWA; 590 mg/m3 TWA (Listed under 'Ketones')

NIOSH - Health Standards - Health Effects and Precautions
Irritation; liver, kidney, and nervous system effects (Urinalysis required) (Listed under 'Ketones')

NTP Chemical Status Reports - Testing Status and NTIS Number
Technical reports printed (PB91185975)

OSHA - Vacated PELs - Time Weighted Averages
750 ppm TWA; 1800 mg/m3 TWA

OSHA - Vacated PELs - Short Term Exposure Limits
1000 ppm STEL; 2400 mg/m3 STEL (The acetone STEL does not apply to the cellulose acetate fiber industry. It is in effect for all other sectors)

OSHA - Final PELs - Time Weighted Averages
1000 ppm TWA; 2400 mg/m3 TWA

ENVIRONMENTAL LISTS

ATSDR Priority List
Rank (of 275): 144

CERCLA/SARA - Section 313 - Emission Reporting
form R reporting required for 1.0% de minimus concentration

CERCLA/SARA - Hazardous Substances and their Reportable Quantities
final RQ = 5000 pounds (2270 kg)

CAA - HON Rule - SOCMI Chemicals
compliance by Oct. 24, 1994

EPA - Master Testing List
[present]

List of Pesticide Product Inert Ingredients
[present]

RCRA - U Series Wastes
waste number U002 (Ignitable waste)

RCRA - Hazardous Constituents-Appendix VIII
waste number U002 (Ignitable waste)

RCRA - Basis for Listing - Appendix VII
Included in waste stream: F039

RCRA - Substances Banned From Land Disposal
[present]

RCRA - TSD Facilities Ground Water Monitoring
TM 8240 = 100 ug/L PQL

RCRA - Universal Treatment Standards (LDR)
WW: 0.28 mg/l; NWW: 160 mg/kg

TSCA - Multichemical Test Rules - Neurotoxicity
administrative stay for neurotoxicity tests effective June 27, 1994

INTERNATIONAL LISTS

Australian Exposure Standards - Time Weighted Averages
500 ppm TWA; 1185 mg/m3 TWA

Australian Exposure Standards - Short Term Exposure Limits
1000 ppm STEL; 2375 mg/m3 STEL

Canada - WHMIS: Ingredient Disclosure
1% item 10 (41)

Canada - NPRI (National Pollutant Release Inventory)
[present]

Canada - Alberta - 8 Hour Occupational Exposure Limit
750 ppm TWA; 1782 mg/m3 TWA

Canada - Alberta - 15 Minute Occupational Exposure Limit
1000 ppm STEL; 2375 mg/m3 STEL

Canada - British Columbia - 8 Hour Exposure Limits
250 ppm TWA

Canada - British Columbia - 15 Minute Exposure Limits
500 ppm STEL

Canada - Ontario - OHSA - TWAEVs
750 ppm TWAEV; 1780 mg/m3 TWAEV

Canada - Ontario - OHSA - STEVs
1000 ppm STEV; 2375 mg/m3 STEV

Canada - Quebec - Time-Weighted Average Exposure Values
750 ppm TWAEV; 1780 mg/m3 TWAEV

Canada - Quebec - Short-term Exposure Values
1000 ppm STEV; 2380 mg/m3 STEV

United Kingdom - Occupational Exposure Standards - TWAs
750 ppm TWA; 1780 mg/m3 TWA

United Kingdom - Occupational Exposure Standards - STELs
1500 ppm STEL; 3560 mg/m3 STEL

German (DFG) - MAK Values
500 ppm MAK; 1200 mg/m3 MAK

German (DFG) - Peak Limitations
2 x normal MAK (30 minute average value); don't exceed 4 times per shift

Israel - Time Weighted Averages
750 ppm TWA; 1780 mg/m3 TWA

Israel - Short Term Exposure Limits
1000 ppm STEL; 2380 mg/m3 STEL

Israel - Action Levels
375 ppm AL; 890 mg/m3 AL

Mexico - Instruction No. 10 - TWAs
1000 ppm TWA; 2400 mg/m3 TWA

Mexico - Instruction No. 10 - STELs
1260 ppm STEL; 3000 mg/m3 STEL

STATE LISTS

California - Air Bill 2588 Appendix A-I
6/91

California - Exposure Limits - PELs
750 ppm PEL; 1780 mg/m3 PEL

California - Exposure Limits - STELs
1000 ppm STEL; 2400 mg/m3 STEL

California - Exposure Limits - Ceilings
C 3000 ppm

California - Directors List of Hazardous Substances (8 CCR 339)
[present]

Florida Hazardous Substance List
[present]

Massachusetts Right To Know List
[present]

Minnesota Hazardous Substance List
[present]

NJ Right to Know List (Total)
sn 0006

NJ Special Hazardous Substances
(flammable - third degree)

Pennsylvania Right to Know List
environmental hazard

PROPOSED REGULATIONS

TSCA - ITC 33rd Report Priority Testing List
designated for testing

TSCA - ITC 34th Report Priority Testing List
designated for testing

Canada - Ontario - Proposed Occupational TWAEVs
125 ppm TWAEV; 295 mg/m3 TWAEV

ACETONE ALCOHOL   116-09-6

HEALTH AND SAFETY LISTS

NIOSH - Selected LD50s and LC50s
Oral, rat: LD50 = 2200 mg/kg

ACETONE CYANOHYDRIN   75-86-5

HEALTH AND SAFETY LISTS

ACGIH 1995 - Ceiling Limits
C 4.7 ppm; C 5 mg/m3

ACGIH 1995 - Skin Designations
skin - potential for cutaneous absorption

AIHA - WEEL - Time Weighted Averages
2 ppm TWA; 7.1 mg/m3 TWA

AIHA - WEEL - Ceilings or Short Term Time Weighted Averages
5 ppm STEL; 17.7 mg/m3 STEL

AIHA - WEEL - Skin Absorption Designations
skin absorber

U.S. DOT - Substances From 49 CFR 172.101
regulated by DOT (UN1541)

U.S. DOT - Hazard Classes
DOT hazard class = 6.1

U.S. DOT - Substances Which Are Poisonous by Inhalation
liquid hazardous material poisonous by inhalation (stabilized form) (UN1541)

U.S. DOT - Appendix B - Marine Pollutants
stabilized form: DOT regulated marine pollutant

U.S. DOT - Appendix A Table 1 - Hazardous Substances
final RQ = 10 pounds (4.54 kg)

NFPA - Flash Points
flash point = 165 degrees F (74 degrees C)

NFPA - Hazard Identification Ratings
health-4; flammability-2; reactivity-2

NIOSH - Selected LD50s and LC50s
Oral, rat: LD50 = 17800 ug/kg Skin, guinea pgi: LD50 = 150 mg/kg

NIOSH - Health Standards - Exposure Limits
C (15 min) 1 ppm; C (15 min) 4 mg/m3 (Listed under 'Nitriles')

NIOSH - Health Standards - Health Effects and Precautions
Hepatic, renal, respiratory, cardiovascular, gastrointestinal, and nervous system effects (Periodic chest X-ray and pulmonary function testing required; prevent skin and eye contact; make first-aid kits and personnel available during use) (Listed under 'Nitriles')

ENVIRONMENTAL LISTS

CERCLA/SARA - Section 302 Extremely Hazardous Substances and TPQs
TPQ = 1000 pounds

CERCLA/SARA - Section 313 - Emission Reporting
form R reporting required

CERCLA/SARA - Hazardous Substances and their Reportable Quantities
final RQ = 10 pounds (4.54 kg)

CAA - HON Rule - SOCMI Chemicals
compliance by Oct. 23, 1995

Clean Water Act - Hazardous Substances
[present]

EPA - Master Testing List
[present]

RCRA - P Series Wastes
waste number P069

RCRA - Hazardous Constituents-Appendix VIII
waste number P069

RCRA - Substances Banned From Land Disposal
[present]

TSCA - Code of Federal Regulations Citations
40 CFR 716.120(a)

TSCA - Health and Safety Reporting List
Effective Date: March 7, 1986

INTERNATIONAL LISTS

Canada - WHMIS: Ingredient Disclosure
1% item 11 (42)

STATE LISTS

California - Directors List of Hazardous Substances (8 CCR 339)
[present]

Florida Hazardous Substance List
[present]

Massachusetts Right To Know List
extraordinarily hazardous

Minnesota Hazardous Substance List
[present]

NJ Right to Know List (Total)
sn 0007

NJ Special Hazardous Substances
(reactive - second degree)

Pennsylvania Right to Know List
environmental hazard

ACETONE OIL   RR-00099-7

HEALTH AND SAFETY LISTS

U.S. DOT - Substances From 49 CFR 172.101
regulated by DOT (UN1091)

U.S. DOT - Hazard Classes
DOT hazard class = 3

STATE LISTS

NJ Right to Know List (Total)
sn 2058

ACETONE THIOSEMICARBAZIDE   1752-30-3

ENVIRONMENTAL LISTS

CERCLA/SARA - Section 302 Extremely Hazardous Substances and TPQs
TPQ = 1000/10,000 pounds

STATE LISTS

Florida Hazardous Substance List
effective March 13, 1992

Massachusetts Right To Know List
extraordinarily hazardous

Pennsylvania Right to Know List
environmental hazard

PROPOSED REGULATIONS

CERCLA/SARA - Proposed Hazardous Substance Additions
proposed RQ = 1 pound (.454 kg)

CERCLA/SARA - 1989 Proposed RQ Adjustments
proposed RQ = 100 pounds (45.4 kg)

ACETONITRILE   75-05-8

SEE ALSO:

K011-HAZARDOUS WASTES
F039-HAZARDOUS WASTES
K014-HAZARDOUS WASTES
K013-HAZARDOUS WASTES

HEALTH AND SAFETY LISTS

ACGIH 1995 - Time Weighted Averages
40 ppm TWA; 67 mg/m3 TWA

ACGIH 1995 - Short Term Exposure Limits
60 ppm STEL; 101 mg/m3 STEL

AIHA - Odor Threshold Values
geometric mean air odor threshold = 1160 ppm (detectable)

U.S. DOT - Substances From 49 CFR 172.101
regulated by DOT (UN1648)

U.S. DOT - Hazard Classes
DOT hazard class = 3

U.S. DOT - Appendix A Table 1 - Hazardous Substances
final RQ = 5000 pounds (2270 kg)

NFPA - Flash Points
flash point = 42 degrees F (6 degrees C)

NFPA - Hazard Identification Ratings
health-2; flammability-3; reactivity-0

NIOSH - Selected LD50s and LC50s
Inhalation, rat: LC50 = 7551 ppm 8 hr Oral, rat: LD50 = 2730 mg/kg Skin, rabbit: LD50 = 1250 mg/kg

NIOSH 1990 - Pocket Guide - RELs
20 ppm TWA; 34 mg/m3 TWA

NIOSH 1990 - Pocket Guide - IDLHs
4000 ppm IDLH

NIOSH 1990 - Pocket Guide - Target organs
kidneys, liver, CVS, CNS, lungs, skin ,eyes

NIOSH - Health Standards - Exposure Limits
20 ppm TWA; 34 mg/m3 TWA (Listed under 'Nitriles')

NIOSH - Health Standards - Health Effects and Precautions
Hepatic, renal, respiratory, cardiovascular, gastrointestinal, and nervous system effects (Periodic chest X-ray and pulmonary function testing required; prevent skin and eye contact; make first-aid kits and personnel available during use) (Listed under 'Nitriles')

NTP Chemical Status Reports - Testing Status and NTIS Number
Post peer review technical reports in progress; prechronic studies for which toxicity technical reports were not prepared

OSHA - Vacated PELs - Time Weighted Averages
40 ppm TWA; 70 mg/m3 TWA

OSHA - Vacated PELs - Short Term Exposure Limits
60 ppm STEL; 105 mg/m3 STEL

OSHA - Final PELs - Time Weighted Averages
40 ppm TWA; 70 mg/m3 TWA

ENVIRONMENTAL LISTS

CERCLA/SARA - Section 313 - Emission Reporting
form R reporting required for 1.0% de minimus concentration

CERCLA/SARA - Hazardous Substances and their Reportable Quantities
final RQ = 5000 pounds (2270 kg)

Clean Air Act (1990) - List of Hazardous Air Contaminants
[present]

CAA - HON Rule - SOCMI Chemicals
compliance by Oct. 24, 1994

CAA - HON Rule - Organic HAPs
[present]

List of Pesticide Product Inert Ingredients
[present]

RCRA - U Series Wastes
waste number U003 (Ignitable waste; Toxic waste)

RCRA - Hazardous Constituents-Appendix VIII
waste number U003

RCRA - Basis for Listing - Appendix VII
Included in waste streams: F039, K011, K013, K014

RCRA - Substances Banned From Land Disposal
[present]

RCRA - TSD Facilities Ground Water Monitoring
TM 8015 = 100 ug/L PQL

RCRA - Universal Treatment Standards (LDR)
WW: 5.6 mg/l; NWW: 1.8 mg/kg

TSCA - Code of Federal Regulations Citations
40 CFR 712.30(d); 40 CFR 716.120(a)

TSCA - PAIR - Reporting List
Reporting Date: November 19, 1982

TSCA - Health and Safety Reporting List
Effective Date: October 4, 1982

INTERNATIONAL LISTS

Australian Exposure Standards - Time Weighted Averages
40 ppm TWA; 70 mg/m3 TWA

Australian Exposure Standards - Short Term Exposure Limits
60 ppm STEL; 101 mg/m3 STEL

Australian Exposure Standards - Skin Effects
skin absorption

Canada - WHMIS: Ingredient Disclosure
0.1% item 12 (43)

Canada - NPRI (National Pollutant Release Inventory)
[present]

Canada - Alberta - 8 Hour Occupational Exposure Limit
40 ppm TWA; 67 mg/m3 TWA

Canada - Alberta - 15 Minute Occupational Exposure Limit
60 ppm STEL; 100 mg/m3 STEL

Canada - Alberta - Skin Designation
can be absorbed through the intact skin

Canada - British Columbia - 8 Hour Exposure Limits
40 ppm TWA; 70 mg/m3 TWA

Canada - British Columbia - 15 Minute Exposure Limits
60 ppm STEL; 105 mg/m3 STEL

Canada - Ontario - OHSA - TWAEVs
40 ppm TWAEV; 67 mg/m3 TWAEV

Canada - Ontario - OHSA - STEVs
60 ppm STEV; 100 mg/m3 STEV

Canada - Ontario - OHSA - Skin Notations
absorption through skin, eyes, or mucous membranes

Canada - Quebec - Time-Weighted Average Exposure Values
40 ppm TWAEV; 67 mg/m3 TWAEV

Canada - Quebec - Short-term Exposure Values
60 ppm STEV; 101 mg/m3 STEV

United Kingdom - Occupational Exposure Standards - TWAs
40 ppm TWA; 70 mg/m3 TWA

United Kingdom - Occupational Exposure Standards - STELs
60 ppm STEL; 105 mg/m3 STEL

German (DFG) - MAK Values
40 ppm MAK; 70 mg/m3 MAK

German (DFG) - Peak Limitations
2 x normal MAK (30 min. average value); don't exceed 4 times during shift

Israel - Time Weighted Averages
40 ppm TWA; 67 mg/m3 TWA

Israel - Short Term Exposure Limits
60 ppm STEL; 101 mg/m3 STEL

Israel - Action Levels
20 ppm AL; 33.5 mg/m3 AL

Mexico - Instruction No. 10 - TWAs
40 ppm TWA; 70 mg/m3 TWA

Mexico - Instruction No. 10 - STELs
60 ppm STEL; 105 mg/m3 STEL

Mexico - Instruction No. 10 - Skin designation
skin - potential for cutaneous absorption

STATE LISTS

California - Air Bill 2588 Appendix A-I
6/91

California - Exposure Limits - PELs
40 ppm PEL; 70 mg/m3 PEL

California - Exposure Limits - STELs
60 ppm STEL; 105 mg/m3 STEL

California - Exposure Limits - Skin Notation
material may be absorbed through the skin, eyes or mucous membrane

California - Directors List of Hazardous Substances (8 CCR 339)
[present]

Florida Hazardous Substance List
[present]

Massachusetts Right To Know List
[present]

Minnesota Hazardous Substance List
skin

NJ Right to Know List (Total)
sn 0008

NJ Special Hazardous Substances
(flammable - third degree)

Pennsylvania Right to Know List
environmental hazard

ACETOPHENONE   98-86-2
SEE ALSO:
F039-HAZARDOUS WASTES

HEALTH AND SAFETY LISTS

ACGIH 1995 - Time Weighted Averages
10 ppm TWA; 49 mg/m3 TWA

AIHA - WEEL - Time Weighted Averages
10 ppm TWA; 50 mg/m3 TWA

U.S. DOT - Appendix A Table 1 - Hazardous Substances
final RQ = 5000 pounds (2270 kg)

NFPA - Flash Points
flash point = 170 degrees F (77 degrees C)

NFPA - Hazard Identification Ratings
health-1; flammability-2; reactivity-0

NIOSH - Selected LD50s and LC50s
Oral, rat: LD50 = 815 mg/kg

ENVIRONMENTAL LISTS

CERCLA/SARA - Section 313 - Emission Reporting
form R reporting required

CERCLA/SARA - Hazardous Substances and their Reportable Quantities
final RQ = 5000 pounds (2270 kg)

Clean Air Act (1990) - List of Hazardous Air Contaminants
[present]

CAA - HON Rule - SOCMI Chemicals
compliance by Oct. 24, 1994

CAA - HON Rule - Organic HAPs
[present]

EPA - Master Testing List
[present]

List of Pesticide Product Inert Ingredients
[present]

RCRA - U Series Wastes
waste number U004

RCRA - Hazardous Constituents-Appendix VIII
waste number U004

RCRA - Basis for Listing - Appendix VII
Included in waste stream: F039

RCRA - Substances Banned From Land Disposal
[present]

RCRA - TSD Facilities Ground Water Monitoring
TM 8270 = 10 ug/L PQL

RCRA - Universal Treatment Standards (LDR)
WW: 0.010 mg/l; NWW: 9.7 mg/kg

INTERNATIONAL LISTS

Canada - WHMIS: Ingredient Disclosure
1% item 13 (44)

STATE LISTS

California - Air Bill 2588 Appendix A-I
6/91

Massachusetts Right To Know List
[present]

Minnesota Hazardous Substance List
[present]

Pennsylvania Right to Know List
environmental hazard

PROPOSED REGULATIONS

TSCA - ITC 33rd Report Priority Testing List
designated for testing

P-ACETOTOLUIDIDE 103-89-9

HEALTH AND SAFETY LISTS

NFPA - Flash Points
flash point = 334 degrees F (168 degrees C)

NFPA - Hazard Identification Ratings
health-2; flammability-1

STATE LISTS

Florida Hazardous Substance List
[present]

Massachusetts Right To Know List
[present]

Pennsylvania Right to Know List
[present]

ACETYL ACETONE PEROXIDE 37187-22-7

HEALTH AND SAFETY LISTS

U.S. DOT - Hazard Classes
Forbidden from transport by the DOT

U.S. DOT - Organic Peroxides Table
Organic peroxide UN3105

STATE LISTS

NJ Right to Know List (Total)
sn 0009

4-(ACETYLAMINO) BENZENESULFONYL CHLORIDE 121-60-8

ENVIRONMENTAL LISTS

TSCA - Code of Federal Regulations Citations
40 CFR 712.30(x); 40 CFR 716.120(d)

TSCA - PAIR - Reporting List
Reporting Date: November 27, 1991

TSCA - Health and Safety Reporting List
Effective Date: September 30, 1991

4-ACETYLAMINOBIPHENYL 4075-79-0

INTERNATIONAL LISTS

Canada - WHMIS: Ingredient Disclosure
1% item 14 (46)

2-ACETYLAMINOFLUORENE 53-96-3

SEE ALSO:
F039-HAZARDOUS WASTES

HEALTH AND SAFETY LISTS

U.S. DOT - Appendix A Table 1 - Hazardous Substances
final RQ = 1 pound (0.454 kg)

NIOSH - Selected LD50s and LC50s
Oral, mouse: LD50 = 1020 mg/kg

NIOSH 1990 - Pocket Guide - Carcinogens
occupational carcinogen

NIOSH 1990 - Pocket Guide - Target organs
liver, bladder, kidneys, pancreas, skin, lungs

NIOSH - Health Standards - Exposure Limits
use 29 CFR 1910.1014

NIOSH - Health Standards - Health Effects and Precautions
has produced tumors of the liver, bladder, lungs, pancreas, and skin in animals

NIOSH - Health Standards - Carcinogenic Chemicals
potential human carcinogen

NTP Seventh Report - Suspect Carcinogens
suspect carcinogen

OSHA - 29 CFR 1910 Specifically Regulated Chemicals
Cancer suspect agent (see 29 CFR 1910.1014)

OSHA - Select Carcinogens
[present]

OSHA - Possible Select Carcinogens
[present]

ENVIRONMENTAL LISTS

CERCLA/SARA - Section 313 - Emission Reporting
form R reporting required for 0.1% de minimus concentration

CERCLA/SARA - Hazardous Substances and their Reportable Quantities
final RQ = 1 pound (0.454 kg)

Clean Air Act (1990) - List of Hazardous Air Contaminants
[present]

EPA - Carcinogen Hazard Ranking for RQ Adjustment
Hazard ranking = High

RCRA - U Series Wastes
waste number U005

RCRA - Hazardous Constituents-Appendix VIII
waste number U005

RCRA - Basis for Listing - Appendix VII
Included in waste stream: F039

RCRA - Substances Banned From Land Disposal
[present]

RCRA - TSD Facilities Ground Water Monitoring
TM 8270 = 10 ug/L PQL

RCRA - Universal Treatment Standards (LDR)
WW: 0.059 mg/l; NWW: 140 mg/kg

INTERNATIONAL LISTS

Canada - WHMIS: Ingredient Disclosure
0.1% item 15 (47)

STATE LISTS

California - Air Bill 2588 Appendix A-I
known or potential carcinogen

California - Air Bill 2588 Appendix A-II
[present]

California - Prop. 65 - Cancer list
carcinogen - initial date 7/1/87

California - Prop. 65 - No Significant Risk Levels
no significant risk level = 0.2 ug/day

California - Exposure Limits - Skin Notation
material may be absorbed through the skin, eyes or mucous membrane

California - Exposure Limits - Carcinogens
cancer-suspect agent (at a concentration >= 1.0%)

California - Directors List of Hazardous Substances (8 CCR 339)
[present]

Florida Hazardous Substance List
[present]

Massachusetts Right To Know List
carcinogen; extraordinarily hazardous

Minnesota Hazardous Substance List
carcinogen

NJ Right to Know List (Total)
sn 0010

NJ Special Hazardous Substances
(carcinogen; mutagen)

Pennsylvania Right to Know List
environmental hazard; special hazardous substance

Pennsylvania RTK - Special Hazardous Substances
[present]

N-ACETYLANTHRANILIC ACID   89-52-1

HEALTH AND SAFETY LISTS

FDA - Controlled Substances Act - Precursor chemicals
Threshold by base weight = 40 kilograms

N-ACETYLANTHRANILIC ACID SALTS   RR-01776-5

HEALTH AND SAFETY LISTS

FDA - Controlled Substances Act - Precursor chemicals
Threshold by base weight = 40 kilograms

ACETYLATED LANOLIN   61788-48-5

ENVIRONMENTAL LISTS

List of Pesticide Product Inert Ingredients
[present]

ACETYLATED LANOLIN ALCOHOL   6178-49-0

ENVIRONMENTAL LISTS

List of Pesticide Product Inert Ingredients
[present]

ACETYL BENZOYL PEROXIDE   644-31-5

HEALTH AND SAFETY LISTS

U.S. DOT - Hazard Classes
Forbidden from transport by the DOT

U.S. DOT - Organic Peroxides Table
Organic peroxide UN3105

STATE LISTS

NJ Right to Know List (Total)
sn 0011

ACETYL BROMIDE   506-96-7

HEALTH AND SAFETY LISTS

U.S. DOT - Substances From 49 CFR 172.101
regulated by DOT (UN1716)

U.S. DOT - Hazard Classes
DOT hazard class = 8

U.S. DOT - Appendix A Table 1 - Hazardous Substances
final RQ = 5000 pounds (2270 kg)

NIOSH - Selected LD50s and LC50s
Inhalation, mammal: LC50 = 48 gm/m3 (8 hr)

ENVIRONMENTAL LISTS

CERCLA/SARA - Hazardous Substances and their Reportable Quantities
final RQ = 5000 pounds (2270 kg)

Clean Water Act - Hazardous Substances
[present]

TSCA - Code of Federal Regulations Citations
40 CFR 716.120(a)

TSCA - Health and Safety Reporting List
Effective Date: June 1, 1987

INTERNATIONAL LISTS

Canada - WHMIS: Ingredient Disclosure
1% item 16 (328)

STATE LISTS

California - Directors List of Hazardous Substances (8 CCR 339)
[present]

Massachusetts Right To Know List
[present]

NJ Right to Know List (Total)
sn 0012

NJ Special Hazardous Substances
(corrosive)

Pennsylvania Right to Know List
environmental hazard

ACETYL CHLORIDE   75-36-5

HEALTH AND SAFETY LISTS

U.S. DOT - Substances From 49 CFR 172.101
regulated by DOT (UN1717)

U.S. DOT - Hazard Classes
DOT hazard class = 3

U.S. DOT - Appendix A Table 1 - Hazardous Substances
final RQ = 5000 pounds (2270 kg)

NFPA - Flash Points
flash point = 40 degrees F (4 degrees C)

NFPA - Hazard Identification Ratings
health-3; flammability-3; reactivity-2 (avoid use of water)

ENVIRONMENTAL LISTS

CERCLA/SARA - Hazardous Substances and their Reportable Quantities
final RQ = 5000 pounds (2270 kg)

RCRA - U Series Wastes
waste number U006 (Corrosive waste; Reactive waste; Toxic waste)

RCRA - Hazardous Constituents-Appendix VIII
waste number U006

RCRA - Substances Banned From Land Disposal
[present]

STATE LISTS

Florida Hazardous Substance List
[present]

Massachusetts Right To Know List
[present]

NJ Right to Know List (Total)
sn 0013

NJ Special Hazardous Substances
(corrosive; flammable - third degree; reactive - second degree)

Pennsylvania Right to Know List
environmental hazard

ACETYL CYCLOHEXANE SULFONYL PEROXIDE   3179-56-4

HEALTH AND SAFETY LISTS

U.S. DOT - Hazard Classes
Forbidden from transport by the DOT

U.S. DOT - Organic Peroxides Table
Organic peroxide UN3112; UN3115

STATE LISTS

NJ Right to Know List (Total)
sn 0014

ACETYLENE   74-86-2

HEALTH AND SAFETY LISTS

ACGIH 1995 - Time Weighted Averages
simple asphyxiant

AIHA - Odor Threshold Values
no geometric mean air odor threshold

U.S. DOT - Substances From 49 CFR 172.101
regulated by DOT (UN1001)

U.S. DOT - Hazard Classes
Forbidden from transport by the DOT

NFPA - Flash Points
gas (no flash point given)

NFPA - Hazard Identification Ratings
health-0; flammability-4; reactivity-3

NIOSH - Health Standards - Exposure Limits
no exposure > 2500 ppm or 2662 mg/m3

NIOSH - Health Standards - Health Effects and Precautions
Asphyxia (check for and inform workers of contaminants such as Arsine and Phosphine)

ENVIRONMENTAL LISTS

CAA - Flammable Substances for Accidental Release Prevention
threshold quantity = 10,000 lbs

List of Pesticide Product Inert Ingredients
[present]

INTERNATIONAL LISTS

Australian Exposure Standards - Time Weighted Averages
Asphyxiant at < 18% oxygen by volume; explosion hazard

Canada - British Columbia - 8 Hour Exposure Limits
asphyxiant substance

Canada - Ontario - OHSA - TWAEVs
simple asphyxiant

Canada - Quebec - Time-Weighted Average Exposure Values
simple asphyxiant

Israel - Time Weighted Averages
Asphyxiant

Mexico - Instruction No. 10 - TWAs
simple asphyxiant

STATE LISTS

California - Exposure Limits - PELs
asphyxiant (limit depends on level of oxygen)

California - Directors List of Hazardous Substances (8 CCR 339)
[present]

Florida Hazardous Substance List
[present]

Massachusetts Right To Know List
[present]

Minnesota Hazardous Substance List
[present]

NJ Right to Know List (Total)
sn 0015

NJ Special Hazardous Substances
(flammable - fourth degree; reactive - third degree)

Pennsylvania Right to Know List
[present]

ACETYLENE SILVER NITRATE   RR-01390-1

HEALTH AND SAFETY LISTS

U.S. DOT - Hazard Classes
Forbidden from transport by the DOT

ACETYLENE TETRABROMIDE   79-27-6

HEALTH AND SAFETY LISTS

ACGIH 1995 - Time Weighted Averages
1 ppm TWA; 14 mg/m3 TWA

U.S. DOT - Appendix B - Marine Pollutants
DOT regulated marine pollutant

NFPA - Hazard Identification Ratings
health-3; flammability-0; reactivity-1

NIOSH - Selected LD50s and LC50s
Inhalation, rat: LC50 = 549 mg/m3 4 hr Oral, rat: LD50 = 1100 mg/kg Skin, rat: LD50 = 5250 mg/kg

NIOSH 1990 - Pocket Guide - RELs
See Appendix D

NIOSH 1990 - Pocket Guide - IDLHs
10 ppm IDLH

NIOSH 1990 - Pocket Guide - Target organs
eyes, upper respiratory system, liver

NTP Chemical Status Reports - Testing Status and NTIS Number
Prechronic studies completed: in review for further evaluation

OSHA - Vacated PELs - Time Weighted Averages
1 ppm TWA; 14 mg/m3 TWA

OSHA - Final PELs - Time Weighted Averages
1 ppm TWA; 14 mg/m3 TWA

ENVIRONMENTAL LISTS

TSCA - Code of Federal Regulations Citations
40 CFR 704.225(a)

TSCA - CAIR - Reporting List
reporting required by: manufacturer; distributor; importer; processor

INTERNATIONAL LISTS

Australian Exposure Standards - Time Weighted Averages
1 ppm TWA; 14 mg/m3 TWA

Canada - WHMIS: Ingredient Disclosure
1% item 17 (327)

Canada - Alberta - 8 Hour Occupational Exposure Limit
1 ppm TWA; 14 mg/m3 TWA

Canada - Alberta - 15 Minute Occupational Exposure Limit
1.5 ppm STEL; 21 mg/m3 STEL

Canada - British Columbia - 8 Hour Exposure Limits
1 ppm TWA; 14 mg/m3 TWA

Canada - British Columbia - 15 Minute Exposure Limits
1.5 ppm STEL; 18 mg/m3 STEL

Canada - Ontario - OHSA - TWAEVs
1 ppm TWAEV; 14 mg/m3 TWAEV

Canada - Quebec - Time-Weighted Average Exposure Values
1 ppm TWAEV; 14 mg/m3 TWAEV

United Kingdom - Occupational Exposure Standards - TWAs
0.5 ppm TWA; 7 mg/m3 TWA

United Kingdom - Occupational Exposure Standards - Notes
can be absorbed through skin

German (DFG) - MAK Values
1 ppm MAK; 14 mg/m3 MAK

German (DFG) - Peak Limitations
2 x normal MAK (30 min. average value); don't exceed 4 times during shift

Israel - Time Weighted Averages
1 ppm TWA;   14 mg/m3 TWA
Israel - Action Levels
0.5 ppm AL;   7 mg/m3 AL
Mexico - Instruction No. 10 - TWAs
1 ppm TWA;   15 mg/m3 TWA
Mexico - Instruction No. 10 - STELs
1.5 ppm STEL;   20 mg/m3 STEL

STATE LISTS

California - Exposure Limits - PELs
1 ppm PEL;   14 mg/m3 PEL
California - Directors List of Hazardous Substances (8 CCR 339)
[present]
Florida Hazardous Substance List
[present]
Massachusetts Right To Know List
[present]
Minnesota Hazardous Substance List
[present]
NJ Right to Know List (Total)
sn 0016
Pennsylvania Right to Know List
[present]

PROPOSED REGULATIONS

Canada - Ontario - Proposed Occupational TWAEVs
0.5 ppm TWAEV;   7 mg/m3 TWAEV

N-ACETYL ETHANOLAMINE   142-26-7

HEALTH AND SAFETY LISTS

NFPA - Flash Points
flash point = 355 degrees F (179 degrees C)
NFPA - Hazard Identification Ratings
health-1;   flammability-1;   reactivity-1

ACETYL IODIDE   507-02-8

HEALTH AND SAFETY LISTS

U.S. DOT - Substances From 49 CFR 172.101
regulated by DOT (UN1898)
U.S. DOT - Hazard Classes
DOT hazard class = 8

INTERNATIONAL LISTS

Canada - WHMIS: Ingredient Disclosure
1% item 18 (1022)

STATE LISTS

NJ Right to Know List (Total)
sn 0017
NJ Special Hazardous Substances
(corrosive)

ACETYL METHYL CARBINOL   513-86-0

HEALTH AND SAFETY LISTS

U.S. DOT - Substances From 49 CFR 172.101
regulated by DOT (UN2621)
U.S. DOT - Hazard Classes
DOT hazard class = 3

INTERNATIONAL LISTS

Canada - WHMIS: Ingredient Disclosure
1% item 19 (48)

STATE LISTS

NJ Right to Know List (Total)
sn 0018

3-ACETYL-1-PROPANOL   1071-73-4

HEALTH AND SAFETY LISTS

NIOSH - Selected LD50s and LC50s
Oral, mouse: LD50 = 1960 mg/kg

4-ACETYLRESORCINOL   89-84-9

INTERNATIONAL LISTS

Canada - WHMIS: Ingredient Disclosure
1% item 20 (49)

ACETYLSALICYLIC ACID (ASPIRIN)   50-78-2

HEALTH AND SAFETY LISTS

ACGIH 1995 - Time Weighted Averages
5 mg/m3 TWA
NIOSH - Selected LD50s and LC50s
Oral, rat: LD50 = 200 mg/kg
OSHA - Vacated PELs - Time Weighted Averages
5 mg/m3 TWA

INTERNATIONAL LISTS

Australian Exposure Standards - Time Weighted Averages
5 mg/m3 TWA
Canada - Alberta - 8 Hour Occupational Exposure Limit
5 mg/m3 TWA
Canada - Alberta - 15 Minute Occupational Exposure Limit
10 mg/m3 STEL
Canada - Ontario - OHSA - TWAEVs
5 mg/m3 TWAEV
Canada - Quebec - Time-Weighted Average Exposure Values
5 mg/m3 TWAEV
United Kingdom - Occupational Exposure Standards - TWAs
5 mg/m3 TWA
Israel - Time Weighted Averages
5 mg/m3 TWA
Israel - Action Levels
2.5 mg/m3 AL

STATE LISTS

California - Air Bill 2588 Appendix A-II
6/91
California - Prop. 65 - Developmental Toxicity
developmental toxicity (especially during the last 3 months of pregnancy) - initial date 7/1/90
California - Prop. 65 - Reproductive - Female
female reproductive toxicity (especially during last 3 months of pregnancy) - initial date 7/1/90
California - Exposure Limits - PELs
5 mg/m3 PEL
California - Directors List of Hazardous Substances (8 CCR 339)
[present] (exempt when crystalline powder is being manufactured or used)
Florida Hazardous Substance List
[present]
Massachusetts Right To Know List
[present]
Minnesota Hazardous Substance List
[present]
NJ Right to Know List (Total)
sn 0020
Pennsylvania Right to Know List
[present]

1-ACETYL-2-THIOUREA   591-08-2

HEALTH AND SAFETY LISTS

U.S. DOT - Appendix A Table 1 - Hazardous Substances
final RQ = 1000 pounds (454 kg)

NIOSH - Selected LD50s and LC50s
Oral, rat: LD50 = 50 mg/kg

ENVIRONMENTAL LISTS

CERCLA/SARA - Hazardous Substances and their Reportable Quantities
final RQ = 1000 pounds (454 kg)

RCRA - P Series Wastes
waste number P002

RCRA - Hazardous Constituents-Appendix VIII
waste number P002

RCRA - Substances Banned From Land Disposal
[present]

TSCA - Code of Federal Regulations Citations
40 CFR 716.120(a)

TSCA - Health and Safety Reporting List
Effective Date: March 7, 1986

STATE LISTS

Massachusetts Right To Know List
[present]

Pennsylvania Right to Know List
environmental hazard

ACETYL TRIETHYL CITRATE 77-89-4

HEALTH AND SAFETY LISTS

NIOSH - Selected LD50s and LC50s
Oral, rat: LD50 = 8000 mg/kg

ACID MODIFIED ACRYLATED EPOXIDE RR-00238-0

ENVIRONMENTAL LISTS

TSCA - Chemicals with Significant New Use Rules
PMN numbers: P-85-1169; P-85-1170

ACID ORANGE 7 633-96-5

ENVIRONMENTAL LISTS

List of Pesticide Product Inert Ingredients
[present]

ACID ORANGE 10 1936-15-8

HEALTH AND SAFETY LISTS

IARC - Group 3 (not classifiable)
[present]

NTP Chemical Status Reports - Testing Status and NTIS Number
Technical reports printed (PB88169347/AS)

NTP Chemical Status Reports - Evidence of Carcinogenicity
male rat-negative; female rat-negative; male mice-negative; female mice-negative

ENVIRONMENTAL LISTS

List of Pesticide Product Inert Ingredients
[present]

INTERNATIONAL LISTS

Canada - WHMIS: Ingredient Disclosure
1% item 21 (1286)

ACID RED 87 17372-87-1

HEALTH AND SAFETY LISTS

NIOSH - Selected LD50s and LC50s
Oral, mouse: LD50 = 2344 mg/kg

ENVIRONMENTAL LISTS

List of Pesticide Product Inert Ingredients
[present]

ACID TREATED LIGHT DISTILLATE (PETROLEUM) 64742-14-9

ENVIRONMENTAL LISTS

List of Pesticide Product Inert Ingredients
[present]

ACID TREATED RESIDUAL OIL (PETROLEUM) 64742-17-2

STATE LISTS

Massachusetts Right To Know List
carcinogen; extraordinarily hazardous

ACIFLUORFEN 62476-59-9

ENVIRONMENTAL LISTS

CERCLA/SARA - Section 313 - Emission Reporting
form R reporting required

STATE LISTS

California - Air Bill 2588 Appendix A-II
known or potential carcinogen: 9/90

California - Prop. 65 - Cancer list
carcinogen - initial date 1/1/90

ACRIDINE 260-94-6
SEE ALSO:
COAL TAR PITCHES

HEALTH AND SAFETY LISTS

U.S. DOT - Substances From 49 CFR 172.101
regulated by DOT (UN2713)

U.S. DOT - Hazard Classes
DOT hazard class = 6.1

STATE LISTS

NJ Right to Know List (Total)
sn 2064

ACRIDINE ORANGE 494-38-2

HEALTH AND SAFETY LISTS

IARC - Group 3 (not classifiable)
[present]

ACRIFLAVINIUM CHLORIDE 8048-52-0

HEALTH AND SAFETY LISTS

IARC - Group 3 (not classifiable)
[present]

ACROLEIN 107-02-8

HEALTH AND SAFETY LISTS

ACGIH 1995 - Time Weighted Averages
0.1 ppm TWA; 0.23 mg/m3 TWA

ACGIH 1995 - Short Term Exposure Limits
0.3 ppm STEL; 0.69 mg/m3 STEL

AIHA - Odor Threshold Values
geometric mean air odor threshold = 1.8 ppm (detectable)

U.S. DOT - Substances From 49 CFR 172.101
regulated by DOT (UN2607, UN1092)

U.S. DOT - Hazard Classes
DOT hazard class = 3

U.S. DOT - Substances Which Are Poisonous by Inhalation
liquid hazardous material poisonous by inhalation (inhibited form) (UN1092)

U.S. DOT - Appendix B - Marine Pollutants
inhibited form: DOT regulated marine pollutant

U.S. DOT - Appendix A Table 1 - Hazardous Substances
final RQ = 1 pound (0.454 kg)

IARC - Group 3 (not classifiable)
[present]

NFPA - Flash Points
flash point = -15 degrees F (-26 degrees C)

NFPA - Hazard Identification Ratings
health-4; flammability-3; reactivity-3

NIOSH - Selected LD50s and LC50s
Inhalation, rat: LC50 = 300 mg/m3 30 mn Oral, rat: LC50 = 25900 ug/kg Skin, rabbit: LD50 = 562 mg/kg

NIOSH 1990 - Pocket Guide - RELs
0.1 ppm TWA; 0.25 mg/m3 TWA; 0.3 ppm STEL; 0.8 mg/m3 STEL

NIOSH 1990 - Pocket Guide - IDLHs
5 ppm IDLH

NIOSH 1990 - Pocket Guide - Target organs
heart, eyes, skin, respiratory system

NTP Chemical Status Reports - Testing Status and NTIS Number
Approved for toxicology/carcinogenesis study

OSHA - Vacated PELs - Time Weighted Averages
0.1 ppm TWA; 0.25 mg/m3 TWA

OSHA - Vacated PELs - Short Term Exposure Limits
0.3 ppm STEL; 0.8 mg/m3 STEL

OSHA - Final PELs - Time Weighted Averages
0.1 ppm TWA; 0.25 mg/m3 TWA

OSHA - List of Highly Hazardous Chemicals
threshhold quantity = 150 pounds

ENVIRONMENTAL LISTS

ATSDR Priority List
Rank (of 275): 135

CERCLA/SARA - Section 302 Extremely Hazardous Substances and TPQs
TPQ = 500 pounds

CERCLA/SARA - Section 313 - Emission Reporting
form R reporting required for 1.0% de minimus concentration

CERCLA/SARA - Hazardous Substances and their Reportable Quantities
final RQ = 1 pound (0.454 kg)

Clean Air Act (1990) - List of Hazardous Air Contaminants
[present]

CAA -Toxic Substances for Accidental Release Prevention
threshold quantity = 5,000 lbs

CAA - HON Rule - SOCMI Chemicals
compliance by July 24, 1995

CAA - HON Rule - Organic HAPs
[present]

Clean Water Act - Hazardous Substances
[present]

Clean Water Act - Priority Pollutants
[present]

Clean Water Act - Toxic Pollutants
[present]

RCRA - P Series Wastes
waste number P003

RCRA - Hazardous Constituents-Appendix VIII
waste number P003

RCRA - Substances Banned From Land Disposal
[present]

RCRA - TSD Facilities Ground Water Monitoring
TM 8030 = 5 ug/L PQL; TM 8240 = 5 ug/L PQL

RCRA - Universal Treatment Standards (LDR)
WW: 0.29 mg/l; NWW: Not applicable

TSCA - Code of Federal Regulations Citations
40 CFR 712.30(x); 40 CFR 716.120(d)

TSCA - PAIR - Reporting List
Reporting Date: November 27, 1991

TSCA - Health and Safety Reporting List
Effective Date: September 30, 1991

INTERNATIONAL LISTS

Australian Exposure Standards - Time Weighted Averages
0.1 ppm TWA; 0.23 mg/m3 TWA

Australian Exposure Standards - Short Term Exposure Limits
0.3 ppm STEL; 0.69 mg/m3 STEL

Canada - WHMIS: Ingredient Disclosure
1% item 22 (151)

Canada - Alberta - 8 Hour Occupational Exposure Limit
0.1 ppm TWA; 0.23 mg/m3 TWA

Canada - Alberta - 15 Minute Occupational Exposure Limit
0.3 ppm STEL; 0.69 mg/m3 STEL

Canada - British Columbia - 8 Hour Exposure Limits
0.1 ppm TWA; 0.25 mg/m3 TWA

Canada - British Columbia - 15 Minute Exposure Limits
0.3 ppm STEL; 0.8 mg/m3 STEL

Canada - Ontario - OHSA - TWAEVs
0.1 ppm TWAEV; 0.23 mg/m3 TWAEV

Canada - Ontario - OHSA - STEVs
0.3 ppm STEV; 0.7 mg/m3 STEV

Canada - Quebec - Time-Weighted Average Exposure Values
0.1 ppm TWAEV; 0.23 mg/m3 TWAEV

Canada - Quebec - Short-term Exposure Values
0.3 ppm STEV; 0.69 mg/m3 STEV

United Kingdom - Occupational Exposure Standards - TWAs
0.1 ppm TWA; 0.25 mg/m3 TWA

United Kingdom - Occupational Exposure Standards - STELs
0.3 ppm STEL; 0.8 mg/m3 STEL

German (DFG) - MAK Values
0.1 ppm MAK; 0.25 mg/m3 MAK

German (DFG) - Peak Limitations
2 x normal MAK (5 min momentary value); don't exceed 8 times during shift

German (DFG) - Pregnancy
classification not yet possible

Israel - Time Weighted Averages
0.1 ppm TWA; 0.23 mg/m3 TWA

Israel - Short Term Exposure Limits
0.3 ppm STEL; 0.69 mg/m3 STEL

Israel - Action Levels
0.05 ppm AL; 0.115 mg/m3 AL

Mexico - Instruction No. 10 - TWAs
0.1 ppm TWA; 0.25 mg/m3 TWA

Mexico - Instruction No. 10 - STELs
0.3 ppm STEL; 0.8 mg/m3 STEL

Mexico - Wastewater - Organic Toxic Pollutants and Heavy Metals
Listed under [Organic Toxic Pollutants]

Mexico - Drinking Water - Ecological Criteria
0.3 mg/l

STATE LISTS

California - Air Bill 2588 Appendix A-I
[present]

California - Exposure Limits - PELs
0.1 ppm PEL; 0.25 mg/m3 PEL

California - Exposure Limits - STELs
0.3 ppm STEL; 0.8 mg/m3 STEL

California - Directors List of Hazardous Substances (8 CCR 339)
[present]

Florida Hazardous Substance List
[present]

Massachusetts Right To Know List
extraordinarily hazardous

Minnesota Hazardous Substance List
[present]

NJ Right to Know List (Total)
sn 0021
NJ Special Hazardous Substances
(flammable - third degree; mutagen; reactive - second degree)
Pennsylvania Right to Know List
environmental hazard

ACROLEIN DIMER   100-73-2

HEALTH AND SAFETY LISTS

NFPA - Flash Points
flash point = 118 degrees F (48 degrees C)
NFPA - Hazard Identification Ratings
health-1; flammability-2; reactivity-1

STATE LISTS

Florida Hazardous Substance List
[present]
Massachusetts Right To Know List
[present]
Pennsylvania Right to Know List
[present]

ACRONYCINE   7008-42-6

HEALTH AND SAFETY LISTS

NTP Chemical Status Reports - Testing Status and NTIS Number
Technical reports printed (PB283347/AS)
NTP Chemical Status Reports - Evidence of Carcinogenicity
male rat-positive; female rat-positive; male mice-inadequate; female mice-inadequate

STATE LISTS

Massachusetts Right To Know List
carcinogen; extraordinarily hazardous

ACRYLAMIDE   79-06-1
SEE ALSO:
K014-HAZARDOUS WASTES

HEALTH AND SAFETY LISTS

ACGIH 1995 - Time Weighted Averages
0.03 mg/m3 TWA
ACGIH 1995 - Skin Designations
skin - potential for cutaneous absorption
ACGIH 1995 - Carcinogens
A2-suspected human carcinogen
U.S. DOT - Substances From 49 CFR 172.101
regulated by DOT (UN2074)
U.S. DOT - Hazard Classes
DOT hazard class = 6.1
U.S. DOT - Appendix A Table 1 - Hazardous Substances
final RQ = 5000 pounds (2270 kg)
IARC - Group 2B (sufficient animal data)
[present] (Overall evaluation based only on evidence of carcinogenicity in monograph (39, 1986) or in Supplement 4)
NIOSH - Selected LD50s and LC50s
Oral, rat: LD50 = 124 mg/kg Skin, rat: LD50 = 400 mg/kg
NIOSH 1990 - Pocket Guide - RELs
0.03 mg/m3 TWA
NIOSH 1990 - Pocket Guide - Carcinogens
occupational carcinogen
NIOSH 1990 - Pocket Guide - Target organs
CNS, PNS, skin, eyes
NIOSH 1990 - Pocket Guide - Skin list
Potential for dermal absorption
NIOSH - Health Standards - Exposure Limits
0.3 mg/m3 TWA

NIOSH - Health Standards - Health Effects and Precautions
Skin, eye, and nervous system effects (Prevent skin and eye contact)
NTP Seventh Report - Suspect Carcinogens
suspect carcinogen
OSHA - Vacated PELs - Time Weighted Averages
0.03 mg/m3 TWA
OSHA - Vacated PELs - Skin Designation
Prevent or reduce skin absorption
OSHA - Final PELs - Time Weighted Averages
0.3 mg/m3 TWA
OSHA - Final PELs - Skin Notations
prevent or reduce skin absorption
OSHA - Possible Select Carcinogens
[present]

ENVIRONMENTAL LISTS

CERCLA/SARA - Section 302 Extremely Hazardous Substances and TPQs
TPQ = 1000/10,000 pounds
CERCLA/SARA - Section 313 - Emission Reporting
form R reporting required for 0.1% de minimus concentration
CERCLA/SARA - Hazardous Substances and their Reportable Quantities
final RQ = 5000 pounds (2270 kg)
Clean Air Act (1990) - List of Hazardous Air Contaminants
[present]
CAA - HON Rule - SOCMI Chemicals
compliance by Oct. 24, 1994
CAA - HON Rule - Organic HAPs
[present]
Safe Drinking Water Act - MCLs
MCL = treatment technique (polymer addition practices)
Safe Drinking Water Act - MCLGs
MCLG = Zero
RCRA - U Series Wastes
waste number U007
RCRA - Hazardous Constituents-Appendix VIII
waste number U007
RCRA - Basis for Listing - Appendix VII
Included in waste stream: K014
RCRA - Substances Banned From Land Disposal
[present]
RCRA - Universal Treatment Standards (LDR)
WW: 19 mg/l; NWW: 23 mg/kg
TSCA - Code of Federal Regulations Citations
40 CFR 712.30(d); 40 CFR 716.120(a)
TSCA - PAIR - Reporting List
Reporting Date: November 19, 1982
TSCA - Health and Safety Reporting List
Effective Date: October 4, 1982

INTERNATIONAL LISTS

Australian Exposure Standards - Time Weighted Averages
0.03 mg/m3 TWA
Australian Exposure Standards - Skin Effects
skin absorption
Australian Exposure Standards - Carcinogens
probable carcinogen
Canada - WHMIS: Ingredient Disclosure
0.1% item 23 (152)
Canada - NPRI (National Pollutant Release Inventory)
[present]
Canada - Alberta - 8 Hour Occupational Exposure Limit
0.3 mg/m3 TWA

Canada - Alberta - 15 Minute Occupational Exposure Limit
0.6 mg/m3 STEL
Canada - Alberta - Skin Designation
can be absorbed through the intact skin
Canada - British Columbia - 8 Hour Exposure Limits
0.3 mg/m3 TWA
Canada - British Columbia - 15 Minute Exposure Limits
0.6 mg/m3 STEL
Canada - British Columbia - Skin Notations
skin - potential for skin absorption
Canada - Ontario - OHSA - TWAEVs
0.03 mg/m3 TWAEV
Canada - Ontario - OHSA - Skin Notations
absorption through skin, eyes, or mucous membranes
Canada - Quebec - Time-Weighted Average Exposure Values
0.03 mg/m3 TWAEV
Canada - Quebec - Skin Designations
absorbed through the skin
Canada - Quebec - Carcinogens
C2 carcinogen: effect suspected in humans
United Kingdom - Maximum Exposure Limits - TWAs
0.3 mg/m3 TWA
United Kingdom - Maximum Exposure Limits - Notes
can be absorbed through skin
German (DFG) - Skin/Sensitizers
danger of cutaneous absorption
German (DFG) - Carcinogens
animal evidence of carcinogenicity
Israel - Time Weighted Averages
0.03 mg/m3 TWA
Israel - Action Levels
0.015 mg/m3 AL
Mexico - Instruction No. 10 - TWAs
0.3 mg/m3 TWA
Mexico - Instruction No. 10 - STELs
2.6 mg/m3 STEL
Mexico - Instruction No. 10 - Skin designation
skin - potential for cutaneous absorption

STATE LISTS
California - Air Bill 2588 Appendix A-I
known or potential carcinogen
California - Prop. 65 - Cancer list
carcinogen - initial date 1/1/90
California - Prop. 65 - No Significant Risk Levels
no significant risk level = 0.2 ug/day
California - Exposure Limits - PELs
0.03 mg/m3 PEL
California - Exposure Limits - Skin Notation
material may be absorbed through the skin, eyes or mucous membrane
California - Directors List of Hazardous Substances (8 CCR 339)
[present]
Florida Hazardous Substance List
[present]
Massachusetts Right To Know List
carcinogen; extraordinarily hazardous
Minnesota Hazardous Substance List
carcinogen; skin
NJ Right to Know List (Total)
sn 0022
Pennsylvania Right to Know List
environmental hazard

ACRYLAMIDE-ACRYLIC ACID POLYMER   9003-06-9
ENVIRONMENTAL LISTS
List of Pesticide Product Inert Ingredients
[present]

ACRYLAMIDE, POLYMERS WITH TETRAALKYL AMMONIUM SALT AND POLYALKYL, AMINO ALKYL METHACRYLAMIDE SALT   RR-01205-5
ENVIRONMENTAL LISTS
TSCA - Chemicals with Significant New Use Rules
PMN numbers: P-88-2100; P-88-2169

ACRYLAMIDE, POLYMER WITH SUBSTITUTED ALKYLACRYLAMIDE SALT   RR-00905-2
ENVIRONMENTAL LISTS
TSCA - Chemicals with Significant New Use Rules
PMN number: P-87-794

ACRYLAMIDE-SODIUM ACRYLATE POLYMER   25085-02-3
ENVIRONMENTAL LISTS
List of Pesticide Product Inert Ingredients
[present]

ACRYLAMIDE-SUBSTITUTED EPOXY   RR-01672-8
ENVIRONMENTAL LISTS
TSCA - Chemicals with Significant New Use Rules
PMN number: P-92-660

ACRYLATE ESTERS   RR-01765-2
ENVIRONMENTAL LISTS
TSCA - Section 12(b) - Export Notification
export notification required - Section 5 Acrylate esters are defined as: 1)having a molecular weight of 1,000 amu or less, or, 2)having a molecular weight over 1,000 amu but containing more than 2% by weight acrylate esters with a molecular weight of 500 amu or less.

ACRYLATED EPOXY PHENOLIC RESIN   RR-01669-3
ENVIRONMENTAL LISTS
TSCA - Chemicals with Significant New Use Rules
PMN number: P-92-44

ACRYLATED EPOXY PHENOLIC RESIN   RR-01670-6
ENVIRONMENTAL LISTS
TSCA - Chemicals with Significant New Use Rules
PMN number: P-92-177

ACRYLATES OF ALIPHATIC POLYOL   RR-01575-8
ENVIRONMENTAL LISTS
TSCA - Chemicals with Significant New Use Rules
PMN number: P-91-1077

ACRYLATE SUBSTITUTED SILOXANES AND SILICONES   RR-01258-8
ENVIRONMENTAL LISTS
TSCA - Chemicals with Significant New Use Rules
PMN number: P-91-1153

ACRYLIC ACID   79-10-7
HEALTH AND SAFETY LISTS
ACGIH 1995 - Time Weighted Averages
2 ppm TWA; 5.9 mg/m3 TWA
ACGIH 1995 - Skin Designations
skin - potential for cutaneous absorption

AIHA - Odor Threshold Values
geometric mean air odor threshold = 0.092 ppm (detectable); 1.0 ppm (recognizable)

U.S. DOT - Substances From 49 CFR 172.101
regulated by DOT (UN2218)

U.S. DOT - Hazard Classes
DOT hazard class = 8

U.S. DOT - Appendix A Table 1 - Hazardous Substances
final RQ = 5000 pounds (2270 kg)

IARC - Group 3 (not classifiable)
[present]

NFPA - Flash Points
flash point = 122 degrees F (50 degrees C)

NFPA - Hazard Identification Ratings
health-3; flammability-2; reactivity-2

NIOSH - Selected LD50s and LC50s
Inhalation, mouse: LC50 = 5300 mg/m3 (2 hr) Oral, rat: LD50 = 250 mg/kg Skin, rabbit: LD50 = 280 mg/kg

OSHA - Vacated PELs - Time Weighted Averages
10 ppm TWA; 30 mg/m3 TWA

OSHA - Vacated PELs - Skin Designation
Prevent or reduce skin absorption

ENVIRONMENTAL LISTS

CERCLA/SARA - Section 313 - Emission Reporting
form R reporting required for 1.0% de minimus concentration

CERCLA/SARA - Hazardous Substances and their Reportable Quantities
final RQ = 5000 pounds (2270 kg)

Clean Air Act (1990) - List of Hazardous Air Contaminants
[present]

CAA - HON Rule - SOCMI Chemicals
compliance by July 24, 1995

CAA - HON Rule - Organic HAPs
[present]

EPA - Master Testing List
[present]

RCRA - U Series Wastes
waste number U008 (Ignitable waste)

RCRA - Hazardous Constituents-Appendix VIII
waste number U008 (Ignitable waste)

RCRA - Substances Banned From Land Disposal
[present]

TSCA - Code of Federal Regulations Citations
40 CFR 799.5000

TSCA - Substances Subject to Testing Consent Orders
Test for: Health Effects

TSCA - Section 12(b) - Export Notification
export notification required - Section 4

INTERNATIONAL LISTS

Australian Exposure Standards - Time Weighted Averages
2 ppm TWA; 5.9 mg/m3 TWA

Australian Exposure Standards - Skin Effects
skin absorption

Canada - WHMIS: Ingredient Disclosure
1% item 24 (52)

Canada - NPRI (National Pollutant Release Inventory)
[present]

Canada - Alberta - 8 Hour Occupational Exposure Limit
10 ppm TWA; 30 mg/m3 TWA

Canada - Alberta - 15 Minute Occupational Exposure Limit
20 ppm STEL; 60 mg/m3 STEL

Canada - Ontario - OHSA - TWAEVs
10 ppm TWAEV; 29 mg/m3 TWAEV

Canada - Quebec - Time-Weighted Average Exposure Values
10 ppm TWAEV; 29 mg/m3 TWAEV

Canada - Quebec - Skin Designations
absorbed through the skin

United Kingdom - Occupational Exposure Standards - TWAs
10 ppm TWA; 30 mg/m3 TWA

United Kingdom - Occupational Exposure Standards - STELs
20 ppm STEL; 60 mg/m3 STEL

Israel - Time Weighted Averages
2 ppm TWA; 5.9 mg/m3 TWA

Israel - Action Levels
1 ppm AL; 2.95 mg/m3 AL

STATE LISTS

California - Air Bill 2588 Appendix A-I
6/91

California - Exposure Limits - PELs
10 ppm PEL; 30 mg/m3 PEL

California - Exposure Limits - Skin Notation
material may be absorbed through the skin, eyes or mucous membrane

California - Directors List of Hazardous Substances (8 CCR 339)
[present]

Florida Hazardous Substance List
[present] (includes the Glacial form)

Massachusetts Right To Know List
[present]

Minnesota Hazardous Substance List
[present]

NJ Right to Know List (Total)
sn 0023

NJ Special Hazardous Substances
(corrosive; reactive - second degree)

Pennsylvania Right to Know List
environmental hazard

ACRYLIC ACID, BUTYL ESTER, POLYMER WITH ETHYLENE   25750-84-9

ENVIRONMENTAL LISTS

List of Pesticide Product Inert Ingredients
[present]

ACRYLIC ACID METHYL ESTER, POLYMER WITH ACRYLONITRILE AND 1,3-BUTADIENE   24968-79-4

ENVIRONMENTAL LISTS

List of Pesticide Product Inert Ingredients
[present]

ACRYLIC ACID POLYMER WITH ETHYLENE   9010-77-9

ENVIRONMENTAL LISTS

List of Pesticide Product Inert Ingredients
[present]

ACRYLIC ACID, POLYMER WITH SUBSTITUTED ETHENE   RR-01217-9

ENVIRONMENTAL LISTS

TSCA - Chemicals with Significant New Use Rules
PMN number: P-91-521

ACRYLIC ACID-SUCROSE POLYALLYL ETHER POLYMER   9007-16-3

HEALTH AND SAFETY LISTS

NIOSH - Selected LD50s and LC50s
Oral, rat: LD50 = 4100 mg/kg

ACRYLIC FIBRES   RR-01524-7

HEALTH AND SAFETY LISTS

IARC - Group 3 (not classifiable)
[present]

ACRYLIC RESIN   9003-01-4

HEALTH AND SAFETY LISTS

IARC - Group 3 (not classifiable)
[present]

ENVIRONMENTAL LISTS

List of Pesticide Product Inert Ingredients
[present]

ACRYLONITRILE   107-13-1
SEE ALSO:
K011-HAZARDOUS WASTES
F039-HAZARDOUS WASTES
K013-HAZARDOUS WASTES

HEALTH AND SAFETY LISTS

ACGIH 1995 - Time Weighted Averages
2 ppm TWA; 4.3 mg/m3 TWA

ACGIH 1995 - Skin Designations
skin - potential for cutaneous absorption

ACGIH 1995 - Carcinogens
A2-suspected human carcinogen

AIHA - Odor Threshold Values
geometric mean air odor threshold = 1.6 ppm (detectable)

U.S. DOT - Substances From 49 CFR 172.101
regulated by DOT (UN1093)

U.S. DOT - Hazard Classes
DOT hazard class = 3

U.S. DOT - Appendix A Table 1 - Hazardous Substances
final RQ = 100 pounds (45.4 kg)

IARC - Group 2A (limited human data)
[present]

NFPA - Flash Points
flash point = 32 degrees F (0 degrees C)

NFPA - Hazard Identification Ratings
health-4; flammability-3; reactivity-2

NIOSH - Selected LD50s and LC50s
Oral, rat: LD50 = 78 mg/kg Skin, rat: LD50 = 148 mg/kg

NIOSH 1990 - Pocket Guide - RELs
1 ppm TWA; C 10 ppm (15 min)

NIOSH 1990 - Pocket Guide - IDLHs
500 ppm IDLH (not considering carcinogenic effects)

NIOSH 1990 - Pocket Guide - Carcinogens
occupational carcinogen

NIOSH 1990 - Pocket Guide - Target organs
CVS, liver, kidneys, CNS, skin, brain tumor, lung and bowel cancer

NIOSH 1990 - Pocket Guide - Skin list
Potential for dermal absorption

NIOSH - Health Standards - Exposure Limits
(skin) 1 ppm TWA (8 hr); C (15 min) 10 ppm

NIOSH - Health Standards - Health Effects and Precautions
Brain tumors, lung and bowel cancer (Periodic chest X-ray required; make first-aid and medical kits available during use; prevent skin contact)

NIOSH - Health Standards - Carcinogenic Chemicals
potential human carcinogen

NTP Chemical Status Reports - Testing Status and NTIS Number
Approved for toxicology/carcinogenesis study

NTP Seventh Report - Suspect Carcinogens
suspect carcinogen

OSHA - 29 CFR 1910 Specifically Regulated Chemicals
2 ppm TWA PEL; C 10 ppm (15 min); 1 ppm TWA action level; skin and eye exposure prohibited; cancer hazard (see 29 CFR 1910.1045)

OSHA - Select Carcinogens
[present]

OSHA - Possible Select Carcinogens
[present]

ENVIRONMENTAL LISTS

CERCLA/SARA - Section 302 Extremely Hazardous Substances and TPQs
TPQ = 10,000 pounds

CERCLA/SARA - Section 313 - Emission Reporting
form R reporting required for 0.1% de minimus concentration

CERCLA/SARA - Hazardous Substances and their Reportable Quantities
final RQ = 100 pounds (45.4 kg)

Clean Air Act (1990) - List of Hazardous Air Contaminants
[present]

CAA -Toxic Substances for Accidental Release Prevention
threshold quantity = 20,000 lbs

CAA - HON Rule - SOCMI Chemicals
compliance by Oct. 24, 1994

CAA - HON Rule - Organic HAPs
[present]

Clean Water Act - Hazardous Substances
[present]

Clean Water Act - Priority Pollutants
[present]

Clean Water Act - Toxic Pollutants
[present]

EPA - Carcinogen Hazard Ranking for RQ Adjustment
Hazard ranking = Medium

EPA - Master Testing List
[present]

RCRA - U Series Wastes
waste number U009

RCRA - Hazardous Constituents-Appendix VIII
waste number U009

RCRA - Basis for Listing - Appendix VII
Included in waste streams: F039, K011, K013

RCRA - Substances Banned From Land Disposal
[present]

RCRA - TSD Facilities Ground Water Monitoring
TM 8030 = 5 ug/L PQL; TM 8240 = 5 ug/L PQL

RCRA - Universal Treatment Standards (LDR)
WW: 0.24 mg/l; NWW: 84 mg/kg

INTERNATIONAL LISTS

Australian Exposure Standards - Time Weighted Averages
2 ppm TWA; 4.3 mg/m3 TWA

Australian Exposure Standards - Skin Effects
skin absorption

Australian Exposure Standards - Carcinogens
probable carcinogen

Canada - WHMIS: Ingredient Disclosure
0.1% item 25 (162)

Canada - NPRI (National Pollutant Release Inventory)
[present]

Canada - Alberta - 8 Hour Occupational Exposure Limit
2 ppm TWA; 4.3 mg/m3 TWA

Canada - Alberta - 15 Minute Occupational Exposure Limit
4 ppm STEL; 8.6 mg/m3 STEL

Canada - Alberta - Skin Designation
can be absorbed through the intact skin

Canada - Alberta - Designated Substances
designated substance - requires code of practice

Canada - British Columbia - 8 Hour Exposure Limits
20 ppm TWA;   45 mg/m3 TWA

Canada - British Columbia - 15 Minute Exposure Limits
30 ppm STEL;   65 mg/m3 STEL

Canada - British Columbia - Skin Notations
skin - potential for skin absorption

Canada - British Columbia - Carcinogens
carcinogen - 20 ppm TWA;   45 mg/m3 TWA

Canada - Ontario - OHSA - TWAEVs
2 ppm TWAEV;   4.3 mg/m3 TWAEV (designated subsatnce regulation) (These values apply to work places in which the designated substance regulation does not apply.)

Canada - Ontario - OHSA - CEVs
10 ppm CEV;   21.7 mg/m3 TWAEV (designated substance regulation)

Canada - Ontario - OHSA - Designated Substances
2 ppm TWAEV;   4.3 mg/m3 TWAEV;   See Ontario Reg. 835 for full information.

Canada - Quebec - Time-Weighted Average Exposure Values
2 ppm TWAEV;   4.3 mg/m3 TWAEV (substance of which the recirculation is prohibited)

Canada - Quebec - Skin Designations
absorbed through the skin

Canada - Quebec - Carcinogens
C2 carcinogen: effect suspected in humans

United Kingdom - Maximum Exposure Limits - TWAs
2 ppm TWA;   4 mg/m3 TWA

United Kingdom - Maximum Exposure Limits - Notes
can be absorbed through skin

German (DFG) - Skin/Sensitizers
danger of cutaneous absorption

German (DFG) - Carcinogens
animal evidence of carcinogenicity

Israel - Time Weighted Averages
2 ppm TWA;   4.3 mg/m3 TWA

Israel - Action Levels
1 ppm AL;   2.15 mg/m3 AL

Mexico - Instruction No. 10 - TWAs
2 ppm TWA;   5.4 mg/m3 TWA

Mexico - Instruction No. 10 - Skin designation
skin - potential for cutaneous absorption

Mexico - Instruction No. 10 - Carcinogens
potential carcinogen in humans - limited epidemiological evidence

Mexico - Wastewater - Organic Toxic Pollutants and Heavy Metals
Listed under [Organic Toxic Pollutants]

Mexico - Drinking Water - Ecological Criteria
0.0006 mg/l Substance presents persistence, bioaccumulations or risk of cancer, reduce human exposure to a minimum;   This level has been extrapolated by using a mathematic model

STATE LISTS

California - Air Bill 2588 Appendix A-I
known or potential carcinogen

California - Prop. 65 - Cancer list
carcinogen - initial date 7/1/87

California - Prop. 65 - No Significant Risk Levels
no significant risk level = 0.7 ug/day

California - Exposure Limits - PELs
2 ppm PEL;   4.5 mg/m3 PEL;   1 ppm Action Level;   avoid eye and skin contact;   see also section 5213

California - Exposure Limits - STELs
10 ppm STEL

California - Exposure Limits - Skin Notation
material may be absorbed through the skin, eyes or mucous membrane

California - Directors List of Hazardous Substances (8 CCR 339)
[present]

Florida Hazardous Substance List
[present]

Massachusetts Right To Know List
carcinogen;   extraordinarily hazardous

Minnesota Hazardous Substance List
carcinogen;   skin

NJ Right to Know List (Total)
sn 0024

NJ Special Hazardous Substances
(carcinogen;   flammable - third degree;   mutagen;   reactive - second degree)

Pennsylvania Right to Know List
environmental hazard;   special hazardous substance

Pennsylvania RTK - Special Hazardous Substances
[present]

PROPOSED REGULATIONS

Safe Drinking Water Act - Priority list
[present]

TSCA - Proposed Substances for Developmental/Reproductive Testing
proposed testing for: Developmental Toxicity - oral

ACRYLONITRILE POLYMER   25014-41-9

ENVIRONMENTAL LISTS

List of Pesticide Product Inert Ingredients
[present]

ACRYLYL CHLORIDE   814-68-6

HEALTH AND SAFETY LISTS

NIOSH - Selected LD50s and LC50s
Inhalation, mouse: LC50 = 92 mg/m3 2 hr

OSHA - List of Highly Hazardous Chemicals
threshhold quantity = 250 pounds

ENVIRONMENTAL LISTS

CERCLA/SARA - Section 302 Extremely Hazardous Substances and TPQs
TPQ = 100 pounds

CAA -Toxic Substances for Accidental Release Prevention
threshold quantity = 5,000 lbs

STATE LISTS

Florida Hazardous Substance List
effective March 13, 1992

Massachusetts Right To Know List
extraordinarily hazardous

Pennsylvania Right to Know List
environmental hazard

PROPOSED REGULATIONS

CERCLA/SARA - Proposed Hazardous Substance Additions
proposed RQ = 1 pound (.454 kg)

CERCLA/SARA - 1989 Proposed RQ Adjustments
proposed RQ = 100 pounds (45.4 kg)

ACTINIUM 224   15755-98-3

HEALTH AND SAFETY LISTS

U.S. DOT - Appendix A Table 2 - Radionuclides
final RQ = 100 curies (3.7E 12 Bq)

ENVIRONMENTAL LISTS

CERCLA/SARA List of Radionuclides (Appendix B) and Their Reportable Quantities
final RQ = 100 curies (3.7E 12 Bq)

ACTINIUM 225   14265-85-1

HEALTH AND SAFETY LISTS

U.S. DOT - Appendix A Table 2 - Radionuclides
final RQ = 1 curie (3.7E 10 Bq)

ENVIRONMENTAL LISTS

CERCLA/SARA List of Radionuclides (Appendix B) and Their Reportable Quantities
final RQ = 1 curie (3.7E 10 Bq)

ACTINIUM 226   20379-10-6

HEALTH AND SAFETY LISTS

U.S. DOT - Appendix A Table 2 - Radionuclides
final RQ = 10 curies (3.7E 11 Bq)

ENVIRONMENTAL LISTS

CERCLA/SARA List of Radionuclides (Appendix B) and Their Reportable Quantities
final RQ = 10 curies (3.7E 11 Bq)

ACTINIUM 227   14952-40-0

HEALTH AND SAFETY LISTS

U.S. DOT - Appendix A Table 2 - Radionuclides
final RQ = 0.001 curies (3.7E 7 Bq)

ENVIRONMENTAL LISTS

CERCLA/SARA List of Radionuclides (Appendix B) and Their Reportable Quantities
final RQ = 0.001 curies (3.7E 7 Bq)

ACTINIUM 228   14331-83-0

HEALTH AND SAFETY LISTS

U.S. DOT - Appendix A Table 2 - Radionuclides
final RQ = 10 curies (3.7E 11 Bq)

ENVIRONMENTAL LISTS

CERCLA/SARA List of Radionuclides (Appendix B) and Their Reportable Quantities
final RQ = 10 curies (3.7E 11 Bq)

ACTINOLITE   12172-67-7

INTERNATIONAL LISTS

Canada - Quebec - Carcinogens
C1 carcinogen: effect detected in humans

ACTINOLITE, NON-ASBESTIFORM   13768-00-8

INTERNATIONAL LISTS

Canada - Quebec - Time-Weighted Average Exposure Values
1 fibre/cm3 TWAEV

Canada - Quebec - Short-term Exposure Values
35 fibres/cm3 STEV

ACTINOMYCIN D   50-76-0

HEALTH AND SAFETY LISTS

IARC - Group 3 (not classifiable)
[present]

NIOSH - Selected LD50s and LC50s
Oral, rat: LD50 = 7200 ug/kg

NTP Chemical Status Reports - Testing Status and NTIS Number
Chronic studies exist for which technical reports were not prepared

STATE LISTS

California - Air Bill 2588 Appendix A-II
known or potential carcinogen: 9/90

California - Prop. 65 - Cancer list
carcinogen - initial date 10/1/89

California - Prop. 65 - Developmental Toxicity
developmental toxicity - initial date 10/1/92

California - Prop. 65 - No Significant Risk Levels
no significant risk level = 0.00008 ug/day

California - Directors List of Hazardous Substances (8 CCR 339)
[present]

Florida Hazardous Substance List
[present]

Massachusetts Right To Know List
carcinogen; extraordinarily hazardous

Minnesota Hazardous Substance List
carcinogen

NJ Right to Know List (Total)
sn 0025

NJ Special Hazardous Substances
(carcinogen; mutagen; teratogen)

Pennsylvania Right to Know List
special hazardous substance

Pennsylvania RTK - Special Hazardous Substances
[present]

ACTIVATED ERGOSTEROL   1406-16-2

ENVIRONMENTAL LISTS

List of Pesticide Product Inert Ingredients
[present]

ADENINE (6-AMINOPURINE)   73-24-5

HEALTH AND SAFETY LISTS

NIOSH - Selected LD50s and LC50s
Oral, rat: LD50 = 745 mg/kg

ADHESIVES   RR-01322-9

HEALTH AND SAFETY LISTS

U.S. DOT - Substances From 49 CFR 172.101
regulated by DOT (UN1133)

U.S. DOT - Hazard Classes
DOT hazard class = 3

STATE LISTS

NJ Right to Know List (Total)
sn 3072;   containing flammable liquid: sn 2067

ADIPAMIDE   628-94-4

HEALTH AND SAFETY LISTS

NIOSH - Selected LD50s and LC50s
Oral, mouse: LD50 = 6000 mg/kg

ADIPIC ACID   124-04-9

HEALTH AND SAFETY LISTS

ACGIH 1995 - Time Weighted Averages
5 mg/m3 TWA

U.S. DOT - Appendix A Table 1 - Hazardous Substances
final RQ = 5000 pounds (2270 kg)

NFPA - Flash Points
flash point = 385 degrees F (196 degrees F)

NFPA - Hazard Identification Ratings
flammability-1;   reactivity-0

NIOSH - Selected LD50s and LC50s
Oral, rat: LD50 = 3600 mg/kg

ENVIRONMENTAL LISTS

CERCLA/SARA - Hazardous Substances and their Reportable Quantities
final RQ = 5000 pounds (2270 kg)

Clean Water Act - Hazardous Substances
[present]

List of Pesticide Product Inert Ingredients
[present]

INTERNATIONAL LISTS

Canada - WHMIS: Ingredient Disclosure
1% item 26 (53)

STATE LISTS

California - Directors List of Hazardous Substances (8 CCR 339)
[present]

Massachusetts Right To Know List
[present]

NJ Right to Know List (Total)
sn 0026

Pennsylvania Right to Know List
environmental hazard

ADIPIC ACID, DIOCTYL ESTER   123-79-5

ENVIRONMENTAL LISTS

List of Pesticide Product Inert Ingredients
[present]

ADIPIC ACID MONOMETHYL ESTER   626-86-8

HEALTH AND SAFETY LISTS

NIOSH - Selected LD50s and LC50s
Oral, mammal: LD50 = 4100 mg/kg

ADIPONITRILE   111-69-3

HEALTH AND SAFETY LISTS

ACGIH 1995 - Time Weighted Averages
2 ppm TWA; 8.8 mg/m3 TWA

ACGIH 1995 - Skin Designations
skin - potential for cutaneous absorption

U.S. DOT - Substances From 49 CFR 172.101
regulated by DOT (UN2205)

U.S. DOT - Hazard Classes
DOT hazard class = 6.1

NFPA - Flash Points
flash point = 200 degrees F (93 degrees C)

NFPA - Hazard Identification Ratings
*health-2; flammability-2; reactivity-1; *See list description*

NIOSH - Selected LD50s and LC50s
Inhalation, rat: LC50 = 1710 mg/m3 4 hr Oral, rat: LD50 = 155 mg/kg

NIOSH - Health Standards - Exposure Limits
4 ppm TWA; 18 mg/m3 TWA (Listed under 'Nitriles')

NIOSH - Health Standards - Health Effects and Precautions
Hepatic, renal, respiratory, cardiovascular, gastrointestinal, and nervous system effects (Periodic chest X-ray and pulmonary function testing required; prevent skin and eye contact; make first-aid kits and personnel available during use) (Listed under 'Nitriles')

ENVIRONMENTAL LISTS

CERCLA/SARA - Section 302 Extremely Hazardous Substances and TPQs
TPQ = 1000 pounds

CAA - HON Rule - SOCMI Chemicals
compliance by Oct. 24, 1994

EPA - Master Testing List
[present]

TSCA - Code of Federal Regulations Citations
40 CFR 716.120(a)

TSCA - Health and Safety Reporting List
Effective Date: June 1, 1987

INTERNATIONAL LISTS

Canada - WHMIS: Ingredient Disclosure
1% item 27 (166)

STATE LISTS

Florida Hazardous Substance List
[present]

Massachusetts Right To Know List
extraordinarily hazardous

NJ Right to Know List (Total)
sn 0027

Pennsylvania Right to Know List
environmental hazard

PROPOSED REGULATIONS

CERCLA/SARA - Proposed Hazardous Substance Additions
proposed RQ = 1 pound (.454 kg)

CERCLA/SARA - 1989 Proposed RQ Adjustments
proposed RQ = 1000 pounds (454 kg)

ADIPOYL CHLORIDE   111-50-2

HEALTH AND SAFETY LISTS

NFPA - Flash Points
flash point = 162 degrees F (72 degrees C)

NFPA - Hazard Identification Ratings
health-2; flammability-2; reactivity-0

STATE LISTS

Florida Hazardous Substance List
[present]

Massachusetts Right To Know List
[present]

Pennsylvania Right to Know List
[present]

ADRIAMYCIN   23214-92-8

HEALTH AND SAFETY LISTS

IARC - Group 2A (limited human data)
[present] (Other relevant data, as given in Supplement 7, influenced the making of the overall evaluation)

NIOSH - Selected LD50s and LC50s
Oral, mouse: LD50 = 570 mg/kg

NTP Seventh Report - Suspect Carcinogens
suspect carcinogen

OSHA - Possible Select Carcinogens
[present]

STATE LISTS

California - Air Bill 2588 Appendix A-II
known or potential carcinogen

California - Prop. 65 - Cancer list
carcinogen - initial date 7/1/87

California - Directors List of Hazardous Substances (8 CCR 339)
[present]

Florida Hazardous Substance List
[present]

Massachusetts Right To Know List
carcinogen; extraordinarily hazardous

Minnesota Hazardous Substance List
carcinogen

NJ Right to Know List (Total)
sn 0028

NJ Special Hazardous Substances
(carcinogen; mutagen)

Pennsylvania Right to Know List
special hazardous substance

Pennsylvania RTK - Special Hazardous Substances
[present]

AEROSOL DISPENSERS   **RR-01514-5**

STATE LISTS

NJ Right to Know List (Total)
sn 2068

AEROSOLS   **RR-01323-0**

HEALTH AND SAFETY LISTS

U.S. DOT - Substances From 49 CFR 172.101
regulated by DOT (UN1950)

U.S. DOT - Hazard Classes
DOT hazard class = 2.2

AF 2(FOAMING AGENT)   **51004-61-6**

STATE LISTS

Pennsylvania Right to Know List
special hazardous substance

Pennsylvania RTK - Special Hazardous Substances
[present]

AFLATOXIN B1   **1162-65-8**

HEALTH AND SAFETY LISTS

IARC - Group Unspecified
[present] (Listed under 'Aflatoxins, naturally occurring mixtures of')

NIOSH - Selected LD50s and LC50s
Oral, rat: LD50 = 5 mg/kg

AFLATOXIN B2   **7220-81-7**

HEALTH AND SAFETY LISTS

IARC - Group Unspecified
[present] (Listed under 'Aflatoxins, naturally occurring mixtures of')

NIOSH - Selected LD50s and LC50s
Oral, duck: LD50 = 1700 ug/kg

STATE LISTS

NJ Special Hazardous Substances
(carcinogen; mutagen)

AFLATOXIN G1   **1165-39-5**

HEALTH AND SAFETY LISTS

IARC - Group Unspecified
[present] (Listed under 'Aflatoxins, naturally occurring mixtures of')

NIOSH - Selected LD50s and LC50s
Oral, duck: LD50 = 785 ug/kg

STATE LISTS

NJ Special Hazardous Substances
(carcinogen; mutagen)

AFLATOXIN G2   **7241-98-7**

HEALTH AND SAFETY LISTS

IARC - Group Unspecified
[present] (Listed under 'Aflatoxins, naturally occurring mixtures of')

AFLATOXIN M1   **6795-23-9**

HEALTH AND SAFETY LISTS

IARC - Group 2B (sufficient animal data)
[present]

OSHA - Possible Select Carcinogens
[present]

AFLATOXINS   **1402-68-2**

HEALTH AND SAFETY LISTS

IARC - Group 1 (carcinogenic to humans)
[present]

NIOSH - Selected LD50s and LC50s
Oral, monkey: LD50 = 1750 ug/kg

NTP Seventh Report - Known Carcinogens
known carcinogen

OSHA - Select Carcinogens
[present]

ENVIRONMENTAL LISTS

RCRA - Hazardous Constituents-Appendix VIII
hazardous constituent - no waste number

STATE LISTS

California - Air Bill 2588 Appendix A-II
known or potential carcinogen

California - Prop. 65 - Cancer list
carcinogen - initial date 1/1/88

California - Directors List of Hazardous Substances (8 CCR 339)
[present]

Massachusetts Right To Know List
carcinogen; extraordinarily hazardous

Minnesota Hazardous Substance List
carcinogen

Pennsylvania Right to Know List
environmental hazard; special hazardous substance

Pennsylvania RTK - Special Hazardous Substances
[present]

AGAR   **9002-18-0**

HEALTH AND SAFETY LISTS

NIOSH - Selected LD50s and LC50s
Oral, rat: LD50 = 11 gm/kg

NTP Chemical Status Reports - Testing Status and NTIS Number
Technical reports printed (PB82227588)

NTP Chemical Status Reports - Evidence of Carcinogenicity
male rat-negative; female rat-negative; male mice-negative; female mice-negative

ENVIRONMENTAL LISTS

List of Pesticide Product Inert Ingredients
[present]

AGARITINE   **2757-90-6**

HEALTH AND SAFETY LISTS

IARC - Group 3 (not classifiable)
[present]

NTP Chemical Status Reports - Testing Status and NTIS Number
Chronic studies exist for which technical reports were not prepared

AIR   **RR-01325-2**

HEALTH AND SAFETY LISTS

U.S. DOT - Substances From 49 CFR 172.101
regulated by DOT (UN1002, UN1003)

U.S. DOT - Hazard Classes
DOT hazard class = 2.2

STATE LISTS

NJ Right to Know List (Total)
compressed: sn 2070; refrigerated liquid: sn 2076

ALACHLOR   **15972-60-8**

HEALTH AND SAFETY LISTS

NIOSH - Selected LD50s and LC50s
Oral, rat: LD50 = 1200 mg/kg Skin, rabbit: LD50 = 3500 mg/kg

ENVIRONMENTAL LISTS

CERCLA/SARA - Section 313 - Emission Reporting
form R reporting required

Safe Drinking Water Act - MCLs
MCL = 0.002 mg/L

Safe Drinking Water Act - MCLGs
MCLG = Zero

INTERNATIONAL LISTS

Canada - CEPA Schedule II Part I - Prohibited Substances (Export)
[present]

STATE LISTS

California - Air Bill 2588 Appendix A-II
known or potential carcinogen: 9/89

California - Prop. 65 - Cancer list
carcinogen - initial date 1/1/89

ALANINE, N-(2-CARBOXYETHYL)-N-ALKYL, SALT   RR-00966-5

ENVIRONMENTAL LISTS

TSCA - Chemicals with Significant New Use Rules
PMN number: P-89-336

ALBUMIN EGG   9006-50-2

HEALTH AND SAFETY LISTS

NIOSH - Selected LD50s and LC50s
Oral, rat: LD50 = 101 mg/kg

ENVIRONMENTAL LISTS

List of Pesticide Product Inert Ingredients
[present} (includes egg shells)

ALCOHOL, ALKALI METAL SALT   RR-00964-3

ENVIRONMENTAL LISTS

TSCA - Chemicals with Significant New Use Rules
PMN number: P-91-151

ALCOHOL C-13 - C-15 POLY(1-3) ETHOXYLATE   RR-01271-5

HEALTH AND SAFETY LISTS

U.S. DOT - Appendix B - Marine Pollutants
DOT regulated marine pollutant

ALCOHOL C-6 - C-17 (SECONDARY)POLY(3-6) ETHOXYLATE   RR-01272-6

HEALTH AND SAFETY LISTS

U.S. DOT - Appendix B - Marine Pollutants
DOT regulated marine pollutant

ALCOHOL, DENATURED   RR-00113-8

HEALTH AND SAFETY LISTS

U.S. DOT - Substances From 49 CFR 172.101
regulated by DOT (NA1987)

U.S. DOT - Hazard Classes
DOT hazard class = 3

NFPA - Flash Points
flash point = 60 degrees F (16 degrees C) (flash point varies for various other government formulas)

NFPA - Hazard Identification Ratings
health-0; flammability-3; reactivity-0

STATE LISTS

NJ Right to Know List (Total)
sn 2080

Pennsylvania Right to Know List
[present]

ALCOHOLS   RR-00111-6

HEALTH AND SAFETY LISTS

U.S. DOT - Substances From 49 CFR 172.101
regulated by DOT (NA3065, UN1986, UN1987)

U.S. DOT - Hazard Classes
DOT hazard class = 3

IARC - Group 1 (carcinogenic to humans)
[present]

OSHA - Select Carcinogens
[present]

STATE LISTS

California - Prop. 65 - Cancer list
carcinogen (when associated with alcohol abuse) - initial date 7/1/88

California - Prop. 65 - Developmental Toxicity
develpomental toxicity - initial date 10/1/87

NJ Right to Know List (Total)
sn 2079; beverage: sn 2078

ALCOHOLS, C6-12, ETHOXYLATED   68439-45-2

ENVIRONMENTAL LISTS

List of Pesticide Product Inert Ingredients
[present]

ALCOHOLS, C8-18, ETHOXYLATED PROPOXYLATED   69013-18-9

ENVIRONMENTAL LISTS

List of Pesticide Product Inert Ingredients
[present]

ALCOHOLS, C8-22, ETHOXYLATED   69013-19-0

ENVIRONMENTAL LISTS

List of Pesticide Product Inert Ingredients
[present]

ALCOHOLS, C9-16, ETHOXYLATED   97043-91-9

ENVIRONMENTAL LISTS

List of Pesticide Product Inert Ingredients
[present]

ALCOHOLS, C10-16, ETHOXYLATED PROPOXYLATED   69227-22-1

ENVIRONMENTAL LISTS

List of Pesticide Product Inert Ingredients
[present]

ALCOHOLS, C11-15-SECONDARY, ETHOXYLATED   68131-40-8

ENVIRONMENTAL LISTS

List of Pesticide Product Inert Ingredients
[present]

ALCOHOLS, C12-14, ETHOXYLATED   68439-50-9

ENVIRONMENTAL LISTS

List of Pesticide Product Inert Ingredients
[present]

ALCOHOLS, C12-18, ETHOXYLATED PROPOXYLATED   69227-21-0

ENVIRONMENTAL LISTS

List of Pesticide Product Inert Ingredients
[present]

ALCOHOLS, C12-20, ETHOXYLATED   68526-94-3

ENVIRONMENTAL LISTS

List of Pesticide Product Inert Ingredients
[present]

ALCOHOLS, C16-18, ETHOXYLATED   68439-49-6

ENVIRONMENTAL LISTS

List of Pesticide Product Inert Ingredients
[present]

ALCOHOLS, C-12 - C-15, ETHOXYLATED   68131-39-5

HEALTH AND SAFETY LISTS

U.S. DOT - Appendix B - Marine Pollutants
DOT regulated marine pollutant

ENVIRONMENTAL LISTS

List of Pesticide Product Inert Ingredients
[present]

ALCOHOLS, C12-18, ETHOXYLATED   68213-23-0

ENVIRONMENTAL LISTS

List of Pesticide Product Inert Ingredients
[present]

ALCOHOLS, TALLOW, ETHOXYLATED   61791-28-4

ENVIRONMENTAL LISTS

List of Pesticide Product Inert Ingredients
[present]

ALDEHYDES   RR-00114-9

HEALTH AND SAFETY LISTS

U.S. DOT - Substances From 49 CFR 172.101
regulated by DOT (UN1988, UN1989)

U.S. DOT - Hazard Classes
DOT hazard class = 3

STATE LISTS

NJ Right to Know List (Total)
sn 2081; toxic: sn 2082

PROPOSED REGULATIONS

TSCA - ITC 33rd Report Priority Testing List
recommended with intent-to-designate

TSCA - ITC 34th Report Priority Testing List
recommended with intent-to-designate

ALDICARB   116-06-3

HEALTH AND SAFETY LISTS

AIHA - WEEL - Time Weighted Averages
0.07 mg/m3 TWA

AIHA - WEEL - Skin Absorption Designations
skin absorber

U.S. DOT - Substances Which Are Poisonous by Inhalation
liquid hazardous material poisonous by inhalation (when mixed with Dichloromethane)

U.S. DOT - Appendix B - Marine Pollutants
DOT regulated marine pollutant

U.S. DOT - Appendix A Table 1 - Hazardous Substances
final RQ = 1 pound (0.454 kg)

IARC - Group 3 (not classifiable)
[present]

NIOSH - Selected LD50s and LC50s
Oral, rat: LD50 = 650 ug/kg Skin, rat: LD50 = 2500 ug/kg

NTP Chemical Status Reports - Testing Status and NTIS Number
Technical reports printed (PB298511/AS)

NTP Chemical Status Reports - Evidence of Carcinogenicity
male rat-negative; female rat-negative; male mice-negative; female mice-negative

ENVIRONMENTAL LISTS

CERCLA/SARA - Section 302 Extremely Hazardous Substances and TPQs
TPQ = 100/10,000 pounds

CERCLA/SARA - Section 313 - Emission Reporting
form R reporting required

CERCLA/SARA - Hazardous Substances and their Reportable Quantities
final RQ = 1 pound (0.454 kg)

Safe Drinking Water Act - MCLs
MCL = 0.003 mg/L

Safe Drinking Water Act - MCLGs
MCLG = 0.001 mg/L

RCRA - P Series Wastes
waste number P070

RCRA - Hazardous Constituents-Appendix VIII
waste number P070

RCRA - Substances Banned From Land Disposal
[present]

INTERNATIONAL LISTS

Canada - Drinking Water Quality - MACs
0.009 mg/L MAC

STATE LISTS

California - Directors List of Hazardous Substances (8 CCR 339)
[present]

Florida Hazardous Substance List
effective March 13, 1992

Massachusetts Right To Know List
extraordinarily hazardous; neurotoxin

NJ Right to Know List (Total)
sn 0031

Pennsylvania Right to Know List
environmental hazard

ALDICARB SULFONE   1646-88-4

ENVIRONMENTAL LISTS

Safe Drinking Water Act - MCLs
MCL = 0.003 mg/L

Safe Drinking Water Act - MCLGs
MCLG = 0.001 mg/L

ALDICARB SULFOXIDE   1646-87-3

ENVIRONMENTAL LISTS

Safe Drinking Water Act - MCLs
MCL = 0.004 mg/L

Safe Drinking Water Act - MCLGs
MCLG = 0.001 mg/L

ALDOL   107-89-1

HEALTH AND SAFETY LISTS

U.S. DOT - Substances From 49 CFR 172.101
regulated by DOT (UN2839)

U.S. DOT - Hazard Classes
DOT hazard class = 6.1

NFPA - Flash Points
flash point = 150 degrees F (66 degrees C)

NFPA - Hazard Identification Ratings
health-3; flammability-2; reactivity-2

NIOSH - Selected LD50s and LC50s
Oral, rat: LD50 = 2180 mg/kg Skin, rabbit: LD50 = 140 mg/kg

ENVIRONMENTAL LISTS

CAA - HON Rule - SOCMI Chemicals
compliance by Jan. 23, 1995

INTERNATIONAL LISTS

Canada - WHMIS: Ingredient Disclosure
1% item 28 (192)

STATE LISTS

Florida Hazardous Substance List
[present]

Massachusetts Right To Know List
[present]

NJ Right to Know List (Total)
sn 0032

Pennsylvania Right to Know List
[present]

ALDRIN   309-00-2
SEE ALSO:
F039-HAZARDOUS WASTES

HEALTH AND SAFETY LISTS

ACGIH 1995 - Time Weighted Averages
0.25 mg/m3 TWA

ACGIH 1995 - Skin Designations
skin - potential for cutaneous absorption

U.S. DOT - Substances From 49 CFR 172.101
regulated by DOT (NA2762, NA2761)

U.S. DOT - Hazard Classes
DOT hazard class = 6.1

U.S. DOT - Appendix B - Marine Pollutants
DOT regulated severe marine pollutant

U.S. DOT - Appendix A Table 1 - Hazardous Substances
final RQ = 1 pound (0.454 kg)

IARC - Group 3 (not classifiable)
[present]

NIOSH - Selected LD50s and LC50s
Oral, rat: LD50 = 39 mg/kg Skin, rat: LD50 = 98 mg/kg

NIOSH 1990 - Pocket Guide - RELs
0.25 mg/m3 TWA

NIOSH 1990 - Pocket Guide - IDLHs
100 mg/m3 IDLH (not considering carcinogenic effects)

NIOSH 1990 - Pocket Guide - Carcinogens
occupational carcinogen

NIOSH 1990 - Pocket Guide - Target organs
cancer, CNS, liver, kidneys, skin

NIOSH 1990 - Pocket Guide - Skin list
Potential for dermal absorption

NIOSH - Health Standards - Exposure Limits
reduce exposure to lowest reliably detectable concentration (Listed under 'Aldrin/dieldrin')

NIOSH - Health Standards - Health Effects and Precautions
has produced tumors of the lungs, liiver, thyroid, and adrenal glands in animals (prevent skin contact) (Listed under 'Aldrin/dieldrin')

NIOSH - Health Standards - Carcinogenic Chemicals
potential human carcinogen (Listed under 'Aldrin/dieldrin')

NTP Chemical Status Reports - Testing Status and NTIS Number
Technical reports printed (PB275666/AS)

NTP Chemical Status Reports - Evidence of Carcinogenicity
male rat-equivocal; female rat-equivocal; male mice-positive; female mice-negative

OSHA - Vacated PELs - Time Weighted Averages
0.25 mg/m3 TWA

OSHA - Vacated PELs - Skin Designation
Prevent or reduce skin absorption

OSHA - Final PELs - Time Weighted Averages
0.25 mg/m3 TWA

OSHA - Final PELs - Skin Notations
prevent or reduce skin absorption

ENVIRONMENTAL LISTS

ATSDR Priority List
Rank (of 275): 027

CERCLA/SARA - Section 302 Extremely Hazardous Substances and TPQs
TPQ = 500/10,000 pounds

CERCLA/SARA - Section 313 - Emission Reporting
form R reporting required for 1.0% de minimus concentration

CERCLA/SARA - Hazardous Substances and their Reportable Quantities
final RQ = 1 pound (0.454 kg)

Clean Water Act - Hazardous Substances
[present]

Clean Water Act - Priority Pollutants
[present]

Clean Water Act - Toxic Pollutants
[present] (Listed under 'Aldrin/dieldrin')

EPA - Carcinogen Hazard Ranking for RQ Adjustment
Hazard ranking = High

RCRA - P Series Wastes
waste number P004

RCRA - Hazardous Constituents-Appendix VIII
waste number P004

RCRA - Basis for Listing - Appendix VII
Included in waste stream: F039

RCRA - Substances Banned From Land Disposal
[present]

RCRA - TSD Facilities Ground Water Monitoring
TM 8080 = 0.05 ug/L PQL; TM 8270 = 10 ug/L PQL

RCRA - Universal Treatment Standards (LDR)
WW: 0.021 mg/l; NWW: 0.066 mg/kg

INTERNATIONAL LISTS

Australian Exposure Standards - Time Weighted Averages
0.25 mg/m3 TWA

Australian Exposure Standards - Skin Effects
skin absorption

Australian Exposure Standards - Carcinogens
suspected carcinogen

Canada - CEPA Schedule II Part II - Toxic Substances (Export)
[present]

Canada - Alberta - 8 Hour Occupational Exposure Limit
0.25 mg/m3 TWA

Canada - Alberta - 15 Minute Occupational Exposure Limit
0.75 mg/m3 STEL

Canada - Alberta - Skin Designation
can be absorbed through the intact skin

Canada - British Columbia - 8 Hour Exposure Limits
0.25 mg/m3 TWA

Canada - British Columbia - 15 Minute Exposure Limits
0.75 mg/m3 STEL

Canada - British Columbia - Skin Notations
skin - potential for skin absorption

Canada - Ontario - OHSA - TWAEVs
0.25 mg/m3 TWAEV

Canada - Ontario - OHSA - Skin Notations
absorption through skin, eyes, or mucous membranes

Canada - Quebec - Time-Weighted Average Exposure Values
0.25 mg/m3 TWAEV

Canada - Quebec - Skin Designations
absorbed through the skin

United Kingdom - Occupational Exposure Standards - TWAs
0.25 mg/m3 TWA

United Kingdom - Occupational Exposure Standards - STELs
0.75 mg/m3 STEL

United Kingdom - Occupational Exposure Standards - Notes
can be absorbed through skin
German (DFG) - MAK Values
total dust: 0.25 ppm MAK
German (DFG) - Peak Limitations
10 x normal MAK (30 min average value);   don't exceed during shift
German (DFG) - Skin/Sensitizers
danger of cutaneous absorption
Israel - Time Weighted Averages
0.25 mg/m3 TWA
Israel - Action Levels
0.125 mg/m3 AL
Mexico - Instruction No. 10 - TWAs
0.25 mg/m3 TWA
Mexico - Instruction No. 10 - STELs
0.75 mg/m3 STEL
Mexico - Instruction No. 10 - Skin designation
skin - potential for cutaneous absorption
Mexico - Wastewater - Organic Toxic Pollutants and Heavy Metals
Listed under [Pesticides and Metabolites]
Mexico - Drinking Water - Ecological Criteria
0.00003 mg/l Substance presents persistence, bioaccumulations or risk of cancer, reduce human exposure to a minimum;   This level has been extrapolated by using a mathematic model

STATE LISTS
California - Air Bill 2588 Appendix A-II
known or potential carcinogen: 9/89
California - Prop. 65 - Cancer list
carcinogen - initial date 7/1/88
California - Prop. 65 - No Significant Risk Levels
no significant risk level = 0.04 ug/day
California - Exposure Limits - PELs
0.25 mg/m3 PEL
California - Exposure Limits - Skin Notation
material may be absorbed through the skin, eyes or mucous membrane
California - Directors List of Hazardous Substances (8 CCR 339)
[present]
Florida Hazardous Substance List
[present]
Massachusetts Right To Know List
carcinogen;   extraordinarily hazardous
Minnesota Hazardous Substance List
skin
NJ Right to Know List (Total)
sn 0033
NJ Special Hazardous Substances
(carcinogen;   teratogen)
Pennsylvania Right to Know List
environmental hazard

ALDRIN AND DIELDRIN   RR-01614-8
INTERNATIONAL LISTS
Canada - Drinking Water Quality - MACs
0.0007 mg/L MAC

ALFALFA   RR-01028-6
ENVIRONMENTAL LISTS
List of Pesticide Product Inert Ingredients
[present]

ALGENIC ACID, SODIUM SALT   9005-38-3
ENVIRONMENTAL LISTS
List of Pesticide Product Inert Ingredients
[present]

ALGIN GUM   9049-05-2
ENVIRONMENTAL LISTS
List of Pesticide Product Inert Ingredients
[present]

ALIPHATIC DICARBOXYLIC ACID SALT   RR-01681-9
ENVIRONMENTAL LISTS
TSCA - Chemicals with Significant New Use Rules
PMN number: P-92-1352

ALIPHATIC DIFUNCTIONAL ACRYLIC ACID ESTER   RR-01680-8
ENVIRONMENTAL LISTS
TSCA - Chemicals with Significant New Use Rules
PMN number: 92-1313

ALIPHATIC DIURETHANE ACRYLATE ESTER   RR-00239-1
ENVIRONMENTAL LISTS
TSCA - Chemicals with Significant New Use Rules
PMN number: P-85-1013

ALIPHATIC ETHER   RR-01733-4
ENVIRONMENTAL LISTS
TSCA - Chemicals with Significant New Use Rules
PMN number: P-93-1381

ALIPHATIC POLYGLYCIDYL ETHER   RR-01019-5
ENVIRONMENTAL LISTS
TSCA - Chemicals with Significant New Use Rules
PMN number: P-89-1036

ALKALI METAL ALLOYS   RR-00116-1
HEALTH AND SAFETY LISTS
U.S. DOT - Substances From 49 CFR 172.101
regulated by DOT (UN1421)
U.S. DOT - Hazard Classes
DOT hazard class = 4.3
STATE LISTS
NJ Right to Know List (Total)
sn 2083

ALKALI METAL AMALGAM, N.O.S.   RR-00117-2
HEALTH AND SAFETY LISTS
U.S. DOT - Substances From 49 CFR 172.101
regulated by DOT (UN1389)
U.S. DOT - Hazard Classes
DOT hazard class = 4.3
STATE LISTS
NJ Right to Know List (Total)
sn 2084

ALKALI METAL AMIDES, N.O.S.   RR-00118-3
HEALTH AND SAFETY LISTS
U.S. DOT - Substances From 49 CFR 172.101
regulated by DOT (UN1390)
U.S. DOT - Hazard Classes
DOT hazard class = 4.3
STATE LISTS
NJ Right to Know List (Total)
sn 2085

ALKALI METAL DISPERSIONS, N.O.S. RR-00119-4

 HEALTH AND SAFETY LISTS

  U.S. DOT - Substances From 49 CFR 172.101
   regulated by DOT (UN1391)

  U.S. DOT - Hazard Classes
   DOT hazard class = 4.3

 STATE LISTS

  NJ Right to Know List (Total)
   sn 2086

ALKALI METAL NITRITES RR-01264-6

 ENVIRONMENTAL LISTS

  TSCA - Chemicals with Significant New Use Rules
   [present] (when used as an ingredient in the metalworking fluid industry)

  TSCA - Section 12(b) - Export Notification
   export notification required - Section 5 Alkali metal nitrites are defined as nitrites of the alkali metals lithium, sodium, potassium, rubidium, cesium and francium (40 CFR 721.4740).

ALKALINE CORROSIVE LIQUID, N.O.S. RR-00121-8

 STATE LISTS

  NJ Right to Know List (Total)
   sn 2088

ALKALINE EARTH METAL ALLOYS, N.O.S. RR-00122-9

 HEALTH AND SAFETY LISTS

  U.S. DOT - Substances From 49 CFR 172.101
   regulated by DOT (UN1393)

  U.S. DOT - Hazard Classes
   DOT hazard class = 4.3

 STATE LISTS

  NJ Right to Know List (Total)
   sn 2089

ALKALINE EARTH METAL AMALGAMS, N.O.S. RR-00123-0

 HEALTH AND SAFETY LISTS

  U.S. DOT - Substances From 49 CFR 172.101
   regulated by DOT (UN1392)

  U.S. DOT - Hazard Classes
   DOT hazard class = 4.3

 STATE LISTS

  NJ Right to Know List (Total)
   sn 2090

ALKALINE EARTH METAL DISPERSIONS, N.O.S. RR-00124-1

 STATE LISTS

  NJ Right to Know List (Total)
   sn 2091

ALKALOIDS RR-01339-8

 HEALTH AND SAFETY LISTS

  U.S. DOT - Substances From 49 CFR 172.101
   regulated by DOT (UN1544, UN3140)

  U.S. DOT - Hazard Classes
   DOT hazard class = 6.1

ALKALOID, SALTS, N.O.S. RR-00125-2

 STATE LISTS

  NJ Right to Know List (Total)
   sn 2093

ALKANAMINIUM, POLYALKYL-[(2-METHYL-1-OXO-2-PROPENYL)OXY]SALT, POLYMER WITH ACRYLAMIDE AND SUBSTITUTED ALKYL METHACRYLATE RR-01229-3

 ENVIRONMENTAL LISTS

  TSCA - Chemicals with Significant New Use Rules
   PMN number: P-87-252

ALKANES, C6-18, CHLORO- 68920-70-7
 SEE ALSO:
  CHLORINATED PARAFFINS

 ENVIRONMENTAL LISTS

  TSCA - Code of Federal Regulations Citations
   40 CFR 716.120(c)

ALKANES, C10-13-ISO- 68551-17-7

 STATE LISTS

  Minnesota Hazardous Substance List
   [present]

ALKANES, C(10-18)-BROMOCHLORO- 68955-41-9

 ENVIRONMENTAL LISTS

  TSCA - Code of Federal Regulations Citations
   40 CFR 712.30(x); 40 CFR 716.120(d)

  TSCA - PAIR - Reporting List
   Reporting Date: December 27, 1990

  TSCA - Health and Safety Reporting List
   Effective Date: October 29, 1990

ALKANE SULFONIC ACID 75-75-2

 HEALTH AND SAFETY LISTS

  NIOSH - Selected LD50s and LC50s
   Oral, quail: LD50 = 1000 mg/kg

 STATE LISTS

  NJ Right to Know List (Total)
   sn 0034

  NJ Special Hazardous Substances
   (corrosive)

ALKANOIC ACID, BUTANEDIOL AND CYCLOHEX-ANEALKANOL POLYMER RR-00938-1

 ENVIRONMENTAL LISTS

  TSCA - Chemicals with Significant New Use Rules
   PMN number: P-89-672

ALKEHYLDICARBOXYLIC ACIDS, POLYMERS WITH ALKANEPOLYOL ANDTDI, ALKANOL BLOCKED, ACRYLATE RR-00215-3

 ENVIRONMENTAL LISTS

  TSCA - Chemicals with Significant New Use Rules
   PMN number: P-89-77

ALKENOIC ACID, TRISUBSTITUTED BENZYL-DISUBSTITUTED PHENYLESTER RR-00221-1

 ENVIRONMENTAL LISTS

  TSCA - Chemicals with Significant New Use Rules
   PMN number: P-89-697

ALKENOIC ACID, TRISUBSTITUTED PHENY-LALKYL DISUBSTITUTEDPHENYL ESTER RR-00222-2

 ENVIRONMENTAL LISTS

  TSCA - Chemicals with Significant New Use Rules
   PMN number: P-89-694

ALKENYL ETHER OF ALKANETRIOL POLYMER   RR-01690-0
  ENVIRONMENTAL LISTS
    TSCA - Chemicals with Significant New Use Rules
    PMN number: P-93-458

ALKOXYLATED ALKANE POLYOL, POLYACRY-   RR-00243-7
LATE ESTER
  ENVIRONMENTAL LISTS
    TSCA - Chemicals with Significant New Use Rules
    PMN number: P-84-713

ALKYOXYLATED DIALKYLDIETHYLENETRI-   RR-01212-4
AMINE, ALKYL SULFATE SALT
  ENVIRONMENTAL LISTS
    TSCA - Chemicals with Significant New Use Rules
    PMN number: P-91-288

ALKYL ANTHRAQUINONES   RR-01746-9
  ENVIRONMENTAL LISTS
    CAA - HON Rule - SOCMI Chemicals
    compliance by Oct. 23, 1995

ALKYL ALKENOATE, AZOBIS-   RR-00946-1
  ENVIRONMENTAL LISTS
    TSCA - Chemicals with Significant New Use Rules
    PMN number: P-88-2470

ALKYLAMINES OR POLYALKYLAMINES   RR-00126-3
  HEALTH AND SAFETY LISTS
    U.S. DOT - Substances From 49 CFR 172.101
    regulated by DOT (UN2733, UN2734)
    U.S. DOT - Hazard Classes
    DOT hazard class = 3
  STATE LISTS
    NJ Right to Know List (Total)
    sn 2094; sn 2095; sn 2096

ALKYLAMINE TETRACHLOROPHENATE   RR-01195-0
  ENVIRONMENTAL LISTS
    TSCA - Section 12(b) - Export Notification
    export notification required - Section 4

3-ALKYL-2-(2-ANILINO)VINYLTHIAZOLINIUM SALT   RR-00199-0
  ENVIRONMENTAL LISTS
    TSCA - Chemicals with Significant New Use Rules
    PMN number: P-84-1007

ALKYL, ARYL, OR TOLUENE SULFONIC ACID   RR-00129-6
  STATE LISTS
    NJ Right to Know List (Total)
    sn 2097; sn 2098; sn 2099; sn 2100

ALKYLARYL SUBSTITUTED PHOSPHITE   RR-01178-9
  ENVIRONMENTAL LISTS
    TSCA - Chemicals with Significant New Use Rules
    PMN number: P-91-899

ALKYLATED DIARYLAMINE, SULFURIZED   RR-00179-6
  ENVIRONMENTAL LISTS
    TSCA - Chemicals with Significant New Use Rules
    PMN number: P-89-506

ALKYLATED DIPHENYL OXIDE   RR-00174-1
  ENVIRONMENTAL LISTS

TSCA - Chemicals with Significant New Use Rules
    PMN number: P-84-1079

ALKYLATED DIPHENYLS   RR-01570-3
  ENVIRONMENTAL LISTS
    TSCA - Chemicals with Significant New Use Rules
    PMN numbers: P-90-237; P-90-248; P-90-249

ALKYLATED SULFONATED DIPHENYL OXIDE,   RR-01688-6
ALKALI AND AMINE SALTS
  ENVIRONMENTAL LISTS
    TSCA - Chemicals with Significant New Use Rules
    PMN numbers: P-93-352, P-93-353

ALKYLBENZENE SULFONATE, AMINE SALT   RR-00992-7
  ENVIRONMENTAL LISTS
    TSCA - Chemicals with Significant New Use Rules
    PMN number: P-90-456

ALKYLBENZENESULFONIC ACIDS AND SODIUM   RR-01233-9
SALTS
  ENVIRONMENTAL LISTS
    TSCA - Chemicals with Significant New Use Rules
    PMN numbers: P-88-1783; P-88-2231; P-88-2237; P-88-2530

ALKYLBENZYLDIMETHYLAMMONIUM CHLORIDE   68391-01-5
  ENVIRONMENTAL LISTS
    List of Pesticide Product Inert Ingredients
    [present]

ALKYLBISOXYALKYL (SUBSTITUTED-1,1-   RR-00930-3
DIMETHYLETHYLPHENYL) BENZOTRIAZOLE
  ENVIRONMENTAL LISTS
    TSCA - Chemicals with Significant New Use Rules
    PMN number: P-86-1771

ALKYL-, BROMO-, CHLORO-, HYDROXYMETHYL   RR-01712-9
DIARYL ETHERS
  PROPOSED REGULATIONS
    TSCA - ITC 33rd Report Priority Testing List
    recommended for testing
    TSCA - ITC 34th Report Priority Testing List
    recommended for testing

ALPHA-ALKYL(C8-C10)-OMEGA-HYDROXYPOLY   68891-29-2
(OXYETHYLENE) AMMONIUMSULFATE
  ENVIRONMENTAL LISTS
    List of Pesticide Product Inert Ingredients
    [present]

ALKYL(C8-10) POLYETHOXYPOLYPROPOXYBEN-   RR-01032-2
ZENE ETHER
  ENVIRONMENTAL LISTS
    List of Pesticide Product Inert Ingredients
    [present]

ALPHA-ALKYL(C8-C14)-OMEGA-HYDROXYPOLY   RR-01033-3
(OXYPROPYLENE) BLOCK COPOLYMER WITH
POLYOXYETHYLENE
  ENVIRONMENTAL LISTS
    List of Pesticide Product Inert Ingredients
    [present]

ALKYL(C10-14) OXYPOLY(ETHYLENEOXY)ETHYL   68585-36-4
PHOSPHATE
  ENVIRONMENTAL LISTS

List of Pesticide Product Inert Ingredients
[present]

ALKYL(C10-14) POLY(OXYETHYLENE)POLY (OXYPROPYLENE) CONDENSATE WITH MONOS-TEARYL ACID PHOSPHATE   RR-01030-0

ENVIRONMENTAL LISTS

List of Pesticide Product Inert Ingredients
[present]

ALKYL(C10-16) DIMETHYLAMINE OXIDE   70592-80-2

ENVIRONMENTAL LISTS

List of Pesticide Product Inert Ingredients
[present]

ALPHA-ALKYL(C10-16)-OMEGA-HYDROXYPOLY (OXYETHYLENE) SULFATE,SODIUM SALT   68585-34-2

ENVIRONMENTAL LISTS

List of Pesticide Product Inert Ingredients
[present]

ALKYL(C11-15) PHENOXYPOLY(OXYETHYLENE) ETHANOL   RR-01031-1

ENVIRONMENTAL LISTS

List of Pesticide Product Inert Ingredients
[present]

ALKYL(C12-14) DIMETHYL AMINE OXIDE   68955-55-5

ENVIRONMENTAL LISTS

List of Pesticide Product Inert Ingredients
[present]

ALKYL (C12-16)DIMETHYLBENZYLAMMONIUM CHLORIDE   68424-85-1

ENVIRONMENTAL LISTS

List of Pesticide Product Inert Ingredients
[present]

ALKYL (C12-C14) GLYCIDYL ETHER   68609-97-2

SEE ALSO:
GLYCIDOL (OXIRANEMETHANOL) AND ITS DERIVATIVES

ENVIRONMENTAL LISTS

EPA - Master Testing List
[present]

TSCA - Code of Federal Regulations Citations
40 CFR 716.120(c)

PROPOSED REGULATIONS

TSCA - Proposed Testing Rule for Glycidyl Ethers
subject to screening subcategory and mutagenicity testing (results apply to all members of Glycidyl subcategory II-A)

ALPHA-ALKYL(C12-C15)-OMEGA-HYDROXYPOLYOXYETHYLENE   68908-63-4

ENVIRONMENTAL LISTS

List of Pesticide Product Inert Ingredients
[present]

ALKYL (C8-C10) GLYCIDYL ETHER   68609-96-1

SEE ALSO:
GLYCIDOL (OXIRANEMETHANOL) AND ITS DERIVATIVES

ENVIRONMENTAL LISTS

EPA - Master Testing List
[present]

TSCA - Code of Federal Regulations Citations
40 CFR 716.120(c)

PROPOSED REGULATIONS

TSCA - Proposed Testing Rule for Glycidyl Ethers
subject to subchronic toxicity and reproductive and fertility effects testing; (results apply to all members of Glycidyl subcategory I-A)

ALKYLCARBAMIC ACID, ALKYNYL ESTER   RR-01248-6

ENVIRONMENTAL LISTS

TSCA - Chemicals with Significant New Use Rules
PMN number: P-91-55

N-ALKYLDIMETHYLBENZYL AMMONIUM CHLORIDE   8001-54-5

HEALTH AND SAFETY LISTS

NIOSH - Selected LD50s and LC50s
Oral, rat: LD50 = 240 mg/kg Skin, rat: LD50 = 1560 mg/kg

ALKYLENE GLYCOL TEREPHTHALATE AND SUB-STITUTED BENZOATEESTERS   RR-00178-5

ENVIRONMENTAL LISTS

TSCA - Chemicals with Significant New Use Rules
PMN number: P-89-596

ALKYLENEBIS(SUBSTITUTED CARBOMONOCYCLE), EPICHLOROHYDRIN, DISUBSTITUTED HETER-MONOCYCLE, ACRYLATE POLYMER   RR-00984-7

ENVIRONMENTAL LISTS

TSCA - Chemicals with Significant New Use Rules
PMN number: P-89-626

ALKYLENEDIOALKYL ETHER   RR-01689-7

ENVIRONMENTAL LISTS

TSCA - Chemicals with Significant New Use Rules
PMN number: P-93-362

ALKYL EPOXIDES   RR-00285-7

ENVIRONMENTAL LISTS

TSCA - Health and Safety Reporting List
Effective Date: October 4, 1982 (includes all noncyclic aliphatic hydrocarbons with one or more epoxy functional groups)

ALKYL ESTER   RR-00923-4

ENVIRONMENTAL LISTS

TSCA - Chemicals with Significant New Use Rules
PMN number: P-84-968

ALKYL (FATTY ACIDS OF COCONUT OIL) DIMETHYLAMMONIUM BETAINE   RR-01029-7

ENVIRONMENTAL LISTS

List of Pesticide Product Inert Ingredients
[present]

ALKYL (HETEROCYCLICYL) PHENYLAZOHETERO MONOCYCLIC POLYONE   RR-00925-6

ENVIRONMENTAL LISTS

TSCA - Chemicals with Significant New Use Rules
PMN number: P-85-1370

ALKYL (HETEROCYCLICYL) PHENYLAZOHETERO MONOCYCLIC POLYONE, [(ALKYLIMIDAZOLYL) METHYL] DERIVATIVE   RR-00927-8

ENVIRONMENTAL LISTS

TSCA - Chemicals with Significant New Use Rules
PMN number: P-86-136

ALKYL IMIDAZOLINIUM METHYL SULFATE (DE-RIVED FROM OLEIC ACID) 70206-24-5
ENVIRONMENTAL LISTS
List of Pesticide Product Inert Ingredients
[present]

ALKYL PEROXY-2-ETHYL HEXANOATE RR-00945-0
ENVIRONMENTAL LISTS
TSCA - Chemicals with Significant New Use Rules
PMN number: P-86-1492

ALKYL PHENOL, N.O.S. RR-00133-2
HEALTH AND SAFETY LISTS
U.S. DOT - Substances From 49 CFR 172.101
regulated by DOT (UN2430, UN3145)
U.S. DOT - Hazard Classes
DOT hazard class = 6.1
U.S. DOT - Appendix B - Marine Pollutants
DOT regulated marine pollutant
STATE LISTS
NJ Right to Know List (Total)
sn 2101

ALKYLPHENOXYPOLY(OXYETHYLENE SULFURIC ACID ESTER, SUBSTITUTED AMINE SALT RR-01652-4
ENVIRONMENTAL LISTS
TSCA - Chemicals with Significant New Use Rules
PMN number: P-92-396

ALKYLPHENOXYPOLYALKOXYAMINE RR-00192-3
ENVIRONMENTAL LISTS
TSCA - Chemicals with Significant New Use Rules
PMN number: P-86-1489

ALKYL PHOSPHONATE AMMONIUM SALTS RR-01659-1
ENVIRONMENTAL LISTS
TSCA - Chemicals with Significant New Use Rules
PMN numbers: P-93-725, P-93-726

ALKYL PHTHALATES RR-00284-6
ENVIRONMENTAL LISTS
TSCA - Health and Safety Reporting List
Effective Date: October 4, 1982 (all alkyl esters of 1,2-benzenedicar-boxylic acid)

ALKYL POLYETHYLENE GLYCOL PHOSPHATE, POTASSIUM SALT RR-00973-4
ENVIRONMENTAL LISTS
TSCA - Chemicals with Significant New Use Rules
PMN number: P-90-481

ALKYL SUBSTITUTED DIAROMATIC HYDROCARBONS RR-01219-1
ENVIRONMENTAL LISTS
TSCA - Chemicals with Significant New Use Rules
PMN number: P-91-710

ALKYLSULFONIUM SALT RR-01726-5
ENVIRONMENTAL LISTS
TSCA - Chemicals with Significant New Use Rules
PMN number: P-93-1166

ALKYLTIN COMPOUNDS RR-00283-5
ENVIRONMENTAL LISTS

TSCA - Health and Safety Reporting List
Effective Date: October 4, 1982

ALLANTOIN 97-59-6
ENVIRONMENTAL LISTS
List of Pesticide Product Inert Ingredients
[present]

ALLETHRIN 584-79-2
HEALTH AND SAFETY LISTS
NIOSH - Selected LD50s and LC50s
Oral, rat: LD50 = 310 mg/kg
STATE LISTS
NJ Right to Know List (Total)
sn 2102

D-TRANS-ALLETHRIN [D-TRANS-CHRYSANTHEMIC ACID OF D-ALLETHRONE] 28057-48-9
ENVIRONMENTAL LISTS
CERCLA/SARA - Section 313 - Emission Reporting
form R reporting required

ALLYL ACETATE 591-87-7
HEALTH AND SAFETY LISTS
U.S. DOT - Substances From 49 CFR 172.101
regulated by DOT (UN2333)
U.S. DOT - Hazard Classes
DOT hazard class = 3
NFPA - Flash Points
flash point = 72 degrees F (22 degrees C)
NFPA - Hazard Identification Ratings
health-1; flammability-3; reactivity-0
NIOSH - Selected LD50s and LC50s
Inhalation, rat: LC50 = 1000 ppm 1 hr Oral, rat: LD50 = 130 mg/kg Skin, rabbit: LD50 = 1021 mg/kg
NTP Chemical Status Reports - Testing Status and NTIS Number
Approved for toxicology/carcinogenesis study
INTERNATIONAL LISTS
Canada - WHMIS: Ingredient Disclosure
1% item 29 (5)
STATE LISTS
Florida Hazardous Substance List
[present]
Massachusetts Right To Know List
[present]
NJ Right to Know List (Total)
sn 0035
NJ Special Hazardous Substances
(flammable - third degree)
Pennsylvania Right to Know List
[present]

ALLYL ALCOHOL 107-18-6
HEALTH AND SAFETY LISTS
ACGIH 1995 - Time Weighted Averages
2 ppm TWA; 4.8 mg/m3 TWA
ACGIH 1995 - Short Term Exposure Limits
4 ppm STEL; 9.5 mg/m3 STEL
ACGIH 1995 - Skin Designations
skin - potential for cutaneous absorption
AIHA - Odor Threshold Values
geometric mean air odor threshold = 1.7 ppm (detectable)
U.S. DOT - Substances From 49 CFR 172.101
regulated by DOT (UN1098)

U.S. DOT - Hazard Classes
DOT hazard class = 6.1

U.S. DOT - Substances Which Are Poisonous by Inhalation
liquid hazardous material poisonous by inhalation (UN1098)

U.S. DOT - Appendix A Table 1 - Hazardous Substances
final RQ = 100 pounds (45.4 kg)

NFPA - Flash Points
flash point = 70 degrees F (21 degrees C)

NFPA - Hazard Identification Ratings
health-4; flammability-3; reactivity-1

NIOSH - Selected LD50s and LC50s
Inhalation, rat: LC50 = 76 ppm 8 hr Oral, rat: LD50 = 64 mg/kg Skin, rabbit: LD50 = 45 mg/kg

NIOSH 1990 - Pocket Guide - RELs
2 ppm TWA; 5 mg/m3 TWA; 4 ppm STEL; 10 mg/m3 STEL

NIOSH 1990 - Pocket Guide - IDLHs
150 ppm IDLH

NIOSH 1990 - Pocket Guide - Target organs
eyes, skin, respiratory system

NIOSH 1990 - Pocket Guide - Skin list
Potential for dermal absorption

NTP Chemical Status Reports - Testing Status and NTIS Number
Approved for toxicology/carcinogenesis study

OSHA - Vacated PELs - Time Weighted Averages
2 ppm TWA; 5 mg/m3 TWA

OSHA - Vacated PELs - Short Term Exposure Limits
4 ppm STEL; 10 mg/m3 STEL

OSHA - Vacated PELs - Skin Designation
Prevent or reduce skin absorption

OSHA - Final PELs - Time Weighted Averages
2 ppm TWA; 5 mg/m3 TWA

OSHA - Final PELs - Skin Notations
prevent or reduce skin absorption

ENVIRONMENTAL LISTS

CERCLA/SARA - Section 302 Extremely Hazardous Substances and TPQs
TPQ = 1000 pounds

CERCLA/SARA - Section 313 - Emission Reporting
form R reporting required for 1.0% de minimus concentration

CERCLA/SARA - Hazardous Substances and their Reportable Quantities
final RQ = 100 pounds (45.4 kg)

CAA -Toxic Substances for Accidental Release Prevention
threshold quantity = 15,000 lbs

CAA - HON Rule - SOCMI Chemicals
compliance by Oct. 24, 1994

Clean Water Act - Hazardous Substances
[present]

List of Pesticide Product Inert Ingredients
[present]

RCRA - P Series Wastes
waste number P005

RCRA - Hazardous Constituents-Appendix VIII
waste number P005

RCRA - Substances Banned From Land Disposal
[present]

TSCA - Health and Safety Reporting List
Effective Date: September 30, 1991

INTERNATIONAL LISTS

Australian Exposure Standards - Time Weighted Averages
2 ppm TWA; 4.8 mg/m3 TWA

Australian Exposure Standards - Short Term Exposure Limits
4 ppm STEL; 9.5 mg/m3 STEL

Australian Exposure Standards - Skin Effects
skin absorption

Canada - WHMIS: Ingredient Disclosure
1% item 30 (167)

Canada - NPRI (National Pollutant Release Inventory)
[present]

Canada - CEPA Schedule II Part II - Toxic Substances (Export)
[present]

Canada - Alberta - 8 Hour Occupational Exposure Limit
2 ppm TWA; 4.7 mg/m3 TWA

Canada - Alberta - 15 Minute Occupational Exposure Limit
4 ppm STEL; 9.5 mg/m3 STEL

Canada - Alberta - Skin Designation
can be absorbed through the intact skin

Canada - British Columbia - 8 Hour Exposure Limits
2 ppm TWA; 5 mg/m3 TWA

Canada - British Columbia - 15 Minute Exposure Limits
4 ppm STEL; 10 mg/m3 STEL

Canada - British Columbia - Skin Notations
skin - potential for skin absorption

Canada - Ontario - OHSA - TWAEVs
2 ppm TWAEV; 5 mg/m3 TWAEV

Canada - Ontario - OHSA - STEVs
4 ppm STEV; 10 mg/m3 STEV

Canada - Ontario - OHSA - Skin Notations
absorption through skin, eyes, or mucous membranes

Canada - Quebec - Time-Weighted Average Exposure Values
2 ppm TWAEV; 4.8 mg/m3 TWAEV

Canada - Quebec - Short-term Exposure Values
4 ppm STEV; 9.5 mg/m3 STEV

Canada - Quebec - Skin Designations
absorbed through the skin

United Kingdom - Occupational Exposure Standards - TWAs
2 ppm TWA; 5 mg/m3 TWA

United Kingdom - Occupational Exposure Standards - STELs
4 ppm STEL; 10 mg/m3 STEL

United Kingdom - Occupational Exposure Standards - Notes
can be absorbed through skin

German (DFG) - MAK Values
2 ppm MAK; 5 mg/m3 MAK

German (DFG) - Peak Limitations
2 x normal MAK (30 min. average value); don't exceed 4 times during shift

German (DFG) - Skin/Sensitizers
danger of cutaneous absorption

Israel - Time Weighted Averages
2 ppm TWA; 4.8 mg/m3 TWA

Israel - Short Term Exposure Limits
4 ppm STEL; 9.5 mg/m3 STEL

Israel - Action Levels
1 ppm AL; 2.4 mg/m3 AL

Mexico - Instruction No. 10 - TWAs
2 ppm TWA; 5 mg/m3 TWA

Mexico - Instruction No. 10 - STELs
4 ppm STEL; 10 mg/m3 STEL

Mexico - Instruction No. 10 - Skin designation
skin - potential for cutaneous absorption

STATE LISTS

California - Air Bill 2588 Appendix A-II
6/91

California - Exposure Limits - PELs
2 ppm PEL; 5 mg/m3 PEL

California - Exposure Limits - STELs
4 ppm STEL; 10 mg/m3 STEL

California - Exposure Limits - Skin Notation
material may be absorbed through the skin, eyes or mucous membrane
California - Directors List of Hazardous Substances (8 CCR 339)
[present]
Florida Hazardous Substance List
[present]
Massachusetts Right To Know List
extraordinarily hazardous
Minnesota Hazardous Substance List
skin
NJ Right to Know List (Total)
sn 0036
NJ Special Hazardous Substances
(flammable - third degree)
Pennsylvania Right to Know List
environmental hazard

ALLYLAMINE   107-11-9

HEALTH AND SAFETY LISTS
U.S. DOT - Substances From 49 CFR 172.101
regulated by DOT (UN2334)
U.S. DOT - Hazard Classes
DOT hazard class = 6.1
U.S. DOT - Substances Which Are Poisonous by Inhalation
liquid hazardous material poisonous by inhalation (UN2334)
NFPA - Flash Points
flash point = -20 degrees F (-29 degrees C)
NFPA - Hazard Identification Ratings
health-4; flammability-3; reactivity-1
NIOSH - Selected LD50s and LC50s
Inhalation, rat: LC50 = 177 ppm 8 hr Oral, rat: LD50 = 106 mg/kg Skin, rabbit: LD50 = 35 mg/kg
OSHA - List of Highly Hazardous Chemicals
threshhold quantity = 1000 pounds

ENVIRONMENTAL LISTS
CERCLA/SARA - Section 302 Extremely Hazardous Substances and TPQs
TPQ = 500 pounds
CERCLA/SARA - Section 313 - Emission Reporting
form R reporting required
CAA -Toxic Substances for Accidental Release Prevention
threshold quantity = 10,000 lbs

INTERNATIONAL LISTS
Canada - WHMIS: Ingredient Disclosure
1% item 31 (193)

STATE LISTS
Florida Hazardous Substance List
[present]
Massachusetts Right To Know List
extraordinarily hazardous
NJ Right to Know List (Total)
sn 0037
NJ Special Hazardous Substances
(flammable - third degree)
Pennsylvania Right to Know List
environmental hazard

PROPOSED REGULATIONS
CERCLA/SARA - Proposed Hazardous Substance Additions
proposed RQ = 1 pound (.454 kg)
CERCLA/SARA - 1989 Proposed RQ Adjustments
proposed RQ = 100 pounds (45.4 kg)

ALLYL BROMIDE   106-95-6

HEALTH AND SAFETY LISTS
U.S. DOT - Substances From 49 CFR 172.101
regulated by DOT (UN1099)
U.S. DOT - Hazard Classes
DOT hazard class = 3
U.S. DOT - Appendix B - Marine Pollutants
DOT regulated marine pollutant
NFPA - Flash Points
flash point = 30 degrees F (-1 degrees C)
NFPA - Hazard Identification Ratings
health-3; flammability-3; reactivity-1
NIOSH - Selected LD50s and LC50s
Inhalation, rat: LC50 = 10000 mg/m3 (30 mn) Oral, guinea pig: LD50 = 30 mg/kg

INTERNATIONAL LISTS
Canada - WHMIS: Ingredient Disclosure
1% item 32 (329)

STATE LISTS
Florida Hazardous Substance List
[present]
Massachusetts Right To Know List
[present]
NJ Right to Know List (Total)
sn 0038
NJ Special Hazardous Substances
(flammable - third degree)
Pennsylvania Right to Know List
[present]

ALLYL CAPROATE   123-68-2

HEALTH AND SAFETY LISTS
NFPA - Flash Points
flash point = 150 degrees F (66 degrees C)
NFPA - Hazard Identification Ratings
health-1; flammability-2; reactivity-0

ALLYL CHLORIDE   107-05-1
SEE ALSO:
F039-HAZARDOUS WASTES
F024-HAZARDOUS WASTES
F025-HAZARDOUS WASTES
BIS(2,4-DIMETHYLBUTYL) MALEATE

HEALTH AND SAFETY LISTS
ACGIH 1995 - Time Weighted Averages
1 ppm TWA; 3 mg/m3 TWA
ACGIH 1995 - Short Term Exposure Limits
2 ppm STEL; 6 mg/m3 STEL
AIHA - Odor Threshold Values
no geometric mean air odor threshold
U.S. DOT - Substances From 49 CFR 172.101
regulated by DOT (UN1100)
U.S. DOT - Hazard Classes
DOT hazard class = 3
U.S. DOT - Appendix B - Marine Pollutants
DOT regulated marine pollutant
U.S. DOT - Appendix A Table 1 - Hazardous Substances
final RQ = 1000 pounds (454 kg)
IARC - Group 3 (not classifiable)
[present]
NFPA - Flash Points
flash point = -25 degrees F (-32 degrees)

NFPA - Hazard Identification Ratings
health-3; flammability-3; reactivity-1

NIOSH - Selected LD50s and LC50s
Oral, rat: LD50 = 64 mg/kg Skin, rabbit: LD50 = 2066 mg/kg

NIOSH 1990 - Pocket Guide - RELs
1 ppm TWA; 3 mg/m3 TWA; 2 ppm STEL; 6 mg/m3 STEL

NIOSH 1990 - Pocket Guide - IDLHs
300 ppm IDLH

NIOSH 1990 - Pocket Guide - Target organs
respiratory system, skin, eyes, liver, kidneys

NIOSH - Health Standards - Exposure Limits
1 ppm TWA; 3.1 mg/m3 TWA; C (15 min) 3 ppm; C (15 min) 9.3 mg/m3

NIOSH - Health Standards - Health Effects and Precautions
Liver, kidney, and lung effects (Urine, blood, and pulmonary function testing required)

NTP Chemical Status Reports - Testing Status and NTIS Number
Technical reports printed (PB287516/AS)

NTP Chemical Status Reports - Evidence of Carcinogenicity
male rat-negative; female rat-negative; male mice-equivocal; female mice-equivocal

OSHA - Vacated PELs - Time Weighted Averages
1 ppm TWA; 3 mg/m3 TWA

OSHA - Vacated PELs - Short Term Exposure Limits
2 ppm STEL; 6 mg/m3 STEL

OSHA - Final PELs - Time Weighted Averages
1 ppm TWA; 3 mg/m3 TWA

OSHA - List of Highly Hazardous Chemicals
threshhold quantity = 1000 pounds

ENVIRONMENTAL LISTS

CERCLA/SARA - Section 313 - Emission Reporting
form R reporting required for 1.0% de minimus concentration

CERCLA/SARA - Hazardous Substances and their Reportable Quantities
final RQ = 1000 pounds (454 kg)

Clean Air Act (1990) - List of Hazardous Air Contaminants
[present]

CAA - HON Rule - SOCMI Chemicals
compliance by July 24, 1995

CAA - HON Rule - Organic HAPs
[present]

Clean Water Act - Hazardous Substances
[present]

RCRA - Hazardous Constituents-Appendix VIII
hazardous constituent - no waste number

RCRA - Basis for Listing - Appendix VII
Included in waste streams: F024, F025, F039

RCRA - TSD Facilities Ground Water Monitoring
TM 8010 = 5 ug/L PQL; TM 8240 = 100 ug/L PQL

RCRA - Universal Treatment Standards (LDR)
WW: 0.036 mg/l; NWW: 30 mg/kg

INTERNATIONAL LISTS

Australian Exposure Standards - Time Weighted Averages
1 ppm TWA; 3 mg/m3 TWA

Australian Exposure Standards - Short Term Exposure Limits
2 ppm STEL; 6 mg/m3 STEL

Canada - WHMIS: Ingredient Disclosure
1% item 33 (469)

Canada - NPRI (National Pollutant Release Inventory)
[present]

Canada - Alberta - 8 Hour Occupational Exposure Limit
1 ppm TWA; 3.1 mg/m3 TWA

Canada - Alberta - 15 Minute Occupational Exposure Limit
2 ppm STEL; 6.3 mg/m3 STEL

Canada - British Columbia - 8 Hour Exposure Limits
1 ppm TWA; 3 mg/m3 TWA

Canada - British Columbia - 15 Minute Exposure Limits
2 ppm STEL; 6 mg/m3 STEL

Canada - Ontario - OHSA - TWAEVs
1 ppm TWAEV; 3 mg/m3 TWAEV

Canada - Ontario - OHSA - STEVs
2 ppm STEV; 6 mg/m3 STEV

Canada - Quebec - Time-Weighted Average Exposure Values
1 ppm TWAEV; 3 mg/m3 TWAEV

Canada - Quebec - Short-term Exposure Values
2 ppm STEV; 6 mg/m3 STEV

German (DFG) - MAK Values
1 ppm MAK; 3 mg/m3 MAK

German (DFG) - Peak Limitations
2 x normal MAK (5 min momentary value); don't exceed 8 times during shift

German (DFG) - Carcinogens
suspected carcinogen

Israel - Time Weighted Averages
1 ppm TWA; 3.0 mg/m3 TWA

Israel - Short Term Exposure Limits
2 ppm STEL; 6.0 mg/m3 STEL

Israel - Action Levels
0.5 ppm AL; 1.5 mg/m3 AL

Mexico - Instruction No. 10 - TWAs
1 ppm TWA; 3 mg/m3 TWA

Mexico - Instruction No. 10 - STELs
2 ppm STEL; 10 mg/m3 STEL

STATE LISTS

California - Air Bill 2588 Appendix A-I
known or potential carcinogen

California - Prop. 65 - Cancer list
carcinogen - initial date 1/1/90

California - Prop. 65 - No Significant Risk Levels
no significant risk level = 30 ug/day

California - Exposure Limits - PELs
1 ppm PEL; 3 mg/m3 PEL

California - Exposure Limits - STELs
2 ppm STEL; 6 mg/m3 STEL

California - Directors List of Hazardous Substances (8 CCR 339)
[present]

Florida Hazardous Substance List
[present]

Massachusetts Right To Know List
[present]

Minnesota Hazardous Substance List
[present]

NJ Right to Know List (Total)
sn 0039

NJ Special Hazardous Substances
(flammable - third degree)

Pennsylvania Right to Know List
environmental hazard

ALLYL CHLOROCARBONATE   2937-50-0

HEALTH AND SAFETY LISTS

U.S. DOT - Substances From 49 CFR 172.101
regulated by DOT (UN1722)

U.S. DOT - Hazard Classes
DOT hazard class = 8

U.S. DOT - Substances Which Are Poisonous by Inhalation
liquid hazardous material poisonous by inhalation (UN1722)

NFPA - Flash Points
flash point = 88 degrees F (31 degrees C)

NFPA - Hazard Identification Ratings
health-3; flammability-3; reactivity-1

NIOSH - Selected LD50s and LC50s
Inhalation, rat: LC50 = 32400 ug/m3 (8 hr) Oral, rat: LD50 = 244 mg/kg

INTERNATIONAL LISTS

Canada - WHMIS: Ingredient Disclosure
1% item 34 (430)

STATE LISTS

Florida Hazardous Substance List
[present]

Massachusetts Right To Know List
[present]

NJ Right to Know List (Total)
sn 0040

NJ Special Hazardous Substances
(corrosive; flammable - third degree)

Pennsylvania Right to Know List
[present]

ALLYL CYANIDE   109-75-1

HEALTH AND SAFETY LISTS

NIOSH - Selected LD50s and LC50s
Inhalation, guinea pig: LC50 = 2500 mg/m3 (4 hr) Oral, rat: LD50 = 115 mg/kg Skin, rabbit: LD50 = 1410 mg/kg

ENVIRONMENTAL LISTS

CAA - HON Rule - SOCMI Chemicals
compliance by July 24, 1995

ALLYL CYANOACETATE   13361-32-5

INTERNATIONAL LISTS

Canada - WHMIS: Ingredient Disclosure
1% item 35 (583)

ALLYL ETHER OF TETRABROMOBISPHENOL-A   25327-89-3

ENVIRONMENTAL LISTS

EPA - Master Testing List
[present]

TSCA - Code of Federal Regulations Citations
40 CFR 712.30(x); 40 CFR 716.120(d); 40 CFR 766.35

TSCA - PAIR - Reporting List
Reporting Date: December 27, 1990

TSCA - Health and Safety Reporting List
Effective Date: October 29, 1990; Sunset Date: November 9, 1993

TSCA - HDD/HDF - Chemicals Required for Testing
[present]

TSCA - Section 12(b) - Export Notification
export notification required - Section 4

ALLYL ETHYL ETHER   557-31-3

HEALTH AND SAFETY LISTS

U.S. DOT - Substances From 49 CFR 172.101
regulated by DOT (UN2335)

U.S. DOT - Hazard Classes
DOT hazard class = 3

INTERNATIONAL LISTS

Canada - WHMIS: Ingredient Disclosure
1% item 36 (806)

STATE LISTS

NJ Right to Know List (Total)
sn 0041

NJ Special Hazardous Substances
(flammable - fourth degree; reactive - fourth degree)

ALLYL FORMATE   1838-59-1

HEALTH AND SAFETY LISTS

U.S. DOT - Substances From 49 CFR 172.101
regulated by DOT (UN2336)

U.S. DOT - Hazard Classes
DOT hazard class = 3

NIOSH - Selected LD50s and LC50s
Inhalation, mouse: LC50 = 14 gm/m3 2 hr Oral, rat: LD50 = 124 mg/kg

INTERNATIONAL LISTS

Canada - WHMIS: Ingredient Disclosure
1% item 37 (920)

STATE LISTS

NJ Right to Know List (Total)
sn 0042

ALLYL GLYCIDYL ETHER   106-92-3
SEE ALSO:
GLYCIDOL (OXIRANEMETHANOL) AND ITS DERIVATIVES

HEALTH AND SAFETY LISTS

ACGIH 1995 - Time Weighted Averages
5 ppm TWA; 23 mg/m3 TWA

ACGIH 1995 - Short Term Exposure Limits
10 ppm STEL; 47 mg/m3 STEL

U.S. DOT - Substances From 49 CFR 172.101
regulated by DOT (UN2219)

U.S. DOT - Hazard Classes
DOT hazard class = 3

NIOSH - Selected LD50s and LC50s
Inhalation, rat: LC50 = 860 ppm 4 hr Oral, rat: LD50 = 922 mg/kg Skin, rabbit: LD50 = 2550 mg/kg

NIOSH 1990 - Pocket Guide - RELs
5 ppm TWA; 22 mg/m3 TWA; 10 ppm STEL; 44 mg/m3 STEL

NIOSH 1990 - Pocket Guide - IDLHs
270 ppm IDLH

NIOSH 1990 - Pocket Guide - Target organs
respiratory system, skin

NIOSH 1990 - Pocket Guide - Skin list
Potential for dermal absorption

NIOSH - Health Standards - Exposure Limits
C (15 min) 9.6 ppm; C (15 min) 45 mg/m3 (Listed under 'Glycidyl ethers')

NIOSH - Health Standards - Health Effects and Precautions
Skin and mucous membrane effects; sensitization potential; possible hematopoietic and reproductive system effects (Medical monitoring required) (Listed under 'Glycidyl ethers')

NTP Chemical Status Reports - Testing Status and NTIS Number
Technical reports printed (PB90260027)

NTP Chemical Status Reports - Evidence of Carcinogenicity
male rat-equivocal evidence; female rat-no evidence; male mice-some evidence; female mice-equivocal evidence

OSHA - Vacated PELs - Time Weighted Averages
5 ppm TWA; 22 mg/m3 TWA

OSHA - Vacated PELs - Short Term Exposure Limits
10 ppm STEL; 44 mg/m3 STEL

OSHA - Final PELs - Ceiling Limits
C 10 ppm; C 45 mg/m3

ENVIRONMENTAL LISTS

EPA - Master Testing List
[present]

TSCA - Code of Federal Regulations Citations
40 CFR 712.30(d); 40 CFR 716.120(c)

TSCA - PAIR - Reporting List
Reporting Date: November 19, 1982

INTERNATIONAL LISTS

Australian Exposure Standards - Time Weighted Averages
5 ppm TWA;   23 mg/m3 TWA

Australian Exposure Standards - Short Term Exposure Limits
10 ppm STEL;   47 mg/m3 STEL

Australian Exposure Standards - Skin Effects
skin absorption;   sensitiser

Canada - WHMIS: Ingredient Disclosure
0.1% item 38 (807)

Canada - Alberta - 8 Hour Occupational Exposure Limit
5 ppm TWA;   23 mg/m3 TWA

Canada - Alberta - 15 Minute Occupational Exposure Limit
10 ppm STEL;   47 mg/m3 STEL

Canada - British Columbia - 8 Hour Exposure Limits
5 ppm TWA;   22 mg/m3 TWA

Canada - British Columbia - 15 Minute Exposure Limits
10 ppm STEL;   44 mg/m3 STEL

Canada - British Columbia - Skin Notations
skin - potential for skin absorption

Canada - Ontario - OHSA - TWAEVs
5 ppm TWAEV;   23 mg/m3 TWAEV

Canada - Ontario - OHSA - STEVs
10 ppm STEV;   47 mg/m3 STEV

Canada - Ontario - OHSA - Skin Notations
absorption through skin, eyes, or mucous membranes

Canada - Quebec - Time-Weighted Average Exposure Values
5 ppm TWAEV;   23 mg/m3 TWAEV

Canada - Quebec - Short-term Exposure Values
10 ppm STEV;   47 mg/m3 STEV

United Kingdom - Occupational Exposure Standards - TWAs
5 ppm TWA;   22 mg/m3 TWA

United Kingdom - Occupational Exposure Standards - STELs
10 ppm STEL;   44 mg/m3 STEL

United Kingdom - Occupational Exposure Standards - Notes
can be absorbed through skin

German (DFG) - Skin/Sensitizers
danger of sensitization (skin or respiratory)

German (DFG) - Carcinogens
animal evidence of carcinogenicity

Israel - Time Weighted Averages
5 ppm TWA;   23 mg/m3 TWA

Israel - Short Term Exposure Limits
10 ppm STEL;   47 mg/m3 STEL

Israel - Action Levels
2.5 ppm AL;   11.5 mg/m3 AL

Mexico - Instruction No. 10 - TWAs
5 ppm TWA;   22 mg/m3 TWA

Mexico - Instruction No. 10 - STELs
10 ppm STEL;   44 mg/m3 STEL

Mexico - Instruction No. 10 - Skin designation
skin - potential for cutaneous absorption

STATE LISTS

California - Exposure Limits - PELs
5 ppm PEL;   22 mg/m3 PEL

California - Exposure Limits - STELs
10 ppm STEL;   44 mg/m3 STEL

California - Exposure Limits - Skin Notation
material may be absorbed through the skin, eyes or mucous membrane

California - Directors List of Hazardous Substances (8 CCR 339)
[present] (exempt when part of a cured epoxy or rubber)

Florida Hazardous Substance List
[present]

Massachusetts Right To Know List
[present]

Minnesota Hazardous Substance List
skin

NJ Right to Know List (Total)
sn 0043

Pennsylvania Right to Know List
[present]

PROPOSED REGULATIONS

TSCA - Proposed Testing Rule for Glycidyl Ethers
subject to neurotoxicity, reproductive and fertility effects, mutagenicity, and screening subcategory testing (results apply to all members of Glycidyl subcategory I-C)

ALLYLIDENE DIACETATE   869-29-4

HEALTH AND SAFETY LISTS

NFPA - Flash Points
flash point = 180 degrees F (82 degrees C)

NFPA - Hazard Identification Ratings
health-2;   flammability-2;   reactivity-1

STATE LISTS

Florida Hazardous Substance List
[present]

Massachusetts Right To Know List
[present]

Pennsylvania Right to Know List
[present]

ALLYL IODIDE   556-56-9

HEALTH AND SAFETY LISTS

U.S. DOT - Substances From 49 CFR 172.101
regulated by DOT (UN1723)

U.S. DOT - Hazard Classes
DOT hazard class = 3

STATE LISTS

NJ Right to Know List (Total)
sn 0044

NJ Special Hazardous Substances
(corrosive)

ALLYL ISOTHIOCYANATE   57-06-7

HEALTH AND SAFETY LISTS

AIHA - WEEL - Ceilings or Short Term Time Weighted Averages
1 ppm STEL;   4 mg/m3 STEL

AIHA - WEEL - Skin Absorption Designations
skin absorber

U.S. DOT - Substances From 49 CFR 172.101
regulated by DOT (UN1545)

U.S. DOT - Hazard Classes
DOT hazard class = 6.1

U.S. DOT - Substances Which Are Poisonous by Inhalation
liquid hazardous material poisonous by inhalation (inhibited form) (UN1545)

IARC - Group 3 (not classifiable)
[present]

NFPA - Flash Points
flash point = 115 degrees F (46 degrees C)

NFPA - Hazard Identification Ratings
health-3;   flammability-2;   reactivity-0

NIOSH - Selected LD50s and LC50s
Oral, rat: LD50 = 112 mg/kg   Skin, rabbit: LD50 = 88 mg/kg

NTP Chemical Status Reports - Testing Status and NTIS Number
Technical reports printed (PB83144238)

NTP Chemical Status Reports - Evidence of Carcinogenicity
male rat-positive; female rat-equivocal; male mice-negative; female mice-negative

INTERNATIONAL LISTS

Canada - WHMIS: Ingredient Disclosure
1% item 39 (1057)

STATE LISTS

California - Directors List of Hazardous Substances (8 CCR 339)
[present]

Florida Hazardous Substance List
[present]

Massachusetts Right To Know List
[present]

NJ Right to Know List (Total)
sn 0045

NJ Special Hazardous Substances
(mutagen)

Pennsylvania Right to Know List
[present]

ALLYL ISOVALERATE   2835-39-4

HEALTH AND SAFETY LISTS

IARC - Group 3 (not classifiable)
[present]

NTP Chemical Status Reports - Testing Status and NTIS Number
Technical reports printed (PB83218214)

NTP Chemical Status Reports - Evidence of Carcinogenicity
male rat-positive; female rat-negative; male mice-negative; female mice-positive

STATE LISTS

California - Directors List of Hazardous Substances (8 CCR 339)
[present]

ALLYL METHACRYLATE   96-05-9

HEALTH AND SAFETY LISTS

NIOSH - Selected LD50s and LC50s
Oral, rat: LD50 = 430 mg/kg Skin, rabbit: LD50 = 500 mg/kg

ENVIRONMENTAL LISTS

TSCA - Code of Federal Regulations Citations
40 CFR 712.30(d)

TSCA - PAIR - Reporting List
Reporting Date: November 19, 1982

INTERNATIONAL LISTS

Canada - WHMIS: Ingredient Disclosure
1% item 40 (1088)

ALLYL PALLADIUM CHLORIDE DIMER   12012-95-2
SEE ALSO:
PALLADIUM
ANTIMONY

INTERNATIONAL LISTS

Canada - WHMIS: Ingredient Disclosure
1% item 41 (514)

ALLYL PROPYL DISULFIDE   2179-59-1

HEALTH AND SAFETY LISTS

ACGIH 1995 - Time Weighted Averages
2 ppm TWA; 12 mg/m3 TWA

ACGIH 1995 - Short Term Exposure Limits
3 ppm STEL; 18 mg/m3 STEL

OSHA - Vacated PELs - Time Weighted Averages
2 ppm TWA; 12 mg/m3 TWA

OSHA - Vacated PELs - Short Term Exposure Limits
3 ppm STEL; 18 mg/m3 STEL

OSHA - Final PELs - Time Weighted Averages
2 ppm TWA; 12 mg/m3 TWA

INTERNATIONAL LISTS

Australian Exposure Standards - Time Weighted Averages
2 ppm TWA; 12 mg/m3 TWA

Australian Exposure Standards - Short Term Exposure Limits
3 ppm STEL; 18 mg/m3 STEL

Canada - WHMIS: Ingredient Disclosure
1% item 42 (786)

Canada - Alberta - 8 Hour Occupational Exposure Limit
2 ppm TWA; 12 mg/m3 TWA

Canada - Alberta - 15 Minute Occupational Exposure Limit
3 ppm STEL; 18 mg/m3 STEL

Canada - British Columbia - 8 Hour Exposure Limits
2 ppm TWA; 12 mg/m3 TWA

Canada - British Columbia - 15 Minute Exposure Limits
3 ppm STEL; 18 mg/m3 STEL

Canada - Ontario - OHSA - TWAEVs
2 ppm TWAEV; 12 mg/m3 TWAEV

Canada - Ontario - OHSA - STEVs
3 ppm STEV; 18 mg/m3 STEV

Canada - Quebec - Time-Weighted Average Exposure Values
2 ppm TWAEV; 12 mg/m3 TWAEV

Canada - Quebec - Short-term Exposure Values
3 ppm STEV; 18 mg/m3 STEV

German (DFG) - MAK Values
2 ppm MAK; 12 mg/m3 MAK

Israel - Time Weighted Averages
2 ppm TWA; 12 mg/m3 TWA

Israel - Short Term Exposure Limits
3 ppm STEL; 18 mg/m3 STEL

Israel - Action Levels
1 ppm AL; 6 mg/m3 AL

STATE LISTS

California - Exposure Limits - PELs
2 ppm PEL; 12 mg/m3 PEL

California - Exposure Limits - STELs
3 ppm STEL; 18 mg/m3 STEL

California - Directors List of Hazardous Substances (8 CCR 339)
[present]

Florida Hazardous Substance List
[present]

Massachusetts Right To Know List
[present]

Minnesota Hazardous Substance List
[present]

NJ Right to Know List (Total)
sn 0046

Pennsylvania Right to Know List
[present]

ALLYL TRICHLOROSILANE   107-37-9

HEALTH AND SAFETY LISTS

U.S. DOT - Substances From 49 CFR 172.101
regulated by DOT (UN1724)

U.S. DOT - Hazard Classes
DOT hazard class = 8

NFPA - Flash Points
flash point = 95 degrees F (35 degrees C)

NFPA - Hazard Identification Ratings
health-3; flammability-3; reactivity-2 (avoid use of water)

INTERNATIONAL LISTS

Canada - WHMIS: Ingredient Disclosure
1% item 43 (194)

STATE LISTS

Florida Hazardous Substance List
[present]

Massachusetts Right To Know List
[present]

NJ Right to Know List (Total)
sn 0047

NJ Special Hazardous Substances
(corrosive; flammable - third degree; reactive - second degree)

Pennsylvania Right to Know List
[present]

ALMOND HULLS   RR-01034-4

ENVIRONMENTAL LISTS

List of Pesticide Product Inert Ingredients
[present]

ALOE   8001-97-6

ENVIRONMENTAL LISTS

List of Pesticide Product Inert Ingredients
[present]

A-ALPHA-C   26148-68-5

HEALTH AND SAFETY LISTS

IARC - Group 2B (sufficient animal data)
[present] (Overall evaluation based only on evidence of carcinogenicity in monograph (40, 1986) or in supplement 4)

OSHA - Possible Select Carcinogens
[present]

STATE LISTS

California - Air Bill 2588 Appendix A-II
known or potential carcinogen: 9/89

California - Prop. 65 - Cancer list
carcinogen - initial date 1/1/90

California - Prop. 65 - No Significant Risk Levels
no significant risk level = 2 ug/day

California - Directors List of Hazardous Substances (8 CCR 339)
[present]

Massachusetts Right To Know List
carcinogen; extraordinarily hazardous

Minnesota Hazardous Substance List
carcinogen

ALPHA RADIATION   12587-46-1

ENVIRONMENTAL LISTS

ATSDR Priority List
Rank (of 275): 075

ALPRAZOLAM   28981-97-7

HEALTH AND SAFETY LISTS

NIOSH - Selected LD50s and LC50s
Oral rat: LD50 = 1220 mg/kg

STATE LISTS

California - Air Bill 2588 Appendix A-II
9/90

California - Prop. 65 - Developmental Toxicity
developmental toxicity - initial date 7/1/90

NJ Special Hazardous Substances
(teratogen)

ALUMINUM   7429-90-5

HEALTH AND SAFETY LISTS

ACGIH 1995 - Time Weighted Averages
metal dust, as Al: 10 mg/m3 TWA; pyro powders, as Al: 5 mg/m3 TWA; welding fumes, as Al: 5 mg/m3 TWA; soluble salts, as Al: 2 mg/m3 TWA; alkyls (NOC), as Al: 2 mg/m3 TWA

U.S. DOT - Substances From 49 CFR 172.101
regulated by DOT (UN1309, UN1396)

U.S. DOT - Hazard Classes
DOT hazard class = 4.3

OSHA - Vacated PELs - Time Weighted Averages
total dust, as Al: 15 mg/m3 TWA; respirable fraction, as Al: 5 mg/m3 TWA

OSHA - Final PELs - Time Weighted Averages
total dust, as Al: 15 mg/m3 TWA; respirable fraction, as Al: 5 mg/m3 TWA

ENVIRONMENTAL LISTS

ATSDR Priority List
Rank (of 275): 163

CERCLA/SARA - Section 313 - Emission Reporting
form R reporting required for 1.0% de minimus concentration form only)

Safe Drinking Water Act - SMCLs
SMCL = 0.05 to 0.2 mg/L

List of Pesticide Product Inert Ingredients
[present]

INTERNATIONAL LISTS

Australian Exposure Standards - Time Weighted Averages
dust: 10 mg/m3 TWA; welding fumes, as Al: 5 mg/m3 TWA; alkyls (NOC), as Al: 2 mg/m3 TWA; pyro powders, as Al: 5 mg/m3 TWA; soluble salts, as Al: 2 mg/m3 TWA

Canada - WHMIS: Ingredient Disclosure
1% item 47 (197)

Canada - NPRI (National Pollutant Release Inventory)
[present] (as fume or dust)

Canada - Alberta - 8 Hour Occupational Exposure Limit
10 mg/m3 TWA

Canada - Alberta - 15 Minute Occupational Exposure Limit
20 mg/m3 STEL

Canada - Ontario - OHSA - TWAEVs
powder: 5 mg/m3 TWAEV; alkyl derivatives of: 2 mg/m3 TWAEV; metal and oxide dust: 10 mg/m3 TWAEV; water soluble compounds of: 2 mg/m3 TWAEV; welding fume or particulate, aluminum containing (as Al): 5 mg/m3 TWAEV

German (DFG) - MAK Values
fine dust: 6 mg/m3 MAK

Israel - Time Weighted Averages
metal dust, as Al: 10 mg/m3 TWA; pyro powders, as Al: 5 mg/m3 TWA; welding fumes, as Al: 5 mg/m3 TWA; soluble salts, as Al: 2 mg/m3 TWA; alkyls (NOC), as Al: 2 mg/m3 TWA

Israel - Action Levels
metal dust, as Al: 5 mg/m3 AL; pyro powders, as Al: 2.5 mg/m3 AL; welding fumes, as Al: 2.5 mg/m3 AL; soluble salts, as Al: 1 mg/m3 AL; alkyls (NOC), as Al: 1 mg/m3 AL

Mexico - Instruction No. 10 - TWAs
10 mg/m3 TWA

Mexico - Instruction No. 10 - STELs
20 mg/m3 STEL

Mexico - Wastewater - Organic Toxic Pollutants and Heavy Metals
Listed under [Heavy Metals]

Mexico - Drinking Water - Ecological Criteria
0.02 mg/l

STATE LISTS

California - Air Bill 2588 Appendix A-I
6/91

California - Exposure Limits - PELs
total dust: 10 mg/m3 PEL; respirable fraction: 5 mg/m3 PEL

California - Directors List of Hazardous Substances (8 CCR 339)
[present]

Florida Hazardous Substance List
[present] (includes Aluminum oxide and Aluminum welding fumes)

Massachusetts Right To Know List
[present]

Minnesota Hazardous Substance List
[present] (includes Aluminum pyro powders, welding fumes, soluble salts, metal and oxide, and alkyls, and the inert or nuisance dusts from these)

NJ Right to Know List (Total)
sn 0054

Pennsylvania Right to Know List
environmental hazard Glycidyl subcategory V-A)

PROPOSED REGULATIONS

Safe Drinking Water Act - Priority list
[present]

ALUMINUM 26   14682-66-7

HEALTH AND SAFETY LISTS

U.S. DOT - Appendix A Table 2 - Radionuclides
final RQ = 10 curies (3.7E 11 Bq)

ENVIRONMENTAL LISTS

CERCLA/SARA List of Radionuclides (Appendix B) and Their Reportable Quantities
final RQ = 10 curies (3.7E 11 Bq)

ALUMINUM ALKYL CHLORIDE   RR-00135-4

STATE LISTS

NJ Right to Know List (Total)
sn 2104

ALUMINUM ALKYL HALIDES   RR-00136-5

HEALTH AND SAFETY LISTS

U.S. DOT - Substances From 49 CFR 172.101
regulated by DOT (UN3052)

U.S. DOT - Hazard Classes
DOT hazard class = 4.2

STATE LISTS

NJ Right to Know List (Total)
in solution: sn 2106; pure: sn 2107

ALUMINUM ALKYL HYDRIDES   RR-01361-6

HEALTH AND SAFETY LISTS

U.S. DOT - Substances From 49 CFR 172.101
regulated by DOT (UN3076)

U.S. DOT - Hazard Classes
DOT hazard class = 4.2

ALUMINUM, ALKYLS (NOC)   RR-00022-6

HEALTH AND SAFETY LISTS

U.S. DOT - Substances From 49 CFR 172.101
regulated by DOT (UN3051)

U.S. DOT - Hazard Classes
DOT hazard class = 4.2

OSHA - Vacated PELs - Time Weighted Averages
as Al: 2 mg/m3 TWA (Enforcement indefinitely stayed) (Listed under 'Aluminum')

OSHA - List of Highly Hazardous Chemicals
threshhold quantity = 5000 pounds

INTERNATIONAL LISTS

Canada - WHMIS: Ingredient Disclosure
1% item 45 (196)

Canada - Alberta - 8 Hour Occupational Exposure Limit
2 mg/m3 TWA

Canada - Alberta - 15 Minute Occupational Exposure Limit
4 mg/m3 STEL

Canada - Quebec - Time-Weighted Average Exposure Values
2 mg/m3 TWAEV

United Kingdom - Occupational Exposure Standards - TWAs
2 mg/m3 TWA

Mexico - Instruction No. 10 - TWAs
2 mg/m3 TWA

STATE LISTS

California - Exposure Limits - PELs
2 mg/m3 PEL

California - Directors List of Hazardous Substances (8 CCR 339)
[present]

NJ Right to Know List (Total)
sn 2547

Pennsylvania Right to Know List
[present]

ALUMINUM BOROHYDRIDE   16962-07-5

HEALTH AND SAFETY LISTS

U.S. DOT - Substances From 49 CFR 172.101
regulated by DOT (UN2870)

U.S. DOT - Hazard Classes
DOT hazard class = 4.2

STATE LISTS

NJ Right to Know List (Total)
sn 2108

ALUMINUM BROMIDE   7727-15-3

HEALTH AND SAFETY LISTS

U.S. DOT - Substances From 49 CFR 172.101
regulated by DOT (UN2580, UN1725)

U.S. DOT - Hazard Classes
DOT hazard class = 8

STATE LISTS

NJ Right to Know List (Total)
sn 0055

NJ Special Hazardous Substances
(corrosive)

ALUMINUM CARBIDE   12656-43-8

HEALTH AND SAFETY LISTS

U.S. DOT - Substances From 49 CFR 172.101
regulated by DOT (UN1394)

U.S. DOT - Hazard Classes
DOT hazard class = 4.3

STATE LISTS

NJ Right to Know List (Total)
sn 0056

ALUMINUM CHLORIDE   7446-70-0
SEE ALSO:
ALUMINUM

HEALTH AND SAFETY LISTS

U.S. DOT - Substances From 49 CFR 172.101
regulated by DOT (UN2581, UN1726)

U.S. DOT - Hazard Classes
DOT hazard class = 8
NIOSH - Selected LD50s and LC50s
Oral, rat: LD50 = 3730 mg/kg

ENVIRONMENTAL LISTS
List of Pesticide Product Inert Ingredients
[present]

INTERNATIONAL LISTS
Canada - WHMIS: Ingredient Disclosure
1% item 46 (470)

STATE LISTS
Florida Hazardous Substance List
[present]
Massachusetts Right To Know List
[present]
NJ Right to Know List (Total)
sn 0057
NJ Special Hazardous Substances
(corrosive; reactive - second degree)
Pennsylvania Right to Know List
[present]

ALUMINUM DROSS 69011-71-8

HEALTH AND SAFETY LISTS
U.S. DOT - Hazard Classes
Forbidden from transport by the DOT

ALUMINUM FERROSILICON 12003-41-7

HEALTH AND SAFETY LISTS
U.S. DOT - Substances From 49 CFR 172.101
regulated by DOT (UN1395)
U.S. DOT - Hazard Classes
DOT hazard class = 4.3

STATE LISTS
NJ Right to Know List (Total)
sn 0058

ALUMINUM FLUOROSULFATE, HYDRATE 73680-58-7

INTERNATIONAL LISTS
Canada - WHMIS: Ingredient Disclosure
1% item 48 (899)

ALUMINUM HYDRIDE 7784-21-6

HEALTH AND SAFETY LISTS
U.S. DOT - Substances From 49 CFR 172.101
regulated by DOT (UN2463)
U.S. DOT - Hazard Classes
DOT hazard class = 4.3

STATE LISTS
NJ Right to Know List (Total)
sn 0060

ALUMINUM HYDROXIDE 21645-51-2
SEE ALSO:
ALUMINUM

ENVIRONMENTAL LISTS
List of Pesticide Product Inert Ingredients
[present]

INTERNATIONAL LISTS
German (DFG) - MAK Values
fine dust: 6 mg/m3 MAK (exposure lasting one year) (Listed under 'Aluminum')

ALUMINUM, HYDROXYBIS(STEARATO)- 300-92-5

STATE LISTS
California - Exposure Limits - PELs
10 mg/m3 PEL

ALUMINUM ISO-PROPOXIDE 555-31-7

HEALTH AND SAFETY LISTS
NIOSH - Selected LD50s and LC50s
Oral, rat: LD50 = 11300 mg/kg

ENVIRONMENTAL LISTS
List of Pesticide Product Inert Ingredients
[present]

ALUMINUM MAGNESIUM PHOSPHIDE RR-00621-3

HEALTH AND SAFETY LISTS
U.S. DOT - Substances From 49 CFR 172.101
regulated by DOT (UN1419)
U.S. DOT - Hazard Classes
DOT hazard class = 4.3

STATE LISTS
NJ Right to Know List (Total)
sn 2524

ALUMINUM - MAGNESIUM STEARATE RR-01035-5

ENVIRONMENTAL LISTS
List of Pesticide Product Inert Ingredients
[present]

ALUMINUM NITRATE 13473-90-0

HEALTH AND SAFETY LISTS
U.S. DOT - Substances From 49 CFR 172.101
regulated by DOT (UN1438)
U.S. DOT - Hazard Classes
DOT hazard class = 5.1
NIOSH - Selected LD50s and LC50s
Oral, rat: LD50 = 3654 mg/kg

INTERNATIONAL LISTS
Canada - WHMIS: Ingredient Disclosure
1% item 49 (1196)

STATE LISTS
NJ Right to Know List (Total)
sn 0061

ALUMINUM OCTANOATE 6028-57-5

ENVIRONMENTAL LISTS
List of Pesticide Product Inert Ingredients
[present]

ALUMINUM OXIDE 1344-28-1

HEALTH AND SAFETY LISTS
ACGIH 1995 - Time Weighted Averages
as Al: 10 mg/m3 TWA (The value is for total dust containing no asbestos and < 1% crystalline silica)
OSHA - Vacated PELs - Time Weighted Averages
total dust: 10 mg/m3 TWA; respirable fraction: 5 mg/m3 TWA
OSHA - Final PELs - Time Weighted Averages
total dust: 15 mg/m3 TWA; respirable fraction: 5 mg/m3 TWA

ENVIRONMENTAL LISTS
CERCLA/SARA - Section 313 - Emission Reporting
form R reporting required for 0.1% de minimus concentration (fibrous form only)
List of Pesticide Product Inert Ingredients
[present]

INTERNATIONAL LISTS

Australian Exposure Standards - Time Weighted Averages
10 mg/m3 TWA

Canada - WHMIS: Ingredient Disclosure
1% item 44 (195)

Canada - NPRI (National Pollutant Release Inventory)
[present] (as fibrous forms)

Canada - British Columbia - 8 Hour Exposure Limits
nuisance dusts, mists, and fumes: 10 mg/m3

Canada - British Columbia - 15 Minute Exposure Limits
20 mg/m3 STEL

Canada - Ontario - OHSA - TWAEVs
total dust: 10 mg/m3 TWAEV (Listed as 'Nuisance Particulate')

German (DFG) - MAK Values
fine dust: 6 mg/m3 MAK (exposure lasting one year)

German (DFG) - Peak Limitations
5 x normal MAK (30 min. average value);   don't exceed 2 times during shift

German (DFG) - Carcinogens
as fibrous dust: animal evidence of carcinogenicity

Israel - Time Weighted Averages
as Al: 10 mg/m3 TWA (The value is for total dust containing no asbestos and < 1% crystalline silica)

Israel - Action Levels
as Al: 5 mg/m3 AL

Mexico - Instruction No. 10 - TWAs
as total dust: 10 mg/m3 TWA (nuisance particulate)

STATE LISTS

California - Air Bill 2588 Appendix A-I
6/91

California - Exposure Limits - PELs
total dust: 10 mg/m3 PEL;   respirable fraction: 5 mg/m3 PEL (Listed under 'Aluminum metal and oxide')

Massachusetts Right To Know List
[present]

Minnesota Hazardous Substance List
[present] (includes inert or nuisance dust)

NJ Right to Know List (Total)
sn 2891

Pennsylvania Right to Know List
environmental hazard

ALUMINUM PHOSPHATE   **7784-30-7**
SEE ALSO:
ALUMINUM

STATE LISTS

NJ Right to Know List (Total)
sn 0062

NJ Special Hazardous Substances
(corrosive)

ALUMINUM PHOSPHIDE   **20859-73-8**
SEE ALSO:
ALUMINUM

HEALTH AND SAFETY LISTS

U.S. DOT - Substances From 49 CFR 172.101
regulated by DOT (UN1397)

U.S. DOT - Hazard Classes
DOT hazard class = 4.3

U.S. DOT - Appendix A Table 1 - Hazardous Substances
final RQ = 100 pounds (45.4 kg)

ENVIRONMENTAL LISTS

CERCLA/SARA - Section 302 Extremely Hazardous Substances and TPQs
TPQ = 500 pounds (This material is a reactive solid. The TPQ does not default to 10,000 pounds for non-powder, non-molten, non-solvent form)

CERCLA/SARA - Section 313 - Emission Reporting
form R reporting required

CERCLA/SARA - Hazardous Substances and their Reportable Quantities
final RQ = 100 pounds (45.4 kg)

RCRA - P Series Wastes
waste number P006 (Reactive waste;   Toxic waste)

RCRA - Hazardous Constituents-Appendix VIII
waste number P006

RCRA - Substances Banned From Land Disposal
[present]

INTERNATIONAL LISTS

Canada - WHMIS: Ingredient Disclosure
1% item 51 (1410)

STATE LISTS

California - Directors List of Hazardous Substances (8 CCR 339)
[present]

Florida Hazardous Substance List
effective March 13, 1992

Massachusetts Right To Know List
extraordinarily hazardous

NJ Right to Know List (Total)
sn 0063

Pennsylvania Right to Know List
environmental hazard

ALUMINUM PHOSPHIDE PESTICIDES   **RR-00139-8**

HEALTH AND SAFETY LISTS

U.S. DOT - Substances From 49 CFR 172.101
regulated by DOT (UN3048)

U.S. DOT - Hazard Classes
DOT hazard class = 6.1

ALUMINUM POWDER   **RR-00141-2**

STATE LISTS

NJ Right to Know List (Total)
sn 2110;   sn 2111

ALUMINUM PRODUCTION   **RR-00546-9**

HEALTH AND SAFETY LISTS

IARC - Group 1 (carcinogenic to humans)
[present]

OSHA - Select Carcinogens
[present]

STATE LISTS

Pennsylvania Right to Know List
special hazardous substance

Pennsylvania RTK - Special Hazardous Substances
[present]

ALUMINUM, PYRO POWDERS   **RR-00019-1**

HEALTH AND SAFETY LISTS

OSHA - Vacated PELs - Time Weighted Averages
as Al: 5 mg/m3 TWA (Listed under 'Aluminum')

INTERNATIONAL LISTS

Canada - Alberta - 8 Hour Occupational Exposure Limit
5 mg/m3 TWA

Canada - Alberta - 15 Minute Occupational Exposure Limit
  10 mg/m3 STEL
Canada - Quebec - Time-Weighted Average Exposure Values
  5 mg/m3 TWAEV
Mexico - Instruction No. 10 - TWAs
  5 mg/m3 TWA
STATE LISTS
  California - Exposure Limits - PELs
    5 mg/m3 PEL
  California - Directors List of Hazardous Substances (8 CCR 339)
    [present]
  Pennsylvania Right to Know List
    [present]

ALUMINUM RESINATE   61789-65-9

HEALTH AND SAFETY LISTS
  U.S. DOT - Substances From 49 CFR 172.101
    regulated by DOT (UN2715)
  U.S. DOT - Hazard Classes
    DOT hazard class = 4.1
STATE LISTS
  NJ Right to Know List (Total)
    sn 0064

ALUMINUM SILICATE   1327-36-2
SEE ALSO:
  ALUMINUM

ENVIRONMENTAL LISTS
  List of Pesticide Product Inert Ingredients
    [present]

ALUMINUM SILICATE, HYDRATE   1335-30-4

ENVIRONMENTAL LISTS
  List of Pesticide Product Inert Ingredients
    [present]

ALUMINUM SILICON   57485-31-1

STATE LISTS
  NJ Right to Know List (Total)
    sn 0067

ALUMINUM SILICON (AL SI)   12042-55-6

HEALTH AND SAFETY LISTS
  U.S. DOT - Substances From 49 CFR 172.101
    regulated by DOT (UN1398)
  U.S. DOT - Hazard Classes
    DOT hazard class = 4.3
STATE LISTS
  NJ Right to Know List (Total)
    sn 0066

ALUMINUM SILICON (AL SI5)   50810-25-8

STATE LISTS
  NJ Right to Know List (Total)
    sn 0065

ALUMINUM SODIUM OXIDE   11138-49-1

STATE LISTS
  NJ Right to Know List (Total)
    sn 1675
  NJ Special Hazardous Substances
    (corrosive)

ALUMINUM, SOLUBLE SALTS   RR-00021-5

HEALTH AND SAFETY LISTS
  OSHA - Vacated PELs - Time Weighted Averages
    as Al: 2 mg/m3 TWA (Listed under 'Aluminum')
INTERNATIONAL LISTS
  Canada - WHMIS: Ingredient Disclosure
    1% item 53 (198)
  Canada - Alberta - 8 Hour Occupational Exposure Limit
    2 mg/m3 TWA
  Canada - Alberta - 15 Minute Occupational Exposure Limit
    4 mg/m3 STEL
  Canada - Quebec - Time-Weighted Average Exposure Values
    2 mg/m3 TWAEV
  United Kingdom - Occupational Exposure Standards - TWAs
    2 mg/m3 TWA
  Mexico - Instruction No. 10 - TWAs
    2 mg/m3 TWA
STATE LISTS
  California - Exposure Limits - PELs
    2 mg/m3 PEL
  California - Directors List of Hazardous Substances (8 CCR 339)
    [present] (refers to water-soluble salts only, all others exempt)
  Pennsylvania Right to Know List
    [present]

ALUMINUM STEARATE   637-12-7
SEE ALSO:
  ALUMINUM

ENVIRONMENTAL LISTS
  List of Pesticide Product Inert Ingredients
    [present]
STATE LISTS
  California - Exposure Limits - PELs
    10 mg/m3 PEL

ALUMINUM STEARATE   7047-84-9

STATE LISTS
  California - Exposure Limits - PELs
    10 mg/m3 PEL

ALUMINUM SULFATE   10043-01-3

HEALTH AND SAFETY LISTS
  U.S. DOT - Appendix A Table 1 - Hazardous Substances
    final RQ = 5000 pounds (2270 kg)
  NIOSH - Selected LD50s and LC50s
    Oral, mouse: LD50 = 6207 mg/kg
ENVIRONMENTAL LISTS
  CERCLA/SARA - Hazardous Substances and their Reportable Quantities
    final RQ = 5000 pounds (2270 kg)
  Clean Water Act - Hazardous Substances
    [present]
  List of Pesticide Product Inert Ingredients
    [present]
STATE LISTS
  Massachusetts Right To Know List
    [present]
  NJ Right to Know List (Total)
    sn 0068
  Pennsylvania Right to Know List
    environmental hazard

ALUMINUM, WELDING FUMES   RR-00020-4
HEALTH AND SAFETY LISTS
OSHA - Vacated PELs - Time Weighted Averages
as Al: 5 mg/m3 TWA (Listed under 'Aluminum')
INTERNATIONAL LISTS
Canada - Alberta - 8 Hour Occupational Exposure Limit
5 mg/m3 TWA
Canada - Alberta - 15 Minute Occupational Exposure Limit
10 mg/m3 STEL
Mexico - Instruction No. 10 - TWAs
5 mg/m3 TWA
STATE LISTS
California - Exposure Limits - PELs
5 mg/m3 PEL
Pennsylvania Right to Know List
[present]

AMERICIUM 237   29492-78-2
HEALTH AND SAFETY LISTS
U.S. DOT - Appendix A Table 2 - Radionuclides
final RQ = 1000 curies (3.7E 13 Bq)
ENVIRONMENTAL LISTS
CERCLA/SARA List of Radionuclides (Appendix B) and Their Reportable Quantities
final RQ = 1000 curies (3.7E 13 Bq)

AMERICIUM 238   18233-96-0
HEALTH AND SAFETY LISTS
U.S. DOT - Appendix A Table 2 - Radionuclides
final RQ = 100 curies (3.7E 12 Bq)
ENVIRONMENTAL LISTS
CERCLA/SARA List of Radionuclides (Appendix B) and Their Reportable Quantities
final RQ = 100 curies (3.7E 12 Bq)

AMERICIUM 239   16652-10-1
HEALTH AND SAFETY LISTS
U.S. DOT - Appendix A Table 2 - Radionuclides
final RQ = 100 curies (3.7E 12 Bq)
ENVIRONMENTAL LISTS
CERCLA/SARA List of Radionuclides (Appendix B) and Their Reportable Quantities
final RQ = 100 curies (3.7E 12 Bq)

AMERICIUM 240   15116-95-7
HEALTH AND SAFETY LISTS
U.S. DOT - Appendix A Table 2 - Radionuclides
final RQ = 10 curies (3.7E 11 Bq)
ENVIRONMENTAL LISTS
CERCLA/SARA List of Radionuclides (Appendix B) and Their Reportable Quantities
final RQ = 10 curies (3.7E 11 Bq)

AMERICIUM 241   14596-10-2
HEALTH AND SAFETY LISTS
U.S. DOT - Appendix A Table 2 - Radionuclides
final RQ = 0.01 curies (3.7E 8 Bq)
ENVIRONMENTAL LISTS
CERCLA/SARA List of Radionuclides (Appendix B) and Their Reportable Quantities
final RQ = 0.01 curies (3.7E 8 Bq)

AMERICIUM 242   13981-54-9
HEALTH AND SAFETY LISTS
U.S. DOT - Appendix A Table 2 - Radionuclides
final RQ = 100 curies (3.7E 12 Bq)
ENVIRONMENTAL LISTS
CERCLA/SARA List of Radionuclides (Appendix B) and Their Reportable Quantities
final RQ = 100 curies (3.7E 12 Bq)

AMERICIUM 242M   RR-00495-5
HEALTH AND SAFETY LISTS
U.S. DOT - Appendix A Table 2 - Radionuclides
final RQ = 0.01 curies (3.7E 8 Bq)
ENVIRONMENTAL LISTS
CERCLA/SARA List of Radionuclides (Appendix B) and Their Reportable Quantities
final RQ = 0.01 curies (3.7E 8 Bq)

AMERICIUM 243   14993-75-0
HEALTH AND SAFETY LISTS
U.S. DOT - Appendix A Table 2 - Radionuclides
final RQ = 0.01 curies (3.7E 8 Bq)
ENVIRONMENTAL LISTS
CERCLA/SARA List of Radionuclides (Appendix B) and Their Reportable Quantities
final RQ = 0.01 curies (3.7E 8 Bq)

AMERICIUM 244   15756-26-0
HEALTH AND SAFETY LISTS
U.S. DOT - Appendix A Table 2 - Radionuclides
final RQ = 10 curies (3.7E 11 Bq)
ENVIRONMENTAL LISTS
CERCLA/SARA List of Radionuclides (Appendix B) and Their Reportable Quantities
final RQ = 10 curies (3.7E 11 Bq)

AMERICIUM 244M   RR-00493-3
HEALTH AND SAFETY LISTS
U.S. DOT - Appendix A Table 2 - Radionuclides
final RQ = 1000 curies (3.7E 13 Bq)
ENVIRONMENTAL LISTS
CERCLA/SARA List of Radionuclides (Appendix B) and Their Reportable Quantities
final RQ = 1000 curies (3.7E 13 Bq)

AMERICIUM 245   16415-43-3
HEALTH AND SAFETY LISTS
U.S. DOT - Appendix A Table 2 - Radionuclides
final RQ = 1000 curies (3.7E 13 Bq)
ENVIRONMENTAL LISTS
CERCLA/SARA List of Radionuclides (Appendix B) and Their Reportable Quantities
final RQ = 1000 curies (3.7E 13 Bq)

AMERICIUM 246   15776-16-6
HEALTH AND SAFETY LISTS
U.S. DOT - Appendix A Table 2 - Radionuclides
final RQ = 1000 curies (3.7E 13 Bq)
ENVIRONMENTAL LISTS
CERCLA/SARA List of Radionuclides (Appendix B) and Their Reportable Quantities
final RQ = 1000 curies (3.7E 13 Bq)

AMERICIUM 246M   RR-00492-2

HEALTH AND SAFETY LISTS

U.S. DOT - Appendix A Table 2 - Radionuclides
final RQ = 1000 curies (3.7E 13 Bq)

ENVIRONMENTAL LISTS

CERCLA/SARA List of Radionuclides (Appendix B) and Their Reportable Quantities
final RQ = 1000 curies (3.7E 13 Bq)

AMETRYN (N-ETHYL-N'-(1-METHYLETHYL)-6-   834-12-8
(METHYLTHIO)-1,3,5,-TRIAZINE-2,4-DIAMINE)

ENVIRONMENTAL LISTS

CERCLA/SARA - Section 313 - Emission Reporting
form R reporting required

AMIDES, COCO, N-(HYDROXYETHYL),   68425-44-5
ETHOXYLATED

ENVIRONMENTAL LISTS

List of Pesticide Product Inert Ingredients
[present]

AMIDINODITHIOPROPIONIC ACID   RR-01209-9
HYDROCHLORIDE

ENVIRONMENTAL LISTS

TSCA - Chemicals with Significant New Use Rules
PMN number: P-91-102

AMIDOSULFOSUCCINATE   70904-61-9

ENVIRONMENTAL LISTS

List of Pesticide Product Inert Ingredients
[present]

AMIKACIN SULFATE   39831-55-5

STATE LISTS

California - Air Bill 2588 Appendix A-II
9/90

California - Prop. 65 - Developmental Toxicity
developmental toxicity - initial date 7/1/90

AMINES, C14-18-ALKYL, ETHOXYLATED   68155-33-9

ENVIRONMENTAL LISTS

List of Pesticide Product Inert Ingredients
[present]

AMINES, SOYA ALKYL, ETHOXYLATED   61791-24-0

ENVIRONMENTAL LISTS

List of Pesticide Product Inert Ingredients
[present]

AMINES, N-TALLOW ALKYLTRIMETHYLENEDI-,   68153-99-1
DIOLEATES

ENVIRONMENTAL LISTS

List of Pesticide Product Inert Ingredients
[present]

5-AMINOACENAPHTHENE   4657-93-6

HEALTH AND SAFETY LISTS

IARC - Group 3 (not classifiable)
[present]

M-AMINOACETOPHENONE   99-03-6

HEALTH AND SAFETY LISTS

NIOSH - Selected LD50s and LC50s
Oral, rat: LD50 = 1870 mg/kg Skin, rabbit: LD50 = 4340 mg/kg

P-AMINOACETOPHENONE   99-92-3

HEALTH AND SAFETY LISTS

NIOSH - Selected LD50s and LC50s
Oral, rat: LD50 = 381 mg/kg

9-AMINOACRIDINE HYDROCHLORIDE   134-50-9

HEALTH AND SAFETY LISTS

NIOSH - Selected LD50s and LC50s
Oral, mouse: LD50 = 78 mg/kg

NTP Chemical Status Reports - Testing Status and NTIS Number
Prechronic studies for which toxicity technical reports were not prepared

AMINO ACRYLATE MONOMER   RR-00241-5

ENVIRONMENTAL LISTS

TSCA - Chemicals with Significant New Use Rules
PMN numbers: P-85-296;   P-85-298

1-AMINOANTHRAQUINONE   82-45-1

ENVIRONMENTAL LISTS

EPA - Master Testing List
[present]

INTERNATIONAL LISTS

Canada - WHMIS: Ingredient Disclosure
1% item 54 (203)

2-AMINOANTHRAQUINONE   117-79-3

HEALTH AND SAFETY LISTS

IARC - Group 3 (not classifiable)
[present]

NTP Chemical Status Reports - Testing Status and NTIS Number
Technical reports printed (PB287739/AS)

NTP Chemical Status Reports - Evidence of Carcinogenicity
male rat-positive; female rat-inadequate; male mice-positive; female mice-positive

NTP Seventh Report - Suspect Carcinogens
suspect carcinogen

OSHA - Possible Select Carcinogens
[present]

ENVIRONMENTAL LISTS

CERCLA/SARA - Section 313 - Emission Reporting
form R reporting required for 0.1% de minimus concentration

INTERNATIONAL LISTS

Canada - WHMIS: Ingredient Disclosure
0.1% item 55 (204)

STATE LISTS

California - Air Bill 2588 Appendix A-I
known or potential carcinogen

California - Prop. 65 - Cancer list
carcinogen - initial date 10/1/89

California - Prop. 65 - No Significant Risk Levels
no significant risk level = 20 ug/day

California - Directors List of Hazardous Substances (8 CCR 339)
[present]

Florida Hazardous Substance List
[present]

Massachusetts Right To Know List
carcinogen; extraordinarily hazardous

Minnesota Hazardous Substance List
carcinogen

NJ Right to Know List (Total)
sn 0069

NJ Special Hazardous Substances
  (carcinogen)
Pennsylvania Right to Know List
  environmental hazard; special hazardous substance
Pennsylvania RTK - Special Hazardous Substances
  [present]

4-AMINOANTIPYRINE   83-07-8
HEALTH AND SAFETY LISTS
NIOSH - Selected LD50s and LC50s
  Oral, rat: LD50 = 1700 mg/kg

4-AMINOAZOBENZENE   60-09-3
HEALTH AND SAFETY LISTS
IARC - Group 2B (sufficient animal data)
  *[present] (Degree of evidence in animals revised on the basis of data
  that appeared after the most recent monograph and/or on the basis of
  present criteria)*
OSHA - Possible Select Carcinogens
  [present]
ENVIRONMENTAL LISTS
CERCLA/SARA - Section 313 - Emission Reporting
  form R reporting required for 0.1% de minimus concentration
INTERNATIONAL LISTS
Canada - WHMIS: Ingredient Disclosure
  1% item 56 (205)
STATE LISTS
California - Air Bill 2588 Appendix A-II
  known or potential carcinogen
California - Prop. 65 - Cancer list
  carcinogen - initial date 1/1/90
California - Directors List of Hazardous Substances (8 CCR 339)
  [present]
Massachusetts Right To Know List
  [present]
Minnesota Hazardous Substance List
  carcinogen
NJ Right to Know List (Total)
  sn 0508
Pennsylvania Right to Know List
  environmental hazard

4-AMINOAZOBENZENE-3,4'-DISULFONIC ACID   101-50-8
INTERNATIONAL LISTS
Canada - WHMIS: Ingredient Disclosure
  1% item 57 (54)

O-AMINOAZOTOLUENE   97-56-3
HEALTH AND SAFETY LISTS
IARC - Group 2B (sufficient animal data)
  *[present] (Overall evaluation based only on evidence of carcinogenicity
  in monograph (8, 1975) or in Supplement 4)*
NTP Seventh Report - Suspect Carcinogens
  suspect carcinogen
OSHA - Possible Select Carcinogens
  [present]
ENVIRONMENTAL LISTS
CERCLA/SARA - Section 313 - Emission Reporting
  form R reporting required for 0.1% de minimus concentration
INTERNATIONAL LISTS
Canada - WHMIS: Ingredient Disclosure
  1% item 58 (206)

German (DFG) - Carcinogens
  animal evidence of carcinogenicity
STATE LISTS
California - Air Bill 2588 Appendix A-II
  known or potential carcinogen
California - Prop. 65 - Cancer list
  carcinogen - initial date 7/1/87
California - Prop. 65 - No Significant Risk Levels
  no significant risk level = 0.2 ug/day
California - Directors List of Hazardous Substances (8 CCR 339)
  [present]
Florida Hazardous Substance List
  [present]
Massachusetts Right To Know List
  carcinogen; extraordinarily hazardous
Minnesota Hazardous Substance List
  carcinogen
NJ Right to Know List (Total)
  sn 0507
Pennsylvania Right to Know List
  environmental hazard; special hazardous substance
Pennsylvania RTK - Special Hazardous Substances
  [present]

2-AMINO-P-BENZENEDISULFONIC ACID   98-44-2
INTERNATIONAL LISTS
Canada - WHMIS: Ingredient Disclosure
  1% item 59 (55)

M-AMINOBENZENESULFONIC ACID, SODIUM SALT   1126-34-7
INTERNATIONAL LISTS
Canada - WHMIS: Ingredient Disclosure
  1% item 60 (207)

P-AMINOBENZOIC ACID   150-13-0
HEALTH AND SAFETY LISTS
AIHA - WEEL - Time Weighted Averages
  5 mg/m3 TWA
IARC - Group 3 (not classifiable)
  [present]
NIOSH - Selected LD50s and LC50s
  Oral, mouse: LD50 = 2850 mg/kg
ENVIRONMENTAL LISTS
List of Pesticide Product Inert Ingredients
  [present]
INTERNATIONAL LISTS
Canada - WHMIS: Ingredient Disclosure
  1% item 61 (56)

4-AMINOBENZOPHENONE   1137-41-3
HEALTH AND SAFETY LISTS
NIOSH - Selected LD50s and LC50s
  Oral, bird: LD50 = 562 mg/kg

4-AMINOBIPHENYL   92-67-1
SEE ALSO:
F039-HAZARDOUS WASTES
HEALTH AND SAFETY LISTS
ACGIH 1995 - Skin Designations
  skin - potential for cutaneous absorption
ACGIH 1995 - Carcinogens
  A1-confirmed human carcinogen

IARC - Group 1 (carcinogenic to humans)
[present]

NIOSH - Selected LD50s and LC50s
Oral, rat: LD50 = 500 mg/kg

NIOSH 1990 - Pocket Guide - Carcinogens
occupational carcinogen

NIOSH 1990 - Pocket Guide - Target organs
bladder, skin

NIOSH - Health Standards - Exposure Limits
use 29 CFR 1910.1011

NIOSH - Health Standards - Health Effects and Precautions
Bladder cancer

NIOSH - Health Standards - Carcinogenic Chemicals
potential human carcinogen

NTP Seventh Report - Known Carcinogens
known carcinogen

OSHA - 29 CFR 1910 Specifically Regulated Chemicals
Cancer suspect agent (see 29 CFR 1910.1011)

OSHA - Select Carcinogens
[present]

ENVIRONMENTAL LISTS

CERCLA/SARA - Section 313 - Emission Reporting
form R reporting required for 0.1% de minimus concentration

CERCLA/SARA - Hazardous Substances and their Reportable Quantities
final RQ = 1 pound (.454 kg)

Clean Air Act (1990) - List of Hazardous Air Contaminants
[present]

RCRA - Hazardous Constituents-Appendix VIII
hazardous constituent - no waste number

RCRA - Basis for Listing - Appendix VII
Included in waste stream: F039

RCRA - TSD Facilities Ground Water Monitoring
TM 8270 = 10 ug/L PQL

RCRA - Universal Treatment Standards (LDR)
WW: 0.13 mg/l; NWW: Not applicable

INTERNATIONAL LISTS

Australian Exposure Standards - Time Weighted Averages
prohibition recommended

Australian Exposure Standards - Skin Effects
skin absorption

Australian Exposure Standards - Carcinogens
confirmed carcinogen

Canada - WHMIS: Ingredient Disclosure
0.1% item 63 (210)

Canada - Alberta - Designated Substances
designated substance - requires code of practice

Canada - British Columbia - 8 Hour Exposure Limits
carcinogen with no permitted exposure or contact by any route

Canada - British Columbia - Skin Notations
skin - potential for skin absorption

Canada - British Columbia - Carcinogens
carcinogen with no permitted exposure or contact by any route

Canada - Quebec - Time-Weighted Average Exposure Values
substance of which the recirculation is prohibited

Canada - Quebec - Skin Designations
absorbed through the skin

Canada - Quebec - Carcinogens
C1 carcinogen: effect detected in humans

German (DFG) - Carcinogens
proven carcinogen

Mexico - Instruction No. 10 - Carcinogens
carcinogen in humans

STATE LISTS

California - Air Bill 2588 Appendix A-I
known or potential carcinogen

California - Air Bill 2588 Appendix A-II
[present]

California - Prop. 65 - Cancer list
carcinogen - initial date 2/27/87

California - Prop. 65 - No Significant Risk Levels
no significant risk level = 0.03 ug/day

California - Exposure Limits - Skin Notation
material may be absorbed through the skin, eyes or mucous membrane

California - Exposure Limits - Carcinogens
cancer-suspect agent (at a concentration of >= 0.1%)

California - Directors List of Hazardous Substances (8 CCR 339)
[present] (refers to any mixture greater than 0.1%)

Florida Hazardous Substance List
[present]

Massachusetts Right To Know List
carcinogen; extraordinarily hazardous

Minnesota Hazardous Substance List
carcinogen; skin

NJ Right to Know List (Total)
sn 0072

NJ Special Hazardous Substances
(carcinogen; mutagen)

Pennsylvania Right to Know List
environmental hazard; special hazardous substance

Pennsylvania RTK - Special Hazardous Substances
[present]

2-AMINO-1-BUTANOL   96-20-8

HEALTH AND SAFETY LISTS

NFPA - Flash Points
flash point = 165 degrees F (74 degrees C)

NFPA - Hazard Identification Ratings
health-2; flammability-2; reactivity-0

NIOSH - Selected LD50s and LC50s
Oral, mouse: LD50 = 2300 mg/kg

STATE LISTS

Florida Hazardous Substance List
[present]

Massachusetts Right To Know List
[present]

Pennsylvania Right to Know List
[present]

AMINOCARB   2032-59-9

HEALTH AND SAFETY LISTS

U.S. DOT - Appendix B - Marine Pollutants
DOT regulated marine pollutant

2-AMINO-4-CHLOROPHENOL   95-85-2

HEALTH AND SAFETY LISTS

U.S. DOT - Substances From 49 CFR 172.101
regulated by DOT (UN2673)

U.S. DOT - Hazard Classes
DOT hazard class = 6.1

NIOSH - Selected LD50s and LC50s
Oral, rat: LD50 = 690 mg/kg

INTERNATIONAL LISTS

Canada - WHMIS: Ingredient Disclosure
1% item 62 (208)

STATE LISTS
> NJ Right to Know List (Total)
> *sn 0070*

1-AMINO-2,4-DIBROMOANTHRAQUINONE   81-49-2

HEALTH AND SAFETY LISTS
> NTP Chemical Status Reports - Testing Status and NTIS Number
> *Post peer review technical reports in progress*

2-AMINO-5-DIETHYLAMINOPENTANE   140-80-7

HEALTH AND SAFETY LISTS
> U.S. DOT - Substances From 49 CFR 172.101
> *regulated by DOT (UN2946)*
> U.S. DOT - Hazard Classes
> *DOT hazard class = 6.1*

STATE LISTS
> NJ Right to Know List (Total)
> *sn 0071*

3-[[4-AMINO-9,10-DIHYDRO-9,10-DIOXO-3-[SULFO-4-(1,1,3,3-TETRAMETHYLBUTYL)PHENOXY]-1-ANTHRACENYL]AMINO]-2,4,6-TRIMETHYL BENZENE-SULFONIC ACID DISODIUM SALT   72243-90-4

ENVIRONMENTAL LISTS
> List of Pesticide Product Inert Ingredients
> *[present]*

2-AMINO-4,5-DIHYDRO-6-METHYL-4-PROPYL-5-TRIAZOLO-(1,5-C)-PYRAMIDIN-5-ONE   27277-00-5

ENVIRONMENTAL LISTS
> List of Pesticide Product Inert Ingredients
> *[present]*

2-AMINO-2,3-DIMETHYLBUTYRONITRILE   RR-00503-8

HEALTH AND SAFETY LISTS
> U.S. DOT - Substances Which Are Poisonous by Inhalation
> *liquid hazardous material poisonous by inhalation (with Toluene solutions)*

7-AMINO-2,2-DIMETHYL-2,3-DIHYDROBENZOFURAN   68298-46-4

ENVIRONMENTAL LISTS
> TSCA - Code of Federal Regulations Citations
> *40 CFR 712.30(g);   40 CFR 716.120(a)*
> TSCA - PAIR - Reporting List
> *Reporting Date: Ocotober 8, 1984*
> TSCA - Health and Safety Reporting List
> *Effective Date: February 13, 1984*

3-AMINO-1,4-DIMETHYL-5H-PYRIDO (4,3-B) INDOLE ACETATE   68808-54-8

STATE LISTS
> Pennsylvania Right to Know List
> *special hazardous substance*
> Pennsylvania RTK - Special Hazardous Substances
> *[present]*

P-AMINODIPHENYLAMINE   101-54-2

HEALTH AND SAFETY LISTS
> NTP Chemical Status Reports - Testing Status and NTIS Number
> *Technical reports printed (PB285856/AS)*
> NTP Chemical Status Reports - Evidence of Carcinogenicity
> *male rat-negative; female rat-negative; male mice-negative; female mice-negative*

ENVIRONMENTAL LISTS
> EPA - Master Testing List
> *[present]*

INTERNATIONAL LISTS
> Canada - WHMIS: Ingredient Disclosure
> *0.1% item 64 (209)*

3-AMINO-4-ETHOXYACETANILIDE   17026-81-2

HEALTH AND SAFETY LISTS
> NTP Chemical Status Reports - Testing Status and NTIS Number
> *Technical reports printed (PB285194/AS)*
> NTP Chemical Status Reports - Evidence of Carcinogenicity
> *male rat-negative; female rat-negative; male mice-positive; female mice-negative*

STATE LISTS
> Massachusetts Right To Know List
> *carcinogen; extraordinarily hazardous*

AMINOETHOXYETHANOL   110-76-9

HEALTH AND SAFETY LISTS
> U.S. DOT - Substances From 49 CFR 172.101
> *regulated by DOT (UN3055)*
> U.S. DOT - Hazard Classes
> *DOT hazard class = 8*

2-(2-AMINOETHOXY)ETHANOL   929-06-6

HEALTH AND SAFETY LISTS
> NIOSH - Selected LD50s and LC50s
> *Oral, rat: LD50 = 5660 mg/kg Skin, rabbit: LD50 = 1190 mg/kg*

ENVIRONMENTAL LISTS
> TSCA - Code of Federal Regulations Citations
> *40 CFR 712.30(x);   40 CFR 716.120(d)*
> TSCA - PAIR - Reporting List
> *Reporting Date: November 27, 1991*
> TSCA - Health and Safety Reporting List
> *Effective Date: September 30, 1991*

INTERNATIONAL LISTS
> Canada - WHMIS: Ingredient Disclosure
> *1% item 65 (211)*

STATE LISTS
> Massachusetts Right To Know List
> *[present]*
> NJ Right to Know List (Total)
> *sn 0073*
> NJ Special Hazardous Substances
> *(corrosive)*

5-AMINO-6-ETHOXY-2-NAPHTHALENESULFONIC ACID   118-28-5

INTERNATIONAL LISTS
> Canada - WHMIS: Ingredient Disclosure
> *1% item 66 (57)*

3-AMINO-9-ETHYLCARBAZOLE   132-32-1

HEALTH AND SAFETY LISTS
> NIOSH - Selected LD50s and LC50s
> *Oral, rat: LD50 = 144 mg/kg*

INTERNATIONAL LISTS
> German (DFG) - Carcinogens
> *suspected carcinogen*

3-AMINO-9-ETHYLCARBAZOLE HCL   6109-97-3

HEALTH AND SAFETY LISTS

NTP Chemical Status Reports - Testing Status and NTIS Number
Technical reports printed (PB287126/AS)

NTP Chemical Status Reports - Evidence of Carcinogenicity
male rat-positive; female rat-positive; male mice-positive; female mice-positive

STATE LISTS

California - Air Bill 2588 Appendix A-II
known or potential carcinogen: 9/89

California - Prop. 65 - Cancer list
carcinogen - initial date 7/1/89

California - Prop. 65 - No Significant Risk Levels
no significant risk level = 9 ug/day

Massachusetts Right To Know List
carcinogen; extraordinarily hazardous

AMINOETHYLETHANOLAMINE   111-41-1

HEALTH AND SAFETY LISTS

NFPA - Flash Points
flash point = 270 degrees F (132 degrees C)

NFPA - Hazard Identification Ratings
health-2; flammability-1; reactivity-0

NIOSH - Selected LD50s and LC50s
Oral, rat: LD50 = 3000 mg/kg Skin, rat: LD50 = 2250 mg/kg

INTERNATIONAL LISTS

Canada - WHMIS: Ingredient Disclosure
0.1% item 67 (212)

STATE LISTS

Florida Hazardous Substance List
[present]

Massachusetts Right To Know List
[present]

Pennsylvania Right to Know List
[present]

AMINOETHYLETHYLENE UREA   RR-01642-2
METHACRYLAMIDE

ENVIRONMENTAL LISTS

TSCA - Chemicals with Significant New Use Rules
PMN number: P-89-1038

4-(2-AMINOETHYL)-MORPHOLINE   2038-03-1

HEALTH AND SAFETY LISTS

NFPA - Hazard Identification Ratings
health-2; flammability-2; reactivity-0

STATE LISTS

Florida Hazardous Substance List
[present]

Massachusetts Right To Know List
[present]

Pennsylvania Right to Know List
[present]

1-(2-AMINOETHYL) PIPERAZINE   140-31-8

HEALTH AND SAFETY LISTS

U.S. DOT - Substances From 49 CFR 172.101
regulated by DOT (UN2815)

U.S. DOT - Hazard Classes
DOT hazard class = 8

NFPA - Flash Points
flash point = 200 degrees F (93 degrees C)

NFPA - Hazard Identification Ratings
health-2; flammability-2; reactivity-0

NIOSH - Selected LD50s and LC50s
Oral, rat: LD50 = 2140 mg/kg Skin, rabbit: LD50 = 880 mg/kg

INTERNATIONAL LISTS

Canada - WHMIS: Ingredient Disclosure
1% item 68 (213)

STATE LISTS

Florida Hazardous Substance List
[present]

Massachusetts Right To Know List
[present]

NJ Right to Know List (Total)
sn 0075

NJ Special Hazardous Substances
(corrosive)

Pennsylvania Right to Know List
[present]

AMINOGLUTETHIMIDE   125-84-8

STATE LISTS

California - Air Bill 2588 Appendix A-II
9/90

California - Prop. 65 - Developmental Toxicity
developmental toxicity - initial date 7/1/90

AMINOGLYCOSIDES   RR-01471-1

STATE LISTS

California - Prop. 65 - Developmental Toxicity
developmental toxicity - initial date 10/1/92

2-AMINO-3-HYDROXYBENZOIC ACID   548-93-6

INTERNATIONAL LISTS

Canada - WHMIS: Ingredient Disclosure
1% item 69 (58)

2-AMINO-4-[(2-HYDROXYETHYL)SULFONYL]   17601-96-6
PHENOL

ENVIRONMENTAL LISTS

TSCA - Code of Federal Regulations Citations
40 CFR 712.30(x); 40 CFR 716.120(d)

TSCA - PAIR - Reporting List
Reporting Date: November 27, 1991

TSCA - Health and Safety Reporting List
Effective Date: September 30, 1991

7-AMINO-4-HYDROXY-2-NAPHTHALENESULFONIC   87-02-5
ACID

ENVIRONMENTAL LISTS

TSCA - Code of Federal Regulations Citations
40 CFR 712.30(x); 40 CFR 716.120(d)

TSCA - PAIR - Reporting List
Reporting Date: November 27, 1991

TSCA - Health and Safety Reporting List
Effective Date: September 30, 1991

1-AMINO-2-METHYLANTHRAQUINONE   82-28-0

HEALTH AND SAFETY LISTS

IARC - Group 3 (not classifiable)
[present]

NTP Chemical Status Reports - Testing Status and NTIS Number
Technical reports printed (PB286852/AS)

NTP Chemical Status Reports - Evidence of Carcinogenicity
male rat-positive; female rat-positive; male mice-negative; female mice-positive

NTP Seventh Report - Suspect Carcinogens
suspect carcinogen

OSHA - Possible Select Carcinogens
[present]

ENVIRONMENTAL LISTS

CERCLA/SARA - Section 313 - Emission Reporting
form R reporting required for 0.1% de minimus concentration

INTERNATIONAL LISTS

Canada - WHMIS: Ingredient Disclosure
0.1% item 70 (214)

STATE LISTS

California - Air Bill 2588 Appendix A-II
known or potential carcinogen

California - Prop. 65 - Cancer list
carcinogen - initial date 10/1/89

California - Prop. 65 - No Significant Risk Levels
no significant risk level = 5 ug/day

California - Directors List of Hazardous Substances (8 CCR 339)
[present]

Florida Hazardous Substance List
[present]

Massachusetts Right To Know List
carcinogen; extraordinarily hazardous

Minnesota Hazardous Substance List
carcinogen

NJ Right to Know List (Total)
sn 0076

NJ Special Hazardous Substances
(carcinogen)

Pennsylvania Right to Know List
environmental hazard; special hazardous substance

Pennsylvania RTK - Special Hazardous Substances
[present]

2-AMINO-2-METHYLBUTANENITRILE   4475-95-0

HEALTH AND SAFETY LISTS

U.S. DOT - Substances Which Are Poisonous by Inhalation
liquid hazardous material poisonous by inhalation

2-AMINO-2-METHYLPROPANENITRILE   19355-69-2

HEALTH AND SAFETY LISTS

U.S. DOT - Substances Which Are Poisonous by Inhalation
liquid hazardous material poisonous by inhalation

2-AMINO-2-METHYL-1-PROPANOL   124-68-5

HEALTH AND SAFETY LISTS

NFPA - Flash Points
flash point = 153 degrees F (67 degrees C)

NFPA - Hazard Identification Ratings
health-2; flammability-2; reactivity-0

ENVIRONMENTAL LISTS

List of Pesticide Product Inert Ingredients
[present]

STATE LISTS

Florida Hazardous Substance List
[present]

Massachusetts Right To Know List
[present]

Pennsylvania Right to Know List
[present]

1-AMINO-2-METHYL-2-PROPANOL   2854-16-2

HEALTH AND SAFETY LISTS

NIOSH - Selected LD50s and LC50s
Oral, mouse: LD50 = 2450 mg/kg

2-AMINO-4-METHYLPYRIDINE   3731-51-9

HEALTH AND SAFETY LISTS

NIOSH - Selected LD50s and LC50s
Oral, bird: LD50 = 562 mg/kg

3-AMINO-1-METHYL-5H-PYRIDO (4,3-B) INDOLE ACETATE   72254-58-1

STATE LISTS

Pennsylvania Right to Know List
special hazardous substance

Pennsylvania RTK - Special Hazardous Substances
[present]

2-AMINO-3-METHYL-9H-PYRIDO(2,3,-B)INDOLE (METHYL A-ALPHA-C)   68006-83-7

HEALTH AND SAFETY LISTS

IARC - Group 2B (sufficient animal data)
[present] (Overall evaluation based only on evidence of carcinogenicity in monograph (40, 1986) or in Supplement 4)

OSHA - Possible Select Carcinogens
[present]

STATE LISTS

California - Air Bill 2588 Appendix A-II
known or potential carcinogen: 9/89

California - Prop. 65 - Cancer list
carcinogen - initial date 1/1/90

California - Prop. 65 - No Significant Risk Levels
no significant risk level = 0.6 ug/day

California - Directors List of Hazardous Substances (8 CCR 339)
[present]

Massachusetts Right To Know List
carcinogen; extraordinarily hazardous

Minnesota Hazardous Substance List
carcinogen

2-AMINO-4-(METHYLSULONYL)PHENOL   98-30-6

ENVIRONMENTAL LISTS

TSCA - Code of Federal Regulations Citations
40 CFR 712.30(x); 40 CFR 716.120(d)

TSCA - PAIR - Reporting List
Reporting Date: November 27, 1991

TSCA - Health and Safety Reporting List
Effective Date: September 30, 1991

3-AMINO-1,5-NAPHTHALENEDISULFONIC ACID   131-27-1

INTERNATIONAL LISTS

Canada - WHMIS: Ingredient Disclosure
1% item 72 (60)

6-AMINO-1,3-NAPHTHALENEDISULFONIC ACID   118-33-2

INTERNATIONAL LISTS

Canada - WHMIS: Ingredient Disclosure
1% item 73 (61)

2-AMINO-1-NAPHTHALENESULFONIC ACID   81-16-3

ENVIRONMENTAL LISTS

CAA - HON Rule - SOCMI Chemicals
compliance by Oct. 23, 1995

INTERNATIONAL LISTS
  Canada - WHMIS: Ingredient Disclosure
    1% item 74 (62)

2-[(6-AMINO-2-NAPHTHALENYL)SULFONYL] ETHANOL   52218-35-6

ENVIRONMENTAL LISTS
  TSCA - Code of Federal Regulations Citations
    40 CFR 712.30(x); 40 CFR 716.120(d)
  TSCA - PAIR - Reporting List
    Reporting Date: November 27, 1991
  TSCA - Health and Safety Reporting List
    Effective Date: September 30, 1991

2-AMINO-1,5-NAPHTHALINEDISULFONIC ACID   117-62-4

INTERNATIONAL LISTS
  Canada - WHMIS: Ingredient Disclosure
    1% item 71 (59)

2-AMINO-4-NITROANILINE   99-56-9
SEE ALSO:
  2-AMINO-4-NITROANILINE
HEALTH AND SAFETY LISTS
  IARC - Group 3 (not classifiable)
    [present]
  NIOSH - Selected LD50s and LC50s
    Oral, rat: LD50 = 681 mg/kg
  NTP Chemical Status Reports - Testing Status and NTIS Number
    Technical reports printed (PB290306/AS)
  NTP Chemical Status Reports - Evidence of Carcinogenicity
    male rat-negative; female rat-negative; male mice-negative; female mice-negative

ENVIRONMENTAL LISTS
  TSCA - Code of Federal Regulations Citations
    40 CFR 712.30(d); 40 CFR 716.120(c)
  TSCA - PAIR - Reporting List
    Reporting Date: November 19, 1982

INTERNATIONAL LISTS
  Canada - WHMIS: Ingredient Disclosure
    1% item 75 (215)

2-AMINO-5-(5-NITRO-2-FURYL)-1,3,4-THIADIAZOLE   712-68-5

STATE LISTS
  California - Air Bill 2588 Appendix A-II
    known or potential carcinogen
  California - Prop. 65 - Cancer list
    carcinogen - initial date 7/1/87
  California - Prop. 65 - No Significant Risk Levels
    no significant risk level = 0.04 ug/day
  California - Directors List of Hazardous Substances (8 CCR 339)
    [present]
  Florida Hazardous Substance List
    [present]
  Massachusetts Right To Know List
    carcinogen; extraordinarily hazardous
  Pennsylvania Right to Know List
    special hazardous substance
  Pennsylvania RTK - Special Hazardous Substances
    [present]

2-AMINO-5-(5-NITRO-2-FURYL)-1,3,4-THIADIAZOLE   59716-87-9

HEALTH AND SAFETY LISTS
  IARC - Group 2B (sufficient animal data)
    [present] (Overall evaluation based only on evidence of carcinogenicity in monograph (7, 1974) or in Supplement 4)
  OSHA - Possible Select Carcinogens
    [present]

STATE LISTS
  Minnesota Hazardous Substance List
    carcinogen

2-AMINO-4-NITROPHENOL   99-57-0

HEALTH AND SAFETY LISTS
  IARC - Group 3 (not classifiable)
    [present]
  NIOSH - Selected LD50s and LC50s
    Oral, rat: LD50 = 1030 mg/kg
  NTP Chemical Status Reports - Testing Status and NTIS Number
    Technical reports printed (PB89128623/AS)
  NTP Chemical Status Reports - Evidence of Carcinogenicity
    male rat-some evidence; female rat-no evidence; male mice-no evidence; female mice-no evidence

2-AMINO-5-NITROPHENOL   121-88-0

HEALTH AND SAFETY LISTS
  IARC - Group 3 (not classifiable)
    [present]
  NTP Chemical Status Reports - Testing Status and NTIS Number
    Technical reports printed (PB88184809/AS)
  NTP Chemical Status Reports - Evidence of Carcinogenicity
    male rat-some evidence; female rat-no evidence; male mice-no evidence; female mice-no evidence

2-AMINO-5-NITROTHIAZOLE   121-66-4

HEALTH AND SAFETY LISTS
  IARC - Group 3 (not classifiable)
    [present]
  NTP Chemical Status Reports - Testing Status and NTIS Number
    Technical reports printed (PB283346/AS)
  NTP Chemical Status Reports - Evidence of Carcinogenicity
    male rat-positive; female rat-negative; male mice-negative; female mice-negative

STATE LISTS
  California - Directors List of Hazardous Substances (8 CCR 339)
    [present]
  Massachusetts Right To Know List
    carcinogen; extraordinarily hazardous

O-AMINOPHENOL   95-55-6

HEALTH AND SAFETY LISTS
  NIOSH - Selected LD50s and LC50s
    Oral, rat: LD50 = 1300 mg/kg

INTERNATIONAL LISTS
  Canada - WHMIS: Ingredient Disclosure
    0.1% item 78 (218)

P-AMINOPHENOL   123-30-8

HEALTH AND SAFETY LISTS
  NIOSH - Selected LD50s and LC50s
    Oral, rat: LD50 = 375 mg/kg Skin, mammal: LD50 = 6400 mg/kg

ENVIRONMENTAL LISTS
  CAA - HON Rule - SOCMI Chemicals
    compliance by Oct. 24, 1994

EPA - Master Testing List
[present]
INTERNATIONAL LISTS
Canada - WHMIS: Ingredient Disclosure
1% item 79 (219)
PROPOSED REGULATIONS
TSCA - Proposed Substances for Developmental/ReproductiveTesting
proposed testing for: Developmental Toxicity - oral

M-AMINOPHENOL   591-27-5
HEALTH AND SAFETY LISTS
NIOSH - Selected LD50s and LC50s
Oral, rat: LD50 = 1 gm/kg
INTERNATIONAL LISTS
Canada - WHMIS: Ingredient Disclosure
0.1% item 77 (217)

AMINOPHENOL   27598-85-2
HEALTH AND SAFETY LISTS
U.S. DOT - Substances From 49 CFR 172.101
regulated by DOT (UN2512)
U.S. DOT - Hazard Classes
DOT hazard class = 6.1
STATE LISTS
NJ Right to Know List (Total)
sn 0078

P-AMINOPHENOL HYDROCHLORIDE   51-78-5
INTERNATIONAL LISTS
Canada - WHMIS: Ingredient Disclosure
0.1% item 80 (220)

AMINOPHENOL SULFONIC ACID   RR-01747-0
ENVIRONMENTAL LISTS
CAA - HON Rule - SOCMI Chemicals
compliance by Oct. 23, 1995

**2-(4-AMINOPHENYL)-6-METHYL-7-BENZOTHIA-
ZOLE SULFONIC ACID**   130-17-6
HEALTH AND SAFETY LISTS
NTP Chemical Status Reports - Testing Status and NTIS Number
*Prechronic studies for which toxicity technical reports were not pre-
pared*

3-AMINOPHENYL SULFONE   599-61-1
HEALTH AND SAFETY LISTS
NIOSH - Selected LD50s and LC50s
Oral, rat: LD50 = 4920 mg/kg

2-[(3-AMINOPHENYL)SULFONYL]ETHANOL   5246-57-1
ENVIRONMENTAL LISTS
TSCA - Code of Federal Regulations Citations
40 CFR 712.30(x); 40 CFR 716.120(d)
TSCA - PAIR - Reporting List
Reporting Date: November 27, 1991
TSCA - Health and Safety Reporting List
Effective Date: September 30, 1991

N-(3-AMINOPROPYL) CYCLOHEXYLAMINE   3312-60-5
HEALTH AND SAFETY LISTS
NFPA - Flash Points
flash point = 175 degrees F (79 degrees C)

NFPA - Hazard Identification Ratings
health-2; flammability-2; reactivity-0
STATE LISTS
Florida Hazardous Substance List
[present]
Massachusetts Right To Know List
[present]
Pennsylvania Right to Know List
[present]

AMINOPROPYLDIETHANOLAMINE   4985-85-7
INTERNATIONAL LISTS
Canada - WHMIS: Ingredient Disclosure
1% item 81 (221)
STATE LISTS
NJ Right to Know List (Total)
sn 0079

N-AMINOPROPYLMORPHOLINE   123-00-2
HEALTH AND SAFETY LISTS
NFPA - Flash Points
flash point = 220 degrees F (104 degrees C)
NFPA - Hazard Identification Ratings
health-2; flammability-1; reactivity-0
NIOSH - Selected LD50s and LC50s
Oral, rat: LD50 = 3560 mg/kg Skin, rabbit: LD50 = 1230 mg/kg
INTERNATIONAL LISTS
Canada - WHMIS: Ingredient Disclosure
1% item 82 (222)
STATE LISTS
Florida Hazardous Substance List
[present]
Massachusetts Right To Know List
[present]
NJ Right to Know List (Total)
sn 0080
NJ Special Hazardous Substances
(corrosive)
Pennsylvania Right to Know List
[present]

AMINOPTERIN   54-62-6
ENVIRONMENTAL LISTS
CERCLA/SARA - Section 302 Extremely Hazardous Substances and
TPQs
TPQ = 500/10,000 pounds
STATE LISTS
California - Air Bill 2588 Appendix A-II
[present]
California - Prop. 65 - Developmental Toxicity
developmental toxicity - initial date 7/1/87
California - Prop. 65 - Reproductive - Female
female reproductive toxicity - initial date 7/1/87
Florida Hazardous Substance List
effective March 13, 1992
Massachusetts Right To Know List
extraordinarily hazardous; teratogen
NJ Right to Know List (Total)
sn 2112
Pennsylvania Right to Know List
environmental hazard

PROPOSED REGULATIONS

CERCLA/SARA - Proposed Hazardous Substance Additions
proposed RQ = 1 pound (.454 kg)
CERCLA/SARA - 1989 Proposed RQ Adjustments
proposed RQ = 10 pounds (4.54 kg)

1-AMINO PYRENE   1606-67-3

HEALTH AND SAFETY LISTS

NIOSH - Selected LD50s and LC50s
Oral, rat: LD50 = 1070 mg/kg

3-AMINO PYRIDINE   462-08-8

HEALTH AND SAFETY LISTS

NIOSH - Selected LD50s and LC50s
Oral, bird: LD50 = 13300 ug/kg

STATE LISTS

NJ Right to Know List (Total)
sn 0082

4-AMINOPYRIDINE   504-24-5

HEALTH AND SAFETY LISTS

U.S. DOT - Appendix A Table 1 - Hazardous Substances
final RQ = 1000 pounds (454 kg)
NIOSH - Selected LD50s and LC50s
Oral, rat: LD50 = 20 mg/kg

ENVIRONMENTAL LISTS

CERCLA/SARA - Section 302 Extremely Hazardous Substances and TPQs
TPQ = 500/10,000 pounds
CERCLA/SARA - Hazardous Substances and their Reportable Quantities
final RQ = 1000 pounds (454 kg)
RCRA - P Series Wastes
waste number P008
RCRA - Hazardous Constituents-Appendix VIII
waste number P008
RCRA - Substances Banned From Land Disposal
[present]

STATE LISTS

California - Directors List of Hazardous Substances (8 CCR 339)
[present]
Florida Hazardous Substance List
effective March 13, 1992
Massachusetts Right To Know List
extraordinarily hazardous
NJ Right to Know List (Total)
sn 0172
Pennsylvania Right to Know List
environmental hazard

2-AMINOPYRIDINE   504-29-0

HEALTH AND SAFETY LISTS

ACGIH 1995 - Time Weighted Averages
0.5 ppm TWA; 1.9 mg/m3 TWA
NIOSH - Selected LD50s and LC50s
Oral, bird: LD50 = 31600 ug/kg
NIOSH 1990 - Pocket Guide - RELs
0.5 ppm TWA; 2 mg/m3 TWA
NIOSH 1990 - Pocket Guide - IDLHs
5 ppm IDLH
NIOSH 1990 - Pocket Guide - Target organs
CNS, respiratory system

OSHA - Vacated PELs - Time Weighted Averages
0.5 ppm TWA; 2 mg/m3 TWA
OSHA - Final PELs - Time Weighted Averages
0.5 ppm TWA; 2 mg/m3 TWA

INTERNATIONAL LISTS

Australian Exposure Standards - Time Weighted Averages
0.5 ppm TWA; 2 mg/m3 TWA
Canada - WHMIS: Ingredient Disclosure
1% item 83 (223)
Canada - Alberta - 8 Hour Occupational Exposure Limit
0.5 ppm TWA; 1.9 mg/m3 TWA
Canada - Alberta - 15 Minute Occupational Exposure Limit
2 ppm STEL; 7.7 mg/m3 STEL
Canada - British Columbia - 8 Hour Exposure Limits
0.5 ppm TWA; 2 mg/m3 TWA
Canada - British Columbia - 15 Minute Exposure Limits
2 ppm STEL; 4 mg/m3 STEL
Canada - Ontario - OHSA - TWAEVs
0.5 ppm TWAEV; 2 mg/m3 TWAEV
Canada - Quebec - Time-Weighted Average Exposure Values
0.5 ppm TWAEV; 2 mg/m3 TWAEV
United Kingdom - Occupational Exposure Standards - TWAs
0.5 ppm TWA; 2 mg/m3 TWA
United Kingdom - Occupational Exposure Standards - STELs
2 ppm STEL; 8 mg/m3 STEL
German (DFG) - MAK Values
0.5 ppm MAK; 2 mg/m3 MAK
Israel - Time Weighted Averages
0.5 ppm TWA; 2 mg/m3 TWA
Israel - Action Levels
0.25 ppm AL; 1.0 mg/m3 AL

STATE LISTS

California - Exposure Limits - PELs
0.5 ppm PEL; 2 mg/m3 PEL
California - Directors List of Hazardous Substances (8 CCR 339)
[present]
Florida Hazardous Substance List
[present]
Massachusetts Right To Know List
[present]
Minnesota Hazardous Substance List
[present]
NJ Right to Know List (Total)
sn 0081
Pennsylvania Right to Know List
[present]

AMINOPYRIDINES   RR-01354-7

HEALTH AND SAFETY LISTS

U.S. DOT - Substances From 49 CFR 172.101
regulated by DOT (UN2671)
U.S. DOT - Hazard Classes
DOT hazard class = 6.1

1-AMINOPYRINE   58-15-1

HEALTH AND SAFETY LISTS

NIOSH - Selected LD50s and LC50s
Oral, rat: LD50 = 685 mg/kg

4-AMINOSALICYLIC ACID   65-49-6

INTERNATIONAL LISTS

Canada - WHMIS: Ingredient Disclosure
1% item 84 (63)

4-AMINO-2,2,6,6-TETRAMETHYLPIPERIDINE   36768-62-4

HEALTH AND SAFETY LISTS
NIOSH - Selected LD50s and LC50s
Oral, rat: LD50 = 906 mg/kg

AMINOTRI(METHYLENEPHOSPHONIC ACID)   6419-19-8

HEALTH AND SAFETY LISTS
NIOSH - Selected LD50s and LC50s
Oral, rat: LD50 = 2100 mg/kg

ENVIRONMENTAL LISTS
EPA - Master Testing List
[present]

List of Pesticide Product Inert Ingredients
[present]

11-AMINOUNDECANOIC ACID   2432-99-7

HEALTH AND SAFETY LISTS
IARC - Group 3 (not classifiable)
[present]

NTP Chemical Status Reports - Testing Status and NTIS Number
Technical reports printed (PB82225640)

NTP Chemical Status Reports - Evidence of Carcinogenicity
male rat-positive; female rat-negative; male mice-equivocal; female mice-negative

ENVIRONMENTAL LISTS
TSCA - Code of Federal Regulations Citations
40 CFR 704.25; 40 CFR 721.350

TSCA - Chemicals with Significant New Use Rules
[present]

TSCA - Section 12(b) - Export Notification
export notification required - Section 5

STATE LISTS
California - Directors List of Hazardous Substances (8 CCR 339)
[present]

AMITON   78-53-5

HEALTH AND SAFETY LISTS
NIOSH - Selected LD50s and LC50s
Oral, rat: LD50 = 3300 ug/kg

ENVIRONMENTAL LISTS
CERCLA/SARA - Section 302 Extremely Hazardous Substances and TPQs
TPQ = 500 pounds

STATE LISTS
Florida Hazardous Substance List
effective March 13, 1992

Massachusetts Right To Know List
extraordinarily hazardous

NJ Right to Know List (Total)
sn 2113

Pennsylvania Right to Know List
environmental hazard

PROPOSED REGULATIONS
CERCLA/SARA - Proposed Hazardous Substance Additions
proposed RQ = 1 pound (.454 kg)

CERCLA/SARA - 1989 Proposed RQ Adjustments
proposed RQ = 100 pounds (45.4 kg)

AMITON OXALATE   3734-97-2

HEALTH AND SAFETY LISTS
NIOSH - Selected LD50s and LC50s
Oral, rat: LD50 = 3 mg/kg

ENVIRONMENTAL LISTS
CERCLA/SARA - Section 302 Extremely Hazardous Substances and TPQs
TPQ = 100/10,000 pounds

STATE LISTS
Florida Hazardous Substance List
effective March 13, 1992

Massachusetts Right To Know List
extraordinarily hazardous

NJ Right to Know List (Total)
sn 2114

Pennsylvania Right to Know List
environmental hazard

PROPOSED REGULATIONS
CERCLA/SARA - Proposed Hazardous Substance Additions
proposed RQ = 1 pound (.454 kg)

CERCLA/SARA - 1989 Proposed RQ Adjustments
proposed RQ = 100 pounds (45.4 kg)

AMITRAZ   33089-61-1

ENVIRONMENTAL LISTS
CERCLA/SARA - Section 313 - Emission Reporting
form R reporting required

STATE LISTS
Massachusetts Right To Know List
[present]

AMITROLE   61-82-5

HEALTH AND SAFETY LISTS
ACGIH 1995 - Time Weighted Averages
0.2 mg/m3 TWA

U.S. DOT - Appendix A Table 1 - Hazardous Substances
final RQ = 10 pounds (4.54 kg)

IARC - Group 2B (sufficient animal data)
[present]

NIOSH - Selected LD50s and LC50s
Oral, rat: LD50 = 1100 mg/kg

NTP Seventh Report - Suspect Carcinogens
suspect carcinogen

OSHA - Vacated PELs - Time Weighted Averages
0.2 mg/m3 TWA

OSHA - Possible Select Carcinogens
[present]

ENVIRONMENTAL LISTS
CERCLA/SARA - Section 313 - Emission Reporting
form R reporting required

CERCLA/SARA - Hazardous Substances and their Reportable Quantities
final RQ = 10 pounds (4.54 kg)

EPA - Carcinogen Hazard Ranking for RQ Adjustment
Hazard ranking = Medium

RCRA - U Series Wastes
waste number U011

RCRA - Hazardous Constituents-Appendix VIII
waste number U011

RCRA - Substances Banned From Land Disposal
[present]

TSCA - Health and Safety Reporting List
Effective Date: March 11, 1994; Sunset Date: March 11, 2004

INTERNATIONAL LISTS
Australian Exposure Standards - Time Weighted Averages
0.2 mg/m3 TWA

Canada - Alberta - 8 Hour Occupational Exposure Limit
0.2 mg/m3 TWA

Canada - Alberta - 15 Minute Occupational Exposure Limit
0.5 mg/m3 STEL

Canada - Ontario - OHSA - TWAEVs
0.2 mg/m3 TWAEV

Canada - Quebec - Time-Weighted Average Exposure Values
0.2 mg/m3 TWAEV

Canada - Quebec - Carcinogens
C3 carcinogen: effect detected in animals

German (DFG) - MAK Values
total dust: 0.2 mg/m3 MAK

Israel - Time Weighted Averages
0.2 mg/m3 TWA

Israel - Action Levels
0.1 mg/m3 AL

STATE LISTS

California - Air Bill 2588 Appendix A-I
known or potential carcinogen

California - Prop. 65 - Cancer list
carcinogen - initial date 7/1/87

California - Prop. 65 - No Significant Risk Levels
no significant risk level = 0.7 ug/day

California - Exposure Limits - PELs
0.2 mg/m3 PEL

California - Directors List of Hazardous Substances (8 CCR 339)
[present]

Florida Hazardous Substance List
[present]

Massachusetts Right To Know List
carcinogen; extraordinarily hazardous

Minnesota Hazardous Substance List
carcinogen

NJ Right to Know List (Total)
sn 0083

NJ Special Hazardous Substances
(carcinogen; mutagen)

Pennsylvania Right to Know List
environmental hazard; special hazardous substance

Pennsylvania RTK - Special Hazardous Substances
[present]

PROPOSED REGULATIONS

TSCA - ITC 32nd Report Priority Testing List
designated for dermal absorption testing

AMMONIA   7664-41-7

HEALTH AND SAFETY LISTS

ACGIH 1995 - Time Weighted Averages
25 ppm TWA; 17 mg/m3 TWA

ACGIH 1995 - Short Term Exposure Limits
35 ppm STEL; 24 mg/m3 STEL

AIHA - Odor Threshold Values
geometric mean air odor threshold = 17 ppm (detectable)

U.S. DOT - Substances From 49 CFR 172.101
regulated by DOT (UN1005, UN2073, UN2672)

U.S. DOT - Hazard Classes
DOT hazard class = 2.3

U.S. DOT - Substances Which Are Poisonous by Inhalation
gaseous hazardous material poisonous by inhalation (anhydrous form) (UN1005)

U.S. DOT - Appendix A Table 1 - Hazardous Substances
final RQ = 100 pounds (45.4 kg)

NFPA - Flash Points
gas (no flash point given)

NFPA - Hazard Identification Ratings
health-3; flammability-1; reactivity-0

NIOSH - Selected LD50s and LC50s
Inhalation, rat: LC50 = 2000 ppm 4 hr Oral, rat: LD50 = 350 mg/kg

NIOSH 1990 - Pocket Guide - RELs
25 ppm TWA; 18 mg/m3 TWA; 35 ppm STEL; 27 mg/m3 STEL

NIOSH 1990 - Pocket Guide - IDLHs
500 ppm IDLH

NIOSH 1990 - Pocket Guide - Target organs
respiratory system, eyes

NIOSH - Health Standards - Exposure Limits
C (5 min) 50 ppm; C (5 min) 34.8 mg/m3

NIOSH - Health Standards - Health Effects and Precautions
Respiratory and eye irritation (Prevent eye contact)

OSHA - Vacated PELs - Short Term Exposure Limits
35 ppm STEL; 27 mg/m3 STEL

OSHA - Final PELs - Time Weighted Averages
50 ppm TWA; 35 mg/m3 TWA

OSHA - List of Highly Hazardous Chemicals
anhydrous: threshhold quantity = 10,000 pounds; solutions > 44% ammonia by weight: threshhold quantity = 15,000 pounds

ENVIRONMENTAL LISTS

ATSDR Priority List
Rank (of 275): 151

CERCLA/SARA - Section 302 Extremely Hazardous Substances and TPQs
TPQ = 500 pounds

CERCLA/SARA - Section 313 - Emission Reporting
form R reporting required for 1.0% de minimus concentration

CERCLA/SARA - Hazardous Substances and their Reportable Quantities
final RQ = 100 pounds (45.4 kg)

CAA -Toxic Substances for Accidental Release Prevention
anhydrous: threshold quantity = 10,000 lbs; 20% or greater: threshold quantity = 20,000 lbs

Clean Water Act - Hazardous Substances
[present]

INTERNATIONAL LISTS

Australian Exposure Standards - Time Weighted Averages
25 ppm TWA; 17 mg/m3 TWA

Australian Exposure Standards - Short Term Exposure Limits
35 ppm STEL; 24 mg/m3 STEL

Canada - WHMIS: Ingredient Disclosure
1% item 86 (225)

Canada - NPRI (National Pollutant Release Inventory)
[present]

Canada - Alberta - 8 Hour Occupational Exposure Limit
25 ppm TWA; 17 mg/m3 TWA

Canada - Alberta - 15 Minute Occupational Exposure Limit
35 ppm STEL; 24 mg/m3 STEL

Canada - British Columbia - 8 Hour Exposure Limits
25 ppm TWA; 18 mg/m3 TWA

Canada - British Columbia - 15 Minute Exposure Limits
35 ppm STEL; 27 mg/m3 STEL

Canada - Ontario - OHSA - TWAEVs
25 ppm TWAEV; 17 mg/m3 TWAEV

Canada - Ontario - OHSA - STEVs
35 ppm STEV; 24 mg/m3 STEV

Canada - Quebec - Time-Weighted Average Exposure Values
25 ppm TWAEV; 17 mg/m3 TWAEV

Canada - Quebec - Short-term Exposure Values
35 ppm STEV; 24 mg/m3 STEV

United Kingdom - Occupational Exposure Standards - TWAs
25 ppm TWA; 17 mg/m3 TWA

United Kingdom - Occupational Exposure Standards - STELs
35 ppm STEL;   24 mg/m3 STEL
German (DFG) - MAK Values
50 ppm MAK;   35 mg/m3 MAK
German (DFG) - Peak Limitations
2 x normal MAK (5 min momentary value);   don't exceed 8 times during shift
German (DFG) - Pregnancy
no risk to embryo/fetus if exposure limits adhered to
Israel - Time Weighted Averages
25 ppm TWA;   17 mg/m3 TWA
Israel - Short Term Exposure Limits
35 ppm STEL;   24 mg/m3 STEL
Israel - Action Levels
12.5 ppm AL;   8.5 mg/m3 AL
Mexico - Instruction No. 10 - TWAs
25 ppm TWA;   18 mg/m3 TWA
Mexico - Instruction No. 10 - STELs
35 ppm STEL;   27 mg/m3 STEL

STATE LISTS
California - Air Bill 2588 Appendix A-I
[present]
California - Exposure Limits - PELs
25 ppm PEL;   18 mg/m3 PEL
California - Exposure Limits - STELs
35 ppm STEL;   27 mg/m3 STEL
California - Directors List of Hazardous Substances (8 CCR 339)
[present]
Florida Hazardous Substance List
[present]
Massachusetts Right To Know List
extraordinarily hazardous
Minnesota Hazardous Substance List
[present]
NJ Right to Know List (Total)
sn 0084
NJ Special Hazardous Substances
(corrosive)
Pennsylvania Right to Know List
environmental hazard

AMMONIUM ACETATE   631-61-8

HEALTH AND SAFETY LISTS
U.S. DOT - Appendix A Table 1 - Hazardous Substances
final RQ = 5000 pounds (2270 kg)

ENVIRONMENTAL LISTS
CERCLA/SARA - Hazardous Substances and their Reportable Quantities
final RQ = 5000 pounds (2270 kg)
Clean Water Act - Hazardous Substances
[present]
List of Pesticide Product Inert Ingredients
[present]

STATE LISTS
California - Directors List of Hazardous Substances (8 CCR 339)
[present]
Massachusetts Right To Know List
[present]
NJ Right to Know List (Total)
sn 0085
Pennsylvania Right to Know List
environmental hazard

AMMONIUM ALUM   7784-25-0

ENVIRONMENTAL LISTS
List of Pesticide Product Inert Ingredients
[present]

AMMONIUM ARSENATE   7784-44-3
SEE ALSO:
ARSENIC

HEALTH AND SAFETY LISTS
U.S. DOT - Substances From 49 CFR 172.101
regulated by DOT (UN1546)
U.S. DOT - Hazard Classes
DOT hazard class = 6.1
U.S. DOT - Appendix B - Marine Pollutants
DOT regulated marine pollutant

STATE LISTS
NJ Right to Know List (Total)
sn 0086

AMMONIUM AZIDE   12164-94-2

HEALTH AND SAFETY LISTS
U.S. DOT - Hazard Classes
Forbidden from transport by the DOT

AMMONIUM BENZOATE   1863-63-4

HEALTH AND SAFETY LISTS
U.S. DOT - Appendix A Table 1 - Hazardous Substances
final RQ = 5000 pounds (2270 kg)

ENVIRONMENTAL LISTS
CERCLA/SARA - Hazardous Substances and their Reportable Quantities
final RQ = 5000 pounds (2270 kg)
Clean Water Act - Hazardous Substances
[present]

STATE LISTS
California - Directors List of Hazardous Substances (8 CCR 339)
[present]
Massachusetts Right To Know List
[present]
NJ Right to Know List (Total)
sn 0087
Pennsylvania Right to Know List
environmental hazard

AMMONIUM BICARBONATE   1066-33-7

HEALTH AND SAFETY LISTS
U.S. DOT - Appendix A Table 1 - Hazardous Substances
final RQ = 5000 pounds (2270 kg)

ENVIRONMENTAL LISTS
CERCLA/SARA - Hazardous Substances and their Reportable Quantities
final RQ = 5000 pounds (2270 kg)
Clean Water Act - Hazardous Substances
[present]
List of Pesticide Product Inert Ingredients
[present]

STATE LISTS
California - Directors List of Hazardous Substances (8 CCR 339)
[present]
Massachusetts Right To Know List
[present]

NJ Right to Know List (Total)
sn 0088

Pennsylvania Right to Know List
environmental hazard

AMMONIUM BICHROMATE   7789-09-5
SEE ALSO:
    CHROMIUM
    CHROMIUM (VI) COMPOUNDS- WATER SOLUBLE
    CHROMIUM COMPOUNDS
    CHROMIUM (VI) COMPOUNDS

HEALTH AND SAFETY LISTS
    U.S. DOT - Substances From 49 CFR 172.101
        regulated by DOT (UN1439)
    U.S. DOT - Hazard Classes
        DOT hazard class = 5.1
    U.S. DOT - Appendix A Table 1 - Hazardous Substances
        final RQ = 10 pounds (4.54 kg)

ENVIRONMENTAL LISTS
    CERCLA/SARA - Hazardous Substances and their Reportable Quantities
        final RQ = 10 pounds (4.54 kg)
    Clean Water Act - Hazardous Substances
        [present]
    EPA - Carcinogen Hazard Ranking for RQ Adjustment
        Hazard ranking = High

INTERNATIONAL LISTS
    Canada - WHMIS: Ingredient Disclosure
        1% item 93 (686)

STATE LISTS
    Florida Hazardous Substance List
        [present]
    Massachusetts Right To Know List
        [present]
    NJ Right to Know List (Total)
        sn 0097
    Pennsylvania Right to Know List
        environmental hazard

AMMONIUM BIFLUORIDE   1341-49-7
HEALTH AND SAFETY LISTS
    U.S. DOT - Substances From 49 CFR 172.101
        regulated by DOT (UN2817, UN1727)
    U.S. DOT - Hazard Classes
        DOT hazard class = 8
    U.S. DOT - Appendix A Table 1 - Hazardous Substances
        final RQ = 100 pounds (45.4 kg)

ENVIRONMENTAL LISTS
    CERCLA/SARA - Hazardous Substances and their Reportable Quantities
        final RQ = 100 pounds (45.4 kg)
    Clean Water Act - Hazardous Substances
        [present]

STATE LISTS
    Massachusetts Right To Know List
        [present]
    NJ Right to Know List (Total)
        sn 0089
    Pennsylvania Right to Know List
        environmental hazard

AMMONIUM BISULFITE   10192-30-0
HEALTH AND SAFETY LISTS
    U.S. DOT - Appendix A Table 1 - Hazardous Substances
        final RQ = 5000 pounds (2270 kg)

ENVIRONMENTAL LISTS
    CERCLA/SARA - Hazardous Substances and their Reportable Quantities
        final RQ = 5000 pounds (2270 kg)
    Clean Water Act - Hazardous Substances
        [present] (Listed under 'Ammonium sulfite')

INTERNATIONAL LISTS
    Canada - WHMIS: Ingredient Disclosure
        1% item 87 (306)

STATE LISTS
    California - Directors List of Hazardous Substances (8 CCR 339)
        [present]
    Massachusetts Right To Know List
        [present]
    NJ Right to Know List (Total)
        sn 0090
    Pennsylvania Right to Know List
        environmental hazard

AMMONIUM BROMATE   13843-59-9
HEALTH AND SAFETY LISTS
    U.S. DOT - Hazard Classes
        Forbidden from transport by the DOT

AMMONIUM BROMIDE   12124-97-9
ENVIRONMENTAL LISTS
    List of Pesticide Product Inert Ingredients
        [present]

STATE LISTS
    Florida Hazardous Substance List
        [present]
    Massachusetts Right To Know List
        [present]
    Pennsylvania Right to Know List
        [present]

AMMONIUM C6-10-ALKYL POLYOXYETHYLENE   68037-05-8
SULFATE
ENVIRONMENTAL LISTS
    List of Pesticide Product Inert Ingredients
        [present]

AMMONIUM CARBAMATE   1111-78-0
HEALTH AND SAFETY LISTS
    U.S. DOT - Appendix A Table 1 - Hazardous Substances
        final RQ = 5000 pounds (2270 kg)

ENVIRONMENTAL LISTS
    CERCLA/SARA - Hazardous Substances and their Reportable Quantities
        final RQ = 5000 pounds (2270 kg)
    Clean Water Act - Hazardous Substances
        [present]
    List of Pesticide Product Inert Ingredients
        [present]
    TSCA - Code of Federal Regulations Citations
        40 CFR 712.30(x); 40 CFR 716.120(d)
    TSCA - PAIR - Reporting List
        Reporting Date: November 27, 1991

TSCA - Health and Safety Reporting List
Effective Date: September 30, 1991

STATE LISTS

California - Directors List of Hazardous Substances (8 CCR 339)
[present]

Massachusetts Right To Know List
[present]

NJ Right to Know List (Total)
sn 0091

Pennsylvania Right to Know List
environmental hazard

AMMONIUM CARBONATE   506-87-6

HEALTH AND SAFETY LISTS

U.S. DOT - Appendix A Table 1 - Hazardous Substances
final RQ = 5000 pounds (2270 kg)

ENVIRONMENTAL LISTS

CERCLA/SARA - Hazardous Substances and their Reportable Quantities
final RQ = 5000 pounds (2270 kg)

Clean Water Act - Hazardous Substances
[present]

List of Pesticide Product Inert Ingredients
[present]

STATE LISTS

California - Directors List of Hazardous Substances (8 CCR 339)
[present]

Massachusetts Right To Know List
[present]

NJ Right to Know List (Total)
sn 0092

Pennsylvania Right to Know List
environmental hazard

AMMONIUM CARBONATE   10361-29-2

STATE LISTS

NJ Right to Know List (Total)
sn 0092

Pennsylvania Right to Know List
environmental hazard

AMMONIUM CASEINATE   9005-42-9

ENVIRONMENTAL LISTS

List of Pesticide Product Inert Ingredients
[present]

AMMONIUM CHLORATE   10192-29-7

HEALTH AND SAFETY LISTS

U.S. DOT - Hazard Classes
Forbidden from transport by the DOT

AMMONIUM CHLORIDE   12125-02-9

HEALTH AND SAFETY LISTS

ACGIH 1995 - Time Weighted Averages
10 mg/m3 TWA

ACGIH 1995 - Short Term Exposure Limits
20 mg/m3 STEL

U.S. DOT - Appendix A Table 1 - Hazardous Substances
final RQ = 5000 pounds (2270 kg)

NIOSH - Selected LD50s and LC50s
Oral, rat: LD50 = 1650 mg/kg

OSHA - Vacated PELs - Time Weighted Averages
10 mg/m3 TWA

OSHA - Vacated PELs - Short Term Exposure Limits
20 mg/m3 STEL

ENVIRONMENTAL LISTS

CERCLA/SARA - Hazardous Substances and their Reportable Quantities
final RQ = 5000 pounds (2270 kg)

Clean Water Act - Hazardous Substances
[present]

List of Pesticide Product Inert Ingredients
[present]

INTERNATIONAL LISTS

Australian Exposure Standards - Time Weighted Averages
fume: 10 mg/m3 TWA

Australian Exposure Standards - Short Term Exposure Limits
fume: 20 mg/m3 STEL

Canada - WHMIS: Ingredient Disclosure
1% item 88 (471)

Canada - Alberta - 8 Hour Occupational Exposure Limit
as fume: 10 mg/m3 TWA

Canada - Alberta - 15 Minute Occupational Exposure Limit
as fume: 20 mg/m3 STEL

Canada - British Columbia - 8 Hour Exposure Limits
fume: 10 mg/m3 TWA

Canada - British Columbia - 15 Minute Exposure Limits
20 mg/m3 STEL

Canada - Ontario - OHSA - TWAEVs
10 mg/m3 TWAEV

Canada - Ontario - OHSA - STEVs
20 mg/m3 STEV

Canada - Quebec - Time-Weighted Average Exposure Values
10 mg/m3 TWAEV

Canada - Quebec - Short-term Exposure Values
20 mg/m3 STEV

United Kingdom - Occupational Exposure Standards - TWAs
fume: 10 mg/m3 TWA

United Kingdom - Occupational Exposure Standards - STELs
fume: 20 mg/m3 STEL

Israel - Time Weighted Averages
10 mg/m3 TWA

Israel - Short Term Exposure Limits
20 mg/m3 STEL

Israel - Action Levels
5 mg/m3 AL

Mexico - Instruction No. 10 - TWAs
10 mg/m3 TWA

Mexico - Instruction No. 10 - STELs
20 mg/m3 STEL

STATE LISTS

California - Exposure Limits - PELs
fume: 10 mg/m3 PEL

California - Exposure Limits - STELs
fume: 20 mg/m3 STEL

California - Directors List of Hazardous Substances (8 CCR 339)
[present]

Florida Hazardous Substance List
[present]

Massachusetts Right To Know List
[present]

Minnesota Hazardous Substance List
[present]

NJ Right to Know List (Total)
sn 0093

Pennsylvania Right to Know List
environmental hazard

AMMONIUM CHLOROPALLADATE   19168-23-1
INTERNATIONAL LISTS
Canada - WHMIS: Ingredient Disclosure
1% item 89 (447)

AMMONIUM CHLOROPALLADITE   13820-40-1
INTERNATIONAL LISTS
Canada - WHMIS: Ingredient Disclosure
1% item 90 (449)

AMMONIUM CHLOROPLATINATE   16919-58-7
SEE ALSO:
PLATINUM
HEALTH AND SAFETY LISTS
NIOSH - Selected LD50s and LC50s
Oral, rat: LD50 = 195 mg/kg
INTERNATIONAL LISTS
Canada - WHMIS: Ingredient Disclosure
0.1% item 91 (456)
STATE LISTS
Massachusetts Right To Know List
[present]

AMMONIUM CHROMATE   7788-98-9
SEE ALSO:
CHROMIUM
HEALTH AND SAFETY LISTS
U.S. DOT - Appendix A Table 1 - Hazardous Substances
final RQ = 10 pounds (4.54 kg)
ENVIRONMENTAL LISTS
CERCLA/SARA - Hazardous Substances and their Reportable Quantities
final RQ = 10 pounds (4.54 kg)
Clean Water Act - Hazardous Substances
[present]
EPA - Carcinogen Hazard Ranking for RQ Adjustment
Hazard ranking = High
INTERNATIONAL LISTS
Canada - WHMIS: Ingredient Disclosure
1% item 92 (546)
STATE LISTS
Massachusetts Right To Know List
[present]
NJ Right to Know List (Total)
sn 0095
Pennsylvania Right to Know List
environmental hazard

AMMONIUM CITRATE, DIBASIC   3012-65-5
HEALTH AND SAFETY LISTS
U.S. DOT - Appendix A Table 1 - Hazardous Substances
final RQ = 5000 pounds (2270 kg)
ENVIRONMENTAL LISTS
CERCLA/SARA - Hazardous Substances and their Reportable Quantities
final RQ = 5000 pounds (2270 kg)
Clean Water Act - Hazardous Substances
[present]
List of Pesticide Product Inert Ingredients
[present]

STATE LISTS
California - Directors List of Hazardous Substances (8 CCR 339)
[present]
Massachusetts Right To Know List
[present]
NJ Right to Know List (Total)
sn 0096
Pennsylvania Right to Know List
environmental hazard

AMMONIUM, (4-(ALPHA-(P-(DIETHYLAMINO)   129-17-9
PHENYL)-2,4-DISULFOBENZYLIDENE)-2,5-CYCLO-
HEXADIEN-1-YLIDENE)DIETHYL-, HYDROXIDE,
MONOSODIUM SALT
HEALTH AND SAFETY LISTS
IARC - Group 3 (not classifiable)
[present]
ENVIRONMENTAL LISTS
List of Pesticide Product Inert Ingredients
[present]
STATE LISTS
California - Directors List of Hazardous Substances (8 CCR 339)
[present]

AMMONIUM DIISODECYL SULFOSUCCINATE   94313-89-0
ENVIRONMENTAL LISTS
List of Pesticide Product Inert Ingredients
[present]

AMMONIUM DINITRO-O-CRESOLATE   2980-64-5
HEALTH AND SAFETY LISTS
U.S. DOT - Substances From 49 CFR 172.101
regulated by DOT (UN1843)
U.S. DOT - Hazard Classes
DOT hazard class = 6.1
U.S. DOT - Appendix B - Marine Pollutants
DOT regulated marine pollutant
STATE LISTS
NJ Right to Know List (Total)
sn 0098

AMMONIUM DINITRO-O-CRESOLATE   29595-25-3
INTERNATIONAL LISTS
Canada - WHMIS: Ingredient Disclosure
1% item 94 (758)

AMMONIUM DODECYL ALCOHOL POLYOXYETHY-   RR-01036-6
LENE PHOSPHATE
ENVIRONMENTAL LISTS
List of Pesticide Product Inert Ingredients
[present]

AMMONIUM ETHYL CARBAMOYLPHOSPHONATE   25954-13-6
HEALTH AND SAFETY LISTS
NIOSH - Selected LD50s and LC50s
Oral, rat: LD50 = 24000 mg/kg

AMMONIUM FLUOBORATE   13826-83-0
HEALTH AND SAFETY LISTS
U.S. DOT - Appendix A Table 1 - Hazardous Substances
final RQ = 5000 pounds (2270 kg)

ENVIRONMENTAL LISTS

CERCLA/SARA - Hazardous Substances and their Reportable Quantities
final RQ = 5000 pounds (2270 kg)

Clean Water Act - Hazardous Substances
[present]

STATE LISTS

California - Directors List of Hazardous Substances (8 CCR 339)
[present]

Massachusetts Right To Know List
[present]

NJ Right to Know List (Total)
sn 0100

Pennsylvania Right to Know List
environmental hazard

AMMONIUM FLUORIDE 12125-01-8

HEALTH AND SAFETY LISTS

U.S. DOT - Substances From 49 CFR 172.101
regulated by DOT (UN2505)

U.S. DOT - Hazard Classes
DOT hazard class = 6.1

U.S. DOT - Appendix A Table 1 - Hazardous Substances
final RQ = 100 pounds (45.4 kg)

ENVIRONMENTAL LISTS

CERCLA/SARA - Hazardous Substances and their Reportable Quantities
final RQ = 100 pounds (45.4 kg)

Clean Water Act - Hazardous Substances
[present]

STATE LISTS

Florida Hazardous Substance List
[present]

Massachusetts Right To Know List
[present]

NJ Right to Know List (Total)
sn 0099

Pennsylvania Right to Know List
environmental hazard

AMMONIUM FLUOSILICATE 1309-32-6

HEALTH AND SAFETY LISTS

U.S. DOT - Substances From 49 CFR 172.101
regulated by DOT (UN2854)

U.S. DOT - Hazard Classes
DOT hazard class = 6.1

STATE LISTS

NJ Right to Know List (Total)
sn 0101

AMMONIUM FORMATE 540-69-2

HEALTH AND SAFETY LISTS

NIOSH - Selected LD50s and LC50s
Oral, mouse: LD50 = 2250 mg/kg

ENVIRONMENTAL LISTS

List of Pesticide Product Inert Ingredients
[present]

AMMONIUM FULMINATE 134282-14-7

HEALTH AND SAFETY LISTS

U.S. DOT - Hazard Classes
Forbidden from transport by the DOT

AMMONIUM FULMINATE RR-01391-2

STATE LISTS

NJ Right to Know List (Total)
sn 2226

AMMONIUM HYDROGEN SULFATE 7803-63-6

HEALTH AND SAFETY LISTS

U.S. DOT - Substances From 49 CFR 172.101
regulated by DOT (UN2506)

U.S. DOT - Hazard Classes
DOT hazard class = 8

ENVIRONMENTAL LISTS

List of Pesticide Product Inert Ingredients
[present]

STATE LISTS

NJ Right to Know List (Total)
sn 0102

NJ Special Hazardous Substances
(corrosive)

AMMONIUM HYDROSULFIDE SOLUTION 12124-99-1

STATE LISTS

NJ Right to Know List (Total)
sn 2115

AMMONIUM HYDROXIDE 1336-21-6

HEALTH AND SAFETY LISTS

U.S. DOT - Appendix A Table 1 - Hazardous Substances
final RQ = 1000 pounds (454 kg)

NIOSH - Selected LD50s and LC50s
Oral, rat: LD50 = 350 mg/kg

ENVIRONMENTAL LISTS

CERCLA/SARA - Hazardous Substances and their Reportable Quantities
final RQ = 1000 pounds (454 kg)

Clean Water Act - Hazardous Substances
[present]

List of Pesticide Product Inert Ingredients
[present]

INTERNATIONAL LISTS

Canada - WHMIS: Ingredient Disclosure
1% item 96 (989)

STATE LISTS

California - Directors List of Hazardous Substances (8 CCR 339)
[present] (refers to sulutions greater than or equal to 4%)

Massachusetts Right To Know List
[present]

NJ Right to Know List (Total)
sn 0103

Pennsylvania Right to Know List
environmental hazard

AMMONIUM LACTATE 515-98-0

ENVIRONMENTAL LISTS

List of Pesticide Product Inert Ingredients
[present]

AMMONIUM LAURYL SULFATE 2235-54-3

ENVIRONMENTAL LISTS

List of Pesticide Product Inert Ingredients
[present]

AMMONIUM MOLYBDATE   13106-76-8

HEALTH AND SAFETY LISTS

  NIOSH - Selected LD50s and LC50s
    Oral, rat: LD50 = 333 mg/kg

INTERNATIONAL LISTS

  Canada - WHMIS: Ingredient Disclosure
    1% item 98 (1161)

AMMONIUM NITRATE   6484-52-2

HEALTH AND SAFETY LISTS

  U.S. DOT - Substances From 49 CFR 172.101
    regulated by DOT (UN2426, UN1942, UN0222)

  U.S. DOT - Hazard Classes
    DOT hazard class = 5.1

  NIOSH - Selected LD50s and LC50s
    Oral, rat: LD50 = 4820 mg/kg

ENVIRONMENTAL LISTS

  CERCLA/SARA - Section 313 - Emission Reporting
    form R reporting required for 1.0% de minimus concentration (only in solution)

  List of Pesticide Product Inert Ingredients
    [present]

INTERNATIONAL LISTS

  Canada - NPRI (National Pollutant Release Inventory)
    [present] (as solution)

STATE LISTS

  California - Air Bill 2588 Appendix A-I
    6/91

  Florida Hazardous Substance List
    [present]

  Massachusetts Right To Know List
    [present]

  NJ Right to Know List (Total)
    sn 0106

  NJ Special Hazardous Substances
    (reactive - third degree)

  Pennsylvania Right to Know List
    environmental hazard

AMMONIUM NITRATE FERTILIZERS   RR-00142-3

HEALTH AND SAFETY LISTS

  U.S. DOT - Substances From 49 CFR 172.101
    regulated by DOT (NA2069, NA2072, UN2071)

  U.S. DOT - Hazard Classes
    DOT hazard class = 5.1; certain non-oxidizing types: DOT hazard class = 9

STATE LISTS

  NJ Right to Know List (Total)
    sn 2116; sn 2117; sn 2118; sn 2119; sn 2120; sn 2121

AMMONIUM NITRATE-FUEL OIL MIXTURES   RR-00148-9

STATE LISTS

  NJ Right to Know List (Total)
    sn 2122

AMMONIUM NITRATE PHOSPHATE   57608-40-9

STATE LISTS

  NJ Right to Know List (Total)
    sn 0107

AMMONIUM NITRATE-SULFATE MIXTURES   RR-00149-0

STATE LISTS

  NJ Right to Know List (Total)
    sn 2126

AMMONIUM NONYLPHENYL POLYOXYETHYLENE SULFATE   9051-57-4

ENVIRONMENTAL LISTS

  List of Pesticide Product Inert Ingredients
    [present]

AMMONIUM OLEATE   544-60-5

ENVIRONMENTAL LISTS

  List of Pesticide Product Inert Ingredients
    [present]

AMMONIUM OXALATE   1113-38-8

STATE LISTS

  NJ Right to Know List (Total)
    sn 0108

AMMONIUM OXALATE, MONOHYDRATE   6009-70-7

HEALTH AND SAFETY LISTS

  U.S. DOT - Appendix A Table 1 - Hazardous Substances
    final RQ = 5000 pounds (2270 kg) (Listed under 'Ammonium oxalate')

ENVIRONMENTAL LISTS

  CERCLA/SARA - Hazardous Substances and their Reportable Quantities
    final RQ = 5000 pounds (2270 kg) (Listed under 'Ammonium oxalate')

  Clean Water Act - Hazardous Substances
    [present] (Listed under 'Ammonium oxalate')

STATE LISTS

  California - Directors List of Hazardous Substances (8 CCR 339)
    [present]

  Massachusetts Right To Know List
    [present]

  Pennsylvania Right to Know List
    environmental hazard

AMMONIUM OXALATE, UNSPECIFIED HYDRATE   5972-73-6

HEALTH AND SAFETY LISTS

  U.S. DOT - Appendix A Table 1 - Hazardous Substances
    final RQ = 5000 pounds (2270 kg) (Listed under 'Ammonium oxalate')

ENVIRONMENTAL LISTS

  CERCLA/SARA - Hazardous Substances and their Reportable Quantities
    final RQ = 5000 pounds (2270 kg) (Listed under 'Ammonium oxalate')

  Clean Water Act - Hazardous Substances
    [present] (Listed under 'Ammonium oxalate')

STATE LISTS

  Massachusetts Right To Know List
    [present]

  Pennsylvania Right to Know List
    environmental hazard

AMMONIUM PERCHLORATE   7790-98-9

HEALTH AND SAFETY LISTS

  U.S. DOT - Substances From 49 CFR 172.101
    regulated by DOT (UN0402, UN1442)

  U.S. DOT - Hazard Classes
    DOT hazard class = 5.1

  OSHA - List of Highly Hazardous Chemicals
    threshhold quantity = 7500 pounds

STATE LISTS

Florida Hazardous Substance List
[present]

Massachusetts Right To Know List
[present]

NJ Right to Know List (Total)
sn 0109

NJ Special Hazardous Substances
(reactive - fourth degree)

Pennsylvania Right to Know List
[present]

AMMONIUM PERFLUOROOCTANOATE   3825-26-1

HEALTH AND SAFETY LISTS

ACGIH 1995 - Time Weighted Averages
0.01 mg/m3 TWA

ACGIH 1995 - Skin Designations
skin - potential for cutaneous absorption

ACGIH 1995 - Carcinogens
A3-animal carcinogen

NIOSH - Selected LD50s and LC50s
Oral, rat: LD50 = 430 mg/kg Skin, rat: LD50 = 7 gm/kg

INTERNATIONAL LISTS

Australian Exposure Standards - Time Weighted Averages
0.1 mg/m3 TWA

Canada - Ontario - OHSA - TWAEVs
0.1 mg/m3 TWAEV

Canada - Quebec - Time-Weighted Average Exposure Values
0.1 mg/m3 TWAEV

Israel - Time Weighted Averages
0.1 mg/m3 TWA

Israel - Action Levels
0.05 mg/m3 AL

STATE LISTS

California - Exposure Limits - PELs
0.1 mg/m3 PEL

California - Exposure Limits - Skin Notation
material may be absorbed through the skin, eyes or mucous membrane

Massachusetts Right To Know List
[present]

Minnesota Hazardous Substance List
skin

AMMONIUM PERMANGANATE   13446-10-1
SEE ALSO:
MANGANESE

STATE LISTS

Florida Hazardous Substance List
[present]

Massachusetts Right To Know List
[present]

NJ Right to Know List (Total)
sn 0110

NJ Special Hazardous Substances
(reactive - third degree)

Pennsylvania Right to Know List
[present]

AMMONIUM PERSULFATE   7727-54-0

HEALTH AND SAFETY LISTS

U.S. DOT - Substances From 49 CFR 172.101
regulated by DOT (UN1444)

U.S. DOT - Hazard Classes
DOT hazard class = 5.1

NIOSH - Selected LD50s and LC50s
Oral, rat: LD50 = 820 mg/kg

INTERNATIONAL LISTS

Canada - Ontario - OHSA - TWAEVs
as S2O8: 5 mg/m3 TWAEV (Listed under 'persulfates')

United Kingdom - Occupational Exposure Standards - TWAs
measured as S2O8: 1 mg/m3 TWA

STATE LISTS

NJ Right to Know List (Total)
sn 0111

PROPOSED REGULATIONS

Canada - Ontario - Proposed Occupational TWAEVs
2 mg/m3 TWAEV

AMMONIUM PICRATE   131-74-8

HEALTH AND SAFETY LISTS

U.S. DOT - Substances From 49 CFR 172.101
regulated by DOT (UN0004, UN1310)

U.S. DOT - Hazard Classes
DOT hazard class = 1.1D

U.S. DOT - Appendix A Table 1 - Hazardous Substances
final RQ = 10 pounds (4.54 kg)

ENVIRONMENTAL LISTS

CERCLA/SARA - Hazardous Substances and their Reportable Quantities
final RQ = 10 pounds (4.54 kg)

RCRA - P Series Wastes
waste number P009 (Reactive waste)

RCRA - Hazardous Constituents-Appendix VIII
waste number P009 (Reactive waste)

RCRA - Substances Banned From Land Disposal
[present]

STATE LISTS

Massachusetts Right To Know List
[present]

NJ Right to Know List (Total)
sn 0112

Pennsylvania Right to Know List
environmental hazard

AMMONIUM POLYSULFIDE   9080-17-5

HEALTH AND SAFETY LISTS

U.S. DOT - Substances From 49 CFR 172.101
regulated by DOT (UN2818)

U.S. DOT - Hazard Classes
DOT hazard class = 8

AMMONIUM POLYSULFIDE   12259-92-6

INTERNATIONAL LISTS

Canada - WHMIS: Ingredient Disclosure
1% item 99 (1441)

STATE LISTS

NJ Right to Know List (Total)
sn 0113

NJ Special Hazardous Substances
(corrosive)

AMMONIUM POLYVANADATE   RR-00150-3

HEALTH AND SAFETY LISTS

U.S. DOT - Substances From 49 CFR 172.101
regulated by DOT (UN2861)

U.S. DOT - Hazard Classes
DOT hazard class = 6.1

STATE LISTS

NJ Right to Know List (Total)
sn 2127

AMMONIUM SILICOFLUORIDE   16919-19-0

HEALTH AND SAFETY LISTS

U.S. DOT - Appendix A Table 1 - Hazardous Substances
final RQ = 1000 pounds (454 kg)

ENVIRONMENTAL LISTS

CERCLA/SARA - Hazardous Substances and their Reportable Quantities
final RQ = 1000 pounds (454 kg)

Clean Water Act - Hazardous Substances
[present]

List of Pesticide Product Inert Ingredients
[present]

STATE LISTS

California - Directors List of Hazardous Substances (8 CCR 339)
[present]

Massachusetts Right To Know List
[present]

NJ Right to Know List (Total)
sn 2128

Pennsylvania Right to Know List
environmental hazard

AMMONIUM STEARATE   1002-89-7

ENVIRONMENTAL LISTS

List of Pesticide Product Inert Ingredients
[present]

STATE LISTS

California - Exposure Limits - PELs
10 mg/m3 PEL

AMMONIUM SULFAMATE   7773-06-0

HEALTH AND SAFETY LISTS

ACGIH 1995 - Time Weighted Averages
10 mg/m3 TWA

U.S. DOT - Appendix A Table 1 - Hazardous Substances
final RQ = 5000 pounds (2270 kg)

NIOSH - Selected LD50s and LC50s
Oral, rat: LD50 = 2 gm/kg

NIOSH 1990 - Pocket Guide - RELs
total: 10 mg/m3 TWA;   respirable dust: 5 mg/m3 TWA

NIOSH 1990 - Pocket Guide - IDLHs
5000 mg/m3 IDLH

NIOSH 1990 - Pocket Guide - Target organs
upper respiratory system, eyes

OSHA - Vacated PELs - Time Weighted Averages
total dust: 10 mg/m3 TWA;   respirable fraction: 5 mg/m3 TWA

OSHA - Final PELs - Time Weighted Averages
total dust: 15 mg/m3 TWA;   respirable fraction: 5 mg/m3 TWA

ENVIRONMENTAL LISTS

CERCLA/SARA - Hazardous Substances and their Reportable Quantities
final RQ = 5000 pounds (2270 kg)

Clean Water Act - Hazardous Substances
[present]

INTERNATIONAL LISTS

Australian Exposure Standards - Time Weighted Averages
10 mg/m3 TWA

Canada - WHMIS: Ingredient Disclosure
1% item 100 (1511)

Canada - Alberta - 8 Hour Occupational Exposure Limit
10 mg/m3 TWA

Canada - Alberta - 15 Minute Occupational Exposure Limit
20 mg/m3 STEL

Canada - British Columbia - 8 Hour Exposure Limits
10 mg/m3 TWA

Canada - British Columbia - 15 Minute Exposure Limits
20 mg/m3 STEL

Canada - Ontario - OHSA - TWAEVs
10 mg/m3 TWAEV

Canada - Quebec - Time-Weighted Average Exposure Values
10 mg/m3 TWAEV

United Kingdom - Occupational Exposure Standards - TWAs
10 mg/m3 TWA

United Kingdom - Occupational Exposure Standards - STELs
20 mg/m3 STEL

German (DFG) - MAK Values
total dust: 15 mg/m3 MAK

Israel - Time Weighted Averages
10 mg/m3 TWA

Israel - Action Levels
5 mg/m3 AL

STATE LISTS

California - Exposure Limits - PELs
total dust: 10 mg/m3 PEL;   respirable fraction: 5 mg/m3 PEL

California - Directors List of Hazardous Substances (8 CCR 339)
[present]

Florida Hazardous Substance List
[present]

Massachusetts Right To Know List
[present]

Minnesota Hazardous Substance List
[present]

NJ Right to Know List (Total)
sn 0114

Pennsylvania Right to Know List
environmental hazard

AMMONIUM SULFATE   7783-20-2

HEALTH AND SAFETY LISTS

NIOSH - Selected LD50s and LC50s
Oral, rat: LD50 = 3000 mg/kg

ENVIRONMENTAL LISTS

CERCLA/SARA - Section 313 - Emission Reporting
form R reporting required for 1.0% de minimus concentration (only in solution)

List of Pesticide Product Inert Ingredients
[present]

INTERNATIONAL LISTS

Canada - WHMIS: Ingredient Disclosure
1% item 95 (1512)

Canada - NPRI (National Pollutant Release Inventory)
[present] (as solution)

STATE LISTS

California - Air Bill 2588 Appendix A-I
6/91

Florida Hazardous Substance List
[present]

Massachusetts Right To Know List
[present]

NJ Right to Know List (Total)
sn 2892

Pennsylvania Right to Know List
environmental hazard

AMMONIUM SULFIDE   12135-76-1

HEALTH AND SAFETY LISTS

U.S. DOT - Substances From 49 CFR 172.101
regulated by DOT (UN2683)

U.S. DOT - Hazard Classes
DOT hazard class = 8

U.S. DOT - Appendix A Table 1 - Hazardous Substances
final RQ = 100 pounds (45.4 kg)

ENVIRONMENTAL LISTS

CERCLA/SARA - Hazardous Substances and their Reportable Quantities
final RQ = 100 pounds (45.4 kg)

Clean Water Act - Hazardous Substances
[present]

STATE LISTS

California - Directors List of Hazardous Substances (8 CCR 339)
[present]

Massachusetts Right To Know List
[present]

NJ Right to Know List (Total)
sn 0115

NJ Special Hazardous Substances
(corrosive)

Pennsylvania Right to Know List
environmental hazard

AMMONIUM SULFITE   10196-04-0

HEALTH AND SAFETY LISTS

U.S. DOT - Appendix A Table 1 - Hazardous Substances
final RQ = 5000 pounds (2270 kg)

ENVIRONMENTAL LISTS

CERCLA/SARA - Hazardous Substances and their Reportable Quantities
final RQ = 5000 pounds (2270 kg)

Clean Water Act - Hazardous Substances
[present]

STATE LISTS

California - Directors List of Hazardous Substances (8 CCR 339)
[present]

Massachusetts Right To Know List
[present]

NJ Right to Know List (Total)
sn 0116

Pennsylvania Right to Know List
environmental hazard

AMMONIUM TARTRATE   14307-43-8

HEALTH AND SAFETY LISTS

U.S. DOT - Appendix A Table 1 - Hazardous Substances
final RQ = 5000 pounds (2270 kg)

ENVIRONMENTAL LISTS

CERCLA/SARA - Hazardous Substances and their Reportable Quantities
final RQ = 5000 pounds (2270 kg)

Clean Water Act - Hazardous Substances
[present]

STATE LISTS

Massachusetts Right To Know List
[present]

Pennsylvania Right to Know List
environmental hazard

AMMONIUM TARTRATE, DIAMMONIUM SALT   3164-29-2

HEALTH AND SAFETY LISTS

U.S. DOT - Appendix A Table 1 - Hazardous Substances
final RQ = 5000 pounds (2270 kg)

ENVIRONMENTAL LISTS

CERCLA/SARA - Hazardous Substances and their Reportable Quantities
final RQ = 5000 pounds (2270 kg)

Clean Water Act - Hazardous Substances
[present] (Listed under 'Ammonium tartrate')

STATE LISTS

California - Directors List of Hazardous Substances (8 CCR 339)
[present]

Massachusetts Right To Know List
[present]

NJ Right to Know List (Total)
sn 0117

Pennsylvania Right to Know List
environmental hazard

AMMONIUM TELLURATE   13453-06-0

INTERNATIONAL LISTS

Canada - WHMIS: Ingredient Disclosure
1% item 101 (1555)

AMMONIUM THIOCYANATE   1762-95-4

HEALTH AND SAFETY LISTS

U.S. DOT - Appendix A Table 1 - Hazardous Substances
final RQ = 5000 pounds (2270 kg)

ENVIRONMENTAL LISTS

CERCLA/SARA - Hazardous Substances and their Reportable Quantities
final RQ = 5000 pounds (2270 kg)

Clean Water Act - Hazardous Substances
[present]

List of Pesticide Product Inert Ingredients
[present]

STATE LISTS

California - Directors List of Hazardous Substances (8 CCR 339)
[present]

Massachusetts Right To Know List
[present]

NJ Right to Know List (Total)
sn 0119

Pennsylvania Right to Know List
environmental hazard

AMMONIUM THIOSULFATE   7783-18-8

HEALTH AND SAFETY LISTS

NIOSH - Selected LD50s and LC50s
Oral, rat: LD50 = 2890 mg/kg

STATE LISTS

Massachusetts Right To Know List
[present]

NJ Right to Know List (Total)
sn 0120

Pennsylvania Right to Know List
environmental hazard

AMMONIUM VANADATE   7803-55-6

HEALTH AND SAFETY LISTS

U.S. DOT - Substances From 49 CFR 172.101
regulated by DOT (UN2859)

U.S. DOT - Hazard Classes
DOT hazard class = 6.1

U.S. DOT - Appendix A Table 1 - Hazardous Substances
final RQ = 1000 pounds (454 kg)

NIOSH - Selected LD50s and LC50s
Oral, rat: LD50 = 160 mg/kg

ENVIRONMENTAL LISTS

CERCLA/SARA - Hazardous Substances and their Reportable Quantities
final RQ = 1000 pounds (454 kg)

RCRA - P Series Wastes
waste number P119

RCRA - Hazardous Constituents-Appendix VIII
waste number P119

RCRA - Substances Banned From Land Disposal
[present]

INTERNATIONAL LISTS

Canada - WHMIS: Ingredient Disclosure
1% item 97 (1085)

STATE LISTS

Massachusetts Right To Know List
[present]

NJ Right to Know List (Total)
sn 0104

Pennsylvania Right to Know List
environmental hazard

AMMONIUM VANADATE   11115-67-6

INTERNATIONAL LISTS

Canada - WHMIS: Ingredient Disclosure
1% item 102 (1710)

AMPHETAMINE   300-62-9

HEALTH AND SAFETY LISTS

NIOSH - Selected LD50s and LC50s
Oral, rat: LD50 = 30 mg/kg

ENVIRONMENTAL LISTS

CERCLA/SARA - Section 302 Extremely Hazardous Substances and TPQs
TPQ = 1000 pounds

STATE LISTS

Florida Hazardous Substance List
effective March 13, 1992

Massachusetts Right To Know List
extraordinarily hazardous

NJ Right to Know List (Total)
sn 2130

Pennsylvania Right to Know List
environmental hazard

PROPOSED REGULATIONS

CERCLA/SARA - Proposed Hazardous Substance Additions
proposed RQ = 1 pound (.454 kg)

CERCLA/SARA - 1989 Proposed RQ Adjustments
proposed RQ = 100 pounds (45.4 kg)

DL-AMPHETAMINE SULFATE   60-13-9

HEALTH AND SAFETY LISTS

NTP Chemical Status Reports - Testing Status and NTIS Number
Technical reports printed (PB921107978)

AMPICILLIN   69-53-4

HEALTH AND SAFETY LISTS

IARC - Group 3 (not classifiable)
[present]

STATE LISTS

NJ Right to Know List (Total)
sn 0121

AMPICILLIN TRIHYDRATE   7177-48-2

HEALTH AND SAFETY LISTS

NTP Chemical Status Reports - Testing Status and NTIS Number
Technical reports printed (PB87204160/AS)

NTP Chemical Status Reports - Evidence of Carcinogenicity
male rat-equivocal evidence; female rat-no evidence; male mice-no evidence; female mice-no evidence

TERT-AMYL ACETATE   625-16-1

HEALTH AND SAFETY LISTS

U.S. DOT - Appendix A Table 1 - Hazardous Substances
final RQ = 5000 pounds (2270 kg) (Listed under 'Amyl acetate')

ENVIRONMENTAL LISTS

CERCLA/SARA - Hazardous Substances and their Reportable Quantities
final RQ = 5000 pounds (2270 kg) (Listed under 'Amyl acetate')

Clean Water Act - Hazardous Substances
[present] (Listed under 'Amyl acetate')

STATE LISTS

California - Directors List of Hazardous Substances (8 CCR 339)
[present] (Listed under 'Amyl acetate, all isomers')

Massachusetts Right To Know List
[present]

Pennsylvania Right to Know List
environmental hazard

SEC-AMYL ACETATE   626-38-0

HEALTH AND SAFETY LISTS

ACGIH 1995 - Time Weighted Averages
125 ppm TWA; 665 mg/m3 TWA

U.S. DOT - Appendix A Table 1 - Hazardous Substances
final RQ = 5000 pounds (2270 kg) (Listed under 'Amyl acetate')

NFPA - Flash Points
flash point = 89 degrees F (32 degrees C)

NFPA - Hazard Identification Ratings
health-1; flammability-3; reactivity-0

NIOSH 1990 - Pocket Guide - RELs
125 ppm TWA; 650 mg/m3 TWA

NIOSH 1990 - Pocket Guide - IDLHs
9000 ppm IDLH

NIOSH 1990 - Pocket Guide - Target organs
respiratory system, eyes, skin

OSHA - Vacated PELs - Time Weighted Averages
125 ppm TWA; 650 mg/m3 TWA

OSHA - Final PELs - Time Weighted Averages
125 ppm TWA; 650 mg/m3 TWA

ENVIRONMENTAL LISTS

CERCLA/SARA - Hazardous Substances and their Reportable Quantities
final RQ = 5000 pounds (2270 kg) (Listed under 'Amyl acetate')

Clean Water Act - Hazardous Substances
[present] (Listed under 'Amyl acetate')

INTERNATIONAL LISTS

Australian Exposure Standards - Time Weighted Averages
125 ppm TWA; 665 mg/m3 TWA

Canada - WHMIS: Ingredient Disclosure
1% item 104 (6)

Canada - Alberta - 8 Hour Occupational Exposure Limit
125 ppm TWA; 665 mg/m3 TWA

Canada - Alberta - 15 Minute Occupational Exposure Limit
150 ppm STEL; 800 mg/m3 STEL

Canada - British Columbia - 8 Hour Exposure Limits
125 ppm TWA; 670 mg/m3 TWA

Canada - British Columbia - 15 Minute Exposure Limits
150 ppm STEL; 800 mg/m3 STEL

Canada - Ontario - OHSA - TWAEVs
125 ppm TWAEV; 660 mg/m3 TWAEV

Canada - Quebec - Time-Weighted Average Exposure Values
125 ppm TWAEV; 665 mg/m3 TWAEV

United Kingdom - Occupational Exposure Standards - STELs
150 ppm STEL; 800 mg/m3 STEL

Israel - Time Weighted Averages
125 ppm TWA; 665 mg/m3 TWA

Israel - Action Levels
62.5 ppm AL; 332.5 mg/m3 AL

Mexico - Instruction No. 10 - TWAs
125 ppm TWA; 670 mg/m3 TWA

Mexico - Instruction No. 10 - STELs
150 ppm STEL; 800 mg/m3 STEL

STATE LISTS

California - Exposure Limits - PELs
125 ppm PEL; 665 mg/m3 PEL (includes all isomers and mixtures)

California - Directors List of Hazardous Substances (8 CCR 339)
[present] (Listed under 'Amyl acetate, all isomers')

Florida Hazardous Substance List
[present]

Massachusetts Right To Know List
[present]

Minnesota Hazardous Substance List
[present]

Pennsylvania Right to Know List
environmental hazard

N-AMYL ACETATE   628-63-7

HEALTH AND SAFETY LISTS

ACGIH 1995 - Time Weighted Averages
100 ppm TWA; 532 mg/m3 TWA

AIHA - Odor Threshold Values
geometric mean air odor threshold = 0.052 ppm (detectable)

U.S. DOT - Substances From 49 CFR 172.101
regulated by DOT (UN1104)

U.S. DOT - Hazard Classes
DOT hazard class = 3

U.S. DOT - Appendix A Table 1 - Hazardous Substances
final RQ = 5000 pounds (2270 kg)

NFPA - Flash Points
flash point = 60 degrees F (16 degrees C); commercial: flash point = 70 degrees F (21 degrees C)

NFPA - Hazard Identification Ratings
health-1; flammability-3; reactivity-0; mixture of isomeric amyl acetates and amyl alcohols: health-2; flammability-3; reactivity-0

NIOSH - Selected LD50s and LC50s
Oral, rat: LD50 = 6500 mg/kg

NIOSH 1990 - Pocket Guide - RELs
100 ppm TWA; 525 mg/m3 TWA

NIOSH 1990 - Pocket Guide - IDLHs
4000 ppm IDLH

NIOSH 1990 - Pocket Guide - Target organs
eyes, skin, respiratory system

OSHA - Vacated PELs - Time Weighted Averages
100 ppm TWA; 525 mg/m3 TWA

OSHA - Final PELs - Time Weighted Averages
100 ppm TWA; 525 mg/m3 TWA

ENVIRONMENTAL LISTS

CERCLA/SARA - Hazardous Substances and their Reportable Quantities
final RQ = 5000 pounds (2270 kg)

Clean Water Act - Hazardous Substances
[present] (Listed under 'Amyl acetate')

EPA - Master Testing List
[present]

List of Pesticide Product Inert Ingredients
[present]

TSCA - PAIR - Reporting List
Effective Date: January 26, 1994; Reporting Date: March 28, 1994

TSCA - Health and Safety Reporting List
Effective Date: January 26, 1994; Sunset Date: January 26, 2004

TSCA - Multichemical Test Rules - Neurotoxicity
administrative stay for neurotoxicity tests effective June 27, 1994

INTERNATIONAL LISTS

Australian Exposure Standards - Time Weighted Averages
100 ppm TWA; 532 mg/m3 TWA

Canada - WHMIS: Ingredient Disclosure
1% item 1245 (28)

Canada - Alberta - 8 Hour Occupational Exposure Limit
100 ppm TWA; 530 mg/m3 TWA

Canada - Alberta - 15 Minute Occupational Exposure Limit
150 ppm STEL; 800 mg/m3 STEL

Canada - British Columbia - 8 Hour Exposure Limits
100 ppm TWA; 530 mg/m3 TWA

Canada - British Columbia - 15 Minute Exposure Limits
150 ppm STEL; 800 mg/m3 STEL

Canada - Ontario - OHSA - TWAEVs
100 ppm TWAEV; 530 mg/m3 TWAEV

Canada - Quebec - Time-Weighted Average Exposure Values
100 ppm TWAEV; 532 mg/m3 TWAEV

United Kingdom - Occupational Exposure Standards - TWAs
100 ppm TWA; 530 mg/m3 TWA

United Kingdom - Occupational Exposure Standards - STELs
150 ppm STEL; 800 mg/m3 STEL

German (DFG) - MAK Values
100 ppm MAK; 525 mg/m3 MAK

Israel - Time Weighted Averages
100 ppm TWA; 532 mg/m3 TWA

Israel - Action Levels
50 ppm AL; 266 mg/m3 AL

Mexico - Instruction No. 10 - TWAs
100 ppm TWA; 530 mg/m3 TWA

Mexico - Instruction No. 10 - STELs
150 ppm STEL; 800 mg/m3 STEL

STATE LISTS

California - Exposure Limits - PELs
100 ppm PEL; 532 mg/m3 PEL

California - Directors List of Hazardous Substances (8 CCR 339)
[present] (Listed under 'Amyl acetate, all isomers')

Florida Hazardous Substance List
[present]

Massachusetts Right To Know List
[present]

Minnesota Hazardous Substance List
[present]

NJ Right to Know List (Total)
sn 1321

NJ Special Hazardous Substances
(flammable - third degree)

Pennsylvania Right to Know List
environmental hazard

PROPOSED REGULATIONS
TSCA - ITC 31st Report Priority Testing List
designated to be tested

AMYL ACID PHOSPHATE   12789-46-7
HEALTH AND SAFETY LISTS
U.S. DOT - Substances From 49 CFR 172.101
regulated by DOT (UN2819)

U.S. DOT - Hazard Classes
DOT hazard class = 8

INTERNATIONAL LISTS
Canada - WHMIS: Ingredient Disclosure
1% item 105 (64)

STATE LISTS
NJ Right to Know List (Total)
sn 0123

NJ Special Hazardous Substances
(corrosive)

AMYL ALCOHOL   71-41-0
HEALTH AND SAFETY LISTS
NFPA - Flash Points
flash point = 91 degrees F (33 degrees C)

NFPA - Hazard Identification Ratings
health-1; flammability-3; reactivity-0

NIOSH - Selected LD50s and LC50s
Oral, rat: LD50 = 2200 mg/kg Skin, rabbit: LD50 = 3600 mg/kg

ENVIRONMENTAL LISTS
List of Pesticide Product Inert Ingredients
[present]

INTERNATIONAL LISTS
Canada - WHMIS: Ingredient Disclosure
1% item 106 (168)

STATE LISTS
Florida Hazardous Substance List
[present]

Massachusetts Right To Know List
[present]

NJ Right to Know List (Total)
sn 0124

Pennsylvania Right to Know List
[present]

SEC-AMYL ALCOHOL   6032-29-7
HEALTH AND SAFETY LISTS
NFPA - Flash Points
flash point = 94 degrees F (34 degrees C); see list description

NFPA - Hazard Identification Ratings
health-1; flammability-3; reactivity-0

NIOSH - Selected LD50s and LC50s
Oral, rat: LD50 = 1470 mg/kg

INTERNATIONAL LISTS
Canada - WHMIS: Ingredient Disclosure
1% item 107 (169)

STATE LISTS
Florida Hazardous Substance List
[present]

Massachusetts Right To Know List
[present]

Pennsylvania Right to Know List
[present]

AMYL ALCOHOLS   RR-00108-1
HEALTH AND SAFETY LISTS
U.S. DOT - Substances From 49 CFR 172.101
regulated by DOT (UN1105)

U.S. DOT - Hazard Classes
DOT hazard class = 3

AMYL AMINE   110-58-7
HEALTH AND SAFETY LISTS
U.S. DOT - Substances From 49 CFR 172.101
regulated by DOT (UN1106)

U.S. DOT - Hazard Classes
DOT hazard class = 3

NFPA - Flash Points
flash point = 30 degrees F (-1 degrees C)

NFPA - Hazard Identification Ratings
health-2; flammability-3; reactivity-0

STATE LISTS
Florida Hazardous Substance List
[present]

Massachusetts Right To Know List
[present]

NJ Right to Know List (Total)
sn 0125

NJ Special Hazardous Substances
(flammable - third degree)

Pennsylvania Right to Know List
[present]

SEC-AMYLAMINE   625-30-9
HEALTH AND SAFETY LISTS
NFPA - Flash Points
flash point = 20 degrees F (-7 degrees C)

NFPA - Hazard Identification Ratings
health-2; flammability-3; reactivity-0

STATE LISTS
Florida Hazardous Substance List
[present]

Massachusetts Right To Know List
[present]

Pennsylvania Right to Know List
[present]

P-TERT-AMYLANILINE   2049-92-5
HEALTH AND SAFETY LISTS
NFPA - Flash Points
flash point = 215 degrees F (102 degrees C)

NFPA - Hazard Identification Ratings
health-3; flammability-1; reactivity-0

STATE LISTS

Florida Hazardous Substance List
[present]

Massachusetts Right To Know List
[present]

Pennsylvania Right to Know List
[present]

AMYL BUTYRATE   540-18-1

HEALTH AND SAFETY LISTS

NFPA - Flash Points
flash point = 135 degrees F (57 degrees C)

NFPA - Hazard Identification Ratings
health-1; flammability-2; reactivity-0

NIOSH - Selected LD50s and LC50s
Oral, rat: LD50 = 12210 mg/kg

STATE LISTS

NJ Right to Know List (Total)
sn 0126

AMYL BUTYRATES   RR-00145-6

HEALTH AND SAFETY LISTS

U.S. DOT - Substances From 49 CFR 172.101
regulated by DOT (UN2620)

U.S. DOT - Hazard Classes
DOT hazard class = 3

AMYL CHLORIDE   543-59-9

HEALTH AND SAFETY LISTS

NFPA - Flash Points
flash point = 55 degrees F (13 degrees C)

NFPA - Hazard Identification Ratings
health-1; flammability-3; reactivity-0

STATE LISTS

Florida Hazardous Substance List
[present]

Massachusetts Right To Know List
[present]

NJ Right to Know List (Total)
sn 0127

NJ Special Hazardous Substances
(flammable - third degree)

Pennsylvania Right to Know List
[present]

TERT-AMYL CHLORIDE   594-36-5

HEALTH AND SAFETY LISTS

NFPA - Hazard Identification Ratings
health-1; flammability-3; reactivity-0

STATE LISTS

Florida Hazardous Substance List
[present]

Massachusetts Right To Know List
[present]

Pennsylvania Right to Know List
[present]

AMYL CHLORIDES (MIXED)   RR-00063-5

HEALTH AND SAFETY LISTS

U.S. DOT - Substances From 49 CFR 172.101
regulated by DOT (UN1107)

U.S. DOT - Hazard Classes
DOT hazard class = 3

NFPA - Flash Points
flash point = 38 degrees F (3 degrees C)

NFPA - Hazard Identification Ratings
health-1; flammability-3; reactivity-0

AMYLCYCLOHEXANE   4292-92-6

HEALTH AND SAFETY LISTS

NFPA - Hazard Identification Ratings
health-1; reactivity-0

BETA-AMYLENE-CIS   627-20-3

HEALTH AND SAFETY LISTS

NFPA - Flash Points
flash point < -4 degrees F (-20 degrees C)

NFPA - Hazard Identification Ratings
health-0; flammability-4

ENVIRONMENTAL LISTS

CAA - Flammable Substances for Accidental Release Prevention
threshold quantity = 10,000 lbs

STATE LISTS

Florida Hazardous Substance List
[present]

Massachusetts Right To Know List
[present]

Pennsylvania Right to Know List
[present]

BETA-AMYLENE-TRANS   646-04-8

HEALTH AND SAFETY LISTS

NFPA - Flash Points
flash point < -4 degrees F (-20 degrees C)

NFPA - Hazard Identification Ratings
health-0; flammability-4

ENVIRONMENTAL LISTS

CAA - Flammable Substances for Accidental Release Prevention
threshold quantity = 10,000 lbs

STATE LISTS

Florida Hazardous Substance List
[present]

Massachusetts Right To Know List
[present]

Pennsylvania Right to Know List
[present]

AMYLENE, NORMAL   25377-72-4

HEALTH AND SAFETY LISTS

U.S. DOT - Substances From 49 CFR 172.101
regulated by DOT (UN1108)

U.S. DOT - Hazard Classes
DOT hazard class = 3

STATE LISTS

NJ Right to Know List (Total)
sn 0128

NJ Special Hazardous Substances
(flammable - fourth degree)

AMYL ETHER   693-65-2

HEALTH AND SAFETY LISTS

NFPA - Flash Points
flash point = 135 degrees F (57 degrees C)

NFPA - Hazard Identification Ratings
health-1; flammability-2; reactivity-0

AMYL FORMATE   638-49-3

HEALTH AND SAFETY LISTS

NFPA - Flash Points
flash point = 79 degrees F (26 degrees C)

NFPA - Hazard Identification Ratings
health-1; flammability-3; reactivity-0

STATE LISTS

Florida Hazardous Substance List
[present]

Massachusetts Right To Know List
[present]

NJ Right to Know List (Total)
sn 0129

NJ Special Hazardous Substances
(flammable - third degree)

Pennsylvania Right to Know List
[present]

AMYL FORMATES   RR-00144-5

HEALTH AND SAFETY LISTS

U.S. DOT - Substances From 49 CFR 172.101
regulated by DOT (UN1109)

U.S. DOT - Hazard Classes
DOT hazard class = 3

TERT-AMYL HYDROPEROXIDE   3425-61-4

HEALTH AND SAFETY LISTS

U.S. DOT - Organic Peroxides Table
Organic peroxide UN3107

STATE LISTS

NJ Right to Know List (Total)
sn 3099

AMYL LACTATE   RR-00620-2

HEALTH AND SAFETY LISTS

NFPA - Flash Points
flash point = 175 degrees F (79 degrees C)

NFPA - Hazard Identification Ratings
health-1; flammability-2; reactivity-0

AMYL LAURATE   RR-00807-1

HEALTH AND SAFETY LISTS

NFPA - Flash Points
flash point = 300 degrees F (149 degrees C)

NFPA - Hazard Identification Ratings
health-0; flammability-1; reactivity-0

AMYL MALEATE   RR-00513-0

HEALTH AND SAFETY LISTS

NFPA - Flash Points
flash point = 270 degrees F (132 degrees C)

NFPA - Hazard Identification Ratings
health-0; flammability-1; reactivity-0

AMYL MERCAPTAN   110-66-7

HEALTH AND SAFETY LISTS

NFPA - Flash Points
(n) and mixed: flashpoint = 65 degrees F (18 degrees C)

NFPA - Hazard Identification Ratings
(n): health-2; flammability-3; (mixed): health-2; flammability-3; reactivity-0

NIOSH - Health Standards - Exposure Limits
C (15 min) 0.5 ppm; C (15 min) 2.1 mg/m3 (Listed under 'Thiols')

NIOSH - Health Standards - Health Effects and Precautions
Irritation; eye, skin, blood, and nervous system effects (Blood and urine monitoring required; prevent skin contact) (Listed under 'Thiols')

STATE LISTS

Florida Hazardous Substance List
[present]

Massachusetts Right To Know List
[present]

NJ Right to Know List (Total)
sn 0130

NJ Special Hazardous Substances
(flammable - third degree)

Pennsylvania Right to Know List
[present]

AMYL MERCAPTANS   RR-01196-1

HEALTH AND SAFETY LISTS

U.S. DOT - Substances From 49 CFR 172.101
regulated by DOT (UN1111)

U.S. DOT - Hazard Classes
DOT hazard class = 3

U.S. DOT - Appendix B - Marine Pollutants
DOT regulated marine pollutant

TERT-AMYL METHYL ETHER   994-05-8

PROPOSED REGULATIONS

TSCA - ITC 34th Report Priority Testing List
recommended for health effects testing

AMYL NAPHTHALENE   RR-00336-1

HEALTH AND SAFETY LISTS

NFPA - Flash Points
flash point = 255 degrees F (124 degrees C)

NFPA - Hazard Identification Ratings
health-0; flammability-1; reactivity-0

AMYL NITRATE   1002-16-0

HEALTH AND SAFETY LISTS

U.S. DOT - Substances From 49 CFR 172.101
regulated by DOT (UN1112)

U.S. DOT - Hazard Classes
DOT hazard class = 3

NFPA - Flash Points
flash point = 118 degrees F (48 degrees C)

NFPA - Hazard Identification Ratings
health-2; flammability-2; reactivity-0 (oxidizing properties)

STATE LISTS

Florida Hazardous Substance List
[present]

Massachusetts Right To Know List
[present]

NJ Right to Know List (Total)
sn 0131

NJ Special Hazardous Substances
(reactive - second degree)

Pennsylvania Right to Know List
[present]

AMYL NITRITE   110-46-3

HEALTH AND SAFETY LISTS

U.S. DOT - Substances From 49 CFR 172.101
regulated by DOT (UN1113)

U.S. DOT - Hazard Classes
DOT hazard class = 3

NFPA - Hazard Identification Ratings
health-1; reactivity-2
NIOSH - Selected LD50s and LC50s
Inhalation, rat: LC50 = 1274 ppm 1 hr Oral, rat: LD50 = 505 mg/kg
STATE LISTS
NJ Right to Know List (Total)
sn 0132
NJ Special Hazardous Substances
(reactive - second degree)

AMYLODEXTRIN   9005-84-9
INTERNATIONAL LISTS
United Kingdom - Occupational Exposure Standards - TWAs
total inhalable dust: 10 mg/m3 TWA; respirable dust: 5 mg/m3 TWA

AMYL OLEATE   RR-00601-9
HEALTH AND SAFETY LISTS
NFPA - Flash Points
flash point = 366 degrees F (186 degrees C)
NFPA - Hazard Identification Ratings
health-0; flammability-1; reactivity-0

AMYL OXALATE   RR-00824-2
HEALTH AND SAFETY LISTS
NFPA - Flash Points
flash point = 245 degrees F (118 degrees C)
NFPA - Hazard Identification Ratings
health-0; flammability-1; reactivity-0

TERT-AMYL PEROXY ALKYLENE ESTER   RR-00924-5
ENVIRONMENTAL LISTS
TSCA - Chemicals with Significant New Use Rules
PMN number: P-85-1180

TERT-AMYL PEROXY-2-EHTYLHEXANOATE   RR-01749-2
HEALTH AND SAFETY LISTS
U.S. DOT - Organic Peroxides Table
Organic peroxide UN3115

TERT-AMYLPEROXY-3,5,5-TRIMETHYLHEXANOATE   RR-01750-5
HEALTH AND SAFETY LISTS
U.S. DOT - Organic Peroxides Table
Organic peroxide UN3101

O-AMYL PHENOL   136-81-2
HEALTH AND SAFETY LISTS
NFPA - Flash Points
flash point = 219 degrees F (104 degrees C)
NFPA - Hazard Identification Ratings
health-2; flammability-1; reactivity-0
STATE LISTS
Florida Hazardous Substance List
[present]
Massachusetts Right To Know List
[present]
Pennsylvania Right to Know List
[present]

P-SEC-AMYLPHENOL   25735-67-5
HEALTH AND SAFETY LISTS
NFPA - Flash Points
flash point = 270 degrees F (132 degrees C)

NFPA - Hazard Identification Ratings
health-1; flammability-1; reactivity-0

2-(P-TERT-AMYLPHENOXY) ETHANOL   6382-07-6
HEALTH AND SAFETY LISTS
NFPA - Flash Points
flash point = 280 degrees F (138 degrees C)
NFPA - Hazard Identification Ratings
health-1; flammability-1; reactivity-0

2-(P-TERT-AMYLPHENOXY) ETHYL LAURATE   RR-00762-5
HEALTH AND SAFETY LISTS
NFPA - Flash Points
flash point = 410 degrees F (210 degrees C)
NFPA - Hazard Identification Ratings
health-0; flammability-1; reactivity-0

P-TERT-AMYLPHENYL ACETATE   5137-52-0
HEALTH AND SAFETY LISTS
NFPA - Flash Points
flash point = 240 degrees F (116 degrees C)
NFPA - Hazard Identification Ratings
health-0; flammability-1; reactivity-0

P-TERT-AMYLPHENYL BUTYL ETHER   RR-00876-4
HEALTH AND SAFETY LISTS
NFPA - Flash Points
flash point = 275 degrees F (135 degrees C)
NFPA - Hazard Identification Ratings
health-0; flammability-1; reactivity-0

AMYL PHENYL ETHER   2050-04-6
HEALTH AND SAFETY LISTS
NFPA - Flash Points
flash point = 185 degrees F (85 degrees C)
NFPA - Hazard Identification Ratings
health-0; flammability-2; reactivity-0

P-TERT-AMYLPHENYL METHYL ETHER   1320-05-4
HEALTH AND SAFETY LISTS
NFPA - Flash Points
flash point = 210 degrees F (99 degrees C)
NFPA - Hazard Identification Ratings
health-0; flammability-1; reactivity-0

AMYL PROPIONATE   624-54-4
HEALTH AND SAFETY LISTS
NFPA - Flash Points
flash point = 106 degrees F (41 degrees C)
NFPA - Hazard Identification Ratings
health-0; flammability-2; reactivity-0

AMYL SALICYLATE   2050-08-0
HEALTH AND SAFETY LISTS
NFPA - Flash Points
flash point = 270 degrees F (132 degrees C)
NFPA - Hazard Identification Ratings
health-0; flammability-1; reactivity-0

AMYL STEARATE   6382-13-4
HEALTH AND SAFETY LISTS
NFPA - Flash Points
flash point = 365 degrees F (185 degrees C)

NFPA - Hazard Identification Ratings
health-0;   flammability-1;   reactivity-0

AMYL SULFIDES, MIXED   RR-00083-9

HEALTH AND SAFETY LISTS

NFPA - Flash Points
flash point = 185 degrees F (85 degrees C)

NFPA - Hazard Identification Ratings
health-2;   flammability-2;   reactivity-0

AMYL TRICHLOROSILANE   107-72-2

HEALTH AND SAFETY LISTS

U.S. DOT - Substances From 49 CFR 172.101
regulated by DOT (UN1728)

U.S. DOT - Hazard Classes
DOT hazard class = 8

NFPA - Flash Points
flash point = 145 degrees F (63 degrees C)

NFPA - Hazard Identification Ratings
health-3;   flammability-2;   reactivity-2 (avoid use of water)

NIOSH - Selected LD50s and LC50s
Oral, rat: LD50 = 2340 mg/kg Skin, rabbit: LD50 = 780 mg/kg

INTERNATIONAL LISTS

Canada - WHMIS: Ingredient Disclosure
1% item 108 (226)

STATE LISTS

Florida Hazardous Substance List
[present]

Massachusetts Right To Know List
[present]

NJ Right to Know List (Total)
sn 0134

NJ Special Hazardous Substances
(corrosive;   reactive - second degree)

Pennsylvania Right to Know List
[present]

ANAESTHETICS, VOLATILE   RR-01525-8

HEALTH AND SAFETY LISTS

IARC - Group 3 (not classifiable)
[present]

ANALGESIC MIXTURES CONTAINING PHENACETIN   RR-00055-5

HEALTH AND SAFETY LISTS

IARC - Group 1 (carcinogenic to humans)
[present] (Listed under 'Phenacetin')

NTP Seventh Report - Known Carcinogens
known carcinogen

OSHA - Select Carcinogens
[present]

STATE LISTS

California - Air Bill 2588 Appendix A-II
known or potential carcinogen

California - Prop. 65 - Cancer list
carcinogen - initial date 2/27/87

Minnesota Hazardous Substance List
carcinogen

Pennsylvania Right to Know List
special hazardous substance

Pennsylvania RTK - Special Hazardous Substances
[present]

ANATASE (TIO2)   1317-70-0

STATE LISTS

Pennsylvania Right to Know List
[present]

ANDROGENIC (ANABOLIC) STEROIDS   RR-00057-7

HEALTH AND SAFETY LISTS

IARC - Group 2A (limited human data)
[present]

OSHA - Possible Select Carcinogens
[present]

STATE LISTS

California - Air Bill 2588 Appendix A-II
known or potential carcinogen

California - Prop. 65 - Reproductive - Female
female reproductive toxicity - initial date 4/1/90

California - Prop. 65 - Reproductive - Male
male reproductive toxicity - initial date 4/1/90

California - Directors List of Hazardous Substances (8 CCR 339)
[present]

ANGELICIN PLUS ULTRAVIOLET A RADIATION   RR-01526-9

HEALTH AND SAFETY LISTS

IARC - Group 3 (not classifiable)
[present]

ANGIOTENSIN CONVERTING ENZYME (ACE)   RR-01472-2

STATE LISTS

California - Prop. 65 - Developmental Toxicity
developmental toxicity - initial date 10/1/92

ANILAZINE   101-05-3

HEALTH AND SAFETY LISTS

NTP Chemical Status Reports - Testing Status and NTIS Number
Technical reports printed (PB287141/AS)

NTP Chemical Status Reports - Evidence of Carcinogenicity
male rat-negative;   female rat-negative;   male mice-negative;   female mice-negative

ENVIRONMENTAL LISTS

CERCLA/SARA - Section 313 - Emission Reporting
form R reporting required

ANILINE   62-53-3

SEE ALSO:
COCO SHELL FLOUR
K083-HAZARDOUS WASTES
F039-HAZARDOUS WASTES
ANILINE AND CHLORO-, BROMO-, AND/OR NITROANILINES
K103-HAZARDOUS WASTES
K113-HAZARDOUS WASTES
K112-HAZARDOUS WASTES
K104-HAZARDOUS WASTES

HEALTH AND SAFETY LISTS

ACGIH 1995 - Time Weighted Averages
2 ppm TWA;   7.6 mg/m3 TWA

ACGIH 1995 - Skin Designations
skin - potential for cutaneous absorption

ACGIH 1995 - Biological Exposure Indices
Total p-aminophenol in urine: 50 mg/g creatinine, end of shift (Ns); Methemoglobin in blood: 1.5% of hemoglobin, during or end of shift (B, Ns, Sq)

AIHA - Odor Threshold Values
geometric mean air odor threshold = 2.4 ppm (detectable)

U.S. DOT - Substances From 49 CFR 172.101
regulated by DOT (UN1547)

U.S. DOT - Hazard Classes
DOT hazard class = 6.1

U.S. DOT - Appendix A Table 1 - Hazardous Substances
final RQ = 5000 pounds (2270 kg)

IARC - Group 3 (not classifiable)
[present]

NFPA - Flash Points
flash point = 158 degrees F (70 degrees C)

NFPA - Hazard Identification Ratings
health-3; flammability-2; reactivity-0

NIOSH - Selected LD50s and LC50s
Inhalation, mouse: LC50 = 175 ppm 7 hr Oral, rat: LD50 = 250 mg/kg Skin, rat: LD50 = 1400 mg/kg

NIOSH 1990 - Pocket Guide - RELs
2 ppm TWA; 8 mg/m3 TWA

NIOSH 1990 - Pocket Guide - IDLHs
100 ppm IDLH (not considering carcinogenic effects)

NIOSH 1990 - Pocket Guide - Carcinogens
occupational carcinogen

NIOSH 1990 - Pocket Guide - Target organs
blood, CVS, liver, kidneys

NIOSH 1990 - Pocket Guide - Skin list
Potential for dermal absorption

OSHA - Vacated PELs - Time Weighted Averages
2 ppm TWA; 8 mg/m3 TWA

OSHA - Vacated PELs - Skin Designation
Prevent or reduce skin absorption

OSHA - Final PELs - Time Weighted Averages
5 ppm TWA; 19 mg/m3 TWA

OSHA - Final PELs - Skin Notations
prevent or reduce skin absorption

ENVIRONMENTAL LISTS

CERCLA/SARA - Section 302 Extremely Hazardous Substances and TPQs
TPQ = 1000 pounds

CERCLA/SARA - Section 313 - Emission Reporting
form R reporting required for 1.0% de minimus concentration

CERCLA/SARA - Hazardous Substances and their Reportable Quantities
final RQ = 5000 pounds (2270 kg)

Clean Air Act (1990) - List of Hazardous Air Contaminants
[present]

CAA - HON Rule - SOCMI Chemicals
compliance by Oct. 24, 1994

CAA - HON Rule - Organic HAPs
[present]

Clean Water Act - Hazardous Substances
[present]

RCRA - U Series Wastes
waste number U012 (Ignitable waste; Toxic waste)

RCRA - Hazardous Constituents-Appendix VIII
waste number U012

RCRA - Basis for Listing - Appendix VII
Included in waste streams: F039, K083, K103, K104, K112, K113

RCRA - Substances Banned From Land Disposal
[present]

RCRA - TSD Facilities Ground Water Monitoring
TM 8270 = 10 ug/L PQL

RCRA - Universal Treatment Standards (LDR)
WW: 0.81 mg/l; NWW: 14 mg/kg

TSCA - Code of Federal Regulations Citations
40 CFR 716.120(c); 40 CFR 799.5000

TSCA - Substances Subject to Testing Consent Orders
Test for: Health and Environmental Effects

TSCA - Section 12(b) - Export Notification
export notification required - Section 4

INTERNATIONAL LISTS

Australian Exposure Standards - Time Weighted Averages
2 ppm TWA; 7.6 mg/m3 TWA

Australian Exposure Standards - Skin Effects
skin absorption

Canada - WHMIS: Ingredient Disclosure
1% item 109 (240)

Canada - NPRI (National Pollutant Release Inventory)
[present]

Canada - CEPA - Priority Substances List
estimated time for completion of assessment reports: 5 years

Canada - Alberta - 8 Hour Occupational Exposure Limit
2 ppm TWA; 7.6 mg/m3 TWA

Canada - Alberta - 15 Minute Occupational Exposure Limit
5 ppm STEL; 19 mg/m3 STEL

Canada - Alberta - Skin Designation
can be absorbed through the intact skin

Canada - British Columbia - 8 Hour Exposure Limits
5 ppm TWA; 19 mg/m3 TWA

Canada - British Columbia - Skin Notations
skin - potential for skin absorption

Canada - Ontario - OHSA - TWAEVs
2 ppm TWAEV; 8 mg/m3 TWAEV

Canada - Ontario - OHSA - Skin Notations
absorption through skin, eyes, or mucous membranes

Canada - Quebec - Time-Weighted Average Exposure Values
2 ppm TWAEV; 7.6 mg/m3 TWAEV

Canada - Quebec - Skin Designations
absorbed through the skin

German (DFG) - MAK Values
2 ppm MAK; 8 mg/m3 MAK

German (DFG) - Peak Limitations
5 x normal MAK (30 min. average value); don't exceed 2 times during shift

German (DFG) - Skin/Sensitizers
danger of cutaneous absorption

German (DFG) - Carcinogens
suspected carcinogen

German (DFG) - Pregnancy
classification not yet possible

Israel - Time Weighted Averages
2 ppm TWA; 7.6 mg/m3 TWA

Israel - Action Levels
1 ppm AL; 3.8 mg/m3 AL

Mexico - Instruction No. 10 - TWAs
2 ppm TWA; 10 mg/m3 TWA

Mexico - Instruction No. 10 - STELs
5 ppm STEL; 20 mg/m3 STEL

Mexico - Instruction No. 10 - Skin designation
skin - potential for cutaneous absorption

STATE LISTS

California - Air Bill 2588 Appendix A-I
known or potential carcinogen: 9/90

California - Prop. 65 - Cancer list
carcinogen - initial date 1/1/90

California - Prop. 65 - No Significant Risk Levels
no significant risk level = 100 ug/day

California - Exposure Limits - PELs
2 ppm PEL; 7.6 mg/m3 PEL

California - Exposure Limits - Skin Notation
material may be absorbed through the skin, eyes or mucous membrane

California - Directors List of Hazardous Substances (8 CCR 339)
[present]

Florida Hazardous Substance List
[present]

Massachusetts Right To Know List
extraordinarily hazardous

Minnesota Hazardous Substance List
skin

NJ Right to Know List (Total)
sn 0135

NJ Special Hazardous Substances
(mutagen)

Pennsylvania Right to Know List
environmental hazard

PROPOSED REGULATIONS

Canada - Ontario - Proposed Occupational TWAEVs
1 ppm TWAEV; 4 mg/m3 TWAEV

Canada - Ontario - Proposed Occupational STEVs
2 ppm STEV; 8 mg/m3 STEV

ANILINE AND CHLORO-, BROMO-, AND/OR NITROANILINES   RR-00282-4

ENVIRONMENTAL LISTS

TSCA - Health and Safety Reporting List
Effective Date: October 4, 1982

ANILINE HYDROCHLORIDE   142-04-1

HEALTH AND SAFETY LISTS

U.S. DOT - Substances From 49 CFR 172.101
regulated by DOT (UN1548)

U.S. DOT - Hazard Classes
DOT hazard class = 6.1

NFPA - Flash Points
flash point = 380 degrees F (193 degrees C)

NFPA - Hazard Identification Ratings
health-3; flammability-1

NIOSH - Selected LD50s and LC50s
Oral, rat: LD50 = 1072 mg/kg

NTP Chemical Status Reports - Testing Status and NTIS Number
Technical reports printed (PB287539/AS)

NTP Chemical Status Reports - Evidence of Carcinogenicity
male rat-positive; female rat-positive; male mice-negative; female mice-negative

ENVIRONMENTAL LISTS

CAA - HON Rule - SOCMI Chemicals
compliance by April 24, 1995

TSCA - Code of Federal Regulations Citations
40 CFR 712.30(d)

TSCA - PAIR - Reporting List
Reporting Date: November 19, 1982

INTERNATIONAL LISTS

Canada - WHMIS: Ingredient Disclosure
1% item 110 (241)

STATE LISTS

Florida Hazardous Substance List
[present]

Massachusetts Right To Know List
carcinogen; extraordinarily hazardous

NJ Right to Know List (Total)
sn 0136

Pennsylvania Right to Know List
[present]

ANILINE, 2,4,6-TRIMETHYL-   88-05-1

HEALTH AND SAFETY LISTS

IARC - Group 3 (not classifiable)
[present]

NIOSH - Selected LD50s and LC50s
Inhalation, mouse: LC50 = 290 mg/m3 (2 hr) Oral, rat: LD50 = 743 mg/kg

ENVIRONMENTAL LISTS

CERCLA/SARA - Section 302 Extremely Hazardous Substances and TPQs
TPQ = 500 pounds

STATE LISTS

Florida Hazardous Substance List
effective March 13, 1992

Massachusetts Right To Know List
extraordinarily hazardous

NJ Right to Know List (Total)
sn 2841

Pennsylvania Right to Know List
environmental hazard

PROPOSED REGULATIONS

CERCLA/SARA - Proposed Hazardous Substance Additions
proposed RQ = 1 pound (.454 kg)

CERCLA/SARA - 1989 Proposed RQ Adjustments
proposed RQ = 100 pounds (45.4 kg)

4-ANILINOPHENOL   122-37-2

HEALTH AND SAFETY LISTS

NIOSH - Selected LD50s and LC50s
Oral, rat: LD50 = 1220 mg/kg

INTERNATIONAL LISTS

Canada - WHMIS: Ingredient Disclosure
1% item 111 (242)

P-ANISALDEHYDE   123-11-5

HEALTH AND SAFETY LISTS

NIOSH - Selected LD50s and LC50s
Oral, rat: LD50 = 1510 mg/kg

ENVIRONMENTAL LISTS

TSCA - Code of Federal Regulations Citations
40 CFR 712.30(x); 40 CFR 716.120(d)

TSCA - PAIR - Reporting List
Reporting Date: November 27, 1991

TSCA - Health and Safety Reporting List
Effective Date: September 30, 1991

INTERNATIONAL LISTS

Canada - WHMIS: Ingredient Disclosure
1% item 113 (244)

O-ANISALDEHYDE   135-02-4

HEALTH AND SAFETY LISTS

NFPA - Flash Points
flash point = 104 degrees F (40 degrees C)

NFPA - Hazard Identification Ratings
health-2; flammability-1; reactivity-0

ENVIRONMENTAL LISTS

TSCA - Code of Federal Regulations Citations
40 CFR 712.30(x); 40 CFR 716.120(d)

TSCA - PAIR - Reporting List
Reporting Date: November 27, 1991

TSCA - Health and Safety Reporting List
Effective Date: September 30, 1991

INTERNATIONAL LISTS

Canada - WHMIS: Ingredient Disclosure
1% item 112 (243)

STATE LISTS

Florida Hazardous Substance List
[present]

Massachusetts Right To Know List
[present]

Pennsylvania Right to Know List
[present]

O-ANISIDINE   90-04-0
SEE ALSO:
ANISIDINE (O-, P- ISOMERS)

HEALTH AND SAFETY LISTS

U.S. DOT - Appendix B - Marine Pollutants
DOT regulated marine pollutant

IARC - Group 2B (sufficient animal data)
[present] (Overall evaluation based only on evidence of carcinogenicity in monograph (27, 1982) or in Supplement 4)

NFPA - Flash Points
flash point = 244 degrees F (118 degrees C)

NFPA - Hazard Identification Ratings
health-2; flammability-1; reactivity-0

NIOSH - Selected LD50s and LC50s
Oral, rat: LD50 = 2000 mg/kg

NIOSH 1990 - Pocket Guide - RELs
0.5 mg/m3 TWA (Listed under 'Anisidine (o-, p- isomers)')

NIOSH 1990 - Pocket Guide - IDLHs
50 mg/m3 IDLH (not considering carcinogenic effects) (Listed under

NIOSH 1990 - Pocket Guide - Carcinogens
occupational carcinogen (Listed under 'Anisidine (o-, p- isomers)')

NIOSH 1990 - Pocket Guide - Target organs
blood, kidneys, liver, CVS (Listed under 'Anisidine (o-, p- isomers)')

NIOSH 1990 - Pocket Guide - Skin list
Potential for dermal absorption (Listed under 'Anisidine (o-, p- isomers)')

OSHA - Possible Select Carcinogens
[present]

ENVIRONMENTAL LISTS

CERCLA/SARA - Section 313 - Emission Reporting
form R reporting required for 0.1% de minimus concentration

CERCLA/SARA - Hazardous Substances and their Reportable Quantities
final RQ = 1 pound (.454 kg)

Clean Air Act (1990) - List of Hazardous Air Contaminants
[present]

CAA - HON Rule - SOCMI Chemicals
compliance by Jan. 23, 1995

CAA - HON Rule - Organic HAPs
[present]

TSCA - Health and Safety Reporting List
Effective Date: March 11, 1994;   Sunset Date: March 11, 2004

INTERNATIONAL LISTS

Canada - WHMIS: Ingredient Disclosure
0.1% item 114 (245)

Canada - Quebec - Carcinogens
C3 carcinogen: effect detected in animals

German (DFG) - MAK Values
0.1 ppm MAK; 0.5 mg/m3 MAK

German (DFG) - Peak Limitations
2 x normal MAK (30 min. average value);   don't exceed 4 times during shift

German (DFG) - Skin/Sensitizers
danger of cutaneous absorption

STATE LISTS

California - Air Bill 2588 Appendix A-I
known or potential carcinogen

California - Prop. 65 - Cancer list
carcinogen - initial date 7/1/87

California - Prop. 65 - No Significant Risk Levels
no significant risk level = 5 ug/day

California - Directors List of Hazardous Substances (8 CCR 339)
[present]

Massachusetts Right To Know List
[present]

NJ Right to Know List (Total)
sn 1421

NJ Special Hazardous Substances
(carcinogen)

Pennsylvania Right to Know List
environmental hazard;   special hazardous substance

Pennsylvania RTK - Special Hazardous Substances
[present]

PROPOSED REGULATIONS

TSCA - ITC 32nd Report Priority Testing List
designated for dermal absorption testing

P-ANISIDINE   104-94-9
SEE ALSO:
ANISIDINE (O-, P- ISOMERS)

HEALTH AND SAFETY LISTS

IARC - Group 3 (not classifiable)
[present]

NIOSH - Selected LD50s and LC50s
Oral, rat: LD50 = 1400 mg/kg Skin, rat: LD50 = 3200 mg/kg

NIOSH 1990 - Pocket Guide - RELs
0.5 mg/m3 TWA (Listed under 'Anisidine (o-, p- isomers)')

NIOSH 1990 - Pocket Guide - IDLHs
50 mg/m3 IDLH (not considering carcinogenic effects) (Listed under 'Anisidine (o-, p- isomers)')

NIOSH 1990 - Pocket Guide - Carcinogens
occupational carcinogen (Listed under 'Anisidine (o-, p- isomers)')

NIOSH 1990 - Pocket Guide - Target organs
blood, kidneys, liver, CVS (Listed under 'Anisidine (o-, p- isomers)')

NIOSH 1990 - Pocket Guide - Skin list
Potential for dermal absorption (Listed under 'Anisidine (o-, p- isomers)')

ENVIRONMENTAL LISTS

CERCLA/SARA - Section 313 - Emission Reporting
form R reporting required for 1.0% de minimus concentration

EPA - Master Testing List
[present]

INTERNATIONAL LISTS

Canada - WHMIS: Ingredient Disclosure
1% item 115 (246)

German (DFG) - MAK Values
0.1 ppm MAK; 0.5 mg/m3 MAK

German (DFG) - Peak Limitations
2 x normal MAK (30 min. average value);   don't exceed 4 times during shift

German (DFG) - Skin/Sensitizers
danger of cutaneous absorption

STATE LISTS
California - Air Bill 2588 Appendix A-II
6/91
California - Directors List of Hazardous Substances (8 CCR 339)
[present]
Florida Hazardous Substance List
[present]
Massachusetts Right To Know List
[present]
NJ Right to Know List (Total)
sn 2893
Pennsylvania Right to Know List
environmental hazard

M-ANISIDINE   536-90-3
SEE ALSO:
ANISIDINE (O-, P- ISOMERS)
HEALTH AND SAFETY LISTS
NIOSH - Selected LD50s and LC50s
Oral, bird: LD50 = 562 mg/kg
ENVIRONMENTAL LISTS
EPA - Master Testing List
[present]

O-ANISIDINE ANTIMONYL TARTRATE   64070-14-0
INTERNATIONAL LISTS
Canada - WHMIS: Ingredient Disclosure
1% item 116 (254)

P-ANISIDINE HYDROCHLORIDE   20265-97-8
HEALTH AND SAFETY LISTS
NTP Chemical Status Reports - Testing Status and NTIS Number
Technical reports printed (PB286951/AS)
NTP Chemical Status Reports - Evidence of Carcinogenicity
male rat-equivocal; female rat-negative; male mice-negative; female mice-negative
STATE LISTS
Massachusetts Right To Know List
[present]

O-ANISIDINE HYDROCHLORIDE   134-29-2
HEALTH AND SAFETY LISTS
NTP Chemical Status Reports - Testing Status and NTIS Number
Technical reports printed (PB285879/AS)
NTP Chemical Status Reports - Evidence of Carcinogenicity
male rat-positive; female rat-positive; male mice-positive; female mice-positive
NTP Seventh Report - Suspect Carcinogens
suspect carcinogen
OSHA - Possible Select Carcinogens
[present]
ENVIRONMENTAL LISTS
CERCLA/SARA - Section 313 - Emission Reporting
form R reporting required for 0.1% de minimus concentration
STATE LISTS
California - Air Bill 2588 Appendix A-II
known or potential carcinogen
California - Prop. 65 - Cancer list
carcinogen - initial date 7/1/87
California - Prop. 65 - No Significant Risk Levels
no significant risk level = 7 ug/day
Florida Hazardous Substance List
[present]

Massachusetts Right To Know List
carcinogen; extraordinarily hazardous
Minnesota Hazardous Substance List
carcinogen
NJ Right to Know List (Total)
sn 1422
NJ Special Hazardous Substances
(carcinogen)
Pennsylvania Right to Know List
environmental hazard; special hazardous substance
Pennsylvania RTK - Special Hazardous Substances
[present]

ANISIDINE (O-, P- ISOMERS)   29191-52-4
HEALTH AND SAFETY LISTS
ACGIH 1995 - Time Weighted Averages
0.1 ppm TWA; 0.50 mg/m3 TWA
ACGIH 1995 - Skin Designations
skin - potential for cutaneous absorption
U.S. DOT - Substances From 49 CFR 172.101
regulated by DOT (UN2431)
U.S. DOT - Hazard Classes
DOT hazard class = 6.1
NIOSH 1990 - Pocket Guide - RELs
0.5 mg/m3 TWA
NIOSH 1990 - Pocket Guide - IDLHs
50 mg/m3 IDLH (not considering carcinogenic effects)
NIOSH 1990 - Pocket Guide - Carcinogens
occupational carcinogen
NIOSH 1990 - Pocket Guide - Target organs
blood, kidneys, liver, CVS
NIOSH 1990 - Pocket Guide - Skin list
Potential for dermal absorption
OSHA - Vacated PELs - Time Weighted Averages
0.5 mg/m3 TWA
OSHA - Vacated PELs - Skin Designation
Prevent or reduce skin absorption
OSHA - Final PELs - Time Weighted Averages
0.5 mg/m3 TWA
OSHA - Final PELs - Skin Notations
prevent or reduce skin absorption
INTERNATIONAL LISTS
Australian Exposure Standards - Time Weighted Averages
0.1 ppm TWA; 0.5 mg/m3 TWA
Australian Exposure Standards - Skin Effects
skin absorption
Canada - Alberta - 8 Hour Occupational Exposure Limit
0.1 ppm TWA; 0.5 mg/m3 TWA
Canada - Alberta - 15 Minute Occupational Exposure Limit
0.3 ppm STEL; 1.5 mg/m3 STEL
Canada - Alberta - Skin Designation
can be absorbed through the intact skin
Canada - British Columbia - 8 Hour Exposure Limits
0.1 ppm TWA; 0.5 mg/m3 TWA
Canada - British Columbia - Skin Notations
skin - potential for skin absorption
Canada - Ontario - OHSA - TWAEVs
0.1 ppm TWAEV; 0.5 mg/m3 TWAEV
Canada - Ontario - OHSA - Skin Notations
absorption through skin, eyes, or mucous membranes
Canada - Quebec - Time-Weighted Average Exposure Values
0.1 ppm TWAEV; 0.5 mg/m3 TWAEV

Canada - Quebec - Skin Designations
absorbed through the skin
United Kingdom - Occupational Exposure Standards - TWAs
0.1 ppm TWA; 0.5 mg/m3 TWA
United Kingdom - Occupational Exposure Standards - Notes
can be absorbed through skin
Israel - Time Weighted Averages
0.1 ppm TWA; 0.50 mg/m3 TWA
Israel - Action Levels
0.05 ppm AL; 0.25 mg/m3 AL
Mexico - Instruction No. 10 - TWAs
0.1 ppm TWA; 0.5 mg/m3 TWA
Mexico - Instruction No. 10 - Skin designation
skin - potential for cutaneous absorption
STATE LISTS
California - Exposure Limits - PELs
0.1 ppm PEL; 0.5 mg/m3 PEL
California - Exposure Limits - Skin Notation
material may be absorbed through the skin, eyes or mucous membrane
Florida Hazardous Substance List
[present]
Massachusetts Right To Know List
carcinogen; extraordinarily hazardous
Minnesota Hazardous Substance List
carcinogen; skin
Pennsylvania Right to Know List
[present]

ORTHO-ANISIDINES   RR-01197-2
HEALTH AND SAFETY LISTS
U.S. DOT - Appendix B - Marine Pollutants
DOT regulated marine pollutant

ANISINDIONE   117-37-3
STATE LISTS
California - Prop. 65 - Developmental Toxicity
developmental toxicity - initial date 10/1/92

ANISOLE   100-66-3
HEALTH AND SAFETY LISTS
U.S. DOT - Substances From 49 CFR 172.101
regulated by DOT (UN2222)
U.S. DOT - Hazard Classes
DOT hazard class = 3
NFPA - Flash Points
flash point = 125 degrees F (52 degrees C)
NFPA - Hazard Identification Ratings
health-1; flammability-2; reactivity-0
NIOSH - Selected LD50s and LC50s
Oral, rat: LD50 = 3700 mg/kg
INTERNATIONAL LISTS
Canada - WHMIS: Ingredient Disclosure
1% item 117 (247)
STATE LISTS
NJ Right to Know List (Total)
sn 0137

ANISOYL CHLORIDE   100-07-2
INTERNATIONAL LISTS
Canada - WHMIS: Ingredient Disclosure
1% item 118 (472)

ANISOYL CHLORIDE   1300-64-7
HEALTH AND SAFETY LISTS
U.S. DOT - Substances From 49 CFR 172.101
regulated by DOT (UN1729)
U.S. DOT - Hazard Classes
DOT hazard class = 8
STATE LISTS
NJ Right to Know List (Total)
sn 0138
NJ Special Hazardous Substances
(corrosive)

ANTHANTHRENE   191-26-4
HEALTH AND SAFETY LISTS
IARC - Group 3 (not classifiable)
[present]
INTERNATIONAL LISTS
Canada - WHMIS: Ingredient Disclosure
1% item 119 (248)
STATE LISTS
California - Directors List of Hazardous Substances (8 CCR 339)
[present]

ANTHOPHYLLITE, NON-ASBESTIFORM   17068-78-9
ENVIRONMENTAL LISTS
TSCA - Code of Federal Regulations Citations
40 CFR 763; 40 CFR 716.120(c)
INTERNATIONAL LISTS
Canada - Quebec - Carcinogens
C1 carcinogen: effect detected in humans
STATE LISTS
Pennsylvania Right to Know List
special hazardous substance
Pennsylvania RTK - Special Hazardous Substances
[present]

ANTHRACENE   120-12-7
SEE ALSO:
F039-HAZARDOUS WASTES
COAL TAR PITCHES
HEALTH AND SAFETY LISTS
U.S. DOT - Appendix A Table 1 - Hazardous Substances
final RQ = 5000 pounds (2270 kg)
IARC - Group 3 (not classifiable)
[present]
NFPA - Flash Points
flash point = 250 degrees F (121 degrees C)
NFPA - Hazard Identification Ratings
health-0; flammability-1
ENVIRONMENTAL LISTS
ATSDR Priority List
Rank (of 275): 235
CERCLA/SARA - Section 313 - Emission Reporting
form R reporting required for 1.0% de minimus concentration
CERCLA/SARA - Hazardous Substances and their Reportable Quantities
final RQ = 5000 pounds (2270 kg)
CAA - HON Rule - SOCMI Chemicals
compliance by Oct. 23, 1995
Clean Water Act - Priority Pollutants
[present]

RCRA - Basis for Listing - Appendix VII
Included in waste stream: F039

RCRA - TSD Facilities Ground Water Monitoring
TM 8100 = 200 ug/L PQL;   TM 8270 = 10 ug/L PQL

RCRA - Universal Treatment Standards (LDR)
WW: 0.059 mg/l;   NWW: 3.4 mg/kg

TSCA - Code of Federal Regulations Citations
40 CFR 716.120(a)

INTERNATIONAL LISTS

Canada - WHMIS: Ingredient Disclosure
1% item 120 (249)

Canada - NPRI (National Pollutant Release Inventory)
[present]

Mexico - Wastewater - Organic Toxic Pollutants and Heavy Metals
Listed under [Aromatic Hydrocarbons]

STATE LISTS

California - Air Bill 2588 Appendix A-I
6/91

California - Directors List of Hazardous Substances (8 CCR 339)
[present]

Massachusetts Right To Know List
[present]

NJ Right to Know List (Total)
sn 0139

Pennsylvania Right to Know List
environmental hazard

ANTHRACENE   120-20-7

ENVIRONMENTAL LISTS

TSCA - Health and Safety Reporting List
Effective Date: June 1, 1987

9,10-ANTHRACENEDIONE, 1-AMINO-4-HYDROXY-2-   17418-58-5
PHENOXY-

ENVIRONMENTAL LISTS

TSCA - Code of Federal Regulations Citations
40 CFR 712.30(v);   40 CFR 716.120(a)

TSCA - PAIR - Reporting List
Reporting Date: February 18, 1988

TSCA - Health and Safety Reporting List
Effective Date: December 21, 1987

9,10-ANTHRACENEDIONE, 1,5-DIAMINOCHLORO-4,   12217-79-7
8-DIHYDROXY-

ENVIRONMENTAL LISTS

TSCA - Code of Federal Regulations Citations
40 CFR 712.30(v);   40 CFR 716.120(a)

TSCA - PAIR - Reporting List
Reporting Date: February 18, 1988

TSCA - Health and Safety Reporting List
Effective Date: December 21, 1987;   Sunset Date: November 9, 1993

2,6-ANTHRACENEDISULFONIC ACID, 4,8-DIAMINO-   128-86-9
9,10-DIHYDRO-1,5-DIHYDROXY-9,10-DIOXO-

ENVIRONMENTAL LISTS

TSCA - Code of Federal Regulations Citations
40 CFR 712.30(v);   40 CFR 716.120(a)

TSCA - PAIR - Reporting List
Reporting Date: February 18, 1988

TSCA - Health and Safety Reporting List
Effective Date: December 21, 1987;   Sunset Date: November 9, 1993

ANTHRACENE OILS (COAL TAR DERIVED PROD-   RR-00781-8
UCT)

STATE LISTS

Minnesota Hazardous Substance List
carcinogen

Pennsylvania Right to Know List
special hazardous substance

Pennsylvania RTK - Special Hazardous Substances
[present]

2-ANTHRACENESULFONIC ACID, 4-((4-(ACETY-   6247-34-3
LAMINO)PHENYL)AMINO)-1-AMINO-9,10-DIHYDRO-
9,10-DIOXO-

ENVIRONMENTAL LISTS

TSCA - Code of Federal Regulations Citations
40 CFR 712.30(v);   40 CFR 716.120(a)

TSCA - PAIR - Reporting List
Reporting Date: February 18, 1988

TSCA - Health and Safety Reporting List
Effective Date: December 21, 1987;   Sunset Date: November 9, 1993

2-ANTHRACENESULFONIC ACID, 4-[[4-(ACETY-   6424-85-7
LAMINO)PHENYL]AMINO]-1-AMINO-9,10-DIHYDRO-
9,10-DIOXO-, MONOSODIUM SALT

ENVIRONMENTAL LISTS

List of Pesticide Product Inert Ingredients
[present]

TSCA - Code of Federal Regulations Citations
40 CFR 712.30(v);   40 CFR 716.120(a)

TSCA - PAIR - Reporting List
Reporting Date: February 18, 1988

TSCA - Health and Safety Reporting List
Effective Date: December 21, 1987

O-ANTHRANILIC ACID   118-92-3

HEALTH AND SAFETY LISTS

FDA - Controlled Substances Act - Precursor chemicals
Threshold by base weight = 30 kilograms

IARC - Group 3 (not classifiable)
[present]

NIOSH - Selected LD50s and LC50s
Oral, rat: LD50 = 4549 mg/kg

NTP Chemical Status Reports - Testing Status and NTIS Number
Technical reports printed (PB278883/AS)

NTP Chemical Status Reports - Evidence of Carcinogenicity
male rat-negative;   female rat-negative;   male mice-negative;   female mice-negative

ANTHRANILIC ACID, ETHYL ESTER   87-25-2

ENVIRONMENTAL LISTS

List of Pesticide Product Inert Ingredients
[present]

ANTHRANILIC ACID SALTS   RR-01772-1

HEALTH AND SAFETY LISTS

FDA - Controlled Substances Act - Precursor chemicals
Threshold by base weight = 30 kilograms

ANTHRAQUINONE   84-65-1

HEALTH AND SAFETY LISTS

NFPA - Flash Points
flash point = 365 degrees F (185 degrees C)

NFPA - Hazard Identification Ratings
health-0;   flammability-1

NTP Chemical Status Reports - Testing Status and NTIS Number
Prechronic studies completed: chemicals in review for further evaluation

ENVIRONMENTAL LISTS

CAA - HON Rule - SOCMI Chemicals
compliance by April 24, 1995

TSCA - Code of Federal Regulations Citations
40 CFR 712.30(m);   40 CFR 716.120(a)

TSCA - PAIR - Reporting List
Reporting Date: February 26, 1985

TSCA - Health and Safety Reporting List
Effective Date: December 28, 1984;   Sunset Date: November 9, 1993

TSCA - Chemical Test Rules
Testing required by: manufacturers; importers; processors (40 CFR 799.500)

TSCA - Section 12(b) - Export Notification
export notification required - Section 4

STATE LISTS

NJ Right to Know List (Total)
sn 0140

ANTHRAQUINONE BLUE   2861-02-1

ENVIRONMENTAL LISTS

TSCA - Code of Federal Regulations Citations
40 CFR 712.30(v);   40 CFR 716.120(a)(1)

TSCA - PAIR - Reporting List
Reporting Date: February 18, 1988

TSCA - Health and Safety Reporting List
Effective Date: December 21, 1987

2-ANTHRAQUINONE SULFONIC ACID, SODIUM SALT   131-08-8

HEALTH AND SAFETY LISTS

NIOSH - Selected LD50s and LC50s
Oral, guinea pig: LD50 = 21 gm/kg

ANTIMONY   7440-36-0

SEE ALSO:
K021-HAZARDOUS WASTES
F039-HAZARDOUS WASTES

HEALTH AND SAFETY LISTS

ACGIH 1995 - Time Weighted Averages
as Sb: 0.5 mg/m3 TWA

U.S. DOT - Substances From 49 CFR 172.101
regulated by DOT (UN2871)

U.S. DOT - Hazard Classes
DOT hazard class = 6.1

U.S. DOT - Appendix A Table 1 - Hazardous Substances
final RQ = 5000 pounds (2270 kg) (no reporting of releases of this hazardous substance is required if the diameter of the pieces of solid metal released is equal to or exceeds 0.004 inches)

NIOSH - Selected LD50s and LC50s
Oral, rat: LD50 = 7 gm/kg

NIOSH 1990 - Pocket Guide - RELs
as Sb: 0.5 mg/m3 TWA

NIOSH 1990 - Pocket Guide - IDLHs
as Sb: 80 mg/m3 IDLH

NIOSH 1990 - Pocket Guide - Target organs
respiratory system, CVS, skin, eyes

NIOSH - Health Standards - Exposure Limits
0.5 mg/m3 TWA

NIOSH - Health Standards - Health Effects and Precautions
Irritation, cardiovascular and lung effects (Periodic chest X-ray, pulmonary function testing, and electrocardiogram required)

OSHA - Vacated PELs - Time Weighted Averages
as Sb: 0.5 mg/m3 TWA

OSHA - Final PELs - Time Weighted Averages
as Sb: 0.5 mg/m3 TWA

ENVIRONMENTAL LISTS

ATSDR Priority List
Rank (of 275): 212

CERCLA/SARA - Section 313 - Emission Reporting
form R reporting required for 1.0% de minimus concentration

CERCLA/SARA - Hazardous Substances and their Reportable Quantities
final RQ = 5000 pounds (2270 kg) (no reporting of releases of this hazardous substance is required if the diameter of the pieces of solid metal released is equal to or exceeds 0.004 inches)

Clean Water Act - Priority Pollutants
[present]

Clean Water Act - Toxic Pollutants
[present] (Listed under 'Antimony and compounds')

Safe Drinking Water Act - MCLs
MCL = 0.006 mg/L

Safe Drinking Water Act - MCLGs
MCLG = 0.006 mg/L

RCRA - Hazardous Constituents-Appendix VIII
hazardous constituent - no waste number

RCRA - Basis for Listing - Appendix VII
Included in waste streams: F039, K021

RCRA - TSD Facilities Ground Water Monitoring
TM 6010 = 300 ug/L PQL;   TM 7040 = 2000 ug/L PQL;   TM 7041 = 30 ug/L PQL (all species in the ground water that contain this element are included)

RCRA - Universal Treatment Standards (LDR)
WW: 1.9 mg/l; NWW: 2.1 mg/l TCLP

TSCA - Code of Federal Regulations Citations
40 CFR 712.30(d);   40 CFR 716.120(a)

TSCA - PAIR - Reporting List
Reporting Date: November 19, 1982

TSCA - Health and Safety Reporting List
Effective Date: October 4, 1982

INTERNATIONAL LISTS

Australian Exposure Standards - Time Weighted Averages
as Sb: 0.5 mg/m3 TWA

Canada - WHMIS: Ingredient Disclosure
1% item 122 (251)

Canada - NPRI (National Pollutant Release Inventory)
[present]

Canada - Alberta - 8 Hour Occupational Exposure Limit
as Sb: 0.5 mg/m3 TWA

Canada - Alberta - 15 Minute Occupational Exposure Limit
as Sb: 1.5 mg/m3 STEL

Canada - British Columbia - 8 Hour Exposure Limits
as Sb: 0.5 mg/m3 TWA

Canada - Ontario - OHSA - TWAEVs
0.5 mg/m3 TWAEV

Canada - Quebec - Time-Weighted Average Exposure Values
as Sb: 0.5 mg/m3 TWAEV

German (DFG) - MAK Values
total dust: 0.5 mg/m3 MAK

German (DFG) - Peak Limitations
10 x normal MAK (30 min average value);   don't exceed during shift

Israel - Time Weighted Averages
as Sb: 0.5 mg/m3 TWA

Israel - Action Levels
as Sb: 0.25 mg/m3 AL

Mexico - Instruction No. 10 - TWAs
0.5 mg/m3 TWA

Mexico - Wastewater - Organic Toxic Pollutants and Heavy Metals
Listed under [Heavy Metals]

Mexico - Drinking Water - Ecological Criteria
0.1 mg/l

STATE LISTS

California - Air Bill 2588 Appendix A-I
6/91

California - Directors List of Hazardous Substances (8 CCR 339)
[present]

Florida Hazardous Substance List
[present]

Massachusetts Right To Know List
[present]

Minnesota Hazardous Substance List
as Sb

NJ Right to Know List (Total)
sn 0141

Pennsylvania Right to Know List
environmental hazard (any compound of this substance is also an environmental hazard)

ANTIMONY 115   17620-10-9

HEALTH AND SAFETY LISTS

U.S. DOT - Appendix A Table 2 - Radionuclides
final RQ = 1000 curies (3.7E 13 Bq)

ENVIRONMENTAL LISTS

CERCLA/SARA List of Radionuclides (Appendix B) and Their Reportable Quantities
final RQ = 1000 curies (3.7E 13 Bq)

ANTIMONY 116   15755-27-8

HEALTH AND SAFETY LISTS

U.S. DOT - Appendix A Table 2 - Radionuclides
final RQ = 1000 curies (3.7E 13 Bq)

ENVIRONMENTAL LISTS

CERCLA/SARA List of Radionuclides (Appendix B) and Their Reportable Quantities
final RQ = 1000 curies (3.7E 13 Bq)

ANTIMONY 116M   RR-00488-6

HEALTH AND SAFETY LISTS

U.S. DOT - Appendix A Table 2 - Radionuclides
final RQ = 100 curies (3.7E 12 Bq)

ENVIRONMENTAL LISTS

CERCLA/SARA List of Radionuclides (Appendix B) and Their Reportable Quantities
final RQ = 100 curies (3.7E 12 Bq)

ANTIMONY 117   15755-18-7

HEALTH AND SAFETY LISTS

U.S. DOT - Appendix A Table 2 - Radionuclides
final RQ = 1000 curies (3.7E 13 Bq)

ENVIRONMENTAL LISTS

CERCLA/SARA List of Radionuclides (Appendix B) and Their Reportable Quantities
final RQ = 1000 curies (3.7E 13 Bq)

ANTIMONY 118M   RR-00487-5

HEALTH AND SAFETY LISTS

U.S. DOT - Appendix A Table 2 - Radionuclides
final RQ = 10 curies (3.7E 11 Bq)

ENVIRONMENTAL LISTS

CERCLA/SARA List of Radionuclides (Appendix B) and Their Reportable Quantities
final RQ = 10 curies (3.7E 11 Bq)

ANTIMONY 119   14914-68-2

HEALTH AND SAFETY LISTS

U.S. DOT - Appendix A Table 2 - Radionuclides
final RQ = 1000 curies (3.7E 13 Bq)

ENVIRONMENTAL LISTS

CERCLA/SARA List of Radionuclides (Appendix B) and Their Reportable Quantities
final RQ = 1000 curies (3.7E 13 Bq)

ANTIMONY 120   14391-68-5

HEALTH AND SAFETY LISTS

U.S. DOT - Appendix A Table 2 - Radionuclides
16 minute half-life: final RQ = 1000 curies (3.7E 13 Bq); 5.76 day half-life: Final RQ = 10 curies (3.7E 11 Bq)

ENVIRONMENTAL LISTS

CERCLA/SARA List of Radionuclides (Appendix B) and Their Reportable Quantities
16 minute half-life: final RQ = 1000 curies (3.7E 13 Bq); 5.76 day half-life: Final RQ = 10 curies (3.7E 11 Bq)

ANTIMONY 122   14374-79-9

HEALTH AND SAFETY LISTS

U.S. DOT - Appendix A Table 2 - Radionuclides
final RQ = 10 curies (3.7E 11 Bq)

ENVIRONMENTAL LISTS

CERCLA/SARA List of Radionuclides (Appendix B) and Their Reportable Quantities
final RQ = 10 curies (3.7E 11 Bq)

ANTIMONY 124   14683-10-4

HEALTH AND SAFETY LISTS

U.S. DOT - Appendix A Table 2 - Radionuclides
final RQ = 10 curies (3.7E 11 Bq)

ENVIRONMENTAL LISTS

CERCLA/SARA List of Radionuclides (Appendix B) and Their Reportable Quantities
final RQ = 10 curies (3.7E 11 Bq)

ANTIMONY 124M   RR-00486-4

HEALTH AND SAFETY LISTS

U.S. DOT - Appendix A Table 2 - Radionuclides
final RQ = 1000 curies (3.7E 13 Bq)

ENVIRONMENTAL LISTS

CERCLA/SARA List of Radionuclides (Appendix B) and Their Reportable Quantities
final RQ = 1000 curies (3.7E 13 Bq)

ANTIMONY 125   14234-35-6

HEALTH AND SAFETY LISTS

U.S. DOT - Appendix A Table 2 - Radionuclides
final RQ = 10 curies (3.7E 11 Bq)

ENVIRONMENTAL LISTS

CERCLA/SARA List of Radionuclides (Appendix B) and Their Reportable Quantities
final RQ = 10 curies (3.7E 11 Bq)

ANTIMONY 126   15756-32-8

HEALTH AND SAFETY LISTS

U.S. DOT - Appendix A Table 2 - Radionuclides
final RQ = 10 curies (3.7E 11 Bq)

ENVIRONMENTAL LISTS

CERCLA/SARA List of Radionuclides (Appendix B) and Their Reportable Quantities
final RQ = 10 curies (3.7E 11 Bq)

ANTIMONY 126M   RR-00484-2

HEALTH AND SAFETY LISTS

U.S. DOT - Appendix A Table 2 - Radionuclides
final RQ = 1000 curies (3.7E 13 Bq)

ENVIRONMENTAL LISTS

CERCLA/SARA List of Radionuclides (Appendix B) and Their Reportable Quantities
final RQ = 1000 curies (3.7E 13 Bq)

ANTIMONY 127   13968-50-8

HEALTH AND SAFETY LISTS

U.S. DOT - Appendix A Table 2 - Radionuclides
final RQ = 10 curies (3.7E 11 Bq)

ENVIRONMENTAL LISTS

CERCLA/SARA List of Radionuclides (Appendix B) and Their Reportable Quantities
final RQ = 10 curies (3.7E 11 Bq)

ANTIMONY 128   15756-34-0

HEALTH AND SAFETY LISTS

U.S. DOT - Appendix A Table 2 - Radionuclides
10.4 minute half-life: final RQ = 1000 curies (3.7E 13 Bq);   9.01 hour half-life: final RQ = 10 curies (3.7E 11 Bq)

ENVIRONMENTAL LISTS

CERCLA/SARA List of Radionuclides (Appendix B) and Their Reportable Quantities
10.4 minute half-life: final RQ = 1000 curies (3.7E 13 Bq);   9.01 hour half-life: final RQ = 10 curies (3.7E 11 Bq)

ANTIMONY 129   14331-88-5

HEALTH AND SAFETY LISTS

U.S. DOT - Appendix A Table 2 - Radionuclides
final RQ = 100 curies (3.7E 12 Bq)

ENVIRONMENTAL LISTS

CERCLA/SARA List of Radionuclides (Appendix B) and Their Reportable Quantities
final RQ = 100 curies (3.7E 12 Bq)

ANTIMONY 130   15756-35-1

HEALTH AND SAFETY LISTS

U.S. DOT - Appendix A Table 2 - Radionuclides
final RQ = 100 curies (3.7E 12 Bq)

ENVIRONMENTAL LISTS

CERCLA/SARA List of Radionuclides (Appendix B) and Their Reportable Quantities
final RQ = 100 curies (3.7E 12 Bq)

ANTIMONY 131   15756-29-3

HEALTH AND SAFETY LISTS

U.S. DOT - Appendix A Table 2 - Radionuclides
final RQ = 1000 curies (3.7E 13 Bq)

ENVIRONMENTAL LISTS

CERCLA/SARA List of Radionuclides (Appendix B) and Their Reportable Quantities
final RQ = 1000 curies (3.7E 13 Bq)

ANTIMONY(III) ACETATE   6923-52-0

SEE ALSO:
ANTIMONY

INTERNATIONAL LISTS

Canada - WHMIS: Ingredient Disclosure
1% item 123 (7)

ANTIMONY COMPOUNDS   RR-00585-6

ENVIRONMENTAL LISTS

CERCLA/SARA - Section 313 - Emission Reporting
form R reporting required for 1.0% de minimus concentration

Clean Air Act (1990) - List of Hazardous Air Contaminants
[present] (includes any unique chemical substance that contains Antimony as part of that chemical's infrastructure)

Clean Water Act - Toxic Pollutants
[present] (Listed under 'Antimony and compounds')

RCRA - Hazardous Constituents-Appendix VIII
hazardous constituent - no waste number

INTERNATIONAL LISTS

Canada - WHMIS: Ingredient Disclosure
1% item 121 (250)

Canada - NPRI (National Pollutant Release Inventory)
[present]

Canada - Ontario - OHSA - TWAEVs
excluding stibine, as Sb: 0.5 mg/m3 TWAEV

Canada - Quebec - Time-Weighted Average Exposure Values
as Sb: 0.5 mg/m3 TWAEV

United Kingdom - Occupational Exposure Standards - TWAs
as Sb: 0.5 mg/m3 TWA

STATE LISTS

California - Air Bill 2588 Appendix A-I
6/91

California - Exposure Limits - PELs
as Sb: 0.5 mg/m3 PEL

California - Directors List of Hazardous Substances (8 CCR 339)
[present] (exempt when in bonded form or when it cannot be released)

NJ Right to Know List (Total)
sn 2223

ANTIMONY, INORGANIC COMPOUNDS   RR-00554-9

HEALTH AND SAFETY LISTS

U.S. DOT - Substances From 49 CFR 172.101
regulated by DOT (UN1549, UN3141)

U.S. DOT - Hazard Classes
DOT hazard class = 6.1

STATE LISTS

NJ Right to Know List (Total)
sn 2866

ANTIMONY LACTATE   58164-88-8

SEE ALSO:
ANTIMONY

HEALTH AND SAFETY LISTS

U.S. DOT - Substances From 49 CFR 172.101
regulated by DOT (UN1550)

U.S. DOT - Hazard Classes
DOT hazard class = 6.1

STATE LISTS
  NJ Right to Know List (Total)
  sn 0142

ANTIMONY OXIDE   1327-33-9
  SEE ALSO:
    ANTIMONY

  INTERNATIONAL LISTS
    German (DFG) - Carcinogens
    animal evidence of carcinogenicity (Listed under 'Antimony trioxide')

ANTIMONY PENTACHLORIDE   7647-18-9
  SEE ALSO:
    ANTIMONY

  HEALTH AND SAFETY LISTS
    U.S. DOT - Substances From 49 CFR 172.101
    regulated by DOT (UN1730, UN1731)
    U.S. DOT - Hazard Classes
    DOT hazard class - 8
    U.S. DOT - Appendix A Table 1 - Hazardous Substances
    final RQ - 1000 pounds (454 kg)
    NIOSH - Selected LD50s and LC50s
    Inhalation, mouse: LC50 - 620 mg/m3 (8 hr) Oral, rat: LD50 - 1115 mg/kg

  ENVIRONMENTAL LISTS
    CERCLA/SARA - Hazardous Substances and their Reportable Quantities
    final RQ - 1000 pounds (454 kg)
    Clean Water Act - Hazardous Substances
    [present]

  STATE LISTS
    Florida Hazardous Substance List
    [present]
    Massachusetts Right To Know List
    [present]
    NJ Right to Know List (Total)
    sn 0143
    NJ Special Hazardous Substances
    (corrosive)
    Pennsylvania Right to Know List
    environmental hazard

ANTIMONY PENTAFLUORIDE   7783-70-2
  SEE ALSO:
    ANTIMONY

  HEALTH AND SAFETY LISTS
    U.S. DOT - Substances From 49 CFR 172.101
    regulated by DOT (UN1732)
    U.S. DOT - Hazard Classes
    DOT hazard class - 8
    NIOSH - Selected LD50s and LC50s
    Inhalation, mouse: LC50 - 270 mg/m3

  ENVIRONMENTAL LISTS
    CERCLA/SARA - Section 302 Extremely Hazardous Substances and TPQs
    TPQ - 500 pounds

  INTERNATIONAL LISTS
    Canada - WHMIS: Ingredient Disclosure
    1% item 124 (1343)

  STATE LISTS
    Florida Hazardous Substance List
    [present]
    Massachusetts Right To Know List
    extraordinarily hazardous

NJ Right to Know List (Total)
sn 0144
NJ Special Hazardous Substances
(corrosive)
Pennsylvania Right to Know List
environmental hazard

  PROPOSED REGULATIONS
    CERCLA/SARA - Proposed Hazardous Substance Additions
    proposed RQ - 1 pound (.454 kg)
    CERCLA/SARA - 1989 Proposed RQ Adjustments
    proposed RQ - 100 pounds (45.4 kg)

ANTIMONY PENTASULFIDE   1315-04-4
  SEE ALSO:
    ANTIMONY

  STATE LISTS
    Florida Hazardous Substance List
    [present]
    Massachusetts Right To Know List
    [present]
    Pennsylvania Right to Know List
    environmental hazard

ANTIMONY POTASSIUM TARTRATE   28300-74-5
  SEE ALSO:
    ANTIMONY

  HEALTH AND SAFETY LISTS
    U.S. DOT - Substances From 49 CFR 172.101
    regulated by DOT (UN1551)
    U.S. DOT - Hazard Classes
    DOT hazard class - 6.1
    U.S. DOT - Appendix A Table 1 - Hazardous Substances
    final RQ - 100 pounds (45.4 kg)
    NIOSH - Selected LD50s and LC50s
    Oral, rat: LD50 - 115 mg/kg
    NTP Chemical Status Reports - Testing Status and NTIS Number
    Technical reports printed (PB93-149714)

  ENVIRONMENTAL LISTS
    CERCLA/SARA - Hazardous Substances and their Reportable Quantities
    final RQ - 100 pounds (45.4 kg)
    Clean Water Act - Hazardous Substances
    [present]
    List of Pesticide Product Inert Ingredients
    [present]

  STATE LISTS
    Massachusetts Right To Know List
    [present]
    NJ Right to Know List (Total)
    sn 0145
    Pennsylvania Right to Know List
    environmental hazard

ANTIMONY SULFIDE   1345-04-6
  SEE ALSO:
    ANTIMONY

  ENVIRONMENTAL LISTS
    TSCA - Code of Federal Regulations Citations
    40 CFR 712.30(d); 40 CFR 716.120(a)
    TSCA - PAIR - Reporting List
    Reporting Date: November 19, 1982
    TSCA - Health and Safety Reporting List
    Effective Date: October 4, 1982

ANTIMONY TRIBROMIDE   7789-61-9
SEE ALSO:
  ANTIMONY

HEALTH AND SAFETY LISTS
  U.S. DOT - Substances From 49 CFR 172.101
  regulated by DOT (NA1549)
  U.S. DOT - Hazard Classes
  DOT hazard class = 8
  U.S. DOT - Appendix A Table 1 - Hazardous Substances
  final RQ = 1000 pounds (454 kg)

ENVIRONMENTAL LISTS
  CERCLA/SARA - Hazardous Substances and their Reportable Quantities
  final RQ = 1000 pounds (454 kg)
  Clean Water Act - Hazardous Substances
  [present]

STATE LISTS
  Massachusetts Right To Know List
  [present]
  NJ Right to Know List (Total)
  sn 0146
  NJ Special Hazardous Substances
  (corrosive)
  Pennsylvania Right to Know List
  environmental hazard

ANTIMONY TRICHLORIDE   10025-91-9
SEE ALSO:
  ANTIMONY

HEALTH AND SAFETY LISTS
  U.S. DOT - Substances From 49 CFR 172.101
  regulated by DOT (UN1733)
  U.S. DOT - Hazard Classes
  DOT hazard class = 8
  U.S. DOT - Appendix A Table 1 - Hazardous Substances
  final RQ = 1000 pounds (454 kg)
  NIOSH - Selected LD50s and LC50s
  Oral, rat: LD50 = 525 mg/kg

ENVIRONMENTAL LISTS
  CERCLA/SARA - Hazardous Substances and their Reportable Quantities
  final RQ = 1000 pounds (454 kg)
  Clean Water Act - Hazardous Substances
  [present]

INTERNATIONAL LISTS
  Canada - WHMIS: Ingredient Disclosure
  1% item 125 (1652)

STATE LISTS
  Massachusetts Right To Know List
  [present]
  NJ Right to Know List (Total)
  sn 0147
  NJ Special Hazardous Substances
  (corrosive)
  Pennsylvania Right to Know List
  environmental hazard

ANTIMONY TRIFLUORIDE   7783-56-4
SEE ALSO:
  ANTIMONY

HEALTH AND SAFETY LISTS
  U.S. DOT - Substances From 49 CFR 172.101
  regulated by DOT (NA1549)

  U.S. DOT - Hazard Classes
  DOT hazard class = 8
  U.S. DOT - Appendix A Table 1 - Hazardous Substances
  final RQ = 1000 pounds (454 kg)
  NIOSH - Selected LD50s and LC50s
  Oral, mouse: LD50 = 804 mg/kg

ENVIRONMENTAL LISTS
  CERCLA/SARA - Hazardous Substances and their Reportable Quantities
  final RQ = 1000 pounds (454 kg)
  Clean Water Act - Hazardous Substances
  [present]

STATE LISTS
  Massachusetts Right To Know List
  [present]
  NJ Right to Know List (Total)
  sn 0148
  NJ Special Hazardous Substances
  (corrosive)
  Pennsylvania Right to Know List
  environmental hazard

ANTIMONY TRIOXIDE   1309-64-4
SEE ALSO:
  ANTIMONY
  ANTIMONY COMPOUNDS
  ANTIMONY

HEALTH AND SAFETY LISTS
  ACGIH 1995 - Time Weighted Averages
  handling and use, as Sb: 0.5 mg/m3 TWA
  ACGIH 1995 - Carcinogens
  production: A2-suspected human carcinogen
  U.S. DOT - Appendix A Table 1 - Hazardous Substances
  final RQ = 1000 pounds (454 kg)
  IARC - Group 2B (sufficient animal data)
  [present]
  IARC - Group 3 (not classifiable)
  [present]
  OSHA - Possible Select Carcinogens
  [present]

ENVIRONMENTAL LISTS
  CERCLA/SARA - Hazardous Substances and their Reportable Quantities
  final RQ = 1000 pounds (454 kg)
  Clean Water Act - Hazardous Substances
  [present]
  EPA - Master Testing List
  [present]
  List of Pesticide Product Inert Ingredients
  [present]
  TSCA - Code of Federal Regulations Citations
  40 CFR 712.30(d);   40 CFR 716.120(a)
  TSCA - PAIR - Reporting List
  Reporting Date: November 19, 1982
  TSCA - Health and Safety Reporting List
  Effective Date: October 4, 1982

INTERNATIONAL LISTS
  Australian Exposure Standards - Time Weighted Averages
  handling and use, as Sb: 0.5 mg/m3 TWA;   production: control to lowest practical level
  Australian Exposure Standards - Carcinogens
  production: probable carcinogen
  Canada - WHMIS: Ingredient Disclosure
  1% item 126 (1691)

Canada - Alberta - 8 Hour Occupational Exposure Limit
0.5 mg/m3 TWA

Canada - Alberta - 15 Minute Occupational Exposure Limit
as Sb: 1.5 mg/m3 STEL

Canada - British Columbia - 8 Hour Exposure Limits
handling and use, as Sb: 0.5 mg/m3 TWA

Canada - Quebec - Time-Weighted Average Exposure Values
as Sb: 0.5 mg/m3 TWAEV

Canada - Quebec - Carcinogens
as Sb: C3 carcinogen: effect detected in animals

German (DFG) - Carcinogens
animal evidence of carcinogenicity

Israel - Time Weighted Averages
handling and use, as Sb: 0.5 mg/m3 TWA

Israel - Action Levels
handling and use, as Sb: 0.25 mg/m3 AL

STATE LISTS

California - Air Bill 2588 Appendix A-I
known or potential carcinogen: 9/90

California - Prop. 65 - Cancer list
carcinogen - initial date 10/1/90

California - Directors List of Hazardous Substances (8 CCR 339)
[present]

Florida Hazardous Substance List
[present]

Massachusetts Right To Know List
[present]

Minnesota Hazardous Substance List
as Sb: carcinogen

NJ Right to Know List (Total)
sn 0149

Pennsylvania Right to Know List
environmental hazard

ANTIMONY TRIOXIDE PRODUCTION   RR-01639-7

INTERNATIONAL LISTS

Canada - Alberta - 8 Hour Occupational Exposure Limit
0.5 mg/m3 TWA

Canada - Alberta - 15 Minute Occupational Exposure Limit
as Sb: 1.5 mg/m3 STEL

Canada - Alberta - Designated Substances
designated substance - requires code of practice

Canada - British Columbia - 8 Hour Exposure Limits
as Sb: 0.5 mg/m3 TWA

Canada - British Columbia - Carcinogens
carcinogen - as Sb: 0.5 mg/m3 TWA

Canada - Quebec - Time-Weighted Average Exposure Values
as Sb: substance of which the recirculation is prohibited

Canada - Quebec - Carcinogens
C2 carcinogen: effect suspected in humans

Mexico - Instruction No. 10 - TWAs
1 mg/m3 TWA

Mexico - Instruction No. 10 - Carcinogens
potential carcinogen in humans - limited epidemiological evidence

ANTIMYCIN A   1397-94-0

HEALTH AND SAFETY LISTS

NIOSH - Selected LD50s and LC50s
Oral, rat: LD50 = 28 mg/kg

ENVIRONMENTAL LISTS

CERCLA/SARA - Section 302 Extremely Hazardous Substances and TPQs
TPQ = 1000/10,000 pounds

STATE LISTS

Florida Hazardous Substance List
effective March 13, 1992

Massachusetts Right To Know List
extraordinarily hazardous

NJ Right to Know List (Total)
sn 2132

Pennsylvania Right to Know List
environmental hazard

PROPOSED REGULATIONS

CERCLA/SARA - Proposed Hazardous Substance Additions
proposed RQ = 1 pound (.454 kg)

CERCLA/SARA - 1989 Proposed RQ Adjustments
proposed RQ = 1000 pounds (454 kg)

ANTU   86-88-4

HEALTH AND SAFETY LISTS

ACGIH 1995 - Time Weighted Averages
0.3 mg/m3 TWA

U.S. DOT - Appendix A Table 1 - Hazardous Substances
final RQ = 100 pounds (45.4 kg)

IARC - Group 3 (not classifiable)
[present]

NIOSH - Selected LD50s and LC50s
Oral, rat: LD50 = 6 mg/kg

NIOSH 1990 - Pocket Guide - RELs
0.3 mg/m3 TWA

NIOSH 1990 - Pocket Guide - IDLHs
100 mg/m3 IDLH

NIOSH 1990 - Pocket Guide - Target organs
respiratory system

OSHA - Vacated PELs - Time Weighted Averages
0.3 mg/m3 TWA

OSHA - Final PELs - Time Weighted Averages
0.3 mg/m3 TWA

ENVIRONMENTAL LISTS

CERCLA/SARA - Section 302 Extremely Hazardous Substances and TPQs
TPQ = 500/10,000 pounds

CERCLA/SARA - Hazardous Substances and their Reportable Quantities
final RQ = 100 pounds (45.4 kg)

RCRA - P Series Wastes
waste number P072

RCRA - Hazardous Constituents-Appendix VIII
waste number P072

RCRA - Substances Banned From Land Disposal
[present]

INTERNATIONAL LISTS

Australian Exposure Standards - Time Weighted Averages
0.3 mg/m3 TWA

Canada - Alberta - 8 Hour Occupational Exposure Limit
0.3 mg/m3 TWA

Canada - Alberta - 15 Minute Occupational Exposure Limit
0.9 mg/m3 STEL

Canada - British Columbia - 8 Hour Exposure Limits
0.3 mg/m3 TWA

Canada - British Columbia - 15 Minute Exposure Limits
0.9 mg/m3 STEL

Canada - Ontario - OHSA - TWAEVs
0.3 mg/m3 TWAEV

Canada - Quebec - Time-Weighted Average Exposure Values
0.3 mg/m3 TWAEV

German (DFG) - MAK Values
total dust: 0.3 mg/m3 MAK
German (DFG) - Peak Limitations
5 x normal MAK (30 min. average value); don't exceed 2 times during shift
Israel - Time Weighted Averages
0.3 mg/m3 TWA
Israel - Action Levels
0.15 mg/m3 AL
Mexico - Instruction No. 10 - TWAs
0.3 mg/m3 TWA
Mexico - Instruction No. 10 - STELs
0.9 mg/m3 STEL

STATE LISTS
California - Exposure Limits - PELs
0.3 mg/m3 PEL
California - Directors List of Hazardous Substances (8 CCR 339)
[present]
Florida Hazardous Substance List
[present]
Massachusetts Right To Know List
extraordinarily hazardous
Minnesota Hazardous Substance List
[present]
Pennsylvania Right to Know List
environmental hazard

APHOLATE   52-46-0

HEALTH AND SAFETY LISTS
IARC - Group 3 (not classifiable)
[present]

APPLE POMACE   RR-01037-7

ENVIRONMENTAL LISTS
List of Pesticide Product Inert Ingredients
[present]

P-ARAMIDE   26125-61-1

INTERNATIONAL LISTS
German (DFG) - Carcinogens
animal evidence of carcinogenicity

ARAMITE   140-57-8

HEALTH AND SAFETY LISTS
IARC - Group 2B (sufficient animal data)
[present] (Overall evaluation based only on evidence of carcinogenicity in monograph (5, 1974) or in Supplement 4)
NIOSH - Selected LD50s and LC50s
Oral, rat: LD50 = 3900 mg/kg
OSHA - Possible Select Carcinogens
[present]

ENVIRONMENTAL LISTS
RCRA - Hazardous Constituents-Appendix VIII
hazardous constituent - no waste number
RCRA - TSD Facilities Ground Water Monitoring
TM 8270 = 10 ug/L PQL
RCRA - Universal Treatment Standards (LDR)
WW: 0.36 mg/l; NWW: Not applicable

STATE LISTS
California - Air Bill 2588 Appendix A-II
known or potential carcinogen
California - Prop. 65 - Cancer list
carcinogen - initial date 7/1/87

California - Prop. 65 - No Significant Risk Levels
no significant risk level = 20 ug/day
California - Directors List of Hazardous Substances (8 CCR 339)
[present]
Florida Hazardous Substance List
[present]
Massachusetts Right To Know List
carcinogen; extraordinarily hazardous
Minnesota Hazardous Substance List
carcinogen
NJ Right to Know List (Total)
sn 0150
NJ Special Hazardous Substances
(corrosive)
Pennsylvania Right to Know List
environmental hazard; special hazardous substance
Pennsylvania RTK - Special Hazardous Substances
[present]

L-ARGININE HCL   1119-34-2

HEALTH AND SAFETY LISTS
NIOSH - Selected LD50s and LC50s
Oral, rat: LD50 = 12 gm/kg

ARGON   7440-37-1

HEALTH AND SAFETY LISTS
ACGIH 1995 - Time Weighted Averages
simple asphyxiant
U.S. DOT - Substances From 49 CFR 172.101
regulated by DOT (UN1006, UN1951)
U.S. DOT - Hazard Classes
DOT hazard class = 2.2

ENVIRONMENTAL LISTS
List of Pesticide Product Inert Ingredients
[present]

INTERNATIONAL LISTS
Australian Exposure Standards - Time Weighted Averages
Asphyxiant at < 18% oxygen by volume
Canada - British Columbia - 8 Hour Exposure Limits
asphyxiant substance
Canada - Ontario - OHSA - TWAEVs
simple asphyxiant
Canada - Quebec - Time-Weighted Average Exposure Values
simple asphyxiant
Israel - Time Weighted Averages
Asphyxiant
Mexico - Instruction No. 10 - TWAs
simple asphyxiant

STATE LISTS
California - Exposure Limits - PELs
asphyxiant (limit depends on level of oxygen)
Florida Hazardous Substance List
[present]
Massachusetts Right To Know List
[present]
Minnesota Hazardous Substance List
[present]
NJ Right to Know List (Total)
sn 0151
Pennsylvania Right to Know List
[present]

© Van Nostrand Reinhold 1995

ARGON 39   25729-41-3

HEALTH AND SAFETY LISTS

U.S. DOT - Appendix A Table 2 - Radionuclides
final RQ = 1000 curies (3.7E 13 Bq)

ENVIRONMENTAL LISTS

CERCLA/SARA List of Radionuclides (Appendix B) and Their Reportable Quantities
final RQ = 1000 curies (3.7E 13 Bq)

ARGON 41   14163-25-8

HEALTH AND SAFETY LISTS

U.S. DOT - Appendix A Table 2 - Radionuclides
final RQ = 10 curies (3.7E 11 Bq)

ENVIRONMENTAL LISTS

CERCLA/SARA List of Radionuclides (Appendix B) and Their Reportable Quantities
final RQ = 10 curies (3.7E 11 Bq)

AROCHLOR 1262   37324-23-5

HEALTH AND SAFETY LISTS

NIOSH - Selected LD50s and LC50s
Oral, rat: LD50 = 11300 mg/kg

AROCHLOR 1268   11100-14-4

HEALTH AND SAFETY LISTS

NIOSH - Selected LD50s and LC50s
Oral, rat: LD50 = 10900 mg/kg

AROCHLOR 4465   11120-29-9

HEALTH AND SAFETY LISTS

NIOSH - Selected LD50s and LC50s
Oral, rat: LD50 = 16 gm/kg

AROCLOR 1016   12674-11-2

SEE ALSO:
F039-HAZARDOUS WASTES
POLYCHLORINATED BIPHENYLS

HEALTH AND SAFETY LISTS

U.S. DOT - Appendix A Table 1 - Hazardous Substances
final RQ = 1 pound (0.454 kg)

ENVIRONMENTAL LISTS

ATSDR Priority List
Rank (of 275): 042

CERCLA/SARA - Hazardous Substances and their Reportable Quantities
final RQ = 1 pound (0.454 kg)

Clean Water Act - Priority Pollutants
[present]

EPA - Carcinogen Hazard Ranking for RQ Adjustment
Hazard ranking = Medium

RCRA - Basis for Listing - Appendix VII
Included in waste stream: F039

INTERNATIONAL LISTS

Mexico - Wastewater - Organic Toxic Pollutants and Heavy Metals
Listed under [Polychlorinated Byphenyls]

STATE LISTS

Massachusetts Right To Know List
[present]

NJ Right to Know List (Total)
sn 1554

Pennsylvania Right to Know List
environmental hazard

AROCLOR 1221   11104-28-2

SEE ALSO:
F039-HAZARDOUS WASTES
POLYCHLORINATED BIPHENYLS

HEALTH AND SAFETY LISTS

U.S. DOT - Appendix A Table 1 - Hazardous Substances
final RQ = 1 pound (0.454 kg)

NIOSH - Selected LD50s and LC50s
Oral, rat: LD50 = 3980 mg/kg

ENVIRONMENTAL LISTS

ATSDR Priority List
Rank (of 275): 038

CERCLA/SARA - Hazardous Substances and their Reportable Quantities
final RQ = 1 pound (0.454 kg)

Clean Water Act - Priority Pollutants
[present]

EPA - Carcinogen Hazard Ranking for RQ Adjustment
Hazard ranking = Medium

RCRA - Basis for Listing - Appendix VII
Included in waste stream: F039

INTERNATIONAL LISTS

Mexico - Wastewater - Organic Toxic Pollutants and Heavy Metals
Listed under [Polychlorinated Byphenyls]

STATE LISTS

Massachusetts Right To Know List
[present]

NJ Right to Know List (Total)
sn 1554

Pennsylvania Right to Know List
environmental hazard

AROCLOR 1232   11141-16-5

SEE ALSO:
F039-HAZARDOUS WASTES
POLYCHLORINATED BIPHENYLS

HEALTH AND SAFETY LISTS

U.S. DOT - Appendix A Table 1 - Hazardous Substances
final RQ = 1 pound (0.454 kg)

NIOSH - Selected LD50s and LC50s
Oral, rat: LD50 = 4470 mg/kg

ENVIRONMENTAL LISTS

ATSDR Priority List
Rank (of 275): 062

CERCLA/SARA - Hazardous Substances and their Reportable Quantities
final RQ = 1 pound (0.454 kg)

Clean Water Act - Priority Pollutants
[present]

EPA - Carcinogen Hazard Ranking for RQ Adjustment
Hazard ranking = Medium

RCRA - Basis for Listing - Appendix VII
Included in waste stream: F039

INTERNATIONAL LISTS

Mexico - Wastewater - Organic Toxic Pollutants and Heavy Metals
Listed under [Polychlorinated Byphenyls]

STATE LISTS

Massachusetts Right To Know List
[present]

NJ Right to Know List (Total)
sn 1554

Pennsylvania Right to Know List
environmental hazard

AROCLOR 1240   71328-89-7

ENVIRONMENTAL LISTS

ATSDR Priority List
Rank (of 275): 123

AROCLOR 1242   53469-21-9
SEE ALSO:
F039-HAZARDOUS WASTES
POLYCHLORINATED BIPHENYLS

HEALTH AND SAFETY LISTS

ACGIH 1995 - Time Weighted Averages
1 mg/m3 TWA

ACGIH 1995 - Skin Designations
skin - potential for cutaneous absorption

U.S. DOT - Appendix A Table 1 - Hazardous Substances
final RQ = 1 pound (0.454 kg)

NIOSH - Selected LD50s and LC50s
Oral, rat: LD50 = 4250 mg/kg

NIOSH 1990 - Pocket Guide - RELs
0.001 mg/m3 TWA

NIOSH 1990 - Pocket Guide - IDLHs
10 mg/m3 IDLH (not considering carcinogenic effects)

NIOSH 1990 - Pocket Guide - Carcinogens
occupational carcinogen

NIOSH 1990 - Pocket Guide - Target organs
skin, eyes, liver

OSHA - Vacated PELs - Time Weighted Averages
1 mg/m3 TWA

OSHA - Vacated PELs - Skin Designation
Prevent or reduce skin absorption

OSHA - Final PELs - Time Weighted Averages
1 mg/m3 TWA

OSHA - Final PELs - Skin Notations
prevent or reduce skin absorption

ENVIRONMENTAL LISTS

ATSDR Priority List
Rank (of 275): 017

CERCLA/SARA - Hazardous Substances and their Reportable Quantities
final RQ = 1 pound (0.454 kg)

Clean Water Act - Priority Pollutants
[present]

EPA - Carcinogen Hazard Ranking for RQ Adjustment
Hazard ranking = Medium

RCRA - Basis for Listing - Appendix VII
Included in waste stream: F039

INTERNATIONAL LISTS

Australian Exposure Standards - Time Weighted Averages
1 mg/m3 TWA

Australian Exposure Standards - Short Term Exposure Limits
2 mg/m3 STEL

Australian Exposure Standards - Skin Effects
skin absorption

Australian Exposure Standards - Carcinogens
probable carcinogen

Australian Exposure Standards - Under Review
exposure limits under review

Canada - WHMIS: Ingredient Disclosure
0.1% item 361 (417)

Canada - Alberta - 8 Hour Occupational Exposure Limit
1 mg/m3 TWA

Canada - Alberta - 15 Minute Occupational Exposure Limit
2 mg/m3 STEL

Canada - Alberta - Skin Designation
can be absorbed through the intact skin

Canada - British Columbia - 8 Hour Exposure Limits
1 mg/m3 TWA

Canada - British Columbia - 15 Minute Exposure Limits
2 mg/m3 STEL

Canada - British Columbia - Skin Notations
skin - potential for skin absorption

Canada - Quebec - Time-Weighted Average Exposure Values
1 mg/m3 TWAEV

Canada - Quebec - Skin Designations
absorbed through the skin

Canada - Quebec - Carcinogens
C2 carcinogen: effect suspected in humans

German (DFG) - MAK Values
0.1 ppm MAK;   1 mg/m3 MAK

German (DFG) - Peak Limitations
10 x normal MAK (30 min average value);   don't exceed during shift

German (DFG) - Skin/Sensitizers
danger of cutaneous absorption

German (DFG) - Carcinogens
suspected carcinogen

German (DFG) - Pregnancy
risk to embryo/fetus probable

Israel - Time Weighted Averages
1 mg/m3 TWA

Israel - Action Levels
0.5 mg/m3 AL

Mexico - Instruction No. 10 - TWAs
1 mg/m3 TWA

Mexico - Instruction No. 10 - STELs
2 mg/m3 STEL

Mexico - Instruction No. 10 - Skin designation
skin - potential for cutaneous absorption

Mexico - Wastewater - Organic Toxic Pollutants and Heavy Metals
Listed under [Polychlorinated Byphenyls]

STATE LISTS

California - Exposure Limits - PELs
1 mg/m3 PEL

California - Exposure Limits - Skin Notation
material may be absorbed through the skin, eyes or mucous membrane

California - Directors List of Hazardous Substances (8 CCR 339)
[present] (Listed under 'Polychlorobiphenyls')

Florida Hazardous Substance List
[present]

Massachusetts Right To Know List
carcinogen;   extraordinarily hazardous;   teratogen

Minnesota Hazardous Substance List
carcinogen;   skin

NJ Right to Know List (Total)
sn 1554

Pennsylvania Right to Know List
environmental hazard

AROCLOR 1248   12672-29-6
SEE ALSO:
F039-HAZARDOUS WASTES
POLYCHLORINATED BIPHENYLS

HEALTH AND SAFETY LISTS

U.S. DOT - Appendix A Table 1 - Hazardous Substances
final RQ = 1 pound (0.454 kg)

NIOSH - Selected LD50s and LC50s
Oral, rat: LD50 = 11 gm/kg

ENVIRONMENTAL LISTS

ATSDR Priority List
Rank (of 275): 024

CERCLA/SARA - Hazardous Substances and their Reportable Quantities
final RQ = 1 pound (0.454 kg)

Clean Water Act - Priority Pollutants
[present]

EPA - Carcinogen Hazard Ranking for RQ Adjustment
Hazard ranking = Medium

RCRA - Basis for Listing - Appendix VII
Included in waste stream: F039

INTERNATIONAL LISTS

Mexico - Wastewater - Organic Toxic Pollutants and Heavy Metals
Listed under [Polychlorinated Byphenyls]

STATE LISTS

Massachusetts Right To Know List
[present]

NJ Right to Know List (Total)
sn 1554

Pennsylvania Right to Know List
environmental hazard

AROCLOR 1260   **11096-82-5**
SEE ALSO:
F039-HAZARDOUS WASTES
POLYCHLORINATED BIPHENYLS

HEALTH AND SAFETY LISTS

U.S. DOT - Appendix A Table 1 - Hazardous Substances
final RQ = 1 pound (0.454 kg)

NIOSH - Selected LD50s and LC50s
Oral, rat: LD50 = 1315 mg/kg

NTP Seventh Report - Suspect Carcinogens
suspect carcinogen (Listed under 'Polychlorinated biphenyls')

ENVIRONMENTAL LISTS

ATSDR Priority List
Rank (of 275): 015

CERCLA/SARA - Hazardous Substances and their Reportable Quantities
final RQ = 1 pound (0.454 kg)

Clean Water Act - Priority Pollutants
[present]

EPA - Carcinogen Hazard Ranking for RQ Adjustment
Hazard ranking = Medium

RCRA - Basis for Listing - Appendix VII
Included in waste stream: F039

INTERNATIONAL LISTS

Canada - WHMIS: Ingredient Disclosure
0.1% item 128 (258)

Mexico - Wastewater - Organic Toxic Pollutants and Heavy Metals
Listed under [Polychlorinated Byphenyls]

STATE LISTS

California - Prop. 65 - Cancer list
carcinogen - initial date 1/1/88

Massachusetts Right To Know List
[present]

NJ Right to Know List (Total)
sn 1554

Pennsylvania Right to Know List
environmental hazard; special hazardous substance

Pennsylvania RTK - Special Hazardous Substances
[present]

AROMATIC AMINE COMPOUND   **RR-01009-3**

ENVIRONMENTAL LISTS

TSCA - Chemicals with Significant New Use Rules
PMN number: P-86-334

AROMATIC AMINE POLYOLS   **RR-01686-4**

ENVIRONMENTAL LISTS

TSCA - Chemicals with Significant New Use Rules
PMN numbers: P-93-212, P-93-213

AROMATIC AMINO ETHER   **RR-01668-2**

ENVIRONMENTAL LISTS

TSCA - Chemicals with Significant New Use Rules
PMN number: P-90-1840

AROMATIC C9 FRACTION FROM PETROLEUM REFINING   **RR-00286-8**

ENVIRONMENTAL LISTS

TSCA - Health and Safety Reporting List
Effective Date: February 13, 1984 (The C9 fraction is primarily composed of o-, m-, p-, and mixed isomers of Ethyltoluene, and 1, 2,3-, 1,2,4-, 1,3,5-, and mixed isomers of Trimethylbenzene)

AROMATIC C9 FRACTION FROM PETROLEUM REFINING   **RR-01183-6**

ENVIRONMENTAL LISTS

TSCA - Health and Safety Reporting List
Effective Date: October 12, 1993; Sunset Date: October 12, 2003

AROMATIC C9 FRACTION FROM PETROLEUM REFINING   **RR-01184-7**

ENVIRONMENTAL LISTS

TSCA - Health and Safety Reporting List
Effective Date: October 12, 1993; Sunset Date: October 12, 2003

AROMATIC C9 FRACTION FROM PETROLEUM REFINING   **RR-01185-8**

ENVIRONMENTAL LISTS

TSCA - Health and Safety Reporting List
Effective Date: October 12, 1993; Sunset Date: October 12, 2003

AROMATIC C9 FRACTION FROM PETROLEUM REFINING   **RR-01186-9**

ENVIRONMENTAL LISTS

TSCA - Health and Safety Reporting List
Effective Date: October 12, 1993; Sunset Date: October 12, 2003

AROMATIC C9 FRACTION FROM PETROLEUM REFINING   **RR-01187-0**

ENVIRONMENTAL LISTS

TSCA - Health and Safety Reporting List
Effective Date: October 12, 1993; Sunset Date: October 12, 2003

AROMATIC C9 FRACTION FROM PETROLEUM REFINING   **RR-01188-1**

ENVIRONMENTAL LISTS

TSCA - Health and Safety Reporting List
Effective Date: October 12, 1993; Sunset Date: October 12, 2003

AROMATIC C9 FRACTION FROM PETROLEUM REFINING   **RR-01189-2**

ENVIRONMENTAL LISTS

TSCA - Health and Safety Reporting List
Effective Date: October 12, 1993; Sunset Date: October 12, 2003

AROMATIC DIAMINES   RR-00164-9

ENVIRONMENTAL LISTS

TSCA - Chemicals with Significant New Use Rules
PMN numbers: P-86-501; P-86-503

C9 AROMATIC HYDROCARBON FRACTION   RR-01766-3

ENVIRONMENTAL LISTS

TSCA - Section 12(b) - Export Notification
*export notification required - Section 4 This substance consists of o-
m-, and p-ethyltoluene (min. 22%), and 1,2,4-, 1,2,3-, and 1,3,
5-trimethylbenzene (min. 15%), and represents a typical C9 fraction
obtained from reforming crude petroleum and used as a solvent end
product.*

AROMATIC NITRO COMPOUND   RR-01010-6

ENVIRONMENTAL LISTS

TSCA - Chemicals with Significant New Use Rules
PMN number: P-86-335

AROMATIC SULFONIC ACID COMPOUND WITH   RR-01660-4
AMINE

ENVIRONMENTAL LISTS

TSCA - Chemicals with Significant New Use Rules
PMN number: P-93-832

ARSENATES, N.O.S.   RR-01200-0

HEALTH AND SAFETY LISTS

U.S. DOT - Appendix B - Marine Pollutants
DOT regulated marine pollutant

ARSENIC   7440-38-2
SEE ALSO:
ARSENIC COMPOUNDS, N.O.S.
K101-HAZARDOUS WASTES
K031-HAZARDOUS WASTES
K102-HAZARDOUS WASTES
K084-HAZARDOUS WASTES
F039-HAZARDOUS WASTES
K060-HAZARDOUS WASTES

HEALTH AND SAFETY LISTS

ACGIH 1995 - Time Weighted Averages
elemental and inorganic compounds, as As: 0.01 mg/m3 TWA

ACGIH 1995 - Carcinogens
*elemental and inorganic compounds (except arsine), as As: A1-con-
firmed human carcinogen*

U.S. DOT - Substances From 49 CFR 172.101
regulated by DOT (UN1558)

U.S. DOT - Hazard Classes
DOT hazard class = 6.1

U.S. DOT - Appendix A Table 1 - Hazardous Substances
*final RQ = 1 pound (0.454 kg) (no reporting of releases of this
hazardous substance is required if the diameter of the pieces of the
solid metal release is equal to or exceeds 0.004 inches)*

NIOSH 1990 - Pocket Guide - RELs
as As: C 0.002 mg/m3 (15 min)

NIOSH 1990 - Pocket Guide - IDLHs
as As: 100 mg/m3 IDLH (not considering carcinogenic effects)

NIOSH 1990 - Pocket Guide - Carcinogens
occupational carcinogen

NIOSH 1990 - Pocket Guide - Target organs
liver, kidneys, skin, lungs, lymphatic system

NTP Seventh Report - Known Carcinogens
*known carcinogen (Listed under 'Arsenic and certain arsenic com-
pounds')*

OSHA - Vacated PELs - Time Weighted Averages
organic compounds, as As: 0.5 mg/m3 TWA

OSHA - Final PELs - Time Weighted Averages
organic compounds, as As: 0.5 mg/m3 TWA

OSHA - Select Carcinogens
[present]

ENVIRONMENTAL LISTS

ATSDR Priority List
Rank (of 275): 002

CERCLA/SARA - Section 313 - Emission Reporting
form R reporting required for 0.1% de minimus concentration

CERCLA/SARA - Hazardous Substances and their Reportable Quantities
*final RQ = 1 pound (0.454 kg) (no reporting of releases of this
hazardous substance is required if the diameter of the pieces of the
solid metal release is equal to or exceeds 0.004 inches)*

Clean Water Act - Priority Pollutants
[present]

Clean Water Act - Toxic Pollutants
[present] (Listed under 'Arsenic and compounds')

Safe Drinking Water Act - MCLs
MCL = 0.05 mg/L

EPA - Carcinogen Hazard Ranking for RQ Adjustment
Hazard ranking = High

RCRA - D Series - Maximum Concentration of Contaminants
waste number D004; regulatory level = 5.0 mg/L

RCRA - D Series - Chronic Toxicity Reference Levels
chronic toxicity reference level = 0.05 mg/L

RCRA - Hazardous Constituents-Appendix VIII
hazardous constituent - no waste number

RCRA - Basis for Listing - Appendix VII
Included in waste streams: F039, K031, K060, K084, K101, K102

RCRA - Substances Banned From Land Disposal
[present]

RCRA - TSD Facilities Ground Water Monitoring
*TM 6010 = 500 ug/L PQL; TM 7060 = 10 ug/L PQL; TM 7061 =
20 ug/L PQL (all species in the ground water that contain this element
are included)*

RCRA - Universal Treatment Standards (LDR)
WW: 1.4 mg/l; NWW: 5.0 mg/l TCLP

INTERNATIONAL LISTS

Australian Exposure Standards - Time Weighted Averages
as As: 0.05 mg/m3 TWA

Australian Exposure Standards - Carcinogens
as As: confirmed carcinogen

Canada - WHMIS: Ingredient Disclosure
0.1% item 130 (266)

Canada - NPRI (National Pollutant Release Inventory)
[present]

Canada - CEPA - Priority Substances List
estimated time for completion of assessment reports: 3 years

Canada - CEPA Schedule III Part II - Restricted Substances (Ocean
Dumping)
[present]

Canada - Drinking Water Quality - IMACs
0.025 mg/L IMAC

Canada - Alberta - 8 Hour Occupational Exposure Limit
as As: 0.2 mg/m3 TWA

Canada - Alberta - 15 Minute Occupational Exposure Limit
0.6 mg/m3 STEL

Canada - British Columbia - 8 Hour Exposure Limits
as As: 0.5 mg/m3 TWA
Canada - Ontario - OHSA - TWAEVs
0.01 mg/m3 TWAEV (designated substance regulation)
Canada - Ontario - OHSA - STEVs
0.05 mg/m3 STEV (designated substance regulation)
Canada - Ontario - OHSA - Designated Substances
0.01 mg/m3; See Ontario Reg. 836 for full information.
Canada - Quebec - Time-Weighted Average Exposure Values
as As: 0.2 mg/m3 TWAEV
Israel - Time Weighted Averages
organic: 0.2 mg/m3 TWA; inorganic: 0.01 mg/m3 TWA
Israel - Action Levels
organic: 0.1 mg/m3 AL; inorganic: 0.005 mg/m3 AL
Mexico - Instruction No. 10 - TWAs
0.2 mg/m3 TWA
Mexico - Wastewater - Organic Toxic Pollutants and Heavy Metals
Listed under [Heavy Metals]
Mexico - Drinking Water - Ecological Criteria
0.5 mg/l Substance presents persistence, bioaccumulations or risk of cancer, reduce human exposure to a minimum; This level has been extrapolated by using a mathematic model

STATE LISTS
California - Air Bill 2588 Appendix A-I
known or potential carcinogen
California - Prop. 65 - No Significant Risk Levels
inhalation: no significant risk level = 0.06 ug/day
California - Exposure Limits - PELs
metal and inorganic arsenic compounds, as As: 0.01 mg/m3 PEL; 0.005 mg/m3 Action Level; harmful if inhaled or swallowed; see also section 5214; organic compounds, as As: 0.2 mg/m3 PEL
California - Directors List of Hazardous Substances (8 CCR 339)
[present] (refers to mixtures with 0.02% or greater inorganic arsenic)
Florida Hazardous Substance List
[present]
Massachusetts Right To Know List
carcinogen; extraordinarily hazardous
Minnesota Hazardous Substance List
carcinogen
NJ Right to Know List (Total)
sn 0152
NJ Special Hazardous Substances
(carcinogen)
Pennsylvania Right to Know List
environmental hazard; special hazardous substance (any compound of this substance is also an environmental hazard)
Pennsylvania RTK - Special Hazardous Substances
[present]

ARSENIC 69   14809-44-0

HEALTH AND SAFETY LISTS
U.S. DOT - Appendix A Table 2 - Radionuclides
final RQ = 1000 curies (3.7E 13 Bq)

ENVIRONMENTAL LISTS
CERCLA/SARA List of Radionuclides (Appendix B) and Their Reportable Quantities
final RQ = 1000 curies (3.7E 13 Bq)

ARSENIC 70   14809-45-1

HEALTH AND SAFETY LISTS
U.S. DOT - Appendix A Table 2 - Radionuclides
final RQ = 100 curies (3.7E 12 Bq)

ENVIRONMENTAL LISTS
CERCLA/SARA List of Radionuclides (Appendix B) and Their Reportable Quantities
final RQ = 100 curies (3.7E 12 Bq)

ARSENIC 71   16685-55-5

HEALTH AND SAFETY LISTS
U.S. DOT - Appendix A Table 2 - Radionuclides
final RQ = 100 curies (3.7E 12 Bq)

ENVIRONMENTAL LISTS
CERCLA/SARA List of Radionuclides (Appendix B) and Their Reportable Quantities
final RQ = 100 curies (3.7E 12 Bq)

ARSENIC 72   15755-33-6

HEALTH AND SAFETY LISTS
U.S. DOT - Appendix A Table 2 - Radionuclides
final RQ = 10 curies (3.7E 11 Bq)

ENVIRONMENTAL LISTS
CERCLA/SARA List of Radionuclides (Appendix B) and Their Reportable Quantities
final RQ = 10 curies (3.7E 11 Bq)

ARSENIC 73   15422-59-0

HEALTH AND SAFETY LISTS
U.S. DOT - Appendix A Table 2 - Radionuclides
final RQ = 100 curies (3.7E 12 Bq)

ENVIRONMENTAL LISTS
CERCLA/SARA List of Radionuclides (Appendix B) and Their Reportable Quantities
final RQ = 100 curies (3.7E 12 Bq)

ARSENIC 74   14304-78-0

HEALTH AND SAFETY LISTS
U.S. DOT - Appendix A Table 2 - Radionuclides
final RQ = 10 curies (3.7E 11 Bq)

ENVIRONMENTAL LISTS
CERCLA/SARA List of Radionuclides (Appendix B) and Their Reportable Quantities
final RQ = 10 curies (3.7E 11 Bq)

ARSENIC 76   15575-20-9

HEALTH AND SAFETY LISTS
U.S. DOT - Appendix A Table 2 - Radionuclides
final RQ = 100 curies (3.7E 12 Bq)

ENVIRONMENTAL LISTS
CERCLA/SARA List of Radionuclides (Appendix B) and Their Reportable Quantities
final RQ = 100 curies (3.7E 12 Bq)

ARSENIC 77   14687-61-7

HEALTH AND SAFETY LISTS
U.S. DOT - Appendix A Table 2 - Radionuclides
final RQ = 1000 curies (3.7E 13 Bq)

ENVIRONMENTAL LISTS
CERCLA/SARA List of Radionuclides (Appendix B) and Their Reportable Quantities
final RQ = 1000 curies (3.7E 13 Bq)

ARSENIC 78   15755-35-8

HEALTH AND SAFETY LISTS
U.S. DOT - Appendix A Table 2 - Radionuclides
final RQ = 100 curies (3.7E 12 Bq)

ENVIRONMENTAL LISTS

CERCLA/SARA List of Radionuclides (Appendix B) and Their Reportable Quantities
final RQ = 100 curies (3.7E 12 Bq)

ARSENIC ACID   1327-52-2

HEALTH AND SAFETY LISTS

U.S. DOT - Substances From 49 CFR 172.101
regulated by DOT (UN1554)

U.S. DOT - Hazard Classes
DOT hazard class = 6.1

U.S. DOT - Appendix A Table 1 - Hazardous Substances
final RQ = 1 pound (0.454 kg)

ENVIRONMENTAL LISTS

CERCLA/SARA - Hazardous Substances and their Reportable Quantities
final RQ = 1 pound (0.454 kg)

EPA - Carcinogen Hazard Ranking for RQ Adjustment
Hazard ranking = High

STATE LISTS

Massachusetts Right To Know List
[present]

Pennsylvania Right to Know List
environmental hazard

ARSENIC ACID   7778-39-4
SEE ALSO:
INORGANIC ARSENIC
ARSENIC

HEALTH AND SAFETY LISTS

U.S. DOT - Appendix A Table 1 - Hazardous Substances
final RQ = 1 pound (0.454 kg)

NIOSH - Selected LD50s and LC50s
Oral, rat: LD50 = 48 mg/kg

ENVIRONMENTAL LISTS

ATSDR Priority List
Rank (of 275): 184

CERCLA/SARA - Hazardous Substances and their Reportable Quantities
final RQ = 1 pound (0.454 kg)

EPA - Carcinogen Hazard Ranking for RQ Adjustment
Hazard ranking = High

RCRA - P Series Wastes
waste number P010

RCRA - Hazardous Constituents-Appendix VIII
waste number P010

RCRA - Substances Banned From Land Disposal
[present]

INTERNATIONAL LISTS

Canada - WHMIS: Ingredient Disclosure
0.1% item 129 (65)

German (DFG) - Carcinogens
proven carcinogen (results apply to all members of Glycidyl subcategory V-A)

STATE LISTS

Massachusetts Right To Know List
[present]

NJ Right to Know List (Total)
sn 0153

Pennsylvania Right to Know List
environmental hazard

ARSENIC ACID, LEAD (4+) SALT   53404-12-9

STATE LISTS

Massachusetts Right To Know List
[present]

ARSENIC ACID, TRISODIUM SALT   13464-38-5

HEALTH AND SAFETY LISTS

U.S. DOT - Appendix B - Marine Pollutants
DOT regulated marine pollutant

STATE LISTS

Massachusetts Right To Know List
[present]

ARSENICAL DUST OR FLUE DUST   RR-00153-6

HEALTH AND SAFETY LISTS

U.S. DOT - Substances From 49 CFR 172.101
regulated by DOT (UN1562)

U.S. DOT - Hazard Classes
DOT hazard class = 6.1

STATE LISTS

NJ Right to Know List (Total)
sn 2133

ARSENICAL PESTICIDES, N.O.S.   RR-00154-7

HEALTH AND SAFETY LISTS

U.S. DOT - Substances From 49 CFR 172.101
regulated by DOT (UN2759, UN2760, UN2993, UN2994)

U.S. DOT - Hazard Classes
toxic or toxic, flammable: DOT hazard class = 6.1; flammable, toxic: DOT hazard class = 3

U.S. DOT - Appendix B - Marine Pollutants
liquid, toxic, flammable: DOT regulated marine pollutant

STATE LISTS

NJ Right to Know List (Total)
sn 2861; sn 2135; sn 2136; sn 2137

ARSENIC BROMIDE   7784-33-0
SEE ALSO:
ARSENIC

HEALTH AND SAFETY LISTS

U.S. DOT - Substances From 49 CFR 172.101
regulated by DOT (UN1555)

U.S. DOT - Hazard Classes
DOT hazard class = 6.1

U.S. DOT - Appendix B - Marine Pollutants
DOT regulated marine pollutant

STATE LISTS

NJ Right to Know List (Total)
sn 0154

ARSENIC COMPOUNDS, N.O.S.   RR-00625-7

HEALTH AND SAFETY LISTS

U.S. DOT - Substances From 49 CFR 172.101
regulated by DOT (UN1556, UN1557)

U.S. DOT - Hazard Classes
DOT hazard class = 6.1

IARC - Group 1 (carcinogenic to humans)
[present] This evaluation applies to the group of chemicals as a whole and not necessarily to all individual chemicals within the group.

OSHA - Select Carcinogens
[present]

ENVIRONMENTAL LISTS

CERCLA/SARA - Section 313 - Emission Reporting
form R reporting reqired for 0.1% (inorganic), 1.0% (organic) de minimus concentration

Clean Air Act (1990) - List of Hazardous Air Contaminants
[present] (includes any unique chemical substance that contains Arsenic as part of that chemical's infrastructure)

Clean Water Act - Toxic Pollutants
[present] (Listed under 'Arsenic and compounds')

RCRA - Hazardous Constituents-Appendix VIII
hazardous constituent - no waste number

INTERNATIONAL LISTS

Canada - NPRI (National Pollutant Release Inventory)
[present]

Canada - CEPA - Priority Substances List
estimated time for completion of assessment reports: 3 years

Canada - CEPA Schedule III Part II - Restricted Substances (Ocean Dumping)
[present]

Canada - British Columbia - 8 Hour Exposure Limits
as As: 0.5 mg/m3 TWA

Canada - Quebec - Time-Weighted Average Exposure Values
as As: 0.5 mg/m3 TWAEV

United Kingdom - Maximum Exposure Limits - TWAs
as As: 0.1 mg/m3 TWA (does not include Arsine)

STATE LISTS

California - Air Bill 2588 Appendix A-I
6/91

California - Directors List of Hazardous Substances (8 CCR 339)
[present] (for any mixture containing 0.02% or greater inorganic arsenic)

NJ Right to Know List (Total)
sn 2138; sn 2139

ARSENIC DISULFIDE   1303-32-8

HEALTH AND SAFETY LISTS

U.S. DOT - Appendix A Table 1 - Hazardous Substances
final RQ = 1 pound (0.454 kg)

ENVIRONMENTAL LISTS

CERCLA/SARA - Hazardous Substances and their Reportable Quantities
final RQ = 1 pound (0.454 kg)

Clean Water Act - Hazardous Substances
[present]

EPA - Carcinogen Hazard Ranking for RQ Adjustment
Hazard ranking = High

STATE LISTS

Massachusetts Right To Know List
[present]

Pennsylvania Right to Know List
environmental hazard

ARSENIC DISULFIDE   56320-22-0
SEE ALSO:
ARSENIC

STATE LISTS

NJ Right to Know List (Total)
sn 0156

ARSENIC ORGANIC COMPOUNDS   RR-00035-1

STATE LISTS

Minnesota Hazardous Substance List
as As

ARSENIC PENTOXIDE   1303-28-2
SEE ALSO:
INORGANIC ARSENIC
ARSENIC

HEALTH AND SAFETY LISTS

U.S. DOT - Substances From 49 CFR 172.101
regulated by DOT (UN1559)

U.S. DOT - Hazard Classes
DOT hazard class = 6.1

U.S. DOT - Appendix A Table 1 - Hazardous Substances
final RQ = 1 pound (0.454 kg)

NIOSH - Selected LD50s and LC50s
Oral, rat: LD50 = 8 mg/kg

NTP Seventh Report - Known Carcinogens
known carcinogen (Listed under 'Arsenic and certain arsenic compounds')

OSHA - Select Carcinogens
[present]

ENVIRONMENTAL LISTS

ATSDR Priority List
Rank (of 275): 174

CERCLA/SARA - Section 302 Extremely Hazardous Substances and TPQs
TPQ = 100/10,000 pounds

CERCLA/SARA - Hazardous Substances and their Reportable Quantities
final RQ = 1 pound (0.454 kg)

Clean Water Act - Hazardous Substances
[present]

EPA - Carcinogen Hazard Ranking for RQ Adjustment
Hazard ranking = High

RCRA - P Series Wastes
waste number P011

RCRA - Hazardous Constituents-Appendix VIII
waste number P011

RCRA - Substances Banned From Land Disposal
[present]

INTERNATIONAL LISTS

Canada - WHMIS: Ingredient Disclosure
1% item 131 (1351)

German (DFG) - Carcinogens
proven carcinogen (Listed under 'Arsenic trioxide')

STATE LISTS

Massachusetts Right To Know List
extraordinarily hazardous

NJ Right to Know List (Total)
sn 0158

NJ Special Hazardous Substances
(carcinogen)

Pennsylvania Right to Know List
environmental hazard; special hazardous substance

Pennsylvania RTK - Special Hazardous Substances
[present]

ARSENIC TRICHLORIDE   60646-36-8

STATE LISTS

NJ Right to Know List (Total)
sn 0160

ARSENIC TRIOXIDE   1327-53-3
SEE ALSO:
  INORGANIC ARSENIC
  ARSENIC

HEALTH AND SAFETY LISTS

U.S. DOT - Substances From 49 CFR 172.101
  regulated by DOT (UN1561)

U.S. DOT - Hazard Classes
  DOT hazard class = 6.1

U.S. DOT - Appendix B - Marine Pollutants
  DOT regulated marine pollutant

U.S. DOT - Appendix A Table 1 - Hazardous Substances
  final RQ = 1 pound (0.454 kg)

NIOSH - Selected LD50s and LC50s
  Oral, rat: LD500 = 40 mg/kg

NTP Seventh Report - Known Carcinogens
  known carcinogen (Listed under 'Arsenic and certain arsenic compounds')

OSHA - Select Carcinogens
  [present]

ENVIRONMENTAL LISTS

ATSDR Priority List
  Rank (of 275): 177

CERCLA/SARA - Section 302 Extremely Hazardous Substances and TPQs
  TPQ = 100/10,000 pounds

CERCLA/SARA - Hazardous Substances and their Reportable Quantities
  final RQ = 1 pound (0.454 kg)

Clean Water Act - Hazardous Substances
  [present]

EPA - Carcinogen Hazard Ranking for RQ Adjustment
  Hazard ranking = High

RCRA - P Series Wastes
  waste number P012

RCRA - Hazardous Constituents-Appendix VIII
  waste number P012

RCRA - Substances Banned From Land Disposal
  [present]

INTERNATIONAL LISTS

Australian Exposure Standards - Time Weighted Averages
  production, as As: control to lowest practical level

Australian Exposure Standards - Carcinogens
  production, as As: confirmed carcinogen

Canada - WHMIS: Ingredient Disclosure
  0.1% item 134 (1692)

Canada - Alberta - 8 Hour Occupational Exposure Limit
  0.05 mg/m3 TWA

Canada - Alberta - 15 Minute Occupational Exposure Limit
  0.15 mg/m3 STEL

Canada - Alberta - Designated Substances
  designated substance - requires code of practice

Canada - British Columbia - 8 Hour Exposure Limits
  as As: 0.05 mg/m3 TWA

Canada - British Columbia - Ceiling Exposure Limits
  SO2: C 5 ppm

Canada - British Columbia - Carcinogens
  carcinogen - As2O3, as As: 0.05 mg/m3 TWA; SO2: C 5 ppm

Canada - Quebec - Time-Weighted Average Exposure Values
  0.2 mg/m3 TWAEV (substance of which the recirculation is prohibited)

Canada - Quebec - Carcinogens
  C2 carcinogen: effect suspected in humans

German (DFG) - Carcinogens
  proven carcinogen

Mexico - Instruction No. 10 - TWAs
  0.5 mg/m3 TWA

Mexico - Instruction No. 10 - Carcinogens
  potential carcinogen in humans - limited epidemiological evidence

STATE LISTS

Florida Hazardous Substance List
  effective March 13, 1992

Massachusetts Right To Know List
  extraordinarily hazardous

Minnesota Hazardous Substance List
  carcinogen

NJ Right to Know List (Total)
  sn 0161

NJ Special Hazardous Substances
  (carcinogen)

Pennsylvania Right to Know List
  environmental hazard; special hazardous substance

Pennsylvania RTK - Special Hazardous Substances
  [present]

ARSENIC TRISULFIDE   1303-33-9
SEE ALSO:
  ARSENIC
  INORGANIC ARSENIC

HEALTH AND SAFETY LISTS

U.S. DOT - Substances From 49 CFR 172.101
  regulated by DOT (NA1557)

U.S. DOT - Hazard Classes
  DOT hazard class = 6.1

U.S. DOT - Appendix A Table 1 - Hazardous Substances
  final RQ = 1 pound (0.454 kg)

ENVIRONMENTAL LISTS

CERCLA/SARA - Hazardous Substances and their Reportable Quantities
  final RQ = 1 pound (0.454 kg)

Clean Water Act - Hazardous Substances
  [present]

EPA - Carcinogen Hazard Ranking for RQ Adjustment
  Hazard ranking = High

INTERNATIONAL LISTS

Canada - WHMIS: Ingredient Disclosure
  1% item 135 (1700)

STATE LISTS

Florida Hazardous Substance List
  [present]

Massachusetts Right To Know List
  [present]

NJ Right to Know List (Total)
  sn 0162

Pennsylvania Right to Know List
  environmental hazard

ARSENIC, WATER-SOLUBLE COMPOUNDS, N.O.S.   RR-00586-7

INTERNATIONAL LISTS

Canada - WHMIS: Ingredient Disclosure
  1% item 132 (265)

ARSENOUS ACID, TRISODIUM SALT   13464-37-4

STATE LISTS

Pennsylvania Right to Know List
  environmental hazard; special hazardous substance

Pennsylvania RTK - Special Hazardous Substances
[present]

ARSENOUS TRICHLORIDE   7784-34-1

SEE ALSO:
ARSENIC

HEALTH AND SAFETY LISTS

U.S. DOT - Substances From 49 CFR 172.101
regulated by DOT (UN1560)

U.S. DOT - Hazard Classes
DOT hazard class = 6.1

U.S. DOT - Substances Which Are Poisonous by Inhalation
liquid hazardous material poisonous by inhalation (UN1560)

U.S. DOT - Appendix B - Marine Pollutants
DOT regulated marine pollutant

U.S. DOT - Appendix A Table 1 - Hazardous Substances
final RQ = 1 pound (0.454 kg)

ENVIRONMENTAL LISTS

CERCLA/SARA - Section 302 Extremely Hazardous Substances and TPQs
TPQ = 500 pounds

CERCLA/SARA - Hazardous Substances and their Reportable Quantities
final RQ = 1 pound (0.454 kg)

CAA -Toxic Substances for Accidental Release Prevention
threshold quantity = 15,000 lbs

Clean Water Act - Hazardous Substances
[present]

EPA - Carcinogen Hazard Ranking for RQ Adjustment
Hazard ranking = High

INTERNATIONAL LISTS

Canada - WHMIS: Ingredient Disclosure
1% item 133 (1653)

STATE LISTS

Florida Hazardous Substance List
[present]

Massachusetts Right To Know List
extraordinarily hazardous

NJ Right to Know List (Total)
sn 0159

Pennsylvania Right to Know List
environmental hazard

ARSENOUS TRIIODIDE   7784-45-4

STATE LISTS

NJ Right to Know List (Total)
sn 0157

ARSINE   7784-42-1

SEE ALSO:
ARSENIC
INORGANIC ARSENIC

HEALTH AND SAFETY LISTS

ACGIH 1995 - Time Weighted Averages
0.05 ppm TWA; 0.16 mg/m3 TWA

ACGIH 1995 - Biological Exposure Indices
Inorganic arsenic metabolites in urine: 50 ug/g creatinine, end of workweek (B)

AIHA - Odor Threshold Values
no geometric mean air odor threshold

U.S. DOT - Substances From 49 CFR 172.101
regulated by DOT (UN2188)

U.S. DOT - Hazard Classes
DOT hazard class = 2.3

U.S. DOT - Substances Which Are Poisonous by Inhalation
gaseous hazardous material poisonous by inhalation (UN2188)

NIOSH - Selected LD50s and LC50s
Inhalation, rat: LC50 = 390 mg/m3 10 mn

NIOSH 1990 - Pocket Guide - RELs
C 0.002 mg/m3 (15 min)

NIOSH 1990 - Pocket Guide - IDLHs
6 ppm IDLH (not considering carcinogenic effects)

NIOSH 1990 - Pocket Guide - Carcinogens
occupational carcinogen

NIOSH 1990 - Pocket Guide - Target organs
blood, kidneys, liver

NIOSH - Health Standards - Exposure Limits
C (15 min) 2 ug/m3;   C (15 min) 0.002 mg/m3

NIOSH - Health Standards - Health Effects and Precautions
sudden extensive hemolysis (Warn workers about working with arsenic compounds in presence of freshly formed hydrogen)

NIOSH - Health Standards - Carcinogenic Chemicals
potential human carcinogen

NTP Chemical Status Reports - Testing Status and NTIS Number
Prechronic studies for which toxicity technical reports were not prepared

OSHA - Vacated PELs - Time Weighted Averages
0.05 ppm TWA; 0.2 mg/m3 TWA

OSHA - Final PELs - Time Weighted Averages
0.05 ppm TWA; 0.2 mg/m3 TWA

OSHA - List of Highly Hazardous Chemicals
threshhold quantity = 100 pounds

ENVIRONMENTAL LISTS

ATSDR Priority List
Rank (of 275): 186

CERCLA/SARA - Section 302 Extremely Hazardous Substances and TPQs
TPQ = 100 pounds

CAA -Toxic Substances for Accidental Release Prevention
threshold quantity = 1,000 lbs

INTERNATIONAL LISTS

Australian Exposure Standards - Time Weighted Averages
0.05 ppm TWA; 0.16 mg/m3 TWA

Canada - WHMIS: Ingredient Disclosure
0.1% item 136 (268)

Canada - Alberta - 8 Hour Occupational Exposure Limit
0.05 ppm TWA; 0.16 mg/m3 TWA

Canada - Alberta - 15 Minute Occupational Exposure Limit
0.15 ppm STEL; 0.48 mg/m3 STEL

Canada - British Columbia - 8 Hour Exposure Limits
0.05 ppm TWA; 0.2 mg/m3 TWA

Canada - Ontario - OHSA - TWAEVs
0.05 ppm TWAEV; 0.16 mg/m3 TWAEV

Canada - Quebec - Time-Weighted Average Exposure Values
0.05 ppm TWAEV; 0.16 mg/m3 TWAEV

United Kingdom - Occupational Exposure Standards - TWAs
0.05 ppm TWA; 0.2 mg/m3 TWA

German (DFG) - MAK Values
0.05 ppm MAK; 0.2 mg/m3 MAK

German (DFG) - Peak Limitations
5 x normal MAK (30 min. average value); don't exceed 2 times during shift

Israel - Time Weighted Averages
0.05 mg/m3 TWA

Israel - Action Levels
0.025 mg/m3 AL

Mexico - Instruction No. 10 - TWAs
0.05 ppm TWA; 0.2 mg/m3 TWA

STATE LISTS

California - Air Bill 2588 Appendix A-I
[present]

California - Exposure Limits - PELs
0.05 ppm PEL; 0.2 mg/m3 PEL

California - Directors List of Hazardous Substances (8 CCR 339)
[present]

Florida Hazardous Substance List
[present]

Massachusetts Right To Know List
extraordinarily hazardous

Minnesota Hazardous Substance List
[present]

NJ Right to Know List (Total)
sn 0163

Pennsylvania Right to Know List
environmental hazard

PROPOSED REGULATIONS

CERCLA/SARA - Proposed Hazardous Substance Additions
proposed RQ = 1 pound (.454 kg)

CERCLA/SARA - 1989 Proposed RQ Adjustments
proposed RQ = 1 pounds (.454 kg)

Canada - Ontario - Proposed Occupational TWAEVs
0.003 ppm TWAEV; 0.01 mg/m3 TWAEV

ARSINE, DIETHYL-   692-42-2

HEALTH AND SAFETY LISTS

U.S. DOT - Appendix A Table 1 - Hazardous Substances
final RQ = 1 pound (0.454 kg)

ENVIRONMENTAL LISTS

CERCLA/SARA - Hazardous Substances and their Reportable Quantities
final RQ = 1 pound (0.454 kg)

RCRA - P Series Wastes
waste number P038

RCRA - Hazardous Constituents-Appendix VIII
waste number P038

RCRA - Substances Banned From Land Disposal
[present]

TSCA - Code of Federal Regulations Citations
40 CFR 716.120(a)

TSCA - Health and Safety Reporting List
Effective Date: March 7, 1986

STATE LISTS

Massachusetts Right To Know List
[present]

NJ Right to Know List (Total)
sn 2962

Pennsylvania Right to Know List
environmental hazard

ARSINO, THIOXO-   12044-79-0

STATE LISTS

Pennsylvania Right to Know List
environmental hazard

ARYL PHOSPHATES   RR-00281-3

ENVIRONMENTAL LISTS

TSCA - Health and Safety Reporting List
Effective Date: October 4, 1982 (phosphate esters of phenol or of alkyl-substituted phenols. Triaryl and mixed alkyl and aryl esters are included but trialkyl esters are excluded)

ARYL SULFONATE OF A FATTY ACID MIXTURE,   RR-01218-0
POLYAMINE CONDENSATE

ENVIRONMENTAL LISTS

TSCA - Chemicals with Significant New Use Rules
PMN number: P-91-584

ASBESTOS   1332-21-4

HEALTH AND SAFETY LISTS

ACGIH 1995 - Time Weighted Averages
mean dust level = 2 fibers/cc (Fibers longer than 5 microns and with an aspect ratio equal to or greater than 3:1)

ACGIH 1995 - Carcinogens
(A1)-confirmed human carcinogen

U.S. DOT - Substances From 49 CFR 172.101
regulated by DOT (UN2590)

U.S. DOT - Hazard Classes
DOT hazard class = 9

U.S. DOT - Appendix A Table 1 - Hazardous Substances
final RQ = 1 pound (0.454 kg) (for friable forms of only)

IARC - Group 1 (carcinogenic to humans)
[present]

NIOSH 1990 - Pocket Guide - RELs
100,000 fibers/m3 TWA ; 0.1 fiber/cm3 TWA (asbestos fibers > 5 micrometers long)

NIOSH 1990 - Pocket Guide - Carcinogens
occupational carcinogen

NIOSH 1990 - Pocket Guide - Target organs
lungs

NIOSH - Health Standards - Exposure Limits
100,000 fibers/m3 (fibers > 5 um long), 8 hour TWA in a 400 liter air sample

NIOSH - Health Standards - Health Effects and Precautions
Lung cancer, mesothelioma, asbestosis (Periodic chest X-ray and pulmonary function testing required)

NIOSH - Health Standards - Carcinogenic Chemicals
potential human carcinogen

NTP Seventh Report - Known Carcinogens
known carcinogen

OSHA - 29 CFR 1910 Specifically Regulated Chemicals
fibers shorter than 5 micrometers: 0.1 fiber/cc action level; 0.2 fiber/cc TWA; 1.0 fiber/cc excursion limit (30 min); Cancer and lung disease hazard (see 29 CFR 1910.1001) (does not include non-asbestiform actinolite, tremolite or anthophylite)

OSHA - Select Carcinogens
[present]

ENVIRONMENTAL LISTS

ATSDR Priority List
Rank (of 275): 074

CERCLA/SARA - Section 313 - Emission Reporting
form R reporting required for 0.1% de minimus concentration (friable form)

CERCLA/SARA - Hazardous Substances and their Reportable Quantities
final RQ = 1 pound (0.454 kg) (for friable forms of only)

Clean Air Act (1990) - List of Hazardous Air Contaminants
[present]

Clean Water Act - Priority Pollutants
[present]

Clean Water Act - Toxic Pollutants
[present]

Safe Drinking Water Act - MCLs
MCL = 7 million fibers/L (fibers longer than 10 micrometers)

Safe Drinking Water Act - MCLGs
MCLG = 7 million fibers/L (fibers longer than 10 micrometers)

EPA - Carcinogen Hazard Ranking for RQ Adjustment
Hazard ranking = High

TSCA - Code of Federal Regulations Citations
40 CFR 763;   40 CFR 716.120(c)

TSCA - Health and Safety Reporting List
Effective Date: October 4, 1982 (asbestiform varieties of chrysolite (serpentine); crocidolite (riebeckite); amosite (cummingtonite-grunerite); anthophyllite; tremolite; and actinolite)

TSCA - Section 12(b) - Export Notification
export notification required - Section 6

INTERNATIONAL LISTS

Australian Exposure Standards - Time Weighted Averages
0.1 fibres per ml of air TWA

Australian Exposure Standards - Carcinogens
confirmed carcinogen

Canada - WHMIS: Ingredient Disclosure
0.1% item 137 (199)

Canada - NPRI (National Pollutant Release Inventory)
[present]

Canada - CEPA Schedule I - Toxic Substances
limited atmospheric releases from asbestos mines and mills

Canada - Alberta - Designated Substances
designated substance - requires code of practice (See additional requirements in Part 3)

Canada - British Columbia - Carcinogens
carcinogen - 0.1 fibres per ml TWA

Canada - Ontario - OHSA - TWAEVs
1.0 fibres/cm3 TWAEV (designated substance regulation)

Canada - Ontario - OHSA - STEVs
5.0 fibres/cm3 STEV (designated substance regulations)

Canada - Ontario - OHSA - Designated Substances
1.0 fibres/cm3; See Ontario Reg. 837 for full information.

Canada - Quebec - Time-Weighted Average Exposure Values
permissible recirculation concentration of respirable dusts: 0.1 mg/m3 TWAEV

United Kingdom - Maximum Exposure Limits - TWAs
containing crocidolite or amosite: 0.2 fibres per milliliter TWA (4 hour) ; other forms: 0.5 fibres per milliliter TWA (4 hour)

United Kingdom - Maximum Exposure Limits - STELs
containing crocidolite and amosite: 0.6 fibres per milliliter STEL; other forms: 1.5 fibres per milliliter STEL

German (DFG) - Carcinogens
proven carcinogen

Israel - Time Weighted Averages
mean dust level = 0.4 fibers/cc (Fibers longer than 5 microns and with an aspect ratio equal to or greater than 3:1)

Israel - Action Levels
mean dust level = 0.1 fibers/cc AL

Mexico - Instruction No. 10 - Carcinogens
potentially carcinogenic contaminant

Mexico - Drinking Water - Ecological Criteria
3000 mg/l Substance presents persistence, bioaccumulations or risk of cancer, reduce human exposure to a minimum; This level has been extrapolated by using a mathematic model

STATE LISTS

California - Air Bill 2588 Appendix A-I
known or potential carcinogen

California - Prop. 65 - Cancer list
carcinogen - initial date 2/27/87

California - Prop. 65 - No Significant Risk Levels
no significant risk level = 100 fibers inhaled/day (fibers equal to or greater than 5 micrometers in length and 0.3 micrometers in width, with a length to width ratio of greater than or equal to 3:1 as measured by phase contrast microscopy)

California - Exposure Limits - PELs
0.2 fibers/cc PEL; see also section 5208 for respiratory protection requirements

California - Directors List of Hazardous Substances (8 CCR 339)
[present] (exemptions include: exterior and interior coatings and laminating resins with encapsulated asbestos fibers, cold process asphalt roof coatings, and non-friable encapsulated products)

Florida Hazardous Substance List
[present]

Massachusetts Right To Know List
carcinogen; extraordinarily hazardous (exempt only when in a non-friable encapsulated product unless processing results in generation of asbestos dust)

Minnesota Hazardous Substance List
carcinogen

NJ Right to Know List (Total)
sn 0164

NJ Special Hazardous Substances
(carcinogen)

Pennsylvania Right to Know List
environmental hazard; special hazardous substance

Pennsylvania RTK - Special Hazardous Substances
[present]

PROPOSED REGULATIONS

ACGIH 1995 - Notice of Intended Changes
0.2 f/cc TWA; A1-confirmed human carcinogen (fibers longer than 5 um and with an aspect ratio equal to or greater than 3:1)

Canada - Ontario - Proposed Occupational TWAEVs
0.1 fibres/cm3 TWAEV

ASBESTOS, ACTINOLITE   77536-66-4

HEALTH AND SAFETY LISTS

IARC - Group 1 (carcinogenic to humans)
[present]

OSHA - Select Carcinogens
[present]

STATE LISTS

Massachusetts Right To Know List
carcinogen; extraordinarily hazardous (exempt only when in a non-friable encapsulated product unless processing results in generation of asbestos dust)

ASBESTOS, AMOSITE   12172-73-5
SEE ALSO:
   ASBESTOS

HEALTH AND SAFETY LISTS

ACGIH 1995 - Time Weighted Averages
mean dust level = 0.5 fiber/cc (Fibers longer than 5 microns and with an aspect ratio equal to or greater than 3:1)

ACGIH 1995 - Carcinogens
(A1)-confirmed human carcinogen

IARC - Group 1 (carcinogenic to humans)
[present]

NTP Chemical Status Reports - Testing Status and NTIS Number
Technical reports printed (PB91172312) (PB87133278/AS); with dimethyl hydrazine: (PB91172312)

NTP Chemical Status Reports - Evidence of Carcinogenicity
PB91172312:   male   rat-negative;   female   rat-negative; PB87133278/AS: hamster-negative

NTP Seventh Report - Known Carcinogens
known carcinogen (Listed under 'Asbestos')

OSHA - Select Carcinogens
[present]

ENVIRONMENTAL LISTS

TSCA - Code of Federal Regulations Citations
40 CFR 763;   40 CFR 716.120

EPA - Carcinogen Hazard Ranking for RQ Adjustment
Hazard ranking = High

TSCA - Code of Federal Regulations Citations
40 CFR 763; 40 CFR 716.120(c)

TSCA - Health and Safety Reporting List
Effective Date: October 4, 1982 (asbestiform varieties of chrysolite (serpentine); crocidolite (riebeckite); amosite (cummingtonite-grunerite); anthophyllite; tremolite; and actinolite)

TSCA - Section 12(b) - Export Notification
export notification required - Section 6

INTERNATIONAL LISTS

Australian Exposure Standards - Time Weighted Averages
0.1 fibres per ml of air TWA

Australian Exposure Standards - Carcinogens
confirmed carcinogen

Canada - WHMIS: Ingredient Disclosure
0.1% item 137 (199)

Canada - NPRI (National Pollutant Release Inventory)
[present]

Canada - CEPA Schedule I - Toxic Substances
limited atmospheric releases from asbestos mines and mills

Canada - Alberta - Designated Substances
designated substance - requires code of practice (See additional requirements in Part 3)

Canada - British Columbia - Carcinogens
carcinogen - 0.1 fibres per ml TWA

Canada - Ontario - OHSA - TWAEVs
1.0 fibres/cm3 TWAEV (designated substance regulation)

Canada - Ontario - OHSA - STEVs
5.0 fibres/cm3 STEV (designated substance regulations)

Canada - Ontario - OHSA - Designated Substances
1.0 fibres/cm3; See Ontario Reg. 837 for full information.

Canada - Quebec - Time-Weighted Average Exposure Values
permissible recirculation concentration of respirable dusts: 0.1 mg/m3 TWAEV

United Kingdom - Maximum Exposure Limits - TWAs
containing crocidolite or amosite: 0.2 fibres per milliliter TWA (4 hour); other forms: 0.5 fibres per milliliter TWA (4 hour)

United Kingdom - Maximum Exposure Limits - STELs
containing crocidolite and amosite: 0.6 fibres per milliliter STEL; other forms: 1.5 fibres per milliliter STEL

German (DFG) - Carcinogens
proven carcinogen

Israel - Time Weighted Averages
mean dust level = 0.4 fibers/cc (Fibers longer than 5 microns and with an aspect ratio equal to or greater than 3:1)

Israel - Action Levels
mean dust level = 0.1 fibers/cc AL

Mexico - Instruction No. 10 - Carcinogens
potentially carcinogenic contaminant

Mexico - Drinking Water - Ecological Criteria
3000 mg/l Substance presents persistence, bioaccumulations or risk of cancer, reduce human exposure to a minimum; This level has been extrapolated by using a mathematic model

STATE LISTS

California - Air Bill 2588 Appendix A-I
known or potential carcinogen

California - Prop. 65 - Cancer list
carcinogen - initial date 2/27/87

California - Prop. 65 - No Significant Risk Levels
no significant risk level = 100 fibers inhaled/day (fibers equal to or greater than 5 micrometers in length and 0.3 micrometers in width, with a length to width ratio of greater than or equal to 3:1 as measured by phase contrast microscopy)

California - Exposure Limits - PELs
0.2 fibers/cc PEL; see also section 5208 for respiratory protection requirements

California - Directors List of Hazardous Substances (8 CCR 339)
[present] (exemptions include: exterior and interior coatings and laminating resins with encapsulated asbestos fibers, cold process asphalt roof coatings, and non-friable encapsulated products)

Florida Hazardous Substance List
[present]

Massachusetts Right To Know List
carcinogen; extraordinarily hazardous (exempt only when in a non-friable encapsulated product unless processing results in generation of asbestos dust)

Minnesota Hazardous Substance List
carcinogen

NJ Right to Know List (Total)
sn 0164

NJ Special Hazardous Substances
(carcinogen)

Pennsylvania Right to Know List
environmental hazard; special hazardous substance

Pennsylvania RTK - Special Hazardous Substances
[present]

PROPOSED REGULATIONS

ACGIH 1995 - Notice of Intended Changes
0.2 f/cc TWA; A1-confirmed human carcinogen (fibers longer than 5 um and with an aspect ratio equal to or greater than 3:1)

Canada - Ontario - Proposed Occupational TWAEVs
0.1 fibres/cm3 TWAEV

ASBESTOS, ACTINOLITE   77536-66-4

HEALTH AND SAFETY LISTS

IARC - Group 1 (carcinogenic to humans)
[present]

OSHA - Select Carcinogens
[present]

STATE LISTS

Massachusetts Right To Know List
carcinogen; extraordinarily hazardous (exempt only when in a non-friable encapsulated product unless processing results in generation of asbestos dust)

ASBESTOS, AMOSITE   12172-73-5
SEE ALSO:
ASBESTOS

HEALTH AND SAFETY LISTS

ACGIH 1995 - Time Weighted Averages
mean dust level = 0.5 fiber/cc (Fibers longer than 5 microns and with an aspect ratio equal to or greater than 3:1)

ACGIH 1995 - Carcinogens
(A1)-confirmed human carcinogen

IARC - Group 1 (carcinogenic to humans)
[present]

NTP Chemical Status Reports - Testing Status and NTIS Number
Technical reports printed (PB91172312) (PB87133278/AS); with dimethyl hydrazine: (PB91172312)

NTP Chemical Status Reports - Evidence of Carcinogenicity
PB91172312: male rat-negative; female rat-negative; PB87133278/AS: hamster-negative

NTP Seventh Report - Known Carcinogens
known carcinogen (Listed under 'Asbestos')

OSHA - Select Carcinogens
[present]

ENVIRONMENTAL LISTS

TSCA - Code of Federal Regulations Citations
40 CFR 763; 40 CFR 716.120

Israel - Action Levels
mean dust level = 0.1 fibers/cc AL

STATE LISTS

Florida Hazardous Substance List
[present]

Massachusetts Right To Know List
carcinogen; extraordinarily hazardous

Minnesota Hazardous Substance List
carcinogen

NJ Right to Know List (Total)
sn 0167

NJ Special Hazardous Substances
(corrosive)

Pennsylvania Right to Know List
special hazardous substance

Pennsylvania RTK - Special Hazardous Substances
[present]

ASBESTOS, CROCIDOLITE   12001-28-4
SEE ALSO:
   ASBESTOS

HEALTH AND SAFETY LISTS

ACGIH 1995 - Time Weighted Averages
mean dust level = 0.2 fiber/cc (Fibers longer than 5 microns and with an aspect ratio equal to or greater than 3:1)

ACGIH 1995 - Carcinogens
(A1)-confirmed human carcinogen

U.S. DOT - Substances From 49 CFR 172.101
regulated by DOT (UN2212)

U.S. DOT - Hazard Classes
DOT hazard class = 9

IARC - Group 1 (carcinogenic to humans)
[present]

NTP Chemical Status Reports - Testing Status and NTIS Number
Technical reports printed (PB89178529/AS)

NTP Chemical Status Reports - Evidence of Carcinogenicity
male rat-negative; female rat-negative

NTP Seventh Report - Known Carcinogens
known carcinogen (Listed under 'Asbestos')

OSHA - Select Carcinogens
[present]

INTERNATIONAL LISTS

Australian Exposure Standards - Time Weighted Averages
0.1 fibres per ml of air TWA (Listed under 'Asbestos')

Australian Exposure Standards - Carcinogens
confirmed carcinogen

Canada - WHMIS: Ingredient Disclosure
0.1% item 447 (202)

Canada - Alberta - 8 Hour Occupational Exposure Limit
0.2 f/cm3 TWA; See additional requirements in Part 3

Canada - Alberta - 15 Minute Occupational Exposure Limit
1 f/cm3 STEL

Canada - British Columbia - 8 Hour Exposure Limits
0.2 fibres/mL TWA

Canada - Ontario - OHSA - TWAEVs
0.2 fibres/cm3 TWAEV (designated substance regulation)

Canada - Ontario - OHSA - STEVs
1.0 fibres/cm3 STEV (designated substance regulations)

Canada - Ontario - OHSA - Designated Substances
0.2 fibres/cm3; See Ontario Reg. 837 for full information.

Canada - Quebec - Time-Weighted Average Exposure Values
0.2 fibres/cm3 TWAEV (where the use of these products is permitted)

Canada - Quebec - Short-term Exposure Values
31 fibre/cm3 STEV (where the use of these products is permitted)

Canada - Quebec - Carcinogens
C1 carcinogen: effect detected in humans

Israel - Time Weighted Averages
mean dust level = 0.4 fiber/cc (Fibers longer than 5 microns and with an aspect ratio equal to or greater than 3:1)

Israel - Action Levels
mean dust level = 0.1 fiber/cc AL

STATE LISTS

Florida Hazardous Substance List
[present]

Massachusetts Right To Know List
carcinogen; extraordinarily hazardous (exempt only when in a non-friable encapsulated product unless processing results in generation of asbestos dust)

Minnesota Hazardous Substance List
carcinogen

NJ Right to Know List (Total)
sn 0168

NJ Special Hazardous Substances
(carcinogen)

Pennsylvania Right to Know List
special hazardous substance

Pennsylvania RTK - Special Hazardous Substances
[present]

PROPOSED REGULATIONS

Canada - Ontario - Proposed Occupational TWAEVs
0.1 fibres/cm3 TWAEV

ASBESTOS, TREMOLITE   77536-68-6
SEE ALSO:
   ASBESTOS

STATE LISTS

Massachusetts Right To Know List
carcinogen; extraordinarily hazardous (exempt only when in a non-friable encapsulated product unless processing results in generation of asbestos dust)

Pennsylvania Right to Know List
[present]

ASCARIDOLE   512-85-6

HEALTH AND SAFETY LISTS

U.S. DOT - Hazard Classes
Forbidden from transport by the DOT

L-ASCORBIC ACID   50-81-7

HEALTH AND SAFETY LISTS

NIOSH - Selected LD50s and LC50s
Oral, rat: LD50 = 11900 mg/kg

NTP Chemical Status Reports - Testing Status and NTIS Number
Technical reports printed (PB83201194)

NTP Chemical Status Reports - Evidence of Carcinogenicity
male rat-negative; female rat-negative; male mice-negative; female mice-negative

ENVIRONMENTAL LISTS

EPA - Master Testing List
[present]

List of Pesticide Product Inert Ingredients
[present]

ASCORBYL PALMITATE   137-66-6

ENVIRONMENTAL LISTS

List of Pesticide Product Inert Ingredients
[present]

ASPARAGINASE   9015-68-3

STATE LISTS

NJ Right to Know List (Total)
sn 0169

DL-ASPARTIC ACID, N-(3-CARBOXY-1-OXO-3-SUL-FOPROPYL)-N-OCTADECYL-, TETRASODIUM SALT   38916-42-6

HEALTH AND SAFETY LISTS

NIOSH - Selected LD50s and LC50s
Oral, rat: LD50 = 6500 mg/kg

ENVIRONMENTAL LISTS

List of Pesticide Product Inert Ingredients
[present]

ASPHALT   8052-42-4

HEALTH AND SAFETY LISTS

ACGIH 1995 - Time Weighted Averages
5 mg/m3 TWA

U.S. DOT - Substances From 49 CFR 172.101
regulated by DOT (NA1999)

U.S. DOT - Hazard Classes
DOT hazard class = 3

NFPA - Flash Points
(fp) cutback: < 50 deg F (10 deg C); liq-med curing: 100 deg F (38 deg C) (MC -30, 70), 150 deg F (66 deg C) (MC-250, 800, 3000); slow curing: 150+ deg F (66 deg C) (SC-70), 175+ deg F (79 deg C) (SC-250), 200+ deg F (93 deg C) (SC-800), 225+ deg F (107 deg C) (SC-3000); typical: 400+ deg F (204 deg C)

NFPA - Hazard Identification Ratings
cutback: health-0; flammability-3; reactivity-0; liquid-medium curing: health-0; flammability-2; reactivity-0; liquid-rapid curing: health-0; flammability-3; reactivity-0; liquid slow curing: health-0; flammability-2 (SC-70 or SC-250) or 1 (SC-800 or SC-3000); health-0; typical: he-0; flam-1; re-0

NIOSH - Health Standards - Exposure Limits
C (15 min) 5 mg/m3 (measured as total particulates)

NIOSH - Health Standards - Health Effects and Precautions
Eye and respiratory irritation (Medical monitoring required; prevent skin contact)

ENVIRONMENTAL LISTS

List of Pesticide Product Inert Ingredients
[present]

INTERNATIONAL LISTS

Australian Exposure Standards - Time Weighted Averages
5 mg/m3 TWA

Canada - Alberta - 8 Hour Occupational Exposure Limit
5 mg/m3 TWA

Canada - Alberta - 15 Minute Occupational Exposure Limit
10 mg/m3 STEL

Canada - British Columbia - 8 Hour Exposure Limits
5 mg/m3 TWA

Canada - British Columbia - 15 Minute Exposure Limits
10 mg/m3 STEL

Canada - Ontario - OHSA - TWAEVs
as total benzene soluble compounds: 5 mg/m3 TWAEV (Listed under 'Agents of variable composition')

Canada - Quebec - Time-Weighted Average Exposure Values
5 mg/m3 TWAEV

United Kingdom - Occupational Exposure Standards - TWAs
petroleum fumes: 5 mg/m3 TWA

United Kingdom - Occupational Exposure Standards - STELs
petroleum fumes: 10 mg/m3 STEL

German (DFG) - Carcinogens
suspected carcinogen

Israel - Time Weighted Averages
5 mg/m3 TWA

Israel - Action Levels
2.5 mg/m3 AL

Mexico - Instruction No. 10 - TWAs
5 mg/m3 TWA

Mexico - Instruction No. 10 - STELs
10 mg/m3 STEL

STATE LISTS

California - Exposure Limits - PELs
5 mg/m3 PEL

California - Directors List of Hazardous Substances (8 CCR 339)
[present] (includes any liquids or products that could give rise to fumes)

Florida Hazardous Substance List
[present]

Massachusetts Right To Know List
[present]

Minnesota Hazardous Substance List
[present]

NJ Right to Know List (Total)
sn 0170

Pennsylvania Right to Know List
[present]

ASPIRIN, PHENACETIN, AND CAFFIENE   8003-03-0

HEALTH AND SAFETY LISTS

NTP Chemical Status Reports - Testing Status and NTIS Number
Technical reports printed (PB284684/AS)

NTP Chemical Status Reports - Evidence of Carcinogenicity
male rat-negative; female rat-equivocal; male mice-negative; female mice-negative

ASTATINE 207   20601-76-7

HEALTH AND SAFETY LISTS

U.S. DOT - Appendix A Table 2 - Radionuclides
final RQ = 100 curies (3.7E 12 Bq)

ENVIRONMENTAL LISTS

CERCLA/SARA List of Radionuclides (Appendix B) and Their Reportable Quantities
final RQ = 100 curies (3.7E 12 Bq)

ASTATINE 211   15755-39-2

HEALTH AND SAFETY LISTS

U.S. DOT - Appendix A Table 2 - Radionuclides
final RQ = 100 curies (3.7E 12 Bq)

ENVIRONMENTAL LISTS

CERCLA/SARA List of Radionuclides (Appendix B) and Their Reportable Quantities
final RQ = 100 curies (3.7E 12 Bq)

ASULAM   3337-71-1

PROPOSED REGULATIONS

Safe Drinking Water Act - Priority list
[present]

ATRAZINE   1912-24-9

HEALTH AND SAFETY LISTS

ACGIH 1995 - Time Weighted Averages
5 mg/m3 TWA

IARC - Group 2B (sufficient animal data)
[present]

NIOSH - Selected LD50s and LC50s
*Inhalation, rat: LC50 = 5200 mg/m3 (4 hr) Oral, rat: LD50 = 672
mg/kg Skin, rabbit: LD50 = 7500 mg/kg*
OSHA - Vacated PELs - Time Weighted Averages
5 mg/m3 TWA
OSHA - Possible Select Carcinogens
[present]

ENVIRONMENTAL LISTS
CERCLA/SARA - Section 313 - Emission Reporting
form R reporting required
Safe Drinking Water Act - MCLs
MCL = 0.003 mg/L
Safe Drinking Water Act - MCLGs
MCLG = 0.003 mg/L
EPA - Master Testing List
[present]

INTERNATIONAL LISTS
Australian Exposure Standards - Time Weighted Averages
5 mg/m3 TWA
Australian Exposure Standards - Under Review
exposure limits under review
Canada - Drinking Water Quality - IMACs
0.06 mg/L IMAC
Canada - Alberta - 8 Hour Occupational Exposure Limit
5 mg/m3 TWA
Canada - Alberta - 15 Minute Occupational Exposure Limit
10 mg/m3 STEL
Canada - British Columbia - 8 Hour Exposure Limits
10 mg/m3 TWA
Canada - Ontario - OHSA - TWAEVs
5 mg/m3 TWAEV
Canada - Quebec - Time-Weighted Average Exposure Values
5 mg/m3 TWAEV
German (DFG) - MAK Values
total dust: 2 mg/m3 MAK
Israel - Time Weighted Averages
5 mg/m3 TWA
Israel - Action Levels
2.5 mg/m3 AL
Mexico - Instruction No. 10 - TWAs
10 mg/m3 TWA

STATE LISTS
California - Exposure Limits - PELs
5 mg/m3 PEL
California - Directors List of Hazardous Substances (8 CCR 339)
[present]
Florida Hazardous Substance List
[present]
Massachusetts Right To Know List
[present]
Minnesota Hazardous Substance List
[present]
NJ Right to Know List (Total)
sn 0171
NJ Special Hazardous Substances
(mutagen)
Pennsylvania Right to Know List
[present]

ATTAPULGITE   1337-76-4

HEALTH AND SAFETY LISTS
IARC - Group 3 (not classifiable)
[present]

ATTAPULGITE   12174-11-7

ENVIRONMENTAL LISTS
List of Pesticide Product Inert Ingredients
[present]

INTERNATIONAL LISTS
Canada - Quebec - Time-Weighted Average Exposure Values
1 fibre/cm3 TWAEV
Canada - Quebec - Carcinogens
C1 carcinogen: effect detected in humans

STATE LISTS
California - Directors List of Hazardous Substances (8 CCR 339)
[present]

AURAMINE   492-80-8

HEALTH AND SAFETY LISTS
U.S. DOT - Appendix A Table 1 - Hazardous Substances
final RQ = 100 pounds (45.4 kg)

ENVIRONMENTAL LISTS
CERCLA/SARA - Section 313 - Emission Reporting
form R reporting required for 0.1% de minimus concentration
CERCLA/SARA - Hazardous Substances and their Reportable Quantities
final RQ = 100 pounds (45.4 kg)
EPA - Carcinogen Hazard Ranking for RQ Adjustment
Hazard ranking = Low
RCRA - U Series Wastes
waste number U014
RCRA - Hazardous Constituents-Appendix VIII
waste number U014
RCRA - Substances Banned From Land Disposal
[present]

INTERNATIONAL LISTS
German (DFG) - Carcinogens
animal evidence of carcinogenicity

STATE LISTS
California - Air Bill 2588 Appendix A-II
known or potential carcinogen
California - Prop. 65 - Cancer list
carcinogen - initial date 7/1/87
California - Prop. 65 - No Significant Risk Levels
no significant risk level = 0.8 ug/day
California - Directors List of Hazardous Substances (8 CCR 339)
[present]
Florida Hazardous Substance List
[present] (includes technical grade)
Massachusetts Right To Know List
carcinogen; extraordinarily hazardous
Minnesota Hazardous Substance List
carcinogen
NJ Right to Know List (Total)
sn 2894
Pennsylvania Right to Know List
environmental hazard

AURAMINE   2465-27-2

HEALTH AND SAFETY LISTS
IARC - Group 2B (sufficient animal data)
[present]
NIOSH - Selected LD50s and LC50s
Oral, mouse: LD50 = 480 mg/kg Skin, mouse: LD50 = 300 mg/kg
OSHA - Possible Select Carcinogens
[present]

ENVIRONMENTAL LISTS

List of Pesticide Product Inert Ingredients
[present]

INTERNATIONAL LISTS

German (DFG) - Carcinogens
animal evidence of carcinogenicity

STATE LISTS

NJ Right to Know List (Total)
sn 0450

NJ Special Hazardous Substances
(carcinogen)

Pennsylvania Right to Know List
environmental hazard; special hazardous substance

Pennsylvania RTK - Special Hazardous Substances
[present]

AURAMINE, MANUFACTURE OF   RR-00525-4

HEALTH AND SAFETY LISTS

IARC - Group 1 (carcinogenic to humans)
[present]

OSHA - Select Carcinogens
[present]

STATE LISTS

Pennsylvania Right to Know List
special hazardous substance

Pennsylvania RTK - Special Hazardous Substances
[present]

AUROTHIOGLUCOSE   12192-57-3

HEALTH AND SAFETY LISTS

IARC - Group 3 (not classifiable)
[present]

STATE LISTS

California - Directors List of Hazardous Substances (8 CCR 339)
[present]

5-AZACYTIDINE   320-67-2

HEALTH AND SAFETY LISTS

IARC - Group 2A (limited human data)
[present]

NTP Chemical Status Reports - Testing Status and NTIS Number
Technical reports printed (PB279526/AS)

NTP Chemical Status Reports - Evidence of Carcinogenicity
male rat-inadequate; female rat-inadequate; male mice-inadequate; female mice-positive

OSHA - Possible Select Carcinogens
[present]

STATE LISTS

California - Prop. 65 - Cancer list
carcinogen - initial date 1/1/92

California - Directors List of Hazardous Substances (8 CCR 339)
[present]

Massachusetts Right To Know List
carcinogen; extraordinarily hazardous

PROPOSED REGULATIONS

NTP - Proposed Additions to Annual Report on Carcinogens
proposed as a suspect carcinogen for the NTP 9th reports

AZASERINE   115-02-6

HEALTH AND SAFETY LISTS

U.S. DOT - Appendix A Table 1 - Hazardous Substances
final RQ = 1 pound (0.454 kg)

IARC - Group 2B (sufficient animal data)
[present] (Overall evaluation based only on evidence of carcinogenicity in monograph (10, 1976) or in Supplement 4)

NIOSH - Selected LD50s and LC50s
Oral, rat: LD50 = 170 mg/kg

OSHA - Possible Select Carcinogens
[present]

ENVIRONMENTAL LISTS

CERCLA/SARA - Hazardous Substances and their Reportable Quantities
final RQ = 1 pound (0.454 kg)

EPA - Carcinogen Hazard Ranking for RQ Adjustment
Hazard ranking = High

RCRA - U Series Wastes
waste number U015

RCRA - Hazardous Constituents-Appendix VIII
waste number U015

RCRA - Substances Banned From Land Disposal
[present]

STATE LISTS

California - Air Bill 2588 Appendix A-II
known or potential carcinogen

California - Prop. 65 - Cancer list
carcinogen - initial date 7/1/87

California - Prop. 65 - No Significant Risk Levels
no significant risk level = 0.06 ug/day

California - Directors List of Hazardous Substances (8 CCR 339)
[present]

Florida Hazardous Substance List
[present]

Massachusetts Right To Know List
carcinogen; extraordinarily hazardous

Minnesota Hazardous Substance List
carcinogen

NJ Right to Know List (Total)
sn 0173

NJ Special Hazardous Substances
(carcinogen)

Pennsylvania Right to Know List
environmental hazard; special hazardous substance

Pennsylvania RTK - Special Hazardous Substances
[present]

AZATHIOPRINE   446-86-6

HEALTH AND SAFETY LISTS

IARC - Group 1 (carcinogenic to humans)
[present]

NIOSH - Selected LD50s and LC50s
Oral, rat: LD50 = 535 mg/kg

NTP Chemical Status Reports - Testing Status and NTIS Number
Chronic studies exist for which technical reports were not prepared

NTP Seventh Report - Known Carcinogens
known carcinogen

OSHA - Select Carcinogens
[present]

STATE LISTS

California - Air Bill 2588 Appendix A-II
known or potential carcinogen

California - Prop. 65 - Cancer list
carcinogen - initial date 2/27/87

California - Prop. 65 - No Significant Risk Levels
no significant risk level = 0.4 ug/day

California - Directors List of Hazardous Substances (8 CCR 339)
[present]
Florida Hazardous Substance List
[present]
Massachusetts Right To Know List
carcinogen; extraordinarily hazardous
Minnesota Hazardous Substance List
carcinogen
NJ Right to Know List (Total)
sn 0174
NJ Special Hazardous Substances
(carcinogen; mutagen)
Pennsylvania Right to Know List
special hazardous substance
Pennsylvania RTK - Special Hazardous Substances
[present]

6-AZAURIDINE   54-25-1
STATE LISTS
Massachusetts Right To Know List
teratogen

AZAUROLIC ACID   134191-17-6
HEALTH AND SAFETY LISTS
U.S. DOT - Hazard Classes
Forbidden from transport by the DOT

AZAUROLIC ACID   RR-01392-3
STATE LISTS
NJ Right to Know List (Total)
with irritant: sn 2227; with poisonous material: sn 2228

3'-AZIDO-3'-DEOXYTHYMIDINE + 2',3'-DIDEOXYI-   RR-01633-1
NOSINE (AIDS INITIATIVE)
HEALTH AND SAFETY LISTS
NTP Chemical Status Reports - Testing Status and NTIS Number
*Prechronic studies for which toxicity technical reports were not pre-
pared*

3'-AZIDO-3'-DEOXYTHYMIDINE (AIDS INITIATIVE)   30516-87-1
HEALTH AND SAFETY LISTS
NTP Chemical Status Reports - Testing Status and NTIS Number
*Two year studies: pathology quality assessment in progress;
prechronic studies completed: in review for further evaluation*

3'-AZIDO-3'-DEOXYTHYMIDINE/2',3'-DIDEOXYCY-   RR-01632-0
TIDINE (AIDS INITIATIVE)
HEALTH AND SAFETY LISTS
NTP Chemical Status Reports - Testing Status and NTIS Number
*Prechronic studies for which toxicity technical reports were not pre-
pared*

AZIDODITHIOCARBONIC ACID   4472-06-4
HEALTH AND SAFETY LISTS
U.S. DOT - Hazard Classes
Forbidden from transport by the DOT

AZIDOETHYL NITRATE   53422-49-4
HEALTH AND SAFETY LISTS
U.S. DOT - Hazard Classes
Forbidden from transport by the DOT

AZIDO GUANIDINE PICRATE   RR-01393-4
HEALTH AND SAFETY LISTS

U.S. DOT - Hazard Classes
Forbidden from transport by the DOT

5-AZIDO-1-HYDROXY TETRAZOLE   RR-01388-7
HEALTH AND SAFETY LISTS
U.S. DOT - Hazard Classes
Forbidden from transport by the DOT

AZIDO HYDROXY TETRAZOLE   RR-01394-5
HEALTH AND SAFETY LISTS
U.S. DOT - Hazard Classes
Forbidden from transport by the DOT

3-AZIDO-1,2-PROPYLENE GLYCOL DINITRATE   RR-01386-5
HEALTH AND SAFETY LISTS
U.S. DOT - Hazard Classes
Forbidden from transport by the DOT

AZINPHOS-ETHYL   2642-71-9
HEALTH AND SAFETY LISTS
U.S. DOT - Appendix B - Marine Pollutants
DOT regulated severe marine pollutant
NIOSH - Selected LD50s and LC50s
*Inhalation, rat: LC50 = 390 mg/m3 8 hr Oral, rat: LD50 = 7 mg/kg
Skin, rat: LD50 = 250 mg/kg*
ENVIRONMENTAL LISTS
CERCLA/SARA - Section 302 Extremely Hazardous Substances and
TPQs
TPQ = 100/10,000 pounds
STATE LISTS
Florida Hazardous Substance List
effective March 13, 1992
Massachusetts Right To Know List
extraordinarily hazardous
NJ Right to Know List (Total)
sn 2140
Pennsylvania Right to Know List
environmental hazard
PROPOSED REGULATIONS
CERCLA/SARA - Proposed Hazardous Substance Additions
proposed RQ = 1 pound (.454 kg)
CERCLA/SARA - 1989 Proposed RQ Adjustments
proposed RQ = 100 pounds (45.4 kg)

AZINPHOS-METHYL   86-50-0
HEALTH AND SAFETY LISTS
ACGIH 1995 - Time Weighted Averages
0.2 mg/m3 TWA
ACGIH 1995 - Skin Designations
skin - potential for cutaneous absorption
U.S. DOT - Appendix B - Marine Pollutants
DOT regulated severe marine pollutant
U.S. DOT - Appendix A Table 1 - Hazardous Substances
final RQ = 1 pound (0.454 kg)
NIOSH - Selected LD50s and LC50s
*Inhalation, rat: LC50 = 69 mg/m3 1 hr Oral, rat: LD50 = 7 mg/kg
Skin, rat: LD50 = 220 mg/kg*
NIOSH 1990 - Pocket Guide - RELs
0.2 mg/m3 TWA
NIOSH 1990 - Pocket Guide - IDLHs
20 mg/m3 IDLH
NIOSH 1990 - Pocket Guide - Target organs
respiratory system, CNS, CVS, blood cholinesterase

NIOSH 1990 - Pocket Guide - Skin list
  Potential for dermal absorption
NTP Chemical Status Reports - Testing Status and NTIS Number
  Technical reports printed (PB286371/AS)
NTP Chemical Status Reports - Evidence of Carcinogenicity
  male rat-equivocal; female rat-negative; male mice-negative; female mice-negative
OSHA - Vacated PELs - Time Weighted Averages
  0.2 mg/m3 TWA
OSHA - Vacated PELs - Skin Designation
  Prevent or reduce skin absorption
OSHA - Final PELs - Time Weighted Averages
  0.2 mg/m3 TWA
OSHA - Final PELs - Skin Notations
  prevent or reduce skin absorption

ENVIRONMENTAL LISTS
  ATSDR Priority List
    Rank (of 275): 180
  CERCLA/SARA - Section 302 Extremely Hazardous Substances and TPQs
    TPQ = 10/10,000 pounds
  CERCLA/SARA - Hazardous Substances and their Reportable Quantities
    final RQ = 1 pound (0.454 kg)
  Clean Water Act - Hazardous Substances
    [present]

INTERNATIONAL LISTS
  Australian Exposure Standards - Time Weighted Averages
    0.2 mg/m3 TWA
  Australian Exposure Standards - Skin Effects
    skin absorption
  Canada - Drinking Water Quality - MACs
    0.02 mg/L MAC
  Canada - Alberta - 8 Hour Occupational Exposure Limit
    0.2 mg/m3 TWA
  Canada - Alberta - 15 Minute Occupational Exposure Limit
    0.6 mg/m3 STEL
  Canada - Alberta - Skin Designation
    can be absorbed through the intact skin
  Canada - British Columbia - 8 Hour Exposure Limits
    0.2 mg/m3 TWA
  Canada - British Columbia - 15 Minute Exposure Limits
    0.6 mg/m3 STEL
  Canada - British Columbia - Skin Notations
    skin - potential for skin absorption
  Canada - Ontario - OHSA - TWAEVs
    0.2 mg/m3 TWAEV
  Canada - Ontario - OHSA - Skin Notations
    absorption through skin, eyes, or mucous membranes
  Canada - Quebec - Time-Weighted Average Exposure Values
    0.2 mg/m3 TWAEV
  Canada - Quebec - Skin Designations
    absorbed through the skin
  United Kingdom - Occupational Exposure Standards - TWAs
    0.2 mg/m3 TWA
  United Kingdom - Occupational Exposure Standards - STELs
    0.6 mg/m3 STEL
  United Kingdom - Occupational Exposure Standards - Notes
    can be absorbed through skin
  German (DFG) - MAK Values
    total dust: 0.2 mg/m3 MAK
  German (DFG) - Peak Limitations
    10 x normal MAK (30 min average value); don't exceed during shift

German (DFG) - Skin/Sensitizers
  danger of cutaneous absorption
Israel - Time Weighted Averages
  0.2 mg/m3 TWA
Israel - Action Levels
  0.1 mg/m3 AL
Mexico - Instruction No. 10 - TWAs
  0.2 mg/m3 TWA
Mexico - Instruction No. 10 - STELs
  0.6 mg/m3 STEL
Mexico - Instruction No. 10 - Skin designation
  skin - potential for cutaneous absorption

STATE LISTS
  California - Exposure Limits - PELs
    0.2 mg/m3 PEL
  California - Exposure Limits - Skin Notation
    material may be absorbed through the skin, eyes or mucous membrane
  California - Directors List of Hazardous Substances (8 CCR 339)
    [present]
  Florida Hazardous Substance List
    [present]
  Massachusetts Right To Know List
    extraordinarily hazardous; neurotoxin
  Minnesota Hazardous Substance List
    skin
  NJ Right to Know List (Total)
    sn 0966
  Pennsylvania Right to Know List
    environmental hazard

1-AZIRIDINEETHANOL   1072-52-2
HEALTH AND SAFETY LISTS
  IARC - Group 3 (not classifiable)
    [present]
  NIOSH - Selected LD50s and LC50s
    Oral, rat: LD50 = 74 mg/kg Skin, rabbit: LD50 = 280 mg/kg
INTERNATIONAL LISTS
  Canada - WHMIS: Ingredient Disclosure
    1% item 138 (270)
STATE LISTS
  California - Directors List of Hazardous Substances (8 CCR 339)
    [present]

AZIRIDYL BENZOQUINONE   800-24-8
HEALTH AND SAFETY LISTS
  IARC - Group 3 (not classifiable)
    [present]
STATE LISTS
  California - Directors List of Hazardous Substances (8 CCR 339)
    [present]

AZOBENZENE   103-33-3
HEALTH AND SAFETY LISTS
  IARC - Group 3 (not classifiable)
    [present]
  NIOSH - Selected LD50s and LC50s
    Oral, rat: LD50 = 1000 mg/kg
  NTP Chemical Status Reports - Testing Status and NTIS Number
    Technical reports printed (PB293835/AS)
  NTP Chemical Status Reports - Evidence of Carcinogenicity
    male rat-positive; female rat-positive; male mice-negative; female mice-negative

ENVIRONMENTAL LISTS

CAA - HON Rule - SOCMI Chemicals
compliance by Oct. 24, 1994

INTERNATIONAL LISTS

Canada - WHMIS: Ingredient Disclosure
1% item 139 (271)

STATE LISTS

California - Air Bill 2588 Appendix A-II
known or potential carcinogen: 9/90

California - Prop. 65 - Cancer list
carcinogen - initial date 1/1/90

California - Prop. 65 - No Significant Risk Levels
no significant risk level = 6 ug/day

California - Directors List of Hazardous Substances (8 CCR 339)
[present]

Massachusetts Right To Know List
carcinogen; extraordinarily hazardous

AZOBISISOBUTYRONITRILE 78-67-1

HEALTH AND SAFETY LISTS

U.S. DOT - Substances From 49 CFR 172.101
regulated by DOT (UN2952)

U.S. DOT - Hazard Classes
DOT hazard class = 4.1

NFPA - Hazard Identification Ratings
health-3; reactivity-2

NIOSH - Selected LD50s and LC50s
Oral, mouse: LD50 = 700 mg/kg

STATE LISTS

Florida Hazardous Substance List
[present]

Massachusetts Right To Know List
[present]

NJ Right to Know List (Total)
sn 0179

Pennsylvania Right to Know List
[present]

AZODICARBONAMIDE 123-77-3

HEALTH AND SAFETY LISTS

NTP Chemical Status Reports - Testing Status and NTIS Number
Prechronic studies for which toxicity technical reports were not prepared

ENVIRONMENTAL LISTS

EPA - Master Testing List
[present]

List of Pesticide Product Inert Ingredients
[present]

2,2'-AZODI-(2,4-DIMETHYL-4-METHOXYVALERONI-TRILE) 15545-97-8

HEALTH AND SAFETY LISTS

U.S. DOT - Substances From 49 CFR 172.101
regulated by DOT (UN2955)

U.S. DOT - Hazard Classes
DOT hazard class = 4.1

STATE LISTS

NJ Right to Know List (Total)
sn 0176

2,2'-AZODI-(2,4-DIMETHYLVALERONITRILE) 28604-91-3

HEALTH AND SAFETY LISTS

U.S. DOT - Substances From 49 CFR 172.101
regulated by DOT (UN2953)

U.S. DOT - Hazard Classes
DOT hazard class = 4.1

STATE LISTS

NJ Right to Know List (Total)
sn 0177

AZODI-(1,1'-HEXAHYDROBENZONITRILE) 25551-14-8

HEALTH AND SAFETY LISTS

U.S. DOT - Substances From 49 CFR 172.101
regulated by DOT (UN2954)

U.S. DOT - Hazard Classes
DOT hazard class = 4.1

STATE LISTS

NJ Right to Know List (Total)
sn 0178

AZODI(2-METHYLBUTYRONITRITE) RR-00768-1

HEALTH AND SAFETY LISTS

U.S. DOT - Substances From 49 CFR 172.101
regulated by DOT (UN3030)

U.S. DOT - Hazard Classes
DOT hazard class = 4.1

AZOTETRAZOLE RR-01395-6

HEALTH AND SAFETY LISTS

U.S. DOT - Hazard Classes
Forbidden from transport by the DOT

AZOXYBENZENE 495-48-7

HEALTH AND SAFETY LISTS

NIOSH - Selected LD50s and LC50s
Oral, rat: LD50 = 620 mg/kg Skin, rabbit: LD50 = 1090 mg/kg

BACITRACIN 1405-87-4

HEALTH AND SAFETY LISTS

NIOSH - Selected LD50s and LC50s
Oral, guinea pig: LD50 = 2 gm/kg

STATE LISTS

Massachusetts Right To Know List
[present]

PROPOSED REGULATIONS

CERCLA/SARA - Proposed Hazardous Substance Additions
proposed RQ = 1 pound (.454 kg)

CERCLA/SARA - 1989 Proposed RQ Adjustments
proposed RQ = 1 pounds (.454 kg)

BACTOPEPTONE 51142-18-8

ENVIRONMENTAL LISTS

List of Pesticide Product Inert Ingredients
[present]

BANCROFT CLAY RR-01038-8

ENVIRONMENTAL LISTS

List of Pesticide Product Inert Ingredients
[present]

BARBAN   101-27-9

STATE LISTS

California - Directors List of Hazardous Substances (8 CCR 339)
[present]

BARBITURATES   RR-01473-3

STATE LISTS

California - Prop. 65 - Developmental Toxicity
developmental toxicity - initial date 10/1/92

BARIUM   7440-39-3

SEE ALSO:
F039-HAZARDOUS WASTES
BARIUM COMPOUNDS, N.O.S.

HEALTH AND SAFETY LISTS

ACGIH 1995 - Time Weighted Averages
soluble compounds, as Ba: 0.5 mg/m3 TWA

U.S. DOT - Substances From 49 CFR 172.101
regulated by DOT (UN1400)

U.S. DOT - Hazard Classes
DOT hazard class = 4.3

OSHA - Vacated PELs - Time Weighted Averages
as Ba: 0.5 mg/m3 TWA

OSHA - Final PELs - Time Weighted Averages
as Ba: 0.5 mg/m3 TWA

ENVIRONMENTAL LISTS

ATSDR Priority List
Rank (of 275): 101

CERCLA/SARA - Section 313 - Emission Reporting
form R reporting required for 1.0% de minimus concentration

Safe Drinking Water Act - MCLs
MCL = 2 mg/L

Safe Drinking Water Act - MCLGs
MCLG = 2 mg/L

RCRA - D Series - Maximum Concentration of Contaminants
waste number D005; regulatory level = 100.0 mg/L

RCRA - D Series - Chronic Toxicity Reference Levels
chronic toxicity reference level = 1.0 mg/L

RCRA - Hazardous Constituents-Appendix VIII
hazardous constituent - no waste number

RCRA - Basis for Listing - Appendix VII
Included in waste stream: F039

RCRA - Substances Banned From Land Disposal
[present]

RCRA - TSD Facilities Ground Water Monitoring
TM 6010 = 20 ug/L PQL; TM 7080 = 1000 ug/L PQL (all species in the ground water that contain this element are included)

RCRA - Universal Treatment Standards (LDR)
WW: 1.2 mg/l; NWW: 7.6 mg/l TCLP

INTERNATIONAL LISTS

Australian Exposure Standards - Time Weighted Averages
soluble compounds, as Ba: 0.5 mg/m3 TWA

Canada - Drinking Water Quality - MACs
1.0 mg/L MAC

Canada - Alberta - 8 Hour Occupational Exposure Limit
0.5 mg/m3 TWA

Canada - Alberta - 15 Minute Occupational Exposure Limit
as Ba: 1.5 mg/m3 STEL

German (DFG) - MAK Values
total dust, as Ba: 0.5 mg/m3 MAK

German (DFG) - Peak Limitations
2 x normal MAK (5 min momentary value); don't exceed 8 times during shift

Israel - Time Weighted Averages
as Ba: 0.5 mg/m3 TWA

Israel - Action Levels
as Ba: 0.25 mg/m3 AL

Mexico - Instruction No. 10 - TWAs
0.5 mg/m3 TWA

Mexico - Drinking Water - Ecological Criteria
1.0 mg/l

STATE LISTS

California - Air Bill 2588 Appendix A-I
6/91

California - Exposure Limits - PELs
soluble compounds, as Ba: 0.5 mg/m3 PEL

Florida Hazardous Substance List
[present]

Massachusetts Right To Know List
[present]

Minnesota Hazardous Substance List
[present] as Ba

NJ Right to Know List (Total)
sn 0180

Pennsylvania Right to Know List
environmental hazard (any compound of this substance is also an environmental hazard)

BARIUM 126   15229-36-4

HEALTH AND SAFETY LISTS

U.S. DOT - Appendix A Table 2 - Radionuclides
final RQ = 1000 curies (3.7E 13 Bq)

ENVIRONMENTAL LISTS

CERCLA/SARA List of Radionuclides (Appendix B) and Their Reportable Quantities
final RQ = 1000 curies (3.7E 13 Bq)

BARIUM 128   15741-25-0

HEALTH AND SAFETY LISTS

U.S. DOT - Appendix A Table 2 - Radionuclides
final RQ = 10 curies (3.7E 11 Bq)

ENVIRONMENTAL LISTS

CERCLA/SARA List of Radionuclides (Appendix B) and Their Reportable Quantities
final RQ = 10 curies (3.7E 11 Bq)

BARIUM 131   14914-75-1

HEALTH AND SAFETY LISTS

U.S. DOT - Appendix A Table 2 - Radionuclides
final RQ = 10 curies (3.7E 11 Bq)

ENVIRONMENTAL LISTS

CERCLA/SARA List of Radionuclides (Appendix B) and Their Reportable Quantities
final RQ = 10 curies (3.7E 11 Bq)

BARIUM 131M   RR-00483-1

HEALTH AND SAFETY LISTS

U.S. DOT - Appendix A Table 2 - Radionuclides
final RQ = 1000 curies (3.7E 13 Bq)

ENVIRONMENTAL LISTS

CERCLA/SARA List of Radionuclides (Appendix B) and Their Reportable Quantities
final RQ = 1000 curies (3.7E 13 Bq)

BARIUM 133   13981-41-4

HEALTH AND SAFETY LISTS

  U.S. DOT - Appendix A Table 2 - Radionuclides
  final RQ = 10 curies (3.7E 11 Bq)

ENVIRONMENTAL LISTS

  CERCLA/SARA List of Radionuclides (Appendix B) and Their Reportable Quantities
  final RQ = 10 curies (3.7E 11 Bq)

BARIUM 133M   RR-00482-0

HEALTH AND SAFETY LISTS

  U.S. DOT - Appendix A Table 2 - Radionuclides
  final RQ = 100 curies (3.7E 12 Bq)

ENVIRONMENTAL LISTS

  CERCLA/SARA List of Radionuclides (Appendix B) and Their Reportable Quantities
  final RQ = 100 curies (3.7E 12 Bq)

BARIUM 135M   RR-00481-9

HEALTH AND SAFETY LISTS

  U.S. DOT - Appendix A Table 2 - Radionuclides
  final RQ = 1000 curies (3.7E 13 Bq)

ENVIRONMENTAL LISTS

  CERCLA/SARA List of Radionuclides (Appendix B) and Their Reportable Quantities
  final RQ = 1000 curies (3.7E 13 Bq)

BARIUM 139   14378-25-7

HEALTH AND SAFETY LISTS

  U.S. DOT - Appendix A Table 2 - Radionuclides
  final RQ = 1000 curies (3.7E 13 Bq)

ENVIRONMENTAL LISTS

  CERCLA/SARA List of Radionuclides (Appendix B) and Their Reportable Quantities
  final RQ = 1000 curies (3.7E 13 Bq)

BARIUM 140   14798-08-4

HEALTH AND SAFETY LISTS

  U.S. DOT - Appendix A Table 2 - Radionuclides
  final RQ = 10 curies (3.7E 11 Bq)

ENVIRONMENTAL LISTS

  CERCLA/SARA List of Radionuclides (Appendix B) and Their Reportable Quantities
  final RQ = 10 curies (3.7E 11 Bq)

BARIUM 141   15741-29-4

HEALTH AND SAFETY LISTS

  U.S. DOT - Appendix A Table 2 - Radionuclides
  final RQ = 1000 curies (3.7E 13 Bq)

ENVIRONMENTAL LISTS

  CERCLA/SARA List of Radionuclides (Appendix B) and Their Reportable Quantities
  final RQ = 1000 curies (3.7E 13 Bq)

BARIUM 142   18879-37-3

HEALTH AND SAFETY LISTS

  U.S. DOT - Appendix A Table 2 - Radionuclides
  final RQ = 1000 curies (3.7E 13 Bq)

ENVIRONMENTAL LISTS

  CERCLA/SARA List of Radionuclides (Appendix B) and Their Reportable Quantities
  final RQ = 1000 curies (3.7E 13 Bq)

BARIUM ACETATE   543-80-6
SEE ALSO:
  BARIUM

HEALTH AND SAFETY LISTS

  NIOSH - Selected LD50s and LC50s
  Oral, rat: LD50 = 921 mg/kg

INTERNATIONAL LISTS

  Canada - WHMIS: Ingredient Disclosure
  1% item 140 (8)

BARIUM ALLOYS   RR-00161-6

HEALTH AND SAFETY LISTS

  U.S. DOT - Substances From 49 CFR 172.101
  regulated by DOT (UN1854)

  U.S. DOT - Hazard Classes
  DOT hazard class = 4.2

STATE LISTS

  NJ Right to Know List (Total)
  sn 2144;   sn 2145

BARIUM AZIDE   18810-58-7
SEE ALSO:
  BARIUM

HEALTH AND SAFETY LISTS

  U.S. DOT - Substances From 49 CFR 172.101
  regulated by DOT (UN1571, UN0224)

  U.S. DOT - Hazard Classes
  DOT hazard class = 1.1A

STATE LISTS

  NJ Right to Know List (Total)
  sn 0181

BARIUM BROMATE   13967-90-3
SEE ALSO:
  BARIUM

HEALTH AND SAFETY LISTS

  U.S. DOT - Substances From 49 CFR 172.101
  regulated by DOT (UN2719)

  U.S. DOT - Hazard Classes
  DOT hazard class = 5.1

STATE LISTS

  NJ Right to Know List (Total)
  sn 0182

BARIUM CARBONATE   513-77-9
SEE ALSO:
  BARIUM

HEALTH AND SAFETY LISTS

  NIOSH - Selected LD50s and LC50s
  Oral, rat: LD50 = 418 mg/kg

ENVIRONMENTAL LISTS

  List of Pesticide Product Inert Ingredients
  [present]

INTERNATIONAL LISTS

  Canada - WHMIS: Ingredient Disclosure
  0.1% item 141 (384)

  Canada - Ontario - OHSA - TWAEVs
  as Ba: 0.5 mg/m3 TWAEV

BARIUM CHLORATE   **13477-00-4**
SEE ALSO:
BARIUM

HEALTH AND SAFETY LISTS
U.S. DOT - Substances From 49 CFR 172.101
regulated by DOT (UN1445)
U.S. DOT - Hazard Classes
DOT hazard class = 5.1

STATE LISTS
Florida Hazardous Substance List
[present]
Massachusetts Right To Know List
[present]
NJ Right to Know List (Total)
sn 0183
NJ Special Hazardous Substances
(reactive - second degree)
Pennsylvania Right to Know List
environmental hazard

BARIUM CHLORIDE   **10361-37-2**
SEE ALSO:
BARIUM

HEALTH AND SAFETY LISTS
NIOSH - Selected LD50s and LC50s
Oral, rat: LD50 = 118 mg/kg
NIOSH 1990 - Pocket Guide - RELs
as Ba: 0.5 mg/m3 TWA (Listed under 'Barium (soluble compounds)')
NIOSH 1990 - Pocket Guide - IDLHs
1100 mg/m3 IDLH (Listed under 'Barium (soluble compounds)')
NIOSH 1990 - Pocket Guide - Target organs
heart, CNS, skin, respiratory system, eyes (Listed under 'Barium (soluble compounds)')

ENVIRONMENTAL LISTS
List of Pesticide Product Inert Ingredients
[present]
TSCA - Code of Federal Regulations Citations
40 CFR 712.30(w)
TSCA - PAIR - Reporting List
Reporting Date: July 13, 1988

INTERNATIONAL LISTS
Canada - WHMIS: Ingredient Disclosure
1% item 142 (473)
Canada - Ontario - OHSA - TWAEVs
as Ba: 0.5 mg/m3 TWAEV

BARIUM CHLORIDE DIHYDRATE   **10326-27-9**
HEALTH AND SAFETY LISTS
NTP Chemical Status Reports - Testing Status and NTIS Number
Technical reports printed (no NTIS number given); prechronic studies for toxicity technical reports were not prepared

BARIUM CHROMATE   **10294-40-3**
SEE ALSO:
CHROMIUM (VI) COMPOUNDS
CHROMIUM
CHROMIUM COMPOUNDS
BARIUM

HEALTH AND SAFETY LISTS
OSHA - Select Carcinogens
[present]

INTERNATIONAL LISTS
Canada - WHMIS: Ingredient Disclosure
0.1% item 143 (547)

STATE LISTS
California - Air Bill 2588 Appendix A-I
known or potential carcinogen: 6/91
Massachusetts Right To Know List
[present]

BARIUM COMPOUNDS, N.O.S.   **RR-00555-0**
HEALTH AND SAFETY LISTS
U.S. DOT - Substances From 49 CFR 172.101
regulated by DOT (UN1564)
U.S. DOT - Hazard Classes
DOT hazard class = 6.1
U.S. DOT - Appendix B - Marine Pollutants
DOT regulated marine pollutant

ENVIRONMENTAL LISTS
CERCLA/SARA - Section 313 - Emission Reporting
form R reporting required for 1.0% de minimus concentration
RCRA - Hazardous Constituents-Appendix VIII
hazardous constituent - no waste number

STATE LISTS
California - Air Bill 2588 Appendix A-I
6/91
NJ Right to Know List (Total)
sn 2146

BARIUM CYANIDE   **542-62-1**
SEE ALSO:
BARIUM
CYANIDE ANION

HEALTH AND SAFETY LISTS
U.S. DOT - Substances From 49 CFR 172.101
regulated by DOT (UN1565)
U.S. DOT - Hazard Classes
DOT hazard class = 6.1
U.S. DOT - Appendix B - Marine Pollutants
DOT regulated marine pollutant
U.S. DOT - Appendix A Table 1 - Hazardous Substances
final RQ = 10 pounds (4.54 kg)

ENVIRONMENTAL LISTS
CERCLA/SARA - Hazardous Substances and their Reportable Quantities
final RQ = 10 pounds (4.54 kg)
Clean Water Act - Hazardous Substances
[present]
RCRA - P Series Wastes
waste number P013
RCRA - Hazardous Constituents-Appendix VIII
waste number P013
RCRA - Substances Banned From Land Disposal
[present]

STATE LISTS
Massachusetts Right To Know List
[present]
NJ Right to Know List (Total)
sn 0184
Pennsylvania Right to Know List
environmental hazard

BARIUM DICHROMATE   **10031-16-0**
STATE LISTS
Massachusetts Right To Know List
[present]

BARIUM DICHROMATE   10031-22-8

STATE LISTS

Massachusetts Right To Know List
[present]

BARIUM FLUORIDE   7787-32-8

SEE ALSO:
BARIUM

HEALTH AND SAFETY LISTS

NIOSH - Selected LD50s and LC50s
Oral, rat: LD50 = 250 mg/kg

INTERNATIONAL LISTS

Canada - WHMIS: Ingredient Disclosure
1% item 144 (901)

BARIUM HYPOCHLORITE   13477-10-6

SEE ALSO:
BARIUM

HEALTH AND SAFETY LISTS

U.S. DOT - Substances From 49 CFR 172.101
regulated by DOT (UN2741)
U.S. DOT - Hazard Classes
DOT hazard class = 5.1

STATE LISTS

NJ Right to Know List (Total)
sn 0185

BARIUM METABORATE   13701-59-2

SEE ALSO:
BARIUM
BARIUM
BARIUM COMPOUNDS, N.O.S.

ENVIRONMENTAL LISTS

List of Pesticide Product Inert Ingredients
[present]

BARIUM NITRATE   10022-31-8

SEE ALSO:
BARIUM

HEALTH AND SAFETY LISTS

U.S. DOT - Substances From 49 CFR 172.101
regulated by DOT (UN1446)
U.S. DOT - Hazard Classes
DOT hazard class = 5.1
NIOSH - Selected LD50s and LC50s
Oral, rat: LD50 = 355 mg/kg
NIOSH 1990 - Pocket Guide - RELs
as Ba: 0.5 mg/m3 TWA (Listed under 'Barium (soluble compounds)')
NIOSH 1990 - Pocket Guide - IDLHs
1100 mg/m3 IDLH (Listed under 'Barium (soluble compounds)')
NIOSH 1990 - Pocket Guide - Target organs
heart, CNS, skin, respiratory system, eyes (Listed under 'Barium (soluble compounds)')

INTERNATIONAL LISTS

Canada - WHMIS: Ingredient Disclosure
1% item 145 (1199)
Canada - Ontario - OHSA - TWAEVs
as Ba: 0.5 mg/m3 TWAEV

STATE LISTS

Florida Hazardous Substance List
[present]
Massachusetts Right To Know List
[present]

NJ Right to Know List (Total)
sn 0186
Pennsylvania Right to Know List
environmental hazard

BARIUM NONYLPHENATE   28987-17-9

ENVIRONMENTAL LISTS

List of Pesticide Product Inert Ingredients
[present]

BARIUM OXIDE   1304-28-5

SEE ALSO:
BARIUM

HEALTH AND SAFETY LISTS

U.S. DOT - Substances From 49 CFR 172.101
regulated by DOT (UN1884)
U.S. DOT - Hazard Classes
DOT hazard class = 6.1

INTERNATIONAL LISTS

Canada - Ontario - OHSA - TWAEVs
as Ba: 0.5 mg/m3 TWAEV

STATE LISTS

NJ Right to Know List (Total)
sn 0187

BARIUM PERCHLORATE   13465-95-7

SEE ALSO:
BARIUM

HEALTH AND SAFETY LISTS

U.S. DOT - Substances From 49 CFR 172.101
regulated by DOT (UN1447)
U.S. DOT - Hazard Classes
DOT hazard class = 5.1

STATE LISTS

NJ Right to Know List (Total)
sn 0188

BARIUM PERMANGANATE   7787-36-2

SEE ALSO:
BARIUM
MANGANESE

HEALTH AND SAFETY LISTS

U.S. DOT - Substances From 49 CFR 172.101
regulated by DOT (UN1448)
U.S. DOT - Hazard Classes
DOT hazard class = 5.1
OSHA - List of Highly Hazardous Chemicals
threshhold quantity = 7500 pounds

STATE LISTS

NJ Right to Know List (Total)
sn 0189

BARIUM PEROXIDE   1304-29-6

SEE ALSO:
BARIUM

HEALTH AND SAFETY LISTS

U.S. DOT - Substances From 49 CFR 172.101
regulated by DOT (UN1449)
U.S. DOT - Hazard Classes
DOT hazard class = 5.1

STATE LISTS

Florida Hazardous Substance List
[present]

Massachusetts Right To Know List
[present]
NJ Right to Know List (Total)
sn 0190

Pennsylvania Right to Know List
environmental hazard

BARIUM SELENATE   **7787-41-9**

STATE LISTS

NJ Right to Know List (Total)
sn 2148

BARIUM SELENITE   **13718-59-7**

STATE LISTS

NJ Right to Know List (Total)
sn 2149

BARIUM SILICOFLUORIDE   **17125-80-3**
SEE ALSO:
BARIUM

INTERNATIONAL LISTS

Canada - WHMIS: Ingredient Disclosure
1% item 146 (1495)

BARIUM SOLUBLE COMPOUNDS   **RR-00049-7**

HEALTH AND SAFETY LISTS

NIOSH 1990 - Pocket Guide - RELs
as Ba: 0.5 mg/m3 TWA

NIOSH 1990 - Pocket Guide - IDLHs
as Ba: 1100 mg/m3 IDLH

NIOSH 1990 - Pocket Guide - Target organs
heart, CNS, skin, respiratory system, eyes

INTERNATIONAL LISTS

Canada - WHMIS: Ingredient Disclosure
1% item 147 (273)

Canada - British Columbia - 8 Hour Exposure Limits
as Ba: 0.5 mg/m3 TWA

Canada - Quebec - Time-Weighted Average Exposure Values
as Ba: 0.5 mg/m3 TWAEV

United Kingdom - Occupational Exposure Standards - TWAs
as Ba: 0.5 mg/m3 TWA

STATE LISTS

California - Directors List of Hazardous Substances (8 CCR 339)
[present] (refers to water-soluble salts only; all other salts are exempt)

BARIUM STEARATE   **6865-35-6**
SEE ALSO:
BARIUM
BARIUM COMPOUNDS, N.O.S.
BARIUM

HEALTH AND SAFETY LISTS

NIOSH - Selected LD50s and LC50s
Oral, rat: LD50 = 4 gm/kg

BARIUM SULFATE   **7727-43-7**
SEE ALSO:
BARIUM

HEALTH AND SAFETY LISTS

ACGIH 1995 - Time Weighted Averages
10 mg/m3 TWA (The value is for the total dust containing no asbestos and <1% crystalline silica)

OSHA - Vacated PELs - Time Weighted Averages
total dust: 10 mg/m3 TWA; respirable fraction: 5 mg/m3 TWA

OSHA - Final PELs - Time Weighted Averages
total dust: 15 mg/m3 TWA; respirable fraction: 5 mg/m3 TWA

ENVIRONMENTAL LISTS

List of Pesticide Product Inert Ingredients
[present]

INTERNATIONAL LISTS

Australian Exposure Standards - Time Weighted Averages
10 mg/m3 TWA

Canada - Ontario - OHSA - TWAEVs
total dust: 10 mg/m3 TWAEV (Listed as nuisance particulate)

Canada - Quebec - Time-Weighted Average Exposure Values
total dust: 10 ppm TWAEV; respirable dust: 5 ppm TWAEV

United Kingdom - Occupational Exposure Standards - TWAs
respirable dust: 2 mg/m3 TWA

Israel - Time Weighted Averages
10 mg/m3 TWA (The value is for the total dust containing no asbestos and <1% crystalline silica)

Israel - Action Levels
5 mg/m3 AL

STATE LISTS

Minnesota Hazardous Substance List
[present]

Pennsylvania Right to Know List
[present]

BARIUM ZINC METHYLBENZOATE-2-   **RR-01039-9**
ETHYLHEXANOATE

ENVIRONMENTAL LISTS

List of Pesticide Product Inert Ingredients
[present]

BARIUM ZIRCONATE   **12009-21-1**
SEE ALSO:
BARIUM
ZIRCONIUM

HEALTH AND SAFETY LISTS

NIOSH - Selected LD50s and LC50s
Oral, rat: LD50 = 1980 mg/kg

BARLEY, MALT   **8002-48-0**

ENVIRONMENTAL LISTS

List of Pesticide Product Inert Ingredients
[present]

BASIC ALUMINUM ACETATE   **142-03-0**

ENVIRONMENTAL LISTS

List of Pesticide Product Inert Ingredients
[present]

BASIC ZINC CHROMATE   **50922-29-7**
SEE ALSO:
CHROMIUM
ZINC
CHROMIUM (VI) COMPOUNDS
ZINC
ZINC COMPOUNDS
CHROMIUM COMPOUNDS

INTERNATIONAL LISTS

Canada - WHMIS: Ingredient Disclosure
0.1% item 148 (555)

BATTERY FLUID, ACID   **RR-00166-1**

HEALTH AND SAFETY LISTS

U.S. DOT - Substances From 49 CFR 172.101
regulated by DOT (UN2796)

U.S. DOT - Hazard Classes
DOT hazard class = 8

STATE LISTS

NJ Right to Know List (Total)
sn 2153

BATTERY FLUID, ALKALI   RR-00167-2

HEALTH AND SAFETY LISTS

U.S. DOT - Substances From 49 CFR 172.101
regulated by DOT (UN2797)

U.S. DOT - Hazard Classes
DOT hazard class = 8

STATE LISTS

NJ Right to Know List (Total)
sn 2154

BAUXITE (AL2O3.XH2O)   1318-16-7

INTERNATIONAL LISTS

Canada - WHMIS: Ingredient Disclosure
1% item 149 (274)

BELLADONNALEAF   8007-93-0

ENVIRONMENTAL LISTS

List of Pesticide Product Inert Ingredients
[present]

BENDIOCARB   22781-23-3

HEALTH AND SAFETY LISTS

U.S. DOT - Appendix B - Marine Pollutants
DOT regulated marine pollutant

NIOSH - Selected LD50s and LC50s
Oral, rat: LD50 = 40 mg/kg Skin, rat: LD50 = 1000 mg/kg

ENVIRONMENTAL LISTS

CERCLA/SARA - Section 313 - Emission Reporting
form R reporting required

INTERNATIONAL LISTS

Canada - Drinking Water Quality - MACs
0.04 mg/L MAC

STATE LISTS

NJ Right to Know List (Total)
sn 0191

BENOMYL   17804-35-2

HEALTH AND SAFETY LISTS

ACGIH 1995 - Time Weighted Averages
0.84 ppm TWA; 10 mg/m3 TWA

NIOSH - Selected LD50s and LC50s
Oral, rat: LD50 = 10 gm/kg

OSHA - Vacated PELs - Time Weighted Averages
total dust: 10 mg/m3 TWA; respirable fraction: 5 mg/m3 TWA

OSHA - Final PELs - Time Weighted Averages
total dust: 15 mg/m3 TWA; respirable fraction: 5 mg/m3 TWA

ENVIRONMENTAL LISTS

CERCLA/SARA - Section 313 - Emission Reporting
form R reporting required

INTERNATIONAL LISTS

Australian Exposure Standards - Time Weighted Averages
0.84 ppm TWA; 10 mg/m3 TWA

Canada - Alberta - 8 Hour Occupational Exposure Limit
0.8 ppm TWA; 10 mg/m3 TWA

Canada - Alberta - 15 Minute Occupational Exposure Limit
1.3 ppm STEL; 15 mg/m3 STEL

Canada - Ontario - OHSA - TWAEVs
0.8 ppm TWAEV; 9 mg/m3 TWAEV

Canada - Quebec - Time-Weighted Average Exposure Values
0.84 ppm TWAEV; 10 mg/m3 TWAEV

United Kingdom - Occupational Exposure Standards - TWAs
10 mg/m3 TWA

United Kingdom - Occupational Exposure Standards - STELs
15 mg/m3 STEL

Israel - Time Weighted Averages
0.84 ppm TWA; 10 mg/m3 TWA

Israel - Action Levels
0.42 ppm AL; 5 mg/m3 AL

Mexico - Instruction No. 10 - TWAs
as total dust: 0.8 ppm TWA; 10 mg/m3 TWA

Mexico - Instruction No. 10 - STELs
as total dust: 1.3 ppm STEL; 15 mg/m3 STEL

STATE LISTS

California - Prop. 65 - Developmental Toxicity
developmental toxicity - initial date 7/1/91

California - Prop. 65 - Reproductive - Male
male reproductive toxicity - initial date 7/1/91

California - Exposure Limits - PELs
total dust: 10 mg/m3 PEL; respirable fraction: 5 mg/m3 PEL

California - Directors List of Hazardous Substances (8 CCR 339)
[present]

Florida Hazardous Substance List
[present]

Massachusetts Right To Know List
neurotoxin

Minnesota Hazardous Substance List
[present]

NJ Right to Know List (Total)
sn 0192

Pennsylvania Right to Know List
[present]

BENQUINOX   495-73-8

HEALTH AND SAFETY LISTS

U.S. DOT - Appendix B - Marine Pollutants
DOT regulated marine pollutant

BENTAZON   25057-89-0

STATE LISTS

California - Directors List of Hazardous Substances (8 CCR 339)
[present]

PROPOSED REGULATIONS

Safe Drinking Water Act - Priority list
[present]

BENTONITE   1302-78-9

ENVIRONMENTAL LISTS

List of Pesticide Product Inert Ingredients
[present]

BENTONITE, ACID-LEACHED   70131-50-9

ENVIRONMENTAL LISTS

List of Pesticide Product Inert Ingredients
[present]

BENZ[A]ACRIDINE   225-11-6

HEALTH AND SAFETY LISTS

IARC - Group 3 (not classifiable)
[present]

BENZ(A)ANTHRACENE   56-55-3
SEE ALSO:
K001-HAZARDOUS WASTES
F039-HAZARDOUS WASTES
K143-HAZARDOUS WASTES
K035-HAZARDOUS WASTES
K147-HAZARDOUS WASTES
K148-HAZARDOUS WASTES
K144-HAZARDOUS WASTES
K141-HAZARDOUS WASTES
K145-HAZARDOUS WASTES
K142-HAZARDOUS WASTES

HEALTH AND SAFETY LISTS
ACGIH 1995 - Carcinogens
A2-suspected human carcinogen

U.S. DOT - Appendix A Table 1 - Hazardous Substances
final RQ = 10 pounds (4.54 kg)

NTP Seventh Report - Suspect Carcinogens
suspect carcinogen (Listed under 'Polycyclic aromatic hydrocarbons')

OSHA - Possible Select Carcinogens
[present]

ENVIRONMENTAL LISTS
ATSDR Priority List
Rank (of 275): 035

CERCLA/SARA - Section 313 - Emission Reporting
form R reporting required; (Listed under 'Polycyclic aromatic compounds')

CERCLA/SARA - Hazardous Substances and their Reportable Quantities
final RQ = 10 pounds (4.54 kg)

Clean Water Act - Priority Pollutants
[present]

EPA - Carcinogen Hazard Ranking for RQ Adjustment
Hazard ranking = Medium

RCRA - U Series Wastes
waste number U018

RCRA - Hazardous Constituents-Appendix VIII
waste number U018

RCRA - Basis for Listing - Appendix VII
Included in waste streams: F039, K001, K035, K141, K142, K143, K144, K145, K147, K148

RCRA - Substances Banned From Land Disposal
[present]

RCRA - TSD Facilities Ground Water Monitoring
TM 8100 = 200 ug/L PQL; TM 8270 = 10 ug/L PQL

RCRA - Universal Treatment Standards (LDR)
WW: 0.059 mg/l; NWW: 3.4mg/kg

INTERNATIONAL LISTS
Canada - WHMIS: Ingredient Disclosure
0.1% item 151 (281)

German (DFG) - Carcinogens
animal evidence of carcinogenicity

Mexico - Wastewater - Organic Toxic Pollutants and Heavy Metals
Listed under [Aromatic Hydrocarbons]

STATE LISTS
California - Air Bill 2588 Appendix A-I
known or potential carcinogen

California - Prop. 65 - Cancer list
carcinogen - initial date 7/1/87

California - Directors List of Hazardous Substances (8 CCR 339)
[present]

Florida Hazardous Substance List
[present]

Massachusetts Right To Know List
carcinogen; extraordinarily hazardous

Minnesota Hazardous Substance List
carcinogen

NJ Special Hazardous Substances
(carcinogen; mutagen)

Pennsylvania Right to Know List
environmental hazard;   special hazardous substance

Pennsylvania RTK - Special Hazardous Substances
[present]

BENZ(C)ACRIDINE   225-51-4
HEALTH AND SAFETY LISTS
U.S. DOT - Appendix A Table 1 - Hazardous Substances
final RQ = 100 pounds (45.4 kg)

IARC - Group 3 (not classifiable)
[present]

ENVIRONMENTAL LISTS
CERCLA/SARA - Hazardous Substances and their Reportable Quantities
final RQ = 100 pounds (45.4 kg)

EPA - Carcinogen Hazard Ranking for RQ Adjustment
Hazard ranking = Low

RCRA - U Series Wastes
waste number U016

RCRA - Hazardous Constituents-Appendix VIII
waste number U016

RCRA - Substances Banned From Land Disposal
[present]

STATE LISTS
California - Directors List of Hazardous Substances (8 CCR 339)
[present]

Massachusetts Right To Know List
[present]

Pennsylvania Right to Know List
environmental hazard

BENZ(C)ACRIDINE, 7,8,9,11-TETRAMETHYL-   51787-44-1
STATE LISTS
Florida Hazardous Substance List
[present]

Massachusetts Right To Know List
carcinogen; extraordinarily hazardous

Pennsylvania Right to Know List
environmental hazard

BENZAL CHLORIDE   98-87-3
HEALTH AND SAFETY LISTS
U.S. DOT - Substances From 49 CFR 172.101
regulated by DOT (UN1886)

U.S. DOT - Hazard Classes
DOT hazard class = 6.1

U.S. DOT - Appendix A Table 1 - Hazardous Substances
final RQ = 5000 pounds (2270 kg)

IARC - Group Unspecified
[present] (Listed under 'alpha-chlorinated toluenes')

NIOSH - Selected LD50s and LC50s
Inhalation, rat: LC50 = 61 ppm 2 hr Oral, rat: LD50 = 3249 mg/kg

ENVIRONMENTAL LISTS
CERCLA/SARA - Section 302 Extremely Hazardous Substances and TPQs
TPQ = 500 pounds

CERCLA/SARA - Section 313 - Emission Reporting
form R reporting required for 1.0% de minimus concentration

CERCLA/SARA - Hazardous Substances and their Reportable Quantities
final RQ = 5000 pounds (2270 kg)

CAA - HON Rule - SOCMI Chemicals
compliance by April 24, 1995

RCRA - U Series Wastes
waste number U017

RCRA - Hazardous Constituents-Appendix VIII
waste number U017

RCRA - Substances Banned From Land Disposal
[present]

RCRA - Universal Treatment Standards (LDR)
WW: 0.055 mg/l; NWW: 6.0 mg/kg

TSCA - Code of Federal Regulations Citations
40 CFR 712.30(d)

TSCA - PAIR - Reporting List
Reporting Date: November 19, 1982

INTERNATIONAL LISTS

Canada - WHMIS: Ingredient Disclosure
1% item 173 (681)

German (DFG) - Carcinogens
proven carcinogen (Listed under 'alpha-Chlorinated toluenes')

STATE LISTS

California - Air Bill 2588 Appendix A-II
6/91

Florida Hazardous Substance List
effective March 13, 1992

Massachusetts Right To Know List
extraordinarily hazardous

NJ Right to Know List (Total)
sn 0195

Pennsylvania Right to Know List
environmental hazard

BENZALDEHYDE   100-52-7

HEALTH AND SAFETY LISTS

AIHA - WEEL - Time Weighted Averages
2 ppm TWA; 8.7 mg/m3 TWA

AIHA - WEEL - Ceilings or Short Term Time Weighted Averages
4 ppm STEL

NFPA - Flash Points
flash point = 145 degrees F (63 degrees C)

NFPA - Hazard Identification Ratings
health-2; flammability-2; reactivity-0

NIOSH - Selected LD50s and LC50s
Oral, rat: LD50 = 1300 mg/kg

NTP Chemical Status Reports - Testing Status and NTIS Number
Technical reports printed (PB902538782/AS)

NTP Chemical Status Reports - Evidence of Carcinogenicity
male rat-no evidence; female rat-no evidence; male mice-some evidence; female mice-some evidence

ENVIRONMENTAL LISTS

CAA - HON Rule - SOCMI Chemicals
compliance by April 24, 1995

EPA - Master Testing List
[present]

List of Pesticide Product Inert Ingredients
[present]

TSCA - Code of Federal Regulations Citations
40 CFR 712.30(x); 40 CFR 716.120(d)

TSCA - PAIR - Reporting List
Reporting Date: November 27, 1991

TSCA - Health and Safety Reporting List
Effective Date: September 30, 1991

INTERNATIONAL LISTS

Canada - WHMIS: Ingredient Disclosure
1% item 150 (275)

STATE LISTS

Florida Hazardous Substance List
[present]

Massachusetts Right To Know List
[present]

Minnesota Hazardous Substance List
[present]

NJ Right to Know List (Total)
sn 0196

Pennsylvania Right to Know List
[present]

BENZALDEHYDE, 3-BROMO-   3132-99-8

ENVIRONMENTAL LISTS

TSCA - Code of Federal Regulations Citations
40 CFR 712.30(x); 40 CFR 716.120(d)

TSCA - PAIR - Reporting List
Reporting Date: November 27, 1991

TSCA - Health and Safety Reporting List
Effective Date: September 30, 1991

BENZALDEHYDE, 4-BUTYL-   1200-14-2

ENVIRONMENTAL LISTS

TSCA - Code of Federal Regulations Citations
40 CFR 712.30(x); 40 CFR 716.120(d)

TSCA - PAIR - Reporting List
Reporting Date: November 27, 1991

TSCA - Health and Safety Reporting List
Effective Date: September 30, 1991

BENZALDEHYDE, 2-CHLORO-   89-98-5

ENVIRONMENTAL LISTS

TSCA - Code of Federal Regulations Citations
40 CFR 712.30(x); 40 CFR 716.120(d)

TSCA - PAIR - Reporting List
Reporting Date: November 27, 1991

TSCA - Health and Safety Reporting List
Effective Date: September 30, 1991

BENZALDEHYDE, 4-(DIETHYLAMINO)-   120-21-8

ENVIRONMENTAL LISTS

TSCA - Code of Federal Regulations Citations
40 CFR 712.30(x); 40 CFR 716.120(d)

TSCA - PAIR - Reporting List
Reporting Date: November 27, 1991

TSCA - Health and Safety Reporting List
Effective Date: September 30, 1991

BENZALDEHYDE, 4-(DIETHYLAMINO)-2-HYDROXY-   17754-90-4

ENVIRONMENTAL LISTS

TSCA - Code of Federal Regulations Citations
40 CFR 712.30(x); 40 CFR 716.120(d)

TSCA - PAIR - Reporting List
Reporting Date: November 27, 1991

TSCA - Health and Safety Reporting List
Effective Date: September 30, 1991

BENZALDEHYDE, 2,4-DIHYDROXY-   95-01-2

ENVIRONMENTAL LISTS

    TSCA - Code of Federal Regulations Citations
    40 CFR 712.30(x); 40 CFR 716.120(d)

    TSCA - PAIR - Reporting List
    Reporting Date: November 27, 1991

    TSCA - Health and Safety Reporting List
    Effective Date: September 30, 1991

BENZALDEHYDE, 4-(DIMETHYLAMINO)-   100-10-7

ENVIRONMENTAL LISTS

    TSCA - Code of Federal Regulations Citations
    40 CFR 712.30(x); 40 CFR 716.120(d)

    TSCA - PAIR - Reporting List
    Reporting Date: November 27, 1991

    TSCA - Health and Safety Reporting List
    Effective Date: September 30, 1991

BENZALDEHYDE, 2,5-DIMETHOXY-   93-02-7

ENVIRONMENTAL LISTS

    TSCA - Code of Federal Regulations Citations
    40 CFR 712.30(x); 40 CFR 716.120(d)

    TSCA - PAIR - Reporting List
    Reporting Date: November 27, 1991

    TSCA - Health and Safety Reporting List
    Effective Date: September 30, 1991

BENZALDEHYDE, (DIMETHYLAMINO)-   28602-27-9

ENVIRONMENTAL LISTS

    TSCA - Code of Federal Regulations Citations
    40 CFR 712.30(x); 40 CFR 716.120(d)

    TSCA - PAIR - Reporting List
    Reporting Date: November 27, 1991

    TSCA - Health and Safety Reporting List
    Effective Date: September 30, 1991

BENZALDEHYDE, 4-(1,1-DIMETHYLETHYL)-   939-97-9

ENVIRONMENTAL LISTS

    TSCA - Code of Federal Regulations Citations
    40 CFR 704.33; 40 CFR 716.120(a); 40 CFR 712.30(x)

    TSCA - PAIR - Reporting List
    Reporting Date: November 27, 1991

    TSCA - Health and Safety Reporting List
    Effective Date: June 25, 1986; Sunset Date: November 9, 1993

BENZALDEHYDE, 4-ETHOXY-   10031-82-0

ENVIRONMENTAL LISTS

    TSCA - Code of Federal Regulations Citations
    40 CFR 712.30(x); 40 CFR 716.120(d)

    TSCA - PAIR - Reporting List
    Reporting Date: November 27, 1991

    TSCA - Health and Safety Reporting List
    Effective Date: September 30, 1991; Sunset Date: November 9, 1993

BENZALDEHYDE, 3-ETHOXY-4-HYDROXY-   121-32-4

ENVIRONMENTAL LISTS

    TSCA - Code of Federal Regulations Citations
    40 CFR 712.30(x); 40 CFR 716.120(d)

    TSCA - PAIR - Reporting List
    Reporting Date: November 27, 1991

    TSCA - Health and Safety Reporting List
    Effective Date: September 30, 1991

BENZALDEHYDE, 4-HYDROXY-   123-08-

ENVIRONMENTAL LISTS

    TSCA - Code of Federal Regulations Citations
    40 CFR 712.30(x); 40 CFR 716.120(d)

    TSCA - PAIR - Reporting List
    Reporting Date: November 27, 1991

    TSCA - Health and Safety Reporting List
    Effective Date: September 30, 1991

BENZALDEHYDE, 2-HYDROXY-5-NITRO-   97-51-

ENVIRONMENTAL LISTS

    TSCA - Code of Federal Regulations Citations
    40 CFR 712.30(x); 40 CFR 716.120(d)

    TSCA - PAIR - Reporting List
    Reporting Date: November 27, 1991

    TSCA - Health and Safety Reporting List
    Effective Date: September 30, 1991

BENZALDEHYDE, 4-METHYL-   104-87-

ENVIRONMENTAL LISTS

    TSCA - Code of Federal Regulations Citations
    40 CFR 712.30(x); 40 CFR 716.120(d)

    TSCA - PAIR - Reporting List
    Reporting Date: November 27, 1991

    TSCA - Health and Safety Reporting List
    Effective Date: September 30, 1991

BENZALDEHYDE, 2-NITRO-   552-89-6

HEALTH AND SAFETY LISTS

    NIOSH - Selected LD50s and LC50s
    Oral, mouse: LD50 = 600 mg/kg

ENVIRONMENTAL LISTS

    TSCA - Code of Federal Regulations Citations
    40 CFR 712.30(x); 40 CFR 716.120(d)

    TSCA - PAIR - Reporting List
    Reporting Date: November 27, 1991

    TSCA - Health and Safety Reporting List
    Effective Date: September 30, 1991

BENZALDEHYDE, 3-PHENOXY-   39515-51-0

ENVIRONMENTAL LISTS

    TSCA - Code of Federal Regulations Citations
    40 CFR 712.30(x); 40 CFR 716.120(d)

    TSCA - PAIR - Reporting List
    Reporting Date: November 27, 1991

    TSCA - Health and Safety Reporting List
    Effective Date: September 30, 1991

BENZALDEHYDE, 4-(TRIFLUOROMETHYL)-   455-19-6

ENVIRONMENTAL LISTS

    TSCA - Code of Federal Regulations Citations
    40 CFR 712.30(x); 40 CFR 716.120(d)

    TSCA - PAIR - Reporting List
    Reporting Date: November 27, 1991

    TSCA - Health and Safety Reporting List
    Effective Date: September 30, 1991

BENZAMIDE   55-21-0

HEALTH AND SAFETY LISTS

    NIOSH - Selected LD50s and LC50s
    Oral, mouse: LD50 = 1160 mg/kg

ENVIRONMENTAL LISTS

CERCLA/SARA - Section 313 - Emission Reporting
form R reporting required for 1.0% de minimus concentration

STATE LISTS

California - Air Bill 2588 Appendix A-II
6/91

Massachusetts Right To Know List
[present]

NJ Right to Know List (Total)
sn 2895

Pennsylvania Right to Know List
environmental hazard

BENZANTHRONE   82-05-3

INTERNATIONAL LISTS

Canada - WHMIS: Ingredient Disclosure
1% item 152 (276)

BENZENAMINE, 2-[(4-AMINOPHENYL)METHYL]-   1208-52-2

ENVIRONMENTAL LISTS

TSCA - Code of Federal Regulations Citations
40 CFR 716.120(a)

TSCA - Health and Safety Reporting List
Effective Date: June 1, 1987; Sunset Date: November 9, 1993

BENZENAMINE, 4-BROMO-   106-40-1
SEE ALSO:
ANILINE AND CHLORO-, BROMO-, AND/OR NITROANILINES

HEALTH AND SAFETY LISTS

NIOSH - Selected LD50s and LC50s
Oral, rat: LD50 = 456 mg/kg

ENVIRONMENTAL LISTS

TSCA - Code of Federal Regulations Citations
40 CFR 712.30(d); 40 CFR 716.120(c)

TSCA - PAIR - Reporting List
Reporting Date: November 19, 1982

BENZENAMINE, 2-BROMO-6-CHLORO-4-NITRO-   99-29-6
SEE ALSO:
ANILINE AND CHLORO-, BROMO-, AND/OR NITROANILINES

ENVIRONMENTAL LISTS

TSCA - Code of Federal Regulations Citations
40 CFR 716.120(c)

BENZENAMINE, 2-BROMO-4,6-DINITRO-   1817-73-8
SEE ALSO:
ANILINE AND CHLORO-, BROMO-, AND/OR NITROANILINES

ENVIRONMENTAL LISTS

TSCA - Code of Federal Regulations Citations
40 CFR 712.30(d); 40 CFR 716.120(c)

TSCA - PAIR - Reporting List
Reporting Date: November 19, 1982

BENZENAMINE, 2-CHLORO-4,6-DINITRO-   3531-19-9
SEE ALSO:
ANILINE AND CHLORO-, BROMO-, AND/OR NITROANILINES

ENVIRONMENTAL LISTS

TSCA - Code of Federal Regulations Citations
40 CFR 716.120(c)

BENZENAMINE, 4-CHLORO-2,6-DINITRO-   5388-62-5
SEE ALSO:
ANILINE AND CHLORO-, BROMO-, AND/OR NITROANILINES

ENVIRONMENTAL LISTS

TSCA - Code of Federal Regulations Citations
40 CFR 712.30(d); 40 CFR 716.120(c)

TSCA - PAIR - Reporting List
Reporting Date: November 19, 1982

BENZENAMINE, N-(2-CHLOROETHYL)-N-ETHYL-   92-49-9

HEALTH AND SAFETY LISTS

NIOSH - Selected LD50s and LC50s
Oral, rat: LD50 = 616 mg/kg Skin, rabbit: LD50 = 200 mg/kg

ENVIRONMENTAL LISTS

TSCA - Code of Federal Regulations Citations
40 CFR 712.30(d)

TSCA - PAIR - Reporting List
Reporting Date: November 19, 1982

BENZENAMINE, 3-CHLORO-, HYDROCHLORIDE   141-85-5
SEE ALSO:
ANILINE AND CHLORO-, BROMO-, AND/OR NITROANILINES

ENVIRONMENTAL LISTS

TSCA - Code of Federal Regulations Citations
40 CFR 712.30(d); 40 CFR 716.120(c)

TSCA - PAIR - Reporting List
Reporting Date: November 19, 1982

BENZENAMINE, 4-CHLORO-2-METHYL-,   3165-93-3
HYDROCHLORIDE

HEALTH AND SAFETY LISTS

U.S. DOT - Substances From 49 CFR 172.101
regulated by DOT (UN1579)

U.S. DOT - Hazard Classes
DOT hazard class = 6.1

U.S. DOT - Appendix A Table 1 - Hazardous Substances
final RQ = 100 pounds (45.4 kg)

NTP Chemical Status Reports - Testing Status and NTIS Number
Technical reports printed (PB295864/AS)

NTP Chemical Status Reports - Evidence of Carcinogenicity
male rat-negative; female rat-negative; male mice-positive; female mice-positive

ENVIRONMENTAL LISTS

CERCLA/SARA - Hazardous Substances and their Reportable Quantities
final RQ = 100 pounds (45.4 kg)

EPA - Carcinogen Hazard Ranking for RQ Adjustment
Hazard ranking = Low

RCRA - U Series Wastes
waste number U049

RCRA - Hazardous Constituents-Appendix VIII
waste number U049

RCRA - Substances Banned From Land Disposal
[present]

TSCA - Code of Federal Regulations Citations
40 CFR 721.462

TSCA - Chemicals with Significant New Use Rules
[present]

TSCA - Section 12(b) - Export Notification
export notification required - Section 5

INTERNATIONAL LISTS

Canada - WHMIS: Ingredient Disclosure
1% item 389 (466)

STATE LISTS

Massachusetts Right To Know List
carcinogen; extraordinarily hazardous

NJ Right to Know List (Total)
sn 0397

Pennsylvania Right to Know List
environmental hazard

BENZENAMINE, 4-CHLORO-2-NITRO-   89-63-4
SEE ALSO:
ANILINE AND CHLORO-, BROMO-, AND/OR NITROANILINES
HEALTH AND SAFETY LISTS
NTP Chemical Status Reports - Testing Status and NTIS Number
Prechronic studies for which toxicity technical reports were not prepared
ENVIRONMENTAL LISTS
TSCA - Code of Federal Regulations Citations
40 CFR 712.30(d); 40 CFR 716.120(c)
TSCA - PAIR - Reporting List
Reporting Date: November 19, 1982

BENZENAMINE, 2-CHLORO-4-NITRO-   121-87-9
SEE ALSO:
ANILINE AND CHLORO-, BROMO-, AND/OR NITROANILINES
ENVIRONMENTAL LISTS
TSCA - Code of Federal Regulations Citations
40 CFR 712.30(d); 40 CFR 716.120(c)
TSCA - PAIR - Reporting List
Reporting Date: November 19, 1982

BENZENAMINE, 4-CHLORO-3-NITRO-   635-22-3
SEE ALSO:
ANILINE AND CHLORO-, BROMO-, AND/OR NITROANILINES
ENVIRONMENTAL LISTS
TSCA - Code of Federal Regulations Citations
40 CFR 712.30(d); 40 CFR 716.120(c)
TSCA - PAIR - Reporting List
Reporting Date: November 19, 1982

BENZENAMINE, 2-CHLORO-5-NITRO-   6283-25-6
SEE ALSO:
ANILINE AND CHLORO-, BROMO-, AND/OR NITROANILINES
ENVIRONMENTAL LISTS
TSCA - Code of Federal Regulations Citations
40 CFR 712.30(d); 40 CFR 716.120
TSCA - PAIR - Reporting List
Reporting Date: November 19, 1982

BENZENAMINE, 2,6-DIBROMO-4-NITRO-   827-94-1
SEE ALSO:
ANILINE AND CHLORO-, BROMO-, AND/OR NITROANILINES
ENVIRONMENTAL LISTS
TSCA - Code of Federal Regulations Citations
40 CFR 712.30(d); 40 CFR 716.120(c)
TSCA - PAIR - Reporting List
Reporting Date: November 19, 1982
TSCA - HDD/HDF - Precursors Required for Reporting
[present]

BENZENAMINE, 2,5-DIBUTOXY-4-(4-MORPHOLINYL)-, SULFATE   130169-66-3
ENVIRONMENTAL LISTS
TSCA - Chemicals with Significant New Use Rules
PMN number: P-90-1809
TSCA - Section 12(b) - Export Notification
P-90-1809; export notification required - Section 5

BENZENAMINE, N,N-DIBUTYL-   613-29-6
HEALTH AND SAFETY LISTS
NFPA - Flash Points
flash point = 230 degrees F (110 degrees C)
NFPA - Hazard Identification Ratings
health-3; flammability-1; reactivity-0
STATE LISTS
Florida Hazardous Substance List
[present]
Massachusetts Right To Know List
[present]
Pennsylvania Right to Know List
[present]

BENZENAMINE, 2,5-DICHLORO-   95-82-9
SEE ALSO:
ANILINE AND CHLORO-, BROMO-, AND/OR NITROANILINES
HEALTH AND SAFETY LISTS
NIOSH - Selected LD50s and LC50s
Oral, rat: LD50 = 2900 mg/kg
ENVIRONMENTAL LISTS
TSCA - Code of Federal Regulations Citations
40 CFR 712.30(d); 40 CFR 716.120(c)
TSCA - PAIR - Reporting List
Reporting Date: November 19, 1982

BENZENAMINE, 2,4-DICHLORO-   554-00-7
SEE ALSO:
ANILINE AND CHLORO-, BROMO-, AND/OR NITROANILINES
ENVIRONMENTAL LISTS
TSCA - Code of Federal Regulations Citations
40 CFR 712.30(d); 40 CFR 716.120(c)
TSCA - PAIR - Reporting List
Reporting Date: November 19, 1982

BENZENAMINE, 2,3-DICHLORO-   608-27-5
SEE ALSO:
ANILINE AND CHLORO-, BROMO-, AND/OR NITROANILINES
ENVIRONMENTAL LISTS
TSCA - Code of Federal Regulations Citations
40 CFR 712.30(d); 40 CFR 716.120(c)
TSCA - PAIR - Reporting List
Reporting Date: November 19, 1982

BENZENAMINE, 3,5-DICHLORO-   626-43-7
SEE ALSO:
ANILINE AND CHLORO-, BROMO-, AND/OR NITROANILINES
ENVIRONMENTAL LISTS
TSCA - Code of Federal Regulations Citations
40 CFR 712.30(d); 40 CFR 716.120(c)
TSCA - PAIR - Reporting List
Reporting Date: November 19, 1982

BENZENAMINE, 2,3-DIMETHYL-   87-59-2
HEALTH AND SAFETY LISTS
NIOSH - Selected LD50s and LC50s
Oral, rat: LD50 = 933 mg/kg
STATE LISTS
Pennsylvania Right to Know List
[present]

BENZENAMINE, 4-ETHOXY-N-[(5-NITRO-2-FURANYL)METHYLENE]-   13410-72-5
STATE LISTS

Pennsylvania Right to Know List
special hazardous substance

Pennsylvania RTK - Special Hazardous Substances
[present]

BENZENAMINE, N-(2-ETHYLHEXYL)-   10137-80-1

HEALTH AND SAFETY LISTS

NFPA - Flash Points
flash point = 325 degrees F (163 degrees C)

NFPA - Hazard Identification Ratings
health-3; flammability-1; reactivity-0

STATE LISTS

Florida Hazardous Substance List
[present]

Massachusetts Right To Know List
[present]

Pennsylvania Right to Know List
[present]

BENZENAMINE, 4-(1-METHYLBUTOXY)-, HYDROCHLORIDE   RR-01237-3

ENVIRONMENTAL LISTS

TSCA - Chemicals with Significant New Use Rules
PMN number: P-90-559

BENZENAMINE, 4,4'-[1,4-PHENYLENEBIS[1-METHYLETHYLIDENE]BIS[2,6-DIMETHYL-,   2716-10-1

ENVIRONMENTAL LISTS

TSCA - Chemicals with Significant New Use Rules
PMN number: P-86-503

TSCA - Section 12(b) - Export Notification
P-86-503; export notification required - Section 5

BENZENAMINE, 4,4'-[1,3-PHENYLENEBIS[1-METHYLETHYLIDENE]BIS[2,6-DIMETHYL-,   2716-12-3

ENVIRONMENTAL LISTS

TSCA - Chemicals with Significant New Use Rules
PMN number: P-86-501

TSCA - Section 12(b) - Export Notification
P-86-501; export notification required - Section 5

BENZENAMINE, 2,4,6-TRIBROMO-   147-82-0

SEE ALSO:
ANILINE AND CHLORO-, BROMO-, AND/OR NITROANILINES

ENVIRONMENTAL LISTS

TSCA - Code of Federal Regulations Citations
40 CFR 712.30(d); 40 CFR 716.120(c)

TSCA - PAIR - Reporting List
Reporting Date: November 19, 1982

BENZENAMINE, 2,4,6-TRICHLORO-   634-93-5

SEE ALSO:
ANILINE AND CHLORO-, BROMO-, AND/OR NITROANILINES

ENVIRONMENTAL LISTS

CAA - HON Rule - SOCMI Chemicals
compliance by April 24, 1995

TSCA - Code of Federal Regulations Citations
40 CFR 712.30(d); 40 CFR 716.120(c)

TSCA - PAIR - Reporting List
Reporting Date: November 19, 1982

BENZENAMINE, 3-(TRIFLUOROMETHYL)-   98-16-8

HEALTH AND SAFETY LISTS

U.S. DOT - Substances From 49 CFR 172.101
regulated by DOT (UN2948)

U.S. DOT - Hazard Classes
DOT hazard class = 6.1

NIOSH - Selected LD50s and LC50s
Inhalation, rat: LC50 = 440 mg/m3 4 hr Oral, rat: LD50 = 480 mg/kg

ENVIRONMENTAL LISTS

CERCLA/SARA - Section 302 Extremely Hazardous Substances and TPQs
TPQ = 500 pounds

INTERNATIONAL LISTS

Canada - WHMIS: Ingredient Disclosure
1% item 1634 (1672)

STATE LISTS

Florida Hazardous Substance List
effective March 13, 1992

Massachusetts Right To Know List
extraordinarily hazardous

NJ Right to Know List (Total)
sn 1916

Pennsylvania Right to Know List
environmental hazard

PROPOSED REGULATIONS

CERCLA/SARA - Proposed Hazardous Substance Additions
proposed RQ = 1 pound (.454 kg)

CERCLA/SARA - 1989 Proposed RQ Adjustments
proposed RQ = 100 pounds (45.4 kg)

BENZENE   71-43-2

SEE ALSO:
K085-HAZARDOUS WASTES
F024-HAZARDOUS WASTES
F005-HAZARDOUS WASTES
F039-HAZARDOUS WASTES
K145-HAZARDOUS WASTES
K147-HAZARDOUS WASTES
K151-HAZARDOUS WASTES
K142-HAZARDOUS WASTES
K143-HAZARDOUS WASTES
K144-HAZARDOUS WASTES
BIS(2,4-DIMETHYLBUTYL) MALEATE
K105-HAZARDOUS WASTES
K104-HAZARDOUS WASTES
GLUCOSE PENTAPROPIONATE
K141-HAZARDOUS WASTES
F025-HAZARDOUS WASTES

HEALTH AND SAFETY LISTS

ACGIH 1995 - Time Weighted Averages
(10) ppm TWA; (32) mg/m3 TWA

ACGIH 1995 - Carcinogens
(A2)-suspected human carcinogen

ACGIH 1995 - Biological Exposure Indices
Total phenol in urine: 50 mg/g creatinine, end of shift (B, Ns); Benzene in exhaled air: 0.08 ppm (mixed exhaled), 0.12 ppm (end-exhaled), prior to next shift (Sq)

AIHA - Odor Threshold Values
geometric mean air odor threshold = 61 ppm (detectable); 97 ppm (recognizable)

U.S. DOT - Substances From 49 CFR 172.101
regulated by DOT (UN1114)

U.S. DOT - Hazard Classes
DOT hazard class = 3

U.S. DOT - Appendix A Table 1 - Hazardous Substances
final RQ = 10 pounds (4.54 kg)

IARC - Group 1 (carcinogenic to humans)
[present]

NFPA - Flash Points
flash point = 12 degrees F (-11 degrees C); Benzol diluent: -25 degrees F (-32 degrees C)

NFPA - Hazard Identification Ratings
health-2; flammability-3; reactivity 0

NIOSH - Selected LD50s and LC50s
Inhalation, rat: LC50 = 10000 ppm 7 hr Oral, rat: LD50 = 3306 mg/kg Skin, mouse: LD50 = 48 mg/kg

NIOSH 1990 - Pocket Guide - RELs
0.1 ppm TWA; 1 ppm STEL

NIOSH 1990 - Pocket Guide - IDLHs
3000 ppm IDLH (not considering carcinogenic effects)

NIOSH 1990 - Pocket Guide - Carcinogens
occupational carcinogen

NIOSH 1990 - Pocket Guide - Target organs
blood, CNS, skin, bone marrow, eyes, respiratory system

NIOSH - Health Standards - Exposure Limits
0.1 ppm TWA (8 hr); 0.32 mg/m3 TWA (8 hr); C (15 min) 1 ppm; C (15 min) 3.2 mg/m3

NIOSH - Health Standards - Health Effects and Precautions
Leukemia (Prevent skin contact)

NIOSH - Health Standards - Carcinogenic Chemicals
potential human carcinogen

NTP Chemical Status Reports - Testing Status and NTIS Number
Technical reports printed (PB86216967/AS)

NTP Chemical Status Reports - Evidence of Carcinogenicity
male rat-clear evidence; female rat-clear evidence; male mice-clear evidence; female mice-clear evidence

NTP Seventh Report - Known Carcinogens
known carcinogen

OSHA - 29 CFR 1910 Specifically Regulated Chemicals
1 ppm TWA; 5 ppm STEL; 0.5 ppm TWA action limit; Cancer hazard; Flammable (see 29 CFR 1910.1028)

OSHA - Vacated PELs - Time Weighted Averages
10 ppm TWA (unless specified in 1910.1028)

OSHA - Vacated PELs - Short Term Exposure Limits
50 ppm STEL (10 min) (unless specified in 1910.1028)

OSHA - Vacated PELs - Ceiling Limits
C 25 ppm (unless specified in 1910.1028)

OSHA - Final PELs - Time Weighted Averages
10 ppm TWA (apply only to exempt industry segments)

OSHA - Select Carcinogens
[present]

ENVIRONMENTAL LISTS

ATSDR Priority List
Rank (of 275): 004

CERCLA/SARA - Section 313 - Emission Reporting
form R reporting required for 0.1% de minimus concentration

CERCLA/SARA - Hazardous Substances and their Reportable Quantities
final RQ = 10 pounds (4.54 kg)

Clean Air Act (1990) - List of Hazardous Air Contaminants
[present] (includes Benzene from gasoline)

CAA - HON Rule - SOCMI Chemicals
compliance by Oct. 24, 1994

CAA - HON Rule - Organic HAPs
[present]

Clean Water Act - Hazardous Substances
[present]

Clean Water Act - Priority Pollutants
[present]

Clean Water Act - Toxic Pollutants
[present]

Safe Drinking Water Act - MCLs
MCL = 0.005 mg/L

Safe Drinking Water Act - MCLGs
MCLG = Zero

EPA - Carcinogen Hazard Ranking for RQ Adjustment
Hazard ranking = Medium

RCRA - D Series - Maximum Concentration of Contaminants
waste number D018; regulatory level = 0.5 mg/L

RCRA - D Series - Chronic Toxicity Reference Levels
chronic toxicity reference level = 0.005 mg/L

RCRA - U Series Wastes
waste number U019 (Ignitable waste; Toxic waste)

RCRA - Hazardous Constituents-Appendix VIII
waste number U019

RCRA - Basis for Listing - Appendix VII
Included in waste streams: F005, F024, F025, F039, K085, K104, K105, K141, K142, K143, K144, K145, K147, K151

RCRA - Substances Banned From Land Disposal
[present]

RCRA - TSD Facilities Ground Water Monitoring
TM 8020 = 2 ug/L PQL; TM 8240 = 5 ug/L PQL

RCRA - Universal Treatment Standards (LDR)
WW: 0.14 mg/l; NWW: 10 mg/kg

INTERNATIONAL LISTS

Australian Exposure Standards - Time Weighted Averages
5 ppm TWA; 16 mg/m3 TWA

Australian Exposure Standards - Carcinogens
confirmed carcinogen

Australian Exposure Standards - Under Review
exposure limits under review

Canada - WHMIS: Ingredient Disclosure
0.1% item 153 (277)

Canada - NPRI (National Pollutant Release Inventory)
[present]

Canada - CEPA - Priority Substances List
estimated time for completion of assessment reports: 3 years

Canada - Drinking Water Quality - MACs
0.005 mg/L MAC

Canada - Alberta - 8 Hour Occupational Exposure Limit
1.0 ppm TWA; 3.2 mg/m3 TWA

Canada - Alberta - 15 Minute Occupational Exposure Limit
5 ppm STEL; 16 mg/m3 STEL

Canada - Alberta - Skin Designation
can be absorbed through the intact skin

Canada - Alberta - Designated Substances
designated substance - requires code of practice

Canada - British Columbia - 8 Hour Exposure Limits
10 ppm TWA; 30 mg/m3 TWA

Canada - British Columbia - Ceiling Exposure Limits
C 10 ppm; C 30 mg/m3

Canada - British Columbia - Carcinogens
carcinogen - C 10 ppm; C 32 mg/m3

Canada - Ontario - OHSA - TWAEVs
1 ppm TWAEV; 3.2 mg/m3 TWAEV (designated substance regulation); 5 ppm TWAEV; 16 mg/m3 TWAEV (These values apply to work places in which the designated substance regulation does not apply.)

Canada - Ontario - OHSA - STEVs
15 ppm STEV; 48 mg/m3 STEV (designated substance regulations)

Canada - Ontario - OHSA - CEVs
5 ppm CEV; 16 mg/m3 CEV (designated substance regulation)

Canada - Ontario - OHSA - Designated Substances
1 ppm TWAEV; 3.2 mg/m3 TWAEV; See Ontario Reg. 839 for full information. (designated substance regulation for which DSR does not apply: 5 ppm TWAEV; 16 mg/m3 TWAEV)

Canada - Quebec - Time-Weighted Average Exposure Values
1 ppm TWAEV; 3 mg/m3 TWAEV (substance of which the recirculation is prohibited)

Canada - Quebec - Short-term Exposure Values
5 ppm STEV; 15.5 mg/m3 STEV

Canada - Quebec - Carcinogens
C1 carcinogen: effect detected in humans

United Kingdom - Maximum Exposure Limits - TWAs
5 ppm TWA; 16 mg/m3 TWA

German (DFG) - Skin/Sensitizers
danger of cutaneous absorption

German (DFG) - Carcinogens
proven carcinogen

Israel - Time Weighted Averages
1 ppm TWA

Israel - Action Levels
0.5 ppm AL; 16 mg/m3 AL

Mexico - Instruction No. 10 - TWAs
10 ppm TWA; 30 mg/m3 TWA

Mexico - Instruction No. 10 - STELs
25 ppm STEL; 75 mg/m3 STEL

Mexico - Instruction No. 10 - Carcinogens
potential carcinogen in humans - limited epidemiological evidence

Mexico - Wastewater - Organic Toxic Pollutants and Heavy Metals
Listed under [Organic Toxic Pollutants]

Mexico - Drinking Water - Ecological Criteria
0.01 mg/l Substance presents persistence, bioaccumulations or risk of cancer, reduce human exposure to a minimum; This level has been extrapolated by using a mathematic model

STATE LISTS

California - Air Bill 2588 Appendix A-I
known or potential carcinogen

California - Prop. 65 - Cancer list
carcinogen - initial date 2/27/87

California - Prop. 65 - No Significant Risk Levels
no significant risk level = 7 ug/day

California - Exposure Limits - PELs
1 ppm PEL; prevent eye and skin contact; see also section 5218

California - Exposure Limits - STELs
5 ppm STEL

California - Exposure Limits - Skin Notation
material may be absorbed through the skin, eyes or mucous membrane

California - Directors List of Hazardous Substances (8 CCR 339)
[present]

Florida Hazardous Substance List
[present]

Massachusetts Right To Know List
carcinogen; extraordinarily hazardous

Minnesota Hazardous Substance List
carcinogen

NJ Right to Know List (Total)
sn 0197

NJ Special Hazardous Substances
(carcinogen; flammable - third degree; mutagen)

Pennsylvania Right to Know List
environmental hazard; special hazardous substance

Pennsylvania RTK - Special Hazardous Substances
[present]

PROPOSED REGULATIONS

ACGIH 1995 - Notice of Intended Changes
(skin) 0.3 ppm TWA; 0.96 mg/m3 TWA; A1-confirmed human carcinogen

Canada - Ontario - Proposed Occupational TWAEVs
0.5 ppm TWAEV; 1.5 mg/m3 TWAEV

Canada - Ontario - Proposed Occupational STEVs
3 ppm STEV; 9 mg/m3 STEV

BENZENEACETALDEHYDE   122-78-1

HEALTH AND SAFETY LISTS

NFPA - Flash Points
flash point = 160 degrees F (71 degrees C)

NFPA - Hazard Identification Ratings
health-1; flammability-2; reactivity-0

NIOSH - Selected LD50s and LC50s
Oral, rat: LD50 = 1550 mg/kg

ENVIRONMENTAL LISTS

TSCA - Code of Federal Regulations Citations
40 CFR 712.30(x); 40 CFR 716.120(d)

TSCA - PAIR - Reporting List
Reporting Date: November 27, 1991

TSCA - Health and Safety Reporting List
Effective Date: September 30, 1991

BENZENEACETALDEHYDE, 4-METHYL-   99-72-9

ENVIRONMENTAL LISTS

TSCA - Code of Federal Regulations Citations
40 CFR 712.30(x); 40 CFR 716.120(d)

TSCA - PAIR - Reporting List
Reporting Date: November 27, 1991

BENZENEACETALDEHYDE, 4-METHYL-   104-09-6

ENVIRONMENTAL LISTS

TSCA - PAIR - Reporting List
Effective Date: September 30, 1991 Reporting Date: November 27, 1991

TSCA - Health and Safety Reporting List
Effective Date: September 30, 1991; Sunset Date: September 30, 2001

BENZENEACETALDEHYDE, .ALPHA.-METHYL-   93-53-8

ENVIRONMENTAL LISTS

TSCA - Code of Federal Regulations Citations
40 CFR 712.30(x); 40 CFR 716.120(d)

TSCA - PAIR - Reporting List
Reporting Date: November 27, 1991

TSCA - Health and Safety Reporting List
Effective Date: September 30, 1991

BENZENEACETIC ACID, ALPHA-(HYDROX-   6106-46-3
YMETHYL)-, 9-METHYL-3-OXA-9-AZATRICYCLO
[3.3.1.02,4]NON-7-YL ESTER, [7(S)-(1-ALPHA,2-
BETA,4-BETA,5-ALPHA,7-BETA)]-, COMPOUND
WITH METHYL NITRATE (1:1)

ENVIRONMENTAL LISTS

TSCA - Code of Federal Regulations Citations
40 CFR 712.30(d)

TSCA - PAIR - Reporting List
Reporting Date: November 19, 1982

BENZENEACETIC ACID, ALPHA-(HYDROX-   6106-81-6
YMETHYL)-, 9-METHYL-3-OXA-9-AZATRICYCLO
[3.3.1.02,4]NON-7-YL ESTER, N-OXIDE, HYDRO-
BROMI DE, [7(S)-(1-ALPHA,2-BETA,4-BETA,5-ALPHA,
7-BETA)]-

ENVIRONMENTAL LISTS

TSCA - Code of Federal Regulations Citations
40 CFR 712.30(d)

TSCA - PAIR - Reporting List
Reporting Date: November 19, 1982

BENZENEARSONIC ACID   98-05-5
SEE ALSO:
ARSENIC

HEALTH AND SAFETY LISTS
NIOSH - Selected LD50s and LC50s
Oral, mouse: LD50 = 270 ug/kg

ENVIRONMENTAL LISTS
CERCLA/SARA - Section 302 Extremely Hazardous Substances and TPQs
TPQ = 10/10,000 pounds

RCRA - Hazardous Constituents-Appendix VIII
hazardous constituent - no waste number

INTERNATIONAL LISTS
Canada - WHMIS: Ingredient Disclosure
1% item 1266 (125)

STATE LISTS
Florida Hazardous Substance List
effective March 13, 1992

Massachusetts Right To Know List
extraordinarily hazardous

NJ Right to Know List (Total)
sn 2155

Pennsylvania Right to Know List
environmental hazard

PROPOSED REGULATIONS
CERCLA/SARA - Proposed Hazardous Substance Additions
proposed RQ = 1 pound (.454 kg)

CERCLA/SARA - 1989 Proposed RQ Adjustments
proposed RQ = 10 pounds (4.54 kg)

BENZENE, BIS(ISOCYANATOMETHYL)-   25854-16-4
ENVIRONMENTAL LISTS
TSCA - Code of Federal Regulations Citations
40 CFR 712.30(x); 40 CFR 716.120(d)

TSCA - PAIR - Reporting List
Reporting Date: December 27, 1990

TSCA - Health and Safety Reporting List
Effective Date: October 29, 1990; Sunset Date: November 9, 1993

BENZENE, 1,3-BIS(1-ISOCYANATO-1-   2778-42-9
METHYLETHYL-
ENVIRONMENTAL LISTS
TSCA - Code of Federal Regulations Citations
40 CFR 716.120(a)

TSCA - Health and Safety Reporting List
Effective Date: June 1, 1987; Sunset Date: November 9, 1993

BENZENE, BROMODIMETHYL   35884-77-6
HEALTH AND SAFETY LISTS
U.S. DOT - Substances From 49 CFR 172.101
regulated by DOT (UN1701)

U.S. DOT - Hazard Classes
DOT hazard class = 6.1

BENZENE, 1-BROMO-4-ISOCYANATO-   2493-02-9
ENVIRONMENTAL LISTS
TSCA - Code of Federal Regulations Citations
40 CFR 712.30(x); 40 CFR 716.120(d)

TSCA - PAIR - Reporting List
Reporting Date: December 27, 1990

TSCA - Health and Safety Reporting List
Effective Date: October 29, 1990; Sunset Date: November 9, 1993

BENZENE, 1-(BROMOMETHYL)-3-PHENOXY-   51632-16-7
ENVIRONMENTAL LISTS
TSCA - PAIR - Reporting List
Reporting Date: June 10, 1993

TSCA - Health and Safety Reporting List
Effective Date: April 12, 1993

BENZENE, 1-BROMO-4-PHENOXY   101-55-3
SEE ALSO:
F039-HAZARDOUS WASTES

HEALTH AND SAFETY LISTS
U.S. DOT - Appendix A Table 1 - Hazardous Substances
final RQ = 100 pounds (45.4 kg)

ENVIRONMENTAL LISTS
CERCLA/SARA - Hazardous Substances and their Reportable Quantities
final RQ = 100 pounds (45.4 kg)

Clean Water Act - Priority Pollutants
[present]

RCRA - U Series Wastes
waste number U030

RCRA - Hazardous Constituents-Appendix VIII
waste number U030

RCRA - Basis for Listing - Appendix VII
Included in waste stream: F039

RCRA - Substances Banned From Land Disposal
[present]

RCRA - TSD Facilities Ground Water Monitoring
TM 8270 = 10 ug/L PQL

RCRA - Universal Treatment Standards (LDR)
WW: 0.055 mg/l; NWW: 15 mg/kg

TSCA - Chemicals with Significant New Use Rules
[present]

TSCA - Section 12(b) - Export Notification
export notification required - Section 5

INTERNATIONAL LISTS
Mexico - Wastewater - Organic Toxic Pollutants and Heavy Metals
Listed under [Haloethers]

Mexico - Drinking Water - Ecological Criteria
None given.

STATE LISTS
California - Directors List of Hazardous Substances (8 CCR 339)
[present]

Massachusetts Right To Know List
[present]

Pennsylvania Right to Know List
environmental hazard

BENZENE, (2-CHLOROETHOXY)-   622-86-6
ENVIRONMENTAL LISTS
TSCA - Chemicals with Significant New Use Rules
PMN number: P-87-1471

TSCA - Section 12(b) - Export Notification
P-87-1471; export notification required - Section 5

BENZENE, 1-CHLORO-4-ISOCYANATO-   104-12-1
HEALTH AND SAFETY LISTS
U.S. DOT - Substances Which Are Poisonous by Inhalation
liquid hazardous material poisonous by inhalation

ENVIRONMENTAL LISTS
CERCLA/SARA - Section 313 - Emission Reporting
form R reporting required

TSCA - Code of Federal Regulations Citations
40 CFR 712.30(x); 40 CFR 716.120(d)
TSCA - PAIR - Reporting List
Reporting Date: December 27, 1990
TSCA - Health and Safety Reporting List
Effective Date: October 29, 1990

BENZENE, 1-CHLORO-3-ISOCYANATO-   2909-38-8

ENVIRONMENTAL LISTS
TSCA - Code of Federal Regulations Citations
40 CFR 712.30(x); 40 CFR 716.120(d)
TSCA - PAIR - Reporting List
Reporting Date: December 27, 1990
TSCA - Health and Safety Reporting List
Effective Date: October 29, 1990; Sunset Date: November 9, 1993

BENZENE, (CHLOROMETHYL)ETHENYL-   30030-25-2

HEALTH AND SAFETY LISTS
NFPA - Flash Points
flash point = 220 degrees F (104 degrees C)
NFPA - Hazard Identification Ratings
health-2; flammability-1

STATE LISTS
Florida Hazardous Substance List
[present]
Massachusetts Right To Know List
[present]
Pennsylvania Right to Know List
[present]

BENZENE, 1-(CHLOROMETHYL)-4-NITRO-   100-14-1

ENVIRONMENTAL LISTS
CERCLA/SARA - Section 302 Extremely Hazardous Substances and TPQs
TPQ = 500/10,000 pounds

STATE LISTS
Florida Hazardous Substance List
effective March 13, 1992
Massachusetts Right To Know List
extraordinarily hazardous
NJ Right to Know List (Total)
sn 2156
Pennsylvania Right to Know List
environmental hazard

PROPOSED REGULATIONS
CERCLA/SARA - Proposed Hazardous Substance Additions
proposed RQ = 1 pound (.454 kg)
CERCLA/SARA - 1989 Proposed RQ Adjustments
proposed RQ = 100 pounds (45.4 kg)

BENZENE, 1-CHLORO-2-(TRIFLUOROMETHYL)-   88-16-4

HEALTH AND SAFETY LISTS
NFPA - Flash Points
flash point = 138 degrees F (59 degrees C)
NFPA - Hazard Identification Ratings
health-2; flammability-2; reactivity-1

STATE LISTS
Florida Hazardous Substance List
[present]
Massachusetts Right To Know List
[present]
Pennsylvania Right to Know List
environmental hazard

1,4-BENZENEDIAMINE, N,N'-BIS(1,4-DIMETHYLPENTYL)-   3081-14-9

HEALTH AND SAFETY LISTS
NFPA - Flash Points
flash point = 347 degrees F (175 degrees C)
NFPA - Hazard Identification Ratings
health-2; flammability-1; reactivity-0

STATE LISTS
Florida Hazardous Substance List
[present]
Massachusetts Right To Know List
[present]
Pennsylvania Right to Know List
[present]

1,4-BENZENEDIAMINE, N,N'-BIS(1-METHYL-PROPYL)-   101-96-2

HEALTH AND SAFETY LISTS
NFPA - Flash Points
flash point = 270 degrees F (132 degrees C)
NFPA - Hazard Identification Ratings
health-2; flammability-1; reactivity-0

STATE LISTS
Florida Hazardous Substance List
[present]
Massachusetts Right To Know List
[present]
Pennsylvania Right to Know List
[present]

1,2-BENZENEDIAMINE, 4-BUTYL-   3663-23-8
SEE ALSO:
1,2-BENZENEDIAMINE, 4-BUTYL-

ENVIRONMENTAL LISTS
TSCA - Code of Federal Regulations Citations
40 CFR 712.30(f); 40 CFR 716.120(c)
TSCA - PAIR - Reporting List
Reporting Date: August 17, 1983

1,4-BENZENEDIAMINE, 2-CHLORO-, DIHYDROCHLORIDE   615-46-3
SEE ALSO:
1,4-BENZENEDIAMINE, 2-CHLORO-, DIHYDROCHLORIDE

ENVIRONMENTAL LISTS
TSCA - Code of Federal Regulations Citations
40 CFR 712.30(f); 40 CFR 716.120(c)
TSCA - PAIR - Reporting List
Reporting Date: August 17, 1983

1,2-BENZENEDIAMINE, 5-CHLORO-3-NITRO-   42389-30-0
SEE ALSO:
1,2-BENZENEDIAMINE, 5-CHLORO-3-NITRO-

ENVIRONMENTAL LISTS
TSCA - Code of Federal Regulations Citations
40 CFR 712.30(f); 40 CFR 716.120(c)
TSCA - PAIR - Reporting List
Reporting Date: August 17, 1983

1,4-BENZENEDIAMINE, 2-CHLORO-, SULFATE   6219-71-2
SEE ALSO:
1,4-BENZENEDIAMINE, 2-CHLORO-, SULFATE

ENVIRONMENTAL LISTS
TSCA - Code of Federal Regulations Citations
40 CFR 712.30(f); 40 CFR 716.120

TSCA - PAIR - Reporting List
Reporting Date: August 17, 1983

1,3-BENZENEDIAMINE, 4-CHLORO-, SULFATE (1:1)   68239-80-5
SEE ALSO:
1,3-BENZENEDIAMINE, 4-CHLORO-, SULFATE (1:1)

ENVIRONMENTAL LISTS

TSCA - Code of Federal Regulations Citations
40 CFR 712.30(f);   40 CFR 716.120(c)

TSCA - PAIR - Reporting List
Reporting Date: August 17, 1983

1,2-BENZENEDIAMINE, 4-CHLORO-, SULFATE (1:1)   68459-98-3
SEE ALSO:
1,2-BENZENEDIAMINE, 4-CHLORO-, SULFATE (1:1)

ENVIRONMENTAL LISTS

TSCA - Code of Federal Regulations Citations
40 CFR 712.30(f);   40 CFR 716.120(c)

TSCA - PAIR - Reporting List
Reporting Date: August 17, 1983

1,4-BENZENEDIAMINE, 2,5-DICHLORO-   20103-09-7
SEE ALSO:
1,4-BENZENEDIAMINE, 2,5-DICHLORO-

ENVIRONMENTAL LISTS

TSCA - Code of Federal Regulations Citations
40 CFR 712.30(f);   40 CFR 716.120(c)

TSCA - PAIR - Reporting List
Reporting Date: August 17, 1983

1,3-BENZENEDIAMINE, DIHYDROCHLORIDE   541-69-5
SEE ALSO:
1,3-BENZENEDIAMINE, DIHYDROCHLORIDE

ENVIRONMENTAL LISTS

TSCA - Code of Federal Regulations Citations
40 CFR 712.30(f);   40 CFR 716.120(c)

TSCA - PAIR - Reporting List
Reporting Date: August 17, 1983

1,2-BENZENEDIAMINE, DIHYDROCHLORIDE   615-28-1
SEE ALSO:
1,2-BENZENEDIAMINE, DIHYDROCHLORIDE

ENVIRONMENTAL LISTS

CERCLA/SARA - Section 313 - Emission Reporting
form R reporting required

TSCA - Code of Federal Regulations Citations
40 CFR 712.30(f);   40 CFR 716.120(c)

TSCA - PAIR - Reporting List
Reporting Date: August 17, 1983

1,4-BENZENEDIAMINE, N-(1,3-DIMETHYLBUTYL-   793-24-8

ENVIRONMENTAL LISTS

EPA - Master Testing List
[present]

1,3-BENZENEDIAMINE, 4-(1,1-DIMETHYLETHYL)-   RR-00246-0
AR-METHYL

ENVIRONMENTAL LISTS

TSCA - Chemicals with Significant New Use Rules
PMN number: P-85-929

1,4-BENZENEDIAMINE, ETHANEDIOATE (1:1)   62654-17-5
SEE ALSO:
1,4-BENZENEDIAMINE, ETHANEDIOATE (1:1)

ENVIRONMENTAL LISTS

TSCA - Code of Federal Regulations Citations
40 CFR 712.30(f);   40 CFR 716.120

TSCA - PAIR - Reporting List
Reporting Date: August 17, 1983

1,2-BENZENEDIAMINE, 4-ETHOXY-   1197-37-1
SEE ALSO:
1,2-BENZENEDIAMINE, 4-ETHOXY-

ENVIRONMENTAL LISTS

TSCA - Code of Federal Regulations Citations
40 CFR 712.30(f);   40 CFR 716.120(c)

TSCA - PAIR - Reporting List
Reporting Date: August 17, 1983

1,3-BENZENEDIAMINE, 4-ETHOXY-,   67801-06-3
DIHYDROCHLORIDE
SEE ALSO:
1,3-BENZENEDIAMINE, 4-ETHOXY-, DIHYDROCHLORIDE

ENVIRONMENTAL LISTS

TSCA - Code of Federal Regulations Citations
40 CFR 712.30(f);   40 CFR 716.120

TSCA - PAIR - Reporting List
Reporting Date: August 17, 1983

1,2-BENZENEDIAMINE, 4-ETHOXY-, SULFATE   RR-00259-5

ENVIRONMENTAL LISTS

TSCA - Chemicals with Significant New Use Rules
PMN number: P-83-105

1,3-BENZENEDIAMINE, 4-ETHOXY-, SULFATE (1:1)   68015-98-5
SEE ALSO:
1,3-BENZENEDIAMINE, 4-ETHOXY-, SULFATE (1:1)

ENVIRONMENTAL LISTS

TSCA - Code of Federal Regulations Citations
40 CFR 712.30(f);   40 CFR 716.120

TSCA - PAIR - Reporting List
Reporting Date: August 17, 1983

1,3-BENZENEDIAMINE, AR-ETHYL-AR-METHYL-   68966-84-7
SEE ALSO:
1,3-BENZENEDIAMINE, AR-ETHYL-AR-METHYL-

ENVIRONMENTAL LISTS

TSCA - Code of Federal Regulations Citations
40 CFR 712.30(f);   40 CFR 716.120(c)

TSCA - PAIR - Reporting List
Reporting Date: August 17, 1983

1,3-BENZENEDIAMINE, 4-METHOXY-,   614-94-8
DIHYDROCHLORIDE
SEE ALSO:
1,3-BENZENEDIAMINE, 4-METHOXY-, DIHYDROCHLORIDE

ENVIRONMENTAL LISTS

TSCA - Code of Federal Regulations Citations
40 CFR 712.30(f);   40 CFR 716.120(c)

TSCA - PAIR - Reporting List
Reporting Date: August 17, 1983

1,3-BENZENEDIAMINE, 4-METHOXY-, SULFATE   6219-67-6
SEE ALSO:
1,3-BENZENEDIAMINE, 4-METHOXY-, SULFATE

ENVIRONMENTAL LISTS

TSCA - Code of Federal Regulations Citations
40 CFR 712.30(f);   40 CFR 716.120

TSCA - PAIR - Reporting List
Reporting Date: August 17, 1983

1,2-BENZENEDIAMINE, 3-METHYL-   **2687-25-4**
SEE ALSO:
  1,2-BENZENEDIAMINE, 3-METHYL-

ENVIRONMENTAL LISTS

  TSCA - Code of Federal Regulations Citations
  40 CFR 712.30(d); 40 CFR 716.120(c)

  TSCA - PAIR - Reporting List
  Reporting Date: November 19, 1982

1,3-BENZENEDIAMINE, 2-METHYL-4,6-BIS   **104983-85-9**
(METHYLTHIO)-

ENVIRONMENTAL LISTS

  TSCA - Code of Federal Regulations Citations
  40 CFR 721.557

1,4-BENZENEDIAMINE, 2-METHYL-,   **615-45-2**
DIHYDROCHLORIDE
SEE ALSO:
  1,4-BENZENEDIAMINE, 2-METHYL-, DIHYDROCHLORIDE

ENVIRONMENTAL LISTS

  TSCA - Code of Federal Regulations Citations
  40 CFR 712.30(f); 40 CFR 716.120(c)

  TSCA - PAIR - Reporting List
  Reporting Date: August 17, 1983

1,4-BENZENEDIAMINE, 2-METHYL-, SULFATE (1:1)   **615-50-9**
SEE ALSO:
  1,4-BENZENEDIAMINE, 2-METHYL-, SULFATE (1:1)

ENVIRONMENTAL LISTS

  TSCA - Code of Federal Regulations Citations
  40 CFR 712.30(f); 40 CFR 716.120(c)

  TSCA - PAIR - Reporting List
  Reporting Date: August 17, 1983

1,3-BENZENEDIAMINE, 5-NITRO-   **5042-55-7**
SEE ALSO:
  1,3-BENZENEDIAMINE, 5-NITRO-

ENVIRONMENTAL LISTS

  TSCA - Code of Federal Regulations Citations
  40 CFR 712.30(f); 40 CFR 716.120

  TSCA - PAIR - Reporting List
  Reporting Date: August 17, 1983

1,3-BENZENEDIAMINE, 4-NITRO-   **5131-58-8**
SEE ALSO:
  1,3-BENZENEDIAMINE, 4-NITRO-

ENVIRONMENTAL LISTS

  TSCA - Code of Federal Regulations Citations
  40 CFR 712.30(f); 40 CFR 716.120(a)

  TSCA - PAIR - Reporting List
  Reporting Date: August 17, 1983

1,2-BENZENEDIAMINE, 4-NITRO-,   **6219-77-8**
DIHYDROCHLORIDE
SEE ALSO:
  1,2-BENZENEDIAMINE, 4-NITRO-, DIHYDROCHLORIDE

ENVIRONMENTAL LISTS

  TSCA - Code of Federal Regulations Citations
  40 CFR 712.30(f); 40 CFR 716.120

  TSCA - PAIR - Reporting List
  Reporting Date: August 17, 1983

1,4-BENZENEDIAMINE, 2-NITRO-,   **18266-52-9**
DIHYDROCHLORIDE
SEE ALSO:
  1,4-BENZENEDIAMINE, 2-NITRO-, DIHYDROCHLORIDE

ENVIRONMENTAL LISTS

  TSCA - Code of Federal Regulations Citations
  40 CFR 712.30(f); 40 CFR 716.120

  TSCA - PAIR - Reporting List
  Reporting Date: August 17, 1983

1,2-BENZENEDIAMINE, 4-NITRO-, SULFATE (1:1)   **68239-82-7**
SEE ALSO:
  1,2-BENZENEDIAMINE, 4-NITRO-, SULFATE (1:1)

ENVIRONMENTAL LISTS

  TSCA - Code of Federal Regulations Citations
  40 CFR 712.30(f); 40 CFR 716.120(c)

  TSCA - PAIR - Reporting List
  Reporting Date: August 17, 1983

1,4-BENZENEDIAMINE, 2-NITRO-, SULFATE (1:1)   **68239-83-8**
SEE ALSO:
  1,4-BENZENEDIAMINE, 2-NITRO-, SULFATE (1:1)

ENVIRONMENTAL LISTS

  TSCA - Code of Federal Regulations Citations
  40 CFR 712.30(f); 40 CFR 716.120(c)

  TSCA - PAIR - Reporting List
  Reporting Date: August 17, 1983

1,3-BENZENEDIAMINE, SULFATE (1:1)   **541-70-8**
SEE ALSO:
  1,3-BENZENEDIAMINE, SULFATE (1:1)

ENVIRONMENTAL LISTS

  TSCA - Code of Federal Regulations Citations
  40 CFR 712.30(f); 40 CFR 716.120(c)

  TSCA - PAIR - Reporting List
  Reporting Date: August 17, 1983

  TSCA - Chemical Test Rules
  Testing required by: manufacturers; importers; processors (40 CFR 799.3300) (Listed under 'Unsubstituted phenylenediamines')

  TSCA - Section 12(b) - Export Notification
  export notification required - Section 4

1,4-BENZENEDIAMINE, SULFATE (1:1)   **16245-77-5**
SEE ALSO:
  1,4-BENZENEDIAMINE, SULFATE (1:1)

ENVIRONMENTAL LISTS

  TSCA - Code of Federal Regulations Citations
  40 CFR 712.30(f); 40 CFR 716.120(c); 40 CFR 799.3300

  TSCA - PAIR - Reporting List
  Reporting Date: August 17, 1983

  TSCA - Chemical Test Rules
  Testing required by: manufacturers; importers; processors (40 CFR 799.3300) (Listed under 'Unsusbstituted phenylenediamines')

  TSCA - Section 12(b) - Export Notification
  export notification required - Section 4

BENZENEDIAZONIUM, 4-(DIMETHYLAMINO)-,   **124737-31-1**
SALT WITH 2-HYDROXY-5-SULFOBENZOIC ACID
(1:1)

ENVIRONMENTAL LISTS

  TSCA - Code of Federal Regulations Citations
  40 CFR 721.490

  TSCA - Chemicals with Significant New Use Rules
  PMN number: P-90-1366

TSCA - Section 12(b) - Export Notification
P-90-1366;   export notification required - Section 5

BENZENE DIAZONIUM NITRATE   619-97-6

HEALTH AND SAFETY LISTS

U.S. DOT - Hazard Classes
Forbidden from transport by the DOT

1,3-BENZENEDICARBOXYLIC ACID, POLYMER   68584-15-6
WITH 5-AMINO-1,3,5-TRIMETHYLCYCLOHEXY-
LAMINE, MODIFIED

ENVIRONMENTAL LISTS

List of Pesticide Product Inert Ingredients
[present]

1,4-BENZENEDICARBOXYLIC ACID, BIS(2-ETHYL-   6422-86-2
HEXYL) ESTER

ENVIRONMENTAL LISTS

TSCA - Code of Federal Regulations Citations
40 CFR 712.30(g);   40 CFR 716.120(a)

TSCA - PAIR - Reporting List
Reporting Date: October 8, 1984

TSCA - Health and Safety Reporting List
Effective Date: January 3, 1983

1,2-BENZENEDICARBOXYLIC ACID, BIS(1-METHYL-   131-15-7
HEPTYL) ESTER
SEE ALSO:
ALKYL PHTHALATES

HEALTH AND SAFETY LISTS

NFPA - Flash Points
flash point = 395 degrees F (202 degrees C)

NFPA - Hazard Identification Ratings
health-0;   flammability-1;   reactivity-0

ENVIRONMENTAL LISTS

TSCA - Code of Federal Regulations Citations
40 CFR 716.120(c)

1,2-BENZENEDICARBOXYLIC ACID, BIS(OXIRANYL-   7195-45-1
METHYL) ESTER

ENVIRONMENTAL LISTS

EPA - Master Testing List
[present]

TSCA - Code of Federal Regulations Citations
40 CFR 716.120(c)

PROPOSED REGULATIONS

TSCA - Proposed Testing Rule for Glycidyl Ethers
member of Glycidyl subcategory VII-D

1,2-BENZENEDICARBOXYLIC ACID, BIS(2-METHYL-   84-69-5
PROPYL) ESTER
SEE ALSO:
ALKYL PHTHALATES

HEALTH AND SAFETY LISTS

NFPA - Flash Points
flash point = 365 degrees F (185 degrees C)

NFPA - Hazard Identification Ratings
health-0;   flammability-1;   reactivity-0

NIOSH - Selected LD50s and LC50s
Oral, rat: LD50 = 15 gm/kg Skin, guinea pig: LD50 = 10 gm/kg

ENVIRONMENTAL LISTS

TSCA - Code of Federal Regulations Citations
40 CFR 716.120(c)

INTERNATIONAL LISTS

United Kingdom - Occupational Exposure Standards - TWAs
5 mg/m3 TWA

1,2-BENZENEDICARBOXYLIC ACID, 2-BUTOXY-2-   85-70-1
OXYETHYL BUTYLESTER
SEE ALSO:
ALKYL PHTHALATES

HEALTH AND SAFETY LISTS

NFPA - Flash Points
flash point = 390 degrees F (199 degrees C)

NFPA - Hazard Identification Ratings
health-1;   flammability-1;   reactivity-0

NIOSH - Selected LD50s and LC50s
Oral, rat: LD50 = 7 gm/kg

ENVIRONMENTAL LISTS

TSCA - Code of Federal Regulations Citations
40 CFR 712.30(d)

TSCA - PAIR - Reporting List
Reporting Date: November 19, 1982

1,2-BENZENEDICARBOXYLIC ACID, BUTYL CYCLO-   84-64-0
HEXYL ESTER
SEE ALSO:
ALKYL PHTHALATES

ENVIRONMENTAL LISTS

TSCA - Code of Federal Regulations Citations
40 CFR 712.30(d);   40 CFR 716.120(c)

1,2-BENZENEDICARBOXYLIC ACID, BUTYL 2-   85-69-8
ETHYLHEXYL ESTER
SEE ALSO:
ALKYL PHTHALATES

ENVIRONMENTAL LISTS

TSCA - Code of Federal Regulations Citations
40 CFR 716.120(c)

1,2-BENZENEDICARBOXYLIC ACID, BUTYL OCTYL   84-78-6
ESTER
SEE ALSO:
ALKYL PHTHALATES

ENVIRONMENTAL LISTS

TSCA - Code of Federal Regulations Citations
40 CFR 716.120(c)

1,2-BENZENEDICARBOXYLIC ACID, DECYL HEXYL   25724-58-7
ESTER
SEE ALSO:
ALKYL PHTHALATES

HEALTH AND SAFETY LISTS

NIOSH - Selected LD50s and LC50s
Oral, rat: LD50 = 49 gm/kg

ENVIRONMENTAL LISTS

TSCA - Code of Federal Regulations Citations
40 CFR 716.120(c)

1,2-BENZENEDICARBOXYLIC ACID, DECYL OCTYL   119-07-3
ESTER
SEE ALSO:
ALKYL PHTHALATES

HEALTH AND SAFETY LISTS

NIOSH - Selected LD50s and LC50s
Oral, rat: LD50 = 45 gm/kg

ENVIRONMENTAL LISTS

TSCA - Code of Federal Regulations Citations
40 CFR 712.30(d);   40 CFR 716.120(c)

TSCA - PAIR - Reporting List
Reporting Date: November 19, 1982

1,2-BENZENEDICARBOXYLIC ACID, DICYCLO-HEXYL ESTER   84-61-7
SEE ALSO:
ALKYL PHTHALATES

HEALTH AND SAFETY LISTS
NIOSH - Selected LD50s and LC50s
Oral, rat: LD50 = 30 gm/kg

ENVIRONMENTAL LISTS
TSCA - Code of Federal Regulations Citations
40 CFR 712.30(d);   40 CFR 716.120(c)
TSCA - PAIR - Reporting List
Reporting Date: November 19, 1982

INTERNATIONAL LISTS
United Kingdom - Occupational Exposure Standards - TWAs
5 mg/m3 TWA

1,2-BENZENEDICARBOXYLIC ACID, DIHEXYL ESTER   84-75-3
SEE ALSO:
ALKYL PHTHALATES

HEALTH AND SAFETY LISTS
NIOSH - Selected LD50s and LC50s
Oral, rat: LD50 = 29600 mg/kg Skin, rabbit: LD50 = 20 gm/kg

ENVIRONMENTAL LISTS
TSCA - Code of Federal Regulations Citations
40 CFR 716.120(c);   40 CFR 799.5000
TSCA - Substances Subject to Testing Consent Orders
Test for: Environmental Effects and Chemical Fate
TSCA - Section 12(b) - Export Notification
export notification required - Section 4

1,2-BENZENEDICARBOXYLIC ACID, DINONYL ESTER   84-76-4
SEE ALSO:
ALKYL PHTHALATES

HEALTH AND SAFETY LISTS
NIOSH - Selected LD50s and LC50s
Oral, rat: LD50 = 2000 mg/kg

ENVIRONMENTAL LISTS
TSCA - Code of Federal Regulations Citations
40 CFR 716.120(c)

INTERNATIONAL LISTS
United Kingdom - Occupational Exposure Standards - TWAs
5 mg/m3 TWA

1,2-BENZENEDICARBOXYLIC ACID, DITRIDECYL ESTER   119-06-2
SEE ALSO:
ALKYL PHTHALATES

HEALTH AND SAFETY LISTS
NFPA - Flash Points
flash point = 470 degrees F (243 degrees C)
NFPA - Hazard Identification Ratings
health-0; flammability-1; reactivity-0

ENVIRONMENTAL LISTS
TSCA - Code of Federal Regulations Citations
40 CFR 712.30(d);   40 CFR 716.120(c);   40 CFR 799.5000
TSCA - PAIR - Reporting List
Reporting Date: November 19, 1982
TSCA - Substances Subject to Testing Consent Orders
Test for: Chemical Fate

TSCA - Section 12(b) - Export Notification
export notification required - Section 4

1,2-BENZENEDICARBOXYLIC ACID, DIUNDECYL ESTER   3648-20-2
SEE ALSO:
ALKYL PHTHALATES

ENVIRONMENTAL LISTS
TSCA - Code of Federal Regulations Citations
40 CFR 716.120(c);   40 CFR 799.5000
TSCA - Substances Subject to Testing Consent Orders
Test for: Environmental Effects
TSCA - Section 12(b) - Export Notification
export notification required - Section 4

1,2-BENZENEDICARBOXYLIC ACID, 2-ETHOXY-2-OXOETHYL ETHYL ESTER   84-72-0
HEALTH AND SAFETY LISTS
NFPA - Flash Points
flash point = 365 degrees F (185 degrees C)
NFPA - Hazard Identification Ratings
health-0; flammability-1; reactivity-0

STATE LISTS
Pennsylvania Right to Know List
[present]

1,2-BENZENEDICARBOXYLIC ACID, 2-ETHYL-HEXYL-8-METHYLNONYLESTER   89-13-4
SEE ALSO:
ALKYL PHTHALATES

ENVIRONMENTAL LISTS
TSCA - Code of Federal Regulations Citations
40 CFR 716.120(c)

1,2-BENZENEDICARBOXYLIC ACID, HEXYL ISODE-CYL ESTER   61702-81-6
SEE ALSO:
ALKYL PHTHALATES

ENVIRONMENTAL LISTS
TSCA - Code of Federal Regulations Citations
40 CFR 716.120(c)

1,2-BENZENEDICARBOXYLIC ACID, ISODECYL TRIDECYL ESTER   61886-60-0
SEE ALSO:
ALKYL PHTHALATES

ENVIRONMENTAL LISTS
TSCA - Code of Federal Regulations Citations
40 CFR 716.120(c)

BENZENE, (1,2-DICHLOROETHENYL)-   6607-45-0
HEALTH AND SAFETY LISTS
NFPA - Flash Points
flash point = 225 degrees F (107 degrees C)
NFPA - Hazard Identification Ratings
health-2; flammability-1; reactivity-2

STATE LISTS
Florida Hazardous Substance List
[present]
Massachusetts Right To Know List
[present]
Pennsylvania Right to Know List
[present]

BENZENE, 2,4-DICHLORO-1-NITRO-   611-06-3

ENVIRONMENTAL LISTS

EPA - Master Testing List
[present]

BENZENE, 1,2-DICHLORO-3-NITRO-   3209-22-1

ENVIRONMENTAL LISTS

EPA - Master Testing List
[present]

BENZENE, 1,4-DIISOCYANATO-   104-49-4

ENVIRONMENTAL LISTS

CERCLA/SARA - Section 313 - Emission Reporting
form R reporting required; (Listed under 'Diisocyanate')

TSCA - Code of Federal Regulations Citations
40 CFR 712.30(x); 40 CFR 716.120(a)

TSCA - PAIR - Reporting List
Reporting Date: December 27, 1990

TSCA - Health and Safety Reporting List
Effective Date: June 1, 1987; Sunset Date: November 9, 1993

BENZENE, 1,3-DIISOCYANATO-   123-61-5

ENVIRONMENTAL LISTS

CERCLA/SARA - Section 313 - Emission Reporting
form R reporting required; (Listed under 'Diisocyanates')

TSCA - Code of Federal Regulations Citations
40 CFR 712.30(x); 40 CFR 716.120(a)

TSCA - PAIR - Reporting List
Reporting Date: December 27, 1990

TSCA - Health and Safety Reporting List
Effective Date: June 1, 1987; Sunset Date: November 9, 1993

BENZENE, DIISOCYANATOMETHYL-   1321-38-6

ENVIRONMENTAL LISTS

TSCA - Code of Federal Regulations Citations
40 CFR 716.120(a); 40 CFR 704.225(a)

TSCA - CAIR - Reporting List
reporting required by: manufacturer; importer; distributor; processor

TSCA - Health and Safety Reporting List
Effective Date: June 1, 1987; Sunset Date: November 9, 1993

BENZENE, 1,1'-(DIISOCYANATOMETHYLENE)BIS-   10031-75-1

ENVIRONMENTAL LISTS

TSCA - Code of Federal Regulations Citations
40 CFR 712.30(x); 40 CFR 716.120(d)

TSCA - PAIR - Reporting List
Reporting Date: December 27, 1990

TSCA - Health and Safety Reporting List
Effective Date: October 29, 1990

BENZENE, DIMETHYL(PENTYLOXY)-   1320-21-4

HEALTH AND SAFETY LISTS

NFPA - Flash Points
flash point = 205 degrees F (96 degrees C)

NFPA - Hazard Identification Ratings
health-2; flammability-1; reactivity-0

STATE LISTS

Florida Hazardous Substance List
[present]

Massachusetts Right To Know List
[present]

Pennsylvania Right to Know List
[present]

BENZENE, 1,2-DIMETHYL-, POLYPROPENE DERIVATIVES, SULFONATED,POTASSIUM SALTS   RR-00967-6

ENVIRONMENTAL LISTS

TSCA - Chemicals with Significant New Use Rules
PMN number: P-89-711

BENZENEDIOL   12385-08-9

STATE LISTS

Pennsylvania Right to Know List
[present]

1,2-BENZENEDIOL, 4-[1-HYDROXY-2-(METHY-LAMINO)ETHYL]   51-43-4

HEALTH AND SAFETY LISTS

U.S. DOT - Appendix A Table 1 - Hazardous Substances
final RQ = 1000 pounds (454 kg)

ENVIRONMENTAL LISTS

CERCLA/SARA - Hazardous Substances and their Reportable Quantities
final RQ = 1000 pounds (454 kg)

RCRA - P Series Wastes
waste number P042 (Reactive waste)

RCRA - Hazardous Constituents-Appendix VIII
waste number P042

RCRA - Substances Banned From Land Disposal
[present]

STATE LISTS

Massachusetts Right To Know List
[present]

Pennsylvania Right to Know List
environmental hazard

1,3-BENZENEDIOL, 2,4,6-TRINITRO-, LEAD SALT   15245-44-0

SEE ALSO:
LEAD

STATE LISTS

Massachusetts Right To Know List
[present]

BENZENE-1,3-DISULFOHYDRAZIDE   RR-01358-1

HEALTH AND SAFETY LISTS

U.S. DOT - Substances From 49 CFR 172.101
regulated by DOT (UN2971)

U.S. DOT - Hazard Classes
DOT hazard class = 4.1

1,3-BENZENEDISULFONIC ACID   98-48-6

ENVIRONMENTAL LISTS

CAA - HON Rule - SOCMI Chemicals
compliance by Oct. 24, 1994

TSCA - Code of Federal Regulations Citations
40 CFR 712.30(x); 40 CFR 716.120(d)

TSCA - PAIR - Reporting List
Reporting Date: November 27, 1991

TSCA - Health and Safety Reporting List
Effective Date: September 30, 1991

1,4-BENZENEDISULFONIC ACID, 2,2'-[1,2-ETHENEDIYLBIS[(3-SULFO-4,1-PHENYLENE)IMINO [6-(DIETHYLAMINO)-1,3,5-TRIAZINE-4,2-DIYL] IMINO]BIS-, HEXASODIUM SALT   41098-56-0

HEALTH AND SAFETY LISTS

NIOSH - Selected LD50s and LC50s
Oral, rat: LD50 = 11320 mg/kg

BENZENEETHANAMINE, .ALPHA.-METHYL-   60-15-1

HEALTH AND SAFETY LISTS

NIOSH - Selected LD50s and LC50s
Oral, mouse: LD50 = 22 mg/kg

BENZENE, ETHENYL-, HOMOPOLYMER,   88497-56-7
BROMINATED

ENVIRONMENTAL LISTS

TSCA - PAIR - Reporting List
Reporting Date: December 27, 1990

TSCA - Health and Safety Reporting List
Effective Date: October 29, 1990; Sunset Date: November 9, 1993

BENZENE, ETHENYL-, MONOMETHYL DERIV.   1319-73-9

STATE LISTS

Pennsylvania Right to Know List
[present]

BENZENE, 1-ETHYL-2-METHYL-   611-14-3
SEE ALSO:
AROMATIC C9 FRACTION FROM PETROLEUM REFINING

HEALTH AND SAFETY LISTS

NFPA - Hazard Identification Ratings
flammability-2; reactivity-0

ENVIRONMENTAL LISTS

TSCA - Code of Federal Regulations Citations
40 CFR 716.120(b)

BENZENE, ETHYLMETHYL-   25550-14-5
SEE ALSO:
AROMATIC C9 FRACTION FROM PETROLEUM REFINING

ENVIRONMENTAL LISTS

TSCA - Code of Federal Regulations Citations
40 CFR 712.30(d); 40 CFR 716.120(a); 40 CFR 799.2175

TSCA - PAIR - Reporting List
Reporting Date: November 19, 1982

TSCA - Health and Safety Reporting List
Effective Date: April 29, 1983

BENZENE, 2-ISOCYANATO-1,3-BIS(1-   28178-42-9
METHYLETHYL)-

ENVIRONMENTAL LISTS

TSCA - Code of Federal Regulations Citations
40 CFR 712.30(x); 40 CFR 716.120(d)

TSCA - PAIR - Reporting List
Reporting Date: December 27, 1990

TSCA - Health and Safety Reporting List
Effective Date: October 29, 1990; Sunset Date: November 9, 1993

BENZENE, 2-ISOCYANATO-1,3-DIMETHYL-   28556-81-2

ENVIRONMENTAL LISTS

TSCA - Code of Federal Regulations Citations
40 CFR 712.30(x); 40 CFR 716.120(d)

TSCA - PAIR - Reporting List
Reporting Date: December 27, 1990

TSCA - Health and Safety Reporting List
Effective Date: October 29, 1990; Sunset Date: November 9, 1993

BENZENE, 2-ISOCYANATO-4-[(4-ISOCYANATO   75790-84-0
PHENYL)METHYL]-1-METHYL-

ENVIRONMENTAL LISTS

CERCLA/SARA - Section 313 - Emission Reporting
form R reporting required; (Listed under 'Diisocyanates')

TSCA - Code of Federal Regulations Citations
40 CFR 716.120(a)

TSCA - Health and Safety Reporting List
Effective Date: June 1, 1987; Sunset Date: November 9, 1993

BENZENE, 1-ISOCYANATO-2-[(4-ISO-   75790-87-3
CYANATOPHENYL)THIO]-

ENVIRONMENTAL LISTS

CERCLA/SARA - Section 313 - Emission Reporting
form R reporting required; (Listed under 'Diisocyanates')

TSCA - Code of Federal Regulations Citations
40 CFR 716.120(a)

TSCA - Health and Safety Reporting List
Effective Date: June 1, 1987; Sunset Date: November 9, 1993

BENZENE, 1-ISOCYANATO-2-METHYL-   614-68-6

ENVIRONMENTAL LISTS

TSCA - Code of Federal Regulations Citations
40 CFR 712.30(x); 40 CFR 716.120(d)

TSCA - PAIR - Reporting List
Reporting Date: December 27, 1990

TSCA - Health and Safety Reporting List
Effective Date: October 29, 1990; Sunset Date: November 9, 1993

BENZENE, 1-ISOCYANATO-4-METHYL-   622-58-2

ENVIRONMENTAL LISTS

TSCA - Code of Federal Regulations Citations
40 CFR 712.30(x); 40 CFR 716.120(d)

TSCA - PAIR - Reporting List
Reporting Date: December 27, 1990

TSCA - Health and Safety Reporting List
Effective Date: October 29, 1990; Sunset Date: November 9, 1993

BENZENE, 1-ISOCYANATO-4-NITRO-   100-28-7

ENVIRONMENTAL LISTS

TSCA - Code of Federal Regulations Citations
40 CFR 712.30(x); 40 CFR 716.120(d)

TSCA - PAIR - Reporting List
Reporting Date: December 27, 1990

TSCA - Health and Safety Reporting List
Effective Date: October 29, 1990; Sunset Date: November 9, 1993

BENZENE, 1-ISOCYANATO-3-(TRIFLUOROMETHYL)   329-01-1
-

ENVIRONMENTAL LISTS

TSCA - Code of Federal Regulations Citations
40 CFR 712.30(x); 40 CFR 716.120(d)

TSCA - PAIR - Reporting List
Reporting Date: December 27, 1990

TSCA - Health and Safety Reporting List
Effective Date: October 29, 1990

BENZENEMETHANAMINE, N,N-DIETHYL-   772-54-3

HEALTH AND SAFETY LISTS

NFPA - Flash Points
flash point = 170 degrees F (77 degrees C)

NFPA - Hazard Identification Ratings
health-2; flammability-2; reactivity-0

STATE LISTS

Florida Hazardous Substance List
[present]

Massachusetts Right To Know List
[present]

Pennsylvania Right to Know List
[present]

BENZENEMETHANOL,.ALPHA.-(AMINOMETHYL)-   7568-93-6
STATE LISTS

Pennsylvania Right to Know List
[present]

BENZENEMETHANOL, 3-AMINO-.ALPHA.-METHYL-   2454-37-7
HEALTH AND SAFETY LISTS

NFPA - Flash Points
flash point = 315 degrees F (157 degrees C)
NFPA - Hazard Identification Ratings
health-2; flammability-1; reactivity-0

STATE LISTS

Florida Hazardous Substance List
[present]
Massachusetts Right To Know List
[present]
Pennsylvania Right to Know List
[present]

BENZENEMETHANOL, 3-PHENOXY-   13826-35-2
ENVIRONMENTAL LISTS

TSCA - PAIR - Reporting List
Reporting Date: June 10, 1993
TSCA - Health and Safety Reporting List
Effective Date: April 12, 1993

BENZENEMETHANOL, 3-PHENOXY-, ACETATE   50789-44-1
ENVIRONMENTAL LISTS

TSCA - PAIR - Reporting List
Reporting Date: June 10, 1993
TSCA - Health and Safety Reporting List
Effective Date: April 12, 1993

BENZENE, 1-METHOXY-4-(1-PROPENYL)-   104-46-1
HEALTH AND SAFETY LISTS

NIOSH - Selected LD50s and LC50s
Oral, rat: LD50 = 2090 mg/kg

ENVIRONMENTAL LISTS

List of Pesticide Product Inert Ingredients
[present]

BENZENE, 1-(1-METHYLBUTOXY)-4-NITRO-,   RR-01238-4
ENVIRONMENTAL LISTS

TSCA - Chemicals with Significant New Use Rules
PMN number: P-90-560

BENZENE, 1,1'-METHYLENEBIS(2-ISOCYANATO-   2536-05-2
ENVIRONMENTAL LISTS

TSCA - Code of Federal Regulations Citations
40 CFR 716.120(a)
TSCA - Health and Safety Reporting List
Effective Date: June 1, 1987; Sunset Date: November 9, 1993

BENZENE, 1,1'-METHYLENEBIS[4-ISOCYANATO-3-   139-25-3
METHYL-
ENVIRONMENTAL LISTS

CERCLA/SARA - Section 313 - Emission Reporting
form R reporting required; (Listed under 'Diisocyanates')
TSCA - Code of Federal Regulations Citations
40 CFR 716.120(a)

TSCA - Health and Safety Reporting List
Effective Date: June 1, 1987; Sunset Date: November 9, 1993

BENZENE, (1-METHYLETHYL)(2-PHENYLETHYL)-   77851-17-3
ENVIRONMENTAL LISTS

TSCA - Code of Federal Regulations Citations
40 CFR 721.466
TSCA - Chemicals with Significant New Use Rules
PMN number: P-88-894
TSCA - Section 12(b) - Export Notification
P-88-894; export notification required - Section 5

BENZENE, 1,1',1"-METHYLIDYENETRIS(4-ISO-   2422-91-5
CYANATO-
ENVIRONMENTAL LISTS

TSCA - Code of Federal Regulations Citations
40 CFR 712.30(x); 40 CFR 716.120(d)
TSCA - PAIR - Reporting List
Reporting Date: December 27, 1990
TSCA - Health and Safety Reporting List
Effective Date: October 29, 1990; Sunset Date: November 9, 1993

BENZENE, METHYLPENTYL-   1320-01-0
HEALTH AND SAFETY LISTS

NFPA - Flash Points
flash point = 180 degrees F (82 degrees C)
NFPA - Hazard Identification Ratings
health-2; flammability-2; reactivity-0

STATE LISTS

Florida Hazardous Substance List
[present]
Massachusetts Right To Know List
[present]
Pennsylvania Right to Know List
[present]

BENZENE, 1-METHYL-3-PHENOXY-   3586-14-9
ENVIRONMENTAL LISTS

TSCA - PAIR - Reporting List
Reporting Date: June 10, 1993
TSCA - Health and Safety Reporting List
Effective Date: April 12, 1993

BENZENE, 1-[(2-METHYL-2-PROPENYL)OXY]-2-   13414-54-5
NITRO-
ENVIRONMENTAL LISTS

TSCA - Code of Federal Regulations Citations
40 CFR 712.30(g); 40 CFR 716.120(a)
TSCA - PAIR - Reporting List
Reporting Date: October 8, 1984
TSCA - Health and Safety Reporting List
Effective Date: February 13, 1984; Sunset Date: November 9, 1993

BENZENE, 1,1'-OXYBIS[DODECYL-   69834-19-1
ENVIRONMENTAL LISTS

TSCA - PAIR - Reporting List
Reporting Date: June 10, 1993
TSCA - Health and Safety Reporting List
Effective Date: April 12, 1993

BENZENE, 1,1'-OXYBIS[METHYL-   28299-41-4
ENVIRONMENTAL LISTS

TSCA - PAIR - Reporting List
Reporting Date: June 10, 1993

TSCA - Health and Safety Reporting List
Effective Date: April 12, 1993

BENZENE, 1,1'-OXYBIS[(1,1,3,3-TETRAMETHYL-BUTYL)-   61702-88-3

ENVIRONMENTAL LISTS

TSCA - PAIR - Reporting List
Reporting Date: June 10, 1993

TSCA - Health and Safety Reporting List
Effective Date: April 12, 1993

BENZENE, 1,1'-(OXYDIETHYLIDENE)BIS-   93-96-9

HEALTH AND SAFETY LISTS

NFPA - Flash Points
flash point = 275 degrees F (135 degrees C)

NFPA - Hazard Identification Ratings
health-2; flammability-1; reactivity-0

NIOSH - Selected LD50s and LC50s
Oral, rat: LD50 = 9800 mg/kg

STATE LISTS

Florida Hazardous Substance List
[present]

Massachusetts Right To Know List
[present]

Pennsylvania Right to Know List
[present]

BENZENE, PENTABROMOMETHYL-   87-83-2

ENVIRONMENTAL LISTS

TSCA - Code of Federal Regulations Citations
40 CFR 712.30(x); 40 CFR 716.120(d)

TSCA - PAIR - Reporting List
Reporting Date: December 27, 1990

TSCA - Health and Safety Reporting List
Effective Date: October 29, 1990

BENZENE, PENTYL-   538-68-1

HEALTH AND SAFETY LISTS

NFPA - Flash Points
flash point = 150 degrees F (66 degrees C)

NFPA - Hazard Identification Ratings
health-1; flammability-2; reactivity-0

STATE LISTS

Pennsylvania Right to Know List
[present]

BENZENE PHOSPHORUS THIODICHLORIDE   14684-25-4

INTERNATIONAL LISTS

Canada - WHMIS: Ingredient Disclosure
1% item 154 (475)

STATE LISTS

NJ Right to Know List (Total)
sn 0201

NJ Special Hazardous Substances
(corrosive)

BENZENEPROPANAL, 4-(1,1-DIMETHYLETHYL)-.ALPHA.-METHYL-   80-54-6

ENVIRONMENTAL LISTS

TSCA - Code of Federal Regulations Citations
40 CFR 712.30(x); 40 CFR 716.120(d)

TSCA - PAIR - Reporting List
Reporting Date: November 27, 1991

TSCA - Health and Safety Reporting List
Effective Date: September 30, 1991

BENZENEPROPANAL, .ALPHA.-METHYL-4-(1-METHYLETHYL)-   103-95-7

HEALTH AND SAFETY LISTS

NFPA - Flash Points
flash point = 190 degrees F (88 degrees C)

NFPA - Hazard Identification Ratings
flammability-2; reactivity-0

ENVIRONMENTAL LISTS

TSCA - Code of Federal Regulations Citations
40 CFR 712.30(x); 40 CFR 716.120(d)

TSCA - PAIR - Reporting List
Reporting Date: November 27, 1991

TSCA - Health and Safety Reporting List
Effective Date: September 30, 1991

BENZENEPROPANOIC ACID, 3-(2H-BENZOTRIA-ZOL-2-YL)-5-(1,1-DIMETHYLETHYL)-4-HYDROXY-, C (7-9) BRANCHED AND LINEAR ALKYL ESTERS   127519-17-9

ENVIRONMENTAL LISTS

TSCA - Code of Federal Regulations Citations
40 CFR 721.500

TSCA - Chemicals with Significant New Use Rules
PMN number: P-90-1635

TSCA - Section 12(b) - Export Notification
P-90-1635; export notification required - Section 5

BENZENEPROPANOIC ACID, 3,5-BIS(1,1-DIMETHYLETHYL)-4-HYDROXY-, METHYL ESTER   6386-38-5

ENVIRONMENTAL LISTS

EPA - Master Testing List
[present]

BENZENE, SUBSTITUTED, ALKYL ACRYLATE DERIVATIVE   RR-00903-0

ENVIRONMENTAL LISTS

TSCA - Chemicals with Significant New Use Rules
PMN number: P-86-1692

BENZENESULFONAMIDE   98-10-2

HEALTH AND SAFETY LISTS

NIOSH - Selected LD50s and LC50s
Oral, rat: LD50 = 991 mg/kg

BENZENESULFONAMIDE, N-(2-MER-CAPTOETHYL)-, S-ESTER WITH O,O-DIISOPROPYLPHOSPHORODITHIOATE   741-58-2

HEALTH AND SAFETY LISTS

NIOSH - Selected LD50s and LC50s
Oral, rat: LD50 = 271 mg/kg Skin, rat: LD50 = 3950 mg/kg

BENZENESULFONAMIDE, 4-METHYL-   70-55-3

HEALTH AND SAFETY LISTS

NIOSH - Selected LD50s and LC50s
Oral, bird: LD50 = 75 mg/kg

ENVIRONMENTAL LISTS

EPA - Master Testing List
[present]

List of Pesticide Product Inert Ingredients
[present]

BENZENESULFONIC ACID, 4-METHYL-, REACTION PRODUCTS WITH OXIRANE MONO[(C10,16-ALKYLOXY)METHYL] DERIVATIVES AND 2,2,4 (OR 2,4,4)-TRIMETHYL-1,6-HEXANEDIAMINE   147170-38-5

ENVIRONMENTAL LISTS

TSCA - Chemicals with Significant New Use Rules
PMN number: P-93-1047

BENZENESULFONIC ACID, 4-AMINO-   121-57-3

ENVIRONMENTAL LISTS

CAA - HON Rule - SOCMI Chemicals
compliance by April 24, 1995

BENZENESULFONIC ACID, C8-24-ALKYL DERIVATIVES   RR-01040-2

ENVIRONMENTAL LISTS

List of Pesticide Product Inert Ingredients
[present]

BENZENESULFONIC ACID, C8-24-ALKYL DERIVATIVES, AMMONIUM, MAGNESIUM, POTASSIUM OR ZINC SALT   RR-01041-3

ENVIRONMENTAL LISTS

List of Pesticide Product Inert Ingredients
[present]

BENZENESULFONIC ACID, C10-13-ALKYL DERIVATIVES SODIUM SALT   68411-30-3

INTERNATIONAL LISTS

Canada - WHMIS: Ingredient Disclosure
1% item 155 (278)

BENZENESULFONIC ACID, C10-16-ALKYL DERIVATIVES   68584-22-5

ENVIRONMENTAL LISTS

List of Pesticide Product Inert Ingredients
[present]

BENZENESULFONIC ACID, C10-16-ALKYL DERIVATIVES, CALCIUM SALTS   68584-23-6

ENVIRONMENTAL LISTS

List of Pesticide Product Inert Ingredients
[present]

BENZENESULFONIC ACID, C10-16-ALKYL DERIVATIVES, COMPOUNDS WITH 2-PROPANAMINE   68584-24-7

ENVIRONMENTAL LISTS

List of Pesticide Product Inert Ingredients
[present]

BENZENESULFONIC ACID, C10-16-ALKYL DERIVATIVES, COMPOUNDS WITH TRIETHANOLAMINE   68584-25-8

ENVIRONMENTAL LISTS

List of Pesticide Product Inert Ingredients
[present]

BENZENESULFONIC ACID, C10-16-ALKYL DERIVATIVES, MAGNESIUM SALTS   68584-26-9

ENVIRONMENTAL LISTS

List of Pesticide Product Inert Ingredients
[present]

BENZENESULFONIC ACID, C10-16-ALKYL DERIVATIVES, POTASSIUM SALTS   68584-27-0

ENVIRONMENTAL LISTS

List of Pesticide Product Inert Ingredients
[present]

BENZENESULFONIC ACID, C10-16-ALKYL DERIVATIVES, SODIUM SALTS   68081-81-2

ENVIRONMENTAL LISTS

List of Pesticide Product Inert Ingredients
[present]

BENZENESULFONIC ACID, 4-[(4,6-DICHLORO-1,3,5-TRIAZIN-2-YL)AMINO]-   16110-89-7

ENVIRONMENTAL LISTS

TSCA - Code of Federal Regulations Citations
40 CFR 712.30(d)

TSCA - PAIR - Reporting List
Reporting Date: November 19, 1982

BENZENESULFONIC ACID, HYDRAZIDE   80-17-1

HEALTH AND SAFETY LISTS

U.S. DOT - Substances From 49 CFR 172.101
regulated by DOT (UN2970)

U.S. DOT - Hazard Classes
DOT hazard class = 4.1

BENZENESULFONIC ACID, MONO-C9-17-BRANCHED ALKYL DERIVATIVES   68648-98-6

ENVIRONMENTAL LISTS

List of Pesticide Product Inert Ingredients
[present]

BENZENESULFONIC ACID, MONO-C9-17-BRANCHED ALKYL DERIVATIVES, ISOPROPYLAMINE SALTS   68649-00-3

ENVIRONMENTAL LISTS

List of Pesticide Product Inert Ingredients
[present]

BENZENESULFONIC ACID, OXYBIS[DODECYL-, DISODIUM SALT]   25167-32-2

ENVIRONMENTAL LISTS

List of Pesticide Product Inert Ingredients
[present]

BENZENESULFONYL CHLORIDE   98-09-9

HEALTH AND SAFETY LISTS

U.S. DOT - Substances From 49 CFR 172.101
regulated by DOT (UN2225)

U.S. DOT - Hazard Classes
DOT hazard class = 8

U.S. DOT - Appendix A Table 1 - Hazardous Substances
final RQ = 100 pounds (45.4 kg)

NIOSH - Selected LD50s and LC50s
Oral, rat: LD50 = 1960 mg/kg

ENVIRONMENTAL LISTS

CERCLA/SARA - Hazardous Substances and their Reportable Quantities
final RQ = 100 pounds (45.4 kg)

RCRA - U Series Wastes
waste number U020 (Corrosive waste; Reactive waste)

RCRA - Hazardous Constituents-Appendix VIII
waste number U020 (Corrosive waste; Reactive waste)

RCRA - Substances Banned From Land Disposal
[present]

TSCA - Code of Federal Regulations Citations
40 CFR 716.120(a)

TSCA - Health and Safety Reporting List
Effective Date: March 7, 1986

INTERNATIONAL LISTS

Canada - WHMIS: Ingredient Disclosure
1% item 156 (474)

STATE LISTS

Massachusetts Right To Know List
[present]

NJ Right to Know List (Total)
sn 0202

NJ Special Hazardous Substances
(corrosive)

Pennsylvania Right to Know List
environmental hazard

BENZENE, 1,2,3,5-TETRACHLORO-   634-90-2

ENVIRONMENTAL LISTS

TSCA - Code of Federal Regulations Citations
40 CFR 712.30(d);   40 CFR 716.120(c)

TSCA - PAIR - Reporting List
Reporting Date: November 19, 1982

BENZENETHIOL   108-98-5

HEALTH AND SAFETY LISTS

ACGIH 1995 - Time Weighted Averages
0.5 ppm TWA;   2.3 mg/m3 TWA

AIHA - Odor Threshold Values
geometric mean air odor threshold = 0.00030 ppm (detectable)

U.S. DOT - Substances From 49 CFR 172.101
regulated by DOT (UN2337)

U.S. DOT - Hazard Classes
DOT hazard class = 6.1

U.S. DOT - Substances Which Are Poisonous by Inhalation
liquid hazardous material poisonous by inhalation (UN2337)

U.S. DOT - Appendix A Table 1 - Hazardous Substances
final RQ = 100 pounds (45.4 kg)

NIOSH - Selected LD50s and LC50s
*Inhalation, rat: LC50 = 33 ppm 4 hr Oral, rat: LD50 = 46 mg/kg
Skin, rat: LD50 = 300 mg/kg*

NIOSH - Health Standards - Exposure Limits
C (15 min) 0.1 ppm;   C (15 min) 0.5 mg/m3 (Listed under 'Thiols')

NIOSH - Health Standards - Health Effects and Precautions
Irritation;   eye, skin, blood, and nervous system effects (Blood and urine monitoring required;   prevent skin contact) (Listed under 'Thiols')

OSHA - Vacated PELs - Time Weighted Averages
0.5 ppm TWA;   2 mg/m3 TWA

ENVIRONMENTAL LISTS

CERCLA/SARA - Section 302 Extremely Hazardous Substances and TPQs
TPQ = 500 pounds

CERCLA/SARA - Hazardous Substances and their Reportable Quantities
final RQ = 100 pounds (45.4 kg)

EPA - Master Testing List
[present]

RCRA - P Series Wastes
waste number P014

RCRA - Hazardous Constituents-Appendix VIII
waste number P014

RCRA - Substances Banned From Land Disposal
[present]

TSCA - Code of Federal Regulations Citations
40 CFR 716.120(a)

TSCA - PAIR - Reporting List
Effective Date: January 26, 1994;   Reporting Date: March 28, 1994

TSCA - Health and Safety Reporting List
Effective Date: January 26, 1994;   Sunset Date: January 26, 2004

INTERNATIONAL LISTS

Australian Exposure Standards - Time Weighted Averages
0.5 ppm TWA;   2.3 mg/m3 TWA

Canada - WHMIS: Ingredient Disclosure
1% item 1278 (1384)

Canada - Alberta - 8 Hour Occupational Exposure Limit
0.5 ppm TWA;   2.3 mg/m3 TWA

Canada - Alberta - 15 Minute Occupational Exposure Limit
1.5 ppm STEL;   6.8 mg/m3 STEL

Canada - British Columbia - 8 Hour Exposure Limits
0.5 ppm TWA;   2 mg/m3 TWA

Canada - Ontario - OHSA - TWAEVs
0.5 ppm TWAEV;   2.2 mg/m3 TWAEV

Canada - Quebec - Time-Weighted Average Exposure Values
0.5 ppm TWAEV;   2.3 mg/m3 TWAEV

United Kingdom - Occupational Exposure Standards - TWAs
0.5 ppm TWA;   2 mg/m3 TWA

Israel - Time Weighted Averages
0.5 ppm TWA;   2.3 mg/m3 TWA

Israel - Action Levels
0.25 ppm AL;   1.15 mg/m3 AL

Mexico - Instruction No. 10 - TWAs
0.5 ppm TWA;   2 mg/m3 TWA

STATE LISTS

California - Exposure Limits - PELs
0.5 ppm PEL;   2 mg/m3 PEL

California - Directors List of Hazardous Substances (8 CCR 339)
[present]

Florida Hazardous Substance List
[present]

Massachusetts Right To Know List
extraordinarily hazardous

Minnesota Hazardous Substance List
[present]

NJ Right to Know List (Total)
sn 0203

Pennsylvania Right to Know List
environmental hazard

PROPOSED REGULATIONS

TSCA - ITC 33rd Report Priority Testing List
designated for testing

TSCA - ITC 34th Report Priority Testing List
designated for testing

1,2,4-BENZENETRICARBOXYLIC ACID, TRIS (2-ETHYLHEXYL) ESTER   3319-31-1

ENVIRONMENTAL LISTS

TSCA - Code of Federal Regulations Citations
40 CFR 712.30(g);   40 CFR 716.120(a)

TSCA - PAIR - Reporting List
Reporting Date: October 8, 1984

TSCA - Health and Safety Reporting List
Effective Date: Jaunary 3, 1983

BENZENE TRIOZONIDE   RR-01396-7

HEALTH AND SAFETY LISTS

U.S. DOT - Hazard Classes
Forbidden from transport by the DOT

BENZENE, UNDECYL-   6742-54-7
ENVIRONMENTAL LISTS
EPA - Master Testing List
[present]

BENZETHONIUM CHLORIDE   121-54-0
HEALTH AND SAFETY LISTS
NIOSH - Selected LD50s and LC50s
Oral, rat: LD50 = 368 mg/kg

NTP Chemical Status Reports - Testing Status and NTIS Number
Post peer review technical reports in progress;   prechronic studies for which technical reports were not prepared

ENVIRONMENTAL LISTS
List of Pesticide Product Inert Ingredients
[present]

BENZHYDROL   91-01-0
HEALTH AND SAFETY LISTS
NIOSH - Selected LD50s and LC50s
Oral, rat: LD50 = 5000 mg/kg

BENZIDINE   92-87-5
HEALTH AND SAFETY LISTS
ACGIH 1995 - Skin Designations
skin - potential for cutaneous absorption

ACGIH 1995 - Carcinogens
A1-confirmed human carcinogen

U.S. DOT - Substances From 49 CFR 172.101
regulated by DOT (UN1885)

U.S. DOT - Hazard Classes
DOT hazard class = 6.1

U.S. DOT - Appendix A Table 1 - Hazardous Substances
final RQ = 1 pound (0.454 kg)

IARC - Group 1 (carcinogenic to humans)
[present]

NIOSH - Selected LD50s and LC50s
Oral, rat: LD50 = 309 mg/kg

NIOSH 1990 - Pocket Guide - Carcinogens
occupational carcinogen

NIOSH 1990 - Pocket Guide - Target organs
bladder, kidneys, liver, skin, blood

NIOSH - Health Standards - Exposure Limits
use 29 CFR 1910.1010

NIOSH - Health Standards - Health Effects and Precautions
Bladder, liver, and kidney cancer

NIOSH - Health Standards - Carcinogenic Chemicals
potential human carcinogen

NTP Seventh Report - Known Carcinogens
known carcinogen

OSHA - 29 CFR 1910 Specifically Regulated Chemicals
Cancer suspect agent (see 29 CFR 1910.1010)

OSHA - Select Carcinogens
[present]

ENVIRONMENTAL LISTS
ATSDR Priority List
Rank (of 275): 049

CERCLA/SARA - Section 313 - Emission Reporting
form R reporting required for 0.1% de minimus concentration

CERCLA/SARA - Hazardous Substances and their Reportable Quantities
final RQ = 1 pound (0.454 kg)

Clean Air Act (1990) - List of Hazardous Air Contaminants
[present]

Clean Water Act - Priority Pollutants
[present]

Clean Water Act - Toxic Pollutants
[present]

EPA - Carcinogen Hazard Ranking for RQ Adjustment
Hazard ranking = High

RCRA - U Series Wastes
waste number U021

RCRA - Hazardous Constituents-Appendix VIII
waste number U021

RCRA - Substances Banned From Land Disposal
[present]

TSCA - Code of Federal Regulations Citations
40 CFR 716.120(a)

TSCA - Health and Safety Reporting List
Effective Date: June 1, 1987

INTERNATIONAL LISTS
Australian Exposure Standards - Time Weighted Averages
prohibition recommended

Australian Exposure Standards - Skin Effects
skin absorption

Australian Exposure Standards - Carcinogens
confirmed carcinogen

Canada - WHMIS: Ingredient Disclosure
0.1% item 157 (279)

Canada - CEPA - Priority Substances List
estimated time for completion of assessment reports: 5 years

Canada - Alberta - Skin Designation
can be absorbed through the intact skin

Canada - Alberta - Designated Substances
designated substance - requires code of practice

Canada - British Columbia - 8 Hour Exposure Limits
carcinogen with no permitted exposure or contact by any route

Canada - British Columbia - Skin Notations
skin - potential for skin absorption

Canada - British Columbia - Carcinogens
carcinogen with no permitted exposure or contact by any route

Canada - Quebec - Time-Weighted Average Exposure Values
substance of which the recirculation is prohibited

Canada - Quebec - Skin Designations
absorbed through the skin

Canada - Quebec - Carcinogens
C1 carcinogen: effect detected in humans

German (DFG) - Skin/Sensitizers
danger of cutaneous absorption

German (DFG) - Carcinogens
proven carcinogen

Mexico - Instruction No. 10 - Skin designation
skin - potential for cutaneous absorption

Mexico - Instruction No. 10 - Carcinogens
carcinogen in humans

Mexico - Wastewater - Organic Toxic Pollutants and Heavy Metals
Listed under [Organic Toxic Pollutants]

Mexico - Drinking Water - Ecological Criteria
0.000001 mg/l Substance presents persistence, bioaccumulations or risk of cancer, reduce human exposure to a minimum;   This level has been extrapolated by using a mathematic model

STATE LISTS
California - Air Bill 2588 Appendix A-I
and salts: known or potential carcinogen

California - Prop. 65 - Cancer list
carcinogen (and its salts) - initial date 2/27/87

California - Prop. 65 - No Significant Risk Levels
no significant risk level = 0.001 ug/day

California - Exposure Limits - Skin Notation
material may be absorbed through the skin, eyes or mucous membrane
California - Exposure Limits - Carcinogens
cancer-suspect agent (at a concentration >= 0.1%) (includes benzidine salts)
California - Directors List of Hazardous Substances (8 CCR 339)
[present] (any mixture containing 0.1% or greater benzidine)
Florida Hazardous Substance List
[present]
Massachusetts Right To Know List
carcinogen; extraordinarily hazardous
Minnesota Hazardous Substance List
carcinogen; skin
NJ Right to Know List (Total)
sn 0204
NJ Special Hazardous Substances
(carcinogen; mutagen)
Pennsylvania Right to Know List
environmental hazard; special hazardous substance
Pennsylvania RTK - Special Hazardous Substances
[present]

BENZIDINE BASED DYES   RR-00532-3

HEALTH AND SAFETY LISTS

IARC - Group 2A (limited human data)
[present] (Other relevant data, as given in Supplement 7, influenced the making of the overall evaluation)
NIOSH - Health Standards - Exposure Limits
reduce exposure to lowest feasible concentration
NIOSH - Health Standards - Health Effects and Precautions
Bladder cancer (Stringent workplace controls and medical monitoring required)
NIOSH - Health Standards - Carcinogenic Chemicals
potential human carcinogen
OSHA - Possible Select Carcinogens
[present]

STATE LISTS

California - Air Bill 2588 Appendix A-I
known or potential carcinogen
California - Prop. 65 - Cancer list
carcinogen - initial date 10/1/92
California - Directors List of Hazardous Substances (8 CCR 339)
[present]
Minnesota Hazardous Substance List
carcinogen

BENZIDINE SALT   531-86-2

STATE LISTS

California - Directors List of Hazardous Substances (8 CCR 339)
[present] (any mixture containing 0.1% or greater benzidine) (Listed under 'Benzidine (and its salts)')

BENZIL   134-81-6

HEALTH AND SAFETY LISTS

NIOSH - Selected LD50s and LC50s
Oral, rat: LD50 = 2710 mg/kg

ENVIRONMENTAL LISTS

CAA - HON Rule - SOCMI Chemicals
compliance by April 24, 1995

INTERNATIONAL LISTS

Canada - WHMIS: Ingredient Disclosure
1% item 158 (280)

BENZILIC ACID   76-93-7

HEALTH AND SAFETY LISTS

NIOSH - Selected LD50s and LC50s
Oral, mouse: LD50 = 2 gm/kg

ENVIRONMENTAL LISTS

CAA - HON Rule - SOCMI Chemicals
compliance by April 24, 1995

1-BENZIMIDAZOLE   51-17-2

HEALTH AND SAFETY LISTS

NIOSH - Selected LD50s and LC50s
Oral, mouse: LD50 = 2910 mg/kg

BENZIMIDAZOLE, 4,5-DICHLORO-2-(TRIFLUO-ROMETHYL)-   3615-21-2

ENVIRONMENTAL LISTS

CERCLA/SARA - Section 302 Extremely Hazardous Substances and TPQs
TPQ = 500/10,000 pounds

STATE LISTS

Florida Hazardous Substance List
effective March 13, 1992
Massachusetts Right To Know List
extraordinarily hazardous
NJ Right to Know List (Total)
sn 2908
Pennsylvania Right to Know List
environmental hazard

PROPOSED REGULATIONS

CERCLA/SARA - Proposed Hazardous Substance Additions
proposed RQ = 1 pound (.454 kg)
CERCLA/SARA - 1989 Proposed RQ Adjustments
proposed RQ = 100 pounds (45.4 kg)

BENZIN   8030-30-6

HEALTH AND SAFETY LISTS

ACGIH 1995 - Time Weighted Averages
400 ppm TWA; 1590 mg/m3 TWA
U.S. DOT - Substances From 49 CFR 172.101
regulated by DOT (UN2553, UN1255, UN1256)
U.S. DOT - Hazard Classes
DOT hazard class = 3
U.S. DOT - Appendix B - Marine Pollutants
DOT regulated marine pollutant
NFPA - Flash Points
50 degree flash: flash point = 50 degrees F (10 degrees C); high flash: flash point = 85 degrees F (29 degrees C); regular: flash point = 28 degrees F (-2 degrees C); (all flash points will vary depending on manufacturer)
NFPA - Hazard Identification Ratings
all forms except petroleum: health-1; flammability-3; reactivity-0; petroleum: health-1; flammability-4; reactivity-0
NIOSH 1990 - Pocket Guide - RELs
100 ppm TWA; 400 mg/m3 TWA
NIOSH 1990 - Pocket Guide - IDLHs
10,000 ppm IDLH (lower explosive level)
NIOSH 1990 - Pocket Guide - Target organs
respiratory system, eyes, skin
OSHA - Vacated PELs - Time Weighted Averages
100 ppm TWA; 400 mg/m3 TWA
OSHA - Final PELs - Time Weighted Averages
100 ppm TWA; 400 mg/m3 TWA

ENVIRONMENTAL LISTS

List of Pesticide Product Inert Ingredients
[present]

INTERNATIONAL LISTS

Canada - Ontario - OHSA - TWAEVs
1350 mg/m3 TWAEV (listed as an agent of variable composition)
As sum of components assayed by chromatographic procedure with reference to the bulk sample.

Canada - Quebec - Time-Weighted Average Exposure Values
300 ppm TWAEV; 1370 mg/m3 TWAEV

STATE LISTS

California - Exposure Limits - PELs
300 ppm PEL; 1350 mg/m3 PEL

California - Exposure Limits - STELs
400 ppm STEL; 1800 mg/m3 STEL

California - Directors List of Hazardous Substances (8 CCR 339)
[present]

Florida Hazardous Substance List
[present]

Massachusetts Right To Know List
[present]

Minnesota Hazardous Substance List
[present]

NJ Right to Know List (Total)
sn 0518

Pennsylvania Right to Know List
[present] (includes 50 degree flash form)

**1H-BENZ(DE)ISOQUINOLINE-1,3(2H)-DIONE, 2-   52821-24-6
(3-HYDROXYPROPYL)-6-[(3-HYDROXYPROPYL)
AMINO]-**

ENVIRONMENTAL LISTS

List of Pesticide Product Inert Ingredients
[present]

1,2-BENZISOTHIAZOLIN-3-ONE   2634-33-5

HEALTH AND SAFETY LISTS

NIOSH - Selected LD50s and LC50s
Oral, rat: LD50 = 1020 mg/kg

ENVIRONMENTAL LISTS

List of Pesticide Product Inert Ingredients
[present]

BENZO(A)FLUORANTHENE   203-33-8

ENVIRONMENTAL LISTS

ATSDR Priority List
Rank (of 275): 060

BENZO(A)FLUORENE   238-84-6

HEALTH AND SAFETY LISTS

IARC - Group 3 (not classifiable)
[present]

BENZO(A)PYRENE   50-32-8

HEALTH AND SAFETY LISTS

ACGIH 1995 - Carcinogens
A2-suspected human carcinogen

U.S. DOT - Appendix A Table 1 - Hazardous Substances
final RQ = 1 pound (0.454 kg)

IARC - Group 2A (limited human data)
[present] (Overall evaluation based only on evidence of carcinogenicity in monograph (32, 1983) or in Supplement 4) (Other relevant data, as given in Supplement 7 or in the monograph influenced the making of the overall evaluation)

NTP Seventh Report - Suspect Carcinogens
suspect carcinogen (Listed under 'Nickel and certain nickel compounds')

OSHA - Possible Select Carcinogens
[present]

ENVIRONMENTAL LISTS

ATSDR Priority List
Rank (of 275): 008

CERCLA/SARA - Section 313 - Emission Reporting
form R reporting required; (Listed under 'Polycyclic aromatic compounds')

CERCLA/SARA - Hazardous Substances and their Reportable Quantities
final RQ = 1 pound (0.454 kg)

Clean Water Act - Priority Pollutants
[present]

Safe Drinking Water Act - MCLs
MCL = 0.0002 mg/L

Safe Drinking Water Act - MCLGs
MCLG = Zero

EPA - Carcinogen Hazard Ranking for RQ Adjustment
Hazard ranking = High

RCRA - U Series Wastes
waste number U022

RCRA - Hazardous Constituents-Appendix VIII
waste number U022

RCRA - Basis for Listing - Appendix VII
Included in waste streams: F039, K001, K035, K141, K142, K143, K144, K145, K147, K148

RCRA - Substances Banned From Land Disposal
[present]

RCRA - TSD Facilities Ground Water Monitoring
TM 8100 = 200 ug/L PQL; TM 8270 = 10 ug/L PQL

RCRA - Universal Treatment Standards (LDR)
WW: 0.061 mg/l; NWW: 3.4 mg/kg

INTERNATIONAL LISTS

Australian Exposure Standards - Time Weighted Averages
control to lowest practical level

Australian Exposure Standards - Carcinogens
probable carcinogen

Canada - WHMIS: Ingredient Disclosure
0.1% item 163 (287)

Canada - Drinking Water Quality - MACs
0.00001 mg/L MAC

Canada - Alberta - Designated Substances
designated substance - requires code of practice

Canada - British Columbia - 8 Hour Exposure Limits
carcinogen with no established permitted concentration

Canada - British Columbia - Carcinogens
carcinogen with no established permitted concentration

Canada - Quebec - Time-Weighted Average Exposure Values
0.005 mg/m3 TWAEV; substance of which the recirculation is prohibited

Canada - Quebec - Carcinogens
C2 carcinogen: effect suspected in humans

German (DFG) - Carcinogens
animal evidence of carcinogenicity

Mexico - Wastewater - Organic Toxic Pollutants and Heavy Metals
Listed under [Aromatic Hydrocarbons]

STATE LISTS

California - Air Bill 2588 Appendix A-I
known or potential carcinogen

California - Prop. 65 - Cancer list
carcinogen - initial date 7/1/87

California - Prop. 65 - No Significant Risk Levels
no significant risk level = 0.06 ug/day

California - Directors List of Hazardous Substances (8 CCR 339)
[present]

Florida Hazardous Substance List
[present]

Massachusetts Right To Know List
carcinogen; extraordinarily hazardous

Minnesota Hazardous Substance List
carcinogen

NJ Special Hazardous Substances
(carcinogen; mutagen)

Pennsylvania Right to Know List
environmental hazard; special hazardous substance

Pennsylvania RTK - Special Hazardous Substances
[present]

BENZO(B)FLUORANTHENE   205-99-2

HEALTH AND SAFETY LISTS

ACGIH 1995 - Carcinogens
A2-suspected human carcinogen

U.S. DOT - Appendix A Table 1 - Hazardous Substances
final RQ = 1 pound (0.454 kg)

IARC - Group 2B (sufficient animal data)
[present] (Overall evaluation based only on evidence of carcinogenicity in monograph (32, 1983) or in Supplement 4)

NTP Seventh Report - Suspect Carcinogens
suspect carcinogen (Listed under 'Polycyclic aromatic hydrocarbons')

OSHA - Possible Select Carcinogens
[present]

ENVIRONMENTAL LISTS

ATSDR Priority List
Rank (of 275): 010

CERCLA/SARA - Section 313 - Emission Reporting
form R reporting required; (Listed under 'Polycyclic aromatic compounds')

CERCLA/SARA - Hazardous Substances and their Reportable Quantities
final RQ = 1 pound (0.454 kg)

Clean Water Act - Priority Pollutants
[present]

EPA - Carcinogen Hazard Ranking for RQ Adjustment
Hazard ranking = High

RCRA - Hazardous Constituents-Appendix VIII
hazardous constituent - no waste number

RCRA - Basis for Listing - Appendix VII
Included in waste streams: F039, K001, K035, K141, K142, K143, K144, K145, K147, K148

RCRA - TSD Facilities Ground Water Monitoring
TM 8100 = 200 ug/L PQL; TM 8270 = 10 ug/L PQL

RCRA - Universal Treatment Standards (LDR)
WW: 0.11 mg/l; NWW: 6.8 mg/kg; (Difficult to distinguish from k isomer)

INTERNATIONAL LISTS

Canada - WHMIS: Ingredient Disclosure
0.1% item 159 (284)

Mexico - Wastewater - Organic Toxic Pollutants and Heavy Metals
Listed under [Aromatic Hydrocarbons]

STATE LISTS

California - Air Bill 2588 Appendix A-I
known or potential carcinogen

California - Prop. 65 - Cancer list
carcinogen - initial date 7/1/87

California - Directors List of Hazardous Substances (8 CCR 339)
[present]

Florida Hazardous Substance List
[present]

Massachusetts Right To Know List
carcinogen; extraordinarily hazardous

Minnesota Hazardous Substance List
carcinogen

NJ Special Hazardous Substances
(carcinogen)

Pennsylvania Right to Know List
environmental hazard; special hazardous substance

Pennsylvania RTK - Special Hazardous Substances
[present]

1H,3H-BENZO[1,2-C:4,5-C']DIFURAN-1,3,5,7-TETRONE   89-32-7

HEALTH AND SAFETY LISTS

NIOSH - Selected LD50s and LC50s
Oral, mouse: LD50 = 2400 mg/kg

BENZO[C]FLUORENE   RR-01527-0

HEALTH AND SAFETY LISTS

IARC - Group 3 (not classifiable)
[present]

BENZO[C]PHENANTHRENE   195-19-7

HEALTH AND SAFETY LISTS

IARC - Group 3 (not classifiable)
[present]

BENZO(E)PYRENE   192-97-2

HEALTH AND SAFETY LISTS

IARC - Group 3 (not classifiable)
[present]

ENVIRONMENTAL LISTS

ATSDR Priority List
Rank (of 275): 067

BENZO[GHI]FLUORANTHENE   203-12-3

HEALTH AND SAFETY LISTS

IARC - Group 3 (not classifiable)
[present]

BENZO(GHI)PERYLENE   191-24-2

SEE ALSO:
F039-HAZARDOUS WASTES

HEALTH AND SAFETY LISTS

U.S. DOT - Appendix A Table 1 - Hazardous Substances
final RQ = 5000 pounds (2270 kg)

IARC - Group 3 (not classifiable)
[present]

ENVIRONMENTAL LISTS

ATSDR Priority List
Rank (of 275): 262

CERCLA/SARA - Hazardous Substances and their Reportable Quantities
final RQ = 5000 pounds (2270 kg)

Clean Water Act - Priority Pollutants
[present]

RCRA - Basis for Listing - Appendix VII
Included in waste stream: F039

RCRA - TSD Facilities Ground Water Monitoring
TM 8100 = 200 ug/L PQL; TM 8270 = 10 ug/L PQL

RCRA - Universal Treatment Standards (LDR)
WW: 0.0055 mg/l;   NWW: 1.8 mg/kg

INTERNATIONAL LISTS

Mexico - Wastewater - Organic Toxic Pollutants and Heavy Metals
Listed under [Aromatic Hydrocarbons]

STATE LISTS

California - Directors List of Hazardous Substances (8 CCR 339)
[present]

Massachusetts Right To Know List
[present]

Pennsylvania Right to Know List
environmental hazard

BENZO(J)FLUORANTHENE   205-82-3

HEALTH AND SAFETY LISTS

IARC - Group 2B (sufficient animal data)
[present] (Overall evaluation based only on evidence of carcinogenicity in monograph (32, 1983) or in Supplement 4)

NTP Seventh Report - Suspect Carcinogens
suspect carcinogen (Listed under 'Polycyclic aromatic hydrocarbons')

OSHA - Possible Select Carcinogens
[present]

ENVIRONMENTAL LISTS

CERCLA/SARA - Section 313 - Emission Reporting
form R reporting required;   (Listed under 'Polycyclic aromatic compounds')

RCRA - Hazardous Constituents-Appendix VIII
hazardous constituent - no waste number

INTERNATIONAL LISTS

German (DFG) - Carcinogens
animal evidence of carcinogenicity

STATE LISTS

California - Air Bill 2588 Appendix A-I
known or potential carcinogen

California - Prop. 65 - Cancer list
carcinogen - initial date 7/1/87

California - Directors List of Hazardous Substances (8 CCR 339)
[present]

Florida Hazardous Substance List
[present]

Massachusetts Right To Know List
carcinogen;   extraordinarily hazardous

Minnesota Hazardous Substance List
carcinogen

NJ Special Hazardous Substances
(carcinogen)

Pennsylvania Right to Know List
environmental hazard;   special hazardous substance

Pennsylvania RTK - Special Hazardous Substances
[present]

BENZO(K)FLUORANTHENE   207-08-9

HEALTH AND SAFETY LISTS

U.S. DOT - Appendix A Table 1 - Hazardous Substances
final RQ = 1 pound (0.454 kg)

IARC - Group 2B (sufficient animal data)
[present] (Overall evaluation based only on evidence of carcinogenicity in monograph (32, 1983) or in Supplement 4)

NTP Seventh Report - Suspect Carcinogens
suspect carcinogen (Listed under 'Polycyclic aromatic hydrocarbons')

OSHA - Possible Select Carcinogens
[present]

ENVIRONMENTAL LISTS

ATSDR Priority List
Rank (of 275): 247

CERCLA/SARA - Section 313 - Emission Reporting
form R reporting required;   (Listed under 'Polycyclic aromatic compounds')

CERCLA/SARA - Hazardous Substances and their Reportable Quantities
final RQ = 1 pound (0.454 kg)

Clean Water Act - Priority Pollutants
[present]

RCRA - Basis for Listing - Appendix VII
Included in waste stream: F039, K141, K142, K143, K144, K145, K147, K148

RCRA - TSD Facilities Ground Water Monitoring
TM 8100 = 200 ug/L PQL;   TM 8270 = 10 ug/L PQL

RCRA - Universal Treatment Standards (LDR)
WW: 0.11 mg/l;   NWW: 6.8 mg/kg;   (Difficult to distinguish from b isomer)

INTERNATIONAL LISTS

German (DFG) - Carcinogens
animal evidence of carcinogenicity

Mexico - Wastewater - Organic Toxic Pollutants and Heavy Metals
Listed under [Aromatic Hydrocarbons]

STATE LISTS

California - Air Bill 2588 Appendix A-I
known or potential carcinogen

California - Prop. 65 - Cancer list
carcinogen - initial date 7/1/87

California - Directors List of Hazardous Substances (8 CCR 339)
[present]

Florida Hazardous Substance List
effective March 13, 1992

Massachusetts Right To Know List
carcinogen;   extraordinarily hazardous

Minnesota Hazardous Substance List
carcinogen

Pennsylvania Right to Know List
environmental hazard;   special hazardous substance

Pennsylvania RTK - Special Hazardous Substances
[present]

BENZOATE ESTER   RR-00974-5

ENVIRONMENTAL LISTS

TSCA - Chemicals with Significant New Use Rules
PMN number: P-90-549

BENZOCYCLOBUTENE   RR-01161-0

HEALTH AND SAFETY LISTS

NFPA - Flash Points
flash point = 95 degrees F (35 degrees C)

BENZODIAZEPINES   RR-01474-4

STATE LISTS

California - Prop. 65 - Developmental Toxicity
developmental toxicity - initial date 10/1/92

1,3-BENZODIOXOLE-5-CARBOXALDEHYDE   120-57-0

HEALTH AND SAFETY LISTS

FDA - Controlled Substances Act - Precursor chemicals
Threshold by base weight = 4 kilograms

ENVIRONMENTAL LISTS

TSCA - Code of Federal Regulations Citations
40 CFR 712.30(x);   40 CFR 716.120(d)

TSCA - PAIR - Reporting List
Reporting Date: November 27, 1991

TSCA - Health and Safety Reporting List
Effective Date: September 30, 1991

BENZOFLUORANTHENE   56832-73-6

ENVIRONMENTAL LISTS

ATSDR Priority List
Rank (of 275): 150

BENZOFURAN   271-89-6

HEALTH AND SAFETY LISTS

NTP Chemical Status Reports - Testing Status and NTIS Number
Technical reports printed (PB90231127/AS)

NTP Chemical Status Reports - Evidence of Carcinogenicity
male rat-no evidence; female rat-some evidence; male mice-clear evidence; female mice-clear evidence

STATE LISTS

California - Air Bill 2588 Appendix A-I
known or potential carcinogen: 6/91

California - Prop. 65 - Cancer list
carcinogen - initial date 10/1/90

BENZOFURAN, 2,3-DIHYDRO-2,2-DIMETHYL-7-NITRO-   13414-55-6

ENVIRONMENTAL LISTS

TSCA - Code of Federal Regulations Citations
40 CFR 712.30(g); 40 CFR 716.120(a)

TSCA - PAIR - Reporting List
Reporting Date: October 8, 1984

TSCA - Health and Safety Reporting List
Effective Date: February 13, 1984

BENZOIC ACID   65-85-0

HEALTH AND SAFETY LISTS

U.S. DOT - Appendix A Table 1 - Hazardous Substances
final RQ = 5000 pounds (2270 kg)

NFPA - Flash Points
flash point = 250 degrees F (121 degrees C)

NFPA - Hazard Identification Ratings
health-2; flammability-1

NIOSH - Selected LD50s and LC50s
Oral, rat: LD50 = 2530 mg/kg

ENVIRONMENTAL LISTS

ATSDR Priority List
Rank (of 275): 250

CERCLA/SARA - Hazardous Substances and their Reportable Quantities
final RQ = 5000 pounds (2270 kg)

CAA - HON Rule - SOCMI Chemicals
compliance by April 24, 1995

Clean Water Act - Hazardous Substances
[present]

List of Pesticide Product Inert Ingredients
[present]

INTERNATIONAL LISTS

Canada - WHMIS: Ingredient Disclosure
1% item 160 (66)

STATE LISTS

California - Directors List of Hazardous Substances (8 CCR 339)
[present] (exempt when used in foods and feeds as a preservative)

Florida Hazardous Substance List
[present]

Massachusetts Right To Know List
[present]

NJ Right to Know List (Total)
sn 0209

Pennsylvania Right to Know List
environmental hazard

BENZOIC ACID, 5-[[4'-[1-AMINO-4-SULFO-2-NAPHTHALENYL]AZO][1,1'-BIPHENYL]-4-YL]AZO]-2-HYDROXY-, DISODIUM SALT   2429-79-0

SEE ALSO:
BISAZOBIPHENYL DYES

ENVIRONMENTAL LISTS

TSCA - Code of Federal Regulations Citations
40 CFR 716.120(c)

BENZOIC ACID, 3-CHLORO-, METHYL ESTER   2905-65-9

ENVIRONMENTAL LISTS

TSCA - Code of Federal Regulations Citations
40 CFR 712.30(d)

TSCA - PAIR - Reporting List
Reporting Date: November 19, 1982

BENZOIC ACID, 3-[2-CHLORO-4-(TRIFLUOROMETHYL)PHENOXY]-   63734-62-3

ENVIRONMENTAL LISTS

TSCA - PAIR - Reporting List
Reporting Date: June 10, 1993

TSCA - Health and Safety Reporting List
Effective Date: April 12, 1993

BENZOIC ACID, 3-[2-CHLORO-4-(TRIFLUOROMETHYL)PHENOXY]-, POTASSIUM SALT   72252-48-3

ENVIRONMENTAL LISTS

TSCA - PAIR - Reporting List
Reporting Date: June 10, 1993

TSCA - Health and Safety Reporting List
Effective Date: April 12, 1993

BENZOIC ACID, 5-[[4'-[[2,6-DIAMINO-3-METHYL-5[(SULFOPHENYL)AZO]PHENYL]AZO][1,1'-BIPHENYL]-4-YL]AZO]-2-HYDROXY-, DISODIUM SALT   2586-58-5

SEE ALSO:
BISAZOBIPHENYL DYES

ENVIRONMENTAL LISTS

TSCA - Code of Federal Regulations Citations
40 CFR 716.120(c)

BENZOIC ACID, ETHYL ESTER   93-89-0

HEALTH AND SAFETY LISTS

NFPA - Flash Points
flash point = 190 degrees F (88 degrees C)

NFPA - Hazard Identification Ratings
health-1; flammability-1; reactivity-0

NIOSH - Selected LD50s and LC50s
Oral, rat: LD50 = 2100 mg/kg

STATE LISTS

Pennsylvania Right to Know List
[present]

BENZOIC ACID, MAGNESIUM SALT   553-70-8

ENVIRONMENTAL LISTS

List of Pesticide Product Inert Ingredients
[present]

BENZOIC ACID, 2-METHYL-   **118-90-1**

HEALTH AND SAFETY LISTS

NIOSH - Selected LD50s and LC50s
Oral, rat: LD50 = 400 mg/kg

BENZOIC ACID, 3-METHYL-, BARIUM SALT   **68092-47-7**

ENVIRONMENTAL LISTS

List of Pesticide Product Inert Ingredients
[present]

BENZOIC ACID, 3,3'-METHYLENEBIS[6-AMINO-DI-   **61386-02-5**
2-PROPENYL] ESTER

ENVIRONMENTAL LISTS

TSCA - Code of Federal Regulations Citations
40 CFR 721.575

TSCA - Section 12(b) - Export Notification
P-82-438; export notification required - Section 5

BENZOIC ACID, 2-(3-PHENYLBUTYLIDENE)AMINO-,   **RR-01663-7**
METHYL ESTER

ENVIRONMENTAL LISTS

TSCA - Chemicals with Significant New Use Rules
PMN number: P-85-1211

BENZOIC DERIVATIVE PESTICIDES, N.O.S.   **RR-00168-3**

HEALTH AND SAFETY LISTS

U.S. DOT - Substances From 49 CFR 172.101
regulated by DOT (UN2769, UN2770, UN3003, UN3004)

U.S. DOT - Hazard Classes
toxic or toxic, flammable: DOT hazard class = 6.1; flammable, toxic: DOT hazard = 3

STATE LISTS

NJ Right to Know List (Total)
sn 2157; sn 2158; sn 2159; sn 2160

BENZOIN   **119-53-9**

HEALTH AND SAFETY LISTS

NTP Chemical Status Reports - Testing Status and NTIS Number
Technical reports printed (PB80217953)

NTP Chemical Status Reports - Evidence of Carcinogenicity
male rat-negative; female rat-negative; male mice-negative; female mice-negative

ENVIRONMENTAL LISTS

CAA - HON Rule - SOCMI Chemicals
compliance by April 24, 1995

ALPHA-BENZOIN OXIME   **441-38-3**

HEALTH AND SAFETY LISTS

NIOSH - Selected LD50s and LC50s
Oral, rat: LD50 = 5960 mg/kg

INTERNATIONAL LISTS

Canada - WHMIS: Ingredient Disclosure
1% item 161 (285)

BENZOL DILUENT   **RR-01762-9**

HEALTH AND SAFETY LISTS

NFPA - Flash Points
flash point = -25 degrees F (-32 degrees C)

NFPA - Hazard Identification Ratings
health-2; flammability-3; reactivity-0

BENZONITRILE   **100-47-0**

HEALTH AND SAFETY LISTS

U.S. DOT - Substances From 49 CFR 172.101
regulated by DOT (UN2224)

U.S. DOT - Hazard Classes
DOT hazard class = 6.1

U.S. DOT - Appendix A Table 1 - Hazardous Substances
final RQ = 5000 pounds (2270 kg)

NIOSH - Selected LD50s and LC50s
Inhalation, mouse: LC50 = 6000 mg/m3 (8 hr) Oral, mouse: LD50 = 971 mg/kg Skin, rat: LD50 = 1200 mg/kg

NTP Chemical Status Reports - Testing Status and NTIS Number
Prechronic studies for which toxicity technical reports were not prepared

ENVIRONMENTAL LISTS

CERCLA/SARA - Hazardous Substances and their Reportable Quantities
final RQ = 5000 pounds (2270 kg)

CAA - HON Rule - SOCMI Chemicals
compliance by April 24, 1995

Clean Water Act - Hazardous Substances
[present]

INTERNATIONAL LISTS

Canada - WHMIS: Ingredient Disclosure
1% item 162 (286)

STATE LISTS

California - Directors List of Hazardous Substances (8 CCR 339)
[present]

Massachusetts Right To Know List
[present]

NJ Right to Know List (Total)
sn 0211

Pennsylvania Right to Know List
environmental hazard

BENZOPHENONE   **119-61-9**

HEALTH AND SAFETY LISTS

AIHA - WEEL - Time Weighted Averages
5 mg/m3 TWA

NIOSH - Selected LD50s and LC50s
Oral, mouse: LD50 = 2895 mg/kg

NTP Chemical Status Reports - Testing Status and NTIS Number
Prechronic studies completed: chemicals in review for further evaluation; Project leader assigned/study in design

ENVIRONMENTAL LISTS

CAA - HON Rule - SOCMI Chemicals
compliance by Oct. 24, 1994

List of Pesticide Product Inert Ingredients
[present]

STATE LISTS

Minnesota Hazardous Substance List
[present]

BENZOPHETAMINE HYDROCHLORIDE   **5411-22-3**

STATE LISTS

California - Air Bill 2588 Appendix A-II
9/90

California - Prop. 65 - Developmental Toxicity
developmental toxicity - initial date 4/1/90

4H-BENZOPYRAN-4-ONE, 6-(2,3-DIHYDROXY-3-   **91681-63-9**
METHYLBUTYL)-3-(2,4-DIHYDROXYPHENYL)-7-
HYDROXY-

STATE LISTS

Pennsylvania Right to Know List
special hazardous substance

Pennsylvania RTK - Special Hazardous Substances
[present]

2H-1-BENZOPYRAN-3,5,7-TRIOL, 2-(3,4-DIHYDROX-YPHENYL)-3,4-DIHYDRO-, (2R-TRANS)-   154-23-4

STATE LISTS

Pennsylvania Right to Know List
[present]

O-BENZOQUINONE   583-63-1

INTERNATIONAL LISTS

Canada - WHMIS: Ingredient Disclosure
1% item 164 (288)

P-BENZOQUINONE DIOXIME   105-11-3

HEALTH AND SAFETY LISTS

IARC - Group 3 (not classifiable)
[present]
NIOSH - Selected LD50s and LC50s
Oral, rat: LD50 = 464 mg/kg
NTP Chemical Status Reports - Testing Status and NTIS Number
Technical reports printed (PB291501/AS)
NTP Chemical Status Reports - Evidence of Carcinogenicity
male rat-negative; female rat-positive; male mice-negative; female mice-negative

STATE LISTS

California - Directors List of Hazardous Substances (8 CCR 339)
[present]
Massachusetts Right To Know List
carcinogen; extraordinarily hazardous

BENZOTHIAZOLE   95-16-9

HEALTH AND SAFETY LISTS

NIOSH - Selected LD50s and LC50s
Oral, mouse: LD50 = 900 mg/kg

1,2,3-BENZOTRIAZOLE   95-14-7

HEALTH AND SAFETY LISTS

NTP Chemical Status Reports - Testing Status and NTIS Number
Technical reports printed (PB285202/AS)
NTP Chemical Status Reports - Evidence of Carcinogenicity
male rat-equivocal; female rat-equivocal; male mice-negative; female mice-equivocal

ENVIRONMENTAL LISTS

List of Pesticide Product Inert Ingredients
[present]
TSCA - Code of Federal Regulations Citations
40 CFR 712.30(w); 40 CFR 716.120(a)
TSCA - PAIR - Reporting List
Reporting Date: June 13, 1989
TSCA - Health and Safety Reporting List
Effective Date: April 13, 1989

STATE LISTS

Massachusetts Right To Know List
[present]

1H-BENZOTRIAZOLE, 5-(PENTYLOXY)-   133145-29-6

ENVIRONMENTAL LISTS

TSCA - Chemicals with Significant New Use Rules
PMN number: P-92-34
TSCA - Section 12(b) - Export Notification
P-92-34; export notification required - Section 5

1H-BENZOTRIAZOLE, 5-(PENTYLOXY)-, POTAS-SIUM SALT   RR-01561-2

ENVIRONMENTAL LISTS

TSCA - Chemicals with Significant New Use Rules
PMN number: P-92-36

1H-BENZOTRIAZOLE, 5-(PENTYLOXY)-, SODIUM SALT   RR-01560-1

ENVIRONMENTAL LISTS

TSCA - Chemicals with Significant New Use Rules
PMN number: P-92-35

2-BENZOTRIAZOLYL-4-METHYLPHENOL   2440-22-4

HEALTH AND SAFETY LISTS

NIOSH - Selected LD50s and LC50s
Oral, mouse: LD50 = 6500 mg/kg

ENVIRONMENTAL LISTS

List of Pesticide Product Inert Ingredients
[present]

BENZOTRICHLORIDE   98-07-7

SEE ALSO:
K149-HAZARDOUS WASTES
K015-HAZARDOUS WASTES

HEALTH AND SAFETY LISTS

U.S. DOT - Substances From 49 CFR 172.101
regulated by DOT (UN2226)
U.S. DOT - Hazard Classes
DOT hazard class = 8
U.S. DOT - Appendix A Table 1 - Hazardous Substances
final RQ = 10 pounds (4.54 kg)
IARC - Group Unspecified
[present] (Listed under 'alpha-chlorinated toluenes')
NFPA - Flash Points
flash point = 260 degrees F (127 degrees C)
NFPA - Hazard Identification Ratings
health-3; flammability-1; reactivity-0
NIOSH - Selected LD50s and LC50s
Inhalation, rat: LC50 = 19 ppm 2 hr Oral, rat: LD50 = 6 gm/kg
NTP Seventh Report - Suspect Carcinogens
suspect carcinogen
OSHA - Possible Select Carcinogens
[present]

ENVIRONMENTAL LISTS

CERCLA/SARA - Section 302 Extremely Hazardous Substances and TPQs
TPQ = 500 pounds
CERCLA/SARA - Section 313 - Emission Reporting
form R reporting required for 0.1% de minimus concentration
CERCLA/SARA - Hazardous Substances and their Reportable Quantities
final RQ = 10 pounds (4.54 kg)
Clean Air Act (1990) - List of Hazardous Air Contaminants
[present]
CAA - HON Rule - SOCMI Chemicals
compliance by April 24, 1995
CAA - HON Rule - Organic HAPs
[present]
EPA - Carcinogen Hazard Ranking for RQ Adjustment
Hazard ranking = Medium
RCRA - U Series Wastes
waste number U023 (Corrosive waste; Reactive waste; Toxic waste)
RCRA - Hazardous Constituents-Appendix VIII
waste number U023

RCRA - Basis for Listing - Appendix VII
Included in waste streams: K015, K149

RCRA - Substances Banned From Land Disposal
[present]

TSCA - Code of Federal Regulations Citations
40 CFR 712.30(d)

TSCA - PAIR - Reporting List
Reporting Date: November 19, 1982

INTERNATIONAL LISTS

Canada - WHMIS: Ingredient Disclosure
0.1% item 166 (478)

German (DFG) - Carcinogens
proven carcinogen (Listed under 'alpha-Chlorinated toluenes')

STATE LISTS

California - Air Bill 2588 Appendix A-I
known or potential carcinogen

California - Prop. 65 - Cancer list
carcinogen - initial date 7/1/87

California - Directors List of Hazardous Substances (8 CCR 339)
[present]

Florida Hazardous Substance List
[present]

Massachusetts Right To Know List
carcinogen; extraordinarily hazardous

Minnesota Hazardous Substance List
carcinogen

NJ Right to Know List (Total)
sn 0212

NJ Special Hazardous Substances
(carcinogen; corrosive)

Pennsylvania Right to Know List
environmental hazard; special hazardous substance

Pennsylvania RTK - Special Hazardous Substances
[present]

BENZOTRIFLUORIDE   98-08-8

HEALTH AND SAFETY LISTS

U.S. DOT - Substances From 49 CFR 172.101
regulated by DOT (UN2338)

U.S. DOT - Hazard Classes
DOT hazard class = 3

NFPA - Flash Points
flash point = 54 degrees F (12 degrees C)

NFPA - Hazard Identification Ratings
health-3; flammability-3; reactivity-1

NIOSH - Selected LD50s and LC50s
Inhalation, rat: LC50 = 70810 mg/m3 (4 hr) Oral, rat: LD50 = 15000 mg/kg

STATE LISTS

Florida Hazardous Substance List
[present]

Massachusetts Right To Know List
[present]

NJ Right to Know List (Total)
sn 0213

NJ Special Hazardous Substances
(flammable - third degree)

Pennsylvania Right to Know List
[present]

2H-1,4-BENZOXAZINE, 4-(DICHLOROACETYL)-3,4-DIHYDRO-3-METHYL-   98730-04-2

ENVIRONMENTAL LISTS

List of Pesticide Product Inert Ingredients
[present]

BENZOXAZOLE   273-53-0

HEALTH AND SAFETY LISTS

NIOSH - Selected LD50s and LC50s
Oral, mouse: LD50 = 750 mg/kg

BENZOXIDIAZOLES   RR-01397-8

HEALTH AND SAFETY LISTS

U.S. DOT - Hazard Classes
Forbidden from transport by the DOT

BENZOYL AZIDE   582-61-6

HEALTH AND SAFETY LISTS

U.S. DOT - Hazard Classes
Forbidden from transport by the DOT

BENZOYL CHLORIDE   98-88-4

HEALTH AND SAFETY LISTS

AIHA - WEEL - Ceilings or Short Term Time Weighted Averages
1 ppm STEL

U.S. DOT - Substances From 49 CFR 172.101
regulated by DOT (UN1736)

U.S. DOT - Hazard Classes
DOT hazard class = 8

U.S. DOT - Appendix A Table 1 - Hazardous Substances
final RQ = 1000 pounds (454 kg)

IARC - Group 3 (not classifiable)
[present]

NFPA - Flash Points
flash point = 162 degrees F (72 degrees C)

NFPA - Hazard Identification Ratings
health-3; flammability-2; reactivity-2 (decomposes in water)

ENVIRONMENTAL LISTS

CERCLA/SARA - Section 313 - Emission Reporting
form R reporting required for 1.0% de minimus concentration

CERCLA/SARA - Hazardous Substances and their Reportable Quantities
final RQ = 1000 pounds (454 kg)

CAA - HON Rule - SOCMI Chemicals
compliance by April 24, 1995

Clean Water Act - Hazardous Substances
[present]

TSCA - Code of Federal Regulations Citations
40 CFR 712.30(d)

TSCA - PAIR - Reporting List
Reporting Date: November 19, 1982

INTERNATIONAL LISTS

Canada - WHMIS: Ingredient Disclosure
1% item 167 (476)

Canada - NPRI (National Pollutant Release Inventory)
[present]

Canada - Alberta - 8 Hour Occupational Exposure Limit
1 ppm TWA; 5.2 mg/m3 TWA

Canada - Alberta - 15 Minute Occupational Exposure Limit
3 ppm STEL; 16 mg/m3 STEL

German (DFG) - Carcinogens
proven carcinogen (Listed under 'alpha-Chlorinated toluenes')

STATE LISTS

California - Air Bill 2588 Appendix A-I
6/91

California - Directors List of Hazardous Substances (8 CCR 339)
[present]

Florida Hazardous Substance List
[present]

Massachusetts Right To Know List
[present]

Minnesota Hazardous Substance List
[present]

NJ Right to Know List (Total)
sn 0214

NJ Special Hazardous Substances
(corrosive)

Pennsylvania Right to Know List
environmental hazard

PROPOSED REGULATIONS

ACGIH 1995 - Notice of Intended Changes
C 0.5 ppm; C 2.8 mg/m3

BENZOYL PEROXIDE   94-36-0

HEALTH AND SAFETY LISTS

ACGIH 1995 - Time Weighted Averages
5 mg/m3 TWA

U.S. DOT - Organic Peroxides Table
Organic peroxide UN3102; UN3104; UN3106; UN3108; exempt at less than or equal to 35%

IARC - Group 3 (not classifiable)
[present]

NIOSH - Selected LD50s and LC50s
Oral, rat: LD50 = 7710 mg/kg

NIOSH 1990 - Pocket Guide - RELs
5 mg/m3 TWA

NIOSH 1990 - Pocket Guide - IDLHs
7000 mg/m3 IDLH

NIOSH 1990 - Pocket Guide - Target organs
skin, respiratory system, eyes

NIOSH - Health Standards - Exposure Limits
5 mg/m3 TWA

NIOSH - Health Standards - Health Effects and Precautions
Respiratory and eye irritation, skin effects

OSHA - Vacated PELs - Time Weighted Averages
5 mg/m3 TWA

OSHA - Final PELs - Time Weighted Averages
5 mg/m3 TWA

OSHA - List of Highly Hazardous Chemicals
threshhold quantity = 7500 pounds

ENVIRONMENTAL LISTS

CERCLA/SARA - Section 313 - Emission Reporting
form R reporting required for 1.0% de minimus concentration

List of Pesticide Product Inert Ingredients
[present]

INTERNATIONAL LISTS

Australian Exposure Standards - Time Weighted Averages
5 mg/m3 TWA

Canada - WHMIS: Ingredient Disclosure
1% item 168 (1362)

Canada - NPRI (National Pollutant Release Inventory)
[present]

Canada - Alberta - 8 Hour Occupational Exposure Limit
5 mg/m3 TWA

Canada - Alberta - 15 Minute Occupational Exposure Limit
10 mg/m3 STEL

Canada - British Columbia - 8 Hour Exposure Limits
5 mg/m3 TWA

Canada - Ontario - OHSA - TWAEVs
5 mg/m3 TWAEV

Canada - Quebec - Time-Weighted Average Exposure Values
5 mg/m3 TWAEV

United Kingdom - Occupational Exposure Standards - TWAs
5 mg/m3 TWA

German (DFG) - MAK Values
total dust: 5 mg/m3 MAK

German (DFG) - Peak Limitations
2 x normal MAK (5 min momentary value); don't exceed 8 times during shift

German (DFG) - Skin/Sensitizers
negligible effects on skin

Israel - Time Weighted Averages
5 mg/m3 TWA

Israel - Action Levels
2.5 mg/m3 AL

Mexico - Instruction No. 10 - TWAs
5 mg/m3 TWA

STATE LISTS

California - Air Bill 2588 Appendix A-I
6/91

California - Exposure Limits - PELs
5 mg/m3 PEL

California - Directors List of Hazardous Substances (8 CCR 339)
[present]

Florida Hazardous Substance List
[present]

Massachusetts Right To Know List
[present]

Minnesota Hazardous Substance List
[present]

NJ Right to Know List (Total)
sn 0215

NJ Special Hazardous Substances
(flammable - fourth degree; reactive - fourth degree)

Pennsylvania Right to Know List
environmental hazard

BENZYL ACETATE   140-11-4

HEALTH AND SAFETY LISTS

IARC - Group 3 (not classifiable)
[present]

NFPA - Flash Points
flash point = 195 degrees F (90 degrees C)

NFPA - Hazard Identification Ratings
health-1; flammability-1; reactivity-0

NIOSH - Selected LD50s and LC50s
Inhalation, cat: LC50 = 245 ppm 8 hr Oral, rat: LD50 = 2490 mg/kg

NTP Chemical Status Reports - Testing Status and NTIS Number
Technical reports printed (PB87115044)

NTP Chemical Status Reports - Evidence of Carcinogenicity
male rat-equivocal evidence; female rat-no evidence; male mice-some evidence; female mice-some evidence

ENVIRONMENTAL LISTS

CAA - HON Rule - SOCMI Chemicals
compliance by April 24, 1995

List of Pesticide Product Inert Ingredients
[present]

STATE LISTS

California - Directors List of Hazardous Substances (8 CCR 339)
[present]

PROPOSED REGULATIONS

ACGIH 1995 - Notice of Intended Changes
10 ppm TWA;   61 mg/m3 TWA;   A4-not classifiable as a human carcinogen

BENZYL ALCOHOL   100-51-6

HEALTH AND SAFETY LISTS

AIHA - WEEL - Time Weighted Averages
10 ppm TWA;   44.2 mg/m3 TWA

NFPA - Flash Points
flash point = 200 degrees F (93 degrees C)

NFPA - Hazard Identification Ratings
health-2;   flammability-1;   reactivity-0

NIOSH - Selected LD50s and LC50s
Oral, rat: LD50 = 1230 mg/kg Skin, rabbit: LD50 = 2000 mg/kg

NTP Chemical Status Reports - Testing Status and NTIS Number
Technical reports printed (PB90110206)

NTP Chemical Status Reports - Evidence of Carcinogenicity
male rat-no evidence;   female rat-no evidence;   male mice-no evidence;   female mice-no evidence

ENVIRONMENTAL LISTS

CAA - HON Rule - SOCMI Chemicals
compliance by April 24, 1995

List of Pesticide Product Inert Ingredients
[present]

RCRA - TSD Facilities Ground Water Monitoring
TM 8270 = 20 ug/L PQL

INTERNATIONAL LISTS

Canada - WHMIS: Ingredient Disclosure
1% item 169 (170)

STATE LISTS

Florida Hazardous Substance List
[present]

Massachusetts Right To Know List
[present]

Pennsylvania Right to Know List
[present]

BENZYL BENZOATE   120-51-4

HEALTH AND SAFETY LISTS

NFPA - Flash Points
flash point = 298 degrees F (148 degrees C)

NFPA - Hazard Identification Ratings
health-1;   flammability-1;   reactivity-0

NIOSH - Selected LD50s and LC50s
Oral, rat: LD50 = 500 mg/kg Skin, rat: LD50 = 4000 mg/kg

ENVIRONMENTAL LISTS

CAA - HON Rule - SOCMI Chemicals
compliance by April 24, 1995

List of Pesticide Product Inert Ingredients
[present]

BENZYL BIS(HYDROGENATED TALLOW ALKYL)   61789-73-9
METHYL AMMONIUM CHLORIDE

ENVIRONMENTAL LISTS

List of Pesticide Product Inert Ingredients
[present]

BENZYL BROMIDE   100-39-0

HEALTH AND SAFETY LISTS

U.S. DOT - Substances From 49 CFR 172.101
regulated by DOT (UN1737)

U.S. DOT - Hazard Classes
DOT hazard class = 6.1

INTERNATIONAL LISTS

Canada - WHMIS: Ingredient Disclosure
1% item 170 (330)

STATE LISTS

Massachusetts Right To Know List
[present]

NJ Right to Know List (Total)
sn 0216

NJ Special Hazardous Substances
(corrosive)

BENZYL CHLORIDE   100-44-7
SEE ALSO:
K015-HAZARDOUS WASTES
K085-HAZARDOUS WASTES
K149-HAZARDOUS WASTES

HEALTH AND SAFETY LISTS

ACGIH 1995 - Time Weighted Averages
1 ppm TWA;   5.2 mg/m3 TWA

AIHA - Odor Threshold Values
geometric mean air odor threshold = 0.041 ppm (detectable)

U.S. DOT - Substances From 49 CFR 172.101
regulated by DOT (UN1738)

U.S. DOT - Hazard Classes
DOT hazard class = 6.1

U.S. DOT - Appendix A Table 1 - Hazardous Substances
final RQ = 100 pounds (45.4 kg)

FDA - Controlled Substances Act - Essential chemicals
Import/Export threshold volume = N/A;   weight = 4 kilograms;   Domestic threshold volume = N/A;   weight = 1 kilogram

IARC - Group Unspecified
[present] (Listed under 'alpha-chlorinated toluenes')

NFPA - Flash Points
flash point = 153 degrees F (67 degrees C)

NFPA - Hazard Identification Ratings
health-3;   flammability-2;   reactivity-1

NIOSH - Selected LD50s and LC50s
Inhalation, rat: LC50 = 150 ppm 2 hr Oral, rat: LD50 = 1231 mg/kg

NIOSH 1990 - Pocket Guide - RELs
C 1 ppm (15 min);   C 5 mg/m3 (15 min)

NIOSH 1990 - Pocket Guide - IDLHs
10 ppm IDLH

NIOSH 1990 - Pocket Guide - Target organs
eyes, respiratory system, skin

NIOSH - Health Standards - Exposure Limits
C (15 min) 5 mg/m3

NIOSH - Health Standards - Health Effects and Precautions
Irritation, skin and eye effects (Periodic chest X-ray and pulmonary function testing required)

NTP Chemical Status Reports - Testing Status and NTIS Number
Chronic studies exist for which technical reports were not prepared

OSHA - Vacated PELs - Time Weighted Averages
1 ppm TWA;   5 mg/m3 TWA

OSHA - Final PELs - Time Weighted Averages
1 ppm TWA;   5 mg/m3 TWA

ENVIRONMENTAL LISTS

CERCLA/SARA - Section 302 Extremely Hazardous Substances and TPQs
TPQ = 500 pounds

CERCLA/SARA - Section 313 - Emission Reporting
form R reporting required for 1.0% de minimus concentration

CERCLA/SARA - Hazardous Substances and their Reportable Quantities
final RQ = 100 pounds (45.4 kg)

Clean Air Act (1990) - List of Hazardous Air Contaminants
[present]

CAA - HON Rule - SOCMI Chemicals
compliance by April 24, 1995

CAA - HON Rule - Organic HAPs
[present]

Clean Water Act - Hazardous Substances
[present]

EPA - Carcinogen Hazard Ranking for RQ Adjustment
Hazard ranking = Low

RCRA - P Series Wastes
waste number P028

RCRA - Hazardous Constituents-Appendix VIII
waste number P028

RCRA - Basis for Listing - Appendix VII
Included in waste streams: K015, K085, K149

RCRA - Substances Banned From Land Disposal
[present]

TSCA - Code of Federal Regulations Citations
40 CFR 712.30(d)

TSCA - PAIR - Reporting List
Reporting Date: November 19, 1982

TSCA - Health and Safety Reporting List
Effective Date: March 11, 1994;   Sunset Date: March 11, 2004

INTERNATIONAL LISTS

Australian Exposure Standards - Time Weighted Averages
1 ppm TWA; 5.2 mg/m3 TWA

Canada - WHMIS: Ingredient Disclosure
1% item 171 (477)

Canada - NPRI (National Pollutant Release Inventory)
[present]

Canada - British Columbia - 8 Hour Exposure Limits
1 ppm TWA; 5 mg/m3 TWA

Canada - Ontario - OHSA - TWAEVs
1 ppm TWAEV; 5 mg/m3 TWAEV

Canada - Quebec - Time-Weighted Average Exposure Values
1 ppm TWAEV; 5.2 mg/m3 TWAEV

German (DFG) - Carcinogens
proven carcinogen (Listed under 'alpha-Chlorinated toluenes')

Israel - Time Weighted Averages
1 ppm TWA; 5.2 mg/m3 TWA

Israel - Action Levels
0.5 ppm AL; 2.6 mg/m3 AL

Mexico - Instruction No. 10 - TWAs
1 ppm TWA; 5 mg/m3 TWA

STATE LISTS

California - Air Bill 2588 Appendix A-I
known or potential carcinogen

California - Prop. 65 - Cancer list
carcinogen - initial date 1/1/90

California - Prop. 65 - No Significant Risk Levels
no significant risk level = 4 ug/day

California - Exposure Limits - PELs
1 ppm PEL; 5 mg/m3 PEL

California - Directors List of Hazardous Substances (8 CCR 339)
[present]

Florida Hazardous Substance List
[present]

Massachusetts Right To Know List
extraordinarily hazardous

Minnesota Hazardous Substance List
[present]

NJ Right to Know List (Total)
sn 0217

NJ Special Hazardous Substances
(corrosive; mutagen)

Pennsylvania Right to Know List
environmental hazard

PROPOSED REGULATIONS

TSCA - ITC 32nd Report Priority Testing List
designated for dermal absorption testing

BENZYL CHLOROFORMATE   501-53-1

HEALTH AND SAFETY LISTS

U.S. DOT - Substances From 49 CFR 172.101
regulated by DOT (UN1739)

U.S. DOT - Hazard Classes
DOT hazard class = 8

U.S. DOT - Appendix B - Marine Pollutants
DOT regulated marine pollutant

INTERNATIONAL LISTS

Canada - WHMIS: Ingredient Disclosure
1% item 172 (431)

STATE LISTS

NJ Right to Know List (Total)
sn 0218

NJ Special Hazardous Substances
(corrosive)

O-BENZYL-P-CHLOROPHENOL   120-32-1

HEALTH AND SAFETY LISTS

NIOSH - Selected LD50s and LC50s
Oral, rat: LD50 = 1700 mg/kg

NTP Chemical Status Reports - Testing Status and NTIS Number
Galley or camera copy technical reports in progress; technical reports printed; prechronic studies for which toxicity technical reports were not prepared

ENVIRONMENTAL LISTS

List of Pesticide Product Inert Ingredients
[present]

TSCA - Code of Federal Regulations Citations
40 CFR 712.30(w); 40 CFR 716.120(a)

TSCA - PAIR - Reporting List
Reporting Date: June 13, 1989

TSCA - Health and Safety Reporting List
Effective Date: April 13, 1989

BENZYL CINNAMATE   103-41-3

ENVIRONMENTAL LISTS

List of Pesticide Product Inert Ingredients
[present]

BENZYL CYANIDE   140-29-4

HEALTH AND SAFETY LISTS

U.S. DOT - Substances From 49 CFR 172.101
regulated by DOT (UN2470)

U.S. DOT - Hazard Classes
DOT hazard class = 6.1

FDA - Controlled Substances Act - Precursor chemicals
Threshold by base weight = 1 kilogram

NFPA - Flash Points
flash point = 235 degrees F (113 degrees C)

NFPA - Hazard Identification Ratings
health-2; flammability-1; reactivity-0

NIOSH - Selected LD50s and LC50s
Inhalation, rat: LC50 = 430 mg/m3 2 hr Oral, rat: LD50 = 270 mg/kg Skin, rat: LD50 = 2 gm/kg

ENVIRONMENTAL LISTS

CERCLA/SARA - Section 302 Extremely Hazardous Substances and TPQs
TPQ = 500 pounds

INTERNATIONAL LISTS

Canada - WHMIS: Ingredient Disclosure
1% item 1264 (1377)

STATE LISTS

Florida Hazardous Substance List
[present]

Massachusetts Right To Know List
extraordinarily hazardous

NJ Right to Know List (Total)
sn 1490

Pennsylvania Right to Know List
environmental hazard

PROPOSED REGULATIONS

CERCLA/SARA - Proposed Hazardous Substance Additions
proposed RQ = 1 pound (.454 kg)

CERCLA/SARA - 1989 Proposed RQ Adjustments
proposed RQ = 100 pounds (45.4 kg)

BENZYL DIBROMOACETATE   64503-07-7

ENVIRONMENTAL LISTS

List of Pesticide Product Inert Ingredients
[present]

BENZYL DIMETHYLAMINE   103-83-3

HEALTH AND SAFETY LISTS

U.S. DOT - Substances From 49 CFR 172.101
regulated by DOT (UN2619)

U.S. DOT - Hazard Classes
DOT hazard class = 8

NIOSH - Selected LD50s and LC50s
Inhalation, rat: LC50 = 2062 mg/m3 4 hr Oral, rat: LD50 = 265 mg/kg Skin, rabbit: LD50 = 1660 mg/kg

INTERNATIONAL LISTS

Canada - WHMIS: Ingredient Disclosure
1% item 174 (290)

STATE LISTS

NJ Right to Know List (Total)
sn 0219

NJ Special Hazardous Substances
(corrosive)

BENZYL DIMETHYL (TALLOW ALKYL) AMMO-   61789-75-1
NIUM CHLORIDE

ENVIRONMENTAL LISTS

List of Pesticide Product Inert Ingredients
[present]

BENZYLDIMETHYLTETRADECYLAMMONIUM   139-08-2
CHLORIDE

ENVIRONMENTAL LISTS

List of Pesticide Product Inert Ingredients
[present]

BENZYL ETHER   103-50-4

HEALTH AND SAFETY LISTS

NFPA - Flash Points
flash point = 275 degrees F (135 degrees C)

NFPA - Hazard Identification Ratings
health-0; flammability-1; reactivity-0

NIOSH - Selected LD50s and LC50s
Oral, rat: LD50 = 2500 mg/kg

INTERNATIONAL LISTS

Canada - WHMIS: Ingredient Disclosure
1% item 175 (808)

4-(BENZYL(ETHYL)AMINO)-3-ETHOXYBENZENEDI-   RR-00769-2
AZONIUM ZINCCHLORIDE

HEALTH AND SAFETY LISTS

U.S. DOT - Substances From 49 CFR 172.101
regulated by DOT (UN3037)

U.S. DOT - Hazard Classes
DOT hazard class = 4.1

BENZYLIDENE ACETONE   122-57-6

INTERNATIONAL LISTS

Canada - WHMIS: Ingredient Disclosure
1% item 176 (291)

BENZYL IODIDE   620-05-3

HEALTH AND SAFETY LISTS

U.S. DOT - Substances From 49 CFR 172.101
regulated by DOT (UN2653)

U.S. DOT - Hazard Classes
DOT hazard class = 6.1

INTERNATIONAL LISTS

Canada - WHMIS: Ingredient Disclosure
1% item 177 (1023)

STATE LISTS

NJ Right to Know List (Total)
sn 0220

BENZYL MERCAPTAN   100-53-8

HEALTH AND SAFETY LISTS

NFPA - Flash Points
flash point = 158 degrees F (70 degrees C)

NFPA - Hazard Identification Ratings
health-2; flammability-2

STATE LISTS

Florida Hazardous Substance List
[present]

Massachusetts Right To Know List
[present]

Pennsylvania Right to Know List
[present]

4-(BENZYL(METHYL)AMINO)-3-ETHOXYBENZENE-   RR-00770-5
DIAZONIUM ZINCCHLORIDE

HEALTH AND SAFETY LISTS

U.S. DOT - Substances From 49 CFR 172.101
regulated by DOT (UN3038)

U.S. DOT - Hazard Classes
DOT hazard class = 4.1

BENZYLPENICILLIN   61-33-6

HEALTH AND SAFETY LISTS

NIOSH - Selected LD50s and LC50s
Oral, hamster: LD50 = 24 mg/kg

STATE LISTS

NJ Right to Know List (Total)
sn 0221

BENZYL SALICILATE   118-58-1

HEALTH AND SAFETY LISTS

NFPA - Flash Points
flash point > 212 degrees F (100 degrees C)

NFPA - Hazard Identification Ratings
health-1; flammability-1; reactivity-0

BENZYLTRIMETHYL AMMONIUM CHLORIDE   56-93-9

HEALTH AND SAFETY LISTS

NTP Chemical Status Reports - Testing Status and NTIS Number
Prechronic studies for which toxicity technical reports were not prepared; prechronic studies completed: in review for further evaluation

BENZYL VIOLET 4B   1694-09-3

HEALTH AND SAFETY LISTS

IARC - Group 2B (sufficient animal data)
[present] (Overall evaluation based only on evidence of carcinogenicity in monograph (16, 1978) or in Supplement 4)
OSHA - Possible Select Carcinogens
[present]

ENVIRONMENTAL LISTS

List of Pesticide Product Inert Ingredients
[present]

STATE LISTS

California - Air Bill 2588 Appendix A-II
known or potential carcinogen
California - Prop. 65 - Cancer list
carcinogen - initial date 7/1/87
California - Prop. 65 - No Significant Risk Levels
no significant risk level = 30 ug/day
California - Directors List of Hazardous Substances (8 CCR 339)
[present]
Florida Hazardous Substance List
[present]
Massachusetts Right To Know List
carcinogen; extraordinarily hazardous
Minnesota Hazardous Substance List
carcinogen
Pennsylvania Right to Know List
special hazardous substance
Pennsylvania RTK - Special Hazardous Substances
[present]

BERGAMOT OIL   8007-75-8

HEALTH AND SAFETY LISTS

NIOSH - Selected LD50s and LC50s
Oral, rat: LD50 = 11520 mg/kg

BERKELIUM 245   16652-07-6

HEALTH AND SAFETY LISTS

U.S. DOT - Appendix A Table 2 - Radionuclides
final RQ = 100 curies (3.7E 12 Bq)

ENVIRONMENTAL LISTS

CERCLA/SARA List of Radionuclides (Appendix B) and Their Reportable Quantities
final RQ = 100 curies (3.7E 12 Bq)

BERKELIUM 246   15715-02-3

HEALTH AND SAFETY LISTS

U.S. DOT - Appendix A Table 2 - Radionuclides
final RQ = 10 curies (3.7E 11 Bq)

ENVIRONMENTAL LISTS

CERCLA/SARA List of Radionuclides (Appendix B) and Their Reportable Quantities
final RQ = 10 curies (3.7E 11 Bq)

BERKELIUM 247   15752-38-2

HEALTH AND SAFETY LISTS

U.S. DOT - Appendix A Table 2 - Radionuclides
final RQ = 0.01 curies (3.7E 8 Bq)

ENVIRONMENTAL LISTS

CERCLA/SARA List of Radionuclides (Appendix B) and Their Reportable Quantities
final RQ = 0.01 curies (3.7E 8 Bq)

BERKELIUM 249   14900-25-5

HEALTH AND SAFETY LISTS

U.S. DOT - Appendix A Table 2 - Radionuclides
final RQ = 1 curie (3.7E 10 Bq)

ENVIRONMENTAL LISTS

CERCLA/SARA List of Radionuclides (Appendix B) and Their Reportable Quantities
final RQ = 1 curie (3.7E 10 Bq)

BERKELIUM 250   15755-53-0

HEALTH AND SAFETY LISTS

U.S. DOT - Appendix A Table 2 - Radionuclides
final RQ = 100 curies (3.7E 12 Bq)

ENVIRONMENTAL LISTS

CERCLA/SARA List of Radionuclides (Appendix B) and Their Reportable Quantities
final RQ = 100 curies (3.7E 12 Bq)

BERTRONDITE   12161-82-9
SEE ALSO:
BERYLLIUM

INTERNATIONAL LISTS

Canada - WHMIS: Ingredient Disclosure
0.1% item 178 (292)

BERYL   1302-52-9
SEE ALSO:
BERYLLIUM

HEALTH AND SAFETY LISTS

NTP Seventh Report - Suspect Carcinogens
suspect carcinogen (Listed under 'Beryllium and certain beryllium compounds')
OSHA - Possible Select Carcinogens
[present]

INTERNATIONAL LISTS

Canada - WHMIS: Ingredient Disclosure
0.1% item 179 (293)

STATE LISTS

Florida Hazardous Substance List
[present]

Massachusetts Right To Know List
carcinogen; extraordinarily hazardous

Pennsylvania Right to Know List
special hazardous substance

Pennsylvania RTK - Special Hazardous Substances
[present]

BERYLLIUM   7440-41-7
SEE ALSO:
F039-HAZARDOUS WASTES

HEALTH AND SAFETY LISTS

ACGIH 1995 - Time Weighted Averages
as Be: 0.002 mg/m3 TWA

ACGIH 1995 - Carcinogens
as Be: A2-suspected human carcinogen

U.S. DOT - Substances From 49 CFR 172.101
regulated by DOT (UN1567)

U.S. DOT - Hazard Classes
DOT hazard class = 6.1

U.S. DOT - Appendix A Table 1 - Hazardous Substances
final RQ = 10 pounds (4.54 kg) (no reporting of releases of this hazardous substance is required if the diameter of the pieces of the solid metal released is equal to or exceeds 0.004 inches)

NIOSH 1990 - Pocket Guide - RELs
as Be: Not to exceed 0.0005 mg/m3

NIOSH 1990 - Pocket Guide - IDLHs
as Be: 10 mg/m3 IDLH (not considering carcinogenic effects)

NIOSH 1990 - Pocket Guide - Carcinogens
occupational carcinogen

NIOSH 1990 - Pocket Guide - Target organs
lungs, skin, eyes, mucous membrane

NIOSH - Health Standards - Exposure Limits
do not exceed 0.5 ug Be/m3

NIOSH - Health Standards - Health Effects and Precautions
Lung cancer, berylliosis (Periodic chest X-ray and pulmonary function testing required)

NIOSH - Health Standards - Carcinogenic Chemicals
potential human carcinogen

NTP Seventh Report - Suspect Carcinogens
suspect carcinogen (Listed under 'Beryllium and certain beryllium compounds')

OSHA - Vacated PELs - Time Weighted Averages
as Be: 2 ug/m3 TWA

OSHA - Vacated PELs - Short Term Exposure Limits
25 ug/m3 STEL (30 min)

OSHA - Vacated PELs - Ceiling Limits
C 5 ug/m3

OSHA - Final PELs - Time Weighted Averages
2 ug/m3 TWA; C 25 ppm

OSHA - Final PELs - Ceiling Limits
C 25 ppm

OSHA - Possible Select Carcinogens
[present]

ENVIRONMENTAL LISTS

ATSDR Priority List
Rank (of 275): 039

CERCLA/SARA - Section 313 - Emission Reporting
form R reporting required for 0.1% de minimus concentration

CERCLA/SARA - Hazardous Substances and their Reportable Quantities
final RQ = 10 pounds (4.54 kg) (no reporting of releases of this hazardous substance is required if the diameter of the pieces of the solid metal released is equal to or exceeds 0.004 inches)

Clean Water Act - Priority Pollutants
[present]

Clean Water Act - Toxic Pollutants
[present] (Listed under 'Beryllium and compounds')

Safe Drinking Water Act - MCLs
MCL = 0.004 mg/L

Safe Drinking Water Act - MCLGs
MCLG = 0.004 mg/L

EPA - Carcinogen Hazard Ranking for RQ Adjustment
Hazard ranking = Medium

RCRA - P Series Wastes
waste number P015

RCRA - Hazardous Constituents-Appendix VIII
waste number P015

RCRA - Basis for Listing - Appendix VII
Included in waste stream: F039

RCRA - Substances Banned From Land Disposal
(powder)

RCRA - TSD Facilities Ground Water Monitoring
TM 6010 = 3 ug/L PQL; TM 7090 = 50 ug/L PQL; TM 7091 = 2 ug/L PQL (all species in the ground water that contain this element are included)

RCRA - Universal Treatment Standards (LDR)
WW: 0.82 mg/l; NWW: 0.014 mg/l TCLP

INTERNATIONAL LISTS

Australian Exposure Standards - Time Weighted Averages
0.002 mg/m3 TWA

Australian Exposure Standards - Carcinogens
probable carcinogen

Canada - WHMIS: Ingredient Disclosure
0.1% item 182 (295)

Canada - CEPA Schedule III Part II - Restricted Substances (Ocean Dumping)
[present]

Canada - Alberta - 8 Hour Occupational Exposure Limit
0.002 mg/m3 TWA

Canada - Alberta - 15 Minute Occupational Exposure Limit
0.006 mg/m3 STEL

Canada - Alberta - Designated Substances
designated substance - requires code of practice

Canada - British Columbia - 8 Hour Exposure Limits
0.002 mg/m3 TWA

Canada - British Columbia - 15 Minute Exposure Limits
0.025 mg/m3 STEL

Canada - British Columbia - Carcinogens
carcinogen - 0.002 mg/m3 TWA

Canada - Ontario - OHSA - TWAEVs
as Be: 0.002 mg/m3 TWAEV

Canada - Quebec - Time-Weighted Average Exposure Values
0.002 mg/m3 TWAEV (substance of which the recirculation is prohibited)

Canada - Quebec - Carcinogens
C2 carcinogen: effect suspected in humans

German (DFG) - Carcinogens
animal evidence of carcinogenicity

Israel - Time Weighted Averages
as Be: 0.002 mg/m3 TWA

Israel - Action Levels
as Be: 0.001 mg/m3 AL

Mexico - Instruction No. 10 - TWAs
0.002 mg/m3 TWA

Mexico - Instruction No. 10 - Carcinogens
potential carcinogen in humans - limited epidemiological evidence

Mexico - Wastewater - Organic Toxic Pollutants and Heavy Metals
Listed under [Heavy Metals]

Mexico - Drinking Water - Ecological Criteria
0.00007 mg/l Substance presents persistence, bioaccumulations or risk of cancer, reduce human exposure to a minimum; This level has been extrapolated by using a mathematic model

STATE LISTS

California - Air Bill 2588 Appendix A-I
known or potential carcinogen

California - Prop. 65 - No Significant Risk Levels
no significant risk level = 0.1 ug/day

California - Directors List of Hazardous Substances (8 CCR 339)
[present] (exempt when encapsulated as a Special Form Material, 40 CFR 173.403)

Florida Hazardous Substance List
[present]

Massachusetts Right To Know List
carcinogen; extraordinarily hazardous

Minnesota Hazardous Substance List
carcinogen (includes its compounds)

NJ Right to Know List (Total)
sn 0222

NJ Special Hazardous Substances
(carcinogen)

Pennsylvania Right to Know List
environmental hazard; special hazardous substance (any compound of this substance is also an environmental hazard)

Pennsylvania RTK - Special Hazardous Substances
[present]

BERYLLIUM 7   13966-02-4

HEALTH AND SAFETY LISTS

U.S. DOT - Appendix A Table 2 - Radionuclides
final RQ = 100 curies (3.7E 12 Bq)

ENVIRONMENTAL LISTS

CERCLA/SARA List of Radionuclides (Appendix B) and Their Reportable Quantities
final RQ = 100 curies (3.7E 12 Bq)

BERYLLIUM 10   14390-89-7

HEALTH AND SAFETY LISTS

U.S. DOT - Appendix A Table 2 - Radionuclides
final RQ = 1 curie (3.7E 10 Bq)

ENVIRONMENTAL LISTS

CERCLA/SARA List of Radionuclides (Appendix B) and Their Reportable Quantities
final RQ = 1 curie (3.7E 10 Bq)

BERYLLIUM ALUMINUM ALLOY   12770-50-2

HEALTH AND SAFETY LISTS

NTP Seventh Report - Suspect Carcinogens
suspect carcinogen (Listed under 'Beryllium and certain beryllium compounds')

OSHA - Possible Select Carcinogens
[present]

STATE LISTS

Florida Hazardous Substance List
[present]

Massachusetts Right To Know List
carcinogen; extraordinarily hazardous

Pennsylvania Right to Know List
environmental hazard; special hazardous substance

Pennsylvania RTK - Special Hazardous Substances
[present]

BERYLLIUM CARBONATE   13106-47-3
SEE ALSO:
BERYLLIUM

STATE LISTS

Florida Hazardous Substance List
[present]

Massachusetts Right To Know List
carcinogen; extraordinarily hazardous

Pennsylvania Right to Know List
environmental hazard; special hazardous substance

Pennsylvania RTK - Special Hazardous Substances
[present]

BERYLLIUM CARBONATE   66104-24-3

HEALTH AND SAFETY LISTS

OSHA - Possible Select Carcinogens
[present]

BERYLLIUM CHLORIDE   7787-47-5
SEE ALSO:
BERYLLIUM

HEALTH AND SAFETY LISTS

NIOSH - Selected LD50s and LC50s
Oral, rat: LD50 = 86 mg/kg

NTP Seventh Report - Suspect Carcinogens
suspect carcinogen (Listed under 'Beryllium and certain beryllium compounds')

OSHA - Possible Select Carcinogens
[present]

ENVIRONMENTAL LISTS

Clean Water Act - Hazardous Substances
[present]

EPA - Carcinogen Hazard Ranking for RQ Adjustment
Hazard ranking = High

INTERNATIONAL LISTS

Canada - WHMIS: Ingredient Disclosure
0.1% item 180 (479)

STATE LISTS

Florida Hazardous Substance List
[present]

Massachusetts Right To Know List
carcinogen; extraordinarily hazardous

NJ Right to Know List (Total)
sn 0223

NJ Special Hazardous Substances
(carcinogen)

Pennsylvania Right to Know List
environmental hazard; special hazardous substance

Pennsylvania RTK - Special Hazardous Substances
[present]

BERYLLIUM COMPOUNDS, N.O.S.   RR-00557-2

HEALTH AND SAFETY LISTS

U.S. DOT - Substances From 49 CFR 172.101
regulated by DOT (UN1566)

U.S. DOT - Hazard Classes
DOT hazard class = 6.1

IARC - Group 2A (limited human data)
[present] (This evaluation applies to the group of chemicals as a whole and not necessarily to all individual chemicals within the group)

OSHA - Possible Select Carcinogens
[present]

ENVIRONMENTAL LISTS

CERCLA/SARA - Section 313 - Emission Reporting
form R reporting required for 0.1% (inorganic), 1.0% (organic) de minimus concentration

Clean Air Act (1990) - List of Hazardous Air Contaminants
[present] (includes any unique chemical substance that contains Beryllium as part of that chemical's infrastructure)

Clean Water Act - Toxic Pollutants
[present] (Listed under 'Beryllium and compounds')

RCRA - Hazardous Constituents-Appendix VIII
hazardous constituent - no waste number

INTERNATIONAL LISTS

Canada - WHMIS: Ingredient Disclosure
0.1% item 181 (294)

Canada - CEPA Schedule III Part II - Restricted Substances (Ocean Dumping)
[present]

Canada - Ontario - OHSA - TWAEVs
as Be: 0.002 mg/m3 TWAEV

STATE LISTS

California - Air Bill 2588 Appendix A-I
known or potential carcinogen: 9/89

California - Prop. 65 - Cancer list
carcinogen - initial date 10/1/87

California - Exposure Limits - PELs
0.002 mg/m3 PEL

California - Exposure Limits - STELs
0.005 mg/m3 30-minute STEL

California - Exposure Limits - Ceilings
C 0.025 mg/m3

California - Directors List of Hazardous Substances (8 CCR 339)
[present]

NJ Right to Know List (Total)
sn 2163; sn 2868

BERYLLIUM FLUORIDE   7787-49-7
SEE ALSO:
BERYLLIUM

HEALTH AND SAFETY LISTS

U.S. DOT - Appendix A Table 1 - Hazardous Substances
final RQ = 1 pound (0.454 kg)

NIOSH - Selected LD50s and LC50s
Oral, rat: LD50 = 98 mg/kg

NTP Seventh Report - Suspect Carcinogens
suspect carcinogen (Listed under 'Beryllium and certain beryllium compounds')

OSHA - Possible Select Carcinogens
[present]

ENVIRONMENTAL LISTS

CERCLA/SARA - Hazardous Substances and their Reportable Quantities
final RQ = 1 pound (0.454 kg)

Clean Water Act - Hazardous Substances
[present]

EPA - Carcinogen Hazard Ranking for RQ Adjustment
Hazard ranking = High

INTERNATIONAL LISTS

Canada - WHMIS: Ingredient Disclosure
0.1% item 183 (902)

STATE LISTS

Florida Hazardous Substance List
[present]

Massachusetts Right To Know List
carcinogen; extraordinarily hazardous

NJ Right to Know List (Total)
sn 0224

NJ Special Hazardous Substances
(carcinogen)

Pennsylvania Right to Know List
environmental hazard; special hazardous substance

Pennsylvania RTK - Special Hazardous Substances
[present]

BERYLLIUM HYDROXIDE   13327-32-7
SEE ALSO:
BERYLLIUM

HEALTH AND SAFETY LISTS

NTP Seventh Report - Suspect Carcinogens
suspect carcinogen (Listed under 'Beryllium and certain beryllium compounds')

OSHA - Possible Select Carcinogens
[present]

INTERNATIONAL LISTS

Canada - WHMIS: Ingredient Disclosure
0.1% item 184 (990)

STATE LISTS

Florida Hazardous Substance List
[present]

Massachusetts Right To Know List
carcinogen; extraordinarily hazardous

Pennsylvania Right to Know List
environmental hazard; special hazardous substance

Pennsylvania RTK - Special Hazardous Substances
[present]

BERYLLIUM NITRATE   13597-99-4
SEE ALSO:
BERYLLIUM

HEALTH AND SAFETY LISTS

U.S. DOT - Substances From 49 CFR 172.101
regulated by DOT (UN2464)

U.S. DOT - Hazard Classes
DOT hazard class = 5.1

U.S. DOT - Appendix A Table 1 - Hazardous Substances
final RQ = 1 pound (0.454 kg)

ENVIRONMENTAL LISTS

CERCLA/SARA - Hazardous Substances and their Reportable Quantities
final RQ = 1 pound (0.454 kg)

Clean Water Act - Hazardous Substances
[present]

EPA - Carcinogen Hazard Ranking for RQ Adjustment
Hazard ranking = High

STATE LISTS

Massachusetts Right To Know List
[present]

NJ Right to Know List (Total)
sn 0225

Pennsylvania Right to Know List
environmental hazard

BERYLLIUM NITRATE TRIHYDRATE   7787-55-5

HEALTH AND SAFETY LISTS

U.S. DOT - Appendix A Table 1 - Hazardous Substances
final RQ = 1 pound (0.454 kg) (Listed under 'Beryllium nitrate')

ENVIRONMENTAL LISTS
  CERCLA/SARA - Hazardous Substances and their Reportable Quantities
    final RQ = 1 pound (0.454 kg) (Listed under 'Beryllium nitrate')
  Clean Water Act - Hazardous Substances
    [present] (Listed under 'Beryllium nitrate')
STATE LISTS
  Massachusetts Right To Know List
    [present]
  Pennsylvania Right to Know List
    environmental hazard

BERYLLIUM OXIDE   1304-56-9
  SEE ALSO:
    BERYLLIUM
  HEALTH AND SAFETY LISTS
    NTP Seventh Report - Suspect Carcinogens
      suspect carcinogen (Listed under 'Beryllium and certain beryllium compounds')
    OSHA - Possible Select Carcinogens
      [present]
  INTERNATIONAL LISTS
    Canada - WHMIS: Ingredient Disclosure
      0.1% item 185 (1299)
  STATE LISTS
    California - Prop. 65 - No Significant Risk Levels
      no significant risk level = 0.1 ug/day
    Florida Hazardous Substance List
      [present]
    Massachusetts Right To Know List
      carcinogen; extraordinarily hazardous
    NJ Right to Know List (Total)
      sn 0226
    NJ Special Hazardous Substances
      (carcinogen)
    Pennsylvania Right to Know List
      environmental hazard;   special hazardous substance
    Pennsylvania RTK - Special Hazardous Substances
      [present]

BERYLLIUM PHOSPHATE   13598-15-7
  HEALTH AND SAFETY LISTS
    NTP Seventh Report - Suspect Carcinogens
      suspect carcinogen (Listed under 'Beryllium and certain beryllium compounds')
    OSHA - Possible Select Carcinogens
      [present]
  STATE LISTS
    Florida Hazardous Substance List
      [present]
    Massachusetts Right To Know List
      carcinogen; extraordinarily hazardous
    Pennsylvania Right to Know List
      environmental hazard;   special hazardous substance
    Pennsylvania RTK - Special Hazardous Substances
      [present]

BERYLLIUM PHOSPHATE   35089-00-0
  STATE LISTS
    Pennsylvania Right to Know List
      environmental hazard;   special hazardous substance
    Pennsylvania RTK - Special Hazardous Substances
      [present]

BERYLLIUM SULFATE   13510-49-1
  SEE ALSO:
    BERYLLIUM
  HEALTH AND SAFETY LISTS
    NIOSH - Selected LD50s and LC50s
      Inhalation, guinea pig: LC50 = 47 mg/m3 (8 hr) Oral, rat: LD50 = 82 mg/kg
    NTP Seventh Report - Suspect Carcinogens
      suspect carcinogen (Listed under 'Beryllium and certain beryllium compounds')
    OSHA - Possible Select Carcinogens
      [present]
  INTERNATIONAL LISTS
    Canada - WHMIS: Ingredient Disclosure
      0.1% item 186 (1514)
  STATE LISTS
    California - Prop. 65 - No Significant Risk Levels
      no significant risk level = 0.0002 ug/day
    Florida Hazardous Substance List
      [present]
    Massachusetts Right To Know List
      carcinogen; extraordinarily hazardous
    Pennsylvania Right to Know List
      environmental hazard; special hazardous substance
    Pennsylvania RTK - Special Hazardous Substances
      [present]

BERYLLIUM SULFATE TETRAHYDRATE   7787-56-6
  HEALTH AND SAFETY LISTS
    NTP Seventh Report - Suspect Carcinogens
      suspect carcinogen (Listed under 'Beryllium and certain beryllium compounds')
    OSHA - Possible Select Carcinogens
      [present]
  STATE LISTS
    Massachusetts Right To Know List
      carcinogen; extraordinarily hazardous

BETA RADIATION   12587-47-2
  ENVIRONMENTAL LISTS
    ATSDR Priority List
      Rank (of 275): 112

BETEL QUID WITH TOBACCO   RR-00027-1
  HEALTH AND SAFETY LISTS
    IARC - Group 1 (carcinogenic to humans)
      [present] (with tobacco)
    IARC - Group 3 (not classifiable)
      [present] (without tobacco)
    OSHA - Select Carcinogens
      [present]
  STATE LISTS
    California - Air Bill 2588 Appendix A-II
      known or potential carcinogen
    California - Prop. 65 - Cancer list
      carcinogen - initial date 1/1/90

ALPHA-BHC   319-84-6
  SEE ALSO:
    HEXACHLOROCYCLOHEXANE (MIXED ISOMERS)
    F039-HAZARDOUS WASTES
  HEALTH AND SAFETY LISTS
    U.S. DOT - Appendix A Table 1 - Hazardous Substances
      final RQ = 10 pounds (4.54 kg)
    IARC - Group Unspecified
      [present] (Listed under 'Hexachlorocyclohexanes (HCH)')
    NIOSH - Selected LD50s and LC50s
      Oral, rat: LD50 = 177 mg/kg
    NTP Seventh Report - Suspect Carcinogens
      suspect carcinogen (Listed under 'Lindane and other hexachlorocyclo-
      hexane isomers')
    OSHA - Possible Select Carcinogens
      [present]
  ENVIRONMENTAL LISTS
    ATSDR Priority List
      Rank (of 275): 099
    CERCLA/SARA - Section 313 - Emission Reporting
      form R reporting required
    CERCLA/SARA - Hazardous Substances and their Reportable Quan-
    tities
      final RQ = 10 pounds (4.54 kg)
    Clean Water Act - Priority Pollutants
      [present]
    EPA - Carcinogen Hazard Ranking for RQ Adjustment
      Hazard ranking = Medium
    RCRA - Basis for Listing - Appendix VII
      Included in waste stream: F039
    RCRA - TSD Facilities Ground Water Monitoring
      TM 8080 = 0.05 ug/L PQL;   TM 8250 = 10 ug/L PQL
    RCRA - Universal Treatment Standards (LDR)
      WW: 0.00014 mg/l;   NWW: 0.066 mg/kg
  INTERNATIONAL LISTS
    Mexico - Wastewater - Organic Toxic Pollutants and Heavy Metals
      Listed under [Hexachlorocyclohexane]
  STATE LISTS
    California - Prop. 65 - No Significant Risk Levels
      no significant risk level = 0.3 ug/day
    California - Directors List of Hazardous Substances (8 CCR 339)
      [present]
    Florida Hazardous Substance List
      [present]
    Massachusetts Right To Know List
      carcinogen;   extraordinarily hazardous
    NJ Special Hazardous Substances
      (carcinogen)
    Pennsylvania Right to Know List
      environmental hazard;   special hazardous substance
    Pennsylvania RTK - Special Hazardous Substances
      [present]

BETA-BHC   319-85-7
  SEE ALSO:
    HEXACHLOROCYCLOHEXANE (MIXED ISOMERS)
    F039-HAZARDOUS WASTES
  HEALTH AND SAFETY LISTS
    U.S. DOT - Appendix A Table 1 - Hazardous Substances
      final RQ = 1 pound (0.454 kg)
    IARC - Group Unspecified
      [present] (Listed under 'Hexachlorocyclohexanes (HCH)')

    NIOSH - Selected LD50s and LC50s
      Oral, rat: LD50 = 6000 mg/kg
    NTP Seventh Report - Suspect Carcinogens
      suspect carcinogen (Listed under 'Lindane and other hexachlorocyclo-
      hexane isomers')
    OSHA - Possible Select Carcinogens
      [present]
  ENVIRONMENTAL LISTS
    ATSDR Priority List
      Rank (of 275): 028
    CERCLA/SARA - Hazardous Substances and their Reportable Quan-
    tities
      final RQ = 1 pound (0.454 kg)
    Clean Water Act - Priority Pollutants
      [present]
    EPA - Carcinogen Hazard Ranking for RQ Adjustment
      Hazard ranking = Low
    RCRA - Basis for Listing - Appendix VII
      Included in waste stream: F039
    RCRA - TSD Facilities Ground Water Monitoring
      TM 8080 = 0.05 ug/L PQL;   TM 8250 = 40 ug/L PQL
    RCRA - Universal Treatment Standards (LDR)
      WW: 0.00014 mg/l;   NWW: 0.066 mg/kg
  INTERNATIONAL LISTS
    Mexico - Wastewater - Organic Toxic Pollutants and Heavy Metals
      Listed under [Hexachlorocyclohexane]
  STATE LISTS
    California - Prop. 65 - No Significant Risk Levels
      no significant risk level = 0.5 ug/day
    California - Directors List of Hazardous Substances (8 CCR 339)
      [present]
    Florida Hazardous Substance List
      [present]
    Massachusetts Right To Know List
      carcinogen;   extraordinarily hazardous
    NJ Special Hazardous Substances
      (carcinogen)
    Pennsylvania Right to Know List
      environmental hazard;   special hazardous substance
    Pennsylvania RTK - Special Hazardous Substances
      [present]

DELTA-BHC   319-86-8
  SEE ALSO:
    F039-HAZARDOUS WASTES
  HEALTH AND SAFETY LISTS
    U.S. DOT - Appendix A Table 1 - Hazardous Substances
      final RQ = 1 pound (0.454 kg)
    NIOSH - Selected LD50s and LC50s
      Oral, rat: LD50 = 1000 mg/kg
  ENVIRONMENTAL LISTS
    ATSDR Priority List
      Rank (of 275): 037
    CERCLA/SARA - Hazardous Substances and their Reportable Quan-
    tities
      final RQ = 1 pound (0.454 kg)
    Clean Water Act - Priority Pollutants
      [present]
    RCRA - Basis for Listing - Appendix VII
      Included in waste stream: F039
    RCRA - TSD Facilities Ground Water Monitoring
      TM 8080 = 0.1 ug/L PQL;   TM 8250 = 30 ug/L PQL

RCRA - Universal Treatment Standards (LDR)
WW: 0.023 mg/l; NWW: 0.066 mg/kg

INTERNATIONAL LISTS
Mexico - Wastewater - Organic Toxic Pollutants and Heavy Metals
Listed under [Hexachlorocyclochexane]

STATE LISTS
California - Directors List of Hazardous Substances (8 CCR 339)
[present]

Massachusetts Right To Know List
[present]

Pennsylvania Right to Know List
environmental hazard

BHUSA   RR-00172-9
STATE LISTS
NJ Right to Know List (Total)
sn 2164

BICHROMATES   RR-00782-9
STATE LISTS
Pennsylvania Right to Know List
[present]

BICYCLO(2.2.1)HEPTA-2,5-DIENE   121-46-0
HEALTH AND SAFETY LISTS
U.S. DOT - Substances From 49 CFR 172.101
regulated by DOT (UN2251)
U.S. DOT - Hazard Classes
DOT hazard class = 3
NFPA - Flash Points
flash point = -6 degrees F (-21 degrees C)
NFPA - Hazard Identification Ratings
flammability-3; reactivity-1

STATE LISTS
Florida Hazardous Substance List
[present]
Massachusetts Right To Know List
[present]
Pennsylvania Right to Know List
[present]

BICYCLO[2.2.1]HEPTANE-2-CARBONITRILE, 5-CHLORO-6-((((METHYLAMINO)CARBONYL)OXY)IMINO)-,(1ST-(1-ALPHA, 2-BETA, 4-ALPHA, 5-ALPHA, 6E))-   15271-41-7
HEALTH AND SAFETY LISTS
NIOSH - Selected LD50s and LC50s
Oral, rat: LD50 = 19 mg/kg Skin, rat: LD50 = 303 mg/kg

ENVIRONMENTAL LISTS
CERCLA/SARA - Section 302 Extremely Hazardous Substances and TPQs
TPQ = 500/10,000 pounds

STATE LISTS
Florida Hazardous Substance List
effective March 13, 1992
Massachusetts Right To Know List
extraordinarily hazardous
NJ Right to Know List (Total)
sn 2856
Pennsylvania Right to Know List
environmental hazard

PROPOSED REGULATIONS
CERCLA/SARA - Proposed Hazardous Substance Additions
proposed RQ = 1 pound (.454 kg)
CERCLA/SARA - 1989 Proposed RQ Adjustments
proposed RQ = 100 pounds (45.4 kg)

BICYCLO [2,2,1]5-HEPTENE-2,3-DICARBOXYLIC ANHYDRIDE   826-62-0
HEALTH AND SAFETY LISTS
NIOSH - Selected LD50s and LC50s
Oral, rat: LD50 = 3250 mg/kg

BICYCLOHEXYL   92-51-3
HEALTH AND SAFETY LISTS
NFPA - Flash Points
flash point = 165 degrees F (74 degrees C)
NFPA - Hazard Identification Ratings
health-1; flammability-2; reactivity-0

[1,1'-BICYCLOHEXYL]-2-ONE   90-42-6
ENVIRONMENTAL LISTS
TSCA - Code of Federal Regulations Citations
40 CFR 716.120(a)
TSCA - Health and Safety Reporting List
Effective Date: June 1, 1987; Sunset Date: November 9, 1993

BIFENTHRIN   82657-04-3
ENVIRONMENTAL LISTS
CERCLA/SARA - Section 313 - Emission Reporting
form R reporting required

BIFLUORIDE, N.O.S.   18130-74-0
STATE LISTS
NJ Right to Know List (Total)
sn 2165

BIFLUORIDES, N.O.S.   RR-01343-4
HEALTH AND SAFETY LISTS
U.S. DOT - Substances From 49 CFR 172.101
regulated by DOT (UN1740)
U.S. DOT - Hazard Classes
DOT hazard class = 8

BINAPACRYL   485-31-4
HEALTH AND SAFETY LISTS
U.S. DOT - Appendix B - Marine Pollutants
DOT regulated severe marine pollutant

BIOBAN-S   137-40-6
HEALTH AND SAFETY LISTS
NIOSH - Selected LD50s and LC50s
Skin, rabbit: LD50 = 1640 mg/kg
ENVIRONMENTAL LISTS
List of Pesticide Product Inert Ingredients
[present]

BIOLOGICAL WARFARE SUBSTANCES   RR-01607-9
INTERNATIONAL LISTS
Canada - CEPA Schedule III Part I - Prohibited Substances (Ocean Dumping)
[present]

BIPHENYL   92-52-4

HEALTH AND SAFETY LISTS

ACGIH 1995 - Time Weighted Averages
0.2 ppm TWA; 1.3 mg/m3 TWA

AIHA - Odor Threshold Values
no geometric mean air odor threshold

U.S. DOT - Appendix B - Marine Pollutants
DOT regulated marine pollutant

NFPA - Flash Points
flash point = 235 degrees F (113 degrees C)

NFPA - Hazard Identification Ratings
health-2; flammability-1; reactivity-0

NIOSH - Selected LD50s and LC50s
Oral, rat: LD50 = 3280 mg/kg

NIOSH 1990 - Pocket Guide - RELs
1 mg/m3 TWA; 0.2 ppm TWA

NIOSH 1990 - Pocket Guide - IDLHs
300 mg/m3 IDLH

NIOSH 1990 - Pocket Guide - Target organs
liver, skin, CNS, eyes, upper respiratory system

OSHA - Vacated PELs - Time Weighted Averages
0.2 ppm TWA; 1 mg/m3 TWA

OSHA - Final PELs - Time Weighted Averages
0.2 ppm TWA; 1 mg/m3 TWA

ENVIRONMENTAL LISTS

ATSDR Priority List
Rank (of 275): 244

CERCLA/SARA - Section 313 - Emission Reporting
form R reporting required for 1.0% de minimus concentration

CERCLA/SARA - Hazardous Substances and their Reportable Quantities
final RQ = 1 pound (.454 kg)

Clean Air Act (1990) - List of Hazardous Air Contaminants
[present]

CAA - HON Rule - SOCMI Chemicals
compliance by Oct. 24, 1994

CAA - HON Rule - Organic HAPs
[present]

TSCA - Code of Federal Regulations Citations
40 CFR 712.30(d); 40 CFR 716.120(a); 40 CFR 799.925

TSCA - PAIR - Reporting List
Reporting Date: November 19, 1982

TSCA - Health and Safety Reporting List
Effective Date: April 29, 1983

TSCA - Chemical Test Rules
Testing required by: manufacturers; processors (40 CFR 799.925)

TSCA - Section 12(b) - Export Notification
export notification required - Section 4

INTERNATIONAL LISTS

Australian Exposure Standards - Time Weighted Averages
0.2 ppm TWA; 1.3 mg/m3 TWA

Canada - WHMIS: Ingredient Disclosure
1% item 658 (780)

Canada - NPRI (National Pollutant Release Inventory)
[present]

Canada - Alberta - 8 Hour Occupational Exposure Limit
0.2 ppm TWA; 1.3 mg/m3 TWA

Canada - Alberta - 15 Minute Occupational Exposure Limit
0.6 ppm STEL; 3.8 mg/m3 STEL

Canada - British Columbia - 8 Hour Exposure Limits
0.2 ppm TWA; 1.5 mg/m3 TWA

Canada - British Columbia - 15 Minute Exposure Limits
0.6 ppm STEL; 4 mg/m3 STEL

Canada - Ontario - OHSA - TWAEVs
0.2 ppm TWAEV; 1.3 mg/m3 TWAEV

Canada - Quebec - Time-Weighted Average Exposure Values
0.2 ppm TWAEV; 1.3 mg/m3 TWAEV

United Kingdom - Occupational Exposure Standards - TWAs
0.2 ppm TWA; 1.5 mg/m3 TWA

United Kingdom - Occupational Exposure Standards - STELs
0.6 ppm STEL; 4 mg/m3 STEL

German (DFG) - MAK Values
0.2 ppm MAK; 1 mg/m3 MAK

German (DFG) - Pregnancy
classification not yet possible

Israel - Time Weighted Averages
0.2 ppm TWA; 1.3 mg/m3 TWA

Israel - Action Levels
0.1 ppm AL; 0.65 mg/m3 AL

Mexico - Instruction No. 10 - TWAs
0.2 ppm TWA; 1.5 mg/m3 TWA

Mexico - Instruction No. 10 - STELs
0.6 ppm STEL; 4 mg/m3 STEL

STATE LISTS

California - Air Bill 2588 Appendix A-I
6/91

California - Exposure Limits - PELs
0.2 ppm PEL; 1.5 mg/m3 PEL

California - Directors List of Hazardous Substances (8 CCR 339)
[present]

Florida Hazardous Substance List
[present]

Massachusetts Right To Know List
[present]

Minnesota Hazardous Substance List
[present]

NJ Right to Know List (Total)
sn 0795

Pennsylvania Right to Know List
environmental hazard

2-BIPHENYLAMINE   90-41-5

HEALTH AND SAFETY LISTS

NFPA - Hazard Identification Ratings
health-2; flammability-1; reactivity-0

NIOSH - Selected LD50s and LC50s
Oral, rat: LD50 = 2340 mg/kg

STATE LISTS

Florida Hazardous Substance List
[present]

Massachusetts Right To Know List
[present]

Pennsylvania Right to Know List
[present]

2-BIPHENYLAMINE HYDROCHLORIDE   2185-92-4

HEALTH AND SAFETY LISTS

NTP Chemical Status Reports - Testing Status and NTIS Number
Technical reports printed (PB83138842)

NTP Chemical Status Reports - Evidence of Carcinogenicity
male rat-negative; female rat-negative; male mice-equivocal; female mice-positive

[1,1'-BIPHENYL]-4,4'-BIS(DIAZONIUM), 3,3'-DIMETHOXY-   20282-70-6
   SEE ALSO:
      BISAZOBIPHENYL DYES
   ENVIRONMENTAL LISTS
      TSCA - Code of Federal Regulations Citations
         40 CFR 716.120(c)

1,1'-BIPHENYL, 2-BROMO-   2052-07-5
   ENVIRONMENTAL LISTS
      TSCA - Code of Federal Regulations Citations
         40 CFR 721.600
      TSCA - Chemicals with Significant New Use Rules
         [present]
      TSCA - Section 12(b) - Export Notification
         export notification required - Section 5

1,1'-BIPHENYL, 3-BROMO-   2113-57-7
   ENVIRONMENTAL LISTS
      TSCA - Code of Federal Regulations Citations
         40 CFR 721.600
      TSCA - Chemicals with Significant New Use Rules
         [present]
      TSCA - Section 12(b) - Export Notification
         export notification required - Section 5

[1,1'-BIPHENYL]-4,4'-DIAMINE, DIHYDROCHLORIDE   531-85-1
   HEALTH AND SAFETY LISTS
      NTP Chemical Status Reports - Testing Status and NTIS Number
         Prechronic studies for which toxicity technical reports were not prepared
   STATE LISTS
      California - Directors List of Hazardous Substances (8 CCR 339)
         [present] (any mixture containing 0.1% or greater benzidine) (Listed under 'Benzidine (and its salts)')

[1,1'-BIPHENYL]-4,4'-DIAMINO, DICHLORO-   1331-47-1
   ENVIRONMENTAL LISTS
      TSCA - Code of Federal Regulations Citations
         40 CFR 716.120(a)
      TSCA - Health and Safety Reporting List
         Effective Date: June 1, 1987;   Sunset Date: November 9, 1993

1,1'-BIPHENYL, 4,4'-DIBROMO-   92-86-4
   ENVIRONMENTAL LISTS
      TSCA - Code of Federal Regulations Citations
         40 CFR 721.600
      TSCA - Chemicals with Significant New Use Rules
         [present]
      TSCA - Section 12(b) - Export Notification
         export notification required - Section 5

1,1'-BIPHENYL, 4,4'-DIISOCYANATO-3,3'-DIMETHYL-   91-97-4
   ENVIRONMENTAL LISTS
      CERCLA/SARA - Section 313 - Emission Reporting
         form R reporting required; (Listed under 'Diisocyanates')
      TSCA - Code of Federal Regulations Citations
         40 CFR 712.30(x); 40 CFR 716.120(c)
      TSCA - PAIR - Reporting List
         Reporting Date: December 27, 1990

1,1'-BIPHENYL, NONABROMO-   27753-52-2
   ENVIRONMENTAL LISTS

      TSCA - Code of Federal Regulations Citations
         40 CFR 721.600
      TSCA - Chemicals with Significant New Use Rules
         [present]
      TSCA - Section 12(b) - Export Notification
         export notification required - Section 5

(1,1'-BIPHENYL)-4-OL   92-69-3
   ENVIRONMENTAL LISTS
      TSCA - Code of Federal Regulations Citations
         40 CFR 716.120(a)
      TSCA - Health and Safety Reporting List
         Effective Date: June 1, 1987

[1,1'-BIPHENYL]-3-OL   580-51-8
   ENVIRONMENTAL LISTS
      TSCA - Code of Federal Regulations Citations
         40 CFR 716.120(a)
      TSCA - Health and Safety Reporting List
         Effective Date: June 1, 1987;   Sunset Date: November 9, 1993

[1,1'-BIPHENYL]-4-OL, 3-CHLORO-   92-04-6
   HEALTH AND SAFETY LISTS
      NFPA - Flash Points
         flash point = 345 degrees F (174 degrees C)
      NFPA - Hazard Identification Ratings
         health-2; flammability-1; reactivity-0
      NIOSH - Selected LD50s and LC50s
         Oral, rat: LD50 = 4220 mg/kg
   ENVIRONMENTAL LISTS
      TSCA - HDD/HDF - Precursors Required for Reporting
         [present]
   STATE LISTS
      Florida Hazardous Substance List
         [present]
      Massachusetts Right To Know List
         [present].
      Pennsylvania Right to Know List
         [present]

1,1'-BIPHENYL, PHENOXY-   28984-89-6
   ENVIRONMENTAL LISTS
      TSCA - PAIR - Reporting List
         Reporting Date: June 10, 1993
      TSCA - Health and Safety Reporting List
         Effective Date: April 12, 1993

BIPHENYL PHENYL ETHER AND DIPHENYL OXIDE, MIXTURES   RR-01273-7
   HEALTH AND SAFETY LISTS
      U.S. DOT - Appendix B - Marine Pollutants
         DOT regulated marine pollutant

BIPHENYL TRIOZONIDE   RR-01398-9
   HEALTH AND SAFETY LISTS
      U.S. DOT - Hazard Classes
         Forbidden from transport by the DOT

BIPYRIDILIUM PESTICIDES, N.O.S.   RR-00173-0
   HEALTH AND SAFETY LISTS
      U.S. DOT - Substances From 49 CFR 172.101
         regulated by DOT (UN2781, UN2782, UN3015, UN3016)

U.S. DOT - Hazard Classes
toxic or toxic, flammable: DOT hazard class = 6.1; flammable, toxic: DOT hazard class = 3
STATE LISTS
NJ Right to Know List (Total)
sn 2166; sn 2167; sn 2168; sn 2169

2,2'-BIPYRIDINE   366-18-7

HEALTH AND SAFETY LISTS
NIOSH - Selected LD50s and LC50s
Oral, rat: LD50 = 100 mg/kg

2,2-BIS[(ACETYLOXY)METHYL]-1,3-PROPANEDIOL DIACETATE   597-71-7

HEALTH AND SAFETY LISTS
NIOSH - Selected LD50s and LC50s
Oral, mouse: LD50 = 3500 mg/kg

BISALKYLATED FATTY ALKYL AMINE OXIDE   RR-00994-9

ENVIRONMENTAL LISTS
TSCA - Chemicals with Significant New Use Rules
PMN number: P-90-643

N,N'-BIS[2-[2-(3-ALKYL)THIAZOLINE]VINYL]-1,4-PHENYLENEDIAMINE METHYL SULFATE, DOUBLE SALT   RR-00211-9

ENVIRONMENTAL LISTS
TSCA - Chemicals with Significant New Use Rules
PMN number: P-84-913

BIS(4-AMINO-1-ANTHRAQUINONYL)AMINE   128-87-0

INTERNATIONAL LISTS
Canada - WHMIS: Ingredient Disclosure
1% item 187 (296)

1,4-BIS(AMINOMETHYL)CYCLOHEXANE   2549-93-1

INTERNATIONAL LISTS
Canada - WHMIS: Ingredient Disclosure
1% item 188 (297)

N,N-BIS(3-AMINOPROPYL)METHYLAMINE   105-83-9

HEALTH AND SAFETY LISTS
NIOSH - Selected LD50s and LC50s
Oral, rat: LD50 = 1540 mg/kg Skin, rabbit: LD50 = 140 mg/kg
INTERNATIONAL LISTS
Canada - WHMIS: Ingredient Disclosure
1% item 189 (298)

BIS(AMINOPROPYL)PIPERAZINE   7209-38-3

STATE LISTS
NJ Right to Know List (Total)
sn 0231
NJ Special Hazardous Substances
(corrosive)

BIS(1-AZIRIDINYL)MORPHOLINOPHOSPHINE SULPHIDE   2168-68-5

HEALTH AND SAFETY LISTS
IARC - Group 3 (not classifiable)
[present]
STATE LISTS
California - Directors List of Hazardous Substances (8 CCR 339)
[present]

BISAZOBIPHENYL DYES   RR-00280-2

ENVIRONMENTAL LISTS
TSCA - Health and Safety Reporting List
Effective Date: October 4, 1991 (derived from benzidine and its congeners, ortho-toluidine, and dianisidine)

2,2-BIS(BROMOMETHYL)-1,3-PROPANEDIOL   3296-90-0

HEALTH AND SAFETY LISTS
NIOSH - Selected LD50s and LC50s
Oral, rat: LD50 = 3458 mg/kg
NTP Chemical Status Reports - Testing Status and NTIS Number
Two year studies scheduled for peer review; prechronic studies for which for which technical reports were not prepared; assigned to laboratory for toxicolgy/carcinogenesis study
ENVIRONMENTAL LISTS
TSCA - Code of Federal Regulations Citations
40 CFR 712.30(w); 40 CFR 716.120(a)
TSCA - PAIR - Reporting List
Reporting Date: March 12, 1990
TSCA - Health and Safety Reporting List
Effective Date: June 1, 1987

BIS[2-N-BUTOXYETHYL] PHTHALATE   117-83-9

HEALTH AND SAFETY LISTS
NFPA - Flash Points
flash point = 407 degrees F (208 degrees C)
NFPA - Hazard Identification Ratings
health-0; flammability-1; reactivity-0
NIOSH - Selected LD50s and LC50s
Oral, rat: LD50 = 8380 mg/kg

1,4-BIS(BUTYLAMINO)-9,10-ANTHRAQUINONE   17354-14-2

ENVIRONMENTAL LISTS
List of Pesticide Product Inert Ingredients
[present]

BIS(P-TERT-BUTYLPHENYL) PHENYL PHOSPHATE RR-00820-8

HEALTH AND SAFETY LISTS
NFPA - Flash Points
flash point = 482 degrees F (250 degrees C)
NFPA - Hazard Identification Ratings
flammability-1; reactivity-0

BIS(2-(2-CHLOROETHOXY)ETHYL)ETHER   638-56-2

HEALTH AND SAFETY LISTS
NFPA - Flash Points
flash point > 250 degrees F (121 degrees C)
NFPA - Hazard Identification Ratings
health-2; flammability-1; reactivity-0
STATE LISTS
Florida Hazardous Substance List
[present]
Massachusetts Right To Know List
[present]
Pennsylvania Right to Know List
[present]

BIS(2-CHLOROETHOXY)METHANE   111-91-1
SEE ALSO:
F039-HAZARDOUS WASTES

HEALTH AND SAFETY LISTS
U.S. DOT - Appendix A Table 1 - Hazardous Substances
final RQ = 1000 pounds (454 kg)

NFPA - Flash Points
flash point = 230 degrees F (110 degrees C)

NFPA - Hazard Identification Ratings
health-2; flammability-1; reactivity-0

NIOSH - Selected LD50s and LC50s
Oral, rat: LD50 = 65 mg/kg Skin, guinea pig: LD50 = 170 mg/kg

ENVIRONMENTAL LISTS

CERCLA/SARA - Section 313 - Emission Reporting
form R reporting required

CERCLA/SARA - Hazardous Substances and their Reportable Quantities
final RQ = 1000 pounds (454 kg)

Clean Water Act - Priority Pollutants
[present]

RCRA - U Series Wastes
waste number U024

RCRA - Hazardous Constituents-Appendix VIII
waste number U024

RCRA - Basis for Listing - Appendix VII
Included in waste stream: F039

RCRA - Substances Banned From Land Disposal
[present]

RCRA - TSD Facilities Ground Water Monitoring
TM 8270 = 10 ug/L PQL

RCRA - Universal Treatment Standards (LDR)
WW: 0.036 mg/l; NWW: 7.2 mg/kg

TSCA - Code of Federal Regulations Citations
40 CFR 716.120(a)

TSCA - Health and Safety Reporting List
Effective Date: March 7, 1986

TSCA - Multichemical Test Rules - Waste Constituents
hydrolysis testing for Chemical Fate; subchronic toxicity testing for Health Effects

TSCA - Section 12(b) - Export Notification
export notification required - Section 4

INTERNATIONAL LISTS

Mexico - Wastewater - Organic Toxic Pollutants and Heavy Metals
Listed under [Haloethers]

STATE LISTS

California - Directors List of Hazardous Substances (8 CCR 339)
[present]

Florida Hazardous Substance List
[present]

Massachusetts Right To Know List
[present]

NJ Right to Know List (Total)
sn 2971

Pennsylvania Right to Know List
environmental hazard

BIS(2-CHLOROETHYL) ETHER   111-44-4
SEE ALSO:
K017-HAZARDOUS WASTES
CHLOROALKYL ETHERS
F039-HAZARDOUS WASTES

HEALTH AND SAFETY LISTS

ACGIH 1995 - Time Weighted Averages
5 ppm TWA; 29 mg/m3 TWA

ACGIH 1995 - Short Term Exposure Limits
10 ppm STEL; 58 mg/m3 STEL

ACGIH 1995 - Skin Designations
skin - potential for cutaneous absorption

U.S. DOT - Substances From 49 CFR 172.101
regulated by DOT (UN1916)

U.S. DOT - Hazard Classes
DOT hazard class = 6.1

U.S. DOT - Appendix B - Marine Pollutants
DOT regulated marine pollutant

U.S. DOT - Appendix A Table 1 - Hazardous Substances
final RQ = 10 pounds (4.54 kg)

IARC - Group 3 (not classifiable)
[present]

NFPA - Flash Points
flash point = 131 degrees F (55 degrees C)

NFPA - Hazard Identification Ratings
health-3; flammability-2; reactivity-1

NIOSH - Selected LD50s and LC50s
Inhalation, rat: LC50 = 330 mg/m3 4 hr Oral, rat: LD50 = 75 mg/kg Skin, guinea pig: LD50 = 300 mg/kg

NIOSH 1990 - Pocket Guide - RELs
5 ppm TWA; 30 mg/m3 TWA; 10 ppm STEL; 60 mg/m3 STEL

NIOSH 1990 - Pocket Guide - IDLHs
250 ppm IDLH (not considering carcinogenic effects)

NIOSH 1990 - Pocket Guide - Carcinogens
occupational carcinogen

NIOSH 1990 - Pocket Guide - Target organs
skin, eyes, respiratory system

NIOSH 1990 - Pocket Guide - Skin list
Potential for dermal absorption

OSHA - Vacated PELs - Time Weighted Averages
5 ppm TWA; 30 mg/m3 TWA

OSHA - Vacated PELs - Short Term Exposure Limits
10 ppm STEL; 60 mg/m3 STEL

OSHA - Vacated PELs - Skin Designation
Prevent or reduce skin absorption

OSHA - Final PELs - Ceiling Limits
C 15 ppm; C 90 mg/m3

OSHA - Final PELs - Skin Notations
prevent or reduce skin absorption

ENVIRONMENTAL LISTS

ATSDR Priority List
Rank (of 275): 077

CERCLA/SARA - Section 302 Extremely Hazardous Substances and TPQs
TPQ = 10,000 pounds

CERCLA/SARA - Section 313 - Emission Reporting
form R reporting required for 1.0% de minimus concentration

CERCLA/SARA - Hazardous Substances and their Reportable Quantities
final RQ = 10 pounds (4.54 kg)

Clean Air Act (1990) - List of Hazardous Air Contaminants
[present]

CAA - HON Rule - SOCMI Chemicals
compliance by Oct. 24, 1994

CAA - HON Rule - Organic HAPs
[present]

Clean Water Act - Priority Pollutants
[present]

EPA - Carcinogen Hazard Ranking for RQ Adjustment
Hazard ranking = Medium

RCRA - U Series Wastes
waste number U025

RCRA - Hazardous Constituents-Appendix VIII
waste number U025

RCRA - Basis for Listing - Appendix VII
Included in waste stream: F039, K017

RCRA - Substances Banned From Land Disposal
[present]

RCRA - TSD Facilities Ground Water Monitoring
TM 8270 = 10 ug/L PQL

RCRA - Universal Treatment Standards (LDR)
WW: 0.033 mg/l; NWW: 6.0 mg/kg

INTERNATIONAL LISTS

Australian Exposure Standards - Time Weighted Averages
5 ppm TWA; 29 mg/m3 TWA

Australian Exposure Standards - Short Term Exposure Limits
10 ppm STEL; 58 mg/m3 STEL

Australian Exposure Standards - Skin Effects
skin absorption

Canada - WHMIS: Ingredient Disclosure
1% item 547 (838)

Canada - Alberta - 8 Hour Occupational Exposure Limit
5 ppm TWA; 29 mg/m3 TWA

Canada - Alberta - 15 Minute Occupational Exposure Limit
10 ppm STEL; 59 mg/m3 STEL

Canada - Alberta - Skin Designation
can be absorbed through the intact skin

Canada - British Columbia - 8 Hour Exposure Limits
5 ppm TWA; 30 mg/m3 TWA

Canada - British Columbia - 15 Minute Exposure Limits
10 ppm STEL; 60 mg/m3 STEL

Canada - British Columbia - Skin Notations
skin - potential for skin absorption

Canada - Ontario - OHSA - TWAEVs
5 ppm TWAEV; 29 mg/m3 TWAEV

Canada - Ontario - OHSA - STEVs
10 ppm STEV; 58 mg/m3 STEV

Canada - Ontario - OHSA - Skin Notations
absorption through skin, mucous membranes, or eyes

Canada - Quebec - Time-Weighted Average Exposure Values
5 ppm TWAEV; 29 mg/m3 TWAEV

Canada - Quebec - Short-term Exposure Values
10 ppm STEV; 58 mg/m3 STEV

Canada - Quebec - Skin Designations
absorbed through the skin

German (DFG) - MAK Values
10 ppm MAK; 60 mg/m3 MAK

German (DFG) - Peak Limitations
5 x normal MAK (30 min. average value); don't exceed 2 times during shift

German (DFG) - Skin/Sensitizers
danger of cutaneous absorption

Israel - Time Weighted Averages
5 ppm TWA; 29 mg/m3 TWA

Israel - Short Term Exposure Limits
10 ppm STEL; 58 mg/m3 STEL

Israel - Action Levels
2.5 ppm AL; 14.5 mg/m3 AL

Mexico - Wastewater - Organic Toxic Pollutants and Heavy Metals
Listed under [Chloroalkyl Ethers]

Mexico - Drinking Water - Ecological Criteria
0.0003 mg/l This level has been extrapolated by using a mathematic model

STATE LISTS

California - Air Bill 2588 Appendix A-I
known or potential carcinogen: 9/89

California - Prop. 65 - Cancer list
carcinogen - initial date 4/1/88

California - Prop. 65 - No Significant Risk Levels
no significant risk level = 0.3 ug/day

California - Exposure Limits - PELs
5 ppm PEL; 30 mg/m3 PEL

California - Exposure Limits - STELs
10 ppm STEL; 60 mg/m3 STEL

California - Exposure Limits - Skin Notation
material may be absorbed through the skin, eyes or mucous membrane

Florida Hazardous Substance List
effective March 13, 1992

Massachusetts Right To Know List
carcinogen; extraordinarily hazardous

Minnesota Hazardous Substance List
skin

NJ Right to Know List (Total)
sn 0232

NJ Special Hazardous Substances
(carcinogen)

Pennsylvania Right to Know List
environmental hazard

N,N-BIS(2-CHLOROETHYL)-2-NAPHTHYLAMINE   494-03-1

HEALTH AND SAFETY LISTS

U.S. DOT - Appendix A Table 1 - Hazardous Substances
final RQ = 100 pounds (45.4 kg)

IARC - Group 1 (carcinogenic to humans)
[present]

OSHA - Select Carcinogens
[present]

ENVIRONMENTAL LISTS

CERCLA/SARA - Hazardous Substances and their Reportable Quantities
final RQ = 100 pounds (45.4 kg)

EPA - Carcinogen Hazard Ranking for RQ Adjustment
Hazard ranking = Low

RCRA - U Series Wastes
waste number U026

RCRA - Hazardous Constituents-Appendix VIII
waste number U026

RCRA - Substances Banned From Land Disposal
[present]

STATE LISTS

California - Air Bill 2588 Appendix A-II
known or potential carcinogen

California - Prop. 65 - Cancer list
carcinogen - initial date 2/27/87

California - Directors List of Hazardous Substances (8 CCR 339)
[present]

Florida Hazardous Substance List
[present]

Massachusetts Right To Know List
carcinogen; extraordinarily hazardous

Minnesota Hazardous Substance List
carcinogen

NJ Right to Know List (Total)
sn 0371

NJ Special Hazardous Substances
(carcinogen)

Pennsylvania Right to Know List
environmental hazard; special hazardous substance

Pennsylvania RTK - Special Hazardous Substances
[present]

BISCHLOROETHYL NITROSOUREA   154-93-8

HEALTH AND SAFETY LISTS

IARC - Group 2A (limited human data)
[present] (Listed under 'Chloroethyl nitrosourea')

NIOSH - Selected LD50s and LC50s
Oral, rat: LD50 = 20 mg/kg

NTP Chemical Status Reports - Testing Status and NTIS Number
Chronic studies exist for which technical reports were not prepared

NTP Seventh Report - Suspect Carcinogens
suspect carcinogen

OSHA - Possible Select Carcinogens
[present]

STATE LISTS

California - Air Bill 2588 Appendix A-II
known or potential carcinogen

California - Prop. 65 - Cancer list
carcinogen - initial date 7/1/87

California - Prop. 65 - Developmental Toxicity
developmental toxicity - initial date 7/1/90

California - Directors List of Hazardous Substances (8 CCR 339)
[present]

Florida Hazardous Substance List
[present]

Massachusetts Right To Know List
carcinogen; extraordinarily hazardous

Minnesota Hazardous Substance List
carcinogen

NJ Right to Know List (Total)
sn 0233

NJ Special Hazardous Substances
(carcinogen; mutagen)

Pennsylvania Right to Know List
special hazardous substance

Pennsylvania RTK - Special Hazardous Substances
[present]

BIS(2-CHLOROISOPROPYL) ETHER   108-60-1

HEALTH AND SAFETY LISTS

AIHA - WEEL - Time Weighted Averages
3 ppm TWA; 20 mg/m3 TWA

U.S. DOT - Substances From 49 CFR 172.101
regulated by DOT (UN2490)

U.S. DOT - Hazard Classes
DOT hazard class = 6.1

U.S. DOT - Appendix A Table 1 - Hazardous Substances
final RQ = 1000 pounds (454 kg)

IARC - Group 3 (not classifiable)
[present]

NFPA - Flash Points
flash point = 185 degrees F (85 degrees C)

NFPA - Hazard Identification Ratings
health-2; flammability-2; reactivity-0

NIOSH - Selected LD50s and LC50s
Oral, rat: LD50 = 240 mg/kg Skin, rabbit: LD50 = 3000 mg/kg

NTP Chemical Status Reports - Testing Status and NTIS Number
Technical reports printed (PB299741/AS) (PB83169615)

NTP Chemical Status Reports - Evidence of Carcinogenicity
PB299741/AS:
male rat-negative; female rat-negative; PB83169615: male mice-positive; female mice-positive

ENVIRONMENTAL LISTS

CERCLA/SARA - Section 313 - Emission Reporting
form R reporting required for 1.0% de minimus concentration

CERCLA/SARA - Hazardous Substances and their Reportable Quantities
final RQ = 1000 pounds (454 kg)

Clean Water Act - Priority Pollutants
[present]

RCRA - U Series Wastes
waste number U027

RCRA - Hazardous Constituents-Appendix VIII
waste number U027

RCRA - Substances Banned From Land Disposal
[present]

RCRA - TSD Facilities Ground Water Monitoring
TM 8010 = 100 ug/L PQL; TM 8270 = 10 ug/L PQL

RCRA - Universal Treatment Standards (LDR)
WW: 0.055 mg/l; NWW: 7.2 mg/kg

TSCA - Code of Federal Regulations Citations
40 CFR 716.120(a); 40 CFR 799.5055(c), (d)(2)

TSCA - Health and Safety Reporting List
Effective Date: June 1, 1987

TSCA - Multichemical Test Rules - Waste Constituents
hydrolysis testing for Chemical Fate

TSCA - Section 12(b) - Export Notification
export notification required - Section 4

INTERNATIONAL LISTS

Canada - WHMIS: Ingredient Disclosure
1% item 550 (839)

STATE LISTS

California - Air Bill 2588 Appendix A-II
6/91

Florida Hazardous Substance List
[present]

Massachusetts Right To Know List
[present]

Minnesota Hazardous Substance List
[present]

NJ Right to Know List (Total)
sn 0235

Pennsylvania Right to Know List
environmental hazard

BIS(2-CHLOROISOPROPYL)ETHER   39638-32-9
SEE ALSO:
F039-HAZARDOUS WASTES

ENVIRONMENTAL LISTS

RCRA - Basis for Listing - Appendix VII
Included in waste stream: F039

INTERNATIONAL LISTS

Canada - WHMIS: Ingredient Disclosure
1% item 190 (840)

Mexico - Wastewater - Organic Toxic Pollutants and Heavy Metals
Listed under [Haloethers]

Mexico - Drinking Water - Ecological Criteria
0.03 mg/l This level has been extrapolated by using a mathematic model

STATE LISTS

Pennsylvania Right to Know List
environmental hazard

1,2-BIS(CHLOROMETHOXY)ETHANE   13483-18-6

HEALTH AND SAFETY LISTS

IARC - Group 3 (not classifiable)
[present]

STATE LISTS

California - Directors List of Hazardous Substances (8 CCR 339)
[present]

1,4-BIS(CHLOROMETHOXYMETHYL)BENZENE   56894-91-8

HEALTH AND SAFETY LISTS
IARC - Group 3 (not classifiable)
[present]

STATE LISTS
California - Directors List of Hazardous Substances (8 CCR 339)
[present]

BIS(CHLOROMETHYL)KETONE   534-07-6

HEALTH AND SAFETY LISTS
U.S. DOT - Substances From 49 CFR 172.101
regulated by DOT (UN2649)
U.S. DOT - Hazard Classes
DOT hazard class = 6.1
NIOSH - Selected LD50s and LC50s
Inhalation, rat: LC50 = 29 mg/m3 2 hr

ENVIRONMENTAL LISTS
CERCLA/SARA - Section 302 Extremely Hazardous Substances and TPQs
TPQ = 10/10,000 pounds
TSCA - Code of Federal Regulations Citations
40 CFR 716.120(a)
TSCA - Health and Safety Reporting List
Effective Date: June 1, 1987

INTERNATIONAL LISTS
Canada - WHMIS: Ingredient Disclosure
1% item 530 (650)

STATE LISTS
Florida Hazardous Substance List
effective March 13, 1992
Massachusetts Right To Know List
extraordinarily hazardous
NJ Right to Know List (Total)
sn 2170
Pennsylvania Right to Know List
environmental hazard

PROPOSED REGULATIONS
CERCLA/SARA - Proposed Hazardous Substance Additions
proposed RQ = 1 pound (.454 kg)
CERCLA/SARA - 1989 Proposed RQ Adjustments
proposed RQ = 10 pounds (4.54 kg)

2,2-BIS(CHLOROMETHYL)-1,3-PROPANEDIYL TE-TRAKIS(2-CHLOROETHYL) PHOSPHATE   38051-10-4

ENVIRONMENTAL LISTS
TSCA - PAIR - Reporting List
Effective Date: June 14, 1993
TSCA - Health and Safety Reporting List
Effective Date: June 14, 1993

BIS(DI-(BETA-CHLOROETHYL)SULFIDE)PAL-LADOUS CHLORIDE   64047-28-5

INTERNATIONAL LISTS
Canada - WHMIS: Ingredient Disclosure
1% item 192 (515)

BIS-DIETHYLENE GLYCOL MONOETHYL ETHER PHTHALATE   RR-00821-9

HEALTH AND SAFETY LISTS
NFPA - Flash Points
flash point = 405 degrees F (207 degrees C)
NFPA - Hazard Identification Ratings
health-1; flammability-1; reactivity-0

1,2-BIS(DIMETHYLAMINO)ETHANE   110-18-9

HEALTH AND SAFETY LISTS
U.S. DOT - Substances From 49 CFR 172.101
regulated by DOT (UN2372)
U.S. DOT - Hazard Classes
DOT hazard class = 3
NIOSH - Selected LD50s and LC50s
Oral, rat: LD50 = 1580 mg/kg Skin, rabbit: LD50 = 5390 mg/kg

ENVIRONMENTAL LISTS
CAA - HON Rule - SOCMI Chemicals
compliance by Oct. 23, 1995

INTERNATIONAL LISTS
Canada - WHMIS: Ingredient Disclosure
1% item 193 (299)

STATE LISTS
NJ Right to Know List (Total)
sn 2329

BIS(2,4-DIMETHYLBUTYL) MALEATE   RR-00818-4

HEALTH AND SAFETY LISTS
NFPA - Flash Points
flash point = 290 degrees F (143 degrees C)
NFPA - Hazard Identification Ratings
health-1; flammability-1; reactivity-0

BIS[(3,4-EPOXYCYCLOHEXYL)METHYL] ADIPATE   3130-19-6

HEALTH AND SAFETY LISTS
NIOSH - Selected LD50s and LC50s
Oral, rat: LD50 = 4390 mg/kg

BIS(2,3-EPOXYCYCLOPENTYL)ETHER   2386-90-5

HEALTH AND SAFETY LISTS
IARC - Group 3 (not classifiable)
[present]

BIS (2-ETHYLHEXYL) ADIPATE   103-23-1

HEALTH AND SAFETY LISTS
IARC - Group 3 (not classifiable)
[present]
NFPA - Flash Points
*flash point = 402 degrees F (206 degrees C); *See list description*
NFPA - Hazard Identification Ratings
health-0; flammability-1; reactivity-0
NIOSH - Selected LD50s and LC50s
Oral, rat: LD50 = 9110 mg/kg Skin, rabbit: LD50 = 8410 mg/kg
NTP Chemical Status Reports - Testing Status and NTIS Number
Technical reports printed (PB82185927)
NTP Chemical Status Reports - Evidence of Carcinogenicity
male rat-negative; female rat-negative; male mice-positive; female mice-positive

ENVIRONMENTAL LISTS
CERCLA/SARA - Section 313 - Emission Reporting
form R reporting required for 1.0% de minimus concentration
Safe Drinking Water Act - MCLs
MCL = 0.4 mg/L
Safe Drinking Water Act - MCLGs
MCLG = 0.4 mg/L
EPA - Master Testing List
[present]
List of Pesticide Product Inert Ingredients
[present]

INTERNATIONAL LISTS

Canada - WHMIS: Ingredient Disclosure
0.1% item 194 (164)

Canada - NPRI (National Pollutant Release Inventory)
[present]

STATE LISTS

California - Air Bill 2588 Appendix A-I
6/91

California - Directors List of Hazardous Substances (8 CCR 339)
[present]

Massachusetts Right To Know List
[present]

NJ Right to Know List (Total)
sn 0237

Pennsylvania Right to Know List
environmental hazard

BIS(2-ETHYLHEXYL)-2-BUTENEDIOATE   142-16-5

HEALTH AND SAFETY LISTS

NFPA - Flash Points
flash point = 365 degrees F (185 degrees C)

NFPA - Hazard Identification Ratings
health-0; flammability-1; reactivity-0

NIOSH - Selected LD50s and LC50s
Oral, rat: LD50 = 14 gm/kg Skin, rabbit: LD50 = 15 gm/kg

ENVIRONMENTAL LISTS

TSCA - Code of Federal Regulations Citations
40 CFR 712.30(x); 40 CFR 716.120(d)

TSCA - PAIR - Reporting List
Reporting Date: November 27, 1991

TSCA - Health and Safety Reporting List
Effective Date: September 30, 1991

BIS(2-ETHYLHEXYL) DODECANEDIOATE   19074-24-9

HEALTH AND SAFETY LISTS

NIOSH - Selected LD50s and LC50s
Oral, rat: LD50 = 21500 mg/kg

BIS(2-ETHYLHEXYL)-ETHANOLAMINE   RR-00819-5

HEALTH AND SAFETY LISTS

NFPA - Flash Points
flash point = 280 degrees F (138 degrees C)

NFPA - Hazard Identification Ratings
health-1; flammability-1; reactivity-0

BIS(2-ETHYLHEXYL) ISOPHTHALATE   137-89-3

HEALTH AND SAFETY LISTS

NIOSH - Selected LD50s and LC50s
Oral, rat: LD50 = 17300 mg/kg skin, rabbit: LD50 = 7940 mg/kg

BIS((2-ETHYLHEXYLOXY)MALEOLOXY)   15546-12-0
DIBUTYLSTANNANE
SEE ALSO:
TIN

INTERNATIONAL LISTS

Canada - WHMIS: Ingredient Disclosure
1% item 196 (301)

BIS(2-ETHYLHEXYL) SUCCINATE   2915-57-3

HEALTH AND SAFETY LISTS

NFPA - Flash Points
flash point = 315 degrees F (157 degrees C)

NFPA - Hazard Identification Ratings
health-0; flammability-1; reactivity-0

BIS(HYDROGENATED TALLOW ALKYL)DIMETHYL QUATERNARY COMPOUNDS, METHYL SULFATES   61789-81-9

ENVIRONMENTAL LISTS

List of Pesticide Product Inert Ingredients
[present]

N,N-BIS(2-HYDROXYETHYL)-C12-18-ALKYLAMINE   68155-06-6

ENVIRONMENTAL LISTS

List of Pesticide Product Inert Ingredients
[present]

N,N-BIS(2-HYDROXYETHYL)-C12-18-ALKYLAMINE   71786-60-2

ENVIRONMENTAL LISTS

List of Pesticide Product Inert Ingredients
[present]

BIS(2-HYDROXYETHYL) COCOAMINE OXIDE   61791-47-7

ENVIRONMENTAL LISTS

List of Pesticide Product Inert Ingredients
[present]

BIS(2-HYDROXYETHYL)-3-(DECYLOXY)PROPY-LAMINE OXIDE   77500-13-1

ENVIRONMENTAL LISTS

List of Pesticide Product Inert Ingredients
[present]

BIS(2-METHOXYETHYL)PHTHALATE   117-82-8

HEALTH AND SAFETY LISTS

NFPA - Flash Points
flash point = 410 degrees F (210 degrees C)

NFPA - Hazard Identification Ratings
health-0; flammability-1; reactivity-0

NIOSH - Selected LD50s and LC50s
Oral, guinea pig: LD50 = 1600 mg/kg

INTERNATIONAL LISTS

Canada - WHMIS: Ingredient Disclosure
0.1% item 197 (1421)

BISMETHYLETHER OF TETRABROMOBISPHENOL-A   37853-61-5

ENVIRONMENTAL LISTS

TSCA - HDD/HDF - Chemicals Required for Testing
[present]

TSCA - Section 12(b) - Export Notification
export notification required - Section 4

2,6-BIS(1-METHYLHEPTADECYL)-P-CRESOL   5012-62-4

ENVIRONMENTAL LISTS

List of Pesticide Product Inert Ingredients
[present]

BIS(1-METHYLHEPTYL)-2-BUTENEDIOATE   1330-76-3

ENVIRONMENTAL LISTS

List of Pesticide Product Inert Ingredients
[present]

N,N-BIS(1-METHYLHEPTYL) ETHYLENEDIAMINE   RR-00871-9

HEALTH AND SAFETY LISTS

NFPA - Flash Points
flash point > 400 degrees F (204 degrees C)

NFPA - Hazard Identification Ratings
health-0; flammability-1; reactivity-0

BISMUTH   **7440-69-9**

INTERNATIONAL LISTS

Mexico - Wastewater - Organic Toxic Pollutants and Heavy Metals
Listed under [Heavy Metals]

BISMUTH 200   **17239-85-9**

HEALTH AND SAFETY LISTS

U.S. DOT - Appendix A Table 2 - Radionuclides
final RQ = 100 curies (3.7E 12 Bq)

ENVIRONMENTAL LISTS

CERCLA/SARA List of Radionuclides (Appendix B) and Their Reportable Quantities
final RQ = 100 curies (3.7E 12 Bq)

BISMUTH 201   **14280-38-7**

HEALTH AND SAFETY LISTS

U.S. DOT - Appendix A Table 2 - Radionuclides
final RQ = 100 curies (3.7E 12 Bq)

ENVIRONMENTAL LISTS

CERCLA/SARA List of Radionuclides (Appendix B) and Their Reportable Quantities
final RQ = 100 curies (3.7E 12 Bq)

BISMUTH 202   **14687-50-4**

HEALTH AND SAFETY LISTS

U.S. DOT - Appendix A Table 2 - Radionuclides
final RQ = 1000 curies (3.7E 13 Bq)

ENVIRONMENTAL LISTS

CERCLA/SARA List of Radionuclides (Appendix B) and Their Reportable Quantities
final RQ = 1000 curies (3.7E 13 Bq)

BISMUTH 203   **24383-94-6**

HEALTH AND SAFETY LISTS

U.S. DOT - Appendix A Table 2 - Radionuclides
final RQ = 10 curies (3.7E 11 Bq)

ENVIRONMENTAL LISTS

CERCLA/SARA List of Radionuclides (Appendix B) and Their Reportable Quantities
final RQ = 10 curies (3.7E 11 Bq)

BISMUTH 205   **14333-38-1**

HEALTH AND SAFETY LISTS

U.S. DOT - Appendix A Table 2 - Radionuclides
final RQ = 10 curies (3.7E 11 Bq)

ENVIRONMENTAL LISTS

CERCLA/SARA List of Radionuclides (Appendix B) and Their Reportable Quantities
final RQ = 10 curies (3.7E 11 Bq)

BISMUTH 206   **15776-19-9**

HEALTH AND SAFETY LISTS

U.S. DOT - Appendix A Table 2 - Radionuclides
final RQ = 10 curies (3.7E 11 Bq)

ENVIRONMENTAL LISTS

CERCLA/SARA List of Radionuclides (Appendix B) and Their Reportable Quantities
final RQ = 10 curies (3.7E 11 Bq)

BISMUTH 207   **13982-38-2**

HEALTH AND SAFETY LISTS

U.S. DOT - Appendix A Table 2 - Radionuclides
final RQ = 10 curies (3.7E 11 Bq)

ENVIRONMENTAL LISTS

CERCLA/SARA List of Radionuclides (Appendix B) and Their Reportable Quantities
final RQ = 10 curies (3.7E 11 Bq)

BISMUTH 210   **14331-79-4**

HEALTH AND SAFETY LISTS

U.S. DOT - Appendix A Table 2 - Radionuclides
final RQ = 10 curies (3.7E 11 Bq)

ENVIRONMENTAL LISTS

CERCLA/SARA List of Radionuclides (Appendix B) and Their Reportable Quantities
final RQ = 10 curies (3.7E 11 Bq)

BISMUTH 210M   **RR-00480-8**

HEALTH AND SAFETY LISTS

U.S. DOT - Appendix A Table 2 - Radionuclides
final RQ = 0.1 curies (3.7E 9 Bq)

ENVIRONMENTAL LISTS

CERCLA/SARA List of Radionuclides (Appendix B) and Their Reportable Quantities
final RQ = 0.1 curies (3.7E 9 Bq)

BISMUTH 212   **14913-49-6**

HEALTH AND SAFETY LISTS

U.S. DOT - Appendix A Table 2 - Radionuclides
final RQ = 100 curies (3.7E 12 Bq)

ENVIRONMENTAL LISTS

CERCLA/SARA List of Radionuclides (Appendix B) and Their Reportable Quantities
final RQ = 100 curies (3.7E 12 Bq)

BISMUTH 213   **15776-20-2**

HEALTH AND SAFETY LISTS

U.S. DOT - Appendix A Table 2 - Radionuclides
final RQ = 100 curies (3.7E 12 Bq)

ENVIRONMENTAL LISTS

CERCLA/SARA List of Radionuclides (Appendix B) and Their Reportable Quantities
final RQ = 100 curies (3.7E 12 Bq)

BISMUTH 214   **14733-03-0**

HEALTH AND SAFETY LISTS

U.S. DOT - Appendix A Table 2 - Radionuclides
final RQ = 100 curies (3.7E 12 Bq)

ENVIRONMENTAL LISTS

CERCLA/SARA List of Radionuclides (Appendix B) and Their Reportable Quantities
final RQ = 100 curies (3.7E 12 Bq)

BISMUTH CHROMATE   **61204-26-0**

STATE LISTS

Massachusetts Right To Know List
[present]

BISMUTH TELLURIDE   1304-82-1

HEALTH AND SAFETY LISTS

ACGIH 1995 - Time Weighted Averages
undoped: 10 mg/m3 TWA; Se-doped: 5 mg/m3 TWA

OSHA - Vacated PELs - Time Weighted Averages
total dust: 15 mg/m3 TWA; respirable fraction: 5 mg/m3 TWA

OSHA - Final PELs - Time Weighted Averages
total dust: 15 mg/m3 TWA; respirable fraction: 5 mg/m3 TWA

INTERNATIONAL LISTS

Australian Exposure Standards - Time Weighted Averages
10 mg/m3 TWA; Se-doped: 5 mg/m3 TWA

Canada - WHMIS: Ingredient Disclosure
1% item 198 (1559)

Canada - Alberta - 8 Hour Occupational Exposure Limit
10 mg/m3 TWA

Canada - Alberta - 15 Minute Occupational Exposure Limit
20 mg/m3 STEL

Canada - British Columbia - 8 Hour Exposure Limits
10 mg/m3 TWA

Canada - British Columbia - 15 Minute Exposure Limits
20 mg/m3 STEL

Canada - Ontario - OHSA - TWAEVs
10 mg/m3 TWAEV; selenium-doped: 5 mg/m3 TWAEV

Canada - Quebec - Time-Weighted Average Exposure Values
Un-doped: 10 mg/m3 TWAEV; Se-doped: 5 mg/m3 TWAEV

United Kingdom - Occupational Exposure Standards - TWAs
10 mg/m3 TWA; selenium doped: 5 mg/m3 TWA

United Kingdom - Occupational Exposure Standards - STELs
20 mg/m3 STEL; selenium doped: 10 mg/m3 STEL

Israel - Time Weighted Averages
10 mg/m3 TWA; Se-doped: 5 mg/m3 TWA

Israel - Action Levels
5 mg/m3 AL; Se-doped: 2.5 mg/m3 AL

Mexico - Instruction No. 10 - TWAs
as undoped: 10 mg/m3 TWA

Mexico - Instruction No. 10 - STELs
as undoped: 20 mg/m3 STEL

STATE LISTS

California - Exposure Limits - PELs
total dust: 10 mg/m3 PEL; respirable fraction: 5 mg/m3 PEL; selenium-doped: 5 mg/m3 PEL

Florida Hazardous Substance List
[present]

Massachusetts Right To Know List
[present]

Minnesota Hazardous Substance List
[present] fertility effects, mutagenicity, and screening subcategory testing; (results apply to all members of Glycidyl subcategory I-A)

Pennsylvania Right to Know List
[present]

BISMUTH TELLURIDE   37293-14-4

HEALTH AND SAFETY LISTS

OSHA - Vacated PELs - Time Weighted Averages
5 mg/m3 TWA

INTERNATIONAL LISTS

Canada - Alberta - 8 Hour Occupational Exposure Limit
5 mg/m3 TWA

Canada - Alberta - 15 Minute Occupational Exposure Limit
10 mg/m3 STEL

Canada - British Columbia - 8 Hour Exposure Limits
5 mg/m3 TWA

Canada - British Columbia - 15 Minute Exposure Limits
10 mg/m3 STEL

Mexico - Instruction No. 10 - TWAs
as Se-doped: 5 mg/m3 TWA

Mexico - Instruction No. 10 - STELs
as Se-doped: 10 mg/m3 STEL

STATE LISTS

Pennsylvania Right to Know List
[present]

BIS (1,2,2,6,6-PENTAMETHYL-4-PIPERIDINYL) SEBACATE   41556-26-7

ENVIRONMENTAL LISTS

List of Pesticide Product Inert Ingredients
[present]

BIS(1,2,2,6,6-PENTAMETHYL-4-PIPERIDIN-4-OL) ESTER OF CYCLOALIPHATIC SPIROKETAL   RR-01576-9

ENVIRONMENTAL LISTS

TSCA - Chemicals with Significant New Use Rules
PMN number: P-91-1361

BISPHENOL A   80-05-7

HEALTH AND SAFETY LISTS

NIOSH - Selected LD50s and LC50s
Oral, rat: LD50 = 3250 mg/kg 8 hr Skin, rabbit: LD50 = 3000 mg/kg

NTP Chemical Status Reports - Testing Status and NTIS Number
Technical reports printed (PB82184060)

NTP Chemical Status Reports - Evidence of Carcinogenicity
male rat-equivocal; female rat-negative; male mice-negative; female mice-negative

ENVIRONMENTAL LISTS

CERCLA/SARA - Section 313 - Emission Reporting
form R reporting required for 1.0% de minimus concentration

CAA - HON Rule - SOCMI Chemicals
compliance by April 24, 1995

EPA - Master Testing List
[present]

List of Pesticide Product Inert Ingredients
[present]

TSCA - Code of Federal Regulations Citations
40 CFR 712.30(k); 40 CFR 716.120(a); 40 CFR 799.940

TSCA - PAIR - Reporting List
Reporting Date: August 27, 1984

TSCA - Health and Safety Reporting List
Effective Date: June 28, 1984

TSCA - Chemical Test Rules
Testing required by: manufacturers; processors (40 CFR 799.940)

TSCA - Section 12(b) - Export Notification
export notification required - Section 4

INTERNATIONAL LISTS

Canada - WHMIS: Ingredient Disclosure
1% item 199 (302)

Canada - NPRI (National Pollutant Release Inventory)
[present]

STATE LISTS

California - Air Bill 2588 Appendix A-I
6/91

Massachusetts Right To Know List
[present]

NJ Right to Know List (Total)
sn 2898

Pennsylvania Right to Know List
environmental hazard

BISPHENOL A DIGLYCIDYL ETHER   1675-54-3
SEE ALSO:
GLYCIDOL (OXIRANEMETHANOL) AND ITS DERIVATIVES

HEALTH AND SAFETY LISTS
IARC - Group 3 (not classifiable)
[present]
NIOSH - Selected LD50s and LC50s
Oral, rat: LD50 = 11 gm/kg

ENVIRONMENTAL LISTS
EPA - Master Testing List
[present]
List of Pesticide Product Inert Ingredients
[present]
TSCA - Code of Federal Regulations Citations
40 CFR 716.120(c)
TSCA - Substances Subject to Testing Consent Orders
Test for: Health Effects and Exposure Evaluation

INTERNATIONAL LISTS
Canada - WHMIS: Ingredient Disclosure
1% item 200 (809)

PROPOSED REGULATIONS
TSCA - Proposed Testing Rule for Glycidyl Ethers
subject to comprehensive subcategory testing (results apply to all members of Glycidyl subcategory VI-A)

BISPHENOL A, EPICHLOROHYDRIN,   RR-00231-3
METHYLENEBIS(SUBSTITUTEDCARBOMONO-
CYCLE), POLYALKYLENE GLYCOL, ALKANOL,
METHACRYLATE POLYMER
ENVIRONMENTAL LISTS
TSCA - Chemicals with Significant New Use Rules
PMN number: P-88-2380

BISPHENOL A-EPICHLOROHYDRIN POLYMER   25068-38-6
HEALTH AND SAFETY LISTS
NIOSH - Selected LD50s and LC50s
Oral, rat: LD50 = 13600 mg/kg Skin, rabbit: LD50 = 2000 mg/kg
ENVIRONMENTAL LISTS
List of Pesticide Product Inert Ingredients
[present]

BISPHENOL A, EPICHLOROHYDRIN,   RR-00947-2
POLYALKYLENEPOLYOL AND POLYISOCYANATO
DERIVATIVE
ENVIRONMENTAL LISTS
TSCA - Chemicals with Significant New Use Rules
PMN number: P-89-750

BISPHENOL DERIVATIVE   RR-01566-7
ENVIRONMENTAL LISTS
TSCA - Chemicals with Significant New Use Rules
PMN number: P-92-509

BIS(SUBSTITUTED)CARBOMONOCYCLIC   RR-00913-2
AZOCARBOMONOCYLICOL
ENVIRONMENTAL LISTS
TSCA - Chemicals with Significant New Use Rules
PMN number: P-88-1753

BIS(2,2,6,6-TETRAMETHYLPIPERIDINYL) ESTER OF   RR-00953-0
CYCLOALKYL SPIROKETAL
ENVIRONMENTAL LISTS

TSCA - Chemicals with Significant New Use Rules
PMN number: P-88-0083

1,4-BIS(P-TOLYLAMINO)ANTHRAQUINONE   128-80-3
HEALTH AND SAFETY LISTS
NIOSH - Selected LD50s and LC50s
Oral, rat: LD50 = 3660 mg/kg

ENVIRONMENTAL LISTS
List of Pesticide Product Inert Ingredients
[present]

INTERNATIONAL LISTS
Canada - WHMIS: Ingredient Disclosure
1% item 201 (304)

BIS(TRIBUTYLTIN)OXIDE   56-35-9
SEE ALSO:
TIN
TIN ORGANIC COMPOUNDS
TIN

HEALTH AND SAFETY LISTS
NIOSH - Selected LD50s and LC50s
Oral, rat: LD50 = 87 mg/m3 Skin, rat: LD50 = 11700 mg/kg

ENVIRONMENTAL LISTS
CERCLA/SARA - Section 313 - Emission Reporting
form R reporting required

INTERNATIONAL LISTS
Canada - WHMIS: Ingredient Disclosure
1% item 202 (1300)
German (DFG) - MAK Values
as TBTO: 0.002 ppm MAK; 0.05 mg/m3 MAK (Listed under 'Tri-n-butyltin compounds')
German (DFG) - Peak Limitations
5 x normal MAK (30 min. average value); don't exceed 2 times during shift (Listed under 'Tri-n-butyltin compounds')
German (DFG) - Pregnancy
no risk to embryo/fetus if exposure limits adhered to (Listed under 'Tri-n-butyltin compounds')

BIS(TRICHLOROMETHYL)SULFONE   3064-70-8
HEALTH AND SAFETY LISTS
NIOSH - Selected LD50s and LC50s
Oral, rat: LD50 = 691 mg/kg

BIS(TRIDECYL) SODIUM SULFOSUCCINATE   2673-22-5
ENVIRONMENTAL LISTS
List of Pesticide Product Inert Ingredients
[present]

BIS(2,2,4-TRIMETHYLPENTANEDIOLISOBU-   RR-00817-3
TYRATE) DIGLYCOLATE
HEALTH AND SAFETY LISTS
NFPA - Flash Points
flash point = 383 degrees F (195 degrees C)
NFPA - Hazard Identification Ratings
health-0; flammability-1; reactivity-0

BIS(TRIS(2-METHYL-2-PHENYLPROPYL)TIN)OXIDE   13356-08-6
SEE ALSO:
TIN

ENVIRONMENTAL LISTS
CERCLA/SARA - Section 313 - Emission Reporting
form R reporting required

INTERNATIONAL LISTS

Canada - WHMIS: Ingredient Disclosure
1% item 203 (1301)

BISULFITES   RR-00177-4

HEALTH AND SAFETY LISTS

U.S. DOT - Substances From 49 CFR 172.101
regulated by DOT (UN2693)

U.S. DOT - Hazard Classes
DOT hazard class = 8

IARC - Group 3 (not classifiable)
[present]

STATE LISTS

NJ Right to Know List (Total)
sn 2171

BITOSCANATE   4044-65-9

HEALTH AND SAFETY LISTS

NIOSH - Selected LD50s and LC50s
Oral, rat: LD50 = 21 mg/kg

ENVIRONMENTAL LISTS

CERCLA/SARA - Section 302 Extremely Hazardous Substances and
TPQs
TPQ = 500/10,000 pounds

STATE LISTS

Florida Hazardous Substance List
effective March 13, 1992

Massachusetts Right To Know List
extraordinarily hazardous

NJ Right to Know List (Total)
sn 2172

Pennsylvania Right to Know List
environmental hazard

PROPOSED REGULATIONS

CERCLA/SARA - Proposed Hazardous Substance Additions
proposed RQ = 1 pound (.454 kg)

CERCLA/SARA - 1989 Proposed RQ Adjustments
proposed RQ = 100 pounds (45.4 kg)

BITREX   3734-33-6

HEALTH AND SAFETY LISTS

NIOSH - Selected LD50s and LC50s
Oral, rat: LD50 = 584 mg/kg

ENVIRONMENTAL LISTS

List of Pesticide Product Inert Ingredients
[present]

BITUMENS, EXTRACTS OF STEAM-REFINED AND   RR-00061-3
AIR-REFINED

HEALTH AND SAFETY LISTS

IARC - Group 2B (sufficient animal data)
[present] (extracts of steam and air refined bitumens)

IARC - Group 3 (not classifiable)
[present] (except for air and steam refined bitumens)

OSHA - Possible Select Carcinogens
[present]

STATE LISTS

California - Air Bill 2588 Appendix A-II
known or potential carcinogen

California - Prop. 65 - Cancer list
carcinogen - initial date 1/1/90

Minnesota Hazardous Substance List
*carcinogen (extracts of steam-refined, air-refined, and pooled mixtures
of steam and air refined)*

Pennsylvania Right to Know List
special hazardous substance

Pennsylvania RTK - Special Hazardous Substances
[present]

BLACK NEWSPAPER INK   RR-00340-7

HEALTH AND SAFETY LISTS

NTP Chemical Status Reports - Testing Status and NTIS Number
Technical reports printed (PB93-131910)

BLEOMYCIN   11056-06-7

HEALTH AND SAFETY LISTS

IARC - Group 2B (sufficient animal data)
*[present] (Other relevant data, as given in Supplement 7, influenced the
making of the overall evaluation)*

OSHA - Possible Select Carcinogens
[present]

STATE LISTS

California - Air Bill 2588 Appendix A-II
known or potential carcinogen

California - Directors List of Hazardous Substances (8 CCR 339)
[present]

Minnesota Hazardous Substance List
carcinogen

BOLERO (THIOBENCARB)   28249-77-6

ENVIRONMENTAL LISTS

CERCLA/SARA - Section 313 - Emission Reporting
form R reporting required

STATE LISTS

California - Directors List of Hazardous Substances (8 CCR 339)
[present]

BONE MEAL   68409-75-6

ENVIRONMENTAL LISTS

List of Pesticide Product Inert Ingredients
[present]

BOOT AND SHOE MANUFACTURE AND REPAIR   RR-00539-0

HEALTH AND SAFETY LISTS

IARC - Group 1 (carcinogenic to humans)
[present] (Listed under 'Leather industries')

OSHA - Select Carcinogens
[present]

STATE LISTS

Pennsylvania Right to Know List
special hazardous substance

Pennsylvania RTK - Special Hazardous Substances
[present]

BORATE AND CHLORATE MIXTURES   RR-00180-9

HEALTH AND SAFETY LISTS

U.S. DOT - Substances From 49 CFR 172.101
regulated by DOT (UN1458)

U.S. DOT - Hazard Classes
DOT hazard class = 5.1

STATE LISTS

NJ Right to Know List (Total)
sn 2176

BORATES, TETRA, SODIUM SALTS, ANHYDROUS   1330-43-4

HEALTH AND SAFETY LISTS

OSHA - Vacated PELs - Time Weighted Averages
10 mg/m3 TWA

ENVIRONMENTAL LISTS

List of Pesticide Product Inert Ingredients
[present]

INTERNATIONAL LISTS

Australian Exposure Standards - Time Weighted Averages
1 mg/m3 TWA

Canada - WHMIS: Ingredient Disclosure
1% item 1428 (315)

Canada - Alberta - 8 Hour Occupational Exposure Limit
1 mg/m3 TWA

Canada - Alberta - 15 Minute Occupational Exposure Limit
3 mg/m3 STEL

Canada - British Columbia - 8 Hour Exposure Limits
1 mg/m3 TWA

Canada - Ontario - OHSA - TWAEVs
1 mg/m3 TWAEV

Canada - Quebec - Time-Weighted Average Exposure Values
1 mg/m3 TWAEV

United Kingdom - Occupational Exposure Standards - TWAs
anhydrous: 1 mg/m3 TWA; decahydrate: 5 mg/m3 TWA; pentahydrate: 1 mg/m3 TWA

Mexico - Instruction No. 10 - TWAs
1 mg/m3 TWA

STATE LISTS

California - Exposure Limits - PELs
5 mg/m3 PEL

California - Directors List of Hazardous Substances (8 CCR 339)
[present] (exempt except when present as free crystal/powder)

Florida Hazardous Substance List
effective March 13, 1992

Minnesota Hazardous Substance List
[present]

Pennsylvania Right to Know List
[present]

BORATES, TETRA, SODIUM SALTS, DECAHYDRATE   1303-96-4

HEALTH AND SAFETY LISTS

ACGIH 1995 - Time Weighted Averages
anhydrous: 1 mg/m3 TWA; decahydrate: 5 mg/m3 TWA; pentahydrate: 1 mg/m3 TWA

NIOSH - Selected LD50s and LC50s
Oral, rat: LD50 = 2660 mg/kg

OSHA - Vacated PELs - Time Weighted Averages
10 mg/m3 TWA

ENVIRONMENTAL LISTS

List of Pesticide Product Inert Ingredients
[present]

INTERNATIONAL LISTS

Australian Exposure Standards - Time Weighted Averages
5 mg/m3 TWA

Canada - WHMIS: Ingredient Disclosure
1% item 1458 (1566)

Canada - Alberta - 8 Hour Occupational Exposure Limit
5 mg/m3 TWA

Canada - Alberta - 15 Minute Occupational Exposure Limit
10 mg/m3 STEL

Canada - British Columbia - 8 Hour Exposure Limits
5 mg/m3 TWA

Canada - Ontario - OHSA - TWAEVs
5 mg/m3 TWAEV

Canada - Quebec - Time-Weighted Average Exposure Values
5 mg/m3 TWAEV

Israel - Time Weighted Averages
anhydrous: 1 mg/m3 TWA; decahydrate: 5 mg/m3 TWA; pentahydrate: 1 mg/m3 TWA

Israel - Action Levels
anhydrous: .5 mg/m3 AL; decahydrate: 2.5 mg/m3 AL; pentahydrate: 0.5 mg/m3 AL

Mexico - Instruction No. 10 - TWAs
5 mg/m3 TWA

STATE LISTS

California - Exposure Limits - PELs
5 mg/m3 PEL

California - Directors List of Hazardous Substances (8 CCR 339)
[present] (exempt except when present as free crystal/powder)

Florida Hazardous Substance List
effective March 13, 1992

Massachusetts Right To Know List
[present]

Minnesota Hazardous Substance List
[present]

NJ Right to Know List (Total)
sn 0241

Pennsylvania Right to Know List
[present]

PROPOSED REGULATIONS

Canada - Ontario - Proposed Occupational TWAEVs
2 mg/m3 TWAEV

Canada - Ontario - Proposed Occupational STEVs
5 mg/m3 STEV

BORATES, TETRA, SODIUM SALTS, PENTAHYDRATE   12179-04-3

HEALTH AND SAFETY LISTS

OSHA - Vacated PELs - Time Weighted Averages
10 mg/m3 TWA

INTERNATIONAL LISTS

Australian Exposure Standards - Time Weighted Averages
1 mg/m3 TWA

Canada - Alberta - 8 Hour Occupational Exposure Limit
1 mg/m3 TWA

Canada - Alberta - 15 Minute Occupational Exposure Limit
3 mg/m3 STEL

Canada - British Columbia - 8 Hour Exposure Limits
1 mg/m3 TWA

Canada - Quebec - Time-Weighted Average Exposure Values
1 mg/m3 TWAEV

Mexico - Instruction No. 10 - TWAs
1 mg/m3 TWA

STATE LISTS

California - Exposure Limits - PELs
5 mg/m3 PEL

California - Directors List of Hazardous Substances (8 CCR 339)
[present] (exempt except when present as free crystal/powder)

Florida Hazardous Substance List
effective March 13, 1992

BORATES, TETRA, SODIUM SALTS, PENTAHYDRATE   25481-93-0

INTERNATIONAL LISTS

Canada - Ontario - OHSA - TWAEVs
1 mg/m3 TWAEV

STATE LISTS

California - Directors List of Hazardous Substances (8 CCR 339)
[present] (exempt except when present as free crystal/powder)

Minnesota Hazardous Substance List
[present]

BORATRAN   283-56-7

HEALTH AND SAFETY LISTS

NIOSH - Selected LD50s and LC50s
Oral, mouse: LD50 = 6200 mg/kg

BORAX PENTAHYDRATE   RR-01042-4

ENVIRONMENTAL LISTS

List of Pesticide Product Inert Ingredients
[present]

BORDEAUX ARSENITE, LIQUID OR SOLID   RR-00181-0

STATE LISTS

NJ Right to Know List (Total)
sn 2177

BORIC ACID   11113-50-1

ENVIRONMENTAL LISTS

List of Pesticide Product Inert Ingredients
[present]

BORIC ACID (H3BO3)   10043-35-3

HEALTH AND SAFETY LISTS

NIOSH - Selected LD50s and LC50s
Oral, rat: LD50 = 2660 mg/kg

NTP Chemical Status Reports - Testing Status and NTIS Number
Technical reports printed (PB88213475)

NTP Chemical Status Reports - Evidence of Carcinogenicity
male mice-no evidence; female mice-no evidence

ENVIRONMENTAL LISTS

List of Pesticide Product Inert Ingredients
[present]

INTERNATIONAL LISTS

Canada - WHMIS: Ingredient Disclosure
1% item 204 (67)

BORIC ACID, ALKYL AND SUBSTITUTED ALKYL ESTERS   RR-00160-5

ENVIRONMENTAL LISTS

TSCA - Chemicals with Significant New Use Rules
PMN number: P-86-1252

BORIC ACID, ETHYL ESTER   51845-86-4

HEALTH AND SAFETY LISTS

U.S. DOT - Substances From 49 CFR 172.101
regulated by DOT (UN1176)

U.S. DOT - Hazard Classes
DOT hazard class = 3

STATE LISTS

Pennsylvania Right to Know List
[present]

BORIC ACID (H3BO3), TRIS(2-METHYLPROPYL) ESTER   13195-76-1

HEALTH AND SAFETY LISTS

NFPA - Flash Points
flash point = 185 degrees F (85 degrees C)

NFPA - Hazard Identification Ratings
health-3; flammability-2; reactivity-1

STATE LISTS

Florida Hazardous Substance List
[present]

Massachusetts Right To Know List
[present]

Pennsylvania Right to Know List
[present]

BORNEOL   507-70-0

HEALTH AND SAFETY LISTS

U.S. DOT - Substances From 49 CFR 172.101
regulated by DOT (UN1312)

U.S. DOT - Hazard Classes
DOT hazard class = 4.1

NFPA - Flash Points
flash point = 150 degrees F (66 degrees C)

NFPA - Hazard Identification Ratings
health-2; flammability-2; reactivity-0

NIOSH - Selected LD50s and LC50s
Oral, rat: LD50 = 500 mg/kg

STATE LISTS

Florida Hazardous Substance List
[present]

Massachusetts Right To Know List
[present]

NJ Right to Know List (Total)
sn 0242

Pennsylvania Right to Know List
[present]

BORON   7440-42-8

HEALTH AND SAFETY LISTS

NIOSH - Selected LD50s and LC50s
Oral, mouse: LD50 = 2000 mg/kg

INTERNATIONAL LISTS

Canada - Drinking Water Quality - IMACs
5.0 mg/L IMAC

Mexico - Wastewater - Organic Toxic Pollutants and Heavy Metals
Listed under [Heavy Metals]

Mexico - Drinking Water - Ecological Criteria
1.0 mg/l Substance presents persistence, bioaccumulations or risk of cancer, reduce human exposure to a minimum

PROPOSED REGULATIONS

Safe Drinking Water Act - Priority list
[present]

BORON OXIDE   1303-86-2

HEALTH AND SAFETY LISTS

ACGIH 1995 - Time Weighted Averages
10 mg/m3 TWA

NIOSH - Selected LD50s and LC50s
Oral, mouse: LD50 = 3163 mg/kg

NIOSH 1990 - Pocket Guide - RELs
10 mg/m3 TWA

NIOSH 1990 - Pocket Guide - Target organs
skin, eyes

OSHA - Vacated PELs - Time Weighted Averages
total dust: 10 mg/m3 TWA

OSHA - Final PELs - Time Weighted Averages
total dust: 15 mg/m3 TWA

INTERNATIONAL LISTS

Australian Exposure Standards - Time Weighted Averages
10 mg/m3 TWA

Canada - WHMIS: Ingredient Disclosure
1% item 205 (228)

Canada - Alberta - 8 Hour Occupational Exposure Limit
10 mg/m3 TWA

Canada - Alberta - 15 Minute Occupational Exposure Limit
20 mg/m3 STEL

Canada - British Columbia - 8 Hour Exposure Limits
10 mg/m3 TWA

Canada - British Columbia - 15 Minute Exposure Limits
20 mg/m3 STEL

Canada - Ontario - OHSA - TWAEVs
10 ppm TWAEV

Canada - Quebec - Time-Weighted Average Exposure Values
10 mg/m3 TWAEV

United Kingdom - Occupational Exposure Standards - TWAs
10 mg/m3 TWA

United Kingdom - Occupational Exposure Standards - STELs
20 mg/m3 STEL

German (DFG) - MAK Values
total dust: 15 mg/m3 MAK

German (DFG) - Peak Limitations
5 x normal MAK (30 min. average value);   don't exceed 2 times during shift

Israel - Time Weighted Averages
10 mg/m3 TWA

Israel - Action Levels
5 mg/m3 AL

Mexico - Instruction No. 10 - TWAs
10 mg/m3 TWA

Mexico - Instruction No. 10 - STELs
20 mg/m3 STEL

STATE LISTS

California - Exposure Limits - PELs
10 mg/m3 PEL

California - Directors List of Hazardous Substances (8 CCR 339)
[present] (exempt except when present as free crystal/powder)

Florida Hazardous Substance List
[present]

Massachusetts Right To Know List
[present]

Minnesota Hazardous Substance List
[present]

Pennsylvania Right to Know List
[present]

BORON OXIDE   54566-73-3

STATE LISTS

Pennsylvania Right to Know List
[present]

BORON SODIUM OXIDE (B8NA2O13)   12008-41-2

HEALTH AND SAFETY LISTS

NIOSH - Selected LD50s and LC50s
Oral, rat: LD50 = 2000 mg/kg

ENVIRONMENTAL LISTS

List of Pesticide Product Inert Ingredients
[present]

BORON TRIBROMIDE   10294-33-4

HEALTH AND SAFETY LISTS

ACGIH 1995 - Ceiling Limits
C 1 ppm;   C 10 mg/m3

U.S. DOT - Substances From 49 CFR 172.101
regulated by DOT (UN2692)

U.S. DOT - Hazard Classes
DOT hazard class = 8

U.S. DOT - Substances Which Are Poisonous by Inhalation
liquid hazardous material poisonous by inhalation (UN2692)

OSHA - Vacated PELs - Ceiling Limits
C 1 ppm;   C 10 mg/m3

INTERNATIONAL LISTS

Australian Exposure Standards - Time Weighted Averages
Peak Limitation: 1 ppm;   10 mg/m3

Canada - WHMIS: Ingredient Disclosure
1% item 206 (1636)

Canada - Alberta - Ceiling Occupational Exposure Limit
C 1 ppm;   C 10 mg/m3

Canada - British Columbia - 8 Hour Exposure Limits
1 ppm TWA;   10 mg/m3 TWA

Canada - British Columbia - 15 Minute Exposure Limits
3 ppm STEL;   30 mg/m3 STEL

Canada - Ontario - OHSA - CEVs
1 ppm CEV;   10 mg/m3 CEV

Canada - Quebec - Ceiling Limits
P 1 ppm;   P 10 mg/m3

United Kingdom - Occupational Exposure Standards - STELs
1 ppm STEL;   10 mg/m3 STEL

Israel - Ceiling Exposure Limits
C 1 ppm;   C 10 mg/m3

Mexico - Instruction No. 10 - TWAs
1 ppm TWA;   10 mg/m3 TWA

Mexico - Instruction No. 10 - STELs
3 ppm STEL;   30 mg/m3 STEL

STATE LISTS

California - Exposure Limits - Ceilings
C 1 ppm;   C 10 mg/m3

California - Directors List of Hazardous Substances (8 CCR 339)
[present]

Florida Hazardous Substance List
[present]

Massachusetts Right To Know List
[present]

Minnesota Hazardous Substance List
[present]

NJ Right to Know List (Total)
sn 0244

NJ Special Hazardous Substances
(corrosive)

Pennsylvania Right to Know List
[present]

BORON TRICHLORIDE   10294-34-5

HEALTH AND SAFETY LISTS

U.S. DOT - Substances From 49 CFR 172.101
regulated by DOT (UN1741)

U.S. DOT - Hazard Classes
DOT hazard class = 2.3

U.S. DOT - Substances Which Are Poisonous by Inhalation
gaseous hazardous material poisonous by inhalation (UN1741)

OSHA - List of Highly Hazardous Chemicals
threshhold quantity = 2500 pounds

ENVIRONMENTAL LISTS

CERCLA/SARA - Section 302 Extremely Hazardous Substances and TPQs
TPQ = 500 pounds

CERCLA/SARA - Section 313 - Emission Reporting
form R reporting required

CAA -Toxic Substances for Accidental Release Prevention
threshold quantity = 5,000 lbs

INTERNATIONAL LISTS

Canada - WHMIS: Ingredient Disclosure
1% item 207 (1654)

STATE LISTS

Florida Hazardous Substance List
effective March 13, 1992

Massachusetts Right To Know List
extraordinarily hazardous

NJ Right to Know List (Total)
sn 0245

NJ Special Hazardous Substances
(corrosive)

Pennsylvania Right to Know List
environmental hazard

PROPOSED REGULATIONS

CERCLA/SARA - Proposed Hazardous Substance Additions
proposed RQ = 1 pound (.454 kg)

CERCLA/SARA - 1989 Proposed RQ Adjustments
proposed RQ = 100 pounds (45.4 kg)

BORON TRIFLUORIDE   7637-07-2

HEALTH AND SAFETY LISTS

ACGIH 1995 - Ceiling Limits
C 1 ppm; C 2.8 mg/m3

AIHA - Odor Threshold Values
no geometric mean air odor threshold

U.S. DOT - Substances From 49 CFR 172.101
regulated by DOT (UN1008)

U.S. DOT - Hazard Classes
DOT hazard class = 2.3

U.S. DOT - Substances Which Are Poisonous by Inhalation
gaseous hazardous material poisonous by inhalation (UN1008)

NIOSH - Selected LD50s and LC50s
Inhalation, rat: LC50 = 1180 mg/m3 4 hr

NIOSH 1990 - Pocket Guide - RELs
C 1 ppm; C 3 mg/m3

NIOSH 1990 - Pocket Guide - IDLHs
100 ppm IDLH

NIOSH 1990 - Pocket Guide - Target organs
respiratory system, kidneys, eyes, skin,

NIOSH - Health Standards - Exposure Limits
reduce exposure to lowest feasible concentration

NIOSH - Health Standards - Health Effects and Precautions
Respiratory effects (Pulmonary function testing required)

OSHA - Vacated PELs - Ceiling Limits
C 1 ppm; C 3 mg/m3

OSHA - Final PELs - Ceiling Limits
C 1 ppm; C 3 mg/m3

OSHA - List of Highly Hazardous Chemicals
threshhold quantity = 250 pounds

ENVIRONMENTAL LISTS

CERCLA/SARA - Section 302 Extremely Hazardous Substances and TPQs
TPQ = 500 pounds

CERCLA/SARA - Section 313 - Emission Reporting
form R reporting required

CAA -Toxic Substances for Accidental Release Prevention
threshold quantity = 5,000 lbs

INTERNATIONAL LISTS

Australian Exposure Standards - Time Weighted Averages
Peak Limitation: 1 ppm; 2.8 mg/m3

Canada - WHMIS: Ingredient Disclosure
1% item 208 (1674)

Canada - Alberta - Ceiling Occupational Exposure Limit
C 1 ppm; C 2.8 mg/m3

Canada - British Columbia - Ceiling Exposure Limits
C 1 ppm; C 3 mg/m3

Canada - Ontario - OHSA - CEVs
1 ppm CEV; 2.8 mg/m3 CEV

Canada - Quebec - Ceiling Limits
P 1 ppm; P 2.8 mg/m3

United Kingdom - Occupational Exposure Standards - STELs
1 ppm STEL; 3 mg/m3 STEL

German (DFG) - MAK Values
1 ppm MAK; 3 mg/m3 MAK

German (DFG) - Peak Limitations
2 x normal MAK (5 min momentary value); don't exceed 8 times during shift

Israel - Ceiling Exposure Limits
C 1 ppm; C 2.8 mg/m3

Mexico - Instruction No. 10 - TWAs
1 ppm TWA; 3 mg/m3 TWA

STATE LISTS

California - Exposure Limits - Ceilings
C 1 ppm; C 3 mg/m3

California - Directors List of Hazardous Substances (8 CCR 339)
[present]

Florida Hazardous Substance List
[present]

Massachusetts Right To Know List
extraordinarily hazardous

Minnesota Hazardous Substance List
[present]

NJ Right to Know List (Total)
sn 0246

Pennsylvania Right to Know List
environmental hazard

PROPOSED REGULATIONS

CERCLA/SARA - Proposed Hazardous Substance Additions
proposed RQ = 1 pound (.454 kg)

CERCLA/SARA - 1989 Proposed RQ Adjustments
proposed RQ = 100 pounds (45.4 kg)

BORON TRIFLUORIDE ACETIC ACID   753-53-7

HEALTH AND SAFETY LISTS

U.S. DOT - Substances From 49 CFR 172.101
regulated by DOT (UN1742)

U.S. DOT - Hazard Classes
DOT hazard class = 8

STATE LISTS

NJ Right to Know List (Total)
sn 0247

BORON TRIFLUORIDE COMPOUND WITH METHYL   353-42-4
ETHER (1:1)

HEALTH AND SAFETY LISTS

U.S. DOT - Substances From 49 CFR 172.101
regulated by DOT (UN2965)

U.S. DOT - Hazard Classes
DOT hazard class = 4.3

ENVIRONMENTAL LISTS

CERCLA/SARA - Section 302 Extremely Hazardous Substances and TPQs
TPQ = 1000 pounds

CAA -Toxic Substances for Accidental Release Prevention
threshold quantity = 15,000 lbs

STATE LISTS

Florida Hazardous Substance List
effective March 13, 1992

Massachusetts Right To Know List
extraordinarily hazardous

NJ Right to Know List (Total)
sn 0250

Pennsylvania Right to Know List
environmental hazard

PROPOSED REGULATIONS

CERCLA/SARA - Proposed Hazardous Substance Additions
proposed RQ = 1 pound (.454 kg)

CERCLA/SARA - 1989 Proposed RQ Adjustments
proposed RQ = 1000 pounds (454 kg)

BORON TRIFLUORIDE DIETHYLETHERATE   109-63-7

HEALTH AND SAFETY LISTS

U.S. DOT - Substances From 49 CFR 172.101
regulated by DOT (UN2604)

U.S. DOT - Hazard Classes
DOT hazard class = 8

NFPA - Flash Points
flash point = 147 degrees F (64 degrees C)

NFPA - Hazard Identification Ratings
health-3; flammability-2; reactivity-1 (decomposes in water)

INTERNATIONAL LISTS

Canada - WHMIS: Ingredient Disclosure
1% item 209 (1675)

STATE LISTS

Florida Hazardous Substance List
[present]

Massachusetts Right To Know List
[present]

NJ Right to Know List (Total)
sn 0248

NJ Special Hazardous Substances
(corrosive)

Pennsylvania Right to Know List
[present]

BORON TRIFLUORIDE DIHYDRATE   13319-75-0

HEALTH AND SAFETY LISTS

U.S. DOT - Substances From 49 CFR 172.101
regulated by DOT (UN2851)

U.S. DOT - Hazard Classes
DOT hazard class = 8

INTERNATIONAL LISTS

Canada - WHMIS: Ingredient Disclosure
1% item 210 (1676)

STATE LISTS

NJ Right to Know List (Total)
sn 0249

NJ Special Hazardous Substances
(corrosive)

BORON TRIFLUORIDE PROPIONIC ACID COMPLEX   RR-01344-5

HEALTH AND SAFETY LISTS

U.S. DOT - Substances From 49 CFR 172.101
regulated by DOT (UN1743)

U.S. DOT - Hazard Classes
DOT hazard class = 8

BOROXIN, TRIS((2-ETHYLHEXYL)OXY)-   67859-60-3

ENVIRONMENTAL LISTS

List of Pesticide Product Inert Ingredients
[present]

BRACKEN FERN   RR-00032-8

HEALTH AND SAFETY LISTS

IARC - Group 2B (sufficient animal data)
[present]

OSHA - Possible Select Carcinogens
[present]

STATE LISTS

California - Prop. 65 - Cancer list
carcinogen - initial date 1/1/90

Massachusetts Right To Know List
carcinogen; extraordinarily hazardous

BRAN   RR-01043-5

ENVIRONMENTAL LISTS

List of Pesticide Product Inert Ingredients
[present]

BREAD CRUMBS   RR-01044-6

ENVIRONMENTAL LISTS

List of Pesticide Product Inert Ingredients
[present]

BRILLIANT BLUE G (ACID BLUE 90)   6104-58-1

ENVIRONMENTAL LISTS

List of Pesticide Product Inert Ingredients
[present]

BRILLIANT BLUE R (ACID BLUE 83)   6104-59-2

ENVIRONMENTAL LISTS

List of Pesticide Product Inert Ingredients
[present]

BRODIFACOUM   56073-10-0

HEALTH AND SAFETY LISTS

U.S. DOT - Appendix B - Marine Pollutants
DOT regulated severe marine pollutant

STATE LISTS

Massachusetts Right To Know List
[present]

BROMACIL   314-40-9

HEALTH AND SAFETY LISTS

ACGIH 1995 - Time Weighted Averages
10 mg/m3 TWA

NIOSH - Selected LD50s and LC50s
Oral, rat: LD50 = 641 mg/kg

OSHA - Vacated PELs - Time Weighted Averages
1 ppm TWA; 10 mg/m3 TWA

ENVIRONMENTAL LISTS

CERCLA/SARA - Section 313 - Emission Reporting
form R reporting required

INTERNATIONAL LISTS

Australian Exposure Standards - Time Weighted Averages
1 ppm TWA; 11 mg/m3 TWA

Canada - Alberta - 8 Hour Occupational Exposure Limit
1 ppm TWA; 10 mg/m3 TWA

Canada - Alberta - 15 Minute Occupational Exposure Limit
2 ppm STEL; 21 mg/m3 STEL

Canada - Ontario - OHSA - TWAEVs
1 ppm TWAEV; 11 mg/m3 TWAEV

Canada - Quebec - Time-Weighted Average Exposure Values
10 mg/m3 TWAEV

United Kingdom - Occupational Exposure Standards - TWAs
1 ppm TWA; 10 mg/m3 TWA

United Kingdom - Occupational Exposure Standards - STELs
2 ppm STEL; 20 mg/m3 STEL

Israel - Time Weighted Averages
1 ppm TWA; 11 mg/m3 TWA

Israel - Action Levels
0.5 ppm AL; 5.5 mg/m3 AL

Mexico - Instruction No. 10 - TWAs
1 ppm TWA; 10 mg/m3 TWA

Mexico - Instruction No. 10 - STELs
2 ppm STEL; 20 mg/m3 STEL

STATE LISTS

California - Exposure Limits - PELs
1 ppm PEL; 10 mg/m3 PEL

California - Directors List of Hazardous Substances (8 CCR 339)
[present]

Florida Hazardous Substance List
[present]

Massachusetts Right To Know List
[present]

Minnesota Hazardous Substance List
[present]

Pennsylvania Right to Know List
[present]

PROPOSED REGULATIONS

Safe Drinking Water Act - Priority list
[present]

BROMACIL, LITHIUM SALT [2,4-(1H,3H)-PYRIMIDINEDIONE, 5-BROMO-6-METHYL-3-(1-METHYLPROPYL), LITHIUM SALT]   53404-19-6

ENVIRONMENTAL LISTS

CERCLA/SARA - Section 313 - Emission Reporting
form R reporting required

BROMADIOLONE   28772-56-7

HEALTH AND SAFETY LISTS

NIOSH - Selected LD50s and LC50s
Oral, rat: LD50 = 490 ug/kg

ENVIRONMENTAL LISTS

CERCLA/SARA - Section 302 Extremely Hazardous Substances and TPQs
TPQ = 100/10,000 pounds

STATE LISTS

Florida Hazardous Substance List
effective March 13, 1992

Massachusetts Right To Know List
extraordinarily hazardous

Pennsylvania Right to Know List
environmental hazard

PROPOSED REGULATIONS

CERCLA/SARA - Proposed Hazardous Substance Additions
proposed RQ = 1 pound (.454 kg)

CERCLA/SARA - 1989 Proposed RQ Adjustments
proposed RQ = 100 pounds (45.4 kg)

BROMAMINE ACID   116-81-4

ENVIRONMENTAL LISTS

TSCA - Code of Federal Regulations Citations
40 CFR 712.30(x); 40 CFR 716.120(d)

TSCA - PAIR - Reporting List
Reporting Date: November 27, 1991

TSCA - Health and Safety Reporting List
Effective Date: September 30, 1991

BROMATES, INORGANIC, N.O.S.   RR-00183-2

HEALTH AND SAFETY LISTS

U.S. DOT - Substances From 49 CFR 172.101
regulated by DOT (UN1450)

U.S. DOT - Hazard Classes
DOT hazard class = 5.1

STATE LISTS

NJ Right to Know List (Total)
sn 2180

BROMINATED AROMATIC COMPOUND   RR-00212-0

ENVIRONMENTAL LISTS

TSCA - Chemicals with Significant New Use Rules
PMN number: P-84-824

BROMINATED FLUOROCARBONS   RR-00489-7

STATE LISTS

California - Air Bill 2588 Appendix A-I
[present]

BROMINATED TRIAZINE DERIVATIVE   RR-01253-3

ENVIRONMENTAL LISTS

TSCA - Chemicals with Significant New Use Rules
PMN number: P-91-403

BROMINE   7726-95-6

HEALTH AND SAFETY LISTS

ACGIH 1995 - Time Weighted Averages
0.1 ppm TWA; 0.66 mg/m3 TWA

ACGIH 1995 - Short Term Exposure Limits
0.2 ppm STEL; 1.3 mg/m3 STEL

AIHA - Odor Threshold Values
no geometric mean air odor threshold

U.S. DOT - Substances From 49 CFR 172.101
regulated by DOT (UN1744)

U.S. DOT - Hazard Classes
DOT hazard class = 8

U.S. DOT - Substances Which Are Poisonous by Inhalation
liquid hazardous material poisonous by inhalation (including solutions) (UN1744)

NIOSH - Selected LD50s and LC50s
Inhalation, mouse: LC50 = 750 ppm 9 mn

NIOSH 1990 - Pocket Guide - RELs
0.1 ppm TWA; 0.7 mg/m3 TWA; 0.3 ppm STEL; 2 mg/m3 STEL

NIOSH 1990 - Pocket Guide - IDLHs
10 ppm IDLH

NIOSH 1990 - Pocket Guide - Target organs
respiratory system, eyes, CNS

OSHA - Vacated PELs - Time Weighted Averages
0.1 ppm TWA; 0.7 mg/m3 TWA

OSHA - Vacated PELs - Short Term Exposure Limits
0.3 ppm STEL; 2 mg/m3 STEL

OSHA - Final PELs - Time Weighted Averages
0.1 ppm TWA; 0.7 mg/m3 TWA

OSHA - List of Highly Hazardous Chemicals
threshhold quantity = 1500 pounds

ENVIRONMENTAL LISTS

CERCLA/SARA - Section 302 Extremely Hazardous Substances and TPQs
TPQ = 500 pounds

CERCLA/SARA - Section 313 - Emission Reporting
form R reporting required

CAA -Toxic Substances for Accidental Release Prevention
threshold quantity = 10,000 lbs

TSCA - Code of Federal Regulations Citations
40 CFR 712.30(w)

TSCA - PAIR - Reporting List
Reporting Date: July 13, 1988

INTERNATIONAL LISTS

Australian Exposure Standards - Time Weighted Averages
0.1 ppm TWA; 0.66 mg/m3 TWA

Australian Exposure Standards - Short Term Exposure Limits
0.3 ppm STEL; 2 mg/m3 STEL

Canada - WHMIS: Ingredient Disclosure
1% item 211 (321)

Canada - Alberta - 8 Hour Occupational Exposure Limit
0.1 ppm TWA; 0.65 mg/m3 TWA

Canada - Alberta - 15 Minute Occupational Exposure Limit
0.3 ppm STEL; 2 mg/m3 STEL

Canada - British Columbia - 8 Hour Exposure Limits
0.1 ppm TWA; 0.7 mg/m3 TWA

Canada - British Columbia - 15 Minute Exposure Limits
0.3 ppm STEL; 2 mg/m3 STEL

Canada - Ontario - OHSA - TWAEVs
0.1 ppm TWAEV; 0.7 mg/m3 TWAEV

Canada - Ontario - OHSA - STEVs
0.3 ppm STEV; 2 mg/m3 STEV

Canada - Quebec - Time-Weighted Average Exposure Values
0.1 ppm TWAEV; 0.66 mg/m3 TWAEV

Canada - Quebec - Short-term Exposure Values
0.3 ppm STEV; 2 mg/m3 STEV

United Kingdom - Occupational Exposure Standards - TWAs
0.1 ppm TWA; 0.7 mg/m3 TWA

United Kingdom - Occupational Exposure Standards - STELs
0.3 ppm STEL; 2 mg/m3 STEL

German (DFG) - MAK Values
0.1 ppm MAK; 0.7 mg/m3 MAK

German (DFG) - Peak Limitations
2 x normal MAK (5 min momentary value); don't exceed 8 times during shift

Israel - Time Weighted Averages
0.1 ppm TWA; 0.66 mg/m3 TWA

Israel - Short Term Exposure Limits
0.3 ppm STEL; 2.0 mg/m3 STEL

Israel - Action Levels
0.05 ppm AL; 0.33 mg/m3 AL

Mexico - Instruction No. 10 - TWAs
0.1 ppm TWA; 0.7 mg/m3 TWA

Mexico - Instruction No. 10 - STELs
0.3 ppm STEL; 2 mg/m3 STEL

STATE LISTS

California - Air Bill 2588 Appendix A-I
[present]

California - Exposure Limits - PELs
0.1 ppm PEL; 0.7 mg/m3 PEL

California - Exposure Limits - STELs
0.3 ppm STEL; 2 mg/m3 STEL

California - Directors List of Hazardous Substances (8 CCR 339)
[present]

Florida Hazardous Substance List
[present]

Massachusetts Right To Know List
extraordinarily hazardous

Minnesota Hazardous Substance List
[present]

NJ Right to Know List (Total)
sn 0252

NJ Special Hazardous Substances
(corrosive)

Pennsylvania Right to Know List
environmental hazard

PROPOSED REGULATIONS

CERCLA/SARA - Proposed Hazardous Substance Additions
proposed RQ = 1 pound (.454 kg)

CERCLA/SARA - 1989 Proposed RQ Adjustments
proposed RQ = 100 pounds (45.4 kg)

BROMINE 74   15720-26-0

HEALTH AND SAFETY LISTS

U.S. DOT - Appendix A Table 2 - Radionuclides
final RQ = 100 curies (3.7E 12 Bq)

ENVIRONMENTAL LISTS

CERCLA/SARA List of Radionuclides (Appendix B) and Their Reportable Quantities
final RQ = 100 curies (3.7E 12 Bq)

BROMINE 74M   RR-00478-4

HEALTH AND SAFETY LISTS

U.S. DOT - Appendix A Table 2 - Radionuclides
final RQ = 100 curies (3.7E 12 Bq)

ENVIRONMENTAL LISTS

CERCLA/SARA List of Radionuclides (Appendix B) and Their Reportable Quantities
final RQ = 100 curies (3.7E 12 Bq)

BROMINE 75   14809-47-3

HEALTH AND SAFETY LISTS

U.S. DOT - Appendix A Table 2 - Radionuclides
final RQ = 100 curies (3.7E 12 Bq)

ENVIRONMENTAL LISTS

CERCLA/SARA List of Radionuclides (Appendix B) and Their Reportable Quantities
final RQ = 100 curies (3.7E 12 Bq)

BROMINE 76   15765-38-5

HEALTH AND SAFETY LISTS

U.S. DOT - Appendix A Table 2 - Radionuclides
final RQ = 10 curies (3.7E 11 Bq)

ENVIRONMENTAL LISTS

CERCLA/SARA List of Radionuclides (Appendix B) and Their Reportable Quantities
final RQ = 10 curies (3.7E 11 Bq)

BROMINE 77   15765-39-6

HEALTH AND SAFETY LISTS
U.S. DOT - Appendix A Table 2 - Radionuclides
final RQ = 100 curies (3.7E 12 Bq)

ENVIRONMENTAL LISTS
CERCLA/SARA List of Radionuclides (Appendix B) and Their Reportable Quantities
final RQ = 100 curies (3.7E 12 Bq)

BROMINE 80   14391-61-8

HEALTH AND SAFETY LISTS
U.S. DOT - Appendix A Table 2 - Radionuclides
final RQ = 1000 curies (3.7E 13 Bq)

ENVIRONMENTAL LISTS
CERCLA/SARA List of Radionuclides (Appendix B) and Their Reportable Quantities
final RQ = 1000 curies (3.7E 13 Bq)

BROMINE 80M   RR-00477-3

HEALTH AND SAFETY LISTS
U.S. DOT - Appendix A Table 2 - Radionuclides
final RQ = 1000 curies (3.7E 13 Bq)

ENVIRONMENTAL LISTS
CERCLA/SARA List of Radionuclides (Appendix B) and Their Reportable Quantities
final RQ = 1000 curies (3.7E 13 Bq)

BROMINE 82   14686-69-2

HEALTH AND SAFETY LISTS
U.S. DOT - Appendix A Table 2 - Radionuclides
final RQ = 10 curies (3.7E 11 Bq)

ENVIRONMENTAL LISTS
CERCLA/SARA List of Radionuclides (Appendix B) and Their Reportable Quantities
final RQ = 10 curies (3.7E 11 Bq)

BROMINE 83   14687-62-8

HEALTH AND SAFETY LISTS
U.S. DOT - Appendix A Table 2 - Radionuclides
final RQ = 1000 curies (3.7E 13 Bq)

ENVIRONMENTAL LISTS
CERCLA/SARA List of Radionuclides (Appendix B) and Their Reportable Quantities
final RQ = 1000 curies (3.7E 13 Bq)

BROMINE 84   14331-90-9

HEALTH AND SAFETY LISTS
U.S. DOT - Appendix A Table 2 - Radionuclides
final RQ = 100 curies (3.7E 12 Bq)

ENVIRONMENTAL LISTS
CERCLA/SARA List of Radionuclides (Appendix B) and Their Reportable Quantities
final RQ = 100 curies (3.7E 12 Bq)

BROMINE AZIDE   13973-87-0

HEALTH AND SAFETY LISTS
U.S. DOT - Hazard Classes
Forbidden from transport by the DOT

BROMINE AZIDE   RR-01399-0

STATE LISTS
NJ Right to Know List (Total)
sn 2253;   liquid, corrosive: sn 2254

BROMINE CHLORIDE   13863-41-7

HEALTH AND SAFETY LISTS
U.S. DOT - Substances From 49 CFR 172.101
regulated by DOT (UN2901)

U.S. DOT - Hazard Classes
DOT hazard class = 2.3

U.S. DOT - Substances Which Are Poisonous by Inhalation
gaseous hazardous material poisonous by inhalation (UN2901)

OSHA - List of Highly Hazardous Chemicals
threshhold quantity = 1500 pounds

INTERNATIONAL LISTS
Canada - WHMIS: Ingredient Disclosure
1% item 212 (480)

STATE LISTS
NJ Right to Know List (Total)
sn 0253

NJ Special Hazardous Substances
(corrosive)

BROMINE COMPOUNDS (INORGANIC)   RR-00331-6

STATE LISTS
California - Air Bill 2588 Appendix A-I
[present]

BROMINE PENTAFLUORIDE   7789-30-2

HEALTH AND SAFETY LISTS
ACGIH 1995 - Time Weighted Averages
0.1 ppm TWA;   0.72 mg/m3 TWA

U.S. DOT - Substances From 49 CFR 172.101
regulated by DOT (UN1745)

U.S. DOT - Hazard Classes
DOT hazard class = 5.1

U.S. DOT - Substances Which Are Poisonous by Inhalation
liquid hazardous material poisonous by inhalation (UN1745)

OSHA - Vacated PELs - Time Weighted Averages
0.1 ppm TWA;   0.7 mg/m3 TWA

OSHA - List of Highly Hazardous Chemicals
threshhold quantity = 2500 pounds

INTERNATIONAL LISTS
Australian Exposure Standards - Time Weighted Averages
0.1 ppm TWA;   0.72 mg/m3 TWA

Canada - WHMIS: Ingredient Disclosure
1% item 213 (1344)

Canada - Alberta - 8 Hour Occupational Exposure Limit
0.1 ppm TWA;   0.72 mg/m3 TWA

Canada - Alberta - 15 Minute Occupational Exposure Limit
0.3 ppm STEL;   2.1 mg/m3 STEL

Canada - British Columbia - 8 Hour Exposure Limits
0.1 ppm TWA;   0.7 mg/m3 TWA

Canada - British Columbia - 15 Minute Exposure Limits
0.3 ppm STEL;   2 mg/m3 STEL

Canada - Ontario - OHSA - TWAEVs
0.1 ppm TWAEV;   0.7 mg/m3 TWAEV

Canada - Quebec - Time-Weighted Average Exposure Values
0.1 ppm TWAEV;   0.72 mg/m3 TWAEV

United Kingdom - Occupational Exposure Standards - TWAs
0.1 ppm TWA;   0.7 mg/m3 TWA

United Kingdom - Occupational Exposure Standards - STELs
0.3 ppm STEL;   2 mg/m3 STEL

Israel - Time Weighted Averages
0.1 ppm TWA;   0.72 mg/m3 TWA

Israel - Action Levels
0.05 ppm AL;   0.36 mg/m3 AL

Mexico - Instruction No. 10 - TWAs
0.1 ppm TWA;   0.7 mg/m3 TWA

Mexico - Instruction No. 10 - STELs
0.3 ppm STEL;   2 mg/m3 STEL

STATE LISTS

California - Exposure Limits - PELs
0.1 ppm PEL;   0.7 mg/m3 PEL

California - Directors List of Hazardous Substances (8 CCR 339)
[present]

Florida Hazardous Substance List
[present]

Massachusetts Right To Know List
[present]

Minnesota Hazardous Substance List
[present]

NJ Right to Know List (Total)
sn 0254

NJ Special Hazardous Substances
(corrosive;   reactive - third degree)

Pennsylvania Right to Know List
[present]

BROMINE TRIFLUORIDE   7787-71-5

HEALTH AND SAFETY LISTS

U.S. DOT - Substances From 49 CFR 172.101
regulated by DOT (UN1746)

U.S. DOT - Hazard Classes
DOT hazard class = 5.1

U.S. DOT - Substances Which Are Poisonous by Inhalation
liquid hazardous material poisonous by inhalation (UN1746)

OSHA - List of Highly Hazardous Chemicals
thresshold quantity = 15,000 pounds

INTERNATIONAL LISTS

Canada - WHMIS: Ingredient Disclosure
1% item 214 (1677)

STATE LISTS

Florida Hazardous Substance List
[present]

Massachusetts Right To Know List
[present]

NJ Right to Know List (Total)
sn 0255

NJ Special Hazardous Substances
(corrosive;   reactive - third degree)

Pennsylvania Right to Know List
[present]

BROMOACETIC ACID   79-08-3

HEALTH AND SAFETY LISTS

U.S. DOT - Substances From 49 CFR 172.101
regulated by DOT (UN1938)

U.S. DOT - Hazard Classes
DOT hazard class = 8

NIOSH - Selected LD50s and LC50s
Oral, mouse: LD50 = 100 mg/kg

INTERNATIONAL LISTS

Canada - WHMIS: Ingredient Disclosure
1% item 215 (68)

STATE LISTS

NJ Right to Know List (Total)
sn 2182

BROMOACETONE   598-31-2

HEALTH AND SAFETY LISTS

U.S. DOT - Substances From 49 CFR 172.101
regulated by DOT (UN1569)

U.S. DOT - Hazard Classes
DOT hazard class = 6.1

U.S. DOT - Substances Which Are Poisonous by Inhalation
liquid hazardous material poisonous by inhalation (UN1569)

U.S. DOT - Appendix A Table 1 - Hazardous Substances
final RQ = 1000 pounds (454 kg)

ENVIRONMENTAL LISTS

CERCLA/SARA - Hazardous Substances and their Reportable Quantities
final RQ = 1000 pounds (454 kg)

RCRA - P Series Wastes
waste number P017

RCRA - Hazardous Constituents-Appendix VIII
waste number P017

RCRA - Substances Banned From Land Disposal
[present]

TSCA - Code of Federal Regulations Citations
40 CFR 716.120(a)

TSCA - Health and Safety Reporting List
Effective Date: March 7, 1986

INTERNATIONAL LISTS

Canada - WHMIS: Ingredient Disclosure
1% item 216 (324)

STATE LISTS

Massachusetts Right To Know List
[present]

NJ Right to Know List (Total)
sn 0256

Pennsylvania Right to Know List
environmental hazard

BROMOACETYL BROMIDE   598-21-0

HEALTH AND SAFETY LISTS

U.S. DOT - Substances From 49 CFR 172.101
regulated by DOT (UN2513)

U.S. DOT - Hazard Classes
DOT hazard class = 8

ENVIRONMENTAL LISTS

TSCA - Code of Federal Regulations Citations
40 CFR 716.120(a)

TSCA - Health and Safety Reporting List
Effective Date: June 1, 1987

INTERNATIONAL LISTS

Canada - WHMIS: Ingredient Disclosure
1% item 217 (331)

STATE LISTS

NJ Right to Know List (Total)
sn 0257

NJ Special Hazardous Substances
(corrosive)

BROMODICHLOROETHANE 683-53-4

ENVIRONMENTAL LISTS

ATSDR Priority List
Rank (of 275): 196

4-BROMO-2,5-DICHLOROPHENOL 1940-42-7

ENVIRONMENTAL LISTS

TSCA - HDD/HDF - Chemicals Required for Testing
[present]

TSCA - Section 12(b) - Export Notification
export notification required - Section 4

4-BROMO-1,2-DINITROBENZENE 610-38-8

HEALTH AND SAFETY LISTS

U.S. DOT - Hazard Classes
Forbidden from transport by the DOT

4-BROMODIPHENYL 92-66-0

HEALTH AND SAFETY LISTS

NFPA - Flash Points
flash point = 291 degrees F (144 degrees C)

NFPA - Hazard Identification Ratings
health-2; flammability-1; reactivity-0

ENVIRONMENTAL LISTS

TSCA - Code of Federal Regulations Citations
40 CFR 721.600

TSCA - Chemicals with Significant New Use Rules
[present]

TSCA - Section 12(b) - Export Notification
export notification required - Section 5

STATE LISTS

Florida Hazardous Substance List
[present]

Massachusetts Right To Know List
[present]

Pennsylvania Right to Know List
[present]

BROMOETHYL ETHYL ETHER 592-55-2

HEALTH AND SAFETY LISTS

U.S. DOT - Substances From 49 CFR 172.101
regulated by DOT (UN2340)

U.S. DOT - Hazard Classes
DOT hazard class = 3

STATE LISTS

NJ Right to Know List (Total)
sn 0261

BROMOETHYLPROPANES RR-01349-0

HEALTH AND SAFETY LISTS

U.S. DOT - Substances From 49 CFR 172.101
regulated by DOT (UN2342)

U.S. DOT - Hazard Classes
DOT hazard class = 3

BROMOFORM 75-25-2

SEE ALSO:
F039-HAZARDOUS WASTES

HEALTH AND SAFETY LISTS

ACGIH 1995 - Time Weighted Averages
0.5 ppm TWA; 5.2 mg/m3 TWA

ACGIH 1995 - Skin Designations
skin - potential for cutaneous absorption

AIHA - Odor Threshold Values
no geometric mean air odor threshold

U.S. DOT - Substances From 49 CFR 172.101
regulated by DOT (UN2515)

U.S. DOT - Hazard Classes
DOT hazard class = 6.1

U.S. DOT - Appendix B - Marine Pollutants
DOT regulated marine pollutant

U.S. DOT - Appendix A Table 1 - Hazardous Substances
final RQ = 100 pounds (45.4 kg)

IARC - Group 3 (not classifiable)
[present]

NIOSH - Selected LD50s and LC50s
Inhalation, mammal: LC50 = 12100 mg/m3 (8 hr) Oral, rat: LD50 = 1147 mg/kg

NIOSH 1990 - Pocket Guide - RELs
0.5 ppm TWA; 5 mg/m3 TWA

NIOSH 1990 - Pocket Guide - Target organs
skin, liver, kidneys, respiratory system, CNS

NIOSH 1990 - Pocket Guide - Skin list
Potential for dermal absorption

NTP Chemical Status Reports - Testing Status and NTIS Number
Technical reports printed (PB90110149/AS)

NTP Chemical Status Reports - Evidence of Carcinogenicity
male rat-some evidence; female rat-clear evidence; male mice-no evidence; female mice-no evidence

OSHA - Vacated PELs - Time Weighted Averages
0.5 ppm TWA; 5 mg/m3 TWA

OSHA - Vacated PELs - Skin Designation
Prevent or reduce skin absorption

OSHA - Final PELs - Time Weighted Averages
0.5 ppm TWA; 5 mg/m3 TWA

OSHA - Final PELs - Skin Notations
prevent or reduce skin absorption

ENVIRONMENTAL LISTS

ATSDR Priority List
Rank (of 275): 209

CERCLA/SARA - Section 313 - Emission Reporting
form R reporting required for 1.0% de minimus concentration

CERCLA/SARA - Hazardous Substances and their Reportable Quantities
final RQ = 100 pounds (45.4 kg)

Clean Air Act (1990) - List of Hazardous Air Contaminants
[present]

CAA - HON Rule - SOCMI Chemicals
compliance by Oct. 23, 1995

CAA - HON Rule - Organic HAPs
[present]

Clean Water Act - Priority Pollutants
[present]

Safe Drinking Water Act - Monitoring
monitoring required

RCRA - U Series Wastes
waste number U225

RCRA - Hazardous Constituents-Appendix VIII
waste number U225

RCRA - Basis for Listing - Appendix VII
Included in waste stream: F039

RCRA - Substances Banned From Land Disposal
[present]

RCRA - TSD Facilities Ground Water Monitoring
TM 8010 = 2 ug/L PQL; TM 8240 = 5 ug/L PQL

RCRA - Universal Treatment Standards (LDR)
WW: 0.63 mg/l; NWW: 15 mg/kg

TSCA - Code of Federal Regulations Citations
40 CFR 716.120(a); 40 CFR 799.5055(c), (d)(2)

TSCA - Health and Safety Reporting List
Effective Date: June 1, 1987

TSCA - Multichemical Test Rules - Waste Constituents
hydrolysis testing for Chemical Fate

TSCA - Section 12(b) - Export Notification
export notification required - Section 4

INTERNATIONAL LISTS

Australian Exposure Standards - Time Weighted Averages
0.5 ppm TWA; 5.2 mg/m3 TWA

Australian Exposure Standards - Skin Effects
skin absorption

Canada - WHMIS: Ingredient Disclosure
1% item 220 (326)

Canada - Alberta - 8 Hour Occupational Exposure Limit
0.5 ppm TWA; 5.2 mg/m3 TWA

Canada - Alberta - 15 Minute Occupational Exposure Limit
1.5 ppm STEL; 16 mg/m3 STEL

Canada - Alberta - Skin Designation
can be absorbed through the intact skin

Canada - British Columbia - 8 Hour Exposure Limits
0.5 ppm TWA; 5 mg/m3 TWA

Canada - British Columbia - Skin Notations
skin - potential for skin absorption

Canada - Ontario - OHSA - TWAEVs
0.5 ppm TWAEV; 5 mg/m3 TWAEV

Canada - Ontario - OHSA - Skin Notations
absorption through skin, eyes, or mucous membranes

Canada - Quebec - Time-Weighted Average Exposure Values
0.5 ppm TWAEV; 5.2 mg/m3 TWAEV

Canada - Quebec - Skin Designations
absorbed through the skin

United Kingdom - Occupational Exposure Standards - TWAs
0.5 ppm TWA; 5 mg/m3 TWA

United Kingdom - Occupational Exposure Standards - Notes
can be absorbed through skin

German (DFG) - Carcinogens
suspected carcinogen

Israel - Time Weighted Averages
0.5 ppm TWA; 5.2 mg/m3 TWA

Israel - Action Levels
0.25 ppm AL; 2.6 mg/m3 AL

Mexico - Instruction No. 10 - TWAs
0.5 ppm TWA; 5 mg/m3 TWA

Mexico - Instruction No. 10 - Skin designation
skin - potential for cutaneous absorption

Mexico - Wastewater - Organic Toxic Pollutants and Heavy Metals
Listed under [Halomethanes]

Mexico - Drinking Water - Ecological Criteria
0.002 mg/l Substance presents persistence, bioaccumulations or risk of cancer, reduce human exposure to a minimum; This level has been extrapolated by using a mathematic model

STATE LISTS

California - Air Bill 2588 Appendix A-I
6/91

California - Prop. 65 - Cancer list
carcinogen - initial date 4/1/91

California - Exposure Limits - PELs
0.5 ppm PEL; 5 mg/m3 PEL

California - Exposure Limits - Skin Notation
material may be absorbed through the skin, eyes or mucous membrane

California - Directors List of Hazardous Substances (8 CCR 339)
[present]

Florida Hazardous Substance List
[present]

Massachusetts Right To Know List
[present]

Minnesota Hazardous Substance List
skin

NJ Right to Know List (Total)
sn 0262

Pennsylvania Right to Know List
environmental hazard

PROPOSED REGULATIONS

Safe Drinking Water Act - Priority list
[present]

TSCA - ITC 32nd Report Priority Testing List
designated for dermal absorption testing

1-BROMOHEXANE   111-25-1

HEALTH AND SAFETY LISTS

NIOSH - Selected LD50s and LC50s
Inhalation, rat: LC50 = 550000 mg/m3 (30 mn)

BROMOMETHYLBUTANE   107-82-4

HEALTH AND SAFETY LISTS

U.S. DOT - Substances From 49 CFR 172.101
regulated by DOT (UN2341)

U.S. DOT - Hazard Classes
DOT hazard class = 3

NIOSH - Selected LD50s and LC50s
Inhalation, mammal: LC50 = 21300 mg/m3 (8 hr)

STATE LISTS

NJ Right to Know List (Total)
sn 0263

5-BROMO-6-METHYL-3-(1-METHYLETHYL)-2,4(1H, 3H)-PYRIMIDINEDIONE   314-42-1

HEALTH AND SAFETY LISTS

NIOSH - Selected LD50s and LC50s
Oral, rat: LD50 = 3400 mg/kg

1-BROMO-2-METHYLPROPANE   78-77-3

HEALTH AND SAFETY LISTS

NIOSH - Selected LD50s and LC50s
Inhalation, mammal: LC50 = 50500 mg/m3 (8 hr)

STATE LISTS

NJ Right to Know List (Total)
sn 0265

2-BROMO-2-METHYLPROPANE   507-19-7

STATE LISTS

NJ Right to Know List (Total)
sn 0264

5-BROMO-5-NITRO-1,3-DIOXANE   30007-47-7

HEALTH AND SAFETY LISTS

NIOSH - Selected LD50s and LC50s
Oral, rat: LD50 = 455 mg/kg

2-BROMO-2-NITRO-1,3-PROPANEDIOL   52-51-7

HEALTH AND SAFETY LISTS

NIOSH - Selected LD50s and LC50s
Oral, rat: LD50 = 180 mg/kg Skin, rat: LD50 = 3500 mg/kg

ENVIRONMENTAL LISTS
   CERCLA/SARA - Section 313 - Emission Reporting
   form R reporting required

BETA-BROMO-BETA-NITROSTYRENE   **7166-19-0**
   HEALTH AND SAFETY LISTS
   NTP Chemical Status Reports - Testing Status and NTIS Number
   Post peer review technical reports in progress

1-BROMOOCTANE   **111-83-1**

   HEALTH AND SAFETY LISTS
   NIOSH - Selected LD50s and LC50s
   Oral, rat: LD50 = 5020 mg/kg Skin, rabbit: LD50 = 8944 mg/kg

2-BROMOPENTANE   **107-81-3**
   HEALTH AND SAFETY LISTS
   U.S. DOT - Substances From 49 CFR 172.101
   regulated by DOT (UN2343)
   U.S. DOT - Hazard Classes
   DOT hazard class = 3

1-BROMOPENTANE   **110-53-2**

   HEALTH AND SAFETY LISTS
   NFPA - Flash Points
   flash point = 90 degrees F (32 degrees C)
   NFPA - Hazard Identification Ratings
   health-1; flammability-3; reactivity-0
   NIOSH - Selected LD50s and LC50s
   Inhalation, mammal: LC50 = 26800 mg/m3 (8 hr)

   STATE LISTS
   Florida Hazardous Substance List
   [present]
   Massachusetts Right To Know List
   [present]
   Pennsylvania Right to Know List
   [present]

BROMOPENTANE   **29756-38-5**
   STATE LISTS
   NJ Right to Know List (Total)
   sn 0266

O-BROMOPHENOL   **95-56-7**
   ENVIRONMENTAL LISTS
   TSCA - HDD/HDF - Precursors Required for Reporting
   [present]

BROMOPHOS-ETHYL   **4824-78-6**

   HEALTH AND SAFETY LISTS
   U.S. DOT - Appendix B - Marine Pollutants
   DOT regulated severe marine pollutant

2-BROMOPROPANE   **75-26-3**

   HEALTH AND SAFETY LISTS
   NIOSH - Selected LD50s and LC50s
   Inhalation, mammal: LC50 = 36 gm/m3 (8 hr)

BROMOPROPANE   **26446-77-5**

   HEALTH AND SAFETY LISTS
   U.S. DOT - Substances From 49 CFR 172.101
   regulated by DOT (UN2344)
   U.S. DOT - Hazard Classes
   DOT hazard class = 3

STATE LISTS
   NJ Right to Know List (Total)
   sn 0267

3-BROMOPROPYNE   **106-96-7**
   HEALTH AND SAFETY LISTS
   U.S. DOT - Substances From 49 CFR 172.101
   regulated by DOT (UN2345)
   U.S. DOT - Hazard Classes
   DOT hazard class = 3
   NFPA - Flash Points
   flash point = 50 degrees F (10 degrees C)
   NFPA - Hazard Identification Ratings
   health-3; flammability-3; reactivity-4
   NIOSH - Selected LD50s and LC50s
   Oral, guinea pig: LD50 = 29 ug/kg
   OSHA - List of Highly Hazardous Chemicals
   threshhold quantity = 100 pounds

   ENVIRONMENTAL LISTS
   CERCLA/SARA - Section 302 Extremely Hazardous Substances and
   TPQs
   TPQ = 10 pounds

   STATE LISTS
   Florida Hazardous Substance List
   [present]
   Massachusetts Right To Know List
   extraordinarily hazardous
   NJ Right to Know List (Total)
   sn 0268
   NJ Special Hazardous Substances
   (flammable - third degree; reactive - fourth degree)
   Pennsylvania Right to Know List
   environmental hazard

   PROPOSED REGULATIONS
   CERCLA/SARA - Proposed Hazardous Substance Additions
   proposed RQ = 1 pound (.454 kg)
   CERCLA/SARA - 1989 Proposed RQ Adjustments
   proposed RQ = 1 pound (.454 kg)

BROMOSILANE   **13465-73-1**

   HEALTH AND SAFETY LISTS
   U.S. DOT - Hazard Classes
   Forbidden from transport by the DOT

O-BROMOTOLUENE   **95-46-5**

   HEALTH AND SAFETY LISTS
   NFPA - Flash Points
   flash point = 174 degrees F (79 degrees C)
   NFPA - Hazard Identification Ratings
   health-2; flammability-2; reactivity-0

   STATE LISTS
   Florida Hazardous Substance List
   [present]
   Massachusetts Right To Know List
   [present]
   Pennsylvania Right to Know List
   [present]

P-BROMOTOLUENE   **106-38-7**

   HEALTH AND SAFETY LISTS
   NFPA - Flash Points
   flash point = 185 degrees F (85 degrees C)

NFPA - Hazard Identification Ratings
health-2; flammability-2; reactivity-0
NIOSH - Selected LD50s and LC50s
Inhalation, mammal: LC50 = 1300 mg/m3

STATE LISTS

Florida Hazardous Substance List
[present]

Massachusetts Right To Know List
[present]

Pennsylvania Right to Know List
[present]

BROMOTRIFLUOROETHYLENE   598-73-2

HEALTH AND SAFETY LISTS

U.S. DOT - Substances From 49 CFR 172.101
regulated by DOT (UN2419)
U.S. DOT - Hazard Classes
DOT hazard class = 2.1

ENVIRONMENTAL LISTS

CAA - Flammable Substances for Accidental Release Prevention
threshold quantity = 10,000 lbs

STATE LISTS

NJ Right to Know List (Total)
sn 0269

BROMOXYNIL   1689-84-5

HEALTH AND SAFETY LISTS

U.S. DOT - Appendix B - Marine Pollutants
DOT regulated marine pollutant

ENVIRONMENTAL LISTS

CERCLA/SARA - Section 313 - Emission Reporting
form R reporting required

INTERNATIONAL LISTS

Canada - Drinking Water Quality - IMACs
0.005 mg/L IMAC

STATE LISTS

California - Air Bill 2588 Appendix A-II
6/91

California - Prop. 65 - Developmental Toxicity
developmental toxicity - initial date 10/1/90

California - Directors List of Hazardous Substances (8 CCR 339)
[present]

BROMOXYNIL OCTANOATE (OCTANOIC ACID,2,6-   1689-99-2
DIBROMO-4-CYANOPHENYL ESTER)

ENVIRONMENTAL LISTS

CERCLA/SARA - Section 313 - Emission Reporting
form R reporting required

BRONZING LIQUID   RR-00783-0

HEALTH AND SAFETY LISTS

NFPA - Flash Points
flash point may be below 80 degrees F (27 degrees C)

STATE LISTS

Pennsylvania Right to Know List
[present]

BROWN COAL TAR   RR-00515-2

INTERNATIONAL LISTS

German (DFG) - Carcinogens
proven carcinogen

BRUCINE   357-57-3

HEALTH AND SAFETY LISTS

U.S. DOT - Substances From 49 CFR 172.101
regulated by DOT (UN1570)
U.S. DOT - Hazard Classes
DOT hazard class = 6.1
U.S. DOT - Appendix A Table 1 - Hazardous Substances
final RQ = 100 pounds (45.4 kg)

ENVIRONMENTAL LISTS

CERCLA/SARA - Section 313 - Emission Reporting
form R reporting required
CERCLA/SARA - Hazardous Substances and their Reportable Quantities
final RQ = 100 pounds (45.4 kg)
RCRA - P Series Wastes
waste number P018
RCRA - Hazardous Constituents-Appendix VIII
waste number P018
RCRA - Substances Banned From Land Disposal
[present] (wastewaters)
TSCA - Code of Federal Regulations Citations
40 CFR 716.120(a)
TSCA - Health and Safety Reporting List
Effective Date: March 7, 1986

INTERNATIONAL LISTS

Canada - WHMIS: Ingredient Disclosure
1% item 221 (346)

STATE LISTS

Massachusetts Right To Know List
[present]

NJ Right to Know List (Total)
sn 0270

Pennsylvania Right to Know List
environmental hazard

BUSULFAN   55-98-1

HEALTH AND SAFETY LISTS

IARC - Group 1 (carcinogenic to humans)
[present]
NIOSH - Selected LD50s and LC50s
Oral, rat: LD50 = 1860 ug/kg
NTP Seventh Report - Known Carcinogens
known carcinogen
OSHA - Select Carcinogens
[present]

STATE LISTS

California - Air Bill 2588 Appendix A-II
known or potential carcinogen

California - Prop. 65 - Cancer list
carcinogen - initial date 2/27/87

California - Prop. 65 - Developmental Toxicity
developmental toxicity - initial date 1/1/89

California - Directors List of Hazardous Substances (8 CCR 339)
[present]

Florida Hazardous Substance List
[present]

Massachusetts Right To Know List
carcinogen; extraordinarily hazardous; teratogen

Minnesota Hazardous Substance List
carcinogen

NJ Right to Know List (Total)
sn 0271

NJ Special Hazardous Substances
(carcinogen; mutagen; teratogen)

Pennsylvania Right to Know List
special hazardous substance

Pennsylvania RTK - Special Hazardous Substances
[present]

BUTABARBITAL SODIUM   143-81-7

STATE LISTS

California - Prop. 65 - Developmental Toxicity
developmental toxicity - initial date 10/1/92

1,3-BUTADIENE   106-99-0

HEALTH AND SAFETY LISTS

ACGIH 1995 - Time Weighted Averages
2 ppm TWA; 4.4 mg/m3 TWA

ACGIH 1995 - Carcinogens
A2-suspected human carcinogen

AIHA - Odor Threshold Values
geometric mean air odor threshold = 0.45 ppm (detectable); 1.1 ppm (recognizable)

U.S. DOT - Substances From 49 CFR 172.101
regulated by DOT (UN1010)

U.S. DOT - Hazard Classes
DOT hazard class = 2.1

IARC - Group 2A (limited human data)
[present]

IARC - Group 2B (sufficient animal data)
[present]

NFPA - Flash Points
gas (no flash point given)

NFPA - Hazard Identification Ratings
health-2; flammability-4; reactivity-2

NIOSH - Selected LD50s and LC50s
Inhalation, rat: LC50 = 285 gm/m3 4 hr

NIOSH 1990 - Pocket Guide - RELs
Reduce exposure to lowest feasible concentration

NIOSH 1990 - Pocket Guide - IDLHs
20,000 ppm IDLH (lower explosive limit) (not conisidering carcinogenic effects)

NIOSH 1990 - Pocket Guide - Carcinogens
occupational carcinogen

NIOSH 1990 - Pocket Guide - Target organs
eyes, respiratory system, CNS

NIOSH - Health Standards - Exposure Limits
reduce exposure to lowest feasible concentration

NIOSH - Health Standards - Health Effects and Precautions
Hematopoietic cancer, teratogenicity, reproductive system effects (Restrict access to areas where 1,3-Butadiene is used)

NIOSH - Health Standards - Carcinogenic Chemicals
potential human carcinogen

NTP Chemical Status Reports - Testing Status and NTIS Number
Technical reports printed (PB85179646/AS); prechronic studies for which toxicity technical reports were not prepared

NTP Chemical Status Reports - Evidence of Carcinogenicity
male mice-clear evidence; female mice-clear evidence

NTP Seventh Report - Suspect Carcinogens
suspect carcinogen

OSHA - Vacated PELs - Time Weighted Averages
1000 ppm TWA; 2200 mg/m3 TWA

OSHA - Final PELs - Time Weighted Averages
1000 ppm TWA; 2200 mg/m3 TWA

OSHA - Possible Select Carcinogens
[present]

ENVIRONMENTAL LISTS

CERCLA/SARA - Section 313 - Emission Reporting
form R reporting required for 0.1% de minimus concentration

CERCLA/SARA - Hazardous Substances and their Reportable Quantities
final RQ = 1 pound (.454 kg)

Clean Air Act (1990) - List of Hazardous Air Contaminants
[present]

CAA - Flammable Substances for Accidental Release Prevention
threshold quantity = 10,000 lbs

CAA - HON Rule - SOCMI Chemicals
compliance by Jan. 23, 1995

CAA - HON Rule - Organic HAPs
[present]

INTERNATIONAL LISTS

Australian Exposure Standards - Time Weighted Averages
10 ppm TWA; 22 mg/m3 TWA

Australian Exposure Standards - Carcinogens
probable carcinogen

Canada - WHMIS: Ingredient Disclosure
0.1% item 222 (349)

Canada - NPRI (National Pollutant Release Inventory)
[present]

Canada - Alberta - 8 Hour Occupational Exposure Limit
10 ppm TWA; 22.1 mg/m3 TWA

Canada - Alberta - 15 Minute Occupational Exposure Limit
25 ppm STEL; 55.3 mg/m3 STEL

Canada - Alberta - Designated Substances
designated substance - requires code of practice

Canada - British Columbia - 8 Hour Exposure Limits
1000 ppm TWA; 2200 mg/m3 TWA

Canada - British Columbia - 15 Minute Exposure Limits
1250 ppm STEL; 2750 mg/m3 STEL

Canada - Ontario - OHSA - TWAEVs
10 ppm TWAEV; 22 mg/m3 TWAEV

Canada - Quebec - Time-Weighted Average Exposure Values
10 ppm TWAEV; 22 mg/m3 TWAEV

Canada - Quebec - Carcinogens
C2 carcinogen: effect suspected in humans

United Kingdom - Maximum Exposure Limits - TWAs
10 ppm TWA; 22 mg/m3 TWA

German (DFG) - Carcinogens
animal evidence of carcinogenicity

Israel - Time Weighted Averages
10 ppm TWA; 22 mg/m3 TWA

Israel - Action Levels
5 ppm AL; 11 mg/m3 AL

Mexico - Instruction No. 10 - TWAs
1000 ppm TWA; 2200 mg/m3 TWA

Mexico - Instruction No. 10 - STELs
1250 ppm STEL; 2750 mg/m3 STEL

STATE LISTS

California - Air Bill 2588 Appendix A-I
known or potential carcinogen

California - Prop. 65 - Cancer list
carcinogen - initial date 4/1/88

California - Prop. 65 - No Significant Risk Levels
no significant risk level = 0.4 ug/day

California - Exposure Limits - PELs
2 ppm PEL; 4.4 mg/m3 PEL

California - Exposure Limits - STELs
10 ppm STEL; 22 mg/m3 STEL

California - Directors List of Hazardous Substances (8 CCR 339)
[present]

Florida Hazardous Substance List
[present]

Massachusetts Right To Know List
carcinogen; extraordinarily hazardous

Minnesota Hazardous Substance List
carcinogen

NJ Right to Know List (Total)
sn 0272

NJ Special Hazardous Substances
(flammable - fourth degree; reactive - second degree)

Pennsylvania Right to Know List
environmental hazard; special hazardous substance

Pennsylvania RTK - Special Hazardous Substances
[present]

PROPOSED REGULATIONS

Canada - Ontario - Proposed Occupational TWAEVs
1 ppm TWAEV; 2.2 mg/m3 TWAEV

BUTADIENE MONOXIDE   930-22-3

HEALTH AND SAFETY LISTS

NFPA - Flash Points
flash point < -58 degrees F (-50 degrees C)

NFPA - Hazard Identification Ratings
health-2; flammability-3; reactivity-2

ENVIRONMENTAL LISTS

TSCA - Code of Federal Regulations Citations
40 CFR 716.120(a)

TSCA - Health and Safety Reporting List
Effective Date: October 4, 1982

STATE LISTS

Florida Hazardous Substance List
[present]

Massachusetts Right To Know List
[present]

Pennsylvania Right to Know List
[present]

2,3-BUTADIONE   431-03-8

HEALTH AND SAFETY LISTS

U.S. DOT - Substances From 49 CFR 172.101
regulated by DOT (UN2346)

U.S. DOT - Hazard Classes
DOT hazard class = 3

NFPA - Flash Points
flash point = 80 degrees F (27 degrees C)

NFPA - Hazard Identification Ratings
health-1; flammability-3; reactivity-0

NIOSH - Selected LD50s and LC50s
Oral, rat: LD50 = 1580 mg/kg

ENVIRONMENTAL LISTS

List of Pesticide Product Inert Ingredients
[present]

INTERNATIONAL LISTS

Canada - WHMIS: Ingredient Disclosure
1% item 226 (352)

STATE LISTS

Florida Hazardous Substance List
[present]

Massachusetts Right To Know List
[present]

NJ Right to Know List (Total)
sn 0274

Pennsylvania Right to Know List
[present]

2,3-BUTADIONE   434-03-7

STATE LISTS

Massachusetts Right To Know List
teratogen

BUTANAL, 2-METHYL-   96-17-3

HEALTH AND SAFETY LISTS

NFPA - Flash Points
flash point = 49 degrees F (9 degrees C)

NFPA - Hazard Identification Ratings
health-2; flammability-3; reactivity-0

STATE LISTS

Florida Hazardous Substance List
[present]

Massachusetts Right To Know List
[present]

Pennsylvania Right to Know List
[present]

BUTANAMIDE, N-(4-ETHOXYPHENYL)-3-OXO-   122-82-7

HEALTH AND SAFETY LISTS

NFPA - Flash Points
flash point = 325 degrees F (163 degrees C)

NFPA - Hazard Identification Ratings
health-2; flammability-1; reactivity-1

STATE LISTS

Florida Hazardous Substance List
[present]

Massachusetts Right To Know List
[present]

Pennsylvania Right to Know List
[present]

BUTANE   106-97-8

HEALTH AND SAFETY LISTS

ACGIH 1995 - Time Weighted Averages
800 ppm TWA; 1900 mg/m3 TWA

AIHA - Odor Threshold Values
no geometric mean air odor threshold

U.S. DOT - Substances From 49 CFR 172.101
regulated by DOT (UN1011)

U.S. DOT - Hazard Classes
DOT hazard class = 2.1

NFPA - Flash Points
flashpoint = -76 degrees F (-60 degrees C)

NFPA - Hazard Identification Ratings
health-1; flammability-4; reactivity-0

NIOSH - Selected LD50s and LC50s
Inhalation, rat: LC50 = 658 gm/m3 4 hr

OSHA - Vacated PELs - Time Weighted Averages
800 ppm TWA; 1900 mg/m3 TWA

ENVIRONMENTAL LISTS

CAA - Flammable Substances for Accidental Release Prevention
threshold quantity = 10,000 lbs

List of Pesticide Product Inert Ingredients
[present]

INTERNATIONAL LISTS

Australian Exposure Standards - Time Weighted Averages
800 ppm TWA; 1900 mg/m3 TWA

Canada - WHMIS: Ingredient Disclosure
1% item 223 (350)

Canada - Alberta - 8 Hour Occupational Exposure Limit
800 ppm TWA; 1901 mg/m3 TWA

Canada - Alberta - 15 Minute Occupational Exposure Limit
1000 ppm STEL; 2576 mg/m3 STEL

Canada - British Columbia - 8 Hour Exposure Limits
600 ppm TWA; 1430 mg/m3 TWA

Canada - British Columbia - 15 Minute Exposure Limits
750 ppm STEL; 1780 mg/m3 STEL

Canada - Ontario - OHSA - TWAEVs
800 ppm TWAEV; 1900 mg/m3 TWAEV

Canada - Quebec - Time-Weighted Average Exposure Values
800 ppm TWAEV; 1900 mg/m3 TWAEV

United Kingdom - Occupational Exposure Standards - TWAs
600 ppm TWA; 1430 mg/m3 TWA

United Kingdom - Occupational Exposure Standards - STELs
750 ppm STEL; 1780 mg/m3 STEL

German (DFG) - MAK Values
1000 ppm MAK; 2350 mg/m3 MAK (Listed under 'Butane')

German (DFG) - Peak Limitations
2 x normal MAK (1 hour momentary value); don't exceed 3 times per shift (Listed under 'Butane')

Israel - Time Weighted Averages
800 ppm TWA; 1900 mg/m3 TWA

Israel - Action Levels
400 ppm AL; 950 mg/m3 AL

Mexico - Instruction No. 10 - TWAs
800 ppm TWA; 1900 mg/m3 TWA

STATE LISTS

California - Exposure Limits - PELs
800 ppm PEL; 1900 mg/m3 PEL

California - Directors List of Hazardous Substances (8 CCR 339)
[present]

Massachusetts Right To Know List
[present]

Minnesota Hazardous Substance List
[present]

NJ Right to Know List (Total)
sn 0273

NJ Special Hazardous Substances
(flammable - fourth degree)

Pennsylvania Right to Know List
[present]

BUTANE, 2-CHLORO-   **78-86-4**

HEALTH AND SAFETY LISTS

NFPA - Flash Points
flash point < 32 degrees F (0 degrees C)

NFPA - Hazard Identification Ratings
health-2; flammability-3

NIOSH - Selected LD50s and LC50s
Oral, rat: LD50 = 17460 mg/kg Skin, rabbit: LD50 = 20 gm/kg

STATE LISTS

Florida Hazardous Substance List
[present]

Massachusetts Right To Know List
[present]

Pennsylvania Right to Know List
[present]

1,3-BUTANEDIAMINE   **590-88-5**

HEALTH AND SAFETY LISTS

NFPA - Flash Points
flash point = 125 degrees F (52 degrees C)

NFPA - Hazard Identification Ratings
health-3; flammability-2; reactivity-0

STATE LISTS

Florida Hazardous Substance List
[present]

Massachusetts Right To Know List
[present]

Pennsylvania Right to Know List
[present]

BUTANEDIOIC ACID, METHYLENE-   **97-65-4**

ENVIRONMENTAL LISTS

EPA - Master Testing List
[present]

1,3-BUTANEDIOL   **107-88-0**

HEALTH AND SAFETY LISTS

NFPA - Flash Points
flash point = 250 degrees F (121 degrees C)

NFPA - Hazard Identification Ratings
health-1; flammability-1; reactivity-0

NIOSH - Selected LD50s and LC50s
Oral, rat: LD50 = 18610 mg/kg

ENVIRONMENTAL LISTS

CAA - HON Rule - SOCMI Chemicals
compliance by Jan. 23, 1995

List of Pesticide Product Inert Ingredients
[present]

INTERNATIONAL LISTS

Canada - WHMIS: Ingredient Disclosure
0.1% item 224 (351)

1,4-BUTANEDIOL   **110-63-4**

HEALTH AND SAFETY LISTS

NFPA - Flash Points
flash point = 250 degrees F (121 degrees C)

NFPA - Hazard Identification Ratings
health-1; flammability-1; reactivity-0

NIOSH - Selected LD50s and LC50s
Oral, rat: LD50 = 1525 mg/kg

NTP Chemical Status Reports - Testing Status and NTIS Number
Approved for toxicology/carcinogenesis study

ENVIRONMENTAL LISTS

CAA - HON Rule - SOCMI Chemicals
compliance by Oct. 24, 1994

List of Pesticide Product Inert Ingredients
[present]

2,3-BUTANEDIOL   **513-85-9**

HEALTH AND SAFETY LISTS

NFPA - Flash Points
flash point = 185 degrees F (85 degrees C)

NFPA - Hazard Identification Ratings
health-0; flammability-2

NIOSH - Selected LD50s and LC50s
Oral, mouse: LD50 = 5462 mg/kg

1,2-BUTANEDIOL 584-03-2

HEALTH AND SAFETY LISTS

NFPA - Flash Points
*flash point = 194 degrees F (90 degrees C); *See list description*

NFPA - Hazard Identification Ratings
health-0; flammability-2

NIOSH - Selected LD50s and LC50s
Oral, mouse: LD50 = 3720 mg/kg

ENVIRONMENTAL LISTS

EPA - Master Testing List
[present]

1,4-BUTANEDIOL DIMETHACRYLATE 2082-81-7

ENVIRONMENTAL LISTS

TSCA - Code of Federal Regulations Citations
40 CFR 712.30(d)

TSCA - PAIR - Reporting List
Reporting Date: November 19, 1982

INTERNATIONAL LISTS

Canada - WHMIS: Ingredient Disclosure
1% item 225 (724)

2,4-BUTANE SULTONE 1121-03-5

INTERNATIONAL LISTS

German (DFG) - Carcinogens
animal evidence of carcinogenicity

1,4-BUTANESULTONE 1633-83-6

HEALTH AND SAFETY LISTS

NIOSH - Selected LD50s and LC50s
Oral, rat: LD50 = 500 mg/kg

INTERNATIONAL LISTS

Canada - WHMIS: Ingredient Disclosure
1% item 227 (353)

German (DFG) - Carcinogens
suspected carcinogen

2-BUTANETHIOL 513-53-1

HEALTH AND SAFETY LISTS

NFPA - Flash Points
flash point = -10 degrees F (-23 degrees C)

NFPA - Hazard Identification Ratings
health-2; flammability-3; reactivity-0

STATE LISTS

Florida Hazardous Substance List
[present]

Massachusetts Right To Know List
[present]

Pennsylvania Right to Know List
[present]

BUTANE, 2,2,3-TRIMETHYL- 464-06-2

HEALTH AND SAFETY LISTS

NFPA - Flash Points
flash point < 32 degrees F (0 degrees C)

NFPA - Hazard Identification Ratings
health-0; flammability-3; reactivity-0

STATE LISTS

Florida Hazardous Substance List
[present]

Massachusetts Right To Know List
[present]

Pennsylvania Right to Know List
[present]

1,2,4-BUTANETRIOL 3068-00-6

HEALTH AND SAFETY LISTS

U.S. DOT - Hazard Classes
Forbidden from transport by the DOT

NIOSH - Selected LD50s and LC50s
Oral, mouse: LD50 = 23 gm/kg

BUTANOIC ACID, PROPYL ESTER 105-66-8

HEALTH AND SAFETY LISTS

NFPA - Flash Points
flash point = 99 degrees F (37 degrees C)

NFPA - Hazard Identification Ratings
health-0; flammability-3; reactivity-0

STATE LISTS

Florida Hazardous Substance List
[present]

Massachusetts Right To Know List
[present]

Pennsylvania Right to Know List
[present]

2-BUTANOL, ALUMINUM SALT 2269-22-9

ENVIRONMENTAL LISTS

List of Pesticide Product Inert Ingredients
[present]

1-BUTANOL, 2,2-BIS[(2-PROPENYLOXY)METHYL]- 682-09-7

HEALTH AND SAFETY LISTS

NIOSH - Selected LD50s and LC50s
Oral, rat: LD50 = 6500 mg/kg

2-BUTENE 107-01-7

ENVIRONMENTAL LISTS

CAA - Flammable Substances for Accidental Release Prevention
threshold quantity = 10,000 lbs

EPA - Master Testing List
[present]

STATE LISTS

Pennsylvania Right to Know List
[present]

1-BUTENE 106-98-9

HEALTH AND SAFETY LISTS

NFPA - Flash Points
gas (no flash point given)

NFPA - Hazard Identification Ratings
health-1; flammability-4; reactivity-0

ENVIRONMENTAL LISTS

CAA - Flammable Substances for Accidental Release Prevention
threshold quantity = 10,000 lbs

EPA - Master Testing List
[present]

STATE LISTS

California - Air Bill 2588 Appendix A-I
[present]

Florida Hazardous Substance List
[present]

Massachusetts Right To Know List
[present]

Pennsylvania Right to Know List
[present]

2-BUTENE, (E)-   624-64-6

HEALTH AND SAFETY LISTS

NFPA - Flash Points
gas (no flash point given)

NFPA - Hazard Identification Ratings
health-1; flammability-4; reactivity-0

ENVIRONMENTAL LISTS

CAA - Flammable Substances for Accidental Release Prevention
threshold quantity = 10,000 lbs

STATE LISTS

Florida Hazardous Substance List
[present]

Massachusetts Right To Know List
[present]

Pennsylvania Right to Know List
[present]

2-BUTENE-CIS   590-18-1

HEALTH AND SAFETY LISTS

NFPA - Flash Points
gas (no flash point given)

NFPA - Hazard Identification Ratings
health-1; flammability-4; reactivity-0

ENVIRONMENTAL LISTS

CAA - Flammable Substances for Accidental Release Prevention
threshold quantity = 10,000 lbs

STATE LISTS

Florida Hazardous Substance List
[present]

Massachusetts Right To Know List
[present]

Pennsylvania Right to Know List
[present]

2-BUTENEDIOIC ACID (Z), MONO(2-((1-OXO-PROPENYLOXY)ETHYL)ESTER   RR-01227-1

ENVIRONMENTAL LISTS

TSCA - Chemicals with Significant New Use Rules
PMN number: P-85-543

2-BUTENEDIOIC ACID (Z-)-, MONO [2-[(1-OXO-2-PROPENYL)OXY]ETHYL]-ESTER   19201-36-6

ENVIRONMENTAL LISTS

TSCA - Section 12(b) - Export Notification
P-85-543; export notification required - Section 5

2-BUTENE-1,4-DIOL   110-64-5

ENVIRONMENTAL LISTS

List of Pesticide Product Inert Ingredients
[present]

BUTENEDIOL   29733-86-6

HEALTH AND SAFETY LISTS

NFPA - Flash Points
flash point = 263 degrees F (128 degrees C)

NFPA - Hazard Identification Ratings
health-1; flammability-1; reactivity-0

BUTENE OXIDE   26249-20-7

HEALTH AND SAFETY LISTS

NIOSH - Selected LD50s and LC50s
Oral, rat: LD50 = 1410 mg/kg Skin, rabbit: LD50 = 2100 mg/kg

1-BUTENE, 2,3,4-TRICHLORO-   2431-50-7

ENVIRONMENTAL LISTS

EPA - Master Testing List
[present]

INTERNATIONAL LISTS

German (DFG) - Carcinogens
animal evidence of carcinogenicity

1-BUTENE, 2,3,3-TRIMETHYL-   594-56-9

HEALTH AND SAFETY LISTS

NFPA - Flash Points
flash point < 32 degrees F (0 degrees C)

NFPA - Hazard Identification Ratings
health-0; flammability-3; reactivity-0

STATE LISTS

Florida Hazardous Substance List
[present]

Massachusetts Right To Know List
[present]

Pennsylvania Right to Know List
[present]

2-BUTENOIC ACID, ETHENYL ESTER   14861-06-4

HEALTH AND SAFETY LISTS

NFPA - Flash Points
flash point = 78 degrees F (26 degrees C)

NFPA - Hazard Identification Ratings
health-2; flammability-3; reactivity-2

STATE LISTS

Florida Hazardous Substance List
[present]

Massachusetts Right To Know List
[present]

Pennsylvania Right to Know List
[present]

3-BUTEN-2-OL, 2-METHYL-   115-18-4

ENVIRONMENTAL LISTS

EPA - Master Testing List
[present]

1-BUTEN-3-YNE   689-97-4

HEALTH AND SAFETY LISTS

NFPA - Hazard Identification Ratings
health-2; flammibility-4; reactivity-3

NIOSH - Selected LD50s and LC50s
Inhalation, mouse: LC50 = 97200 mg/m3 (2 hr)

ENVIRONMENTAL LISTS

CAA - Flammable Substances for Accidental Release Prevention
threshold quantity = 10,000 lbs

STATE LISTS

Florida Hazardous Substance List
[present]

Massachusetts Right To Know List
[present]

Pennsylvania Right to Know List
[present]

TERT-BUTOXYCARBONYL AZIDE   1070-19-5

HEALTH AND SAFETY LISTS
U.S. DOT - Hazard Classes
Forbidden from transport by the DOT

2-BUTOXYETHANOL   111-76-2
SEE ALSO:
GLYCOL ETHERS

HEALTH AND SAFETY LISTS
ACGIH 1995 - Time Weighted Averages
25 ppm TWA; 121 mg/m3 TWA

ACGIH 1995 - Skin Designations
skin - potential for cutaneous absorption

AIHA - Odor Threshold Values
geometric mean air odor threshold = 0.10 ppm (detectable); 0.35 ppm (recognizable)

U.S. DOT - Substances From 49 CFR 172.101
regulated by DOT (UN2369)

U.S. DOT - Hazard Classes
DOT hazard class = 6.1

NFPA - Flash Points
flash point = 150 degrees F (66 degrees C);

NFPA - Hazard Identification Ratings
health-1; flammability-2; reactivity-0;

NIOSH - Selected LD50s and LC50s
Inhalation, rat: LC50 = 450 ppm 4 hr Oral, rat: LD50 = 530 mg/kg Skin, guinea pig: LD50 = 230 mg/kg

NIOSH 1990 - Pocket Guide - RELs
25 ppm TWA; 120 mg/m3 TWA

NIOSH 1990 - Pocket Guide - IDLHs
700 ppm IDLH

NIOSH 1990 - Pocket Guide - Target organs
liver, kidneys, lymphoid system, skin, blood, eyes, respiratory system

NIOSH 1990 - Pocket Guide - Skin list
Potential for dermal absorption

NTP Chemical Status Reports - Testing Status and NTIS Number
Two year studies in progress; technical reports printed (PB94-118106)

OSHA - Vacated PELs - Time Weighted Averages
25 ppm TWA; 120 mg/m3 TWA

OSHA - Vacated PELs - Skin Designation
Prevent or reduce skin absorption

OSHA - Final PELs - Time Weighted Averages
50 ppm TWA; 240 mg/m3 TWA

OSHA - Final PELs - Skin Notations
prevent or reduce skin absorption

ENVIRONMENTAL LISTS
CAA - HON Rule - SOCMI Chemicals
compliance by Oct. 24, 1994

List of Pesticide Product Inert Ingredients
[present]

TSCA - Code of Federal Regulations Citations
40 CFR 712.30(w); 40 CFR 716.120(a)

TSCA - PAIR - Reporting List
Reporting Date: June 13, 1989

TSCA - Health and Safety Reporting List
Effective Date: April 13, 1989

INTERNATIONAL LISTS
Australian Exposure Standards - Time Weighted Averages
25 ppm TWA; 121 mg/m3 TWA

Australian Exposure Standards - Skin Effects
skin absorption

Australian Exposure Standards - Under Review
exposure limits under review

Canada - WHMIS: Ingredient Disclosure
1% item 721 (824)

Canada - Alberta - 8 Hour Occupational Exposure Limit
25 ppm TWA; 120 mg/m3 TWA

Canada - Alberta - 15 Minute Occupational Exposure Limit
75 ppm STEL; 360 mg/m3 STEL

Canada - Alberta - Skin Designation
can be absorbed through the intact skin

Canada - British Columbia - 8 Hour Exposure Limits
50 ppm TWA; 240 mg/m3 TWA

Canada - British Columbia - 15 Minute Exposure Limits
150 ppm STEL; 720 mg/m3 STEL

Canada - British Columbia - Skin Notations
skin - potential for skin absorption

Canada - Ontario - OHSA - TWAEVs
25 ppm TWAEV; 120 mg/m3 TWAEV

Canada - Ontario - OHSA - Skin Notations
absorption through skin, eyes, or mucous membranes

Canada - Quebec - Time-Weighted Average Exposure Values
25 ppm TWAEV; 121 mg/m3 TWAEV

Canada - Quebec - Skin Designations
absorbed through the skin

United Kingdom - Maximum Exposure Limits - TWAs
25 ppm TWA; 120 mg/m3 TWA

United Kingdom - Maximum Exposure Limits - Notes
can be absorbed through skin

German (DFG) - MAK Values
20 ppm MAK; 100 mg/m3 MAK

German (DFG) - Peak Limitations
2 x normal MAK (30 min. average value); don't exceed 4 times during shift

German (DFG) - Skin/Sensitizers
danger of cutaneous absorption

German (DFG) - Pregnancy
no risk to embryo/fetus if exposure limits adhered to

Israel - Time Weighted Averages
25 ppm TWA; 121 mg/m3 TWA

Israel - Action Levels
12.5 ppm AL; 60.5 mg/m3 AL

Mexico - Instruction No. 10 - TWAs
25 ppm TWA; 120 mg/m3 TWA

Mexico - Instruction No. 10 - STELs
75 ppm STEL; 360 mg/m3 STEL

Mexico - Instruction No. 10 - Skin designation
skin - potential for cutaneous absorption

STATE LISTS
California - Air Bill 2588 Appendix A-I
9/90

California - Exposure Limits - PELs
25 ppm PEL; 120 mg/m3 PEL

California - Exposure Limits - Skin Notation
material may be absorbed through the skin, eyes or mucous membrane

California - Directors List of Hazardous Substances (8 CCR 339)
[present]

Florida Hazardous Substance List
[present]

Massachusetts Right To Know List
[present]

Minnesota Hazardous Substance List
skin

NJ Right to Know List (Total)
sn 0275

Pennsylvania Right to Know List
[present]

2-(2-BUTOXYETHOXY)ETHYL ACETATE   124-17-4
SEE ALSO:
GLYCOL ETHERS

HEALTH AND SAFETY LISTS
NFPA - Flash Points
flash point = 240 degrees F (116 degrees C)

NFPA - Hazard Identification Ratings
health-1;   flammability-1;   reactivity-0

NIOSH - Selected LD50s and LC50s
Oral, rat: LD50 = 6500 mg/kg Skin, rabbit: LD50 = 14500 mg/kg

ENVIRONMENTAL LISTS
CAA - HON Rule - SOCMI Chemicals
compliance by Oct. 24, 1994

TSCA - Code of Federal Regulations Citations
40 CFR 712.30(j);   40 CFR 716.120(a)

TSCA - PAIR - Reporting List
Reporting Date: March 13, 1984

TSCA - Health and Safety Reporting List
Effective Date: January 13, 1982

TSCA - Chemical Test Rules
Testing required by: manufacturers; importers; processors (40 CFR 799.1560)

TSCA - Section 12(b) - Export Notification
export notification required - Section 4

2-BETA-BUTOXYETHOXYETHYL CHLORIDE   1120-23-6

HEALTH AND SAFETY LISTS
NFPA - Flash Points
flash point = 190 degrees F (88 degrees C)

NFPA - Hazard Identification Ratings
health-2;   flammability-2;   reactivity-0

STATE LISTS
Florida Hazardous Substance List
[present]

Massachusetts Right To Know List
[present]

Pennsylvania Right to Know List
[present]

2-BUTOXYETHYL DIHYDROGENPHOSPHATE, DIETHYLAMINE SALT   64051-23-6

ENVIRONMENTAL LISTS
List of Pesticide Product Inert Ingredients
[present]

BETA-BUTOXYETHYL SALICYLATE   RR-00791-0

HEALTH AND SAFETY LISTS
NFPA - Flash Points
flash point = 315 degrees F (157 degrees C)

NFPA - Hazard Identification Ratings
health-0;   flammability-1;   reactivity-0

BUTOXYL   4435-53-4

HEALTH AND SAFETY LISTS
U.S. DOT - Substances From 49 CFR 172.101
regulated by DOT (UN2708)

U.S. DOT - Hazard Classes
DOT hazard class = 3

NFPA - Flash Points
flash point = 170 degrees F (77 degrees C)

NFPA - Hazard Identification Ratings
health-1;   flammability-2;   reactivity-0

NIOSH - Selected LD50s and LC50s
Oral, rat: LD50 = 4210 mg/kg

STATE LISTS
NJ Right to Know List (Total)
sn 0276

1-BUTOXY-2-PROPANOL   5131-66-8

HEALTH AND SAFETY LISTS
NIOSH - Selected LD50s and LC50s
Oral, rat: LD50 = 2200 mg/kg Skin, rabbit: LD50 = 3100 mg/kg

ENVIRONMENTAL LISTS
List of Pesticide Product Inert Ingredients
[present]

TSCA - Code of Federal Regulations Citations
40 CFR 712.30(w);   40 CFR 716.120

TSCA - PAIR - Reporting List
Reporting Date: June 13, 1989

TSCA - Health and Safety Reporting List
Effective Date: April 13, 1989

BUTOXY-SUBSTITUTED ETHER ALKANE   RR-01674-0

ENVIRONMENTAL LISTS
TSCA - Chemicals with Significant New-Use Rules
PMN number: P-92-755

BUTOXYTRIETHYLENE GLYCOL PHOSPHATE   RR-01045-7

ENVIRONMENTAL LISTS
List of Pesticide Product Inert Ingredients
[present]

N-BUTYL ACETAMIDE   1119-49-9

HEALTH AND SAFETY LISTS
NFPA - Flash Points
flash point = 240 degrees F (116 degrees C)

NFPA - Hazard Identification Ratings
health-2;   flammability-1;   reactivity-0

STATE LISTS
Florida Hazardous Substance List
[present]

Massachusetts Right To Know List
[present]

Pennsylvania Right to Know List
[present]

N-BUTYLACETANILIDE   91-49-6

HEALTH AND SAFETY LISTS
NFPA - Flash Points
flash point = 286 degrees F (141 degrees C)

NFPA - Hazard Identification Ratings
health-2;   flammability-1;   reactivity-0

STATE LISTS
Florida Hazardous Substance List
[present]

Massachusetts Right To Know List
[present]

Pennsylvania Right to Know List
[present]

SEC-BUTYL ACETATE   105-46-4

HEALTH AND SAFETY LISTS
ACGIH 1995 - Time Weighted Averages
200 ppm TWA;   950 mg/m3 TWA

U.S. DOT - Appendix A Table 1 - Hazardous Substances
final RQ = 5000 pounds (2270 kg) (Listed under 'Butyl acetate')

NFPA - Flash Points
flash point = 88 degrees F (31 degrees C)

NFPA - Hazard Identification Ratings
health-1; flammability-3; reactivity-0

NIOSH 1990 - Pocket Guide - RELs
200 ppm TWA; 950 mg/m3 TWA

NIOSH 1990 - Pocket Guide - IDLHs
10,000 ppm IDLH

NIOSH 1990 - Pocket Guide - Target organs
eyes, skin, respiratory system

OSHA - Vacated PELs - Time Weighted Averages
200 ppm TWA; 950 mg/m3 TWA

OSHA - Final PELs - Time Weighted Averages
200 ppm TWA; 950 mg/m3 TWA

ENVIRONMENTAL LISTS

CERCLA/SARA - Hazardous Substances and their Reportable Quantities
final RQ = 5000 pounds (2270 kg) (Listed under 'Butyl acetate')

Clean Water Act - Hazardous Substances
[present] (Listed under 'Butyl acetate')

TSCA - PAIR - Reporting List
Effective Date: January 26, 1994; Reporting Date: March 28, 1994

TSCA - Health and Safety Reporting List
Effective Date: January 26, 1994; Sunset Date: January 26, 2004

INTERNATIONAL LISTS

Australian Exposure Standards - Time Weighted Averages
200 ppm TWA; 950 mg/m3 TWA

Canada - WHMIS: Ingredient Disclosure
1% item 231 (10)

Canada - Alberta - 8 Hour Occupational Exposure Limit
200 ppm TWA; 950 mg/m3 TWA

Canada - Alberta - 15 Minute Occupational Exposure Limit
250 ppm STEL; 1187 mg/m3 STEL

Canada - British Columbia - 8 Hour Exposure Limits
200 ppm TWA; 950 mg/m3 TWA

Canada - British Columbia - 15 Minute Exposure Limits
250 ppm STEL; 1190 mg/m3 STEL

Canada - Ontario - OHSA - TWAEVs
200 ppm TWAEV; 950 mg/m3 TWAEV

Canada - Quebec - Time-Weighted Average Exposure Values
200 ppm TWAEV; 950 mg/m3 TWAEV

United Kingdom - Occupational Exposure Standards - TWAs
200 ppm TWA; 950 mg/m3 TWA

United Kingdom - Occupational Exposure Standards - STELs
250 ppm STEL; 1190 mg/m3 STEL

German (DFG) - MAK Values
200 ppm MAK; 950 mg/m3 MAK (Listed under 'Butyl acetate')

German (DFG) - Peak Limitations
2 x normal MAK (5 min momentary value); don't exceed 8 times during shift (Listed under 'Butyl acetate')

Israel - Time Weighted Averages
200 ppm TWA; 950 mg/m3 TWA

Israel - Action Levels
100 ppm AL; 475 mg/m3 AL

Mexico - Instruction No. 10 - TWAs
200 ppm TWA; 950 mg/m3 TWA

Mexico - Instruction No. 10 - STELs
250 ppm STEL; 1190 mg/m3 STEL

STATE LISTS

California - Exposure Limits - PELs
200 ppm PEL; 950 mg/m3 PEL

California - Directors List of Hazardous Substances (8 CCR 339)
[present] (Listed under 'Butyl acetate, all isomers')

Florida Hazardous Substance List
[present]

Massachusetts Right To Know List
[present]

Minnesota Hazardous Substance List
[present]

Pennsylvania Right to Know List
environmental hazard

PROPOSED REGULATIONS

TSCA - ITC 31st Report Priority Testing List
designated to be tested

N-BUTYL ACETATE   123-86-4

HEALTH AND SAFETY LISTS

ACGIH 1995 - Time Weighted Averages
(150) ppm TWA; (713) mg/m3 TWA

ACGIH 1995 - Short Term Exposure Limits
(200) ppm STEL; (950) mg/m3 STEL

AIHA - Odor Threshold Values
geometric mean air odor threshold = 0.31 ppm (detectable); 0.68 ppm (recognizable)

U.S. DOT - Appendix A Table 1 - Hazardous Substances
final RQ = 5000 pounds (2270 kg)

NFPA - Flash Points
flash point = 72 degrees F (22 degrees C)

NFPA - Hazard Identification Ratings
health-1; flammability-3; reactivity-0

NIOSH - Selected LD50s and LC50s
Inhalation, rat: LC50 = 2000 ppm 4 hr Oral, rat: LD50 = 14 gm/kg

NIOSH 1990 - Pocket Guide - RELs
150 ppm TWA; 710 mg/m3 TWA; 200 ppm STEL; 950 mg/m3 STEL

NIOSH 1990 - Pocket Guide - IDLHs
10,000 ppm IDLH

NIOSH 1990 - Pocket Guide - Target organs
eyes, skin, respiratory system

OSHA - Vacated PELs - Time Weighted Averages
150 ppm TWA; 710 mg/m3 TWA

OSHA - Vacated PELs - Short Term Exposure Limits
200 ppm STEL; 950 mg/m3 STEL

OSHA - Final PELs - Time Weighted Averages
150 ppm TWA; 710 mg/m3 TWA

ENVIRONMENTAL LISTS

CERCLA/SARA - Hazardous Substances and their Reportable Quantities
final RQ = 5000 pounds (2270 kg)

Clean Water Act - Hazardous Substances
[present] (Listed under 'Butyl acetate')

EPA - Master Testing List
[present]

List of Pesticide Product Inert Ingredients
[present]

TSCA - Multichemical Test Rules - Neurotoxicity
administrative stay for neurotoxicity tests effective June 27, 1994

INTERNATIONAL LISTS

Australian Exposure Standards - Time Weighted Averages
150 ppm TWA; 713 mg/m3 TWA

Australian Exposure Standards - Short Term Exposure Limits
200 ppm STEL; 950 mg/m3 STEL

Canada - WHMIS: Ingredient Disclosure
1% item 230 (9)

Canada - Alberta - 8 Hour Occupational Exposure Limit
150 ppm TWA; 713 mg/m3 TWA

Canada - Alberta - 15 Minute Occupational Exposure Limit
200 ppm STEL; 950 mg/m3 STEL

Canada - British Columbia - 8 Hour Exposure Limits
150 ppm TWA; 710 mg/m3 TWA

Canada - British Columbia - 15 Minute Exposure Limits
200 ppm STEL; 950 mg/m3 STEL

Canada - Ontario - OHSA - TWAEVs
150 ppm TWAEV; 710 mg/m3 TWAEV

Canada - Ontario - OHSA - STEVs
200 ppm STEV; 950 mg/m3 STEV

Canada - Quebec - Time-Weighted Average Exposure Values
150 ppm TWAEV; 713 mg/m3 TWAEV

Canada - Quebec - Short-term Exposure Values
200 ppm STEV; 950 mg/m3 STEV

United Kingdom - Occupational Exposure Standards - TWAs
150 ppm TWA; 710 mg/m3 TWA

United Kingdom - Occupational Exposure Standards - STELs
200 ppm STEL; 950 mg/m3 STEL

German (DFG) - MAK Values
200 ppm MAK; 950 mg/m3 MAK (Listed under 'Butyl acetate')

German (DFG) - Peak Limitations
2 x normal MAK (5 min momentary value); don't exceed 8 times during shift (Listed under 'Butyl acetate')

Israel - Time Weighted Averages
150 ppm TWA; 713 mg/m3 TWA

Israel - Short Term Exposure Limits
200 ppm STEL; 950 mg/m3 STEL

Israel - Action Levels
75 ppm AL; 356.5 mg/m3 AL

Mexico - Instruction No. 10 - TWAs
150 ppm TWA; 710 mg/m3 TWA

Mexico - Instruction No. 10 - STELs
200 ppm STEL; 950 mg/m3 STEL

STATE LISTS

California - Exposure Limits - PELs
150 ppm PEL; 710 mg/m3 PEL

California - Exposure Limits - STELs
200 ppm STEL; 950 mg/m3 STEL

California - Directors List of Hazardous Substances (8 CCR 339)
[present] (Listed under 'Butyl acetate, all isomers')

Florida Hazardous Substance List
[present]

Massachusetts Right To Know List
[present]

Minnesota Hazardous Substance List
[present]

NJ Right to Know List (Total)
sn 1329

NJ Special Hazardous Substances
(flammable - third degree)

Pennsylvania Right to Know List
environmental hazard

PROPOSED REGULATIONS

ACGIH 1995 - Notice of Intended Changes
20 ppm TWA; 95 mg/m3 TWA

TERT-BUTYL ACETATE   540-88-5

HEALTH AND SAFETY LISTS

ACGIH 1995 - Time Weighted Averages
200 ppm TWA; 950 mg/m3 TWA

U.S. DOT - Appendix A Table 1 - Hazardous Substances
final RQ = 5000 pounds (2270 kg) (Listed under 'Butyl acetate')

NIOSH 1990 - Pocket Guide - RELs
200 ppm TWA; 950 mg/m3 TWA

NIOSH 1990 - Pocket Guide - IDLHs
10,000 ppm IDLH

NIOSH 1990 - Pocket Guide - Target organs
respiratory system, eyes, skin

OSHA - Vacated PELs - Time Weighted Averages
200 ppm TWA; 950 mg/m3 TWA

OSHA - Final PELs - Time Weighted Averages
200 ppm TWA; 950 mg/m3 TWA

ENVIRONMENTAL LISTS

CERCLA/SARA - Hazardous Substances and their Reportable Quantities
final RQ = 5000 pounds (2270 kg) (Listed under 'Butyl acetate')

Clean Water Act - Hazardous Substances
[present] (Listed under 'Butyl acetate')

TSCA - PAIR - Reporting List
Effective Date: January 26, 1994;   Reporting Date: March 28, 1994

TSCA - Health and Safety Reporting List
Effective Date: January 26, 1994;   Sunset Date: January 26, 2004

INTERNATIONAL LISTS

Australian Exposure Standards - Time Weighted Averages
200 ppm TWA; 950 mg/m3 TWA

Canada - WHMIS: Ingredient Disclosure
1% item 232 (11)

Canada - Alberta - 8 Hour Occupational Exposure Limit
200 ppm TWA; 950 mg/m3 TWA

Canada - Alberta - 15 Minute Occupational Exposure Limit
250 ppm STEL; 1187 mg/m3 STEL

Canada - British Columbia - 8 Hour Exposure Limits
200 ppm TWA; 950 mg/m3 TWA

Canada - British Columbia - 15 Minute Exposure Limits
250 ppm STEL; 1190 mg/m3 STEL

Canada - Ontario - OHSA - TWAEVs
200 ppm TWAEV; 950 mg/m3 TWAEV

Canada - Quebec - Time-Weighted Average Exposure Values
200 ppm TWAEV; 950 mg/m3 TWAEV

United Kingdom - Occupational Exposure Standards - TWAs
200 ppm TWA; 950 mg/m3 TWA

United Kingdom - Occupational Exposure Standards - STELs
250 ppm STEL; 1190 mg/m3 STEL

German (DFG) - MAK Values
200 ppm MAK; 950 mg/m3 MAK (Listed under 'Butyl acetate')

German (DFG) - Peak Limitations
2 x normal MAK (5 min momentary value); don't exceed 8 times during shift (Listed under 'Butyl acetate')

Israel - Time Weighted Averages
200 ppm TWA; 950 mg/m3 TWA

Israel - Action Levels
100 ppm AL; 475 mg/m3 AL

Mexico - Instruction No. 10 - TWAs
200 ppm TWA; 950 mg/m3 TWA

Mexico - Instruction No. 10 - STELs
250 ppm STEL; 1190 mg/m3 STEL

STATE LISTS

California - Exposure Limits - PELs
200 ppm PEL; 950 mg/m3 PEL

California - Directors List of Hazardous Substances (8 CCR 339)
[present] (Listed under 'Butyl acetate, all isomers')

Florida Hazardous Substance List
[present]

Massachusetts Right To Know List
[present]

Minnesota Hazardous Substance List
[present]

NJ Right to Know List (Total)
sn 1786

Pennsylvania Right to Know List
environmental hazard

PROPOSED REGULATIONS

TSCA - ITC 31st Report Priority Testing List
designated to be tested

BUTYL ACETATES RR-00147-8

HEALTH AND SAFETY LISTS

U.S. DOT - Substances From 49 CFR 172.101
regulated by DOT (UN1123)

U.S. DOT - Hazard Classes
DOT hazard class = 3

BUTYL ACETOACETATE 591-60-6

HEALTH AND SAFETY LISTS

NFPA - Flash Points
flash point = 185 degrees F (85 degrees C)

NFPA - Hazard Identification Ratings
health-1; flammability-2; reactivity-0

BUTYL ACETYL RICINOLEATE 140-04-5

HEALTH AND SAFETY LISTS

NFPA - Flash Points
flash point = 230 degrees F (110 degrees C)

NFPA - Hazard Identification Ratings
health-2; flammability-1; reactivity-0

STATE LISTS

Florida Hazardous Substance List
[present]

Massachusetts Right To Know List
[present]

Pennsylvania Right to Know List
[present]

N-TERT-BUTYLACRYLAMIDE 107-58-4

HEALTH AND SAFETY LISTS

NIOSH - Selected LD50s and LC50s
Oral, mouse: LD50 = 941 mg/kg

BUTYL ACRYLATE 141-32-2

HEALTH AND SAFETY LISTS

ACGIH 1995 - Time Weighted Averages
10 ppm TWA; 52 mg/m3 TWA

AIHA - Odor Threshold Values
no geometric mean air odor threshold

U.S. DOT - Substances From 49 CFR 172.101
regulated by DOT (UN2348)

U.S. DOT - Hazard Classes
DOT hazard class = 3

IARC - Group 3 (not classifiable)
[present]

NFPA - Flash Points
flash point = 84 degrees F (29 degrees C)

NFPA - Hazard Identification Ratings
health-2; flammability-2; reactivity-2

NIOSH - Selected LD50s and LC50s
Inhalation, rat: LC50 = 2730 ppm 4 hr Oral, rat: LD50 = 900 mg/kg
Skin, rabbit: LD50 = 2000 mg/kg

OSHA - Vacated PELs - Time Weighted Averages
10 ppm TWA; 55 mg/m3 TWA

ENVIRONMENTAL LISTS

CERCLA/SARA - Section 313 - Emission Reporting
form R reporting required for 1.0% de minimus concentration

CAA - HON Rule - SOCMI Chemicals
compliance by Oct. 23, 1995

List of Pesticide Product Inert Ingredients
[present]

TSCA - Code of Federal Regulations Citations
40 CFR 712.30(d)

TSCA - PAIR - Reporting List
Reporting Date: November 19, 1982

INTERNATIONAL LISTS

Australian Exposure Standards - Time Weighted Averages
10 ppm TWA; 55 mg/m3 TWA

Australian Exposure Standards - Skin Effects
sensitiser

Canada - WHMIS: Ingredient Disclosure
1% item 233 (153)

Canada - NPRI (National Pollutant Release Inventory)
[present]

Canada - Alberta - 8 Hour Occupational Exposure Limit
10 ppm TWA; 52 mg/m3 TWA

Canada - Alberta - 15 Minute Occupational Exposure Limit
20 ppm STEL; 105 mg/m3 STEL

Canada - British Columbia - 8 Hour Exposure Limits
10 ppm TWA; 55 mg/m3 TWA

Canada - Ontario - OHSA - TWAEVs
10 ppm TWAEV; 52 mg/m3 TWAEV

Canada - Quebec - Time-Weighted Average Exposure Values
10 ppm TWAEV; 52 mg/m3 TWAEV

United Kingdom - Occupational Exposure Standards - TWAs
10 ppm TWA; 55 mg/m3 TWA

German (DFG) - MAK Values
10 ppm MAK; 55 mg/m3 MAK

German (DFG) - Peak Limitations
2 x normal MAK (5 min momentary value); don't exceed 8 times during shift

German (DFG) - Skin/Sensitizers
danger of sensitization (skin or respiratory)

German (DFG) - Pregnancy
classification not yet possible

Israel - Time Weighted Averages
10 ppm TWA; 52 mg/m3 TWA

Israel - Action Levels
5 ppm AL; 26 mg/m3 AL

Mexico - Instruction No. 10 - TWAs
10 ppm TWA; 55 mg/m3 TWA

STATE LISTS

California - Air Bill 2588 Appendix A-I
6/91

California - Exposure Limits - PELs
10 ppm PEL; 55 mg/m3 PEL

California - Directors List of Hazardous Substances (8 CCR 339)
[present]

Florida Hazardous Substance List
[present]

Massachusetts Right To Know List
[present]

Minnesota Hazardous Substance List
[present]

NJ Right to Know List (Total)
sn 0278

NJ Special Hazardous Substances
(reactive - second degree)

Pennsylvania Right to Know List
environmental hazard

BUTYL ACRYLATE-HYDROXYETHYL ACRYLATE-METHYL METHACRYLATE COPOLYMER   25951-38-6
ENVIRONMENTAL LISTS
List of Pesticide Product Inert Ingredients
[present]

BUTYL ACRYLATE, POLYMER WITH SUBSTITUTED METHYL STYRENE, METHACRYLATE, AND SUBSTITUTED SILANE   RR-01251-1
ENVIRONMENTAL LISTS
TSCA - Chemicals with Significant New Use Rules
PMN number: P-91-272

N-BUTYL ALCOHOL   71-36-3
SEE ALSO:
F003-HAZARDOUS WASTES
F039-HAZARDOUS WASTES

HEALTH AND SAFETY LISTS
ACGIH 1995 - Ceiling Limits
C 50 ppm;   C 152 mg/m3

ACGIH 1995 - Skin Designations
skin - potential for cutaneous absorption

AIHA - Odor Threshold Values
geometric mean air odor threshold = 1.2 ppm (detectable);   5.8 ppm (recognizable)

U.S. DOT - Substances From 49 CFR 172.101
regulated by DOT (UN1120)

U.S. DOT - Hazard Classes
DOT hazard class = 3

U.S. DOT - Appendix A Table 1 - Hazardous Substances
final RQ = 5000 pounds (2270 kg)

NFPA - Flash Points
flash point = 98 degrees F (37 degrees C)

NFPA - Hazard Identification Ratings
health-1;   flammability-3;   reactivity-0

NIOSH - Selected LD50s and LC50s
Inhalation, rat: LC50 = 8000 ppm 4 hr Oral, rat: LD50 = 790 mg/kg Skin, rabbit: LD50 = 3400 mg/kg

NIOSH 1990 - Pocket Guide - RELs
C 50 ppm;   C 150 mg/m3

NIOSH 1990 - Pocket Guide - IDLHs
8000 ppm IDLH

NIOSH 1990 - Pocket Guide - Target organs
skin, eyes, respiratory system

NIOSH 1990 - Pocket Guide - Skin list
Potential for dermal absorption

OSHA - Vacated PELs - Ceiling Limits
C 50 ppm;   C 150 mg/m3

OSHA - Vacated PELs - Skin Designation
Prevent or reduce skin absorption

OSHA - Final PELs - Time Weighted Averages
100 ppm TWA;   300 mg/m3 TWA

ENVIRONMENTAL LISTS
CERCLA/SARA - Section 313 - Emission Reporting
form R reporting required for 1.0% de minimus concentration

CERCLA/SARA - Hazardous Substances and their Reportable Quantities
final RQ = 5000 pounds (2270 kg)

EPA - Master Testing List
[present]

List of Pesticide Product Inert Ingredients
[present]

RCRA - U Series Wastes
waste number U031 (Ignitable waste)

RCRA - Hazardous Constituents-Appendix VIII
waste number U031 (Ignitable waste)

RCRA - Basis for Listing - Appendix VII
Included in waste stream: F039

RCRA - Substances Banned From Land Disposal
[present]

RCRA - Universal Treatment Standards (LDR)
WW: 5.6 mg/l;   NWW: 2.6 mg/kg

TSCA - Multichemical Test Rules - Neurotoxicity
administrative stay for neurotoxicity tests effective June 27, 1994

INTERNATIONAL LISTS
Australian Exposure Standards - Time Weighted Averages
Peak Limitation: 50 ppm;   152 mg/m3

Australian Exposure Standards - Skin Effects
skin absorption

Canada - WHMIS: Ingredient Disclosure
1% item 234 (354)

Canada - NPRI (National Pollutant Release Inventory)
[present]

Canada - Alberta - Ceiling Occupational Exposure Limit
C 50 ppm;   C 152 mg/m3

Canada - Alberta - Skin Designation
can be absorbed through the intact skin

Canada - British Columbia - Ceiling Exposure Limits
C 50 ppm;   C 150 mg/m3

Canada - British Columbia - Skin Notations
skin - potential for skin absorption

Canada - Ontario - OHSA - CEVs
50 ppm CEV;   150 mg/m3 CEV

Canada - Ontario - OHSA - Skin Notations
absorption through skin, eyes, or mucous membranes

Canada - Quebec - Ceiling Limits
P 50 ppm;   P 152 mg/m3

Canada - Quebec - Skin Designations
absorbed through the skin

United Kingdom - Occupational Exposure Standards - STELs
50 ppm STEL;   150 mg/m3 STEL

United Kingdom - Occupational Exposure Standards - Notes
can be absorbed through skin

German (DFG) - MAK Values
100 ppm MAK;   300 mg/m3 MAK (Listed under 'Butyl alcohol')

German (DFG) - Peak Limitations
2 x normal MAK (30 min. average value);   don't exceed 4 times during shift (Listed under 'Butyl alcohol')

German (DFG) - Pregnancy
classification not yet possible

Israel - Ceiling Exposure Limits
C 50 ppm;   C 152 mg/m3

Mexico - Instruction No. 10 - TWAs
50 ppm TWA;   150 mg/m3 TWA

Mexico - Instruction No. 10 - Skin designation
skin - potential for cutaneous absorption

STATE LISTS
California - Air Bill 2588 Appendix A-I
6/91

California - Exposure Limits - Ceilings
C 50 ppm; C 150 mg/m3

California - Exposure Limits - Skin Notation
material may be absorbed through the skin, eyes or mucous membrane

California - Directors List of Hazardous Substances (8 CCR 339)
[present] (Listed under 'Butyl alcohol')

Florida Hazardous Substance List
[present]

Massachusetts Right To Know List
[present]

Minnesota Hazardous Substance List
skin

NJ Right to Know List (Total)
sn 1330

NJ Special Hazardous Substances
(flammable - third degree)

Pennsylvania Right to Know List
environmental hazard

TERT-BUTYL ALCOHOL   75-65-0

HEALTH AND SAFETY LISTS

ACGIH 1995 - Time Weighted Averages
(100) ppm TWA; (303) mg/m3 TWA

AIHA - Odor Threshold Values
geometric mean air odor threshold = 960 ppm (detectable)

NFPA - Flash Points
flash point = 52 degrees F (11 degrees C)

NFPA - Hazard Identification Ratings
health-1; flammability-3; reactivity-0

NIOSH - Selected LD50s and LC50s
Oral, rat: LD50 = 3500 mg/kg

NIOSH 1990 - Pocket Guide - RELs
100 ppm TWA; 300 mg/m3 TWA; 150 ppm STEL; 450 mg/m3 STEL

NIOSH 1990 - Pocket Guide - IDLHs
8000 ppm IDLH

NIOSH 1990 - Pocket Guide - Target organs
eyes, skin

NTP Chemical Status Reports - Testing Status and NTIS Number
prechronic studies for which toxicity technical reports were not prepared; post peer review technical reports in progress

OSHA - Vacated PELs - Time Weighted Averages
100 ppm TWA; 300 mg/m3 TWA

OSHA - Vacated PELs - Short Term Exposure Limits
150 ppm STEL; 450 mg/m3 STEL

OSHA - Final PELs - Time Weighted Averages
100 ppm TWA; 300 mg/m3 TWA

ENVIRONMENTAL LISTS

CERCLA/SARA - Section 313 - Emission Reporting
form R reporting required for 1.0% de minimus concentration

List of Pesticide Product Inert Ingredients
[present]

TSCA - PAIR - Reporting List
Effective Date: January 26, 1994; Reporting Date: March 28, 1994

TSCA - Health and Safety Reporting List
Effective Date: January 26, 1994; Sunset Date: January 26, 2004

INTERNATIONAL LISTS

Australian Exposure Standards - Time Weighted Averages
100 ppm TWA; 303 mg/m3 TWA

Australian Exposure Standards - Short Term Exposure Limits
150 ppm STEL; 455 mg/m3 STEL

Canada - WHMIS: Ingredient Disclosure
1% item 229 (356)

Canada - NPRI (National Pollutant Release Inventory)
[present]

Canada - Alberta - 8 Hour Occupational Exposure Limit
100 ppm TWA; 303 mg/m3 TWA

Canada - Alberta - 15 Minute Occupational Exposure Limit
150 ppm STEL; 455 mg/m3 STEL

Canada - British Columbia - 8 Hour Exposure Limits
100 ppm TWA; 300 mg/m3 TWA

Canada - British Columbia - 15 Minute Exposure Limits
150 ppm STEL; 450 mg/m3 STEL

Canada - Ontario - OHSA - TWAEVs
100 ppm TWAEV; 303 mg/m3 TWAEV

Canada - Ontario - OHSA - STEVs
150 ppm STEV; 454 mg/m3 STEV

Canada - Quebec - Time-Weighted Average Exposure Values
100 ppm TWAEV; 303 mg/m3 TWAEV

Canada - Quebec - Short-term Exposure Values
150 ppm STEV; 455 mg/m3 STEV

United Kingdom - Occupational Exposure Standards - TWAs
100 ppm TWA; 300 mg/m3 TWA

United Kingdom - Occupational Exposure Standards - STELs
150 ppm STEL; 450 mg/m3 STEL

German (DFG) - MAK Values
100 ppm MAK; 300 mg/m3 MAK (Listed under 'Butyl alcohol')

German (DFG) - Peak Limitations
2 x normal MAK (30 min. average value); don't exceed 4 times during shift (Listed under 'Butyl alcohol')

Israel - Time Weighted Averages
100 ppm TWA; 303 mg/m3 TWA

Israel - Short Term Exposure Limits
150 ppm STEL; 455 mg/m3 STEL

Israel - Action Levels
50 ppm AL; 151.5 mg/m3 AL

Mexico - Instruction No. 10 - TWAs
100 ppm TWA; 300 mg/m3 TWA

Mexico - Instruction No. 10 - STELs
150 ppm STEL; 450 mg/m3 STEL

STATE LISTS

California - Air Bill 2588 Appendix A-I
6/91

California - Exposure Limits - PELs
100 ppm PEL; 300 mg/m3 PEL

California - Exposure Limits - STELs
150 ppm STEL; 450 mg/m3 STEL

California - Directors List of Hazardous Substances (8 CCR 339)
[present] (Listed under 'Butyl alcohol')

Florida Hazardous Substance List
[present]

Massachusetts Right To Know List
[present]

Minnesota Hazardous Substance List
[present]

NJ Right to Know List (Total)
sn 1787

Pennsylvania Right to Know List
environmental hazard

PROPOSED REGULATIONS

ACGIH 1995 - Notice of Intended Changes
100 ppm TWA; 303 mg/m3 TWA; A4-not classifiable as a human carcinogen

TSCA - ITC 31st Report Priority Testing List
designated to be tested

SEC-BUTYL ALCOHOL   78-92-2

HEALTH AND SAFETY LISTS

ACGIH 1995 - Time Weighted Averages
100 ppm TWA; 303 mg/m3 TWA

AIHA - Odor Threshold Values
geometric mean air odor threshold = 3.2 ppm (detectable); 0.41 ppm (recognizable)

NFPA - Flash Points
flash point = 75 degrees F (24 degrees C)

NFPA - Hazard Identification Ratings
health-1; flammability-3; reactivity-0

NIOSH - Selected LD50s and LC50s
Oral, rat: LD50 = 6480 mg/kg

NIOSH 1990 - Pocket Guide - RELs
100 ppm TWA; 305 mg/m3 TWA; 150 ppm STEL; 455 mg/m3 STEL

NIOSH 1990 - Pocket Guide - IDLHs
10,000 ppm IDLH

NIOSH 1990 - Pocket Guide - Target organs
eyes, CNS

OSHA - Vacated PELs - Time Weighted Averages
100 ppm TWA; 305 mg/m3 TWA

OSHA - Final PELs - Time Weighted Averages
150 ppm TWA; 450 mg/m3 TWA

ENVIRONMENTAL LISTS

CERCLA/SARA - Section 313 - Emission Reporting
form R reporting required for 1.0% de minimus concentration

List of Pesticide Product Inert Ingredients
[present]

TSCA - PAIR - Reporting List
Effective Date: January 26, 1994; Reporting Date: March 28, 1994

TSCA - Health and Safety Reporting List
Effective Date: January 26, 1994; Sunset Date: January 26, 2004

INTERNATIONAL LISTS

Australian Exposure Standards - Time Weighted Averages
100 ppm TWA; 303 mg/m3 TWA

Canada - WHMIS: Ingredient Disclosure
1% item 228 (355)

Canada - NPRI (National Pollutant Release Inventory)
[present]

Canada - Alberta - 8 Hour Occupational Exposure Limit
100 ppm TWA; 303 mg/m3 TWA

Canada - Alberta - 15 Minute Occupational Exposure Limit
150 ppm STEL; 455 mg/m3 STEL

Canada - British Columbia - 8 Hour Exposure Limits
150 ppm TWA; 450 mg/m3 TWA

Canada - Ontario - OHSA - TWAEVs
100 ppm TWAEV; 303 mg/m3 TWAEV

Canada - Ontario - OHSA - STEVs
150 ppm STEV; 454 mg/m3 STEV

Canada - Quebec - Time-Weighted Average Exposure Values
100 ppm TWAEV; 303 mg/m3 TWAEV

United Kingdom - Occupational Exposure Standards - TWAs
100 ppm TWA; 300 mg/m3 TWA

United Kingdom - Occupational Exposure Standards - STELs
150 ppm STEL; 450 mg/m3 STEL

German (DFG) - MAK Values
100 ppm MAK; 300 mg/m3 MAK (Listed under 'Butyl alcohol')

German (DFG) - Peak Limitations
2 x normal MAK (30 min. average value); don't exceed 4 times during shift (Listed under 'Butyl alcohol')

Israel - Time Weighted Averages
100 ppm TWA; 303 mg/m3 TWA

Israel - Action Levels
50 ppm AL; 151.5 mg/m3 AL

Mexico - Instruction No. 10 - TWAs
100 ppm TWA; 305 mg/m3 TWA

Mexico - Instruction No. 10 - STELs
150 ppm STEL; 455 mg/m3 STEL

STATE LISTS

California - Air Bill 2588 Appendix A-I
6/91

California - Exposure Limits - PELs
100 ppm PEL; 305 mg/m3 PEL

California - Directors List of Hazardous Substances (8 CCR 339)
[present] (Listed under 'Butyl alcohol')

Florida Hazardous Substance List
[present]

Massachusetts Right To Know List
[present]

Minnesota Hazardous Substance List
[present]

NJ Right to Know List (Total)
sn 1645

Pennsylvania Right to Know List
environmental hazard

PROPOSED REGULATIONS

TSCA - ITC 31st Report Priority Testing List
designated to be tested

TERT-BUTYLAMINE   75-64-9

HEALTH AND SAFETY LISTS

U.S. DOT - Appendix A Table 1 - Hazardous Substances
final RQ = 1000 pounds (454 kg) (Listed under 'Butylamine')

NFPA - Hazard Identification Ratings
health-2; flammability-4; reactivity-0

NIOSH - Selected LD50s and LC50s
Oral, rat: LD50 = 78 mg/kg

ENVIRONMENTAL LISTS

CERCLA/SARA - Hazardous Substances and their Reportable Quantities
final RQ = 1000 pounds (454 kg) (Listed under 'Butylamine')

Clean Water Act - Hazardous Substances
[present] (Listed under 'Butylamine')

INTERNATIONAL LISTS

German (DFG) - MAK Values
5 ppm MAK; 15 mg/m3 MAK (Listed under 'Butylamine')

German (DFG) - Peak Limitations
5 x normal MAK (30 min. average value); don't exceed 2 times during shift (Listed under 'Butylamine')

German (DFG) - Skin/Sensitizers
danger of cutaneous absorption (Listed under 'Butylamine')

STATE LISTS

California - Directors List of Hazardous Substances (8 CCR 339)
[present] (Listed under 'Butylamine, all isomers')

Florida Hazardous Substance List
[present]

Massachusetts Right To Know List
[present]

Pennsylvania Right to Know List
environmental hazard

SEC-BUTYLAMINE   513-49-5

HEALTH AND SAFETY LISTS

U.S. DOT - Appendix A Table 1 - Hazardous Substances
final RQ = 1000 pounds (454 kg) (Listed under 'Butylamine')

ENVIRONMENTAL LISTS

CERCLA/SARA - Hazardous Substances and their Reportable Quantities
final RQ = 1000 pounds (454 kg) (Listed under 'Butylamine')

Clean Water Act - Hazardous Substances
[present] (Listed under 'Butylamine')

STATE LISTS

California - Directors List of Hazardous Substances (8 CCR 339)
[present] (Listed under 'Butylamine, all isomers')

Massachusetts Right To Know List
[present]

Pennsylvania Right to Know List
environmental hazard

SEC-BUTYLAMINE 13952-84-6

HEALTH AND SAFETY LISTS

U.S. DOT - Appendix A Table 1 - Hazardous Substances
final RQ = 1000 pounds (454 kg) (Listed under 'Butylamine')

NFPA - Flash Points
flash point = 16 degrees F (-9 degrees C)

NFPA - Hazard Identification Ratings
health-3; flammability-3

NIOSH - Selected LD50s and LC50s
Oral, rat: LD50 = 152 mg/kg Skin, rabbit: LD50 = 2500 mg/kg

ENVIRONMENTAL LISTS

CERCLA/SARA - Hazardous Substances and their Reportable Quantities
final RQ = 1000 pounds (454 kg) (Listed under 'Butylamine')

Clean Water Act - Hazardous Substances
[present] (Listed under 'Butylamine')

INTERNATIONAL LISTS

German (DFG) - MAK Values
5 ppm MAK; 15 mg/m3 MAK (Listed under 'Butylamine')

German (DFG) - Peak Limitations
5 x normal MAK (30 min. average value); don't exceed 2 times during shift (Listed under 'Butylamine')

German (DFG) - Skin/Sensitizers
danger of cutaneous absorption (Listed under 'Butylamine')

STATE LISTS

California - Directors List of Hazardous Substances (8 CCR 339)
[present] (Listed under 'Butylamine, all isomers')

Florida Hazardous Substance List
[present]

Massachusetts Right To Know List
[present]

Pennsylvania Right to Know List
environmental hazard

BUTYLAMINE (N-) 109-73-9

HEALTH AND SAFETY LISTS

ACGIH 1995 - Ceiling Limits
C 5 ppm; C 15 mg/m3

ACGIH 1995 - Skin Designations
skin - potential for cutaneous absorption

AIHA - Odor Threshold Values
geometric mean air odor threshold = 0.080 ppm (detectable); 1.8 ppm (recognizable)

U.S. DOT - Substances From 49 CFR 172.101
regulated by DOT (UN1125)

U.S. DOT - Hazard Classes
DOT hazard class = 3

U.S. DOT - Appendix A Table 1 - Hazardous Substances
final RQ = 1000 pounds (454 kg)

NFPA - Flash Points
flash point = 10 degrees F (-12 degrees C)

NFPA - Hazard Identification Ratings
health-3; flammability-3; reactivity-0

NIOSH - Selected LD50s and LC50s
Inhalation, mouse: LC50 = 800 mg/m3 (2 hr) Oral, rat: LD50 = 366 mg/kg Skin, guinea pig: LD50 = 366 mg/kg

NIOSH 1990 - Pocket Guide - RELs
C 5 ppm; C 15 mg/m3

NIOSH 1990 - Pocket Guide - IDLHs
2000 ppm IDLH

NIOSH 1990 - Pocket Guide - Target organs
respiratory system, skin, eyes

NIOSH 1990 - Pocket Guide - Skin list
Potential for dermal absorption

OSHA - Vacated PELs - Ceiling Limits
C 5 ppm; C 15 mg/m3

OSHA - Vacated PELs - Skin Designation
Prevent or reduce skin absorption

OSHA - Final PELs - Ceiling Limits
C 5 ppm; C 15 mg/m3

OSHA - Final PELs - Skin Notations
prevent or reduce skin absorption

ENVIRONMENTAL LISTS

CERCLA/SARA - Hazardous Substances and their Reportable Quantities
final RQ = 1000 pounds (454 kg)

Clean Water Act - Hazardous Substances
[present] (Listed under 'Butylamine')

INTERNATIONAL LISTS

Australian Exposure Standards - Time Weighted Averages
Peak Limitation: 5 ppm; 15 mg/m3

Australian Exposure Standards - Skin Effects
skin absorption

Canada - WHMIS: Ingredient Disclosure
1% item 235 (357)

Canada - Alberta - Ceiling Occupational Exposure Limit
C 5 ppm; C 15 mg/m3

Canada - Alberta - Skin Designation
can be absorbed through the intact skin

Canada - British Columbia - Ceiling Exposure Limits
C 5 ppm; C 15 mg/m3

Canada - British Columbia - Skin Notations
skin - potential for skin absorption

Canada - Ontario - OHSA - CEVs
5 ppm CEV; 15 mg/m3 CEV

Canada - Ontario - OHSA - Skin Notations
absorption through skin, eyes, or mucous membranes

Canada - Quebec - Ceiling Limits
P 5 ppm; P 15 mg/m3

Canada - Quebec - Skin Designations
absorbed through the skin

United Kingdom - Occupational Exposure Standards - STELs
5 ppm STEL; 15 mg/m3 STEL

United Kingdom - Occupational Exposure Standards - Notes
can be absorbed through skin

German (DFG) - MAK Values
5 ppm MAK; 15 mg/m3 MAK (Listed under 'Butylamine')

German (DFG) - Peak Limitations
5 x normal MAK (30 min. average value); don't exceed 2 times during shift (Listed under 'Butylamine')

German (DFG) - Skin/Sensitizers
danger of cutaneous absorption (Listed under 'Butylamine')

Israel - Ceiling Exposure Limits
C 5 ppm;   C 15 mg/m3
Mexico - Instruction No. 10 - TWAs
5 ppm TWA;   15 mg/m3 TWA
Mexico - Instruction No. 10 - Skin designation
skin - potential for cutaneous absorption
STATE LISTS
California - Exposure Limits - Ceilings
C 5 ppm;   C 15 mg/m3
California - Exposure Limits - Skin Notation
material may be absorbed through the skin, eyes or mucous membrane
California - Directors List of Hazardous Substances (8 CCR 339)
[present] (Listed under 'Butylamine, all isomers')
Florida Hazardous Substance List
[present]
Massachusetts Right To Know List
[present]
Minnesota Hazardous Substance List
skin
NJ Right to Know List (Total)
sn 0280
NJ Special Hazardous Substances
(flammable - third degree)
Pennsylvania Right to Know List
environmental hazard

BUTYLAMINE OLEATE   26094-13-3
HEALTH AND SAFETY LISTS
NFPA - Flash Points
flash point = 150 degrees F (66 degrees C)
NFPA - Hazard Identification Ratings
health-3; flammability-2; reactivity-0
STATE LISTS
Florida Hazardous Substance List
[present]
Massachusetts Right To Know List
[present]
Pennsylvania Right to Know List
[present]

TERT-BUTYLAMINOETHYL METHACRYLATE   3775-90-4
HEALTH AND SAFETY LISTS
NFPA - Flash Points
flash point = 205 degrees F (96 degrees C)
NFPA - Hazard Identification Ratings
health-2; flammability-1; reactivity-0
ENVIRONMENTAL LISTS
TSCA - Code of Federal Regulations Citations
40 CFR 712.30(d)
TSCA - PAIR - Reporting List
Reporting Date: November 19, 1982
STATE LISTS
Florida Hazardous Substance List
[present]
Massachusetts Right To Know List
[present]
Pennsylvania Right to Know List
[present]

N-BUTYLANILINE   1126-78-9
HEALTH AND SAFETY LISTS
U.S. DOT - Substances From 49 CFR 172.101
regulated by DOT (UN2738)

U.S. DOT - Hazard Classes
DOT hazard class = 6.1
NFPA - Flash Points
flash point = 225 degrees F (107 degrees C)
NFPA - Hazard Identification Ratings
health-3; flammability-1; reactivity-0
NIOSH - Selected LD50s and LC50s
Oral, rat: LD50 = 1620 mg/kg Skin, rabbit: LD50 = 5990 mg/kg
INTERNATIONAL LISTS
Canada - WHMIS: Ingredient Disclosure
1% item 236 (358)
STATE LISTS
Florida Hazardous Substance List
[present]
Massachusetts Right To Know List
[present]
NJ Right to Know List (Total)
sn 0281
Pennsylvania Right to Know List
[present]

TERT-BUTYLARSINE   RR-01577-
HEALTH AND SAFETY LISTS
U.S. DOT - Substances Which Are Poisonous by Inhalation
liquid hazardous material poisonous by inhalation

BUTYLATED HYDROXYANISOLE   25013-16-
HEALTH AND SAFETY LISTS
IARC - Group 2B (sufficient animal data)
[present] (Overall evaluation based only on evidence of carcinogenicity in monograph (40, 1986) or in Supplement 4)
NTP Seventh Report - Suspect Carcinogens
suspect carcinogen
OSHA - Possible Select Carcinogens
[present]
INTERNATIONAL LISTS
Canada - WHMIS: Ingredient Disclosure
1% item 237 (984)
STATE LISTS
California - Air Bill 2588 Appendix A-II
known or potential carcinogen
California - Prop. 65 - Cancer list
carcinogen - initial date 1/1/90
California - Directors List of Hazardous Substances (8 CCR 339)
[present]
Massachusetts Right To Know List
carcinogen; extraordinarily hazardous
Minnesota Hazardous Substance List
carcinogen

BUTYLATED HYDROXYANISOLE   489-01-0
ENVIRONMENTAL LISTS
List of Pesticide Product Inert Ingredients
[present]
STATE LISTS
California - Prop. 65 - No Significant Risk Levels
no significant risk level = 4000 ug/day

BUTYLATED POLYVINYLPYRROLIDONE   26160-96-3
ENVIRONMENTAL LISTS
List of Pesticide Product Inert Ingredients
[present]

TERT-BUTYLBENZENE   98-06-6

HEALTH AND SAFETY LISTS

U.S. DOT - Appendix B - Marine Pollutants
DOT regulated marine pollutant

NFPA - Flash Points
flash point = 140 degrees F (60 degrees C)

NFPA - Hazard Identification Ratings
health-2; flammability-2; reactivity-0

ENVIRONMENTAL LISTS

Safe Drinking Water Act - Monitoring
monitoring required at discretion of the state

TSCA - Code of Federal Regulations Citations
40 CFR 716.120(a)

TSCA - Health and Safety Reporting List
Effective Date: June 1, 1987

STATE LISTS

Florida Hazardous Substance List
[present]

Massachusetts Right To Know List
[present]

Pennsylvania Right to Know List
[present]

BUTYL BENZENE   104-51-8

HEALTH AND SAFETY LISTS

U.S. DOT - Appendix B - Marine Pollutants
DOT regulated marine pollutant

NFPA - Flash Points
flash point = 160 degrees F (71 degrees C)

NFPA - Hazard Identification Ratings
health-2; flammability-2; reactivity-0

ENVIRONMENTAL LISTS

Safe Drinking Water Act - Monitoring
monitoring required at discretion of the state

TSCA - Code of Federal Regulations Citations
40 CFR 716.120(a)

TSCA - Health and Safety Reporting List
Effective Date: June 1, 1987

STATE LISTS

Florida Hazardous Substance List
[present]

Massachusetts Right To Know List
[present]

NJ Right to Know List (Total)
sn 0282

Pennsylvania Right to Know List
[present]

SEC-BUTYLBENZENE   135-98-8

HEALTH AND SAFETY LISTS

U.S. DOT - Appendix B - Marine Pollutants
DOT regulated marine pollutant

NFPA - Flash Points
flash point = 126 degrees F (52 degrees C)

NFPA - Hazard Identification Ratings
health-2; flammability-2; reactivity-0

NIOSH - Selected LD50s and LC50s
Oral, rat: LD50 = 2240 mg/kg

ENVIRONMENTAL LISTS

Safe Drinking Water Act - Monitoring
monitoring required at discretion of the state

TSCA - Code of Federal Regulations Citations
40 CFR 716.120(a)

TSCA - Health and Safety Reporting List
Effective Date: June 1, 1987

STATE LISTS

Florida Hazardous Substance List
[present]

Massachusetts Right To Know List
[present]

Pennsylvania Right to Know List
[present]

BUTYL BENZENES   RR-00151-4

HEALTH AND SAFETY LISTS

U.S. DOT - Substances From 49 CFR 172.101
regulated by DOT (UN2709)

U.S. DOT - Hazard Classes
DOT hazard class = 3

U.S. DOT - Appendix B - Marine Pollutants
DOT regulated marine pollutant

N-BUTYLBENZENESULFONAMIDE   3622-84-2

HEALTH AND SAFETY LISTS

NIOSH - Selected LD50s and LC50s
Oral, rat: LD50 = 2050 mg/kg

BUTYL BENZOATE   136-60-7

HEALTH AND SAFETY LISTS

NFPA - Flash Points
flash point = 225 degrees F (107 degrees C)

NFPA - Hazard Identification Ratings
health-1; flammability-1; reactivity-0

NIOSH - Selected LD50s and LC50s
Oral, rat: LD50 = 5140 mg/kg Skin, rabbit: LD50 = 4000 mg/kg

INTERNATIONAL LISTS

Canada - WHMIS: Ingredient Disclosure
1% item 239 (282)

STATE LISTS

Pennsylvania Right to Know List
[present]

P-TERT-BUTYLBENZOIC ACID   98-73-7

HEALTH AND SAFETY LISTS

NIOSH - Selected LD50s and LC50s
Oral, rat: LD50 = 735 mg/kg

ENVIRONMENTAL LISTS

List of Pesticide Product Inert Ingredients
[present]

TSCA - Code of Federal Regulations Citations
40 CFR 704.33; 40 CFR 716.120(a)

TSCA - Health and Safety Reporting List
Effective Date: June 25, 1986

INTERNATIONAL LISTS

Canada - WHMIS: Ingredient Disclosure
0.1% item 240 (69)

ALPHA-BUTYL-BENZYL ALCOHOL   583-03-9

HEALTH AND SAFETY LISTS

NIOSH - Selected LD50s and LC50s
Oral, rat: LD50 = 5432 mg/kg

BUTYL BENZYL PHTHALATE   **85-68-7**
SEE ALSO:
  F039-HAZARDOUS WASTES
  HEALTH AND SAFETY LISTS
    U.S. DOT - Appendix B - Marine Pollutants
      DOT regulated marine pollutant
    U.S. DOT - Appendix A Table 1 - Hazardous Substances
      final RQ = 100 pounds (45.4 kg)
    IARC - Group 3 (not classifiable)
      [present]
    NFPA - Flash Points
      flash point = 390 degrees F (199 degrees C)
    NFPA - Hazard Identification Ratings
      health-1; flammability-1; reactivity-0
    NIOSH - Selected LD50s and LC50s
      Oral, rat: LD50 = 2330 mg/kg
    NTP Chemical Status Reports - Testing Status and NTIS Number
      Technical reports printed (PB83118398); two year studies: pathology quality assessment in progress; prechronic studies for which toxicity technical reports were not prepared
    NTP Chemical Status Reports - Evidence of Carcinogenicity
      male rat-inadequate; female rat-positive; male mice-negative; female mice-negative
  ENVIRONMENTAL LISTS
    ATSDR Priority List
      Rank (of 275): 156
    CERCLA/SARA - Section 313 - Emission Reporting
      form R reporting required for 1.0% de minimus concentration
    CERCLA/SARA - Hazardous Substances and their Reportable Quantities
      final RQ = 100 pounds (45.4 kg)
    Clean Water Act - Priority Pollutants
      [present]
    List of Pesticide Product Inert Ingredients
      [present]
    RCRA - Hazardous Constituents-Appendix VIII
      hazardous constituent - no waste number
    RCRA - Basis for Listing - Appendix VII
      Included in waste stream: F039
    RCRA - TSD Facilities Ground Water Monitoring
      TM 8060 = 5 ug/L PQL; TM 8270 = 10 ug/L PQL
    RCRA - Universal Treatment Standards (LDR)
      WW: 0.017 mg/l; NWW: 28 mg/kg
    TSCA - Code of Federal Regulations Citations
      40 CFR 712.30(d); 40 CFR 716.120(c)
    TSCA - PAIR - Reporting List
      Reporting Date: November 19, 1982
    TSCA - Health and Safety Reporting List
      Effective Date: April 29, 1983
  INTERNATIONAL LISTS
    Canada - WHMIS: Ingredient Disclosure
      1% item 241 (1415)
    Canada - NPRI (National Pollutant Release Inventory)
      [present]
    United Kingdom - Occupational Exposure Standards - TWAs
      5 mg/m3 TWA
    Mexico - Wastewater - Organic Toxic Pollutants and Heavy Metals
      Listed under [Phthalate Esters]
  STATE LISTS
    California - Air Bill 2588 Appendix A-I
      6/91
    California - Directors List of Hazardous Substances (8 CCR 339)
      [present]

    Massachusetts Right To Know List
      [present]
    NJ Right to Know List (Total)
      sn 2896
    Pennsylvania Right to Know List
      environmental hazard

2-BUTYLBIPHENYL   **54532-97-7**
  HEALTH AND SAFETY LISTS
    NFPA - Flash Points
      flash point > 212 degrees F (100 degrees C)
    NFPA - Hazard Identification Ratings
      health-0; flammability-1

BUTYL BROMIDE   **109-65-9**
  HEALTH AND SAFETY LISTS
    U.S. DOT - Substances From 49 CFR 172.101
      regulated by DOT (UN1126)
    U.S. DOT - Hazard Classes
      DOT hazard class = 3
    NFPA - Flash Points
      flash point = 65 degrees F (18 degrees C)
    NFPA - Hazard Identification Ratings
      health-2; flammability-3; reactivity-0
    NIOSH - Selected LD50s and LC50s
      Inhalation, rat: LC50 = 237000 mg/m3 (30 mn)
  INTERNATIONAL LISTS
    Canada - WHMIS: Ingredient Disclosure
      0.1% item 242 (332)
  STATE LISTS
    Florida Hazardous Substance List
      [present]
    Massachusetts Right To Know List
      [present]
    NJ Right to Know List (Total)
      sn 0283
    NJ Special Hazardous Substances
      (flammable - third degree)
    Pennsylvania Right to Know List
      [present]

BUTYL BUTYRATE   **109-21-7**
  HEALTH AND SAFETY LISTS
    NFPA - Flash Points
      flash point = 128 degrees F (53 degrees C)
    NFPA - Hazard Identification Ratings
      health-2; flammability-2; reactivity-0
    NIOSH - Selected LD50s and LC50s
      Oral, rabbit: LD50 = 9520 mg/kg
  ENVIRONMENTAL LISTS
    List of Pesticide Product Inert Ingredients
      [present]
  STATE LISTS
    Florida Hazardous Substance List
      [present]
    Massachusetts Right To Know List
      [present]
    Pennsylvania Right to Know List
      [present]

4-TERT-BUTYL CATECHOL   98-29-3

HEALTH AND SAFETY LISTS

AIHA - WEEL - Time Weighted Averages
5 mg/m3 TWA

AIHA - WEEL - Skin Absorption Designations
skin absorber

NFPA - Flash Points
flash point = 266 degrees F (130 degrees C)

NFPA - Hazard Identification Ratings
health-2; flammability-1; reactivity-0

NIOSH - Selected LD50s and LC50s
Oral, rat: LD50 = 2820 mg/kg Skin, rabbit: LD50 = 630 mg/kg

NTP Chemical Status Reports - Testing Status and NTIS Number
Approved for toxicology/carcinogenesis study

INTERNATIONAL LISTS

Canada - WHMIS: Ingredient Disclosure
1% item 243 (359)

STATE LISTS

Florida Hazardous Substance List
[present]

Massachusetts Right To Know List
[present]

Pennsylvania Right to Know List
[present]

BUTYL CHLORIDE   109-69-3

HEALTH AND SAFETY LISTS

NFPA - Flash Points
flash point = 15 degrees F (-9 degrees C)

NFPA - Hazard Identification Ratings
health-2; flammability-3; reactivity-0

NIOSH - Selected LD50s and LC50s
Oral, rat: LD50 = 2670 mg/kg

NTP Chemical Status Reports - Testing Status and NTIS Number
Technical reports printed (PB86218526/AS)

NTP Chemical Status Reports - Evidence of Carcinogenicity
male rat-no evidence; female rat-no evidence; male mice-no evidence; female mice-no evidence

ENVIRONMENTAL LISTS

EPA - Master Testing List
[present]

STATE LISTS

Florida Hazardous Substance List
[present]

Massachusetts Right To Know List
[present]

NJ Right to Know List (Total)
sn 0284

NJ Special Hazardous Substances
(flammable - third degree)

Pennsylvania Right to Know List
[present]

TERT-BUTYL CHLORIDE   507-20-0

HEALTH AND SAFETY LISTS

NFPA - Flash Points
flash point < 32 degrees F (0 degrees C)

NFPA - Hazard Identification Ratings
health-2; flammability-3

STATE LISTS

Florida Hazardous Substance List
[present]

Massachusetts Right To Know List
[present]

Pennsylvania Right to Know List
[present]

N-BUTYL CHLOROFORMATE   592-34-7

HEALTH AND SAFETY LISTS

U.S. DOT - Substances From 49 CFR 172.101
regulated by DOT (UN2743)

U.S. DOT - Hazard Classes
DOT hazard class = 6.1

U.S. DOT - Substances Which Are Poisonous by Inhalation
liquid hazardous material poisonous by inhalation (UN2743)

INTERNATIONAL LISTS

Canada - WHMIS: Ingredient Disclosure
1% item 244 (433)

United Kingdom - Occupational Exposure Standards - TWAs
1 ppm TWA; 5.6 mg/m3 TWA

STATE LISTS

NJ Right to Know List (Total)
sn 0285

NJ Special Hazardous Substances
(corrosive)

SEC-BUTYL CHLOROFORMATE   17462-58-7

HEALTH AND SAFETY LISTS

U.S. DOT - Substances From 49 CFR 172.101
regulated by DOT (NA2743, NA2742)

U.S. DOT - Hazard Classes
DOT hazard class = 6.1

U.S. DOT - Substances Which Are Poisonous by Inhalation
liquid hazardous material poisonous by inhalation (NA2742)

4-TERT-BUTYL-2-CHLOROPHENOL   98-28-2

HEALTH AND SAFETY LISTS

NFPA - Flash Points
flash point = 225 degrees F (107 degrees C)

NFPA - Hazard Identification Ratings
health-2; flammability-1; reactivity-0

STATE LISTS

Florida Hazardous Substance List
[present]

Massachusetts Right To Know List
[present]

Pennsylvania Right to Know List
[present]

TERT-BUTYL CHROMATE   1189-85-1
SEE ALSO:
CHROMIUM

HEALTH AND SAFETY LISTS

ACGIH 1995 - Ceiling Limits
as CrO3: C 0.1 mg/m3

ACGIH 1995 - Skin Designations
as CrO3: skin - potential for cutaneous absorption

NIOSH 1990 - Pocket Guide - RELs
as CrO3: 0.001 mg/m3 TWA

NIOSH 1990 - Pocket Guide - IDLHs
as CrO3: 30 mg/m3 IDLH (not considering carcinogenic effects)

NIOSH 1990 - Pocket Guide - Carcinogens
occupational carcinogen

NIOSH 1990 - Pocket Guide - Target organs
respiratory system, skin, eyes, CNS

OSHA - Vacated PELs - Ceiling Limits
as CrO3: C 0.1 mg/m3

OSHA - Vacated PELs - Skin Designation
Prevent or reduce skin absorption

OSHA - Final PELs - Ceiling Limits
as CrO3: C 0.1 mg/m3

OSHA - Final PELs - Skin Notations
prevent or reduce skin absorption

INTERNATIONAL LISTS

Australian Exposure Standards - Time Weighted Averages
as CrO3: Peak Limitation: 0.1 mg/m3

Australian Exposure Standards - Skin Effects
as CrO3: skin absorption

Canada - Alberta - Ceiling Occupational Exposure Limit
C 0.1 mg/m3

Canada - Alberta - Skin Designation
as CrO3: can be absorbed through the intact skin

Canada - British Columbia - Ceiling Exposure Limits
as CrO3: C 0.1 mg/m3

Canada - British Columbia - Skin Notations
as CrO3: skin - potential for skin absorption

Canada - Ontario - OHSA - CEVs
as chromate: 0.1 mg/m3 CEV

Canada - Ontario - OHSA - Skin Notations
absorption through skin, eyes, or mucous membranes

Canada - Quebec - Ceiling Limits
as CrO: P 0.1 mg/m3

Canada - Quebec - Skin Designations
as CrO: absorbed through the skin

Israel - Ceiling Exposure Limits
as CrO3: C 0.1 mg/m3

Mexico - Instruction No. 10 - TWAs
0.1 mg/m3 TWA

Mexico - Instruction No. 10 - Skin designation
skin - potential for cutaneous absorption

STATE LISTS

California - Exposure Limits - Ceilings
C 0.1 mg/m3

California - Exposure Limits - Skin Notation
material may be absorbed through the skin, eyes or mucous membrane

Florida Hazardous Substance List
[present]

Massachusetts Right To Know List
[present]

Minnesota Hazardous Substance List
as CrO3: skin

Pennsylvania Right to Know List
[present]

P-TERT-BUTYL-O-CRESOL   98-27-1

HEALTH AND SAFETY LISTS

NFPA - Flash Points
flash point = 244 degrees F (118 degrees C)

NFPA - Hazard Identification Ratings
health-2; flammability-1; reactivity-0

STATE LISTS

Florida Hazardous Substance List
[present]

Massachusetts Right To Know List
[present]

Pennsylvania Right to Know List
[present]

TERT-BUTYL-M-CRESOL   1333-13-7

HEALTH AND SAFETY LISTS

NFPA - Flash Points
flash point = 116 degrees F (47 degrees C)

NFPA - Hazard Identification Ratings
health-2; flammability-2; reactivity-0

STATE LISTS

Florida Hazardous Substance List
[present]

Massachusetts Right To Know List
[present]

Pennsylvania Right to Know List
[present]

TERT-BUTYL CUMYL PEROXIDE   3457-61-2

STATE LISTS

NJ Right to Know List (Total)
sn 2184

BUTYLCYCLOHEXANE   1678-93-9

HEALTH AND SAFETY LISTS

NFPA - Hazard Identification Ratings
health-0; reactivity-0

TERT-BUTYLCYCLOHEXANE   3178-22-1

HEALTH AND SAFETY LISTS

NFPA - Hazard Identification Ratings
health-0; reactivity-0

SEC-BUTYLCYCLOHEXANE   RR-00882-2

HEALTH AND SAFETY LISTS

NFPA - Hazard Identification Ratings
health-0; reactivity-0

4-TERT-BUTYLCYCLOHEXANOL   98-52-2

HEALTH AND SAFETY LISTS

NIOSH - Selected LD50s and LC50s
Oral, rat: LD50 = 4200 mg/kg

N-BUTYLCYCLOHEXYLAMINE   10108-56-2

HEALTH AND SAFETY LISTS

NFPA - Flash Points
flash point = 200 degrees F (93 degrees C)

NFPA - Hazard Identification Ratings
health-2; flammability-1; reactivity-0

STATE LISTS

Florida Hazardous Substance List
[present]

Massachusetts Right To Know List
[present]

Pennsylvania Right to Know List
[present]

TERT-BUTYLCYCLOHEXYL CHLOROFORMATE   70042-58-9

HEALTH AND SAFETY LISTS

U.S. DOT - Substances From 49 CFR 172.101
regulated by DOT (UN2747)

U.S. DOT - Hazard Classes
DOT hazard class = 6.1

INTERNATIONAL LISTS

Canada - WHMIS: Ingredient Disclosure
1% item 245 (432)

STATE LISTS
NJ Right to Know List (Total)
sn 1789

BUTYLCYCLOPENTANE   RR-00822-0
HEALTH AND SAFETY LISTS
NFPA - Hazard Identification Ratings
health-0; reactivity-0

BUTYLDECALIN   RR-00823-1
HEALTH AND SAFETY LISTS
NFPA - Flash Points
flash point = 500 degrees F (260 degrees C)
NFPA - Hazard Identification Ratings
health-1; flammability-1; reactivity-0

TERT-BUTYLDECALIN   RR-00884-4
HEALTH AND SAFETY LISTS
NFPA - Flash Points
flash point = 640 degrees F (338 degrees C)
NFPA - Hazard Identification Ratings
health-1; flammability-1; reactivity-0

SEC-BUTYL 2,4-D ESTER   94-79-1
HEALTH AND SAFETY LISTS
U.S. DOT - Appendix A Table 1 - Hazardous Substances
final RQ = 100 pounds (45.4 kg) (Listed under '2,4-D esters')
ENVIRONMENTAL LISTS
CERCLA/SARA - Hazardous Substances and their Reportable Quantities
final RQ = 100 pounds (45.4 kg) (Listed under '2,4-D esters')
Clean Water Act - Hazardous Substances
[present]
STATE LISTS
California - Directors List of Hazardous Substances (8 CCR 339)
[present] (Listed under '2,4-D esters...')
Massachusetts Right To Know List
[present]
NJ Right to Know List (Total)
sn 2942
Pennsylvania Right to Know List
environmental hazard

N-BUTYL 2,4-D ESTER   94-80-4
HEALTH AND SAFETY LISTS
U.S. DOT - Appendix A Table 1 - Hazardous Substances
final RQ = 100 pounds (45.4 kg) (Listed under '2,4-D esters')
NIOSH - Selected LD50s and LC50s
Oral, rat: LD50 = 995 mg/kg
ENVIRONMENTAL LISTS
CERCLA/SARA - Section 313 - Emission Reporting
form R reporting required
CERCLA/SARA - Hazardous Substances and their Reportable Quantities
final RQ = 100 pounds (45.4 kg) (Listed under '2,4-D esters')
Clean Water Act - Hazardous Substances
[present] (Listed under '2,4-D ester')
STATE LISTS
California - Directors List of Hazardous Substances (8 CCR 339)
[present] (Listed under '2,4-D esters...')
Massachusetts Right To Know List
[present]

NJ Right to Know List (Total)
sn 2943
Pennsylvania Right to Know List
environmental hazard

N-BUTYL-4,4-DI(TERT-BUTYL-PEROXY)VALERATE   995-33-5
HEALTH AND SAFETY LISTS
U.S. DOT - Organic Peroxides Table
Organic peroxide UN3103; UN3106
STATE LISTS
NJ Right to Know List (Total)
pure: sn 2186; maximum 52%: sn 2185

N-BUTYLDIETHANOLAMINE   102-79-4
HEALTH AND SAFETY LISTS
NFPA - Flash Points
flash point = 245 degrees F (118 degrees C)
NFPA - Hazard Identification Ratings
health-2; flammability-1; reactivity-0
NIOSH - Selected LD50s and LC50s
Oral, rat: LD50 = 4250 mg/kg
STATE LISTS
Florida Hazardous Substance List
[present]
Massachusetts Right To Know List
[present]
Pennsylvania Right to Know List
[present]

TERT-BUTYLDIETHANOLAMINE   2160-93-2
HEALTH AND SAFETY LISTS
NFPA - Flash Points
flash point = 285 degrees F (141 degrees C)
NFPA - Hazard Identification Ratings
health-2; flammability-1; reactivity-0
STATE LISTS
Florida Hazardous Substance List
[present]
Massachusetts Right To Know List
[present]
Pennsylvania Right to Know List
[present]

BUTYLENE   25167-67-3
HEALTH AND SAFETY LISTS
U.S. DOT - Substances From 49 CFR 172.101
regulated by DOT (UN1012)
U.S. DOT - Hazard Classes
DOT hazard class = 2.1
ENVIRONMENTAL LISTS
CAA - Flammable Substances for Accidental Release Prevention
threshold quantity = 10,000 lbs
STATE LISTS
NJ Right to Know List (Total)
sn 0286
NJ Special Hazardous Substances
(flammable - fourth degree)
Pennsylvania Right to Know List
[present]

1,2-BUTYLENE OXIDE   106-88-7

HEALTH AND SAFETY LISTS

AIHA - WEEL - Time Weighted Averages
2 ppm TWA; 5.9 mg/m3 TWA

U.S. DOT - Substances From 49 CFR 172.101
regulated by DOT (UN3022)

U.S. DOT - Hazard Classes
DOT hazard class = 3

IARC - Group 3 (not classifiable)
[present]

NFPA - Flash Points
flash point = -7 degrees F (-22 degrees C)

NFPA - Hazard Identification Ratings
health-2; flammability-3; reactivity-2

NIOSH - Selected LD50s and LC50s
Oral, rat: LD50 = 500 mg/kg Skin, rabbit: LD50 = 2100 mg/kg

NTP Chemical Status Reports - Testing Status and NTIS Number
Technical reports printed (PB88216262/AS)

NTP Chemical Status Reports - Evidence of Carcinogenicity
male rat-clear evidence; female rat-equivocal evidence; male mice-no evidence; female mice-no evidence

ENVIRONMENTAL LISTS

CERCLA/SARA - Section 313 - Emission Reporting
form R reporting required for 1.0% de minimus concentration

CERCLA/SARA - Hazardous Substances and their Reportable Quantities
final RQ = 1 pound (.454 kg)

Clean Air Act (1990) - List of Hazardous Air Contaminants
[present]

List of Pesticide Product Inert Ingredients
[present]

TSCA - Code of Federal Regulations Citations
40 CFR 712.30(d); 40 CFR 716.120(a)

TSCA - PAIR - Reporting List
Reporting Date: November 19, 1982

TSCA - Health and Safety Reporting List
Effective Date: October 4, 1982

INTERNATIONAL LISTS

Canada - NPRI (National Pollutant Release Inventory)
[present]

German (DFG) - Skin/Sensitizers
danger of cutaneous absorption

German (DFG) - Carcinogens
animal evidence of carcinogenicity

STATE LISTS

California - Air Bill 2588 Appendix A-I
6/91

Florida Hazardous Substance List
[present]

Massachusetts Right To Know List
[present]

NJ Right to Know List (Total)
sn 0287

NJ Special Hazardous Substances
(flammable - third degree; mutagen)

Pennsylvania Right to Know List
environmental hazard

2,3,4,5-BIS(2-BUTYLENE)TETRAHYDRO-2-FURFURAL   RR-01610-4

INTERNATIONAL LISTS

Canada - CEPA Schedule II Part I - Prohibited Substances (Export)
[present]

N-BUTYLETHANOLAMINE   111-75-1

HEALTH AND SAFETY LISTS

NFPA - Flash Points
flash point = 170 degrees F (77 degrees C)

NFPA - Hazard Identification Ratings
health-1; flammability-2; reactivity-0

NIOSH - Selected LD50s and LC50s
Oral, rat: LD50 = 1150 mg/kg

BUTYL ETHER   142-96-1

HEALTH AND SAFETY LISTS

U.S. DOT - Substances From 49 CFR 172.101
regulated by DOT (UN1149)

U.S. DOT - Hazard Classes
DOT hazard class = 3

NFPA - Flash Points
flash point = 77 degrees F (25 degrees C)

NFPA - Hazard Identification Ratings
health-2; flammability-3; reactivity-1

NIOSH - Selected LD50s and LC50s
Oral, rat: LD50 = 7400 mg/kg Skin, rabbit: LD50 = 10 gm/kg

STATE LISTS

Florida Hazardous Substance List
[present]

Massachusetts Right To Know List
[present]

NJ Right to Know List (Total)
sn 0288

NJ Special Hazardous Substances
(flammable - third degree)

Pennsylvania Right to Know List
[present]

N-BUTYL-N-ETHYL-2,6-DINITRO-4-(TRIFLUO-   1861-40-1
ROMETHYL)BENZENAMINE

HEALTH AND SAFETY LISTS

NIOSH - Selected LD50s and LC50s
Oral, rat: LD50 = 10 gm/kg

ENVIRONMENTAL LISTS

CERCLA/SARA - Section 313 - Emission Reporting
form R reporting required

BUTYL FORMATE   592-84-7

HEALTH AND SAFETY LISTS

U.S. DOT - Substances From 49 CFR 172.101
regulated by DOT (UN1128)

U.S. DOT - Hazard Classes
DOT hazard class = 3

NFPA - Flash Points
flash point = 64 degrees F (18 degrees C)

NFPA - Hazard Identification Ratings
health-2; flammability-3; reactivity-0

NIOSH - Selected LD50s and LC50s
Oral, rabbit: LD50 = 2656 mg/kg

STATE LISTS

Florida Hazardous Substance List
[present]

Massachusetts Right To Know List
[present]

NJ Right to Know List (Total)
sn 0289

Pennsylvania Right to Know List
[present]

TERT-BUTYL GLYCIDYL ETHER   **7665-72-7**
SEE ALSO:
GLYCIDOL (OXIRANEMETHANOL) AND ITS DERIVATIVES

ENVIRONMENTAL LISTS

EPA - Master Testing List
[present]

TSCA - Code of Federal Regulations Citations
40 CFR 716.120(c)

INTERNATIONAL LISTS

German (DFG) - Skin/Sensitizers
danger of cutaneous absorption; danger of sensitization (skin or respiratory)
German (DFG) - Carcinogens
suspected carcinogen

PROPOSED REGULATIONS

TSCA - Proposed Testing Rule for Glycidyl Ethers
subject to screening subcategory testing (results apply to all members of Glycidyl subcategory I-A)

N-BUTYL GLYCIDYL ETHER (BGE)   **2426-08-6**
SEE ALSO:
GLYCIDOL (OXIRANEMETHANOL) AND ITS DERIVATIVES

HEALTH AND SAFETY LISTS

ACGIH 1995 - Time Weighted Averages
25 ppm TWA; 133 mg/m3 TWA

NIOSH - Selected LD50s and LC50s
Oral, rat: LD50 = 2050 mg/kg Skin, rabbit: LD50 = 2520 mg/kg

NIOSH 1990 - Pocket Guide - RELs
C 5.6 ppm (15 min); C 30 mg/m3 (15 min)

NIOSH 1990 - Pocket Guide - IDLHs
3500 ppm IDLH

NIOSH 1990 - Pocket Guide - Target organs
eyes, skin, respiratory system, CNS

NIOSH - Health Standards - Exposure Limits
C (15 min) 5.6 ppm; C (15 min) 30 mg/m3 (Listed under 'Glycidyl ethers')

NIOSH - Health Standards - Health Effects and Precautions
Skin and mucous membrane effects; sensitization potential; possible hematopoietic and reproductive system effects (Medical monitoring required) (Listed under 'Glycidyl ethers')

OSHA - Vacated PELs - Time Weighted Averages
25 ppm TWA; 135 mg/m3 TWA

OSHA - Final PELs - Time Weighted Averages
50 ppm TWA; 270 mg/m3 TWA

ENVIRONMENTAL LISTS

EPA - Master Testing List
[present]

TSCA - Code of Federal Regulations Citations
40 CFR 712.30(d); 40 CFR 716.120(c)

TSCA - PAIR - Reporting List
Reporting Date: November 19, 1982

INTERNATIONAL LISTS

Australian Exposure Standards - Time Weighted Averages
25 ppm TWA; 133 mg/m3 TWA

Australian Exposure Standards - Skin Effects
sensitiser

Canada - WHMIS: Ingredient Disclosure
0.1% item 246 (810)

Canada - Alberta - 8 Hour Occupational Exposure Limit
25 ppm TWA; 133 mg/m3 TWA

Canada - Alberta - 15 Minute Occupational Exposure Limit
38 ppm STEL; 200 mg/m3 STEL

Canada - British Columbia - 8 Hour Exposure Limits
50 ppm TWA; 270 mg/m3 TWA

Canada - Ontario - OHSA - TWAEVs
25 ppm TWAEV; 133 mg/m3 TWAEV

Canada - Quebec - Time-Weighted Average Exposure Values
25 ppm TWAEV; 133 mg/m3 TWAEV

United Kingdom - Occupational Exposure Standards - TWAs
25 ppm TWA; 135 mg/m3 TWA

German (DFG) - Skin/Sensitizers
danger of cutaneous absorption; danger of sensitization (skin or respiratory)

German (DFG) - Carcinogens
suspected carcinogen

Israel - Time Weighted Averages
25 ppm TWA; 133 mg/m3 TWA

Israel - Action Levels
12.5 ppm AL; 66.5 mg/m3 AL

Mexico - Instruction No. 10 - TWAs
25 ppm TWA; 135 mg/m3 TWA

STATE LISTS

California - Exposure Limits - PELs
25 ppm PEL; 135 mg/m3 PEL

California - Directors List of Hazardous Substances (8 CCR 339)
[present] (exempt when part of a cured epoxy or rubber)

Florida Hazardous Substance List
[present]

Massachusetts Right To Know List
[present]

Minnesota Hazardous Substance List
[present]

Pennsylvania Right to Know List
[present]

PROPOSED REGULATIONS

TSCA - Proposed Testing Rule for Glycidyl Ethers
subject to subchronic toxicity, neurotoxicity, reproductive and fertility testing, screening subcategory and mutagenicity testing (results apply to all members of Glycidyl subcategory I-A)

Canada - Ontario - Proposed Occupational TWAEVs
10 ppm TWAEV; 50 mg/m3 TWAEV

Canada - Ontario - Proposed Occupational STEVs
15 ppm STEV; 80 mg/m3 STEV

BUTYL GLYCOLATE   **7397-62-8**

HEALTH AND SAFETY LISTS

NFPA - Flash Points
flash point = 142 degrees F (61 degrees C)

NFPA - Hazard Identification Ratings
health-0; flammability-2

TERT-BUTYL HYDROPEROXIDE   **75-91-2**

HEALTH AND SAFETY LISTS

U.S. DOT - Hazard Classes
Forbidden from transport by the DOT

U.S. DOT - Organic Peroxides Table
Organic peroxide UN3103; UN3105; UN3109

NFPA - Flash Points
flash point < 80 degrees F (27 degrees C) or higher

NFPA - Hazard Identification Ratings
health-1; flammability-4; reactivity-4 (oxidizing properties)

NIOSH - Selected LD50s and LC50s
Inhalation, rat: LC50 = 500 ppm 4 hr Oral, rat: LD50 = 406 mg/kg Skin, rat: LD50 = 790 mg/kg

OSHA - List of Highly Hazardous Chemicals
threshhold quantity = 5000 pounds

ENVIRONMENTAL LISTS

EPA - Master Testing List
[present]

INTERNATIONAL LISTS

Canada - WHMIS: Ingredient Disclosure
1% item 247 (981)

German (DFG) - Skin/Sensitizers
mg/L IMAC

STATE LISTS

Florida Hazardous Substance List
[present]

Massachusetts Right To Know List
[present]

NJ Right to Know List (Total)
sn 1790;   in solution: sn 2187

NJ Special Hazardous Substances
(mutagen)

Pennsylvania Right to Know List
[present]

T-BUTYLHYDROQUINONE   **1948-33-0**

HEALTH AND SAFETY LISTS

NIOSH - Selected LD50s and LC50s
Oral, rat: LD50 = 700 mg/kg

NTP Chemical Status Reports - Testing Status and NTIS Number
Two year studies: pathology quality assessment in progress

ENVIRONMENTAL LISTS

List of Pesticide Product Inert Ingredients
[present]

N-(N-BUTYL)IMIDAZOLE   **4316-42-1**

HEALTH AND SAFETY LISTS

U.S. DOT - Substances From 49 CFR 172.101
regulated by DOT (UN2690)

U.S. DOT - Hazard Classes
DOT hazard class = 6.1

INTERNATIONAL LISTS

Canada - WHMIS: Ingredient Disclosure
1% item 248 (360)

STATE LISTS

NJ Right to Know List (Total)
sn 1402

SEC-BUTYL IODIDE   **513-48-4**

HEALTH AND SAFETY LISTS

U.S. DOT - Substances From 49 CFR 172.101
regulated by DOT (UN2390)

U.S. DOT - Hazard Classes
DOT hazard class = 3

STATE LISTS

NJ Right to Know List (Total)
sn 1029

N-BUTYL ISOCYANATE   **111-36-4**

HEALTH AND SAFETY LISTS

U.S. DOT - Substances From 49 CFR 172.101
regulated by DOT (UN2485)

U.S. DOT - Hazard Classes
DOT hazard class = 3

U.S. DOT - Substances Which Are Poisonous by Inhalation
liquid hazardous material poisonous by inhalation (UN2485)

NFPA - Flash Points
flash point = 66 degrees F (19 degrees C)

NFPA - Hazard Identification Ratings
health-3; flammability-2; reactivity-2

NIOSH - Selected LD50s and LC50s
Inhalation, rat: LC50 = 3000 mg/m3 Skin, rat: LD50 = 600 mg/kg

ENVIRONMENTAL LISTS

TSCA - Code of Federal Regulations Citations
40 CFR 712.30(x);   716.120(d)

TSCA - PAIR - Reporting List
Reporting Date: December 27, 1990

TSCA - Health and Safety Reporting List
Effective Date: October 29, 1990

INTERNATIONAL LISTS

Canada - WHMIS: Ingredient Disclosure
0.1% item 249 (1035)

STATE LISTS

Florida Hazardous Substance List
[present]

Massachusetts Right To Know List
[present]

NJ Right to Know List (Total)
sn 1332

Pennsylvania Right to Know List
[present]

TERT-BUTYL ISOCYANATE   **1609-86-5**

HEALTH AND SAFETY LISTS

U.S. DOT - Substances From 49 CFR 172.101
regulated by DOT (UN2484)

U.S. DOT - Hazard Classes
DOT hazard class = 3

U.S. DOT - Substances Which Are Poisonous by Inhalation
liquid hazardous material poisonous by inhalation (UN2484)

INTERNATIONAL LISTS

Canada - WHMIS: Ingredient Disclosure
1% item 250 (1036)

STATE LISTS

NJ Right to Know List (Total)
sn 1791

TERT-BUTYL ISOPROPYL BENZENE   **30026-92-7**
HYDROPEROXIDE

STATE LISTS

NJ Right to Know List (Total)
sn 2188

BUTYL ISOVALERATE   **109-19-3**

HEALTH AND SAFETY LISTS

NFPA - Flash Points
flash point = 127 degrees F (53 degrees C)

NFPA - Hazard Identification Ratings
health-0

NIOSH - Selected LD50s and LC50s
Oral, rabbit: LD50 = 8200 mg/kg

STATE LISTS

Massachusetts Right To Know List
[present]

N-BUTYL LACTATE — 138-22-7

HEALTH AND SAFETY LISTS

ACGIH 1995 - Time Weighted Averages
5 ppm TWA; 30 mg/m3 TWA

NFPA - Flash Points
flash point = 160 degrees F (71 degrees C)

NFPA - Hazard Identification Ratings
health-1; flammability-2; reactivity-0

OSHA - Vacated PELs - Time Weighted Averages
5 ppm TWA; 25 mg/m3 TWA

INTERNATIONAL LISTS

Australian Exposure Standards - Time Weighted Averages
5 ppm TWA; 30 mg/m3 TWA

Canada - WHMIS: Ingredient Disclosure
1% item 251 (1062)

Canada - Alberta - 8 Hour Occupational Exposure Limit
5 ppm TWA; 30 mg/m3 TWA

Canada - Alberta - 15 Minute Occupational Exposure Limit
10 ppm STEL; 60 mg/m3 STEL

Canada - British Columbia - 8 Hour Exposure Limits
5 ppm TWA; 25 mg/m3 TWA

Canada - Ontario - OHSA - TWAEVs
5 ppm TWAEV; 30 mg/m3 TWAEV

Canada - Quebec - Time-Weighted Average Exposure Values
5 ppm TWAEV; 30 mg/m3 TWAEV

United Kingdom - Occupational Exposure Standards - TWAs
5 ppm TWA; 25 mg/m3 TWA

Israel - Time Weighted Averages
5 ppm TWA; 30 mg/m3 TWA

Israel - Action Levels
2.5 ppm AL; 15 mg/m3 AL

Mexico - Instruction No. 10 - TWAs
5 ppm TWA; 25 mg/m3 TWA

STATE LISTS

California - Exposure Limits - PELs
5 ppm PEL; 25 mg/m3 PEL

California - Directors List of Hazardous Substances (8 CCR 339)
[present]

Florida Hazardous Substance List
[present]

Massachusetts Right To Know List
[present]

Minnesota Hazardous Substance List
[present]

Pennsylvania Right to Know List
[present]

BUTYL LITHIUM — 109-72-8

STATE LISTS

NJ Right to Know List (Total)
sn 2190

Pennsylvania Right to Know List
[present] (in hydrocarbon solvents)

BUTYL MERCAPTAN — 109-79-5

HEALTH AND SAFETY LISTS

ACGIH 1995 - Time Weighted Averages
0.5 ppm TWA; 1.8 mg/m3 TWA

AIHA - Odor Threshold Values
geometric mean air odor threshold = 0.0010 ppm (detectable); 0.00073 ppm (recognizable)

U.S. DOT - Substances From 49 CFR 172.101
regulated by DOT (UN2347)

U.S. DOT - Hazard Classes
DOT hazard class = 3

NFPA - Flash Points
flash point = 35 degrees F (2 degrees C)

NFPA - Hazard Identification Ratings
health-2; flammability-3; reactivity-0

NIOSH - Selected LD50s and LC50s
Inhalation, rat: LC50 = 4020 ppm 4 hr Oral, rat: LD50 = 1500 mg/kg

NIOSH 1990 - Pocket Guide - RELs
C 0.5 ppm (15 min); C 1.8 mg/m3 (15 min)

NIOSH 1990 - Pocket Guide - IDLHs
2500 ppm IDLH

NIOSH 1990 - Pocket Guide - Target organs
respiratory system; in animals: CNS, liver, kidneys

NIOSH - Health Standards - Exposure Limits
C (15 min) 0.5 ppm; C (15 min) 1.8 mg/m3 (Listed under 'Thiols')

NIOSH - Health Standards - Health Effects and Precautions
Irritation; eye, skin, blood, and nervous system effects (Blood and urine monitoring required; prevent skin contact) (Listed under 'Thiols')

OSHA - Vacated PELs - Time Weighted Averages
0.5 ppm TWA; 1.5 mg/m3 TWA

OSHA - Final PELs - Time Weighted Averages
10 ppm TWA; 35 mg/m3 TWA

INTERNATIONAL LISTS

Australian Exposure Standards - Time Weighted Averages
0.5 ppm TWA; 1.8 mg/m3 TWA

Canada - WHMIS: Ingredient Disclosure
1% item 252 (361)

Canada - Alberta - 8 Hour Occupational Exposure Limit
0.5 ppm TWA; 1.8 mg/m3 TWA

Canada - Alberta - 15 Minute Occupational Exposure Limit
1.5 ppm STEL; 5.5 mg/m3 STEL

Canada - British Columbia - Ceiling Exposure Limits
C 3 ppm; C 9.3 mg/m3

Canada - Ontario - OHSA - TWAEVs
0.5 ppm TWAEV; 1.8 mg/m3 TWAEV

Canada - Quebec - Time-Weighted Average Exposure Values
0.5 ppm TWAEV; 1.8 mg/m3 TWAEV

German (DFG) - MAK Values
0.5 ppm MAK; 1.5 mg/m3 MAK

German (DFG) - Peak Limitations
2 x normal MAK (10 min momentary value); don't exceed 4 times per shift

Israel - Time Weighted Averages
0.5 ppm TWA; 1.8 mg/m3 TWA

Israel - Action Levels
0.25 ppm AL; 0.9 mg/m3 AL

Mexico - Instruction No. 10 - TWAs
0.5 ppm TWA; 1.5 mg/m3 TWA

STATE LISTS

California - Exposure Limits - PELs
0.5 ppm PEL; 1.5 mg/m3 PEL

California - Directors List of Hazardous Substances (8 CCR 339)
[present]

Florida Hazardous Substance List
[present]

Massachusetts Right To Know List
[present]

Minnesota Hazardous Substance List
[present]

NJ Right to Know List (Total)
sn 0290

NJ Special Hazardous Substances
(flammable - third degree)

Pennsylvania Right to Know List
[present]

BUTYL METHACRYLATE   97-88-1

HEALTH AND SAFETY LISTS

U.S. DOT - Substances From 49 CFR 172.101
regulated by DOT (UN2227)

U.S. DOT - Hazard Classes
DOT hazard class = 3

NFPA - Flash Points
flash point = 126 degrees F (52 degrees C)

NFPA - Hazard Identification Ratings
health-2; flammability-2; reactivity-0

NIOSH - Selected LD50s and LC50s
Inhalation, rat: LC50 = 4910 ppm 4 hr Oral, rat: LD50 = 22600 mg/kg Skin, rabbit: LD50 = 11300 mg/kg

ENVIRONMENTAL LISTS

List of Pesticide Product Inert Ingredients
[present]

TSCA - Code of Federal Regulations Citations
40 CFR 712.30(d),(x); 40 CFR 716.120(a)

TSCA - PAIR - Reporting List
Reporting Dates: November 19, 1982, November 27, 1991

TSCA - Health and Safety Reporting List
Effective Date: April 13, 1989

STATE LISTS

Florida Hazardous Substance List
[present]

Massachusetts Right To Know List
[present]

NJ Right to Know List (Total)
sn 0291

Pennsylvania Right to Know List
[present]

BUTYL METHACRYLATE, METHYL METHACRY-   32458-06-3
LATE, 2-HYDROXYETHYL METHACRYLATE AND
STYRENE COPOLYMER

ENVIRONMENTAL LISTS

List of Pesticide Product Inert Ingredients
[present]

BUTYL METHYL ETHER   628-28-4

HEALTH AND SAFETY LISTS

U.S. DOT - Substances From 49 CFR 172.101
regulated by DOT (UN2350)

U.S. DOT - Hazard Classes
DOT hazard class = 3

STATE LISTS

NJ Right to Know List (Total)
sn 3104

TERT-BUTYL MONOPEROXYMALEATE   1931-62-0

HEALTH AND SAFETY LISTS

U.S. DOT - Organic Peroxides Table
Organic peroxide UN3102; UN3103; UN3108

STATE LISTS

NJ Right to Know List (Total)
sn 1792

BUTYL NAPHTHALENE   31711-50-9

HEALTH AND SAFETY LISTS

NFPA - Flash Points
flash point = 680 degrees F (360 degrees C)

NFPA - Hazard Identification Ratings
health-1; flammability-1; reactivity-0

ENVIRONMENTAL LISTS

List of Pesticide Product Inert Ingredients
[present]

BUTYLNAPHTHALENESULFONIC ACID, SODIUM   25638-17-9
SALT

ENVIRONMENTAL LISTS

List of Pesticide Product Inert Ingredients
[present]

BUTYL NITRATE   928-45-0

HEALTH AND SAFETY LISTS

NFPA - Flash Points
flash point = 97 degrees F (36 degrees C)

NFPA - Hazard Identification Ratings
health-1; flammability-3; reactivity-3

STATE LISTS

Florida Hazardous Substance List
[present]

Massachusetts Right To Know List
[present]

Pennsylvania Right to Know List
[present]

BUTYL NITRITE   544-16-1

HEALTH AND SAFETY LISTS

U.S. DOT - Substances From 49 CFR 172.101
regulated by DOT (UN2351)

U.S. DOT - Hazard Classes
DOT hazard class = 3

NIOSH - Selected LD50s and LC50s
Inhalation, rat: LC50 = 918 ppm 1 hr Oral, rat: LD50 = 83 mg/kg

STATE LISTS

NJ Right to Know List (Total)
sn 0292

2-BUTYLOCTANOL   3913-02-8

HEALTH AND SAFETY LISTS

NFPA - Flash Points
flash point = 230 degrees F (110 degrees C)

NFPA - Hazard Identification Ratings
health-1; flammability-1; reactivity-0

BUTYL OLEATE   142-77-8

HEALTH AND SAFETY LISTS

NFPA - Flash Points
flash point = 356 degrees F (180 degrees C)

NFPA - Hazard Identification Ratings
health-0; flammability-1; reactivity-0

INTERNATIONAL LISTS

Canada - WHMIS: Ingredient Disclosure
1% item 253 (1285)

BUTYL OXALATE   2050-60-4

HEALTH AND SAFETY LISTS

NFPA - Flash Points
*flash point = 265 degrees F (129 degrees C); *See list description*

NFPA - Hazard Identification Ratings
health-0; flammability-1; reactivity-0

ENVIRONMENTAL LISTS

List of Pesticide Product Inert Ingredients
[present]

BUTYL PARABAN   94-26-8

ENVIRONMENTAL LISTS

List of Pesticide Product Inert Ingredients
[present]

TERT-BUTYL PEROXYACETATE   107-71-1

HEALTH AND SAFETY LISTS

U.S. DOT - Hazard Classes
Forbidden from transport by the DOT

U.S. DOT - Organic Peroxides Table
Organic peroxide UN3101; UN3103

NFPA - Flash Points
diluted with 25% of benzene: flash point < 80 degrees F (27 degrees C)

NFPA - Hazard Identification Ratings
diluted with 25% of benzene: health-2; flammability-3; reactivity-4

NIOSH - Selected LD50s and LC50s
Oral, rat: LD50 = 675 mg/kg

INTERNATIONAL LISTS

Canada - WHMIS: Ingredient Disclosure
1% item 254 (1361)

German (DFG) - Skin/Sensitizers
moderate effects on skin

STATE LISTS

Florida Hazardous Substance List
[present]

Massachusetts Right To Know List
[present]

NJ Right to Know List (Total)
sn 1793

NJ Special Hazardous Substances
(flammable - third degree; reactive - fourth degree)

Pennsylvania Right to Know List
[present]

TERT-BUTYL PEROXYBENZOATE   614-45-9

HEALTH AND SAFETY LISTS

U.S. DOT - Organic Peroxides Table
Organic peroxide UN3103; UN3105; UN3106

NFPA - Flash Points
flash point > 190 degrees F (88 degrees C)

NFPA - Hazard Identification Ratings
health-1; flammability-3; reactivity-4 (oxidizing properties)

NIOSH - Selected LD50s and LC50s
Oral, rat: LD50 = 1012 mg/kg

NTP Chemical Status Reports - Testing Status and NTIS Number
Technical reports printed (PB93-105690/AS)

OSHA - List of Highly Hazardous Chemicals
threshhold quantity = 7500 pounds

ENVIRONMENTAL LISTS

List of Pesticide Product Inert Ingredients
[present]

STATE LISTS

Florida Hazardous Substance List
[present]

Massachusetts Right To Know List
[present]

NJ Right to Know List (Total)
sn 1794

NJ Special Hazardous Substances
(flammable - third degree; reactive - fourth degree)

Pennsylvania Right to Know List
[present]

TERT-BUTYL PEROXYCROTONATE   23474-91-1

HEALTH AND SAFETY LISTS

U.S. DOT - Organic Peroxides Table
Organic peroxide UN3105

STATE LISTS

NJ Right to Know List (Total)
sn 1795

BUTYL PEROXYDICARBONATE   16215-49-9

HEALTH AND SAFETY LISTS

U.S. DOT - Hazard Classes
Forbidden from transport by the DOT

U.S. DOT - Organic Peroxides Table
Organic peroxide UN3115; UN3117

STATE LISTS

NJ Right to Know List (Total)
sn 0293

TERT-BUTYL PEROXYDIETHYL-ACETATE   2550-33-6

HEALTH AND SAFETY LISTS

U.S. DOT - Organic Peroxides Table
Organic peroxide UN3113

STATE LISTS

NJ Right to Know List (Total)
sn 2191

T-BUTYL PEROXYDIETHYL-ACETATE WITH T-   RR-00185-4
BUTYL PEROXYBENZOATE

STATE LISTS

NJ Right to Know List (Total)
sn 2192

TERT-BUTYL PEROXY(2-ETHYL)-HEXANOATE   3006-82-4

HEALTH AND SAFETY LISTS

U.S. DOT - Organic Peroxides Table
Organic peroxide UN3106; UN3113; UN3115; UN3117

STATE LISTS

NJ Right to Know List (Total)
sn 1796

TERT-BUTYL PEROXYISOBUTYRATE   109-13-7

HEALTH AND SAFETY LISTS

U.S. DOT - Hazard Classes
Forbidden from transport by the DOT

U.S. DOT - Organic Peroxides Table
Organic peroxide UN3111; UN3115

STATE LISTS

NJ Right to Know List (Total)
sn 1797

TERT-BUTYL PEROXYISONONANOATE   RR-00771-6

STATE LISTS

NJ Right to Know List (Total)
sn 2193

TERT-BUTYL PEROXYISOPROPYL CARBONATE   2372-21-6
  HEALTH AND SAFETY LISTS
    U.S. DOT - Organic Peroxides Table
    Organic peroxide UN3103
  STATE LISTS
    NJ Right to Know List (Total)
    sn 1798

TERT-BUTYL PEROXYNEODECANOATE   26748-41-4
  HEALTH AND SAFETY LISTS
    U.S. DOT - Organic Peroxides Table
    Organic peroxide UN3115
  STATE LISTS
    NJ Right to Know List (Total)
    sn 1799

3-TERT-BUTYL PEROXY-3-PHENYLPHTHALIDE   25251-51-8
  HEALTH AND SAFETY LISTS
    U.S. DOT - Organic Peroxides Table
    Organic peroxide UN3106
  STATE LISTS
    NJ Right to Know List (Total)
    sn 2194

TERT-BUTYL PEROXYPHTHALATE   RR-00772-7
  STATE LISTS
    NJ Right to Know List (Total)
    sn 2195

TERT-BUTYL PEROXYPIVALATE   927-07-1
  HEALTH AND SAFETY LISTS
    U.S. DOT - Organic Peroxides Table
    Organic peroxide UN3113;   UN3115
    NFPA - Flash Points
    diluted with 25% of mineral spirits: flash point > 155 degrees F (68 degrees C)
    NFPA - Hazard Identification Ratings
    diluted with 25% mineral spirits: health-0;   flammability-3;   reactivity-4 (oxidizing properties)
    NIOSH - Selected LD50s and LC50s
    Oral, rat: LD50 = 4300 mg/kg
  STATE LISTS
    Florida Hazardous Substance List
    [present]
    Massachusetts Right To Know List
    [present]
    NJ Right to Know List (Total)
    sn 1800
    NJ Special Hazardous Substances
    (flammable - third degree;   reactive - fourth degree)
    Pennsylvania Right to Know List
    [present]

TERT-BUTYLPEROXY STEARYLCARBONATE   RR-01751-6
  HEALTH AND SAFETY LISTS
    U.S. DOT - Organic Peroxides Table
    Organic peroxide UN3106

TERT-BUTYL PEROXY-3,5,5-TRIMETHYLHEXANOATE   13122-18-4
  HEALTH AND SAFETY LISTS
    U.S. DOT - Organic Peroxides Table
    Organic peroxide UN3105

  STATE LISTS
    NJ Right to Know List (Total)
    sn 2196

O-SEC-BUTYLPHENOL   89-72-5
  HEALTH AND SAFETY LISTS
    ACGIH 1995 - Time Weighted Averages
    5 ppm TWA;   31 mg/m3 TWA
    ACGIH 1995 - Skin Designations
    skin - potential for cutaneous absorption
    NIOSH - Selected LD50s and LC50s
    Oral, rat: LD50 = 2700 mg/kg Skin, guinea pig: LD50 = 600 mg/kg
    OSHA - Vacated PELs - Time Weighted Averages
    5 ppm TWA;   30 mg/m3 TWA
    OSHA - Vacated PELs - Skin Designation
    Prevent or reduce skin absorption
  ENVIRONMENTAL LISTS
    TSCA - Health and Safety Reporting List
    Effective Date: March 11, 1994;   Sunset Date: March 11, 2004
  INTERNATIONAL LISTS
    Australian Exposure Standards - Time Weighted Averages
    5 ppm TWA;   31 mg/m3 TWA
    Australian Exposure Standards - Skin Effects
    skin absorption
    Canada - WHMIS: Ingredient Disclosure
    1% item 257 (364)
    Canada - Alberta - 8 Hour Occupational Exposure Limit
    5 ppm TWA;   31 mg/m3 TWA
    Canada - Alberta - 15 Minute Occupational Exposure Limit
    10 ppm STEL;   62 mg/m3 STEL
    Canada - Alberta - Skin Designation
    can be absorbed through the intact skin
    Canada - Ontario - OHSA - TWAEVs
    5 ppm TWAEV;   31 mg/m3 TWAEV
    Canada - Ontario - OHSA - Skin Notations
    absorption through skin, eyes, or mucous membranes
    Canada - Quebec - Time-Weighted Average Exposure Values
    5 ppm TWAEV;   31 mg/m3 TWAEV
    Canada - Quebec - Skin Designations
    absorbed through the skin
    United Kingdom - Occupational Exposure Standards - TWAs
    5 ppm TWA;   30 mg/m3 TWA
    United Kingdom - Occupational Exposure Standards - Notes
    can be absorbed through skin
    Israel - Time Weighted Averages
    5 ppm TWA;   31 mg/m3 TWA
    Israel - Action Levels
    2.5 ppm AL;   15.5 mg/m3 AL
  STATE LISTS
    California - Exposure Limits - PELs
    5 ppm PEL;   30 mg/m3 PEL
    California - Exposure Limits - Skin Notation
    material may be absorbed through the skin, eyes or mucous membrane
    California - Directors List of Hazardous Substances (8 CCR 339)
    [present]
    Florida Hazardous Substance List
    [present]
    Massachusetts Right To Know List
    [present]
    Minnesota Hazardous Substance List
    skin
    NJ Right to Know List (Total)
    sn 1440

Pennsylvania Right to Know List
[present]

PROPOSED REGULATIONS
TSCA - ITC 32nd Report Priority Testing List
designated for dermal absorption testing

P-TERT-BUTYL PHENOL   98-54-4

HEALTH AND SAFETY LISTS
NIOSH - Selected LD50s and LC50s
Oral, rat: LD50 = 2951 mg/kg Skin, rabbit: LD50 = 2288 mg/kg

ENVIRONMENTAL LISTS
List of Pesticide Product Inert Ingredients
[present]

INTERNATIONAL LISTS
Canada - WHMIS: Ingredient Disclosure
1% item 258 (366)
German (DFG) - MAK Values
0.08 ppm MAK; 0.5 mg/m3 MAK
German (DFG) - Peak Limitations
5 x normal MAK (30 min. average value); don't exceed 2 times during shift
German (DFG) - Skin/Sensitizers
danger of cutaneous absorption
Mexico - Instruction No. 10 - TWAs
10 ppm TWA; 60 mg/m3 TWA
Mexico - Instruction No. 10 - STELs
20 ppm STEL; 120 mg/m3 STEL

P-BUTYLPHENOL   1638-22-8

INTERNATIONAL LISTS
Canada - WHMIS: Ingredient Disclosure
1% item 259 (365)

O-BUTYLPHENOL   3180-09-4

INTERNATIONAL LISTS
Canada - WHMIS: Ingredient Disclosure
1% item 256 (363)

M-BUTYLPHENOL   4074-43-5

INTERNATIONAL LISTS
Canada - WHMIS: Ingredient Disclosure
1% item 255 (362)

BUTYL PHENOL   28805-86-9

HEALTH AND SAFETY LISTS
U.S. DOT - Substances From 49 CFR 172.101
regulated by DOT (UN2228, UN2229)
U.S. DOT - Hazard Classes
DOT hazard class = 6.1
U.S. DOT - Appendix B - Marine Pollutants
DOT regulated marine pollutant

STATE LISTS
NJ Right to Know List (Total)
sn 0294

P-TERT-BUTYLPHENOL, ETHOXYLATED,   68332-64-9
PHOSPHATED

ENVIRONMENTAL LISTS
List of Pesticide Product Inert Ingredients
[present]

BETA(P-TERT-BUTYLPHENOXY) ETHANOL   713-46-2

HEALTH AND SAFETY LISTS
NFPA - Flash Points
flash point = 248 degrees F (120 degrees C)
NFPA - Hazard Identification Ratings
health-0; flammability-1; reactivity-0

BETA-(P-TERT-BUTYLPHENOXY) ETHYL ACETATE   RR-00794-3

HEALTH AND SAFETY LISTS
NFPA - Flash Points
flash point = 324 degrees F (162 degrees C)
NFPA - Hazard Identification Ratings
health-0; flammability-1; reactivity-0

TERT-BUTYLPHENYL DIPHENYL PHOSPHATE   56803-37-3
SEE ALSO:
BISAZOBIPHENYL DYES

ENVIRONMENTAL LISTS
EPA - Master Testing List
[present]
TSCA - Code of Federal Regulations Citations
40 CFR 712.30(d); 40 CFR 716.120
TSCA - PAIR - Reporting List
Reporting Date: November 19, 1982

BUTYL PHENYL ETHER   1126-79-0

HEALTH AND SAFETY LISTS
NFPA - Flash Points
flash point = 180 degrees F (82 degrees C)
NFPA - Hazard Identification Ratings
health-1; flammability-2; reactivity-0

P-TERT-BUTYL PHENYL GLYCIDYL ETHER   3101-60-8
SEE ALSO:
GLYCIDOL (OXIRANEMETHANOL) AND ITS DERIVATIVES

ENVIRONMENTAL LISTS
EPA - Master Testing List
[present]
TSCA - Code of Federal Regulations Citations
40 CFR 712.30(d); 40 CFR 716.120(c)

PROPOSED REGULATIONS
TSCA - Proposed Testing Rule for Glycidyl Ethers
member of Glycidyl subcategory IV-A

4-TERT-BUTYL-2-PHENYLPHENOL   42479-87-8

HEALTH AND SAFETY LISTS
NFPA - Flash Points
flash point = 320 degrees F (160 degrees C)
NFPA - Hazard Identification Ratings
health-1; flammability-1; reactivity-0

BUTYL PHOSPHORIC ACID   12788-93-1

HEALTH AND SAFETY LISTS
U.S. DOT - Substances From 49 CFR 172.101
regulated by DOT (UN1718)
U.S. DOT - Hazard Classes
DOT hazard class = 8

ENVIRONMENTAL LISTS
List of Pesticide Product Inert Ingredients
[present]

INTERNATIONAL LISTS
Canada - WHMIS: Ingredient Disclosure
1% item 260 (70)

STATE LISTS
NJ Right to Know List (Total)
sn 0277

NJ Special Hazardous Substances
(corrosive)

BUTYLPOLYETHOXYETHANOL ESTERS OF PHOS-PHORIC ACID   50769-39-6
ENVIRONMENTAL LISTS
List of Pesticide Product Inert Ingredients
[present]

BUTYL PROPIONATE   590-01-2
HEALTH AND SAFETY LISTS
U.S. DOT - Substances From 49 CFR 172.101
regulated by DOT (UN1914)

U.S. DOT - Hazard Classes
DOT hazard class = 3

NFPA - Flash Points
flash point = 90 degrees F (32 degrees C)

NFPA - Hazard Identification Ratings
health-2; flammability-3; reactivity-0

NIOSH - Selected LD50s and LC50s
Oral, rat: LD50 = 5000 mg/kg

INTERNATIONAL LISTS
Canada - WHMIS: Ingredient Disclosure
1% item 261 (1447)

STATE LISTS
Florida Hazardous Substance List
[present]

Massachusetts Right To Know List
[present]

NJ Right to Know List (Total)
sn 0295

NJ Special Hazardous Substances
(flammable - third degree)

Pennsylvania Right to Know List
[present]

BUTYL RICINOLEATE   151-13-3
HEALTH AND SAFETY LISTS
NFPA - Flash Points
flash point = 230 degrees F (110 degrees C)

NFPA - Hazard Identification Ratings
health-1; flammability-1; reactivity-0

BUTYL STEARATE   123-95-5
HEALTH AND SAFETY LISTS
NFPA - Flash Points
flash point = 320 degrees F (160 degrees C)

NFPA - Hazard Identification Ratings
health-1; flammability-1; reactivity-0

ENVIRONMENTAL LISTS
List of Pesticide Product Inert Ingredients
[present]

TERT-BUTYL STYRENE   25338-51-6
HEALTH AND SAFETY LISTS
NFPA - Flash Points
flash point = 177 degrees F (81 degrees C)

NFPA - Hazard Identification Ratings
health-2; flammability-2; reactivity-2

STATE LISTS
Pennsylvania Right to Know List
[present]

TERT-BUTYL TETRALIN   73090-68-3
HEALTH AND SAFETY LISTS
NFPA - Flash Points
flash point = 680 degrees F (360 degrees C)

NFPA - Hazard Identification Ratings
health-2; flammability-1; reactivity-0

STATE LISTS
Florida Hazardous Substance List
[present]

Massachusetts Right To Know List
[present]

Pennsylvania Right to Know List
[present]

BUTYL TITANATE   5593-70-4
HEALTH AND SAFETY LISTS
NIOSH - Selected LD50s and LC50s
Oral, rat: LD50 = 3122 mg/kg

P-TERT-BUTYLTOLUENE   98-51-1
HEALTH AND SAFETY LISTS
ACGIH 1995 - Time Weighted Averages
1 ppm TWA; 6.1 mg/m3 TWA

U.S. DOT - Appendix B - Marine Pollutants
DOT regulated marine pollutant

NIOSH - Selected LD50s and LC50s
Inhalation, rat: LC50 = 165 ppm 8 hr Oral, rat: LD50 = 1800 mg/kg Skin, rabbit: LD50 = 19600 mg/kg

NIOSH 1990 - Pocket Guide - RELs
10 ppm TWA; 60 mg/m3 TWA; 20 ppm STEL; 120 mg/m3 STEL

NIOSH 1990 - Pocket Guide - IDLHs
1000 ppm IDLH

NIOSH 1990 - Pocket Guide - Target organs
CVS, CNS, skin, bone marrow, eyes, upper respiratory system

OSHA - Vacated PELs - Time Weighted Averages
10 ppm TWA; 60 mg/m3 TWA

OSHA - Vacated PELs - Short Term Exposure Limits
20 ppm STEL; 120 mg/m3 STEL

OSHA - Final PELs - Time Weighted Averages
10 ppm TWA; 60 mg/m3 TWA

ENVIRONMENTAL LISTS
CAA - HON Rule - SOCMI Chemicals
compliance by April 24, 1995

TSCA - Code of Federal Regulations Citations
40 CFR 704.33; 40 CFR 716.120(a)

TSCA - Health and Safety Reporting List
Effective Date: June 25, 1986

INTERNATIONAL LISTS
Australian Exposure Standards - Time Weighted Averages
10 ppm TWA; 61 mg/m3 TWA

Australian Exposure Standards - Short Term Exposure Limits
20 ppm STEL; 121 mg/m3 STEL

Canada - WHMIS: Ingredient Disclosure
1% item 265 (370)

Canada - Alberta - 8 Hour Occupational Exposure Limit
10 ppm TWA; 61 mg/m3 TWA

Canada - Alberta - 15 Minute Occupational Exposure Limit
20 ppm STEL; 121 mg/m3 STEL

Canada - British Columbia - 8 Hour Exposure Limits
  10 ppm TWA; 60 mg/m3 TWA
Canada - British Columbia - 15 Minute Exposure Limits
  20 ppm STEL; 120 mg/m3 STEL
Canada - Ontario - OHSA - TWAEVs
  10 ppm TWAEV; 60 mg/m3 TWAEV
Canada - Ontario - OHSA - STEVs
  20 ppm STEV; 120 mg/m3 STEV
Canada - Quebec - Time-Weighted Average Exposure Values
  10 ppm TWAEV; 61 mg/m3 TWAEV
Canada - Quebec - Short-term Exposure Values
  20 ppm STEV; 122 mg/m3 STEV
German (DFG) - MAK Values
  10 ppm MAK; 60 mg/m3 MAK
German (DFG) - Peak Limitations
  2 x normal MAK (5 min momentary value); don't exceed 8 times during shift
Israel - Time Weighted Averages
  10 ppm TWA; 61 mg/m3 TWA
Israel - Short Term Exposure Limits
  20 ppm STEL; 121 mg/m3 STEL
Israel - Action Levels
  5 ppm AL; 30.5 mg/m3 AL

STATE LISTS

California - Exposure Limits - PELs
  10 ppm PEL; 60 mg/m3 PEL
California - Exposure Limits - STELs
  20 ppm STEL; 120 mg/m3 STEL
California - Directors List of Hazardous Substances (8 CCR 339)
  [present]
Florida Hazardous Substance List
  [present]
Massachusetts Right To Know List
  [present]
Minnesota Hazardous Substance List
  [present]
Pennsylvania Right to Know List
  [present]

M-BUTYLTOLUENE   1595-04-6

INTERNATIONAL LISTS

Canada - WHMIS: Ingredient Disclosure
  1% item 262 (367)

P-BUTYLTOLUENE   1595-05-7

INTERNATIONAL LISTS

Canada - WHMIS: Ingredient Disclosure
  1% item 264 (369)

O-BUTYLTOLUENE   1595-11-5

INTERNATIONAL LISTS

Canada - WHMIS: Ingredient Disclosure
  1% item 263 (368)

BUTYLTOLUENES   RR-01353-6

HEALTH AND SAFETY LISTS

U.S. DOT - Substances From 49 CFR 172.101
  regulated by DOT (UN2667)
U.S. DOT - Hazard Classes
  DOT hazard class = 6.1

BUTYLTRICHLOROSILANE   7521-80-4

HEALTH AND SAFETY LISTS

U.S. DOT - Substances From 49 CFR 172.101
  regulated by DOT (UN1747)
U.S. DOT - Hazard Classes
  DOT hazard class = 8
NFPA - Flash Points
  flash point = 130 degrees F (54 degrees C)
NFPA - Hazard Identification Ratings
  health-2; flammability-2; reactivity-0

INTERNATIONAL LISTS

Canada - WHMIS: Ingredient Disclosure
  1% item 266 (371)

STATE LISTS

Florida Hazardous Substance List
  [present]
Massachusetts Right To Know List
  [present]
NJ Right to Know List (Total)
  sn 0296
NJ Special Hazardous Substances
  (corrosive)
Pennsylvania Right to Know List
  [present]

BUTYLTRICHLOROSTANNANE   1118-46-3
SEE ALSO:
TIN

HEALTH AND SAFETY LISTS

NIOSH - Selected LD50s and LC50s
  Oral, rat: LD50 = 2140 mg/kg

INTERNATIONAL LISTS

Canada - WHMIS: Ingredient Disclosure
  1% item 1094 (1656)

N-BUTYLURETHANE   591-62-8

HEALTH AND SAFETY LISTS

NFPA - Flash Points
  flash point = 197 degrees F (92 degrees C)
NFPA - Hazard Identification Ratings
  flammability-2; reactivity-0

BUTYL VINYL ETHER   111-34-2

HEALTH AND SAFETY LISTS

U.S. DOT - Substances From 49 CFR 172.101
  regulated by DOT (UN2352)
U.S. DOT - Hazard Classes
  DOT hazard class = 3
NFPA - Flash Points
  flash point = 15 degrees F (-9 degrees C)
NFPA - Hazard Identification Ratings
  health-2; flammability-3; reactivity-2
NIOSH - Selected LD50s and LC50s
  Inhalation, mouse: LC50 = 62 gm/m3 2 hr Oral, rat: LD50 = 10 gm/kg Skin, rabbit: LD50 = 4240 mg/kg

STATE LISTS

Florida Hazardous Substance List
  [present]
Massachusetts Right To Know List
  [present]
NJ Right to Know List (Total)
  sn 0297
NJ Special Hazardous Substances
  (flammable - third degree; reactive - second degree)

Pennsylvania Right to Know List
[present]

2-BUTYNE 503-17-3

HEALTH AND SAFETY LISTS

U.S. DOT - Substances From 49 CFR 172.101
regulated by DOT (UN1144)

U.S. DOT - Hazard Classes
DOT hazard class = 3

NFPA - Flash Points
flash point < -4 degrees F (-20 degrees C)

NFPA - Hazard Identification Ratings
flammability-4

STATE LISTS

Florida Hazardous Substance List
[present]

Massachusetts Right To Know List
[present]

NJ Right to Know List (Total)
sn 0540

Pennsylvania Right to Know List
[present]

1,4-BUTYNEDIOL 110-65-6

HEALTH AND SAFETY LISTS

U.S. DOT - Substances From 49 CFR 172.101
regulated by DOT (UN2716)

U.S. DOT - Hazard Classes
DOT hazard class = 6.1

NIOSH - Selected LD50s and LC50s
Oral, rat: LD50 = 104 mg/kg

STATE LISTS

NJ Right to Know List (Total)
sn 0298

3-BUTYN-2-ONE 1423-60-5

HEALTH AND SAFETY LISTS

NIOSH - Selected LD50s and LC50s
Oral, rat: LD50 = 6300 ug/kg

BUTYRALDEHYDE 123-72-8

HEALTH AND SAFETY LISTS

AIHA - WEEL - Time Weighted Averages
25 ppm TWA

U.S. DOT - Substances From 49 CFR 172.101
regulated by DOT (UN1129)

U.S. DOT - Hazard Classes
DOT hazard class = 3

NFPA - Flash Points
flash point = -8 degrees F (-22 degrees C)

NFPA - Hazard Identification Ratings
health-3; flammability-3; reactivity-0

NIOSH - Selected LD50s and LC50s
Inhalation, rat: LC50 = 174 gm/m3 30 mn Oral, rat: LD50 = 2490 mg/kg Skin, rabbit: LD50 = 3560 mg/kg

NTP Chemical Status Reports - Testing Status and NTIS Number
Prechronic studies for which toxicity technical reports were not prepared

ENVIRONMENTAL LISTS

CERCLA/SARA - Section 313 - Emission Reporting
form R reporting required for 1.0% de minimus concentration

EPA - Master Testing List
[present]

TSCA - Code of Federal Regulations Citations
40 CFR 712.30(w)

TSCA - PAIR - Reporting List
Reporting Date: February 14, 1989

TSCA - Health and Safety Reporting List
Effective Date: December 16, 1988

INTERNATIONAL LISTS

Canada - WHMIS: Ingredient Disclosure
1% item 267 (372)

Canada - NPRI (National Pollutant Release Inventory)
[present]

STATE LISTS

California - Air Bill 2588 Appendix A-II
6/91

Florida Hazardous Substance List
[present]

Massachusetts Right To Know List
[present]

NJ Right to Know List (Total)
sn 0299

NJ Special Hazardous Substances
(flammable - third degree)

Pennsylvania Right to Know List
environmental hazard

PROPOSED REGULATIONS

TSCA - ITC 33rd Report Priority Testing List
recommended to be tested

TSCA - ITC 34th Report Priority Testing List
recommended to be tested

BUTYRALDOL 496-03-7

HEALTH AND SAFETY LISTS

NFPA - Flash Points
flash point = 165 degrees F (74 degrees C)

NFPA - Hazard Identification Ratings
health-2; flammability-2; reactivity-0

STATE LISTS

Florida Hazardous Substance List
[present]

Massachusetts Right To Know List
[present]

Pennsylvania Right to Know List
[present]

BUTYRALDOXIME 110-69-0

HEALTH AND SAFETY LISTS

U.S. DOT - Substances From 49 CFR 172.101
regulated by DOT (UN2840)

U.S. DOT - Hazard Classes
DOT hazard class = 3

NFPA - Flash Points
flash point = 136 degrees F (58 degrees C)

NFPA - Hazard Identification Ratings
health-2; flammability-2; reactivity-0

NTP Chemical Status Reports - Testing Status and NTIS Number
Project leader assigned/study in design

ENVIRONMENTAL LISTS

List of Pesticide Product Inert Ingredients
[present]

STATE LISTS

Florida Hazardous Substance List
[present]

Massachusetts Right To Know List
[present]

NJ Right to Know List (Total)
sn 0279

Pennsylvania Right to Know List
[present]

BUTYRIC ACID   107-92-6

HEALTH AND SAFETY LISTS

U.S. DOT - Substances From 49 CFR 172.101
regulated by DOT (UN2820)

U.S. DOT - Hazard Classes
DOT hazard class = 8

U.S. DOT - Appendix A Table 1 - Hazardous Substances
final RQ = 5000 pounds (2270 kg)

NFPA - Flash Points
flash point = 161 degrees F (72 degrees C)

NFPA - Hazard Identification Ratings
health-3; flammability-2; reactivity-0

NIOSH - Selected LD50s and LC50s
Oral, rat: LD50 = 2940 mg/kg Skin, rabbit: LD50 = 530 mg/kg

ENVIRONMENTAL LISTS

CERCLA/SARA - Hazardous Substances and their Reportable Quantities
final RQ = 5000 pounds (2270 kg)

Clean Water Act - Hazardous Substances
[present] (Listed under 'Butyric acid')

List of Pesticide Product Inert Ingredients
[present]

INTERNATIONAL LISTS

Canada - WHMIS: Ingredient Disclosure
1% item 268 (71)

STATE LISTS

California - Directors List of Hazardous Substances (8 CCR 339)
[present]

Florida Hazardous Substance List
[present]

Massachusetts Right To Know List
[present]

NJ Right to Know List (Total)
sn 0300

NJ Special Hazardous Substances
(corrosive)

Pennsylvania Right to Know List
environmental hazard

BUTYRIC ANHYDRIDE   106-31-0

HEALTH AND SAFETY LISTS

U.S. DOT - Substances From 49 CFR 172.101
regulated by DOT (UN2739)

U.S. DOT - Hazard Classes
DOT hazard class = 8

NFPA - Flash Points
flash point = 180 degrees F (54 degrees C)

NFPA - Hazard Identification Ratings
health-1; flammability-2; reactivity-1 (avoid use of water)

ENVIRONMENTAL LISTS

TSCA - Code of Federal Regulations Citations
40 CFR 712.30(x); 40 CFR 716.120(d)

TSCA - PAIR - Reporting List
Reporting Date: November 27, 1991

TSCA - Health and Safety Reporting List
Effective Date: September 30, 1991

INTERNATIONAL LISTS

Canada - WHMIS: Ingredient Disclosure
1% item 269 (229)

STATE LISTS

NJ Right to Know List (Total)
sn 0301

NJ Special Hazardous Substances
(corrosive)

GAMMA-BUTYROLACTONE   96-48-0

HEALTH AND SAFETY LISTS

IARC - Group 3 (not classifiable)
[present]

NFPA - Flash Points
flash point = 209 degrees F (98 degrees C)

NFPA - Hazard Identification Ratings
health-0; flammability-1; reactivity-0

NIOSH - Selected LD50s and LC50s
Oral, rat: LD50 = 1800 mg/kg

NTP Chemical Status Reports - Testing Status and NTIS Number
Technical reports printed (PB92-189323)

NTP Chemical Status Reports - Evidence of Carcinogenicity
male rat: no evidence; female rat: no evidence; male mice: equivocal evidence; female mice: no evidence

ENVIRONMENTAL LISTS

CAA - HON Rule - SOCMI Chemicals
compliance by Oct. 24, 1994

List of Pesticide Product Inert Ingredients
[present]

BETA-BUTYROLACTONE   3068-88-0

HEALTH AND SAFETY LISTS

IARC - Group 2B (sufficient animal data)
[present] (Overall evaluation based only on evidence of carcinogenicity in monograph (11, 1976) or in Supplement 4)

NIOSH - Selected LD50s and LC50s
Oral, rat: LD50 = 17 gm/kg

OSHA - Possible Select Carcinogens
[present]

INTERNATIONAL LISTS

Canada - WHMIS: Ingredient Disclosure
1% item 270 (376)

STATE LISTS

California - Air Bill 2588 Appendix A-II
known or potential carcinogen

California - Prop. 65 - Cancer list
carcinogen - initial date 7/1/87

California - Prop. 65 - No Significant Risk Levels
no significant risk level = 0.7 ug/day

California - Directors List of Hazardous Substances (8 CCR 339)
[present]

Florida Hazardous Substance List
[present]

Massachusetts Right To Know List
carcinogen; extraordinarily hazardous

Minnesota Hazardous Substance List
carcinogen

Pennsylvania Right to Know List
special hazardous substance

Pennsylvania RTK - Special Hazardous Substances
[present]

BUTYRONITRILE   109-74-0

HEALTH AND SAFETY LISTS

U.S. DOT - Substances From 49 CFR 172.101
regulated by DOT (UN2411)

U.S. DOT - Hazard Classes
DOT hazard class = 3

NFPA - Flash Points
flash point = 76 degrees F (24 degrees C)

NFPA - Hazard Identification Ratings
health-3; flammability-3; reactivity-0

NIOSH - Selected LD50s and LC50s
Inhalation, mouse: LC50 = 249 ppm 1 hr Oral, rat: LD50 = 140 mg/kg Skin, rabbit: LD50 = 500 mg/kg

NIOSH - Health Standards - Exposure Limits
8 ppm TWA; 22 mg/m3 TWA (Listed under 'Nitriles')

NIOSH - Health Standards - Health Effects and Precautions
Hepatic, renal, respiratory, cardiovascular, gastrointestinal, and nervous sytem effects (Periodic chest X-ray and pulmonary function testing required; prevent skin and eye contact; make first-aid kits and personnel available during use) (Listed under 'Nitriles')

INTERNATIONAL LISTS

Canada - WHMIS: Ingredient Disclosure
1% item 271 (377)

STATE LISTS

Florida Hazardous Substance List
[present]

Massachusetts Right To Know List
[present]

Minnesota Hazardous Substance List
[present]

NJ Right to Know List (Total)
sn 0302

NJ Special Hazardous Substances
(flammable - third degree)

Pennsylvania Right to Know List
[present]

BUTYRYL CHLORIDE   141-75-3

HEALTH AND SAFETY LISTS

U.S. DOT - Substances From 49 CFR 172.101
regulated by DOT (UN2353)

U.S. DOT - Hazard Classes
DOT hazard class = 3

INTERNATIONAL LISTS

Canada - WHMIS: Ingredient Disclosure
1% item 272 (481)

STATE LISTS

NJ Right to Know List (Total)
sn 0303

NJ Special Hazardous Substances
(corrosive)

C.I. 74180   1330-38-7
SEE ALSO:
COPPER

ENVIRONMENTAL LISTS

List of Pesticide Product Inert Ingredients
[present]

C.I. ACID BLACK 1   1064-48-8

ENVIRONMENTAL LISTS

List of Pesticide Product Inert Ingredients
[present]

C.I. ACID BLACK 2   8005-03-6

ENVIRONMENTAL LISTS

List of Pesticide Product Inert Ingredients
[present]

C.I. ACID BLUE 7, SODIUM SALT   3486-30-4

ENVIRONMENTAL LISTS

List of Pesticide Product Inert Ingredients
[present]

C.I. ACID BLUE 9, DIAMMONIUM SALT   2650-18-2

ENVIRONMENTAL LISTS

List of Pesticide Product Inert Ingredients
[present]

STATE LISTS

Massachusetts Right To Know List
[present]

NJ Right to Know List (Total)
sn 2897

C.I. ACID BLUE 9, DISODIUM SALT   3844-45-9

HEALTH AND SAFETY LISTS

IARC - Group 3 (not classifiable)
[present]

ENVIRONMENTAL LISTS

List of Pesticide Product Inert Ingredients
[present]

STATE LISTS

California - Directors List of Hazardous Substances (8 CCR 339)
[present]

Massachusetts Right To Know List
[present]

C.I. ACID BLUE 22   28631-66-5

ENVIRONMENTAL LISTS

List of Pesticide Product Inert Ingredients
[present]

C.I. ACID BLUE 80   4474-24-2

ENVIRONMENTAL LISTS

List of Pesticide Product Inert Ingredients
[present]

C.I. ACID BLUE 93   28983-56-4

ENVIRONMENTAL LISTS

List of Pesticide Product Inert Ingredients
[present]

C.I. ACID BLUE 145   6408-80-6

ENVIRONMENTAL LISTS

List of Pesticide Product Inert Ingredients
[present]

C.I. ACID BLUE 182   72152-54-6

ENVIRONMENTAL LISTS

List of Pesticide Product Inert Ingredients
[present]

C.I. ACID GREEN 1   19381-50-1

ENVIRONMENTAL LISTS

List of Pesticide Product Inert Ingredients
[present]

C.I. ACID GREEN 3   4680-78-8

HEALTH AND SAFETY LISTS

  IARC - Group 3 (not classifiable)
  [present]

ENVIRONMENTAL LISTS

  CERCLA/SARA - Section 313 - Emission Reporting
  form R reporting required for 1.0% de minimus concentration

INTERNATIONAL LISTS

  Canada - NPRI (National Pollutant Release Inventory)
  [present]

STATE LISTS

  California - Air Bill 2588 Appendix A-II
  6/91

  California - Directors List of Hazardous Substances (8 CCR 339)
  [present]

  Massachusetts Right To Know List
  [present]

  NJ Right to Know List (Total)
  sn 0442

  Pennsylvania Right to Know List
  environmental hazard

C.I. ACID GREEN 16   12768-78-4

ENVIRONMENTAL LISTS

  List of Pesticide Product Inert Ingredients
  [present]

C.I. ACID GREEN 28   71927-89-4

ENVIRONMENTAL LISTS

  List of Pesticide Product Inert Ingredients
  [present]

C.I. ACID ORANGE 3   6373-74-6

HEALTH AND SAFETY LISTS

  IARC - Group 3 (not classifiable)
  [present]

  NTP Chemical Status Reports - Testing Status and NTIS Number
  Technical reports printed (PB89216550/AS)

  NTP Chemical Status Reports - Evidence of Carcinogenicity
  male rat-no evidence; female rat-clear evidence; male mice-no evidence; female mice-no evidence

C.I. ACID ORANGE 45   2429-80-3

STATE LISTS

  NJ Right to Know List (Total)
  sn 0443

  NJ Special Hazardous Substances
  (carcinogen)

C.I. ACID ORANGE 52   547-58-0

INTERNATIONAL LISTS

  Canada - WHMIS: Ingredient Disclosure
  1% item 273 (1287)

C.I. ACID RED 1   3734-67-6

ENVIRONMENTAL LISTS

  List of Pesticide Product Inert Ingredients
  [present]

C.I. ACID RED 14   3567-69-9

HEALTH AND SAFETY LISTS

  IARC - Group 3 (not classifiable)
  [present]

  NTP Chemical Status Reports - Testing Status and NTIS Number
  Technical reports printed (PB82201468)

  NTP Chemical Status Reports - Evidence of Carcinogenicity
  male rat-negative; female rat-negative; male mice-negative; female mice-negative

ENVIRONMENTAL LISTS

  List of Pesticide Product Inert Ingredients
  [present]

C.I. ACID RED 17, DISODIUM SALT   5858-33-3

ENVIRONMENTAL LISTS

  List of Pesticide Product Inert Ingredients
  [present]

C.I. ACID RED 27, TRISODIUM SALT   915-67-3

HEALTH AND SAFETY LISTS

  IARC - Group 3 (not classifiable)
  [present]

ENVIRONMENTAL LISTS

  List of Pesticide Product Inert Ingredients
  [present]

C.I. ACID RED 33, DISODIUM SALT   3567-66-6

ENVIRONMENTAL LISTS

  List of Pesticide Product Inert Ingredients
  [present]

C.I. ACID RED 52   3520-42-1

ENVIRONMENTAL LISTS

  List of Pesticide Product Inert Ingredients
  [present]

C.I. ACID RED 73, DISODIUM SALT   5413-75-2

ENVIRONMENTAL LISTS

  List of Pesticide Product Inert Ingredients
  [present]

C.I. ACID RED 85   3567-65-5

SEE ALSO:
  BISAZOBIPHENYL DYES

ENVIRONMENTAL LISTS

  TSCA - Code of Federal Regulations Citations
  40 CFR 716.120(c)

STATE LISTS

  NJ Right to Know List (Total)
  sn 0444

  NJ Special Hazardous Substances
  (carcinogen)

C.I. ACID RED 101, DISODIUM SALT   6844-74-2

ENVIRONMENTAL LISTS

  List of Pesticide Product Inert Ingredients
  [present]

C.I. ACID RED 114, DISODIUM SALT   6459-94-5

HEALTH AND SAFETY LISTS

  IARC - Group 2B (sufficient animal data)
  [present]

  NTP Chemical Status Reports - Testing Status and NTIS Number
  Technical reports printed (PB92-189380)

  NTP Chemical Status Reports - Evidence of Carcinogenicity
  male rat: clear evidence; female rat: clear evidence

  OSHA - Possible Select Carcinogens
  [present]

ENVIRONMENTAL LISTS

CERCLA/SARA - Section 313 - Emission Reporting
form R reporting required

STATE LISTS

California - Prop. 65 - Cancer list
carcinogen - initial date 7/1/92

NJ Right to Know List (Total)
sn 0445

C.I. ACID VIOLET 7, DISODIUM SALT   4321-69-1

ENVIRONMENTAL LISTS

List of Pesticide Product Inert Ingredients
[present]

C.I. ACID VIOLET 12, DISODIUM SALT   6625-46-3

ENVIRONMENTAL LISTS

List of Pesticide Product Inert Ingredients
[present]

C.I. ACID YELLOW 11, SODIUM SALT   6359-82-6

ENVIRONMENTAL LISTS

List of Pesticide Product Inert Ingredients
[present]

C.I. ACID YELLOW 17, DISODIUM SALT   6359-98-4

ENVIRONMENTAL LISTS

List of Pesticide Product Inert Ingredients
[present]

C.I. ACID YELLOW 34   6359-97-3

ENVIRONMENTAL LISTS

List of Pesticide Product Inert Ingredients
[present]

C.I. ACID YELLOW 218   71873-51-3

ENVIRONMENTAL LISTS

List of Pesticide Product Inert Ingredients
[present]

C.I. AZOIC COUPLING COMPONENT   91-92-9
SEE ALSO:
BISAZOBIPHENYL DYES

ENVIRONMENTAL LISTS

TSCA - Code of Federal Regulations Citations
40 CFR 716.120(c)

STATE LISTS

NJ Right to Know List (Total)
sn 0447

C.I. AZOIC COUPLING COMPONENT 5   91-96-3
SEE ALSO:
BISAZOBIPHENYL DYES

ENVIRONMENTAL LISTS

TSCA - Code of Federal Regulations Citations
40 CFR 716.120(c)

STATE LISTS

NJ Right to Know List (Total)
sn 0446

C.I. BASIC GREEN 1   633-03-4

STATE LISTS

Massachusetts Right To Know List
[present]

NJ Right to Know List (Total)
sn 2249

C.I. BASIC GREEN 4, OXALATE   18015-76-4

ENVIRONMENTAL LISTS

List of Pesticide Product Inert Ingredients
[present]

C.I. BASIC ORANGE 15   12768-82-0

STATE LISTS

Pennsylvania Right to Know List
environmental hazard

C.I. BASIC RED 1   989-38-8

HEALTH AND SAFETY LISTS

IARC - Group 3 (not classifiable)
[present]

NTP Chemical Status Reports - Testing Status and NTIS Number
Technical reports printed (PB90219460/AS)

NTP Chemical Status Reports - Evidence of Carcinogenicity
male rat-equivocal evidence; female rat-equivocal evidence; male mice-no evidence; female mice-no evidence

ENVIRONMENTAL LISTS

CERCLA/SARA - Section 313 - Emission Reporting
form R reporting required for 0.1% de minimus concentration

List of Pesticide Product Inert Ingredients
[present]

INTERNATIONAL LISTS

Canada - NPRI (National Pollutant Release Inventory)
[present]

STATE LISTS

California - Air Bill 2588 Appendix A-II
6/91

California - Directors List of Hazardous Substances (8 CCR 339)
[present]

Massachusetts Right To Know List
[present]

NJ Right to Know List (Total)
sn 0449

Pennsylvania Right to Know List
environmental hazard

C.I. BASIC RED 9 MONOHYDROCHLORIDE   569-61-9

HEALTH AND SAFETY LISTS

IARC - Group 2B (sufficient animal data)
[present]

NIOSH - Selected LD50s and LC50s
Oral, mouse: LD50 = 5000 mg/kg

NTP Chemical Status Reports - Testing Status and NTIS Number
Technical reports printed (PB86186509/AS)

NTP Chemical Status Reports - Evidence of Carcinogenicity
male rat-clear evidence; female rat-clear evidence; male mice-clear evidence; female mice-clear evidence

NTP Seventh Report - Suspect Carcinogens
suspect carcinogen

OSHA - Possible Select Carcinogens
[present]

INTERNATIONAL LISTS

Canada - WHMIS: Ingredient Disclosure
1% item 1237 (1335)

STATE LISTS

California - Air Bill 2588 Appendix A-II
known or potential carcinogen: 9/89

California - Prop. 65 - Cancer list
carcinogen - initial date 7/1/89
California - Prop. 65 - No Significant Risk Levels
no significant risk level = 3 ug/day
Florida Hazardous Substance List
effective March 13, 1992

C.I. BASIC RED 14 65122-06-7
ENVIRONMENTAL LISTS
List of Pesticide Product Inert Ingredients
[present]

C.I. BASIC VIOLET 1 548-62-9
HEALTH AND SAFETY LISTS
NIOSH - Selected LD50s and LC50s
Oral, rat: LD50 = 420 mg/kg
NTP Chemical Status Reports - Testing Status and NTIS Number
Technical reports printed (call NCTR for documents)
INTERNATIONAL LISTS
Canada - WHMIS: Ingredient Disclosure
0.1% item 274 (1719)

C.I. BASIC VIOLET 11 2390-63-8
ENVIRONMENTAL LISTS
List of Pesticide Product Inert Ingredients
[present]

C.I. BASIC VIOLET 16 6359-45-1
ENVIRONMENTAL LISTS
List of Pesticide Product Inert Ingredients
[present]

C.I. DIRECT BLACK 4 25156-49-4
STATE LISTS
NJ Right to Know List (Total)
sn 0452
NJ Special Hazardous Substances
(carcinogen)

C.I. DIRECT BLACK 4, DISODIUM SALT 2429-83-6
SEE ALSO:
BISAZOBIPHENYL DYES
ENVIRONMENTAL LISTS
TSCA - Code of Federal Regulations Citations
40 CFR 716.120(c)
STATE LISTS
NJ Right to Know List (Total)
sn 0451
NJ Special Hazardous Substances
(carcinogen)

C.I. DIRECT BLACK 38 1937-37-7
SEE ALSO:
BISAZOBIPHENYL DYES
HEALTH AND SAFETY LISTS
IARC - Group Unspecified
[present] (Listed under 'Benzidine-based dyes')
NTP Chemical Status Reports - Testing Status and NTIS Number
Technical reports printed (PB280204/AS)
NTP Chemical Status Reports - Evidence of Carcinogenicity
male rat-positive; female rat-positive (toxicity study)
NTP Seventh Report - Suspect Carcinogens
suspect carcinogen

OSHA - Possible Select Carcinogens
[present]
ENVIRONMENTAL LISTS
CERCLA/SARA - Section 313 - Emission Reporting
form R reporting required for 0.1% de minimus concentration
TSCA - Code of Federal Regulations Citations
40 CFR 712.30(d); 40 CFR 716.120(c)
TSCA - PAIR - Reporting List
Reporting Date: November 19, 1982
INTERNATIONAL LISTS
Canada - WHMIS: Ingredient Disclosure
0.1% item 671 (1273)
STATE LISTS
California - Air Bill 2588 Appendix A-I
known or potential carcinogen
California - Prop. 65 - Cancer list
carcinogen - initial date 1/1/88
California - Prop. 65 - No Significant Risk Levels
no significant risk level = 0.09 ug/day
California - Directors List of Hazardous Substances (8 CCR 339)
[present]
Florida Hazardous Substance List
[present]
Massachusetts Right To Know List
carcinogen; extraordinarily hazardous
Minnesota Hazardous Substance List
carcinogen
NJ Right to Know List (Total)
sn 0453
NJ Special Hazardous Substances
(carcinogen)
Pennsylvania Right to Know List
environmental hazard; special hazardous substance
Pennsylvania RTK - Special Hazardous Substances
[present]

C.I. DIRECT BLACK 91, TRISODIUM SALT 6739-62-4
SEE ALSO:
BISAZOBIPHENYL DYES
ENVIRONMENTAL LISTS
TSCA - Code of Federal Regulations Citations
40 CFR 716.120(c)
STATE LISTS
NJ Right to Know List (Total)
sn 0455

C.I. DIRECT BLACK 114 61703-05-7
STATE LISTS
NJ Right to Know List (Total)
sn 0454

C.I. DIRECT BLUE 1 2610-05-1
SEE ALSO:
BISAZOBIPHENYL DYES
ENVIRONMENTAL LISTS
TSCA - Code of Federal Regulations Citations
40 CFR 712.30(d); 40 CFR 716.120(c)
TSCA - PAIR - Reporting List
Reporting Date: November 19, 1982
STATE LISTS
NJ Right to Know List (Total)
sn 0458

C.I. DIRECT BLUE 2   25180-19-2

STATE LISTS

NJ Right to Know List (Total)
sn 0457

NJ Special Hazardous Substances
(carcinogen)

C.I. DIRECT BLUE 2, TRISODIUM SALT   2429-73-4

SEE ALSO:
BISAZOBIPHENYL DYES

ENVIRONMENTAL LISTS

TSCA - Code of Federal Regulations Citations
40 CFR 716.120(c)

STATE LISTS

NJ Right to Know List (Total)
sn 0461

NJ Special Hazardous Substances
(carcinogen)

C.I. DIRECT BLUE 6   2602-46-2

SEE ALSO:
BISAZOBIPHENYL DYES

HEALTH AND SAFETY LISTS

IARC - Group Unspecified
[present] (Listed under 'Benzidine-based dyes')

NTP Chemical Status Reports - Testing Status and NTIS Number
Technical reports printed (PB280204/AS);   prechronic studies for which toxicity technical reports were not prepared

NTP Chemical Status Reports - Evidence of Carcinogenicity
male rat-positive; female rat-positive (toxicity study)

NTP Seventh Report - Suspect Carcinogens
suspect carcinogen

OSHA - Possible Select Carcinogens
[present]

ENVIRONMENTAL LISTS

CERCLA/SARA - Section 313 - Emission Reporting
form R reporting required for 0.1% de minimus concentration

TSCA - Code of Federal Regulations Citations
40 CFR 712.30(d);   40 CFR 716.120(c)

TSCA - PAIR - Reporting List
Reporting Date: November 19, 1982

INTERNATIONAL LISTS

Canada - WHMIS: Ingredient Disclosure
0.1% item 673 (312)

STATE LISTS

California - Air Bill 2588 Appendix A-I
known or potential carcinogen

California - Prop. 65 - Cancer list
carcinogen - initial date 1/1/88

California - Prop. 65 - No Significant Risk Levels
no significant risk level = 0.09 ug/day

California - Directors List of Hazardous Substances (8 CCR 339)
[present]

Florida Hazardous Substance List
[present]

Massachusetts Right To Know List
carcinogen; extraordinarily hazardous

Minnesota Hazardous Substance List
carcinogen

NJ Right to Know List (Total)
sn 0462

NJ Special Hazardous Substances
(carcinogen)

Pennsylvania Right to Know List
environmental hazard; special hazardous substance

Pennsylvania RTK - Special Hazardous Substances
[present]

C.I. DIRECT BLUE 8, DISODIUM SALT   2429-71-2

SEE ALSO:
BISAZOBIPHENYL DYES

ENVIRONMENTAL LISTS

TSCA - Code of Federal Regulations Citations
40 CFR 716.120(c)

STATE LISTS

NJ Right to Know List (Total)
sn 0463

C.I. DIRECT BLUE 15   2429-74-5

SEE ALSO:
BISAZOBIPHENYL DYES

HEALTH AND SAFETY LISTS

IARC - Group 2B (sufficient animal data)
[present]

NTP Chemical Status Reports - Testing Status and NTIS Number
Technical reports printed (PB93-126373)

NTP Chemical Status Reports - Evidence of Carcinogenicity
male rat: clear evidence; female rat: clear evidence

OSHA - Possible Select Carcinogens
[present]

ENVIRONMENTAL LISTS

TSCA - Code of Federal Regulations Citations
40 CFR 716.120(c)

STATE LISTS

NJ Right to Know List (Total)
sn 0459

C.I. DIRECT BLUE 22, DISODIUM SALT   2586-57-4

SEE ALSO:
BISAZOBIPHENYL DYES

ENVIRONMENTAL LISTS

TSCA - Code of Federal Regulations Citations
40 CFR 716.120(c)

STATE LISTS

NJ Right to Know List (Total)
sn 0464

C.I. DIRECT BLUE 25   25180-27-2

STATE LISTS

NJ Right to Know List (Total)
sn 0460

C.I. DIRECT BLUE 25, TETRASODIUM SALT   2150-54-1

SEE ALSO:
BISAZOBIPHENYL DYES

ENVIRONMENTAL LISTS

TSCA - Code of Federal Regulations Citations
40 CFR 716.120(c)

STATE LISTS

NJ Right to Know List (Total)
sn 0456

C.I. DIRECT BLUE 199   12222-04-7

ENVIRONMENTAL LISTS

List of Pesticide Product Inert Ingredients
[present]

C.I. DIRECT BLUE 218   10401-50-0
  SEE ALSO:
    BISAZOBIPHENYL DYES

  ENVIRONMENTAL LISTS
    TSCA - Code of Federal Regulations Citations
    40 CFR 716.120(c)

C.I. DIRECT BLUE 218   28407-37-6

  HEALTH AND SAFETY LISTS
    NTP Chemical Status Reports - Testing Status and NTIS Number
    Technical reports printed; prechronic studies for which toxicity technical reports were not prepared

  ENVIRONMENTAL LISTS
    CERCLA/SARA - Section 313 - Emission Reporting
    form R reporting required

C.I. DIRECT BROWN   8014-91-3
  SEE ALSO:
    BISAZOBIPHENYL DYES

  ENVIRONMENTAL LISTS
    TSCA - Code of Federal Regulations Citations
    40 CFR 716.120(c)

  STATE LISTS
    NJ Right to Know List (Total)
    sn 0469

    NJ Special Hazardous Substances
    (carcinogen)

C.I. DIRECT BROWN 1   3811-71-0

  STATE LISTS
    NJ Right to Know List (Total)
    sn 0474

    NJ Special Hazardous Substances
    (carcinogen)

C.I. DIRECT BROWN 2   25255-06-5

  STATE LISTS
    NJ Right to Know List (Total)
    sn 0477

    NJ Special Hazardous Substances
    (carcinogen)

C.I. DIRECT BROWN 2, DISODIUM SALT   2429-82-5
  SEE ALSO:
    BISAZOBIPHENYL DYES

  ENVIRONMENTAL LISTS
    TSCA - Code of Federal Regulations Citations
    40 CFR 716.120(c)

  STATE LISTS
    NJ Right to Know List (Total)
    sn 0471

    NJ Special Hazardous Substances
    (carcinogen)

C.I. DIRECT BROWN 6   25180-39-6

  STATE LISTS
    NJ Right to Know List (Total)
    sn 0468

    NJ Special Hazardous Substances
    (carcinogen)

C.I. DIRECT BROWN 6, DISODIUM SALT   2893-80-3
  SEE ALSO:
    BISAZOBIPHENYL DYES

  ENVIRONMENTAL LISTS
    TSCA - Code of Federal Regulations Citations
    40 CFR 716.120(c)

  STATE LISTS
    NJ Right to Know List (Total)
    sn 0467

    NJ Special Hazardous Substances
    (carcinogen)

C.I. DIRECT BROWN 31   25180-41-0

  STATE LISTS
    NJ Right to Know List (Total)
    sn 0476

    NJ Special Hazardous Substances
    (carcinogen)

C.I. DIRECT BROWN 31, TETRASODIUM SALT   2429-81-4
  SEE ALSO:
    BISAZOBIPHENYL DYES

  ENVIRONMENTAL LISTS
    TSCA - Code of Federal Regulations Citations
    40 CFR 716.120(c)

  STATE LISTS
    NJ Right to Know List (Total)
    sn 0475

    NJ Special Hazardous Substances
    (carcinogen)

C.I. DIRECT BROWN 59   6247-51-4

  STATE LISTS
    NJ Right to Know List (Total)
    sn 0466

    NJ Special Hazardous Substances
    (carcinogen)

C.I. DIRECT BROWN 59, DISODIUM SALT   3476-90-2

  STATE LISTS
    NJ Right to Know List (Total)
    sn 0472

    NJ Special Hazardous Substances
    (carcinogen)

C.I. DIRECT BROWN 95   16071-86-6
  SEE ALSO:
    COPPER
    BISAZOBIPHENYL DYES

  HEALTH AND SAFETY LISTS
    IARC - Group Unspecified
    [present] (Listed under 'Benzidine-based dyes')

    NTP Chemical Status Reports - Testing Status and NTIS Number
    Technical reports printed (PB280204/AS)

    NTP Chemical Status Reports - Evidence of Carcinogenicity
    male rat-negative; female rat-positive (toxicity study)

  ENVIRONMENTAL LISTS
    CERCLA/SARA - Section 313 - Emission Reporting
    form R reporting required for 0.1% de minimus concentration

    TSCA - Code of Federal Regulations Citations
    40 CFR 716.120(c)

INTERNATIONAL LISTS
  Canada - WHMIS: Ingredient Disclosure
    0.1% item 674 (348)
STATE LISTS
  California - Air Bill 2588 Appendix A-I
    known or potential carcinogen: 9/89
  California - Prop. 65 - Cancer list
    carcinogen - initial date 10/1/88
  California - Prop. 65 - No Significant Risk Levels
    no significant risk level = 0.1 ug/day
  California - Directors List of Hazardous Substances (8 CCR 339)
    [present]
  Florida Hazardous Substance List
    [present]
  Massachusetts Right To Know List
    carcinogen; extraordinarily hazardous
  Minnesota Hazardous Substance List
    carcinogen
  NJ Right to Know List (Total)
    sn 0478
  NJ Special Hazardous Substances
    (carcinogen)
  Pennsylvania Right to Know List
    environmental hazard; special hazardous substance
  Pennsylvania RTK - Special Hazardous Substances
    [present]

C.I. DIRECT BROWN 111   12222-20-7
STATE LISTS
  NJ Right to Know List (Total)
    sn 0470
  NJ Special Hazardous Substances
    (carcinogen)

C.I. DIRECT BROWN 154   6360-54-9
SEE ALSO:
  BISAZOBIPHENYL DYES
ENVIRONMENTAL LISTS
  TSCA - Code of Federal Regulations Citations
    40 CFR 716.120(c)
STATE LISTS
  NJ Right to Know List (Total)
    sn 0473
  NJ Special Hazardous Substances
    (carcinogen)

C.I. DIRECT GREEN 1   25180-45-4
STATE LISTS
  NJ Right to Know List (Total)
    sn 0479
  NJ Special Hazardous Substances
    (carcinogen)

C.I. DIRECT GREEN 1, DISODIUM SALT   3626-28-6
SEE ALSO:
  BISAZOBIPHENYL DYES
ENVIRONMENTAL LISTS
  TSCA - Code of Federal Regulations Citations
    40 CFR 716.120(c)
STATE LISTS
  NJ Right to Know List (Total)
    sn 0480

NJ Special Hazardous Substances
  (carcinogen)

C.I. DIRECT GREEN 6   25180-46-5
STATE LISTS
  NJ Right to Know List (Total)
    sn 0484
  NJ Special Hazardous Substances
    (carcinogen)

C.I. DIRECT GREEN 6, DISODIUM SALT   4335-09-5
SEE ALSO:
  BISAZOBIPHENYL DYES
ENVIRONMENTAL LISTS
  TSCA - Code of Federal Regulations Citations
    40 CFR 716.120(c)
STATE LISTS
  NJ Right to Know List (Total)
    sn 0481
  NJ Special Hazardous Substances
    (carcinogen)

C.I. DIRECT GREEN 8   25180-47-6
STATE LISTS
  NJ Right to Know List (Total)
    sn 0483
  NJ Special Hazardous Substances
    (carcinogen)

C.I. DIRECT GREEN 8, TRISODIUM SALT   5422-17-3
STATE LISTS
  NJ Right to Know List (Total)
    sn 0482
  NJ Special Hazardous Substances
    (carcinogen)

C.I. DIRECT ORANGE 1   54579-28-1
STATE LISTS
  NJ Right to Know List (Total)
    sn 0485
  NJ Special Hazardous Substances
    (carcinogen)

C.I. DIRECT ORANGE 6, DISODIUM SALT   6637-88-3
SEE ALSO:
  BISAZOBIPHENYL DYES
ENVIRONMENTAL LISTS
  TSCA - Code of Federal Regulations Citations
    40 CFR 716.120(c)
STATE LISTS
  NJ Right to Know List (Total)
    sn 0487

C.I. DIRECT ORANGE 8   64083-59-6
STATE LISTS
  NJ Right to Know List (Total)
    sn 0486
  NJ Special Hazardous Substances
    (carcinogen)

C.I. DIRECT RED 1   25188-24-3
STATE LISTS
  NJ Right to Know List (Total)
    sn 0495

NJ Special Hazardous Substances
(carcinogen)

C.I. DIRECT RED 1, DISODIUM SALT   2429-84-7
  SEE ALSO:
    BISAZOBIPHENYL DYES

  ENVIRONMENTAL LISTS
    TSCA - Code of Federal Regulations Citations
      40 CFR 716.120(c)

  STATE LISTS
    NJ Right to Know List (Total)
      sn 0492

    NJ Special Hazardous Substances
      (carcinogen)

C.I. DIRECT RED 2, DISODIUM SALT   992-59-6
  SEE ALSO:
    BISAZOBIPHENYL DYES

  ENVIRONMENTAL LISTS
    TSCA - Code of Federal Regulations Citations
      40 CFR 716.120(c)

  STATE LISTS
    NJ Right to Know List (Total)
      sn 0498

C.I. DIRECT RED 10   25188-29-8
  STATE LISTS
    NJ Right to Know List (Total)
      sn 0494

    NJ Special Hazardous Substances
      (carcinogen)

C.I. DIRECT RED 10, DISODIUM SALT   2429-70-1
  STATE LISTS
    NJ Right to Know List (Total)
      sn 0493

    NJ Special Hazardous Substances
      (carcinogen)

C.I. DIRECT RED 13   25188-30-1
  STATE LISTS
    NJ Right to Know List (Total)
      sn 0491

    NJ Special Hazardous Substances
      (carcinogen)

C.I. DIRECT RED 13, DISODIUM SALT   1937-35-5
  STATE LISTS
    NJ Right to Know List (Total)
      sn 0489

    NJ Special Hazardous Substances
      (carcinogen)

C.I. DIRECT RED 28   573-58-0
  SEE ALSO:
    BISAZOBIPHENYL DYES

  ENVIRONMENTAL LISTS
    TSCA - Code of Federal Regulations Citations
      40 CFR 716.120(c)

  STATE LISTS
    NJ Right to Know List (Total)
      sn 0488

    NJ Special Hazardous Substances
      (carcinogen)

C.I. DIRECT RED 37   3530-19-6
  SEE ALSO:
    BISAZOBIPHENYL DYES

  ENVIRONMENTAL LISTS
    TSCA - Code of Federal Regulations Citations
      40 CFR 716.120(c)

  STATE LISTS
    NJ Right to Know List (Total)
      sn 0490

    NJ Special Hazardous Substances
      (carcinogen)

C.I. DIRECT RED 39, DISODIUM SALT   6358-29-8
  SEE ALSO:
    BISAZOBIPHENYL DYES

  ENVIRONMENTAL LISTS
    TSCA - Code of Federal Regulations Citations
      40 CFR 716.120(c)

  STATE LISTS
    NJ Right to Know List (Total)
      sn 0497

C.I. DIRECT RED 81   25188-42-5
  ENVIRONMENTAL LISTS
    List of Pesticide Product Inert Ingredients
      [present]

C.I. DIRECT RED 81, DISODIUM SALT   2610-11-9
  ENVIRONMENTAL LISTS
    List of Pesticide Product Inert Ingredients
      [present]

C.I. DIRECT RED 81, TRIETHANOLAMINE SALT   75768-93-3
  ENVIRONMENTAL LISTS
    List of Pesticide Product Inert Ingredients
      [present]

C.I. DIRECT VIOLET 1   25188-44-7
  STATE LISTS
    NJ Right to Know List (Total)
      sn 0501

    NJ Special Hazardous Substances
      (carcinogen)

C.I. DIRECT VIOLET 1, DISODIUM SALT   2586-60-9
  STATE LISTS
    NJ Right to Know List (Total)
      sn 0498

    NJ Special Hazardous Substances
      (carcinogen)

C.I. DIRECT VIOLET 9, DISODIUM SALT   6227-14-1
  ENVIRONMENTAL LISTS
    List of Pesticide Product Inert Ingredients
      [present]

C.I. DIRECT VIOLET 22   25329-82-2
  STATE LISTS
    NJ Right to Know List (Total)
      sn 0499

    NJ Special Hazardous Substances
      (carcinogen)

C.I. DIRECT VIOLET 22, TRISODIUM SALT   6426-67-1

STATE LISTS

NJ Right to Know List (Total)
sn 0500

NJ Special Hazardous Substances
(carcinogen)

C.I. DIRECT YELLOW 20   6426-62-6

STATE LISTS

NJ Right to Know List (Total)
sn 0502

NJ Special Hazardous Substances
(carcinogen)

C.I. DIRECT YELLOW 50, TETRASODIUM SALT   3214-47-9

ENVIRONMENTAL LISTS

List of Pesticide Product Inert Ingredients
[present]

C.I. DISPERSE BLUE 1   2475-45-8

HEALTH AND SAFETY LISTS

IARC - Group 2B (sufficient animal data)
[present]

NTP Chemical Status Reports - Testing Status and NTIS Number
Technical reports printed (PB86248051/AS)

NTP Chemical Status Reports - Evidence of Carcinogenicity
male rat-clear evidence; female rat-clear evidence; male mice-equiv-ocal evidence; female mice-no evidence

OSHA - Possible Select Carcinogens
[present]

STATE LISTS

California - Air Bill 2588 Appendix A-II
known or potential carcinogen: 6/91

California - Prop. 65 - Cancer list
carcinogen - initial date 10/1/90

California - Prop. 65 - No Significant Risk Levels
no significant risk level = 200 ug/day

California - Directors List of Hazardous Substances (8 CCR 339)
[present]

PROPOSED REGULATIONS

NTP - Proposed Additions to Annual Report on Carcinogens
proposed as a suspect carcinogen for NTP 8th report

C.I. DISPERSE BLUE 79:1 ACETAMIDE, N-[5-[BIS[2-(ACETYLOXYL)ETHYL]AMINO]-2-[(2-BROMO-4,6-DINITROPHENYL)AZO]-4-METHOXY- PHENYL]-   3618-72-2

ENVIRONMENTAL LISTS

TSCA - Code of Federal Regulations Citations
40 CFR 712.30(u);   40 CFR 799.5000

TSCA - PAIR - Reporting List
Reporting Date: August 18, 1987

TSCA - Health and Safety Reporting List
Effective Date: June 19, 1987;   Sunset Date: November 9, 1993

TSCA - Substances Subject to Testing Consent Orders
Test for: Health and Environmental Effects

TSCA - Section 12(b) - Export Notification
export notification required - Section 4

C.I. DISPERSE YELLOW 3   2832-40-8

HEALTH AND SAFETY LISTS

IARC - Group 3 (not classifiable)
[present]

NTP Chemical Status Reports - Testing Status and NTIS Number
Technical reports printed (PB2230061)

NTP Chemical Status Reports - Evidence of Carcinogenicity
male rat-positive; female rat-negative; male mice-negative; female mice-positive

ENVIRONMENTAL LISTS

CERCLA/SARA - Section 313 - Emission Reporting
form R reporting required for 1.0% de minimus concentration

INTERNATIONAL LISTS

Canada - WHMIS: Ingredient Disclosure
1% item 275 (1060)

Canada - NPRI (National Pollutant Release Inventory)
[present]

STATE LISTS

California - Air Bill 2588 Appendix A-II
6/91

Massachusetts Right To Know List
[present]

NJ Right to Know List (Total)
sn 0503

Pennsylvania Right to Know List
environmental hazard

C.I. FLUORESCENT BRIGHTENER 46   6416-68-

INTERNATIONAL LISTS

Canada - WHMIS: Ingredient Disclosure
1% item 276 (269)

C.I. FOOD RED 1   4548-53-

HEALTH AND SAFETY LISTS

IARC - Group 3 (not classifiable)
[present]

ENVIRONMENTAL LISTS

List of Pesticide Product Inert Ingredients
[present]

C.I. FOOD RED 15   81-88-

HEALTH AND SAFETY LISTS

IARC - Group 3 (not classifiable)
[present]

NIOSH - Selected LD50s and LC50s
Oral, mouse: LD50 = 887 mg/kg

ENVIRONMENTAL LISTS

CERCLA/SARA - Section 313 - Emission Reporting
form R reporting required for 0.1% de minimus concentration

List of Pesticide Product Inert Ingredients
[present]

INTERNATIONAL LISTS

Canada - WHMIS: Ingredient Disclosure
1% item 1381 (1466)

Canada - NPRI (National Pollutant Release Inventory)
[present]

STATE LISTS

California - Air Bill 2588 Appendix A-II
known or potential carcinogen: 9/90

California - Prop. 65 - Cancer list
carcinogen - initial date 7/1/90

California - Directors List of Hazardous Substances (8 CCR 339)
[present]

Massachusetts Right To Know List
[present]

NJ Right to Know List (Total)
sn 0505
Pennsylvania Right to Know List
environmental hazard

C.I. FOOD YELLOW 6 — 104-23-4
ENVIRONMENTAL LISTS
  List of Pesticide Product Inert Ingredients
  [present]

C.I. PIGMENT RED 3 — 2425-85-6
HEALTH AND SAFETY LISTS
  IARC - Group 3 (not classifiable)
  [present]
  NTP Chemical Status Reports - Testing Status and NTIS Number
  Technical reports printed (PB92191634/AS)
  NTP Chemical Status Reports - Evidence of Carcinogenicity
  male rat: some evidence; female rat: some evidence; male mice: some evidence; female mice: no evidence
ENVIRONMENTAL LISTS
  List of Pesticide Product Inert Ingredients
  [present]

C.I. PIGMENT RED 5 — 6410-41-9
ENVIRONMENTAL LISTS
  List of Pesticide Product Inert Ingredients
  [present]

C.I. PIGMENT RED 23 — 6471-49-4
HEALTH AND SAFETY LISTS
  NTP Chemical Status Reports - Testing Status and NTIS Number
  Technical reports printed (no NTIS number)
  NTP Chemical Status Reports - Evidence of Carcinogenicity
  male rat: equivocal evidence; female rat: no evidence; male mice: no evidence; female mice: no evidence
ENVIRONMENTAL LISTS
  List of Pesticide Product Inert Ingredients
  [present]

C.I. PIGMENT RED 48, CALCIUM SALT — 7023-61-2
ENVIRONMENTAL LISTS
  List of Pesticide Product Inert Ingredients
  [present]

C.I. PIGMENT RED 52 — 27757-95-5
ENVIRONMENTAL LISTS
  List of Pesticide Product Inert Ingredients
  [present]

C.I. PIGMENT YELLOW 73 — 13515-40-7
ENVIRONMENTAL LISTS
  List of Pesticide Product Inert Ingredients
  [present]

C.I. PIGMENT YELLOW 104 — 15790-07-5
ENVIRONMENTAL LISTS
  List of Pesticide Product Inert Ingredients
  [present]

C.I. PIGMENT YELLOW 138 — 30125-47-4
ENVIRONMENTAL LISTS
  List of Pesticide Product Inert Ingredients
  [present]

C.I. REACTIVE BLUE 52 — 74204-30-1
ENVIRONMENTAL LISTS
  List of Pesticide Product Inert Ingredients
  [present]

C.I. REACTIVE GREEN 12 — 72152-45-5
ENVIRONMENTAL LISTS
  List of Pesticide Product Inert Ingredients
  [present]

C.I. REACTIVE RED 56 — 72139-15-2
ENVIRONMENTAL LISTS
  List of Pesticide Product Inert Ingredients
  [present]

C.I. SOLVENT BLUE 36 — 14233-37-5
ENVIRONMENTAL LISTS
  List of Pesticide Product Inert Ingredients
  [present]

C.I. SOLVENT BLUE 53 — 61969-42-4
SEE ALSO:
  COBALT
INTERNATIONAL LISTS
  Canada - WHMIS: Ingredient Disclosure
  1% item 277 (314)

C.I. SOLVENT BLUE 68 — 4395-65-7
ENVIRONMENTAL LISTS
  List of Pesticide Product Inert Ingredients
  [present]

C.I. SOLVENT BLUE 98 — 71819-49-3
ENVIRONMENTAL LISTS
  List of Pesticide Product Inert Ingredients
  [present]

C.I. SOLVENT ORANGE 3 — 495-54-5
HEALTH AND SAFETY LISTS
  NIOSH - Selected LD50s and LC50s
  Oral, rat: LD50 = 1650 mg/kg

C.I. SOLVENT ORANGE 7 — 3118-97-6
HEALTH AND SAFETY LISTS
  IARC - Group 3 (not classifiable)
  [present]
ENVIRONMENTAL LISTS
  CERCLA/SARA - Section 313 - Emission Reporting
  form R reporting required for 1.0% de minimus concentration
  List of Pesticide Product Inert Ingredients
  [present]
INTERNATIONAL LISTS
  Canada - NPRI (National Pollutant Release Inventory)
  [present]
STATE LISTS
  California - Directors List of Hazardous Substances (8 CCR 339)
  [present]
  Massachusetts Right To Know List
  [present]
  NJ Right to Know List (Total)
  sn 0506
  Pennsylvania Right to Know List
  environmental hazard

C.I. SOLVENT RED 164   **71819-51-7**
  ENVIRONMENTAL LISTS
    List of Pesticide Product Inert Ingredients
     [present]

C.I. SOLVENT RED 169   **27354-18-3**
  ENVIRONMENTAL LISTS
    List of Pesticide Product Inert Ingredients
     [present]

C.I. SOLVENT VIOLET 13   **81-48-1**
  ENVIRONMENTAL LISTS
    List of Pesticide Product Inert Ingredients
     [present]

C.I. SOLVENT YELLOW 14   **842-07-9**
  HEALTH AND SAFETY LISTS
    IARC - Group 3 (not classifiable)
     [present]
    NTP Chemical Status Reports - Testing Status and NTIS Number
     Technical reports printed (PB83126474)
    NTP Chemical Status Reports - Evidence of Carcinogenicity
     male rat-positive; female rat-positive; male mice-negative; female mice-negative
  ENVIRONMENTAL LISTS
    CERCLA/SARA - Section 313 - Emission Reporting
     form R reporting required for 0.1% de minimus concentration
  INTERNATIONAL LISTS
    Canada - NPRI (National Pollutant Release Inventory)
     [present]
  STATE LISTS
    California - Directors List of Hazardous Substances (8 CCR 339)
     [present]
    Massachusetts Right To Know List
     [present]
    NJ Right to Know List (Total)
     sn 0509
    Pennsylvania Right to Know List
     environmental hazard

C.I. SOLVENT YELLOW 107   **67990-27-6**
  ENVIRONMENTAL LISTS
    List of Pesticide Product Inert Ingredients
     [present]

C.I. VAT BLACK 27   **2379-81-9**
  INTERNATIONAL LISTS
    Canada - WHMIS: Ingredient Disclosure
     1% item 278 (1272)

C.I. VAT BLUE 16   **6424-76-6**
  INTERNATIONAL LISTS
    Canada - WHMIS: Ingredient Disclosure
     1% item 279 (309)

C.I. VAT BLUE 22   **6373-20-2**
  INTERNATIONAL LISTS
    Canada - WHMIS: Ingredient Disclosure
     1% item 280 (310)

C.I. VAT BROWN 1   **2475-33-4**
  INTERNATIONAL LISTS
    Canada - WHMIS: Ingredient Disclosure
     1% item 281 (347)

C.I. VAT GREEN 2   **25704-81-8**
  INTERNATIONAL LISTS
    Canada - WHMIS: Ingredient Disclosure
     1% item 282 (1714)

C.I. VAT YELLOW 4   **128-66-5**
  HEALTH AND SAFETY LISTS
    IARC - Group 3 (not classifiable)
     [present]
    NTP Chemical Status Reports - Testing Status and NTIS Number
     Technical reports printed (PB288821/AS)
    NTP Chemical Status Reports - Evidence of Carcinogenicity
     male rat-negative; female rat-negative; male mice-positive; female mice-negative
  ENVIRONMENTAL LISTS
    CERCLA/SARA - Section 313 - Emission Reporting
     form R reporting required for 1.0% de minimus concentration
  STATE LISTS
    Massachusetts Right To Know List
     carcinogen; extraordinarily hazardous
    NJ Right to Know List (Total)
     sn 0512
    Pennsylvania Right to Know List
     environmental hazard

C7-C35 REFINED PETROLEUM OIL   **68783-08-4**
  INTERNATIONAL LISTS
    Canada - CEPA Schedule III Part I - Prohibited Substances (Ocean Dumping)
     [present]

C9 AROMATIC HYDROCARBON FRACTION   **RR-00502-7**
  ENVIRONMENTAL LISTS
    TSCA - Chemical Test Rules
     Testing required by: manufacturers; processors (40 CFR 799.2175)

C9-19 FATTY ACID ESTER PHTHALIC ALKYL   **68459-31-4**
  ENVIRONMENTAL LISTS
    List of Pesticide Product Inert Ingredients
     [present]

CACODYLIC ACID   **75-60-5**
  SEE ALSO:
    ARSENIC
  HEALTH AND SAFETY LISTS
    U.S. DOT - Substances From 49 CFR 172.101
     regulated by DOT (UN1572)
    U.S. DOT - Hazard Classes
     DOT hazard class = 6.1
    U.S. DOT - Appendix B - Marine Pollutants
     DOT regulated marine pollutant
    U.S. DOT - Appendix A Table 1 - Hazardous Substances
     final RQ = 1 pound (0.454 kg)
    NIOSH - Selected LD50s and LC50s
     Oral, rat: LD50 = 644 mg/kg
  ENVIRONMENTAL LISTS
    ATSDR Priority List
     Rank (of 275): 083
    CERCLA/SARA - Hazardous Substances and their Reportable Quantities
     final RQ = 1 pound (0.454 kg)

RCRA - U Series Wastes
waste number U136

RCRA - Hazardous Constituents-Appendix VIII
waste number U136

RCRA - Substances Banned From Land Disposal
[present]

STATE LISTS

Massachusetts Right To Know List
[present]

NJ Right to Know List (Total)
sn 0304

Pennsylvania Right to Know List
environmental hazard

CADMATE (6-), [[[1,2-ETHANEDIYLBIS[NITRILOBIS (METHYLENE)]TETRAKIS[PHOSPHONATO](8-)-N,N', O,O",O"",O"""]-, PENTAPOTASSIUM HYDROGEN, (OC-6-21)-   68309-98-8

ENVIRONMENTAL LISTS

TSCA - Code of Federal Regulations Citations
40 CFR 704.95

CADMIUM   7440-43-9

SEE ALSO:
F006-HAZARDOUS WASTES
F039-HAZARDOUS WASTES
K100-HAZARDOUS WASTES
K069-HAZARDOUS WASTES
CADMIUM COMPOUNDS
ISOBUTYLCYCLOHEXANE
CHROMIUM COMPOUNDS
K061-HAZARDOUS WASTES
K064-HAZARDOUS WASTES
K066-HAZARDOUS WASTES
K065-HAZARDOUS WASTES

HEALTH AND SAFETY LISTS

ACGIH 1995 - Time Weighted Averages
elemental and compounds, total dust, as Cd: 0.01 mg/m3 TWA; respirable fraction: 0.002 mg/m3 TWA

ACGIH 1995 - Carcinogens
elemental and compounds, as Cd: A2-suspected human carcinogen

ACGIH 1995 - Biological Exposure Indices
Cadmium in urine: 5 ug/g creatinine, not critical (B); Cadmium in blood: 5 ug/L creatinine, not critical (B)

U.S. DOT - Appendix A Table 1 - Hazardous Substances
final RQ = 10 pounds (4.54 kg) (No reporting of releases of this hazardous substance is required if the diameter of the pieces of the solid metal released is equal to or exceeds 0.004 inches)

NIOSH - Selected LD50s and LC50s
Oral, rat: LD50 = 225 mg/kg

NIOSH 1990 - Pocket Guide - RELs
as Cd: Reduce exposure to lowest feasible concentration

NIOSH 1990 - Pocket Guide - IDLHs
as Cd: 50 mg/m3 IDLH (not considering carcinogenic effects)

NIOSH 1990 - Pocket Guide - Carcinogens
occupational carcinogen

NIOSH 1990 - Pocket Guide - Target organs
respiratory system, kidneys, prostate, blood

NIOSH - Health Standards - Exposure Limits
reduce exposure to lowest feasible concentration

NIOSH - Health Standards - Health Effects and Precautions
Lung cancer, prostatic cancer, renal system effects

NIOSH - Health Standards - Carcinogenic Chemicals
potential human carcinogen

NTP Seventh Report - Suspect Carcinogens
suspect carcinogen (Listed under 'Cadmium and certain cadmium compounds')

OSHA - 29 CFR 1910 Specifically Regulated Chemicals
2.5 ug/m3 TWA action level; 5 ug/m3 TWA; do not eat, drink or chew tobacco or gum or apply cosmetics in regulated areas; carcinogen; dust can cause lung and kidney disease (see 29 CFR 1910.1027)

OSHA - Vacated PELs - Time Weighted Averages
see 1910.1027

OSHA - Vacated PELs - Short Term Exposure Limits
see 1910.1027

OSHA - Vacated PELs - Ceiling Limits
see 1910.1027

OSHA - Final PELs - Time Weighted Averages
dust: 0.2 mg/m3 TWA; C 0.6 mg/m3; apply only to exempt operations

OSHA - Final PELs - Ceiling Limits
C 0.6 mg/m3; apply only to exempt operations

OSHA - Select Carcinogens
[present]

OSHA - Possible Select Carcinogens
[present]

ENVIRONMENTAL LISTS

ATSDR Priority List
Rank (of 275): 006

CERCLA/SARA - Section 313 - Emission Reporting
form R reporting required for 0.1% de minimus concentration

CERCLA/SARA - Hazardous Substances and their Reportable Quantities
final RQ = 10 pounds (4.54 kg) (No reporting of releases of this hazardous substance is required if the diameter of the pieces of the solid metal released is equal to or exceeds 0.004 inches)

Clean Water Act - Priority Pollutants
[present]

Clean Water Act - Toxic Pollutants
[present] (Listed under 'Cadmium and compounds')

Safe Drinking Water Act - MCLs
MCL = 0.005 mg/L

Safe Drinking Water Act - MCLGs
MCLG = 0.005 mg/L

EPA - Carcinogen Hazard Ranking for RQ Adjustment
Hazard ranking = Medium

RCRA - D Series - Maximum Concentration of Contaminants
waste number D006; regulatory level = 1.0 mg/L

RCRA - D Series - Chronic Toxicity Reference Levels
chronic toxicity reference level = 0.01 mg/L

RCRA - Hazardous Constituents-Appendix VIII
hazardous constituent - no waste number

RCRA - Basis for Listing - Appendix VII
Included in waste streams: F006, F039, K061, K064, K065, K066, K069, K100

RCRA - Substances Banned From Land Disposal
[present]

RCRA - TSD Facilities Ground Water Monitoring
TM 6010 = 40 ug/L PQL; TM 7130 = 50 ug/L PQL; TM 7031 = 1 ug/L PQL (all species in the ground water that contain this element are included)

RCRA - Universal Treatment Standards (LDR)
WW: 0.69 mg/l; NWW: 0.19 mg/l TCLP

INTERNATIONAL LISTS

Australian Exposure Standards - Time Weighted Averages
dusts and salts, as Cd: 0.05 mg/m3 TWA

Australian Exposure Standards - Carcinogens
dusts and salts, as Cd: probable carcinogen

Canada - WHMIS: Ingredient Disclosure
0.1% item 287 (378)

Canada - NPRI (National Pollutant Release Inventory)
[present]

Canada - CEPA - Priority Substances List
estimated time for completion of assessment reports: 4 years

Canada - CEPA Schedule III Part I - Prohibited Substances (Ocean Dumping)
[present]

Canada - Drinking Water Quality - MACs
0.005 mg/L MAC

Canada - Alberta - 8 Hour Occupational Exposure Limit
0.05 mg/m3 TWA

Canada - Alberta - 15 Minute Occupational Exposure Limit
0.2 mg/m3 STEL

Canada - British Columbia - 8 Hour Exposure Limits
as Cd: 0.05 mg/m3 TWA

Canada - British Columbia - 15 Minute Exposure Limits
as Cd: 0.2 mg/m3 STEL

Canada - Ontario - OHSA - TWAEVs
0.02 mg/m3 TWAEV

Canada - Quebec - Time-Weighted Average Exposure Values
as Cd: 0.05 mg/m3 TWAEV

Canada - Quebec - Carcinogens
C2 carcinogen: effect suspected in humans

German (DFG) - Carcinogens
animal evidence of carcinogenicity

Israel - Time Weighted Averages
dusts and salts, as Cd: (0.05) mg/m3 TWA

Israel - Action Levels
dust and salts, as Cd: 0.025 mg/m3 AL

Mexico - Instruction No. 10 - STELs
0.2 mg/m3 STEL

Mexico - Instruction No. 10 - Ceiling Limits
P 0.05 mg/m3

Mexico - Wastewater - Organic Toxic Pollutants and Heavy Metals
Listed under [Heavy Metals]

Mexico - Drinking Water - Ecological Criteria
0.01 mg/l Substance presents persistence, bioaccumulations or risk of cancer, reduce human exposure to a minimum

STATE LISTS

California - Air Bill 2588 Appendix A-I
known or potential carcinogen

California - Prop. 65 - Cancer list
carcinogen - initial date 10/1/87

California - Prop. 65 - No Significant Risk Levels
inhalation: no significant risk level = 0.05 ug/day

California - Exposure Limits - PELs
metal dust, as Cd: 0.05 mg/m3 PEL; soluble salts, as Cd: 0.05 mg/m3 PEL

California - Exposure Limits - Ceilings
metal dust, as Cd: C 0.6 mg/m3; soluble salts, as Cd: C 0.6 mg/m3

California - Directors List of Hazardous Substances (8 CCR 339)
[present]

Florida Hazardous Substance List
[present]

Massachusetts Right To Know List
carcinogen; extraordinarily hazardous

Minnesota Hazardous Substance List
as Cd: carcinogen (includes compounds and dusts and salts)

NJ Right to Know List (Total)
sn 0305

NJ Special Hazardous Substances
(carcinogen; teratogen)

Pennsylvania Right to Know List
environmental hazard; special hazardous substance (any compound of this substance is also an environmental hazard)

Pennsylvania RTK - Special Hazardous Substances
[present]

CADMIUM 104   30905-38-9

HEALTH AND SAFETY LISTS

U.S. DOT - Appendix A Table 2 - Radionuclides
final RQ = 1000 curies (3.7E 13 Bq)

ENVIRONMENTAL LISTS

CERCLA/SARA List of Radionuclides (Appendix B) and Their Reportable Quantities
final RQ = 1000 curies (3.7E 13 Bq)

CADMIUM 107   14709-52-5

HEALTH AND SAFETY LISTS

U.S. DOT - Appendix A Table 2 - Radionuclides
final RQ = 1000 curies (3.7E 13 Bq)

ENVIRONMENTAL LISTS

CERCLA/SARA List of Radionuclides (Appendix B) and Their Reportable Quantities
final RQ = 1000 curies (3.7E 13 Bq)

CADMIUM 109   14109-32-1

HEALTH AND SAFETY LISTS

U.S. DOT - Appendix A Table 2 - Radionuclides
final RQ = 1 curie (3.7E 10 Bq)

ENVIRONMENTAL LISTS

CERCLA/SARA List of Radionuclides (Appendix B) and Their Reportable Quantities
final RQ = 1 curie (3.7E 10 Bq)

CADMIUM 113   14336-66-4

HEALTH AND SAFETY LISTS

U.S. DOT - Appendix A Table 2 - Radionuclides
final RQ = 0.1 curies (3.7E 9 Bq)

ENVIRONMENTAL LISTS

CERCLA/SARA List of Radionuclides (Appendix B) and Their Reportable Quantities
final RQ = 0.1 curies (3.7E 9 Bq)

CADMIUM 113M   RR-00474-0

HEALTH AND SAFETY LISTS

U.S. DOT - Appendix A Table 2 - Radionuclides
final RQ = 0.1 curies (3.7E 9 Bq)

ENVIRONMENTAL LISTS

CERCLA/SARA List of Radionuclides (Appendix B) and Their Reportable Quantities
final RQ = 0.1 curies (3.7E 9 Bq)

CADMIUM 115   14336-68-6

HEALTH AND SAFETY LISTS

U.S. DOT - Appendix A Table 2 - Radionuclides
final RQ = 100 curies (3.7E 12 Bq)

ENVIRONMENTAL LISTS

CERCLA/SARA List of Radionuclides (Appendix B) and Their Reportable Quantities
final RQ = 100 curies (3.7E 12 Bq)

CADMIUM 115M   RR-00473-9

HEALTH AND SAFETY LISTS

U.S. DOT - Appendix A Table 2 - Radionuclides
final RQ = 10 curies (3.7E 11 Bq)

ENVIRONMENTAL LISTS

CERCLA/SARA List of Radionuclides (Appendix B) and Their Reportable Quantities
final RQ = 10 curies (3.7E 11 Bq)

CADMIUM 117   15139-70-5

HEALTH AND SAFETY LISTS

U.S. DOT - Appendix A Table 2 - Radionuclides
final RQ = 100 curies (3.7E 12 Bq)

ENVIRONMENTAL LISTS

CERCLA/SARA List of Radionuclides (Appendix B) and Their Reportable Quantities
final RQ = 100 curies (3.7E 12 Bq)

CADMIUM 117M   RR-00471-7

HEALTH AND SAFETY LISTS

U.S. DOT - Appendix A Table 2 - Radionuclides
final RQ = 10 curies (3.7E 11 Bq)

ENVIRONMENTAL LISTS

CERCLA/SARA List of Radionuclides (Appendix B) and Their Reportable Quantities
final RQ = 10 curies (3.7E 11 Bq)

CADMIUM ACETATE   543-90-8

SEE ALSO:
CADMIUM

HEALTH AND SAFETY LISTS

U.S. DOT - Appendix A Table 1 - Hazardous Substances
final RQ = 10 pounds (4.54 kg)

ENVIRONMENTAL LISTS

CERCLA/SARA - Hazardous Substances and their Reportable Quantities
final RQ = 10 pounds (4.54 kg)

Clean Water Act - Hazardous Substances
[present]

EPA - Carcinogen Hazard Ranking for RQ Adjustment
Hazard ranking = Medium

INTERNATIONAL LISTS

Canada - WHMIS: Ingredient Disclosure
1% item 283 (12)

STATE LISTS

Massachusetts Right To Know List
[present]

NJ Right to Know List (Total)
sn 0306

NJ Special Hazardous Substances
(carcinogen)

Pennsylvania Right to Know List
environmental hazard

CADMIUM BROMIDE   7789-42-6

SEE ALSO:
CADMIUM

HEALTH AND SAFETY LISTS

U.S. DOT - Appendix A Table 1 - Hazardous Substances
final RQ = 10 pounds (4.54 kg)

ENVIRONMENTAL LISTS

CERCLA/SARA - Hazardous Substances and their Reportable Quantities
final RQ = 10 pounds (4.54 kg)

Clean Water Act - Hazardous Substances
[present]

EPA - Carcinogen Hazard Ranking for RQ Adjustment
Hazard ranking = Medium

STATE LISTS

Massachusetts Right To Know List
[present]

NJ Right to Know List (Total)
sn 0307

Pennsylvania Right to Know List
environmental hazard

CADMIUM CARBONATE   513-78-0

SEE ALSO:
CADMIUM

HEALTH AND SAFETY LISTS

NIOSH - Selected LD50s and LC50s
Oral, mouse: LD50 = 310 mg/kg

INTERNATIONAL LISTS

Canada - WHMIS: Ingredient Disclosure
1% item 284 (385)

CADMIUM CHLORIDE   10108-64-2

SEE ALSO:
CADMIUM

HEALTH AND SAFETY LISTS

U.S. DOT - Appendix A Table 1 - Hazardous Substances
final RQ = 10 pounds (4.54 kg)

NIOSH - Selected LD50s and LC50s
Inhalation, mouse: LC50 = 2300 mg/m3 (8 hr) Oral, rat: LD50 = 88 mg/kg

NTP Seventh Report - Suspect Carcinogens
suspect carcinogen (Listed under 'Cadmium and certain cadmium compounds')

OSHA - Possible Select Carcinogens
[present]

ENVIRONMENTAL LISTS

CERCLA/SARA - Hazardous Substances and their Reportable Quantities
final RQ = 10 pounds (4.54 kg)

Clean Water Act - Hazardous Substances
[present]

EPA - Carcinogen Hazard Ranking for RQ Adjustment
Hazard ranking = Medium

INTERNATIONAL LISTS

Canada - WHMIS: Ingredient Disclosure
0.1% item 285 (482)

German (DFG) - Carcinogens
animal evidence of carcinogenicity (Listed under 'Cadmium and its compounds')

STATE LISTS

Florida Hazardous Substance List
[present]

Massachusetts Right To Know List
carcinogen; extraordinarily hazardous

NJ Right to Know List (Total)
sn 0308

NJ Special Hazardous Substances
(carcinogen; mutagen)

Pennsylvania Right to Know List
environmental hazard; special hazardous substance

Pennsylvania RTK - Special Hazardous Substances
[present]

CADMIUM COMPOUNDS RR-00559-4

HEALTH AND SAFETY LISTS

ACGIH 1995 - Biological Exposure Indices
Cadmium in urine: 5 ug/g creatinine, not critical; Cadmium in blood: 5 ug/L, not critical (B)

U.S. DOT - Substances From 49 CFR 172.101
regulated by DOT (UN2570)

U.S. DOT - Hazard Classes
DOT hazard class = 6.1

U.S. DOT - Appendix B - Marine Pollutants
DOT regulated severe marine pollutant

IARC - Group 2A (limited human data)
[present] (This evaluation applies to the group of chemicals as a whole and not necessarily to all individual chemicals within the group)

OSHA - Possible Select Carcinogens
[present]

ENVIRONMENTAL LISTS

CERCLA/SARA - Section 313 - Emission Reporting
form R reporting required for 0.1% (inorganic), 1.0% (organic) de minimus concentration

Clean Air Act (1990) - List of Hazardous Air Contaminants
[present] (includes any unique chemical substance that contains Cadmium as part of that chemical's infrastructure)

Clean Water Act - Toxic Pollutants
[present] (Listed under 'Cadmium and compounds')

RCRA - Hazardous Constituents-Appendix VIII
hazardous constituent - no waste number

INTERNATIONAL LISTS

Canada - WHMIS: Ingredient Disclosure
1% item 292 (379)

Canada - NPRI (National Pollutant Release Inventory)
[present]

Canada - CEPA - Priority Substances List
estimated time for completion of assessment reports: 4 years

Canada - CEPA Schedule III Part I - Prohibited Substances (Ocean Dumping)
[present]

Canada - Ontario - OHSA - TWAEVs
0.02 mg/m3 TWAEV

United Kingdom - Maximum Exposure Limits - TWAs
as Cd: 0.05 mg/m3 TWA (does not include cadmium oxide fume or cadmium sulphide pigments)

STATE LISTS

California - Air Bill 2588 Appendix A-I
known or potential carcinogen

California - Prop. 65 - Cancer list
carcinogen - initial date 10/1/87

California - Directors List of Hazardous Substances (8 CCR 339)
[present]

NJ Right to Know List (Total)
sn 2199; sn 2869

CADMIUM DIETHYLDITHIOCARBAMATE 14239-68-0
SEE ALSO:
CADMIUM

INTERNATIONAL LISTS

Canada - WHMIS: Ingredient Disclosure
1% item 286 (702)

CADMIUM FLUOBORATE 14486-19-2
SEE ALSO:
CADMIUM

INTERNATIONAL LISTS

Canada - WHMIS: Ingredient Disclosure
1% item 288 (886)

CADMIUM IODIDE 7790-80-9
SEE ALSO:
CADMIUM

HEALTH AND SAFETY LISTS

NIOSH - Selected LD50s and LC50s
Oral, mouse: LD50 = 166 mg/kg

INTERNATIONAL LISTS

Canada - WHMIS: Ingredient Disclosure
1% item 289 (1024)

CADMIUM NITRATE 10325-94-7
SEE ALSO:
CADMIUM

HEALTH AND SAFETY LISTS

NIOSH - Selected LD50s and LC50s
Inhalation, mouse: LC50 = 3850 mg/m3 (8 hr) Oral, mouse: LD50 = 100 mg/kg

INTERNATIONAL LISTS

Canada - WHMIS: Ingredient Disclosure
1% item 290 (1200)

CADMIUM ORANGE PIGMENT 12656-57-4
SEE ALSO:
SELENIUM COMPOUNDS
CADMIUM
CADMIUM
CADMIUM COMPOUNDS

ENVIRONMENTAL LISTS

List of Pesticide Product Inert Ingredients
[present]

CADMIUM OXIDE 1306-19-0
SEE ALSO:
CADMIUM

HEALTH AND SAFETY LISTS

NIOSH - Selected LD50s and LC50s
Inhalation, rat: LC50 = 780 mg/m3 10 mn Oral, mouse: LD50 = 72 mg/kg

NIOSH 1990 - Pocket Guide - RELs
as Cd: Reduce exposure to lowest feasible concentration

NIOSH 1990 - Pocket Guide - IDLHs
as Cd: 9 mg/m3 IDLH (not considering carcinogenic effects)

NIOSH 1990 - Pocket Guide - Carcinogens
occupational carcinogen

NIOSH 1990 - Pocket Guide - Target organs
respiratory system, kidneys, blood

NTP Chemical Status Reports - Testing Status and NTIS Number
Post peer review technical reports in progress

NTP Seventh Report - Suspect Carcinogens
suspect carcinogen (Listed under 'Cadmium and certain cadmium compounds')

OSHA - Vacated PELs - Short Term Exposure Limits
as Cd: 0.3 ppm STEL

OSHA - Final PELs - Time Weighted Averages
0.1 mg/m3 TWA; apply only to exempt operations

OSHA - Final PELs - Ceiling Limits
C 0.3 mg/m3; apply only to exempt operations

OSHA - Possible Select Carcinogens
[present]

ENVIRONMENTAL LISTS

CERCLA/SARA - Section 302 Extremely Hazardous Substances and TPQs
TPQ = 10/10,000 pounds

INTERNATIONAL LISTS

Australian Exposure Standards - Time Weighted Averages
fume, as Cd: Peak Limitation: 0.05 mg/m3; production: 0.05 mg/m3 TWA

Australian Exposure Standards - Carcinogens
fume, as Cd: probable carcinogen; production: probable carcinogen

Canada - WHMIS: Ingredient Disclosure
0.1% item 291 (1302)

Canada - Alberta - Ceiling Occupational Exposure Limit
C 0.05 mg/m3

Canada - Alberta - Designated Substances
designated substance - requires code of practice

Canada - British Columbia - 8 Hour Exposure Limits
as Cd: 0.05 mg/m3 TWA

Canada - British Columbia - Ceiling Exposure Limits
as Cd: C 0.05 mg/m3

Canada - Ontario - OHSA - TWAEVs
0.05 mg/m3 TWAEV

Canada - Ontario - OHSA - CEVs
as Cd: 0.05 mg/m3 CEV

Canada - Quebec - Ceiling Limits
as Cd: P 0.05 mg/m3

Canada - Quebec - Carcinogens
as Cd: C2 carcinogen: effect suspected in humans

United Kingdom - Maximum Exposure Limits - TWAs
fume, as Cd: 0.05 mg/m3 TWA

United Kingdom - Maximum Exposure Limits - STELs
fume, as Cd: 0.05 mg/m3 STEL

German (DFG) - Carcinogens
animal evidence of carcinogenicity (Listed under 'Cadmium and its compounds')

Israel - Time Weighted Averages
production: (0.05) mg/m3 TWA

Israel - Ceiling Exposure Limits
fume, as Cd: C (0.05) mg/m3

Israel - Action Levels
production: 0.025 mg/m3 AL; fume, as Cd: 0.025 mg/m3

Mexico - Instruction No. 10 - Ceiling Limits
P 0.05 mg/m3

STATE LISTS

California - Exposure Limits - Ceilings
fume, as Cd: C 0.05 mg/m3

California - Directors List of Hazardous Substances (8 CCR 339)
[present]

Florida Hazardous Substance List
[present]

Massachusetts Right To Know List
carcinogen; extraordinarily hazardous

Minnesota Hazardous Substance List
as Cd: carcinogen (includes oxide production)

NJ Right to Know List (Total)
sn 2200

Pennsylvania Right to Know List
environmental hazard; special hazardous substance

Pennsylvania RTK - Special Hazardous Substances
[present]

PROPOSED REGULATIONS

CERCLA/SARA - Proposed Hazardous Substance Additions
proposed RQ = 1 pound (.454 kg)

CERCLA/SARA - 1989 Proposed RQ Adjustments
proposed RQ = 10 pounds (4.54 kg)

CADMIUM OXIDE PRODUCTION   RR-01638-6

INTERNATIONAL LISTS

Canada - Alberta - Ceiling Occupational Exposure Limit
as Cd: C 0.05 mg/m3

Canada - Alberta - Designated Substances
designated substance - requires code of practice

Canada - British Columbia - 8 Hour Exposure Limits
as Cd: 0.05 mg/m3 TWA

Canada - British Columbia - Carcinogens
carcinogen - as Cd: 0.05 mg/m3 TWA

Canada - Quebec - Time-Weighted Average Exposure Values
0.05 mg/m3 TWAEV; substance of which the recirculation is prohibited

Canada - Quebec - Carcinogens
C2 carcinogen: effect suspected in humans

Mexico - Instruction No. 10 - Carcinogens
potential carcinogen in humans - limited epidemiological evidence

CADMIUM STEARATE   2223-93-0
SEE ALSO:
CADMIUM

HEALTH AND SAFETY LISTS

NIOSH - Selected LD50s and LC50s
Inhalation, rat: LC50 = 130 mg/m3 2 hr Oral, rat: LD50 = 1125 mg/kg

ENVIRONMENTAL LISTS

CERCLA/SARA - Section 302 Extremely Hazardous Substances and TPQs
TPQ = 1000/10,000 pounds

INTERNATIONAL LISTS

Canada - WHMIS: Ingredient Disclosure
1% item 293 (1502)

STATE LISTS

Florida Hazardous Substance List
effective March 13, 1992

Massachusetts Right To Know List
extraordinarily hazardous

NJ Right to Know List (Total)
sn 2201

Pennsylvania Right to Know List
environmental hazard

PROPOSED REGULATIONS

CERCLA/SARA - Proposed Hazardous Substance Additions
proposed RQ = 1 pound (.454 kg)

CERCLA/SARA - 1989 Proposed RQ Adjustments
proposed RQ = 10 pounds (4.54 kg)

CADMIUM SUCCINATE   141-00-4
SEE ALSO:
CADMIUM

INTERNATIONAL LISTS

Canada - WHMIS: Ingredient Disclosure
1% item 294 (1510)

CADMIUM SULFATE   10124-36-4
  SEE ALSO:
    CADMIUM
  HEALTH AND SAFETY LISTS
    NIOSH - Selected LD50s and LC50s
      Oral, rat: LD50 = 280 mg/kg
    NTP Seventh Report - Suspect Carcinogens
      suspect carcinogen (Listed under 'Cadmium and certain cadmium compounds')
    OSHA - Possible Select Carcinogens
      [present]
  INTERNATIONAL LISTS
    Canada - WHMIS: Ingredient Disclosure
      0.1% item 295 (1515)
    German (DFG) - Carcinogens
      animal evidence of carcinogenicity (Listed under 'Cadmium and its compounds')
  STATE LISTS
    Florida Hazardous Substance List
      [present]
    Massachusetts Right To Know List
      carcinogen; extraordinarily hazardous
    Pennsylvania Right to Know List
      environmental hazard; special hazardous substance
    Pennsylvania RTK - Special Hazardous Substances
      [present]

CADMIUM SULFIDE   1306-23-6
  SEE ALSO:
    CADMIUM
    CADMIUM COMPOUNDS
    CADMIUM
  HEALTH AND SAFETY LISTS
    U.S. DOT - Appendix B - Marine Pollutants
      DOT regulated marine pollutant
    NIOSH - Selected LD50s and LC50s
      Oral, rat: LD50 = 7080 mg/kg
    NTP Seventh Report - Suspect Carcinogens
      suspect carcinogen (Listed under 'Cadmium and certain cadmium compounds')
    OSHA - Possible Select Carcinogens
      [present]
  INTERNATIONAL LISTS
    Canada - WHMIS: Ingredient Disclosure
      1% item 296 (1543)
    United Kingdom - Maximum Exposure Limits - TWAs
      respirable dust of pigments, as Cd: 0.04 mg/m3 TWA
    German (DFG) - Carcinogens
      animal evidence of carcinogenicity (Listed under 'Cadmium and its compounds')
  STATE LISTS
    Florida Hazardous Substance List
      [present]
    Massachusetts Right To Know List
      carcinogen; extraordinarily hazardous
    Pennsylvania Right to Know List
      environmental hazard; special hazardous substance
    Pennsylvania RTK - Special Hazardous Substances
      [present]

CAFFEIC ACID   331-39-5
  HEALTH AND SAFETY LISTS
    IARC - Group 2B (sufficient animal data)
      [present]

    OSHA - Possible Select Carcinogens
      [present]
  STATE LISTS
    California - Prop. 65 - Cancer list
      carcinogen - initial date 10/01/94

CAFFEINE   58-08-2
  HEALTH AND SAFETY LISTS
    IARC - Group 3 (not classifiable)
      [present]
    NTP Chemical Status Reports - Testing Status and NTIS Number
      Prechronic studies for which toxicity technical reports were not prepared

CALCIUM   7440-70-2
  HEALTH AND SAFETY LISTS
    U.S. DOT - Substances From 49 CFR 172.101
      regulated by DOT (UN1401, UN1855)
    U.S. DOT - Hazard Classes
      DOT hazard class = 4.3; pyrophoric metal or alloys: DOT hazard class = 4.2
  STATE LISTS
    California - Directors List of Hazardous Substances (8 CCR 339)
      [present]
    Florida Hazardous Substance List
      [present]
    Massachusetts Right To Know List
      [present]
    NJ Right to Know List (Total)
      sn 0309
    NJ Special Hazardous Substances
      (reactive - second degree)
    Pennsylvania Right to Know List
      [present]

CALCIUM 41   14092-95-6
  HEALTH AND SAFETY LISTS
    U.S. DOT - Appendix A Table 2 - Radionuclides
      final RQ = 10 curies (3.7E 11 Bq)
  ENVIRONMENTAL LISTS
    CERCLA/SARA List of Radionuclides (Appendix B) and Their Reportable Quantities
      final RQ = 10 curies (3.7E 11 Bq)

CALCIUM 45   13966-05-7
  HEALTH AND SAFETY LISTS
    U.S. DOT - Appendix A Table 2 - Radionuclides
      final RQ = 10 curies (3.7E 11 Bq)
  ENVIRONMENTAL LISTS
    CERCLA/SARA List of Radionuclides (Appendix B) and Their Reportable Quantities
      final RQ = 10 curies (3.7E 11 Bq)

CALCIUM 47   14391-99-2
  HEALTH AND SAFETY LISTS
    U.S. DOT - Appendix A Table 2 - Radionuclides
      final RQ = 10 curies (3.7E 11 Bq)
  ENVIRONMENTAL LISTS
    CERCLA/SARA List of Radionuclides (Appendix B) and Their Reportable Quantities
      final RQ = 10 curies (3.7E 11 Bq)

CALCIUM ABIETATE   13463-98-4

ENVIRONMENTAL LISTS

List of Pesticide Product Inert Ingredients
[present]

CALCIUM ACETATE, MONOHYDRATE   5743-26-0

ENVIRONMENTAL LISTS

List of Pesticide Product Inert Ingredients
[present]

CALCIUM ALKYL(C8-24)BENZENESULFONATE   RR-01046-8

ENVIRONMENTAL LISTS

List of Pesticide Product Inert Ingredients
[present]

**CALCIUM AND SODIUM SALTS OF SUGAR DE-   RR-01047-9
RIVED ACIDS**

ENVIRONMENTAL LISTS

List of Pesticide Product Inert Ingredients
[present]

CALCIUM ARSENATE   7778-44-1
SEE ALSO:
INORGANIC ARSENIC
ARSENIC

HEALTH AND SAFETY LISTS

U.S. DOT - Substances From 49 CFR 172.101
regulated by DOT (UN1573)

U.S. DOT - Hazard Classes
DOT hazard class = 6.1

U.S. DOT - Appendix B - Marine Pollutants
DOT regulated marine pollutant

U.S. DOT - Appendix A Table 1 - Hazardous Substances
final RQ = 1 pound (0.454 kg)

NIOSH - Selected LD50s and LC50s
Oral, rat: LD50 = 20 mg/kg Skin, rat: LD50 = 2400 mg/kg

NIOSH 1990 - Pocket Guide - RELs
as As: C 0.002 mg/m3 (15 min)

NIOSH 1990 - Pocket Guide - IDLHs
as As: 100 mg/m3 IDLH (not considering carcinogenic effects)

NIOSH 1990 - Pocket Guide - Carcinogens
occupational carcinogen

NIOSH 1990 - Pocket Guide - Target organs
eyes, respiratory system, liver, skin, lymphatics, CNS

ENVIRONMENTAL LISTS

CERCLA/SARA - Section 302 Extremely Hazardous Substances and
TPQs
TPQ = 500/10,000 pounds

CERCLA/SARA - Hazardous Substances and their Reportable Quantities
final RQ = 1 pound (0.454 kg)

Clean Water Act - Hazardous Substances
[present]

EPA - Carcinogen Hazard Ranking for RQ Adjustment
Hazard ranking = High

INTERNATIONAL LISTS

Canada - WHMIS: Ingredient Disclosure
0.1% item 297 (260)

Canada - Alberta - 8 Hour Occupational Exposure Limit
as As: 0.2 mg/m3 TWA

Canada - Alberta - 15 Minute Occupational Exposure Limit
0.6 mg/m3 STEL

Canada - British Columbia - 8 Hour Exposure Limits
as As: 1 mg/m3 TWA

German (DFG) - Carcinogens
proven carcinogen (Listed under 'Arsenic trioxide')

STATE LISTS

Massachusetts Right To Know List
extraordinarily hazardous

NJ Right to Know List (Total)
sn 0310

NJ Special Hazardous Substances
(carcinogen)

Pennsylvania Right to Know List
environmental hazard

CALCIUM ARSENATE   10103-62-5

HEALTH AND SAFETY LISTS

NTP Seventh Report - Known Carcinogens
*known carcinogen (Listed under 'Arsenic and certain arsenic com-
pounds')*

OSHA - Select Carcinogens
[present]

**CALCIUM ARSENATE AND CALCIUM ARSENITE,   RR-01275-9
MIXTURES, SOLID**

HEALTH AND SAFETY LISTS

U.S. DOT - Appendix B - Marine Pollutants
DOT regulated marine pollutant

CALCIUM ARSENITE   52740-16-6

HEALTH AND SAFETY LISTS

U.S. DOT - Substances From 49 CFR 172.101
regulated by DOT (NA1574, UN1574))

U.S. DOT - Hazard Classes
DOT hazard class = 6.1

U.S. DOT - Appendix A Table 1 - Hazardous Substances
final RQ = 1 pound (0.454 kg)

NTP Seventh Report - Known Carcinogens
*known carcinogen (Listed under 'Arsenic and certain arsenic com-
pounds')*

OSHA - Select Carcinogens
[present]

ENVIRONMENTAL LISTS

CERCLA/SARA - Hazardous Substances and their Reportable Quantities
final RQ = 1 pound (0.454 kg)

Clean Water Act - Hazardous Substances
[present]

EPA - Carcinogen Hazard Ranking for RQ Adjustment
Hazard ranking = High

STATE LISTS

Massachusetts Right To Know List
[present]

NJ Right to Know List (Total)
sn 0311

Pennsylvania Right to Know List
environmental hazard

CALCIUM ARSENITE (2:1)   15194-98-6

HEALTH AND SAFETY LISTS

NTP Seventh Report - Known Carcinogens
*known carcinogen (Listed under 'Arsenic and certain arsenic com-
pounds')*

OSHA - Select Carcinogens
[present]

CALCIUM ARSENITE (2:3)   27152-57-4
HEALTH AND SAFETY LISTS
NTP Seventh Report - Known Carcinogens
known carcinogen (Listed under 'Arsenic and certain arsenic compounds')
OSHA - Select Carcinogens
[present]

CALCIUM BENZOATE   2090-05-3
ENVIRONMENTAL LISTS
List of Pesticide Product Inert Ingredients
[present]

CALCIUM, BIS(2,4-PENTANEDIONATO-O,O')   19372-44-2
ENVIRONMENTAL LISTS
TSCA - Chemicals with Significant New Use Rules
PMN number: P-93-214
TSCA - Section 12(b) - Export Notification
P-93-214; export notification required - Section 5

CALCIUM BISULFITE   13780-03-5
INTERNATIONAL LISTS
Canada - WHMIS: Ingredient Disclosure
1% item 298 (307)
STATE LISTS
NJ Right to Know List (Total)
sn 0321
NJ Special Hazardous Substances
(corrosive)

CALCIUM BORATE   12007-56-6
HEALTH AND SAFETY LISTS
NIOSH - Selected LD50s and LC50s
Oral, rat: LD50 = 5600 mg/kg

CALCIUM CARBIDE   75-20-7
HEALTH AND SAFETY LISTS
U.S. DOT - Substances From 49 CFR 172.101
regulated by DOT (UN1402)
U.S. DOT - Hazard Classes
DOT hazard class = 4.3
U.S. DOT - Appendix A Table 1 - Hazardous Substances
final RQ = 10 pounds (4.54 kg)
ENVIRONMENTAL LISTS
CERCLA/SARA - Hazardous Substances and their Reportable Quantities
final RQ = 10 pounds (4.54 kg)
Clean Water Act - Hazardous Substances
[present]
STATE LISTS
California - Directors List of Hazardous Substances (8 CCR 339)
[present]
Florida Hazardous Substance List
[present]
Massachusetts Right To Know List
[present]
NJ Right to Know List (Total)
sn 0312
NJ Special Hazardous Substances
(flammable - fourth degree; reactive - second degree)
Pennsylvania Right to Know List
environmental hazard

CALCIUM CARBONATE   1317-65-
HEALTH AND SAFETY LISTS
ACGIH 1995 - Time Weighted Averages
10 mg/m3 TWA (The value is for total dust containing no asbesto and <1% crystalline silica)
OSHA - Vacated PELs - Time Weighted Averages
total dust: 15 mg/m3 TWA; respirable fraction: 5 mg/m3 TWA
OSHA - Final PELs - Time Weighted Averages
total dust: 15 mg/m3 TWA; respirable fraction: 5 mg/m3 TWA
ENVIRONMENTAL LISTS
List of Pesticide Product Inert Ingredients
[present]
INTERNATIONAL LISTS
Australian Exposure Standards - Time Weighted Averages
10 mg/m3 TWA
Canada - Alberta - 8 Hour Occupational Exposure Limit
respirable mass: 5 mg/m3 TWA; total mass: 10 mg/m3 TWA
Canada - British Columbia - 8 Hour Exposure Limits
nuisance dust: 10 mg/m3 TWA
Canada - British Columbia - 15 Minute Exposure Limits
20 mg/m3 STEL
Canada - Ontario - OHSA - TWAEVs
total dust: 10 mg/m3 TWAEV (listed as a nuisance particulate)
Canada - Quebec - Time-Weighted Average Exposure Values
total dust: 10 mg/m3 TWAEV; respirable dust: 5 mg/m3 TWAEV
United Kingdom - Occupational Exposure Standards - TWAs
total inhalable dust: 10 mg/m3 TWA; respirable dust: 5 mg/m3 TWA
Israel - Time Weighted Averages
10 mg/m3 TWA (The value is for total dust containing no asbestos and <1% crystalline silica)
Israel - Action Levels
5 mg/m3 AL
Mexico - Instruction No. 10 - TWAs
10 mg/m3 TWA; (nuisance particulate)
STATE LISTS
Massachusetts Right To Know List
[present]
Minnesota Hazardous Substance List
[present] (includes inert or nuisance dust)
Pennsylvania Right to Know List
[present]

CALCIUM CARBONATE DIMETHYLHEXANOATE   68442-82-0
ENVIRONMENTAL LISTS
List of Pesticide Product Inert Ingredients
[present]

CALCIUM CHLORATE   10137-74-3
HEALTH AND SAFETY LISTS
U.S. DOT - Substances From 49 CFR 172.101
regulated by DOT (UN1452, UN2429)
U.S. DOT - Hazard Classes
DOT hazard class = 5.1
ENVIRONMENTAL LISTS
List of Pesticide Product Inert Ingredients
[present]
STATE LISTS
Florida Hazardous Substance List
[present]
Massachusetts Right To Know List
[present]

NJ Right to Know List (Total)
sn 0313

Pennsylvania Right to Know List
[present]

CALCIUM CHLORIDE   10043-52-4
HEALTH AND SAFETY LISTS
NIOSH - Selected LD50s and LC50s
Oral, rat: LD50 = 1000 mg/kg

ENVIRONMENTAL LISTS
List of Pesticide Product Inert Ingredients
[present]

CALCIUM CHLORIDE HYDROXIDE HYPOCHLO-   64175-94-6
RITE, DIHYDRATE
ENVIRONMENTAL LISTS
List of Pesticide Product Inert Ingredients
[present]

CALCIUM CHLORITE   14674-72-7
HEALTH AND SAFETY LISTS
U.S. DOT - Substances From 49 CFR 172.101
regulated by DOT (UN1453)
U.S. DOT - Hazard Classes
DOT hazard class = 5.1

STATE LISTS
NJ Right to Know List (Total)
sn 0314

CALCIUM CHROMATE   13765-19-0
SEE ALSO:
CHROMIUM (VI) COMPOUNDS- WATER SOLUBLE
CHROMIUM
HEALTH AND SAFETY LISTS
ACGIH 1995 - Time Weighted Averages
as Cr: 0.001 mg/m3 TWA
ACGIH 1995 - Carcinogens
as Cr: A2-suspected human carcinogen
U.S. DOT - Appendix A Table 1 - Hazardous Substances
final RQ = 10 pounds (4.54 kg)
NTP Seventh Report - Known Carcinogens
known carcinogen (Listed under 'Chromium and certain chromium compounds')
OSHA - Select Carcinogens
[present]

ENVIRONMENTAL LISTS
CERCLA/SARA - Hazardous Substances and their Reportable Quantities
final RQ = 10 pounds (4.54 kg)
Clean Water Act - Hazardous Substances
[present]
EPA - Carcinogen Hazard Ranking for RQ Adjustment
Hazard ranking = High
RCRA - U Series Wastes
waste number U032
RCRA - Hazardous Constituents-Appendix VIII
waste number U032
RCRA - Substances Banned From Land Disposal
[present]

INTERNATIONAL LISTS
Canada - WHMIS: Ingredient Disclosure
0.1% item 299 (548)

STATE LISTS
California - Air Bill 2588 Appendix A-I
known or potential carcinogen: 6/91
Florida Hazardous Substance List
[present]
Massachusetts Right To Know List
carcinogen; extraordinarily hazardous
Minnesota Hazardous Substance List
carcinogen
NJ Right to Know List (Total)
sn 0315
NJ Special Hazardous Substances
(carcinogen; mutagen)
Pennsylvania Right to Know List
environmental hazard; special hazardous substance
Pennsylvania RTK - Special Hazardous Substances
[present]

CALCIUM CITRATE   813-94-5
ENVIRONMENTAL LISTS
List of Pesticide Product Inert Ingredients
[present]

CALCIUM CYANAMIDE   156-62-7
HEALTH AND SAFETY LISTS
ACGIH 1995 - Time Weighted Averages
0.5 mg/m3 TWA
U.S. DOT - Substances From 49 CFR 172.101
regulated by DOT (UN1403)
U.S. DOT - Hazard Classes
DOT hazard class = 4.3
NIOSH - Selected LD50s and LC50s
Oral, rat: LD50 = 158 mg/kg Skin, rat: LD50 = 84 mg/kg
NTP Chemical Status Reports - Testing Status and NTIS Number
Technical reports printed (PB293625/AS)
NTP Chemical Status Reports - Evidence of Carcinogenicity
male rat-negative; female rat-negative; male mice-negative; female mice-negative
OSHA - Vacated PELs - Time Weighted Averages
0.5 mg/m3 TWA

ENVIRONMENTAL LISTS
CERCLA/SARA - Section 313 - Emission Reporting
form R reporting required for 1.0% de minimus concentration
CERCLA/SARA - Hazardous Substances and their Reportable Quantities
final RQ = 1 pound (.454 kg)
Clean Air Act (1990) - List of Hazardous Air Contaminants
[present]

INTERNATIONAL LISTS
Australian Exposure Standards - Time Weighted Averages
0.5 mg/m3 TWA
Canada - WHMIS: Ingredient Disclosure
1% item 300 (582)
Canada - NPRI (National Pollutant Release Inventory)
[present]
Canada - Alberta - 8 Hour Occupational Exposure Limit
0.5 mg/m3 TWA
Canada - Alberta - 15 Minute Occupational Exposure Limit
1 mg/m3 STEL
Canada - British Columbia - 8 Hour Exposure Limits
0.5 mg/m3 TWA
Canada - British Columbia - 15 Minute Exposure Limits
1 mg/m3 STEL

Canada - Ontario - OHSA - TWAEVs
0.5 mg/m3 TWAEV

Canada - Quebec - Time-Weighted Average Exposure Values
0.5 mg/m3 TWAEV

United Kingdom - Occupational Exposure Standards - TWAs
0.5 mg/m3 TWA

United Kingdom - Occupational Exposure Standards - STELs
1 mg/m3 STEL

German (DFG) - MAK Values
total dust: 1 mg/m3 MAK

German (DFG) - Peak Limitations
10 x normal MAK (30 min average value);   don't exceed during shift

German (DFG) - Skin/Sensitizers
danger of cutaneous absorption

Israel - Time Weighted Averages
0.5 mg/m3 TWA

Israel - Action Levels
0.25 mg/m3 AL

Mexico - Instruction No. 10 - TWAs
0.5 mg/m3 TWA

Mexico - Instruction No. 10 - STELs
1 mg/m3 STEL

STATE LISTS

California - Air Bill 2588 Appendix A-I
6/91

California - Exposure Limits - PELs
0.5 mg/m3 PEL

California - Directors List of Hazardous Substances (8 CCR 339)
[present]

Florida Hazardous Substance List
[present]

Massachusetts Right To Know List
[present]

Minnesota Hazardous Substance List
[present]

NJ Right to Know List (Total)
sn 0316

Pennsylvania Right to Know List
environmental hazard

CALCIUM CYANIDE   592-01-8
SEE ALSO:
  CYANIDE ANION

HEALTH AND SAFETY LISTS

ACGIH 1995 - Ceiling Limits
C 5 mg/m3

ACGIH 1995 - Skin Designations
skin - potential for cutaneous absorption

U.S. DOT - Substances From 49 CFR 172.101
regulated by DOT (UN1575)

U.S. DOT - Hazard Classes
DOT hazard class = 6.1

U.S. DOT - Appendix B - Marine Pollutants
DOT regulated marine pollutant

U.S. DOT - Appendix A Table 1 - Hazardous Substances
final RQ = 10 pounds (4.54 kg)

NIOSH - Selected LD50s and LC50s
Oral, rat: LD50 = 39 mg/kg

ENVIRONMENTAL LISTS

CERCLA/SARA - Hazardous Substances and their Reportable Quantities
final RQ = 10 pounds (4.54 kg)

Clean Water Act - Hazardous Substances
[present]

RCRA - P Series Wastes
waste number P021

RCRA - Hazardous Constituents-Appendix VIII
waste number P021

RCRA - Substances Banned From Land Disposal
[present]

STATE LISTS

Florida Hazardous Substance List
[present]

Massachusetts Right To Know List
[present]

NJ Right to Know List (Total)
sn 0317

Pennsylvania Right to Know List
environmental hazard

CALCIUM DITHIONITE   13846-18-9

HEALTH AND SAFETY LISTS

U.S. DOT - Substances From 49 CFR 172.101
regulated by DOT (UN1923)

U.S. DOT - Hazard Classes
DOT hazard class = 4.2

STATE LISTS

NJ Right to Know List (Total)
sn 2203

CALCIUM DODECYLBENZENE SULFONATE   26264-06-2

HEALTH AND SAFETY LISTS

U.S. DOT - Appendix A Table 1 - Hazardous Substances
final RQ = 1000 pounds (454 kg)

ENVIRONMENTAL LISTS

CERCLA/SARA - Hazardous Substances and their Reportable Quantities
final RQ = 1000 pounds (454 kg)

Clean Water Act - Hazardous Substances
[present]

List of Pesticide Product Inert Ingredients
[present]

STATE LISTS

California - Directors List of Hazardous Substances (8 CCR 339)
[present] (exempt when in solution)

Massachusetts Right To Know List
[present]

NJ Right to Know List (Total)
sn 0318

Pennsylvania Right to Know List
environmental hazard

CALCIUM 2-ETHYLHEXANOATE   136-51-6

ENVIRONMENTAL LISTS

List of Pesticide Product Inert Ingredients
[present]

CALCIUM 2-ETHYLHEXANOATE/ISONONANOATE   68478-54-6
COMPLEXES

ENVIRONMENTAL LISTS

List of Pesticide Product Inert Ingredients
[present]

CALCIUM FLUORIDE (CAF2)   7789-75-5

HEALTH AND SAFETY LISTS

NIOSH - Selected LD50s and LC50s
Oral, rat: LD50 = 4250 mg/kg

CALCIUM FORMATE   544-17-2

INTERNATIONAL LISTS

Canada - WHMIS: Ingredient Disclosure
1% item 301 (921)

CALCIUM HYDRIDE   57308-10-8

HEALTH AND SAFETY LISTS

U.S. DOT - Substances From 49 CFR 172.101
regulated by DOT (UN1404)

U.S. DOT - Hazard Classes
DOT hazard class = 4.3

STATE LISTS

NJ Right to Know List (Total)
sn 0320

CALCIUM HYDROXIDE   1305-62-0

HEALTH AND SAFETY LISTS

ACGIH 1995 - Time Weighted Averages
5 mg/m3 TWA

NIOSH - Selected LD50s and LC50s
Oral, rat: LD50 = 7340 mg/kg

OSHA - Vacated PELs - Time Weighted Averages
5 mg/m3 TWA (not in effect as a result of reconsideration)

ENVIRONMENTAL LISTS

List of Pesticide Product Inert Ingredients
[present]

INTERNATIONAL LISTS

Australian Exposure Standards - Time Weighted Averages
5 mg/m3 TWA

Canada - WHMIS: Ingredient Disclosure
1% item 302 (991)

Canada - Alberta - 8 Hour Occupational Exposure Limit
5 mg/m3 TWA

Canada - Alberta - 15 Minute Occupational Exposure Limit
10 mg/m3 STEL

Canada - British Columbia - 8 Hour Exposure Limits
5 mg/m3 TWA

Canada - Ontario - OHSA - TWAEVs
5 mg/m3 TWAEV

Canada - Quebec - Time-Weighted Average Exposure Values
5 mg/m3 TWAEV

United Kingdom - Occupational Exposure Standards - TWAs
5 mg/m3 TWA

Israel - Time Weighted Averages
5 mg/m3 TWA

Israel - Action Levels
2.5 mg/m3 AL

Mexico - Instruction No. 10 - TWAs
5 mg/m3 TWA

STATE LISTS

California - Exposure Limits - PELs
5 mg/m3 PEL

California - Directors List of Hazardous Substances (8 CCR 339)
[present]

Florida Hazardous Substance List
[present]

Massachusetts Right To Know List
[present]

Minnesota Hazardous Substance List
[present]

NJ Right to Know List (Total)
sn 0322

Pennsylvania Right to Know List
[present]

CALCIUM HYPOCHLORITE   7778-54-3

HEALTH AND SAFETY LISTS

U.S. DOT - Substances From 49 CFR 172.101
regulated by DOT (UN2208, UN2880, UN1748)

U.S. DOT - Hazard Classes
DOT hazard class = 5.1

U.S. DOT - Appendix A Table 1 - Hazardous Substances
final RQ = 10 pounds (4.54 kg)

NIOSH - Selected LD50s and LC50s
Oral, rat: LD50 = 850 mg/kg

ENVIRONMENTAL LISTS

CERCLA/SARA - Hazardous Substances and their Reportable Quantities
final RQ = 10 pounds (4.54 kg)

Clean Water Act - Hazardous Substances
[present]

List of Pesticide Product Inert Ingredients
[present]

STATE LISTS

California - Directors List of Hazardous Substances (8 CCR 339)
[present]

Florida Hazardous Substance List
[present]

Massachusetts Right To Know List
[present]

NJ Right to Know List (Total)
sn 0323

NJ Special Hazardous Substances
(reactive - second degree)

Pennsylvania Right to Know List
environmental hazard

CALCIUM IODATE   7789-80-2

ENVIRONMENTAL LISTS

List of Pesticide Product Inert Ingredients
[present]

CALCIUM ISONONANOATE   53988-05-9

ENVIRONMENTAL LISTS

List of Pesticide Product Inert Ingredients
[present]

CALCIUM LIGNOSULFONATE   8061-52-7

ENVIRONMENTAL LISTS

List of Pesticide Product Inert Ingredients
[present]

CALCIUM MANGANESE SILICON   13573-15-4

HEALTH AND SAFETY LISTS

U.S. DOT - Substances From 49 CFR 172.101
regulated by DOT (UN2844)

U.S. DOT - Hazard Classes
DOT hazard class = 4.3

STATE LISTS

NJ Right to Know List (Total)
sn 2204

CALCIUM MOLYBDATE   7789-82-4

HEALTH AND SAFETY LISTS

NIOSH - Selected LD50s and LC50s
Oral, rat: LD50 = 101 mg/kg

INTERNATIONAL LISTS

Canada - WHMIS: Ingredient Disclosure
1% item 303 (1162)

CALCIUM NAPHTHENATE   61789-36-4

HEALTH AND SAFETY LISTS

U.S. DOT - Appendix B - Marine Pollutants
DOT regulated marine pollutant

ENVIRONMENTAL LISTS

List of Pesticide Product Inert Ingredients
[present]

TSCA - Code of Federal Regulations Citations
40 CFR 712.30(h);   40 CFR 716.120(a)

TSCA - PAIR - Reporting List
Reporting Date: September 20, 1983

TSCA - Health and Safety Reporting List
Effective Date: July 1, 1983

CALCIUM NEODECANOATE   27253-33-4

ENVIRONMENTAL LISTS

List of Pesticide Product Inert Ingredients
[present]

CALCIUM NITRATE   10124-37-5

HEALTH AND SAFETY LISTS

U.S. DOT - Substances From 49 CFR 172.101
regulated by DOT (UN1454)

U.S. DOT - Hazard Classes
DOT hazard class = 5.1

STATE LISTS

NJ Right to Know List (Total)
sn 0324

CALCIUM NITRATE.4H20   13477-34-4

HEALTH AND SAFETY LISTS

NIOSH - Selected LD50s and LC50s
Oral, rat: LD50 = 3900 mg/kg

CALCIUM OCTANOATE   6107-56-8

ENVIRONMENTAL LISTS

List of Pesticide Product Inert Ingredients
[present]

CALCIUM OXIDE   1305-78-8

HEALTH AND SAFETY LISTS

ACGIH 1995 - Time Weighted Averages
2 mg/m3 TWA

U.S. DOT - Substances From 49 CFR 172.101
regulated by DOT (UN1910)

U.S. DOT - Hazard Classes
DOT hazard class = 8

NIOSH 1990 - Pocket Guide - RELs
2 mg/m3 TWA

NIOSH 1990 - Pocket Guide - Target organs
respiratory system, skin, eyes

OSHA - Vacated PELs - Time Weighted Averages
5 mg/m3 TWA (not in effect as a result of reconsideration)

OSHA - Final PELs - Time Weighted Averages
5 mg/m3 TWA

ENVIRONMENTAL LISTS

List of Pesticide Product Inert Ingredients
[present]

INTERNATIONAL LISTS

Australian Exposure Standards - Time Weighted Averages
2 mg/m3 TWA

Canada - WHMIS: Ingredient Disclosure
1% item 304 (1303)

Canada - Alberta - 8 Hour Occupational Exposure Limit
2 mg/m3 TWA

Canada - Alberta - 15 Minute Occupational Exposure Limit
4 mg/m3 STEL

Canada - British Columbia - 8 Hour Exposure Limits
2 mg/m3 TWA

Canada - Ontario - OHSA - TWAEVs
2 mg/m3 TWAEV

Canada - Quebec - Time-Weighted Average Exposure Values
2 mg/m3 TWAEV

United Kingdom - Occupational Exposure Standards - TWAs
2 mg/m3 TWA

German (DFG) - MAK Values
total dust: 5 mg/m3 MAK

German (DFG) - Peak Limitations
2 x̄ normal MAK (5 min momentary value);   don't exceed 8 times during shift

Israel - Time Weighted Averages
2 mg/m3 TWA

Israel - Action Levels
1 mg/m3 AL

Mexico - Instruction No. 10 - TWAs
2 mg/m3 TWA

STATE LISTS

California - Exposure Limits - PELs
2 mg/m3 PEL

California - Directors List of Hazardous Substances (8 CCR 339)
[present]

Florida Hazardous Substance List
[present]

Massachusetts Right To Know List
[present]

Minnesota Hazardous Substance List
[present]

NJ Right to Know List (Total)
sn 0325

Pennsylvania Right to Know List
[present]

CALCIUM PERCHLORATE   13477-36-6

HEALTH AND SAFETY LISTS

U.S. DOT - Substances From 49 CFR 172.101
regulated by DOT (UN1455)

U.S. DOT - Hazard Classes
DOT hazard class = 5.1

ENVIRONMENTAL LISTS

List of Pesticide Product Inert Ingredients
[present]

STATE LISTS

NJ Right to Know List (Total)
sn 0326

CALCIUM PERMANGANATE 10118-76-0
SEE ALSO:
MANGANESE

HEALTH AND SAFETY LISTS
U.S. DOT - Substances From 49 CFR 172.101
regulated by DOT (UN1456)
U.S. DOT - Hazard Classes
DOT hazard class = 5.1

STATE LISTS
NJ Right to Know List (Total)
sn 0327

CALCIUM PEROXIDE 1305-79-9

HEALTH AND SAFETY LISTS
U.S. DOT - Substances From 49 CFR 172.101
regulated by DOT (UN1457)
U.S. DOT - Hazard Classes
DOT hazard class = 5.1

STATE LISTS
NJ Right to Know List (Total)
sn 0328

CALCIUM PETROLEUM SULFONATE 61789-86-4

ENVIRONMENTAL LISTS
List of Pesticide Product Inert Ingredients
[present]

CALCIUM PHOSPHATE 7758-87-4

ENVIRONMENTAL LISTS
List of Pesticide Product Inert Ingredients
[present]

CALCIUM PHOSPHATE 10103-46-5

ENVIRONMENTAL LISTS
List of Pesticide Product Inert Ingredients
[present]

CALCIUM PHOSPHIDE 1305-99-3

HEALTH AND SAFETY LISTS
U.S. DOT - Substances From 49 CFR 172.101
regulated by DOT (UN1360)
U.S. DOT - Hazard Classes
DOT hazard class = 4.3

STATE LISTS
NJ Right to Know List (Total)
sn 0329

CALCIUM POLYACRYLATE 25987-55-7

ENVIRONMENTAL LISTS
List of Pesticide Product Inert Ingredients
[present]

CALCIUM PROPIONATE 4075-81-4

ENVIRONMENTAL LISTS
List of Pesticide Product Inert Ingredients
[present]

CALCIUM RESINATE 9007-13-0

HEALTH AND SAFETY LISTS
U.S. DOT - Substances From 49 CFR 172.101
regulated by DOT (UN1313, UN1314)
U.S. DOT - Hazard Classes
DOT hazard class = 4.1

ENVIRONMENTAL LISTS
List of Pesticide Product Inert Ingredients
[present]

STATE LISTS
NJ Right to Know List (Total)
sn 0330

CALCIUM SALTS OF TALL-OIL FATTY ACIDS 68187-71-3

ENVIRONMENTAL LISTS
List of Pesticide Product Inert Ingredients
[present]

CALCIUM SELENATE 14019-91-1

STATE LISTS
NJ Right to Know List (Total)
sn 0331

CALCIUM SILICATE 1344-95-2

HEALTH AND SAFETY LISTS
ACGIH 1995 - Time Weighted Averages
10 mg/m3 TWA (The value is for total dust containing no asbestos and < 1%crystalline silica)
OSHA - Vacated PELs - Time Weighted Averages
total dust: 15 mg/m3 TWA; respirable fraction: 5 mg/m3 TWA
OSHA - Final PELs - Time Weighted Averages
total dust: 15 mg/m3 TWA; respirable fraction: 5 mg/m3 TWA

ENVIRONMENTAL LISTS
List of Pesticide Product Inert Ingredients
[present]

INTERNATIONAL LISTS
Australian Exposure Standards - Time Weighted Averages
10 mg/m3 TWA
Canada - Alberta - 8 Hour Occupational Exposure Limit
respirable mass: 5 mg/m3 TWA; total mass: 10 mg/m3 TWA
Canada - British Columbia - 8 Hour Exposure Limits
nuisance dust, mists, and fumes: 10 mg/m3 TWA
Canada - Ontario - OHSA - TWAEVs
total dust: 10 mg/m3 TWAEV (listed as nuisance particulate)
Canada - Quebec - Time-Weighted Average Exposure Values
10 mg/m3 TWAEV
United Kingdom - Occupational Exposure Standards - TWAs
total inhalable dust: 10 mg/m3 TWA; respirable dust: 5 mg/m3 TWA
Israel - Time Weighted Averages
10 mg/m3 TWA (The value is for total dust containing no asbestos and <1% crystalline silica)
Israel - Action Levels
5 mg/m3 AL
Mexico - Instruction No. 10 - TWAs
10 mg/m3 TWA (nuisance particulate)

STATE LISTS
Florida Hazardous Substance List
effective March 13, 1992
Minnesota Hazardous Substance List
[present] (includes inert or nuisance dust)
Pennsylvania Right to Know List
[present]

CALCIUM SILICIDE 12737-18-7

HEALTH AND SAFETY LISTS
U.S. DOT - Substances From 49 CFR 172.101
regulated by DOT (UN1405)
U.S. DOT - Hazard Classes
DOT hazard class = 4.3

STATE LISTS
NJ Right to Know List (Total)
sn 0332

CALCIUM SILICON   12013-56-8

STATE LISTS
NJ Right to Know List (Total)
sn 2205

CALCIUM SODIUM METAPHOSPHATE   23209-59-8

INTERNATIONAL LISTS
German (DFG) - Carcinogens
as fibrous dust: suspected carcinogen

CALCIUM STEARATE   1592-23-0

ENVIRONMENTAL LISTS
List of Pesticide Product Inert Ingredients
[present]

STATE LISTS
California - Exposure Limits - PELs
10 mg/m3 PEL

CALCIUM SULFATE   7778-18-9

HEALTH AND SAFETY LISTS
ACGIH 1995 - Time Weighted Averages
10 mg/m3 TWA (The value is for total dust containing no asbestos and < 1%crystalline silica)
OSHA - Vacated PELs - Time Weighted Averages
total dust: 15 mg/m3 TWA; respirable fraction: 5 mg/m3 TWA
OSHA - Final PELs - Time Weighted Averages
total dust: 15 mg/m3 TWA; respirable fraction: 5 mg/m3 TWA

ENVIRONMENTAL LISTS
List of Pesticide Product Inert Ingredients
[present]

INTERNATIONAL LISTS
Australian Exposure Standards - Time Weighted Averages
10 mg/m3 TWA
Canada - Quebec - Time-Weighted Average Exposure Values
total dust: 10 mg/m3 TWAEV; respirable dust: 5 mg/m3 TWAEV
German (DFG) - MAK Values
fine dust: 6 mg/m3 MAK
Israel - Time Weighted Averages
10 mg/m3 TWA (The value is for total dust containing no asbestos and <1% crystalline silica)
Israel - Action Levels
5 mg/m3 AL

STATE LISTS
Minnesota Hazardous Substance List
[present] (includes inert or nuisance dust)
Pennsylvania Right to Know List
[present]

CALCIUM SULFATE DIHYDRATE   10101-41-4

HEALTH AND SAFETY LISTS
OSHA - Final PELs - Time Weighted Averages
total dust: 15 mg/m3 TWA; respirable fraction: 5 mg/m3 TWA

ENVIRONMENTAL LISTS
List of Pesticide Product Inert Ingredients
[present]

INTERNATIONAL LISTS
Canada - Alberta - 8 Hour Occupational Exposure Limit
respirable mass: 5 mg/m3 TWA; total mass: 10 mg/m3 TWA

Canada - British Columbia - 8 Hour Exposure Limits
nuisance dust: 10 mg/m3 TWA
Canada - British Columbia - 15 Minute Exposure Limits
20 mg/m3 STEL
Canada - Ontario - OHSA - TWAEVs
total dust: 10 mg/m3 TWAEV (listed as nuisance particulate)
United Kingdom - Occupational Exposure Standards - TWAs
total inhalable dust: 10 mg/m3 TWA; respirable dust: 5 mg/m3 TWA
Mexico - Instruction No. 10 - TWAs
10 mg/m3 TWA; (nuisance particulate)

CALCIUM SULFATE HEMIHYDRATE   10034-76-1

ENVIRONMENTAL LISTS
List of Pesticide Product Inert Ingredients
[present]

CALCIUM THIOSULFATE   10124-41-1

ENVIRONMENTAL LISTS
List of Pesticide Product Inert Ingredients
[present]

CALCO OIL BLUE   29887-08-9

ENVIRONMENTAL LISTS
List of Pesticide Product Inert Ingredients
[present]

CALIFORNIUM 244   16044-16-9

HEALTH AND SAFETY LISTS
U.S. DOT - Appendix A Table 2 - Radionuclides
final RQ = 1000 curies (3.7E 13 Bq)

ENVIRONMENTAL LISTS
CERCLA/SARA List of Radionuclides (Appendix B) and Their Reportable Quantities
final RQ = 1000 curies (3.7E 13 Bq)

CALIFORNIUM 246   15117-45-0

HEALTH AND SAFETY LISTS
U.S. DOT - Appendix A Table 2 - Radionuclides
final RQ = 10 curies (3.7E 11 Bq)

ENVIRONMENTAL LISTS
CERCLA/SARA List of Radionuclides (Appendix B) and Their Reportable Quantities
final RQ = 10 curies (3.7E 11 Bq)

CALIFORNIUM 248   15758-24-4

HEALTH AND SAFETY LISTS
U.S. DOT - Appendix A Table 2 - Radionuclides
final RQ = 0.1 curies (3.7E 9 Bq)

ENVIRONMENTAL LISTS
CERCLA/SARA List of Radionuclides (Appendix B) and Their Reportable Quantities
final RQ = 0.1 curies (3.7E 9 Bq)

CALIFORNIUM 249   15237-97-5

HEALTH AND SAFETY LISTS
U.S. DOT - Appendix A Table 2 - Radionuclides
final RQ = 0.01 curies (3.7E 8 Bq)

ENVIRONMENTAL LISTS
CERCLA/SARA List of Radionuclides (Appendix B) and Their Reportable Quantities
final RQ = 0.01 curies (3.7E 8 Bq)

CALIFORNIUM 250   13982-11-1

HEALTH AND SAFETY LISTS

U.S. DOT - Appendix A Table 2 - Radionuclides
final RQ = 0.01 curies (3.7E 8 Bq)

ENVIRONMENTAL LISTS

CERCLA/SARA List of Radionuclides (Appendix B) and Their Reportable Quantities
final RQ = 0.01 curies (3.7E 8 Bq)

CALIFORNIUM 251   15765-19-2

HEALTH AND SAFETY LISTS

U.S. DOT - Appendix A Table 2 - Radionuclides
final RQ = 0.01 curies (3.7E 8 Bq)

ENVIRONMENTAL LISTS

CERCLA/SARA List of Radionuclides (Appendix B) and Their Reportable Quantities
final RQ = 0.01 curies (3.7E 8 Bq)

CALIFORNIUM 252   13981-17-4

HEALTH AND SAFETY LISTS

U.S. DOT - Appendix A Table 2 - Radionuclides
final RQ = 0.1 curies (3.7E 9 Bq)

ENVIRONMENTAL LISTS

CERCLA/SARA List of Radionuclides (Appendix B) and Their Reportable Quantities
final RQ = 0.1 curies (3.7E 9 Bq)

CALIFORNIUM 253   15720-29-3

HEALTH AND SAFETY LISTS

U.S. DOT - Appendix A Table 2 - Radionuclides
final RQ = 10 curies (3.7E 11 Bq)

ENVIRONMENTAL LISTS

CERCLA/SARA List of Radionuclides (Appendix B) and Their Reportable Quantities
final RQ = 10 curies (3.7E 11 Bq)

CALIFORNIUM 254   22095-76-7

HEALTH AND SAFETY LISTS

U.S. DOT - Appendix A Table 2 - Radionuclides
final RQ = 0.1 curies (3.7E 9 Bq)

ENVIRONMENTAL LISTS

CERCLA/SARA List of Radionuclides (Appendix B) and Their Reportable Quantities
final RQ = 0.1 curies (3.7E 9 Bq)

CAMPHENE   79-92-5

ENVIRONMENTAL LISTS

EPA - Master Testing List
[present]

STATE LISTS

NJ Right to Know List (Total)
sn 0333

CAMPHOR   76-22-2

HEALTH AND SAFETY LISTS

ACGIH 1995 - Time Weighted Averages
2 ppm TWA;   12 mg/m3 TWA

ACGIH 1995 - Short Term Exposure Limits
3 ppm STEL;   19 mg/m3 STEL

AIHA - Odor Threshold Values
geometric mean air odor threshold = 0.079 ppm (detectable)

U.S. DOT - Substances From 49 CFR 172.101
regulated by DOT (UN2717)

U.S. DOT - Hazard Classes
DOT hazard class = 4.1

NFPA - Flash Points
flash point = 150 degrees F (66 degrees C)

NFPA - Hazard Identification Ratings
health-0; flammability-2; reactivity-0

NIOSH - Selected LD50s and LC50s
Oral, mouse: LD50 = 1310 mg/kg

NIOSH 1990 - Pocket Guide - RELs
2 mg/m3 TWA

NIOSH 1990 - Pocket Guide - IDLHs
200 mg/m3 IDLH

NIOSH 1990 - Pocket Guide - Target organs
CNS, eyes, skin, respiratory system

OSHA - Vacated PELs - Time Weighted Averages
2 mg/m3 TWA

OSHA - Final PELs - Time Weighted Averages
2 mg/m3 TWA

ENVIRONMENTAL LISTS

List of Pesticide Product Inert Ingredients
[present]

TSCA - PAIR - Reporting List
Effective Date: January 26, 1994;   Reporting Date: March 28, 1994

TSCA - Health and Safety Reporting List
Effective Date: January 26, 1994;   Sunset Date: January 26, 2004

INTERNATIONAL LISTS

Australian Exposure Standards - Time Weighted Averages
2 ppm TWA;   12 mg/m3 TWA

Australian Exposure Standards - Short Term Exposure Limits
3 ppm STEL;   19 mg/m3 STEL

Canada - Alberta - 8 Hour Occupational Exposure Limit
2 ppm TWA;   12 mg/m3 TWA

Canada - Alberta - 15 Minute Occupational Exposure Limit
3 ppm STEL;   19 mg/m3 STEL

Canada - British Columbia - 8 Hour Exposure Limits
2 ppm TWA;   12 mg/m3 TWA

Canada - British Columbia - 15 Minute Exposure Limits
3 ppm STEL;   18 mg/m3 STEL

Canada - Ontario - OHSA - TWAEVs
2 ppm TWAEV;   12 mg/m3 TWAEV

Canada - Ontario - OHSA - STEVs
3 ppm STEV;   19 mg/m3 STEV

Canada - Quebec - Time-Weighted Average Exposure Values
2 ppm TWAEV;   12 mg/m3 TWAEV

Canada - Quebec - Short-term Exposure Values
3 ppm STEV;   19 mg/m3 STEV

United Kingdom - Occupational Exposure Standards - TWAs
2 ppm TWA;   12 mg/m3 TWA

United Kingdom - Occupational Exposure Standards - STELs
3 ppm STEL;   18 mg/m3 STEL

German (DFG) - MAK Values
2 ppm MAK;   13 mg/m3 MAK

Israel - Time Weighted Averages
2 ppm TWA;   12 mg/m3 TWA

Israel - Short Term Exposure Limits
3 ppm STEL;   19 mg/m3 STEL

Israel - Action Levels
1 ppm AL;   6 mg/m3 AL

Mexico - Instruction No. 10 - TWAs
2 ppm TWA;   12 mg/m3 TWA

STATE LISTS

California - Exposure Limits - PELs
2 mg/m3 PEL

California - Directors List of Hazardous Substances (8 CCR 339)
[present]

Florida Hazardous Substance List
[present] (includes the synthetic form)

Massachusetts Right To Know List
[present]

Minnesota Hazardous Substance List
[present]

NJ Right to Know List (Total)
sn 0334

Pennsylvania Right to Know List
[present]

PROPOSED REGULATIONS

TSCA - ITC 31st Report Priority Testing List
designated to be tested

(+)-CAMPHOR   464-49-3

HEALTH AND SAFETY LISTS

NIOSH - Selected LD50s and LC50s
Oral, mouse: LD50 = 1310 mg/kg

INTERNATIONAL LISTS

Canada - WHMIS: Ingredient Disclosure
1% item 305 (380)

DL-CAMPHOR   21368-68-3

HEALTH AND SAFETY LISTS

NTP Chemical Status Reports - Testing Status and NTIS Number
Approved for toxicology/carcinogenesis study

CAMPHOR OIL   8008-51-3

HEALTH AND SAFETY LISTS

U.S. DOT - Substances From 49 CFR 172.101
regulated by DOT (UN1130)

U.S. DOT - Hazard Classes
DOT hazard class = 3

NFPA - Flash Points
flash point = 117 degrees F (47 degrees C)

NFPA - Hazard Identification Ratings
health-2; flammability-2; reactivity-0

NIOSH - Selected LD50s and LC50s
Oral, rat: LD50 = 3730 mg/m3

STATE LISTS

Florida Hazardous Substance List
[present]

Massachusetts Right To Know List
[present]

NJ Right to Know List (Total)
sn 0335

Pennsylvania Right to Know List
[present]

CANARY SEED   RR-01048-0

ENVIRONMENTAL LISTS

List of Pesticide Product Inert Ingredients
[present]

CANTHARIDIN   56-25-7

HEALTH AND SAFETY LISTS

IARC - Group 3 (not classifiable)
[present]

ENVIRONMENTAL LISTS

CERCLA/SARA - Section 302 Extremely Hazardous Substances and TPQs
TPQ = 100/10,000 pounds

STATE LISTS

California - Directors List of Hazardous Substances (8 CCR 339)
[present]

Florida Hazardous Substance List
effective March 13, 1992

Massachusetts Right To Know List
extraordinarily hazardous

Pennsylvania Right to Know List
environmental hazard

PROPOSED REGULATIONS

CERCLA/SARA - Proposed Hazardous Substance Additions
proposed RQ = 1 pound (.454 kg)

CERCLA/SARA - 1989 Proposed RQ Adjustments
proposed RQ = 100 pounds (45.4 kg)

CAPPED ALIPHATIC ISOCYANATE   RR-01664-

ENVIRONMENTAL LISTS

TSCA - Chemicals with Significant New Use Rules
PMN number: P-86-1146

CAPRIC ACID   334-48-

ENVIRONMENTAL LISTS

List of Pesticide Product Inert Ingredients
[present]

INTERNATIONAL LISTS

Canada - WHMIS: Ingredient Disclosure
1% item 306 (72)

CAPROIC ACID   142-62-

HEALTH AND SAFETY LISTS

U.S. DOT - Substances From 49 CFR 172.101
regulated by DOT (UN2829)

U.S. DOT - Hazard Classes
DOT hazard class = 8

NFPA - Flash Points
flash point = 215 degrees F (102 degrees C)

NFPA - Hazard Identification Ratings
health-2; flammability-1; reactivity-0

NIOSH - Selected LD50s and LC50s
Inhalation, mouse: LD50 = 4100 mg/m3 (2 hr) ORal, rat: LD50 = 3000 mg/kg Skin, guinea pgi: LD50 = 4635 mg/kg

ENVIRONMENTAL LISTS

List of Pesticide Product Inert Ingredients
[present]

INTERNATIONAL LISTS

Canada - WHMIS: Ingredient Disclosure
1% item 830 (92)

STATE LISTS

Florida Hazardous Substance List
[present]

Massachusetts Right To Know List
[present]

NJ Right to Know List (Total)
sn 0336

NJ Special Hazardous Substances
(corrosive)

Pennsylvania Right to Know List
[present]

CAPROLACTAM   105-60-2

HEALTH AND SAFETY LISTS

ACGIH 1995 - Time Weighted Averages
dust: 1 mg/m3 TWA; vapor: 5 ppm TWA; 23 mg/m3 TWA

ACGIH 1995 - Short Term Exposure Limits
dust: 3 mg/m3 STEL; vapor: 10 ppm STEL; 46 mg/m3 STEL

IARC - Group 4 (probably not carcinogenic)
[present]

NIOSH - Selected LD50s and LC50s
Oral, rat: LD50 = 1370 mg/kg Skin, rabbit: LD50 = 1438 mg/kg

NTP Chemical Status Reports - Testing Status and NTIS Number
Technical reports printed (PB82190182)

NTP Chemical Status Reports - Evidence of Carcinogenicity
male rat-negative; female rat-negative; male mice-negative; female mice-negative

OSHA - Vacated PELs - Time Weighted Averages
dust: 1 mg/m3 TWA; vapor: 5 ppm TWA; 20 mg/m3 TWA

OSHA - Vacated PELs - Short Term Exposure Limits
dust: 3 mg/m3 STEL; vapor: 10 ppm STEL; 40 mg/m3 STEL

ENVIRONMENTAL LISTS

CERCLA/SARA - Hazardous Substances and their Reportable Quantities
final RQ = 1 pound (.454 kg)

Clean Air Act (1990) - List of Hazardous Air Contaminants
[present]

CAA - HON Rule - SOCMI Chemicals
compliance by Jan. 23, 1995

CAA - HON Rule - Organic HAPs
[present]

TSCA - Code of Federal Regulations Citations
40 CFR 716.120(a)

TSCA - Health and Safety Reporting List
Effective Date: June 1, 1987

INTERNATIONAL LISTS

Australian Exposure Standards - Time Weighted Averages
dust: 1 mg/m3 TWA; vapour: 4.3 ppm TWA; 20 mg/m3 TWA

Australian Exposure Standards - Short Term Exposure Limits
dust: 3 mg/m3 STEL; vapour: 8.6 ppm STEL; 40 mg/m3 STEL

Canada - WHMIS: Ingredient Disclosure
1% item 307 (381)

Canada - Alberta - 8 Hour Occupational Exposure Limit
as dust: 1 mg/m3 TWA; as vapor: 5 ppm TWA; 23 mg/m3 TWA

Canada - Alberta - 15 Minute Occupational Exposure Limit
as dust: 3 mg/m3 STEL; as vapour: 10 ppm STEL; 46 mg/m3

Canada - British Columbia - 8 Hour Exposure Limits
dust: 1 mg/m3 TWA; vapor: 5 ppm TWA; 20 mg/m3 TWA

Canada - British Columbia - 15 Minute Exposure Limits
dust: 3 mg/m3 STEL; vapour: 10 ppm STEL; 40 mg/m3 STEL

Canada - Ontario - OHSA - TWAEVs
0.25 ppm TWAEV; 1 mg/m3 TWAEV

Canada - Quebec - Time-Weighted Average Exposure Values
vapor: 4.3 ppm TWAEV; 20 mg/m3 TWAEV dust: 1 mg/m3 TWAEV

Canada - Quebec - Short-term Exposure Values
vapor: 8.6 ppm STEV, 40 mg/m3 STEV; dust: 3 mg/m3 STEV

United Kingdom - Occupational Exposure Standards - TWAs
dust: 1 mg/m3 TWA; vapour: 5 ppm TWA; 20 mg/m3 TWA

United Kingdom - Occupational Exposure Standards - STELs
dust: 3 mg/m3 STEL; vapour: 10 ppm STEL; 40 mg/m3 STEL

German (DFG) - MAK Values
vapor and dust: 5 mg/m3 MAK

German (DFG) - Pregnancy
no risk to embryo/fetus if exposure limits adhered to

Israel - Time Weighted Averages
dust: (1) mg/m3 TWA; vapor: (4.3) ppm TWA; (20) mg/m3 TWA

Israel - Short Term Exposure Limits
dust: (3) mg/m3 STEL; vapor: (8.6) ppm STEL; (40) mg/m3 STEL

Israel - Action Levels
dust: 0.5 mg/m3 AL; vapor: 2.15 ppm AL; 10 mg/m3 AL

Mexico - Instruction No. 10 - TWAs
as vapor: 5 ppm TWA; 20 mg/m3 TWA as dust: 1 mg/m3

Mexico - Instruction No. 10 - STELs
as dust: 3 mg/m3 STEL as vapor: 10 ppm STEL; 40 mg/m3 STEL

STATE LISTS

California - Air Bill 2588 Appendix A-I
6/91

California - Exposure Limits - PELs
dust: 1 mg/m3 PEL; vapor: 5 ppm PEL; 20 mg/m3 PEL

California - Exposure Limits - STELs
dust: 3 mg/m3 STEL; vapor: 10 ppm STEL; 40 mg/m3 STEL

California - Directors List of Hazardous Substances (8 CCR 339)
[present]

Florida Hazardous Substance List
[present]

Massachusetts Right To Know List
[present]

Minnesota Hazardous Substance List
[present]

Pennsylvania Right to Know List
[present]

CAPROLACTONE   502-44-3

HEALTH AND SAFETY LISTS

NIOSH - Selected LD50s and LC50s
Oral, rat: LD50 = 4290 mg/kg Skin, rabbit: LD50 = 5990 mg/kg

INTERNATIONAL LISTS

Canada - WHMIS: Ingredient Disclosure
1% item 308 (382)

CAPROLACTONE MODIFIED ACRYLATE   RR-00993-8
MONOMER

ENVIRONMENTAL LISTS

TSCA - Chemicals with Significant New Use Rules
PMN number: P-90-584

CAPROLACTONE, POLYMER WITH HEXAM-   RR-00217-5
ETHYLENE DIISOCYANATE, HYDROXYALKYL
ACRYLATE ESTER, REACTION PRODUCTS WITH
SUBSTITUTED ALKANOIC ACID AND METAL
HETEROMONOCYCLE

ENVIRONMENTAL LISTS

TSCA - Chemicals with Significant New Use Rules
PMN number: P-89-946

CAPRYLALDEHYDE   124-13-0

HEALTH AND SAFETY LISTS

NFPA - Flash Points
flash point = 125 degrees F (52 degrees C)

NFPA - Hazard Identification Ratings
health-2; flammability-2; reactivity-0

NIOSH - Selected LD50s and LC50s
Oral, rat: LD50 = 5630 mg/kg Skin, rabbit: LD50 = 6350 mg/kg

ENVIRONMENTAL LISTS

TSCA - Code of Federal Regulations Citations
40 CFR 712.30(x); 40 CFR 716.120(d)

TSCA - PAIR - Reporting List
Reporting Date: November 27, 1991

TSCA - Health and Safety Reporting List
Effective Date: September 30, 1991

STATE LISTS

Florida Hazardous Substance List
[present]

Massachusetts Right To Know List
[present]

CAPRYLIC ACID MONOGLYCERIDE   26402-26-6

ENVIRONMENTAL LISTS

List of Pesticide Product Inert Ingredients
[present]

CAPRYLIC/CAPRIC TRIGLYCERIDE   37332-31-3

ENVIRONMENTAL LISTS

List of Pesticide Product Inert Ingredients
[present]

CAPRYLYL CHLORIDE   111-64-8

HEALTH AND SAFETY LISTS

NFPA - Flash Points
flash point = 180 degrees F (82 degrees C)

NFPA - Hazard Identification Ratings
health-3; flammability-2; reactivity-1

STATE LISTS

Florida Hazardous Substance List
[present]

Massachusetts Right To Know List
[present]

Pennsylvania Right to Know List
[present]

CAPRYLYL PEROXIDE (N-OCTANOYL PEROXIDE)   7530-07-6

STATE LISTS

NJ Right to Know List (Total)
sn 2208

CAPTAFOL   2425-06-1

HEALTH AND SAFETY LISTS

ACGIH 1995 - Time Weighted Averages
0.1 mg/m3 TWA

ACGIH 1995 - Skin Designations
skin - potential for cutaneous absorption

IARC - Group 2A (limited human data)
[present]

NIOSH - Selected LD50s and LC50s
Oral, rat: LD50 = 2500 mg/kg

OSHA - Vacated PELs - Time Weighted Averages
0.1 mg/m3 TWA

OSHA - Possible Select Carcinogens
[present]

INTERNATIONAL LISTS

Australian Exposure Standards - Time Weighted Averages
0.1 mg/m3 TWA

Australian Exposure Standards - Skin Effects
skin absorption

Canada - Alberta - 8 Hour Occupational Exposure Limit
0.1 mg/m3 TWA

Canada - Alberta - 15 Minute Occupational Exposure Limit
0.3 mg/m3 STEL

Canada - Alberta - Skin Designation
can be absorbed through the intact skin

Canada - British Columbia - 8 Hour Exposure Limits
0.1 mg/m3 TWA

Canada - British Columbia - Skin Notations
skin - potential for skin absorption

Canada - Ontario - OHSA - TWAEVs
0.1 mg/m3 TWAEV

Canada - Ontario - OHSA - Skin Notations
absorption through skin, eyes, or mucous membranes

Canada - Quebec - Time-Weighted Average Exposure Values
0.1 mg/m3 TWAEV

Canada - Quebec - Skin Designations
absorbed through the skin

United Kingdom - Occupational Exposure Standards - TWAs
0.1 mg/m3 TWA

United Kingdom - Occupational Exposure Standards - Notes
can be absorbed through skin

Israel - Time Weighted Averages
0.1 mg/m3 TWA

Israel - Action Levels
0.05 mg/m3 AL

Mexico - Instruction No. 10 - TWAs
0.1 mg/m3 TWA

Mexico - Instruction No. 10 - Skin designation
skin - potential for cutaneous absorption

STATE LISTS

California - Air Bill 2588 Appendix A-I
known or potential carcinogen: 9/89

California - Prop. 65 - Cancer list
carcinogen - initial date 10/1/88

California - Prop. 65 - No Significant Risk Levels
no significant risk level = 5 ug/day

California - Exposure Limits - PELs
0.1 mg/m3 PEL

California - Exposure Limits - Skin Notation
material may be absorbed through the skin, eyes or mucous membrane

California - Directors List of Hazardous Substances (8 CCR 339)
[present]

Florida Hazardous Substance List
[present]

Massachusetts Right To Know List
[present]

Minnesota Hazardous Substance List
skin

NJ Right to Know List (Total)
sn 0338

Pennsylvania Right to Know List
[present]

CAPTAFOL, CIS-   2939-80-2

STATE LISTS

Massachusetts Right To Know List
[present]

CAPTAN   133-06-2

HEALTH AND SAFETY LISTS

ACGIH 1995 - Time Weighted Averages
5 mg/m3 TWA

U.S. DOT - Appendix A Table 1 - Hazardous Substances
final RQ = 10 pounds (4.54 kg)

IARC - Group 3 (not classifiable)
[present]

NIOSH - Selected LD50s and LC50s
Inhalation, mouse: LC50 = 5000 mg/m3 (2 hr) Oral, rat: LD50 = 9 gm/kg

NTP Chemical Status Reports - Testing Status and NTIS Number
Technical reports printed (PB273475/AS)

NTP Chemical Status Reports - Evidence of Carcinogenicity
male rat-negative; female rat-negative; male mice-positive; female mice-positive

OSHA - Vacated PELs - Time Weighted Averages
5 mg/m3 TWA

ENVIRONMENTAL LISTS

CERCLA/SARA - Section 313 - Emission Reporting
form R reporting required for 1.0% de minimus concentration

CERCLA/SARA - Hazardous Substances and their Reportable Quantities
final RQ = 10 pounds (4.54 kg)

Clean Air Act (1990) - List of Hazardous Air Contaminants
[present]

Clean Water Act - Hazardous Substances
[present]

INTERNATIONAL LISTS

Australian Exposure Standards - Time Weighted Averages
5 mg/m3 TWA

Canada - Alberta - 8 Hour Occupational Exposure Limit
5 mg/m3 TWA

Canada - Alberta - 15 Minute Occupational Exposure Limit
15 mg/m3 STEL

Canada - British Columbia - 8 Hour Exposure Limits
5 mg/m3 TWA

Canada - British Columbia - 15 Minute Exposure Limits
15 mg/m3 STEL

Canada - Ontario - OHSA - TWAEVs
5 mg/m3 TWAEV

Canada - Quebec - Time-Weighted Average Exposure Values
5 mg/m3 TWAEV

United Kingdom - Occupational Exposure Standards - TWAs
5 mg/m3 TWA

United Kingdom - Occupational Exposure Standards - STELs
15 mg/m3 STEL

Israel - Time Weighted Averages
5 mg/m3 TWA

Israel - Action Levels
2.5 mg/m3 AL

Mexico - Instruction No. 10 - TWAs
5 mg/m3 TWA

Mexico - Instruction No. 10 - STELs
15 mg/m3 STEL

STATE LISTS

California - Air Bill 2588 Appendix A-I
known or potential carcinogen: 9/90

California - Prop. 65 - Cancer list
carcinogen - initial date 1/1/90

California - Prop. 65 - No Significant Risk Levels
no significant risk level = 300 ug/day

California - Exposure Limits - PELs
5 mg/m3 PEL

California - Directors List of Hazardous Substances (8 CCR 339)
[present]

Florida Hazardous Substance List
[present]

Massachusetts Right To Know List
carcinogen; extraordinarily hazardous

Minnesota Hazardous Substance List
[present]

NJ Right to Know List (Total)
sn 0339

NJ Special Hazardous Substances
(mutagen)

Pennsylvania Right to Know List
environmental hazard

CARAMEL   8028-89-5

ENVIRONMENTAL LISTS

List of Pesticide Product Inert Ingredients
[present]

CARBACHOL CHLORIDE   51-83-2

HEALTH AND SAFETY LISTS

NIOSH - Selected LD50s and LC50s
Oral, rat: LD50 = 40 mg/kg

ENVIRONMENTAL LISTS

CERCLA/SARA - Section 302 Extremely Hazardous Substances and TPQs
TPQ = 500/10,000 pounds

STATE LISTS

Florida Hazardous Substance List
effective March 13, 1992

Massachusetts Right To Know List
extraordinarily hazardous

Pennsylvania Right to Know List
environmental hazard

PROPOSED REGULATIONS

CERCLA/SARA - Proposed Hazardous Substance Additions
proposed RQ = 1 pound (.454 kg)

CERCLA/SARA - 1989 Proposed RQ Adjustments
proposed RQ = 100 pounds (45.4 kg)

CARBAMATE PESTICIDES, N.O.S.   RR-00186-5

HEALTH AND SAFETY LISTS

U.S. DOT - Substances From 49 CFR 172.101
regulated by DOT (UN2757, UN2758, UN2991, UN2992)

U.S. DOT - Hazard Classes
toxic or toxic, flammable: DOT hazard class = 6.1; flammable, toxic: DOT hazard class = 3

STATE LISTS

NJ Right to Know List (Total)
sn 2210; sn 2211; sn 2212; sn 2213

CARBAMIC ACID, MANGANESE SALT   78812-39-2

ENVIRONMENTAL LISTS

List of Pesticide Product Inert Ingredients
[present]

CARBAMIC ACID, METHYL-, O-(((2,4-DIMETHYL-1,   26419-73-8
3-DITHIOLAN-2-YL)METHYLENE)AMINO)-

HEALTH AND SAFETY LISTS

NIOSH - Selected LD50s and LC50s
Oral, rat: LD50 = 1 mg/kg Skin, rat: LD50 = 300 mg/kg

ENVIRONMENTAL LISTS

CERCLA/SARA - Section 302 Extremely Hazardous Substances and TPQs
TPQ = 100/10,000 pounds

STATE LISTS

Florida Hazardous Substance List
effective March 13, 1992

Massachusetts Right To Know List
extraordinarily hazardous

Pennsylvania Right to Know List
environmental hazard

PROPOSED REGULATIONS
CERCLA/SARA - Proposed Hazardous Substance Additions
proposed RQ = 1 pound (.454 kg)
CERCLA/SARA - 1989 Proposed RQ Adjustments
proposed RQ = 100 pounds (45.4 kg)

CARBAMIC ACID, (TRIALKYLOXYSILYALKYL)-   RR-00225-5
SUBSTITUTED ACRYLATEESTER
ENVIRONMENTAL LISTS
TSCA - Chemicals with Significant New Use Rules
PMN number: P-89-424

CARBAMIMIDOSELENOIC ACID   630-10-4
SEE ALSO:
SELENIUM
HEALTH AND SAFETY LISTS
U.S. DOT - Appendix A Table 1 - Hazardous Substances
final RQ = 1000 pounds (454 kg)
ENVIRONMENTAL LISTS
CERCLA/SARA - Hazardous Substances and their Reportable Quantities
final RQ = 1000 pounds (454 kg)
RCRA - P Series Wastes
waste number P103
RCRA - Hazardous Constituents-Appendix VIII
waste number P103
RCRA - Substances Banned From Land Disposal
[present]
STATE LISTS
Massachusetts Right To Know List
[present]
Pennsylvania Right to Know List
environmental hazard

CARBAMODITHIOIC ACID, 1,2-ETHANEDIYL   34731-32-3
ESTER
STATE LISTS
Pennsylvania Right to Know List
environmental hazard; special hazardous substance
Pennsylvania RTK - Special Hazardous Substances
[present]

CARBARYL   63-25-2
HEALTH AND SAFETY LISTS
ACGIH 1995 - Time Weighted Averages
5 mg/m3 TWA
U.S. DOT - Appendix B - Marine Pollutants
DOT regulated marine pollutant
U.S. DOT - Appendix A Table 1 - Hazardous Substances
final RQ = 100 pounds (45.4 kg)
IARC - Group 3 (not classifiable)
[present]
NIOSH - Selected LD50s and LC50s
Oral, rat: LD50 = 250 mg/kg Skin, rat: LD50 = 4000 mg/kg
NIOSH 1990 - Pocket Guide - RELs
5 mg/m3 TWA
NIOSH 1990 - Pocket Guide - IDLHs
600 mg/m3 IDLH
NIOSH 1990 - Pocket Guide - Target organs
respiratory system, CNS, CVS, skin
NIOSH - Health Standards - Exposure Limits
5 mg/m3 TWA

NIOSH - Health Standards - Health Effects and Precautions
Central nervous system and reproductive system effects (Permit only minimum exposure during pregnancy; prevent skin and eye contact)
OSHA - Vacated PELs - Time Weighted Averages
5 mg/m3 TWA
OSHA - Final PELs - Time Weighted Averages
5 mg/m3 TWA
ENVIRONMENTAL LISTS
CERCLA/SARA - Section 313 - Emission Reporting
form R reporting required for 1.0% de minimus concentration
CERCLA/SARA - Hazardous Substances and their Reportable Quantities
final RQ = 100 pounds (45.4 kg)
Clean Air Act (1990) - List of Hazardous Air Contaminants
[present]
CAA - HON Rule - SOCMI Chemicals
compliance by Oct. 23, 1995
Clean Water Act - Hazardous Substances
[present]
INTERNATIONAL LISTS
Australian Exposure Standards - Time Weighted Averages
5 mg/m3 TWA
Australian Exposure Standards - Under Review
exposure limits under review
Canada - Drinking Water Quality - MACs
0.09 mg/L MAC
Canada - Alberta - 8 Hour Occupational Exposure Limit
5 mg/m3 TWA
Canada - Alberta - 15 Minute Occupational Exposure Limit
10 mg/m3 STEL
Canada - British Columbia - 8 Hour Exposure Limits
5 mg/m3 TWA
Canada - British Columbia - 15 Minute Exposure Limits
10 mg/m3 STEL
Canada - Ontario - OHSA - TWAEVs
5 mg/m3 TWAEV
Canada - Quebec - Time-Weighted Average Exposure Values
5 mg/m3 TWAEV
United Kingdom - Occupational Exposure Standards - TWAs
5 mg/m3 TWA
United Kingdom - Occupational Exposure Standards - STELs
10 mg/m3 STEL
German (DFG) - MAK Values
total dust: 5 mg/m3 MAK
German (DFG) - Skin/Sensitizers
danger of cutaneous absorption
Israel - Time Weighted Averages
5 mg/m3 TWA
Israel - Action Levels
2.5 mg/m3 AL
Mexico - Instruction No. 10 - TWAs
5 mg/m3 TWA
Mexico - Instruction No. 10 - STELs
10 mg/m3 STEL
STATE LISTS
California - Air Bill 2588 Appendix A-I
6/91
California - Exposure Limits - PELs
5 mg/m3 PEL
California - Directors List of Hazardous Substances (8 CCR 339)
[present]
Florida Hazardous Substance List
[present]

Massachusetts Right To Know List
neurotoxin

Minnesota Hazardous Substance List
[present]

NJ Right to Know List (Total)
sn 0340

Pennsylvania Right to Know List
environmental hazard

9H-CARBAZOLE   86-74-8

HEALTH AND SAFETY LISTS

IARC - Group 3 (not classifiable)
[present]

ENVIRONMENTAL LISTS

ATSDR Priority List
Rank (of 275): 246

CAA - HON Rule - SOCMI Chemicals
compliance by Oct. 23, 1995

TSCA - Code of Federal Regulations Citations
40 CFR 716.120(a)

TSCA - Health and Safety Reporting List
Effective Date: March 7, 1986

STATE LISTS

California - Directors List of Hazardous Substances (8 CCR 339)
[present]

3-CARBETHOXYPSORALEN   RR-01528-1

HEALTH AND SAFETY LISTS

IARC - Group 3 (not classifiable)
[present]

CARBINOL ACETATE   112-15-2
SEE ALSO:
GLYCOL ETHERS

HEALTH AND SAFETY LISTS

NIOSH - Selected LD50s and LC50s
Oral, rat: LD50 = 11 gm/kg Skin, rabbit: LD50 = 15 gm/kg

ENVIRONMENTAL LISTS

CAA - HON Rule - SOCMI Chemicals
compliance by Oct. 24, 1994

TSCA - Code of Federal Regulations Citations
40 CFR 712.30(x); 40 CFR 716.120(d)

TSCA - PAIR - Reporting List
Effective Date: September 30, 1991; Reporting Date: November 27, 1991

TSCA - Health and Safety Reporting List
Effective Date: September 30, 1991; Sunset Date: September 30, 2001

CARBOFURAN   1563-66-2

HEALTH AND SAFETY LISTS

ACGIH 1995 - Time Weighted Averages
0.1 mg/m3 TWA

U.S. DOT - Appendix B - Marine Pollutants
DOT regulated marine pollutant

U.S. DOT - Appendix A Table 1 - Hazardous Substances
final RQ = 10 pounds (4.54 kg)

NIOSH - Selected LD50s and LC50s
Inhalation, rat: LC50 = 85 mg/m3 Oral, rat: LD50 = 5300 ug/kg Skin, rat: LD50 = 120 mg/kg

OSHA - Vacated PELs - Time Weighted Averages
0.1 mg/m3 TWA

ENVIRONMENTAL LISTS

CERCLA/SARA - Section 302 Extremely Hazardous Substances and TPQs
TPQ = 10/10,000 pounds

CERCLA/SARA - Section 313 - Emission Reporting
form R reporting required

CERCLA/SARA - Hazardous Substances and their Reportable Quantities
final RQ = 10 pounds (4.54 kg)

Clean Water Act - Hazardous Substances
[present]

Safe Drinking Water Act - MCLs
MCL = 0.04 mg/L

Safe Drinking Water Act - MCLGs
MCLG = 0.04 mg/L

INTERNATIONAL LISTS

Australian Exposure Standards - Time Weighted Averages
0.1 mg/m3 TWA

Canada - Drinking Water Quality - MACs
0.09 mg/L MAC

Canada - Alberta - 8 Hour Occupational Exposure Limit
0.1 mg/m3 TWA

Canada - Alberta - 15 Minute Occupational Exposure Limit
0.3 mg/m3 STEL

Canada - British Columbia - 8 Hour Exposure Limits
0.1 mg/m3 TWA

Canada - Ontario - OHSA - TWAEVs
0.1 mg/m3 TWAEV

Canada - Quebec - Time-Weighted Average Exposure Values
0.1 mg/m3 TWAEV

United Kingdom - Occupational Exposure Standards - TWAs
0.1 mg/m3 TWA

Israel - Time Weighted Averages
0.1 mg/m3 TWA

Israel - Action Levels
0.05 mg/m3 AL

Mexico - Instruction No. 10 - TWAs
0.1 mg/m3 TWA

STATE LISTS

California - Exposure Limits - PELs
0.1 mg/m3 PEL

California - Directors List of Hazardous Substances (8 CCR 339)
[present]

Florida Hazardous Substance List
[present]

Massachusetts Right To Know List
extraordinarily hazardous; neurotoxin

Minnesota Hazardous Substance List
[present]

NJ Right to Know List (Total)
sn 0341

Pennsylvania Right to Know List
environmental hazard

CARBON   7440-44-0

HEALTH AND SAFETY LISTS

U.S. DOT - Substances From 49 CFR 172.101
regulated by DOT (UN1362, UN1361)

U.S. DOT - Hazard Classes
DOT hazard class = 4.2

NIOSH 1990 - Pocket Guide - Target organs
respiratory system, CVS

INTERNATIONAL LISTS
German (DFG) - MAK Values
fine dust: 6 mg/m3 MAK

Mexico - Drinking Water - Ecological Criteria
Extractable in alcohol: 1.5 mg/l; Extractable in chloroform: 3.0 mg/l

STATE LISTS
California - Exposure Limits - PELs
3.5 mg/m3 PEL

CARBON 11   14333-33-6

HEALTH AND SAFETY LISTS
U.S. DOT - Appendix A Table 2 - Radionuclides
final RQ = 1000 curies (3.7E 13 Bq)

ENVIRONMENTAL LISTS
CERCLA/SARA List of Radionuclides (Appendix B) and Their Reportable Quantities
final RQ = 1000 curies (3.7E 13 Bq)

CARBON 14   14762-75-5

HEALTH AND SAFETY LISTS
U.S. DOT - Appendix A Table 2 - Radionuclides
final RQ = 10 curies (3.7E 11 Bq)

ENVIRONMENTAL LISTS
ATSDR Priority List
Rank (of 275): 198

CERCLA/SARA List of Radionuclides (Appendix B) and Their Reportable Quantities
final RQ = 10 curies (3.7E 11 Bq)

CARBON BLACK   1333-86-4

HEALTH AND SAFETY LISTS
ACGIH 1995 - Time Weighted Averages
3.5 mg/m3 TWA

IARC - Group 3 (not classifiable)
[present]

NIOSH 1990 - Pocket Guide - RELs
3.5 mg/m3 TWA

NIOSH 1990 - Pocket Guide - Target organs
none known

NIOSH - Health Standards - Exposure Limits
3.5 mg/m3 TWA; in presence of polycyclic aromatic hydrocarbons: 0.1 mg/m3 TWA

NIOSH - Health Standards - Health Effects and Precautions
Lung, cardiovascular, and skin effects (Periodic chest X-ray, pulmonary function testing, and ECG required)

NIOSH - Health Standards - Carcinogenic Chemicals
in presence of polycyclic aromatic hydrocarbons: potential human carcinogen

OSHA - Vacated PELs - Time Weighted Averages
3.5 mg/m3 TWA

OSHA - Final PELs - Time Weighted Averages
3.5 mg/m3 TWA

ENVIRONMENTAL LISTS
List of Pesticide Product Inert Ingredients
[present]

INTERNATIONAL LISTS
Australian Exposure Standards - Time Weighted Averages
3 mg/m3 TWA

Canada - WHMIS: Ingredient Disclosure
1% item 309 (1271)

Canada - Alberta - 8 Hour Occupational Exposure Limit
3.5 mg/m3 TWA

Canada - Alberta - 15 Minute Occupational Exposure Limit
7 mg/m3 STEL

Canada - British Columbia - 8 Hour Exposure Limits
3.5 mg/m3 TWA

Canada - British Columbia - 15 Minute Exposure Limits
7 mg/m3 STEL

Canada - Ontario - OHSA - TWAEVs
3.5 mg/m3 TWAEV

Canada - Quebec - Time-Weighted Average Exposure Values
3.5 mg/m3 TWAEV

United Kingdom - Occupational Exposure Standards - TWAs
3.5 mg/m3 TWA

United Kingdom - Occupational Exposure Standards - STELs
7 mg/m3 STEL

Israel - Time Weighted Averages
3.5 mg/m3 TWA

Israel - Action Levels
1.75 mg/m3 AL

Mexico - Instruction No. 10 - TWAs
3.5 mg/m3 TWA

Mexico - Instruction No. 10 - STELs
7 mg/m3 STEL

STATE LISTS
California - Directors List of Hazardous Substances (8 CCR 339)
[present] (exempt when in form where exposure to dust cannot occur)

Massachusetts Right To Know List
[present] Exempt when encapsulated or if particulates are not present and cannot be substantially generated through use of the product.

Minnesota Hazardous Substance List
[present]

Pennsylvania Right to Know List
[present]

CARBON BLACK EXTRACTS   RR-00060-2

HEALTH AND SAFETY LISTS
IARC - Group 2B (sufficient animal data)
[present] (Listed under 'Carbon blacks')

OSHA - Possible Select Carcinogens
[present]

STATE LISTS
California - Air Bill 2588 Appendix A-I
known or potential carcinogen

California - Prop. 65 - Cancer list
carcinogen - initial date 1/1/90

Pennsylvania Right to Know List
special hazardous substance

Pennsylvania RTK - Special Hazardous Substances
[present]

CARBON DIOXIDE   124-38-9

HEALTH AND SAFETY LISTS
ACGIH 1995 - Time Weighted Averages
5000 ppm TWA; 9000 mg/m3 TWA

ACGIH 1995 - Short Term Exposure Limits
30,000 ppm STEL; 54,000 mg/m3 STEL

U.S. DOT - Substances From 49 CFR 172.101
regulated by DOT (UN1041, UN1952, UN2187, UN1845, UN1015, UN1013)

U.S. DOT - Hazard Classes
DOT hazard class = 2.2

U.S. DOT - Substances Which Are Poisonous by Inhalation
gaseous hazardous material poisonous by inhalation (when mixed with Ethylene oxide) (UN1041)

NIOSH 1990 - Pocket Guide - RELs
5000 ppm TWA; 9000 mg/m3 TWA; 30,000 ppm STEL; 54,000 mg/m3 STEL

NIOSH 1990 - Pocket Guide - IDLHs
50,000 ppm IDLH

NIOSH 1990 - Pocket Guide - Target organs
lungs, skin, CVS

NIOSH - Health Standards - Exposure Limits
10,000 ppm TWA; 18,000 mg/m3 TWA; C (10 min) 30,000 ppm; C (10 min) 54,000 mg/m3

NIOSH - Health Standards - Health Effects and Precautions
Respiratory effects

OSHA - Vacated PELs - Time Weighted Averages
10,000 ppm TWA; 18,000 mg/m3 TWA

OSHA - Vacated PELs - Short Term Exposure Limits
30,000 ppm STEL; 54,000 mg/m3 STEL

OSHA - Final PELs - Time Weighted Averages
5000 ppm TWA (exposures < 10,000 ppm to be cited de minimus); 9000 mg/m3 TWA

ENVIRONMENTAL LISTS

List of Pesticide Product Inert Ingredients
[present]

INTERNATIONAL LISTS

Australian Exposure Standards - Time Weighted Averages
5000 ppm TWA; 9000 mg/m3 TWA; in coal mines: 12,500 ppm TWA; 22,500 mg/m3 TWA

Australian Exposure Standards - Short Term Exposure Limits
30,000 ppm STEL; 54,000 mg/m3 STEL; in coal mines: 30,000 ppm STEL; 54,000 mg/m3 STEL

Canada - WHMIS: Ingredient Disclosure
1% item 310 (770)

Canada - Alberta - 8 Hour Occupational Exposure Limit
5000 ppm TWA; 9000 mg/m3 TWA

Canada - Alberta - 15 Minute Occupational Exposure Limit
15000 ppm STEL; 27000 mg/m3 STEL

Canada - British Columbia - 8 Hour Exposure Limits
5000 ppm TWA; 9000 mg/m3 TWA

Canada - British Columbia - 15 Minute Exposure Limits
15000 ppm STEL; 27000 mg/m3 STEL

Canada - Ontario - OHSA - TWAEVs
5000 ppm TWAEV; 9000 mg/m3 TWAEV

Canada - Ontario - OHSA - STEVs
30000 ppm STEV; 54000 mg/m3 STEV

Canada - Quebec - Time-Weighted Average Exposure Values
5000 ppm TWAEV; 9000 mg/m3 TWAEV

Canada - Quebec - Short-term Exposure Values
30000 ppm STEV; 54000 mg/m3 STEV

United Kingdom - Occupational Exposure Standards - TWAs
5000 ppm TWA; 9000 mg/m3 TWA

United Kingdom - Occupational Exposure Standards - STELs
15,000 ppm STEL; 27,000 mg/m3 STEL

German (DFG) - MAK Values
5000 ppm MAK; 9000 mg/m3 MAK

German (DFG) - Peak Limitations
2 x normal MAK (1 hour momentary value); don't exceed 3 times per shift

Israel - Time Weighted Averages
5000 ppm TWA; 9000 mg/m3 TWA

Israel - Short Term Exposure Limits
30,000 ppm STEL; 54,000 mg/m3 STEL

Israel - Action Levels
2500 ppm AL; 4500 mg/m3 AL

Mexico - Instruction No. 10 - TWAs
5000 ppm TWA; 9000 mg/m3 TWA

Mexico - Instruction No. 10 - STELs
15000 ppm STEL; 27000 mg/m3 STEL

STATE LISTS

California - Exposure Limits - PELs
5000 ppm PEL; 9000 mg/m3 PEL

California - Exposure Limits - STELs
30,000 ppm STEL; 54,000 mg/m3 STEL

California - Directors List of Hazardous Substances (8 CCR 339)
[present]

Florida Hazardous Substance List
[present]

Massachusetts Right To Know List
[present]

Minnesota Hazardous Substance List
[present]

NJ Right to Know List (Total)
sn 0343

Pennsylvania Right to Know List
[present]

PROPOSED REGULATIONS

Canada - Ontario - Proposed Occupational STEVs
10000 ppm STEV; 18000 mg/m3 STEV

CARBON DIOXIDE AND ETHYLENE OXIDE MIXTURES   8070-50-6

HEALTH AND SAFETY LISTS

U.S. DOT - Substances From 49 CFR 172.101
regulated by DOT (UN1041, UN1952)

U.S. DOT - Hazard Classes
between 6% and 25% Ethylene oxide: DOT hazard class = 2.1; with more than 25% Ethylene oxide: DOT hazard class = 2.3

CARBON DISULFIDE   75-15-0
SEE ALSO:
F039-HAZARDOUS WASTES
F005-HAZARDOUS WASTES

HEALTH AND SAFETY LISTS

ACGIH 1995 - Time Weighted Averages
10 ppm TWA; 31 mg/m3 TWA

ACGIH 1995 - Skin Designations
skin - potential for cutaneous absorption

ACGIH 1995 - Biological Exposure Indices
2-Thiothiazolidine-4-carboxylic acid in urine: 5 mg/g creatinine, end of shift

AIHA - Odor Threshold Values
no geometric mean air odor threshold

U.S. DOT - Substances From 49 CFR 172.101
regulated by DOT (UN1131)

U.S. DOT - Hazard Classes
DOT hazard class = 3

U.S. DOT - Appendix B - Marine Pollutants
DOT regulated marine pollutant

U.S. DOT - Appendix A Table 1 - Hazardous Substances
final RQ = 100 pounds (45.4 kg)

NFPA - Flash Points
flash point = -22 degrees F (-30 degrees C)

NFPA - Hazard Identification Ratings
health-3; flammability-3; reactivity-0

NIOSH - Selected LD50s and LC50s
Oral, rat: LD50 = 3188 mg/kg

NIOSH 1990 - Pocket Guide - RELs
1 ppm TWA; 3 mg/m3 TWA; 10 ppm STEL; 30 mg/m3 STEL

NIOSH 1990 - Pocket Guide - IDLHs
500 ppm IDLH

NIOSH 1990 - Pocket Guide - Target organs
CNS, PNS, CVS, eyes, kidneys, liver, skin

NIOSH 1990 - Pocket Guide - Skin list
Potential for dermal absorption

NIOSH - Health Standards - Exposure Limits
1 ppm TWA; 3 mg/m3 TWA; C (15 min) 10 ppm; C (15 min) 30 mg/m3

NIOSH - Health Standards - Health Effects and Precautions
Cardiovascular, central nervous system, and reproductive system effects (Warn workers of potential reproductive system effects)

NTP Chemical Status Reports - Testing Status and NTIS Number
Prechronic studies for which toxicity technical reports were not prepared; prechronic studies completed: in review for further evaluation

OSHA - Vacated PELs - Time Weighted Averages
4 ppm TWA; 12 mg/m3 TWA

OSHA - Vacated PELs - Short Term Exposure Limits
12 ppm STEL; 36 mg/m3 STEL

OSHA - Vacated PELs - Skin Designation
Prevent or reduce skin absorption

OSHA - Final PELs - Time Weighted Averages
20 ppm TWA; C 30 ppm

OSHA - Final PELs - Ceiling Limits
C 30 ppm

ENVIRONMENTAL LISTS

ATSDR Priority List
Rank (of 275): 158

CERCLA/SARA - Section 302 Extremely Hazardous Substances and TPQs
TPQ = 10,000 pounds

CERCLA/SARA - Section 313 - Emission Reporting
form R reporting required for 1.0% de minimus concentration

CERCLA/SARA - Hazardous Substances and their Reportable Quantities
final RQ = 100 pounds (45.4 kg)

Clean Air Act (1990) - List of Hazardous Air Contaminants
[present]

CAA -Toxic Substances for Accidental Release Prevention
threshold quantity = 20,000 lbs

CAA - HON Rule - SOCMI Chemicals
compliance by July 24, 1995

CAA - HON Rule - Organic HAPs
[present]

Clean Water Act - Hazardous Substances
[present]

EPA - Master Testing List
[present]

RCRA - P Series Wastes
waste number P022

RCRA - Hazardous Constituents-Appendix VIII
waste number P022

RCRA - Basis for Listing - Appendix VII
Included in waste stream: F005, F039

RCRA - Substances Banned From Land Disposal
[present]

RCRA - TSD Facilities Ground Water Monitoring
TM 8240 = 5 ug/L PQL

RCRA - Universal Treatment Standards (LDR)
WW: 3.8 mg/l; NWW: 4.8 mg/l TCLP

TSCA - Health and Safety Reporting List
Effective Date: March 11, 1994;   Sunset Date: March 11, 2004

INTERNATIONAL LISTS

Australian Exposure Standards - Time Weighted Averages
10 ppm TWA;   31 mg/m3 TWA

Australian Exposure Standards - Skin Effects
skin absorption

Australian Exposure Standards - Under Review
exposure limits under review

Canada - WHMIS: Ingredient Disclosure
0.1% item 311 (787)

Canada - NPRI (National Pollutant Release Inventory)
[present]

Canada - Alberta - 8 Hour Occupational Exposure Limit
10 ppm TWA;   31 mg/m3 TWA

Canada - Alberta - 15 Minute Occupational Exposure Limit
20 ppm STEL;   62 mg/m3 STEL

Canada - Alberta - Skin Designation
can be absorbed through the intact skin

Canada - British Columbia - 8 Hour Exposure Limits
20 ppm TWA;   60 mg/m3 TWA

Canada - British Columbia - 15 Minute Exposure Limits
30 ppm STEL;   90 mg/m3 STEL

Canada - British Columbia - Skin Notations
skin - potential for skin absorption

Canada - Ontario - OHSA - TWAEVs
10 ppm TWAEV;   31 mg/m3 TWAEV

Canada - Ontario - OHSA - Skin Notations
absorption through skin, eyes, or mucous membranes

Canada - Quebec - Time-Weighted Average Exposure Values
4 ppm TWAEV;   12 mg/m3 TWAEV

Canada - Quebec - Short-term Exposure Values
12 ppm STEV;   36 mg/m3 STEV

Canada - Quebec - Skin Designations
absorbed through the skin

United Kingdom - Maximum Exposure Limits - TWAs
10 ppm TWA;   30 mg/m3 TWA

United Kingdom - Maximum Exposure Limits - Notes
can be absorbed through skin

German (DFG) - MAK Values
10 ppm MAK;   30 mg/m3 MAK

German (DFG) - Peak Limitations
2 x normal MAK (30 min. average value);   don't exceed 4 times during shift

German (DFG) - Skin/Sensitizers
danger of cutaneous absorption

German (DFG) - Pregnancy
risk to embryo/fetus probable

Israel - Time Weighted Averages
10 ppm TWA;   31 mg/m3 TWA

Israel - Action Levels
5 ppm AL;   15.5 mg/m3 AL

Mexico - Instruction No. 10 - TWAs
10 ppm TWA;   30 mg/m3 TWA

STATE LISTS

California - Air Bill 2588 Appendix A-I
9/89

California - Prop. 65 - Developmental Toxicity
developmental toxicity - initial date 7/1/89

California - Prop. 65 - Reproductive - Female
female reproductive toxicity - initial date 7/1/89

California - Prop. 65 - Reproductive - Male
male reproductive toxicity - initial date 7/1/89

California - Exposure Limits - PELs
4 ppm PEL;   12 mg/m3 PEL

California - Exposure Limits - STELs
12 ppm STEL;   36 mg/m3 STEL

California - Exposure Limits - Ceilings
C 30 ppm

California - Exposure Limits - Skin Notation
material may be absorbed through the skin, eyes or mucous membrane

California - Directors List of Hazardous Substances (8 CCR 339)
[present]

Florida Hazardous Substance List
[present]

Massachusetts Right To Know List
extraordinarily hazardous

Minnesota Hazardous Substance List
skin

NJ Right to Know List (Total)
sn 0344

NJ Special Hazardous Substances
(flammable - third degree)

Pennsylvania Right to Know List
environmental hazard

PROPOSED REGULATIONS

TSCA - Proposed Substances for Developmental/ReproductiveTesting
proposed testing for: Developmental Toxicity - inhalation; Reproductive Toxicity - inhalation

TSCA - ITC 32nd Report Priority Testing List
designated for dermal absorption testing

Canada - Ontario - Proposed Occupational TWAEVs
5 ppm TWAEV; 16 mg/m3 TWAEV

Canada - Ontario - Proposed Occupational STEVs
8 ppm STEV; 25 mg/m3 STEV

CARBONIC ACID, CALCIUM SALT (1:1)   471-34-1

HEALTH AND SAFETY LISTS

NIOSH - Selected LD50s and LC50s
Oral, rat: LD50 = 6450 mg/kg

ENVIRONMENTAL LISTS

List of Pesticide Product Inert Ingredients
[present]

INTERNATIONAL LISTS

Canada - Alberta - 8 Hour Occupational Exposure Limit
respirable mass: 5 mg/m3 TWA; total mass: 10 mg/m3 TWA

Canada - British Columbia - 8 Hour Exposure Limits
nuisance dust, mists, and fumes: 10 mg/m3 TWA

Canada - British Columbia - 15 Minute Exposure Limits
20 mg/m3 STEL

Canada - Ontario - OHSA - TWAEVs
total dust: 10 mg/m3 TWAEV (listed as a nuisance particulate)

Canada - Quebec - Time-Weighted Average Exposure Values
total dust: 10 mg/m3 TWAEV; respirable dust: 5 mg/m3 TWAEV

United Kingdom - Occupational Exposure Standards - TWAs
total inhalable dust: 10 mg/m3 TWA; respirable dust: 5 mg/m3 TWA

Mexico - Instruction No. 10 - TWAs
10 mg/m3 TWA (nuisance particulate)

Mexico - Instruction No. 10 - STELs
20 mg/m3 STEL

CARBONIC ACID, NICKEL SALT   16337-84-1

STATE LISTS

Pennsylvania Right to Know List
environmental hazard; special hazardous substance

Pennsylvania RTK - Special Hazardous Substances
[present]

CARBONIC ACID, NICKEL(2+) SALT (2:1)   17237-93-3

STATE LISTS

Pennsylvania Right to Know List
environmental hazard; special hazardous substance

Pennsylvania RTK - Special Hazardous Substances
[present]

CARBONIC DIHYDRAZIDE   497-18-7

HEALTH AND SAFETY LISTS

U.S. DOT - Hazard Classes
Forbidden from transport by the DOT

CARBON MONOXIDE   630-08-0

HEALTH AND SAFETY LISTS

ACGIH 1995 - Time Weighted Averages
25 ppm TWA; 29 mg/m3 TWA

ACGIH 1995 - Biological Exposure Indices
Carboxyhemoglobin in blood: 3.5% of hemoglobin, end of shift (B, Ns); Carbon monoxide in end-exhaled air: 20 ppm, end of shift (B, Ns)

U.S. DOT - Substances From 49 CFR 172.101
regulated by DOT (UN1016, NA9202)

U.S. DOT - Hazard Classes
DOT hazard class = 2.3

U.S. DOT - Substances Which Are Poisonous by Inhalation
gaseous hazardous material poisonous by inhalation (regular, refrigerated liquid or > mixed with Hydrogen) (UN1016, NA9202, UN2600)

NFPA - Flash Points
gas (no flash point given)

NFPA - Hazard Identification Ratings
health-3; flammability-4; reactivity-0

NIOSH - Selected LD50s and LC50s
Inhalation, rat: LC50 = 1807 ppm 4 hr

NIOSH 1990 - Pocket Guide - RELs
35 ppm TWA; 40 mg/m3 TWA; C 200 ppm; C 229 mg/m3

NIOSH 1990 - Pocket Guide - IDLHs
1500 ppm IDLH

NIOSH 1990 - Pocket Guide - Target organs
CVS, lungs, blood, CNS

NIOSH - Health Standards - Exposure Limits
35 ppm TWA (8 hr); 40 mg/m3 TWA (8 hr) C 200 ppm; C 229 mg/m3

NIOSH - Health Standards - Health Effects and Precautions
Cardiovascular effects

OSHA - Vacated PELs - Time Weighted Averages
35 ppm TWA; 40 mg/m3 TWA

OSHA - Vacated PELs - Ceiling Limits
C 200 ppm; C 229 mg/m3

OSHA - Final PELs - Time Weighted Averages
50 ppm TWA; 55 mg/m3 TWA

ENVIRONMENTAL LISTS

List of Pesticide Product Inert Ingredients
[present]

INTERNATIONAL LISTS

Australian Exposure Standards - Time Weighted Averages
50 ppm TWA; 57 mg/m3 TWA

Australian Exposure Standards - Short Term Exposure Limits
400 ppm STEL; 458 mg/m3 STEL

Canada - WHMIS: Ingredient Disclosure
0.1% item 312 (1174)

Canada - National Air Quality Objectives - Schedule I
desirable limits: 0-6 mg/m3 8 hours; 0-15 mg/m3 hour; acceptable limits: 6-15 mg/m3 8 hours; 15-35 mg/m3 hour; tolerable limits: 15-20 mg/m3 8 hours

Canada - Alberta - 8 Hour Occupational Exposure Limit
25 ppm TWA; 29 mg/m3 TWA

Canada - Alberta - 15 Minute Occupational Exposure Limit
200 ppm STEL; 229 mg/m3 STEL

Canada - British Columbia - 8 Hour Exposure Limits
  25 ppm TWA

Canada - British Columbia - 15 Minute Exposure Limits
  100 ppm STEL

Canada - Ontario - OHSA - TWAEVs
  35 ppm TWAEV;   40 mg/m3 TWAEV

Canada - Ontario - OHSA - STEVs
  400 ppm STEV;   460 mg/m3 STEV

Canada - Quebec - Time-Weighted Average Exposure Values
  35 ppm TWAEV;   40 mg/m3 TWAEV

Canada - Quebec - Short-term Exposure Values
  200 ppm STEV;   230 mg/m3 STEV

United Kingdom - Occupational Exposure Standards - TWAs
  50 ppm TWA;   55 mg/m3 TWA

United Kingdom - Occupational Exposure Standards - STELs
  300 ppm STEL;   330 mg/m3 STEL

German (DFG) - MAK Values
  30 ppm MAK;   33 mg/m3 MAK

German (DFG) - Peak Limitations
  2 x normal MAK (30 min. average value);   don't exceed 4 times during shift

German (DFG) - Pregnancy
  risk to embryo/fetus probable

Israel - Time Weighted Averages
  50 ppm TWA;   57 mg/m3 TWA

Israel - Short Term Exposure Limits
  400 ppm STEL;   458 mg/m3 STEL

Israel - Action Levels
  25 ppm AL;   28.5 mg/m3 AL

Mexico - Instruction No. 10 - TWAs
  50 ppm TWA;   55 mg/m3 TWA

Mexico - Instruction No. 10 - STELs
  400 ppm STEL;   440 mg/m3 STEL

STATE LISTS

California - Air Bill 2588 Appendix A-II
  9/89

California - Prop. 65 - Developmental Toxicity
  developmental toxicity - initial date 7/1/89

California - Exposure Limits - PELs
  35 ppm PEL;   40 mg/m3 PEL

California - Exposure Limits - Ceilings
  C 200 ppm

California - Directors List of Hazardous Substances (8 CCR 339)
  [present]

Florida Hazardous Substance List
  [present]

Massachusetts Right To Know List
  teratogen

Minnesota Hazardous Substance List
  [present]

NJ Right to Know List (Total)
  sn 0345

NJ Special Hazardous Substances
  (flammable - fourth degree)

Pennsylvania Right to Know List
  environmental hazard

PROPOSED REGULATIONS

Canada - Ontario - Proposed Occupational TWAEVs
  25 ppm TWAEV;   29 mg/m3 TWAEV

CARBONOCHLORIDOTHIOIC ACID, S-PROPYL ESTER   13889-92-

HEALTH AND SAFETY LISTS

NFPA - Flash Points
  flash point = 145 degrees F (63 degrees C)

NFPA - Hazard Identification Ratings
  health-2;   flammability-2;   reactivity-0

STATE LISTS

Florida Hazardous Substance List
  [present]

Massachusetts Right To Know List
  [present]

Pennsylvania Right to Know List
  [present]

CARBONODITHIOIC ACID, O-ETHYL ESTER, POTASSIUM SALT   140-89-6

HEALTH AND SAFETY LISTS

NFPA - Flash Points
  flash point = 205 degrees F (96 degrees C)

NFPA - Hazard Identification Ratings
  health-2;   flammability-1;   reactivity-0

ENVIRONMENTAL LISTS

CAA - HON Rule - SOCMI Chemicals
  compliance by Oct. 23, 1995

STATE LISTS

Florida Hazardous Substance List
  [present]

Massachusetts Right To Know List
  [present]

Pennsylvania Right to Know List
  [present]

CARBON TETRABROMIDE   558-13-4

HEALTH AND SAFETY LISTS

ACGIH 1995 - Time Weighted Averages
  0.1 ppm TWA;   1.4 mg/m3 TWA

ACGIH 1995 - Short Term Exposure Limits
  0.3 ppm STEL;   4.1 mg/m3 STEL

U.S. DOT - Substances From 49 CFR 172.101
  regulated by DOT (UN2516)

U.S. DOT - Hazard Classes
  DOT hazard class = 6.1

U.S. DOT - Appendix B - Marine Pollutants
  DOT regulated marine pollutant

OSHA - Vacated PELs - Time Weighted Averages
  0.1 ppm TWA;   1.4 mg/m3 TWA

OSHA - Vacated PELs - Short Term Exposure Limits
  0.3 ppm STEL;   4 mg/m3 STEL

ENVIRONMENTAL LISTS

CAA - HON Rule - SOCMI Chemicals
  compliance by Jan. 23, 1995

INTERNATIONAL LISTS

Australian Exposure Standards - Time Weighted Averages
  0.1 ppm TWA;   1.4 mg/m3 TWA

Australian Exposure Standards - Short Term Exposure Limits
  0.3 ppm STEL;   4.1 mg/m3 STEL

Canada - WHMIS: Ingredient Disclosure
  1% item 313 (1569)

Canada - Alberta - 8 Hour Occupational Exposure Limit
  0.1 ppm TWA;   1.4 mg/m3 TWA

Canada - Alberta - 15 Minute Occupational Exposure Limit
0.3 ppm STEL;   4.1 mg/m3 STEL

Canada - British Columbia - 8 Hour Exposure Limits
0.1 ppm TWA;   1.4 mg/m3 TWA

Canada - British Columbia - 15 Minute Exposure Limits
0.3 ppm STEL;   4 mg/m3 STEL

Canada - Ontario - OHSA - TWAEVs
0.1 ppm TWAEV;   1.4 mg/m3 TWAEV

Canada - Ontario - OHSA - STEVs
0.3 ppm STEV;   4.1 mg/m3 STEV

Canada - Quebec - Time-Weighted Average Exposure Values
0.1 ppm TWAEV;   1.4 mg/m3 TWAEV

Canada - Quebec - Short-term Exposure Values
0.3 ppm STEV;   4.1 mg/m3 STEV

United Kingdom - Occupational Exposure Standards - TWAs
0.1 ppm TWA;   1.4 mg/m3 TWA

United Kingdom - Occupational Exposure Standards - STELs
0.3 ppm STEL;   4 mg/m3 STEL

Israel - Time Weighted Averages
0.1 ppm TWA;   1.4 mg/m3 TWA

Israel - Short Term Exposure Limits
0.3 ppm STEL;   4.1 mg/m3 STEL

Israel - Action Levels
0.05 ppm AL;   0.7 mg/m3 AL

Mexico - Instruction No. 10 - TWAs
0.1 ppm TWA;   1.4 mg/m3 TWA

Mexico - Instruction No. 10 - STELs
0.3 ppm STEL;   4 mg/m3 STEL

STATE LISTS

California - Exposure Limits - PELs
0.1 ppm PEL;   1.4 mg/m3 PEL

California - Exposure Limits - STELs
0.3 ppm STEL;   4 mg/m3 STEL

California - Directors List of Hazardous Substances (8 CCR 339)
[present]

Florida Hazardous Substance List
[present]

Massachusetts Right To Know List
[present]

Minnesota Hazardous Substance List
[present]

NJ Right to Know List (Total)
sn 0346

Pennsylvania Right to Know List
[present]

CARBON TETRACHLORIDE   56-23-5
SEE ALSO:
BIS(2,4-DIMETHYLBUTYL) MALEATE
K150-HAZARDOUS WASTES
K116-HAZARDOUS WASTES
ISODECALDEHYDE
K020-HAZARDOUS WASTES
F025-HAZARDOUS WASTES
K019-HAZARDOUS WASTES
F039-HAZARDOUS WASTES
K021-HAZARDOUS WASTES
K151-HAZARDOUS WASTES
K016-HAZARDOUS WASTES
F001-HAZARDOUS WASTES
K073-HAZARDOUS WASTES
F024-HAZARDOUS WASTES

HEALTH AND SAFETY LISTS

ACGIH 1995 - Time Weighted Averages
5 ppm TWA;   31 mg/m3 TWA

ACGIH 1995 - Short Term Exposure Limits
10 ppm STEL;   63 mg/m3 STEL

ACGIH 1995 - Skin Designations
skin - potential for cutaneous absorption

ACGIH 1995 - Carcinogens
A3-animal carcinogen

AIHA - Odor Threshold Values
geometric mean air odor threshold = 252 ppm (detectable);   250 ppm (recognizable)

U.S. DOT - Substances From 49 CFR 172.101
regulated by DOT (UN1846)

U.S. DOT - Hazard Classes
DOT hazard class = 6.1

U.S. DOT - Appendix B - Marine Pollutants
DOT regulated marine pollutant

U.S. DOT - Appendix A Table 1 - Hazardous Substances
final RQ = 10 pounds (4.54 kg)

IARC - Group 2B (sufficient animal data)
[present]

NIOSH - Selected LD50s and LC50s
Inhalation, rat: LC50 = 8000 ppm 4 hr Oral, rat: LD50 = 2800 mg/kg Skin, rat: LD50 = 5070 mg/kg

NIOSH 1990 - Pocket Guide - RELs
2 ppm STEL (60 min);   12.6 mg/m3 STEL (60 min)

NIOSH 1990 - Pocket Guide - IDLHs
300 ppm IDLH (not considering carcinogenic effects)

NIOSH 1990 - Pocket Guide - Carcinogens
occupational carcinogen

NIOSH 1990 - Pocket Guide - Target organs
CNS, eyes, lungs, liver, kidneys, skin

NIOSH - Health Standards - Exposure Limits
C 2 ppm;   C 16.6 mg/m3 (45 liter, 60 min sample)

NIOSH - Health Standards - Health Effects and Precautions
Liver cancer

NIOSH - Health Standards - Carcinogenic Chemicals
potential human carcinogen

NTP Seventh Report - Suspect Carcinogens
suspect carcinogen

OSHA - Vacated PELs - Time Weighted Averages
2 ppm TWA;   12.6 mg/m3 TWA

OSHA - Final PELs - Time Weighted Averages
10 ppm TWA

OSHA - Final PELs - Ceiling Limits
C 25 ppm

OSHA - Possible Select Carcinogens
[present]

ENVIRONMENTAL LISTS

ATSDR Priority List
Rank (of 275): 043

CERCLA/SARA - Section 313 - Emission Reporting
form R reporting required for 0.1% de minimus concentration

CERCLA/SARA - Hazardous Substances and their Reportable Quantities
final RQ = 10 pounds (4.54 kg)

Clean Air Act (1990) - List of Hazardous Air Contaminants
[present]

Class 1 Ozone Depletors
ozone depletion potential = 1.1

CAA - HON Rule - SOCMI Chemicals
compliance by Oct. 24, 1994

CAA - HON Rule - Organic HAPs
[present]

Clean Water Act - Hazardous Substances
[present]

Clean Water Act - Priority Pollutants
[present]

Clean Water Act - Toxic Pollutants
[present]

Safe Drinking Water Act - MCLs
MCL = 0.005 mg/L

Safe Drinking Water Act - MCLGs
MCLG = Zero

EPA - Carcinogen Hazard Ranking for RQ Adjustment
Hazard ranking = Medium

RCRA - D Series - Maximum Concentration of Contaminants
waste number D019; regulatory level = 0.5 mg/L

RCRA - D Series - Chronic Toxicity Reference Levels
chronic toxicity reference level = 0.005 mg/L

RCRA - U Series Wastes
waste number U211

RCRA - Hazardous Constituents-Appendix VIII
waste number U211

RCRA - Basis for Listing - Appendix VII
Included in waste streams: F001, F024, F025, F039, K016, K019, K020, K021, K073, K116, K150, K151

RCRA - Substances Banned From Land Disposal
[present]

RCRA - TSD Facilities Ground Water Monitoring
TM 8010 = 1 ug/L PQL; TM 8240 = 5 ug/L PQL

RCRA - Universal Treatment Standards (LDR)
WW: 0.057 mg/l; NWW: 6.0 mg/kg

INTERNATIONAL LISTS

Australian Exposure Standards - Time Weighted Averages
5 ppm TWA; 31 mg/m3 TWA

Australian Exposure Standards - Skin Effects
skin absorption

Australian Exposure Standards - Carcinogens
probable carcinogen

Australian Exposure Standards - Under Review
exposure limits under review

Canada - WHMIS: Ingredient Disclosure
0.1% item 314 (1579)

Canada - NPRI (National Pollutant Release Inventory)
[present]

Canada - CEPA Schedule II Part II - Toxic Substances (Export)
[present]

Canada - CEPA Schedule I - Toxic Substances
quantities that may be manufactured or imported

Canada - Drinking Water Quality - MACs
0.005 mg/L MAC

Canada - Alberta - 8 Hour Occupational Exposure Limit
5 ppm TWA; 32 mg/m3 TWA

Canada - Alberta - 15 Minute Occupational Exposure Limit
20 ppm STEL; 126 mg/m3 STEL

Canada - Alberta - Skin Designation
can be absorbed through the intact skin

Canada - British Columbia - 8 Hour Exposure Limits
10 ppm TWA; 65 mg/m3 TWA

Canada - British Columbia - 15 Minute Exposure Limits
20 ppm STEL; 130 mg/m3 STEL

Canada - British Columbia - Skin Notations
skin - potential for skin absorption

Canada - Ontario - OHSA - TWAEVs
2 ppm TWAEV; 13 mg/m3 TWAEV

Canada - Ontario - OHSA - Skin Notations
absorption through skin, eyes, or mucous membranes

Canada - Quebec - Time-Weighted Average Exposure Values
5 ppm TWAEV; 31 mg/m3 TWAEV

Canada - Quebec - Skin Designations
absorbed through the skin

Canada - Quebec - Carcinogens
C2 carcinogen: effect suspected in humans

German (DFG) - MAK Values
10 ppm MAK; 65 mg/m3 MAK

German (DFG) - Peak Limitations
2 x normal MAK (30 min. average value); don't exceed 4 times during shift

German (DFG) - Skin/Sensitizers
danger of cutaneous absorption

German (DFG) - Carcinogens
suspected carcinogen

German (DFG) - Pregnancy
classification not yet possible

Israel - Time Weighted Averages
5 ppm TWA; 31 mg/m3 TWA

Israel - Action Levels
2.5 ppm AL; 15.5 mg/m3 AL

Mexico - Instruction No. 10 - TWAs
5 ppm TWA; 30 mg/m3 TWA

Mexico - Instruction No. 10 - STELs
20 ppm STEL; 125 mg/m3 STEL

Mexico - Instruction No. 10 - Skin designation
skin - potential for cutaneous absorption

Mexico - Instruction No. 10 - Carcinogens
potential carcinogen in humans - limited epidemiological evidence

Mexico - Wastewater - Organic Toxic Pollutants and Heavy Metals
Listed under [Organic Toxic Pollutants]

Mexico - Drinking Water - Ecological Criteria
0.004 mg/l Substance presents persistence, bioaccumulations or risk of cancer, reduce human exposure to a minimum; This level has been extrapolated by using a mathematic model

STATE LISTS

California - Air Bill 2588 Appendix A-I
known or potential carcinogen

California - Prop. 65 - Cancer list
carcinogen - initial date 10/1/87

California - Prop. 65 - No Significant Risk Levels
no significant risk level = 5 ug/day

California - Exposure Limits - PELs
2 ppm PEL; 12.6 mg/m3 PEL

California - Exposure Limits - Ceilings
C 200 ppm

California - Exposure Limits - Skin Notation
material may be absorbed through the skin, eyes or mucous membrane

California - Directors List of Hazardous Substances (8 CCR 339)
[present]

Florida Hazardous Substance List
[present]

Massachusetts Right To Know List
carcinogen; extraordinarily hazardous

Minnesota Hazardous Substance List
carcinogen; skin

NJ Right to Know List (Total)
sn 0347

NJ Special Hazardous Substances
(carcinogen)

Pennsylvania Right to Know List
environmental hazard; special hazardous substance

Pennsylvania RTK - Special Hazardous Substances
[present]

CARBONYL FLUORIDE   353-50-4

HEALTH AND SAFETY LISTS

ACGIH 1995 - Time Weighted Averages
2 ppm TWA; 5.4 mg/m3 TWA

ACGIH 1995 - Short Term Exposure Limits
5 ppm STEL; 13 mg/m3 STEL

U.S. DOT - Substances From 49 CFR 172.101
regulated by DOT (UN2417)

U.S. DOT - Hazard Classes
DOT hazard class = 2.3

U.S. DOT - Substances Which Are Poisonous by Inhalation
gaseous hazardous material poisonous by inhalation (UN2417)

U.S. DOT - Appendix A Table 1 - Hazardous Substances
final RQ = 1000 pounds (454 kg)

NIOSH - Selected LD50s and LC50s
Inhalation, rat: LC50 = 360 ppm 1 hr

OSHA - Vacated PELs - Time Weighted Averages
2 ppm TWA; 5 mg/m3 TWA

OSHA - Vacated PELs - Short Term Exposure Limits
5 ppm STEL; 15 mg/m3 STEL

OSHA - List of Highly Hazardous Chemicals
threshhold quantity = 2500 pounds

ENVIRONMENTAL LISTS

CERCLA/SARA - Hazardous Substances and their Reportable Quantities
final RQ = 1000 pounds (454 kg)

RCRA - U Series Wastes
waste number U033 (Reactive waste; Toxic waste)

RCRA - Hazardous Constituents-Appendix VIII
waste number U033

RCRA - Substances Banned From Land Disposal
[present]

TSCA - Chemicals with Significant New Use Rules
[present]

TSCA - Section 12(b) - Export Notification
export notification required - Section 5

INTERNATIONAL LISTS

Australian Exposure Standards - Time Weighted Averages
2 ppm TWA; 5.4 mg/m3 TWA

Australian Exposure Standards - Short Term Exposure Limits
5 ppm STEL; 13 mg/m3 STEL

Canada - WHMIS: Ingredient Disclosure
1% item 315 (903)

Canada - Alberta - 8 Hour Occupational Exposure Limit
2 ppm TWA; 5.4 mg/m3 TWA

Canada - Alberta - 15 Minute Occupational Exposure Limit
5 ppm STEL; 13.5 mg/m3 STEL

Canada - British Columbia - 8 Hour Exposure Limits
5 ppm TWA; 15 mg/m3 TWA

Canada - Ontario - OHSA - TWAEVs
2 ppm TWAEV; 5.4 mg/m3 TWAEV

Canada - Ontario - OHSA - STEVs
5 ppm STEV; 13 mg/m3 STEV

Canada - Quebec - Time-Weighted Average Exposure Values
2 ppm TWAEV; 5.4 mg/m3 TWAEV

Canada - Quebec - Short-term Exposure Values
5 ppm STEV; 13 mg/m3 STEV

Israel - Time Weighted Averages
2 ppm TWA; 5.4 mg/m3 TWA

Israel - Short Term Exposure Limits
5 ppm STEL; 13 mg/m3 STEL

Israel - Action Levels
1 ppm AL; 2.7 mg/m3 AL

STATE LISTS

California - Exposure Limits - PELs
2 ppm PEL; 5 mg/m3 PEL

California - Exposure Limits - STELs
5 ppm STEL; 15 mg/m3 STEL

California - Directors List of Hazardous Substances (8 CCR 339)
[present]

Florida Hazardous Substance List
[present]

Massachusetts Right To Know List
[present]

Minnesota Hazardous Substance List
[present]

NJ Right to Know List (Total)
sn 0348

Pennsylvania Right to Know List
environmental hazard

CARBONYL SULFIDE   463-58-1

HEALTH AND SAFETY LISTS

U.S. DOT - Substances From 49 CFR 172.101
regulated by DOT (UN2204)

U.S. DOT - Hazard Classes
DOT hazard class = 2.3

U.S. DOT - Substances Which Are Poisonous by Inhalation
gaseous hazardous material poisonous by inhalation (UN2204)

NFPA - Flash Points
gas (no flash point given)

NFPA - Hazard Identification Ratings
health-3; flammability-4; reactivity-1

ENVIRONMENTAL LISTS

CERCLA/SARA - Section 313 - Emission Reporting
form R reporting required for 1.0% de minimus concentration

CERCLA/SARA - Hazardous Substances and their Reportable Quantities
final RQ = 1 pound (.454 kg)

Clean Air Act (1990) - List of Hazardous Air Contaminants
[present]

CAA - Flammable Substances for Accidental Release Prevention
threshold quantity = 10,000 lbs

INTERNATIONAL LISTS

Canada - WHMIS: Ingredient Disclosure
1% item 316 (1544)

STATE LISTS

California - Air Bill 2588 Appendix A-I
6/91

Florida Hazardous Substance List
[present]

Massachusetts Right To Know List
[present]

NJ Right to Know List (Total)
sn 0349

Pennsylvania Right to Know List
environmental hazard

CARBOPHENOTHION   786-19-6

HEALTH AND SAFETY LISTS

U.S. DOT - Appendix B - Marine Pollutants
DOT regulated severe marine pollutant

NIOSH - Selected LD50s and LC50s
Oral, rat: LD50 = 6800 ug/kg Skin, rat: LD50 = 27 mg/kg

ENVIRONMENTAL LISTS
  ATSDR Priority List
    Rank (of 275): 189
  CERCLA/SARA - Section 302 Extremely Hazardous Substances and TPQs
    TPQ = 500 pounds
STATE LISTS
  California - Directors List of Hazardous Substances (8 CCR 339)
    [present]
  Florida Hazardous Substance List
    effective March 13, 1992
  Massachusetts Right To Know List
    extraordinarily hazardous; neurotoxin
  NJ Right to Know List (Total)
    sn 2218
  Pennsylvania Right to Know List
    environmental hazard
PROPOSED REGULATIONS
  CERCLA/SARA - Proposed Hazardous Substance Additions
    proposed RQ = 1 pound (.454 kg)
  CERCLA/SARA - 1989 Proposed RQ Adjustments
    proposed RQ = 100 pounds (45.4 kg)

CARBOPLATIN   41575-94-4
STATE LISTS
  California - Air Bill 2588 Appendix A-II
    9/90
  California - Prop. 65 - Developmental Toxicity
    developmental toxicity - initial date 7/1/90

CARBOPOLYCYCLICOL AZOALKYLAMINOALKYL-   RR-01232-8
CARBOMONOCYCLIC ESTER, HALOGEN ACID
SALTS
ENVIRONMENTAL LISTS
  TSCA - Chemicals with Significant New Use Rules
    PMN number: P-88-1682

CARBOXIN (5,6-DIHYDRO-2-METHYL-N-PHENYL-1,   5234-68-4
4-OXATHIIN-3-CARBOXAMIDE)
ENVIRONMENTAL LISTS
  CERCLA/SARA - Section 313 - Emission Reporting
    form R reporting required

6-CARBOXY-4-HEXYL-2-CYCLOHEXENE-1-OC-   68630-89-7
TANOIC ACID, MONOPOTASSIUM SALT
ENVIRONMENTAL LISTS
  List of Pesticide Product Inert Ingredients
    [present]

CARBOXYLIC ACID GLYCIDYL ESTERS   RR-01721-0
ENVIRONMENTAL LISTS
  TSCA - Chemicals with Significant New Use Rules
    PMN numbers: P-92-776 and P-92-777

CARBOXYMETHYL CELLULOSE   9000-11-7
ENVIRONMENTAL LISTS
  List of Pesticide Product Inert Ingredients
    [present]

CARBOXYPOLYMETHYLENE RESIN   68649-45-6
ENVIRONMENTAL LISTS
  List of Pesticide Product Inert Ingredients
    [present]

CARBROMAL   77-65-
HEALTH AND SAFETY LISTS
  NTP Chemical Status Reports - Testing Status and NTIS Number
    Technical reports printed (PB290130/AS)
  NTP Chemical Status Reports - Evidence of Carcinogenicity
    male rat-negative; female rat-negative; male mice-negative; femal mice-negative

CARDBOARD   RR-01049-
ENVIRONMENTAL LISTS
  List of Pesticide Product Inert Ingredients
    [present]

CARISOPRODOL   78-44-
HEALTH AND SAFETY LISTS
  NTP Chemical Status Reports - Testing Status and NTIS Number
    Prechronic studies completed: in review for further evaluation, prechronic studies for which toxicity technical reports were not prepared

CARNAUBA WAX   8015-86-9
HEALTH AND SAFETY LISTS
  NFPA - Flash Points
    flash point = 540 degrees F (282 degrees C)
  NFPA - Hazard Identification Ratings
    health-0; flammability-1; reactivity-0
ENVIRONMENTAL LISTS
  List of Pesticide Product Inert Ingredients
    [present]

CARPENTRY AND JOINERY   RR-00544-7
HEALTH AND SAFETY LISTS
  IARC - Group 2B (sufficient animal data)
    [present] (Listed under 'Wood industries')
  OSHA - Possible Select Carcinogens
    [present]

CARRAGEENAN   9000-07-1
HEALTH AND SAFETY LISTS
  IARC - Group 2B (sufficient animal data)
    [present] (Overall evaluation based only on evidence of carcinogenicity in monograph (31, 1983) or in supplement 4)
  OSHA - Possible Select Carcinogens
    [present]
ENVIRONMENTAL LISTS
  List of Pesticide Product Inert Ingredients
    [present]
STATE LISTS
  California - Air Bill 2588 Appendix A-I
    known or potential carcinogen
  Massachusetts Right To Know List
    carcinogen; extraordinarily hazardous
  Minnesota Hazardous Substance List
    carcinogen
  Pennsylvania Right to Know List
    special hazardous substance
  Pennsylvania RTK - Special Hazardous Substances
    [present]

CARRAGEENAN, NATIVE   RR-01529-2
HEALTH AND SAFETY LISTS
  IARC - Group 3 (not classifiable)
    [present]

CARROTS   RR-01050-4

ENVIRONMENTAL LISTS

List of Pesticide Product Inert Ingredients
[present]

CARTAP HYDROCHLORIDE   15263-52-2

HEALTH AND SAFETY LISTS

U.S. DOT - Appendix B - Marine Pollutants
DOT regulated marine pollutant

CARVACROL   499-75-2

HEALTH AND SAFETY LISTS

NIOSH - Selected LD50s and LC50s
Oral, rat: LD50 - 810 mg/kg

INTERNATIONAL LISTS

Canada - WHMIS: Ingredient Disclosure
1% item 317 (394)

CARVONE   99-49-0

HEALTH AND SAFETY LISTS

NIOSH - Selected LD50s and LC50s
Oral, rat: LD50 - 1640 mg/kg

CARVONE   2244-16-8

HEALTH AND SAFETY LISTS

NIOSH - Selected LD50s and LC50s
Oral, rat: LD50 - 3710 ug/kg Skin, rabbit: LD50 - 4 mg/kg
NTP Chemical Status Reports - Testing Status and NTIS Number
Technical reports printed (PB90241100)
NTP Chemical Status Reports - Evidence of Carcinogenicity
male mice-no evidence; female mice-no evidence

STATE LISTS

Massachusetts Right To Know List
[present]
NJ Right to Know List (Total)
sn 2219

(-)-CARVONE   6485-40-1

HEALTH AND SAFETY LISTS

NIOSH - Selected LD50s and LC50s
Oral, rat: LD50 - 1640 mg/kg

CASEIN   9000-71-9

ENVIRONMENTAL LISTS

List of Pesticide Product Inert Ingredients
[present]

CASTOR OIL   8001-79-4

HEALTH AND SAFETY LISTS

NFPA - Flash Points
flash point - 445 degrees F (229 degrees C); hydrogenated: flash point - 401 degrees F (205 degrees C)
NFPA - Hazard Identification Ratings
health-0; flammability-1; reactivity-0
NTP Chemical Status Reports - Testing Status and NTIS Number
Technical reports printed (PB93-151439)

ENVIRONMENTAL LISTS

List of Pesticide Product Inert Ingredients
[present]

INTERNATIONAL LISTS

Canada - WHMIS: Ingredient Disclosure
1% item 318 (976)

CASTOR OIL, DEHYDRATED, POLYMER WITH   68071-54-5
P-TERT-BUTYLBENZOIC ACID, GLYCEROL AND
PHTHALIC ANHYDRIDE

ENVIRONMENTAL LISTS

List of Pesticide Product Inert Ingredients
[present]

CASTOR OIL, EPOXIDIZED   105839-17-6

ENVIRONMENTAL LISTS

List of Pesticide Product Inert Ingredients
[present]

CASTOR OIL, MALEIC ANHYDRIDE, AND   RR-01051-5
POLYETHYLENE GLYCOL COPOLYMER

ENVIRONMENTAL LISTS

List of Pesticide Product Inert Ingredients
[present]

CASTOR OIL, OXIDIZED   68187-84-8

ENVIRONMENTAL LISTS

List of Pesticide Product Inert Ingredients
[present]

CASTOR OIL, SULFATED   8002-33-3

HEALTH AND SAFETY LISTS

NFPA - Flash Points
flash point - 476 degrees F (247 degrees C)
NFPA - Hazard Identification Ratings
health-0; flammability-1; reactivity-0

ENVIRONMENTAL LISTS

List of Pesticide Product Inert Ingredients
[present]

CATECHOL   120-80-9

HEALTH AND SAFETY LISTS

ACGIH 1995 - Time Weighted Averages
5 ppm TWA; 23 mg/m3 TWA
ACGIH 1995 - Skin Designations
skin - potential for cutaneous absorption
IARC - Group 3 (not classifiable)
[present]
NFPA - Flash Points
flash point - 260 degrees F (127 degrees C)
NFPA - Hazard Identification Ratings
flammability-1; reactivity-0
NIOSH - Selected LD50s and LC50s
Oral, rat: LD50 - 260 mg/kg Skin, rabbit: LD50 - 800 mg/kg
OSHA - Vacated PELs - Time Weighted Averages
5 ppm TWA; 20 mg/m3 TWA
OSHA - Vacated PELs - Skin Designation
Prevent or reduce skin absorption

ENVIRONMENTAL LISTS

CERCLA/SARA - Section 313 - Emission Reporting
form R reporting required for 1.0% de minimus concentration
CERCLA/SARA - Hazardous Substances and their Reportable Quantities
final RQ - 1 pound (.454 kg)
Clean Air Act (1990) - List of Hazardous Air Contaminants
[present]
EPA - Master Testing List
[present]
List of Pesticide Product Inert Ingredients
[present]

INTERNATIONAL LISTS

Australian Exposure Standards - Time Weighted Averages
5 ppm TWA;   23 mg/m3 TWA

Canada - WHMIS: Ingredient Disclosure
1% item 319 (395)

Canada - NPRI (National Pollutant Release Inventory)
[present]

Canada - Alberta - 8 Hour Occupational Exposure Limit
5 ppm TWA;   23 mg/m3 TWA

Canada - Alberta - 15 Minute Occupational Exposure Limit
10 ppm STEL;   45 mg/m3 STEL

Canada - British Columbia - 8 Hour Exposure Limits
5 ppm TWA;   20 mg/m3 TWA

Canada - Ontario - OHSA - TWAEVs
5 ppm TWAEV;   22 mg/m3 TWAEV

Canada - Quebec - Time-Weighted Average Exposure Values
5 ppm TWAEV;   23 mg/m3 TWAEV

Canada - Quebec - Skin Designations
absorbed through the skin

United Kingdom - Occupational Exposure Standards - TWAs
5 ppm TWA;   20 mg/m3 TWA

Israel - Time Weighted Averages
5 ppm TWA;   23 mg/m3 TWA

Israel - Action Levels
2.5 ppm AL;   11.5 mg/m3 AL

Mexico - Instruction No. 10 - TWAs
5 ppm TWA;   20 mg/m3 TWA

STATE LISTS

California - Air Bill 2588 Appendix A-I
6/91

California - Exposure Limits - PELs
5 ppm PEL;   20 mg/m3 PEL

California - Exposure Limits - Skin Notation
material may be absorbed through the skin, eyes or mucous membrane

California - Directors List of Hazardous Substances (8 CCR 339)
[present]

Florida Hazardous Substance List
[present]

Massachusetts Right To Know List
[present]

Minnesota Hazardous Substance List
[present]

NJ Right to Know List (Total)
sn 0722

Pennsylvania Right to Know List
environmental hazard

PROPOSED REGULATIONS

TSCA - Proposed Substances for Developmental/Reproductive Testing
proposed testing for: Developmental Toxicity - oral

CAT FOOD   RR-01052-6

ENVIRONMENTAL LISTS

List of Pesticide Product Inert Ingredients
[present]

CAUSTIC ALKALI LIQUIDS, N.O.S.   RR-00193-4

HEALTH AND SAFETY LISTS

U.S. DOT - Substances From 49 CFR 172.101
regulated by DOT (UN1719)

U.S. DOT - Hazard Classes
DOT hazard class = 8

STATE LISTS

NJ Right to Know List (Total)
sn 2221

CAUSTIC ARSENIC OIL   RR-00801-5

STATE LISTS

Pennsylvania Right to Know List
environmental hazard

CEDAR WOOD OIL   8000-27-9

ENVIRONMENTAL LISTS

List of Pesticide Product Inert Ingredients
[present]

INTERNATIONAL LISTS

Canada - WHMIS: Ingredient Disclosure
1% item 320 (973)

ALPHA-CEDRENE   469-61-4

INTERNATIONAL LISTS

Canada - WHMIS: Ingredient Disclosure
1% item 321 (396)

CEDROL METHYL ETHER   67874-81-1

INTERNATIONAL LISTS

Canada - WHMIS: Ingredient Disclosure
1% item 322 (811)

CELLULOID   RR-00195-6

HEALTH AND SAFETY LISTS

U.S. DOT - Substances From 49 CFR 172.101
regulated by DOT (UN2000, UN2002)

U.S. DOT - Hazard Classes
scrap: DOT hazard class = 4.2;   except for scrap: DOT hazard class = 4.1

STATE LISTS

NJ Right to Know List (Total)
blocks, rods, rolls, sheets, tubes: sn 2224;   scrap: sn 2225

CELLULOSE   9004-34-6

HEALTH AND SAFETY LISTS

ACGIH 1995 - Time Weighted Averages
10 mg/m3 TWA

OSHA - Vacated PELs - Time Weighted Averages
total dust: 15 mg/m3 TWA;   respirable fraction: 5 mg/m3 TWA

OSHA - Final PELs - Time Weighted Averages
total dust: 15 mg/m3 TWA;   respirable fraction: 5 mg/m3 TWA

ENVIRONMENTAL LISTS

List of Pesticide Product Inert Ingredients
[present]

INTERNATIONAL LISTS

Australian Exposure Standards - Time Weighted Averages
10 mg/m3 TWA

Canada - Alberta - 8 Hour Occupational Exposure Limit
respirable mass: 5 mg/m3 TWA;   total mass: 10 mg/m3 TWA

Canada - British Columbia - 8 Hour Exposure Limits
nuisance dust, mist, and fumes: 10 mg/m3 TWA

Canada - British Columbia - 15 Minute Exposure Limits
20 mg/m3 STEL

Canada - Ontario - OHSA - TWAEVs
paper fibre, total dust: 10 mg/m3 TWAEV (listed as nuisance particulate)

Canada - Quebec - Time-Weighted Average Exposure Values
total dust: 10 mg/m3 TWAEV

United Kingdom - Occupational Exposure Standards - TWAs
total inhalable dust: 10 mg/m3 TWA; respirable dust: 5 mg/m3 TWA

United Kingdom - Occupational Exposure Standards - STELs
total inhalable dust: 20 mg/m3 STEL

Israel - Time Weighted Averages
10 mg/m3 TWA

Israel - Action Levels
5 mg/m3 AL

Mexico - Instruction No. 10 - TWAs
10 mg/m3 TWA (nuisance particulate)

Mexico - Instruction No. 10 - STELs
20 mg/m3 STEL

STATE LISTS

Minnesota Hazardous Substance List
[present] (includes inert or nuisance dust)

Pennsylvania Right to Know List
[present]

CELLULOSE ACETATE BUTYRATE   9004-36-8

ENVIRONMENTAL LISTS

List of Pesticide Product Inert Ingredients
[present]

CELLULOSE, OMEGA-ETHER WITH ETHOXY-LATED 2-HYDROXY-3-(TRIMETHYLAMMONIO) PROPANOL, CHLORIDE   68610-92-4

ENVIRONMENTAL LISTS

List of Pesticide Product Inert Ingredients
[present]

CELLULOSE HYDROXYPROPYL METHYL ETHER   9004-65-3

ENVIRONMENTAL LISTS

List of Pesticide Product Inert Ingredients
[present]

CELLULOSE NITRATE, WET WITH ALCOHOL   RR-01760-7

HEALTH AND SAFETY LISTS

NFPA - Flash Points
flash point = 55 degrees F (13 degrees C)

NFPA - Hazard Identification Ratings
health-2; flammability-3; reactivity-3

CELLULOSE, REGENERATED   68442-85-3

ENVIRONMENTAL LISTS

List of Pesticide Product Inert Ingredients
[present]

CERESIN WAX   8001-75-0

ENVIRONMENTAL LISTS

List of Pesticide Product Inert Ingredients
[present]

CERIUM   7440-45-1

HEALTH AND SAFETY LISTS

U.S. DOT - Substances From 49 CFR 172.101
turnings or gritty powder: regulated by DOT (UN3078)

U.S. DOT - Hazard Classes
DOT hazard class = 4.3

STATE LISTS

NJ Right to Know List (Total)
sn 0352

CERIUM 134   15055-11-5

HEALTH AND SAFETY LISTS

U.S. DOT - Appendix A Table 2 - Radionuclides
final RQ = 10 curies (3.7E 11 Bq)

ENVIRONMENTAL LISTS

CERCLA/SARA List of Radionuclides (Appendix B) and Their Reportable Quantities
final RQ = 10 curies (3.7E 11 Bq)

CERIUM 135   15757-94-5

HEALTH AND SAFETY LISTS

U.S. DOT - Appendix A Table 2 - Radionuclides
final RQ = 10 curies (3.7E 11 Bq)

ENVIRONMENTAL LISTS

CERCLA/SARA List of Radionuclides (Appendix B) and Their Reportable Quantities
final RQ = 10 curies (3.7E 11 Bq)

CERIUM 137   13968-49-5

HEALTH AND SAFETY LISTS

U.S. DOT - Appendix A Table 2 - Radionuclides
final RQ = 1000 curies (3.7E 13 Bq)

ENVIRONMENTAL LISTS

CERCLA/SARA List of Radionuclides (Appendix B) and Their Reportable Quantities
final RQ = 1000 curies (3.7E 13 Bq)

CERIUM 137M   RR-00470-6

HEALTH AND SAFETY LISTS

U.S. DOT - Appendix A Table 2 - Radionuclides
final RQ = 100 curies (3.7E 12 Bq)

ENVIRONMENTAL LISTS

CERCLA/SARA List of Radionuclides (Appendix B) and Their Reportable Quantities
final RQ = 100 curies (3.7E 12 Bq)

CERIUM 139   13982-30-4

HEALTH AND SAFETY LISTS

U.S. DOT - Appendix A Table 2 - Radionuclides
final RQ = 100 curies (3.7E 12 Bq)

ENVIRONMENTAL LISTS

CERCLA/SARA List of Radionuclides (Appendix B) and Their Reportable Quantities
final RQ = 100 curies (3.7E 12 Bq)

CERIUM 141   13967-74-3

HEALTH AND SAFETY LISTS

U.S. DOT - Appendix A Table 2 - Radionuclides
final RQ = 10 curies (3.7E 11 Bq)

ENVIRONMENTAL LISTS

CERCLA/SARA List of Radionuclides (Appendix B) and Their Reportable Quantities
final RQ = 10 curies (3.7E 11 Bq)

CERIUM 143   14119-19-8

HEALTH AND SAFETY LISTS

U.S. DOT - Appendix A Table 2 - Radionuclides
final RQ = 100 curies (3.7E 12 Bq)

ENVIRONMENTAL LISTS

CERCLA/SARA List of Radionuclides (Appendix B) and Their Reportable Quantities
final RQ = 100 curies (3.7E 12 Bq)

CERIUM 144   14762-78-8

HEALTH AND SAFETY LISTS

U.S. DOT - Appendix A Table 2 - Radionuclides
final RQ = 1 curie (3.7E 10 Bq)

ENVIRONMENTAL LISTS

CERCLA/SARA List of Radionuclides (Appendix B) and Their Reportable Quantities
final RQ = 1 curie (3.7E 10 Bq)

CERIUM 2-ETHYLHEXOATE   56797-01-4

ENVIRONMENTAL LISTS

List of Pesticide Product Inert Ingredients
[present]

CERIUM NITRATE   10108-73-3

HEALTH AND SAFETY LISTS

NIOSH - Selected LD50s and LC50s
Oral, rat: LD50 = 3154 mg/kg

CESIUM   7440-46-2

HEALTH AND SAFETY LISTS

U.S. DOT - Substances From 49 CFR 172.101
regulated by DOT (UN1407, UN3078, UN1333)
U.S. DOT - Hazard Classes
DOT hazard class = 4.3

STATE LISTS

NJ Right to Know List (Total)
sn 0353

CESIUM 125   15758-27-7

HEALTH AND SAFETY LISTS

U.S. DOT - Appendix A Table 2 - Radionuclides
final RQ = 1000 curies (3.7E 13 Bq)

ENVIRONMENTAL LISTS

CERCLA/SARA List of Radionuclides (Appendix B) and Their Reportable Quantities
final RQ = 1000 curies (3.7E 13 Bq)

CESIUM 127   15720-35-1

HEALTH AND SAFETY LISTS

U.S. DOT - Appendix A Table 2 - Radionuclides
final RQ = 100 curies (3.7E 12 Bq)

ENVIRONMENTAL LISTS

CERCLA/SARA List of Radionuclides (Appendix B) and Their Reportable Quantities
final RQ = 100 curies (3.7E 12 Bq)

CESIUM 129   15047-05-9

HEALTH AND SAFETY LISTS

U.S. DOT - Appendix A Table 2 - Radionuclides
final RQ = 100 curies (3.7E 12 Bq)

ENVIRONMENTAL LISTS

CERCLA/SARA List of Radionuclides (Appendix B) and Their Reportable Quantities
final RQ = 100 curies (3.7E 12 Bq)

CESIUM 130   15066-92-9

HEALTH AND SAFETY LISTS

U.S. DOT - Appendix A Table 2 - Radionuclides
final RQ = 1000 curies (3.7E 13 Bq)

ENVIRONMENTAL LISTS

CERCLA/SARA List of Radionuclides (Appendix B) and Their Reportable Quantities
final RQ = 1000 curies (3.7E 13 Bq)

CESIUM 131   14914-76-2

HEALTH AND SAFETY LISTS

U.S. DOT - Appendix A Table 2 - Radionuclides
final RQ = 1000 curies (3.7E 13 Bq)

ENVIRONMENTAL LISTS

CERCLA/SARA List of Radionuclides (Appendix B) and Their Reportable Quantities
final RQ = 1000 curies (3.7E 13 Bq)

CESIUM 132   15758-03-9

HEALTH AND SAFETY LISTS

U.S. DOT - Appendix A Table 2 - Radionuclides
final RQ = 10 curies (3.7E 11 Bq)

ENVIRONMENTAL LISTS

CERCLA/SARA List of Radionuclides (Appendix B) and Their Reportable Quantities
final RQ = 10 curies (3.7E 11 Bq)

CESIUM 134   13967-70-9

HEALTH AND SAFETY LISTS

U.S. DOT - Appendix A Table 2 - Radionuclides
final RQ = 1 curie (3.7E 10 Bq)

ENVIRONMENTAL LISTS

CERCLA/SARA List of Radionuclides (Appendix B) and Their Reportable Quantities
final RQ = 1 curie (3.7E 10 Bq)

CESIUM 134M   RR-00469-3

HEALTH AND SAFETY LISTS

U.S. DOT - Appendix A Table 2 - Radionuclides
final RQ = 1000 curies (3.7E 13 Bq)

ENVIRONMENTAL LISTS

CERCLA/SARA List of Radionuclides (Appendix B) and Their Reportable Quantities
final RQ = 1000 curies (3.7E 13 Bq)

CESIUM 135   15726-30-4

HEALTH AND SAFETY LISTS

U.S. DOT - Appendix A Table 2 - Radionuclides
final RQ = 10 curies (3.7E 11 Bq)

ENVIRONMENTAL LISTS

CERCLA/SARA List of Radionuclides (Appendix B) and Their Reportable Quantities
final RQ = 10 curies (3.7E 11 Bq)

CESIUM 135M   RR-00468-2

HEALTH AND SAFETY LISTS

U.S. DOT - Appendix A Table 2 - Radionuclides
final RQ = 100 curies (3.7E 12 Bq)

ENVIRONMENTAL LISTS

CERCLA/SARA List of Radionuclides (Appendix B) and Their Reportable Quantities
final RQ = 100 curies (3.7E 12 Bq)

CESIUM 136   14234-29-8

HEALTH AND SAFETY LISTS

U.S. DOT - Appendix A Table 2 - Radionuclides
final RQ = 10 curies (3.7E 11 Bq)

ENVIRONMENTAL LISTS
  CERCLA/SARA List of Radionuclides (Appendix B) and Their Reportable Quantities
    final RQ = 10 curies (3.7E 11 Bq)

CESIUM 137   10045-97-3
  HEALTH AND SAFETY LISTS
    U.S. DOT - Appendix A Table 2 - Radionuclides
      final RQ = 1 curie (3.7E 10 Bq)
  ENVIRONMENTAL LISTS
    ATSDR Priority List
      Rank (of 275): 119
    CERCLA/SARA List of Radionuclides (Appendix B) and Their Reportable Quantities
      final RQ = 1 curie (3.7E 10 Bq)

CESIUM 138   15758-29-9
  HEALTH AND SAFETY LISTS
    U.S. DOT - Appendix A Table 2 - Radionuclides
      final RQ = 100 curies (3.7E 12 Bq)
  ENVIRONMENTAL LISTS
    CERCLA/SARA List of Radionuclides (Appendix B) and Their Reportable Quantities
      final RQ = 100 curies (3.7E 12 Bq)

CESIUM CARBONATE   534-17-8
  HEALTH AND SAFETY LISTS
    NIOSH - Selected LD50s and LC50s
      Oral, rat: LD50 = 2333 mg/kg
  INTERNATIONAL LISTS
    Canada - WHMIS: Ingredient Disclosure
      1% item 323 (386)

CESIUM CHLORIDE   7647-17-8
  HEALTH AND SAFETY LISTS
    NIOSH - Selected LD50s and LC50s
      Oral, rat: LD50 = 2600 mg/kg
  INTERNATIONAL LISTS
    Canada - WHMIS: Ingredient Disclosure
      1% item 324 (483)

CESIUM CHROMATE   13454-78-9
  STATE LISTS
    Massachusetts Right To Know List
      [present]

CESIUM HYDROXIDE   21351-79-1
  HEALTH AND SAFETY LISTS
    ACGIH 1995 - Time Weighted Averages
      2 mg/m3 TWA
    U.S. DOT - Substances From 49 CFR 172.101
      regulated by DOT (UN2682, UN2681)
    U.S. DOT - Hazard Classes
      DOT hazard class = 8
    NIOSH - Selected LD50s and LC50s
      Oral, rat: LD50 = 570 mg/kg
    OSHA - Vacated PELs - Time Weighted Averages
      2 mg/m3 TWA
  INTERNATIONAL LISTS
    Australian Exposure Standards - Time Weighted Averages
      2 mg/m3 TWA
    Canada - WHMIS: Ingredient Disclosure
      1% item 325 (992)

Canada - Alberta - 8 Hour Occupational Exposure Limit
  2 mg/m3 TWA
Canada - Alberta - 15 Minute Occupational Exposure Limit
  4 mg/m3 STEL
Canada - British Columbia - 8 Hour Exposure Limits
  2 mg/m3 TWA
Canada - Ontario - OHSA - TWAEVs
  2 mg/m3 TWAEV
Canada - Quebec - Time-Weighted Average Exposure Values
  2 mg/m3 TWAEV
United Kingdom - Occupational Exposure Standards - TWAs
  2 mg/m3 TWA
Israel - Time Weighted Averages
  2 mg/m3 TWA
Israel - Action Levels
  1 mg/m3 AL
Mexico - Instruction No. 10 - TWAs
  2 mg/m3 TWA
STATE LISTS
  California - Exposure Limits - PELs
    2 mg/m3 PEL
  California - Directors List of Hazardous Substances (8 CCR 339)
    [present]
  Florida Hazardous Substance List
    [present]
  Massachusetts Right To Know List
    [present]
  Minnesota Hazardous Substance List
    [present]
  NJ Right to Know List (Total)
    sn 0354
  NJ Special Hazardous Substances
    (corrosive)
  Pennsylvania Right to Know List
    [present]

CESIUM IODIDE   7789-17-5
  HEALTH AND SAFETY LISTS
    NIOSH - Selected LD50s and LC50s
      Oral, rat: LD50 = 2386 mg/kg

CESIUM NITRATE   7789-18-6
  HEALTH AND SAFETY LISTS
    U.S. DOT - Substances From 49 CFR 172.101
      regulated by DOT (UN1451)
    U.S. DOT - Hazard Classes
      DOT hazard class = 5.1
    NIOSH - Selected LD50s and LC50s
      Oral, rat: LD50 = 2390 mg/kg
  INTERNATIONAL LISTS
    Canada - WHMIS: Ingredient Disclosure
      1% item 326 (1202)
  STATE LISTS
    NJ Right to Know List (Total)
      sn 0355

CESIUM SULFATE   10294-54-9
  INTERNATIONAL LISTS
    Canada - WHMIS: Ingredient Disclosure
      1% item 327 (1518)

CETYL ALCOHOL   36653-82-4

HEALTH AND SAFETY LISTS

NIOSH - Selected LD50s and LC50s
Oral, rat: LD50 = 6400 mg/kg

ENVIRONMENTAL LISTS

List of Pesticide Product Inert Ingredients
[present]

CETYLDIMETHYLAMINE OXIDE   7128-91-8

ENVIRONMENTAL LISTS

List of Pesticide Product Inert Ingredients
[present]

CETYL OCTANOATE   29710-31-4

ENVIRONMENTAL LISTS

List of Pesticide Product Inert Ingredients
[present]

CHEESE   RR-01053-7

ENVIRONMENTAL LISTS

List of Pesticide Product Inert Ingredients
[present]

CHEMICAL WARFARE SUBSTANCES   RR-01608-0

INTERNATIONAL LISTS

Canada - CEPA Schedule III Part I - Prohibited Substances (Ocean Dumping)
[present]

CHENODIOL   474-25-9

STATE LISTS

California - Air Bill 2588 Appendix A-II
9/90

California - Prop. 65 - Developmental Toxicity
developmental toxicity - initial date 4/1/90

CHLORINATED HYDROCARBONS (CHORINATED   63449-39-8
PARAFFINS)
SEE ALSO:
CHLORINATED PARAFFINS

ENVIRONMENTAL LISTS

List of Pesticide Product Inert Ingredients
[present]

TSCA - Code of Federal Regulations Citations
40 CFR 716.120

INTERNATIONAL LISTS

Canada - CEPA - Priority Substances List
estimated time for completion of assessment reports: 5 years

German (DFG) - Carcinogens
suspected carcinogen

CHINOMETHIONAT [6-METHYL-1,3-DITHIOLO[4,5-   2439-01-2
B]QUINOXALIN-2-ONE]

ENVIRONMENTAL LISTS

CERCLA/SARA - Section 313 - Emission Reporting
form R reporting required

CHLORAL   75-87-6

HEALTH AND SAFETY LISTS

U.S. DOT - Substances From 49 CFR 172.101
regulated by DOT (UN2075)

U.S. DOT - Hazard Classes
DOT hazard class = 6.1

U.S. DOT - Appendix A Table 1 - Hazardous Substances
final RQ = 5000 pounds (2270 kg)

ENVIRONMENTAL LISTS

CERCLA/SARA - Hazardous Substances and their Reportable Quantities
final RQ = 5000 pounds (2270 kg)

CAA - HON Rule - SOCMI Chemicals
compliance by Jan. 23, 1995

RCRA - U Series Wastes
waste number U034

RCRA - Hazardous Constituents-Appendix VIII
waste number U034

RCRA - Substances Banned From Land Disposal
[present]

TSCA - Code of Federal Regulations Citations
40 CFR 712.30(x); 40 CFR 716.120(d)

TSCA - PAIR - Reporting List
Reporting Date: November 27, 1991

TSCA - Health and Safety Reporting List
Effective Date: September 30, 1991

INTERNATIONAL LISTS

Canada - WHMIS: Ingredient Disclosure
1% item 328 (399)

STATE LISTS

Massachusetts Right To Know List
[present]

NJ Right to Know List (Total)
sn 0356

Pennsylvania Right to Know List
environmental hazard

CHLORAMBEN   133-90-4

HEALTH AND SAFETY LISTS

NIOSH - Selected LD50s and LC50s
Oral, rat: LD50 = 3500 mg/kg Skin, rabbit: LD50 = 3136 mg/kg

NTP Chemical Status Reports - Testing Status and NTIS Number
Technical reports printed (PB273065/AS)

NTP Chemical Status Reports - Evidence of Carcinogenicity
male rat-negative; female rat-negative; male mice-equivocal; female mice-positive

ENVIRONMENTAL LISTS

CERCLA/SARA - Section 313 - Emission Reporting
form R reporting required for 1.0% de minimus concentration

CERCLA/SARA - Hazardous Substances and their Reportable Quantities
final RQ = 1 pound (.454 kg)

Clean Air Act (1990) - List of Hazardous Air Contaminants
[present]

STATE LISTS

California - Air Bill 2588 Appendix A-I
6/91

Massachusetts Right To Know List
carcinogen; extraordinarily hazardous

NJ Right to Know List (Total)
sn 0357

Pennsylvania Right to Know List
environmental hazard

CHLORAMBUCIL   305-03-3

HEALTH AND SAFETY LISTS

U.S. DOT - Appendix A Table 1 - Hazardous Substances
final RQ = 10 pounds (4.54 kg)

IARC - Group 1 (carcinogenic to humans)
[present]

NIOSH - Selected LD50s and LC50s
Oral, rat: LD50 = 76 mg/kg

NTP Chemical Status Reports - Testing Status and NTIS Number
Chronic studies exist for which technical reports were not prepared

NTP Seventh Report - Known Carcinogens
known carcinogen

OSHA - Select Carcinogens
[present]

ENVIRONMENTAL LISTS

CERCLA/SARA - Hazardous Substances and their Reportable Quantities
final RQ = 10 pounds (4.54 kg)

EPA - Carcinogen Hazard Ranking for RQ Adjustment
Hazard ranking = Medium

RCRA - U Series Wastes
waste number U035

RCRA - Hazardous Constituents-Appendix VIII
waste number U035

RCRA - Substances Banned From Land Disposal
[present]

STATE LISTS

California - Air Bill 2588 Appendix A-II
known or potential carcinogen

California - Prop. 65 - Cancer list
carcinogen - initial date 2/27/87

California - Prop. 65 - Developmental Toxicity
developmental toxicity - initial date 1/1/89

California - Prop. 65 - No Significant Risk Levels
no significant risk level = 0.002 ug/day

California - Directors List of Hazardous Substances (8 CCR 339)
[present]

Florida Hazardous Substance List
[present]

Massachusetts Right To Know List
carcinogen; extraordinarily hazardous; teratogen

Minnesota Hazardous Substance List
carcinogen

NJ Right to Know List (Total)
sn 0358

NJ Special Hazardous Substances
(carcinogen; mutagen; teratogen)

Pennsylvania Right to Know List
environmental hazard; special hazardous substance

Pennsylvania RTK - Special Hazardous Substances
[present]

CHLORAMINATED WATER   RR-01022-0

HEALTH AND SAFETY LISTS

NTP Chemical Status Reports - Testing Status and NTIS Number
Technical reports printed (PB92191659)

NTP Chemical Status Reports - Evidence of Carcinogenicity
male rat: no evidence; female rat: equivocal evidence; male mice: no evidence; female mice: no evidence

CHLORAMINE   10599-90-3

PROPOSED REGULATIONS

Safe Drinking Water Act - Priority list
[present]

CHLORAMINE-T   127-65-1

INTERNATIONAL LISTS

Canada - WHMIS: Ingredient Disclosure
0.1% item 329 (400)

CHLORAMPHENICOL   56-75-7

HEALTH AND SAFETY LISTS

IARC - Group 2A (limited human data)
[present]

NIOSH - Selected LD50s and LC50s
Oral, rat: LD50 = 2500 mg/kg

OSHA - Possible Select Carcinogens
[present]

STATE LISTS

California - Air Bill 2588 Appendix A-I
known or potential carcinogen

California - Prop. 65 - Cancer list
carcinogen - initial date 10/1/89

California - Directors List of Hazardous Substances (8 CCR 339)
[present]

Florida Hazardous Substance List
[present]

Massachusetts Right To Know List
carcinogen; extraordinarily hazardous

Minnesota Hazardous Substance List
carcinogen

NJ Right to Know List (Total)
sn 0360

NJ Special Hazardous Substances
(carcinogen; mutagen; teratogen)

Pennsylvania Right to Know List
special hazardous substance

Pennsylvania RTK - Special Hazardous Substances
[present]

CHLORAMPHENICOL SODIUM SUCCINATE   982-57-0

HEALTH AND SAFETY LISTS

NTP Chemical Status Reports - Testing Status and NTIS Number
Prechronic studies for which toxicity technical reports were not prepared

CHLORATE   14866-68-3

PROPOSED REGULATIONS

Safe Drinking Water Act - Priority list
[present]

CHLORATE AND MAGNESIUM CHLORIDE MIXTURES   RR-00200-6

HEALTH AND SAFETY LISTS

U.S. DOT - Substances From 49 CFR 172.101
regulated by DOT (UN1459)

U.S. DOT - Hazard Classes
DOT hazard class = 5.1

STATE LISTS

NJ Right to Know List (Total)
sn 2231

CHLORATES, INORGANIC, N.O.S.   RR-00201-7

HEALTH AND SAFETY LISTS

U.S. DOT - Substances From 49 CFR 172.101
regulated by DOT (UN1461)

U.S. DOT - Hazard Classes
DOT hazard class = 5.1

STATE LISTS

NJ Right to Know List (Total)
sn 2232

CHLORCYCLIZINE HYDROCHLORIDE   **1620-21-9**

STATE LISTS

California - Prop. 65 - Developmental Toxicity
developmental toxicity - initial date 7/1/87

CHLORCYCLIZINE HYDROCHLORIDE   **14362-31-3**

STATE LISTS

California - Air Bill 2588 Appendix A-II
[present]

CHLORDANE   **57-74-9**
SEE ALSO:
  ISOOCTYL NITRATE
  F039-HAZARDOUS WASTES
  K097-HAZARDOUS WASTES

HEALTH AND SAFETY LISTS

ACGIH 1995 - Time Weighted Averages
0.5 mg/m3 TWA

ACGIH 1995 - Skin Designations
skin - potential for cutaneous absorption

U.S. DOT - Appendix B - Marine Pollutants
DOT regulated severe marine pollutant

U.S. DOT - Appendix A Table 1 - Hazardous Substances
final RQ = 1 pound (0.454 kg)

IARC - Group 2B (sufficient animal data)
[present]

NIOSH - Selected LD50s and LC50s
Inhalation, cat: LC50 = 100 mg/m3 4 hr Oral, rat: LD50 = 283 mg/kg Skin, rat: LD50 = 690 mg/kg

NIOSH 1990 - Pocket Guide - RELs
0.5 mg/m3 TWA

NIOSH 1990 - Pocket Guide - IDLHs
500 mg/m3 IDLH (not considering carcinogenic effects)

NIOSH 1990 - Pocket Guide - Carcinogens
occupational carcinogen

NIOSH 1990 - Pocket Guide - Target organs
CNS, eyes, lungs, liver, kidneys, skin

NIOSH 1990 - Pocket Guide - Skin list
Potential for dermal absorption

NTP Chemical Status Reports - Testing Status and NTIS Number
Technical reports printed (PB271977/AS)

NTP Chemical Status Reports - Evidence of Carcinogenicity
male rat-negative; female rat-negative; male mice-positive; female mice-positive

OSHA - Vacated PELs - Time Weighted Averages
0.5 mg/m3 TWA

OSHA - Vacated PELs - Skin Designation
Prevent or reduce skin absorption

OSHA - Final PELs - Time Weighted Averages
0.5 mg/m3 TWA

OSHA - Final PELs - Skin Notations
prevent or reduce skin absorption

OSHA - Possible Select Carcinogens
[present]

ENVIRONMENTAL LISTS

ATSDR Priority List
Rank (of 275): 021

CERCLA/SARA - Section 302 Extremely Hazardous Substances and TPQs
TPQ = 1000 pounds

CERCLA/SARA - Section 313 - Emission Reporting
form R reporting required for 1.0% de minimus concentration

CERCLA/SARA - Hazardous Substances and their Reportable Quantities
final RQ = 1 pound (0.454 kg)

Clean Air Act (1990) - List of Hazardous Air Contaminants
[present]

Clean Water Act - Hazardous Substances
[present]

Clean Water Act - Priority Pollutants
[present]

Clean Water Act - Toxic Pollutants
[present]

Safe Drinking Water Act - MCLs
MCL = 0.002 mg/L

Safe Drinking Water Act - MCLGs
MCLG = Zero

EPA - Carcinogen Hazard Ranking for RQ Adjustment
Hazard ranking = Medium

RCRA - D Series - Maximum Concentration of Contaminants
waste number D020; regulatory level = 0.03 mg/L

RCRA - D Series - Chronic Toxicity Reference Levels
chronic toxicity reference level = 0.0003 mg/L

RCRA - U Series Wastes
waste number U036

RCRA - Hazardous Constituents-Appendix VIII
waste number U036

RCRA - Basis for Listing - Appendix VII
Included in waste streams: F039, K097

RCRA - Substances Banned From Land Disposal
[present]

RCRA - TSD Facilities Ground Water Monitoring
TM 8080 = 0.01 ug/L PQL;   TM 8250 = 10 ug/L PQL

RCRA - Universal Treatment Standards (LDR)
WW: 0.0033 mg/l;   NWW: 0.26 mg/kg

INTERNATIONAL LISTS

Australian Exposure Standards - Time Weighted Averages
0.5 mg/m3 TWA

Australian Exposure Standards - Skin Effects
skin absorption

Australian Exposure Standards - Carcinogens
suspected carcinogen

Canada - CEPA Schedule II Part II - Toxic Substances (Export)
[present]

Canada - Drinking Water Quality - MACs
0.007 mg/L MAC

Canada - Alberta - 8 Hour Occupational Exposure Limit
0.5 mg/m3 TWA

Canada - Alberta - 15 Minute Occupational Exposure Limit
2 mg/m3 STEL

Canada - Alberta - Skin Designation
can be absorbed through the intact skin

Canada - British Columbia - 8 Hour Exposure Limits
0.5 mg/m3 TWA

Canada - British Columbia - 15 Minute Exposure Limits
2 mg/m3 STEL

Canada - British Columbia - Skin Notations
skin - potential for skin absorption

Canada - Ontario - OHSA - TWAEVs
0.5 mg/m3 TWAEV

Canada - Ontario - OHSA - STEVs
2 mg/m3 STEV

Canada - Ontario - OHSA - Skin Notations
absorption through skin, eyes, or mucous membranes

Canada - Quebec - Time-Weighted Average Exposure Values
0.5 mg/m3 TWAEV
Canada - Quebec - Skin Designations
absorbed through the skin
German (DFG) - MAK Values
total dust: 0.5 mg/m3 MAK
German (DFG) - Peak Limitations
10 x normal MAK (30 min average value);   don't exceed during shift
German (DFG) - Skin/Sensitizers
danger of cutaneous absorption
German (DFG) - Carcinogens
suspected carcinogen
Israel - Time Weighted Averages
0.5 mg/m3 TWA
Israel - Action Levels
0.25 mg/m3 AL
Mexico - Instruction No. 10 - TWAs
0.5 mg/m3 TWA
Mexico - Instruction No. 10 - STELs
2 mg/m3 STEL
Mexico - Instruction No. 10 - Skin designation
skin - potential for cutaneous absorption
Mexico - Wastewater - Organic Toxic Pollutants and Heavy Metals
Listed under [Pesticides and Metabolites]
Mexico - Drinking Water - Ecological Criteria
0.003 mg/l Substance presents persistence, bioaccumulations or risk of cancer, reduce human exposure to a minimum;   This level has been extrapolated by using a mathematic model

STATE LISTS
California - Air Bill 2588 Appendix A-I
known or potential carcinogen: 9/89
California - Prop. 65 - Cancer list
carcinogen - initial date 7/1/88
California - Prop. 65 - No Significant Risk Levels
no significant risk level = 0.5 ug/day
California - Exposure Limits - PELs
0.5 mg/m3 PEL
California - Exposure Limits - Skin Notation
material may be absorbed through the skin, eyes or mucous membrane
California - Directors List of Hazardous Substances (8 CCR 339)
[present]
Florida Hazardous Substance List
[present]
Massachusetts Right To Know List
carcinogen;   extraordinarily hazardous
Minnesota Hazardous Substance List
skin
NJ Right to Know List (Total)
sn 0361
NJ Special Hazardous Substances
(carcinogen)
Pennsylvania Right to Know List
environmental hazard (including technical mixture and metabolites)

ALPHA CHLORDANE   5103-71-9
ENVIRONMENTAL LISTS
ATSDR Priority List
Rank (of 275): 048

GAMMA CHLORDANE   5103-74-2
ENVIRONMENTAL LISTS
ATSDR Priority List
Rank (of 275): 054

CHLORDANE/HEPTACHLOR   RR-01530-5
HEALTH AND SAFETY LISTS
IARC - Group 3 (not classifiable)
[present]

CHLORDENE   3734-48-3
ENVIRONMENTAL LISTS
ATSDR Priority List
Rank (of 275): 189

CHLORDIAZEPOXIDE   58-25-3
HEALTH AND SAFETY LISTS
NIOSH - Selected LD50s and LC50s
Oral, rat: LD50 = 548 mg/kg
STATE LISTS
California - Prop. 65 - Developmental Toxicity
developmental toxicity - initial date 1/1/92
NJ Right to Know List (Total)
sn 0362
NJ Special Hazardous Substances
(teratogen)

CHLORDIAZEPOXIDE HYDROCHLORIDE   438-41-5
STATE LISTS
California - Prop. 65 - Developmental Toxicity
developmental toxicity - initial date 1/1/92
NJ Right to Know List (Total)
sn 0363
NJ Special Hazardous Substances
(teratogen)

CHLORDIMEFORM   6164-98-3
HEALTH AND SAFETY LISTS
IARC - Group 3 (not classifiable)
[present]
STATE LISTS
California - Air Bill 2588 Appendix A-II
known or potential carcinogen: 9/89
California - Prop. 65 - Cancer list
carcinogen - initial date 1/1/89
California - Directors List of Hazardous Substances (8 CCR 339)
[present]
Massachusetts Right To Know List
[present]

CHLORENDIC ACID   115-28-6
HEALTH AND SAFETY LISTS
IARC - Group 2B (sufficient animal data)
[present]
NTP Chemical Status Reports - Testing Status and NTIS Number
Technical reports printed (PB87206835/AS)
NTP Chemical Status Reports - Evidence of Carcinogenicity
male rat-clear evidence;   female rat-clear evidence;   male mice-clear evidence;   female mice-no evidence
NTP Seventh Report - Suspect Carcinogens
suspect carcinogen
OSHA - Possible Select Carcinogens
[present]
ENVIRONMENTAL LISTS
CERCLA/SARA - Section 313 - Emission Reporting
form R reporting required
TSCA - Code of Federal Regulations Citations
40 CFR 712.30(d);   40 CFR 716.120(a)

TSCA - PAIR - Reporting List
Reporting Date: November 19, 1982

TSCA - Health and Safety Reporting List
Effective Date: April 29, 1983

STATE LISTS

California - Air Bill 2588 Appendix A-II
known or potential carcinogen: 9/89

California - Prop. 65 - Cancer list
carcinogen - initial date 7/1/89

California - Prop. 65 - No Significant Risk Levels
no significant risk level = 8 ug/day

California - Directors List of Hazardous Substances (8 CCR 339)
[present]

Florida Hazardous Substance List
effective March 13, 1992

CHLORFENVINPHOS   470-90-6

HEALTH AND SAFETY LISTS

U.S. DOT - Appendix B - Marine Pollutants
DOT regulated marine pollutant

NIOSH - Selected LD50s and LC50s
Oral, rat: LD50 = 10 mg/kg Skin, rat: LD50 = 26400 ug/kg

ENVIRONMENTAL LISTS

CERCLA/SARA - Section 302 Extremely Hazardous Substances and TPQs
TPQ = 500 pounds

STATE LISTS

California - Directors List of Hazardous Substances (8 CCR 339)
[present]

Florida Hazardous Substance List
effective March 13, 1992

Massachusetts Right To Know List
extraordinarily hazardous

NJ Right to Know List (Total)
sn 0364

Pennsylvania Right to Know List
environmental hazard

PROPOSED REGULATIONS

CERCLA/SARA - Proposed Hazardous Substance Additions
proposed RQ = 1 pound (.454 kg)

CERCLA/SARA - 1989 Proposed RQ Adjustments
proposed RQ = 100 pounds (45.4 kg)

CHLORHEXIDINE DIACETATE   56-95-1

ENVIRONMENTAL LISTS

List of Pesticide Product Inert Ingredients
[present]

CHLORIC ACID   7790-93-4

HEALTH AND SAFETY LISTS

U.S. DOT - Substances From 49 CFR 172.101
regulated by DOT (UN2626)

U.S. DOT - Hazard Classes
DOT hazard class = 5.1

INTERNATIONAL LISTS

Canada - WHMIS: Ingredient Disclosure
1% item 330 (73)

STATE LISTS

NJ Right to Know List (Total)
frozen: sn 0365

CHLORIDE   16887-00-6

ENVIRONMENTAL LISTS

Safe Drinking Water Act - SMCLs
SMCL = 250 mg/L

INTERNATIONAL LISTS

Canada - Drinking Water Quality - AOs
<= 250 mg/L AO

Mexico - Drinking Water - Ecological Criteria
250 mg/l

CHLORIMURON ETHYL [ETHYL-2-[[[(4-CHLORO-6-   90982-32-4
METHOXYPYRIMIDIN-2-YL) -CARBONYL]-AMINO]
SULFONYL]BENZOATE]

ENVIRONMENTAL LISTS

CERCLA/SARA - Section 313 - Emission Reporting
form R reporting required

CHLORINATED BENZENES   RR-00077-1

ENVIRONMENTAL LISTS

Clean Water Act - Toxic Pollutants
[present]

RCRA - Hazardous Constituents-Appendix VIII
hazardous constituent - no waste number

TSCA - Health and Safety Reporting List
Effective Date: October 4, 1991 (mono-, di-, tri-, tetra-, and penta-)

STATE LISTS

California - Directors List of Hazardous Substances (8 CCR 339)
[present]

Pennsylvania Right to Know List
environmental hazard

CHLORINATED CRESOLS   RR-01024-2

STATE LISTS

California - Directors List of Hazardous Substances (8 CCR 339)
[present]

CHLORINATED DIBENZODIOXINS   RR-01531-6

HEALTH AND SAFETY LISTS

IARC - Group 3 (not classifiable)
[present] (does not include TCDD)

CHLORINATED DIPHENYL OXIDE   31242-93-0

HEALTH AND SAFETY LISTS

NFPA - Hazard Identification Ratings
health-2; flammability-1; reactivity-1

NIOSH 1990 - Pocket Guide - Target organs
skin, liver

OSHA - Final PELs - Time Weighted Averages
0.5 mg/m3 TWA

INTERNATIONAL LISTS

Canada - WHMIS: Ingredient Disclosure
1% item 331 (1309)

Canada - Alberta - 8 Hour Occupational Exposure Limit
0.5 mg/m3 TWA

Canada - Alberta - 15 Minute Occupational Exposure Limit
2 mg/m3 STEL

Canada - British Columbia - 8 Hour Exposure Limits
0.5 mg/m3 TWA

Canada - British Columbia - 15 Minute Exposure Limits
2 mg/m3 STEL

Canada - Quebec - Time-Weighted Average Exposure Values
0.5 mg/m3 TWAEV

Israel - Time Weighted Averages
0.5 mg/m3 TWA
Israel - Action Levels
0.25 mg/m3 AL
STATE LISTS
California - Exposure Limits - PELs
0.5 mg/m3 PEL
Massachusetts Right To Know List
[present]
Pennsylvania Right to Know List
[present]

CHLORINATED DIPHENYL OXIDE   55720-99-5

HEALTH AND SAFETY LISTS
ACGIH 1995 - Time Weighted Averages
0.5 mg/m3 TWA
NIOSH 1990 - Pocket Guide - RELs
0.5 mg/m3 TWA
OSHA - Vacated PELs - Time Weighted Averages
0.5 mg/m3 TWA
INTERNATIONAL LISTS
Australian Exposure Standards - Time Weighted Averages
0.5 mg/m3 TWA
Australian Exposure Standards - Short Term Exposure Limits
2 mg/m3 STEL
Canada - Ontario - OHSA - TWAEVs
0.5 mg/m3 TWAEV (listed as agent of variable composition)
Canada - Ontario - OHSA - STEVs
2 mg/m3 STEV
German (DFG) - MAK Values
total dust: 0.5 mg/m3 MAK
German (DFG) - Skin/Sensitizers
danger of cutaneous absorption
STATE LISTS
California - Directors List of Hazardous Substances (8 CCR 339)
[present]
Florida Hazardous Substance List
[present]
Minnesota Hazardous Substance List
[present]
Pennsylvania Right to Know List
[present]

CHLORINATED DIPHENYL OXIDES   RR-01624-0

INTERNATIONAL LISTS
Canada - Ontario - OHSA - TWAEVs
0.5 mg/m3 TWAEV (listed as agent of variable composition)
Canada - Ontario - OHSA - STEVs
2 mg/m3 STEV

CHLORINATED ETHANES   RR-00202-8

ENVIRONMENTAL LISTS
Clean Water Act - Toxic Pollutants
[present]
RCRA - Hazardous Constituents-Appendix VIII
hazardous constituent - no waste number
STATE LISTS
California - Directors List of Hazardous Substances (8 CCR 339)
[present]
Pennsylvania Right to Know List
environmental hazard

CHLORINATED FLUOROCARBONS   RR-00747-6

ENVIRONMENTAL LISTS
RCRA - Hazardous Constituents-Appendix VIII
hazardous constituent - no waste number
RCRA - Basis for Listing - Appendix VII
Included in waste stream: F001
STATE LISTS
California - Air Bill 2588 Appendix A-I
[present]

CHLORINATED NAPHTHALENES   RR-00279-9

ENVIRONMENTAL LISTS
TSCA - Health and Safety Reporting List
Effective Date: October 4, 1991 (chlorinated derivatives of naphthalene (empirical formula) C10HxCly where x+y=8)

CHLORINATED PARAFFINS   RR-00278-8

ENVIRONMENTAL LISTS
TSCA - Health and Safety Reporting List
Effective Date: October 4, 1991 (chlorinated paraffin oils and chlorinated parffin waxes, with chlorine content of 35 percent through 70 percent by weight)
STATE LISTS
California - Prop. 65 - No Significant Risk Levels
no significant risk level = 8 ug/day (average chain length - C12)

CHLORINATED PARAFFINS: C12, 60% CHLORINE   108171-26-2

HEALTH AND SAFETY LISTS
IARC - Group 2B (sufficient animal data)
[present]
NTP Chemical Status Reports - Testing Status and NTIS Number
Technical reports printed (PB86248101/AS)
NTP Chemical Status Reports - Evidence of Carcinogenicity
male rat-clear evidence; female rat-clear evidence; male mice-clear evidence; female mice-clear evidence
NTP Seventh Report - Suspect Carcinogens
suspect carcinogen
OSHA - Possible Select Carcinogens
[present]
STATE LISTS
California - Air Bill 2588 Appendix A-I
known or potential carcinogen: 9/89
California - Prop. 65 - Cancer list
carcinogen - initial date 7/1/89
California - Directors List of Hazardous Substances (8 CCR 339)
[present]

CHLORINATED PARAFFINS: C23, 43% CHLORINE   108171-27-3

HEALTH AND SAFETY LISTS
NTP Chemical Status Reports - Testing Status and NTIS Number
Technical reports printed (PB86248093/AS)
NTP Chemical Status Reports - Evidence of Carcinogenicity
male rat-no evidence; female rat-equivocal evidence; male mice-clear evidence; female mice-equivocal evidence

CHLORINATED PHENOLS   RR-00204-0

ENVIRONMENTAL LISTS
Clean Water Act - Toxic Pollutants
[present]
RCRA - Hazardous Constituents-Appendix VIII
hazardous constituent - no waste number

STATE LISTS
Pennsylvania Right to Know List
environmental hazard

CHLORINATED POLYETHYLENE   64754-90-1
ENVIRONMENTAL LISTS
List of Pesticide Product Inert Ingredients
[present]

CHLORINATED RUBBER   9006-03-5
ENVIRONMENTAL LISTS
List of Pesticide Product Inert Ingredients
[present]

ALPHA-CHLORINATED TOLUENES   RR-00534-5
HEALTH AND SAFETY LISTS
IARC - Group 2B (sufficient animal data)
[present]
OSHA - Possible Select Carcinogens
[present]
STATE LISTS
California - Air Bill 2588 Appendix A-II
known or potential carcinogen
California - Directors List of Hazardous Substances (8 CCR 339)
[present]
Minnesota Hazardous Substance List
carcinogen

CHLORINATED TRISODIUM PHOSPHATE   56802-99-4
HEALTH AND SAFETY LISTS
NTP Chemical Status Reports - Testing Status and NTIS Number
Technical reports printed (PB87189718/AS)
NTP Chemical Status Reports - Evidence of Carcinogenicity
male rat-inadequate; female rat-inadequate; male mice-no evidence; female mice-no evidence

CHLORINATED WASTEWATER EFFLUENTS   RR-01613-7
INTERNATIONAL LISTS
Canada - CEPA - Priority Substances List
estimated time for completion of assessment reports: 4 years

CHLORINATED WATER   RR-01023-1
HEALTH AND SAFETY LISTS
IARC - Group 3 (not classifiable)
[present]
NTP Chemical Status Reports - Testing Status and NTIS Number
Technical reports printed (no NTIS number given)
NTP Chemical Status Reports - Evidence of Carcinogenicity
male rat: no evidence; female rat: equivocal evidence; male mice: no evidence; female mice: no evidence

CHLORINATION/CHLORAMINATION BY PRODUCTS (MISC.)   RR-00112-7
PROPOSED REGULATIONS
Safe Drinking Water Act - Priority list
[present]

CHLORINE   7782-50-5
HEALTH AND SAFETY LISTS
ACGIH 1995 - Time Weighted Averages
0.5 ppm TWA; 1.5 mg/m3 TWA
ACGIH 1995 - Short Term Exposure Limits
1 ppm STEL; 2.9 mg/m3 STEL

AIHA - Odor Threshold Values
geometric mean air odor threshold = 0.080 ppm (detectable)
U.S. DOT - Substances From 49 CFR 172.101
regulated by DOT (UN1017)
U.S. DOT - Hazard Classes
DOT hazard class = 2.3
U.S. DOT - Substances Which Are Poisonous by Inhalation
gaseous hazardous material poisonous by inhalation (UN1017)
U.S. DOT - Appendix B - Marine Pollutants
DOT regulated marine pollutant
U.S. DOT - Appendix A Table 1 - Hazardous Substances
final RQ = 10 pounds (4.54 kg)
NIOSH - Selected LD50s and LC50s
Inhalation, rat: LC50 = 293 ppm 1 hr
NIOSH 1990 - Pocket Guide - RELs
0.5 ppm TWA; 1.5 mg/m3 TWA; 1 ppm STEL; 3 mg/m3 STEL
NIOSH 1990 - Pocket Guide - IDLHs
30 ppm IDLH
NIOSH 1990 - Pocket Guide - Target organs
respiratory system
NIOSH - Health Standards - Exposure Limits
C (15 min) 0.5 ppm; C (15 min) 1.45 mg/m3
NIOSH - Health Standards - Health Effects and Precautions
Eye and respiratory irritation (Periodic chest X-ray required)
OSHA - Vacated PELs - Time Weighted Averages
0.5 ppm TWA; 1.5 mg/m3 TWA
OSHA - Vacated PELs - Short Term Exposure Limits
1 ppm STEL; 3 mg/m3 STEL
OSHA - Final PELs - Ceiling Limits
C 1 ppm; C 3 mg/m3
OSHA - List of Highly Hazardous Chemicals
threshhold quantity = 1500 pounds
ENVIRONMENTAL LISTS
ATSDR Priority List
Rank (of 275): 208
CERCLA/SARA - Section 302 Extremely Hazardous Substances and TPQs
TPQ = 100 pounds
CERCLA/SARA - Section 313 - Emission Reporting
form R reporting required for 1.0% de minimus concentration
CERCLA/SARA - Hazardous Substances and their Reportable Quantities
final RQ = 10 pounds (4.54 kg)
Clean Air Act (1990) - List of Hazardous Air Contaminants
[present]
CAA -Toxic Substances for Accidental Release Prevention
threshold quantity = 2,500 lbs
Clean Water Act - Hazardous Substances
[present]
TSCA - Code of Federal Regulations Citations
40 CFR 704.225(a)
TSCA - CAIR - Reporting List
reporting required by: manufacturer
INTERNATIONAL LISTS
Australian Exposure Standards - Time Weighted Averages
Peak Limitation: 1 ppm; 3 mg/m3
Australian Exposure Standards - Under Review
exposure limits under review
Canada - WHMIS: Ingredient Disclosure
1% item 332 (401)
Canada - NPRI (National Pollutant Release Inventory)
[present]

Canada - Alberta - 8 Hour Occupational Exposure Limit
0.5 ppm TWA; 1.5 mg/m3 TWA
Canada - Alberta - 15 Minute Occupational Exposure Limit
1 ppm STEL; 3 mg/m3 STEL
Canada - British Columbia - 8 Hour Exposure Limits
0.5 ppm TWA
Canada - British Columbia - 15 Minute Exposure Limits
1 ppm STEL
Canada - Ontario - OHSA - TWAEVs
1 ppm TWAEV; 3 mg/m3 TWAEV
Canada - Ontario - OHSA - STEVs
3 ppm STEV; 9 mg/m3 STEV
Canada - Quebec - Time-Weighted Average Exposure Values
1 ppm TWAEV; 3 mg/m3 TWAEV
Canada - Quebec - Short-term Exposure Values
3 ppm STEV; 9 mg/m3 STEV
United Kingdom - Occupational Exposure Standards - TWAs
0.5 ppm TWA; 1.5 mg/m3 TWA
United Kingdom - Occupational Exposure Standards - STELs
1 ppm STEL; 3 mg/m3 STEL
German (DFG) - MAK Values
0.5 ppm MAK; 1.5 mg/m3 MAK
German (DFG) - Peak Limitations
2 x normal MAK (5 min momentary value);   don't exceed 8 times during shift
German (DFG) - Pregnancy
no risk to embryo/fetus if exposure limits adhered to
Israel - Time Weighted Averages
0.5 ppm TWA; 1.5 mg/m3 TWA
Israel - Short Term Exposure Limits
1 ppm STEL; 2.9 mg/m3 STEL
Israel - Action Levels
0.25 ppm AL; 0.75 mg/m3 AL
Mexico - Instruction No. 10 - TWAs
1 ppm TWA; 3 mg/m3 TWA
Mexico - Instruction No. 10 - STELs
3 ppm STEL; 9 mg/m3 STEL

STATE LISTS

California - Air Bill 2588 Appendix A-I
[present]
California - Exposure Limits - PELs
0.5 ppm PEL; 1.5 mg/m3 PEL
California - Exposure Limits - STELs
1 ppm STEL; 3 mg/m3 STEL
California - Directors List of Hazardous Substances (8 CCR 339)
[present]
Florida Hazardous Substance List
[present]
Massachusetts Right To Know List
extraordinarily hazardous
Minnesota Hazardous Substance List
[present]
NJ Right to Know List (Total)
sn 0367
Pennsylvania Right to Know List
environmental hazard

PROPOSED REGULATIONS

Safe Drinking Water Act - Priority list
[present]

CHLORINE 36   13981-43-6

HEALTH AND SAFETY LISTS

U.S. DOT - Appendix A Table 2 - Radionuclides
final RQ = 10 curies (3.7E 11 Bq)

ENVIRONMENTAL LISTS

CERCLA/SARA List of Radionuclides (Appendix B) and Their Reportable Quantities
final RQ = 10 curies (3.7E 11 Bq)

CHLORINE 38   14158-34-0

HEALTH AND SAFETY LISTS

U.S. DOT - Appendix A Table 2 - Radionuclides
final RQ = 100 curies (3.7E 12 Bq)

ENVIRONMENTAL LISTS

CERCLA/SARA List of Radionuclides (Appendix B) and Their Reportable Quantities
final RQ = 100 curies (3.7E 12 Bq)

CHLORINE 39   15585-26-9

HEALTH AND SAFETY LISTS

U.S. DOT - Appendix A Table 2 - Radionuclides
final RQ = 100 curies (3.7E 12 Bq)

ENVIRONMENTAL LISTS

CERCLA/SARA List of Radionuclides (Appendix B) and Their Reportable Quantities
final RQ = 100 curies (3.7E 12 Bq)

CHLORINE AZIDE   13973-88-1

HEALTH AND SAFETY LISTS

U.S. DOT - Hazard Classes
Forbidden from transport by the DOT

CHLORINE DIOXIDE   10049-04-4

HEALTH AND SAFETY LISTS

ACGIH 1995 - Time Weighted Averages
0.1 ppm TWA; 0.28 mg/m3 TWA
ACGIH 1995 - Short Term Exposure Limits
0.3 ppm STEL; 0.83 mg/m3 STEL
AIHA - Odor Threshold Values
no geometric mean air odor threshold
U.S. DOT - Substances From 49 CFR 172.101
regulated by DOT (NA9191)
U.S. DOT - Hazard Classes
Forbidden from transport by the DOT
NIOSH - Selected LD50s and LC50s
Oral, rat: LD50 = 292 mg/kg
NIOSH 1990 - Pocket Guide - RELs
0.1 ppm TWA; 0.3 mg/m3 TWA; 0.3 ppm STEL; 0.9 mg/m3 STEL
NIOSH 1990 - Pocket Guide - IDLHs
10 ppm IDLH
NIOSH 1990 - Pocket Guide - Target organs
respiratory system, eyes
OSHA - Vacated PELs - Time Weighted Averages
0.1 ppm TWA; 0.3 mg/m3 TWA
OSHA - Vacated PELs - Short Term Exposure Limits
0.3 ppm STEL; 0.9 mg/m3 STEL
OSHA - Final PELs - Time Weighted Averages
0.1 ppm TWA; 0.3 mg/m3 TWA
OSHA - List of Highly Hazardous Chemicals
threshhold quantity = 1000 pounds

ENVIRONMENTAL LISTS

CERCLA/SARA - Section 313 - Emission Reporting
form R reporting required for 1.0% de minimus concentration

CAA -Toxic Substances for Accidental Release Prevention
threshold quantity = 1,000 lbs

List of Pesticide Product Inert Ingredients
[present]

INTERNATIONAL LISTS

Australian Exposure Standards - Time Weighted Averages
0.1 ppm TWA; 0.28 mg/m3 TWA

Australian Exposure Standards - Short Term Exposure Limits
0.3 ppm STEL; 0.83 mg/m3 STEL

Canada - WHMIS: Ingredient Disclosure
1% item 333 (771)

Canada - NPRI (National Pollutant Release Inventory)
[present]

Canada - Alberta - 8 Hour Occupational Exposure Limit
0.1 ppm TWA; 0.27 mg/m3 TWA

Canada - Alberta - 15 Minute Occupational Exposure Limit
0.3 ppm STEL; 0.82 mg/m3 STEL

Canada - British Columbia - 8 Hour Exposure Limits
0.1 ppm TWA; 0.3 mg/m3 TWA

Canada - British Columbia - 15 Minute Exposure Limits
0.3 ppm STEL; 0.9 mg/m3 STEL

Canada - Ontario - OHSA - TWAEVs
0.1 ppm TWAEV; 0.3 mg/m3 TWAEV

Canada - Ontario - OHSA - STEVs
0.3 ppm STEV; 0.9 mg/m3 STEV

Canada - Quebec - Time-Weighted Average Exposure Values
0.1 ppm TWAEV; 0.28 mg/m3 TWAEV

Canada - Quebec - Short-term Exposure Values
0.3 ppm STEV; 0.83 mg/m3 STEV

United Kingdom - Occupational Exposure Standards - TWAs
0.1 ppm TWA; 0.3 mg/m3 TWA

United Kingdom - Occupational Exposure Standards - STELs
0.3 ppm STEL; 0.9 mg/m3 STEL

German (DFG) - MAK Values
0.1 ppm MAK; 0.3 mg/m3 MAK

German (DFG) - Peak Limitations
2 x normal MAK (5 min momentary value);   don't exceed 8 times during shift

Israel - Time Weighted Averages
0.1 ppm TWA; 0.28 mg/m3 TWA

Israel - Short Term Exposure Limits
0.3 ppm STEL; 0.83 mg/m3 STEL

Israel - Action Levels
0.05 ppm AL; 0.14 mg/m3 AL

Mexico - Instruction No. 10 - TWAs
0.1 ppm TWA; 0.3 mg/m3 TWA

Mexico - Instruction No. 10 - STELs
0.3 ppm STEL; 0.9 mg/m3 STEL

STATE LISTS

California - Air Bill 2588 Appendix A-I
6/91

California - Exposure Limits - PELs
0.1 ppm PEL; 0.3 mg/m3 PEL

California - Exposure Limits - STELs
0.3 ppm STEL; 0.9 mg/m3 STEL

California - Directors List of Hazardous Substances (8 CCR 339)
[present]

Florida Hazardous Substance List
[present]

Massachusetts Right To Know List
[present]

Minnesota Hazardous Substance List
[present]

NJ Right to Know List (Total)
sn 0368

Pennsylvania Right to Know List
environmental hazard

PROPOSED REGULATIONS

Safe Drinking Water Act - Priority list
[present]

CHLORINE OXIDE (CL2O)   7791-21-1

HEALTH AND SAFETY LISTS

NFPA - Flash Points
gas (no flash point given)

NFPA - Hazard Identification Ratings
health-3; flammability-4; reactivity-3 (explodes on heating)

ENVIRONMENTAL LISTS

CAA - Flammable Substances for Accidental Release Prevention
threshold quantity = 10,000 lbs

STATE LISTS

Florida Hazardous Substance List
[present]

Massachusetts Right To Know List
[present]

Pennsylvania Right to Know List
[present]

CHLORINE PENTAFLUORIDE   13637-63-3

HEALTH AND SAFETY LISTS

U.S. DOT - Substances From 49 CFR 172.101
regulated by DOT (UN2548)

U.S. DOT - Hazard Classes
DOT hazard class = 2.3

U.S. DOT - Substances Which Are Poisonous by Inhalation
gaseous hazardous material poisonous by inhalation (UN2548)

NIOSH - Selected LD50s and LC50s
Inhalation, rat: LC50 = 122 ppm 1 hr

OSHA - List of Highly Hazardous Chemicals
threshhold quantity = 1000 pounds

INTERNATIONAL LISTS

Canada - WHMIS: Ingredient Disclosure
1% item 334 (1345)

STATE LISTS

NJ Right to Know List (Total)
sn 0369

NJ Special Hazardous Substances
(corrosive)

CHLORINE TRIFLUORIDE   7790-91-2

HEALTH AND SAFETY LISTS

ACGIH 1995 - Ceiling Limits
C 0.1 ppm; C 0.38 mg/m3

U.S. DOT - Substances From 49 CFR 172.101
regulated by DOT (UN1749)

U.S. DOT - Hazard Classes
DOT hazard class = 2.3

U.S. DOT - Substances Which Are Poisonous by Inhalation
gaseous hazardous material poisonous by inhalation (UN1749)

NIOSH - Selected LD50s and LC50s
Inhalation, mouse: LC50 = 178 ppm 1 hr

NIOSH 1990 - Pocket Guide - RELs
C 0.1 ppm; C 0.4 mg/m3

NIOSH 1990 - Pocket Guide - IDLHs
20 ppm IDLH

NIOSH 1990 - Pocket Guide - Target organs
skin, eyes

OSHA - Vacated PELs - Ceiling Limits
C 0.1 ppm; C 0.4 mg/m3

OSHA - Final PELs - Ceiling Limits
C 0.1 ppm; C 0.4 mg/m3

OSHA - List of Highly Hazardous Chemicals
threshhold quantity = 1000 pounds

INTERNATIONAL LISTS

Australian Exposure Standards - Time Weighted Averages
Peak Limitation: 0.1 ppm; 0.38 mg/m3

Canada - WHMIS: Ingredient Disclosure
1% item 335 (1678)

Canada - Alberta - Ceiling Occupational Exposure Limit
C 0.1 ppm; C 0.38 mg/m3

Canada - British Columbia - Ceiling Exposure Limits
C 0.1 ppm; C 0.4 mg/m3

Canada - Ontario - OHSA - CEVs
0.1 ppm CEV; 0.4 mg/m3 CEV

Canada - Quebec - Ceiling Limits
P 0.1 ppm; P 0.38 mg/m3

United Kingdom - Occupational Exposure Standards - STELs
0.1 ppm STEL; 0.4 mg/m3 STEL

German (DFG) - MAK Values
0.1 ppm MAK; 0.4 mg/m3 MAK

German (DFG) - Peak Limitations
2 x normal MAK (5 min momentary value); don't exceed 8 times during shift

Israel - Ceiling Exposure Limits
C 0.1 ppm; C 0.38 mg/m3

Mexico - Instruction No. 10 - TWAs
0.1 ppm TWA; 0.4 mg/m3 TWA

STATE LISTS

California - Exposure Limits - Ceilings
C 0.1 ppm; C 0.4 mg/m3

California - Directors List of Hazardous Substances (8 CCR 339)
[present]

Florida Hazardous Substance List
[present]

Massachusetts Right To Know List
[present]

Minnesota Hazardous Substance List
[present]

NJ Right to Know List (Total)
sn 0370

Pennsylvania Right to Know List
[present]

CHLORITE   14998-27-7

PROPOSED REGULATIONS

Safe Drinking Water Act - Priority list
[present]

CHLORITE, INORGANIC, N.O.S.   RR-00205-1

HEALTH AND SAFETY LISTS

U.S. DOT - Substances From 49 CFR 172.101
regulated by DOT (UN1462)

U.S. DOT - Hazard Classes
DOT hazard class = 5.1

STATE LISTS

NJ Right to Know List (Total)
sn 2234

CHLORMADINONE ACETATE AND OESTROGENS   RR-00101-4

HEALTH AND SAFETY LISTS

IARC - Group Unspecified
[present] (Listed under 'Combined oral contraceptives')

CHLORMEPHOS   24934-91-6

HEALTH AND SAFETY LISTS

U.S. DOT - Appendix B - Marine Pollutants
DOT regulated marine pollutant

NIOSH - Selected LD50s and LC50s
Oral, rat: LD50 = 7 mg/kg Skin, rat: LD50 = 27 mg/kg

ENVIRONMENTAL LISTS

CERCLA/SARA - Section 302 Extremely Hazardous Substances and TPQs
TPQ = 500 pounds

STATE LISTS

Florida Hazardous Substance List
effective March 13, 1992

Massachusetts Right To Know List
extraordinarily hazardous

NJ Right to Know List (Total)
sn 2235

Pennsylvania Right to Know List
environmental hazard

PROPOSED REGULATIONS

CERCLA/SARA - Proposed Hazardous Substance Additions
proposed RQ = 1 pound (.454 kg)

CERCLA/SARA - 1989 Proposed RQ Adjustments
proposed RQ = 100 pounds (45.4 kg)

CHLORMEQUAT CHLORIDE   999-81-5

HEALTH AND SAFETY LISTS

NIOSH - Selected LD50s and LC50s
Oral, rat: LD50 = 398 mg/kg Skin, rat: LD50 = 4000 mg/kg

NTP Chemical Status Reports - Testing Status and NTIS Number
Technical reports printed (PB293627/AS)

NTP Chemical Status Reports - Evidence of Carcinogenicity
male rat-negative; female rat-negative; male mice-negative; female mice-negative

ENVIRONMENTAL LISTS

CERCLA/SARA - Section 302 Extremely Hazardous Substances and TPQs
TPQ = 100/10,000 pounds

STATE LISTS

Florida Hazardous Substance List
effective March 13, 1992

Massachusetts Right To Know List
extraordinarily hazardous

NJ Right to Know List (Total)
sn 2236

Pennsylvania Right to Know List
environmental hazard

PROPOSED REGULATIONS

CERCLA/SARA - Proposed Hazardous Substance Additions
proposed RQ = 1 pound (.454 kg)

CERCLA/SARA - 1989 Proposed RQ Adjustments
proposed RQ = 100 pounds (45.4 kg)

CHLOROACETALDEHYDE   107-20-0
SEE ALSO:
K010-HAZARDOUS WASTES

HEALTH AND SAFETY LISTS
ACGIH 1995 - Ceiling Limits
C 1 ppm; C 3.2 mg/m3

U.S. DOT - Substances From 49 CFR 172.101
regulated by DOT (UN2232)

U.S. DOT - Hazard Classes
DOT hazard class = 6.1

U.S. DOT - Substances Which Are Poisonous by Inhalation
liquid hazardous material poisonous by inhalation (UN2232)

U.S. DOT - Appendix A Table 1 - Hazardous Substances
final RQ = 1000 pounds (454 kg)

NIOSH - Selected LD50s and LC50s
Oral, rat: LD50 = 75 mg/kg Skin, rabbit: LD50 = 224 mg/kg

NIOSH 1990 - Pocket Guide - RELs
C 1 ppm; C 3 mg/m3

NIOSH 1990 - Pocket Guide - IDLHs
100 ppm IDLH

NIOSH 1990 - Pocket Guide - Target organs
eyes, skin, respiratory system

OSHA - Vacated PELs - Ceiling Limits
C 1 ppm; C 3 mg/m3

OSHA - Final PELs - Ceiling Limits
C 1 ppm; C 3 mg/m3

ENVIRONMENTAL LISTS
CERCLA/SARA - Hazardous Substances and their Reportable Quantities
final RQ = 1000 pounds (454 kg)

RCRA - P Series Wastes
waste number P023

RCRA - Hazardous Constituents-Appendix VIII
waste number P023

RCRA - Basis for Listing - Appendix VII
Included in waste stream: K010

RCRA - Substances Banned From Land Disposal
[present]

TSCA - Code of Federal Regulations Citations
40 CFR 712.30(x); 40 CFR 716.120(d)

TSCA - PAIR - Reporting List
Reporting Date: November 27, 1991

TSCA - Health and Safety Reporting List
Effective Date: September 30, 1991

INTERNATIONAL LISTS
Australian Exposure Standards - Time Weighted Averages
Peak Limitation: 1 ppm; 3.2 mg/m3

Canada - WHMIS: Ingredient Disclosure
1% item 336 (404)

Canada - Alberta - Ceiling Occupational Exposure Limit
C 1 ppm; C 3.2 mg/m3

Canada - British Columbia - Ceiling Exposure Limits
C 1 ppm; C 3 mg/m3

Canada - Ontario - OHSA - CEVs
1 ppm CEV; 3 mg/m3 CEV

Canada - Quebec - Ceiling Limits
P 1 ppm; P 3.2 mg/m3

United Kingdom - Occupational Exposure Standards - STELs
1 ppm STEL; 3 mg/m3 STEL

United Kingdom - Occupational Exposure Standards - Notes
can be absorbed through skin

German (DFG) - MAK Values
1 ppm MAK; 3 mg/m3 MAK

German (DFG) - Peak Limitations
2 x normal MAK (5 min momentary value); don't exceed 8 time. during shift

Israel - Ceiling Exposure Limits
C 1 ppm; C 3.2 mg/m3

Mexico - Instruction No. 10 - Ceiling Limits
P 1 ppm; P 3 mg/m3

STATE LISTS
California - Exposure Limits - Ceilings
C 1 ppm; C 3 mg/m3

California - Directors List of Hazardous Substances (8 CCR 339)
[present]

Florida Hazardous Substance List
[present]

Massachusetts Right To Know List
[present]

Minnesota Hazardous Substance List
[present]

NJ Right to Know List (Total)
sn 0372

Pennsylvania Right to Know List
environmental hazard

2-CHLOROACETAMIDE   79-07-2

HEALTH AND SAFETY LISTS
NIOSH - Selected LD50s and LC50s
Oral, rat: LD50 = 70 mg/kg

ENVIRONMENTAL LISTS
List of Pesticide Product Inert Ingredients
[present]

INTERNATIONAL LISTS
Canada - WHMIS: Ingredient Disclosure
1% item 337 (405)

CHLOROACETIC ACID   79-11-8

HEALTH AND SAFETY LISTS
AIHA - WEEL - Time Weighted Averages
0.3 ppm TWA; 1 mg/m3 TWA

AIHA - WEEL - Ceilings or Short Term Time Weighted Averages
1 ppm STEL; 4 mg/m3 STEL

AIHA - WEEL - Skin Absorption Designations
skin absorber

U.S. DOT - Substances From 49 CFR 172.101
regulated by DOT (UN1751, UN1750)

U.S. DOT - Hazard Classes
DOT hazard class = 8

NFPA - Flash Points
flash point = 259 degrees F (126 degrees C)

NFPA - Hazard Identification Ratings
health-3; flammability-1; reactivity-0

NIOSH - Selected LD50s and LC50s
Inhalation, rat: LC50 = 180 mg/m3 8 hr Oral, mouse: LD50 = 165 mg/kg

NTP Chemical Status Reports - Testing Status and NTIS Number
Technical reports printed (PB92-189372)

NTP Chemical Status Reports - Evidence of Carcinogenicity
male rat: no evidence; female rat: no evidence; male mice: no evidence; female mice: no evidence

ENVIRONMENTAL LISTS
CERCLA/SARA - Section 302 Extremely Hazardous Substances and TPQs
TPQ = 100/10,000 pounds

CERCLA/SARA - Section 313 - Emission Reporting
form R reporting required for 1.0% de minimus concentration

CERCLA/SARA - Hazardous Substances and their Reportable Quantities
final RQ = 1 pound (.454 kg)
Clean Air Act (1990) - List of Hazardous Air Contaminants
[present]
CAA - HON Rule - SOCMI Chemicals
compliance by Jan. 23, 1995
CAA - HON Rule - Organic HAPs
[present]
EPA - Master Testing List
[present]
List of Pesticide Product Inert Ingredients
[present]

INTERNATIONAL LISTS
Canada - WHMIS: Ingredient Disclosure
1% item 338 (74)
Canada - NPRI (National Pollutant Release Inventory)
[present]
United Kingdom - Occupational Exposure Standards - TWAs
0.3 ppm TWA; 1 mg/m3 TWA
United Kingdom - Occupational Exposure Standards - Notes
can be absorbed through skin

STATE LISTS
California - Air Bill 2588 Appendix A-I
6/91
Florida Hazardous Substance List
[present]
Massachusetts Right To Know List
extraordinarily hazardous
Minnesota Hazardous Substance List
skin
NJ Right to Know List (Total)
sn 0373
NJ Special Hazardous Substances
(corrosive)
Pennsylvania Right to Know List
environmental hazard

PROPOSED REGULATIONS
CERCLA/SARA - Proposed Hazardous Substance Additions
proposed RQ = 1 pound (.454 kg)
CERCLA/SARA - 1989 Proposed RQ Adjustments
proposed RQ = 100 pounds (45.4 kg)

CHLOROACETONE   78-95-5
HEALTH AND SAFETY LISTS
ACGIH 1995 - Ceiling Limits
C 1 ppm; C 3.8 mg/m3
ACGIH 1995 - Skin Designations
skin - potential for cutaneous absorption
U.S. DOT - Substances From 49 CFR 172.101
regulated by DOT (UN1695)
U.S. DOT - Hazard Classes
Forbidden from transport by the DOT
U.S. DOT - Substances Which Are Poisonous by Inhalation
liquid hazardous material poisonous by inhalation (stabilized form) (UN1695)
NIOSH - Selected LD50s and LC50s
Inhalation, rat: LC50 = 262 ppm 1 hr Oral, rat: LD50 = 100 mg/kg Skin, rabbit: LD50 = 141 mg/kg
INTERNATIONAL LISTS
Australian Exposure Standards - Time Weighted Averages
Peak Limitation: 1 ppm; 3.8 mg/m3

Australian Exposure Standards - Skin Effects
skin absorption
Canada - WHMIS: Ingredient Disclosure
1% item 1095 (1168)
Canada - Quebec - Ceiling Limits
P 1 ppm; P 3.8 mg/m3
Canada - Quebec - Skin Designations
absorbed through the skin
Israel - Ceiling Exposure Limits
C 1 ppm; C 3.8 mg/m3
STATE LISTS
California - Exposure Limits - Ceilings
C 1 ppm; C 3.8 mg/m3
California - Exposure Limits - Skin Notation
material may be absorbed through the skin, eyes or mucous membrane
Florida Hazardous Substance List
effective March 13, 1992
Massachusetts Right To Know List
[present]
Minnesota Hazardous Substance List
skin
NJ Right to Know List (Total)
sn 0374
Pennsylvania Right to Know List
[present]

CHLOROACETONITRILE   107-14-2
HEALTH AND SAFETY LISTS
U.S. DOT - Substances From 49 CFR 172.101
regulated by DOT (UN2668)
U.S. DOT - Hazard Classes
DOT hazard class = 6.1
U.S. DOT - Substances Which Are Poisonous by Inhalation
liquid hazardous material poisonous by inhalation (UN2668)
IARC - Group 3 (not classifiable)
[present]
NIOSH - Selected LD50s and LC50s
Oral, rat: LD50 = 220 mg/kg Skin, rabbit: LD50 = 71 mg/kg
INTERNATIONAL LISTS
Canada - WHMIS: Ingredient Disclosure
1% item 339 (408)
STATE LISTS
NJ Right to Know List (Total)
sn 0375

ALPHA-CHLOROACETOPHENONE   532-27-4
HEALTH AND SAFETY LISTS
ACGIH 1995 - Time Weighted Averages
0.05 ppm TWA; 0.32 mg/m3 TWA
NIOSH - Selected LD50s and LC50s
Oral, rat: LD50 = 50 mg/kg
NIOSH 1990 - Pocket Guide - RELs
0.3 mg/m3 TWA; 0.05 ppm TWA
NIOSH 1990 - Pocket Guide - IDLHs
100 mg/m3 IDLH
NIOSH 1990 - Pocket Guide - Target organs
eyes, skin, respiratory system
NTP Chemical Status Reports - Testing Status and NTIS Number
Technical reports printed (PB90256066)
NTP Chemical Status Reports - Evidence of Carcinogenicity
male rat-no evidence; female rat-equivocal evidence; male mice-no evidence; female mice-no evidence

OSHA - Vacated PELs - Time Weighted Averages
0.05 ppm TWA;   0.3 mg/m3 TWA

OSHA - Final PELs - Time Weighted Averages
0.05 ppm TWA;   0.3 mg/m3 TWA

ENVIRONMENTAL LISTS

CERCLA/SARA - Section 313 - Emission Reporting
form R reporting required for 1.0% de minimus concentration

CERCLA/SARA - Hazardous Substances and their Reportable Quantities
final RQ = 1 pound (.454 kg)

Clean Air Act (1990) - List of Hazardous Air Contaminants
[present]

CAA - HON Rule - SOCMI Chemicals
compliance by Oct. 24, 1994

CAA - HON Rule - Organic HAPs
[present]

TSCA - PAIR - Reporting List
Effective Date: January 26, 1994;   Reporting Date: March 28, 1994

TSCA - Health and Safety Reporting List
Effective Date: January 26, 1994;   Sunset Date: January 26, 2004

INTERNATIONAL LISTS

Australian Exposure Standards - Time Weighted Averages
0.05 ppm TWA;   0.32 mg/m3 TWA

Canada - WHMIS: Ingredient Disclosure
1% item 340 (409)

Canada - Alberta - 8 Hour Occupational Exposure Limit
0.05 ppm TWA;   0.32 mg/m3 TWA

Canada - Alberta - 15 Minute Occupational Exposure Limit
0.15 ppm STEL;   0.95 mg/m3 STEL

Canada - British Columbia - 8 Hour Exposure Limits
0.05 ppm TWA;   0.3 mg/m3 TWA

Canada - Ontario - OHSA - TWAEVs
0.05 ppm TWAEV;   0.32 mg/m3 TWAEV

United Kingdom - Occupational Exposure Standards - TWAs
0.05 ppm TWA;   0.3 mg/m3 TWA

Israel - Time Weighted Averages
0.05 ppm TWA;   0.32 mg/m3 TWA

Israel - Action Levels
0.025 ppm AL;   0.16 mg/m3 AL

Mexico - Instruction No. 10 - TWAs
0.05 ppm TWA;   0.3 mg/m3 TWA

STATE LISTS

California - Air Bill 2588 Appendix A-I
6/91

California - Exposure Limits - PELs
0.05 ppm PEL;   0.3 mg/m3 PEL

California - Directors List of Hazardous Substances (8 CCR 339)
[present]

Florida Hazardous Substance List
[present]

Massachusetts Right To Know List
[present]

Minnesota Hazardous Substance List
[present]

NJ Right to Know List (Total)
sn 0048

Pennsylvania Right to Know List
environmental hazard

PROPOSED REGULATIONS

TSCA - ITC 31st Report Priority Testing List
designated to be tested

CHLOROACETOPHENONE   1341-24-8

HEALTH AND SAFETY LISTS

U.S. DOT - Substances From 49 CFR 172.101
regulated by DOT (UN1697)

U.S. DOT - Hazard Classes
DOT hazard class = 6.1

NFPA - Flash Points
flash point = 244 degrees F (118 degrees C)

NFPA - Hazard Identification Ratings
health-2;   flammability-1;   reactivity-0

STATE LISTS

Florida Hazardous Substance List
[present]

Massachusetts Right To Know List
[present]

Pennsylvania Right to Know List
[present]

4-(CHLOROACETYL)ACETANILIDE   140-49-8

HEALTH AND SAFETY LISTS

NTP Chemical Status Reports - Testing Status and NTIS Number
Technical reports printed (PB288754/AS)

NTP Chemical Status Reports - Evidence of Carcinogenicity
male rat-negative;   female rat-negative;   male mice-negative;   female mice-negative

CHLOROACETYL CHLORIDE   79-04-9

HEALTH AND SAFETY LISTS

ACGIH 1995 - Time Weighted Averages
0.05 ppm TWA;   0.23 mg/m3 TWA

ACGIH 1995 - Short Term Exposure Limits
0.15 ppm STEL;   0.69 mg/m3 STEL

ACGIH 1995 - Skin Designations
skin - potential for cutaneous absorption

U.S. DOT - Substances From 49 CFR 172.101
regulated by DOT (UN1752)

U.S. DOT - Hazard Classes
DOT hazard class = 8

U.S. DOT - Substances Which Are Poisonous by Inhalation
liquid hazardous material poisonous by inhalation (UN1752)

NIOSH - Selected LD50s and LC50s
Inhalation, rat: LC50 = 1000 ppm 4 hr Oral, rat: LD50 = 120 mg/kg

OSHA - Vacated PELs - Time Weighted Averages
0.05 ppm TWA;   0.2 mg/m3 TWA

INTERNATIONAL LISTS

Australian Exposure Standards - Time Weighted Averages
0.05 ppm TWA;   0.23 mg/m3 TWA

Canada - WHMIS: Ingredient Disclosure
1% item 341 (484)

Canada - Alberta - 8 Hour Occupational Exposure Limit
0.05 ppm TWA;   0.23 mg/m3 TWA

Canada - Alberta - 15 Minute Occupational Exposure Limit
0.15 ppm STEL;   0.69 mg/m3 STEL

Canada - Ontario - OHSA - TWAEVs
0.05 ppm TWAEV;   0.23 mg/m3 TWAEV

Canada - Quebec - Time-Weighted Average Exposure Values
0.05 ppm TWAEV;   0.23 mg/m3 TWAEV

Canada - Quebec - Short-term Exposure Values
0.15 ppm STEV;   0.69 mg/m3 STEV

Canada - Quebec - Skin Designations
absorbed through the skin

Israel - Time Weighted Averages
(0.05) ppm TWA;   (0.23) mg/m3 TWA

Israel - Action Levels
0.025 ppm AL; 0.115 mg/m3 AL

STATE LISTS

California - Exposure Limits - PELs
0.05 ppm PEL; 0.2 mg/m3 PEL

California - Directors List of Hazardous Substances (8 CCR 339)
[present]

Florida Hazardous Substance List
[present]

Massachusetts Right To Know List
[present]

Minnesota Hazardous Substance List
[present]

NJ Right to Know List (Total)
sn 0377

NJ Special Hazardous Substances
(corrosive)

Pennsylvania Right to Know List
[present]

CHLOROALKANES   61788-76-9

SEE ALSO:
CHLORINATED PARAFFINS

HEALTH AND SAFETY LISTS

U.S. DOT - Appendix B - Marine Pollutants
DOT regulated marine pollutant

ENVIRONMENTAL LISTS

TSCA - Code of Federal Regulations Citations
40 CFR 716.120(a)

CHLOROALKYL ETHERS   RR-00206-2

ENVIRONMENTAL LISTS

Clean Water Act - Toxic Pollutants
[present]

RCRA - Hazardous Constituents-Appendix VIII
hazardous constituent - no waste number

STATE LISTS

California - Directors List of Hazardous Substances (8 CCR 339)
[present]

Pennsylvania Right to Know List
environmental hazard

CHLOROALKYL PHOSPHATES   RR-01714-1

PROPOSED REGULATIONS

TSCA - ITC 33rd Report Priority Testing List
recommended for testing

TSCA - ITC 34th Report Priority Testing List
recommended for testing

3-CHLORO-4-AMINOANILINE   615-66-7

INTERNATIONAL LISTS

Canada - WHMIS: Ingredient Disclosure
1% item 342 (410)

CHLORO-4-TERT-AMYLPHENOL   73090-69-4

HEALTH AND SAFETY LISTS

NFPA - Flash Points
flash point = 225 degrees F (107 degrees C)

NFPA - Hazard Identification Ratings
health-2; flammability-1; reactivity-0

STATE LISTS

Florida Hazardous Substance List
[present]

Massachusetts Right To Know List
[present]

Pennsylvania Right to Know List
[present]

2-CHLORO-4-TERT-AMYLPHENYL METHYL ETHER   RR-00629-1

HEALTH AND SAFETY LISTS

NFPA - Flash Points
flash point = 230 degrees F (110 degrees C)

NFPA - Hazard Identification Ratings
health-1; flammability-1; reactivity-0

2-CHLOROANILINE   95-51-2

SEE ALSO:
ANILINE AND CHLORO-, BROMO-, AND/OR NITROANILINES

HEALTH AND SAFETY LISTS

NIOSH - Selected LD50s and LC50s
Oral, mouse: LD50 = 256 mg/kg Skin, cat: LD50 = 222 mg/kg

NTP Chemical Status Reports - Testing Status and NTIS Number
Short term toxicity studies scheduled for peer review

ENVIRONMENTAL LISTS

TSCA - Code of Federal Regulations Citations
40 CFR 712.30(d); 40 CFR 716.120(c); 40 CFR 799.5000

TSCA - PAIR - Reporting List
Reporting Date: November 19, 1982

TSCA - Substances Subject to Testing Consent Orders
Test for: Health and Environmental Effects

TSCA - Section 12(b) - Export Notification
export notification required - Section 4

P-CHLOROANILINE   106-47-8

SEE ALSO:
ANILINE AND CHLORO-, BROMO-, AND/OR NITROANILINES
F039-HAZARDOUS WASTES

HEALTH AND SAFETY LISTS

U.S. DOT - Appendix A Table 1 - Hazardous Substances
final RQ = 1000 pounds (454 kg)

IARC - Group 2B (sufficient animal data)
[present]

NIOSH - Selected LD50s and LC50s
Oral, rat: LD50 = 310 mg/kg Skin, rat: LD50 = 3200 mg/kg

NTP Chemical Status Reports - Testing Status and NTIS Number
Technical reports printed (PB295896/AS)

NTP Chemical Status Reports - Evidence of Carcinogenicity
male rat-equivocal; female rat-negative; male mice-equivocal; female mice-equivocal

OSHA - Possible Select Carcinogens
[present]

ENVIRONMENTAL LISTS

CERCLA/SARA - Section 313 - Emission Reporting
form R reporting required

CERCLA/SARA - Hazardous Substances and their Reportable Quantities
final RQ = 1000 pounds (454 kg)

CAA - HON Rule - SOCMI Chemicals
compliance by Jan. 23, 1995

RCRA - P Series Wastes
waste number P024

RCRA - Hazardous Constituents-Appendix VIII
waste number P024

RCRA - Basis for Listing - Appendix VII
Included in waste stream: F039

RCRA - Substances Banned From Land Disposal
[present]

RCRA - TSD Facilities Ground Water Monitoring
TM 8270 = 20 ug/L PQL

RCRA - Universal Treatment Standards (LDR)
WW: 0.46 mg/l;   NWW: 16 mg/kg

TSCA - Code of Federal Regulations Citations
40 CFR 716.120(a);   40 CFR 799.5000

TSCA - PAIR - Reporting List
Reporting Date: November 19, 1982

TSCA - Substances Subject to Testing Consent Orders
Test for: Health Effects

TSCA - Section 12(b) - Export Notification
export notification required - Section 4

INTERNATIONAL LISTS

Canada - WHMIS: Ingredient Disclosure
1% item 344 (412)

German (DFG) - Skin/Sensitizers
danger of cutaneous absorption

German (DFG) - Carcinogens
animal evidence of carcinogenicity

STATE LISTS

California - Prop. 65 - Cancer list
carcinogen - initial date 10/01/94

Massachusetts Right To Know List
[present]

NJ Right to Know List (Total)
sn 2964

Pennsylvania Right to Know List
environmental hazard

M-CHLOROANILINE   108-42-9
SEE ALSO:
ANILINE AND CHLORO-, BROMO-, AND/OR NITROANILINES

HEALTH AND SAFETY LISTS

NIOSH - Selected LD50s and LC50s
Oral, rat: LD50 = 256 mg/kg Skin, cat: LD50 = 223 mg/kg

NTP Chemical Status Reports - Testing Status and NTIS Number
Prechronic studies completed: in review for further evaluation

ENVIRONMENTAL LISTS

TSCA - Code of Federal Regulations Citations
40 CFR 712.30(d);   40 CFR 716.120(a)

TSCA - PAIR - Reporting List
Reporting Date: November 19, 1982

INTERNATIONAL LISTS

Canada - WHMIS: Ingredient Disclosure
1% item 343 (411)

P-CHLOROANILINE HYDROCHLORIDE   20265-96-7

HEALTH AND SAFETY LISTS

NTP Chemical Status Reports - Testing Status and NTIS Number
Technical reports printed (PB90222563/AS)

NTP Chemical Status Reports - Evidence of Carcinogenicity
male rat-clear evidence;   female rat-equivocal evidence;   male mice-some evidence;   female mice-no evidence

CHLOROANILINES   27134-26-5
HEALTH AND SAFETY LISTS

U.S. DOT - Substances From 49 CFR 172.101
regulated by DOT (UN2019, UN2018)

U.S. DOT - Hazard Classes
DOT hazard class = 6.1

STATE LISTS

NJ Right to Know List (Total)
sn 0378

P-CHLORO-O-ANISIDINE   93-50-5
INTERNATIONAL LISTS

Canada - WHMIS: Ingredient Disclosure
1% item 345 (413)

STATE LISTS

NJ Right to Know List (Total)
sn 1466

CHLOROANISIDINES   RR-01347-8
HEALTH AND SAFETY LISTS

U.S. DOT - Substances From 49 CFR 172.101
regulated by DOT (UN2233)

U.S. DOT - Hazard Classes
DOT hazard class = 6.1

P-CHLOROBENZALDEHYDE   104-88-1
HEALTH AND SAFETY LISTS

NFPA - Flash Points
flash point = 190 degrees F (88 degrees C)

NFPA - Hazard Identification Ratings
health-2;   flammability-2;   reactivity-0

ENVIRONMENTAL LISTS

TSCA - Code of Federal Regulations Citations
40 CFR 712.30(x);   40 CFR 716.120(d)

TSCA - PAIR - Reporting List
Reporting Date: November 27, 1991

TSCA - Health and Safety Reporting List
Effective Date: September 30, 1991

STATE LISTS

Florida Hazardous Substance List
[present]

Massachusetts Right To Know List
[present]

Pennsylvania Right to Know List
[present]

CHLOROBENZENE   108-90-7
SEE ALSO:
BIS(2,4-DIMETHYLBUTYL) MALEATE
K105-HAZARDOUS WASTES
K149-HAZARDOUS WASTES
F025-HAZARDOUS WASTES
F024-HAZARDOUS WASTES
K015-HAZARDOUS WASTES
COCONUT OIL, POLYMER WITH ISOPHTHALIC ACID, TRIMELITTIC ANHY
F039-HAZARDOUS WASTES
F002-HAZARDOUS WASTES

HEALTH AND SAFETY LISTS

ACGIH 1995 - Time Weighted Averages
10 ppm TWA;   46 mg/m3 TWA

ACGIH 1995 - Biological Exposure Indices
Total 4-chlorocatechol in urine: 150 mg/g creatinine, end of shift (Ns) ;   Total p-chlorophenol in urine: 25 mg/g creatinine, end of shift (Ns)

AIHA - Odor Threshold Values
geometric mean air odor theshold = 1.3 ppm (detectable)

U.S. DOT - Substances From 49 CFR 172.101
regulated by DOT (UN1134)

U.S. DOT - Hazard Classes
DOT hazard class = 3

U.S. DOT - Appendix A Table 1 - Hazardous Substances
final RQ = 100 pounds (45.4 kg)

NFPA - Flash Points
flash point = 82 degrees F (28 degrees C)

NFPA - Hazard Identification Ratings
health-2; flammability-3; reactivity-0

NIOSH - Selected LD50s and LC50s
Oral, rat: LD50 = 2910 mg/kg

NIOSH 1990 - Pocket Guide - RELs
See Appendix D

NIOSH 1990 - Pocket Guide - IDLHs
2400 ppm IDLH

NIOSH 1990 - Pocket Guide - Target organs
respiratory system, eyes, skin, CNS, liver

NTP Chemical Status Reports - Testing Status and NTIS Number
Technical reports printed (PB86144714/AS)

NTP Chemical Status Reports - Evidence of Carcinogenicity
male rat-equivocal; female rat-negative; male mice-negative; female mice-negative

OSHA - Vacated PELs - Time Weighted Averages
75 ppm TWA; 350 mg/m3 TWA

OSHA - Final PELs - Time Weighted Averages
75 ppm TWA; 350 mg/m3 TWA

ENVIRONMENTAL LISTS

ATSDR Priority List
Rank (of 275): 085

CERCLA/SARA - Section 313 - Emission Reporting
form R reporting required for 1.0% de minimus concentration

CERCLA/SARA - Hazardous Substances and their Reportable Quantities
final RQ = 100 pounds (45.4 kg)

Clean Air Act (1990) - List of Hazardous Air Contaminants
[present]

CAA - HON Rule - SOCMI Chemicals
compliance by Oct. 24, 1994

CAA - HON Rule - Organic HAPs
[present]

Clean Water Act - Hazardous Substances
[present]

Clean Water Act - Priority Pollutants
[present]

Safe Drinking Water Act - MCLs
MCL = 0.1 mg/L

Safe Drinking Water Act - MCLGs
MCLG = 0.1 mg/L

Safe Drinking Water Act - Monitoring
monitoring required

RCRA - D Series - Maximum Concentration of Contaminants
waste number D021; regulatory level = 100.0 mg/L

RCRA - D Series - Chronic Toxicity Reference Levels
chronic toxicity reference level = 1 mg/L

RCRA - U Series Wastes
waste number U037

RCRA - Hazardous Constituents-Appendix VIII
waste number U037

RCRA - Basis for Listing - Appendix VII
Included in waste streams: F002, F024, F025, F039, K015, K105, K149

RCRA - Substances Banned From Land Disposal
[present]

RCRA - TSD Facilities Ground Water Monitoring
TM 8010 = 2 ug/L PQL; TM 8020 = 2 ug/L PQL; TM 8240 = 5 ug/L PQL

RCRA - Universal Treatment Standards (LDR)
WW: 0.057 mg/l; NWW: 6.0 mg/kg

TSCA - Code of Federal Regulations Citations
40 CFR 712.30(d); 40 CFR 716.120(a); 40 CFR 799.1051

TSCA - PAIR - Reporting List
Reporting Date: November 19, 1982

TSCA - Health and Safety Reporting List
Effective Date: March 11, 1994; Sunset Date: March 11, 2004

TSCA - HDD/HDF - Precursors Required for Reporting
[present]

TSCA - Chemical Test Rules
Testing required by: manufacturers; importers; processors (40 CFR 799.1051)

TSCA - Section 12(b) - Export Notification
export notification required - Section 4

INTERNATIONAL LISTS

Australian Exposure Standards - Time Weighted Averages
75 ppm TWA; 345 mg/m3 TWA

Canada - WHMIS: Ingredient Disclosure
1% item 346 (414)

Canada - NPRI (National Pollutant Release Inventory)
[present]

Canada - CEPA - Priority Substances List
estimated time for completion of assessment reports: 4 years

Canada - Drinking Water Quality - MACs
0.08 mg/L MAC

Canada - Drinking Water Quality - AOs
<= 0.03 mg/L AO

Canada - Alberta - 8 Hour Occupational Exposure Limit
75 ppm TWA; 345 mg/m3 TWA

Canada - Alberta - 15 Minute Occupational Exposure Limit
115 ppm STEL; 520 mg/m3 STEL

Canada - British Columbia - 8 Hour Exposure Limits
75 ppm TWA; 350 mg/m3 TWA

Canada - Ontario - OHSA - TWAEVs
75 ppm TWAEV; 345 mg/m3 TWAEV

Canada - Quebec - Time-Weighted Average Exposure Values
75 ppm TWAEV; 345 mg/m3 TWAEV

United Kingdom - Occupational Exposure Standards - TWAs
50 ppm TWA; 230 mg/m3 TWA

German (DFG) - MAK Values
50 ppm MAK; 230 mg/m3 MAK

German (DFG) - Peak Limitations
2 x normal MAK (30 min. average value); don't exceed 4 times during shift

German (DFG) - Pregnancy
no risk to embryo/fetus if exposure limits adhered to

Israel - Time Weighted Averages
(75) ppm TWA; (345) mg/m3 TWA

Israel - Action Levels
37.5 ppm AL; 172.5 mg/m3 AL

Mexico - Instruction No. 10 - TWAs
75 ppm TWA; 350 mg/m3 TWA

Mexico - Wastewater - Organic Toxic Pollutants and Heavy Metals
Listed under [Chlorinated Benzenes]

STATE LISTS

California - Air Bill 2588 Appendix A-I
[present]

California - Exposure Limits - PELs
75 ppm PEL; 350 mg/m3 PEL

Florida Hazardous Substance List
[present]

Massachusetts Right To Know List
[present]

Minnesota Hazardous Substance List
[present]

NJ Right to Know List (Total)
sn 0379

NJ Special Hazardous Substances
(flammable - third degree)

Pennsylvania Right to Know List
environmental hazard

PROPOSED REGULATIONS

TSCA - ITC 32nd Report Priority Testing List
designated for dermal absorption testing

CHLOROBENZENES, N.O.S.   RR-01737-8

HEALTH AND SAFETY LISTS

U.S. DOT - Appendix A Table 1 - Hazardous Substances
final RQ = 10 pounds (4.54 kg)

STATE LISTS

California - Air Bill 2588 Appendix A-I
6/91

P-CHLOROBENZENESULFONIC ACID, SODIUM   5138-90-9
SALT

INTERNATIONAL LISTS

Canada - WHMIS: Ingredient Disclosure
1% item 347 (415)

CHLOROBENZILATE   510-15-6
SEE ALSO:
F039-HAZARDOUS WASTES

HEALTH AND SAFETY LISTS

U.S. DOT - Appendix A Table 1 - Hazardous Substances
final RQ = 10 pounds (4.54 kg)

IARC - Group 3 (not classifiable)
[present]

NIOSH - Selected LD50s and LC50s
Oral, rat: LD50 = 700 mg/kg

NTP Chemical Status Reports - Testing Status and NTIS Number
Technical reports printed (PB287123/AS)

NTP Chemical Status Reports - Evidence of Carcinogenicity
male rat-equivocal; female rat-equivocal; male mice-positive; female mice-positive

ENVIRONMENTAL LISTS

CERCLA/SARA - Section 313 - Emission Reporting
form R reporting required for 1.0% de minimus concentration

CERCLA/SARA - Hazardous Substances and their Reportable Quantities
final RQ = 10 pounds (4.54 kg)

Clean Air Act (1990) - List of Hazardous Air Contaminants
[present]

EPA - Carcinogen Hazard Ranking for RQ Adjustment
Hazard ranking = Medium

RCRA - U Series Wastes
waste number U038

RCRA - Hazardous Constituents-Appendix VIII
waste number U038

RCRA - Basis for Listing - Appendix VII
Included in waste stream: F039

RCRA - Substances Banned From Land Disposal
[present]

RCRA - TSD Facilities Ground Water Monitoring
TM 8270 = 10 ug/L PQL

RCRA - Universal Treatment Standards (LDR)
WW: 0.10 mg/l; NWW: Not applicable

STATE LISTS

California - Air Bill 2588 Appendix A-I
known or potential carcinogen: 9/90

California - Prop. 65 - Cancer list
carcinogen - initial date 1/1/90

California - Prop. 65 - No Significant Risk Levels
no significant risk level = 7 ug/day

California - Directors List of Hazardous Substances (8 CCR 339)
[present]

Florida Hazardous Substance List
[present]

Massachusetts Right To Know List
carcinogen; extraordinarily hazardous

NJ Right to Know List (Total)
sn 0205

NJ Special Hazardous Substances
(carcinogen)

Pennsylvania Right to Know List
environmental hazard

CHLOROBENZILATE   512-26-5
STATE LISTS

Massachusetts Right To Know List
[present]

O-CHLOROBENZOIC ACID   118-91-2
HEALTH AND SAFETY LISTS

NIOSH - Selected LD50s and LC50s
Oral, rat: LD50 = 6460 mg/kg

INTERNATIONAL LISTS

Canada - WHMIS: Ingredient Disclosure
1% item 348 (75)

4-CHLOROBENZOTRICHLORIDE   5216-25-1
ENVIRONMENTAL LISTS

TSCA - Code of Federal Regulations Citations
40 CFR 799.5055(c), (e)(1)

TSCA - Multichemical Test Rules - Waste Constituents
subchronic toxicity testing for Health Effects

TSCA - Section 12(b) - Export Notification
export notification required - Section 4

INTERNATIONAL LISTS

German (DFG) - Skin/Sensitizers
danger of cutaneous absorption

German (DFG) - Carcinogens
animal evidence of carcinogenicity

STATE LISTS

California - Air Bill 2588 Appendix A-II
known or potential carcinogen: 9/90

California - Prop. 65 - Cancer list
carcinogen - initial date 1/1/90

CHLOROBENZOTRIFLUORIDE   52181-51-8
HEALTH AND SAFETY LISTS

U.S. DOT - Substances From 49 CFR 172.101
regulated by DOT (UN2234)

U.S. DOT - Hazard Classes
DOT hazard class = 3

NFPA - Flash Points
flash point = 117 degrees F (47 degrees C)

NFPA - Hazard Identification Ratings
flammability-2; reactivity-0

STATE LISTS

NJ Right to Know List (Total)
sn 0380

P-CHLOROBENZOYL PEROXIDE   94-17-7
STATE LISTS

NJ Right to Know List (Total)
sn 1464

P-CHLOROBENZYL CHLORIDE   104-83-6

HEALTH AND SAFETY LISTS

U.S. DOT - Appendix B - Marine Pollutants
DOT regulated marine pollutant

INTERNATIONAL LISTS

Canada - WHMIS: Ingredient Disclosure
1% item 349 (485)

STATE LISTS

NJ Right to Know List (Total)
sn 1465

CHLOROBENZYLCHLORIDES   RR-01318-3

HEALTH AND SAFETY LISTS

U.S. DOT - Substances From 49 CFR 172.101
regulated by DOT (UN2235)

U.S. DOT - Hazard Classes
DOT hazard class = 6.1

U.S. DOT - Appendix B - Marine Pollutants
DOT regulated marine pollutant

O-CHLOROBENZYLIDENE MALONITRILE   2698-41-1

HEALTH AND SAFETY LISTS

ACGIH 1995 - Ceiling Limits
C 0.05 ppm;   C 0.39 mg/m3

ACGIH 1995 - Skin Designations
skin - potential for cutaneous absorption

NIOSH - Selected LD50s and LC50s
Oral, rat: LD50 = 178 mg/kg

NIOSH 1990 - Pocket Guide - RELs
C 0.05 ppm;   C 0.4 mg/m3

NIOSH 1990 - Pocket Guide - IDLHs
2 mg/m3 IDLH

NIOSH 1990 - Pocket Guide - Target organs
respiratory system, skin, eyes

NIOSH 1990 - Pocket Guide - Skin list
Potential for dermal absorption

NTP Chemical Status Reports - Testing Status and NTIS Number
Technical reports printed (PB90256280)

NTP Chemical Status Reports - Evidence of Carcinogenicity
male rat-no evidence;   female rat-no evidence;   male mice-no evidence;   female mice-no evidence

OSHA - Vacated PELs - Ceiling Limits
C 0.05 ppm;   C 0.4 mg/m3

OSHA - Vacated PELs - Skin Designation
Prevent or reduce skin absorption

OSHA - Final PELs - Time Weighted Averages
0.05 ppm TWA;   0.4 mg/m3 TWA

INTERNATIONAL LISTS

Australian Exposure Standards - Time Weighted Averages
Peak Limitation: 0.05 ppm;   0.39 mg/m3

Australian Exposure Standards - Skin Effects
skin absorption

Canada - WHMIS: Ingredient Disclosure
1% item 350 (416)

Canada - Alberta - Ceiling Occupational Exposure Limit
C 0.05 ppm;   C 0.39 mg/m3

Canada - Alberta - Skin Designation
can be absorbed through the intact skin

Canada - British Columbia - 8 Hour Exposure Limits
0.05 ppm TWA;   0.4 mg/m3 TWA

Canada - British Columbia - 15 Minute Exposure Limits
0.15 ppm STEL;   1.0 mg/m3 STEL

Canada - British Columbia - Skin Notations
skin - potential for skin absorption

Canada - Ontario - OHSA - CEVs
0.05 ppm CEV;   0.4 mg/m3 CEV

Canada - Ontario - OHSA - Skin Notations
absorption through skin, eyes, or mucous membranes

Canada - Quebec - Ceiling Limits
P 0.05 ppm;   P 0.39 mg/m3

Canada - Quebec - Skin Designations
absorbed through the skin

Israel - Ceiling Exposure Limits
C 0.05 ppm;   C 0.39 mg/m3

Mexico - Instruction No. 10 - TWAs
0.05 ppm TWA;   0.4 mg/m3 TWA

Mexico - Instruction No. 10 - Skin designation
skin - potential for cutaneous absorption

STATE LISTS

California - Exposure Limits - Ceilings
C 0.05 ppm;   C 0.4 mg/m3

California - Exposure Limits - Skin Notation
material may be absorbed through the skin, eyes or mucous membrane

California - Directors List of Hazardous Substances (8 CCR 339)
[present]

Florida Hazardous Substance List
[present]

Massachusetts Right To Know List
[present]

Minnesota Hazardous Substance List
skin

Pennsylvania Right to Know List
[present]

CHLOROBIPHENYL   27323-18-8

STATE LISTS

Massachusetts Right To Know List
carcinogen;   extraordinarily hazardous;   teratogen

CHLOROBIPHENYLS   RR-01555-4

INTERNATIONAL LISTS

Canada - CEPA Schedule II Part II - Toxic Substances (Export)
[present] (having the molecular formula C12H10-nCln where 'n' is greater than 2)

Canada - CEPA Schedule I - Toxic Substances
(a) prohibited commercial, manufacturing or processing uses;   (b) maximum concent rations in products;   (c) maximum quantities and concentrations that may be released into the environment;   (d) treatment and destruction of waste;   (e) prohibition of the export of waste;   (f) storage of PCB material

2-CHLORO-4,6-BIS(SUBSTITUTED)-1,3,5-TRIAZINE,   RR-01719-6
DIHYDROCHLORIDE

ENVIRONMENTAL LISTS

TSCA - Chemicals with Significant New Use Rules
PMN number: P-91-659

CHLOROBROMOMETHANE   74-97-5

HEALTH AND SAFETY LISTS

ACGIH 1995 - Time Weighted Averages
200 ppm TWA;   1060 mg/m3 TWA

U.S. DOT - Substances From 49 CFR 172.101
regulated by DOT (UN1887)

U.S. DOT - Hazard Classes
DOT hazard class = 6.1

NIOSH - Selected LD50s and LC50s
Inhalation, mouse: LC50 = 15850 mg/m3 (8 hr) Oral, rat: LD50 = 5000 mg/kg

NIOSH 1990 - Pocket Guide - RELs
200 ppm TWA; 1050 mg/m3 TWA

NIOSH 1990 - Pocket Guide - IDLHs
5000 ppm IDLH

NIOSH 1990 - Pocket Guide - Target organs
skin, liver, kidneys, respiratory system, CNS

OSHA - Vacated PELs - Time Weighted Averages
200 ppm TWA; 1050 mg/m3 TWA

OSHA - Final PELs - Time Weighted Averages
200 ppm TWA; 1050 mg/m3 TWA

ENVIRONMENTAL LISTS

Safe Drinking Water Act - Monitoring
monitoring required at discretion of the state

EPA - Master Testing List
[present]

TSCA - Code of Federal Regulations Citations
40 CFR 712.30(x); 40 CFR 716.120(a)

TSCA - PAIR - Reporting List
Reporting Date: December 27, 1990

TSCA - Health and Safety Reporting List
Effective Date: June 1, 1987

INTERNATIONAL LISTS

Australian Exposure Standards - Time Weighted Averages
200 ppm TWA; 1060 mg/m3 TWA

Australian Exposure Standards - Short Term Exposure Limits
250 ppm STEL; 1300 mg/m3 STEL

Canada - WHMIS: Ingredient Disclosure
1% item 351 (418)

Canada - Alberta - 8 Hour Occupational Exposure Limit
200 ppm TWA; 1060 mg/m3 TWA

Canada - Alberta - 15 Minute Occupational Exposure Limit
250 ppm STEL; 1320 mg/m3 STEL

Canada - British Columbia - 8 Hour Exposure Limits
200 ppm TWA; 1050 mg/m3 TWA

Canada - British Columbia - 15 Minute Exposure Limits
250 ppm STEL; 1300 mg/m3 STEL

Canada - Ontario - OHSA - TWAEVs
200 ppm TWAEV; 1060 mg/m3 TWAEV

Canada - Ontario - OHSA - STEVs
250 ppm STEV; 1320 mg/m3 STEV

Canada - Quebec - Time-Weighted Average Exposure Values
200 ppm TWAEV; 1058 mg/m3 TWAEV

United Kingdom - Occupational Exposure Standards - TWAs
200 ppm TWA; 1050 mg/m3 TWA

United Kingdom - Occupational Exposure Standards - STELs
250 ppm STEL; 1300 mg/m3 STEL

German (DFG) - MAK Values
200 ppm MAK; 1050 mg/m3 MAK

German (DFG) - Peak Limitations
2 x normal MAK (30 min. average value); don't exceed 4 times during shift

Israel - Time Weighted Averages
200 ppm TWA; 1060 mg/m3 TWA

Israel - Action Levels
100 ppm AL; 530 mg/m3 AL

Mexico - Instruction No. 10 - TWAs
200 ppm TWA; 1050 mg/m3 TWA

Mexico - Instruction No. 10 - STELs
250 ppm STEL; 1300 mg/m3 STEL

STATE LISTS

California - Exposure Limits - PELs
200 ppm PEL; 1050 mg/m3 PEL

California - Directors List of Hazardous Substances (8 CCR 339)
[present]

Florida Hazardous Substance List
[present]

Massachusetts Right To Know List
[present]

Minnesota Hazardous Substance List
[present]

NJ Right to Know List (Total)
sn 0381

Pennsylvania Right to Know List
[present]

PROPOSED REGULATIONS

TSCA - Proposed Substances for Developmental/Reproductive Testing
proposed testing for: Reproductive Toxicity - oral

1-CHLORO-3-BROMOPROPANE   109-70-6

HEALTH AND SAFETY LISTS

U.S. DOT - Substances From 49 CFR 172.101
regulated by DOT (UN2688)

U.S. DOT - Hazard Classes
DOT hazard class = 6.1

NIOSH - Selected LD50s and LC50s
Inhalation, rat: LC50 = 5668 mg/m3 8 hr Oral, rat: LD50 = 930 mg/kg

INTERNATIONAL LISTS

Canada - WHMIS: Ingredient Disclosure
1% item 352 (419)

STATE LISTS

NJ Right to Know List (Total)
sn 2237

CHLOROBUTANE   25154-42-1

HEALTH AND SAFETY LISTS

U.S. DOT - Substances From 49 CFR 172.101
regulated by DOT (UN1127)

U.S. DOT - Hazard Classes
DOT hazard class = 3

STATE LISTS

NJ Right to Know List (Total)
sn 0382

2-CHLOROBUTENE-2   4461-41-0

HEALTH AND SAFETY LISTS

NFPA - Flash Points
flash point = -3 degrees F (-19 degrees C)

NFPA - Hazard Identification Ratings
health-2; flammability-3; reactivity-0

STATE LISTS

Florida Hazardous Substance List
[present]

Massachusetts Right To Know List
[present]

Pennsylvania Right to Know List
[present]

4-CHLORO-M-CRESOL   59-50-7
SEE ALSO:
K001-HAZARDOUS WASTES
F039-HAZARDOUS WASTES

HEALTH AND SAFETY LISTS

U.S. DOT - Appendix A Table 1 - Hazardous Substances
final RQ = 5000 pounds (2270 kg)

NIOSH - Selected LD50s and LC50s
Oral, rat: LD50 = 1830 mg/kg

ENVIRONMENTAL LISTS

CERCLA/SARA - Hazardous Substances and their Reportable Quantities
final RQ = 5000 pounds (2270 kg)

Clean Water Act - Priority Pollutants
[present]

RCRA - U Series Wastes
waste number U039

RCRA - Hazardous Constituents-Appendix VIII
waste number U039

RCRA - Basis for Listing - Appendix VII
Included in waste stream: F039, K001

RCRA - Substances Banned From Land Disposal
[present]

RCRA - TSD Facilities Ground Water Monitoring
TM 8040 = 5 ug/L PQL; TM 8270 = 20 ug/L PQL

RCRA - Universal Treatment Standards (LDR)
WW: 0.018 mg/l; NWW: 14 mg/kg

INTERNATIONAL LISTS

Mexico - Wastewater - Organic Toxic Pollutants and Heavy Metals
Listed under [Chlorinated Phenols]

STATE LISTS

California - Directors List of Hazardous Substances (8 CCR 339)
[present]

Massachusetts Right To Know List
[present]

Pennsylvania Right to Know List
environmental hazard

CHLOROCRESOL   1321-10-4
HEALTH AND SAFETY LISTS

U.S. DOT - Substances From 49 CFR 172.101
regulated by DOT (UN2669)

U.S. DOT - Hazard Classes
DOT hazard class = 6.1

STATE LISTS

NJ Right to Know List (Total)
sn 0383

CHLORO-M-CRESOL, ALL ISOMERS   RR-00588-9
INTERNATIONAL LISTS

Canada - WHMIS: Ingredient Disclosure
1% item 353 (421)

CHLORO-O-CRESOL, ALL ISOMERS   RR-00589-0
INTERNATIONAL LISTS

Canada - WHMIS: Ingredient Disclosure
1% item 354 (422)

CHLORO-P-CRESOL, ALL ISOMERS   RR-00590-3
INTERNATIONAL LISTS

Canada - WHMIS: Ingredient Disclosure
1% item 355 (423)

4-CHLORO-2-CYCLOPENTYLPHENOL   13347-42-7
INTERNATIONAL LISTS

Canada - WHMIS: Ingredient Disclosure
1% item 356 (424)

2-CHLORO-4,6-DI-TERT-AMYLPHENOL   42350-99-2
HEALTH AND SAFETY LISTS

NFPA - Flash Points
flash point = 250 degrees F (121 degrees C)

NFPA - Hazard Identification Ratings
health-2; flammability-1; reactivity-0

STATE LISTS

Florida Hazardous Substance List
[present]

Massachusetts Right To Know List
[present]

Pennsylvania Right to Know List
[present]

CHLORODIBROMOMETHANE   124-48-1
SEE ALSO:
F039-HAZARDOUS WASTES

HEALTH AND SAFETY LISTS

U.S. DOT - Appendix A Table 1 - Hazardous Substances
final RQ = 100 pounds (45.4 kg)

IARC - Group 3 (not classifiable)
[present]

NIOSH - Selected LD50s and LC50s
Oral, rat: LD50 = 848 mg/kg

NTP Chemical Status Reports - Testing Status and NTIS Number
Technical reports printed (PB86166675/AS)

NTP Chemical Status Reports - Evidence of Carcinogenicity
male rat-no evidence; female rat-no evidence; male mice-equivocal evidence; female mice-some evidence

ENVIRONMENTAL LISTS

ATSDR Priority List
Rank (of 275): 207

CERCLA/SARA - Hazardous Substances and their Reportable Quantities
final RQ = 100 pounds (45.4 kg)

Clean Water Act - Priority Pollutants
[present]

Safe Drinking Water Act - Monitoring
monitoring required

RCRA - Basis for Listing - Appendix VII
Included in waste stream: F039

RCRA - TSD Facilities Ground Water Monitoring
TM 8010 = 1 ug/L PQL; TM 8240 = 5 ug/L PQL

RCRA - Universal Treatment Standards (LDR)
WW: 0.057 mg/l; NWW: 15 mg/kg

TSCA - Code of Federal Regulations Citations
40 CFR 716.120(a)

TSCA - Health and Safety Reporting List
Effective Date: June 1, 1987

INTERNATIONAL LISTS

Mexico - Wastewater - Organic Toxic Pollutants and Heavy Metals
Listed under [Halomethanes]

STATE LISTS

California - Air Bill 2588 Appendix A II
known or potential carcinogen: 9/90

California - Prop. 65 - Cancer list
carcinogen - initial date 1/1/90

California - Prop. 65 - No Significant Risk Levels
no significant risk level = 7 ug/day

California - Directors List of Hazardous Substances (8 CCR 339)
[present]

Massachusetts Right To Know List
[present]

Pennsylvania Right to Know List
environmental hazard

PROPOSED REGULATIONS

Safe Drinking Water Act - Priority list
[present]

5-CHLORO-2-(2,4-DICHLOROPHENOXY)PHENOL 3380-34-5

HEALTH AND SAFETY LISTS

NIOSH - Selected LD50s and LC50s
Oral, rat: LD50 = 3700 mg/kg Skin, rat: LD50 = 9300 mg/kg

ENVIRONMENTAL LISTS

List of Pesticide Product Inert Ingredients
[present]

3-CHLORO-4-DIETHYLAMINO-BENZENEDIAZO- RR-00775-0
NIUM ZINC CHLORIDE

HEALTH AND SAFETY LISTS

U.S. DOT - Substances From 49 CFR 172.101
regulated by DOT (UN3033)

U.S. DOT - Hazard Classes
DOT hazard class = 4.1

CHLORODIETHYLSILANE 1609-19-4

STATE LISTS

Florida Hazardous Substance List
[present]

Massachusetts Right To Know List
[present]

Pennsylvania Right to Know List
[present]

CHLORODIFLUOROBROMO-METHANE 353-59-3

HEALTH AND SAFETY LISTS

U.S. DOT - Substances From 49 CFR 172.101
regulated by DOT (UN1974)

U.S. DOT - Hazard Classes
DOT hazard class = 2.2

ENVIRONMENTAL LISTS

CERCLA/SARA - Section 313 - Emission Reporting
form R reporting required for 1.0% de minimus concentration

Class 1 Ozone Depletors
ozone depletion potential = 3.0

INTERNATIONAL LISTS

Canada - CEPA Schedule II Part II - Toxic Substances (Export)
[present] (with the molecular formula CF2BrCl)

Canada - CEPA Schedule I - Toxic Substances
that has the molecular formula CF2BrCl: (a) quantities that may be imported; (b) prohibited commercial, manufacturing or processing uses

STATE LISTS

NJ Right to Know List (Total)
sn 0384

1-CHLORO-1,1-DIFLUOROETHANE 75-68-3

HEALTH AND SAFETY LISTS

AIHA - WEEL - Time Weighted Averages
1000 ppm TWA; 4100 mg/m3 TWA

NFPA - Flash Points
gas (no flash point given)

NFPA - Hazard Identification Ratings
flammability-4; reactivity-0

ENVIRONMENTAL LISTS

CERCLA/SARA - Section 313 - Emission Reporting
form R reporting required

Class 2 Ozone Depletors
ozone depletion weight = 0.06

List of Pesticide Product Inert Ingredients
[present]

TSCA - Code of Federal Regulations Citations
40 CFR 712.30(w); 40 CFR 716.120(a)

TSCA - PAIR - Reporting List
Reporting Date: June 13, 1989

TSCA - Health and Safety Reporting List
Effective Date: April 13, 1989

INTERNATIONAL LISTS

Canada - WHMIS: Ingredient Disclosure
1% item 357 (425)

German (DFG) - MAK Values
1000 ppm MAK; 4170 mg/m3 MAK

German (DFG) - Peak Limitations
2 x normal MAK (1 hour momentary value); don't exceed 3 times per shift

STATE LISTS

Florida Hazardous Substance List
[present]

Massachusetts Right To Know List
[present]

NJ Right to Know List (Total)
sn 0385

Pennsylvania Right to Know List
[present]

CHLORODIFLUOROETHANES 25497-29-4

HEALTH AND SAFETY LISTS

U.S. DOT - Substances From 49 CFR 172.101
regulated by DOT (UN2517)

U.S. DOT - Hazard Classes
DOT hazard class = 2.1

ENVIRONMENTAL LISTS

CAA - HON Rule - SOCMI Chemicals
compliance by Oct. 23, 1995

CHLORODIFLUOROETHANES 27497-51-4

ENVIRONMENTAL LISTS

CAA - HON Rule - SOCMI Chemicals
compliance by July 24, 1995

CHLORODIFLUOROMETHANE 75-45-6

HEALTH AND SAFETY LISTS

ACGIH 1995 - Time Weighted Averages
1000 ppm TWA; 3540 mg/m3 TWA

U.S. DOT - Substances From 49 CFR 172.101
regulated by DOT (UN1018)

U.S. DOT - Hazard Classes
DOT hazard class = 2.2

IARC - Group 3 (not classifiable)
[present]

NIOSH - Selected LD50s and LC50s
Inhalation, mouse: LC50 = 28 pph (30 mn)

OSHA - Vacated PELs - Time Weighted Averages
1000 ppm TWA; 3500 mg/m3 TWA

ENVIRONMENTAL LISTS

ATSDR Priority List
Rank (of 275): 236

CERCLA/SARA - Section 313 - Emission Reporting
form R reporting required

Class 2 Ozone Depletors
ozone depletion weight = 0.05

CAA - HON Rule - SOCMI Chemicals
compliance by Oct. 24, 1994

List of Pesticide Product Inert Ingredients
[present]

TSCA - Code of Federal Regulations Citations
40 CFR 712.30(w); 40 CFR 716.120(a)

TSCA - PAIR - Reporting List
Reporting Date: June 13, 1989

TSCA - Health and Safety Reporting List
Effective Date: April 13, 1989

INTERNATIONAL LISTS

Australian Exposure Standards - Time Weighted Averages
1000 ppm TWA; 3540 mg/m3 TWA

Canada - WHMIS: Ingredient Disclosure
1% item 358 (426)

Canada - Alberta - 8 Hour Occupational Exposure Limit
1000 ppm TWA; 3520 mg/m3 TWA

Canada - Alberta - 15 Minute Occupational Exposure Limit
1250 ppm STEL; 4400 mg/m3 STEL

Canada - British Columbia - 8 Hour Exposure Limits
1000 ppm TWA; 3500 mg/m3 TWA

Canada - British Columbia - 15 Minute Exposure Limits
1250 ppm STEL; 4375 mg/m3 STEL

Canada - Ontario - OHSA - TWAEVs
1000 ppm TWAEV; 3535 mg/m3 TWAEV

Canada - Ontario - OHSA - STEVs
1250 ppm STEV; 4415 mg/m3 STEV

Canada - Quebec - Time-Weighted Average Exposure Values
1000 ppm TWAEV; 3540 mg/m3 TWAEV

German (DFG) - MAK Values
500 ppm MAK; 1800 mg/m3 MAK (applies only to pure substance)

German (DFG) - Peak Limitations
2 x normal MAK (1 hour momentary value); don't exceed 3 times per shift

German (DFG) - Pregnancy
no risk to embryo/fetus if exposure limits adhered to

Israel - Time Weighted Averages
1000 ppm TWA; 3540 mg/m3 TWA

Israel - Action Levels
500 ppm AL; 1770 mg/m3 AL

Mexico - Instruction No. 10 - TWAs
1000 ppm TWA; 3500 mg/m3 TWA

Mexico - Instruction No. 10 - STELs
1250 ppm STEL; 4375 mg/m3 STEL

STATE LISTS

California - Exposure Limits - PELs
1000 ppm PEL; 3500 mg/m3 PEL

California - Directors List of Hazardous Substances (8 CCR 339)
[present]

Florida Hazardous Substance List
[present]

Massachusetts Right To Know List
[present]

Minnesota Hazardous Substance List
[present]

NJ Right to Know List (Total)
sn 0386

Pennsylvania Right to Know List
[present]

PROPOSED REGULATIONS

Canada - Ontario - Proposed Occupational TWAEVs
500 ppm TWAEV; 1800 mg/m3 TWAEV

Canada - Ontario - Proposed Occupational STEVs
750 ppm STEV; 2500 mg/m3 STEV

CHLORODIFLUOROPROPANE   134190-53-7

ENVIRONMENTAL LISTS

Class 2 Ozone Depletors
ozone depletion weight reserved

3-CHLORO-1,2-DIHYDROXYPROPANE   96-24-2

HEALTH AND SAFETY LISTS

U.S. DOT - Substances From 49 CFR 172.101
regulated by DOT (UN2689)

U.S. DOT - Hazard Classes
DOT hazard class = 6.1

NIOSH - Selected LD50s and LC50s
Oral, rat: LD50 = 26 mg/kg

INTERNATIONAL LISTS

Canada - WHMIS: Ingredient Disclosure
1% item 359 (427)

STATE LISTS

NJ Right to Know List (Total)
sn 2453

5-CHLORO-2,4-DIMETHOXYANILINE   97-50-7

ENVIRONMENTAL LISTS

TSCA - HDD/HDF - Precursors Required for Reporting
[present]

1-CHLORO-2,4-DINITROBENZENE   97-00-7

HEALTH AND SAFETY LISTS

NIOSH - Selected LD50s and LC50s
Oral, rat: LD50 = 1070 mg/kg Skin, rabbit: LD50 = 130 mg/kg

OSHA - List of Highly Hazardous Chemicals
threshhold quantity = 5000 pounds

STATE LISTS

Florida Hazardous Substance List
[present]

Massachusetts Right To Know List
[present]

Pennsylvania Right to Know List
[present]

CHLORODINITROBENZENE, ALL ISOMERS   RR-00587-8

HEALTH AND SAFETY LISTS

U.S. DOT - Substances From 49 CFR 172.101
regulated by DOT (UN1577)

U.S. DOT - Hazard Classes
DOT hazard class = 6.1

U.S. DOT - Appendix B - Marine Pollutants
DOT regulated marine pollutant

INTERNATIONAL LISTS

Canada - WHMIS: Ingredient Disclosure
1% item 360 (428)

CHLORODIPHENYL (54% CHLORINE)   11097-69-1

SEE ALSO:
F039-HAZARDOUS WASTES
POLYCHLORINATED BIPHENYLS

HEALTH AND SAFETY LISTS

ACGIH 1995 - Time Weighted Averages
0.5 mg/m3 TWA

ACGIH 1995 - Skin Designations
skin - potential for cutaneous absorption

U.S. DOT - Appendix A Table 1 - Hazardous Substances
final RQ = 1 pound (0.454 kg)

NIOSH - Selected LD50s and LC50s
Oral, rat: LD50 = 1010 mg/kg

NIOSH 1990 - Pocket Guide - RELs
0.001 mg/m3 TWA

NIOSH 1990 - Pocket Guide - IDLHs
5 mg/m3 IDLH (not considering carcinogenic effects)

NIOSH 1990 - Pocket Guide - Carcinogens
occupational carcinogen

NIOSH 1990 - Pocket Guide - Target organs
skin, eyes, liver

NTP Chemical Status Reports - Testing Status and NTIS Number
Technical reports printed (PB279624/AS)

NTP Chemical Status Reports - Evidence of Carcinogenicity
male rat-equivocal; female rat-equivocal

NTP Seventh Report - Suspect Carcinogens
suspect carcinogen (Listed under 'Polychlorinated biphenyls')

OSHA - Vacated PELs - Time Weighted Averages
0.5 mg/m3 TWA

OSHA - Vacated PELs - Skin Designation
Prevent or reduce skin absorption

OSHA - Final PELs - Time Weighted Averages
0.5 mg/m3 TWA

OSHA - Final PELs - Skin Notations
prevent or reduce skin absorption

ENVIRONMENTAL LISTS

ATSDR Priority List
Rank (of 275): 013

CERCLA/SARA - Hazardous Substances and their Reportable Quantities
final RQ = 1 pound (0.454 kg)

Clean Water Act - Priority Pollutants
[present]

EPA - Carcinogen Hazard Ranking for RQ Adjustment
Hazard ranking = Medium

RCRA - Basis for Listing - Appendix VII
Included in waste stream: F039

INTERNATIONAL LISTS

Australian Exposure Standards - Time Weighted Averages
0.5 mg/m3 TWA

Australian Exposure Standards - Short Term Exposure Limits
1 mg/m3 STEL

Australian Exposure Standards - Skin Effects
skin absorption

Australian Exposure Standards - Carcinogens
probable carcinogen

Australian Exposure Standards - Under Review
exposure limits under review

Canada - WHMIS: Ingredient Disclosure
0.1% item 127 (257)

Canada - Alberta - 8 Hour Occupational Exposure Limit
0.5 mg/m3 TWA

Canada - Alberta - 15 Minute Occupational Exposure Limit
1 mg/m3 STEL

Canada - Alberta - Skin Designation
can be absorbed through the intact skin

Canada - British Columbia - 8 Hour Exposure Limits
0.5 mg/m3 TWA

Canada - British Columbia - 15 Minute Exposure Limits
1 mg/m3 STEL

Canada - British Columbia - Skin Notations
skin - potential for skin absorption

Canada - Quebec - Time-Weighted Average Exposure Values
0.5 mg/m3 TWAEV

Canada - Quebec - Skin Designations
absorbed through the skin

Canada - Quebec - Carcinogens
C2 carcinogen: effect suspected in humans

German (DFG) - MAK Values
0.05 ppm MAK; 0.5 mg/m3 MAK

German (DFG) - Peak Limitations
10 x normal MAK (30 min average value); don't exceed during shift

German (DFG) - Skin/Sensitizers
danger of cutaneous absorption

German (DFG) - Carcinogens
suspected carcinogen

German (DFG) - Pregnancy
risk to embryo/fetus probable

Israel - Time Weighted Averages
0.5 mg/m3 TWA

Israel - Action Levels
0.25 mg/m3 AL

Mexico - Instruction No. 10 - TWAs
0.5 mg/m3 TWA

Mexico - Instruction No. 10 - STELs
1 mg/m3 STEL

Mexico - Instruction No. 10 - Skin designation
skin - potential for cutaneous absorption

Mexico - Wastewater - Organic Toxic Pollutants and Heavy Metals
Listed under [Polychlorinated Byphenyls]

STATE LISTS

California - Exposure Limits - PELs
0.5 mg/m3 PEL

California - Exposure Limits - Skin Notation
material may be absorbed through the skin, eyes or mucous membrane

California - Directors List of Hazardous Substances (8 CCR 339)
[present] (Listed under 'Polychlorobiphenyls')

Florida Hazardous Substance List
[present]

Massachusetts Right To Know List
carcinogen; extraordinarily hazardous; teratogen

Minnesota Hazardous Substance List
carcinogen; skin

Pennsylvania Right to Know List
environmental hazard; special hazardous substance

Pennsylvania RTK - Special Hazardous Substances
[present]

CHLOROETHYL ACETATE   RR-00826-4

HEALTH AND SAFETY LISTS

NFPA - Flash Points
flash point = 129 degrees F (54 degrees C)

NFPA - Hazard Identification Ratings
health-2; flammability-2; reactivity-0

CHLORO-4-ETHYLBENZENE   RR-00825-3

HEALTH AND SAFETY LISTS

NFPA - Flash Points
flash point = 147 degrees F (64 degrees C)

NFPA - Hazard Identification Ratings
health-1; flammability-2; reactivity-0

CHLOROETHYL CHLOROFORMATE   627-11-2

ENVIRONMENTAL LISTS

CERCLA/SARA - Section 302 Extremely Hazardous Substances and TPQs
TPQ = 1000 pounds

STATE LISTS

Florida Hazardous Substance List
effective March 13, 1992

Massachusetts Right To Know List
extraordinarily hazardous

NJ Right to Know List (Total)
sn 2239

Pennsylvania Right to Know List
environmental hazard

PROPOSED REGULATIONS

CERCLA/SARA - Proposed Hazardous Substance Additions
proposed RQ = 1 pound (.454 kg)

CERCLA/SARA - 1989 Proposed RQ Adjustments
proposed RQ = 100 pounds (45.4 kg)

1-(2-CHLOROETHYL)-3-CYCLOHEXYL-1-   13010-47-4
NITROSOUREA

HEALTH AND SAFETY LISTS

IARC - Group 2A (limited human data)
[present] (Other relevant data, as given in Supplement 7, influenced the making of the overall evaluation) (Listed under 'Chloroethyl nitrosourea')

NIOSH - Selected LD50s and LC50s
Oral, rat: LD50 = 70 mg/kg

NTP Chemical Status Reports - Testing Status and NTIS Number
Chronic studies exist for which technical reports were not prepared

NTP Seventh Report - Suspect Carcinogens
suspect carcinogen

OSHA - Possible Select Carcinogens
[present]

STATE LISTS

California - Air Bill 2588 Appendix A-II
known or potential carcinogen

California - Prop. 65 - Cancer list
carcinogen - initial date 1/1/88

California - Prop. 65 - Developmental Toxicity
developmental toxicity - initial date 7/1/90

California - Directors List of Hazardous Substances (8 CCR 339)
[present]

Florida Hazardous Substance List
[present]

Massachusetts Right To Know List
carcinogen; extraordinarily hazardous

Minnesota Hazardous Substance List
carcinogen

NJ Special Hazardous Substances
(carcinogen)

Pennsylvania Right to Know List
special hazardous substance

Pennsylvania RTK - Special Hazardous Substances
[present]

2-CHLOROETHYL ETHYL SULFIDE   693-07-2

INTERNATIONAL LISTS

Canada - WHMIS: Ingredient Disclosure
1% item 362 (1545)

1-(2-CHLOROETHYL)-3-(4-METHYLCYCLOHEXYL)-   13909-09-6
1-NITROSOUREA

HEALTH AND SAFETY LISTS

IARC - Group 1 (carcinogenic to humans)
[present]

NTP Chemical Status Reports - Testing Status and NTIS Number
Chronic studies exist for which technical reports were not prepared

NTP Seventh Report - Known Carcinogens
known carcinogen

OSHA - Select Carcinogens
[present]

STATE LISTS

California - Air Bill 2588 Appendix A-I
known or potential carcinogen

California - Prop. 65 - Cancer list
carcinogen - initial date 10/1/88

California - Directors List of Hazardous Substances (8 CCR 339)
[present]

Minnesota Hazardous Substance List
carcinogen

2-CHLOROETHYL VINYL ETHER   110-75-8

HEALTH AND SAFETY LISTS

U.S. DOT - Appendix A Table 1 - Hazardous Substances
final RQ = 1000 pounds (454 kg)

NFPA - Flash Points
flash point = 80 degrees F (27 degrees C)

NFPA - Hazard Identification Ratings
health-2; flammability-3; reactivity-2

NIOSH - Selected LD50s and LC50s
Oral, rat: LD50 = 250 mg/kg Skin, rabbit: LD50 = 3354 mg/kg

ENVIRONMENTAL LISTS

CERCLA/SARA - Hazardous Substances and their Reportable Quantities
final RQ = 1000 pounds (454 kg)

Clean Water Act - Priority Pollutants
[present]

RCRA - U Series Wastes
waste number U042

RCRA - Hazardous Constituents-Appendix VIII
waste number U042

RCRA - Basis for Listing - Appendix VII
Included in waste stream: F039

RCRA - Substances Banned From Land Disposal
[present]

RCRA - Universal Treatment Standards (LDR)
WW: 0.062 mg/l; NWW: Not applicable

TSCA - Code of Federal Regulations Citations
40 CFR 716.120(a)

TSCA - Health and Safety Reporting List
Effective Date: March 7, 1986

INTERNATIONAL LISTS

Canada - WHMIS: Ingredient Disclosure
1% item 1687 (836)

Mexico - Wastewater - Organic Toxic Pollutants and Heavy Metals
Listed under [Chloroalkyl Ethers]

Mexico - Drinking Water - Ecological Criteria
None given.

STATE LISTS

California - Directors List of Hazardous Substances (8 CCR 339)
[present]

Florida Hazardous Substance List
[present]

Massachusetts Right To Know List
[present]

Pennsylvania Right to Know List
environmental hazard

2-CHLOROETHYL-2-XENYL ETHER   RR-00558-3

HEALTH AND SAFETY LISTS

NFPA - Flash Points
flash point = 320 degrees F (160 degrees C)

NFPA - Hazard Identification Ratings
flammability-1; reactivity-0

CHLOROFLUOROALKANES, FULLY HALOGENATED   RR-01767-4

ENVIRONMENTAL LISTS

TSCA - Section 12(b) - Export Notification
export notification required - Section 6 These fully halogenated chlorofluoroalkanes include the chlorofluorocarbons.

O-CHLOROFLUOROBENZENE   348-51-6

ENVIRONMENTAL LISTS

TSCA - HDD/HDF - Precursors Required for Reporting
[present]

CHLOROFLUOROCARBON   RR-00078-2

INTERNATIONAL LISTS

Canada - CEPA Schedule II Part II - Toxic Substances (Export)
[present] (totally halogenated with the molecular formula CnClxF (2n+2-x))

Canada - CEPA Schedule I - Toxic Substances
totally halogenated chlorofluorocarbons that have the molecular formula CnClxF(2 n + 2-x): (a) prohibited commercial manufacturing or processing uses; (b) quantities that may be manufactured or imported

CHLOROFLUOROMETHANE   593-70-4

HEALTH AND SAFETY LISTS

IARC - Group 3 (not classifiable)
[present]

ENVIRONMENTAL LISTS

Class 2 Ozone Depletors
ozone depletion weight reserved

List of Pesticide Product Inert Ingredients
[present]

INTERNATIONAL LISTS

German (DFG) - Carcinogens
animal evidence of carcinogenicity

STATE LISTS

California - Directors List of Hazardous Substances (8 CCR 339)
[present]

3-CHLORO-4-FLUORONITROBENZENE   350-30-1

ENVIRONMENTAL LISTS

TSCA - HDD/HDF - Precursors Required for Reporting
[present]

CHLOROFLUOROPROPANE   134190-54-8

ENVIRONMENTAL LISTS

Class 2 Ozone Depletors
ozone depletion weight reserved

CHLOROFORM   67-66-

SEE ALSO:

K009-HAZARDOUS WASTES
K021-HAZARDOUS WASTES
K010-HAZARDOUS WASTES
K151-HAZARDOUS WASTES
F024-HAZARDOUS WASTES
BIS(2,4-DIMETHYLBUTYL) MALEATE
K020-HAZARDOUS WASTES
METHYL DIHYDROABIETATE
F039-HAZARDOUS WASTES
K073-HAZARDOUS WASTES
K029-HAZARDOUS WASTES
K150-HAZARDOUS WASTES
K019-HAZARDOUS WASTES
F025-HAZARDOUS WASTES
K116-HAZARDOUS WASTES
K149-HAZARDOUS WASTES

HEALTH AND SAFETY LISTS

ACGIH 1995 - Time Weighted Averages
10 ppm TWA; 49 mg/m3 TWA

ACGIH 1995 - Carcinogens
A2-suspected human carcinogen

AIHA - Odor Threshold Values
geometric mean air odor threshold = 192 ppm

U.S. DOT - Substances From 49 CFR 172.101
regulated by DOT (UN1888)

U.S. DOT - Hazard Classes
DOT hazard class = 6.1

U.S. DOT - Appendix A Table 1 - Hazardous Substances
final RQ = 10 pounds (4.54 kg)

IARC - Group 2B (sufficient animal data)
[present]

NIOSH - Selected LD50s and LC50s
Inhalation, rat: LC50 = 75 gm/m3 1 hr Oral, rat: LD50 = 908 mg/kg

NIOSH 1990 - Pocket Guide - RELs
2 ppm STEL (60 min); 9.78 mg/m3 STEL (60 min)

NIOSH 1990 - Pocket Guide - IDLHs
1000 ppm IDLH (not considering carcinogenic effects)

NIOSH 1990 - Pocket Guide - Carcinogens
occupational carcinogen

NIOSH 1990 - Pocket Guide - Target organs
liver, kidneys, heart, eyes, skin

NIOSH - Health Standards - Exposure Limits
C 2 ppm; C 9.78 mg/m3 (45 liter, 60 min sample)

NIOSH - Health Standards - Health Effects and Precautions
has produced cancer of the liver and kidneys in animals; central nervous system effects

NIOSH - Health Standards - Carcinogenic Chemicals
potential human carcinogen

NTP Chemical Status Reports - Testing Status and NTIS Number
Technical reports printed

NTP Seventh Report - Suspect Carcinogens
suspect carcinogen

OSHA - Vacated PELs - Time Weighted Averages
2 ppm TWA; 9.78 mg/m3 TWA

OSHA - Final PELs - Ceiling Limits
C 50 ppm; C 240 mg/m3

OSHA - Possible Select Carcinogens
[present]

ENVIRONMENTAL LISTS

ATSDR Priority List
Rank (of 275): 009

CERCLA/SARA - Section 302 Extremely Hazardous Substances and TPQs
TPQ = 10,000 pounds

CERCLA/SARA - Section 313 - Emission Reporting
form R reporting required for 0.1% de minimus concentration

CERCLA/SARA - Hazardous Substances and their Reportable Quantities
final RQ = 10 pounds (4.54 kg)

Clean Air Act (1990) - List of Hazardous Air Contaminants
[present]

CAA -Toxic Substances for Accidental Release Prevention
threshold quantity = 20,000 lbs

CAA - HON Rule - SOCMI Chemicals
compliance by Oct. 24, 1994

CAA - HON Rule - Organic HAPs
[present]

Clean Water Act - Hazardous Substances
[present]

Clean Water Act - Priority Pollutants
[present]

Clean Water Act - Toxic Pollutants
[present]

Safe Drinking Water Act - Monitoring
monitoring required

EPA - Carcinogen Hazard Ranking for RQ Adjustment
Hazard ranking = Medium

RCRA - D Series - Maximum Concentration of Contaminants
waste number D022; regulatory level = 6.0 mg/L

RCRA - D Series - Chronic Toxicity Reference Levels
chronic toxicity reference level = 0.06 mg/L

RCRA - U Series Wastes
waste number U044

RCRA - Hazardous Constituents-Appendix VIII
waste number U044

RCRA - Basis for Listing - Appendix VII
Included in waste streams: F024, F025, F039, K009, K010, K019, K020, K021, K029, K073, K116, K149, K150, K151

RCRA - Substances Banned From Land Disposal
[present]

RCRA - TSD Facilities Ground Water Monitoring
TM 8010 = 0.5 ug/L PQL; TM 8240 = 5 ug/L PQL

RCRA - Universal Treatment Standards (LDR)
WW: 0.046 mg/l; NWW: 6.0 mg/kg

TSCA - Code of Federal Regulations Citations
40 CFR 716.120(a)

TSCA - Health and Safety Reporting List
Effective Date: June 1, 1987

INTERNATIONAL LISTS

Australian Exposure Standards - Time Weighted Averages
10 ppm TWA; 49 mg/m3 TWA

Australian Exposure Standards - Carcinogens
suspected carcinogen

Canada - WHMIS: Ingredient Disclosure
0.1% item 363 (429)

Canada - NPRI (National Pollutant Release Inventory)
[present]

Canada - Alberta - 8 Hour Occupational Exposure Limit
10 ppm TWA; 49 mg/m3 TWA

Canada - Alberta - 15 Minute Occupational Exposure Limit
50 ppm STEL; 225 mg/m3 STEL

Canada - British Columbia - 8 Hour Exposure Limits
10 ppm TWA; 50 mg/m3 TWA

Canada - British Columbia - Carcinogens
carcinogen - 10 ppm TWA; 50 mg/m3 TWA

Canada - Ontario - OHSA - TWAEVs
10 ppm TWAEV; 49 mg/m3 TWAEV

Canada - Quebec - Time-Weighted Average Exposure Values
5 ppm TWAEV; 24.4 mg/m3 TWAEV (substance of which the recirculation is prohibited)

Canada - Quebec - Carcinogens
C2 carcinogen: effect suspected in humans

German (DFG) - MAK Values
10 ppm MAK; 50 mg/m3 MAK

German (DFG) - Peak Limitations
2 x normal MAK (30 min. average value); don't exceed 4 times during shift

German (DFG) - Carcinogens
suspected carcinogen

German (DFG) - Pregnancy
risk to embryo/fetus probable

Israel - Time Weighted Averages
10 ppm TWA; 49 mg/m3 TWA

Israel - Action Levels
5 ppm AL; 24.5 mg/m3 AL

Mexico - Instruction No. 10 - TWAs
10 ppm TWA; 50 mg/m3 TWA

Mexico - Instruction No. 10 - STELs
50 ppm STEL; 225 mg/m3 STEL

Mexico - Instruction No. 10 - Carcinogens
potential carcinogen in humans - limited epidemiological evidence

Mexico - Wastewater - Organic Toxic Pollutants and Heavy Metals
Listed under [Organic Toxic Pollutants]

Mexico - Drinking Water - Ecological Criteria
0.03 mg/l Substance presents persistence, bioaccumulations or risk of cancer, reduce human exposure to a minimum; This level has been extrapolated by using a mathematic model

STATE LISTS

California - Air Bill 2588 Appendix A-I
known or potential carcinogen

California - Prop. 65 - Cancer list
carcinogen - initial date 10/1/87

California - Prop. 65 - No Significant Risk Levels
ingestion: no significant risk level = 20 ug/day; inhalation: no significant risk level = 40 ug/day

California - Exposure Limits - PELs
2 ppm PEL; 9.78 mg/m3 PEL

California - Directors List of Hazardous Substances (8 CCR 339)
[present]

Florida Hazardous Substance List
[present]

Massachusetts Right To Know List
carcinogen; extraordinarily hazardous

Minnesota Hazardous Substance List
carcinogen

NJ Right to Know List (Total)
sn 0388

NJ Special Hazardous Substances
(carcinogen)

Pennsylvania Right to Know List
environmental hazard; special hazardous substance

Pennsylvania RTK - Special Hazardous Substances
[present]

PROPOSED REGULATIONS

Safe Drinking Water Act - Priority list
[present]

Canada - Ontario - Proposed Occupational TWAEVs
2 ppm TWAEV; 10 mg/m3 TWAEV

CHLOROFORMATE, N.O.S.   RR-00207-3

HEALTH AND SAFETY LISTS

U.S. DOT - Substances From 49 CFR 172.101
regulated by DOT (UN2742)

U.S. DOT - Hazard Classes
DOT hazard class = 6.1

U.S. DOT - Substances Which Are Poisonous by Inhalation
liquid hazardous material poisonous by inhalation (UN2742)

INTERNATIONAL LISTS

Canada - WHMIS: Ingredient Disclosure
1% item 364 (441)

STATE LISTS

NJ Right to Know List (Total)
sn 2240

N-CHLOROFORMYL MORPHOLINE   15159-40-7

INTERNATIONAL LISTS

German (DFG) - Carcinogens
animal evidence of carcinogenicity

2-CHLORO-1,1,1,2,3,3,3-HEPTAFLUOROPROPANE   125426-39-3

ENVIRONMENTAL LISTS

Class 1 Ozone Depletors
ozone depletion potential = 1.0

1-CHLOROHEPTANE   629-06-1

HEALTH AND SAFETY LISTS

U.S. DOT - Appendix B - Marine Pollutants
DOT regulated marine pollutant

CHLOROHEXAFLUOROPROPANE   134308-72-8

ENVIRONMENTAL LISTS

Class 2 Ozone Depletors
ozone depletion weight reserved

1-CHLOROHEXANE   544-10-5

HEALTH AND SAFETY LISTS

U.S. DOT - Appendix B - Marine Pollutants
DOT regulated marine pollutant

NFPA - Flash Points
flash point = 95 degrees F (35 degrees C)

NFPA - Hazard Identification Ratings
flammability-3; reactivity-0

STATE LISTS

Florida Hazardous Substance List
[present]

Massachusetts Right To Know List
[present]

Pennsylvania Right to Know List
[present]

CHLOROHEXIDINE DIGLUCONATE   18472-51-0

HEALTH AND SAFETY LISTS

NIOSH - Selected LD50s and LC50s
Oral, rat: LD50 = 2 gm/kg

CHLOROHYDROQUINONE   615-67-8

ENVIRONMENTAL LISTS

TSCA - HDD/HDF - Precursors Required for Reporting
[present]

2-CHLORO-N-HYDROXYMETHYLACETAMIDE   2832-19-1

INTERNATIONAL LISTS

German (DFG) - Skin/Sensitizers
danger of sensitization

German (DFG) - Carcinogens
suspected carcinogen

CHLORO(O-HYDROXYPHENYL)MERCURY   90-03-9
SEE ALSO:
MERCURY

INTERNATIONAL LISTS

Canada - WHMIS: Ingredient Disclosure
1% item 365 (443)

CHLOROISOCYANURIC ACID   13057-78-8

STATE LISTS

Florida Hazardous Substance List
[present]

Massachusetts Right To Know List
[present]

Pennsylvania Right to Know List
[present]

CHLOROMETHYL (OXIRANE)   6806-86-6

STATE LISTS

Florida Hazardous Substance List
[present]

Massachusetts Right To Know List
[present]

3-CHLORO-2-METHYL ANILINE   87-60-5

HEALTH AND SAFETY LISTS

NIOSH - Selected LD50s and LC50s
Oral, rat: LD50 = 574 mg/kg

STATE LISTS

California - Air Bill 2588 Appendix A-II
known or potential carcinogen: 6/91

Massachusetts Right To Know List
carcinogen; extraordinarily hazardous

NJ Right to Know List (Total)
sn 0419

6-CHLORO-2-METHYL ANILINE   87-63-8

ENVIRONMENTAL LISTS

TSCA - Code of Federal Regulations Citations
40 CFR 721.462

TSCA - Chemicals with Significant New Use Rules
[present]

TSCA - Section 12(b) - Export Notification
export notification required - Section 5

STATE LISTS

NJ Right to Know List (Total)
sn 0423

2-CHLORO-5-METHYL ANILINE   95-81-8

STATE LISTS

NJ Right to Know List (Total)
sn 0420

2-CHLORO-4-METHYL ANILINE   615-65-6

HEALTH AND SAFETY LISTS

NIOSH - Selected LD50s and LC50s
Oral, rat: LD50 = 367 mg/kg

STATE LISTS
  NJ Right to Know List (Total)
    sn 0417

4-CHLORO-3-METHYL ANILINE   **7149-75-9**
  STATE LISTS
    NJ Right to Know List (Total)
      sn 0424

2-CHLORO-3-METHYL ANILINE   **29027-17-6**
  STATE LISTS
    NJ Right to Know List (Total)
      sn 0422

CHLOROMETHYLCHLOROFORMATE   **22128-62-7**
  HEALTH AND SAFETY LISTS
    U.S. DOT - Substances From 49 CFR 172.101
      regulated by DOT (UN2745)
    U.S. DOT - Hazard Classes
      DOT hazard class = 6.1
  INTERNATIONAL LISTS
    Canada - WHMIS: Ingredient Disclosure
      1% item 366 (434)
  STATE LISTS
    NJ Right to Know List (Total)
      sn 0389
    NJ Special Hazardous Substances
      (corrosive)

BIS(CHLOROMETHYL) ETHER   **542-88-1**
  SEE ALSO:
    CHLOROALKYL ETHERS
    K017-HAZARDOUS WASTES
  HEALTH AND SAFETY LISTS
    ACGIH 1995 - Time Weighted Averages
      0.001 ppm TWA; 0.0047 mg/m3 TWA
    ACGIH 1995 - Carcinogens
      A1-confirmed human carcinogen
    U.S. DOT - Substances From 49 CFR 172.101
      regulated by DOT (UN2249)
    U.S. DOT - Hazard Classes
      DOT hazard class = 6.1
    U.S. DOT - Appendix A Table 1 - Hazardous Substances
      final RQ = 10 pounds (4.54 kg)
    IARC - Group 1 (carcinogenic to humans)
      [present]
    NIOSH - Selected LD50s and LC50s
      Inhalation, rat: LC50 = 7 ppm 7 hr Oral, rat: LD50 = 210 mg/kg
      Skin, rabbit: LD50 = 280 mg/kg
    NIOSH 1990 - Pocket Guide - Carcinogens
      occupational carcinogen
    NIOSH 1990 - Pocket Guide - Target organs
      lungs, eyes, skin
    NIOSH - Health Standards - Exposure Limits
      use 29 CFR 1910.1008
    NIOSH - Health Standards - Health Effects and Precautions
      Lung cancer
    NIOSH - Health Standards - Carcinogenic Chemicals
      potential human carcinogen
    NTP Seventh Report - Known Carcinogens
      known carcinogen
    OSHA - 29 CFR 1910 Specifically Regulated Chemicals
      Cancer suspect agent (see 29 CFR 1910.1008)

OSHA - Select Carcinogens
  [present]
OSHA - List of Highly Hazardous Chemicals
  threshhold quantity = 100 pounds
ENVIRONMENTAL LISTS
  CERCLA/SARA - Section 302 Extremely Hazardous Substances and TPQs
    TPQ = 100 pounds
  CERCLA/SARA - Section 313 - Emission Reporting
    form R reporting required for 0.1% de minimus concentration
  CERCLA/SARA - Hazardous Substances and their Reportable Quantities
    final RQ = 10 pounds (4.54 kg)
  Clean Air Act (1990) - List of Hazardous Air Contaminants
    [present]
  CAA -Toxic Substances for Accidental Release Prevention
    threshold quantity = 1,000 lbs
  CAA - HON Rule - SOCMI Chemicals
    compliance by Oct. 24, 1994
  CAA - HON Rule - Organic HAPs
    [present]
  EPA - Carcinogen Hazard Ranking for RQ Adjustment
    Hazard ranking = High
  RCRA - P Series Wastes
    waste number P016
  RCRA - Hazardous Constituents-Appendix VIII
    waste number P016
  RCRA - Basis for Listing - Appendix VII
    Included in waste stream: K017
  RCRA - Substances Banned From Land Disposal
    [present]
INTERNATIONAL LISTS
  Australian Exposure Standards - Time Weighted Averages
    0.001 ppm TWA; 0.005 mg/m3 TWA
  Australian Exposure Standards - Carcinogens
    confirmed carcinogen
  Canada - WHMIS: Ingredient Disclosure
    0.1% item 191 (841)
  Canada - Alberta - 8 Hour Occupational Exposure Limit
    0.001 ppm TWA; 0.0047 mg/m3 TWA
  Canada - Alberta - 15 Minute Occupational Exposure Limit
    0.003 ppm STEL; 0.014 mg/m3 STEL
  Canada - Alberta - Designated Substances
    designated substance - requires code of practice
  Canada - British Columbia - 8 Hour Exposure Limits
    0.001 ppm TWA; 0.005 mg/m3 TWA
  Canada - British Columbia - Carcinogens
    carcinogen - 0.001 ppm TWA; 0.005 mg/m3 TWA
  Canada - Ontario - OHSA - TWAEVs
    0.001 ppm TWAEV; 0.005 mg/m3 TWAEV
  Canada - Quebec - Time-Weighted Average Exposure Values
    0.001 ppm TWAEV; 0.0047 mg/m3 TWAEV (substance of which the recirculation is prohibited)
  Canada - Quebec - Carcinogens
    C1 carcinogen: effect detected in humans
  German (DFG) - Carcinogens
    proven carcinogen
  Israel - Time Weighted Averages
    0.001 ppm TWA; 0.0047 mg/m3 TWA
  Israel - Action Levels
    0.0005 ppm AL; 0.00235 mg/m3 AL
  Mexico - Instruction No. 10 - TWAs
    No exposure permitted

Mexico - Instruction No. 10 - STELs
no exposure permitted

Mexico - Wastewater - Organic Toxic Pollutants and Heavy Metals
Listed under [Chloroalkyl Ethers]

STATE LISTS

California - Air Bill 2588 Appendix A-I
known or potential carcinogen

California - Prop. 65 - Cancer list
carcinogen - initial date 2/27/87

California - Prop. 65 - No Significant Risk Levels
no significant risk level = 0.02 ug/day

California - Exposure Limits - PELs
0.001 ppm PEL; 0.005 mg/m3 PEL

California - Exposure Limits - Carcinogens
cancer-suspect agent (at a concentration >= 0.1%)

California - Directors List of Hazardous Substances (8 CCR 339)
[present] (refers to any mixture containing 0.1% or more)

Florida Hazardous Substance List
[present]

Massachusetts Right To Know List
carcinogen; extraordinarily hazardous

Minnesota Hazardous Substance List
carcinogen

NJ Right to Know List (Total)
sn 0234

NJ Special Hazardous Substances
(carcinogen)

Pennsylvania Right to Know List
environmental hazard; special hazardous substance

Pennsylvania RTK - Special Hazardous Substances
[present]

2-CHLORO-1-METHYLETHYL BIS(2-CHLORO-PROPYL) PHOSPHATE   76649-15-5

ENVIRONMENTAL LISTS

TSCA - PAIR - Reporting List
Effective Date: June 14, 1993

TSCA - Health and Safety Reporting List
Effective Date: June 14, 1993

CHLOROMETHYL ETHYL ETHER   3188-13-4

HEALTH AND SAFETY LISTS

U.S. DOT - Substances From 49 CFR 172.101
regulated by DOT (UN2354)

U.S. DOT - Hazard Classes
DOT hazard class = 3

NFPA - Flash Points
flash point = 185 degrees F (85 degrees C)

NFPA - Hazard Identification Ratings
health-2; flammability-2; reactivity-0

INTERNATIONAL LISTS

Canada - WHMIS: Ingredient Disclosure
1% item 367 (812)

STATE LISTS

NJ Right to Know List (Total)
sn 0390

5-CHLORO-2-METHYL-3-ISOTHIAZOLONE   26172-55-4

ENVIRONMENTAL LISTS

List of Pesticide Product Inert Ingredients
[present]

INTERNATIONAL LISTS

German (DFG) - MAK Values
0.05 mg/m3 MAK

German (DFG) - Skin/Sensitizers
danger of sensitization (skin or respiratory)

German (DFG) - Pregnancy
classification not yet possible

CHLOROMETHYL METHYL ETHER   107-30-2
SEE ALSO:
CHLOROALKYL ETHERS

HEALTH AND SAFETY LISTS

ACGIH 1995 - Carcinogens
A2-suspected human carcinogen

U.S. DOT - Substances From 49 CFR 172.101
regulated by DOT (UN1239)

U.S. DOT - Hazard Classes
DOT hazard class = 6.1

U.S. DOT - Substances Which Are Poisonous by Inhalation
liquid hazardous material poisonous by inhalation (UN1239)

U.S. DOT - Appendix A Table 1 - Hazardous Substances
final RQ = 10 pounds (4.54 kg)

IARC - Group 1 (carcinogenic to humans)
[present]

NIOSH - Selected LD50s and LC50s
Inhalation, rat: LC50 = 55 ppm 7 hr Oral, rat: LD50 = 817 mg/kg

NIOSH 1990 - Pocket Guide - Carcinogens
occupational carcinogen

NIOSH 1990 - Pocket Guide - Target organs
respiratory system, skin, eyes, mucous membrane

NIOSH - Health Standards - Exposure Limits
use 29 CFR 1910.1006

NIOSH - Health Standards - Health Effects and Precautions
Lung cancer

NIOSH - Health Standards - Carcinogenic Chemicals
potential human carcinogen

NTP Seventh Report - Known Carcinogens
technical grade: known carcinogen

OSHA - 29 CFR 1910 Specifically Regulated Chemicals
Cancer suspect agent (see 29 CFR 1910.1006)

OSHA - Select Carcinogens
[present]

OSHA - List of Highly Hazardous Chemicals
threshhold quantity = 500 pounds

ENVIRONMENTAL LISTS

CERCLA/SARA - Section 302 Extremely Hazardous Substances and TPQs
TPQ = 100 pounds

CERCLA/SARA - Section 313 - Emission Reporting
form R reporting required for 0.1% de minimus concentration

CERCLA/SARA - Hazardous Substances and their Reportable Quantities
final RQ = 10 pounds (4.54 kg)

Clean Air Act (1990) - List of Hazardous Air Contaminants
[present]

CAA -Toxic Substances for Accidental Release Prevention
threshold quantity = 5,000 lbs

EPA - Carcinogen Hazard Ranking for RQ Adjustment
Hazard ranking = High

RCRA - U Series Wastes
waste number U046

RCRA - Hazardous Constituents-Appendix VIII
waste number U046

© Van Nostrand Reinhold 1995

RCRA - Substances Banned From Land Disposal
[present]

INTERNATIONAL LISTS

Australian Exposure Standards - Time Weighted Averages
control to the lowest practical level

Australian Exposure Standards - Carcinogens
probable carcinogen

Canada - WHMIS: Ingredient Disclosure
0.1% item 1029 (827)

Canada - NPRI (National Pollutant Release Inventory)
[present]

Canada - Alberta - 8 Hour Occupational Exposure Limit
0.005 ppm TWA; 0.02 mg/m3 TWA

Canada - Alberta - 15 Minute Occupational Exposure Limit
0.015 ppm STEL; 0.05 mg/m3 STEL

Canada - Alberta - Designated Substances
designated substance - requires code of practice

Canada - Quebec - Time-Weighted Average Exposure Values
substance of which the recirculation is prohibited

Canada - Quebec - Carcinogens
C1 carcinogen: effect detected in humans

German (DFG) - Carcinogens
proven carcinogen

Mexico - Instruction No. 10 - Carcinogens
potential carcinogen in humans - limited epidemiological evidence

STATE LISTS

California - Air Bill 2588 Appendix A-I
known or potential carcinogen

California - Prop. 65 - Cancer list
carcinogen - initial date 2/27/87

California - Prop. 65 - No Significant Risk Levels
no significant risk level = 0.3 ug/day

California - Exposure Limits - Carcinogens
cancer-suspect agent (at a concentration >= 0.1%)

California - Directors List of Hazardous Substances (8 CCR 339)
[present] (refers to any mixture containing greater than 0.1%)

Florida Hazardous Substance List
[present] (includes technical grade)

Massachusetts Right To Know List
carcinogen; extraordinarily hazardous

Minnesota Hazardous Substance List
carcinogen

NJ Right to Know List (Total)
sn 0391

NJ Special Hazardous Substances
(carcinogen)

Pennsylvania Right to Know List
environmental hazard; special hazardous substance

Pennsylvania RTK - Special Hazardous Substances
[present]

2-CHLORO-1-(3-METHYLPHENOXY)-4-(TRIFLUO-ROMETHYL)BENZENE   42874-96-4

ENVIRONMENTAL LISTS

TSCA - PAIR - Reporting List
Reporting Date: June 10, 1993

TSCA - Health and Safety Reporting List
Effective Date: April 12, 1993

N'-(3-CHLORO-4-METHYLPHENYL)-N,N-DIMETHYLUREA   15545-48-9

HEALTH AND SAFETY LISTS

NIOSH - Selected LD50s and LC50s
Inhalation, rat: LC50 = 1300 mg/m3 8 hr Oral, rat: LD50 = 5800 mg/kg

3-CHLORO-4-METHYLPHENYL ISOCYANATE   28479-22-3

HEALTH AND SAFETY LISTS

U.S. DOT - Substances From 49 CFR 172.101
regulated by DOT (UN2236)

U.S. DOT - Hazard Classes
DOT hazard class = 6.1

INTERNATIONAL LISTS

Canada - WHMIS: Ingredient Disclosure
1% item 368 (1037)

STATE LISTS

NJ Right to Know List (Total)
sn 0392

3-CHLORO-2-METHYLPROPENE   563-47-3

HEALTH AND SAFETY LISTS

U.S. DOT - Substances From 49 CFR 172.101
regulated by DOT (UN2554)

U.S. DOT - Hazard Classes
DOT hazard class = 3

NFPA - Flash Points
flash point = 11 degrees F (-12 degrees C)

NFPA - Hazard Identification Ratings
health-2; flammability-3; reactivity-1

NIOSH - Selected LD50s and LC50s
Inhalation, rat: LC50 = 34000 mg/m3 (30 mn)

NTP Chemical Status Reports - Testing Status and NTIS Number
Technical reports printed (PB86247293/AS)

NTP Chemical Status Reports - Evidence of Carcinogenicity
male rat-clear evidence; female rat-clear evidence; male mice-clear evidence; female mice-clear evidence

NTP Seventh Report - Suspect Carcinogens
suspect carcinogen

OSHA - Possible Select Carcinogens
[present]

ENVIRONMENTAL LISTS

CERCLA/SARA - Section 313 - Emission Reporting
form R reporting required

INTERNATIONAL LISTS

German (DFG) - Carcinogens
suspected carcinogen

STATE LISTS

California - Air Bill 2588 Appendix A-II
known or potential carcinogen: 9/89

California - Prop. 65 - Cancer list
carcinogen - initial date 7/1/89

California - Prop. 65 - No Significant Risk Levels
no significant risk level = 5 ug/day

Florida Hazardous Substance List
[present]

Massachusetts Right To Know List
[present]

NJ Right to Know List (Total)
sn 1223

Pennsylvania Right to Know List
special hazardous substance

Pennsylvania RTK - Special Hazardous Substances
[present]

2-CHLOROMETHYLPYRIDINE HYDROCHLORIDE   6959-47-3

HEALTH AND SAFETY LISTS

NTP Chemical Status Reports - Testing Status and NTIS Number
Technical reports printed (PB295895/AS)

NTP Chemical Status Reports - Evidence of Carcinogenicity
male rat-negative; female rat-negative; male mice-negative; female mice-negative

3-CHLOROMETHYLPYRIDINE HYDROCHLORIDE   6959-48-4

HEALTH AND SAFETY LISTS

NIOSH - Selected LD50s and LC50s
Oral, rat: LD50 = 316 mg/kg

NTP Chemical Status Reports - Testing Status and NTIS Number
Technical reports printed (PB287125/AS)

NTP Chemical Status Reports - Evidence of Carcinogenicity
male rat-positive; female rat-equivocal; male mice-positive; female mice-positive

STATE LISTS

Massachusetts Right To Know List
carcinogen; extraordinarily hazardous

2-CHLORO-N-METHYL-N-SUBSTITUTED ACETAMIDE   RR-00919-8

ENVIRONMENTAL LISTS

TSCA - Chemicals with Significant New Use Rules
PMN number: P-84-393

BETA-CHLORONAPHTHALENE   91-58-7
SEE ALSO:
F039-HAZARDOUS WASTES

HEALTH AND SAFETY LISTS

U.S. DOT - Appendix A Table 1 - Hazardous Substances
final RQ = 5000 pounds (2270 kg)

NIOSH - Selected LD50s and LC50s
Oral, rat: LD50 = 2078 mg/kg

ENVIRONMENTAL LISTS

CERCLA/SARA - Hazardous Substances and their Reportable Quantities
final RQ = 5000 pounds (2270 kg)

Clean Water Act - Priority Pollutants
[present]

RCRA - U Series Wastes
waste number U047

RCRA - Hazardous Constituents-Appendix VIII
waste number U047

RCRA - Basis for Listing - Appendix VII
Included in waste stream: F039

RCRA - Substances Banned From Land Disposal
[present]

RCRA - TSD Facilities Ground Water Monitoring
TM 8120 = 10 ug/L PQL; TM 8270 = 10 ug/L PQL

RCRA - Universal Treatment Standards (LDR)
WW: 0.055 mg/l; NWW: 5.6 mg/kg

TSCA - Code of Federal Regulations Citations
40 CFR 704.83; 40 CFR 716.120(a)

TSCA - Health and Safety Reporting List
Effective Date: October 4, 1982

STATE LISTS

California - Directors List of Hazardous Substances (8 CCR 339)
[present]

Massachusetts Right To Know List
[present]

NJ Right to Know List (Total)
sn 3017

Pennsylvania Right to Know List
environmental hazard

CHLORONITROANILINE   41587-36-4

HEALTH AND SAFETY LISTS

U.S. DOT - Substances From 49 CFR 172.101
regulated by DOT (UN2237)

U.S. DOT - Hazard Classes
DOT hazard class = 6.1

U.S. DOT - Appendix B - Marine Pollutants
DOT regulated marine pollutant

INTERNATIONAL LISTS

Canada - WHMIS: Ingredient Disclosure
1% item 369 (444)

STATE LISTS

NJ Right to Know List (Total)
sn 0393

CHLORONITROBENZENE   25167-93-5

HEALTH AND SAFETY LISTS

U.S. DOT - Substances From 49 CFR 172.101
regulated by DOT (UN1578)

U.S. DOT - Hazard Classes
DOT hazard class = 6.1

NFPA - Flash Points
flash point = 261 degrees F (127 degrees C)

NFPA - Hazard Identification Ratings
health-3; flammability-1; reactivity-1

STATE LISTS

Florida Hazardous Substance List
[present]

Massachusetts Right To Know List
[present]

NJ Right to Know List (Total)
sn 0394

Pennsylvania Right to Know List
[present]

2-CHLORO-4-NITROBENZOIC ACID   99-60-5

INTERNATIONAL LISTS

Canada - WHMIS: Ingredient Disclosure
1% item 370 (76)

1-CHLORO-1-NITROETHANE   598-92-5

HEALTH AND SAFETY LISTS

NFPA - Flash Points
flash point = 133 degrees F (56 degrees C)

NFPA - Hazard Identification Ratings
flammability-2; reactivity-3

STATE LISTS

Florida Hazardous Substance List
[present]

Massachusetts Right To Know List
[present]

Pennsylvania Right to Know List
[present]

4-CHLORO-2-NITROPHENOL   89-64-5

  ENVIRONMENTAL LISTS

    TSCA - HDD/HDF - Precursors Required for Reporting
    [present]

2-CHLORO-2-NITROPROPANE   594-71-8

  HEALTH AND SAFETY LISTS

    NFPA - Flash Points
    flash point = 135 degrees F (57 degrees C)

    NFPA - Hazard Identification Ratings
    flammability-2; reactivity-3 (explodes on heating)

  STATE LISTS

    Florida Hazardous Substance List
    [present]

    Massachusetts Right To Know List
    [present]

    Pennsylvania Right to Know List
    [present]

1-CHLORO-1-NITROPROPANE   600-25-9

  HEALTH AND SAFETY LISTS

    ACGIH 1995 - Time Weighted Averages
    2 ppm TWA; 10 mg/m3 TWA

    NFPA - Flash Points
    flash point = 144 degrees F (62 degrees C)

    NFPA - Hazard Identification Ratings
    flammability-2; reactivity-3

    NIOSH - Selected LD50s and LC50s
    Inhalation, mouse: LC50 = 66 gm/m3 3 hr Oral, mouse: LD50 = 510 mg/kg

    NIOSH 1990 - Pocket Guide - RELs
    2 ppm TWA; 10 mg/m3 TWA

    NIOSH 1990 - Pocket Guide - IDLHs
    2000 ppm IDLH

    NIOSH 1990 - Pocket Guide - Target organs
    in animals: respiratory system, liver, kidneys, CVS

    OSHA - Vacated PELs - Time Weighted Averages
    2 ppm TWA; 10 mg/m3 TWA

    OSHA - Final PELs - Time Weighted Averages
    20 ppm TWA; 100 mg/m3 TWA

  INTERNATIONAL LISTS

    Australian Exposure Standards - Time Weighted Averages
    2 ppm TWA; 10 mg/m3 TWA

    Canada - WHMIS: Ingredient Disclosure
    1% item 371 (445)

    Canada - Alberta - 8 Hour Occupational Exposure Limit
    2 ppm TWA; 10 mg/m3 TWA

    Canada - Alberta - 15 Minute Occupational Exposure Limit
    4 ppm STEL; 20 mg/m3 STEL

    Canada - British Columbia - 8 Hour Exposure Limits
    20 ppm TWA; 100 mg/m3 TWA

    Canada - British Columbia - 15 Minute Exposure Limits
    30 ppm STEL; 150 mg/m3 STEL

    Canada - Ontario - OHSA - TWAEVs
    2 ppm TWAEV; 10 mg/m3 TWAEV

    Canada - Quebec - Time-Weighted Average Exposure Values
    2 ppm TWAEV; 10 mg/m3 TWAEV

    German (DFG) - MAK Values
    20 ppm MAK; 100 mg/m3 MAK

    Israel - Time Weighted Averages
    2 ppm TWA; 10 mg/m3 TWA

    Israel - Action Levels
    1 ppm AL; 5 mg/m3 AL

    Mexico - Instruction No. 10 - TWAs
    20 ppm TWA; 100 mg/m3 TWA

  STATE LISTS

    California - Exposure Limits - PELs
    2 ppm PEL; 10 mg/m3 PEL

    California - Directors List of Hazardous Substances (8 CCR 339)
    [present]

    Florida Hazardous Substance List
    [present]

    Massachusetts Right To Know List
    [present]

    Minnesota Hazardous Substance List
    [present]

    NJ Special Hazardous Substances
    (reactive - third degree)

    Pennsylvania Right to Know List
    [present]

4-CHLORO-3-NITROTOLUENE   89-60-1

  INTERNATIONAL LISTS

    Canada - WHMIS: Ingredient Disclosure
    1% item 372 (446)

CHLORONITROTOLUENE   25567-68-4

  HEALTH AND SAFETY LISTS

    U.S. DOT - Substances From 49 CFR 172.101
    regulated by DOT (UN2433)

    U.S. DOT - Hazard Classes
    DOT hazard class = 6.1

    U.S. DOT - Appendix B - Marine Pollutants
    DOT regulated marine pollutant

  STATE LISTS

    NJ Right to Know List (Total)
    sn 0396

CHLOROPENTAFLUOROETHANE   76-15-3

  HEALTH AND SAFETY LISTS

    ACGIH 1995 - Time Weighted Averages
    1000 ppm TWA; 6320 mg/m3 TWA

    U.S. DOT - Substances From 49 CFR 172.101
    regulated by DOT (UN1020, UN1973)

    U.S. DOT - Hazard Classes
    DOT hazard class = 2.2

    OSHA - Vacated PELs - Time Weighted Averages
    1000 ppm TWA; 6320 mg/m3 TWA

  ENVIRONMENTAL LISTS

    CERCLA/SARA - Section 313 - Emission Reporting
    form R reporting required for 1.0% de minimus concentration

    Class 1 Ozone Depletors
    ozone depletion potential = 0.6

  INTERNATIONAL LISTS

    Australian Exposure Standards - Time Weighted Averages
    1000 ppm TWA; 6320 mg/m3 TWA

    Canada - WHMIS: Ingredient Disclosure
    1% item 373 (450)

    Canada - Alberta - 8 Hour Occupational Exposure Limit
    1000 ppm TWA; 6340 mg/m3 TWA

    Canada - Alberta - 15 Minute Occupational Exposure Limit
    1250 ppm STEL; 7925 mg/m3 STEL

    Canada - Ontario - OHSA - TWAEVs
    1000 ppm TWAEV; 6315 mg/m3 TWAEV

    Canada - Quebec - Time-Weighted Average Exposure Values
    1000 ppm TWAEV; 6320 mg/m3 TWAEV

United Kingdom - Occupational Exposure Standards - TWAs
1000 ppm TWA;   6320 mg/m3 TWA

Israel - Time Weighted Averages
1000 ppm TWA;   6320 mg/m3 TWA

Israel - Action Levels
500 ppm AL;   3160 mg/m3 AL

STATE LISTS

California - Exposure Limits - PELs
1000 ppm PEL;   6320 mg/m3 PEL

California - Directors List of Hazardous Substances (8 CCR 339)
[present]

Florida Hazardous Substance List
[present]

Massachusetts Right To Know List
[present]

Minnesota Hazardous Substance List
[present]

NJ Right to Know List (Total)
sn 0398

Pennsylvania Right to Know List
[present]

CHLOROPENTAFLUOROPROPANE   134237-41-5

ENVIRONMENTAL LISTS

Class 2 Ozone Depletors
ozone depletion weight reserved

3-CHLOROPEROXYBENZOIC ACID   937-14-4

HEALTH AND SAFETY LISTS

U.S. DOT - Organic Peroxides Table
Organic peroxide UN3102;   UN3106

STATE LISTS

NJ Right to Know List (Total)
sn 0399

CHLOROPHACINONE   3691-35-8

HEALTH AND SAFETY LISTS

NIOSH - Selected LD50s and LC50s
Oral, rat: LD50 = 2100 ug/kg Skin, rabbit: LD50 = 200 mg/kg

ENVIRONMENTAL LISTS

CERCLA/SARA - Section 302 Extremely Hazardous Substances and TPQs
TPQ = 100/10,000 pounds

STATE LISTS

Florida Hazardous Substance List
effective March 13, 1992

Massachusetts Right To Know List
extraordinarily hazardous

Pennsylvania Right to Know List
environmental hazard

PROPOSED REGULATIONS

CERCLA/SARA - Proposed Hazardous Substance Additions
proposed RQ = 1 pound (.454 kg)

CERCLA/SARA - 1989 Proposed RQ Adjustments
proposed RQ = 100 pounds (45.4 kg)

CHLOROPHENATES   RR-01276-0

HEALTH AND SAFETY LISTS

U.S. DOT - Substances From 49 CFR 172.101
regulated by DOT (UN2904, UN2905)

U.S. DOT - Hazard Classes
DOT hazard class = 8

U.S. DOT - Appendix B - Marine Pollutants
DOT regulated severe marine pollutant

BETA-CHLOROPHENETOLE   RR-00816-2

HEALTH AND SAFETY LISTS

NFPA - Flash Points
flash point = 225 degrees F (107 degrees C)

NFPA - Hazard Identification Ratings
flammability-1;   reactivity-0

2-CHLOROPHENOL   95-57-8

SEE ALSO:

K001-HAZARDOUS WASTES
TETRACHLORODIBENZO-P-DIOXINS
CHLOROPHENOL (MIXED ISOMERS)
F039-HAZARDOUS WASTES

HEALTH AND SAFETY LISTS

U.S. DOT - Appendix A Table 1 - Hazardous Substances
final RQ = 100 pounds (45.4 kg)

NFPA - Flash Points
flash point = 147 degrees F (64 degrees C)

NFPA - Hazard Identification Ratings
health-3;   flammability-2;   reactivity-0

NIOSH - Selected LD50s and LC50s
Oral, rat: LD50 = 670 mg/kg

ENVIRONMENTAL LISTS

ATSDR Priority List
Rank (of 275): 202

CERCLA/SARA - Hazardous Substances and their Reportable Quantities
final RQ = 100 pounds (45.4 kg)

CAA - HON Rule - SOCMI Chemicals
compliance by Jan. 23, 1995

Clean Water Act - Priority Pollutants
[present]

Clean Water Act - Toxic Pollutants
[present]

RCRA - U Series Wastes
waste number U048

RCRA - Hazardous Constituents-Appendix VIII
waste number U048

RCRA - Basis for Listing - Appendix VII
Included in waste streams: F039, K001

RCRA - Substances Banned From Land Disposal
[present]

RCRA - TSD Facilities Ground Water Monitoring
TM 8040 = 5 ug/L PQL;   TM 8270 = 10 ug/L PQL

RCRA - Universal Treatment Standards (LDR)
WW: 0.044 mg/l;   NWW: 5.7 mg/kg

TSCA - HDD/HDF - Precursors Required for Reporting
[present]

INTERNATIONAL LISTS

Canada - WHMIS: Ingredient Disclosure
1% item 375 (452)

Mexico - Wastewater - Organic Toxic Pollutants and Heavy Metals
Listed under [Chlorinated Phenols]

Mexico - Drinking Water - Ecological Criteria
0.03 mg/l

STATE LISTS

Florida Hazardous Substance List
[present]

Massachusetts Right To Know List
[present]

Minnesota Hazardous Substance List
carcinogen

NJ Right to Know List (Total)
sn 0403

NJ Special Hazardous Substances
(carcinogen)

Pennsylvania Right to Know List
environmental hazard

P-CHLOROPHENOL   106-48-9
SEE ALSO:
CHLOROPHENOL (MIXED ISOMERS)

HEALTH AND SAFETY LISTS

NFPA - Flash Points
flash point = 250 degrees F (121 degrees C)

NFPA - Hazard Identification Ratings
health-3; flammability-1; reactivity-0

NIOSH - Selected LD50s and LC50s
Oral, rat: LD50 = 261 mg/kg Skin, mammal: LD50 = 1000 mg/kg

ENVIRONMENTAL LISTS

CAA - HON Rule - SOCMI Chemicals
compliance by Jan. 23, 1995

INTERNATIONAL LISTS

Canada - WHMIS: Ingredient Disclosure
1% item 376 (453)

STATE LISTS

Florida Hazardous Substance List
[present]

Massachusetts Right To Know List
[present]

Minnesota Hazardous Substance List
carcinogen

NJ Right to Know List (Total)
sn 0401

NJ Special Hazardous Substances
(corrosive)

Pennsylvania Right to Know List
[present]

M-CHLOROPHENOL   108-43-0
SEE ALSO:
CHLOROPHENOL (MIXED ISOMERS)

HEALTH AND SAFETY LISTS

NIOSH - Selected LD50s and LC50s
Oral, rat: LD50 = 570 mg/kg

ENVIRONMENTAL LISTS

CAA - HON Rule - SOCMI Chemicals
compliance by Jan. 23, 1995

INTERNATIONAL LISTS

Canada - WHMIS: Ingredient Disclosure
1% item 374 (451)

STATE LISTS

NJ Right to Know List (Total)
sn 0402

NJ Special Hazardous Substances
(corrosive)

CHLOROPHENOL (MIXED ISOMERS)   26982-03-6
SEE ALSO:
CHLOROPHENOL (MIXED ISOMERS)

HEALTH AND SAFETY LISTS

U.S. DOT - Substances From 49 CFR 172.101
regulated by DOT (UN2020, UN2021)

U.S. DOT - Hazard Classes
DOT hazard class = 6.1

U.S. DOT - Appendix B - Marine Pollutants
DOT regulated marine pollutant

IARC - Group 2B (sufficient animal data)
[present]

OSHA - Possible Select Carcinogens
[present]

ENVIRONMENTAL LISTS

CERCLA/SARA - Section 313 - Emission Reporting
form R reporting required for 0.1% de minimus concentration

STATE LISTS

California - Air Bill 2588 Appendix A-I
known or potential carcinogen

California - Directors List of Hazardous Substances (8 CCR 339)
[present]

NJ Right to Know List (Total)
sn 3133

Pennsylvania Right to Know List
environmental hazard; special hazardous substance

Pennsylvania RTK - Special Hazardous Substances
[present]

P-CHLOROPHENOXYACETIC ACID   122-88-3

HEALTH AND SAFETY LISTS

NIOSH - Selected LD50s and LC50s
Oral, rat: LD50 = 850 mg/kg

CHLOROPHENOXY DERIVATIVE ACIDS, ESTER,   RR-00499-9
ETHERS, AMINES AND OTHER SALTS
SEE ALSO:
F020-HAZARDOUS WASTES

ENVIRONMENTAL LISTS

RCRA - Basis for Listing - Appendix VII
Included in waste streams: F020, F023, F027, F028

CHLOROPHENOXY HERBICIDES   RR-00059-9

HEALTH AND SAFETY LISTS

IARC - Group 2B (sufficient animal data)
[present]

OSHA - Possible Select Carcinogens
[present]

STATE LISTS

California - Air Bill 2588 Appendix A-II
known or potential carcinogen

California - Directors List of Hazardous Substances (8 CCR 339)
[present]

Minnesota Hazardous Substance List
carcinogen

1-(4-CHLOROPHENYL)-3,3-DIMETHYL TRIAZINE   7203-90-9

HEALTH AND SAFETY LISTS

NIOSH - Selected LD50s and LC50s
Oral, rat: LD50 = 362 mg/kg

4-CHLORO-O-PHENYLENEDIAMINE   95-83-0

HEALTH AND SAFETY LISTS

IARC - Group 2B (sufficient animal data)
[present] (Overall evaluation based only on evidence of carcinogenicity in monograph (27, 1982) or in Supplement 4)

NTP Chemical Status Reports - Testing Status and NTIS Number
Technical reports printed (PB283362/AS)

NTP Chemical Status Reports - Evidence of Carcinogenicity
male rat-positive; female rat-positive; male mice-positive; female mice-positive

NTP Seventh Report - Suspect Carcinogens
suspect carcinogen

OSHA - Possible Select Carcinogens
[present]

ENVIRONMENTAL LISTS

TSCA - Code of Federal Regulations Citations
40 CFR 712.30(f); 40 CFR 716.120(c)

TSCA - PAIR - Reporting List
Reporting Date: August 17, 1983

STATE LISTS

California - Air Bill 2588 Appendix A-I
known or potential carcinogen

California - Prop. 65 - Cancer list
carcinogen - initial date 1/1/88

California - Prop. 65 - No Significant Risk Levels
no significant risk level = 40 ug/day

California - Directors List of Hazardous Substances (8 CCR 339)
[present]

Florida Hazardous Substance List
[present]

Massachusetts Right To Know List
carcinogen; extraordinarily hazardous

Minnesota Hazardous Substance List
carcinogen

Pennsylvania Right to Know List
special hazardous substance

Pennsylvania RTK - Special Hazardous Substances
[present]

4-CHLORO-M-PHENYLENEDIAMINE   5131-60-2
SEE ALSO:
4-CHLORO-M-PHENYLENEDIAMINE

HEALTH AND SAFETY LISTS

IARC - Group 3 (not classifiable)
[present]

NTP Chemical Status Reports - Testing Status and NTIS Number
Technical reports printed (PB285201/AS)

NTP Chemical Status Reports - Evidence of Carcinogenicity
male rat-positive; female rat-negative; male mice-negative; female mice-positive

ENVIRONMENTAL LISTS

TSCA - Code of Federal Regulations Citations
40 CFR 712.30(f); 40 CFR 716.120(c)

TSCA - PAIR - Reporting List
Reporting Date: August 17, 1983

STATE LISTS

Massachusetts Right To Know List
carcinogen; extraordinarily hazardous

2-CHLORO-P-PHENYLENEDIAMINE SULFATE   61702-44-1

HEALTH AND SAFETY LISTS

NTP Chemical Status Reports - Testing Status and NTIS Number
Technical reports printed (PB286370/AS)

NTP Chemical Status Reports - Evidence of Carcinogenicity
male rat-negative; female rat-negative; male mice-negative; female mice-negative

STATE LISTS

Massachusetts Right To Know List
[present]

4-CHLOROPHENYL PHENYL ETHER   7005-72-?

HEALTH AND SAFETY LISTS

U.S. DOT - Appendix A Table 1 - Hazardous Substances
final RQ = 5000 pounds (2270 kg)

ENVIRONMENTAL LISTS

CERCLA/SARA - Hazardous Substances and their Reportable Quantities
final RQ = 5000 pounds (2270 kg)

Clean Water Act - Priority Pollutants
[present]

RCRA - TSD Facilities Ground Water Monitoring
TM 8270 = 10 ug/L PQL

INTERNATIONAL LISTS

Mexico - Wastewater - Organic Toxic Pollutants and Heavy Metals
Listed under [Haloethers]

STATE LISTS

California - Directors List of Hazardous Substances (8 CCR 339)
[present]

Massachusetts Right To Know List
[present]

Pennsylvania Right to Know List
environmental hazard

CHLOROPHENYLTRICHLOROSILANE   26571-79-9

HEALTH AND SAFETY LISTS

U.S. DOT - Substances From 49 CFR 172.101
regulated by DOT (UN1753)

U.S. DOT - Hazard Classes
DOT hazard class = 8

U.S. DOT - Appendix B - Marine Pollutants
DOT regulated marine pollutant

INTERNATIONAL LISTS

Canada - WHMIS: Ingredient Disclosure
1% item 377 (454)

STATE LISTS

NJ Right to Know List (Total)
sn 0404

CHLOROPHYLL   479-61-8

ENVIRONMENTAL LISTS

List of Pesticide Product Inert Ingredients
[present]

CHLOROPICRIN   76-06-2

HEALTH AND SAFETY LISTS

ACGIH 1995 - Time Weighted Averages
0.1 ppm TWA; 0.67 mg/m3 TWA

U.S. DOT - Substances From 49 CFR 172.101
regulated by DOT (UN1580, UN1581, UN1582)

U.S. DOT - Hazard Classes
DOT hazard class = 6.1

U.S. DOT - Substances Which Are Poisonous by Inhalation
liquid or gaseous hazardous material poisonous by inhalation (including mixtures) (UN1580, UN1581, UN1582, UN1583, NA1955)

NIOSH - Selected LD50s and LC50s
Inhalation, mouse: LC50 = 1600 mg/m3 (10 mn) Oral, rat: LD50 = 250 mg/kg

NIOSH 1990 - Pocket Guide - RELs
0.1 ppm TWA; 0.7 mg/m3 TWA

NIOSH 1990 - Pocket Guide - IDLHs
4 ppm IDLH

NIOSH 1990 - Pocket Guide - Target organs
respiratory system, skin, eyes

NTP Chemical Status Reports - Testing Status and NTIS Number
Technical reports printed (PB282311/AS)
NTP Chemical Status Reports - Evidence of Carcinogenicity
male rat-inadequate; female rat-inadequate; male mice-negative; female mice-negative
OSHA - Vacated PELs - Time Weighted Averages
0.1 ppm TWA; 0.7 mg/m3 TWA
OSHA - Final PELs - Time Weighted Averages
0.1 ppm TWA; 0.7 mg/m3 TWA
OSHA - List of Highly Hazardous Chemicals
threshhold quantity = 500 pounds

ENVIRONMENTAL LISTS
CERCLA/SARA - Section 313 - Emission Reporting
form R reporting required
List of Pesticide Product Inert Ingredients
[present]

INTERNATIONAL LISTS
Australian Exposure Standards - Time Weighted Averages
0.1 ppm TWA; 0.67 mg/m3 TWA
Canada - WHMIS: Ingredient Disclosure
0.1% item 378 (455)
Canada - Alberta - 8 Hour Occupational Exposure Limit
0.1 ppm TWA; 0.67 mg/m3 TWA
Canada - Alberta - 15 Minute Occupational Exposure Limit
50 ppm STEL; 225 mg/m3 STEL
Canada - British Columbia - 8 Hour Exposure Limits
0.1 ppm TWA; 0.7 mg/m3 TWA
Canada - British Columbia - 15 Minute Exposure Limits
0.3 ppm STEL; 2 mg/m3 STEL
Canada - Ontario - OHSA - TWAEVs
0.1 ppm TWAEV; 0.67 mg/m3 TWAEV
Canada - Ontario - OHSA - STEVs
0.3 ppm STEV; 2 mg/m3 STEV
Canada - Quebec - Time-Weighted Average Exposure Values
total dust: 0.1 ppm Avg; 0.67 mg/m3 TWAEV; respirable dust: 5 mg/m3 TWAEV
United Kingdom - Occupational Exposure Standards - TWAs
0.1 ppm TWA; 0.7 mg/m3 TWA
United Kingdom - Occupational Exposure Standards - STELs
0.3 ppm STEL; 2 mg/m3 STEL
German (DFG) - MAK Values
0.1 ppm MAK; 0.7 mg/m3 MAK
German (DFG) - Peak Limitations
2 x normal MAK (5 min momentary value); don't exceed 8 times during shift
Israel - Time Weighted Averages
0.1 ppm TWA; 0.67 mg/m3 TWA
Israel - Action Levels
0.05 ppm AL; 0.335 mg/m3 AL
Mexico - Instruction No. 10 - TWAs
0.1 ppm TWA; 0.7 mg/m3 TWA
Mexico - Instruction No. 10 - STELs
0.3 ppm STEL; 2 mg/m3 STEL

STATE LISTS
California - Air Bill 2588 Appendix A-I
[present]
California - Exposure Limits - PELs
0.1 ppm PEL; 0.7 mg/m3 PEL
California - Directors List of Hazardous Substances (8 CCR 339)
[present]
Florida Hazardous Substance List
[present]
Massachusetts Right To Know List
[present]

Minnesota Hazardous Substance List
[present]
NJ Right to Know List (Total)
sn 0405
Pennsylvania Right to Know List
[present]
PROPOSED REGULATIONS
Safe Drinking Water Act - Priority list
[present]

CHLOROPICRIN/METHYL BROMIDE   8004-09-9
HEALTH AND SAFETY LISTS
OSHA - List of Highly Hazardous Chemicals
threshhold quantity = 1500 pounds

CHLOROPICRIN/METHYL CHLORIDE   RR-00806-0
HEALTH AND SAFETY LISTS
OSHA - Vacated PELs - Time Weighted Averages
2 mg/m3 TWA
OSHA - List of Highly Hazardous Chemicals
threshhold quantity = 1500 pounds

CHLOROPIVALOYL CHLORIDE   4300-97-4
HEALTH AND SAFETY LISTS
U.S. DOT - Substances From 49 CFR 172.101
regulated by DOT (NA9263)
U.S. DOT - Hazard Classes
DOT hazard class = 6.1
U.S. DOT - Substances Which Are Poisonous by Inhalation
liquid hazardous material poisonous by inhalation (NA9263)

CHLOROPLATINIC ACID   16941-12-1
HEALTH AND SAFETY LISTS
U.S. DOT - Substances From 49 CFR 172.101
regulated by DOT (UN2507)
U.S. DOT - Hazard Classes
DOT hazard class = 8
INTERNATIONAL LISTS
Canada - WHMIS: Ingredient Disclosure
0.1% item 379 (77)
STATE LISTS
NJ Right to Know List (Total)
sn 0406
NJ Special Hazardous Substances
(corrosive)

2-CHLORO-1,3-BUTADIENE (CHLOROPRENE)   126-99-8
SEE ALSO:
BIS(2,4-DIMETHYLBUTYL) MALEATE
F025-HAZARDOUS WASTES
F024-HAZARDOUS WASTES
HEALTH AND SAFETY LISTS
ACGIH 1995 - Time Weighted Averages
10 ppm TWA; 36 mg/m3 TWA
ACGIH 1995 - Skin Designations
skin - potential for cutaneous absorption
AIHA - Odor Threshold Values
no geometric mean air odor threshold
U.S. DOT - Substances From 49 CFR 172.101
regulated by DOT (UN1991)
U.S. DOT - Hazard Classes
Forbidden from transport by the DOT

IARC - Group 3 (not classifiable)
[present]

NFPA - Flash Points
flash point = -4 degrees F (-20 degrees C)

NFPA - Hazard Identification Ratings
health-2; flammability-3; reactivity-0

NIOSH - Selected LD50s and LC50s
Oral, rat: LD50 = 900 mg/kg

NIOSH 1990 - Pocket Guide - RELs
C 1 ppm (15 min); C 3.6 mg/m3 (15 min)

NIOSH 1990 - Pocket Guide - IDLHs
400 ppm IDLH (not considering carcinogenic effects)

NIOSH 1990 - Pocket Guide - Carcinogens
occupational carcinogen

NIOSH 1990 - Pocket Guide - Target organs
respiratory system, skin, eyes

NIOSH - Health Standards - Exposure Limits
C (15 min) 1 ppm; C (15 min) 3.6 mg/m3

NIOSH - Health Standards - Health Effects and Precautions
Lung and skin cancer, reproductive effects (Periodic chest X-ray and pulmonary function testing required; consel pregnant workers about reproductive hazards)

NIOSH - Health Standards - Carcinogenic Chemicals
potential human carcinogen

NTP Chemical Status Reports - Testing Status and NTIS Number
Two year studies: pathology quality assessment in progress; prechronic studies for which toxicity technical reports were not prepared

OSHA - Vacated PELs - Time Weighted Averages
10 ppm TWA; 35 mg/m3 TWA

OSHA - Vacated PELs - Skin Designation
Prevent or reduce skin absorption

OSHA - Final PELs - Time Weighted Averages
25 ppm TWA; 90 mg/m3 TWA

OSHA - Final PELs - Skin Notations
prevent or reduce skin absorption

ENVIRONMENTAL LISTS

CERCLA/SARA - Section 313 - Emission Reporting
form R reporting required for 1.0% de minimus concentration

CERCLA/SARA - Hazardous Substances and their Reportable Quantities
final RQ = 1 pound (.454 kg)

Clean Air Act (1990) - List of Hazardous Air Contaminants
[present]

CAA - HON Rule - SOCMI Chemicals
compliance by Jan. 23, 1995

CAA - HON Rule - Organic HAPs
[present]

EPA - Master Testing List
[present]

RCRA - Hazardous Constituents-Appendix VIII
hazardous constituent - no waste number

RCRA - Basis for Listing - Appendix VII
Included in waste streams: F024, F025

RCRA - TSD Facilities Ground Water Monitoring
TM 8010 = 50 ug/L PQL; TM 8240 = 5 ug/L PQL

RCRA - Universal Treatment Standards (LDR)
WW: 0.057 mg/l; NWW: 0.28 mg/kg

TSCA - Code of Federal Regulations Citations
40 CFR 712.30(d),(m); 40 CFR 716.120(a)

TSCA - PAIR - Reporting List
Reporting Dates: November 19, 1982; February 26, 1985

TSCA - Health and Safety Reporting List
Effective Date: December 28, 1984

INTERNATIONAL LISTS

Australian Exposure Standards - Time Weighted Averages
10 ppm TWA; 36 mg/m3 TWA

Australian Exposure Standards - Skin Effects
skin absorption

Canada - WHMIS: Ingredient Disclosure
0.1% item 380 (457)

Canada - Alberta - 8 Hour Occupational Exposure Limit
10 ppm TWA; 36 mg/m3 TWA

Canada - Alberta - 15 Minute Occupational Exposure Limit
20 ppm STEL; 72 mg/m3 STEL

Canada - Alberta - Skin Designation
can be absorbed through the intact skin

Canada - British Columbia - 8 Hour Exposure Limits
25 ppm TWA; 90 mg/m3 TWA

Canada - British Columbia - 15 Minute Exposure Limits
35 ppm STEL; 135 mg/m3 STEL

Canada - British Columbia - Skin Notations
skin - potential for skin absorption

Canada - Ontario - OHSA - TWAEVs
10 ppm TWAEV; 36 mg/m3 TWAEV

Canada - Ontario - OHSA - Skin Notations
absorption through skin, eyes, or mucous membranes

Canada - Quebec - Time-Weighted Average Exposure Values
10 ppm TWAEV; 36 mg/m3 TWAEV

Canada - Quebec - Skin Designations
absorbed through the skin

United Kingdom - Occupational Exposure Standards - TWAs
10 ppm TWA; 36 mg/m3 TWA

United Kingdom - Occupational Exposure Standards - Notes
can be absorbed through skin

German (DFG) - MAK Values
5 ppm MAK; 18 mg/m3 MAK

German (DFG) - Peak Limitations
2 x normal MAK (30 min. average value); don't exceed 4 times during shift

German (DFG) - Skin/Sensitizers
danger of cutaneous absorption

German (DFG) - Pregnancy
classification not yet possible

Israel - Time Weighted Averages
10 ppm TWA; 36 mg/m3 TWA

Israel - Action Levels
5 ppm AL; 18 mg/m3 AL

Mexico - Instruction No. 10 - TWAs
10 ppm TWA; 45 mg/m3 TWA

Mexico - Instruction No. 10 - Skin designation
skin - potential for cutaneous absorption

STATE LISTS

California - Air Bill 2588 Appendix A-I
[present]

California - Exposure Limits - PELs
10 ppm PEL; 36 mg/m3 PEL

California - Exposure Limits - Skin Notation
material may be absorbed through the skin, eyes or mucous membrane

California - Directors List of Hazardous Substances (8 CCR 339)
[present]

Florida Hazardous Substance List
[present]

Massachusetts Right To Know List
[present]

Minnesota Hazardous Substance List
skin

NJ Right to Know List (Total)
sn 0407

NJ Special Hazardous Substances
(flammable - third degree; mutagen)

Pennsylvania Right to Know List
environmental hazard

PROPOSED REGULATIONS

TSCA - ITC 32nd Report Priority Testing List
designated for dermal absorption testing

Canada - Ontario - Proposed Occupational TWAEVs
1 ppm TWAEV; 3.5 mg/m3 TWAEV

Canada - Ontario - Proposed Occupational STEVs
5 ppm STEV; 18 mg/m3 STEV

2-CHLOROPROPANE   75-29-6

HEALTH AND SAFETY LISTS

U.S. DOT - Substances From 49 CFR 172.101
regulated by DOT (UN2356)

U.S. DOT - Hazard Classes
DOT hazard class = 3

U.S. DOT - Appendix B - Marine Pollutants
DOT regulated marine pollutant

NFPA - Flash Points
flash point = -26 degrees F (-32 degrees C)

NFPA - Hazard Identification Ratings
health-2; flammability-4; reactivity-0

ENVIRONMENTAL LISTS

CAA - Flammable Substances for Accidental Release Prevention
threshold quantity = 10,000 lbs

TSCA - Code of Federal Regulations Citations
40 CFR 716.120(a)

TSCA - Health and Safety Reporting List
Effective Date: June 1, 1987

STATE LISTS

Florida Hazardous Substance List
[present]

Massachusetts Right To Know List
[present]

NJ Right to Know List (Total)
sn 2241

Pennsylvania Right to Know List
[present]

1-CHLORO-2-PROPANOL   127-00-4

HEALTH AND SAFETY LISTS

NFPA - Flash Points
flash point = 125 degrees F (52 degrees C)

NFPA - Hazard Identification Ratings
health-2; flammability-2; reactivity-0

NTP Chemical Status Reports - Testing Status and NTIS Number
Two year studies: pathology quality assessment in progress; prechronic studies for which toxcity technical reports were not prepared

STATE LISTS

Florida Hazardous Substance List
[present]

Massachusetts Right To Know List
[present]

Pennsylvania Right to Know List
[present]

3-CHLORO-1-PROPANOL   627-30-5

HEALTH AND SAFETY LISTS

U.S. DOT - Substances From 49 CFR 172.101
regulated by DOT (UN2849)

U.S. DOT - Hazard Classes
DOT hazard class = 6.1

NIOSH - Selected LD50s and LC50s
Oral, mouse: LD50 = 2300 mg/kg

INTERNATIONAL LISTS

Canada - WHMIS: Ingredient Disclosure
1% item 381 (458)

STATE LISTS

NJ Right to Know List (Total)
sn 0408

2-CHLOROPROPENE   557-98-2

HEALTH AND SAFETY LISTS

U.S. DOT - Substances From 49 CFR 172.101
regulated by DOT (UN2456)

U.S. DOT - Hazard Classes
DOT hazard class = 3

U.S. DOT - Appendix B - Marine Pollutants
DOT regulated marine pollutant

NFPA - Flash Points
flash point < -4 degrees F (-20 degrees C)

NFPA - Hazard Identification Ratings
health-2; flammability-4; reactivity-0

NIOSH - Selected LD50s and LC50s
Inhalation, mouse: LC50 = 267 gm/m3 (8 hr)

ENVIRONMENTAL LISTS

CAA - Flammable Substances for Accidental Release Prevention
threshold quantity = 10,000 lbs

STATE LISTS

Florida Hazardous Substance List
[present]

Massachusetts Right To Know List
[present]

NJ Right to Know List (Total)
sn 0409

Pennsylvania Right to Know List
[present]

2-CHLOROPROPIONIC ACID   598-78-7

HEALTH AND SAFETY LISTS

ACGIH 1995 - Time Weighted Averages
0.1 ppm TWA; 0.44 mg/m3 TWA

ACGIH 1995 - Skin Designations
skin - potential for cutaneous absorption

U.S. DOT - Substances From 49 CFR 172.101
regulated by DOT (UN2511)

U.S. DOT - Hazard Classes
DOT hazard class = 8

NFPA - Flash Points
flash point = 225 degrees F (107 degrees C)

NFPA - Hazard Identification Ratings
flammability-1; reactivity-0

CHLOROPROPIONIC ACID   28554-00-9

STATE LISTS

NJ Right to Know List (Total)
sn 0410

3-CHLOROPROPIONITRILE   542-76-7

HEALTH AND SAFETY LISTS

U.S. DOT - Appendix A Table 1 - Hazardous Substances
final RQ = 1000 pounds (454 kg)

NFPA - Flash Points
flash point = 168 degrees F (76 degrees C)

NFPA - Hazard Identification Ratings
flammability-2;   reactivity-1

NIOSH - Selected LD50s and LC50s
Oral, rat: LD50 = 100 mg/kg

ENVIRONMENTAL LISTS

CERCLA/SARA - Section 302 Extremely Hazardous Substances and
TPQs
TPQ = 1000 pounds

CERCLA/SARA - Section 313 - Emission Reporting
form R reporting required

CERCLA/SARA - Hazardous Substances and their Reportable Quantities
final RQ = 1000 pounds (454 kg)

RCRA - P Series Wastes
waste number P027

RCRA - Hazardous Constituents-Appendix VIII
waste number P027

RCRA - Substances Banned From Land Disposal
[present]

STATE LISTS

Florida Hazardous Substance List
effective March 13, 1992

Massachusetts Right To Know List
extraordinarily hazardous

NJ Right to Know List (Total)
sn 2711

Pennsylvania Right to Know List
environmental hazard

2-CHLOROPROPIONYL CHLORIDE   7623-09-8

HEALTH AND SAFETY LISTS

NFPA - Flash Points
flash point = 88 degrees F (31 degrees C)

1-CHLOROPROPYLENE   590-21-6

HEALTH AND SAFETY LISTS

U.S. DOT - Appendix B - Marine Pollutants
DOT regulated marine pollutant

NFPA - Flash Points
flash point < 21 degrees F (-6 degrees C)

NFPA - Hazard Identification Ratings
health-2;   flammability-4;   reactivity-2

ENVIRONMENTAL LISTS

CAA - Flammable Substances for Accidental Release Prevention
threshold quantity = 10,000 lbs

STATE LISTS

Florida Hazardous Substance List
[present]

Massachusetts Right To Know List
[present]

Pennsylvania Right to Know List
[present]

ALPHA-CHLOROPROPYLENE   RR-01277-1

HEALTH AND SAFETY LISTS

U.S. DOT - Appendix B - Marine Pollutants
DOT regulated marine pollutant

2-CHLOROPYRIDINE   109-09-1

HEALTH AND SAFETY LISTS

U.S. DOT - Substances From 49 CFR 172.101
regulated by DOT (UN2822)

U.S. DOT - Hazard Classes
DOT hazard class = 6.1

NIOSH - Selected LD50s and LC50s
Oral, mouse: LD50 = 110 mg/kg Skin, rabbit: LD50 = 64 mg/kg

STATE LISTS

NJ Right to Know List (Total)
sn 0411

CHLOROQUINE   54-05-7

HEALTH AND SAFETY LISTS

IARC - Group 3 (not classifiable)
[present]

NIOSH - Selected LD50s and LC50s
Oral, rat: LD50 = 330 mg/kg

STATE LISTS

NJ Right to Know List (Total)
sn 0412

NJ Special Hazardous Substances
(teratogen)

4-CHLORORESORCINOL   95-88-5

ENVIRONMENTAL LISTS

TSCA - HDD/HDF - Precursors Required for Reporting
[present]

CHLOROSILANES, N.O.S.   RR-00208-4

HEALTH AND SAFETY LISTS

U.S. DOT - Substances From 49 CFR 172.101
regulated by DOT (UN2985, UN2986, UN2987, UN2988)

U.S. DOT - Hazard Classes
*flashpoint less than 23 degrees C: DOT hazard class = 3;   flashpoint
23 degrees C or more: DOT hazard class = 8;   emit flammable gases
when in contact with water: DOT hazard class = 4.3*

CHLOROSTYRENE   1331-28-8

HEALTH AND SAFETY LISTS

NIOSH - Selected LD50s and LC50s
Oral, rat: LD50 = 5200 mg/kg Skin, rabbit: LD50 = 20 gm/kg

INTERNATIONAL LISTS

Canada - Alberta - 8 Hour Occupational Exposure Limit
50 ppm TWA;   285 mg/m3 TWA

Canada - Alberta - 15 Minute Occupational Exposure Limit
75 ppm STEL;   425 mg/m3 STEL

Canada - Quebec - Time-Weighted Average Exposure Values
50 ppm TWAEV;   283 mg/m3 TWAEV

Canada - Quebec - Short-term Exposure Values
75 ppm STEV;   425 mg/m3 STEV

Mexico - Instruction No. 10 - TWAs
50 ppm TWA;   285 mg/m3 TWA

Mexico - Instruction No. 10 - STELs
75 ppm STEL;   430 mg/m3 STEL

STATE LISTS

California - Exposure Limits - PELs
50 ppm PEL;   285 mg/m3 PEL

California - Exposure Limits - STELs
75 ppm STEL;   428 mg/m3 STEL
California - Directors List of Hazardous Substances (8 CCR 339)
[present]
Florida Hazardous Substance List
[present]
Massachusetts Right To Know List
[present]

O-CHLOROSTYRENE   2039-87-4

HEALTH AND SAFETY LISTS

ACGIH 1995 - Time Weighted Averages
50 ppm TWA;   283 mg/m3 TWA
ACGIH 1995 - Short Term Exposure Limits
75 ppm STEL;   425 mg/m3 STEL
OSHA - Vacated PELs - Time Weighted Averages
50 ppm TWA;   285 mg/m3 TWA
OSHA - Vacated PELs - Short Term Exposure Limits
75 ppm STEL;   428 mg/m3 STEL

INTERNATIONAL LISTS

Australian Exposure Standards - Time Weighted Averages
50 ppm TWA;   283 mg/m3 TWA
Australian Exposure Standards - Short Term Exposure Limits
75 ppm STEL;   425 mg/m3 STEL
Canada - WHMIS: Ingredient Disclosure
1% item 382 (459)
Canada - British Columbia - 8 Hour Exposure Limits
50 ppm TWA;   285 mg/m3 TWA
Canada - British Columbia - 15 Minute Exposure Limits
75 ppm STEL;   430 mg/m3 STEL
Canada - Ontario - OHSA - TWAEVs
50 ppm TWAEV;   283 mg/m3 TWAEV
Canada - Ontario - OHSA - STEVs
75 ppm STEV;   425 mg/m3 STEV
Israel - Time Weighted Averages
50 ppm TWA;   283 mg/m3 TWA
Israel - Short Term Exposure Limits
75 ppm STEL;   425 mg/m3 STEL
Israel - Action Levels
25 ppm AL;   141.5 mg/m3 AL

STATE LISTS

Massachusetts Right To Know List
[present]
Minnesota Hazardous Substance List
[present]
Pennsylvania Right to Know List
[present]

CHLOROSULFONIC ACID   7790-94-5

HEALTH AND SAFETY LISTS

AIHA - WEEL - Time Weighted Averages
0.3 ppm TWA;   1.3 mg/m3 TWA
U.S. DOT - Substances From 49 CFR 172.101
regulated by DOT (UN1754)
U.S. DOT - Hazard Classes
DOT hazard class = 8
U.S. DOT - Substances Which Are Poisonous by Inhalation
liquid hazardous material poisonous by inhalation (UN1754)
U.S. DOT - Appendix A Table 1 - Hazardous Substances
final RQ = 1000 pounds (454 kg)
IARC - Group Unspecified
[present] (Listed under 'Para-chloro- ortho-toluidine and its strong acid salts')

ENVIRONMENTAL LISTS

CERCLA/SARA - Hazardous Substances and their Reportable Quantities
final RQ = 1000 pounds (454 kg)
Clean Water Act - Hazardous Substances
[present]

INTERNATIONAL LISTS

Canada - WHMIS: Ingredient Disclosure
1% item 383 (78)
United Kingdom - Occupational Exposure Standards - TWAs
1 mg/m3 TWA

STATE LISTS

California - Directors List of Hazardous Substances (8 CCR 339)
[present]
Florida Hazardous Substance List
[present]
Massachusetts Right To Know List
[present]
Minnesota Hazardous Substance List
[present]
NJ Right to Know List (Total)
sn 0413
NJ Special Hazardous Substances
(corrosive; reactive - third degree)
Pennsylvania Right to Know List
environmental hazard

1-CHLORO-1,1,2,2-TETRAFLUOROETHANE (HCFC-   354-25-6
124A)

ENVIRONMENTAL LISTS

CERCLA/SARA - Section 313 - Emission Reporting
form R reporting required

CHLOROTETRAFLUOROETHANE   63938-10-3

HEALTH AND SAFETY LISTS

U.S. DOT - Substances From 49 CFR 172.101
regulated by DOT (UN1021)
U.S. DOT - Hazard Classes
DOT hazard class = 2.2

ENVIRONMENTAL LISTS

CERCLA/SARA - Section 313 - Emission Reporting
form R reporting required

STATE LISTS

NJ Right to Know List (Total)
sn 0414

3-CHLORO-1,1,2,2-TETRAFLUOROPROPANE   679-85-6

ENVIRONMENTAL LISTS

Class 2 Ozone Depletors
ozone depletion weight reserved

CHLOROTHALONIL   1897-45-6

HEALTH AND SAFETY LISTS

IARC - Group 3 (not classifiable)
[present]
NIOSH - Selected LD50s and LC50s
Oral, rat: LD50 = 10 gm/kg
NTP Chemical Status Reports - Testing Status and NTIS Number
Technical reports printed (PB286369/AS)
NTP Chemical Status Reports - Evidence of Carcinogenicity
male rat-positive; female rat-positive; male mice-negative; female mice-negative

ENVIRONMENTAL LISTS

   CERCLA/SARA - Section 313 - Emission Reporting
   form R reporting required for 1.0% de minimus concentration

   List of Pesticide Product Inert Ingredients
   [present]

INTERNATIONAL LISTS

   German (DFG) - Skin/Sensitizers
   danger of sensitization

   German (DFG) - Carcinogens
   suspected carcinogen

STATE LISTS

   California - Air Bill 2588 Appendix A-II
   known or potential carcinogen: 9/89

   California - Prop. 65 - Cancer list
   carcinogen - initial date 1/1/89

   California - Prop. 65 - No Significant Risk Levels
   no significant risk level = 200 ug/day

   California - Directors List of Hazardous Substances (8 CCR 339)
   [present]

   Massachusetts Right To Know List
   carcinogen; extraordinarily hazardous

   NJ Right to Know List (Total)
   sn 0415

   Pennsylvania Right to Know List
   environmental hazard

P-CHLOROTHIOPHENOL   106-54-7

INTERNATIONAL LISTS

   Canada - WHMIS: Ingredient Disclosure
   1% item 384 (461)

4-CHLORO-O-TOLOXY ACETIC ACID   94-74-6

HEALTH AND SAFETY LISTS

   IARC - Group Unspecified
   [present] (Listed under 'Chlorophenoxy herbicides')

ENVIRONMENTAL LISTS

   CERCLA/SARA - Section 313 - Emission Reporting
   form R reporting required

   TSCA - HDD/HDF - Precursors Required for Reporting
   [present]

STATE LISTS

   California - Directors List of Hazardous Substances (8 CCR 339)
   [present]

O-CHLOROTOLUENE   95-49-8

HEALTH AND SAFETY LISTS

   ACGIH 1995 - Time Weighted Averages
   50 ppm TWA; 259 mg/m3 TWA

   OSHA - Vacated PELs - Time Weighted Averages
   50 ppm TWA; 250 mg/m3 TWA

ENVIRONMENTAL LISTS

   CAA - HON Rule - SOCMI Chemicals
   compliance by April 24, 1995

   Safe Drinking Water Act - Monitoring
   monitoring required

   List of Pesticide Product Inert Ingredients
   [present]

   TSCA - Code of Federal Regulations Citations
   40 CFR 712.30(d); 40 CFR 716.120(a)

   TSCA - PAIR - Reporting List
   Reporting Date: November 19, 1982

TSCA - Health and Safety Reporting List
Effective Date: April 29, 1983

INTERNATIONAL LISTS

   Australian Exposure Standards - Time Weighted Averages
   50 ppm TWA; 259 mg/m3 TWA

   Canada - WHMIS: Ingredient Disclosure
   1% item 385 (462)

   Canada - Alberta - 8 Hour Occupational Exposure Limit
   50 ppm TWA; 260 mg/m3 TWA

   Canada - Alberta - 15 Minute Occupational Exposure Limit
   75 ppm STEL; 390 mg/m3 STEL

   Canada - Alberta - Skin Designation
   can be absorbed through the intact skin

   Canada - British Columbia - 8 Hour Exposure Limits
   50 ppm TWA; 250 mg/m3 TWA

   Canada - British Columbia - 15 Minute Exposure Limits
   75 ppm STEL; 375 mg/m3 STEL

   Canada - British Columbia - Skin Notations
   skin - potential for skin absorption

   Canada - Ontario - OHSA - TWAEVs
   50 ppm TWAEV; 260 mg/m3 TWAEV

   Canada - Ontario - OHSA - STEVs
   75 ppm STEV; 388 mg/m3 STEV

   Canada - Quebec - Time-Weighted Average Exposure Values
   50 ppm TWAEV; 259 mg/m3 TWAEV

   United Kingdom - Occupational Exposure Standards - TWAs
   50 ppm TWA; 250 mg/m3 TWA

   Israel - Time Weighted Averages
   50 ppm TWA; 259 mg/m3 TWA

   Israel - Action Levels
   25 ppm AL; 129.5 mg/m3 AL

STATE LISTS

   California - Exposure Limits - PELs
   50 ppm PEL; 250 mg/m3 PEL

   California - Exposure Limits - Skin Notation
   material may be absorbed through the skin, eyes or mucous membrane

   California - Directors List of Hazardous Substances (8 CCR 339)
   [present]

   Florida Hazardous Substance List
   [present]

   Massachusetts Right To Know List
   [present]

   Minnesota Hazardous Substance List
   skin

   Pennsylvania Right to Know List
   [present]

PROPOSED REGULATIONS

   Safe Drinking Water Act - Priority list
   [present]

   TSCA - ITC 32nd Report Priority Testing List
   designated for dermal absorption testing

P-CHLOROTOLUENE   106-43-4

HEALTH AND SAFETY LISTS

   NIOSH - Selected LD50s and LC50s
   Inhalation, mouse: LC50 = 34 gm/m3 2 hr Oral, rat: LD50 = 3600 mg/kg

ENVIRONMENTAL LISTS

   CAA - HON Rule - SOCMI Chemicals
   compliance by April 24, 1995

   Safe Drinking Water Act - Monitoring
   monitoring required

TSCA - Code of Federal Regulations Citations
40 CFR 716.120(a)

TSCA - Health and Safety Reporting List
Effective Date: June 1, 1987

STATE LISTS

Massachusetts Right To Know List
[present]

PROPOSED REGULATIONS

Safe Drinking Water Act - Priority list
[present]

M-CHLOROTOLUENE   108-41-8

ENVIRONMENTAL LISTS

CAA - HON Rule - SOCMI Chemicals
compliance by April 24, 1995

CHLOROTOLUENES   25168-05-2

HEALTH AND SAFETY LISTS

U.S. DOT - Substances From 49 CFR 172.101
regulated by DOT (UN2238)

U.S. DOT - Hazard Classes
DOT hazard class = 3

U.S. DOT - Appendix B - Marine Pollutants
DOT regulated marine pollutant

NFPA - Flash Points
flash point = 126 degrees F (52 degrees C)

NFPA - Hazard Identification Ratings
health-2; flammability-2; reactivity-0

ENVIRONMENTAL LISTS

ATSDR Priority List
Rank (of 275): 260

STATE LISTS

NJ Right to Know List (Total)
sn 0416

NJ Special Hazardous Substances
(corrosive)

Pennsylvania Right to Know List
[present]

4-CHLORO-O-TOLUIDINE   95-69-2

HEALTH AND SAFETY LISTS

IARC - Group 2B (sufficient animal data)
[present] (Overall evaluation based only on evidence of carcinogenicity in monograph (30, 1983) or in Supplement 4)

OSHA - Possible Select Carcinogens
[present]

ENVIRONMENTAL LISTS

CERCLA/SARA - Section 313 - Emission Reporting
form R reporting required

TSCA - Code of Federal Regulations Citations
40 CFR 721.462

TSCA - Chemicals with Significant New Use Rules
[present]

TSCA - Section 12(b) - Export Notification
export notification required - Section 5

INTERNATIONAL LISTS

Canada - WHMIS: Ingredient Disclosure
1% item 388 (465)

German (DFG) - Carcinogens
proven carcinogen

STATE LISTS

California - Air Bill 2588 Appendix A-I
known or potential carcinogen

California - Prop. 65 - Cancer list
carcinogen - initial date 1/1/90

California - Prop. 65 - No Significant Risk Levels
no significant risk level = 3 ug/day

California - Directors List of Hazardous Substances (8 CCR 339)
[present]

Minnesota Hazardous Substance List
carcinogen

Pennsylvania Right to Know List
special hazardous substance

Pennsylvania RTK - Special Hazardous Substances
[present]

PROPOSED REGULATIONS

NTP - Proposed Additions to Annual Report on Carcinogens
proposed as a suspect carcinogen for NTP 8th report (including its hydrochloride salt)

3-CHLORO-P-TOLUIDINE   95-74-9

HEALTH AND SAFETY LISTS

NIOSH - Selected LD50s and LC50s
Oral, rat: LD50 = 1500 mg/kg

NTP Chemical Status Reports - Testing Status and NTIS Number
Technical reports printed (PB287401/AS)

NTP Chemical Status Reports - Evidence of Carcinogenicity
male rat-negative; female rat-negative; male mice-negative; female mice-negative

STATE LISTS

NJ Right to Know List (Total)
sn 0418

5-CHLORO-O-TOLUIDINE   95-79-4

HEALTH AND SAFETY LISTS

NIOSH - Selected LD50s and LC50s
Oral, rat: LD50 = 464 mg/kg

NTP Chemical Status Reports - Testing Status and NTIS Number
Technical reports printed (PB291468/AS)

NTP Chemical Status Reports - Evidence of Carcinogenicity
male rat-negative; female rat-negative; male mice-positive; female mice-positive

INTERNATIONAL LISTS

German (DFG) - Carcinogens
suspected carcinogen

STATE LISTS

Massachusetts Right To Know List
carcinogen; extraordinarily hazardous

NJ Right to Know List (Total)
sn 0421

CHLORO-M-TOLUIDINE, ALL ISOMERS   RR-00592-5

INTERNATIONAL LISTS

Canada - WHMIS: Ingredient Disclosure
1% item 386 (463)

CHLORO-O-TOLUIDINE, ALL ISOMERS, N.O.S.   RR-00593-6

HEALTH AND SAFETY LISTS

IARC - Group 2A (limited human data)
[present]

OSHA - Possible Select Carcinogens
[present]

INTERNATIONAL LISTS
Canada - WHMIS: Ingredient Disclosure
1% item 387 (464)

CHLORO-P-TOLUIDINE, ALL ISOMERS   **RR-00594-7**
INTERNATIONAL LISTS
Canada - WHMIS: Ingredient Disclosure
1% item 390 (467)

CHLOROTOLUIDINES   **RR-00115-0**
HEALTH AND SAFETY LISTS
U.S. DOT - Substances From 49 CFR 172.101
regulated by DOT (UN2239)
U.S. DOT - Hazard Classes
DOT hazard class = 6.1

CHLOROTRIANISENE   **569-57-3**
HEALTH AND SAFETY LISTS
IARC - Group Unspecified
[present] (Listed under 'Nonsteroidal oestrogens')

2-CHLORO-1,1,1-TRIFLUOROETHANE   **75-88-7**
HEALTH AND SAFETY LISTS
U.S. DOT - Substances From 49 CFR 172.101
regulated by DOT (UN1983)
U.S. DOT - Hazard Classes
DOT hazard class = 2.2
IARC - Group 3 (not classifiable)
[present]
ENVIRONMENTAL LISTS
CERCLA/SARA - Section 313 - Emission Reporting
form R reporting required
Class 2 Ozone Depletors
ozone depletion weight reserved
TSCA - Code of Federal Regulations Citations
40 CFR 716.120(a)
TSCA - Section 12(b) - Export Notification
export notification required - Section 5
STATE LISTS
California - Directors List of Hazardous Substances (8 CCR 339)
[present]

CHLOROTRIFLUOROETHANE   **1330-45-6**
STATE LISTS
NJ Right to Know List (Total)
sn 0416

CHLOROTRIFLUOROETHYLENE   **79-38-9**
HEALTH AND SAFETY LISTS
AIHA - WEEL - Time Weighted Averages
5 ppm TWA
U.S. DOT - Substances From 49 CFR 172.101
regulated by DOT (UN1082)
U.S. DOT - Hazard Classes
DOT hazard class = 2.1
U.S. DOT - Substances Which Are Poisonous by Inhalation
gaseous hazardous material poisonous by inhalation (inhibited form) (UN1082)
NFPA - Flash Points
gas (no flash point given)
NFPA - Hazard Identification Ratings
flammability-4; reactivity-0

NIOSH - Selected LD50s and LC50s
Inhalation, rat: LC50 = 1000 ppm 4 hr Oral, mouse: LD50 = 268 mg/kg
OSHA - List of Highly Hazardous Chemicals
threshhold quantity = 10,000 pounds
ENVIRONMENTAL LISTS
CAA - Flammable Substances for Accidental Release Prevention
threshold quantity = 10,000 lbs
STATE LISTS
Florida Hazardous Substance List
[present]
Massachusetts Right To Know List
[present]
Minnesota Hazardous Substance List
[present]
NJ Special Hazardous Substances
(flammable - fourth degree)
Pennsylvania Right to Know List
[present]

CHLOROTRIFLUOROMETHANE   **75-72-9**
HEALTH AND SAFETY LISTS
U.S. DOT - Substances From 49 CFR 172.101
regulated by DOT (UN1022, UN2599)
U.S. DOT - Hazard Classes
DOT hazard class = 2.2
ENVIRONMENTAL LISTS
CERCLA/SARA - Section 313 - Emission Reporting
form R reporting required
Class 1 Ozone Depletors
ozone depletion potential = 1.0
CAA - HON Rule - SOCMI Chemicals
compliance by Jan. 23, 1995
INTERNATIONAL LISTS
German (DFG) - MAK Values
1000 ppm MAK; 4330 mg/m3 MAK
German (DFG) - Peak Limitations
2 x normal MAK (1 hour momentary value); don't exceed 3 time per shift
STATE LISTS
NJ Right to Know List (Total)
sn 0425

3-CHLORO-1,1,1-TRIFLUORO-PROPANE (HCFC-253FB)   **460-35-**
ENVIRONMENTAL LISTS
CERCLA/SARA - Section 313 - Emission Reporting
form R reporting required

CHLOROTRIFLUOROPROPANE   **134237-44-**
ENVIRONMENTAL LISTS
Class 2 Ozone Depletors
ozone depletion weight reserved

CHLOROTRIFLUOROPYRIDINE   **RR-01167-**
HEALTH AND SAFETY LISTS
U.S. DOT - Substances Which Are Poisonous by Inhalation
gaseous hazardous material poisonous by inhalation

P-CHLORO-A,A,A-TRIFLUOROTOLUENE   **98-56-**
HEALTH AND SAFETY LISTS
NTP Chemical Status Reports - Testing Status and NTIS Number
Technical reports printed (PB93-105682/AS)

ENVIRONMENTAL LISTS
  EPA - Master Testing List
  [present]
  TSCA - Code of Federal Regulations Citations
  40 CFR 712.30(f); 40 CFR 716.120(a)
  TSCA - PAIR - Reporting List
  Reporting Date: August 17, 1983
  TSCA - Health and Safety Reporting List
  Effective Date: April 29, 1983

CHLOROXURON   1982-47-4

HEALTH AND SAFETY LISTS
  NIOSH - Selected LD50s and LC50s
  Oral, rat: LD50 = 3700 mg/kg

ENVIRONMENTAL LISTS
  CERCLA/SARA - Section 302 Extremely Hazardous Substances and
  TPQs
  TPQ = 500/10,000 pounds

STATE LISTS
  Florida Hazardous Substance List
  effective March 13, 1992
  Massachusetts Right To Know List
  extraordinarily hazardous
  NJ Right to Know List (Total)
  sn 2246
  Pennsylvania Right to Know List
  environmental hazard

PROPOSED REGULATIONS
  CERCLA/SARA - Proposed Hazardous Substance Additions
  proposed RQ = 1 pound (.454 kg)
  CERCLA/SARA - 1989 Proposed RQ Adjustments
  proposed RQ = 100 pounds (45.4 kg)

P-CHLORO-M-XYLENOL   88-04-0

HEALTH AND SAFETY LISTS
  NIOSH - Selected LD50s and LC50s
  Oral, rat: LD50 = 3830 mg/kg

ENVIRONMENTAL LISTS
  List of Pesticide Product Inert Ingredients
  [present]
  TSCA - Code of Federal Regulations Citations
  40 CFR 712.30(w); 40 CFR 716.120(a)
  TSCA - PAIR - Reporting List
  Reporting Date: June 13, 1989
  TSCA - Health and Safety Reporting List
  Effective Date: April 13, 1989

CHLOROZOTOCIN   54749-90-5

HEALTH AND SAFETY LISTS
  IARC - Group 2A (limited human data)
  [present]
  OSHA - Possible Select Carcinogens
  [present]

STATE LISTS
  California - Prop. 65 - Cancer list
  carcinogen - initial date 1/1/92
  California - Prop. 65 - No Significant Risk Levels
  no significant risk level = 0.003 ug/day

PROPOSED REGULATIONS
  NTP - Proposed Additions to Annual Report on Carcinogens
  proposed as a suspect carcinogen for the NTP 9th report

CHLORPHENIRAMINE MALEATE   113-92-8

HEALTH AND SAFETY LISTS
  NTP Chemical Status Reports - Testing Status and NTIS Number
  Technical reports printed (PB87146759/AS)
  NTP Chemical Status Reports - Evidence of Carcinogenicity
  male rat-no evidence; female rat-no evidence; male mice-no evidence; female mice-no evidence

CHLORPROMAZINE (2-CHLORO-10-(3-DIMETHY-   50-53-3
LAMINOPROPYL) PHENOTHIAZINE)

INTERNATIONAL LISTS
  German (DFG) - Skin/Sensitizers
  danger of photo-contact sensitization

CHLORPROPAMIDE   94-20-2

HEALTH AND SAFETY LISTS
  NTP Chemical Status Reports - Testing Status and NTIS Number
  Technical reports printed (PB275178/AS)
  NTP Chemical Status Reports - Evidence of Carcinogenicity
  male rat-negative; female rat-negative; male mice-negative; female mice-negative

CHLORPYRIFOS   2921-88-2

HEALTH AND SAFETY LISTS
  ACGIH 1995 - Time Weighted Averages
  0.2 mg/m3 TWA
  ACGIH 1995 - Skin Designations
  skin - potential for cutaneous absorption
  U.S. DOT - Appendix B - Marine Pollutants
  DOT regulated severe marine pollutant
  U.S. DOT - Appendix A Table 1 - Hazardous Substances
  final RQ = 1 pound (0.454 kg)
  NIOSH - Selected LD50s and LC50s
  *Inhalation, rat: LD50 = 78 mg/kg 8 hr Oral, rat: LD50 = 82 mg/kg
  Skin, rat: LD50 = 202 mg/kg*
  OSHA - Vacated PELs - Time Weighted Averages
  0.2 mg/m3 TWA
  OSHA - Vacated PELs - Skin Designation
  Prevent or reduce skin absorption

ENVIRONMENTAL LISTS
  ATSDR Priority List
  Rank (of 275): 115
  CERCLA/SARA - Hazardous Substances and their Reportable Quantities
  final RQ = 1 pound (0.454 kg)
  Clean Water Act - Hazardous Substances
  [present]

INTERNATIONAL LISTS
  Australian Exposure Standards - Time Weighted Averages
  0.2 mg/m3 TWA
  Australian Exposure Standards - Skin Effects
  skin absorption
  Canada - Drinking Water Quality - MACs
  0.09 mg/L MAC
  Canada - Alberta - 8 Hour Occupational Exposure Limit
  0.2 mg/m3 TWA
  Canada - Alberta - 15 Minute Occupational Exposure Limit
  0.6 mg/m3 STEL
  Canada - Alberta - Skin Designation
  can be absorbed through the intact skin
  Canada - British Columbia - 8 Hour Exposure Limits
  0.2 mg/m3 TWA
  Canada - British Columbia - 15 Minute Exposure Limits
  0.6 mg/m3 STEL

Canada - British Columbia - Skin Notations
skin - potential for skin absorption

Canada - Ontario - OHSA - TWAEVs
0.2 mg/m3 TWAEV

Canada - Ontario - OHSA - STEVs
0.6 mg/m3 STEV

Canada - Ontario - OHSA - Skin Notations
absorption through skin, eyes, or mucous membranes

Canada - Quebec - Time-Weighted Average Exposure Values
0.2 mg/m3 TWAEV

Canada - Quebec - Skin Designations
absorbed through the skin

United Kingdom - Occupational Exposure Standards - TWAs
0.2 mg/m3 TWA

United Kingdom - Occupational Exposure Standards - STELs
0.6 mg/m3 STEL

United Kingdom - Occupational Exposure Standards - Notes
can be absorbed through skin

Israel - Time Weighted Averages
0.2 mg/m3 TWA

Israel - Action Levels
0.1 mg/m3 AL

Mexico - Instruction No. 10 - TWAs
0.2 mg/m3 TWA

Mexico - Instruction No. 10 - STELs
0.6 mg/m3 STEL

Mexico - Instruction No. 10 - Skin designation
skin - potential for cutaneous absorption

STATE LISTS

California - Exposure Limits - PELs
0.2 mg/m3 PEL

California - Exposure Limits - Skin Notation
material may be absorbed through the skin, eyes or mucous membrane

California - Directors List of Hazardous Substances (8 CCR 339)
[present]

Florida Hazardous Substance List
[present]

Massachusetts Right To Know List
neurotoxin

Minnesota Hazardous Substance List
skin

NJ Right to Know List (Total)
sn 0426

Pennsylvania Right to Know List
environmental hazard

CHLORPYRIFOS METHYL [O,O-DIMETHYL-O-(3,5,6-   5598-13-0 TRICHLORO-2-PYRIDYL)PHOSPHOROTHIOATE

ENVIRONMENTAL LISTS

CERCLA/SARA - Section 313 - Emission Reporting
form R reporting required

CHLORSULFURON [2-CHLORO-N-[[4-METHOXY-6-   64902-72-3 METHYL-1,3,5-TRIAZIN-2-YL) AMINO]CARBONYL] BENZENESULFONAMIDE

ENVIRONMENTAL LISTS

CERCLA/SARA - Section 313 - Emission Reporting
form R reporting required

CHLORTHION   500-28-7

HEALTH AND SAFETY LISTS

NIOSH - Selected LD50s and LC50s
Oral, rat: LD50 = 625 mg/kg Skin, rat: LD50 = 1500 mg/kg

CHLORTHIOPHOS   21923-23-9

HEALTH AND SAFETY LISTS

U.S. DOT - Appendix B - Marine Pollutants
DOT regulated severe marine pollutant

NIOSH - Selected LD50s and LC50s
Oral, rat: LD50 = 7800 ug/kg Skin, rat: LD50 = 58 mg/kg

ENVIRONMENTAL LISTS

CERCLA/SARA - Section 302 Extremely Hazardous Substances and TPQs
TPQ = 500 pounds

STATE LISTS

Florida Hazardous Substance List
effective March 13, 1992

Massachusetts Right To Know List
extraordinarily hazardous

NJ Right to Know List (Total)
sn 2247

Pennsylvania Right to Know List
environmental hazard

PROPOSED REGULATIONS

CERCLA/SARA - Proposed Hazardous Substance Additions
proposed RQ = 1 pound (.454 kg)

CERCLA/SARA - 1989 Proposed RQ Adjustments
proposed RQ = 100 pounds (45.4 kg)

CHOLESTEROL   57-88-5

HEALTH AND SAFETY LISTS

IARC - Group 3 (not classifiable)
[present]

CHOLINE   62-49-7

HEALTH AND SAFETY LISTS

NIOSH - Selected LD50s and LC50s
Oral, rat: LD50 = 6640 mg/kg

CHOLINE CHLORIDE   67-48-1

HEALTH AND SAFETY LISTS

NIOSH - Selected LD50s and LC50s
Oral, rat: LD50 = 3400 mg/kg

CHROMATES   RR-01554-3

INTERNATIONAL LISTS

Canada - Alberta - Designated Substances
designated substance - requires code of practice

Canada - Ontario - OHSA - TWAEVs
as Cr: 0.05 mg/m3 TWAEV

Canada - Quebec - Time-Weighted Average Exposure Values
0.05 mg/m3 TWAEV (substance of which the recirculation is prohibited)

PROPOSED REGULATIONS

Canada - Ontario - Proposed Occupational TWAEVs
as Cr: 0.02 mg/m3 TWAEV

CHROMIC ACETATE   1066-30-4
SEE ALSO:
CHROMIUM
CHROMIUM (III) COMPOUNDS

HEALTH AND SAFETY LISTS

U.S. DOT - Appendix A Table 1 - Hazardous Substances
final RQ = 1000 pounds (454 kg)

ENVIRONMENTAL LISTS
 CERCLA/SARA - Hazardous Substances and their Reportable Quantities
 final RQ = 1000 pounds (454 kg)
 Clean Water Act - Hazardous Substances
 [present]
INTERNATIONAL LISTS
 Canada - WHMIS: Ingredient Disclosure
 1% item 392 (13)
STATE LISTS
 Massachusetts Right To Know List
 [present]
 NJ Right to Know List (Total)
 sn 0428
 Pennsylvania Right to Know List
 environmental hazard

CHROMIC ACID   7738-94-5
SEE ALSO:
 CHROMIUM
HEALTH AND SAFETY LISTS
 U.S. DOT - Appendix A Table 1 - Hazardous Substances
 final RQ = 10 pounds (4.54 kg)
 NIOSH 1990 - Pocket Guide - RELs
 as CrO3: 0.001 mg/m3 TWA
 NIOSH 1990 - Pocket Guide - IDLHs
 as CrO3: 30 mg/m3 IDLH (not considering carcinogenic effects)
 NIOSH 1990 - Pocket Guide - Carcinogens
 occupational carcinogen
 NIOSH 1990 - Pocket Guide - Target organs
 blood, respiratory system, liver, kidneys, eyes, skin
 NIOSH - Health Standards - Exposure Limits
 as non-carcinogenic Cr(VI): 25 ug/m3 TWA; C (15 min) 50 ug/m3
 NIOSH - Health Standards - Health Effects and Precautions
 Nasal ulceration
 OSHA - Final PELs - Time Weighted Averages
 and chromates: C 1 mg/10m3
 OSHA - Final PELs - Ceiling Limits
 and chromates: C 1 mg/10m3
ENVIRONMENTAL LISTS
 ATSDR Priority List
 Rank (of 275): 271
 CERCLA/SARA - Hazardous Substances and their Reportable Quantities
 final RQ = 10 pounds (4.54 kg)
INTERNATIONAL LISTS
 Canada - WHMIS: Ingredient Disclosure
 1% item 391 (79)
 Canada - British Columbia - 8 Hour Exposure Limits
 as Cr: 0.05 mg/m3 TWA
STATE LISTS
 Florida Hazardous Substance List
 [present]
 Massachusetts Right To Know List
 [present]
 Minnesota Hazardous Substance List
 [present]
 NJ Right to Know List (Total)
 sn 0429
 NJ Special Hazardous Substances
 (corrosive)
 Pennsylvania Right to Know List
 environmental hazard

CHROMIC ACID   11115-74-5
HEALTH AND SAFETY LISTS
 U.S. DOT - Substances From 49 CFR 172.101
 regulated by DOT (UN1755, NA1463)
 U.S. DOT - Hazard Classes
 DOT hazard class = 8
 U.S. DOT - Appendix A Table 1 - Hazardous Substances
 final RQ = 10 pounds (4.54 kg) (Listed under 'Chromic acid')
ENVIRONMENTAL LISTS
 CERCLA/SARA - Hazardous Substances and their Reportable Quantities
 final RQ = 10 pounds (4.54 kg) (Listed under 'Chromic acid')
 Clean Water Act - Hazardous Substances
 [present]
 EPA - Carcinogen Hazard Ranking for RQ Adjustment
 Hazard ranking = High
STATE LISTS
 Massachusetts Right To Know List
 [present]
 NJ Right to Know List (Total)
 sn 3063
 Pennsylvania Right to Know List
 environmental hazard

CHROMIC ACID (H2CR2O7)   13530-68-2
SEE ALSO:
 CHROMIUM
STATE LISTS
 Pennsylvania Right to Know List
 environmental hazard

CHROMIC ACID (H2CR2O7), CALCIUM SALT (1:1)   14307-33-6
STATE LISTS
 Florida Hazardous Substance List
 [present]
 Massachusetts Right To Know List
 [present]
 Pennsylvania Right to Know List
 [present]

CHROMIC ACID (H2CRO4), MAGNESIUM SALT (1:1)   13423-61-5
STATE LISTS
 Florida Hazardous Substance List
 [present]
 Massachusetts Right To Know List
 [present]
 Pennsylvania Right to Know List
 [present]

CHROMIC CHLORIDE   10025-73-7
SEE ALSO:
 CHROMIUM (III) COMPOUNDS
 CHROMIUM
HEALTH AND SAFETY LISTS
 NIOSH - Selected LD50s and LC50s
 Inhalation, mouse: LC50 = 31500 ug/m3 (2 hr)
ENVIRONMENTAL LISTS
 CERCLA/SARA - Section 302 Extremely Hazardous Substances and TPQs
 TPQ = 1/10,000 pounds
INTERNATIONAL LISTS
 Canada - WHMIS: Ingredient Disclosure
 1% item 394 (486)

STATE LISTS
Florida Hazardous Substance List
effective March 13, 1992
Massachusetts Right To Know List
extraordinarily hazardous
NJ Right to Know List (Total)
sn 2248
Pennsylvania Right to Know List
environmental hazard

PROPOSED REGULATIONS
CERCLA/SARA - Proposed Hazardous Substance Additions
proposed RQ = 1 pound (.454 kg)
CERCLA/SARA - 1989 Proposed RQ Adjustments
proposed RQ = 1 pounds (.454 kg)

CHROMIC(II) CHLORIDE   10049-05-5
SEE ALSO:
CHROMIUM

HEALTH AND SAFETY LISTS
U.S. DOT - Appendix A Table 1 - Hazardous Substances
final RQ = 1000 pounds (454 kg)
NIOSH - Selected LD50s and LC50s
Oral, rat: LD50 = 1870 mg/kg

ENVIRONMENTAL LISTS
CERCLA/SARA - Hazardous Substances and their Reportable Quantities
final RQ = 1000 pounds (454 kg)
Clean Water Act - Hazardous Substances
[present]

STATE LISTS
Massachusetts Right To Know List
[present]
NJ Right to Know List (Total)
sn 0440
Pennsylvania Right to Know List
environmental hazard

CHROMIC FLUORIDE   7788-97-8
SEE ALSO:
CHROMIUM

HEALTH AND SAFETY LISTS
U.S. DOT - Substances From 49 CFR 172.101
regulated by DOT (UN1756, UN1757)
U.S. DOT - Hazard Classes
DOT hazard class = 8

STATE LISTS
NJ Right to Know List (Total)
sn 0430
NJ Special Hazardous Substances
(corrosive)

CHROMIC SULFATE   10101-53-8
SEE ALSO:
CHROMIUM
CHROMIUM (III) COMPOUNDS

HEALTH AND SAFETY LISTS
U.S. DOT - Appendix A Table 1 - Hazardous Substances
final RQ = 1000 pounds (454 kg)

ENVIRONMENTAL LISTS
CERCLA/SARA - Hazardous Substances and their Reportable Quantities
final RQ = 1000 pounds (454 kg)

Clean Water Act - Hazardous Substances
[present]

INTERNATIONAL LISTS
Canada - WHMIS: Ingredient Disclosure
1% item 403 (1516)

STATE LISTS
Massachusetts Right To Know List
[present]
NJ Right to Know List (Total)
sn 0431
Pennsylvania Right to Know List
environmental hazard

CHROMITE   1308-31-2
SEE ALSO:
IRON

STATE LISTS
Massachusetts Right To Know List
[present]

CHROMITE ORE PROCESSING   RR-00070-4
HEALTH AND SAFETY LISTS
ACGIH 1995 - Time Weighted Averages
as Cr: 0.05 mg/m3 TWA
ACGIH 1995 - Carcinogens
as Cr: A1-confirmed human carcinogen

INTERNATIONAL LISTS
Canada - Alberta - 8 Hour Occupational Exposure Limit
chromate (as Cr): 0.05 mg/m3 TWA
Canada - Alberta - 15 Minute Occupational Exposure Limit
0.15 mg/m3 STEL
Canada - Alberta - Designated Substances
designated substance - requires code of practice
Canada - British Columbia - 8 Hour Exposure Limits
chromate, as Cr: 0.05 mg/m3 TWA
Canada - British Columbia - Carcinogens
carcinogen - chromate, as Cr: 0.05 mg/m3 TWA
Canada - Quebec - Time-Weighted Average Exposure Values
as Cr: 0.05 mg/m3 TWAEV (substance of which the recirculation is prohibited)
Canada - Quebec - Carcinogens
as Cr: C1 carcinogen: effect detected in humans
Israel - Time Weighted Averages
as Cr: 0.05 mg/m3 TWA
Israel - Action Levels
as Cr: 0.025 mg/m3 AL
Mexico - Instruction No. 10 - TWAs
0.05 mg/m3 TWA
Mexico - Instruction No. 10 - Carcinogens
potentially carcinogenic contaminant

STATE LISTS
California - Exposure Limits - PELs
as Cr: 0.05 mg/m3 PEL
Minnesota Hazardous Substance List
as Cr: carcinogen
Pennsylvania Right to Know List
[present]

CHROMIUM   7440-47-3
SEE ALSO:
K090-HAZARDOUS WASTES
K091-HAZARDOUS WASTES
K069-HAZARDOUS WASTES
F039-HAZARDOUS WASTES
CHROMIUM COMPOUNDS

HEALTH AND SAFETY LISTS

ACGIH 1995 - Time Weighted Averages
metal and Cr III compounds, as Cr: 0.5 mg/m3 TWA; water-soluble Cr VI compounds NOC: 0.05 mg/m3; insoluble Cr VI compounds, NOC: 0.01 mg/m3 TWA

ACGIH 1995 - Carcinogens
metal and Cr III compounds: A4-not classifiable as a human carcinogen; water soluble Cr VI compounds, NOC: A1-confirmed human carcinogen; insoluble Cr VI compounds: A1-confirmed human carcinogen

ACGIH 1995 - Biological Exposure Indices
Total Chromium in urine: 10 ug/g creatinine, increase during shift, 30 ug/g creatinine, end of shift at end of workweek (B)

U.S. DOT - Appendix A Table 1 - Hazardous Substances
final RQ = 5000 pounds (2270 kg) (no reporting of releases of this hazardous material is required if the diameter of the pieces of the solid metal released is equal to or exceeds 0.004 inches)

IARC - Group 3 (not classifiable)
[present]

NIOSH 1990 - Pocket Guide - RELs
as Cr: 0.5 mg/m3 TWA

NIOSH 1990 - Pocket Guide - Target organs
respiratory system

OSHA - Vacated PELs - Time Weighted Averages
Chromium, sol. chromic, chromous salts (as Cr): 0.5 mg/m3 TWA; Chromium metal and insoluble salts (as Cr): 1 mg/m3 TWA

OSHA - Vacated PELs - Ceiling Limits
Chromic acid and chromates: C 0.1 mg/m3

OSHA - Final PELs - Time Weighted Averages
Chromium, sol. chromic, chromous salts (as Cr): 0.5 mg/m3 TWA; Chromium, metal and insoluble salts (as Cr): 1 mg/m3 TWA

ENVIRONMENTAL LISTS

ATSDR Priority List
Rank (of 275): 061

CERCLA/SARA - Section 313 - Emission Reporting
form R reporting required for 0.1% de minimus concentration

CERCLA/SARA - Hazardous Substances and their Reportable Quantities
final RQ = 5000 pounds (2270 kg) (no reporting of releases of this hazardous material is required if the diameter of the pieces of the solid metal released is equal to or exceeds 0.004 inches)

Clean Water Act - Priority Pollutants
[present]

Clean Water Act - Toxic Pollutants
[present] (Listed under 'Chromium and compounds')

Safe Drinking Water Act - MCLs
MCL = 0.1 mg/L

Safe Drinking Water Act - MCLGs
MCLG = 0.1 mg/L

RCRA - D Series - Maximum Concentration of Contaminants
waste number D007; regulatory level = 5.0 mg/L

RCRA - D Series - Chronic Toxicity Reference Levels
chronic toxicity reference level = 0.05 mg/L

RCRA - Hazardous Constituents-Appendix VIII
hazardous constituent - no waste number

RCRA - Basis for Listing - Appendix VII
Included in waste streams: F039, K090, K091

RCRA - Substances Banned From Land Disposal
[present]

RCRA - TSD Facilities Ground Water Monitoring
TM 6010 = 70 ug/L PQL; TM 7190 = 500 ug/L PQL; TM 7190 = 10 ug/L PQL (all species in the ground water that contain this element are included)

RCRA - Universal Treatment Standards (LDR)
WW: 2.77 mg/l; NWW: 0.86 mg/l TCLP

INTERNATIONAL LISTS

Australian Exposure Standards - Time Weighted Averages
0.5 mg/m3 TWA

Canada - WHMIS: Ingredient Disclosure
0.1% item 399 (561)

Canada - NPRI (National Pollutant Release Inventory)
[present]

Canada - CEPA - Priority Substances List
estimated time for completion of assessment reports: 4 years

Canada - Drinking Water Quality - MACs
0.05 mg/L MAC

Canada - Alberta - 8 Hour Occupational Exposure Limit
0.5 mg/m3 TWA

Canada - Alberta - 15 Minute Occupational Exposure Limit
1.5 mg/m3 STEL

Canada - Ontario - OHSA - TWAEVs
0.5 mg/m3 TWAEV

Canada - Quebec - Time-Weighted Average Exposure Values
as Cr: 0.5 mg/m3 TWAEV

United Kingdom - Occupational Exposure Standards - TWAs
0.5 mg/m3 TWA

Israel - Time Weighted Averages
0.5 mg/m3 TWA

Israel - Action Levels
0.25 mg/m3 AL

Mexico - Instruction No. 10 - TWAs
0.5 mg/m3 TWA

Mexico - Wastewater - Organic Toxic Pollutants and Heavy Metals
Listed under [Heavy Metals]

STATE LISTS

California - Air Bill 2588 Appendix A-I
6/91

California - Exposure Limits - PELs
0.5 mg/m3 PEL

California - Directors List of Hazardous Substances (8 CCR 339)
[present]

Florida Hazardous Substance List
[present]

Massachusetts Right To Know List
carcinogen; extraordinarily hazardous

Minnesota Hazardous Substance List
[present]

NJ Right to Know List (Total)
sn 0432

NJ Special Hazardous Substances
(carcinogen; mutagen)

Pennsylvania Right to Know List
environmental hazard; special hazardous substance (any compound of this substance is also an environmental hazard)

Pennsylvania RTK - Special Hazardous Substances
[present]

CHROMIUM (VI) COMPOUNDS (CERTAIN WATER   RR-01791-4
INSOLUBLE FORMS)

INTERNATIONAL LISTS

Canada - Quebec - Carcinogens
C1 carcinogen: effect detected in humans

CHROMIUM 48   14833-09-1

HEALTH AND SAFETY LISTS

U.S. DOT - Appendix A Table 2 - Radionuclides
final RQ = 100 curies (3.7E 12 Bq)

ENVIRONMENTAL LISTS

CERCLA/SARA List of Radionuclides (Appendix B) and Their Reportable Quantities
final RQ = 100 curies (3.7E 12 Bq)

CHROMIUM 49   15758-14-2

HEALTH AND SAFETY LISTS

U.S. DOT - Appendix A Table 2 - Radionuclides
final RQ = 1000 curies (3.7E 13 Bq)

ENVIRONMENTAL LISTS

CERCLA/SARA List of Radionuclides (Appendix B) and Their Reportable Quantities
final RQ = 1000 curies (3.7E 13 Bq)

CHROMIUM 51   14392-02-0

HEALTH AND SAFETY LISTS

U.S. DOT - Appendix A Table 2 - Radionuclides
final RQ = 1000 curies (3.7E 13 Bq)

ENVIRONMENTAL LISTS

CERCLA/SARA List of Radionuclides (Appendix B) and Their Reportable Quantities
final RQ = 1000 curies (3.7E 13 Bq)

CHROMIUM (II)   22541-79-3

INTERNATIONAL LISTS

Canada - Ontario - OHSA - TWAEVs
0.5 mg/m3 TWAEV

STATE LISTS

NJ Right to Know List (Total)
sn 0433

Pennsylvania Right to Know List
environmental hazard

CHROMIUM (VI)   18540-29-9

HEALTH AND SAFETY LISTS

IARC - Group 1 (carcinogenic to humans)
[present]

NIOSH - Health Standards - Exposure Limits
carcinogenic Cr(VI): 1 ug/m3 TWA; other Cr(VI): 25 ug/m3 TWA; C (15 min) 50 ug/m3

NIOSH - Health Standards - Health Effects and Precautions
Lung cancer, skin ulcers, and lung irritation (periodic chest X-ray required)

NIOSH - Health Standards - Carcinogenic Chemicals
potential human carcinogen

OSHA - Select Carcinogens
[present]

ENVIRONMENTAL LISTS

ATSDR Priority List
Rank (of 275): 019

RCRA - Basis for Listing - Appendix VII
Included in waste streams: F006, F019, K002, K003, K004, K005, K006, K007, K008, K048, K049, K050, K051, K061, K062, K069, K086, K100

INTERNATIONAL LISTS

Canada - Alberta - 15 Minute Occupational Exposure Limit
water insoluble: 0.15 mg/m3 STEL

Mexico - Drinking Water - Ecological Criteria
0.05 mg/l

STATE LISTS

California - Air Bill 2588 Appendix A-I
known or potential carcinogen

Pennsylvania Right to Know List
environmental hazard

PROPOSED REGULATIONS

Safe Drinking Water Act - Priority list
[present]

CHROMIUM ACETYLACETONATE   21679-31-2
SEE ALSO:
CHROMIUM

HEALTH AND SAFETY LISTS

NIOSH - Selected LD50s and LC50s
Oral, rat: LD50 = 3360 mg/kg Skin, rabbit: LD50 = 6350 mg/kg

CHROMIUM CARBONATE   29689-14-3

HEALTH AND SAFETY LISTS

NTP Seventh Report - Known Carcinogens
known carcinogen (Listed under 'Chromium and certain chromium compounds')

OSHA - Select Carcinogens
[present]

STATE LISTS

Florida Hazardous Substance List
[present]

CHROMIUM CARBONYL   13007-92-6
SEE ALSO:
CHROMIUM

INTERNATIONAL LISTS

Canada - WHMIS: Ingredient Disclosure
1% item 393 (557)

German (DFG) - Carcinogens
suspected carcinogen

CHROMIUM(III) CHROMATE   24613-89-6
SEE ALSO:
CHROMIUM

INTERNATIONAL LISTS

Canada - WHMIS: Ingredient Disclosure
0.1% item 395 (549)

STATE LISTS

Massachusetts Right To Know List
[present]

CHROMIUM (II) COMPOUNDS   RR-00023-7

HEALTH AND SAFETY LISTS

NIOSH 1990 - Pocket Guide - RELs
as Cr: 0.5 mg/m3 TWA

NIOSH 1990 - Pocket Guide - Target organs
skin

INTERNATIONAL LISTS

Australian Exposure Standards - Time Weighted Averages
as Cr: 0.5 mg/m3 TWA

Canada - WHMIS: Ingredient Disclosure
1% ITEM 396 (558)

Canada - Alberta - 8 Hour Occupational Exposure Limit
as Cr: 0.5 mg/m3 TWA

Canada - Alberta - 15 Minute Occupational Exposure Limit
1.5 mg/m3 STEL

Canada - British Columbia - 8 Hour Exposure Limits
as Cr: 0.5 mg/m3 TWA

Canada - Quebec - Time-Weighted Average Exposure Values
  as Cr: 0.05 mg/m3 TWAEV
United Kingdom - Occupational Exposure Standards - TWAs
  as Cr: 0.5 mg/m3 TWA
Israel - Time Weighted Averages
  as Cr: 0.5 mg/m3 TWA
Israel - Action Levels
  as Cr: 0.25 mg/m3 AL
Mexico - Instruction No. 10 - TWAs
  0.5 mg/m3 TWA

STATE LISTS

California - Exposure Limits - PELs
  as Cr: 0.5 mg/m3 PEL
Minnesota Hazardous Substance List
  [present] as Cr

CHROMIUM (III) COMPOUNDS   RR-00024-8

HEALTH AND SAFETY LISTS

IARC - Group 3 (not classifiable)
  [present]
NIOSH 1990 - Pocket Guide - RELs
  as Cr: 0.5 mg/m3 TWA
NIOSH 1990 - Pocket Guide - Target organs
  skin

INTERNATIONAL LISTS

Australian Exposure Standards - Time Weighted Averages
  as Cr: 0.5 mg/m3 TWA
Canada - WHMIS: Ingredient Disclosure
  1% item 397 (559)
Canada - Alberta - 8 Hour Occupational Exposure Limit
  as Cr: 0.5 mg/m3 TWA
Canada - Alberta - 15 Minute Occupational Exposure Limit
  1.5 mg/m3 STEL
Canada - British Columbia - 8 Hour Exposure Limits
  as Cr: 0.5 mg/m3 TWA
United Kingdom - Occupational Exposure Standards - TWAs
  as Cr: 0.5 mg/m3 TWA
Israel - Time Weighted Averages
  as Cr: 0.5 mg/m3 TWA
Israel - Action Levels
  as Cr: 0.25 mg/m3 AL
Mexico - Instruction No. 10 - TWAs
  water soluble: 0.5 mg/m3 TWA

STATE LISTS

California - Exposure Limits - PELs
  as Cr: 0.5 mg/m3 PEL
Minnesota Hazardous Substance List
  [present] as Cr

CHROMIUM (VI) COMPOUNDS   RR-00026-0

HEALTH AND SAFETY LISTS

IARC - Group 1 (carcinogenic to humans)
  [present] (This evaluation applies to the group of chemicals as a whole and not necessarily to all individual chemicals within the group) (Listed under 'Chromium and chromium compounds')
OSHA - Select Carcinogens
  [present]

INTERNATIONAL LISTS

Australian Exposure Standards - Time Weighted Averages
  certain water insoluble, as Cr: 0.05 mg/m3 TWA; certain water soluble, as Cr: 0.05 mg/m3 TWA
Australian Exposure Standards - Skin Effects
  as Cr: sensitiser

Australian Exposure Standards - Carcinogens
  certain water insoluble compounds as Cr: confirmed carcinogen
Australian Exposure Standards - Under Review
  exposure limits under review (regular and water soluble)
Canada - WHMIS: Ingredient Disclosure
  1% item 398 (560)
Canada - Alberta - 8 Hour Occupational Exposure Limit
  as Cr, water soluble: 0.05 mg/m3 TWA; water insoluble: 0.05 mg/m3 TWA
Canada - Ontario - OHSA - TWAEVs
  as Cr: 0.05 mg/m3 TWAEV
German (DFG) - Skin/Sensitizers
  danger of sensitization
German (DFG) - Carcinogens
  as inspirable dusts/aerosols: animal evidence of carcinogenicity
Israel - Time Weighted Averages
  water soluble as Cr: 0.05 mg/m3 TWA; certain water insoluble compounds as Cr: 0.05 mg/m3 TWA
Israel - Action Levels
  Certain water insoluble compounds as Cr: 0.025 mg/m3 AL

STATE LISTS

California - Prop. 65 - Cancer list
  carcinogen - initial date 2/27/87
California - Prop. 65 - No Significant Risk Levels
  inhalation: no significant risk level = 0.001 ug/day
California - Exposure Limits - PELs
  water soluble and certain water insoluble, as Cr: 0.05 mg/m3 PEL
California - Exposure Limits - Ceilings
  as Cr: C 0.1 mg/m3
Minnesota Hazardous Substance List
  [present] as Cr
Pennsylvania Right to Know List
  environmental hazard; special hazardous substance
Pennsylvania RTK - Special Hazardous Substances
  [present]

PROPOSED REGULATIONS

Canada - Ontario - Proposed Occupational TWAEVs
  as Cr: 0.02 mg/m3 TWAEV

CHROMIUM COMPOUNDS   RR-00634-8

ENVIRONMENTAL LISTS

CERCLA/SARA - Section 313 - Emission Reporting
  form R reporting required for 0.1% (chromium VI), 1.0% (chromium III) de minimus concentration
Clean Air Act (1990) - List of Hazardous Air Contaminants
  [present] (includes any unique chemical substance that contains Chromium as part of that chemical's infrastructure)
Clean Water Act - Toxic Pollutants
  [present] (Listed under 'Chromium and compounds')
RCRA - Hazardous Constituents-Appendix VIII
  hazardous constituent - no waste number

INTERNATIONAL LISTS

Canada - NPRI (National Pollutant Release Inventory)
  [present]
Canada - CEPA - Priority Substances List
  estimated time for completion of assessment reports: 4 years

STATE LISTS

California - Air Bill 2588 Appendix A-I
  other than hexavalent: 6/91
California - Directors List of Hazardous Substances (8 CCR 339)
  [present]
NJ Right to Know List (Total)
  sn 3134

CHROMIUM (VI) COMPOUNDS- WATER SOLUBLE   RR-00025-9

INTERNATIONAL LISTS

Canada - Alberta - 15 Minute Occupational Exposure Limit
0.15 mg/m3 STEL

CHROMIUM, INORGANIC COMPOUNDS   RR-00560-7

INTERNATIONAL LISTS

United Kingdom - Maximum Exposure Limits - TWAs
0.05 mg/m3 TWA

STATE LISTS

NJ Right to Know List (Total)
sn 2870

CHROMIUM, ION (CR3+)   16065-83-1

INTERNATIONAL LISTS

Canada - Ontario - OHSA - TWAEVs
as Cr: 0.5 mg/m3 TWAEV

STATE LISTS

Pennsylvania Right to Know List
environmental hazard

PROPOSED REGULATIONS

Safe Drinking Water Act - Priority list
[present]

CHROMIUM LEAD OXIDE   11119-70-3

STATE LISTS

Pennsylvania Right to Know List
environmental hazard; special hazardous substance
Pennsylvania RTK - Special Hazardous Substances
[present]

CHROMIUM (III) NITRATE   7789-02-8
SEE ALSO:
CHROMIUM (III) COMPOUNDS
CHROMIUM
CHROMIUM COMPOUNDS

HEALTH AND SAFETY LISTS

U.S. DOT - Substances From 49 CFR 172.101
regulated by DOT (UN2720)
U.S. DOT - Hazard Classes
DOT hazard class = 5.1
NIOSH - Selected LD50s and LC50s
Oral, rat: LD50 = 3250 mg/kg

CHROMIUM NITRATE   13548-38-4
SEE ALSO:
CHROMIUM COMPOUNDS
CHROMIUM (III) COMPOUNDS
CHROMIUM

HEALTH AND SAFETY LISTS

NIOSH - Selected LD50s and LC50s
Oral, mouse: LD50 = 2976 mg/kg

STATE LISTS

NJ Right to Know List (Total)
sn 0435

CHROMIUM (III) OXIDE   1308-38-9
SEE ALSO:
CHROMIUM COMPOUNDS
CHROMIUM (III) COMPOUNDS
CHROMIUM

ENVIRONMENTAL LISTS

List of Pesticide Product Inert Ingredients
[present]

INTERNATIONAL LISTS

Canada - WHMIS: Ingredient Disclosure
1% item 400 (1304)

STATE LISTS

Massachusetts Right To Know List
carcinogen; extraordinarily hazardous
NJ Right to Know List (Total)
sn 0434

CHROMIUM PHOSPHATE   7789-04-0
SEE ALSO:
CHROMIUM

HEALTH AND SAFETY LISTS

NTP Seventh Report - Known Carcinogens
known carcinogen (Listed under 'Chromium and certain chromium compounds')
OSHA - Select Carcinogens
[present]

CHROMIUM(III) POTASSIUM SULFATE   10141-00-1
SEE ALSO:
CHROMIUM

INTERNATIONAL LISTS

Canada - WHMIS: Ingredient Disclosure
1% item 402 (1540)

STATE LISTS

Massachusetts Right To Know List
[present]

CHROMIUM SODIUM OXIDE   12680-48-7

STATE LISTS

Pennsylvania Right to Know List
environmental hazard

CHROMIUM TRIOXIDE (CRO3)   1333-82-0
SEE ALSO:
CHROMIUM COMPOUNDS
CHROMIUM (VI) COMPOUNDS
CHROMIUM (VI) COMPOUNDS- WATER SOLUBLE
CHROMIUM

HEALTH AND SAFETY LISTS

U.S. DOT - Substances From 49 CFR 172.101
regulated by DOT (UN1463)
U.S. DOT - Hazard Classes
DOT hazard class = 5.1
NIOSH - Selected LD50s and LC50s
Oral, rat: LD50 = 80 mg/kg
NTP Seventh Report - Known Carcinogens
known carcinogen (Listed under 'Chromium and certain chromium compounds')
OSHA - Select Carcinogens
[present]

INTERNATIONAL LISTS

Canada - WHMIS: Ingredient Disclosure
0.1% item 401 (1305)

STATE LISTS

California - Air Bill 2588 Appendix A-I
known or potential carcinogen: 6/91
Florida Hazardous Substance List
[present]
Massachusetts Right To Know List
carcinogen; extraordinarily hazardous
NJ Right to Know List (Total)
sn 0437

NJ Special Hazardous Substances
(carcinogen)

Pennsylvania Right to Know List
special hazardous substance

Pennsylvania RTK - Special Hazardous Substances
[present]

CHROMIUM ZINC OXIDE   12018-19-8
SEE ALSO:
  ZINC
  CHROMIUM

STATE LISTS

Massachusetts Right To Know List
[present]

CHROMOMYCIN A3   7059-24-7

HEALTH AND SAFETY LISTS

NIOSH - Selected LD50s and LC50s
Oral, mouse: LD50 = 1431 ug/kg

CHROMOSULFURIC ACID   14489-25-9

STATE LISTS

NJ Right to Know List (Total)
sn 0439

NJ Special Hazardous Substances
(corrosive)

CHROMOSULFURIC ACID   15005-90-0
SEE ALSO:
  CHROMIUM

HEALTH AND SAFETY LISTS

U.S. DOT - Substances From 49 CFR 172.101
regulated by DOT (UN2240)

U.S. DOT - Hazard Classes
DOT hazard class = 8

CHROMOSULFURIC ACID   64093-79-4
SEE ALSO:
  CHROMIUM

STATE LISTS

NJ Right to Know List (Total)
sn 0438

CHROMOUS (II) SULFATE   13825-86-0

STATE LISTS

Massachusetts Right To Know List
[present]

CHROMYL CHLORIDE   14977-61-8
SEE ALSO:
  CHROMIUM

HEALTH AND SAFETY LISTS

ACGIH 1995 - Time Weighted Averages
0.025 ppm TWA; 0.16 mg/m3 TWA

U.S. DOT - Substances From 49 CFR 172.101
regulated by DOT (UN1758)

U.S. DOT - Hazard Classes
DOT hazard class = 8

U.S. DOT - Appendix B - Marine Pollutants
DOT regulated marine pollutant

INTERNATIONAL LISTS

Canada - WHMIS: Ingredient Disclosure
1% item 404 (487)

Canada - Alberta - 8 Hour Occupational Exposure Limit
0.025 ppm TWA; 0.16 mg/m3 TWA

Canada - Alberta - 15 Minute Occupational Exposure Limit
0.075 ppm STEL; 0.48 mg/m3 STEL

Canada - Ontario - OHSA - TWAEVs
0.025 ppm TWAEV; 0.16 mg/m3 TWAEV

Canada - Quebec - Time-Weighted Average Exposure Values
0.025 ppm TWAEV; 0.16 mg/m3 TWAEV

Israel - Time Weighted Averages
0.025 ppm TWA; 0.16 mg/m3 TWA

Israel - Action Levels
0.0125 ppm AL; 0.08 mg/m3 AL

STATE LISTS

California - Exposure Limits - PELs
0.025 ppm PEL; 0.15 mg/m3 PEL

Florida Hazardous Substance List
[present]

Massachusetts Right To Know List
[present]

Minnesota Hazardous Substance List
[present]

NJ Right to Know List (Total)
sn 0436

NJ Special Hazardous Substances
(corrosive)

Pennsylvania Right to Know List
environmental hazard

CHRYSENE   218-01-9
SEE ALSO:
  K148-HAZARDOUS WASTES
  K143-HAZARDOUS WASTES
  F039-HAZARDOUS WASTES
  K141-HAZARDOUS WASTES
  K142-HAZARDOUS WASTES
  K035-HAZARDOUS WASTES
  K145-HAZARDOUS WASTES
  COAL TAR PITCHES
  K001-HAZARDOUS WASTES
  K144-HAZARDOUS WASTES
  K147-HAZARDOUS WASTES

HEALTH AND SAFETY LISTS

ACGIH 1995 - Carcinogens
A2-suspected human carcinogen

U.S. DOT - Appendix A Table 1 - Hazardous Substances
final RQ = 100 pounds (45.4 kg)

IARC - Group 3 (not classifiable)
[present]

NIOSH - Health Standards - Exposure Limits
control as an occupational carcinogen

NIOSH - Health Standards - Health Effects and Precautions
Liver and skin cancer

NIOSH - Health Standards - Carcinogenic Chemicals
potential human carcinogen

ENVIRONMENTAL LISTS

ATSDR Priority List
Rank (of 275): 110

CERCLA/SARA - Section 313 - Emission Reporting
form R reporting required; (Listed under 'Polycyclic aromatic compounds')

CERCLA/SARA - Hazardous Substances and their Reportable Quantities
final RQ = 100 pounds (45.4 kg)

CAA - HON Rule - SOCMI Chemicals
compliance by Oct. 23, 1995

Clean Water Act - Priority Pollutants
[present]

EPA - Carcinogen Hazard Ranking for RQ Adjustment
Hazard ranking = Low

RCRA - U Series Wastes
waste number U050

RCRA - Hazardous Constituents-Appendix VIII
waste number U050

RCRA - Basis for Listing - Appendix VII
Included in waste streams: F039, K001, K035, K141, K142, K143, K144, K145, K147, K148

RCRA - Substances Banned From Land Disposal
[present]

RCRA - TSD Facilities Ground Water Monitoring
TM 8100 = 200 ug/L PQL;   TM 8270 = 10 ug/L PQL

RCRA - Universal Treatment Standards (LDR)
WW: 0.059 mg/l;   NWW: 3.4 mg/kg

INTERNATIONAL LISTS

Australian Exposure Standards - Time Weighted Averages
control to the lowest practical level

Australian Exposure Standards - Carcinogens
suspected carcinogen

Canada - WHMIS: Ingredient Disclosure
0.1% item 405 (562)

Canada - Alberta - Designated Substances
designated substance - requires code of practice

Canada - Quebec - Time-Weighted Average Exposure Values
substance for which the recirculation is prohibited

Canada - Quebec - Carcinogens
C2 carcinogen: effect suspected in humans

Mexico - Wastewater - Organic Toxic Pollutants and Heavy Metals
Listed under [Aromatic Hydrocarbons]

STATE LISTS

California - Air Bill 2588 Appendix A-I
known or potential carcinogen: 9/90

California - Prop. 65 - Cancer list
carcinogen - initial date 1/1/90

Florida Hazardous Substance List
[present]

Massachusetts Right To Know List
carcinogen;   extraordinarily hazardous

Minnesota Hazardous Substance List
carcinogen

NJ Special Hazardous Substances
(carcinogen)

Pennsylvania Right to Know List
environmental hazard

CHRYSOIDINE   **532-82-1**

HEALTH AND SAFETY LISTS

IARC - Group 3 (not classifiable)
[present]

STATE LISTS

California - Directors List of Hazardous Substances (8 CCR 339)
[present]

C.I. DIRECT BLACK 80   **8003-69-8**

HEALTH AND SAFETY LISTS

NTP Chemical Status Reports - Testing Status and NTIS Number
Project leader assigned/study in design

CIMETIDINE   **51481-61-9**

HEALTH AND SAFETY LISTS

IARC - Group 3 (not classifiable)
[present]

1,8-CINEOL   **470-82-6**

HEALTH AND SAFETY LISTS

NTP Chemical Status Reports - Testing Status and NTIS Number
Prechronic studies for which toxicity technical reports were not prepared

CINERIN I   **25402-06-6**

STATE LISTS

Massachusetts Right To Know List
[present]

CINNAMALDEHYDE   **104-55-2**

HEALTH AND SAFETY LISTS

NIOSH - Selected LD50s and LC50s
Oral, rat: LD50 = 2220 mg/kg

NTP Chemical Status Reports - Testing Status and NTIS Number
Approved for toxicology/carcinogenesis study;   Prechronic studies for which toxicity technical reports were not prepared

ENVIRONMENTAL LISTS

List of Pesticide Product Inert Ingredients
[present]

TSCA - Code of Federal Regulations Citations
40 CFR 712.30(x);   40 CFR 716.120(d)

TSCA - PAIR - Reporting List
Reporting Date: November 27, 1991

TSCA - Health and Safety Reporting List
Effective Date: September 30, 1991

INTERNATIONAL LISTS

Canada - WHMIS: Ingredient Disclosure
0.1% item 407 (563)

TRANS-CINNAMIC ACID   **140-10-3**

HEALTH AND SAFETY LISTS

NIOSH - Selected LD50s and LC50s
Oral, bird: LD50 = 100 mg/kg

CINNAMON BARK OIL   **8007-80-5**

HEALTH AND SAFETY LISTS

NIOSH - Selected LD50s and LC50s
Oral, rat: LD50 = 2800 mg/kg Skin, rabbit: LD50 = 320 mg/kg

CINNAMYL ALCOHOL   **104-54-1**

HEALTH AND SAFETY LISTS

NIOSH - Selected LD50s and LC50s
Oral, rat: LD50 = 2000 mg/kg

ENVIRONMENTAL LISTS

List of Pesticide Product Inert Ingredients
[present]

CINNAMYL ANTHRANILATE   **87-29-6**

HEALTH AND SAFETY LISTS

IARC - Group 3 (not classifiable)
[present]

NTP Chemical Status Reports - Testing Status and NTIS Number
Technical reports printed (PB81143141)

NTP Chemical Status Reports - Evidence of Carcinogenicity
male rat-positive;   female rat-negative;   male mice-positive;   female mice-positive

STATE LISTS

California - Air Bill 2588 Appendix A-II
known or potential carcinogen: 9/89

California - Prop. 65 - Cancer list
carcinogen - initial date 7/1/89

California - Prop. 65 - No Significant Risk Levels
no significant risk level = 200 ug/day
California - Directors List of Hazardous Substances (8 CCR 339)
[present]

CIPC (ISOPROPYL N-(3-CHLOROPHENYL) CARBA-   **101-21-3**
MATE)
HEALTH AND SAFETY LISTS
IARC - Group 3 (not classifiable)
[present]

CISPLATIN   **15663-27-1**
HEALTH AND SAFETY LISTS
IARC - Group 2A (limited human data)
[present] (Other relevant data, as given in Supplement 7, influenced the making of the overall evaluation)
NIOSH - Selected LD50s and LC50s
Oral, rat: LD50 = 25800 ug/kg
NTP Seventh Report - Suspect Carcinogens
suspect carcinogen
OSHA - Possible Select Carcinogens
[present]
INTERNATIONAL LISTS
Canada - WHMIS: Ingredient Disclosure
0.1% item 495 (629)
STATE LISTS
California - Air Bill 2588 Appendix A-II
known or potential carcinogen
California - Prop. 65 - Cancer list
carcinogen - initial date 10/1/88
California - Directors List of Hazardous Substances (8 CCR 339)
[present]
Florida Hazardous Substance List
[present]
Massachusetts Right To Know List
carcinogen; extraordinarily hazardous
Minnesota Hazardous Substance List
carcinogen
NJ Right to Know List (Total)
sn 0510
NJ Special Hazardous Substances
(carcinogen; mutagen)
Pennsylvania Right to Know List
special hazardous substance
Pennsylvania RTK - Special Hazardous Substances
[present]

CITRACONIC ANHYDRIDE   **616-02-4**
HEALTH AND SAFETY LISTS
NIOSH - Selected LD50s and LC50s
Oral, rat: LD50 = 2600 mg/kg Skin, guinea pig: LD50 = 1247 mg/kg

CITRAL   **5392-40-5**
HEALTH AND SAFETY LISTS
NFPA - Flash Points
flash point = 195 degrees F (91 degrees C)
NFPA - Hazard Identification Ratings
health-0; flammability-2; reactivity-0
NIOSH - Selected LD50s and LC50s
Oral, rat: LD50 = 4960 mg/kg
NTP Chemical Status Reports - Testing Status and NTIS Number
Approved for toxicology/carcinogenesis study; prechronic studies for which toxicity technical reports were not prepared

ENVIRONMENTAL LISTS
EPA - Master Testing List
[present]
List of Pesticide Product Inert Ingredients
[present]
INTERNATIONAL LISTS
Canada - WHMIS: Ingredient Disclosure
1% item 408 (564)

CITRIC ACID   **77-92-9**
HEALTH AND SAFETY LISTS
NIOSH - Selected LD50s and LC50s
Oral, rat: LD50 = 6730 mg/kg
ENVIRONMENTAL LISTS
List of Pesticide Product Inert Ingredients
[present]
INTERNATIONAL LISTS
Canada - WHMIS: Ingredient Disclosure
1% item 409 (80)

CITRIC ACID, TRIS(DIMETHYLAMINE) SALT   **52217-48-8**
ENVIRONMENTAL LISTS
List of Pesticide Product Inert Ingredients
[present]

CITRININ   **518-75-2**
HEALTH AND SAFETY LISTS
IARC - Group 3 (not classifiable)
[present]
STATE LISTS
California - Directors List of Hazardous Substances (8 CCR 339)
[present]

CITRONELLA OIL   **8000-29-1**
HEALTH AND SAFETY LISTS
NIOSH - Selected LD50s and LC50s
Oral, rat: LD50 = 7200 mg/kg Skin, rabbit: LD50 = 4700 mg/kg
ENVIRONMENTAL LISTS
List of Pesticide Product Inert Ingredients
[present]

CITRONELLEL   **RR-00827-5**
HEALTH AND SAFETY LISTS
NFPA - Flash Points
flash point = 165 degrees F (74 degrees C)
NFPA - Hazard Identification Ratings
health-0; flammability-2; reactivity-0

CITRONELLOL   **106-22-9**
HEALTH AND SAFETY LISTS
NFPA - Flash Points
flash point = 205 degrees F (96 degrees C)
NFPA - Hazard Identification Ratings
health-0; flammability-1; reactivity-0
NIOSH - Selected LD50s and LC50s
Oral, rat: LD50 = 3450 mg/kg Skin, rabbit: LD50 = 2650 mg/kg

CITRUS PULP, ORANGE   **68514-76-1**
ENVIRONMENTAL LISTS
List of Pesticide Product Inert Ingredients
[present]

CITRUS RED NO. 2   6358-53-8

HEALTH AND SAFETY LISTS
IARC - Group 2B (sufficient animal data)
[present] (Overall evaluation based only on evidence of carcinogenicity in monograph (8, 1975) or in Supplement 4)
OSHA - Possible Select Carcinogens
[present]

ENVIRONMENTAL LISTS
RCRA - Hazardous Constituents-Appendix VIII
hazardous constituent - no waste number

STATE LISTS
California - Air Bill 2588 Appendix A-II
known or potential carcinogen
California - Prop. 65 - Cancer list
carcinogen - initial date 10/1/89
California - Directors List of Hazardous Substances (8 CCR 339)
[present]
Florida Hazardous Substance List
[present]
Massachusetts Right To Know List
carcinogen; extraordinarily hazardous
Minnesota Hazardous Substance List
carcinogen
NJ Right to Know List (Total)
sn 0511
NJ Special Hazardous Substances
(carcinogen)
Pennsylvania Right to Know List
environmental hazard; special hazardous substance
Pennsylvania RTK - Special Hazardous Substances
[present]

CLEANING SOLVENTS, 140 (60) CLASS   RR-00103-6

HEALTH AND SAFETY LISTS
NFPA - Flash Points
flash point = 138.2 degrees F (59 degrees C) or higher
NFPA - Hazard Identification Ratings
health-0; flammability-2; reactivity-0

CLOFIBRATE   637-07-0

HEALTH AND SAFETY LISTS
IARC - Group 3 (not classifiable)
[present]

STATE LISTS
California - Directors List of Hazardous Substances (8 CCR 339)
[present]

CLOMIPHENE CITRATE   50-41-9

HEALTH AND SAFETY LISTS
IARC - Group 3 (not classifiable)
[present]
NIOSH - Selected LD50s and LC50s
Oral, rat: LD50 = 5750 mg/kg

STATE LISTS
California - Air Bill 2588 Appendix A-II
9/90
California - Prop. 65 - Developmental Toxicity
developmental toxicity - initial date 4/1/90

CLONITRALID   1420-04-8

HEALTH AND SAFETY LISTS
NTP Chemical Status Reports - Testing Status and NTIS Number
Technical reports printed (PB287124/AS)
NTP Chemical Status Reports - Evidence of Carcinogenicity
male rat-negative; female rat-equivocal; male mice-inadequate; female mice-negative

STATE LISTS
California - Directors List of Hazardous Substances (8 CCR 339)
[present]
Massachusetts Right To Know List
[present]
Pennsylvania Right to Know List
[present]

CLOPIDOL   2971-90-6

HEALTH AND SAFETY LISTS
ACGIH 1995 - Time Weighted Averages
10 mg/m3 TWA
NIOSH - Selected LD50s and LC50s
Oral, rat: LD50 = 18 gm/kg
OSHA - Vacated PELs - Time Weighted Averages
total dust: 15 mg/m3 TWA; respirable fraction: 5 mg/m3 TWA
OSHA - Final PELs - Time Weighted Averages
total dust: 15 mg/m3 TWA; respirable fraction: 5 mg/m3 TWA

INTERNATIONAL LISTS
Australian Exposure Standards - Time Weighted Averages
10 mg/m3 TWA
Canada - Alberta - 8 Hour Occupational Exposure Limit
10 mg/m3 TWA
Canada - Alberta - 15 Minute Occupational Exposure Limit
20 mg/m3 STEL
Canada - British Columbia - 8 Hour Exposure Limits
10 mg/m3 TWA
Canada - British Columbia - 15 Minute Exposure Limits
20 mg/m3 STEL
Canada - Ontario - OHSA - TWAEVs
10 mg/m3 TWAEV
Canada - Ontario - OHSA - STEVs
20 mg/m3 STEV
Canada - Quebec - Time-Weighted Average Exposure Values
10 mg/m3 TWAEV
Israel - Time Weighted Averages
10 mg/m3 TWA
Israel - Action Levels
5 mg/m3 AL
Mexico - Instruction No. 10 - TWAs
10 mg/m3 TWA
Mexico - Instruction No. 10 - STELs
20 mg/m3 STEL

STATE LISTS
California - Exposure Limits - PELs
total dust: 10 mg/m3 PEL; respirable fraction: 5 mg/m3 PEL
California - Directors List of Hazardous Substances (8 CCR 339)
[present]
Florida Hazardous Substance List
[present]
Massachusetts Right To Know List
[present]
Minnesota Hazardous Substance List
[present]
Pennsylvania Right to Know List
[present]

CLORAZEPATE DIPOTASSIUM   **57109-90-7**

STATE LISTS

California - Prop. 65 - Developmental Toxicity
developmental toxicity - initial date 10/1/92

NJ Special Hazardous Substances
(teratogen)

COAL BRIQUETTES, HOT   **RR-01400-6**

HEALTH AND SAFETY LISTS

U.S. DOT - Hazard Classes
Forbidden from transport by the DOT

COAL DUST   **RR-00011-3**

HEALTH AND SAFETY LISTS

ACGIH 1995 - Time Weighted Averages
*respirable fraction: 2 mg/m3 TWA (The value is for dust containing <
5% silica Any dust containing more than 5% silica should be evaluated
using the TWA of 0.1 mg/m3 for respirable quartz)*

INTERNATIONAL LISTS

Australian Exposure Standards - Time Weighted Averages
respirable dust containing < 5% quartz: 3 mg/m3 TWA

Canada - Alberta - 8 Hour Occupational Exposure Limit
(See requirements in Part 4)

Canada - British Columbia - 8 Hour Exposure Limits
respirable mass: 2 mg/m3 TWA

Canada - Ontario - OHSA - TWAEVs
*total dust: 4 mg/m3 TWAEV; respirable dust: 2 mg/m3 TWAEV
(listed as mineral dust)*

Canada - Quebec - Time-Weighted Average Exposure Values
*2 mg/m3 TWAEV (<5% crystalline silica); 0.1 mg/m3 TWAEV
(>5% crystalline silica)*

United Kingdom - Occupational Exposure Standards - TWAs
respirable dust: 2 mg/m3 TWA

Israel - Time Weighted Averages
*respirable fraction: 2 mg/m3 TWA (The value is for dust containing <
5% silica Any dust containing more than 5% silica should be evaluated
using the TWA of 0.1 mg/m3 for respirable quartz)*

Israel - Action Levels
respirable fraction: 1 mg/m3 AL

STATE LISTS

California - Exposure Limits - PELs
*< 5% quartz, repirable fraction: 2 mg/m3 PEL; > 5% quartz,
respirable fraction: 0.1 mg/m3 PEL*

California - Directors List of Hazardous Substances (8 CCR 339)
[present]

Florida Hazardous Substance List
[present]

Massachusetts Right To Know List
*[present] exempt when encapsulated or if particulates are not present
and cannot be substantially generated through use of the product.*

Minnesota Hazardous Substance List
[present]

Pennsylvania Right to Know List
[present]

COAL DUST, > 5% QUARTZ   **RR-00805-9**

HEALTH AND SAFETY LISTS

OSHA - Vacated PELs - Time Weighted Averages
0.1 mg/m3 TWA

OSHA - Final PELs - Time Weighted Averages
see Table Z-3

COAL GAS   **RR-00216-4**

HEALTH AND SAFETY LISTS

U.S. DOT - Substances From 49 CFR 172.101
regulated by DOT (UN1023)

U.S. DOT - Hazard Classes
DOT hazard class = 2.3

U.S. DOT - Substances Which Are Poisonous by Inhalation
gaseous hazardous material poisonous by inhalation (UN1023)

NFPA - Hazard Identification Ratings
health-2; flammability-4; reactivity-0

STATE LISTS

NJ Right to Know List (Total)
sn 2255

Pennsylvania Right to Know List
[present]

COAL GASIFICATION   **RR-00535-6**

HEALTH AND SAFETY LISTS

IARC - Group 1 (carcinogenic to humans)
[present]

OSHA - Select Carcinogens
[present]

STATE LISTS

Pennsylvania Right to Know List
special hazardous substance

Pennsylvania RTK - Special Hazardous Substances
[present]

COAL SOOT   **RR-00786-3**

STATE LISTS

Pennsylvania Right to Know List
special hazardous substance

Pennsylvania RTK - Special Hazardous Substances
[present]

COAL TAR   **8007-45-2**

SEE ALSO:
COAL TAR PITCHES

HEALTH AND SAFETY LISTS

U.S. DOT - Appendix B - Marine Pollutants
DOT regulated marine pollutant

IARC - Group 1 (carcinogenic to humans)
[present]

NIOSH - Health Standards - Exposure Limits
*cyclohexane-extractable fraction: 0.1 mg/m3 TWA (Listed under 'Coal
tar products')*

NIOSH - Health Standards - Health Effects and Precautions
*Lung and skin cancer (Periodic chest X-ray and pulmonary function
testing required) (Listed under 'Coal tar products')*

NIOSH - Health Standards - Carcinogenic Chemicals
potential human carcinogen (Listed under 'Coal tar products')

NTP Seventh Report - Known Carcinogens
known carcinogen (Listed under 'Soots, tars, and mineral oils')

OSHA - Select Carcinogens
[present]

ENVIRONMENTAL LISTS

List of Pesticide Product Inert Ingredients
[present]

RCRA - Hazardous Constituents-Appendix VIII
hazardous constituent - no waste number

INTERNATIONAL LISTS

Canada - WHMIS: Ingredient Disclosure
1% item 410 (937)

Canada - Ontario - OHSA - TWAEVs
as total benzene-soluble compounds: 0.2 mg/m3 TWAEV; listed as agents of variable composition (as sum of components assayed by chromatograhic procedure with reference to the bulk sample)
German (DFG) - Carcinogens
proven carcinogen

STATE LISTS

California - Air Bill 2588 Appendix A-II
known or potential carcinogen: 9/89
California - Directors List of Hazardous Substances (8 CCR 339)
[present] (includes products which could give rise to coal tar pitch volatiles)
Florida Hazardous Substance List
[present]
Massachusetts Right To Know List
carcinogen; extraordinarily hazardous
Minnesota Hazardous Substance List
carcinogen
NJ Right to Know List (Total)
sn 0519
Pennsylvania Right to Know List
environmental hazard; special hazardous substance
Pennsylvania RTK - Special Hazardous Substances
[present]

COAL TAR DISTILLATE   65996-92-1

HEALTH AND SAFETY LISTS

U.S. DOT - Substances From 49 CFR 172.101
regulated by DOT (UN1136)
U.S. DOT - Hazard Classes
DOT hazard class = 3

STATE LISTS

NJ Right to Know List (Total)
combustible liquid: sn 2256; flammable liquid: sn 2257

COAL TAR LIGHT OIL   65996-91-0

HEALTH AND SAFETY LISTS

NFPA - Flash Points
flash point < 80 degrees F (27 degrees C)
NFPA - Hazard Identification Ratings
health-2; flammability-3; reactivity-0

INTERNATIONAL LISTS

German (DFG) - Carcinogens
proven carcinogen

STATE LISTS

Florida Hazardous Substance List
[present]
Massachusetts Right To Know List
[present]
Pennsylvania Right to Know List
[present]

COAL TAR PITCHES   65996-93-2

HEALTH AND SAFETY LISTS

ACGIH 1995 - Time Weighted Averages
as benzene solubles: 0.2 mg/m3
ACGIH 1995 - Carcinogens
Benzene solubles: A1-confirmed human carcinogen
IARC - Group 1 (carcinogenic to humans)
[present]
NFPA - Flash Points
flash point = 405 degrees F (207 degrees C)

NFPA - Hazard Identification Ratings
health-0; flammability-1; reactivity-0
NIOSH 1990 - Pocket Guide - RELs
cyclohexane-extractable fraction: 0.1 mg/m3 TWA
NIOSH 1990 - Pocket Guide - IDLHs
700 mg/m3 IDLH (not considering carcinogenic effects)
NIOSH 1990 - Pocket Guide - Carcinogens
occupational carcinogen
NIOSH 1990 - Pocket Guide - Target organs
respiratory system, bladder, kidneys, skin
NIOSH - Health Standards - Exposure Limits
cyclohexane-extractable fraction: 0.1 mg/m3 TWA (Listed under 'Coal tar products')
NIOSH - Health Standards - Health Effects and Precautions
Lung and skin cancer (Periodic chest X-ray and pulmonary function testing required) (Listed under 'Coal tar products')
NIOSH - Health Standards - Carcinogenic Chemicals
potential human carcinogen (Listed under 'Coal tar products')
NTP Seventh Report - Known Carcinogens
known carcinogen (Listed under 'Soots, tars, and mineral oils')
OSHA - Vacated PELs - Time Weighted Averages
benzene soluble fraction: 0.2 mg/m3 TWA (anthracene, BaP, phenanthrene, acridine, chrysene, and pyrene)
OSHA - Final PELs - Time Weighted Averages
benzene soluble fraction: 0.2 mg/m3 TWA (includes anthracene, BaP, phenanthrene, acridine, chrysene, and pyrene)
OSHA - Select Carcinogens
[present]

ENVIRONMENTAL LISTS

ATSDR Priority List
Rank (of 275): 171

INTERNATIONAL LISTS

Australian Exposure Standards - Time Weighted Averages
as benzene solubles: 0.2 mg/m3 TWA
Australian Exposure Standards - Carcinogens
as benzene solubles: confirmed carcinogen
Canada - WHMIS: Ingredient Disclosure
0.1% item 411 (1721)
Canada - Alberta - 8 Hour Occupational Exposure Limit
as benzene solubles: 0.2 mg/m3 TWA
Canada - Alberta - 15 Minute Occupational Exposure Limit
as benzene solubles: 0.6 mg/m3 STEL
Canada - Alberta - Designated Substances
designated substance - requires code of practice
Canada - British Columbia - 8 Hour Exposure Limits
0.2 mg/m3 TWA
Canada - Quebec - Time-Weighted Average Exposure Values
0.2 mg/m3 TWAEV (substance of which the recirculation is prohibited)
Canada - Quebec - Carcinogens
C1 carcinogen: effect detected in humans
German (DFG) - Carcinogens
proven carcinogen
Israel - Time Weighted Averages
as benzene solubles: 0.2 mg/m3
Israel - Action Levels
as benzene solubles: 0.1 mg/m3 AL
Mexico - Instruction No. 10 - TWAs
0.2 mg/m3 TWA
Mexico - Instruction No. 10 - Carcinogens
potentially carcinogenic contaminant

STATE LISTS

California - Exposure Limits - PELs
0.2 mg/m3 PEL

California - Directors List of Hazardous Substances (8 CCR 339)
[present]

Massachusetts Right To Know List
carcinogen; extraordinarily hazardous

Minnesota Hazardous Substance List
carcinogen (includes coal tar pitch volatiles)

Pennsylvania Right to Know List
special hazardous substance

Pennsylvania RTK - Special Hazardous Substances
[present]

COAL TARS (DURING DESTRUCTIVE DISTILLA-   RR-00787-4
TION)

STATE LISTS

Pennsylvania Right to Know List
special hazardous substance

Pennsylvania RTK - Special Hazardous Substances
[present]

COBALT   7440-48-4

HEALTH AND SAFETY LISTS

ACGIH 1995 - Time Weighted Averages
elemental and inorganic compounds, as Co: 0.02 mg/m3 TWA

ACGIH 1995 - Carcinogens
elemental and inorganic compounds, as Co: A3-animal carcinogen

U.S. DOT - Substances From 49 CFR 172.101
regulated by DOT (UN1318)

U.S. DOT - Hazard Classes
DOT hazard class = 4.1

IARC - Group Unspecified
[present] (Listed under 'Cobalt and cobalt compounds')

NIOSH - Selected LD50s and LC50s
Oral, rat: LD50 = 6170 mg/kg

NIOSH 1990 - Pocket Guide - RELs
as Co: 0.05 mg/m3 TWA

NIOSH 1990 - Pocket Guide - IDLHs
as Co: 20 mg/m3 IDLH

NIOSH 1990 - Pocket Guide - Target organs
respiratory system, skin

NIOSH - Health Standards - Exposure Limits
insufficient evidence for exposure limit

NIOSH - Health Standards - Health Effects and Precautions
Dermatitis, potential for pulmonary fibrosis

OSHA - Vacated PELs - Time Weighted Averages
as Co: 0.05 mg/m3 TWA

OSHA - Final PELs - Time Weighted Averages
0.1 mg/m3 TWA

ENVIRONMENTAL LISTS

ATSDR Priority List
Rank (of 275): 136

CERCLA/SARA - Section 313 - Emission Reporting
form R reporting required for 1.0% de minimus concentration

RCRA - TSD Facilities Ground Water Monitoring
TM 6010 = 70 ug/L PQL; TM 7200 = 500 ug/L PQL; TM 7201 = 10 ug/L PQL (all species in the ground water that contain this element are included)

TSCA - Code of Federal Regulations Citations
40 CFR 716.120(a)

TSCA - Health and Safety Reporting List
Effective Date: June 1, 1987

INTERNATIONAL LISTS

Australian Exposure Standards - Time Weighted Averages
metal dust and fume, as Co: 0.05 mg/m3 TWA

Australian Exposure Standards - Skin Effects
metal dust and fume, as Co: sensitiser

Canada - WHMIS: Ingredient Disclosure
0.1% item 417 (566)

Canada - NPRI (National Pollutant Release Inventory)
[present]

Canada - Alberta - 8 Hour Occupational Exposure Limit
as Co: 0.05 mg/m3 TWA

Canada - Alberta - 15 Minute Occupational Exposure Limit
0.1 mg/m3 STEL

Canada - British Columbia - 8 Hour Exposure Limits
as Co: 0.02 mg/m3 TWA

Canada - British Columbia - Skin Notations
skin - potential for skin absorption

Canada - British Columbia - Carcinogens
carcinogen - 0.02 mg/m3 TWA

Canada - Ontario - OHSA - TWAEVs
metal, dust, and fume: 0.05 mg/m3 TWAEV

Canada - Ontario - OHSA - STEVs
0.1 mg/m3 STEV

Canada - Quebec - Time-Weighted Average Exposure Values
as Co: 0.05 mg/m3 TWAEV

German (DFG) - Skin/Sensitizers
danger of sensitization (skin or respiratory)

German (DFG) - Carcinogens
as inspirable dusts/aerosols: animal evidence of carcinogenicity

Israel - Time Weighted Averages
metal dust & fume, as Co: 0.05 mg/m3 TWA

Israel - Action Levels
metal dust & fumes as Co: 0.025 mg/m3 AL

Mexico - Instruction No. 10 - TWAs
0.1 mg/m3 TWA

Mexico - Wastewater - Organic Toxic Pollutants and Heavy Metals
Listed under [Heavy Metals]

STATE LISTS

California - Air Bill 2588 Appendix A-I
6/91

California - Prop. 65 - Cancer list
carcinogen - initial date 7/1/92

California - Exposure Limits - PELs
metal fume and dust, as Co: 0.05 mg/m3 PEL

California - Directors List of Hazardous Substances (8 CCR 339)
[present]

Florida Hazardous Substance List
[present] (includes metal, dust, and fume)

Massachusetts Right To Know List
[present]

Minnesota Hazardous Substance List
[present] as Co

NJ Right to Know List (Total)
sn 0520

Pennsylvania Right to Know List
environmental hazard (any compound of this substance is also an environmental hazard)

PROPOSED REGULATIONS

ACGIH 1995 - Proposed Biological Exposure Indices
Cobalt in urine: 15 ug/L, end of shift at end of workweek (B); Cobalt in blood: 1 ug/L, end of shift at end of workweek (B)

COBALT 55   13982-25-7

HEALTH AND SAFETY LISTS

U.S. DOT - Appendix A Table 2 - Radionuclides
final RQ = 10 curies (3.7E 11 Bq)

ENVIRONMENTAL LISTS
CERCLA/SARA List of Radionuclides (Appendix B) and Their Reportable Quantities
final RQ = 10 curies (3.7E 11 Bq)

COBALT 56   14093-03-9

HEALTH AND SAFETY LISTS
U.S. DOT - Appendix A Table 2 - Radionuclides
final RQ = 10 curies (3.7E 11 Bq)

ENVIRONMENTAL LISTS
CERCLA/SARA List of Radionuclides (Appendix B) and Their Reportable Quantities
final RQ = 10 curies (3.7E 11 Bq)

COBALT 57   13981-50-5

HEALTH AND SAFETY LISTS
U.S. DOT - Appendix A Table 2 - Radionuclides
final RQ = 100 curies (3.7E 12 Bq)

ENVIRONMENTAL LISTS
CERCLA/SARA List of Radionuclides (Appendix B) and Their Reportable Quantities
final RQ = 100 curies (3.7E 12 Bq)

COBALT 58   13981-38-9

HEALTH AND SAFETY LISTS
U.S. DOT - Appendix A Table 2 - Radionuclides
final RQ = 10 curies (3.7E 11 Bq)

ENVIRONMENTAL LISTS
CERCLA/SARA List of Radionuclides (Appendix B) and Their Reportable Quantities
final RQ = 10 curies (3.7E 11 Bq)

COBALT 58M   RR-00467-1

HEALTH AND SAFETY LISTS
U.S. DOT - Appendix A Table 2 - Radionuclides
final RQ = 1000 curies (3.7E 13 Bq)

ENVIRONMENTAL LISTS
CERCLA/SARA List of Radionuclides (Appendix B) and Their Reportable Quantities
final RQ = 1000 curies (3.7E 13 Bq)

COBALT 60   10198-40-0

HEALTH AND SAFETY LISTS
U.S. DOT - Appendix A Table 2 - Radionuclides
final RQ = 10 curies (3.7E 11 Bq)

ENVIRONMENTAL LISTS
ATSDR Priority List
Rank (of 275): 186
CERCLA/SARA List of Radionuclides (Appendix B) and Their Reportable Quantities
final RQ = 10 curies (3.7E 11 Bq)

COBALT 60M   RR-00466-0

HEALTH AND SAFETY LISTS
U.S. DOT - Appendix A Table 2 - Radionuclides
final RQ = 10 curies (3.7E 11 Bq)

ENVIRONMENTAL LISTS
CERCLA/SARA List of Radionuclides (Appendix B) and Their Reportable Quantities
final RQ = 10 curies (3.7E 11 Bq)

COBALT 61   13981-83-

HEALTH AND SAFETY LISTS
U.S. DOT - Appendix A Table 2 - Radionuclides
final RQ = 1000 curies (3.7E 13 Bq)

ENVIRONMENTAL LISTS
CERCLA/SARA List of Radionuclides (Appendix B) and Their Reportable Quantities
final RQ = 1000 curies (3.7E 13 Bq)

COBALT 62M   RR-00465-

HEALTH AND SAFETY LISTS
U.S. DOT - Appendix A Table 2 - Radionuclides
final RQ = 1000 curies (3.7E 13 Bq)

ENVIRONMENTAL LISTS
CERCLA/SARA List of Radionuclides (Appendix B) and Their Reportable Quantities
final RQ = 1000 curies (3.7E 13 Bq)

COBALT(II) ACETATE   71-48-
SEE ALSO:
COBALT
COBALT COMPOUNDS
COBALT

INTERNATIONAL LISTS
Canada - WHMIS: Ingredient Disclosure
1% item 412 (14)

COBALTATE (6-), [[[1,2-ETHANEDIYLBIS[NITRILO-   68025-39-
BIS(METHYLENE)]TETRAKIS[PHOSPHONATO](6-)-
N,N',O,O'',O''''',O''''''']-, PENTAAMMONIUM HYDRO-
GEN, (OC-6-21)-

ENVIRONMENTAL LISTS
TSCA - Code of Federal Regulations Citations
40 CFR 704.95

COBALTATE (6-), [[[1,2-ETHANEDIYLBIS[NITRILO-   67924-23-
BIS(METHYLENE)]TETRAKIS[PHOSPHONATO](8-)-
N,N',O,O'',O''''',O''''''']-, PENTAPOTASSIUM HYDRO-
GEN, (OC-6-21)-

ENVIRONMENTAL LISTS
TSCA - Code of Federal Regulations Citations
40 CFR 704.95

COBALTATE (6-), [[[1,2-ETHANEDIYLBIS[NITRILO-   67969-67-
BIS(METHYLENE)]TETRAKIS[PHOSPHONATO](8-)-
N,N',O,O',O'''',O''''''']-, PENTASODIUM HYDROGEN,
(OC-6-21)-

ENVIRONMENTAL LISTS
TSCA - Code of Federal Regulations Citations
40 CFR 704.95

COBALT CARBONATE   513-79-
SEE ALSO:
COBALT

HEALTH AND SAFETY LISTS
NIOSH - Selected LD50s and LC50s
Oral, rat: LD50 = 640 mg/kg

ENVIRONMENTAL LISTS
List of Pesticide Product Inert Ingredients
[present]

COBALT CARBONYL   10210-68-1
SEE ALSO:
COBALT

HEALTH AND SAFETY LISTS

ACGIH 1995 - Time Weighted Averages
as Co: 0.1 mg/m3 TWA

NIOSH - Selected LD50s and LC50s
Inhalation, mouse: LC50 = 27 mg/m3 2 hr Oral, rat: LD50 = 754 mg/kg

OSHA - Vacated PELs - Time Weighted Averages
as Co: 0.1 mg/m3 TWA

ENVIRONMENTAL LISTS

CERCLA/SARA - Section 302 Extremely Hazardous Substances and TPQs
TPQ = 10/10,000 pounds

INTERNATIONAL LISTS

Australian Exposure Standards - Time Weighted Averages
as Co: 0.1 mg/m3 TWA

Australian Exposure Standards - Skin Effects
as Co: sensitiser

Canada - WHMIS: Ingredient Disclosure
1% item 414 (565)

Canada - Alberta - 8 Hour Occupational Exposure Limit
0.1 mg/m3 TWA

Canada - Alberta - 15 Minute Occupational Exposure Limit
0.3 mg/m3 STEL

Canada - Ontario - OHSA - TWAEVs
as Co: 0.1 mg/m3 TWAEV

Canada - Quebec - Time-Weighted Average Exposure Values
0.1 mg/m3 TWAEV

Israel - Time Weighted Averages
as Co: 0.1 mg/m3 TWA

Israel - Action Levels
as Co: 0.05 mg/m3 AL

STATE LISTS

California - Directors List of Hazardous Substances (8 CCR 339)
[present]

Florida Hazardous Substance List
effective March 13, 1992

Massachusetts Right To Know List
extraordinarily hazardous

Minnesota Hazardous Substance List
[present] as Co

NJ Right to Know List (Total)
sn 0521

Pennsylvania Right to Know List
environmental hazard

PROPOSED REGULATIONS

CERCLA/SARA - Proposed Hazardous Substance Additions
proposed RQ = 1 pound (.454 kg)

CERCLA/SARA - 1989 Proposed RQ Adjustments
proposed RQ = 10 pounds (4.54 kg)

COBALT CARBONYLS   37264-96-3
HEALTH AND SAFETY LISTS

NIOSH - Selected LD50s and LC50s
Inhalation, rat: LC50 = 1400 mg/m3 8 hr

STATE LISTS

California - Exposure Limits - PELs
as Co: 0.1 mg/m3 PEL

Pennsylvania Right to Know List
[present]

COBALT CHROMATE   13455-25-9
STATE LISTS

Massachusetts Right To Know List
[present]

COBALT CHROMIUM ALLOY   11114-92-4
HEALTH AND SAFETY LISTS

OSHA - Select Carcinogens
[present]

STATE LISTS

Florida Hazardous Substance List
[present]

Massachusetts Right To Know List
carcinogen; extraordinarily hazardous

Pennsylvania Right to Know List
special hazardous substance

Pennsylvania RTK - Special Hazardous Substances
[present]

COBALT COMPOUNDS   RR-00107-0
HEALTH AND SAFETY LISTS

IARC - Group 2B (sufficient animal data)
[present]

OSHA - Possible Select Carcinogens
[present]

ENVIRONMENTAL LISTS

CERCLA/SARA - Section 313 - Emission Reporting
form R reporting required for 1.0% de minimus concentration

Clean Air Act (1990) - List of Hazardous Air Contaminants
[present] (includes any unique chemical substance that contains Cobalt as part of that chemical's infrastructure)

INTERNATIONAL LISTS

Canada - NPRI (National Pollutant Release Inventory)
[present]

Canada - British Columbia - 8 Hour Exposure Limits
as Co: 0.02 mg/m3 TWA

Canada - British Columbia - Carcinogens
carcinogen - 0.02 mg/m3 TWA

STATE LISTS

California - Air Bill 2588 Appendix A-I
6/91

NJ Right to Know List (Total)
sn 3135

COBALT DISULFIDE   12013-10-4
SEE ALSO:
COBALT

INTERNATIONAL LISTS

Canada - WHMIS: Ingredient Disclosure
1% item 416 (788)

COBALT, ((2,2'-(1,2-ETHANEDIYLBIS(NI-TRILOMETHYLIDYNE))BIS(6-FLUOROPHENOLATO))(2)-   62207-76-5
SEE ALSO:
COBALT

ENVIRONMENTAL LISTS

CERCLA/SARA - Section 302 Extremely Hazardous Substances and TPQs
TPQ = 100/10,000 pounds

INTERNATIONAL LISTS

Canada - WHMIS: Ingredient Disclosure
1% item 418 (858)

STATE LISTS

Florida Hazardous Substance List
effective March 13, 1992

Massachusetts Right To Know List
extraordinarily hazardous

Pennsylvania Right to Know List
environmental hazard

PROPOSED REGULATIONS

CERCLA/SARA - Proposed Hazardous Substance Additions
proposed RQ = 1 pound (.454 kg)

CERCLA/SARA - 1989 Proposed RQ Adjustments
proposed RQ = 100 pounds (45.4 kg)

COBALT(II) FLUOBORATE 26490-63-1
SEE ALSO:
COBALT

INTERNATIONAL LISTS

Canada - WHMIS: Ingredient Disclosure
1% item 419 (887)

COBALT(II) FLUORIDE 10026-17-2
SEE ALSO:
COBALT

INTERNATIONAL LISTS

Canada - WHMIS: Ingredient Disclosure
1% item 420 (904)

COBALT HYDROCARBONYL 16842-03-8
SEE ALSO:
COBALT

HEALTH AND SAFETY LISTS

ACGIH 1995 - Time Weighted Averages
as Co: 0.1 mg/m3 TWA

NIOSH - Selected LD50s and LC50s
Inhalation, rat: LC50 = 165 mg/m3 30 mn

OSHA - Vacated PELs - Time Weighted Averages
as Co: 0.1 mg/m3 TWA

INTERNATIONAL LISTS

Australian Exposure Standards - Time Weighted Averages
as Co: 0.1 mg/m3 TWA

Australian Exposure Standards - Skin Effects
as Co: sensitiser

Canada - WHMIS: Ingredient Disclosure
1% item 421 (567)

Canada - Alberta - 8 Hour Occupational Exposure Limit
0.1 mg/m3 TWA

Canada - Alberta - 15 Minute Occupational Exposure Limit
0.3 mg/m3 STEL

Canada - Ontario - OHSA - TWAEVs
as Co: 0.1 mg/m3 TWAEV

Canada - Quebec - Time-Weighted Average Exposure Values
0.1 mg/m3 TWAEV

Israel - Time Weighted Averages
as Co: 0.1 mg/m3 TWA

Israel - Action Levels
as Co: 0.05 mg/m3 AL

STATE LISTS

California - Exposure Limits - PELs
as Co: 0.1 mg/m3 PEL

California - Directors List of Hazardous Substances (8 CCR 339)
[present]

Florida Hazardous Substance List
effective March 13, 1992

Massachusetts Right To Know List
[present]

Minnesota Hazardous Substance List
[present] as Co

Pennsylvania Right to Know List
[present]

COBALT HYDROXIDE 21041-93-0
ENVIRONMENTAL LISTS

List of Pesticide Product Inert Ingredients
[present]

COBALT(II) MOLYBDATE 13762-14-6
SEE ALSO:
MOLYBDENUM
COBALT

INTERNATIONAL LISTS

Canada - WHMIS: Ingredient Disclosure
1% item 422 (1163)

COBALT NAPHTHENATE 61789-51-3
SEE ALSO:
COBALT COMPOUNDS
COBALT

HEALTH AND SAFETY LISTS

U.S. DOT - Substances From 49 CFR 172.101
regulated by DOT (UN2001)

U.S. DOT - Hazard Classes
DOT hazard class = 4.1

NFPA - Flash Points
flash point = 121 degrees F (49 degrees C)

NFPA - Hazard Identification Ratings
health-1; flammability-2; reactivity-0

ENVIRONMENTAL LISTS

List of Pesticide Product Inert Ingredients
[present]

TSCA - Code of Federal Regulations Citations
40 CFR 712.30(h); 40 CFR 716.120(a)

TSCA - PAIR - Reporting List
Reporting Date: September 20, 1983

TSCA - Health and Safety Reporting List
Effective Date: July 1, 1983

STATE LISTS

NJ Right to Know List (Total)
sn 0523

COBALT NEODECANOATE 27253-31-2
SEE ALSO:
COBALT
COBALT
COBALT COMPOUNDS

ENVIRONMENTAL LISTS

List of Pesticide Product Inert Ingredients
[present]

COBALT NITRATE 10026-22-9
SEE ALSO:
COBALT
COBALT
COBALT COMPOUNDS

HEALTH AND SAFETY LISTS

NIOSH - Selected LD50s and LC50s
Oral, rat: LD50 = 691 mg/kg

COBALT(II) NITRATE   10141-05-6
SEE ALSO:
  COBALT
  COBALT
  COBALT COMPOUNDS

HEALTH AND SAFETY LISTS
  NIOSH - Selected LD50s and LC50s
    Oral, rat: LD50 = 434 mg/kg

INTERNATIONAL LISTS
  Canada - WHMIS: Ingredient Disclosure
    1% item 423 (1201)

STATE LISTS
  Florida Hazardous Substance List
    [present]
  Pennsylvania Right to Know List
    [present]

COBALT OCTOATE   13586-82-8
ENVIRONMENTAL LISTS
  List of Pesticide Product Inert Ingredients
    [present]

COBALTOUS BROMIDE   7789-43-7
SEE ALSO:
  COBALT

HEALTH AND SAFETY LISTS
  U.S. DOT - Appendix A Table 1 - Hazardous Substances
    final RQ = 1000 pounds (454 kg)
  NIOSH - Selected LD50s and LC50s
    Oral, rat: LD50 = 406 mg/kg

ENVIRONMENTAL LISTS
  CERCLA/SARA - Hazardous Substances and their Reportable Quantities
    final RQ = 1000 pounds (454 kg)
  Clean Water Act - Hazardous Substances
    [present]

INTERNATIONAL LISTS
  Canada - WHMIS: Ingredient Disclosure
    1% item 413 (333)

STATE LISTS
  California - Directors List of Hazardous Substances (8 CCR 339)
    [present]
  Massachusetts Right To Know List
    [present]
  NJ Right to Know List (Total)
    sn 0524
  Pennsylvania Right to Know List
    environmental hazard

COBALTOUS CHLORIDE   7646-79-9
SEE ALSO:
  COBALT

HEALTH AND SAFETY LISTS
  NIOSH - Selected LD50s and LC50s
    Oral, rat: LD50 = 80 mg/kg

INTERNATIONAL LISTS
  Canada - WHMIS: Ingredient Disclosure
    0.1% item 415 (488)

COBALTOUS FORMATE   544-18-3
SEE ALSO:
  COBALT

HEALTH AND SAFETY LISTS
  U.S. DOT - Appendix A Table 1 - Hazardous Substances
    final RQ = 1000 pounds (454 kg)

ENVIRONMENTAL LISTS
  CERCLA/SARA - Hazardous Substances and their Reportable Quantities
    final RQ = 1000 pounds (454 kg)
  Clean Water Act - Hazardous Substances
    [present]

STATE LISTS
  California - Directors List of Hazardous Substances (8 CCR 339)
    [present]
  Massachusetts Right To Know List
    [present]
  NJ Right to Know List (Total)
    sn 0525
  Pennsylvania Right to Know List
    environmental hazard

COBALTOUS SULFAMATE   14017-41-5
SEE ALSO:
  COBALT

HEALTH AND SAFETY LISTS
  U.S. DOT - Appendix A Table 1 - Hazardous Substances
    final RQ = 1000 pounds (454 kg)

ENVIRONMENTAL LISTS
  CERCLA/SARA - Hazardous Substances and their Reportable Quantities
    final RQ = 1000 pounds (454 kg)
  Clean Water Act - Hazardous Substances
    [present]

STATE LISTS
  California - Directors List of Hazardous Substances (8 CCR 339)
    [present]
  Massachusetts Right To Know List
    [present]
  NJ Right to Know List (Total)
    sn 2261
  Pennsylvania Right to Know List
    environmental hazard

COBALT(II) OXIDE   1307-96-6
SEE ALSO:
  COBALT

HEALTH AND SAFETY LISTS
  IARC - Group Unspecified
    [present] (Listed under 'Cobalt and cobalt compounds')
  NIOSH - Selected LD50s and LC50s
    Oral, rat: LD50 = 202 mg/kg

INTERNATIONAL LISTS
  Canada - WHMIS: Ingredient Disclosure
    1% item 424 (1306)

STATE LISTS
  California - Prop. 65 - Cancer list
    carcinogen - initial date 7/1/92

COBALT(II) PHOSPHATE   13455-36-2
  SEE ALSO:
    COBALT

  INTERNATIONAL LISTS
    Canada - WHMIS: Ingredient Disclosure
    1% item 425 (1391)

COBALT PROPIONATE   1560-69-6

  ENVIRONMENTAL LISTS
    List of Pesticide Product Inert Ingredients
    [present]

COBALT RESINATE, PRECIPITATED   RR-00219-7

  STATE LISTS
    NJ Right to Know List (Total)
    sn 2262

COBALT SULFATE   10124-43-3
  SEE ALSO:
    COBALT

  HEALTH AND SAFETY LISTS
    NIOSH - Selected LD50s and LC50s
    Oral, rat: LD50 = 424 mg/kg

  ENVIRONMENTAL LISTS
    List of Pesticide Product Inert Ingredients
    [present]
    TSCA - Code of Federal Regulations Citations
    40 CFR 712.30(w)
    TSCA - PAIR - Reporting List
    Reporting Date: July 13, 1988

  INTERNATIONAL LISTS
    Canada - WHMIS: Ingredient Disclosure
    1% item 426 (1517)

COBALT(II) SULFATE (1:1), HEPTAHYDRATE   10026-24-1
  SEE ALSO:
    COBALT

  HEALTH AND SAFETY LISTS
    NIOSH - Selected LD50s and LC50s
    Oral, rat: LD50 = 768 mg/kg
    NTP Chemical Status Reports - Testing Status and NTIS Number
    Technical reports printed (PB91185348); Two year studies: pathology quality assessment in progress

COBALT(II) SULFIDE   1317-42-6
  SEE ALSO:
    COBALT

  INTERNATIONAL LISTS
    Canada - WHMIS: Ingredient Disclosure
    1% item 427 (1546)

COBALT TALLATE   61789-52-4
  SEE ALSO:
    COBALT COMPOUNDS
    COBALT

  ENVIRONMENTAL LISTS
    List of Pesticide Product Inert Ingredients
    [present]

COCAINE   50-36-2

  STATE LISTS
    California - Prop. 65 - Developmental Toxicity
    developmental toxicity - initial date 7/1/89
    California - Prop. 65 - Reproductive - Female
    female reproductive toxicity - initial date 7/1/89

COCOA   8002-31-1

  ENVIRONMENTAL LISTS
    List of Pesticide Product Inert Ingredients
    [present]

COCO ACID TRIAMINE CONDENSATE, POLYCAR-BOXYLIC ACID SALTS   RR-01226-0

  ENVIRONMENTAL LISTS
    TSCA - Chemicals with Significant New Use Rules
    PMN number: P-92-446

COCO ALKYLDIMETHYLAMINES   61788-93-0

  ENVIRONMENTAL LISTS
    List of Pesticide Product Inert Ingredients
    [present]

(COCO ALKYL)DIMETHYL BETAINES   68424-94-2

  ENVIRONMENTAL LISTS
    List of Pesticide Product Inert Ingredients
    [present]

COCO ALKYLTRIMETHYL QUATERNARY AMMO-NIUM CHLORIDES   61789-18-2

  ENVIRONMENTAL LISTS
    List of Pesticide Product Inert Ingredients
    [present]

COCOAMIDO(PROPYLBETAINE)   61789-40-0

  ENVIRONMENTAL LISTS
    List of Pesticide Product Inert Ingredients
    [present]

COCOAMINO PROPIONIC ACID   RR-01055-9

  ENVIRONMENTAL LISTS
    List of Pesticide Product Inert Ingredients
    [present]

COCOAMPHOCARBOXYGLYCINATE, DISODIUM SALT   RR-01056-0

  ENVIRONMENTAL LISTS
    List of Pesticide Product Inert Ingredients
    [present]

COCOMONOETHANOLAMIDE   142-78-9

  ENVIRONMENTAL LISTS
    List of Pesticide Product Inert Ingredients
    [present]

COCONUT DIETHANOLAMIDE   68603-42-9

  HEALTH AND SAFETY LISTS
    NTP Chemical Status Reports - Testing Status and NTIS Number
    Two year studies in progress

  ENVIRONMENTAL LISTS
    List of Pesticide Product Inert Ingredients
    [present]

COCONUT FATTY ACIDS, POTASSIUM SALT   61789-30-8

  ENVIRONMENTAL LISTS
    List of Pesticide Product Inert Ingredients
    [present]

COCONUT MONOETHANOLAMIDE   68140-00-1
ENVIRONMENTAL LISTS
List of Pesticide Product Inert Ingredients
[present]

COCONUT OIL ACID, DIETHANOLAMINE SALT   61790-63-4
ENVIRONMENTAL LISTS
List of Pesticide Product Inert Ingredients
[present]

COCONUT OIL FATTY ACIDS   61788-47-4
ENVIRONMENTAL LISTS
List of Pesticide Product Inert Ingredients
[present]

COCONUT OIL, POLYMER WITH ISOPH-THALIC ACID, TRIMELITTIC ANHYDRIDE AND TRIMETHYLOLPROPANE   RR-01057-1
ENVIRONMENTAL LISTS
List of Pesticide Product Inert Ingredients
[present]

COCO SHELL FLOUR   RR-01054-8
ENVIRONMENTAL LISTS
List of Pesticide Product Inert Ingredients
[present]

CODEINE   76-57-3
HEALTH AND SAFETY LISTS
NTP Chemical Status Reports - Testing Status and NTIS Number
Prechronic studies for which toxicity technical reports were not prepared; Two year studies: pathology working group in progress

COD-LIVER OIL   8001-69-2
HEALTH AND SAFETY LISTS
NFPA - Flash Points
flash point = 412 degrees F (211 degrees C)
NFPA - Hazard Identification Ratings
health-0; flammability-1; reactivity-0
ENVIRONMENTAL LISTS
List of Pesticide Product Inert Ingredients
[present]
STATE LISTS
Pennsylvania Right to Know List
[present]

COD OIL, SULFONATED   97553-00-9
ENVIRONMENTAL LISTS
List of Pesticide Product Inert Ingredients
[present]

COFFEE GROUNDS   RR-01058-2
HEALTH AND SAFETY LISTS
IARC - Group 2B (sufficient animal data)
[present]
OSHA - Possible Select Carcinogens
[present]
ENVIRONMENTAL LISTS
List of Pesticide Product Inert Ingredients
[present]

COKE (COAL)   65996-77-2
HEALTH AND SAFETY LISTS
U.S. DOT - Hazard Classes
Forbidden from transport by the DOT

COKE OVEN EMISSIONS   RR-00528-7
HEALTH AND SAFETY LISTS
U.S. DOT - Appendix A Table 1 - Hazardous Substances
final RQ = 1 pound (0.454 kg)
NFPA - Hazard Identification Ratings
health-2; flammability-4; reactivity-0
NIOSH - Health Standards - Exposure Limits
0.5-0.7 mg/m3 (total particulates) as screening level
NIOSH - Health Standards - Health Effects and Precautions
Lung cancer, bladder cancer (Periodic chest X-ray required; minimize exposure to emissions)
NIOSH - Health Standards - Carcinogenic Chemicals
potential human carcinogen
NTP Seventh Report - Known Carcinogens
known carcinogen
OSHA - 29 CFR 1910 Specifically Regulated Chemicals
150 ug/m3 TWA PEL; Cancer hazard (see 29 CFR 1910.1029)
OSHA - Select Carcinogens
[present]
ENVIRONMENTAL LISTS
CERCLA/SARA - Hazardous Substances and their Reportable Quantities
final RQ = 1 pound (0.454 kg)
Clean Air Act (1990) - List of Hazardous Air Contaminants
[present]
EPA - Carcinogen Hazard Ranking for RQ Adjustment
Hazard ranking = High
INTERNATIONAL LISTS
Canada - Ontario - OHSA - TWAEVs
0.15 mg/m3 TWAEV (designated substance regulation)
Canada - Ontario - OHSA - Designated Substances
0.15 mg/m3 TWAEV; See Ontario Reg. 840 for full information.
German (DFG) - Carcinogens
proven carcinogen
STATE LISTS
California - Air Bill 2588 Appendix A-I
known or potential carcinogen
California - Prop. 65 - Cancer list
carcinogen - initial date 2/27/87
California - Prop. 65 - No Significant Risk Levels
no significant risk level = 0.3 ug/day
California - Exposure Limits - PELs
0.15 mg/m3 PEL
Minnesota Hazardous Substance List
carcinogen
Pennsylvania Right to Know List
environmental hazard; special hazardous substance
Pennsylvania RTK - Special Hazardous Substances
[present]

COKE PRODUCTION   RR-00536-7
HEALTH AND SAFETY LISTS
IARC - Group 1 (carcinogenic to humans)
[present]
OSHA - Select Carcinogens
[present]

STATE LISTS

Pennsylvania Right to Know List
special hazardous substance

Pennsylvania RTK - Special Hazardous Substances
[present]

COLCHICINE   64-86-8

HEALTH AND SAFETY LISTS

NIOSH - Selected LD50s and LC50s
Oral, mouse: LD50 = 5886 ug/kg

ENVIRONMENTAL LISTS

CERCLA/SARA - Section 302 Extremely Hazardous Substances and TPQs
TPQ = 10/10,000 pounds

List of Pesticide Product Inert Ingredients
[present]

STATE LISTS

California - Prop. 65 - Developmental Toxicity
developmental toxicity initial date 10/1/92

California - Prop. 65 - Reproductive - Male
male reproductive toxicity - initial date 10/1/92

Florida Hazardous Substance List
effective March 13, 1992

Massachusetts Right To Know List
extraordinarily hazardous

NJ Right to Know List (Total)
sn 2263

Pennsylvania Right to Know List
environmental hazard

PROPOSED REGULATIONS

CERCLA/SARA - Proposed Hazardous Substance Additions
proposed RQ = 1 pound (.454 kg)

CERCLA/SARA - 1989 Proposed RQ Adjustments
proposed RQ = 10 pounds (4.54 kg)

COMBINED CHEMOTHERAPY FOR LYMPHOMAS   RR-00081-7

STATE LISTS

California - Prop. 65 - Cancer list
carcinogen - initial date 2/27/87

Minnesota Hazardous Substance List
carcinogen (includes MOPP)

COMBINED ORAL CONTRACEPTIVES   RR-00543-6

HEALTH AND SAFETY LISTS

IARC - Group 1 (carcinogenic to humans)
[present] (There is also conclusive evidence that these agents have a protective effect against cancers of the ovary and endometrium) (Listed under 'Oestrogen-progestin combinations')

OSHA - Select Carcinogens
[present]

STATE LISTS

California - Prop. 65 - Cancer list
carcinogen - initial date 10/1/89

California - Directors List of Hazardous Substances (8 CCR 339)
[present]

Minnesota Hazardous Substance List
carcinogen

Pennsylvania Right to Know List
[present]

COMBUSTIBLE LIQUID, N.O.S.   RR-00220-0

STATE LISTS

NJ Right to Know List (Total)
sn 2267

COMMERCIAL HEXANE   RR-00152-5

ENVIRONMENTAL LISTS

EPA - Master Testing List
[present]

TSCA - Chemical Test Rules
Testing required by: manufacturers; importers; processors (40 CFR 799.2155) (Commercial hexane (as regulated) must have between 40% and 55% n-Hexane and less than 10% Methylcyclopentane, by volume)

TSCA - Section 12(b) - Export Notification
export notification required - Section 4 Commercial hexane is obtained from crude oil, natural gas liquids, or petroleum refinery processing in accordance with ASTM D 1836, consists of 6-carbon alkanes or cycloalkanes, and contains at least 40% n-hexane and 5% MCP.

COMPRESSED OR LIQUIFIED GASES, N.O.S.   RR-00224-4

HEALTH AND SAFETY LISTS

U.S. DOT - Substances From 49 CFR 172.101
regulated by DOT (UN1953, UN1954, UN1955, UN1956)

U.S. DOT - Hazard Classes
toxic or flammable, toxic: DOT hazard class = 2.3; flammable: DOT hazard class = 2.1; nonflammable and non-toxic: DOT hazard class = 2.2

U.S. DOT - Substances Which Are Poisonous by Inhalation
gaseous hazardous material poisonous by inhalation (UN1953, UN1955)

STATE LISTS

NJ Right to Know List (Total)
sn 2271; sn 2272; sn 2273; sn 2274

CONDENSATION PRODUCT OF SORBITOL   RR-01059-3
EPICHLOROHYDRIN WITH THE OLEIC ACID DI-
AMIDE OF DIETHYLENETRIAMINE

ENVIRONMENTAL LISTS

List of Pesticide Product Inert Ingredients
[present]

CONJUGATED ESTROGENS   RR-00082-8

HEALTH AND SAFETY LISTS

IARC - Group Unspecified
[present] (Listed under 'Steroidal oestogens')

STATE LISTS

California - Prop. 65 - Cancer list
carcinogen - initial date 2/27/87

California - Prop. 65 - Developmental Toxicity
developmental toxicity - initial date 4/1/90

Minnesota Hazardous Substance List
carcinogen

Pennsylvania Right to Know List
special hazardous substance

Pennsylvania RTK - Special Hazardous Substances
[present]

CONJUGATED ESTROGENS   RR-01744-7

STATE LISTS

California - Air Bill 2588 Appendix A-II
known or potential carcinogen: 9/90

COPPER   7440-50-8

SEE ALSO:
COPPER COMPOUNDS, N.O.S.
F039-HAZARDOUS WASTES
COPPER

HEALTH AND SAFETY LISTS

ACGIH 1995 - Time Weighted Averages
fume: 0.2 mg/m3 TWA; dusts and mists, as Cu: 1 mg/m3 TWA

U.S. DOT - Appendix A Table 1 - Hazardous Substances
final RQ = 5000 pounds (2270 kg) (no reporting of releases of this hazardous substance is required if the diameter of the pieces of the solid metal released is equal to or exceeds 0.004 inches)

NIOSH 1990 - Pocket Guide - RELs
as Cu: 1 mg/m3 TWA (dusts and mists); 0.1 mg/m3 TWA (fume)

NIOSH 1990 - Pocket Guide - Target organs
dusts and mists: respiratory system, skin, liver, kidneys, (increased risk with Wilson's disease); fume: respiratory system, skin, eyes, (increased risk with Wilson's disease)

OSHA - Vacated PELs - Time Weighted Averages
fume, as Cu: 0.1 mg/m3 TWA; dusts and mists, as Cu: 1 mg/m3 TWA

OSHA - Final PELs - Time Weighted Averages
fume, as Cu: 0.1 mg/m3 TWA; dusts and mists, as Cu: 1 mg/m3 TWA

ENVIRONMENTAL LISTS

ATSDR Priority List
Rank (of 275): 098

CERCLA/SARA - Section 313 - Emission Reporting
form R reporting required for 1.0% de minimus concentration

CERCLA/SARA - Hazardous Substances and their Reportable Quantities
final RQ = 5000 pounds (2270 kg) (no reporting of releases of this hazardous substance is required if the diameter of the pieces of the solid metal released is equal to or exceeds 0.004 inches)

Clean Water Act - Priority Pollutants
[present]

Clean Water Act - Toxic Pollutants
[present] (Listed under 'Copper and compounds')

Safe Drinking Water Act - MCLGs
MCLG = 1.3 mg/L

Safe Drinking Water Act - SMCLs
SMCL = 1 mg/L

RCRA - Basis for Listing - Appendix VII
Included in waste stream: F039

RCRA - TSD Facilities Ground Water Monitoring
TM 6010 = 60 ug/L PQL; TM 7210 = 200 ug/L PQL (all species in the ground water that contain this element are included)

INTERNATIONAL LISTS

Australian Exposure Standards - Time Weighted Averages
dusts and mists, as Cu: 1 mg/m3 TWA; fume, as Cu: 0.2 mg/m3 TWA

Canada - WHMIS: Ingredient Disclosure
1% item 433 (578)

Canada - NPRI (National Pollutant Release Inventory)
[present]

Canada - CEPA Schedule III Part II - Restricted Substances (Ocean Dumping)
[present]

Canada - Drinking Water Quality - AOs
<= 1.0 mg/L AO

Canada - Alberta - 8 Hour Occupational Exposure Limit
Copper fume: 0.2 mg/m3 TWA; Copper dust and mist (as Cu): 1 mg/m3 TWA

Canada - Alberta - 15 Minute Occupational Exposure Limit
fume: 0.6 mg/m3 STEL; dusts and mists: 2 mg/m3

Canada - British Columbia - 8 Hour Exposure Limits
fume: 0.2 mg/m3 TWA; dusts and mists, as Cu: 1 mg/m3 TWA

Canada - British Columbia - 15 Minute Exposure Limits
dusts and mists, as Cu: 2 mg/m3 STEL

Canada - Ontario - OHSA - TWAEVs
fume, as Cu: 0.2 mg/m3 TWAEV; dusts and mists, as Cu: 1 mg/m3 TWAEV

Canada - Quebec - Time-Weighted Average Exposure Values
fume: 0.2 mg/m3 TWAEV; dusts and mists, as Cu: 1 mg/m3 TWAEV

United Kingdom - Occupational Exposure Standards - TWAs
fume: 0.2 ppm TWA; dusts and mists, as Cu: 1 mg/m3 TWA

United Kingdom - Occupational Exposure Standards - STELs
dusts and mists, as Cu: 2 mg/m3 STEL

German (DFG) - MAK Values
total dust: 1 mg/m3 MAK; fine dust (fume): 0.1 mg/m3 MAK

German (DFG) - Peak Limitations
2 x normal MAK (30 min. average value); don't exceed 4 times during shift

Israel - Time Weighted Averages
fume: 0.2 mg/m3 TWA; dusts and mists, as Cu: 1 mg/m3 TWA

Israel - Action Levels
fume: 0.1 mg/m3 AL; dusts and mists, as Cu: 0.5 mg/m3 AL

Mexico - Instruction No. 10 - TWAs
0.2 mg/m3 TWA

Mexico - Instruction No. 10 - STELs
2 mg/m3 STEL

Mexico - Wastewater - Organic Toxic Pollutants and Heavy Metals
Listed under [Heavy Metals]

Mexico - Drinking Water - Ecological Criteria
1.0 mg/l

STATE LISTS

California - Air Bill 2588 Appendix A-I
[present]

California - Exposure Limits - PELs
metal fume, as Cu: 0.1 mg/m3 PEL; salts, dusts and mists, as Cu: 1 mg/m3 PEL

California - Directors List of Hazardous Substances (8 CCR 339)
[present]

Florida Hazardous Substance List
[present] (includes fume, dust, and mist)

Massachusetts Right To Know List
[present]

Minnesota Hazardous Substance List
[present] as Cu

NJ Right to Know List (Total)
sn 0528

Pennsylvania Right to Know List
environmental hazard (any compound of this substance is also an environmental hazard)

COPPER 60   13982-06-4

HEALTH AND SAFETY LISTS

U.S. DOT - Appendix A Table 2 - Radionuclides
final RQ = 100 curies (3.7E 12 Bq)

ENVIRONMENTAL LISTS

CERCLA/SARA List of Radionuclides (Appendix B) and Their Reportable Quantities
final RQ = 100 curies (3.7E 12 Bq)

COPPER 61   15128-03-7

HEALTH AND SAFETY LISTS

U.S. DOT - Appendix A Table 2 - Radionuclides
final RQ = 100 curies (3.7E 12 Bq)

ENVIRONMENTAL LISTS
CERCLA/SARA List of Radionuclides (Appendix B) and Their Reportable Quantities
final RQ = 100 curies (3.7E 12 Bq)

COPPER 64   13981-25-4
HEALTH AND SAFETY LISTS
U.S. DOT - Appendix A Table 2 - Radionuclides
final RQ = 1000 curies (3.7E 13 Bq)

ENVIRONMENTAL LISTS
CERCLA/SARA List of Radionuclides (Appendix B) and Their Reportable Quantities
final RQ = 1000 curies (3.7E 13 Bq)

COPPER 67   15757-86-5
HEALTH AND SAFETY LISTS
U.S. DOT - Appendix A Table 2 - Radionuclides
final RQ = 100 curies (3.7E 12 Bq)

ENVIRONMENTAL LISTS
CERCLA/SARA List of Radionuclides (Appendix B) and Their Reportable Quantities
final RQ = 100 curies (3.7E 12 Bq)

COPPER ACETOARSENITE   12002-03-8
SEE ALSO:
ARSENIC

HEALTH AND SAFETY LISTS
U.S. DOT - Substances From 49 CFR 172.101
regulated by DOT (UN1585)
U.S. DOT - Hazard Classes
DOT hazard class = 6.1
U.S. DOT - Appendix B - Marine Pollutants
DOT regulated marine pollutant
U.S. DOT - Appendix A Table 1 - Hazardous Substances
final RQ = 1 pound (0.454 kg)
NIOSH - Selected LD50s and LC50s
Oral, rat: LD50 = 22 mg/kg Skin, rat: LD50 = 2400 mg/kg

ENVIRONMENTAL LISTS
CERCLA/SARA - Section 302 Extremely Hazardous Substances and TPQs
TPQ = 500/10,000 pounds
CERCLA/SARA - Hazardous Substances and their Reportable Quantities
final RQ = 1 pound (0.454 kg)
Clean Water Act - Hazardous Substances
[present]
EPA - Carcinogen Hazard Ranking for RQ Adjustment
Hazard ranking = High

STATE LISTS
Florida Hazardous Substance List
effective March 13, 1992
Massachusetts Right To Know List
extraordinarily hazardous
NJ Right to Know List (Total)
sn 0529
Pennsylvania Right to Know List
environmental hazard

COPPER ACETYLIDE   12540-13-5
HEALTH AND SAFETY LISTS
U.S. DOT - Hazard Classes
Forbidden from transport by the DOT

COPPER AMINE AZIDE   RR-01401-
HEALTH AND SAFETY LISTS
U.S. DOT - Hazard Classes
Forbidden from transport by the DOT

COPPER ARSENITE   10290-12-
HEALTH AND SAFETY LISTS
U.S. DOT - Substances From 49 CFR 172.101
regulated by DOT (UN1586)
U.S. DOT - Hazard Classes
DOT hazard class = 6.1
U.S. DOT - Appendix B - Marine Pollutants
DOT regulated marine pollutant

STATE LISTS
NJ Right to Know List (Total)
sn 0530

COPPER BASED PESTICIDES, N.O.S.   RR-00228-
HEALTH AND SAFETY LISTS
U.S. DOT - Substances From 49 CFR 172.101
regulated by DOT (UN2775, UN2776, UN3009, UN3010)
U.S. DOT - Hazard Classes
toxic or toxic, flammable: DOT hazard class = 6.1; flammable, toxic:
DOT hazard class = 3
STATE LISTS
NJ Right to Know List (Total)
sn 2275; sn 2276; sn 2277; sn 2278

COPPER(II) CARBONATE HYDROXIDE   12069-69-1
SEE ALSO:
COPPER

HEALTH AND SAFETY LISTS
NIOSH - Selected LD50s and LC50s
Oral, rat: LD50 = 159 mg/kg

INTERNATIONAL LISTS
Canada - WHMIS: Ingredient Disclosure
1% item 428 (985)

COPPER CHLORATE   26506-47-8
SEE ALSO:
COPPER

HEALTH AND SAFETY LISTS
U.S. DOT - Substances From 49 CFR 172.101
regulated by DOT (UN2721)
U.S. DOT - Hazard Classes
DOT hazard class = 5.1

STATE LISTS
NJ Right to Know List (Total)
sn 0531

COPPER CHLORIDE   1344-67-8
SEE ALSO:
COPPER

HEALTH AND SAFETY LISTS
U.S. DOT - Substances From 49 CFR 172.101
regulated by DOT (UN2802)
U.S. DOT - Hazard Classes
DOT hazard class = 8
U.S. DOT - Appendix B - Marine Pollutants
DOT regulated marine pollutant
NIOSH - Selected LD50s and LC50s
Oral, rat: LD50 = 140 mg/kg

INTERNATIONAL LISTS
  Canada - WHMIS: Ingredient Disclosure
  1% item 430 (490)

STATE LISTS
  Massachusetts Right To Know List
  [present]
  NJ Right to Know List (Total)
  sn 0532

COPPER(I) CHLORIDE   7758-89-6
  SEE ALSO:
  COPPER

  HEALTH AND SAFETY LISTS
  U.S. DOT - Appendix B - Marine Pollutants
  DOT regulated marine pollutant
  NIOSH - Selected LD50s and LC50s
  Oral, rat: LD50 = 140 mg/kg

  INTERNATIONAL LISTS
  Canada - WHMIS: Ingredient Disclosure
  1% item 429 (489)

COPPER CHROMATE   13548-42-0
  STATE LISTS
  Massachusetts Right To Know List
  [present]

COPPER COMPOUNDS, N.O.S.   RR-00595-8
  ENVIRONMENTAL LISTS
  CERCLA/SARA - Section 313 - Emission Reporting
  form R reporting required for 1.0% de minimus concentration
  Clean Water Act - Toxic Pollutants
  [present] (Listed under 'Copper and compounds')

  INTERNATIONAL LISTS
  Canada - WHMIS: Ingredient Disclosure
  1% item 431 (577)
  Canada - NPRI (National Pollutant Release Inventory)
  [present]
  Canada - CEPA Schedule III Part II - Restricted Substances (Ocean Dumping)
  [present]

  STATE LISTS
  California - Air Bill 2588 Appendix A-I
  9/89
  California - Directors List of Hazardous Substances (8 CCR 339)
  [present]
  NJ Right to Know List (Total)
  sn 3136

COPPER CYANIDE   544-92-3
  SEE ALSO:
  CYANIDE ANION
  COPPER

  HEALTH AND SAFETY LISTS
  U.S. DOT - Appendix B - Marine Pollutants
  DOT regulated severe marine pollutant
  U.S. DOT - Appendix A Table 1 - Hazardous Substances
  final RQ = 10 pounds (4.54 kg)

  ENVIRONMENTAL LISTS
  CERCLA/SARA - Hazardous Substances and their Reportable Quantities
  final RQ = 10 pounds (4.54 kg)
  RCRA - P Series Wastes
  waste number P029

RCRA - Hazardous Constituents-Appendix VIII
  waste number P029
  RCRA - Substances Banned From Land Disposal
  [present]

  STATE LISTS
  Massachusetts Right To Know List
  [present]
  Pennsylvania Right to Know List
  environmental hazard

COPPER CYANIDE (VAN)   39377-49-6
  HEALTH AND SAFETY LISTS
  U.S. DOT - Substances From 49 CFR 172.101
  regulated by DOT (UN1587)
  U.S. DOT - Hazard Classes
  DOT hazard class = 6.1

COPPER DIACETATE MONOHYDRATE   6046-93-1
  SEE ALSO:
  COPPER

  HEALTH AND SAFETY LISTS
  NIOSH - Selected LD50s and LC50s
  Oral, rat: LD50 = 710 mg/kg

COPPER(II) DIMETHYLDITHIOCARBAMATE   137-29-1
  SEE ALSO:
  COPPER

  INTERNATIONAL LISTS
  Canada - WHMIS: Ingredient Disclosure
  1% item 432 (739)

COPPER HYDROXIDE   20427-59-2
  ENVIRONMENTAL LISTS
  List of Pesticide Product Inert Ingredients
  [present]

COPPER(II) 8-HYDROXYQUINOLINATE   10380-28-6
  SEE ALSO:
  COPPER

  HEALTH AND SAFETY LISTS
  IARC - Group 3 (not classifiable)
  [present]

  INTERNATIONAL LISTS
  Canada - WHMIS: Ingredient Disclosure
  1% item 434 (1005)

COPPER, INORGANIC COMPOUNDS   RR-00561-8
  STATE LISTS
  NJ Right to Know List (Total)
  sn 2871

COPPER NAPHTHENATE   1338-02-9
  SEE ALSO:
  COPPER
  COPPER
  COPPER COMPOUNDS, N.O.S.

  ENVIRONMENTAL LISTS
  List of Pesticide Product Inert Ingredients
  [present]

  INTERNATIONAL LISTS
  Canada - WHMIS: Ingredient Disclosure
  1% item 435 (1182)

COPPER(II) NITRATE, TRIHYDRATE (1:2:3)   10031-43-3
SEE ALSO:
COPPER COMPOUNDS, N.O.S.
COPPER
COPPER

HEALTH AND SAFETY LISTS
NIOSH - Selected LD50s and LC50s
Oral, rat: LD50 = 940 mg/kg

COPPER OXALATE (CUC2O4)   814-91-5
STATE LISTS
NJ Right to Know List (Total)
sn 0548

COPPER(+1) OXIDE   1317-39-1
SEE ALSO:
COPPER

HEALTH AND SAFETY LISTS
NIOSH - Selected LD50s and LC50s
Oral, rat: LD50 = 470 mg/kg

ENVIRONMENTAL LISTS
List of Pesticide Product Inert Ingredients
[present]

INTERNATIONAL LISTS
Canada - WHMIS: Ingredient Disclosure
1% item 437 (1307)

COPPER, [29H, 31H-PHTHALOCYANINATO(2-)-N29,N30,N31,N32-, [[3-(METHYLETHOXY)-, PROPY-LAMINOSULFONYL DERIVATIVES   81457-65-0
ENVIRONMENTAL LISTS
List of Pesticide Product Inert Ingredients
[present]

COPPER SELENATE   15123-69-0
STATE LISTS
NJ Right to Know List (Total)
sn 2279

COPPER SELENITE   10214-40-1
STATE LISTS
NJ Right to Know List (Total)
sn 2280

COPPER SODIUM CYANIDE   14264-31-4
HEALTH AND SAFETY LISTS
U.S. DOT - Substances From 49 CFR 172.101
regulated by DOT (UN2316, UN2317)
U.S. DOT - Hazard Classes
DOT hazard class = 6.1
U.S. DOT - Appendix B - Marine Pollutants
DOT regulated severe marine pollutant
STATE LISTS
NJ Right to Know List (Total)
sn 2774

COPPER (II) SULFATE PENTAHYDRATE (1:1:5)   7758-99-8
SEE ALSO:
COPPER

HEALTH AND SAFETY LISTS
NIOSH - Selected LD50s and LC50s
Oral, rat: LD50 = 300 mg/kg
NTP Chemical Status Reports - Testing Status and NTIS Number
Technical reports printed

COPPER(II) SULFIDE   1317-40-4
SEE ALSO:
COPPER

INTERNATIONAL LISTS
Canada - WHMIS: Ingredient Disclosure
1% item 440 (1548)

COPPER(I) SULFIDE   22205-45-4
SEE ALSO:
COPPER

INTERNATIONAL LISTS
Canada - WHMIS: Ingredient Disclosure
1% item 439 (1547)

COPPER TETRAMINE NITRATE   RR-01402-8
HEALTH AND SAFETY LISTS
U.S. DOT - Hazard Classes
Forbidden from transport by the DOT

COPRA   8001-31-8
HEALTH AND SAFETY LISTS
U.S. DOT - Substances From 49 CFR 172.101
regulated by DOT (UN1363)
U.S. DOT - Hazard Classes
DOT hazard class = 4.2
NFPA - Flash Points
crude: flash point = 420 degrees F (216 degrees C); refined: 548 degrees F (287 degrees C)
NFPA - Hazard Identification Ratings
health-0; flammability-1; reactivity-0

ENVIRONMENTAL LISTS
List of Pesticide Product Inert Ingredients
[present]

CORK   61789-98-8
ENVIRONMENTAL LISTS
List of Pesticide Product Inert Ingredients
[present]

CORN   RR-01060-6
ENVIRONMENTAL LISTS
List of Pesticide Product Inert Ingredients
[present] (includes corn cobs and corn meal)

CORN OIL   8001-30-7
HEALTH AND SAFETY LISTS
NFPA - Flash Points
flash point = 490 degrees F (254 degrees C); cooking: 610 degrees F (321 degrees C)
NFPA - Hazard Identification Ratings
health-0; flammability-1; reactivity-0
NTP Chemical Status Reports - Testing Status and NTIS Number
Post peer review technical reports in progress

ENVIRONMENTAL LISTS
List of Pesticide Product Inert Ingredients
[present]
STATE LISTS
Pennsylvania Right to Know List
[present]

CORN SYRUP   8029-43-4
ENVIRONMENTAL LISTS
List of Pesticide Product Inert Ingredients
[present]

CORONENE 191-07-1
HEALTH AND SAFETY LISTS
IARC - Group 3 (not classifiable)
[present]

CORROSIVE LIQUIDS, N.O.S. RR-00233-5
HEALTH AND SAFETY LISTS
U.S. DOT - Substances From 49 CFR 172.101
regulated by DOT (UN1760, UN2920, UN2922, UN3093, UN3094)
U.S. DOT - Hazard Classes
DOT hazard class = 8
STATE LISTS
NJ Right to Know List (Total)
sn 2281; sn 2282; sn 2283

CORROSIVE SOLIDS, N.O.S. RR-00236-8
HEALTH AND SAFETY LISTS
U.S. DOT - Substances From 49 CFR 172.101
regulated by DOT (UN1759, UN2921, UN2923, UN3084, UN3095, UN3096)
U.S. DOT - Hazard Classes
DOT hazard class = 8
STATE LISTS
NJ Right to Know List (Total)
sn 2284; sn 2285; sn 2286

CORUNDUM 1302-74-5
HEALTH AND SAFETY LISTS
ACGIH 1995 - Time Weighted Averages
10 mg/m3 TWA (The value is for total dust containing no asbestos and <1% crystalline silica)
INTERNATIONAL LISTS
Canada - WHMIS: Ingredient Disclosure
1% item 441 (568)
Canada - Alberta - 8 Hour Occupational Exposure Limit
Al2O3, respirable mass: 5 mg/m3 TWA; total mass: 10 mg/m3 TWA
Canada - Quebec - Time-Weighted Average Exposure Values
10 ppm TWAEV
German (DFG) - MAK Values
fine dust: 6 mg/m3 MAK (exposure lasting one year) (Listed under 'Aluminum')
Israel - Time Weighted Averages
10 mg/m3 TWA (The value is for total dust containing no asbestos and < 1%crystalline silica)
Israel - Action Levels
5 mg/m3 AL

COTTON DUST (RAW) RR-00001-1
HEALTH AND SAFETY LISTS
ACGIH 1995 - Time Weighted Averages
0.2 mg/m3 TWA (Lint-free dust as measured by the vertical elutriator cotton-dust sampler)
U.S. DOT - Substances From 49 CFR 172.101
regulated by DOT (NA1365, UN1364, UN1365)
U.S. DOT - Hazard Classes
DOT hazard class = 9; wet or waste: DOT hazard class = 4.2
NIOSH 1990 - Pocket Guide - RELs
0.200 mg/m3 TWA
NIOSH 1990 - Pocket Guide - Target organs
respiratory system, CVS
NIOSH - Health Standards - Exposure Limits
lint-free cotton dust: 200 ug/m3
NIOSH - Health Standards - Health Effects and Precautions
Pulmonary disease (Pulmonary function testing required)

OSHA - Vacated PELs - Time Weighted Averages
1 mg/m3 TWA (See also 1910.1043 for limits to various sectors)
OSHA - Final PELs - Time Weighted Averages
1 mg/m3 TWA
ENVIRONMENTAL LISTS
List of Pesticide Product Inert Ingredients
[present]
INTERNATIONAL LISTS
Australian Exposure Standards - Time Weighted Averages
0.2 mg/m3 TWA
Canada - Alberta - 8 Hour Occupational Exposure Limit
0.2 mg/m3 TWA
Canada - Alberta - 15 Minute Occupational Exposure Limit
0.6 mg/m3 STEL
Canada - British Columbia - 8 Hour Exposure Limits
0.2 mg/m3 TWA
Canada - British Columbia - 15 Minute Exposure Limits
0.6 mg/m3 STEL
Canada - Ontario - OHSA - TWAEVs
0.2 mg/m3 TWAEV
Canada - Quebec - Time-Weighted Average Exposure Values
0.5 mg/m3 TWAEV
United Kingdom - Occupational Exposure Standards - TWAs
total dust less fly: 0.5 mg/m3 TWA; total inhalable dust: 10 mg/m3 TWA; respirable dust: 5 mg/m3 TWA
German (DFG) - MAK Values
total dust: 1.5 mg/m3 MAK
German (DFG) - Peak Limitations
in preparation
German (DFG) - Pregnancy
no risk to embryo/fetus if exposure limits adhered to
Israel - Time Weighted Averages
0.2 mg/m3 TWA (Lint-free dust as measured by the vertical elutriator cotton-dust sampler)
Israel - Action Levels
0.1 mg/m3 AL
Mexico - Instruction No. 10 - TWAs
0.2 mg/m3 TWA
Mexico - Instruction No. 10 - STELs
0.6 mg/m3 STEL
STATE LISTS
California - Exposure Limits - PELs
1 mg/m3 PEL; prevent eye and skin contact; see also section 5190 for respiratory specifications (applies to cotton waste processing operations of waste recycling)
California - Directors List of Hazardous Substances (8 CCR 339)
[present] (applicable to cotton fiber used as covered by General Industry Safety Order 5219)
Florida Hazardous Substance List
[present]
Massachusetts Right To Know List
[present]
Minnesota Hazardous Substance List
[present]
Pennsylvania Right to Know List
[present]

COTTONSEED MEAL 68424-10-2
ENVIRONMENTAL LISTS
List of Pesticide Product Inert Ingredients
[present]

COTTONSEED OIL   8001-29-4

HEALTH AND SAFETY LISTS

NFPA - Flash Points
refined: flash point = 486 degrees F (252 degrees C);   cooking: flash point = 610 degrees F (321 degrees C)

NFPA - Hazard Identification Ratings
health-0;   flammability-1;   reactivity-0

ENVIRONMENTAL LISTS

List of Pesticide Product Inert Ingredients
[present]

STATE LISTS

Pennsylvania Right to Know List
[present]

COUMACHLOR   81-82-3

HEALTH AND SAFETY LISTS

U.S. DOT - Appendix B - Marine Pollutants
DOT regulated marine pollutant

COUMAFURYL   117-52-2

STATE LISTS

Massachusetts Right To Know List
[present]

NJ Right to Know List (Total)
sn 0950

COUMAPHOS   56-72-4

HEALTH AND SAFETY LISTS

U.S. DOT - Appendix B - Marine Pollutants
DOT regulated severe marine pollutant

U.S. DOT - Appendix A Table 1 - Hazardous Substances
final RQ = 10 pounds (4.54 kg)

NIOSH - Selected LD50s and LC50s
Inhalation, rat: LC50 = 303 mg/m3 8 hr Oral, rat: LD50 = 13 mg/kg Skin, rat: LD50 = 860 mg/kg

NTP Chemical Status Reports - Testing Status and NTIS Number
Technical reports printed (PB290305/AS)

NTP Chemical Status Reports - Evidence of Carcinogenicity
male rat-negative;   female rat-negative;   male mice-negative;   female mice-negative

ENVIRONMENTAL LISTS

CERCLA/SARA - Section 302 Extremely Hazardous Substances and TPQs
TPQ = 100/10,000 pounds

CERCLA/SARA - Hazardous Substances and their Reportable Quantities
final RQ = 10 pounds (4.54 kg)

Clean Water Act - Hazardous Substances
[present]

STATE LISTS

California - Directors List of Hazardous Substances (8 CCR 339)
[present]

Florida Hazardous Substance List
effective March 13, 1992

Massachusetts Right To Know List
extraordinarily hazardous

NJ Right to Know List (Total)
sn 0536

Pennsylvania Right to Know List
environmental hazard

COUMARIN   91-64-

HEALTH AND SAFETY LISTS

IARC - Group 3 (not classifiable)
[present]

NIOSH - Selected LD50s and LC50s
Oral, rat: LD50 = 293 mg/kg

NTP Chemical Status Reports - Testing Status and NTIS Number
Prechronic studies for which toxicity technical reports were not prepared;   Technical reports printed

STATE LISTS

California - Directors List of Hazardous Substances (8 CCR 339)
[present]

COUMARIN DERIVATIVE PESTICIDES, N.O.S.   RR-00244-8

HEALTH AND SAFETY LISTS

U.S. DOT - Substances From 49 CFR 172.101
regulated by DOT (UN3024, UN3025, UN3026, UN3027)

U.S. DOT - Hazard Classes
toxic or toxic, flammable: DOT hazard class = 6.1;   flammable, toxic: DOT hazard class = 3

COUMARIN, 7-(DIETHYLAMINO)-4-METHYL-   91-44-1

HEALTH AND SAFETY LISTS

NIOSH - Selected LD50s and LC50s
Oral, rat: LD50 = 5 gm/kg

COUMARONE-INDENE RESIN   63393-89-5

ENVIRONMENTAL LISTS

List of Pesticide Product Inert Ingredients
[present]

COUMATETRALYL   5836-29-3

HEALTH AND SAFETY LISTS

NIOSH - Selected LD50s and LC50s
Oral, rat: LD50 = 16500 ug/kg

ENVIRONMENTAL LISTS

CERCLA/SARA - Section 302 Extremely Hazardous Substances and TPQs
TPQ = 500/10,000 pounds

STATE LISTS

Massachusetts Right To Know List
extraordinarily hazardous

Pennsylvania Right to Know List
environmental hazard

PROPOSED REGULATIONS

CERCLA/SARA - Proposed Hazardous Substance Additions
proposed RQ = 1 pound (.454 kg)

CERCLA/SARA - 1989 Proposed RQ Adjustments
proposed RQ = 100 pounds (45.4 kg)

CRACKED OATS AND WHEAT   RR-01061-7

ENVIRONMENTAL LISTS

List of Pesticide Product Inert Ingredients
[present]

CREOSOTE   8001-58-9

SEE ALSO:
K035-HAZARDOUS WASTES
K001-HAZARDOUS WASTES

HEALTH AND SAFETY LISTS

U.S. DOT - Appendix B - Marine Pollutants
DOT regulated marine pollutant

U.S. DOT - Appendix A Table 1 - Hazardous Substances
final RQ = 1 pound (0.454 kg)

NFPA - Flash Points
flash point = 165 degrees F (74 degrees C)

NFPA - Hazard Identification Ratings
health-2; flammability-2; reactivity-0

NIOSH - Selected LD50s and LC50s
Oral, rat: LD50 = 725 mg/kg

NIOSH - Health Standards - Exposure Limits
cyclohexane-extractable fraction: 0.1 mg/m3 TWA (Listed under 'Coal tar products')

NIOSH - Health Standards - Health Effects and Precautions
Lung and skin cancer (Periodic chest X-ray and pulmonary function testing required) (Listed under 'Coal tar products')

NIOSH - Health Standards - Carcinogenic Chemicals
potential human carcinogen (Listed under 'Coal tar products')

NTP Seventh Report - Known Carcinogens
known carcinogen (Listed under 'Soots, tars, and mineral oils')

OSHA - Select Carcinogens
[present]

ENVIRONMENTAL LISTS

ATSDR Priority List
Rank (of 275): 026

CERCLA/SARA - Section 313 - Emission Reporting
form R reporting required for 0.1% de minimus concentration

CERCLA/SARA - Hazardous Substances and their Reportable Quantities
final RQ = 1 pound (0.454 kg)

EPA - Carcinogen Hazard Ranking for RQ Adjustment
Hazard ranking = High

RCRA - U Series Wastes
waste number U051

RCRA - Hazardous Constituents-Appendix VIII
waste number U051

RCRA - Basis for Listing - Appendix VII
Included in waste streams: K001, K035

RCRA - Substances Banned From Land Disposal
[present]

STATE LISTS

California - Directors List of Hazardous Substances (8 CCR 339)
[present]

Florida Hazardous Substance List
[present]

Massachusetts Right To Know List
carcinogen; extraordinarily hazardous

NJ Right to Know List (Total)
sn 0517

NJ Special Hazardous Substances
(carcinogen)

Pennsylvania Right to Know List
environmental hazard; special hazardous substance

Pennsylvania RTK - Special Hazardous Substances
[present]

CREOSOTE IMPREGNATED WASTE MATERIALS   RR-01601-3

INTERNATIONAL LISTS

Canada - CEPA - Priority Substances List
estimated time for completion of assessment reports: 4 years

CREOSOTE OIL   61789-28-4

HEALTH AND SAFETY LISTS

IARC - Group 2A (limited human data)
[present]

OSHA - Possible Select Carcinogens
[present]

STATE LISTS

Massachusetts Right To Know List
[present]

CREOSOTES   RR-00091-9

HEALTH AND SAFETY LISTS

IARC - Group 2A (limited human data)
[present]

OSHA - Possible Select Carcinogens
[present]

STATE LISTS

California - Air Bill 2588 Appendix A-I
known or potential carcinogen

California - Prop. 65 - Cancer list
carcinogen - initial date 10/1/88

Minnesota Hazardous Substance List
carcinogen

CREOSOTE, WOOD   8021-39-4

HEALTH AND SAFETY LISTS

U.S. DOT - Appendix B - Marine Pollutants
DOT regulated marine pollutant

NTP Seventh Report - Known Carcinogens
known carcinogen (Listed under 'Soots, tars, and mineral oils')

OSHA - Select Carcinogens
[present]

STATE LISTS

Massachusetts Right To Know List
[present]

M-CRESIDINE   102-50-1

HEALTH AND SAFETY LISTS

IARC - Group 3 (not classifiable)
[present]

NTP Chemical Status Reports - Testing Status and NTIS Number
Technical reports printed (PB286188/AS)

NTP Chemical Status Reports - Evidence of Carcinogenicity
male rat-positive; female rat-positive; male mice-inadequate; female mice-negative

STATE LISTS

Massachusetts Right To Know List
carcinogen; extraordinarily hazardous

P-CRESIDINE   120-71-8

HEALTH AND SAFETY LISTS

IARC - Group 2B (sufficient animal data)
[present] (Overall evaluation based only on evidence of carcinogenicity in monograph (27, 1982) or in Supplement 4)

NIOSH - Selected LD50s and LC50s
Oral, rat: LD50 = 1450 mg/kg

NTP Chemical Status Reports - Testing Status and NTIS Number
Technical reports printed (PB295835/AS); prechronic studies for which toxicity technical reports were not prepared

NTP Chemical Status Reports - Evidence of Carcinogenicity
male rat-positive; female rat-positive; male mice-positive; female mice-positive

NTP Seventh Report - Suspect Carcinogens
suspect carcinogen

OSHA - Possible Select Carcinogens
[present]

ENVIRONMENTAL LISTS

CERCLA/SARA - Section 313 - Emission Reporting
form R reporting required for 0.1% de minimus concentration

INTERNATIONAL LISTS

Canada - WHMIS: Ingredient Disclosure
0.1% item 442 (569)

German (DFG) - Carcinogens
animal evidence of carcinogenicty

STATE LISTS

California - Air Bill 2588 Appendix A-I
known or potential carcinogen

California - Prop. 65 - Cancer list
carcinogen - initial date 1/1/88

California - Prop. 65 - No Significant Risk Levels
no significant risk level = 5 ug/day

California - Directors List of Hazardous Substances (8 CCR 339)
[present]

Florida Hazardous Substance List
[present]

Massachusetts Right To Know List
carcinogen; extraordinarily hazardous

Minnesota Hazardous Substance List
carcinogen

NJ Right to Know List (Total)
sn 1467

NJ Special Hazardous Substances
(carcinogen)

Pennsylvania Right to Know List
environmental hazard; special hazardous substance

Pennsylvania RTK - Special Hazardous Substances
[present]

O-CRESOL   **95-48-7**
SEE ALSO:
F004-HAZARDOUS WASTES
F039-HAZARDOUS WASTES

HEALTH AND SAFETY LISTS

U.S. DOT - Appendix A Table 1 - Hazardous Substances
final RQ = 1000 pounds (454 kg) (Listed under 'Cresols')

NFPA - Flash Points
flash point = 178 degrees F (81 degrees C)

NFPA - Hazard Identification Ratings
health-3; flammability-2; reactivity-0

NIOSH - Selected LD50s and LC50s
Inhalation, mouse: LC50 = 179 mg/m3 (2 hr) Oral, rat: LD50 = 121 mg/kg Skin, rat: LD50 = 620 mg/kg

NTP Chemical Status Reports - Testing Status and NTIS Number
Technical reports printed (PB92-174242); project leader assigned/ study in design

ENVIRONMENTAL LISTS

ATSDR Priority List
Rank (of 275): 266

CERCLA/SARA - Section 302 Extremely Hazardous Substances and TPQs
TPQ = 1000/10,000 pounds

CERCLA/SARA - Section 313 - Emission Reporting
form R reporting required for 1.0% de minimus concentration

CERCLA/SARA - Hazardous Substances and their Reportable Quantities
final RQ = 1000 pounds (454 kg) (Listed under 'Cresols')

Clean Air Act (1990) - List of Hazardous Air Contaminants
[present]

CAA - HON Rule - SOCMI Chemicals
compliance by April 24, 1995

CAA - HON Rule - Organic HAPs
[present]

Clean Water Act - Hazardous Substances
[present] (Listed under 'Cresol')

EPA - Master Testing List
[present]

List of Pesticide Product Inert Ingredients
[present]

RCRA - D Series - Maximum Concentration of Contaminants
waste number D023; regulatory level = 200.0 mg/L

RCRA - D Series - Chronic Toxicity Reference Levels
chronic toxicity reference level = 2.0 mg/L

RCRA - Basis for Listing - Appendix VII
Included in waste stream: F039

RCRA - TSD Facilities Ground Water Monitoring
TM 8270 = 10 ug/L PQL

RCRA - Universal Treatment Standards (LDR)
WW: 0.11 mg/l; NWW: 5.6 mg/kg

TSCA - Code of Federal Regulations Citations
40 CFR 712.30(d); 40 CFR 716.120(a); 40 CFR 799.1250

TSCA - PAIR - Reporting List
Reporting Date: November 19, 1982

TSCA - Health and Safety Reporting List
Effective Date: October 4, 1982

TSCA - Chemical Test Rules
Testing required by: manufacturers; processors (40 CFR 799.1250) (Listed under 'Cresols')

TSCA - Section 12(b) - Export Notification
export notification required - Section 4

INTERNATIONAL LISTS

Canada - WHMIS: Ingredient Disclosure
1% item 445 (572)

Canada - NPRI (National Pollutant Release Inventory)
[present]

STATE LISTS

California - Air Bill 2588 Appendix A-I
6/91

California - Directors List of Hazardous Substances (8 CCR 339)
[present] (Listed under 'Cresol (all isomers)')

Florida Hazardous Substance List
[present]

Massachusetts Right To Know List
extraordinarily hazardous

NJ Right to Know List (Total)
sn 1426

NJ Special Hazardous Substances
(corrosive; reactive - second degree)

Pennsylvania Right to Know List
environmental hazard

P-CRESOL   **106-44-5**
SEE ALSO:
F004-HAZARDOUS WASTES

HEALTH AND SAFETY LISTS

U.S. DOT - Appendix A Table 1 - Hazardous Substances
final RQ = 1000 pounds (454 kg) (Listed under 'Cresols')

NIOSH - Selected LD50s and LC50s
Oral, rat: LD50 = 207 mg/kg Skin, rat: LD50 = 750 mg/kg

NTP Chemical Status Reports - Testing Status and NTIS Number
Technical reports printed (PB92-174242); project leader assigned/ study in design

ENVIRONMENTAL LISTS

ATSDR Priority List
Rank (of 275): 229

CERCLA/SARA - Section 313 - Emission Reporting
form R reporting required for 1.0% de minimus concentration

CERCLA/SARA - Hazardous Substances and their Reportable Quantities
final RQ = 1000 pounds (454 kg) (Listed under 'Cresols')

Clean Air Act (1990) - List of Hazardous Air Contaminants
[present]

CAA - HON Rule - SOCMI Chemicals
compliance by April 24, 1995

CAA - HON Rule - Organic HAPs
[present]

Clean Water Act - Hazardous Substances
[present] (Listed under 'Cresol')

List of Pesticide Product Inert Ingredients
[present]

RCRA - D Series - Maximum Concentration of Contaminants
waste number D025; regulatory level = 200.0 mg/L

RCRA - D Series - Chronic Toxicity Reference Levels
chronic toxicity reference level = 2 mg/L

RCRA - TSD Facilities Ground Water Monitoring
TM 8270 = 10 ug/L PQL

RCRA - Universal Treatment Standards (LDR)
WW: 0.77 mg/l; NWW: 5.6 mg/kg; (difficult to distinguish from m-cresol)

TSCA - Code of Federal Regulations Citations
40 CFR 712.30(d); 40 CFR 716.120(a); 40 CFR 799.1250

TSCA - PAIR - Reporting List
Reporting Date: November 19, 1982

TSCA - Health and Safety Reporting List
Effective Date: October 4, 1982

TSCA - Chemical Test Rules
Testing required by: manufacturers; processors (40 CFR 799.1250) (Listed under 'Cresols')

TSCA - Section 12(b) - Export Notification
export notification required - Section 4

INTERNATIONAL LISTS

Canada - WHMIS: Ingredient Disclosure
1% item 446 (573)

Canada - NPRI (National Pollutant Release Inventory)
[present]

STATE LISTS

California - Air Bill 2588 Appendix A-I
6/91

California - Directors List of Hazardous Substances (8 CCR 339)
[present] (Listed under 'Cresol (all isomers)')

Florida Hazardous Substance List
[present]

Massachusetts Right To Know List
[present]

NJ Right to Know List (Total)
sn 1468

NJ Special Hazardous Substances
(corrosive)

Pennsylvania Right to Know List
environmental hazard

M-CRESOL   108-39-4
SEE ALSO:
F004-HAZARDOUS WASTES

HEALTH AND SAFETY LISTS

AIHA - Odor Threshold Values
geometric mean air odor threshold = 0.00060 ppm (detectable)

U.S. DOT - Appendix A Table 1 - Hazardous Substances
final RQ = 1000 pounds (454 kg) (Listed under 'Cresols')

NIOSH - Selected LD50s and LC50s
Oral, rat: LD50 = 242 mg/kg Skin, rat: LD50 = 1100 mg/kg

NTP Chemical Status Reports - Testing Status and NTIS Number
Technical reports printed (PB92-174242); project leader assigned/ study in design

ENVIRONMENTAL LISTS

CERCLA/SARA - Section 313 - Emission Reporting
form R reporting required for 1.0% de minimus concentration

CERCLA/SARA - Hazardous Substances and their Reportable Quantities
final RQ = 1000 pounds (454 kg) (Listed under 'Cresols')

Clean Air Act (1990) - List of Hazardous Air Contaminants
[present]

CAA - HON Rule - SOCMI Chemicals
compliance by April 24, 1995

CAA - HON Rule - Organic HAPs
[present]

Clean Water Act - Hazardous Substances
[present] (Listed under 'Cresol')

RCRA - D Series - Maximum Concentration of Contaminants
waste number D024; regulatory level = 200.0 mg/L

RCRA - D Series - Chronic Toxicity Reference Levels
chronic toxicity reference level = 2 mg/L

RCRA - TSD Facilities Ground Water Monitoring
TM 8270 = 10 ug/L PQL

RCRA - Universal Treatment Standards (LDR)
WW: 0.77 mg/l; NWW: 5.6 mg/kg; (difficult to distinguish from p-cresol)

TSCA - Code of Federal Regulations Citations
40 CFR 712.30(d); 40 CFR 716.120(a); 40 CFR 799.1250

TSCA - PAIR - Reporting List
Reporting Date: November 19, 1982

TSCA - Health and Safety Reporting List
Effective Date: October 4, 1982

TSCA - Chemical Test Rules
Testing required by: manufacturers; processors (40 CFR 799.1250) (Listed under 'Cresols')

TSCA - Section 12(b) - Export Notification
export notification required - Section 4

INTERNATIONAL LISTS

Canada - WHMIS: Ingredient Disclosure
1% item 444 (571)

Canada - NPRI (National Pollutant Release Inventory)
[present]

STATE LISTS

California - Air Bill 2588 Appendix A-I
6/91

California - Directors List of Hazardous Substances (8 CCR 339)
[present] (Listed under 'Cresol (all isomers)')

Florida Hazardous Substance List
[present]

Massachusetts Right To Know List
[present]

NJ Right to Know List (Total)
sn 1161

NJ Special Hazardous Substances
(corrosive)

Pennsylvania Right to Know List
environmental hazard

CRESOL   1319-77-3

SEE ALSO:
N,N-DIBUTYLTOLUENESULFONAMIDE
F004-HAZARDOUS WASTES
F039-HAZARDOUS WASTES

HEALTH AND SAFETY LISTS

ACGIH 1995 - Time Weighted Averages
5 ppm TWA;   22 mg/m3 TWA

ACGIH 1995 - Skin Designations
skin - potential for cutaneous absorption

U.S. DOT - Substances From 49 CFR 172.101
regulated by DOT (UN2076, UN2022)

U.S. DOT - Hazard Classes
DOT hazard class = 6.1

U.S. DOT - Appendix B - Marine Pollutants
DOT regulated marine pollutant

U.S. DOT - Appendix A Table 1 - Hazardous Substances
final RQ = 1000 pounds (454 kg)

NFPA - Flash Points
flash point = 187 degrees F (86 degrees C)

NFPA - Hazard Identification Ratings
health-3;   flammability-2;   reactivity-0

NIOSH - Selected LD50s and LC50s
Oral, rat: LD50 = 1454 mg/kg Skin, rabbit: LD50 = 2000 mg/kg

NIOSH 1990 - Pocket Guide - RELs
2.3 ppm TWA;   10 mg/m3 TWA

NIOSH 1990 - Pocket Guide - IDLHs
250 ppm IDLH

NIOSH 1990 - Pocket Guide - Target organs
CNS, respiratory system, liver, kidneys, skin, eyes

NIOSH - Health Standards - Exposure Limits
2.3 ppm TWA;   10 mg/m3 TWA

NIOSH - Health Standards - Health Effects and Precautions
Skin, liver, kidney, and pancreas effects (Prevent skin and eye contact; possible delayed effects)

NTP Chemical Status Reports - Testing Status and NTIS Number
Technical reports printed (PB92-174242)

OSHA - Vacated PELs - Time Weighted Averages
5 ppm TWA;   22 mg/m3 TWA

OSHA - Vacated PELs - Skin Designation
Prevent or reduce skin absorption

OSHA - Final PELs - Time Weighted Averages
5 ppm TWA;   22 mg/m3 TWA

OSHA - Final PELs - Skin Notations
prevent or reduce skin absorption

ENVIRONMENTAL LISTS

CERCLA/SARA - Section 313 - Emission Reporting
form R reporting required for 1.0% de minimus concentration

CERCLA/SARA - Hazardous Substances and their Reportable Quantities
final RQ = 1000 pounds (454 kg)

Clean Air Act (1990) - List of Hazardous Air Contaminants
[present]

CAA - HON Rule - SOCMI Chemicals
compliance by April 24, 1995

CAA - HON Rule - Organic HAPs
[present]

Clean Water Act - Hazardous Substances
[present]

List of Pesticide Product Inert Ingredients
[present]

RCRA - D Series - Maximum Concentration of Contaminants
waste number D026;   regulatory level = 200.0 mg/L

RCRA - U Series Wastes
waste number U052

RCRA - Hazardous Constituents-Appendix VIII
waste number U052

RCRA - Basis for Listing - Appendix VII
Included in waste stream: F004

RCRA - Substances Banned From Land Disposal
[present]

TSCA - Code of Federal Regulations Citations
40 CFR 712.30(d)

TSCA - PAIR - Reporting List
Reporting Date: November 19, 1982

INTERNATIONAL LISTS

Australian Exposure Standards - Time Weighted Averages
5 ppm TWA;   22 mg/m3 TWA

Australian Exposure Standards - Skin Effects
skin absorption

Canada - WHMIS: Ingredient Disclosure
1% item 443 (570)

Canada - NPRI (National Pollutant Release Inventory)
[present]

Canada - Alberta - 8 Hour Occupational Exposure Limit
5 ppm TWA;   22 mg/m3 TWA

Canada - Alberta - 15 Minute Occupational Exposure Limit
10 ppm STEL;   44 mg/m3 STEL

Canada - Alberta - Skin Designation
can be absorbed through the intact skin

Canada - British Columbia - 8 Hour Exposure Limits
5 ppm TWA;   22 mg/m3 TWA

Canada - British Columbia - Skin Notations
skin - potential for skin absorption

Canada - Ontario - OHSA - TWAEVs
5 ppm TWAEV;   22 mg/m3 TWAEV

Canada - Ontario - OHSA - Skin Notations
absorption through skin, eyes, or mucous membranes

Canada - Quebec - Time-Weighted Average Exposure Values
5 ppm TWAEV;   22 mg/m3 TWAEV

Canada - Quebec - Skin Designations
absorbed through the skin

United Kingdom - Occupational Exposure Standards - TWAs
5 ppm TWA;   22 mg/m3 TWA

United Kingdom - Occupational Exposure Standards - Notes
can be absorbed through skin

German (DFG) - Peak Limitations
2 x normal MAK (5 min momentary value);   don't exceed 8 times during shift

German (DFG) - Skin/Sensitizers
danger of cutaneous absorption

Israel - Time Weighted Averages
5 ppm TWA;   22 mg/m3 TWA

Israel - Action Levels
2.5 ppm AL;   11 mg/m3 AL

Mexico - Instruction No. 10 - TWAs
5 ppm TWA;   22 mg/m3 TWA

Mexico - Instruction No. 10 - Skin designation
skin - potential for cutaneous absorption

STATE LISTS

California - Air Bill 2588 Appendix A-I
[present]

California - Exposure Limits - PELs
5 ppm PEL;   22 mg/m3 PEL

California - Exposure Limits - Skin Notation
material may be absorbed through the skin, eyes or mucous membrane

California - Directors List of Hazardous Substances (8 CCR 339)
[present]
Florida Hazardous Substance List
[present]
Massachusetts Right To Know List
[present]
Minnesota Hazardous Substance List
skin
NJ Right to Know List (Total)
sn 0537
Pennsylvania Right to Know List
environmental hazard

M-CRESOL, POLYMER WITH FORMALDEHYDE   122436-67-3
AND SULFANILIC ACID
ENVIRONMENTAL LISTS
List of Pesticide Product Inert Ingredients
[present]

CRESYL DIPHENYL PHOSPHATE   10303-47-6
HEALTH AND SAFETY LISTS
U.S. DOT - Appendix B - Marine Pollutants
DOT regulated severe marine pollutant
NFPA - Flash Points
flash point = 450 degrees F (232 degrees C)
NFPA - Hazard Identification Ratings
health-0; flammability-1; reactivity-0

CRESYL GLYCIDYL ETHER   26447-14-3
SEE ALSO:
GLYCIDOL (OXIRANEMETHANOL) AND ITS DERIVATIVES
HEALTH AND SAFETY LISTS
NIOSH - Selected LD50s and LC50s
Inhalation, rat: LC50 = 282 mg/m3 8 hr Oral, rat: LD50 = 5140 mg/kg
ENVIRONMENTAL LISTS
EPA - Master Testing List
[present]
TSCA - Code of Federal Regulations Citations
40 CFR 712.30(d); 40 CFR 716.120(c)
TSCA - PAIR - Reporting List
Reporting Date: November 19, 1982
PROPOSED REGULATIONS
TSCA - Proposed Testing Rule for Glycidyl Ethers
member of Glycidyl subcategory IV-A

CRESYLIC ACID, POTASSIUM SALT   12002-51-6
ENVIRONMENTAL LISTS
List of Pesticide Product Inert Ingredients
[present]

CRESYLIC ACID SODIUM SALT   34689-46-8
HEALTH AND SAFETY LISTS
U.S. DOT - Appendix B - Marine Pollutants
DOT regulated marine pollutant

CRIMIDINE   535-89-7
HEALTH AND SAFETY LISTS
NIOSH - Selected LD50s and LC50s
Oral, rat: LD50 = 1250 ug/kg
ENVIRONMENTAL LISTS
CERCLA/SARA - Section 302 Extremely Hazardous Substances and TPQs
TPQ = 100/10,000 pounds

STATE LISTS
Florida Hazardous Substance List
effective March 13, 1992
Massachusetts Right To Know List
extraordinarily hazardous
NJ Right to Know List (Total)
sn 0351
Pennsylvania Right to Know List
environmental hazard
PROPOSED REGULATIONS
CERCLA/SARA - Proposed Hazardous Substance Additions
proposed RQ = 1 pound (.454 kg)
CERCLA/SARA - 1989 Proposed RQ Adjustments
proposed RQ = 100 pounds (45.4 kg)

CROCIDOLITE (FE2MG3NA2(SIO3)8)   61105-31-5
STATE LISTS
Pennsylvania Right to Know List
special hazardous substance
Pennsylvania RTK - Special Hazardous Substances
[present]

CROCIDOLITE (FE5NA2(SIO3)8)   53799-46-5
STATE LISTS
Pennsylvania Right to Know List
special hazardous substance
Pennsylvania RTK - Special Hazardous Substances
[present]

CROSCARMELLOSE SODIUM   74811-65-7
ENVIRONMENTAL LISTS
List of Pesticide Product Inert Ingredients
[present]

CROSS LINKED POLYMER OF SEBACRYL CHLO-   RR-01062-8
RIDE & POLYMETHYLENEPOLYPHENYL ISO-
CYANATE, WITH ETHYLENEDIAMINE AND DI-
ETHYLENETRIA MINE
ENVIRONMENTAL LISTS
List of Pesticide Product Inert Ingredients
[present]

CROTONALDEHYDE   123-73-9
HEALTH AND SAFETY LISTS
U.S. DOT - Appendix B - Marine Pollutants
inhibited form: DOT regulated marine pollutant
U.S. DOT - Appendix A Table 1 - Hazardous Substances
final RQ = 100 pounds (45.4 kg)
NFPA - Flash Points
flash point = 55 degrees F (13 degrees C)
NFPA - Hazard Identification Ratings
health-4; flammability-3; reactivity-2
NIOSH - Selected LD50s and LC50s
Inhalation, rat: LC50 = 4000 mg/m3 (30 mn) Oral, mouse: LD50 = 240 mg/kg Skin, rabbit: LD50 = 380 mg/kg
NIOSH 1990 - Pocket Guide - RELs
2 ppm TWA; 6 mg/m3 TWA
NIOSH 1990 - Pocket Guide - IDLHs
400 ppm IDLH
NIOSH 1990 - Pocket Guide - Target organs
eyes, skin, respiratory system
OSHA - Vacated PELs - Time Weighted Averages
2 ppm TWA; 6 mg/m3 TWA (Listed under 'Crotonaldehyde')

OSHA - Final PELs - Time Weighted Averages
2 ppm TWA;   6 mg/m3 TWA (listed under 'Crotonaldehyde')

ENVIRONMENTAL LISTS

CERCLA/SARA - Section 302 Extremely Hazardous Substances and
TPQs
TPQ = 1000 pounds

CERCLA/SARA - Hazardous Substances and their Reportable Quantities
final RQ = 100 pounds (45.4 kg)

CAA -Toxic Substances for Accidental Release Prevention
threshold quantity = 20,000 lbs

INTERNATIONAL LISTS

Canada - WHMIS: Ingredient Disclosure
1% item 449 (575)

Canada - Alberta - 8 Hour Occupational Exposure Limit
2 ppm TWA;   5.8 mg/m3 TWA

Canada - Alberta - 15 Minute Occupational Exposure Limit
6 ppm STEL;   17 mg/m3 STEL

Canada - British Columbia - 8 Hour Exposure Limits
2 ppm TWA;   6 mg/m3 TWA

Canada - British Columbia - 15 Minute Exposure Limits
6 ppm STEL;   18 mg/m3 STEL

Canada - Quebec - Time-Weighted Average Exposure Values
2 ppm TWAEV;   5.7 mg/m3 TWAEV

German (DFG) - Skin/Sensitizers
danger of cutaneous absorption

German (DFG) - Carcinogens
suspected carcinogen

Mexico - Instruction No. 10 - TWAs
2 ppm TWA;   6 mg/m3 TWA

Mexico - Instruction No. 10 - STELs
6 ppm STEL;   18 mg/m3 STEL

STATE LISTS

California - Exposure Limits - PELs
2 ppm PEL;   6 mg/m3 PEL

California - Directors List of Hazardous Substances (8 CCR 339)
[present]

Florida Hazardous Substance List
[present]

Massachusetts Right To Know List
extraordinarily hazardous

NJ Right to Know List (Total)
sn 0538

NJ Special Hazardous Substances
(flammable - third degree;   mutagen;   reactive - second degree)

Pennsylvania Right to Know List
environmental hazard

CROTONALDEHYDE   4170-30-3

HEALTH AND SAFETY LISTS

ACGIH 1995 - Time Weighted Averages
2 ppm TWA;   5.7 mg/m3 TWA

AIHA - Odor Threshold Values
geometric mean air odor threshold = 0.11 ppm (detectable)

U.S. DOT - Substances From 49 CFR 172.101
regulated by DOT (UN1143)

U.S. DOT - Hazard Classes
DOT hazard class = 3

U.S. DOT - Substances Which Are Poisonous by Inhalation
liquid hazardous material poisonous by inhalation (stabilized form) (UN1143)

U.S. DOT - Appendix B - Marine Pollutants
DOT regulated marine pollutant

U.S. DOT - Appendix A Table 1 - Hazardous Substances
final RQ = 100 pounds (45.4 kg)

NIOSH - Selected LD50s and LC50s
Inhalation, rat: LC50 = 200 mg/m3 2 hr Oral, rat: LD50 = 200 mg/kg Skin, guinea pig: LD50 = 25590 ug/kg

NTP Chemical Status Reports - Testing Status and NTIS Number
Prechronic studies for which toxicity technical reports were not prepared

OSHA - Vacated PELs - Time Weighted Averages
2 ppm TWA;   6 mg/m3 TWA

OSHA - Final PELs - Time Weighted Averages
2 ppm TWA;   6 mg/m3 TWA

ENVIRONMENTAL LISTS

CERCLA/SARA - Section 302 Extremely Hazardous Substances and
TPQs
TPQ = 1000 pounds

CERCLA/SARA - Section 313 - Emission Reporting
form R reporting required

CERCLA/SARA - Hazardous Substances and their Reportable Quantities
final RQ = 100 pounds (45.4 kg)

CAA -Toxic Substances for Accidental Release Prevention
threshold quantity = 20,000 lbs

Clean Water Act - Hazardous Substances
[present]

EPA - Master Testing List
[present]

RCRA - U Series Wastes
waste number U053

RCRA - Hazardous Constituents-Appendix VIII
waste number U053

RCRA - Substances Banned From Land Disposal
[present]

TSCA - Code of Federal Regulations Citations
40 CFR 716.120(a);   40 CFR 712.30(w);   40 CFR 799.5000

TSCA - PAIR - Reporting List
Reporting Date: August 18, 1988

TSCA - Health and Safety Reporting List
Effective Date: March 7, 1986

TSCA - Substances Subject to Testing Consent Orders
Test for: Environmental Effects and Chemical Fate

TSCA - Section 12(b) - Export Notification
export notification required - Section 4

INTERNATIONAL LISTS

Australian Exposure Standards - Time Weighted Averages
2 ppm TWA;   5.7 mg/m3 TWA

Canada - WHMIS: Ingredient Disclosure
1% item 448 (574)

Canada - Ontario - OHSA - TWAEVs
2 ppm TWAEV;   5.7 mg/m3 TWAEV

Israel - Time Weighted Averages
2 ppm TWA;   5.7 mg/m3 TWA

Israel - Action Levels
1 ppm AL;   2.85 mg/m3 AL

STATE LISTS

Massachusetts Right To Know List
extraordinarily hazardous

Minnesota Hazardous Substance List
[present]

NJ Right to Know List (Total)
sn 2888

Pennsylvania Right to Know List
environmental hazard

CROTONIC ACID — 3724-65-0

HEALTH AND SAFETY LISTS

U.S. DOT - Substances From 49 CFR 172.101
regulated by DOT (UN2833)

U.S. DOT - Hazard Classes
DOT hazard class = 8

NFPA - Flash Points
flash point = 190 degrees F (88 degrees C)

NFPA - Hazard Identification Ratings
health-3; flammability-2; reactivity-0

NIOSH - Selected LD50s and LC50s
Oral, rat: LD50 = 1000 mg/kg Skin, guinea pig: LD50 = 616 mg/kg

STATE LISTS

Florida Hazardous Substance List
[present]

Massachusetts Right To Know List
[present]

NJ Right to Know List (Total)
sn 0539

NJ Special Hazardous Substances
(corrosive)

Pennsylvania Right to Know List
[present]

CROTONONITRILE — 4786-20-3

HEALTH AND SAFETY LISTS

NFPA - Flash Points
flash point < 212 degrees F (100 degrees C)

NFPA - Hazard Identification Ratings
flammability-1; reactivity-0

CROTONYL ALCOHOL — 6117-91-5

HEALTH AND SAFETY LISTS

NFPA - Flash Points
flash point = 81 degrees F (27 degrees C)

NFPA - Hazard Identification Ratings
flammability-3; reactivity-2

STATE LISTS

Florida Hazardous Substance List
[present]

Massachusetts Right To Know List
[present]

Pennsylvania Right to Know List
[present]

CROTOXYPHOS — 7700-17-6

HEALTH AND SAFETY LISTS

U.S. DOT - Appendix B - Marine Pollutants
DOT regulated marine pollutant

1-CROTYL BROMIDE — 4784-77-4

HEALTH AND SAFETY LISTS

NFPA - Hazard Identification Ratings
health-2; flammability-3; reactivity-2

STATE LISTS

Florida Hazardous Substance List
[present]

Massachusetts Right To Know List
[present]

Pennsylvania Right to Know List
[present]

1-CROTYL CHLORIDE — 591-97-9

HEALTH AND SAFETY LISTS

NFPA - Hazard Identification Ratings
health-2; flammability-3; reactivity-2

STATE LISTS

Florida Hazardous Substance List
[present]

Massachusetts Right To Know List
[present]

Pennsylvania Right to Know List
[present]

CRUDE OIL WASTES — RR-01606-8

INTERNATIONAL LISTS

Canada - CEPA Schedule III Part I - Prohibited Substances (Ocean Dumping)
[present]

CRUFOMATE — 299-86-5

HEALTH AND SAFETY LISTS

ACGIH 1995 - Time Weighted Averages
5 mg/m3 TWA

NIOSH - Selected LD50s and LC50s
Oral, rat: LD50 = 460 mg/kg Skin, rabbit: LD50 = 2000 mg/kg

OSHA - Vacated PELs - Time Weighted Averages
5 mg/m3 TWA

INTERNATIONAL LISTS

Australian Exposure Standards - Time Weighted Averages
5 mg/m3 TWA

Canada - Alberta - 8 Hour Occupational Exposure Limit
5 mg/m3 TWA

Canada - Alberta - 15 Minute Occupational Exposure Limit
20 mg/m3 STEL

Canada - British Columbia - 8 Hour Exposure Limits
5 mg/m3 TWA

Canada - British Columbia - 15 Minute Exposure Limits
20 mg/m3 STEL

Canada - Ontario - OHSA - TWAEVs
5 mg/m3 TWAEV

Canada - Ontario - OHSA - STEVs
20 mg/m3 STEV

Canada - Quebec - Time-Weighted Average Exposure Values
5 mg/m3 TWAEV

Israel - Time Weighted Averages
5 mg/m3 TWA

Israel - Action Levels
2.5 mg/m3 AL

Mexico - Instruction No. 10 - TWAs
5 mg/m3 TWA

Mexico - Instruction No. 10 - STELs
20 mg/m3 STEL

STATE LISTS

California - Exposure Limits - PELs
5 mg/m3 PEL

California - Directors List of Hazardous Substances (8 CCR 339)
[present]

Florida Hazardous Substance List
[present]

Massachusetts Right To Know List
[present]

Minnesota Hazardous Substance List
[present]

NJ Right to Know List (Total)
  sn 0541

Pennsylvania Right to Know List
  [present]

CRYPTOSPORIDUM   RR-00109-2

PROPOSED REGULATIONS

Safe Drinking Water Act - Priority list
  [present]

CUMENE   98-82-8

HEALTH AND SAFETY LISTS

ACGIH 1995 - Time Weighted Averages
  50 ppm TWA; 246 mg/m3 TWA

ACGIH 1995 - Skin Designations
  skin - potential for cutaneous absorption

AIHA - Odor Threshold Values
  geometric mean air odor threshold = 0.032 ppm (detectable); 0.047 ppm (recognizable)

U.S. DOT - Substances From 49 CFR 172.101
  regulated by DOT (UN1918)

U.S. DOT - Hazard Classes
  DOT hazard class = 3

U.S. DOT - Appendix B - Marine Pollutants
  DOT regulated marine pollutant

U.S. DOT - Appendix A Table 1 - Hazardous Substances
  final RQ = 5000 pounds (2270 kg)

NFPA - Flash Points
  flash point = 96 degrees F (36 degrees C)

NFPA - Hazard Identification Ratings
  health-2; flammability-3; reactivity-1

NIOSH - Selected LD50s and LC50s
  Inhalation, mouse: LC50 = 24700 mg/m3 (2 hr) Oral, rat: LD50 = 1400 mg/kg Skin, rabbit: LD50 = 12300 mg/kg

NIOSH 1990 - Pocket Guide - RELs
  50 ppm TWA; 245 mg/m3 TWA

NIOSH 1990 - Pocket Guide - IDLHs
  8000 ppm IDLH

NIOSH 1990 - Pocket Guide - Target organs
  eyes, upper respiratory system, skin, CNS

NIOSH 1990 - Pocket Guide - Skin list
  Potential for dermal absorption

OSHA - Vacated PELs - Time Weighted Averages
  50 ppm TWA; 245 mg/m3 TWA

OSHA - Vacated PELs - Skin Designation
  Prevent or reduce skin absorption

OSHA - Final PELs - Time Weighted Averages
  50 ppm TWA; 245 mg/m3 TWA

OSHA - Final PELs - Skin Notations
  prevent or reduce skin absorption

ENVIRONMENTAL LISTS

CERCLA/SARA - Section 313 - Emission Reporting
  form R reporting required for 1.0% de minimus concentration

CERCLA/SARA - Hazardous Substances and their Reportable Quantities
  final RQ = 5000 pounds (2270 kg)

Clean Air Act (1990) - List of Hazardous Air Contaminants
  [present]

CAA - HON Rule - SOCMI Chemicals
  compliance by Oct. 24, 1994

CAA - HON Rule - Organic HAPs
  [present]

Safe Drinking Water Act - Monitoring
  monitoring required at discretion of the state

List of Pesticide Product Inert Ingredients
  [present]

RCRA - U Series Wastes
  waste number U055 (Ignitable waste)

RCRA - Hazardous Constituents-Appendix VIII
  waste number U055 (Ignitable waste)

RCRA - Substances Banned From Land Disposal
  [present]

TSCA - Code of Federal Regulations Citations
  40 CFR 712.30(m); 40 CFR 716.120(a)

TSCA - PAIR - Reporting List
  Reporting Date: February 26, 1985

TSCA - Health and Safety Reporting List
  Effective Date: December 28, 1984

TSCA - Chemical Test Rules
  Testing required by: manufacturers; importers; processors (40 CFR 799.1285)

TSCA - Section 12(b) - Export Notification
  export notification required - Section 4

INTERNATIONAL LISTS

Australian Exposure Standards - Time Weighted Averages
  50 ppm TWA; 246 mg/m3 TWA

Australian Exposure Standards - Skin Effects
  skin absorption

Canada - WHMIS: Ingredient Disclosure
  1% item 909 (1054)

Canada - NPRI (National Pollutant Release Inventory)
  [present]

Canada - Alberta - 8 Hour Occupational Exposure Limit
  50 ppm TWA; 245 mg/m3 TWA

Canada - Alberta - 15 Minute Occupational Exposure Limit
  75 ppm STEL; 370 mg/m3 STEL

Canada - Alberta - Skin Designation
  can be absorbed through the intact skin

Canada - British Columbia - 8 Hour Exposure Limits
  50 ppm TWA; 245 mg/m3 TWA

Canada - British Columbia - 15 Minute Exposure Limits
  75 ppm STEL; 365 mg/m3 STEL

Canada - British Columbia - Skin Notations
  skin - potential for skin absorption

Canada - Ontario - OHSA - TWAEVs
  50 ppm TWAEV; 245 mg/m3 TWAEV

Canada - Ontario - OHSA - Skin Notations
  absorption through skin, eyes, or mucous membranes

Canada - Quebec - Time-Weighted Average Exposure Values
  50 ppm TWAEV; 246 mg/m3 TWAEV

Canada - Quebec - Skin Designations
  absorbed through the skin

German (DFG) - MAK Values
  50 ppm MAK; 245 mg/m3 MAK

German (DFG) - Skin/Sensitizers
  danger of cutaneous absorption

Israel - Time Weighted Averages
  50 ppm TWA; 246 mg/m3 TWA

Israel - Action Levels
  25 ppm AL; 123 mg/m3 AL

Mexico - Instruction No. 10 - TWAs
  50 ppm TWA; 245 mg/m3 TWA

Mexico - Instruction No. 10 - STELs
  75 ppm STEL; 365 mg/m3 STEL

Mexico - Instruction No. 10 - Skin designation
  skin - potential for cutaneous absorption

STATE LISTS

California - Air Bill 2588 Appendix A-I
6/91

California - Exposure Limits - PELs
50 ppm PEL;   245 mg/m3 PEL

California - Exposure Limits - Skin Notation
material may be absorbed through the skin, eyes or mucous membrane

California - Directors List of Hazardous Substances (8 CCR 339)
[present]

Florida Hazardous Substance List
[present]

Massachusetts Right To Know List
[present]

Minnesota Hazardous Substance List
skin

NJ Right to Know List (Total)
sn 0542

Pennsylvania Right to Know List
environmental hazard

PROPOSED REGULATIONS

Canada - Ontario - Proposed Occupational TWAEVs
25 ppm TWAEV;   120 mg/m3 TWAEV

Canada - Ontario - Proposed Occupational STEVs
35 ppm STEV;   190 mg/m3 STEV

CUMENE HYDROPEROXIDE   80-15-9

HEALTH AND SAFETY LISTS

AIHA - WEEL - Time Weighted Averages
1 ppm TWA;   6 mg/m3 TWA

U.S. DOT - Appendix A Table 1 - Hazardous Substances
final RQ = 10 pounds (4.54 kg)

U.S. DOT - Organic Peroxides Table
Organic peroxide UN3109

NFPA - Flash Points
flash point = 175 degrees F (79 degrees C)

NFPA - Hazard Identification Ratings
health-1;   flammability-2;   reactivity-4 (oxidizing properties)

NIOSH - Selected LD50s and LC50s
Inhalation, rat: LC50 = 220 ppm 4hr Oral, rat: LD50 = 382 mg/kg Skin, rat: LD50 = 500 mg/kg

OSHA - List of Highly Hazardous Chemicals
threshhold quantity = 5000 pounds

ENVIRONMENTAL LISTS

CERCLA/SARA - Section 313 - Emission Reporting
form R reporting required for 1.0% de minimus concentration

CERCLA/SARA - Hazardous Substances and their Reportable Quantities
final RQ = 10 pounds (4.54 kg)

CAA - HON Rule - SOCMI Chemicals
compliance by Oct. 24, 1994

RCRA - U Series Wastes
waste number U096 (Reactive waste)

RCRA - Hazardous Constituents-Appendix VIII
waste number U096 (Reactive waste)

RCRA - Substances Banned From Land Disposal
[present]

TSCA - Code of Federal Regulations Citations
40 CFR 716.120(a)

TSCA - Health and Safety Reporting List
Effective Date: March 7, 1986

INTERNATIONAL LISTS

Canada - WHMIS: Ingredient Disclosure
1% item 450 (982)

Canada - NPRI (National Pollutant Release Inventory)
[present]

German (DFG) - Skin/Sensitizers
very strong effects on skin

STATE LISTS

California - Air Bill 2588 Appendix A-I
6/91

Florida Hazardous Substance List
[present]

Massachusetts Right To Know List
[present]

NJ Right to Know List (Total)
sn 0543

NJ Special Hazardous Substances
(reactive - fourth degree)

Pennsylvania Right to Know List
environmental hazard

CUMINIC ALCOHOL   536-60-7

INTERNATIONAL LISTS

Canada - WHMIS: Ingredient Disclosure
1% item 451 (171)

CUMYL PEROXYPIVALATE   23383-59-7

HEALTH AND SAFETY LISTS

U.S. DOT - Organic Peroxides Table
Organic peroxide UN3115

CUMYL PEROXYNEODECANOATE   26748-47-0

HEALTH AND SAFETY LISTS

U.S. DOT - Organic Peroxides Table
Organic peroxide UN3115

STATE LISTS

NJ Right to Know List (Total)
sn 0544

CUMYL PEROXYPIVALATE   RR-00580-1

STATE LISTS

NJ Right to Know List (Total)
sn 2299

CUPFERRON   135-20-6

HEALTH AND SAFETY LISTS

NIOSH - Selected LD50s and LC50s
Oral, rat: LD50 = 257 mg/kg

NTP Chemical Status Reports - Testing Status and NTIS Number
Technical reports printed (PB287409/AS)

NTP Chemical Status Reports - Evidence of Carcinogenicity
male rat-positive;   female rat-positive;   male mice-positive;   female mice-positive

NTP Seventh Report - Suspect Carcinogens
suspect carcinogen

OSHA - Possible Select Carcinogens
[present]

ENVIRONMENTAL LISTS

CERCLA/SARA - Section 313 - Emission Reporting
form R reporting required for 0.1% de minimus concentration

INTERNATIONAL LISTS

Canada - WHMIS: Ingredient Disclosure
0.1% item 452 (579)

STATE LISTS

California - Air Bill 2588 Appendix A-I
known or potential carcinogen

California - Prop. 65 - Cancer list
carcinogen - initial date 1/1/88

California - Prop. 65 - No Significant Risk Levels
no significant risk level = 3 ug/day

Florida Hazardous Substance List
[present]

Massachusetts Right To Know List
carcinogen; extraordinarily hazardous

Minnesota Hazardous Substance List
carcinogen

NJ Right to Know List (Total)
sn 0545

NJ Special Hazardous Substances
(carcinogen)

Pennsylvania Right to Know List
environmental hazard; special hazardous substance

Pennsylvania RTK - Special Hazardous Substances
[present]

CUPRATE(4-), [U-[[6,6'-[3,3'-DIHYDROXY[1,1'-BIPHENYL]-4,4'-DIYL)BIS(AZO)BIS[4-AMINO-5-HY-DROXY-1,3-NAPHTHALENE DISULFONATO](8-)]DI-, TETRASODIUM   16143-79-6
SEE ALSO:
    BISAZOBIPHENYL DYES

ENVIRONMENTAL LISTS
    TSCA - Code of Federal Regulations Citations
    40 CFR 716.120(c)

CUPRATE(3-), [U-[7-[[3,3'-DIHYDROXY-4'-[[1-HY-DROXY-6-(PHENYLAMINO)-3-SULFO-2-NAPH-THALENYL]AZO][1,1'-BIPHENYL]-4-YL]AZO] -8-HY-DROXY-1,6-NAPHTHALENEDISULFONATO(7-)]DI-, TRISODIUM   6656-03-7
SEE ALSO:
    BISAZOBIPHENYL DYES

ENVIRONMENTAL LISTS
    TSCA - Code of Federal Regulations Citations
    40 CFR 716.120(c)

CUPRATE(6-), [[[1,2-ETHANEDIYLBIS[NITRILOBIS(METHYLENE)]TETRAKIS[PHOSPHONATO](8-)N,N', O,O",O"",O"""']-, PENTAPOTASSIUM HYDROGEN, (OC-6-21)-   67989-89-3
ENVIRONMENTAL LISTS
    TSCA - Code of Federal Regulations Citations
    40 CFR 704.95

CUPRATE(2-), ((ETHYLENEDINITRILO)TETRAAC-ETATO)-, DISODIUM   14025-15-1
SEE ALSO:
    COPPER

ENVIRONMENTAL LISTS
    List of Pesticide Product Inert Ingredients
    [present]

CUPRIC ACETATE   142-71-2
SEE ALSO:
    COPPER

HEALTH AND SAFETY LISTS
    U.S. DOT - Appendix A Table 1 - Hazardous Substances
    final RQ = 100 pounds (45.4 kg)
    NIOSH - Selected LD50s and LC50s
    Oral, rat: LD50 = 595 mg/kg

ENVIRONMENTAL LISTS
    CERCLA/SARA - Hazardous Substances and their Reportable Quantities
    final RQ = 100 pounds (45.4 kg)
    Clean Water Act - Hazardous Substances
    [present]
    List of Pesticide Product Inert Ingredients
    [present]

STATE LISTS
    Massachusetts Right To Know List
    [present]
    NJ Right to Know List (Total)
    sn 0546
    Pennsylvania Right to Know List
    environmental hazard

CUPRIC CHLORIDE   7447-39-4
SEE ALSO:
    COPPER

HEALTH AND SAFETY LISTS
    U.S. DOT - Appendix B - Marine Pollutants
    DOT regulated marine pollutant
    U.S. DOT - Appendix A Table 1 - Hazardous Substances
    final RQ = 10 pounds (4.54 kg)

ENVIRONMENTAL LISTS
    CERCLA/SARA - Hazardous Substances and their Reportable Quantities
    final RQ = 10 pounds (4.54 kg)
    Clean Water Act - Hazardous Substances
    [present]

STATE LISTS
    Massachusetts Right To Know List
    [present]
    Pennsylvania Right to Know List
    environmental hazard

CUPRIC CYANIDE   14763-77-0
SEE ALSO:
    CYANIDE ANION
    COPPER

HEALTH AND SAFETY LISTS
    U.S. DOT - Appendix B - Marine Pollutants
    DOT regulated severe marine pollutant

INTERNATIONAL LISTS
    Canada - WHMIS: Ingredient Disclosure
    1% item 453 (591)

STATE LISTS
    NJ Right to Know List (Total)
    sn 0533

CUPRIC GLUCONATE   527-09-3
ENVIRONMENTAL LISTS
    List of Pesticide Product Inert Ingredients
    [present]

CUPRIC NITRATE   3251-23-8
SEE ALSO:
    COPPER

HEALTH AND SAFETY LISTS
    U.S. DOT - Appendix A Table 1 - Hazardous Substances
    final RQ = 100 pounds (45.4 kg)
    NIOSH - Selected LD50s and LC50s
    Oral, rat: LD50 = 940 mg/kg

ENVIRONMENTAL LISTS

CERCLA/SARA - Hazardous Substances and their Reportable Quantities
final RQ = 100 pounds (45.4 kg)

Clean Water Act - Hazardous Substances
[present]

List of Pesticide Product Inert Ingredients
[present]

INTERNATIONAL LISTS

Canada - WHMIS: Ingredient Disclosure
1% item 436 (1203)

STATE LISTS

Florida Hazardous Substance List
[present]

Massachusetts Right To Know List
[present]

NJ Right to Know List (Total)
sn 0547

Pennsylvania Right to Know List
environmental hazard

CUPRIC OXALATE   5893-66-3

HEALTH AND SAFETY LISTS

U.S. DOT - Appendix A Table 1 - Hazardous Substances
final RQ = 100 pounds (45.4 kg)

ENVIRONMENTAL LISTS

CERCLA/SARA - Hazardous Substances and their Reportable Quantities
final RQ = 100 pounds (45.4 kg)

Clean Water Act - Hazardous Substances
[present]

STATE LISTS

Massachusetts Right To Know List
[present]

Pennsylvania Right to Know List
environmental hazard

CUPRIC SULFATE   7758-98-7
SEE ALSO:
COPPER

HEALTH AND SAFETY LISTS

U.S. DOT - Appendix A Table 1 - Hazardous Substances
final RQ = 10 pounds (4.54 kg)

NIOSH - Selected LD50s and LC50s
Oral, rat: LD50 = 300 mg/kg

ENVIRONMENTAL LISTS

CERCLA/SARA - Hazardous Substances and their Reportable Quantities
final RQ = 10 pounds (4.54 kg)

Clean Water Act - Hazardous Substances
[present]

List of Pesticide Product Inert Ingredients
[present]

INTERNATIONAL LISTS

Canada - WHMIS: Ingredient Disclosure
1% item 438 (1519)

STATE LISTS

Massachusetts Right To Know List
[present]

NJ Right to Know List (Total)
sn 0549

Pennsylvania Right to Know List
environmental hazard

CUPRIC SULFATE AMMONIATED   10380-29-7

HEALTH AND SAFETY LISTS

U.S. DOT - Appendix A Table 1 - Hazardous Substances
final RQ = 100 pounds (45.4 kg)

ENVIRONMENTAL LISTS

CERCLA/SARA - Hazardous Substances and their Reportable Quantities
final RQ = 100 pounds (45.4 kg)

Clean Water Act - Hazardous Substances
[present]

STATE LISTS

Massachusetts Right To Know List
[present]

NJ Right to Know List (Total)
sn 2300

Pennsylvania Right to Know List
environmental hazard

CUPRIC TARTRATE   815-82-7
SEE ALSO:
COPPER

HEALTH AND SAFETY LISTS

U.S. DOT - Appendix A Table 1 - Hazardous Substances
final RQ = 100 pounds (45.4 kg)

ENVIRONMENTAL LISTS

CERCLA/SARA - Hazardous Substances and their Reportable Quantities
final RQ = 100 pounds (45.4 kg)

Clean Water Act - Hazardous Substances
[present]

STATE LISTS

Massachusetts Right To Know List
[present]

Pennsylvania Right to Know List
environmental hazard

CUPRIETHYLENEDIAMINE   13426-91-0
SEE ALSO:
COPPER

HEALTH AND SAFETY LISTS

U.S. DOT - Substances From 49 CFR 172.101
regulated by DOT (UN1761)

U.S. DOT - Hazard Classes
DOT hazard class = 8

U.S. DOT - Appendix B - Marine Pollutants
solution: DOT regulated marine pollutant

INTERNATIONAL LISTS

Canada - WHMIS: Ingredient Disclosure
1% item 454 (580)

STATE LISTS

NJ Right to Know List (Total)
sn 0551

NJ Special Hazardous Substances
(corrosive)

CURACRON   41198-08-7

ENVIRONMENTAL LISTS

CERCLA/SARA - Section 313 - Emission Reporting
form R reporting required

STATE LISTS
  Massachusetts Right To Know List
  [present]

CURIUM 238   30989-40-3

  HEALTH AND SAFETY LISTS
    U.S. DOT - Appendix A Table 2 - Radionuclides
    final RQ = 1000 curies (3.7E 13 Bq)

  ENVIRONMENTAL LISTS
    CERCLA/SARA List of Radionuclides (Appendix B) and Their Reportable Quantities
    final RQ = 1000 curies (3.7E 13 Bq)

CURIUM 240   15411-90-2

  HEALTH AND SAFETY LISTS
    U.S. DOT - Appendix A Table 2 - Radionuclides
    final RQ = 1 curie (3.7E 10 Bq)

  ENVIRONMENTAL LISTS
    CERCLA/SARA List of Radionuclides (Appendix B) and Their Reportable Quantities
    final RQ = 1 curie (3.7E 10 Bq)

CURIUM 241   15411-91-3

  HEALTH AND SAFETY LISTS
    U.S. DOT - Appendix A Table 2 - Radionuclides
    final RQ = 10 curies (3.7E 11 Bq)

  ENVIRONMENTAL LISTS
    CERCLA/SARA List of Radionuclides (Appendix B) and Their Reportable Quantities
    final RQ = 10 curies (3.7E 11 Bq)

CURIUM 242   15510-73-3

  HEALTH AND SAFETY LISTS
    U.S. DOT - Appendix A Table 2 - Radionuclides
    final RQ = 1 curie (3.7E 10 Bq)

  ENVIRONMENTAL LISTS
    CERCLA/SARA List of Radionuclides (Appendix B) and Their Reportable Quantities
    final RQ = 1 curie (3.7E 10 Bq)

CURIUM 243   15757-87-6

  HEALTH AND SAFETY LISTS
    U.S. DOT - Appendix A Table 2 - Radionuclides
    final RQ = 0.01 curies (3.7E 8 Bq)

  ENVIRONMENTAL LISTS
    CERCLA/SARA List of Radionuclides (Appendix B) and Their Reportable Quantities
    final RQ = 0.01 curies (3.7E 8 Bq)

CURIUM 244   13981-15-2

  HEALTH AND SAFETY LISTS
    U.S. DOT - Appendix A Table 2 - Radionuclides
    final RQ = 0.01 curies (3.7E 8 Bq)

  ENVIRONMENTAL LISTS
    CERCLA/SARA List of Radionuclides (Appendix B) and Their Reportable Quantities
    final RQ = 0.01 curies (3.7E 8 Bq)

CURIUM 245   15621-76-8

  HEALTH AND SAFETY LISTS
    U.S. DOT - Appendix A Table 2 - Radionuclides
    final RQ = 0.01 curies (3.7E 8 Bq)

ENVIRONMENTAL LISTS
  CERCLA/SARA List of Radionuclides (Appendix B) and Their Reportable Quantities
  final RQ = 0.01 curies (3.7E 8 Bq)

CURIUM 246   15757-90-

  HEALTH AND SAFETY LISTS
    U.S. DOT - Appendix A Table 2 - Radionuclides
    final RQ = 0.01 curies (3.7E 8 Bq)

  ENVIRONMENTAL LISTS
    CERCLA/SARA List of Radionuclides (Appendix B) and Their Reportable Quantities
    final RQ = 0.01 curies (3.7E 8 Bq)

CURIUM 247   15758-32-4

  HEALTH AND SAFETY LISTS
    U.S. DOT - Appendix A Table 2 - Radionuclides
    final RQ = 0.01 curies (3.7E 8 Bq)

  ENVIRONMENTAL LISTS
    CERCLA/SARA List of Radionuclides (Appendix B) and Their Reportable Quantities
    final RQ = 0.01 curies (3.7E 8 Bq)

CURIUM 248   15758-33-5

  HEALTH AND SAFETY LISTS
    U.S. DOT - Appendix A Table 2 - Radionuclides
    final RQ = 0.001 curies (3.7E 7 Bq)

  ENVIRONMENTAL LISTS
    CERCLA/SARA List of Radionuclides (Appendix B) and Their Reportable Quantities
    final RQ = 0.001 curies (3.7E 7 Bq)

CURIUM 249   15701-07-2

  HEALTH AND SAFETY LISTS
    U.S. DOT - Appendix A Table 2 - Radionuclides
    final RQ = 1000 curies (3.7E 13 Bq)

  ENVIRONMENTAL LISTS
    CERCLA/SARA List of Radionuclides (Appendix B) and Their Reportable Quantities
    final RQ = 1000 curies (3.7E 13 Bq)

CYANAMIDE   420-04-2

  HEALTH AND SAFETY LISTS
    ACGIH 1995 - Time Weighted Averages
    2 mg/m3 TWA

    NFPA - Flash Points
    flash point = 286 degrees F (141 degrees C)

    NFPA - Hazard Identification Ratings
    health-4; flammability-1; reactivity-3

    NIOSH - Selected LD50s and LC50s
    Oral, rat: LD50 = 125 mg/kg Skin, rabbit: LD50 = 590 mg/kg

    OSHA - Vacated PELs - Time Weighted Averages
    2 mg/m3 TWA

  ENVIRONMENTAL LISTS
    List of Pesticide Product Inert Ingredients
    [present]

  INTERNATIONAL LISTS
    Australian Exposure Standards - Time Weighted Averages
    2 mg/m3 TWA

    Canada - WHMIS: Ingredient Disclosure
    1% item 455 (581)

    Canada - Alberta - 8 Hour Occupational Exposure Limit
    2 mg/m3 TWA

Canada - Alberta - 15 Minute Occupational Exposure Limit
4 mg/m3 STEL

Canada - British Columbia - 8 Hour Exposure Limits
2 mg/m3 TWA

Canada - Ontario - OHSA - TWAEVs
2 mg/m3 TWAEV

Canada - Quebec - Time-Weighted Average Exposure Values
2 mg/m3 TWAEV

United Kingdom - Occupational Exposure Standards - TWAs
2 mg/m3 TWA

Israel - Time Weighted Averages
2 mg/m3 TWA

Israel - Action Levels
1 mg/m3 AL

Mexico - Instruction No. 10 - TWAs
2 mg/m3 TWA

STATE LISTS

California - Exposure Limits - PELs
2 mg/m3 PEL

California - Directors List of Hazardous Substances (8 CCR 339)
[present]

Florida Hazardous Substance List
[present]

Massachusetts Right To Know List
[present]

Minnesota Hazardous Substance List
[present]

NJ Special Hazardous Substances
(reactive - third degree)

Pennsylvania Right to Know List
[present]

CYANAZINE   21725-46-2

HEALTH AND SAFETY LISTS

NIOSH - Selected LD50s and LC50s
Oral, rat: LD50 = 149 mg/kg Skin, rat: LD50 = 1200 mg/kg

ENVIRONMENTAL LISTS

CERCLA/SARA - Section 313 - Emission Reporting
form R reporting required

INTERNATIONAL LISTS

Canada - Drinking Water Quality - IMACs
0.01 mg/L IMAC

STATE LISTS

California - Air Bill 2588 Appendix A-II
9/90

California - Prop. 65 - Developmental Toxicity
developmental toxicity - initial date 4/1/90

Massachusetts Right To Know List
[present]

NJ Right to Know List (Total)
sn 0240

PROPOSED REGULATIONS

Safe Drinking Water Act - Priority list
[present]

CYANIDE ANION   57-12-5

SEE ALSO:

F006-HAZARDOUS WASTES
K088-HAZARDOUS WASTES
F039-HAZARDOUS WASTES
K060-HAZARDOUS WASTES
K007-HAZARDOUS WASTES
F012-HAZARDOUS WASTES
F019-HAZARDOUS WASTES

HEALTH AND SAFETY LISTS

U.S. DOT - Substances From 49 CFR 172.101
regulated by DOT (UN1935)

U.S. DOT - Hazard Classes
DOT hazard class = 6.1

U.S. DOT - Appendix B - Marine Pollutants
DOT regulated marine pollutant

U.S. DOT - Appendix A Table 1 - Hazardous Substances
soluble salts and complexes (n.o.s.): Final RQ = 10 pounds (4.54 kg)

NIOSH 1990 - Pocket Guide - RELs
as CN: C 4.7 ppm (10 min); C 5 mg/m3 (10 min) (Listed under 'Cyanides')

NIOSH 1990 - Pocket Guide - IDLHs
as CN: 50 mg/m3 IDLH

NIOSH 1990 - Pocket Guide - Target organs
CVS, CNS, liver, kidneys, skin

NIOSH - Health Standards - Exposure Limits
C (2 hr) 0.06 ppm; C (2 hr) 0.15 mg/m3 (Listed under 'Hydrazines')

NIOSH - Health Standards - Health Effects and Precautions
has produced tumors of the lung, liver, blood vessel, and intestines in animals; blood, liver, and skin effects (Blood and urine testing and periodic chest X-ray required; bowel examinations for worker above age 40) (Listed under 'Hydrazines')

NIOSH - Health Standards - Carcinogenic Chemicals
potential human carcinogen (Listed under 'Hydrazines')

OSHA - Vacated PELs - Time Weighted Averages
as CN: 5 mg/m3 TWA

OSHA - Final PELs - Time Weighted Averages
as CN: 5 mg/m3 TWA

ENVIRONMENTAL LISTS

ATSDR Priority List
Rank (of 275): 031

CERCLA/SARA - Hazardous Substances and their Reportable Quantities
soluble salts and complexes (n.o.s.): final RQ = 10 pounds (4.54 kg)

Clean Air Act (1990) - List of Hazardous Air Contaminants
[present] (includes any X'CN where X=H' or any other group where a formal dissociation may occur)

Clean Water Act - Priority Pollutants
[present]

Clean Water Act - Toxic Pollutants
[present]

Safe Drinking Water Act - MCLs
MCL = 0.2 mg/L

Safe Drinking Water Act - MCLGs
MCLG = 0.2 mg/L

RCRA - P Series Wastes
waste number P030

RCRA - Hazardous Constituents-Appendix VIII
waste number P030

RCRA - Basis for Listing - Appendix VII
Included in waste streams: F006, F012, F019, F039, K007, K060, K088

RCRA - TSD Facilities Ground Water Monitoring
TM 9010 = 40 ug/L PQL

RCRA - Universal Treatment Standards (LDR)
*Total: WW: 1.2 mg/l;   NWW: 590 mg/kg Amenable: WW: 0.86 mg/l;
NWW: 30 mg/kg*

INTERNATIONAL LISTS

Australian Exposure Standards - Time Weighted Averages
as CN: 5 mg/m3 TWA

Australian Exposure Standards - Skin Effects
skin absorption

Canada - NPRI (National Pollutant Release Inventory)
[present]

Canada - CEPA Schedule III Part II - Restricted Substances (Ocean Dumping)
[present]

Canada - Drinking Water Quality - MACs
0.2 mg/L MAC

Canada - Alberta - 8 Hour Occupational Exposure Limit
as Cn: 5 mg/m3 TWA

Canada - Alberta - 15 Minute Occupational Exposure Limit
as CN: 10 mg/m3 STEL

Canada - Alberta - Skin Designation
as CN: can be absorbed through the intact skin

Canada - British Columbia - 8 Hour Exposure Limits
5 mg/m3 TWA

Canada - British Columbia - Skin Notations
skin - potential for skin absorption

Canada - Ontario - OHSA - TWAEVs
as Cyanide: 5 mg/m3 TWAEV

Canada - Ontario - OHSA - Skin Notations
as CN: absorption through skin, eyes, or mucous membranes

Canada - Quebec - Time-Weighted Average Exposure Values
as CN: 5 mg/m3 TWAEV

Canada - Quebec - Skin Designations
as CN: absorbed through the skin

United Kingdom - Occupational Exposure Standards - TWAs
as CN: 5 mg/m3 TWA (does not include hydrogen cyanide, cyanogen or cyanogen chloride)

United Kingdom - Occupational Exposure Standards - Notes
can be absorbed through skin

German (DFG) - MAK Values
total dust, as CN: 5 mg/m3 MAK

German (DFG) - Peak Limitations
2 x normal MAK (30 min. average value);   don't exceed 4 times during shift

German (DFG) - Skin/Sensitizers
danger of cutaneous absorption

Mexico - Instruction No. 10 - TWAs
5 mg/m3 TWA

Mexico - Instruction No. 10 - Skin designation
skin - potential for cutaneous absorption

Mexico - Drinking Water - Ecological Criteria
0.2 mg/l

STATE LISTS

California - Exposure Limits - PELs
as CN: 5 mg/m3 PEL

California - Exposure Limits - Skin Notation
material may be absorbed through the skin, eyes or mucous membrane

Massachusetts Right To Know List
[present]

NJ Right to Know List (Total)
sn 0553;   sn 2872

Pennsylvania Right to Know List
environmental hazard (for any compound of this substance)

CYANIDE COMPOUNDS   RR-00812-8

ENVIRONMENTAL LISTS

CERCLA/SARA - Section 313 - Emission Reporting
form R reporting required for 1.0% de minimus concentration

RCRA - Substances Banned From Land Disposal
[present]

STATE LISTS

California - Air Bill 2588 Appendix A-I
6/91

NJ Right to Know List (Total)
sn 2308

CYANIDE, INORGANIC COMPOUNDS   RR-00573-2

HEALTH AND SAFETY LISTS

U.S. DOT - Substances From 49 CFR 172.101
regulated by DOT (UN1588)

U.S. DOT - Hazard Classes
DOT hazard class = 6.1

U.S. DOT - Appendix B - Marine Pollutants
DOT regulated marine pollutant

INTERNATIONAL LISTS

Canada - WHMIS: Ingredient Disclosure
1% item 456 (590)

STATE LISTS

California - Directors List of Hazardous Substances (8 CCR 339)
[present]

NJ Right to Know List (Total)
sn 2872

CYANIDE (SALTS)   RR-00500-5

SEE ALSO:
F009-HAZARDOUS WASTES
F010-HAZARDOUS WASTES
F011-HAZARDOUS WASTES
F008-HAZARDOUS WASTES
F007-HAZARDOUS WASTES

HEALTH AND SAFETY LISTS

NIOSH - Health Standards - Exposure Limits
C (10 min) 4.7 ppm CN;   C (10 min) 5 mg CN/m3 (Listed under 'Hydrogen cyanide and cyanide salts')

NIOSH - Health Standards - Health Effects and Precautions
Thyroid, blood, and respiratory system effect (Prevent skin and eye contact;   make first-aid kits and personnel available during use) (Listed under 'Hydrogen cyanide and cyanide salts')

ENVIRONMENTAL LISTS

RCRA - P Series Wastes
waste number P030

RCRA - Hazardous Constituents-Appendix VIII
waste number P030

RCRA - Basis for Listing - Appendix VII
Included in waste streams: F007, F008, F009, F010, F011

2-CYANOACETAMIDE   107-91-5

HEALTH AND SAFETY LISTS

NIOSH - Selected LD50s and LC50s
Oral, rat: LD50 = 7230 mg/kg

CYANOACRYLATES   RR-01711-8

PROPOSED REGULATIONS

TSCA - ITC 33rd Report Priority Testing List
recommended for testing

TSCA - ITC 34th Report Priority Testing List
recommended for testing

2-CYANOETHYL ACRYLATE   **106-71-8**

HEALTH AND SAFETY LISTS

NFPA - Flash Points
flash point = 255 degrees F (124 degrees C)

NFPA - Hazard Identification Ratings
health-2; flammability-1; reactivity-1

NIOSH - Selected LD50s and LC50s
Oral, rat: LD50 = 180 mg/kg Skin, rabbit: LD50 = 220 mg/kg

ENVIRONMENTAL LISTS

TSCA - Code of Federal Regulations Citations
40 CFR 712.30(d)

TSCA - PAIR - Reporting List
Reporting Date: November 19, 1982

INTERNATIONAL LISTS

Canada - WHMIS: Ingredient Disclosure
1% item 457 (154)

STATE LISTS

Florida Hazardous Substance List
[present]

Massachusetts Right To Know List
[present]

Pennsylvania Right to Know List
[present]

N-(2-CYANOETHYL)CYCLOHEXYLAMINE   **702-03-4**

HEALTH AND SAFETY LISTS

NFPA - Flash Points
flash point = 255 degrees F (124 degrees C)

NFPA - Hazard Identification Ratings
health-2; flammability-1; reactivity-0

STATE LISTS

Florida Hazardous Substance List
[present]

Massachusetts Right To Know List
[present]

Pennsylvania Right to Know List
[present]

CYANOGEN   **460-19-5**

HEALTH AND SAFETY LISTS

ACGIH 1995 - Time Weighted Averages
10 ppm TWA; 21 mg/m3 TWA

U.S. DOT - Substances From 49 CFR 172.101
regulated by DOT (UN1026)

U.S. DOT - Hazard Classes
DOT hazard class = 2.3

U.S. DOT - Substances Which Are Poisonous by Inhalation
gaseous hazardous material poisonous by inhalation (UN1026)

U.S. DOT - Appendix A Table 1 - Hazardous Substances
final RQ = 100 pounds (45.4 kg)

NFPA - Flash Points
gas (no flash point given)

NFPA - Hazard Identification Ratings
health-4; flammability-4; reactivity-2

NIOSH - Selected LD50s and LC50s
Inhalation, rat: LC50 = 350 ppm 1 hr

OSHA - Vacated PELs - Time Weighted Averages
10 ppm TWA; 20 mg/m3 TWA

OSHA - List of Highly Hazardous Chemicals
threshhold quantity = 2500 pounds

ENVIRONMENTAL LISTS

CERCLA/SARA - Hazardous Substances and their Reportable Quantities
final RQ = 100 pounds (45.4 kg)

CAA - Flammable Substances for Accidental Release Prevention
threshold quantity = 10,000 lbs

RCRA - P Series Wastes
waste number P031

RCRA - Hazardous Constituents-Appendix VIII
waste number P031

RCRA - Substances Banned From Land Disposal
[present]

INTERNATIONAL LISTS

Australian Exposure Standards - Time Weighted Averages
10 ppm TWA; 21 mg/m3 TWA

Canada - WHMIS: Ingredient Disclosure
1% item 458 (586)

Canada - Alberta - 8 Hour Occupational Exposure Limit
10 ppm TWA; 21 mg/m3 TWA

Canada - Alberta - 15 Minute Occupational Exposure Limit
20 ppm STEL; 43 mg/m3 STEL

Canada - British Columbia - 8 Hour Exposure Limits
10 ppm TWA; 20 mg/m3 TWA

Canada - Ontario - OHSA - TWAEVs
10 ppm TWAEV; 21 mg/m3 TWAEV

United Kingdom - Occupational Exposure Standards - TWAs
10 ppm TWA; 20 mg/m3 TWA

German (DFG) - MAK Values
10 ppm MAK; 22 mg/m3 MAK

German (DFG) - Peak Limitations
5 x normal MAK (30 min. average value); don't exceed 2 times during shift

German (DFG) - Skin/Sensitizers
danger of cutaneous absorption

Israel - Time Weighted Averages
10 ppm TWA; 21 mg/m3 TWA

Israel - Action Levels
5 ppm AL; 10.5 mg/m3 AL

Mexico - Instruction No. 10 - TWAs
10 ppm TWA; 20 mg/m3 TWA

STATE LISTS

California - Exposure Limits - PELs
10 ppm PEL; 20 mg/m3 PEL

California - Directors List of Hazardous Substances (8 CCR 339)
[present]

Florida Hazardous Substance List
[present]

Massachusetts Right To Know List
[present]

Minnesota Hazardous Substance List
[present]

NJ Right to Know List (Total)
sn 0554

NJ Special Hazardous Substances
(flammable - fourth degree; reactive - third degree)

Pennsylvania Right to Know List
environmental hazard

CYANOGEN   **2074-87-5**

INTERNATIONAL LISTS

Canada - Quebec - Time-Weighted Average Exposure Values
10 ppm TWAEV; 21 mg/m3 TWAEV

STATE LISTS

Pennsylvania Right to Know List
environmental hazard

CYANOGEN BROMIDE   506-68-3

HEALTH AND SAFETY LISTS

U.S. DOT - Substances From 49 CFR 172.101
regulated by DOT (UN1889)

U.S. DOT - Hazard Classes
DOT hazard class = 6.1

U.S. DOT - Substances Which Are Poisonous by Inhalation
liquid hazardous material poisonous by inhalation (UN1889)

U.S. DOT - Appendix B - Marine Pollutants
DOT regulated marine pollutant

U.S. DOT - Appendix A Table 1 - Hazardous Substances
final RQ = 1000 pounds (454 kg)

ENVIRONMENTAL LISTS

CERCLA/SARA - Section 302 Extremely Hazardous Substances and TPQs
TPQ = 500/10,000 pounds

CERCLA/SARA - Hazardous Substances and their Reportable Quantities
final RQ = 1000 pounds (454 kg)

RCRA - U Series Wastes
waste number U246

RCRA - Hazardous Constituents-Appendix VIII
waste number U246

RCRA - Substances Banned From Land Disposal
[present]

INTERNATIONAL LISTS

Canada - WHMIS: Ingredient Disclosure
1% item 459 (334)

STATE LISTS

Florida Hazardous Substance List
[present]

Massachusetts Right To Know List
extraordinarily hazardous

NJ Right to Know List (Total)
sn 2302

Pennsylvania Right to Know List
environmental hazard

CYANOGEN CHLORIDE   506-77-4

HEALTH AND SAFETY LISTS

ACGIH 1995 - Ceiling Limits
C 0.3 ppm;   C 0.75 mg/m3

U.S. DOT - Substances From 49 CFR 172.101
regulated by DOT (UN1589)

U.S. DOT - Hazard Classes
DOT hazard class = 2.3

U.S. DOT - Substances Which Are Poisonous by Inhalation
gaseous hazardous material poisonous by inhalation (inhibited form) (UN1589)

U.S. DOT - Appendix B - Marine Pollutants
inhibited form: DOT regulated marine pollutant

U.S. DOT - Appendix A Table 1 - Hazardous Substances
final RQ = 10 pounds (4.54 kg)

NIOSH - Selected LD50s and LC50s
Inhalation, rat: LC50 = 5400 mg/m3 3 mn Oral, cat: LD50 = 6 mg/kg

OSHA - Vacated PELs - Ceiling Limits
C 0.3 ppm;   C 0.6 mg/m3

OSHA - List of Highly Hazardous Chemicals
threshhold quantity = 500 pounds

ENVIRONMENTAL LISTS

CERCLA/SARA - Hazardous Substances and their Reportable Quantities
final RQ = 10 pounds (4.54 kg)

CAA -Toxic Substances for Accidental Release Prevention
threshold quantity = 10,000 lbs

Clean Water Act - Hazardous Substances
[present]

RCRA - P Series Wastes
waste number P033

RCRA - Hazardous Constituents-Appendix VIII
waste number P033

RCRA - Substances Banned From Land Disposal
[present]

INTERNATIONAL LISTS

Australian Exposure Standards - Time Weighted Averages
Peak Limitation: 0.3 ppm; 0.75 mg/m3

Canada - WHMIS: Ingredient Disclosure
1% item 460 (491)

Canada - Alberta - Ceiling Occupational Exposure Limit
C 0.3 ppm; C 0.75 mg/m3

Canada - Ontario - OHSA - CEVs
0.3 ppm CEV; 0.75 mg/m3 CEV

Canada - Quebec - Ceiling Limits
P 0.3 ppm; P 0.75 mg/m3

United Kingdom - Occupational Exposure Standards - STELs
0.3 ppm STEL; 0.6 mg/m3 STEL

Israel - Ceiling Exposure Limits
C 0.3 ppm; C 0.75 mg/m3

STATE LISTS

California - Exposure Limits - Ceilings
C 0.3 ppm; C 0.6 mg/m3

California - Directors List of Hazardous Substances (8 CCR 339)
[present]

Florida Hazardous Substance List
[present]

Massachusetts Right To Know List
[present]

Minnesota Hazardous Substance List
[present]

NJ Right to Know List (Total)
sn 0556

Pennsylvania Right to Know List
environmental hazard

PROPOSED REGULATIONS

Safe Drinking Water Act - Priority list
[present]

Canada - Ontario - Proposed Occupational CEVs
0.25 ppm CEV; 0.6 mg/m3 CEV

CYANOGEN IODIDE   506-78-5

ENVIRONMENTAL LISTS

CERCLA/SARA - Section 302 Extremely Hazardous Substances and TPQs
TPQ = 1000/10,000 pounds

STATE LISTS

Florida Hazardous Substance List
effective March 13, 1992

Massachusetts Right To Know List
extraordinarily hazardous

NJ Right to Know List (Total)
sn 2303

Pennsylvania Right to Know List
environmental hazard
PROPOSED REGULATIONS
CERCLA/SARA - Proposed Hazardous Substance Additions
proposed RQ = 1 pound (454 kg)
CERCLA/SARA - 1989 Proposed RQ Adjustments
proposed RQ = 1000 pounds (454 kg)

CYANOPHOS   2636-26-2
HEALTH AND SAFETY LISTS
U.S. DOT - Appendix B - Marine Pollutants
DOT regulated marine pollutant
NIOSH - Selected LD50s and LC50s
Oral, rat: LD50 = 25 mg/kg Skin, rat: LD50 = 800 mg/kg
ENVIRONMENTAL LISTS
CERCLA/SARA - Section 302 Extremely Hazardous Substances and
TPQs
TPQ = 1000 pounds
STATE LISTS
Florida Hazardous Substance List
effective March 13, 1992
Massachusetts Right To Know List
extraordinarily hazardous
NJ Right to Know List (Total)
sn 2304
Pennsylvania Right to Know List
environmental hazard
PROPOSED REGULATIONS
CERCLA/SARA - Proposed Hazardous Substance Additions
proposed RQ = 1 pound (.454 kg)
CERCLA/SARA - 1989 Proposed RQ Adjustments
proposed RQ = 1000 pounds (454 kg)

CYANURIC CHLORIDE   108-77-0
HEALTH AND SAFETY LISTS
U.S. DOT - Substances From 49 CFR 172.101
regulated by DOT (UN2670)
U.S. DOT - Hazard Classes
DOT hazard class = 8
NIOSH - Selected LD50s and LC50s
Oral, rat: LD50 = 485 mg/kg
INTERNATIONAL LISTS
Canada - WHMIS: Ingredient Disclosure
1% item 461 (468)
STATE LISTS
NJ Right to Know List (Total)
sn 0557
NJ Special Hazardous Substances
(corrosive)

CYANURIC FLUORIDE   675-14-9
HEALTH AND SAFETY LISTS
NIOSH - Selected LD50s and LC50s
*Inhalation, rat: LC50 = 3100 ppb 4 hr Skin, rabbit: LD50 = 160
mg/kg*
OSHA - List of Highly Hazardous Chemicals
threshhold quantity = 100 pounds
ENVIRONMENTAL LISTS
CERCLA/SARA - Section 302 Extremely Hazardous Substances and
TPQs
TPQ = 100 pounds

STATE LISTS
Florida Hazardous Substance List
effective March 13, 1992
Massachusetts Right To Know List
extraordinarily hazardous
Pennsylvania Right to Know List
environmental hazard
PROPOSED REGULATIONS
CERCLA/SARA - Proposed Hazardous Substance Additions
proposed RQ = 1 pound (.454 kg)
CERCLA/SARA - 1989 Proposed RQ Adjustments
proposed RQ = 100 pounds (45.4 kg)

CYANURIC TRIAZIDE   5637-83-2
HEALTH AND SAFETY LISTS
U.S. DOT - Hazard Classes
Forbidden from transport by the DOT

CYCASIN   14901-08-7
HEALTH AND SAFETY LISTS
IARC - Group 2B (sufficient animal data)
*[present] (Overall evaluation based only on evidence of carcinogenicity
in monograph (10, 1976) or in Supplement 4)*
NIOSH - Selected LD50s and LC50s
Oral, rat: LD50 = 270 mg/kg
OSHA - Possible Select Carcinogens
[present]
ENVIRONMENTAL LISTS
RCRA - Hazardous Constituents-Appendix VIII
hazardous constituent - no waste number
STATE LISTS
California - Air Bill 2588 Appendix A-II
known or potential carcinogen
California - Prop. 65 - Cancer list
carcinogen - initial date 1/1/88
California - Directors List of Hazardous Substances (8 CCR 339)
[present]
Florida Hazardous Substance List
[present]
Massachusetts Right To Know List
carcinogen; extraordinarily hazardous
Minnesota Hazardous Substance List
carcinogen
NJ Special Hazardous Substances
(carcinogen)
Pennsylvania Right to Know List
environmental hazard; special hazardous substance
Pennsylvania RTK - Special Hazardous Substances
[present]

CYCLAMATES   RR-01025-3
HEALTH AND SAFETY LISTS
IARC - Group 3 (not classifiable)
[present]
STATE LISTS
California - Directors List of Hazardous Substances (8 CCR 339)
[present]

CYCLAMIC ACID   100-88-9
HEALTH AND SAFETY LISTS
NIOSH - Selected LD50s and LC50s
Oral, rat: LD50 = 12 gm/kg

CYCLIC AMIDE
RR-01223-7

ENVIRONMENTAL LISTS

TSCA - Chemicals with Significant New Use Rules
PMN number: P-92-131

CYCLIC PHOSPHAZENE, METHACRYLATE DERIVATIVE
RR-01679-5

ENVIRONMENTAL LISTS

TSCA - Chemicals with Significant New Use Rules
PMN number: P-92-1134

CYCLOATE
1134-23-2

ENVIRONMENTAL LISTS

CERCLA/SARA - Section 313 - Emission Reporting
form R reporting required

CYCLOBUTANE
287-23-0

HEALTH AND SAFETY LISTS

U.S. DOT - Substances From 49 CFR 172.101
regulated by DOT (UN2601, UN1146)

U.S. DOT - Hazard Classes
DOT hazard class = 2.1

NFPA - Flash Points
gas (no flash point given)

NFPA - Hazard Identification Ratings
health-1; flammability-4; reactivity-0

STATE LISTS

Florida Hazardous Substance List
[present]

Massachusetts Right To Know List
[present]

NJ Right to Know List (Total)
sn 0559

NJ Special Hazardous Substances
(flammable - fourth degree)

Pennsylvania Right to Know List
[present]

CYCLOBUTYL CHLOROFORMATE
81228-87-7

HEALTH AND SAFETY LISTS

U.S. DOT - Substances From 49 CFR 172.101
regulated by DOT (UN2744)

U.S. DOT - Hazard Classes
DOT hazard class = 6.1

INTERNATIONAL LISTS

Canada - WHMIS: Ingredient Disclosure
1% item 462 (435)

STATE LISTS

NJ Right to Know List (Total)
sn 0560

NJ Special Hazardous Substances
(corrosive)

CYCLOCHLOROTINE
12663-46-6

HEALTH AND SAFETY LISTS

IARC - Group 3 (not classifiable)
[present]

CYCLODODECANE
294-62-2

ENVIRONMENTAL LISTS

EPA - Master Testing List
[present]

CYCLODODECANE, HEXABROMO-
25637-99-

ENVIRONMENTAL LISTS

TSCA - Code of Federal Regulations Citations
40 CFR 712.30(d)

TSCA - PAIR - Reporting List
Reporting Date: November 19, 1982

1,5,9-CYCLODODECATRIENE
4904-61-4

HEALTH AND SAFETY LISTS

U.S. DOT - Substances From 49 CFR 172.101
regulated by DOT (UN2518)

U.S. DOT - Hazard Classes
DOT hazard class = 6.1

NFPA - Flash Points
flash point = 160 degrees F (71 degrees C)

NFPA - Hazard Identification Ratings
flammability-2; reactivity-0

INTERNATIONAL LISTS

Canada - WHMIS: Ingredient Disclosure
1% item 463 (599)

STATE LISTS

NJ Right to Know List (Total)
sn 0561

CYCLOHEPTAAMYLOSE
7585-39-9

HEALTH AND SAFETY LISTS

NIOSH - Selected LD50s and LC50s
Oral, rat: LD50 = 18800 mg/kg

CYCLOHEPTANE
291-64-5

HEALTH AND SAFETY LISTS

U.S. DOT - Substances From 49 CFR 172.101
regulated by DOT (UN2241)

U.S. DOT - Hazard Classes
DOT hazard class = 3

NFPA - Flash Points
flash point < 70 degrees F (21 degrees C)

NFPA - Hazard Identification Ratings
health-0; flammability-3; reactivity-0

STATE LISTS

Florida Hazardous Substance List
[present]

Massachusetts Right To Know List
[present]

NJ Right to Know List (Total)
sn 0562

NJ Special Hazardous Substances
(flammable - third degree)

Pennsylvania Right to Know List
[present]

CYCLOHEPTATRIENE
544-25-2

HEALTH AND SAFETY LISTS

U.S. DOT - Substances From 49 CFR 172.101
regulated by DOT (UN2603)

U.S. DOT - Hazard Classes
DOT hazard class = 3

NIOSH - Selected LD50s and LC50s
Oral, rat: LD50 = 57 mg/kg Skin, rat: LD50 = 442 mg/kg

INTERNATIONAL LISTS

Canada - WHMIS: Ingredient Disclosure
1% item 464 (600)

STATE LISTS
  NJ Right to Know List (Total)
  sn 0563

CYCLOHEPTENE   628-92-2

HEALTH AND SAFETY LISTS
  U.S. DOT - Substances From 49 CFR 172.101
  regulated by DOT (UN2242)
  U.S. DOT - Hazard Classes
  DOT hazard class = 3

STATE LISTS
  NJ Right to Know List (Total)
  sn 0564

CYCLOHEXANE   110-82-7

HEALTH AND SAFETY LISTS
  ACGIH 1995 - Time Weighted Averages
  300 ppm TWA; 1030 mg/m3 TWA
  AIHA - Odor Threshold Values
  geometric mean air odor threshold = 780 ppm (detectable)
  U.S. DOT - Substances From 49 CFR 172.101
  regulated by DOT (UN1145)
  U.S. DOT - Hazard Classes
  DOT hazard class = 3
  U.S. DOT - Appendix A Table 1 - Hazardous Substances
  final RQ = 1000 pounds (454 kg)
  NFPA - Flash Points
  flash point = -4 degrees F (-20 degrees C)
  NFPA - Hazard Identification Ratings
  health-1; flammability-3; reactivity-0
  NIOSH - Selected LD50s and LC50s
  Oral, rat: LD50 = 12705 mg/kg
  NIOSH 1990 - Pocket Guide - RELs
  300 ppm TWA; 1050 mg/m3 TWA
  NIOSH 1990 - Pocket Guide - IDLHs
  10,000 ppm IDLH
  NIOSH 1990 - Pocket Guide - Target organs
  eyes, skin, CNS, respiratory system
  OSHA - Vacated PELs - Time Weighted Averages
  300 ppm TWA; 1050 mg/m3 TWA
  OSHA - Final PELs - Time Weighted Averages
  300 ppm TWA; 1050 mg/m3 TWA

ENVIRONMENTAL LISTS
  CERCLA/SARA - Section 313 - Emission Reporting
  form R reporting required for 1.0% de minimus concentration
  CERCLA/SARA - Hazardous Substances and their Reportable Quantities
  final RQ = 1000 pounds (454 kg)
  CAA - HON Rule - SOCMI Chemicals
  compliance by Oct. 24, 1994
  Clean Water Act - Hazardous Substances
  [present]
  EPA - Master Testing List
  [present]
  List of Pesticide Product Inert Ingredients
  [present]
  RCRA - U Series Wastes
  waste number U056 (Ignitable waste)
  RCRA - Hazardous Constituents-Appendix VIII
  waste number U056 (Ignitable waste)
  RCRA - Substances Banned From Land Disposal
  [present]

TSCA - Code of Federal Regulations Citations
  40 CFR 712.30(p); 40 CFR 716.120(a)
TSCA - PAIR - Reporting List
  Reporting Date: February 18, 1986
TSCA - Health and Safety Reporting List
  Effective Date: December 19, 1985

INTERNATIONAL LISTS
  Australian Exposure Standards - Time Weighted Averages
  300 ppm TWA; 1030 mg/m3 TWA
  Canada - WHMIS: Ingredient Disclosure
  1% item 465 (601)
  Canada - NPRI (National Pollutant Release Inventory)
  [present]
  Canada - Alberta - 8 Hour Occupational Exposure Limit
  300 ppm TWA; 1030 mg/m3 TWA
  Canada - Alberta - 15 Minute Occupational Exposure Limit
  375 ppm STEL; 1290 mg/m3 STEL
  Canada - British Columbia - 8 Hour Exposure Limits
  300 ppm TWA; 1050 mg/m3 TWA
  Canada - British Columbia - 15 Minute Exposure Limits
  375 ppm STEL; 1300 mg/m3 STEL
  Canada - Ontario - OHSA - TWAEVs
  300 ppm TWAEV; 1030 mg/m3 TWAEV
  Canada - Quebec - Time-Weighted Average Exposure Values
  300 ppm TWAEV; 1030 mg/m3 TWAEV
  German (DFG) - MAK Values
  300 ppm MAK; 1050 mg/m3 MAK
  German (DFG) - Peak Limitations
  2 x normal MAK (30 min. average value); don't exceed 4 times during shift
  Israel - Time Weighted Averages
  300 ppm TWA; 1030 mg/m3 TWA
  Israel - Action Levels
  150 ppm AL; 515 mg/m3 AL
  Mexico - Instruction No. 10 - TWAs
  300 ppm TWA; 1050 mg/m3 TWA
  Mexico - Instruction No. 10 - STELs
  375 ppm STEL; 1300 mg/m3 STEL

STATE LISTS
  California - Air Bill 2588 Appendix A-I
  6/91
  California - Exposure Limits - PELs
  300 ppm PEL; 1050 mg/m3 PEL
  California - Directors List of Hazardous Substances (8 CCR 339)
  [present]
  Florida Hazardous Substance List
  [present]
  Massachusetts Right To Know List
  [present]
  Minnesota Hazardous Substance List
  [present]
  NJ Right to Know List (Total)
  sn 0565
  NJ Special Hazardous Substances
  (flammable - third degree)
  Pennsylvania Right to Know List
  environmental hazard

PROPOSED REGULATIONS
  Canada - Ontario - Proposed Occupational TWAEVs
  150 ppm TWAEV; 525 mg/m3 TWAEV

CYCLOHEXANE, 1,4-BIS(ISOCYANATOMETHYL)-   10347-54-3
ENVIRONMENTAL LISTS
CERCLA/SARA - Section 313 - Emission Reporting
form R reporting required; (Listed under 'Diisocyanates')
TSCA - Code of Federal Regulations Citations
40 CFR 716.120(a)
TSCA - Health and Safety Reporting List
Effective Date: June 1, 1987

CYCLOHEXANE, 1,3-BIS(ISOCYANATOMETHYL)-   38661-72-2
ENVIRONMENTAL LISTS
CERCLA/SARA - Section 313 - Emission Reporting
form R reporting required; (Listed under ''Diisocyanates')
TSCA - Code of Federal Regulations Citations
40 CFR 716.120(a)
TSCA - Health and Safety Reporting List
Effective Date: June 1, 1987; Sunset Date: November 9, 1993

CYCLOHEXANECARBONITRILE, 1,3,3-TRIMETHYL-   7027-11-4
5-OXO-,
ENVIRONMENTAL LISTS
CAA - HON Rule - SOCMI Chemicals
compliance by Oct. 23, 1995
TSCA - Chemicals with Significant New Use Rules
PMN number: P-90-1358
TSCA - Section 12(b) - Export Notification
P-90-1358; export notification required - Section 5

CYCLOHEXANE CARBOXYLIC ACID   98-89-5
HEALTH AND SAFETY LISTS
NIOSH - Selected LD50s and LC50s
Oral, rat: LD50 = 3265 mg/kg

CIS- AND TRANS-1,4-CYCLOHEXANEDIAMINE   2615-25-0
ENVIRONMENTAL LISTS
TSCA - Chemicals with Significant New Use Rules
PMN numbers: P-87-1881; P-87-1882
TSCA - Section 12(b) - Export Notification
P-87-1882; export notification required - Section 5

(CIS) 1,4-CYLCOHEXANEDIAMINE   15827-56-2
ENVIRONMENTAL LISTS
TSCA - Section 12(b) - Export Notification
P-87-1881; export notification required - Section 5

1,2-CYCLOHEXANEDICARBOXYLIC ACID, BIS   5493-45-8
(OXIRANYLMETHYL) ESTER
SEE ALSO:
GLYCIDOL (OXIRANEMETHANOL) AND ITS DERIVATIVES
ENVIRONMENTAL LISTS
EPA - Master Testing List
[present]
TSCA - Code of Federal Regulations Citations
40 CFR 716.120(c)
PROPOSED REGULATIONS
TSCA - Proposed Testing Rule for Glycidyl Ethers
subject to mutagenicity testing (results apply to all members of Glycidyl subcategory VII-C)

1,2-CYCLOHEXANEDICARBOXYLIC ACID, 2,   RR-01653-
2-BIS[[[[2-[(OXIRANYLMETHOXY)CARBONYL]
CYCLOHEXYLCARBONYL]OXY]METHYL]-1,3-
PROPANEDIYL BIS(OXIRANYLMETHYL) ESTER
ENVIRONMENTAL LISTS
TSCA - Chemicals with Significant New Use Rules
PMN number: P-92-471

CYCLOHEXANE, 1,4-DIISOCYANATO-   2556-36-7
ENVIRONMENTAL LISTS
CERCLA/SARA - Section 313 - Emission Reporting
form R reporting required; (Listed under 'Diisocyanates')
TSCA - Code of Federal Regulations Citations
40 CFR 716.120(a)
TSCA - Health and Safety Reporting List
Effective Date: June 1, 1987; Sunset Date: November 9, 1993

1,4-CYCLOHEXANE DIMETHANOL   105-08-8
HEALTH AND SAFETY LISTS
NFPA - Flash Points
flash point = 332 degrees F (167 degrees C)
NFPA - Hazard Identification Ratings
flammability-1; reactivity-0

CYCLOHEXANE, 2-HEPTYL-3,4-BIS(9-ISO-   68239-06-5
CYANATONONYL)-1-PENTYL-
ENVIRONMENTAL LISTS
TSCA - Code of Federal Regulations Citations
40 CFR 712.30(x); 40 CFR 716.120(d)
TSCA - PAIR - Reporting List
Reporting Date: December 27, 1990
TSCA - Health and Safety Reporting List
Effective Date: October 29, 1990; Sunset Date: November 9, 1993

CYCLOHEXANE, TETRABROMODICHLORO-   30554-72-4
ENVIRONMENTAL LISTS
TSCA - Code of Federal Regulations Citations
40 CFR 712.30(x); 40 CFR 716.120(d)
TSCA - PAIR - Reporting List
Reporting Date: December 27, 1990
TSCA - Health and Safety Reporting List
Effective Date: October 29, 1990

CYCLOHEXANETHIOL   1569-69-3
HEALTH AND SAFETY LISTS
U.S. DOT - Substances From 49 CFR 172.101
regulated by DOT (UN3054)
U.S. DOT - Hazard Classes
DOT hazard class = 3
NFPA - Flash Points
flash point = 110 degrees F (43 degrees C)
NFPA - Hazard Identification Ratings
flammability-2; reactivity-0
NIOSH - Selected LD50s and LC50s
Oral, rat: LD50 = 558 mg/kg
NIOSH - Health Standards - Exposure Limits
C (15 min) 0.5 ppm; C (15 min) 2.4 mg/m3 (Listed under 'Thiols')
NIOSH - Health Standards - Health Effects and Precautions
Irritation; eye, skin, blood, and nervous system effects (Blood and urine monitoring required; prevent skin contact) (Listed under 'Thiols')
STATE LISTS
Minnesota Hazardous Substance List
[present]

NJ Right to Know List (Total)
sn 2306

CYCLOHEXANE, TRIBROMOCHLORO-   30554-73-5

ENVIRONMENTAL LISTS

TSCA - Code of Federal Regulations Citations
40 CFR 712.30(x); 40 CFR 716.120(d)

TSCA - PAIR - Reporting List
Reporting Date: December 27, 1990

TSCA - Health and Safety Reporting List
Effective Date: October 29, 1990

CYCLOHEXANOL   108-93-0

HEALTH AND SAFETY LISTS

ACGIH 1995 - Time Weighted Averages
50 ppm TWA; 206 mg/m3 TWA

ACGIH 1995 - Skin Designations
skin - potential for cutaneous absorption

AIHA - Odor Threshold Values
geometric mean air odor threshold = 0.16 ppm (detectable)

NFPA - Flash Points
flash point = 154 degrees F (68 degrees C)

NFPA - Hazard Identification Ratings
health-1; flammability-2; reactivity-0

NIOSH - Selected LD50s and LC50s
Oral, rat: LD50 = 2060 mg/kg

NIOSH 1990 - Pocket Guide - RELs
50 ppm TWA; 200 mg/m3 TWA

NIOSH 1990 - Pocket Guide - IDLHs
3500 ppm IDLH

NIOSH 1990 - Pocket Guide - Target organs
eyes, skin, respiratory system

NIOSH 1990 - Pocket Guide - Skin list
Potential for dermal absorption

OSHA - Vacated PELs - Time Weighted Averages
50 ppm TWA; 200 mg/m3 TWA

OSHA - Vacated PELs - Skin Designation
Prevent or reduce skin absorption

OSHA - Final PELs - Time Weighted Averages
50 ppm TWA; 200 mg/m3 TWA

ENVIRONMENTAL LISTS

CERCLA/SARA - Section 313 - Emission Reporting
form R reporting required

CAA - HON Rule - SOCMI Chemicals
compliance by Oct. 24, 1994

List of Pesticide Product Inert Ingredients
[present]

INTERNATIONAL LISTS

Australian Exposure Standards - Time Weighted Averages
50 ppm TWA; 206 mg/m3 TWA

Australian Exposure Standards - Skin Effects
skin absorption

Canada - WHMIS: Ingredient Disclosure
1% item 466 (602)

Canada - Alberta - 8 Hour Occupational Exposure Limit
50 ppm TWA; 205 mg/m3 TWA

Canada - Alberta - 15 Minute Occupational Exposure Limit
75 ppm STEL; 305 mg/m3 STEL

Canada - British Columbia - 8 Hour Exposure Limits
50 ppm TWA; 200 mg/m3 TWA

Canada - Ontario - OHSA - TWAEVs
50 ppm TWAEV; 200 mg/m3 TWAEV

Canada - Ontario - OHSA - Skin Notations
absorption through skin, eyes, or mucous membranes

Canada - Quebec - Time-Weighted Average Exposure Values
50 ppm TWAEV; 206 mg/m3 TWAEV

Canada - Quebec - Skin Designations
absorbed through the skin

United Kingdom - Occupational Exposure Standards - TWAs
50 ppm TWA; 200 mg/m3 TWA

German (DFG) - MAK Values
50 ppm MAK; 200 mg/m3 MAK

German (DFG) - Peak Limitations
2 x normal MAK (30 min. average value); don't exceed 4 times during shift

Israel - Time Weighted Averages
50 ppm TWA; 206 mg/m3 TWA

Israel - Action Levels
25 ppm AL; 103 mg/m3 AL

Mexico - Instruction No. 10 - TWAs
50 ppm TWA; 200 mg/m3 TWA

STATE LISTS

California - Exposure Limits - PELs
50 ppm PEL; 200 mg/m3 PEL

California - Exposure Limits - Skin Notation
material may be absorbed through the skin, eyes or mucous membrane

California - Directors List of Hazardous Substances (8 CCR 339)
[present]

Florida Hazardous Substance List
[present]

Massachusetts Right To Know List
[present]

Minnesota Hazardous Substance List
skin

NJ Right to Know List (Total)
sn 0569

Pennsylvania Right to Know List
[present]

CYCLOHEXANONE   108-94-1

SEE ALSO:

F039-HAZARDOUS WASTES
F003-HAZARDOUS WASTES

HEALTH AND SAFETY LISTS

ACGIH 1995 - Time Weighted Averages
25 ppm TWA; 100 mg/m3 TWA

ACGIH 1995 - Skin Designations
skin - potential for cutaneous absorption

AIHA - Odor Threshold Values
geometric mean air odor threshold = 3.5 ppm (detectable); 0.12 ppm (recognizable)

U.S. DOT - Substances From 49 CFR 172.101
regulated by DOT (UN1915)

U.S. DOT - Hazard Classes
DOT hazard class = 3

U.S. DOT - Appendix A Table 1 - Hazardous Substances
final RQ = 5000 pounds (2270 kg)

IARC - Group 3 (not classifiable)
[present]

NFPA - Flash Points
flash point = 111 degrees F (44 degrees C)

NFPA - Hazard Identification Ratings
health-1; flammability-2; reactivity-0

NIOSH - Selected LD50s and LC50s
Inhalation, rat: LC50 = 8000 ppm 4 hr Oral, rat: LD50 = 1535 mg/kg Skin, rabbit: LD50 = 948 mg/kg

NIOSH 1990 - Pocket Guide - RELs
25 ppm TWA;   100 mg/m3 TWA

NIOSH 1990 - Pocket Guide - IDLHs
5000 ppm IDLH

NIOSH 1990 - Pocket Guide - Target organs
eyes, skin, CNS, respiratory system

NIOSH 1990 - Pocket Guide - Skin list
Potential for dermal absorption

NIOSH - Health Standards - Exposure Limits
25 ppm TWA;   100 mg/m3 TWA (Listed under 'Ketones')

NIOSH - Health Standards - Health Effects and Precautions
Irritation;   liver, kidney, and nervous system effects (Urinalysis required) (Listed under "Ketones")

NTP Chemical Status Reports - Testing Status and NTIS Number
Chronic studies exist for which technical reports were not prepared

OSHA - Vacated PELs - Time Weighted Averages
25 ppm TWA;   100 mg/m3 TWA

OSHA - Vacated PELs - Skin Designation
Prevent or reduce skin absorption

OSHA - Final PELs - Time Weighted Averages
50 ppm TWA;   200 mg/m3 TWA

ENVIRONMENTAL LISTS

CERCLA/SARA - Hazardous Substances and their Reportable Quantities
final RQ = 5000 pounds (2270 kg)

CAA - HON Rule - SOCMI Chemicals
compliance by Oct. 24, 1994

EPA - Master Testing List
[present]

List of Pesticide Product Inert Ingredients
[present]

RCRA - U Series Wastes
waste number U057 (Ignitable waste)

RCRA - Hazardous Constituents-Appendix VIII
waste number U057 (Ignitable waste)

RCRA - Basis for Listing - Appendix VII
Included in waste stream: F039

RCRA - Substances Banned From Land Disposal
[present]

RCRA - Universal Treatment Standards (LDR)
WW: 0.36 mg/l;   NWW: 0.75 mg/l TCLP

TSCA - Code of Federal Regulations Citations
40 CFR 712.30(d);   40 CFR 716.120

TSCA - PAIR - Reporting List
Reporting Date: November 19, 1982

TSCA - Health and Safety Reporting List
Effective Date: October 4, 1982

INTERNATIONAL LISTS

Australian Exposure Standards - Time Weighted Averages
25 ppm TWA;   100 mg/m3 TWA

Australian Exposure Standards - Skin Effects
skin absorption

Canada - WHMIS: Ingredient Disclosure
0.1% item 467 (603)

Canada - Alberta - 8 Hour Occupational Exposure Limit
25 ppm TWA;   100 mg/m3 TWA

Canada - Alberta - 15 Minute Occupational Exposure Limit
100 ppm STEL;   400 mg/m3 STEL

Canada - British Columbia - 8 Hour Exposure Limits
50 ppm TWA;   200 mg/m3 TWA

Canada - Ontario - OHSA - TWAEVs
25 ppm TWAEV;   100 mg/m3 TWAEV

Canada - Ontario - OHSA - Skin Notations
absorption through skin, eyes, or mucous membranes

Canada - Quebec - Time-Weighted Average Exposure Values
25 ppm TWAEV;   100 mg/m3 TWAEV

Canada - Quebec - Skin Designations
absorbed through the skin

United Kingdom - Occupational Exposure Standards - TWAs
25 ppm TWA;   100 mg/m3 TWA

United Kingdom - Occupational Exposure Standards - STELs
100 ppm STEL;   400 mg/m3 STEL

German (DFG) - MAK Values
50 ppm MAK;   200 mg/m3 MAK

German (DFG) - Peak Limitations
2 x normal MAK (30 min. average value);   don't exceed 4 times during shift

German (DFG) - Skin/Sensitizers
danger of cutaneous absorption

German (DFG) - Carcinogens
suspected carcinogen

German (DFG) - Pregnancy
no risk to embryo/fetus if exposure limits adhered to

Israel - Time Weighted Averages
25 ppm TWA;   100 mg/m3 TWA

Israel - Action Levels
12.5 ppm AL;   50 mg/m3 AL

Mexico - Instruction No. 10 - TWAs
50 ppm TWA;   200 mg/m3 TWA

Mexico - Instruction No. 10 - STELs
100 ppm STEL;   400 mg/m3 STEL

STATE LISTS

California - Exposure Limits - PELs
25 ppm PEL;   100 mg/m3 PEL

California - Exposure Limits - Skin Notation
material may be absorbed through the skin, eyes or mucous membrane

California - Directors List of Hazardous Substances (8 CCR 339)
[present]

Florida Hazardous Substance List
[present]

Massachusetts Right To Know List
[present]

Minnesota Hazardous Substance List
skin

NJ Right to Know List (Total)
sn 0570

Pennsylvania Right to Know List
environmental hazard

CYCLOHEXANONE, 2-(O-CHLOROPHENYL)-2-   **1867-66-9**
(METHYLAMINE)-, HYDROCHLORIDE

HEALTH AND SAFETY LISTS

NIOSH - Selected LD50s and LC50s
Oral, rat: LD50 = 447 mg/kg

CYCLOHEXANONE OXIME   **100-64-1**

HEALTH AND SAFETY LISTS

NTP Chemical Status Reports - Testing Status and NTIS Number
Short term toxicity studies scheduled for peer review;   project leader assigned/study in design

CYCLOHEXANONE PEROXIDE   **78-18-2**

INTERNATIONAL LISTS

Canada - WHMIS: Ingredient Disclosure
0.1% item 468 (1363)

German (DFG) - Skin/Sensitizers
very strong effects on skin

STATE LISTS
  NJ Right to Know List (Total)
    sn 2475; sn 2474

CYCLOHEXANONE PEROXIDE   12262-58-7

STATE LISTS
  NJ Right to Know List (Total)
    sn 0571

CYCLOHEXENE   110-83-8

HEALTH AND SAFETY LISTS
  ACGIH 1995 - Time Weighted Averages
    300 ppm TWA; 1010 mg/m3 TWA
  AIHA - Odor Threshold Values
    no geometric mean air odor threshold
  U.S. DOT - Substances From 49 CFR 172.101
    regulated by DOT (UN2256)
  U.S. DOT - Hazard Classes
    DOT hazard class = 3
  NFPA - Flash Points
    flash point < 20 degrees F (-7 degrees C)
  NFPA - Hazard Identification Ratings
    health-1; flammability-3; reactivity-0
  NIOSH 1990 - Pocket Guide - RELs
    300 ppm TWA; 1015 mg/m3 TWA
  NIOSH 1990 - Pocket Guide - IDLHs
    10,000 ppm IDLH
  NIOSH 1990 - Pocket Guide - Target organs
    skin, eyes, respiratory system
  OSHA - Vacated PELs - Time Weighted Averages
    300 ppm TWA; 1015 mg/m3 TWA
  OSHA - Final PELs - Time Weighted Averages
    300 ppm TWA; 1015 mg/m3 TWA

ENVIRONMENTAL LISTS
  TSCA - PAIR - Reporting List
    Effective Date: January 26, 1994; Reporting Date: March 28, 1994
  TSCA - Health and Safety Reporting List
    Effective Date: January 26, 1994; Sunset Date: January 26, 2004

INTERNATIONAL LISTS
  Australian Exposure Standards - Time Weighted Averages
    300 ppm TWA; 1010 mg/m3 TWA
  Canada - WHMIS: Ingredient Disclosure
    1% item 469 (604)
  Canada - Alberta - 8 Hour Occupational Exposure Limit
    300 ppm TWA; 1010 mg/m3 TWA
  Canada - Alberta - 15 Minute Occupational Exposure Limit
    375 ppm STEL; 1260 mg/m3 STEL
  Canada - British Columbia - 8 Hour Exposure Limits
    300 ppm TWA; 1015 mg/m3 TWA
  Canada - Ontario - OHSA - TWAEVs
    300 ppm TWAEV; 1010 mg/m3 TWAEV
  Canada - Quebec - Time-Weighted Average Exposure Values
    300 ppm TWAEV; 1010 mg/m3 TWAEV
  United Kingdom - Occupational Exposure Standards - TWAs
    300 ppm TWA; 1015 mg/m3 TWA
  German (DFG) - MAK Values
    300 ppm MAK; 1015 mg/m3 MAK
  German (DFG) - Peak Limitations
    2 x normal MAK (30 min. average value); don't exceed 4 times during shift
  Israel - Time Weighted Averages
    300 ppm TWA; 1010 mg/m3 TWA

Israel - Action Levels
    150 ppm AL; 505 mg/m3 AL
Mexico - Instruction No. 10 - TWAs
    300 ppm TWA; 1015 mg/m3 TWA

STATE LISTS
  California - Exposure Limits - PELs
    300 ppm PEL; 1015 mg/m3 PEL
  California - Directors List of Hazardous Substances (8 CCR 339)
    [present]
  Florida Hazardous Substance List
    [present]
  Massachusetts Right To Know List
    [present]
  Minnesota Hazardous Substance List
    [present]
  NJ Right to Know List (Total)
    sn 0572
  Pennsylvania Right to Know List
    [present]

PROPOSED REGULATIONS
  TSCA - ITC 31st Report Priority Testing List
    designated to be tested
  Canada - Ontario - Proposed Occupational TWAEVs
    150 ppm TWAEV; 510 mg/m3 TWAEV

3-CYCLOHEXENE-1-CARBOXALDEHYDE, DIMETHYL-   27939-60-2

ENVIRONMENTAL LISTS
  TSCA - Code of Federal Regulations Citations
    40 CFR 712.30(x), 40 CFR 716.120(d)
  TSCA - PAIR - Reporting List
    Reporting Date: November 27, 1991
  TSCA - Health and Safety Reporting List
    Effective Date: September 30, 1991

3-CYCLOHEXENE-1-CARBOXALDEHYDE, 4-(4-HYDROXY-4-METHYLPENTYL)-   31906-04-4

ENVIRONMENTAL LISTS
  TSCA - Code of Federal Regulations Citations
    40 CR 712.30(x); 40 CFR 716.120(d)
  TSCA - PAIR - Reporting List
    Reporting Date: November 27, 1991
  TSCA - Health and Safety Reporting List
    Effective Date: September 30, 1991

3-CYCLOHEXENE-1-CARBOXALDEHYDE, 1-METHYL-4-(4-METHYL-3-PENTENYL)-   52475-86-2

ENVIRONMENTAL LISTS
  TSCA - Code of Federal Regulations Citations
    40 CFR 712.30(x); 40 CFR 716.120(d)
  TSCA - PAIR - Reporting List
    Reporting Date: November 27, 1991
  TSCA - Health and Safety Reporting List
    Effective Date: September 30, 1991

3-CYCLOHEXENE-1-CARBOXALDEHYDE, 1-METHYL-4-(4-METHYLPENTYL)-   66327-54-6

ENVIRONMENTAL LISTS
  TSCA - Code of Federal Regulations Citations
    40 CFR 712.30(x); 40 CFR 716.120(d)
  TSCA - PAIR - Reporting List
    Reporting Date: November 27, 1991
  TSCA - Health and Safety Reporting List
    Effective Date: September 30, 1991

3-CYCLOHEXENE-1-CARBOXALDEHYDE, 4-(4-METHYL-3-PENTENYL)-   37677-14-8

ENVIRONMENTAL LISTS

TSCA - Code of Federal Regulations Citations
40 CFR 712.30(x);   40 CFR 716.120(d)

TSCA - PAIR - Reporting List
Reporting Date: November 27, 1991

TSCA - Health and Safety Reporting List
Effective Date: September 30, 1991

3-CYCLOHEXENE-1-CARBOXALDEHYDE, 2,4,6-TRIMETHYL-   1423-46-7

ENVIRONMENTAL LISTS

TSCA - Code of Federal Regulations Citations
40 CFR 712.30(x);   40 CFR 716.120(d)

TSCA - PAIR - Reporting List
Reporting Date: November 27, 1991

TSCA - Health and Safety Reporting List
Effective Date: September 30, 1991

2-CYCLOHEXENE-1-OCTANOIC ACID, 5(OR 6)-CARBOXY-4-HEXYL-   53980-88-4

ENVIRONMENTAL LISTS

List of Pesticide Product Inert Ingredients
[present]

2-CYCLOHEXENE-1-ONE   930-68-7

HEALTH AND SAFETY LISTS

NFPA - Flash Points
flash point = 93 degrees F (34 degrees C)

NFPA - Hazard Identification Ratings
health-1;   flammability-3;   reactivity-0

NIOSH - Selected LD50s and LC50s
Inhalation, rat: LC50 = 250 ppm 4 hr Oral, rat: LD50 = 220 mg/kg Skin, rabbit: LD50 = 70 mg/kg

NTP Chemical Status Reports - Testing Status and NTIS Number
Project leader assigned/study in design

CYCLOHEXENYL TRICHLOROSILANE   10137-69-6

HEALTH AND SAFETY LISTS

U.S. DOT - Substances From 49 CFR 172.101
regulated by DOT (UN1762)

U.S. DOT - Hazard Classes
DOT hazard class = 8

NIOSH - Selected LD50s and LC50s
Oral, rat: LD50 = 2830 mg/kg Skin, rabbit: LD50 = 630 mg/kg

INTERNATIONAL LISTS

Canada - WHMIS: Ingredient Disclosure
1% item 470 (605)

STATE LISTS

NJ Right to Know List (Total)
sn 0573

NJ Special Hazardous Substances
(corrosive)

CYCLOHEXIMIDE   66-81-9

HEALTH AND SAFETY LISTS

NIOSH - Selected LD50s and LC50s
Oral, rat: LD50 = 2 mg/kg

ENVIRONMENTAL LISTS

CERCLA/SARA - Section 302 Extremely Hazardous Substances and TPQs
TPQ = 100/10,000 pounds

STATE LISTS

California - Air Bill 2588 Appendix A-I
[present]

California - Prop. 65 - Developmental Toxicity
developmental toxicity - initial date 1/1/89

California - Directors List of Hazardous Substances (8 CCR 339)
[present]

Florida Hazardous Substance List
effective March 13, 1992

Massachusetts Right To Know List
extraordinarily hazardous

NJ Right to Know List (Total)
sn 0574

NJ Special Hazardous Substances
(mutagen)

Pennsylvania Right to Know List
environmental hazard

PROPOSED REGULATIONS

CERCLA/SARA - Proposed Hazardous Substance Additions
proposed RQ = 1 pound (.454 kg)

CERCLA/SARA - 1989 Proposed RQ Adjustments
proposed RQ = 100 pounds (45.4 kg)

CYCLOHEXYL ACETATE   622-45-7

HEALTH AND SAFETY LISTS

U.S. DOT - Substances From 49 CFR 172.101
regulated by DOT (UN2243)

U.S. DOT - Hazard Classes
DOT hazard class = 3

NFPA - Flash Points
flash point = 136 degrees F (58 degrees C)

NFPA - Hazard Identification Ratings
health-1;   flammability-2;   reactivity-0

NIOSH - Selected LD50s and LC50s
Oral, rat: LD50 = 6730 mg/kg Skin, rabbit: LD50 = 10 gm/kg

INTERNATIONAL LISTS

Canada - WHMIS: Ingredient Disclosure
1% item 471 (15)

STATE LISTS

NJ Right to Know List (Total)
sn 0575

CYCLOHEXYLAMINE   108-91-8

HEALTH AND SAFETY LISTS

ACGIH 1995 - Time Weighted Averages
10 ppm TWA;   41 mg/m3 TWA

U.S. DOT - Substances From 49 CFR 172.101
regulated by DOT (UN2357)

U.S. DOT - Hazard Classes
DOT hazard class = 8

NFPA - Flash Points
flash point = 88 degrees F (31 degrees C)

NFPA - Hazard Identification Ratings
health-3;   flammability-3;   reactivity-0

NIOSH - Selected LD50s and LC50s
Oral, rat: LD50 = 156 mg/kg Skin, rabbit: LD50 = 277 mg/kg

OSHA - Vacated PELs - Time Weighted Averages
10 ppm TWA;   40 mg/m3 TWA

ENVIRONMENTAL LISTS

CERCLA/SARA - Section 302 Extremely Hazardous Substances and TPQs
TPQ = 10,000 pounds

CAA -Toxic Substances for Accidental Release Prevention
threshold quantity = 15,000 lbs

CAA - HON Rule - SOCMI Chemicals
compliance by April 24, 1995

INTERNATIONAL LISTS

Australian Exposure Standards - Time Weighted Averages
10 ppm TWA; 41 mg/m3 TWA

Canada - WHMIS: Ingredient Disclosure
0.1% item 472 (606)

Canada - Alberta - 8 Hour Occupational Exposure Limit
10 ppm TWA; 41 mg/m3 TWA

Canada - Alberta - 15 Minute Occupational Exposure Limit
20 ppm STEL; 82 mg/m3 STEL

Canada - Alberta - Skin Designation
can be absorbed through the intact skin

Canada - British Columbia - 8 Hour Exposure Limits
10 ppm TWA; 40 mg/m3 TWA

Canada - British Columbia - Skin Notations
skin - potential for skin absorption

Canada - Ontario - OHSA - TWAEVs
10 ppm TWAEV; 40 mg/m3 TWAEV

Canada - Quebec - Time-Weighted Average Exposure Values
10 ppm TWAEV; 40 mg/m3 TWAEV

United Kingdom - Occupational Exposure Standards - TWAs
10 ppm TWA; 40 mg/m3 TWA

United Kingdom - Occupational Exposure Standards - Notes
can be absorbed through skin

German (DFG) - MAK Values
10 ppm MAK; 40 mg/m3 MAK

German (DFG) - Peak Limitations
2 x normal MAK (10 min momentary value); don't exceed 4 times per shift

German (DFG) - Pregnancy
classification not yet possible

Israel - Time Weighted Averages
10 ppm TWA; 41 mg/m3 TWA

Israel - Action Levels
5 ppm AL; 20.5 mg/m3 AL

Mexico - Instruction No. 10 - TWAs
10 ppm TWA; 40 mg/m3 TWA

Mexico - Instruction No. 10 - Skin designation
skin - potential for cutaneous absorption

STATE LISTS

California - Exposure Limits - PELs
10 ppm PEL; 40 mg/m3 PEL

California - Exposure Limits - Skin Notation
material may be absorbed through the skin, eyes or mucous membrane

California - Directors List of Hazardous Substances (8 CCR 339)
[present]

Florida Hazardous Substance List
[present]

Massachusetts Right To Know List
extraordinarily hazardous

Minnesota Hazardous Substance List
[present]

NJ Right to Know List (Total)
sn 0576

NJ Special Hazardous Substances
(corrosive; flammable - third degree; mutagen)

Pennsylvania Right to Know List
environmental hazard

PROPOSED REGULATIONS

CERCLA/SARA - Proposed Hazardous Substance Additions
proposed RQ = 1 pound (.454 kg)

CERCLA/SARA - 1989 Proposed RQ Adjustments
proposed RQ = 1000 pounds (454 kg)

CYCLOHEXYLBENZENE   827-52-1

HEALTH AND SAFETY LISTS

NFPA - Flash Points
flash point = 210 degrees F (99 degrees C)

NFPA - Hazard Identification Ratings
health-2; flammability-1; reactivity-0

STATE LISTS

Florida Hazardous Substance List
[present]

Massachusetts Right To Know List
[present]

Pennsylvania Right to Know List
[present]

N-CYCLOHEXYL-2-BENZOTHIAZOLESULFENAMIDE   95-33-0

INTERNATIONAL LISTS

Canada - WHMIS: Ingredient Disclosure
1% item 473 (607)

CYCLOHEXYL CHLORIDE   542-18-7

HEALTH AND SAFETY LISTS

NFPA - Flash Points
flash point = 90 degrees F (32 degrees C)

NFPA - Hazard Identification Ratings
health-2; flammability-3; reactivity-0

STATE LISTS

Florida Hazardous Substance List
[present]

Massachusetts Right To Know List
[present]

Pennsylvania Right to Know List
[present]

CYCLOHEXYLCYCLOHEXANOL   6531-86-8

HEALTH AND SAFETY LISTS

NFPA - Flash Points
flash point = 270 degrees F (132 degrees C)

NFPA - Hazard Identification Ratings
health-0; flammability-1; reactivity-0

CYCLOHEXYL FORMATE   RR-00828-6

HEALTH AND SAFETY LISTS

NFPA - Flash Points
flash point = 124 degrees F (51 degrees C)

NFPA - Hazard Identification Ratings
flammability-2; reactivity-0

CYCLOHEXYL ISOCYANATE   3173-53-3

HEALTH AND SAFETY LISTS

U.S. DOT - Substances From 49 CFR 172.101
regulated by DOT (UN2488)

U.S. DOT - Hazard Classes
DOT hazard class = 6.1

U.S. DOT - Substances Which Are Poisonous by Inhalation
liquid hazardous material poisonous by inhalation (UN2488)

ENVIRONMENTAL LISTS

TSCA - Code of Federal Regulations Citations
40 CFR 712.30(x); 40 CFR 716.120(d)

TSCA - PAIR - Reporting List
Reporting Date: December 27, 1990

TSCA - Health and Safety Reporting List
Effective Date: October 29, 1990; Sunset Date: November 9, 1993

INTERNATIONAL LISTS

Canada - WHMIS: Ingredient Disclosure
0.1% item 474 (1038)

STATE LISTS

NJ Right to Know List (Total)
sn 0577

CYCLOHEXYL METHACRYLATE, 2-HYDROX-YLETHYL METHACRYLATE, ISODECYL METHACRYLATE, 2-MORPHOLINOETHYL METHACRYLATE COPOLYMER   119239-21-3

ENVIRONMENTAL LISTS

List of Pesticide Product Inert Ingredients
[present]

O-CYCLOHEXYLPHENOL   119-42-6

HEALTH AND SAFETY LISTS

NFPA - Flash Points
flash point = 273 degrees F (134 degrees C)

NFPA - Hazard Identification Ratings
health-2; flammability-1; reactivity-0

STATE LISTS

Florida Hazardous Substance List
[present]

Massachusetts Right To Know List
[present]

Pennsylvania Right to Know List
[present]

CYCLOHEXYLTRICHLOROSILANE   98-12-4

HEALTH AND SAFETY LISTS

U.S. DOT - Substances From 49 CFR 172.101
regulated by DOT (UN1763)

U.S. DOT - Hazard Classes
DOT hazard class = 8

NFPA - Flash Points
flash point = 196 degrees F (91 degrees C)

NFPA - Hazard Identification Ratings
health-2; flammability-2; reactivity-1

INTERNATIONAL LISTS

Canada - WHMIS: Ingredient Disclosure
1% item 475 (608)

STATE LISTS

Florida Hazardous Substance List
[present]

Massachusetts Right To Know List
[present]

NJ Right to Know List (Total)
sn 0578

NJ Special Hazardous Substances
(corrosive)

Pennsylvania Right to Know List
[present]

CYCLONITE   121-82-4

HEALTH AND SAFETY LISTS

ACGIH 1995 - Time Weighted Averages
1.5 mg/m3 TWA

ACGIH 1995 - Skin Designations
skin - potential for cutaneous absorption

NIOSH - Selected LD50s and LC50s
Oral, rat: LD50 = 100 mg/kg

OSHA - Vacated PELs - Time Weighted Averages
1.5 mg/m3 TWA

OSHA - Vacated PELs - Skin Designation
Prevent or reduce skin absorption

ENVIRONMENTAL LISTS

ATSDR Priority List
Rank (of 275): 213

INTERNATIONAL LISTS

Australian Exposure Standards - Time Weighted Averages
1.5 mg/m3 TWA

Australian Exposure Standards - Skin Effects
skin absorption

Canada - Alberta - 8 Hour Occupational Exposure Limit
1.5 mg/m3 TWA

Canada - Alberta - 15 Minute Occupational Exposure Limit
3 mg/m3 STEL

Canada - Alberta - Skin Designation
can be absorbed through the intact skin

Canada - British Columbia - 8 Hour Exposure Limits
1.5 mg/m3 TWA

Canada - British Columbia - 15 Minute Exposure Limits
3 mg/m3 STEL

Canada - British Columbia - Skin Notations
skin - potential for skin absorption

Canada - Ontario - OHSA - TWAEVs
1.5 mg/m3 TWAEV

Canada - Ontario - OHSA - Skin Notations
absorption through skin, eyes, or mucous membranes

Canada - Quebec - Time-Weighted Average Exposure Values
1.5 mg/m3 TWAEV

Canada - Quebec - Skin Designations
absorbed through the skin

United Kingdom - Occupational Exposure Standards - TWAs
1.5 mg/m3 TWA

United Kingdom - Occupational Exposure Standards - STELs
3 mg/m3 STEL

United Kingdom - Occupational Exposure Standards - Notes
can be absorbed through skin

Israel - Time Weighted Averages
1.5 mg/m3 TWA

Israel - Action Levels
0.75 mg/m3 AL

Mexico - Instruction No. 10 - TWAs
1.5 mg/m3 TWA

Mexico - Instruction No. 10 - Skin designation
skin - potential for cutaneous absorption

STATE LISTS

California - Exposure Limits - PELs
1.5 mg/m3 PEL

California - Exposure Limits - Skin Notation
material may be absorbed through the skin, eyes or mucous membrane

California - Directors List of Hazardous Substances (8 CCR 339)
[present]

Florida Hazardous Substance List
  [present]
Massachusetts Right To Know List
  [present]
Minnesota Hazardous Substance List
  skin
NJ Right to Know List (Total)
  sn 0579
Pennsylvania Right to Know List
  [present]

1,5-CYCLOOCTADIENE   111-78-4

HEALTH AND SAFETY LISTS
NFPA - Flash Points
  flash point = 95 degrees F (35 degrees C)
NFPA - Hazard Identification Ratings
  flammability-3; reactivity-0

STATE LISTS
Pennsylvania Right to Know List
  [present]

1,5-CYCLOOCTADIENE   1552-12-1

STATE LISTS
Florida Hazardous Substance List
  [present]
Massachusetts Right To Know List
  [present]

CYCLOOCTADIENE   29965-97-7

HEALTH AND SAFETY LISTS
U.S. DOT - Substances From 49 CFR 172.101
  regulated by DOT (UN2520)
U.S. DOT - Hazard Classes
  DOT hazard class = 3

ENVIRONMENTAL LISTS
CAA - HON Rule - SOCMI Chemicals
  compliance by Jan. 23, 1995

STATE LISTS
NJ Right to Know List (Total)
  sn 0580
NJ Special Hazardous Substances
  (flammable - third degree)

CYCLOOCTATETRAENE   629-20-9

HEALTH AND SAFETY LISTS
U.S. DOT - Substances From 49 CFR 172.101
  regulated by DOT (UN2358)
U.S. DOT - Hazard Classes
  DOT hazard class = 3

STATE LISTS
NJ Right to Know List (Total)
  sn 0581

CYCLOPENTA[C,D]PYRENE   27208-37-3

HEALTH AND SAFETY LISTS
IARC - Group 3 (not classifiable)
  [present]

STATE LISTS
California - Directors List of Hazardous Substances (8 CCR 339)
  [present]

CYCLOPENTADIENE   542-92-7

HEALTH AND SAFETY LISTS
ACGIH 1995 - Time Weighted Averages
  75 ppm TWA; 203 mg/m3 TWA
AIHA - Odor Threshold Values
  no geometric mean air odor threshold
NIOSH 1990 - Pocket Guide - RELs
  75 ppm TWA; 200 mg/m3 TWA
NIOSH 1990 - Pocket Guide - IDLHs
  2000 ppm IDLH
NIOSH 1990 - Pocket Guide - Target organs
  eyes, respiratory system
OSHA - Vacated PELs - Time Weighted Averages
  75 ppm TWA; 200 mg/m3 TWA
OSHA - Final PELs - Time Weighted Averages
  75 ppm TWA; 200 mg/m3 TWA

INTERNATIONAL LISTS
Australian Exposure Standards - Time Weighted Averages
  75 ppm TWA; 203 mg/m3 TWA
Canada - WHMIS: Ingredient Disclosure
  1% item 476 (609)
Canada - Alberta - 8 Hour Occupational Exposure Limit
  75 ppm TWA; 205 mg/m3 TWA
Canada - Alberta - 15 Minute Occupational Exposure Limit
  150 ppm STEL; 405 mg/m3 STEL
Canada - British Columbia - 8 Hour Exposure Limits
  75 ppm TWA; 200 mg/m3 TWA
Canada - British Columbia - 15 Minute Exposure Limits
  150 ppm STEL; 400 mg/m3 STEL
Canada - Ontario - OHSA - TWAEVs
  75 ppm TWAEV; 200 mg/m3 TWAEV
Canada - Quebec - Time-Weighted Average Exposure Values
  75 ppm TWAEV; 203 mg/m3 TWAEV
German (DFG) - MAK Values
  75 ppm MAK; 200 mg/m3 MAK
Israel - Time Weighted Averages
  75 ppm TWA; 203 mg/m3 TWA
Israel - Action Levels
  37.5 ppm AL; 101.5 mg/m3 AL
Mexico - Instruction No. 10 - TWAs
  75 ppm TWA; 200 mg/m3 TWA

STATE LISTS
California - Exposure Limits - PELs
  75 ppm PEL; 200 mg/m3 PEL
California - Directors List of Hazardous Substances (8 CCR 339)
  [present]
Florida Hazardous Substance List
  [present]
Massachusetts Right To Know List
  [present]
Minnesota Hazardous Substance List
  [present]
Pennsylvania Right to Know List
  [present]

CYCLOPENTANE   287-92-3

HEALTH AND SAFETY LISTS
ACGIH 1995 - Time Weighted Averages
  600 ppm TWA; 1720 mg/m3 TWA
U.S. DOT - Substances From 49 CFR 172.101
  regulated by DOT (UN1146)
U.S. DOT - Hazard Classes
  DOT hazard class = 3

NFPA - Flash Points
flash point < 20 degrees F (-7 degrees C)

NFPA - Hazard Identification Ratings
health-1; flammability-3; reactivity-0

OSHA - Vacated PELs - Time Weighted Averages
600 ppm TWA; 1720 mg/m3 TWA

ENVIRONMENTAL LISTS

TSCA - PAIR - Reporting List
Effective Date: January 26, 1994;   Reporting Date: March 28, 1994

TSCA - Health and Safety Reporting List
Effective Date: January 26, 1994; Sunset Date: January 26, 2004

INTERNATIONAL LISTS

Australian Exposure Standards - Time Weighted Averages
600 ppm TWA; 1720 mg/m3 TWA

Canada - WHMIS: Ingredient Disclosure
1% item 477 (610)

Canada - Alberta - 8 Hour Occupational Exposure Limit
600 ppm TWA; 1720 mg/m3 TWA

Canada - Alberta - 15 Minute Occupational Exposure Limit
900 ppm STEL; 2580 mg/m3 STEL

Canada - Ontario - OHSA - TWAEVs
600 ppm TWAEV; 1720 mg/m3 TWAEV

Canada - Quebec - Time-Weighted Average Exposure Values
600 ppm TWAEV; 1720 mg/m3 TWAEV

Israel - Time Weighted Averages
600 ppm TWA; 1720 mg/m3 TWA

Israel - Action Levels
300 ppm AL; 860 mg/m3 AL

STATE LISTS

California - Exposure Limits - PELs
600 ppm PEL; 1720 mg/m3 PEL

California - Directors List of Hazardous Substances (8 CCR 339)
[present]

Florida Hazardous Substance List
[present]

Massachusetts Right To Know List
[present]

Minnesota Hazardous Substance List
[present]

NJ Right to Know List (Total)
sn 0583

NJ Special Hazardous Substances
(flammable - third degree)

Pennsylvania Right to Know List
[present]

PROPOSED REGULATIONS

TSCA - ITC 31st Report Priority Testing List
designated to be tested

CYCLOPENTANEHEPTANOIC ACID, 3-.ALPHA.-HYDROXY-2-(3-HYDROXY-1-OCTENYL)-5-OXO-   745-65-3

HEALTH AND SAFETY LISTS

NIOSH - Selected LD50s and LC50s
Oral, rat: LD50 = 228 mg/kg

CYCLOPENTANOL   96-41-3

HEALTH AND SAFETY LISTS

U.S. DOT - Substances From 49 CFR 172.101
regulated by DOT (UN2244)

U.S. DOT - Hazard Classes
DOT hazard class = 3

NFPA - Flash Points
flash point = 124 degrees F (51 degrees C)

NFPA - Hazard Identification Ratings
health-0; flammability-2; reactivity-0

STATE LISTS

NJ Right to Know List (Total)
sn 0584

CYCLOPENTANONE   120-92-3

HEALTH AND SAFETY LISTS

U.S. DOT - Substances From 49 CFR 172.101
regulated by DOT (UN2245)

U.S. DOT - Hazard Classes
DOT hazard class = 3

NFPA - Flash Points
flash point = 79 degrees F (26 degrees C)

NFPA - Hazard Identification Ratings
health-2; flammability-3; reactivity-0

INTERNATIONAL LISTS

Canada - WHMIS: Ingredient Disclosure
1% item 478 (611)

STATE LISTS

Florida Hazardous Substance List
[present]

Massachusetts Right To Know List
[present]

NJ Right to Know List (Total)
sn 0585

NJ Special Hazardous Substances
(flammable - third degree)

Pennsylvania Right to Know List
[present]

CYCLOPENTENE   142-29-0

HEALTH AND SAFETY LISTS

U.S. DOT - Substances From 49 CFR 172.101
regulated by DOT (UN2246)

U.S. DOT - Hazard Classes
DOT hazard class = 3

NFPA - Flash Points
flash point = -20 degrees F (-29 degrees C)

NFPA - Hazard Identification Ratings
health-1; flammability-3; reactivity-1

NIOSH - Selected LD50s and LC50s
Oral, rat: LD50 = 1656 mg/kg Skin, rabbit: LD50 = 1231 mg/kg

STATE LISTS

Florida Hazardous Substance List
[present]

Massachusetts Right To Know List
[present]

NJ Right to Know List (Total)
sn 0586

NJ Special Hazardous Substances
(flammable - third degree)

Pennsylvania Right to Know List
[present]

3-CYCLOPENTENE-1-ACETALDEHYDE, 2,2,3-TRIMETHYL-   4501-58-0

ENVIRONMENTAL LISTS

TSCA - Code of Federal Regulations Citations
40 CFR 712.30(x); 40 CFR 716.120(d)

TSCA - PAIR - Reporting List
Reporting Date: November 27, 1991

CYCLOPHOSPHAMIDE   50-18-0

HEALTH AND SAFETY LISTS

U.S. DOT - Appendix A Table 1 - Hazardous Substances
final RQ = 10 pounds (4.54 kg)

IARC - Group 1 (carcinogenic to humans)
[present]

NIOSH - Selected LD50s and LC50s
Oral, mouse: LD50 = 137 mg/kg

NTP Chemical Status Reports - Testing Status and NTIS Number
Chronic studies exist for which technical reports were not prepared

NTP Seventh Report - Known Carcinogens
known carcinogen

OSHA - Select Carcinogens
[present]

ENVIRONMENTAL LISTS

CERCLA/SARA - Hazardous Substances and their Reportable Quantities
final RQ = 10 pounds (4.54 kg)

EPA - Carcinogen Hazard Ranking for RQ Adjustment
Hazard ranking = Medium

RCRA - U Series Wastes
waste number U058

RCRA - Hazardous Constituents-Appendix VIII
waste number U058

RCRA - Substances Banned From Land Disposal
[present]

STATE LISTS

California - Air Bill 2588 Appendix A-II
known or potential carcinogen

California - Prop. 65 - Cancer list
carcinogen - initial date 2/27/87

California - Prop. 65 - Developmental Toxicity
developmental toxicity - initial date 1/1/89

California - Prop. 65 - Reproductive - Female
female reproductive toxicity - initial date 1/1/89

California - Prop. 65 - Reproductive - Male
male reproductive toxicity - initial date 1/1/89

California - Prop. 65 - No Significant Risk Levels
no significant risk level = 1 ug/day

California - Directors List of Hazardous Substances (8 CCR 339)
[present]

Florida Hazardous Substance List
[present]

Massachusetts Right To Know List
carcinogen; extraordinarily hazardous; teratogen

Minnesota Hazardous Substance List
carcinogen

NJ Right to Know List (Total)
sn 0587

NJ Special Hazardous Substances
(carcinogen; mutagen; teratogen)

Pennsylvania Right to Know List
environmental hazard; special hazardous substance

Pennsylvania RTK - Special Hazardous Substances
[present]

CYCLOPHOSPHAMIDE C   6055-19-2

STATE LISTS

California - Prop. 65 - Cancer list
carcinogen - initial date 2/27/87

California - Prop. 65 - Developmental Toxicity
developmental toxicity - initial date 1/1/89

California - Prop. 65 - Reproductive - Female
female reproductive toxicity - initial date 1/1/89

California - Prop. 65 - Reproductive - Male
male reproductive toxicity - initial date 1/1/89

California - Prop. 65 - No Significant Risk Levels
no significant risk level = 1 ug/day

California - Directors List of Hazardous Substances (8 CCR 339)
[present]

CYCLOPROPANE   75-19-4

HEALTH AND SAFETY LISTS

U.S. DOT - Substances From 49 CFR 172.101
regulated by DOT (UN1027)

U.S. DOT - Hazard Classes
DOT hazard class = 2.1

NFPA - Flash Points
gas (no flash point given)

NFPA - Hazard Identification Ratings
health-1; flammability-4; reactivity-0

ENVIRONMENTAL LISTS

CAA - Flammable Substances for Accidental Release Prevention
threshold quantity = 10,000 lbs

STATE LISTS

Florida Hazardous Substance List
[present]

Massachusetts Right To Know List
[present]

NJ Right to Know List (Total)
sn 0588

NJ Special Hazardous Substances
(flammable - fourth degree)

Pennsylvania Right to Know List
[present]

CYCLOSILANES, DIMETHYL-   69430-24-6

ENVIRONMENTAL LISTS

List of Pesticide Product Inert Ingredients
[present]

TSCA - PAIR - Reporting List
Effective Date: October 12, 1993;   Reporting Date: February 28, 1994

TSCA - Health and Safety Reporting List
Effective Date: October 12, 1993;   Sunset Date: October 12, 2003

CYCLOSPORIN A   59865-13-3

STATE LISTS

California - Prop. 65 - Cancer list
carcinogen - initial date 1/1/92

PROPOSED REGULATIONS

NTP - Proposed Additions to Annual Report on Carcinogens
proposed as a known carcinogen for the NTP 9th report

CYCLOSPORINE   79217-60-0

HEALTH AND SAFETY LISTS

IARC - Group 1 (carcinogenic to humans)
[present]

OSHA - Select Carcinogens
[present]

STATE LISTS

California - Prop. 65 - Cancer list
carcinogen - initial date 1/1/92

CYCLOTETRAMETHYLENETETRANITRAMINE   **2691-41-0**
HEALTH AND SAFETY LISTS
U.S. DOT - Substances From 49 CFR 172.101
regulated by DOT (UN0226, UN0484, UN0391, UN0483, UN0072)
U.S. DOT - Hazard Classes
DOT hazard class = 1.1D
NIOSH - Selected LD50s and LC50s
Oral, mouse: LD50 = 1500 mg/kg
STATE LISTS
NJ Right to Know List (Total)
sn 0589
NJ Special Hazardous Substances
(corrosive)

CYFLUTHRIN [3-(2,2-DICHLOROETHENYL)-2,2-   **68359-37-5**
DIMETHYLCYCLOPROPANECARBOXYLIC ACID,
CYANO(4-FLUORO-3-PHENOXYPHENYL)METHYL
ESTER]
ENVIRONMENTAL LISTS
CERCLA/SARA - Section 313 - Emission Reporting
form R reporting required

CYHALOTHRIN [3-(2-CHLORO-3,3,3-TRIFLUORO-1-   **68085-85-8**
PROPENYL)-2,2-DIMETHYLCYCLOPROPANECAR-
BOXYLIC ACID CYANO(3-PHENOXYPHENYL)
METHYL ESTER]
ENVIRONMENTAL LISTS
CERCLA/SARA - Section 313 - Emission Reporting
form R reporting required

CYHEXATIN   **13121-70-5**
SEE ALSO:
TIN
HEALTH AND SAFETY LISTS
ACGIH 1995 - Time Weighted Averages
5 mg/m3 TWA
U.S. DOT - Appendix B - Marine Pollutants
DOT regulated severe marine pollutant
NIOSH - Selected LD50s and LC50s
Inhalation, rat: LC50 - 244 mg/m3 8 hr Oral, rat: LD50 = 180 mg/kg
Skin, rat: LD50 = 446 mg/kg
OSHA - Vacated PELs - Time Weighted Averages
5 mg/m3 TWA
INTERNATIONAL LISTS
Australian Exposure Standards - Time Weighted Averages
5 mg/m3 TWA
Canada - CEPA Schedule II Part I - Prohibited Substances (Export)
[present]
Canada - Alberta - 8 Hour Occupational Exposure Limit
5 mg/m3 TWA
Canada - Alberta - 15 Minute Occupational Exposure Limit
10 mg/m3 STEL
Canada - British Columbia - 8 Hour Exposure Limits
5 mg/m3 TWA
Canada - British Columbia - 15 Minute Exposure Limits
10 mg/m3 STEL
Canada - Ontario - OHSA - TWAEVs
5 mg/m3 TWAEV
Canada - Quebec - Time-Weighted Average Exposure Values
5 mg/m3 TWAEV
United Kingdom - Occupational Exposure Standards - TWAs
5 mg/m3 TWA
United Kingdom - Occupational Exposure Standards - STELs
10 mg/m3 STEL

Israel - Time Weighted Averages
5 mg/m3 TWA
Israel - Action Levels
2.5 mg/m3 AL
Mexico - Instruction No. 10 - TWAs
5 mg/m3 TWA
STATE LISTS
California - Air Bill 2588 Appendix A-II
9/89
California - Prop. 65 - Developmental Toxicity
developmental toxicity - initial date 1/1/89
California - Exposure Limits - PELs
5 mg/m3 PEL
California - Directors List of Hazardous Substances (8 CCR 339)
[present]
Florida Hazardous Substance List
[present]
Massachusetts Right To Know List
[present]
Minnesota Hazardous Substance List
[present]
Pennsylvania Right to Know List
[present]

P-CYMENE   **99-87-6**
HEALTH AND SAFETY LISTS
NFPA - Flash Points
flash point = 117 degrees F (47 degrees C); technical grade: flash
point = 127 degrees F (53 degrees C)
NFPA - Hazard Identification Ratings
health-2; flammability-2; reactivity-0
NIOSH - Selected LD50s and LC50s
Oral, rat: LD50 = 4750 mg/kg
ENVIRONMENTAL LISTS
Safe Drinking Water Act - Monitoring
monitoring required at discretion of the state
TSCA - Code of Federal Regulations Citations
40 CFR 716.120(c)
INTERNATIONAL LISTS
Canada - WHMIS: Ingredient Disclosure
1% item 479 (612)
STATE LISTS
Florida Hazardous Substance List
[present]
Massachusetts Right To Know List
[present]
Pennsylvania Right to Know List
[present]

CYMENE   **25155-15-1**
HEALTH AND SAFETY LISTS
U.S. DOT - Substances From 49 CFR 172.101
regulated by DOT (UN2046)
U.S. DOT - Hazard Classes
DOT hazard class = 3
STATE LISTS
NJ Right to Know List (Total)
sn 0591

CYPERMETHRIN   **52315-07-8**
HEALTH AND SAFETY LISTS
U.S. DOT - Appendix B - Marine Pollutants
DOT regulated severe marine pollutant

STATE LISTS
Massachusetts Right To Know List
[present]

CYROMAZINE   66215-27-8
PROPOSED REGULATIONS
Safe Drinking Water Act - Priority list
[present]

L-CYSTEINE   52-90-4
HEALTH AND SAFETY LISTS
NIOSH - Selected LD50s and LC50s
Oral, rat: LD50 = 5580 mg/kg

CYSTOSINE ARABINOSIDE   147-94-4
HEALTH AND SAFETY LISTS
NIOSH - Selected LD50s and LC50s
Oral, mouse: LD50 = 3150 mg/kg

NTP Chemical Status Reports - Testing Status and NTIS Number
Chronic studies exist for which technical reports were not prepared

STATE LISTS
California - Air Bill 2588 Appendix A-II
9/89

California - Prop. 65 - Developmental Toxicity
developmental toxicity - initial date 1/1/89

Massachusetts Right To Know List
teratogen

NJ Special Hazardous Substances
(mutagen; teratogen)

CYTEMBENA   21739-91-3
HEALTH AND SAFETY LISTS
NTP Chemical Status Reports - Testing Status and NTIS Number
Technical reports printed (PB82163312)

NTP Chemical Status Reports - Evidence of Carcinogenicity
male rat-positive; female rat-positive; male mice-negative; female mice-negative

CYTOCHALASIN B   14930-96-2
INTERNATIONAL LISTS
Canada - WHMIS: Ingredient Disclosure
1% item 480 (613)

CYTOXAL ALCOHOL   4465-94-5
HEALTH AND SAFETY LISTS
NTP Chemical Status Reports - Testing Status and NTIS Number
Chronic studies exist for which technical reports were not prepared

2,4-D   94-75-7
SEE ALSO:
N,N-DIETHYLSTEARAMIDE
F039-HAZARDOUS WASTES

HEALTH AND SAFETY LISTS
ACGIH 1995 - Time Weighted Averages
10 mg/m3 TWA

U.S. DOT - Appendix B - Marine Pollutants
DOT regulated marine pollutant

U.S. DOT - Appendix A Table 1 - Hazardous Substances
final RQ = 100 pounds (45.4 kg)

IARC - Group Unspecified
[present] (Listed under 'Chlorophenoxy herbicides')

NIOSH - Selected LD50s and LC50s
Oral, rat: LD50 = 370 mg/kg Skin, rat: LD50 = 1500 mg/kg

NIOSH 1990 - Pocket Guide - RELs
10 mg/m3 TWA

NIOSH 1990 - Pocket Guide - IDLHs
500 mg/m3 IDLH

NIOSH 1990 - Pocket Guide - Target organs
skin, CNS

NTP Chemical Status Reports - Testing Status and NTIS Number
Prechronic studies in progress

OSHA - Vacated PELs - Time Weighted Averages
10 mg/m3 TWA

OSHA - Final PELs - Time Weighted Averages
10 mg/m3 TWA

ENVIRONMENTAL LISTS
ATSDR Priority List
Rank (of 275): 222

CERCLA/SARA - Section 313 - Emission Reporting
form R reporting required for 1.0% de minimus concentration

CERCLA/SARA - Hazardous Substances and their Reportable Quantities
final RQ = 100 pounds (45.4 kg)

Clean Air Act (1990) - List of Hazardous Air Contaminants
[present]

Clean Water Act - Hazardous Substances
[present]

Safe Drinking Water Act - MCLs
MCL = 0.1 mg/L

Safe Drinking Water Act - MCLGs
MCLG = 0.07 mg/L

RCRA - D Series - Maximum Concentration of Contaminants
waste number D016; regulatory level = 10.0 mg/L

RCRA - D Series - Chronic Toxicity Reference Levels
chronic toxicity reference level = 0.1 mg/L

RCRA - U Series Wastes
waste number U240

RCRA - Hazardous Constituents-Appendix VIII
waste number U240

RCRA - Basis for Listing - Appendix VII
Included in waste stream: F039

RCRA - Substances Banned From Land Disposal
[present]

RCRA - TSD Facilities Ground Water Monitoring
TM 8150 = 10 ug/L PQL

RCRA - Universal Treatment Standards (LDR)
WW: 0.72 mg/l; NWW: 10 mg/kg

TSCA - Code of Federal Regulations Citations
40 CFR 799.5055(c), (d)(2)

TSCA - Multichemical Test Rules - Waste Constituents
hydrolysis testing for Chemical Fate

TSCA - Section 12(b) - Export Notification
export notification required - Section 4

INTERNATIONAL LISTS
Australian Exposure Standards - Time Weighted Averages
10 mg/m3 TWA

Canada - Drinking Water Quality - IMACs
0.1 mg/L IMAC

Canada - Alberta - 8 Hour Occupational Exposure Limit
10 mg/m3 TWA

Canada - Alberta - 15 Minute Occupational Exposure Limit
20 mg/m3 STEL

Canada - British Columbia - 8 Hour Exposure Limits
10 mg/m3 TWA

Canada - British Columbia - 15 Minute Exposure Limits
20 mg/m3 STEL

Canada - Ontario - OHSA - TWAEVs
10 mg/m3 TWAEV

Canada - Quebec - Time-Weighted Average Exposure Values
10 mg/m3 TWAEV

Canada - Quebec - Carcinogens
C2 carcinogen: effect suspected in humans

United Kingdom - Occupational Exposure Standards - TWAs
10 mg/m3 TWA

United Kingdom - Occupational Exposure Standards - STELs
20 mg/m3 STEL

German (DFG) - MAK Values
total dust: 1 mg/m3 MAK (includes salts and esters)

German (DFG) - Peak Limitations
5 x normal MAK (30 min. average value);   don't exceed 2 times during shift

German (DFG) - Skin/Sensitizers
danger of cutaneous absorption (including amine and ester forms)

German (DFG) - Pregnancy
no risk to embryo/fetus if exposure limits adhered to

Israel - Time Weighted Averages
10 mg/m3 TWA

Israel - Action Levels
5 mg/m3 AL

Mexico - Instruction No. 10 - TWAs
10 mg/m3 TWA

Mexico - Instruction No. 10 - STELs
20 mg/m3 STEL

Mexico - Drinking Water - Ecological Criteria
0.1 mg/l

STATE LISTS

California - Air Bill 2588 Appendix A-I
6/91

California - Exposure Limits - PELs
10 mg/m3 PEL

California - Directors List of Hazardous Substances (8 CCR 339)
[present]

Florida Hazardous Substance List
[present]

Massachusetts Right To Know List
[present]

Minnesota Hazardous Substance List
[present]

NJ Right to Know List (Total)
sn 0593

Pennsylvania Right to Know List
environmental hazard

2,4-D BUTOXYETHYL ESTER   1929-73-3

HEALTH AND SAFETY LISTS

U.S. DOT - Appendix A Table 1 - Hazardous Substances
final RQ = 100 pounds (45.4 kg) (Listed under '2,4-D esters')

ENVIRONMENTAL LISTS

CERCLA/SARA - Section 313 - Emission Reporting
form R reporting required

CERCLA/SARA - Hazardous Substances and their Reportable Quantities
final RQ = 100 pounds (45.4 kg) (Listed under '2,4-D esters')

Clean Water Act - Hazardous Substances
[present] (Listed under '2,4-D ester')

STATE LISTS

California - Directors List of Hazardous Substances (8 CCR 339)
[present] (Listed under '2,4-D esters...')

Massachusetts Right To Know List
[present]

NJ Right to Know List (Total)
sn 2949

Pennsylvania Right to Know List
environmental hazard

2,4-D CHLOROCROTYL ESTER   2971-38-?

HEALTH AND SAFETY LISTS

U.S. DOT - Appendix A Table 1 - Hazardous Substances
final RQ = 100 pounds (45.4 kg) (Listed under '2,4-D esters')

ENVIRONMENTAL LISTS

CERCLA/SARA - Section 313 - Emission Reporting
form R reporting required

CERCLA/SARA - Hazardous Substances and their Reportable Quantities
final RQ = 100 pounds (45.4 kg) (Listed under '2,4-D esters')

Clean Water Act - Hazardous Substances
[present] (Listed under '2,4-D ester')

STATE LISTS

California - Directors List of Hazardous Substances (8 CCR 339)
[present] (Listed under '2,4-D esters...')

Massachusetts Right To Know List
[present]

NJ Right to Know List (Total)
sn 2947

Pennsylvania Right to Know List
environmental hazard

2,4-D ESTERS   94-11-1

HEALTH AND SAFETY LISTS

U.S. DOT - Appendix A Table 1 - Hazardous Substances
final RQ = 100 pounds (45.4 kg)

NIOSH - Selected LD50s and LC50s
Oral, rat: LD50 = 700 mg/kg

ENVIRONMENTAL LISTS

CERCLA/SARA - Section 313 - Emission Reporting
form R reporting required

CERCLA/SARA - Hazardous Substances and their Reportable Quantities
final RQ = 100 pounds (45.4 kg)

Clean Water Act - Hazardous Substances
[present]

INTERNATIONAL LISTS

Canada - Ontario - OHSA - TWAEVs
10 mg/m3 TWAEV

STATE LISTS

California - Directors List of Hazardous Substances (8 CCR 339)
[present]

Massachusetts Right To Know List
[present]

NJ Right to Know List (Total)
sn 2941

Pennsylvania Right to Know List
environmental hazard

2,4-D ESTERS   53467-11-1

HEALTH AND SAFETY LISTS

U.S. DOT - Appendix A Table 1 - Hazardous Substances
final RQ = 100 pounds (45.4 kg)

ENVIRONMENTAL LISTS

  CERCLA/SARA - Hazardous Substances and their Reportable Quantities
    final RQ = 100 pounds (45.4 kg)

  Clean Water Act - Hazardous Substances
    [present]

STATE LISTS

  California - Directors List of Hazardous Substances (8 CCR 339)
    [present]

  Massachusetts Right To Know List
    [present]

  NJ Right to Know List (Total)
    sn 2948

  Pennsylvania Right to Know List
    environmental hazard

2,4-D ISOOCTYL ESTER   25168-26-7

HEALTH AND SAFETY LISTS

  U.S. DOT - Appendix A Table 1 - Hazardous Substances
    final RQ = 100 pounds (45.4 kg) (Listed under '2,4-D Esters')

  NIOSH - Selected LD50s and LC50s
    Oral, rat: LD50 = 500 mg/kg

ENVIRONMENTAL LISTS

  CERCLA/SARA - Hazardous Substances and their Reportable Quantities
    final RQ = 100 pounds (45.4 kg) (Listed under '2,4-D Esters')

  Clean Water Act - Hazardous Substances
    [present] (Listed under '2,4-D ester')

STATE LISTS

  California - Directors List of Hazardous Substances (8 CCR 339)
    [present] (Listed under '2,4-D esters...')

  Massachusetts Right To Know List
    [present]

  NJ Right to Know List (Total)
    sn 2930

  Pennsylvania Right to Know List
    environmental hazard

2,4-D METHYL ESTER   1928-38-7

HEALTH AND SAFETY LISTS

  U.S. DOT - Appendix A Table 1 - Hazardous Substances
    final RQ = 100 pounds (45.4 kg) (Listed under '2,4-D esters')

ENVIRONMENTAL LISTS

  CERCLA/SARA - Hazardous Substances and their Reportable Quantities
    final RQ = 100 pounds (45.4 kg) (Listed under '2,4-D esters')

  Clean Water Act - Hazardous Substances
    [present] (Listed under '2,4-D ester')

STATE LISTS

  California - Directors List of Hazardous Substances (8 CCR 339)
    [present] (Listed under '2,4-D esters...')

  Massachusetts Right To Know List
    [present]

  NJ Right to Know List (Total)
    sn 2945

  Pennsylvania Right to Know List
    environmental hazard

2,4-D PROPYL ESTER   1928-61-6

HEALTH AND SAFETY LISTS

  U.S. DOT - Appendix A Table 1 - Hazardous Substances
    final RQ = 100 pounds (45.4 kg) (Listed under '2,4-D esters')

ENVIRONMENTAL LISTS

  CERCLA/SARA - Hazardous Substances and their Reportable Quantities
    final RQ = 100 pounds (45.4 kg) (Listed under '2,4-D esters')

  Clean Water Act - Hazardous Substances
    [present] (Listed under '2,4-D ester')

STATE LISTS

  California - Directors List of Hazardous Substances (8 CCR 339)
    [present] (Listed under '2,4-D esters...')

  Massachusetts Right To Know List
    [present]

  NJ Right to Know List (Total)
    sn 2946

  Pennsylvania Right to Know List
    environmental hazard

2,4-D PROPYLENE GLYCOL BUTYL ETHER ESTER   1320-18-9

HEALTH AND SAFETY LISTS

  U.S. DOT - Appendix A Table 1 - Hazardous Substances
    final RQ = 100 pounds (45.4 kg) (Listed under '2,4-D Esters')

ENVIRONMENTAL LISTS

  CERCLA/SARA - Section 313 - Emission Reporting
    form R reporting required

  CERCLA/SARA - Hazardous Substances and their Reportable Quantities
    final RQ = 100 pounds (45.4 kg) (Listed under '2,4-D Esters')

  Clean Water Act - Hazardous Substances
    [present] (Listed under '2,4-D ester')

STATE LISTS

  California - Directors List of Hazardous Substances (8 CCR 339)
    [present] (Listed under '2,4-D esters...')

  Massachusetts Right To Know List
    [present]

  NJ Right to Know List (Total)
    sn 2944

  Pennsylvania Right to Know List
    environmental hazard

D AND C GREEN NO. 5   4403-90-1

ENVIRONMENTAL LISTS

  List of Pesticide Product Inert Ingredients
    [present]

D&C ORANGE NO. 17   3468-63-1

STATE LISTS

  California - Air Bill 2588 Appendix A-II
    known or potential carcinogen: 9/90

  California - Prop. 65 - Cancer list
    carcinogen - initial date 7/1/90

D&C ORANGE NO. 5   596-03-2

ENVIRONMENTAL LISTS

  List of Pesticide Product Inert Ingredients
    [present]

D&C RED NO. 8   2092-56-0

STATE LISTS

  California - Air Bill 2588 Appendix A-II
    known or potential carcinogen: 6/91

  California - Prop. 65 - Cancer list
    carcinogen - initial date 10/1/90

D & C RED NO. 9   5160-02-1
SEE ALSO:
  BARIUM
  BARIUM
  BARIUM COMPOUNDS, N.O.S.
HEALTH AND SAFETY LISTS
  IARC - Group 3 (not classifiable)
  [present]
  NTP Chemical Status Reports - Testing Status and NTIS Number
  Technical reports printed (PB82229592)
  NTP Chemical Status Reports - Evidence of Carcinogenicity
  male rat-positive; female rat-equivocal; male mice-negative; female mice-negative
STATE LISTS
  California - Air Bill 2588 Appendix A-II
  known or potential carcinogen: 9/90
  California - Prop. 65 - Cancer list
  carcinogen - initial date 7/1/90
  California - Prop. 65 - No Significant Risk Levels
  no significant risk level = 100 ug/day

D & C YELLOW NO. 11   8003-22-3
HEALTH AND SAFETY LISTS
  NTP Chemical Status Reports - Testing Status and NTIS Number
  Technical reports printed (PB91185355); two year studies: pathology quality assessment in progress
ENVIRONMENTAL LISTS
  List of Pesticide Product Inert Ingredients
  [present]

D001-IGNITABLE UNLISTED HAZARDOUS WASTES   RR-00631-5
HEALTH AND SAFETY LISTS
  U.S. DOT - Appendix A Table 1 - Hazardous Substances
  final RQ = 100 pounds (45.4 kg)
ENVIRONMENTAL LISTS
  CERCLA/SARA - Hazardous Substances and their Reportable Quantities
  final RQ = 100 pounds (45.4 kg)
  RCRA - Substances Banned From Land Disposal
  [present]

D002-CORROSIVE UNLISTED HAZARDOUS WASTES   RR-00632-6
HEALTH AND SAFETY LISTS
  U.S. DOT - Appendix A Table 1 - Hazardous Substances
  final RQ = 100 pounds (45.4 kg)
ENVIRONMENTAL LISTS
  CERCLA/SARA - Hazardous Substances and their Reportable Quantities
  final RQ = 100 pounds (45.4 kg)
  RCRA - Substances Banned From Land Disposal
  [present]

D003-REACTIVE UNLISTED HAZARDOUS WASTES   RR-00633-7
HEALTH AND SAFETY LISTS
  U.S. DOT - Appendix A Table 1 - Hazardous Substances
  final RQ = 100 pounds (45.4 kg)
ENVIRONMENTAL LISTS
  CERCLA/SARA - Hazardous Substances and their Reportable Quantities
  final RQ = 100 pounds (45.4 kg)
  RCRA - Substances Banned From Land Disposal
  [present]

DACARBAZINE   4342-03-
HEALTH AND SAFETY LISTS
  IARC - Group 2B (sufficient animal data)
  [present]
  NIOSH - Selected LD50s and LC50s
  Oral, rat: LD50 = 2147 mg/kg
  NTP Chemical Status Reports - Testing Status and NTIS Number
  Chronic studies exist for which technical reports were not prepared
  NTP Seventh Report - Suspect Carcinogens
  suspect carcinogen
  OSHA - Possible Select Carcinogens
  [present]
STATE LISTS
  California - Air Bill 2588 Appendix A-II
  known or potential carcinogen
  California - Prop. 65 - Cancer list
  carcinogen - initial date 1/1/88
  California - Prop. 65 - No Significant Risk Levels
  no significant risk level = 0.01 ug/day
  California - Directors List of Hazardous Substances (8 CCR 339)
  [present]
  Florida Hazardous Substance List
  [present]
  Massachusetts Right To Know List
  carcinogen; extraordinarily hazardous
  Minnesota Hazardous Substance List
  carcinogen
  NJ Special Hazardous Substances
  (carcinogen)
  Pennsylvania Right to Know List
  special hazardous substance
  Pennsylvania RTK - Special Hazardous Substances
  [present]

DAMINOZIDE   1596-84-5
HEALTH AND SAFETY LISTS
  NTP Chemical Status Reports - Testing Status and NTIS Number
  Technical reports printed (PB285073/AS)
  NTP Chemical Status Reports - Evidence of Carcinogenicity
  male rat-negative; female rat-positive; male mice-equivocal; female mice-negative
STATE LISTS
  California - Air Bill 2588 Appendix A-II
  known or potential carcinogen: 9/90
  California - Prop. 65 - Cancer list
  carcinogen - initial date 1/1/90
  California - Prop. 65 - No Significant Risk Levels
  no significant risk level = 40 ug/day
  Massachusetts Right To Know List
  carcinogen; extraordinarily hazardous

DANAZOL   17230-88-5
STATE LISTS
  California - Air Bill 2588 Appendix A-II
  9/90
  California - Prop. 65 - Developmental Toxicity
  developmental toxicity - initial date 4/1/90
  Massachusetts Right To Know List
  teratogen

DAPSONE   80-08-0
HEALTH AND SAFETY LISTS
IARC - Group 3 (not classifiable)
[present]
NIOSH - Selected LD50s and LC50s
Oral, mouse: LD50 = 375 mg/kg
NTP Chemical Status Reports - Testing Status and NTIS Number
Technical reports printed (PB274394/AS)
NTP Chemical Status Reports - Evidence of Carcinogenicity
male rat-positive; female rat-negative; male mice-negative; female mice-negative
ENVIRONMENTAL LISTS
TSCA - Code of Federal Regulations Citations
40 CFR 712.30(x); 40 CFR 716.120(d)
TSCA - PAIR - Reporting List
Reporting Date: November 27, 1991
TSCA - Health and Safety Reporting List
Effective Date: September 30, 1991
STATE LISTS
California - Directors List of Hazardous Substances (8 CCR 339)
[present]
Massachusetts Right To Know List
carcinogen; extraordinarily hazardous

DAUNOMYCIN   20830-81-3
HEALTH AND SAFETY LISTS
U.S. DOT - Appendix A Table 1 - Hazardous Substances
final RQ = 10 pounds (4.54 kg)
IARC - Group 2B (sufficient animal data)
[present] (Overall evaluation based only on evidence of carcinogenicity in monograph (10, 1976) or in Supplement 4)
NTP Chemical Status Reports - Testing Status and NTIS Number
Chronic studies exist for which technical reports were not prepared
OSHA - Possible Select Carcinogens
[present]
ENVIRONMENTAL LISTS
CERCLA/SARA - Hazardous Substances and their Reportable Quantities
final RQ = 10 pounds (4.54 kg)
EPA - Carcinogen Hazard Ranking for RQ Adjustment
Hazard ranking = Medium
RCRA - U Series Wastes
waste number U059
RCRA - Hazardous Constituents-Appendix VIII
waste number U059
RCRA - Substances Banned From Land Disposal
[present]
STATE LISTS
California - Air Bill 2588 Appendix A-II
known or potential carcinogen
California - Prop. 65 - Cancer list
carcinogen - initial date 1/1/88
California - Directors List of Hazardous Substances (8 CCR 339)
[present]
Florida Hazardous Substance List
[present]
Massachusetts Right To Know List
carcinogen; extraordinarily hazardous
Minnesota Hazardous Substance List
carcinogen
NJ Right to Know List (Total)
sn 0594

NJ Special Hazardous Substances
(carcinogen; mutagen)
Pennsylvania Right to Know List
environmental hazard; special hazardous substance
Pennsylvania RTK - Special Hazardous Substances
[present]

DAUNORUBICIN HYDROCHLORIDE   23541-50-6
STATE LISTS
California - Air Bill 2588 Appendix A-II
9/90
California - Prop. 65 - Developmental Toxicity
developmental toxicity - initial date 7/1/90

DAWSONITE   12011-76-6
INTERNATIONAL LISTS
German (DFG) - Carcinogens
as fibrous dust: animal evidence of carcinogenicity

DAZOMET, SODIUM SALT [TETRAHYDRO-3,5-   53404-60-7
DIMETHYL-2H-1,3,5-THIADIAZINE-2-THIONE, ION
(1-), SODIUM]
ENVIRONMENTAL LISTS
CERCLA/SARA - Section 313 - Emission Reporting
form R reporting required

2,4-DB (2,4-DICHLORO-PHENOXYBUTYRIC ACID)   94-82-6
ENVIRONMENTAL LISTS
CERCLA/SARA - Section 313 - Emission Reporting
form R reporting required
STATE LISTS
California - Directors List of Hazardous Substances (8 CCR 339)
[present]

O,P'-DDD   53-19-0
SEE ALSO:
F039-HAZARDOUS WASTES
HEALTH AND SAFETY LISTS
NTP Chemical Status Reports - Testing Status and NTIS Number
Chronic studies exist for which technical reports were not prepared
ENVIRONMENTAL LISTS
RCRA - Basis for Listing - Appendix VII
Included in waste stream: F039
RCRA - Universal Treatment Standards (LDR)
WW: 0.023 mg/l; NWW: 0.087 mg/kg

DDD   72-54-8
SEE ALSO:
F039-HAZARDOUS WASTES
HEALTH AND SAFETY LISTS
U.S. DOT - Appendix A Table 1 - Hazardous Substances
final RQ = 1 pound (0.454 kg)
NIOSH - Selected LD50s and LC50s
Oral, rat: LD50 = 113 mg/kg Skin, rabbit: LD50 = 1200 mg/kg
NTP Chemical Status Reports - Testing Status and NTIS Number
Technical reports printed (PB286367/AS)
NTP Chemical Status Reports - Evidence of Carcinogenicity
male rat-equivocal; female rat-negative; male mice-negative; female mice-negative
ENVIRONMENTAL LISTS
ATSDR Priority List
Rank (of 275): 025

CERCLA/SARA - Hazardous Substances and their Reportable Quantities
final RQ = 1 pound (0.454 kg)

Clean Water Act - Hazardous Substances
[present]

Clean Water Act - Priority Pollutants
[present]

EPA - Carcinogen Hazard Ranking for RQ Adjustment
Hazard ranking = Medium

RCRA - U Series Wastes
waste number U060

RCRA - Hazardous Constituents-Appendix VIII
waste number U060

RCRA - Basis for Listing - Appendix VII
Included in waste stream: F039

RCRA - Substances Banned From Land Disposal
[present]

RCRA - TSD Facilities Ground Water Monitoring
TM 8080 = 0.1 ug/L PQL; TM 8270 = 10 ug/L PQL

RCRA - Universal Treatment Standards (LDR)
WW: 0.023 mg/l; NWW: 0.087 mg/kg

INTERNATIONAL LISTS

Mexico - Wastewater - Organic Toxic Pollutants and Heavy Metals
Listed under [DDT and Metabolites]

Mexico - Drinking Water - Ecological Criteria
0.0000002 mg/l Substance presents persistence, bioaccumulations or risk of cancer, reduce human exposure to a minimum; This level has been extrapolated by using a mathematic model

STATE LISTS

California - Air Bill 2588 Appendix A-II
known or potential carcinogen: 9/89

California - Prop. 65 - Cancer list
carcinogen - initial date 1/1/89

California - Directors List of Hazardous Substances (8 CCR 339)
[present]

Massachusetts Right To Know List
[present]

NJ Right to Know List (Total)
sn 0646

Pennsylvania Right to Know List
environmental hazard

DDE   72-55-9
SEE ALSO:
F039-HAZARDOUS WASTES

HEALTH AND SAFETY LISTS

U.S. DOT - Appendix A Table 1 - Hazardous Substances
final RQ = 1 pound (0.454 kg)

NIOSH - Selected LD50s and LC50s
Oral, rat: LD50 = 880 mg/kg

NTP Chemical Status Reports - Testing Status and NTIS Number
Technical reports printed (PB286367/AS)

NTP Chemical Status Reports - Evidence of Carcinogenicity
male rat-negative; female rat-negative; male mice-positive; female mice-positive

ENVIRONMENTAL LISTS

ATSDR Priority List
Rank (of 275): 016

CERCLA/SARA - Hazardous Substances and their Reportable Quantities
final RQ = 1 pound (0.454 kg)

Clean Water Act - Priority Pollutants
[present]

EPA - Carcinogen Hazard Ranking for RQ Adjustment
Hazard ranking = Medium

RCRA - Hazardous Constituents-Appendix VIII
hazardous constituent - no waste number

RCRA - Basis for Listing - Appendix VII
Included in waste stream: F039

RCRA - TSD Facilities Ground Water Monitoring
TM 8080 = 0.05 ug/L PQL; TM 8270 = 10 ug/L PQL

RCRA - Universal Treatment Standards (LDR)
WW: 0.031 mg/l; NWW: 0.087 mg/kg

INTERNATIONAL LISTS

Canada - WHMIS: Ingredient Disclosure
1% item 539 (658)

Mexico - Wastewater - Organic Toxic Pollutants and Heavy Metals
Listed under [DDT and Metabolites]

Mexico - Drinking Water - Ecological Criteria
None given.

STATE LISTS

California - Air Bill 2588 Appendix A-I
known or potential carcinogen: 9/89

California - Prop. 65 - Cancer list
carcinogen - initial date 1/1/89

California - Directors List of Hazardous Substances (8 CCR 339)
[present]

Massachusetts Right To Know List
carcinogen; extraordinarily hazardous

NJ Right to Know List (Total)
sn 2979

Pennsylvania Right to Know List
environmental hazard

O,P'-DDE   3424-82-6
SEE ALSO:
F039-HAZARDOUS WASTES

ENVIRONMENTAL LISTS

RCRA - Basis for Listing - Appendix VII
Included in waste stream: F039

RCRA - Universal Treatment Standards (LDR)
WW: 0.031 mg/l; NWW: 0.087 mg/kg

DDE   3547-04-4

ENVIRONMENTAL LISTS

CERCLA/SARA - Hazardous Substances and their Reportable Quantities
final RQ = 1 pound (.454 kg)

Clean Air Act (1990) - List of Hazardous Air Contaminants
[present]

DDT   50-29-3
SEE ALSO:
F039-HAZARDOUS WASTES

HEALTH AND SAFETY LISTS

ACGIH 1995 - Time Weighted Averages
1 mg/m3 TWA

U.S. DOT - Appendix B - Marine Pollutants
DOT regulated severe marine pollutant

U.S. DOT - Appendix A Table 1 - Hazardous Substances
final RQ = 1 pound (0.454 kg)

IARC - Group 2B (sufficient animal data)
[present]

NIOSH - Selected LD50s and LC50s
Oral, rat: LD50 = 87 mg/kg Skin, rat: LD50 = 1931 mg/kg

NIOSH 1990 - Pocket Guide - RELs
0.5 mg/m3 TWA

NIOSH 1990 - Pocket Guide - Carcinogens
occupational carcinogen

NIOSH 1990 - Pocket Guide - Target organs
CNS, PNS, kidneys, liver, skin

NIOSH - Health Standards - Exposure Limits
lowest reliable detectable concentration: (0.5 mg/m3 TWA by NIOSH)

NIOSH - Health Standards - Health Effects and Precautions
has produced tumors of the liver, lungs, and lymphatic system in animals (Prevent skin contact)

NIOSH - Health Standards - Carcinogenic Chemicals
potential human carcinogen

NTP Chemical Status Reports - Testing Status and NTIS Number
Technical reports printed (PB286367/AS)

NTP Chemical Status Reports - Evidence of Carcinogenicity
male rat-negative; female rat-negative; male mice-negative; female mice-negative

NTP Seventh Report - Suspect Carcinogens
suspect carcinogen

OSHA - Vacated PELs - Time Weighted Averages
1 mg/m3 TWA

OSHA - Vacated PELs - Skin Designation
Prevent or reduce skin absorption

OSHA - Final PELs - Time Weighted Averages
1 mg/m3 TWA

OSHA - Final PELs - Skin Notations
prevent or reduce skin absorption

OSHA - Possible Select Carcinogens
[present]

ENVIRONMENTAL LISTS

ATSDR Priority List
Rank (of 275): 012

CERCLA/SARA - Hazardous Substances and their Reportable Quantities
final RQ = 1 pound (0.454 kg)

Clean Water Act - Hazardous Substances
[present]

Clean Water Act - Priority Pollutants
[present]

Clean Water Act - Toxic Pollutants
[present]

EPA - Carcinogen Hazard Ranking for RQ Adjustment
Hazard ranking = Medium

RCRA - U Series Wastes
waste number U061

RCRA - Hazardous Constituents-Appendix VIII
waste number U061

RCRA - Basis for Listing - Appendix VII
Included in waste stream: F039

RCRA - Substances Banned From Land Disposal
[present]

RCRA - TSD Facilities Ground Water Monitoring
TM 8080 = 0.1 ug/L PQL; TM 8270 = 10 ug/L PQL

RCRA - Universal Treatment Standards (LDR)
WW: 0.0039 mg/l; NWW: 0.087 mg/kg

TSCA - Chemicals with Significant New Use Rules
[present]

TSCA - Section 12(b) - Export Notification
export notification required - Section 5

INTERNATIONAL LISTS

Australian Exposure Standards - Time Weighted Averages
1 mg/m3 TWA

Canada - CEPA Schedule II Part II - Toxic Substances (Export)
[present]

Canada - Drinking Water Quality - MACs
0.03 mg/L MAC

Canada - Alberta - 8 Hour Occupational Exposure Limit
1 mg/m3 TWA

Canada - Alberta - 15 Minute Occupational Exposure Limit
3 mg/m3 STEL

Canada - British Columbia - 8 Hour Exposure Limits
1 mg/m3 TWA

Canada - British Columbia - 15 Minute Exposure Limits
3 mg/m3 STEL

Canada - Ontario - OHSA - TWAEVs
1 mg/m3 TWAEV

Canada - Quebec - Time-Weighted Average Exposure Values
1 mg/m3 TWAEV

Canada - Quebec - Carcinogens
C3 carcinogen: effect detected in animals

United Kingdom - Occupational Exposure Standards - TWAs
1 mg/m3 TWA

United Kingdom - Occupational Exposure Standards - STELs
3 mg/m3 STEL

German (DFG) - MAK Values
total dust: 1 mg/m3 MAK

German (DFG) - Peak Limitations
10 x normal MAK (30 min average value); don't exceed during shift

German (DFG) - Skin/Sensitizers
danger of cutaneous absorption

Israel - Time Weighted Averages
1 mg/m3 TWA

Israel - Action Levels
0.5 mg/m3 AL

Mexico - Instruction No. 10 - TWAs
1 mg/m3 TWA

Mexico - Instruction No. 10 - STELs
3 mg/m3 STEL

Mexico - Wastewater - Organic Toxic Pollutants and Heavy Metals
Listed under [DDT and Metabolites]

Mexico - Drinking Water - Ecological Criteria
0.001 mg/l Substance presents persistence, bioaccumulations or risk of cancer, reduce human exposure to a minimum; This level has been extrapolated by using a mathematic model

STATE LISTS

California - Air Bill 2588 Appendix A-II
known or potential carcinogen

California - Prop. 65 - Cancer list
carcinogen - initial date 10/1/87

California - Exposure Limits - PELs
1 mg/m3 PEL

California - Exposure Limits - Skin Notation
material may be absorbed through the skin, eyes or mucous membrane

California - Directors List of Hazardous Substances (8 CCR 339)
[present]

Florida Hazardous Substance List
[present]

Massachusetts Right To Know List
carcinogen; extraordinarily hazardous

Minnesota Hazardous Substance List
carcinogen

NJ Right to Know List (Total)
sn 0596

NJ Special Hazardous Substances
(carcinogen; mutagen; teratogen)

Pennsylvania Right to Know List
environmental hazard; special hazardous substance

Pennsylvania RTK - Special Hazardous Substances
[present]

O,P'-DDT   **789-02-6**
SEE ALSO:
F039-HAZARDOUS WASTES
ENVIRONMENTAL LISTS
RCRA - Basis for Listing - Appendix VII
Included in waste stream: F039
RCRA - Universal Treatment Standards (LDR)
WW: 0.0039 mg/l; NWW: 0.087 mg/kg

DDT, DDE, AND DDD (IN COMBINATION)   **RR-00080-6**
STATE LISTS
California - Prop. 65 - No Significant Risk Levels
no significant risk level = 2 ug/day

DDT METABOLITES   **RR-00628-0**
STATE LISTS
Pennsylvania Right to Know List
environmental hazard

DECABORANE   **17702-41-9**
HEALTH AND SAFETY LISTS
ACGIH 1995 - Time Weighted Averages
0.05 ppm TWA; 0.25 mg/m3 TWA
ACGIH 1995 - Short Term Exposure Limits
0.15 ppm STEL; 0.75 mg/m3 STEL
ACGIH 1995 - Skin Designations
skin - potential for cutaneous absorption
AIHA - Odor Threshold Values
no geometric mean air odor threshold
U.S. DOT - Substances From 49 CFR 172.101
regulated by DOT (UN1868)
U.S. DOT - Hazard Classes
DOT hazard class = 4.1
NFPA - Flash Points
flash point = 176 degrees F (80 degrees C)
NFPA - Hazard Identification Ratings
health-3; flammability-2; reactivity-1
NIOSH - Selected LD50s and LC50s
Inhalation, rat: LC50 = 46 ppm 4 hr Oral, rat: LD50 = 64 mg/kg Skin, rat: LD50 = 740 mg/kg
NIOSH 1990 - Pocket Guide - RELs
0.3 mg/m3 TWA; 0.05 ppm TWA; 0.9 mg/m3 STEL; 0.15 ppm STEL
NIOSH 1990 - Pocket Guide - IDLHs
100 mg/m3 IDLH
NIOSH 1990 - Pocket Guide - Target organs
CNS
NIOSH 1990 - Pocket Guide - Skin list
Potential for dermal absorption
OSHA - Vacated PELs - Time Weighted Averages
0.05 ppm TWA; 0.3 mg/m3 TWA
OSHA - Vacated PELs - Short Term Exposure Limits
0.15 ppm STEL; 0.9 mg/m3 STEL
OSHA - Vacated PELs - Skin Designation
Prevent or reduce skin absorption
OSHA - Final PELs - Time Weighted Averages
0.05 ppm TWA; 0.3 mg/m3 TWA
OSHA - Final PELs - Skin Notations
prevent or reduce skin absorption

ENVIRONMENTAL LISTS
CERCLA/SARA - Section 302 Extremely Hazardous Substances and TPQs
TPQ = 500/10,000 pounds
INTERNATIONAL LISTS
Australian Exposure Standards - Time Weighted Averages
0.05 ppm TWA; 0.25 mg/m3 TWA
Australian Exposure Standards - Short Term Exposure Limits
0.15 ppm STEL; 0.75 mg/m3 STEL
Australian Exposure Standards - Skin Effects
skin absorption
Canada - WHMIS: Ingredient Disclosure
1% item 481 (614)
Canada - Alberta - 8 Hour Occupational Exposure Limit
0.05 ppm TWA; 0.25 mg/m3 TWA
Canada - Alberta - 15 Minute Occupational Exposure Limit
0.15 ppm STEL; 0.75 mg/m3 STEL
Canada - Alberta - Skin Designation
can be absorbed through the intact skin
Canada - British Columbia - 8 Hour Exposure Limits
0.05 ppm TWA; 0.3 mg/m3 TWA
Canada - British Columbia - 15 Minute Exposure Limits
0.15 ppm STEL; 0.9 mg/m3 STEL
Canada - British Columbia - Skin Notations
skin - potential for skin absorption
Canada - Ontario - OHSA - TWAEVs
0.05 ppm TWAEV; 0.25 mg/m3 TWAEV
Canada - Ontario - OHSA - STEVs
0.15 ppm STEV; 0.75 mg/m3 STEV
Canada - Ontario - OHSA - Skin Notations
absorption through skin, eyes, or mucous membranes
Canada - Quebec - Time-Weighted Average Exposure Values
0.05 ppm TWAEV; 0.25 mg/m3 TWAEV
Canada - Quebec - Short-term Exposure Values
0.15 ppm STEV; 0.75 mg/m3 STEV
Canada - Quebec - Skin Designations
absorbed through the skin
German (DFG) - MAK Values
0.05 ppm MAK; 0.3 mg/m3 MAK
German (DFG) - Peak Limitations
2 x normal MAK (5 min momentary value); don't exceed 8 times during shift
German (DFG) - Skin/Sensitizers
danger of cutaneous absorption
Israel - Time Weighted Averages
0.05 ppm TWA; 0.25 mg/m3 TWA
Israel - Short Term Exposure Limits
0.15 ppm STEL; 0.75 mg/m3 STEL
Israel - Action Levels
0.025 ppm AL; 0.125 mg/m3 AL
Mexico - Instruction No. 10 - TWAs
0.05 ppm TWA; 0.3 mg/m3 TWA
Mexico - Instruction No. 10 - STELs
0.15 ppm STEL; 0.9 mg/m3 STEL
Mexico - Instruction No. 10 - Skin designation
skin - potential for cutaneous absorption
STATE LISTS
California - Exposure Limits - PELs
0.05 ppm PEL; 0.3 mg/m3 PEL
California - Exposure Limits - STELs
0.15 ppm STEL; 0.9 mg/m3 STEL
California - Exposure Limits - Skin Notation
material may be absorbed through the skin, eyes or mucous membrane

California - Directors List of Hazardous Substances (8 CCR 339)
[present]

Florida Hazardous Substance List
[present]

Massachusetts Right To Know List
extraordinarily hazardous

Minnesota Hazardous Substance List
skin

NJ Right to Know List (Total)
sn 0597

Pennsylvania Right to Know List
environmental hazard

PROPOSED REGULATIONS

CERCLA/SARA - Proposed Hazardous Substance Additions
proposed RQ = 1 pound (.454 kg)

CERCLA/SARA - 1989 Proposed RQ Adjustments
proposed RQ = 10 pounds (4.54 kg)

DECABROMODIPHENYL OXIDE   1163-19-5

HEALTH AND SAFETY LISTS

AIHA - WEEL - Time Weighted Averages
5 mg/m3 TWA

IARC - Group 3 (not classifiable)
[present]

NTP Chemical Status Reports - Testing Status and NTIS Number
Technical reports printed (PB86247780/AS)

NTP Chemical Status Reports - Evidence of Carcinogenicity
male rat-some evidence; female rat-some evidence; male mice-equivocal evidence; female mice-no evidence

ENVIRONMENTAL LISTS

CERCLA/SARA - Section 313 - Emission Reporting
form R reporting required for 1.0% de minimus concentration

EPA - Master Testing List
[present]

TSCA - Code of Federal Regulations Citations
40 CFR 712.30; 40 CFR 716.120(a); 40 CFR 766.35

TSCA - PAIR - Reporting List
Reporting Date: March 12, 1990

TSCA - Health and Safety Reporting List
Effective Date: January 11, 1990

TSCA - HDD/HDF - Chemicals Required for Testing
[present]

TSCA - Section 12(b) - Export Notification
export notification required - Section 4

INTERNATIONAL LISTS

Canada - NPRI (National Pollutant Release Inventory)
[present]

STATE LISTS

California - Air Bill 2588 Appendix A-I
6/91

Massachusetts Right To Know List
[present]

Minnesota Hazardous Substance List
[present]

NJ Right to Know List (Total)
sn 0598

Pennsylvania Right to Know List
environmental hazard

2,4-DECADIENAL   25152-84-5

HEALTH AND SAFETY LISTS

NTP Chemical Status Reports - Testing Status and NTIS Number
Project leader assigned/study in design

DECAHYDRONAPHTHALENE   91-17-8

HEALTH AND SAFETY LISTS

U.S. DOT - Substances From 49 CFR 172.101
regulated by DOT (UN1147)

U.S. DOT - Hazard Classes
DOT hazard class = 3

NFPA - Flash Points
flash point = 136 degrees F (58 degrees C); trans-: 129 degrees F (54 degrees C)

NFPA - Hazard Identification Ratings
health-2; flammability-2; reactivity-0; trans-: health-0; flammability-2; reactivity-0

NIOSH - Selected LD50s and LC50s
Oral, rat: LD50 = 4170 mg/kg Skin, rabbit: LD50 = 5900 mg/kg

NTP Chemical Status Reports - Testing Status and NTIS Number
Approved for toxicology/carcinogenesis study

ENVIRONMENTAL LISTS

CAA - HON Rule - SOCMI Chemicals
compliance by July 24, 1995

STATE LISTS

Florida Hazardous Substance List
[present]

Massachusetts Right To Know List
[present]

NJ Right to Know List (Total)
sn 0599

Pennsylvania Right to Know List
[present]

DECAMETHYLCYCLOPENTASILOXANE   541-02-6

ENVIRONMENTAL LISTS

TSCA - PAIR - Reporting List
Effective Date: October 12, 1993; Reporting Date: February 28, 1994

TSCA - Health and Safety Reporting List
Effective Date: October 12, 1993; Sunset Date: October 12, 2003

DECAMETHYLTETRASILOXANE   141-62-8

ENVIRONMENTAL LISTS

TSCA - PAIR - Reporting List
Effective Date: October 12, 1993; Reporting Date: February 28, 1994

TSCA - Health and Safety Reporting List
Effective Date: October 12, 1993; Sunset Date: October 12, 2003

DECANAL   112-31-2

HEALTH AND SAFETY LISTS

NIOSH - Selected LD50s and LC50s
Oral, rat: LD50 = 3730 mg/kg Skin, rabbit: LD50 = 5040 mg/kg

ENVIRONMENTAL LISTS

TSCA - Code of Federal Regulations Citations
40 CFR 712.30(x); 40 CFR 716.120(d)

TSCA - PAIR - Reporting List
Reporting Date: November 27, 1991

TSCA - Health and Safety Reporting List
Effective Date: September 30, 1991

1-DECANAMINE, N-DECYL-N-METHYL-N-OXIDE   100545-50-4

ENVIRONMENTAL LISTS

TSCA - Section 12(b) - Export Notification
P-86-566; export notification required - Section 5

DECANE 124-18-5

HEALTH AND SAFETY LISTS

U.S. DOT - Substances From 49 CFR 172.101
regulated by DOT (UN2247)

U.S. DOT - Hazard Classes
DOT hazard class = 3

NFPA - Flash Points
flash point = 115 degrees F (46 degrees C)

NFPA - Hazard Identification Ratings
health-0; flammability-2; reactivity-0

NIOSH - Selected LD50s and LC50s
Inhalation, mouse: LC50 = 72300 mg/m3 (2 hr)

ENVIRONMENTAL LISTS

EPA - Master Testing List
[present]

STATE LISTS

NJ Right to Know List (Total)
sn 0600

Pennsylvania Right to Know List
[present]

DECANEDIOIC ACID 111-20-6

HEALTH AND SAFETY LISTS

NIOSH - Selected LD50s and LC50s
Oral, rat: LD50 = 3400 mg/kg

1-DECANETHIOL 143-10-2

HEALTH AND SAFETY LISTS

NIOSH - Health Standards - Exposure Limits
C (15 min) 0.5 ppm; C (15 min) 3.6 mg/m3 (Listed under 'Thiols')

NIOSH - Health Standards - Health Effects and Precautions
Irritation; eye, skin, blood, and nervous system effects (Blood and urine monitoring required; prevent skin contact) (Listed under 'Thiols')

TERT-DECANETHIOL 30174-58-4

HEALTH AND SAFETY LISTS

NFPA - Flash Points
flash point = 190 degrees F (88 degrees C)

NFPA - Hazard Identification Ratings
health-2; flammability-2; reactivity-0

STATE LISTS

Florida Hazardous Substance List
[present]

Massachusetts Right To Know List
[present]

Pennsylvania Right to Know List
[present]

DECANOIC ACID, BIS(2-ETHYLHEXYL) ESTER 122-62-3

HEALTH AND SAFETY LISTS

NIOSH - Selected LD50s and LC50s
Oral, rat: LD50 = 12800 mg/kg

1-DECANOL 112-30-1

HEALTH AND SAFETY LISTS

NFPA - Flash Points
flash point = 180 degrees F (82 degrees C)

NFPA - Hazard Identification Ratings
health-0; flammability-2; reactivity-0

NIOSH - Selected LD50s and LC50s
Inhalation, mouse: LC50 = 4 gm/m3 2 hr Oral, rat: LD50 = 4720 mg/kg Skin, rabbit: LD50 = 3560 mg/kg

ENVIRONMENTAL LISTS

List of Pesticide Product Inert Ingredients
[present]

INTERNATIONAL LISTS

Canada - WHMIS: Ingredient Disclosure
1% item 482 (615)

STATE LISTS

Pennsylvania Right to Know List
[present]

2-DECANONE 693-54-9

HEALTH AND SAFETY LISTS

NIOSH - Selected LD50s and LC50s
Oral, mouse: LD50 = 7936 mg/kg

DECANOYL PEROXIDE 762-12-9

HEALTH AND SAFETY LISTS

U.S. DOT - Organic Peroxides Table
Organic peroxide UN3102

STATE LISTS

NJ Right to Know List (Total)
sn 0602

DECARBROMOBIPHENYL 13654-09-6

ENVIRONMENTAL LISTS

TSCA - Code of Federal Regulations Citations
40 CFR 721.600

TSCA - Chemicals with Significant New Use Rules
[present]

TSCA - Section 12(b) - Export Notification
export notification required - Section 5

STATE LISTS

Massachusetts Right To Know List
carcinogen; extraordinarily hazardous

NJ Right to Know List (Total)
sn 3144; sn 3145

1-DECENE 872-05-9

HEALTH AND SAFETY LISTS

NFPA - Flash Points
flash point < 131 degrees F (55 degrees C)

NFPA - Hazard Identification Ratings
health-0; flammability-2; reactivity-0

ENVIRONMENTAL LISTS

EPA - Master Testing List
[present]

DECYL ACRYLATE 2156-96-9

HEALTH AND SAFETY LISTS

U.S. DOT - Appendix B - Marine Pollutants
DOT regulated marine pollutant

NFPA - Flash Points
flash point = 441 degrees F (227 degrees C)

NFPA - Hazard Identification Ratings
health-2; flammability-1; reactivity-0

ENVIRONMENTAL LISTS

TSCA - Code of Federal Regulations Citations
40 CFR 712.30(d)

TSCA - PAIR - Reporting List
Reporting Date: November 19, 1982

STATE LISTS
  Florida Hazardous Substance List
  [present]
  Massachusetts Right To Know List
  [present]
  Pennsylvania Right to Know List
  [present]

DECYL ALCOHOL BOTTOMS   68526-90-9
ENVIRONMENTAL LISTS
  List of Pesticide Product Inert Ingredients
  [present]

DECYL ALCOHOL, ETHOXYLATED   26183-52-8
ENVIRONMENTAL LISTS
  List of Pesticide Product Inert Ingredients
  [present]

DECYLAMINE   2016-57-1
HEALTH AND SAFETY LISTS
  NFPA - Flash Points
  flash point = 210 degrees F (99 degrees C)
  NFPA - Hazard Identification Ratings
  health-2; flammability-1; reactivity-0
  NIOSH - Selected LD50s and LC50s
  Oral, rat: LD50 = 280 mg/kg Skin, rabbit: LD50 = 350 mg/kg
STATE LISTS
  Florida Hazardous Substance List
  [present]
  Massachusetts Right To Know List
  [present]
  Pennsylvania Right to Know List
  [present]

DECYLBENZENE   104-72-3
HEALTH AND SAFETY LISTS
  NFPA - Flash Points
  flash point = 225 degrees F (107 degrees C)
  NFPA - Hazard Identification Ratings
  health-2; flammability-1; reactivity-0
STATE LISTS
  Florida Hazardous Substance List
  [present]
  Massachusetts Right To Know List
  [present]
  Pennsylvania Right to Know List
  [present]

DECYL GLUCOSIDE   41444-55-7
ENVIRONMENTAL LISTS
  List of Pesticide Product Inert Ingredients
  [present]

DECYLNAPHTHALENE   RR-00830-0
HEALTH AND SAFETY LISTS
  NFPA - Flash Points
  flash point = 350 degrees F (177 degrees C)
  NFPA - Hazard Identification Ratings
  health-1; flammability-1; reactivity-0

DECYL NITRATE   RR-00829-7
HEALTH AND SAFETY LISTS

NFPA - Flash Points
  flash point = 235 degrees F (113 degrees C)
NFPA - Hazard Identification Ratings
  flammability-1; reactivity-0

3-(DECYLOXY)TETRAHYDROTHIOPHENE 1,1- 18760-44-6
DIOXIDE
ENVIRONMENTAL LISTS
  TSCA - Code of Federal Regulations Citations
  40 CFR 712.30(x); 40 CFR 716.120(d)
  TSCA - PAIR - Reporting List
  Reporting Date: November 27, 1991
  TSCA - Health and Safety Reporting List
  Effective Date: September 30, 1991

DECYL PHENOXYBENZENEDISULFONIC ACID 70191-75-2
ENVIRONMENTAL LISTS
  List of Pesticide Product Inert Ingredients
  [present]

DECYL PHENOXYBENZENEDISULFONIC ACID, 36445-71-3
DISODIUM SALT
ENVIRONMENTAL LISTS
  List of Pesticide Product Inert Ingredients
  [present]

DECYL PHTHALATE   84-77-5
HEALTH AND SAFETY LISTS
  NIOSH - Selected LD50s and LC50s
  Skin, rabbit: LD50 = 16800 mg/kg

5-DECYNE-4,7-DIOL, 2,4,7,9-TETRAMETHYL- 126-86-3
ENVIRONMENTAL LISTS
  List of Pesticide Product Inert Ingredients
  [present]

DEF   78-48-8
HEALTH AND SAFETY LISTS
  U.S. DOT - Appendix B - Marine Pollutants
  DOT regulated marine pollutant
  NIOSH - Selected LD50s and LC50s
  Oral, rat: LD50 = 150 mg/kg Skin, ra: LD50 = 168 mg/kg
ENVIRONMENTAL LISTS
  CERCLA/SARA - Section 313 - Emission Reporting
  form R reporting required
STATE LISTS
  California - Directors List of Hazardous Substances (8 CCR 339)
  [present]

DEHYDROABIETYLAMINE, ETHOXYLATED 80584-98-1
ENVIRONMENTAL LISTS
  List of Pesticide Product Inert Ingredients
  [present]

DEHYDROABIETYLAMINE-ETHYLENE OXIDE 51344-62-8
ADDUCT
ENVIRONMENTAL LISTS
  List of Pesticide Product Inert Ingredients
  [present]

DEHYDROACETIC ACID   520-45-6
HEALTH AND SAFETY LISTS
  NFPA - Flash Points
  flash point = 315 degrees F (157 degrees C)

NFPA - Hazard Identification Ratings
health-1; flammability-1; reactivity-0
NIOSH - Selected LD50s and LC50s
Oral, rat: LD50 = 500 mg/kg

DELTAMETHRIN   52918-63-5

HEALTH AND SAFETY LISTS
IARC - Group 3 (not classifiable)
[present]

DEMECLOCYCLINE HYDROCHLORIDE   64-73-3

STATE LISTS
California - Prop. 65 - Developmental Toxicity
internal use: developmental toxicity - initial date 1/1/92

DEMETON   8065-48-3

HEALTH AND SAFETY LISTS
ACGIH 1995 - Time Weighted Averages
0.01 ppm TWA; 0.11 mg/m3 TWA
ACGIH 1995 - Skin Designations
skin - potential for cutaneous absorption
NIOSH - Selected LD50s and LC50s
Oral, rat: LD50 = 1700 ug/kg Skin, rat: LD50 = 8200 ug/kg
NIOSH 1990 - Pocket Guide - RELs
0.1 mg/m3 TWA
NIOSH 1990 - Pocket Guide - IDLHs
20 mg/m3 IDLH
NIOSH 1990 - Pocket Guide - Target organs
CVS, CNS, skin, eyes, blood cholinesterase, respiratory system
NIOSH 1990 - Pocket Guide - Skin list
Potential for dermal absorption
OSHA - Vacated PELs - Time Weighted Averages
0.1 mg/m3 TWA
OSHA - Vacated PELs - Skin Designation
Prevent or reduce skin absorption
OSHA - Final PELs - Time Weighted Averages
0.1 mg/m3 TWA
OSHA - Final PELs - Skin Notations
prevent or reduce skin absorption

ENVIRONMENTAL LISTS
CERCLA/SARA - Section 302 Extremely Hazardous Substances and TPQs
TPQ = 500 pounds

INTERNATIONAL LISTS
Australian Exposure Standards - Time Weighted Averages
0.01 ppm TWA; 0.11 mg/m3 TWA
Australian Exposure Standards - Skin Effects
skin absorption
Canada - Alberta - 8 Hour Occupational Exposure Limit
0.01 ppm TWA; 0.11 mg/m3 TWA
Canada - Alberta - 15 Minute Occupational Exposure Limit
0.03 ppm STEL; 0.32 mg/m3 STEL
Canada - Alberta - Skin Designation
can be absorbed through the intact skin
Canada - British Columbia - 8 Hour Exposure Limits
0.01 ppm TWA; 0.1 mg/m3 TWA
Canada - British Columbia - 15 Minute Exposure Limits
0.03 ppm STEL; 0.3 mg/m3 STEL
Canada - British Columbia - Skin Notations
skin - potential for skin absorption
Canada - Ontario - OHSA - TWAEVs
0.01 ppm TWAEV; 0.11 mg/m3 TWAEV

Canada - Ontario - OHSA - Skin Notations
absorption through skin, eyes, or mucous membranes
Canada - Quebec - Time-Weighted Average Exposure Values
0.01 ppm TWAEV; 0.11 mg/m3 TWAEV
Canada - Quebec - Skin Designations
absorbed through the skin
German (DFG) - MAK Values
0.01 ppm MAK; 0.1 mg/m3 MAK
German (DFG) - Peak Limitations
10 x normal MAK (30 min average value); don't exceed during shift
German (DFG) - Skin/Sensitizers
danger of cutaneous absorption
Israel - Time Weighted Averages
0.01 ppm TWA; 0.11 mg/m3 TWA
Israel - Action Levels
0.005 ppm AL; 0.055 mg/m3 AL
Mexico - Instruction No. 10 - TWAs
0.01 ppm TWA; 0.1 mg/m3 TWA
Mexico - Instruction No. 10 - STELs
0.03 ppm STEL; 0.3 mg/m3 STEL

STATE LISTS
California - Exposure Limits - PELs
0.01 ppm PEL; 0.1 mg/m3 PEL
California - Exposure Limits - Skin Notation
material may be absorbed through the skin, eyes or mucous membrane
California - Directors List of Hazardous Substances (8 CCR 339)
[present]
Florida Hazardous Substance List
[present]
Massachusetts Right To Know List
extraordinarily hazardous; neurotoxin
Minnesota Hazardous Substance List
skin
Pennsylvania Right to Know List
environmental hazard

PROPOSED REGULATIONS
CERCLA/SARA - Proposed Hazardous Substance Additions
proposed RQ = 1 pound (.454 kg)
CERCLA/SARA - 1989 Proposed RQ Adjustments
proposed RQ = 100 pounds (45.4 kg)

DEMETON-O   298-03-3

STATE LISTS
California - Directors List of Hazardous Substances (8 CCR 339)
[present]

DEMETON-S   126-75-0

STATE LISTS
California - Directors List of Hazardous Substances (8 CCR 339)
[present]

DEMETON-S-METHYL   919-86-8

HEALTH AND SAFETY LISTS
NIOSH - Selected LD50s and LC50s
Inhalation, rat: LC50 = 500 mg/m3 4 hr Oral, rat: LD50 = 60 mg/kg Skin, rat: LD50 = 85 mg/kg

ENVIRONMENTAL LISTS
CERCLA/SARA - Section 302 Extremely Hazardous Substances and TPQs
TPQ = 500 pounds

STATE LISTS
Florida Hazardous Substance List
effective March 13, 1992

Massachusetts Right To Know List
extraordinarily hazardous

NJ Right to Know List (Total)
sn 2886

Pennsylvania Right to Know List
environmental hazard

PROPOSED REGULATIONS

CERCLA/SARA - Proposed Hazardous Substance Additions
proposed RQ = 1 pound (.454 kg)

CERCLA/SARA - 1989 Proposed RQ Adjustments
proposed RQ = 100 pounds (45.4 kg)

DEOXYCHOLIC ACID   83-44-3

HEALTH AND SAFETY LISTS

NIOSH - Selected LD50s and LC50s
Oral, rat: LD50 = 1 gm/kg

DEOXYNIVALENOL   51481-10-8

HEALTH AND SAFETY LISTS

IARC - Group Unspecified
[present] (Listed under 'Toxins derived from Fusarium graminearum, F. culmorum, and F. crookwellense')

DERIVATIVE OF TETRACHLOROETHYLENE   RR-00249-3

ENVIRONMENTAL LISTS

TSCA - Chemicals with Significant New Use Rules
PMN number: P-82-684

DESMEDIPHAM   13684-56-5

ENVIRONMENTAL LISTS

CERCLA/SARA - Section 313 - Emission Reporting
form R reporting required

2,4-D 2-ETHYLHEXYL ESTER   1928-43-4

ENVIRONMENTAL LISTS

CERCLA/SARA - Section 313 - Emission Reporting
form R reporting required

2,4-D 2-ETHYL-4-METHYLPENTYL ESTER   53404-37-8

ENVIRONMENTAL LISTS

CERCLA/SARA - Section 313 - Emission Reporting
form R reporting required

DEUTERIUM   7782-39-0

HEALTH AND SAFETY LISTS

U.S. DOT - Substances From 49 CFR 172.101
regulated by DOT (UN1957)

U.S. DOT - Hazard Classes
DOT hazard class = 2.1

NFPA - Flash Points
gas (no flash point given)

NFPA - Hazard Identification Ratings
health-0; flammability-4; reactivity-0

STATE LISTS

Florida Hazardous Substance List
[present]

Massachusetts Right To Know List
[present]

NJ Right to Know List (Total)
sn 0605

NJ Special Hazardous Substances
(flammable - fourth degree)

Pennsylvania Right to Know List
[present]

DEXTRIN   9004-53-9

ENVIRONMENTAL LISTS

List of Pesticide Product Inert Ingredients
[present]

DIACETONE ALCOHOL   123-42-2

HEALTH AND SAFETY LISTS

ACGIH 1995 - Time Weighted Averages
50 ppm TWA; 238 mg/m3 TWA

AIHA - Odor Threshold Values
geometric mean air odor threshold = 0.27 ppm (detectable); 1.1 ppm (recognizable)

U.S. DOT - Substances From 49 CFR 172.101
regulated by DOT (UN1148)

U.S. DOT - Hazard Classes
DOT hazard class = 3

NFPA - Flash Points
flash point = 148 degrees F (64 degrees C); Acetone-free: 136 degrees F (58 degrees C); commercial: 148 degrees F (64 degrees C)

NFPA - Hazard Identification Ratings
health-1; flammability-2; reactivity-0

NIOSH - Selected LD50s and LC50s
Oral, rat: LD50 = 4000 mg/kg Skin, rabbit: LD50 = 13500 mg/kg

NIOSH 1990 - Pocket Guide - RELs
50 ppm TWA; 240 mg/m3 TWA

NIOSH 1990 - Pocket Guide - IDLHs
2100 ppm IDLH

NIOSH 1990 - Pocket Guide - Target organs
eyes, skin, respiratory system

NIOSH - Health Standards - Exposure Limits
50 ppm TWA; 240 mg/m3 TWA (Listed under 'Ketones')

NIOSH - Health Standards - Health Effects and Precautions
Irritation; liver, kidney, and nervous system effects (Urinalysis required) (Listed under 'Ketones')

OSHA - Vacated PELs - Time Weighted Averages
50 ppm TWA; 240 mg/m3 TWA

OSHA - Final PELs - Time Weighted Averages
50 ppm TWA; 240 mg/m3 TWA

ENVIRONMENTAL LISTS

List of Pesticide Product Inert Ingredients
[present]

INTERNATIONAL LISTS

Australian Exposure Standards - Time Weighted Averages
50 ppm TWA; 238 mg/m3 TWA

Canada - WHMIS: Ingredient Disclosure
1% item 483 (172)

Canada - Alberta - 8 Hour Occupational Exposure Limit
50 ppm TWA; 235 mg/m3 TWA

Canada - Alberta - 15 Minute Occupational Exposure Limit
75 ppm STEL; 355 mg/m3 STEL

Canada - British Columbia - 8 Hour Exposure Limits
50 ppm TWA; 240 mg/m3 TWA

Canada - British Columbia - 15 Minute Exposure Limits
75 ppm STEL; 360 mg/m3 STEL

Canada - Ontario - OHSA - TWAEVs
50 ppm TWAEV; 240 mg/m3 TWAEV

Canada - Ontario - OHSA - STEVs
75 ppm STEV; 360 mg/m3 STEV

Canada - Quebec - Time-Weighted Average Exposure Values
50 ppm TWAEV; 238 mg/m3 TWAEV

United Kingdom - Occupational Exposure Standards - TWAs
50 ppm TWA; 240 mg/m3 TWA

United Kingdom - Occupational Exposure Standards - STELs
75 ppm STEL; 360 mg/m3 STEL
German (DFG) - MAK Values
50 ppm MAK; 240 mg/m3 MAK
Israel - Time Weighted Averages
50 ppm TWA; 238 mg/m3 TWA
Israel - Action Levels
25 ppm AL; 119 mg/m3 AL
Mexico - Instruction No. 10 - TWAs
50 ppm TWA; 240 mg/m3 TWA
Mexico - Instruction No. 10 - STELs
75 ppm STEL; 360 mg/m3 STEL

STATE LISTS

California - Exposure Limits - PELs
50 ppm PEL; 240 mg/m3 PEL
California - Directors List of Hazardous Substances (8 CCR 339)
[present]
Florida Hazardous Substance List
[present]
Massachusetts Right To Know List
[present]
Minnesota Hazardous Substance List
[present]
NJ Right to Know List (Total)
sn 0606
Pennsylvania Right to Know List
[present]

DIACETONE ALCOHOL PEROXIDE   54693-46-8

HEALTH AND SAFETY LISTS

U.S. DOT - Hazard Classes
Forbidden from transport by the DOT
U.S. DOT - Organic Peroxides Table
Organic peroxide UN3115

STATE LISTS

NJ Right to Know List (Total)
sn 0607

DIACETONE ALCOHOL PEROXIDE   55794-20-2

HEALTH AND SAFETY LISTS

U.S. DOT - Organic Peroxides Table
Organic peroxide UN3103; UN3105; UN3106

DIACETONE ALCOHOL PEROXIDE   67567-23-1

HEALTH AND SAFETY LISTS

U.S. DOT - Organic Peroxides Table
Organic peroxide UN3105

DIACETONE ALCOHOL PEROXIDE   68299-16-1

HEALTH AND SAFETY LISTS

U.S. DOT - Organic Peroxides Table
Organic peroxide UN3115

1,4-DIACETOXYBUT-2-ENE (1,4)   25260-60-0

ENVIRONMENTAL LISTS

CAA - HON Rule - SOCMI Chemicals
compliance by Oct. 23, 1995

DIACETYLAMINOAZOTOLUENE   83-63-6

HEALTH AND SAFETY LISTS

IARC - Group 3 (not classifiable)
[present]

N,N'-DIACETYLBENZIDINE   613-35-

HEALTH AND SAFETY LISTS

IARC - Group 2B (sufficient animal data)
[present] (Overall evaluation based only on evidence of carcinogenici in monograph (16, 1978) or in Supplement 4)
OSHA - Possible Select Carcinogens
[present]

STATE LISTS

California - Air Bill 2588 Appendix A-II
known or potential carcinogen
California - Prop. 65 - Cancer list
carcinogen - initial date 10/1/89
California - Directors List of Hazardous Substances (8 CCR 339)
[present]
Florida Hazardous Substance List
[present]
Massachusetts Right To Know List
carcinogen; extraordinarily hazardous
Minnesota Hazardous Substance List
carcinogen
Pennsylvania Right to Know List
special hazardous substance
Pennsylvania RTK - Special Hazardous Substances
[present]

DIACETYL PEROXIDE   110-22-5

HEALTH AND SAFETY LISTS

U.S. DOT - Hazard Classes
Forbidden from transport by the DOT
U.S. DOT - Organic Peroxides Table
Organic peroxide UN3115
NFPA - Hazard Identification Ratings
25% solution in Dimethyl phthalate: health-1; flammability-2; reactivity-4
OSHA - List of Highly Hazardous Chemicals
concentration > 70%: threshhold quantity = 5000 pounds

INTERNATIONAL LISTS

Canada - WHMIS: Ingredient Disclosure
1% item 484 (1364)
German (DFG) - Skin/Sensitizers
very strong effects on skin

STATE LISTS

Florida Hazardous Substance List
[present]
Massachusetts Right To Know List
[present]
NJ Right to Know List (Total)
sn 0019
NJ Special Hazardous Substances
(reactive - fourth degree)
Pennsylvania Right to Know List
[present]

DIACETYL TARTARIC ACID ESTERS OF MONO AND DIGLYCERIDES OF EDIBLE FATS   RR-01064-0

ENVIRONMENTAL LISTS

List of Pesticide Product Inert Ingredients
[present]

DI ACETYL TARTARIC ESTERS OF MONO AND DIGLYCERIDES OF EDIBLE FATTY ACIDS   RR-01063-9

ENVIRONMENTAL LISTS

List of Pesticide Product Inert Ingredients
[present]

DIALIFOS   10311-84-9

HEALTH AND SAFETY LISTS

U.S. DOT - Appendix B - Marine Pollutants
DOT regulated severe marine pollutant

NIOSH - Selected LD50s and LC50s
Oral, rat: LD50 = 5 mg/kg Skin, rat: LD50 = 28 mg/kg

ENVIRONMENTAL LISTS

CERCLA/SARA - Section 302 Extremely Hazardous Substances and TPQs
TPQ = 100/10,000 pounds

STATE LISTS

California - Directors List of Hazardous Substances (8 CCR 339)
[present]

Massachusetts Right To Know List
extraordinarily hazardous

NJ Right to Know List (Total)
sn 2309

Pennsylvania Right to Know List
environmental hazard

PROPOSED REGULATIONS

CERCLA/SARA - Proposed Hazardous Substance Additions
proposed RQ = 1 pound (.454 kg)

CERCLA/SARA - 1989 Proposed RQ Adjustments
proposed RQ = 100 pounds (45.4 kg)

DI(ALKANEPOLYOL) ETHER, POLYACRYLATE   RR-00978-9

ENVIRONMENTAL LISTS

TSCA - Chemicals with Significant New Use Rules
PMN number: P-85-718

DIALKENYLAMIDE   RR-00191-2

ENVIRONMENTAL LISTS

TSCA - Chemicals with Significant New Use Rules
PMN number: P-87-502

DIALKYL 79 PHTHALATE   83968-18-7

INTERNATIONAL LISTS

United Kingdom - Occupational Exposure Standards - TWAs
5 mg/m3 TWA

DIALKYLAMINO ALKANOATE, METAL SALT   RR-00971-2

ENVIRONMENTAL LISTS

TSCA - Chemicals with Significant New Use Rules
PMN number: P-90-274

DIALKYLDITHIOPHOSPHORIC ACID, ALIPHATIC AMINE SALT   RR-01245-3

ENVIRONMENTAL LISTS

TSCA - Chemicals with Significant New Use Rules
PMN number: P-90-1839

DIALKYLNITROSAMINES   RR-00414-8

STATE LISTS

California - Air Bill 2588 Appendix A-I
[present]

DIALKYL PHOSPHORODITHIOATE PHOSPHATE COMPOUNDS, 2-PROPENAMIDE, N-[3-(DIMETHYLAMINO)PROPYL]-   RR-00961-0

ENVIRONMENTAL LISTS

TSCA - Chemicals with Significant New Use Rules
PMN numbers: P-90-1642 throug P-90-1649

DIALLATE   2303-16-4

HEALTH AND SAFETY LISTS

U.S. DOT - Appendix B - Marine Pollutants
DOT regulated marine pollutant

U.S. DOT - Appendix A Table 1 - Hazardous Substances
final RQ = 100 pounds (45.4 kg)

IARC - Group 3 (not classifiable)
[present]

NIOSH - Selected LD50s and LC50s
Oral, rat: LD50 = 395 mg/kg Skin, rat: LD50 = 2124 mg/kg

ENVIRONMENTAL LISTS

CERCLA/SARA - Section 313 - Emission Reporting
form R reporting required for 1.0% de minimus concentration

CERCLA/SARA - Hazardous Substances and their Reportable Quantities
final RQ = 100 pounds (45.4 kg)

EPA - Carcinogen Hazard Ranking for RQ Adjustment
Hazard ranking = Low

RCRA - U Series Wastes
waste number U062

RCRA - Hazardous Constituents-Appendix VIII
waste number U062

RCRA - Substances Banned From Land Disposal
[present]

RCRA - TSD Facilities Ground Water Monitoring
TM 8270 = 10 ug/L PQL

STATE LISTS

California - Air Bill 2588 Appendix A-II
6/91

California - Directors List of Hazardous Substances (8 CCR 339)
[present]

Florida Hazardous Substance List
[present]

Massachusetts Right To Know List
carcinogen; extraordinarily hazardous

NJ Right to Know List (Total)
sn 0608

NJ Special Hazardous Substances
(carcinogen; mutagen)

Pennsylvania Right to Know List
environmental hazard

DIALLYL ADIPATE   2998-04-1

HEALTH AND SAFETY LISTS

NIOSH - Selected LD50s and LC50s
Oral, mouse: LD50 = 180 mg/kg

DIALLYLAMINE   124-02-7

HEALTH AND SAFETY LISTS

AIHA - WEEL - Time Weighted Averages
1 ppm TWA; 3.97 mg/m3 TWA

U.S. DOT - Substances From 49 CFR 172.101
regulated by DOT (UN2359)

U.S. DOT - Hazard Classes
DOT hazard class = 3

NIOSH - Selected LD50s and LC50s
Inhalation, rat: LC50 = 795 ppm 8 hr Oral, rat: LD50 = 578 mg/kg
Skin, rabbit: LD50 = 280 mg/kg

STATE LISTS

NJ Right to Know List (Total)
sn 0609

N,N-DIALLYL-2,3-DICHLOROACETAMIDE   37764-25-3

ENVIRONMENTAL LISTS

List of Pesticide Product Inert Ingredients
[present]

DIALLYL ETHER   557-40-4

HEALTH AND SAFETY LISTS

U.S. DOT - Substances From 49 CFR 172.101
regulated by DOT (UN2360)
U.S. DOT - Hazard Classes
DOT hazard class = 3
NFPA - Flash Points
flash point = 20 degrees F (-7 degrees C)
NFPA - Hazard Identification Ratings
health-3; flammability-3; reactivity-2
NIOSH - Selected LD50s and LC50s
Oral, rat: LD50 = 320 mg/kg Skin, rabbit: LD50 = 600 mg/kg

INTERNATIONAL LISTS

Canada - WHMIS: Ingredient Disclosure
1% item 485 (837)

STATE LISTS

Florida Hazardous Substance List
[present]
Massachusetts Right To Know List
[present]
NJ Right to Know List (Total)
sn 0610
NJ Special Hazardous Substances
(flammable - third degree; reactive - second degree)
Pennsylvania Right to Know List
[present]

DIALLYL MALEATE   999-21-3

HEALTH AND SAFETY LISTS

NIOSH - Selected LD50s and LC50s
Oral, rat: LD50 = 300 mg/kg Skin, rabbit: LD50 = 1150 mg/kg

DIALLYL PHTHALATE   131-17-9

HEALTH AND SAFETY LISTS

NFPA - Flash Points
flash point = 330 degrees F (166 degrees C)
NFPA - Hazard Identification Ratings
health-2; flammability-1; reactivity-0
NIOSH - Selected LD50s and LC50s
Oral, rat: LD50 = 770 mg/kg
NTP Chemical Status Reports - Testing Status and NTIS Number
Technical reports printed (PB83200824) (PB86203742/AS)
NTP Chemical Status Reports - Evidence of Carcinogenicity
PB83200824:   male mice-equivocal;   female mice-equivocal;
PB86203742/AS: male rat-no evidence; female rat-equivocal evi-
dence

ENVIRONMENTAL LISTS

List of Pesticide Product Inert Ingredients
[present]
TSCA - Code of Federal Regulations Citations
40 CFR 712.30(w); 40 CFR 716.120(a)

TSCA - PAIR - Reporting List
Reporting Date: June 13, 1989
TSCA - Health and Safety Reporting List
Effective Date: April 13, 1989

INTERNATIONAL LISTS

United Kingdom - Occupational Exposure Standards - TWAs
5 mg/m3 TWA

STATE LISTS

Florida Hazardous Substance List
[present]
Massachusetts Right To Know List
[present]
Pennsylvania Right to Know List
[present]

2,4-DIAMINOANISOLE   615-05-4

HEALTH AND SAFETY LISTS

IARC - Group 2B (sufficient animal data)
[present] (Overall evaluation based only on evidence of carcinogenicity in monograph (27, 1982) or in Supplement 4)
NIOSH - Selected LD50s and LC50s
Oral, rat: LD50 = 460 mg/kg
NIOSH - Health Standards - Exposure Limits
reduce exposure to lowest feasible concentration
NIOSH - Health Standards - Health Effects and Precautions
has produced tumors of the thyroid, skin, and lymphatic system in animals (Prevent skin contact)
NIOSH - Health Standards - Carcinogenic Chemicals
potential human carcinogen
OSHA - Possible Select Carcinogens
[present]

ENVIRONMENTAL LISTS

CERCLA/SARA - Section 313 - Emission Reporting
form R reporting required for 0.1% de minimus concentration
TSCA - Code of Federal Regulations Citations
40 CFR 712.30(d);   40 CFR 716.120(c)
TSCA - PAIR - Reporting List
Reporting Date: November 19, 1982

INTERNATIONAL LISTS

Canada - WHMIS: Ingredient Disclosure
1% item 486 (621)
German (DFG) - Carcinogens
animal evidence of carcinogenicity

STATE LISTS

California - Air Bill 2588 Appendix A-I
known or potential carcinogen
California - Prop. 65 - Cancer list
carcinogen - initial date 10/1/90
California - Prop. 65 - No Significant Risk Levels
no significant risk level = 30 ug/day
California - Directors List of Hazardous Substances (8 CCR 339)
[present]
Massachusetts Right To Know List
[present]
Minnesota Hazardous Substance List
carcinogen (includes its salts)
NJ Right to Know List (Total)
sn 0611
Pennsylvania Right to Know List
environmental hazard

2,4-DIAMINOANISOLE SULFATE   39156-41-7
SEE ALSO:
2,4-DIAMINOANISOLE SULFATE

HEALTH AND SAFETY LISTS
NTP Chemical Status Reports - Testing Status and NTIS Number
Technical reports printed (PB279940/AS)
NTP Chemical Status Reports - Evidence of Carcinogenicity
male rat-positive; female rat-positive; male mice-positive; female mice-positive
NTP Seventh Report - Suspect Carcinogens
suspect carcinogen
OSHA - Possible Select Carcinogens
[present]

ENVIRONMENTAL LISTS
CERCLA/SARA - Section 313 - Emission Reporting
form R reporting required for 0.1% de minimus concentration
TSCA - Code of Federal Regulations Citations
40 CFR 712.30(f); 40 CFR 716.120(c)
TSCA - PAIR - Reporting List
Reporting Date: August 17, 1983

INTERNATIONAL LISTS
Canada - WHMIS: Ingredient Disclosure
0.1% item 487 (1520)

STATE LISTS
California - Air Bill 2588 Appendix A-II
known or potential carcinogen
California - Prop. 65 - Cancer list
carcinogen - initial date 1/1/88
California - Prop. 65 - No Significant Risk Levels
no significant risk level = 50 ug/day
California - Directors List of Hazardous Substances (8 CCR 339)
[present]
Florida Hazardous Substance List
[present]
Massachusetts Right To Know List
carcinogen; extraordinarily hazardous
Minnesota Hazardous Substance List
carcinogen
NJ Right to Know List (Total)
sn 2899
Pennsylvania Right to Know List
environmental hazard; special hazardous substance
Pennsylvania RTK - Special Hazardous Substances
[present]

1,4-DIAMINOANTHRAQUINONE   128-95-0

HEALTH AND SAFETY LISTS
NIOSH - Selected LD50s and LC50s
Oral, rat: LD50 = 5790 mg/kg

INTERNATIONAL LISTS
Canada - WHMIS: Ingredient Disclosure
1% item 488 (622)

3,3'-DIAMINOBENZIDINE   91-95-2

INTERNATIONAL LISTS
German (DFG) - Carcinogens
suspected carcinogen

3,3'-DIAMINOBENZIDINE TETRAHYDROCHLORIDE   7411-49-6

INTERNATIONAL LISTS
German (DFG) - Carcinogens
suspected carcinogen

4,4'-DIAMINODIPHENYL ETHER   101-80-4

HEALTH AND SAFETY LISTS
IARC - Group 2B (sufficient animal data)
[present] (Overall evaluation based only on evidence of carcinogenicity in monograph (29, 1982) or in Supplement 4)
NIOSH - Selected LD50s and LC50s
Oral, rat: LD50 = 725 mg/kg
NTP Chemical Status Reports - Testing Status and NTIS Number
Technical reports printed (PB80217938)
NTP Chemical Status Reports - Evidence of Carcinogenicity
male rat-positive; female rat-positive; male mice-positive; female mice-positive
NTP Seventh Report - Suspect Carcinogens
suspect carcinogen
OSHA - Possible Select Carcinogens
[present]

ENVIRONMENTAL LISTS
CERCLA/SARA - Section 313 - Emission Reporting
form R reporting required for 0.1% de minimus concentration

INTERNATIONAL LISTS
German (DFG) - Skin/Sensitizers
danger of sensitization (skin or respiratory)
German (DFG) - Carcinogens
animal evidence of carcinogenicity

STATE LISTS
California - Air Bill 2588 Appendix A-II
known or potential carcinogen
California - Prop. 65 - Cancer list
carcinogen - initial date 1/1/88
California - Prop. 65 - No Significant Risk Levels
no significant risk level = 5 ug/day
California - Directors List of Hazardous Substances (8 CCR 339)
[present]
Florida Hazardous Substance List
[present]
Massachusetts Right To Know List
carcinogen; extraordinarily hazardous
Minnesota Hazardous Substance List
carcinogen
NJ Right to Know List (Total)
sn 0612
Pennsylvania Right to Know List
environmental hazard; special hazardous substance
Pennsylvania RTK - Special Hazardous Substances
[present]

1,2-DIAMINOETHANE, DIHYDROIODIDE   5700-49-2

ENVIRONMENTAL LISTS
List of Pesticide Product Inert Ingredients
[present]

2,4-DIAMINOPHENOL DIHYDROCHLORIDE   137-09-7
SEE ALSO:
2,4-DIAMINOPHENOL DIHYDROCHLORIDE

HEALTH AND SAFETY LISTS
NTP Chemical Status Reports - Testing Status and NTIS Number
Technical reports printed (PB93-1117919)
NTP Chemical Status Reports - Evidence of Carcinogenicity
male rat: no evidence; female rat: no evidence; male mice: some evidence; female mice: no evidence

ENVIRONMENTAL LISTS
CAA - HON Rule - SOCMI Chemicals
compliance by Oct. 23, 1995

TSCA - Code of Federal Regulations Citations
40 CFR 712.30(f);   40 CFR 716.120(c)
TSCA - PAIR - Reporting List
Reporting Date: August 17, 1983
INTERNATIONAL LISTS
Canada - WHMIS: Ingredient Disclosure
0.1% item 490 (624)

1,3-DIAMINO-2-PROPANOL   616-29-5
HEALTH AND SAFETY LISTS
NFPA - Flash Points
flash point = 270 degrees F (132 degrees C)
NFPA - Hazard Identification Ratings
health-2;   flammability-1;   reactivity-0
STATE LISTS
Florida Hazardous Substance List
[present]
Massachusetts Right To Know List
[present]
Pennsylvania Right to Know List
[present]

4,4'-DIAMINO-2,2'-STILBENEDISULFONIC ACID, DISODIUM SALT   81-11-8
HEALTH AND SAFETY LISTS
NIOSH - Selected LD50s and LC50s
Oral, guinea pig: LD50 = 47 gm/kg
ENVIRONMENTAL LISTS
EPA - Master Testing List
[present]

4,4'-DIAMINO-2,2'-STILBENEDISULFONIC ACID, DISODIUM SALT   7336-20-1
HEALTH AND SAFETY LISTS
NTP Chemical Status Reports - Testing Status and NTIS Number
Technical reports printed (PB93-132504)
NTP Chemical Status Reports - Evidence of Carcinogenicity
male rat: no evidence; female rat: no evidence; male mice: no evidence; female mice: no evidence

2,5-DIAMINOTOLUENE   95-70-5
SEE ALSO:
2,5-DIAMINOTOLUENE
HEALTH AND SAFETY LISTS
IARC - Group 3 (not classifiable)
[present]
ENVIRONMENTAL LISTS
TSCA - Code of Federal Regulations Citations
40 CFR 712.30(f);   40 CFR 716.120(c)
TSCA - PAIR - Reporting List
Reporting Date: August 17, 1983
INTERNATIONAL LISTS
Canada - WHMIS: Ingredient Disclosure
1% item 492 (626)

DIAMINOTOLUENE   496-72-0
SEE ALSO:
DIAMINOTOLUENE
HEALTH AND SAFETY LISTS
U.S. DOT - Appendix A Table 1 - Hazardous Substances
final RQ = 10 pounds (4.54 kg) (Listed under 'Toluenediamine')

ENVIRONMENTAL LISTS
CERCLA/SARA - Hazardous Substances and their Reportable Quantities
final RQ = 10 pounds (4.54 kg) (Listed under 'Toluenediamine')
RCRA - Hazardous Constituents-Appendix VIII
hazardous constituent - no waste number
TSCA - Code of Federal Regulations Citations
40 CFR 712.30(d);   40 CFR 716.120(a)
TSCA - PAIR - Reporting List
Reporting Date: November 19, 1982
INTERNATIONAL LISTS
Canada - WHMIS: Ingredient Disclosure
0.1% item 1580 (1625)
STATE LISTS
Massachusetts Right To Know List
[present]
NJ Right to Know List (Total)
sn 2981
Pennsylvania Right to Know List
environmental hazard

2,6-DIAMINOTOLUENE   823-40-5
SEE ALSO:
2,6-DIAMINOTOLUENE
HEALTH AND SAFETY LISTS
U.S. DOT - Appendix A Table 1 - Hazardous Substances
final RQ = 10 pounds (4.54 kg) (Listed under 'Toluenediamine')
ENVIRONMENTAL LISTS
CERCLA/SARA - Hazardous Substances and their Reportable Quantities
final RQ = 10 pounds (4.54 kg) (Listed under 'Toluenediamine')
RCRA - Hazardous Constituents-Appendix VIII
hazardous constituent - no waste number
TSCA - Code of Federal Regulations Citations
40 CFR 712.30(d);   40 CFR 716.120(c)
TSCA - PAIR - Reporting List
Reporting Date: November 19, 1982
STATE LISTS
Massachusetts Right To Know List
[present]
NJ Right to Know List (Total)
sn 2980
Pennsylvania Right to Know List
environmental hazard

DIAMINOTOLUENE (MIXED ISOMERS)   25376-45-8
SEE ALSO:
DIAMINOTOLUENE (MIXED ISOMERS)
ENVIRONMENTAL LISTS
CERCLA/SARA - Section 313 - Emission Reporting
form R reporting required for 0.1% de minimus concentration
RCRA - U Series Wastes
waste number U221
RCRA - Hazardous Constituents-Appendix VIII
waste number U221
RCRA - Substances Banned From Land Disposal
[present]
TSCA - Code of Federal Regulations Citations
40 CFR 712.30(f);   40 CFR 716.120(c)
TSCA - PAIR - Reporting List
Reporting Date: August 17, 1983

INTERNATIONAL LISTS
Canada - WHMIS: Ingredient Disclosure
0.1% item 1582 (1623)

STATE LISTS
California - Air Bill 2588 Appendix A-I
known or potential carcinogen: 9/90
California - Prop. 65 - Cancer list
carcinogen - initial date 1/1/90
California - Directors List of Hazardous Substances (8 CCR 339)
[present]
Massachusetts Right To Know List
[present]
NJ Right to Know List (Total)
sn 2134
Pennsylvania Right to Know List
environmental hazard

DIAMMINEDICHLOROPALLADIUM   14323-43-4

INTERNATIONAL LISTS
Canada - WHMIS: Ingredient Disclosure
1% item 493 (627)

CIS-DIAMMINEDICHLOROPALLADIUM(II)   15684-18-1

INTERNATIONAL LISTS
Canada - WHMIS: Ingredient Disclosure
1% item 494 (628)

DIAMMONIUM ETHYLENEDIAMINETETRAACETATE   20824-56-0

ENVIRONMENTAL LISTS
List of Pesticide Product Inert Ingredients
[present]

DIAMMONIUM PHOSPHATE   7783-28-0

ENVIRONMENTAL LISTS
List of Pesticide Product Inert Ingredients
[present]

DI-N-AMYLAMINE   2050-92-2

HEALTH AND SAFETY LISTS
U.S. DOT - Substances From 49 CFR 172.101
regulated by DOT (UN2841)
U.S. DOT - Hazard Classes
DOT hazard class = 6.1
U.S. DOT - Substances Which Are Poisonous by Inhalation
liquid hazardous material poisonous by inhalation (mixed isomers)
NFPA - Flash Points
flash point = 124 degrees F (51 degrees C)
NFPA - Hazard Identification Ratings
health-3; flammability-2; reactivity-0
NIOSH - Selected LD50s and LC50s
Oral, rat: LD50 = 270 mg/kg Skin, rabbit: LD50 = 350 mg/kg

INTERNATIONAL LISTS
Canada - WHMIS: Ingredient Disclosure
1% item 496 (630)

STATE LISTS
Florida Hazardous Substance List
[present]
Massachusetts Right To Know List
[present]
NJ Right to Know List (Total)
sn 0614

Pennsylvania Right to Know List
[present]

DIAMYLBENZENE   RR-00836-6

HEALTH AND SAFETY LISTS
NFPA - Flash Points
flash point = 225 degrees F (107 degrees C)
NFPA - Hazard Identification Ratings
health-0; flammability-1; reactivity-0

DIAMYLBIPHENYL   RR-00837-7

HEALTH AND SAFETY LISTS
NFPA - Flash Points
flash point = 340 degrees F (171 degrees C)
NFPA - Hazard Identification Ratings
health-0; flammability-1; reactivity-0

DI-TERT-AMYLCYCLOHEXANOL   RR-00833-3

HEALTH AND SAFETY LISTS
NFPA - Flash Points
flash point = 270 degrees F (132 degrees C)
NFPA - Hazard Identification Ratings
health-0; flammability-1; reactivity-0

DIAMYLENE   RR-00839-9

HEALTH AND SAFETY LISTS
NFPA - Flash Points
flash point = 118 degrees F (48 degrees C)
NFPA - Hazard Identification Ratings
health-0; flammability-2; reactivity-0

DIAMYL MALEATE   10099-71-5

HEALTH AND SAFETY LISTS
NFPA - Flash Points
flash point = 270 degrees F (132 degrees C)
NFPA - Hazard Identification Ratings
health-0; flammability-1; reactivity-0

DIAMYL NAPHTHALENE   50696-42-9

HEALTH AND SAFETY LISTS
NFPA - Flash Points
flash point = 315 degrees F (159 degrees C)
NFPA - Hazard Identification Ratings
health-0; flammability-1; reactivity-0

DI-TERT-AMYLPHENOXY ETHANOL   RR-00834-4

HEALTH AND SAFETY LISTS
NFPA - Flash Points
flash point = 300 degrees F (149 degrees C)
NFPA - Hazard Identification Ratings
health-0; flammability-1; reactivity-0

DIAMYL PHTHALATE   131-18-0

HEALTH AND SAFETY LISTS
NFPA - Flash Points
flash point = 245 degrees F (118 degrees C)
NFPA - Hazard Identification Ratings
health-0; flammability-1; reactivity-0

O-DIANISIDINE-BASED DYES   RR-00062-4

HEALTH AND SAFETY LISTS
NIOSH - Health Standards - Exposure Limits
handle with caution in the workplace and minimize exposures

NIOSH - Health Standards - Health Effects and Precautions
*has produced tumors of the bladder, stomach, and mammary glands
in animals (Substitute less toxic dyes wherever possible)*
NIOSH - Health Standards - Carcinogenic Chemicals
potential human carcinogen

1,1-DIANTHRIMIDE   82-22-4
HEALTH AND SAFETY LISTS
NIOSH - Selected LD50s and LC50s
Oral, rat: LD50 = 16200 mg/kg

DIARYLANILIDE YELLOW   6358-85-6
HEALTH AND SAFETY LISTS
NTP Chemical Status Reports - Testing Status and NTIS Number
Technical reports printed (PB278272/AS)
NTP Chemical Status Reports - Evidence of Carcinogenicity
*male rat-negative; female rat-negative; male mice-negative; female
mice-negative*

DIATOMACEOUS EARTH   61790-53-2
HEALTH AND SAFETY LISTS
ACGIH 1995 - Time Weighted Averages
*10 mg/m3 TWA (The value is for total dust containing no asbestos
and <1% free silica)*
OSHA - Vacated PELs - Time Weighted Averages
6 mg/m3 TWA
OSHA - Final PELs - Time Weighted Averages
see Table Z-3
ENVIRONMENTAL LISTS
List of Pesticide Product Inert Ingredients
[present]
INTERNATIONAL LISTS
Australian Exposure Standards - Time Weighted Averages
10 mg/m3 TWA
Canada - Alberta - 8 Hour Occupational Exposure Limit
respirable mass: 2 mg/m3 TWA; total mass: 5 mg/m3 TWA
Canada - British Columbia - 8 Hour Exposure Limits
respirable mass: 1.5 mg/m3 TWA
United Kingdom - Occupational Exposure Standards - TWAs
respirable dust: 1.5 mg/m3 TWA
German (DFG) - MAK Values
total dust: 4 mg/m3 MAK
German (DFG) - Pregnancy
no risk to embryo/fetus if exposure limits adhered to
Israel - Time Weighted Averages
*10 mg/m3 TWA (The value is for total dust containing no asbestos
and <1% free silica)*
Israel - Action Levels
5 mg/m3 AL
STATE LISTS
California - Exposure Limits - PELs
6 mg/m3 PEL
PROPOSED REGULATIONS
ACGIH 1995 - Notice of Intended Changes
*inhalable particulate: 10 mg/m3 TWA; respirable particulate: 3
mg/m3 TWA These values are for total dust containing no asbestos
and <1% crystalline silica.*

1,4-DIAZABICYCLO[2,2,2]OCTANE   280-57-9
HEALTH AND SAFETY LISTS
NIOSH - Selected LD50s and LC50s
Oral, rat: LD50 = 1700 mg/kg

INTERNATIONAL LISTS
Canada - WHMIS: Ingredient Disclosure
1% item 1624 (1666)

DIAZEPAM   439-14-
HEALTH AND SAFETY LISTS
IARC - Group 3 (not classifiable)
[present]
NIOSH - Selected LD50s and LC50s
Oral, rat: LD50 = 352 mg/kg Skin, mouse: LD50 = 800 mg/kg
STATE LISTS
California - Prop. 65 - Developmental Toxicity
developmental toxicity - initial date 1/1/92
Massachusetts Right To Know List
teratogen
NJ Right to Know List (Total)
sn 0617
NJ Special Hazardous Substances
(teratogen)

1,3-DIAZETIDINE-2,4-DIONE, 1,3-BIS(3-ISO-   26747-90-0
CYANATOMETHYLPHENYL)-
ENVIRONMENTAL LISTS
TSCA - Code of Federal Regulations Citations
40 CFR 712.30(x); 40 CFR 716.120(d)
TSCA - PAIR - Reporting List
Reporting Date: December 27, 1990
TSCA - Health and Safety Reporting List
Effective Date: October 29, 1990; Sunset Date: November 9, 1993

P-DIAZIDOBENZENE   2294-47-5
HEALTH AND SAFETY LISTS
U.S. DOT - Hazard Classes
Forbidden from transport by the DOT

1,2-DIAZIDOETHANE   RR-01365-0
HEALTH AND SAFETY LISTS
U.S. DOT - Hazard Classes
Forbidden from transport by the DOT

DIAZINON   333-41-5
HEALTH AND SAFETY LISTS
ACGIH 1995 - Time Weighted Averages
0.1 mg/m3 TWA
ACGIH 1995 - Skin Designations
skin - potential for cutaneous absorption
U.S. DOT - Appendix B - Marine Pollutants
DOT regulated severe marine pollutant
U.S. DOT - Appendix A Table 1 - Hazardous Substances
final RQ = 1 pound (0.454 kg)
NIOSH - Selected LD50s and LC50s
*Inhalation, rat: LC50 = 3500 mg/m3 4 hr Oral, rat: LD50 = 66
mg/kg Skin, rat: LD50 = 180 mg/kg*
NTP Chemical Status Reports - Testing Status and NTIS Number
Technical reports printed (PB293889/AS)
NTP Chemical Status Reports - Evidence of Carcinogenicity
*male rat-negative; female rat-negative; male mice-negative; female
mice-negative*
OSHA - Vacated PELs - Time Weighted Averages
0.1 mg/m3 TWA
OSHA - Vacated PELs - Skin Designation
Prevent or reduce skin absorption

ENVIRONMENTAL LISTS

ATSDR Priority List
Rank (of 275): 086

CERCLA/SARA - Section 313 - Emission Reporting
form R reporting required

CERCLA/SARA - Hazardous Substances and their Reportable Quantities
final RQ = 1 pound (0.454 kg)

Clean Water Act - Hazardous Substances
[present]

INTERNATIONAL LISTS

Australian Exposure Standards - Time Weighted Averages
0.1 mg/m3 TWA

Australian Exposure Standards - Skin Effects
skin absorption

Canada - Drinking Water Quality - MACs
0.02 mg/L MAC

Canada - Alberta - 8 Hour Occupational Exposure Limit
0.1 mg/m3 TWA

Canada - Alberta - 15 Minute Occupational Exposure Limit
0.3 mg/m3 STEL

Canada - Alberta - Skin Designation
can be absorbed through the intact skin

Canada - British Columbia - 8 Hour Exposure Limits
0.1 mg/m3 TWA

Canada - British Columbia - 15 Minute Exposure Limits
0.3 mg/m3 STEL

Canada - British Columbia - Skin Notations
skin - potential for skin absorption

Canada - Ontario - OHSA - TWAEVs
0.1 mg/m3 TWAEV

Canada - Ontario - OHSA - Skin Notations
absorption through skin, eyes, or mucous membranes

Canada - Quebec - Time-Weighted Average Exposure Values
0.1 mg/m3 TWAEV

Canada - Quebec - Skin Designations
absorbed through the skin

United Kingdom - Occupational Exposure Standards - TWAs
0.1 mg/m3 TWA

United Kingdom - Occupational Exposure Standards - STELs
0.3 mg/m3 STEL

United Kingdom - Occupational Exposure Standards - Notes
can be absorbed through skin

German (DFG) - MAK Values
total dust: 1 mg/m3 MAK

German (DFG) - Peak Limitations
10 x normal MAK (30 min average value); don't exceed during shift

German (DFG) - Skin/Sensitizers
danger of cutaneous absorption

German (DFG) - Pregnancy
no risk to embryo/fetus if exposure limits adhered to

Israel - Time Weighted Averages
0.1 mg/m3 TWA

Israel - Action Levels
0.05 mg/m3 AL

Mexico - Instruction No. 10 - TWAs
0.1 mg/m3 TWA

Mexico - Instruction No. 10 - STELs
0.3 mg/m3 STEL

Mexico - Instruction No. 10 - Skin designation
skin - potential for cutaneous absorption

STATE LISTS

California - Exposure Limits - PELs
0.1 mg/m3 PEL

California - Exposure Limits - Skin Notation
material may be absorbed through the skin, eyes or mucous membrane

California - Directors List of Hazardous Substances (8 CCR 339)
[present]

Florida Hazardous Substance List
[present]

Massachusetts Right To Know List
neurotoxin

Minnesota Hazardous Substance List
skin

NJ Right to Know List (Total)
sn 0618

Pennsylvania Right to Know List
environmental hazard

DIAZOAMINOTETRAZOLE   RR-01407-3

HEALTH AND SAFETY LISTS

U.S. DOT - Hazard Classes
Forbidden from transport by the DOT

DIAZODINTIROPHENOL   87-31-0

STATE LISTS

NJ Right to Know List (Total)
sn 0619

DIAZODINITROPHENOL   4682-03-5

HEALTH AND SAFETY LISTS

U.S. DOT - Substances From 49 CFR 172.101
regulated by DOT (UN0074)

U.S. DOT - Hazard Classes
DOT hazard class = 1.1A

DIAZODIPHENYLMETHANE   883-40-9

HEALTH AND SAFETY LISTS

U.S. DOT - Hazard Classes
Forbidden from transport by the DOT

DIAZOLIDINYL UREA   78491-02-8

ENVIRONMENTAL LISTS

List of Pesticide Product Inert Ingredients
[present]

DIAZOMETHANE   334-88-3

HEALTH AND SAFETY LISTS

ACGIH 1995 - Time Weighted Averages
0.2 ppm TWA; 0.34 mg/m3 TWA

IARC - Group 3 (not classifiable)
[present]

NIOSH 1990 - Pocket Guide - RELs
0.2 ppm TWA; 0.4 mg/m3 TWA

NIOSH 1990 - Pocket Guide - IDLHs
2 ppm IDLH

NIOSH 1990 - Pocket Guide - Target organs
eyes, skin, respiratory system

OSHA - Vacated PELs - Time Weighted Averages
0.2 ppm TWA; 0.4 mg/m3 TWA

OSHA - Final PELs - Time Weighted Averages
0.2 ppm TWA; 0.4 mg/m3 TWA

OSHA - List of Highly Hazardous Chemicals
threshhold quantity = 500 pounds

ENVIRONMENTAL LISTS

CERCLA/SARA - Section 313 - Emission Reporting
form R reporting required for 1.0% de minimus concentration

CERCLA/SARA - Hazardous Substances and their Reportable Quantities
final RQ = 1 pound (.454 kg)

Clean Air Act (1990) - List of Hazardous Air Contaminants
[present]

INTERNATIONAL LISTS

Australian Exposure Standards - Time Weighted Averages
0.2 ppm TWA; 0.34 mg/m3 TWA

Australian Exposure Standards - Carcinogens
suspected carcinogen

Canada - WHMIS: Ingredient Disclosure
0.1% item 497 (631)

Canada - Alberta - 8 Hour Occupational Exposure Limit
0.2 ppm TWA; 0.34 mg/m3 TWA

Canada - Alberta - 15 Minute Occupational Exposure Limit
0.6 ppm STEL; 1 mg/m3 STEL

Canada - British Columbia - 8 Hour Exposure Limits
0.2 ppm TWA; 0.4 mg/m3 TWA

Canada - Ontario - OHSA - TWAEVs
0.2 ppm TWAEV; 0.34 mg/m3 TWAEV

Canada - Quebec - Time-Weighted Average Exposure Values
0.2 ppm TWAEV; 0.34 mg/m3 TWAEV

German (DFG) - Carcinogens
animal evidence of carcinogenicity

Israel - Time Weighted Averages
0.2 ppm TWA; 0.34 mg/m3 TWA

Israel - Action Levels
0.1 ppm AL; 0.17 mg/m3 AL

STATE LISTS

California - Air Bill 2588 Appendix A-I
known or potential carcinogen: 6/91

California - Exposure Limits - PELs
0.2 ppm PEL; 0.4 mg/m3 PEL

California - Directors List of Hazardous Substances (8 CCR 339)
[present]

Florida Hazardous Substance List
[present]

Massachusetts Right To Know List
[present]

Minnesota Hazardous Substance List
[present]

NJ Right to Know List (Total)
sn 0620

Pennsylvania Right to Know List
environmental hazard

2-DIAZO-1-NAPHTHOL-4-SULFOCHLORIDE   RR-00763-6

HEALTH AND SAFETY LISTS

U.S. DOT - Substances From 49 CFR 172.101
regulated by DOT (UN3042)

U.S. DOT - Hazard Classes
DOT hazard class = 4.1

2-DIAZO-1-NAPHTHOL-5-SULFOCHLORIDE   RR-00764-7

HEALTH AND SAFETY LISTS

U.S. DOT - Substances From 49 CFR 172.101
regulated by DOT (UN3043)

U.S. DOT - Hazard Classes
DOT hazard class = 4.1

DIAZONIUM NITRATES   RR-01408-

HEALTH AND SAFETY LISTS

U.S. DOT - Hazard Classes
Forbidden from transport by the DOT

DIAZONIUM PERCHLORATES   RR-01409-

HEALTH AND SAFETY LISTS

U.S. DOT - Hazard Classes
Forbidden from transport by the DOT

1,3-DIAZOPROPANE   5239-06-

HEALTH AND SAFETY LISTS

U.S. DOT - Hazard Classes
Forbidden from transport by the DOT

DIBASIC LEAD PHOSPHATE   15845-52-
SEE ALSO:
LEAD

STATE LISTS

Massachusetts Right To Know List
[present]

NJ Right to Know List (Total)
sn 2293

DIBASIC LEAD STEARATE   56189-09-
SEE ALSO:
LEAD STEARATE
LEAD

HEALTH AND SAFETY LISTS

U.S. DOT - Appendix A Table 1 - Hazardous Substances
*final RQ = 5000 pounds (2270 kg) (the RQ is subject to change whe.
the assessment of potential carcinogenicity is completed) (Listed unde
'Lead stearate')*

ENVIRONMENTAL LISTS

CERCLA/SARA - Hazardous Substances and their Reportable Quantities
*final RQ = 5000 pounds (2270 kg) (the RQ is subject to change whe.
the assessment of potential carcinogenicity is completed) (Listed unde
'Lead stearate')*

STATE LISTS

Massachusetts Right To Know List
[present]

Pennsylvania Right to Know List
environmental hazard

DIBENZ[A,C]ANTHRACENE   215-58-

HEALTH AND SAFETY LISTS

IARC - Group 3 (not classifiable)
[present]

STATE LISTS

California - Directors List of Hazardous Substances (8 CCR 339)
[present]

DIBENZ(A,H)ACRIDINE   226-36-

HEALTH AND SAFETY LISTS

IARC - Group 2B (sufficient animal data)
*[present] (Overall evaluation based only on evidence of carcinogenici
in monograph (32, 1983) or in Supplement 4)*

NTP Seventh Report - Suspect Carcinogens
suspect carcinogen (Listed under 'Polycyclic aromatic hydrocarbons

OSHA - Possible Select Carcinogens
[present]

ENVIRONMENTAL LISTS

CERCLA/SARA - Section 313 - Emission Reporting
form R reporting required; (Listed under 'Polycyclic aromatic compounds')

RCRA - Hazardous Constituents-Appendix VIII
hazardous constituent - no waste number

INTERNATIONAL LISTS

Canada - WHMIS: Ingredient Disclosure
0.1% item 498 (634)

STATE LISTS

California - Air Bill 2588 Appendix A-I
known or potential carcinogen

California - Prop. 65 - Cancer list
carcinogen - initial date 1/1/88

California - Directors List of Hazardous Substances (8 CCR 339)
[present]

Florida Hazardous Substance List
[present]

Massachusetts Right To Know List
carcinogen; extraordinarily hazardous

Minnesota Hazardous Substance List
carcinogen

NJ Special Hazardous Substances
(carcinogen)

Pennsylvania Right to Know List
environmental hazard; special hazardous substance

Pennsylvania RTK - Special Hazardous Substances
[present]

DIBENZ(A,J)ACRIDINE   224-42-0

HEALTH AND SAFETY LISTS

IARC - Group 2B (sufficient animal data)
[present] (Overall evaluation based only on evidence of carcinogenicity in monograph (32, 1983) or in Supplement 4)

NTP Seventh Report - Suspect Carcinogens
suspect carcinogen (Listed under 'Polycyclic aromatic hydrocarbons)

OSHA - Possible Select Carcinogens
[present]

ENVIRONMENTAL LISTS

CERCLA/SARA - Section 313 - Emission Reporting
form R reporting required; (Listed under 'Polycylic aromatic compounds')

RCRA - Hazardous Constituents-Appendix VIII
hazardous constituent - no waste number

INTERNATIONAL LISTS

Canada - WHMIS: Ingredient Disclosure
0.1% item 499 (635)

STATE LISTS

California - Air Bill 2588 Appendix A-I
known or potential carcinogen

California - Prop. 65 - Cancer list
carcinogen - initial date 1/1/88

California - Directors List of Hazardous Substances (8 CCR 339)
[present]

Florida Hazardous Substance List
[present]

Massachusetts Right To Know List
carcinogen; extraordinarily hazardous

Minnesota Hazardous Substance List
carcinogen

NJ Special Hazardous Substances
(carcinogen)

Pennsylvania Right to Know List
environmental hazard; special hazardous substance

Pennsylvania RTK - Special Hazardous Substances
[present]

DIBENZ[A,J]ANTHRACENE   224-41-9

HEALTH AND SAFETY LISTS

IARC - Group 3 (not classifiable)
[present]

STATE LISTS

California - Directors List of Hazardous Substances (8 CCR 339)
[present]

4,4'-DIBENZAMIDO-1,1'-DIANTHRIMIDE   128-79-0

INTERNATIONAL LISTS

Canada - WHMIS: Ingredient Disclosure
1% item 500 (632)

4,4'-DIBENZANTHRONYL   116-90-5

INTERNATIONAL LISTS

Canada - WHMIS: Ingredient Disclosure
1% item 502 (633)

DIBENZO[A,E]FLUORANTHENE   5385-75-1

HEALTH AND SAFETY LISTS

IARC - Group 3 (not classifiable)
[present]

ENVIRONMENTAL LISTS

CERCLA/SARA - Section 313 - Emission Reporting
form R reporting required; (Listed under 'Polycyclic aromatic compounds')

STATE LISTS

California - Directors List of Hazardous Substances (8 CCR 339)
[present]

DIBENZO(A,E)PYRENE   192-65-4

HEALTH AND SAFETY LISTS

IARC - Group 2B (sufficient animal data)
[present] (Overall evaluation based only on evidence of carcinogenicity in monograph (32, 1983) or in Supplement 4)

NTP Seventh Report - Suspect Carcinogens
suspect carcinogen (Listed under 'Polycyclic aromatic hydrocarbons')

OSHA - Possible Select Carcinogens
[present]

ENVIRONMENTAL LISTS

CERCLA/SARA - Section 313 - Emission Reporting
form R reporting required; (Listed under 'Polycyclic aromatic compounds')

RCRA - Hazardous Constituents-Appendix VIII
hazardous constituent - no waste number

RCRA - Universal Treatment Standards (LDR)
WW: 0.061 mg/l; NWW: Not applicable

INTERNATIONAL LISTS

German (DFG) - Carcinogens
animal evidence of carcinogenicity

STATE LISTS

California - Air Bill 2588 Appendix A-I
known or potential carcinogen

California - Prop. 65 - Cancer list
carcinogen - initial date 1/1/88

California - Directors List of Hazardous Substances (8 CCR 339)
[present]

Florida Hazardous Substance List
[present]

Massachusetts Right To Know List
carcinogen; extraordinarily hazardous

Minnesota Hazardous Substance List
carcinogen

NJ Special Hazardous Substances
(carcinogen; mutagen)

Pennsylvania Right to Know List
environmental hazard; special hazardous substance

Pennsylvania RTK - Special Hazardous Substances
[present]

DIBENZO(A,H)ANTHRACENE   53-70-3

HEALTH AND SAFETY LISTS

U.S. DOT - Appendix A Table 1 - Hazardous Substances
final RQ = 1 pound (0.454 kg)

IARC - Group 2A (limited human data)
[present] (Overall evaluation based only on evidence of carcinogenicity in monograph (32, 1983) or in Supplement 4) (Other relevant data, as given in Supplement 7 or in the monograph influenced the making of the overall evaluation)

NTP Seventh Report - Suspect Carcinogens
suspect carcinogen (Listed under 'Polycyclic aromatic hydrocarbons')

OSHA - Possible Select Carcinogens
[present]

ENVIRONMENTAL LISTS

ATSDR Priority List
Rank (of 275): 018

CERCLA/SARA - Section 313 - Emission Reporting
form R reporting required; (Listed under 'Polycyclic aromatic compounds')

CERCLA/SARA - Hazardous Substances and their Reportable Quantities
final RQ = 1 pound (0.454 kg)

Clean Water Act - Priority Pollutants
[present]

EPA - Carcinogen Hazard Ranking for RQ Adjustment
Hazard ranking = High

RCRA - U Series Wastes
waste number U063

RCRA - Hazardous Constituents-Appendix VIII
waste number U063

RCRA - Basis for Listing - Appendix VII
Included in waste streams: F039, K001, K035, K141, K142, K144, K145, K147, K148

RCRA - Substances Banned From Land Disposal
[present]

RCRA - TSD Facilities Ground Water Monitoring
TM 8100 = 200 ug/L PQL; TM 8270 = 10 ug/L PQL

RCRA - Universal Treatment Standards (LDR)
WW: 0.055 mg/l; NWW: 8.2 mg/kg

INTERNATIONAL LISTS

Canada - WHMIS: Ingredient Disclosure
0.1% item 501 (636)

German (DFG) - Carcinogens
animal evidence of carcinogenicity

Mexico - Wastewater - Organic Toxic Pollutants and Heavy Metals
Listed under [Aromatic Hydrocarbons]

STATE LISTS

California - Air Bill 2588 Appendix A-I
known or potential carcinogen

California - Prop. 65 - Cancer list
carcinogen - initial date 1/1/88

California - Prop. 65 - No Significant Risk Levels
no significant risk level = 0.2 ug/day

California - Directors List of Hazardous Substances (8 CCR 339)
[present]

Florida Hazardous Substance List
[present]

Massachusetts Right To Know List
carcinogen; extraordinarily hazardous

Minnesota Hazardous Substance List
carcinogen

NJ Special Hazardous Substances
(carcinogen; mutagen)

Pennsylvania Right to Know List
environmental hazard; special hazardous substance

Pennsylvania RTK - Special Hazardous Substances
[present]

DIBENZO(A,H)PYRENE   189-64-

HEALTH AND SAFETY LISTS

IARC - Group 2B (sufficient animal data)
[present] (Overall evaluation based only on evidence of carcinogenicity in monograph (32, 1983) or in Supplement 4)

NTP Seventh Report - Suspect Carcinogens
suspect carcinogen (Listed under 'Polycyclic aromatic hydrocarbons')

OSHA - Possible Select Carcinogens
[present]

ENVIRONMENTAL LISTS

CERCLA/SARA - Section 313 - Emission Reporting
form R reporting required; (Listed under 'Polycyclic aromatic compounds')

RCRA - Hazardous Constituents-Appendix VIII
hazardous constituent - no waste number

INTERNATIONAL LISTS

Canada - WHMIS: Ingredient Disclosure
0.1% item 504 (638)

German (DFG) - Carcinogens
animal evidence of carcinogenicity

STATE LISTS

California - Air Bill 2588 Appendix A-I
known or potential carcinogen

California - Prop. 65 - Cancer list
carcinogen - initial date 1/1/88

California - Directors List of Hazardous Substances (8 CCR 339)
[present]

Florida Hazardous Substance List
[present]

Massachusetts Right To Know List
carcinogen; extraordinarily hazardous

Minnesota Hazardous Substance List
carcinogen

NJ Special Hazardous Substances
(carcinogen)

Pennsylvania Right to Know List
environmental hazard; special hazardous substance

Pennsylvania RTK - Special Hazardous Substances
[present]

DIBENZO(A,I)PYRENE   189-55-9

HEALTH AND SAFETY LISTS

U.S. DOT - Appendix A Table 1 - Hazardous Substances
final RQ = 10 pounds (4.54 kg)

IARC - Group 2B (sufficient animal data)
[present] (Overall evaluation based only on evidence of carcinogenicity in monograph (32, 1983) or in Supplement 4)

NTP Seventh Report - Suspect Carcinogens
suspect carcinogen (Listed under 'Polycyclic aromatic hydrocarbons)
OSHA - Possible Select Carcinogens
[present]

ENVIRONMENTAL LISTS

CERCLA/SARA - Section 313 - Emission Reporting
form R reporting required; (Listed under 'Polycyclic aromatic compounds')
CERCLA/SARA - Hazardous Substances and their Reportable Quantities
final RQ = 10 pounds (4.54 kg)
EPA - Carcinogen Hazard Ranking for RQ Adjustment
Hazard ranking = Medium
RCRA - U Series Wastes
waste number U064
RCRA - Hazardous Constituents-Appendix VIII
waste number U064
RCRA - Substances Banned From Land Disposal
[present]

INTERNATIONAL LISTS

Canada - WHMIS: Ingredient Disclosure
0.1% item 505 (639)
German (DFG) - Carcinogens
animal evidence of carcinogenicity

STATE LISTS

California - Air Bill 2588 Appendix A-I
known or potential carcinogen
California - Prop. 65 - Cancer list
carcinogen - initial date 1/1/88
California - Directors List of Hazardous Substances (8 CCR 339)
[present]
Florida Hazardous Substance List
[present]
Massachusetts Right To Know List
carcinogen; extraordinarily hazardous
Minnesota Hazardous Substance List
carcinogen
NJ Special Hazardous Substances
(carcinogen)
Pennsylvania Right to Know List
environmental hazard; special hazardous substance
Pennsylvania RTK - Special Hazardous Substances
[present]

DIBENZO(A,L)PYRENE   191-30-0

HEALTH AND SAFETY LISTS

IARC - Group 2B (sufficient animal data)
[present] (Overall evaluation based only on evidence of carcinogenicity in monograph (32, 1983) or in Supplement 4)
NTP Seventh Report - Suspect Carcinogens
suspect carcinogen (Listed under 'Polycyclic aromatic hydrocarbons)
OSHA - Possible Select Carcinogens
[present]

ENVIRONMENTAL LISTS

CERCLA/SARA - Section 313 - Emission Reporting
form R reporting required; (Listed under 'Polycyclic aromatic compounds')

INTERNATIONAL LISTS

Canada - WHMIS: Ingredient Disclosure
1% item 506 (640)
German (DFG) - Carcinogens
animal evidence of carcinogenicity

STATE LISTS

California - Air Bill 2588 Appendix A-I
known or potential carcinogen
California - Prop. 65 - Cancer list
carcinogen - initial date 1/1/88
California - Directors List of Hazardous Substances (8 CCR 339)
[present]
Florida Hazardous Substance List
effective March 13, 1992
Massachusetts Right To Know List
carcinogen; extraordinarily hazardous
Minnesota Hazardous Substance List
carcinogen
Pennsylvania Right to Know List
special hazardous substance
Pennsylvania RTK - Special Hazardous Substances
[present]

7H-DIBENZO(C,G)CARBAZOLE   194-59-2

HEALTH AND SAFETY LISTS

IARC - Group 2B (sufficient animal data)
[present] (Overall evaluation based only on evidence of carcinogenicity in monograph (32, 1983) or in Supplement 4)
NTP Seventh Report - Suspect Carcinogens
suspect carcinogen (Listed under 'Polycyclic aromatic hydrocarbons)
OSHA - Possible Select Carcinogens
[present]

ENVIRONMENTAL LISTS

CERCLA/SARA - Section 313 - Emission Reporting
form R reporting required; (Listed under 'Polycyclic aromatic compounds')
RCRA - Hazardous Constituents-Appendix VIII
hazardous constituent - no waste number

INTERNATIONAL LISTS

Canada - WHMIS: Ingredient Disclosure
0.1% item 503 (637)

STATE LISTS

California - Air Bill 2588 Appendix A-I
known or potential carcinogen
California - Prop. 65 - Cancer list
carcinogen - initial date 1/1/88
California - Directors List of Hazardous Substances (8 CCR 339)
[present]
Florida Hazardous Substance List
[present]
Massachusetts Right To Know List
carcinogen; extraordinarily hazardous
Minnesota Hazardous Substance List
carcinogen
NJ Special Hazardous Substances
(carcinogen)
Pennsylvania Right to Know List
environmental hazard; special hazardous substance
Pennsylvania RTK - Special Hazardous Substances
[present]

DIBENZO[H,RST]PENTAPHENE   192-47-2

HEALTH AND SAFETY LISTS

IARC - Group 3 (not classifiable)
[present]

STATE LISTS

California - Directors List of Hazardous Substances (8 CCR 339)
[present]

DIBENZO-18-CROWN-6   14187-32-7

HEALTH AND SAFETY LISTS

NIOSH - Selected LD50s and LC50s
Oral, rat: LD50 = 2600 mg/kg

DIBENZO-P-DIOXIN   262-12-4

HEALTH AND SAFETY LISTS

NTP Chemical Status Reports - Testing Status and NTIS Number
Technical reports printed (PB288475/AS)

NTP Chemical Status Reports - Evidence of Carcinogenicity
male rat-negative; female rat-negative; male mice-negative; female mice-negative

INTERNATIONAL LISTS

Canada - CEPA Schedule I - Toxic Substances
that has the molecular formula C12H8O2: maximum concentrations in products

DIBENZOFURAN   132-64-9

ENVIRONMENTAL LISTS

ATSDR Priority List
Rank (of 275): 263

CERCLA/SARA - Section 313 - Emission Reporting
form R reporting required for 1.0% de minimus concentration

CERCLA/SARA - Hazardous Substances and their Reportable Quantities
final RQ = 1 pound (.454 kg)

Clean Air Act (1990) - List of Hazardous Air Contaminants
[present]

RCRA - TSD Facilities Ground Water Monitoring
TM 8270 = 10 ug/L PQL

INTERNATIONAL LISTS

Canada - CEPA Schedule I - Toxic Substances
that has the molecular formula C12H8O: maximum concentrations in products

STATE LISTS

California - Air Bill 2588 Appendix A-I
6/91

Massachusetts Right To Know List
[present]

NJ Right to Know List (Total)
sn 2230

Pennsylvania Right to Know List
environmental hazard

4A(4H)-DIBENZOFURANCARBOXALDEHYDE, 1,5A,6,   126-15-8
9,9A,9B,-HEXAHYDRO-

ENVIRONMENTAL LISTS

TSCA - Code of Federal Regulations Citations
40 CFR 712.30(x);   40 CFR 716.120(d)

TSCA - PAIR - Reporting List
Reporting Date: November 27, 1991

TSCA - Health and Safety Reporting List
Effective Date: September 30, 1991

2,2'-DIBENZOTHIAZYL DISULFIDE   120-78-5

HEALTH AND SAFETY LISTS

NIOSH - Selected LD50s and LC50s
Oral, rat: LD50 = 7 gm/kg

ENVIRONMENTAL LISTS

EPA - Master Testing List
[present]

INTERNATIONAL LISTS

Canada - WHMIS: Ingredient Disclosure
1% item 507 (789)

DIBENZOTHIOPHENE   132-65-0

ENVIRONMENTAL LISTS

ATSDR Priority List
Rank (of 275): 205

DIBENZOYL CHLORIDE   RR-00563-0

STATE LISTS

Pennsylvania Right to Know List
[present]

DIBENZYLDICHLOROSILANE   18414-36-3

HEALTH AND SAFETY LISTS

U.S. DOT - Substances From 49 CFR 172.101
regulated by DOT (UN2434)

U.S. DOT - Hazard Classes
DOT hazard class = 8

STATE LISTS

NJ Right to Know List (Total)
sn 0627

NJ Special Hazardous Substances
(corrosive)

DIBENZYL DISULFIDE   150-60-7

HEALTH AND SAFETY LISTS

NIOSH - Selected LD50s and LC50s
Oral, rat: LD50 = 3780 mg/kg

INTERNATIONAL LISTS

Canada - WHMIS: Ingredient Disclosure
1% item 508 (790)

DIBENZYL PEROXYDICARBONATE   2144-45-8

HEALTH AND SAFETY LISTS

U.S. DOT - Hazard Classes
Forbidden from transport by the DOT

U.S. DOT - Organic Peroxides Table
Organic peroxide UN3112

STATE LISTS

NJ Right to Know List (Total)
sn 0628

DIBORANE   19287-45-7

HEALTH AND SAFETY LISTS

ACGIH 1995 - Time Weighted Averages
0.1 ppm TWA;   0.11 mg/m3 TWA

AIHA - Odor Threshold Values
no geometric mean air odor threshold

U.S. DOT - Substances From 49 CFR 172.101
regulated by DOT (UN1911, NA1911)

U.S. DOT - Hazard Classes
DOT hazard class = 2.3

U.S. DOT - Substances Which Are Poisonous by Inhalation
gaseous hazardous material poisonous by inhalation (alone or in mixtures) (UN1911, NA1911)

NFPA - Flash Points
gas (no flash point given)

NFPA - Hazard Identification Ratings
health-4;   flammability-4;   reactivity-3 (avoid use of water or halogenated extinguishing agents)

NIOSH - Selected LD50s and LC50s
Inhalation, rat: LC50 = 40 ppm 4 hr

NIOSH 1990 - Pocket Guide - RELs
0.1 ppm TWA; 0.1 mg/m3 TWA

NIOSH 1990 - Pocket Guide - IDLHs
40 ppm IDLH

NIOSH 1990 - Pocket Guide - Target organs
CNS, respiratory system

OSHA - Vacated PELs - Time Weighted Averages
0.1 ppm TWA; 0.1 mg/m3 TWA

OSHA - Final PELs - Time Weighted Averages
0.1 ppm TWA; 0.1 mg/m3 TWA

OSHA - List of Highly Hazardous Chemicals
threshhold quantity = 100 pounds

ENVIRONMENTAL LISTS

CERCLA/SARA - Section 302 Extremely Hazardous Substances and TPQs
TPQ = 100 pounds

CAA -Toxic Substances for Accidental Release Prevention
threshold quantity = 2,500 lbs

INTERNATIONAL LISTS

Australian Exposure Standards - Time Weighted Averages
0.1 ppm TWA; 0.11 mg/m3 TWA

Canada - WHMIS: Ingredient Disclosure
1% item 509 (641)

Canada - Alberta - 8 Hour Occupational Exposure Limit
0.1 ppm TWA; 0.11 mg/m3 TWA

Canada - Alberta - 15 Minute Occupational Exposure Limit
0.3 ppm STEL; 0.34 mg/m3 STEL

Canada - British Columbia - 8 Hour Exposure Limits
0.1 ppm TWA; 0.1 mg/m3 TWA

Canada - Ontario - OHSA - TWAEVs
0.1 ppm TWAEV; 0.11 mg/m3 TWAEV

Canada - Quebec - Time-Weighted Average Exposure Values
0.1 ppm TWAEV; 0.11 mg/m3 TWAEV

United Kingdom - Occupational Exposure Standards - TWAs
0.1 ppm TWA; 0.1 mg/m3 TWA

German (DFG) - MAK Values
0.1 ppm MAK; 0.1 mg/m3 MAK

German (DFG) - Peak Limitations
2 x normal MAK (5 min momentary value); don't exceed 8 times during shift

Israel - Time Weighted Averages
0.1 ppm TWA; 0.11 mg/m3 TWA

Israel - Action Levels
0.05 ppm AL; 0.055 mg/m3 AL

Mexico - Instruction No. 10 - TWAs
0.1 ppm TWA; 0.1 mg/m3 TWA

STATE LISTS

California - Exposure Limits - PELs
0.1 ppm PEL; 0.1 mg/m3 PEL

California - Directors List of Hazardous Substances (8 CCR 339)
[present]

Florida Hazardous Substance List
[present]

Massachusetts Right To Know List
extraordinarily hazardous

Minnesota Hazardous Substance List
[present]

NJ Right to Know List (Total)
sn 0629

NJ Special Hazardous Substances
(flammable - fourth degree; reactive - third degree)

Pennsylvania Right to Know List
environmental hazard

PROPOSED REGULATIONS

CERCLA/SARA - Proposed Hazardous Substance Additions
proposed RQ = 1 pound (.454 kg)

CERCLA/SARA - 1989 Proposed RQ Adjustments
proposed RQ = 10 pounds (4.54 kg)

DIBROMOACETONITRILE   3252-43-5

HEALTH AND SAFETY LISTS

IARC - Group 3 (not classifiable)
[present]

PROPOSED REGULATIONS

Safe Drinking Water Act - Priority list
[present]

DIBROMOACETYLENE   624-61-3

HEALTH AND SAFETY LISTS

U.S. DOT - Hazard Classes
Forbidden from transport by the DOT

P-DIBROMOBENZENE   106-37-6

HEALTH AND SAFETY LISTS

NIOSH - Selected LD50s and LC50s
Oral, mouse: LD50 = 3120 mg/kg

DIBROMOBENZENE   26249-12-7

HEALTH AND SAFETY LISTS

U.S. DOT - Substances From 49 CFR 172.101
regulated by DOT (UN2711)

U.S. DOT - Hazard Classes
DOT hazard class = 3

STATE LISTS

NJ Right to Know List (Total)
sn 0630

DIBROMOBUTANONE   3479-86-5

STATE LISTS

NJ Right to Know List (Total)
sn 0631

1,2-DIBROMO-3-BUTANONE   25109-57-3

HEALTH AND SAFETY LISTS

U.S. DOT - Substances From 49 CFR 172.101
regulated by DOT (UN2648)

U.S. DOT - Hazard Classes
DOT hazard class = 6.1

INTERNATIONAL LISTS

Canada - WHMIS: Ingredient Disclosure
1% item 510 (642)

1,2-DIBROMO-3-CHLOROPROPANE   96-12-8
SEE ALSO:
F039-HAZARDOUS WASTES

HEALTH AND SAFETY LISTS

U.S. DOT - Substances From 49 CFR 172.101
regulated by DOT (UN2872)

U.S. DOT - Hazard Classes
DOT hazard class = 6.1

U.S. DOT - Appendix A Table 1 - Hazardous Substances
final RQ = 1 pound (0.454 kg)

IARC - Group 2B (sufficient animal data)
[present]

NIOSH - Selected LD50s and LC50s
Inhalation, rat: LC50 = 103 ppm 8 hr Oral, rat: LD50 = 170 mg/kg Skin, rabbit: LD50 = 1400 mg/kg
NIOSH 1990 - Pocket Guide - Carcinogens
occupational carcinogen
NIOSH 1990 - Pocket Guide - Target organs
CNS, skin, liver, kidneys, spleen, reproductive system, digestive system
NIOSH - Health Standards - Exposure Limits
10 ppb TWA; 0.1 mg/m3 TWA (superseded by 1978 OSHA standard)
NIOSH - Health Standards - Health Effects and Precautions
Sterility, renal and liver effects
NTP Chemical Status Reports - Testing Status and NTIS Number
Technical reports printed (PB277472/AS) (PB82225632)
NTP Chemical Status Reports - Evidence of Carcinogenicity
PB277472/AS: male rat-positive; female rat-positive; male mice-positive; female mice-positive; PB82225632: male rat-positive; female rat-positive; male mice-positive; female mice-positive
NTP Seventh Report - Suspect Carcinogens
suspect carcinogen
OSHA - 29 CFR 1910 Specifically Regulated Chemicals
1 ppb TWA PEL; Cancer hazard; eye and skin contact prohibited (see 29 CFR 1910.1044)
OSHA - Select Carcinogens
[present]
OSHA - Possible Select Carcinogens
[present]

ENVIRONMENTAL LISTS
ATSDR Priority List
Rank (of 275): 119
CERCLA/SARA - Section 313 - Emission Reporting
form R reporting required for 0.1% de minimus concentration
CERCLA/SARA - Hazardous Substances and their Reportable Quantities
final RQ = 1 pound (0.454 kg)
Clean Air Act (1990) - List of Hazardous Air Contaminants
[present]
Safe Drinking Water Act - MCLs
MCL = 0.0002 mg/L
Safe Drinking Water Act - MCLGs
MCLG = Zero
Safe Drinking Water Act - Monitoring
monitoring required
EPA - Carcinogen Hazard Ranking for RQ Adjustment
Hazard ranking = High
RCRA - U Series Wastes
waste number U066
RCRA - Hazardous Constituents-Appendix VIII
waste number U066
RCRA - Basis for Listing - Appendix VII
Included in waste stream: F039
RCRA - Substances Banned From Land Disposal
[present]
RCRA - TSD Facilities Ground Water Monitoring
TM 8010 = 100 ug/L PQL; TM 8240 = 5 ug/L PQL; TM 8270 = 10 ug/L PQL
RCRA - Universal Treatment Standards (LDR)
WW: 0.11 mg/l; NWW: 15 mg/kg
TSCA - Code of Federal Regulations Citations
40 CFR 712.30(d)
TSCA - PAIR - Reporting List
Reporting Date: November 19, 1982

INTERNATIONAL LISTS
Canada - WHMIS: Ingredient Disclosure
0.1% item 511 (643)

Canada - CEPA Schedule II Part II - Toxic Substances (Export)
[present]
German (DFG) - Carcinogens
animal evidence of carcinogenicity

STATE LISTS
California - Air Bill 2588 Appendix A-I
known or potential carcinogen
California - Prop. 65 - Cancer list
carcinogen - initial date 7/1/87
California - Prop. 65 - Reproductive - Male
male reproductive toxicity - initial date 2/27/87
California - Prop. 65 - No Significant Risk Levels
no significant risk level = 0.1 ug/day
California - Exposure Limits - PELs
0.001 ppm PEL; 0.01 mg/m3 PEL; prevent eye and skin contact see also section 5212 for chronic effects
California - Directors List of Hazardous Substances (8 CCR 339)
[present]
Florida Hazardous Substance List
[present]
Massachusetts Right To Know List
carcinogen; extraordinarily hazardous
Minnesota Hazardous Substance List
carcinogen
NJ Right to Know List (Total)
sn 0595
NJ Special Hazardous Substances
(carcinogen; mutagen)
Pennsylvania Right to Know List
environmental hazard; special hazardous substance
Pennsylvania RTK - Special Hazardous Substances
[present]

DIBROMOCHLOROPROPANE   67708-83-2
ENVIRONMENTAL LISTS
ATSDR Priority List
Rank (of 275): 057

1,2-DIBROMO-4-(1,2-DIBROMOETHYL) CYCLOHEXANE   3322-93-8
ENVIRONMENTAL LISTS
TSCA - Code of Federal Regulations Citations
40 CFR 712.30(k); 40 CFR 716.120(a)
TSCA - PAIR - Reporting List
Reporting Date: August 27, 1984
TSCA - Health and Safety Reporting List
Effective Date: June 28, 1984

1,2-DIBROMO-2,4-DICYANOBUTANE   35691-65-7
ENVIRONMENTAL LISTS
CERCLA/SARA - Section 313 - Emission Reporting
form R reporting required
List of Pesticide Product Inert Ingredients
[present]

DIBROMODULCITOL   10318-26-0
HEALTH AND SAFETY LISTS
NTP Chemical Status Reports - Testing Status and NTIS Number
Chronic studies exist for which technical reports were not prepared

1,2-DIBROMOETHENE   540-49-8
HEALTH AND SAFETY LISTS
U.S. DOT - Appendix B - Marine Pollutants
DOT regulated marine pollutant

© Van Nostrand Reinhold 1995

DIBROMOMANNITOL   488-41-5

HEALTH AND SAFETY LISTS

NTP Chemical Status Reports - Testing Status and NTIS Number
Chronic studies exist for which technical reports were not prepared

2,6-DIBROMO-4-METHYLPHENYL GYCIDYL ETHER   22421-59-6

ENVIRONMENTAL LISTS

EPA - Master Testing List
[present]

PROPOSED REGULATIONS

TSCA - Proposed Testing Rule for Glycidyl Ethers
member of Glycidyl subcategory IV-C

2,4-DIBROMO-6-METHYLPHENYL GYCIDYL ETHER   75150-13-9

ENVIRONMENTAL LISTS

EPA - Master Testing List
[present]

PROPOSED REGULATIONS

TSCA - Proposed Testing Rule for Glycidyl Ethers
member of Glycidyl subcategory IV-C

2,2-DIBROMO-3-NITRILOPROPIONAMIDE   10222-01-2

HEALTH AND SAFETY LISTS

NIOSH - Selected LD50s and LC50s
Oral, mammal: LD50 = 118 mg/kg

ENVIRONMENTAL LISTS

CERCLA/SARA - Section 313 - Emission Reporting
form R reporting required

2,6-DIBROMO-4-NITROPHENOL   99-28-5

ENVIRONMENTAL LISTS

TSCA - HDD/HDF - Chemicals Required for Testing
[present]

TSCA - Section 12(b) - Export Notification
export notification required - Section 4

2,4-DIBROMOPHENOL   615-58-7

ENVIRONMENTAL LISTS

TSCA - Code of Federal Regulations Citations
40 CFR 712.30(x); 40 CFR 716.120(d)

TSCA - PAIR - Reporting List
Reporting Date: December 27, 1990

TSCA - Health and Safety Reporting List
Effective Date: October 29, 1990

TSCA - HDD/HDF - Chemicals Required for Testing
[present]

TSCA - Section 12(b) - Export Notification
export notification required - Section 4

1,2-DIBROMOPROPANE   78-75-1

HEALTH AND SAFETY LISTS

NIOSH - Selected LD50s and LC50s
Inhalation, rat: LC50 = 15344 mg/m3 (8 hr) Oral, rat: LD50 = 1373 mg/kg

INTERNATIONAL LISTS

Canada - WHMIS: Ingredient Disclosure
1% item 513 (645)

2,3-DIBROMO-1-PROPANOL   96-13-9

HEALTH AND SAFETY LISTS

NTP Chemical Status Reports - Testing Status and NTIS Number
Technical reports printed (no NTIS number given)

ENVIRONMENTAL LISTS

TSCA - Code of Federal Regulations Citations
40 CFR 712.30(x); 40 CFR 716.120(d)

TSCA - PAIR - Reporting List
Reporting Date: December 27, 1990

TSCA - Health and Safety Reporting List
Effective Date: October 29, 1990

STATE LISTS

California - Prop. 65 - Cancer list
carcinogen - initial date 10/01/94

3,5-DIBROMOSALICYLANILIDE   2577-72-2

ENVIRONMENTAL LISTS

TSCA - HDD/HDF - Chemicals Required for Testing
[present]

TSCA - Section 12(b) - Export Notification
export notification required - Section 4

DIBROMOTETRAFLUOROETHANE (HALON 2402)   124-73-2

ENVIRONMENTAL LISTS

CERCLA/SARA - Section 313 - Emission Reporting
form R reporting required for 1.0% de minimus concentration

Class 1 Ozone Depletors
ozone depletion potential = 6.0

INTERNATIONAL LISTS

Canada - CEPA Schedule II Part II - Toxic Substances (Export)
[present] (with the molecular formula C2F4Br2)

Canada - CEPA Schedule I - Toxic Substances
that has the molecular formula C2F4Br2: (a) quantities that may be imported; (b) prohibited commercial, manufacturing or processing uses

STATE LISTS

NJ Right to Know List (Total)
sn 3137

DIBUTOXYMETHANE   2568-90-3

HEALTH AND SAFETY LISTS

NFPA - Flash Points
flash point = 140 degrees F (60 degrees C)

NFPA - Hazard Identification Ratings
health-0; flammability-2; reactivity-0

DIBUTOXY TETRAGLYCOL   112-98-1

HEALTH AND SAFETY LISTS

NFPA - Flash Points
flash point = 305 degrees F (152 degrees C)

NFPA - Hazard Identification Ratings
health-2; flammability-1; reactivity-0

STATE LISTS

Florida Hazardous Substance List
[present]

Massachusetts Right To Know List
[present]

Pennsylvania Right to Know List
[present]

N,N-DIBUTYLACETAMIDE   1563-90-2

HEALTH AND SAFETY LISTS

NFPA - Flash Points
flash point = 225 degrees F (107 degrees C)

NFPA - Hazard Identification Ratings
health-0; flammability-1; reactivity-0

DIBUTYL ADIPATE   105-99-7

HEALTH AND SAFETY LISTS

NIOSH - Selected LD50s and LC50s
Oral, rat: LD50 = 12900 mg/kg Skin, rabbit: LD50 = 20000 mg/kg

ENVIRONMENTAL LISTS

EPA - Master Testing List
[present]

DI-(N-BUTYL)AMINE   111-92-2

HEALTH AND SAFETY LISTS

AIHA - WEEL - Ceilings or Short Term Time Weighted Averages
C 5 ppm;   C 26.5 mg/m3

AIHA - WEEL - Skin Absorption Designations
skin absorber

U.S. DOT - Substances From 49 CFR 172.101
regulated by DOT (UN2248)

U.S. DOT - Hazard Classes
DOT hazard class = 8

NFPA - Flash Points
flash point = 117 degrees F (47 degrees C)

NFPA - Hazard Identification Ratings
health-3; flammability-2; reactivity-0

NIOSH - Selected LD50s and LC50s
Oral, rat: LD50 = 220 mg/kg Skin, rabbit: LD50 = 1010 mg/kg

ENVIRONMENTAL LISTS

List of Pesticide Product Inert Ingredients
[present]

TSCA - Code of Federal Regulations Citations
40 CFR 716.120(a)

TSCA - Health and Safety Reporting List
Effective Date: June 1, 1987

INTERNATIONAL LISTS

Canada - WHMIS: Ingredient Disclosure
1% item 514 (646)

STATE LISTS

Florida Hazardous Substance List
[present]

Massachusetts Right To Know List
[present]

NJ Right to Know List (Total)
sn 0632

NJ Special Hazardous Substances
(corrosive)

Pennsylvania Right to Know List
[present]

DI-SEC-BUTYLAMINE   626-23-3

HEALTH AND SAFETY LISTS

NFPA - Flash Points
flash point = 75 degrees F (24 degrees C)

NFPA - Hazard Identification Ratings
health-3; flammability-3; reactivity-0

STATE LISTS

Florida Hazardous Substance List
[present]

Massachusetts Right To Know List
[present]

Pennsylvania Right to Know List
[present]

2-N-DIBUTYLAMINOETHANOL   102-81-8

HEALTH AND SAFETY LISTS

ACGIH 1995 - Time Weighted Averages
0.5 ppm TWA 3.5 mg/m3 TWA

ACGIH 1995 - Skin Designations
skin - potential for cutaneous absorption

U.S. DOT - Substances From 49 CFR 172.101
regulated by DOT (UN2873)

U.S. DOT - Hazard Classes
DOT hazard class = 6.1

NFPA - Flash Points
flash point = 200 degrees F (93 degrees C)

NFPA - Hazard Identification Ratings
health-3; flammability-2; reactivity-0

NIOSH - Selected LD50s and LC50s
Oral, rat: LD50 = 1070 mg/kg Skin, rabbit: LD50 = 1680 mg/kg

OSHA - Vacated PELs - Time Weighted Averages
2 ppm TWA;   14 mg/m3 TWA

INTERNATIONAL LISTS

Australian Exposure Standards - Time Weighted Averages
2 ppm TWA;   14 mg/m3 TWA

Australian Exposure Standards - Skin Effects
skin absorption

Canada - WHMIS: Ingredient Disclosure
1% item 515 (647)

Canada - Alberta - 8 Hour Occupational Exposure Limit
2 ppm TWA;   14 mg/m3 TWA

Canada - Alberta - 15 Minute Occupational Exposure Limit
4 ppm STEL;   28 mg/m3

Canada - Alberta - Skin Designation
can be absorbed through the intact skin

Canada - British Columbia - 8 Hour Exposure Limits
2 ppm TWA;   14 mg/m3 TWA

Canada - British Columbia - 15 Minute Exposure Limits
4 ppm STEL;   28 mg/m3 STEL

Canada - British Columbia - Skin Notations
skin - potential for skin absorption

Canada - Ontario - OHSA - TWAEVs
2 ppm TWAEV;   14 mg/m3 TWAEV

Canada - Ontario - OHSA - Skin Notations
absorption through skin, eyes, or mucous membranes

Canada - Quebec - Time-Weighted Average Exposure Values
2 ppm TWAEV;   14 mg/m3 TWAEV

Canada - Quebec - Skin Designations
absorbed through the skin

Israel - Time Weighted Averages
2 ppm TWA;   14 mg/m3 TWA

Israel - Action Levels
1 ppm AL;   7 mg/m3 AL

Mexico - Instruction No. 10 - TWAs
2 ppm TWA;   14 mg/m3 TWA

Mexico - Instruction No. 10 - STELs
4 ppm STEL;   28 mg/m3 STEL

Mexico - Instruction No. 10 - Skin designation
skin - potential for cutaneous absorption

STATE LISTS

California - Exposure Limits - PELs
2 ppm PEL;   14 mg/m3 PEL

California - Exposure Limits - Skin Notation
material may be absorbed through the skin, eyes or mucous membrane

California - Directors List of Hazardous Substances (8 CCR 339)
[present]

Florida Hazardous Substance List
[present]

Massachusetts Right To Know List
[present]

Minnesota Hazardous Substance List
skin

NJ Right to Know List (Total)
sn 1334

Pennsylvania Right to Know List
[present]

1-DIBUTYLAMINO-2-PROPANOL   RR-00552-7

STATE LISTS

Pennsylvania Right to Know List
[present]

2,6-DI-TERT-BUTYL-P-CRESOL   128-37-0

HEALTH AND SAFETY LISTS

ACGIH 1995 - Time Weighted Averages
10 mg/m3 TWA

IARC - Group 3 (not classifiable)
[present]

NIOSH - Selected LD50s and LC50s
Oral, rat: LD50 = 890 mg/kg

·NTP Chemical Status Reports - Testing Status and NTIS Number
Technical reports printed (PB298539/AS)

NTP Chemical Status Reports - Evidence of Carcinogenicity
male rat-negative; female rat-negative; male mice-negative; female mice-negative

OSHA - Vacated PELs - Time Weighted Averages
10 mg/m3 TWA

ENVIRONMENTAL LISTS

List of Pesticide Product Inert Ingredients
[present]

INTERNATIONAL LISTS

Australian Exposure Standards - Time Weighted Averages
10 mg/m3 TWA

Canada - WHMIS: Ingredient Disclosure
1% item 238 (1007)

Canada - Alberta - 8 Hour Occupational Exposure Limit
10 mg/m3 TWA

Canada - Alberta - 15 Minute Occupational Exposure Limit
20 mg/m3 STEL

Canada - British Columbia - 8 Hour Exposure Limits
10 mg/m3 TWA

Canada - British Columbia - 15 Minute Exposure Limits
20 mg/m3 STEL

Canada - Ontario - OHSA - TWAEVs
10 mg/m3 TWAEV

Canada - Quebec - Time-Weighted Average Exposure Values
10 mg/m3 TWAEV

Israel - Time Weighted Averages
10 mg/m3 TWA

Israel - Action Levels
5 mg/m3 AL

Mexico - Instruction No. 10 - TWAs
10 mg/m3 TWA

Mexico - Instruction No. 10 - STELs
20 mg/m3 STEL

STATE LISTS

California - Exposure Limits - PELs
10 mg/m3 PEL

California - Directors List of Hazardous Substances (8 CCR 339)
[present] (exempt when used in foods and feeds as a preservative)

Florida Hazardous Substance List
[present]

Massachusetts Right To Know List
[present]

Minnesota Hazardous Substance List
[present]

Pennsylvania Right to Know List
[present]

DI-TERT-BUTYL-P-CRESOL   25377-21-3

HEALTH AND SAFETY LISTS

NFPA - Flash Points
flash point = 261 degrees F (127 degrees C)

NFPA - Hazard Identification Ratings
health-0; flammability-1; reactivity-0

INTERNATIONAL LISTS

United Kingdom - Occupational Exposure Standards - TWAs
10 mg/m3 TWA

DI(4-TERT-BUTYLCYCLOHEXYL)-   15520-11-3
PEROXYDICARBONATE

HEALTH AND SAFETY LISTS

U.S. DOT - Organic Peroxides Table
Organic Peroxoide UN3114;   UN3119

STATE LISTS

NJ Right to Know List (Total)
sn 0813

DI-N-BUTYL FUMARATE   105-75-9

HEALTH AND SAFETY LISTS

NIOSH - Selected LD50s and LC50s
Oral, rat: LD50 = 8530 mg/kg Skin, rabbit: LD50 = 16 gm/kg

2,5-DI-TERT-BUTYLHYDROQUINONE   88-58-4

HEALTH AND SAFETY LISTS

NFPA - Flash Points
flash point = 420 degrees F (216 degrees C)

NFPA - Hazard Identification Ratings
health-1; flammability-1; reactivity-0

ENVIRONMENTAL LISTS

List of Pesticide Product Inert Ingredients
[present]

DIBUTYL ISOPHTHALATE   3126-90-7

HEALTH AND SAFETY LISTS

NFPA - Flash Points
flash point = 322 degrees F (161 degrees C)

NFPA - Hazard Identification Ratings
health-0; flammability-1; reactivity-0

DIBUTYLISOPROPANOLAMINE   2109-64-0

HEALTH AND SAFETY LISTS

NFPA - Flash Points
flash point = 205 degrees F (96 degrees C)

NFPA - Hazard Identification Ratings
health-2; flammability-1; reactivity-0

STATE LISTS

Florida Hazardous Substance List
[present]

Massachusetts Right To Know List
[present]

Pennsylvania Right to Know List
[present]

DIBUTYLNAPHTHALENESULFONIC ACID, SODIUM SALT   25417-20-3

HEALTH AND SAFETY LISTS
NIOSH - Selected LD50s and LC50s
Oral, rat: LD50 = 1250 mg/kg

ENVIRONMENTAL LISTS
List of Pesticide Product Inert Ingredients
[present]

DI-TERT-BUTYL PEROXIDE   110-05-4

HEALTH AND SAFETY LISTS
U.S. DOT - Organic Peroxides Table
Organic peroxide UN3103;   UN3107
NFPA - Flash Points
flash point = 65 degrees F (18 degrees C)
NFPA - Hazard Identification Ratings
health-3;   flammability-2;   reactivity-4 (oxidizing properties)
NIOSH - Selected LD50s and LC50s
Oral, rat: LD50 = 10200 mg/kg
OSHA - List of Highly Hazardous Chemicals
threshhold quantity = 5000 pounds

ENVIRONMENTAL LISTS
List of Pesticide Product Inert Ingredients
[present]

INTERNATIONAL LISTS
German (DFG) - Skin/Sensitizers
negligible effects on skin

STATE LISTS
Florida Hazardous Substance List
[present]
Massachusetts Right To Know List
[present]
NJ Right to Know List (Total)
sn 0815
Pennsylvania Right to Know List
[present]

2,2-DI(TERT-BUTYLPEROXY)-BUTANE   2167-23-9

HEALTH AND SAFETY LISTS
U.S. DOT - Hazard Classes
Forbidden from transport by the DOT
U.S. DOT - Organic Peroxides Table
Organic peroxide UN3103;   UN3106;   UN3115

STATE LISTS
NJ Right to Know List (Total)
sn 0816

1,1-DI(TERT-BUTYLPEROXY)-CYCLOHEXANE   3006-86-8

HEALTH AND SAFETY LISTS
U.S. DOT - Organic Peroxides Table
Organic peroxide UN3101;   UN3103;   UN3105;   UN3106;   UN3107

STATE LISTS
NJ Right to Know List (Total)
sn 0817

2,2-DI-(4,4-DI-TERT-BUTYLPEROXYCYCLOHEXYL) PROPANE   RR-01372-

HEALTH AND SAFETY LISTS
U.S. DOT - Hazard Classes
Forbidden from transport by the DOT

DI-N-BUTYL PEROXYDICARBONATE   RR-01406-

HEALTH AND SAFETY LISTS
U.S. DOT - Hazard Classes
Forbidden from transport by the DOT

DI-SEC-BUTYL PEROXYDICARBONATE   19910-65-

HEALTH AND SAFETY LISTS
U.S. DOT - Organic Peroxides Table
Organic peroxide UN3113;   UN3115
NIOSH - Selected LD50s and LC50s
Skin, rabbit: LD50 = 1200 mg/kg

1,4-DI-(2-T-BUTYLPEROXY ISOPROPYL)BENZENE   RR-00250-6

STATE LISTS
NJ Right to Know List (Total)
sn 2314

DI-(2-TERT-BUTYLPEROXYISOPROPYL)-BENZENE   RR-01752-7

HEALTH AND SAFETY LISTS
U.S. DOT - Organic Peroxides Table
Organic peroxide UN3106;   exempt

DI-TERT-BUTYLPEROXYPHTHALATE   2155-71-7

STATE LISTS
NJ Right to Know List (Total)
technically pure: sn 2317;   55% in paste: sn 2315;   55% in solution: sn 2316

DI-(TERT-BUTYLPEROXY) PHTHALATE   RR-01405-1

HEALTH AND SAFETY LISTS
U.S. DOT - Hazard Classes
Forbidden from transport by the DOT
U.S. DOT - Organic Peroxides Table
Organic peroxide UN3105

2,2-DI(T-BUTYLPEROXY)PROPANE   RR-00253-9

HEALTH AND SAFETY LISTS
U.S. DOT - Organic Peroxides Table
Organic peorxide UN3105;   UN3106

STATE LISTS
NJ Right to Know List (Total)
sn 2318;   sn 2319

1,1-DI(TERT-BUTYLPEROXY)-3,3,5-TRIMETHYLCYCLOHEXANE PEROXIDE   6731-36-8

HEALTH AND SAFETY LISTS
U.S. DOT - Organic Peroxides Table
Organic peroxide UN3101;   UN3106;   UN3107

STATE LISTS
NJ Right to Know List (Total)
sn 0818

2,6-DI-T-BUTYLPHENOL   128-39-2

ENVIRONMENTAL LISTS
EPA - Master Testing List
[present]
TSCA - Code of Federal Regulations Citations
40 CFR 712.30(p);   40 CFR 716.120(a)

TSCA - PAIR - Reporting List
Reporting Date: February 18, 1986
TSCA - Health and Safety Reporting List
Effective Date: December 19, 1985; Sunset Date: November 9, 1993

DIBUTYL PHENYL PHOSPHATE   2528-36-1
SEE ALSO:
BISAZOBIPHENYL DYES

HEALTH AND SAFETY LISTS
ACGIH 1995 - Time Weighted Averages
0.3 ppm TWA; 3.5 mg/m3 TWA
ACGIH 1995 - Skin Designations
skin - potential for cutaneous absorption
NIOSH - Selected LD50s and LC50s
Oral, rat: LD50 = 2140 mg/kg

ENVIRONMENTAL LISTS
EPA - Master Testing List
[present]
TSCA - Code of Federal Regulations Citations
40 CFR 712.30(d); 40 CFR 716.120(c)
TSCA - PAIR - Reporting List
Reporting Date: November 19, 1982

INTERNATIONAL LISTS
Australian Exposure Standards - Time Weighted Averages
0.3 ppm TWA; 3.5 mg/m3 TWA
Australian Exposure Standards - Skin Effects
skin absorption
Israel - Time Weighted Averages
0.3 ppm TWA; 3.5 mg/m3 TWA
Israel - Action Levels
0.15 ppm AL; 1.75 mg/m3 AL

STATE LISTS
Florida Hazardous Substance List
effective March 13, 1992
Minnesota Hazardous Substance List
skin

DIBUTYL PHOSPHATE   107-66-4

HEALTH AND SAFETY LISTS
ACGIH 1995 - Time Weighted Averages
1 ppm TWA; 8.6 mg/m3 TWA
ACGIH 1995 - Short Term Exposure Limits
2 ppm STEL; 17 mg/m3 STEL
NIOSH 1990 - Pocket Guide - RELs
1 ppm TWA; 5 mg/m3 TWA; 2 ppm STEL; 10 mg/m3 STEL
NIOSH 1990 - Pocket Guide - IDLHs
125 ppm IDLH
NIOSH 1990 - Pocket Guide - Target organs
skin, respiratory system
OSHA - Vacated PELs - Time Weighted Averages
1 ppm TWA; 5 mg/m3 TWA
OSHA - Vacated PELs - Short Term Exposure Limits
2 ppm STEL; 10 mg/m3 STEL
OSHA - Final PELs - Time Weighted Averages
1 ppm TWA; 5 mg/m3 TWA

ENVIRONMENTAL LISTS
EPA - Master Testing List
[present]
TSCA - PAIR - Reporting List
Effective Date: January 26, 1994; Reporting Date: March 28, 1994
TSCA - Health and Safety Reporting List
Effective Date: January 26, 1994; Sunset Date: January 26, 2004

INTERNATIONAL LISTS
Australian Exposure Standards - Time Weighted Averages
1 ppm TWA; 8.6 mg/m3 TWA
Australian Exposure Standards - Short Term Exposure Limits
2 ppm STEL; 17 mg/m3 STEL
Canada - WHMIS: Ingredient Disclosure
1% item 516 (1392)
Canada - Alberta - 8 Hour Occupational Exposure Limit
1 ppm TWA; 8.6 mg/m3 TWA
Canada - Alberta - 15 Minute Occupational Exposure Limit
2 ppm STEL; 17.2 mg/m3 STEL
Canada - British Columbia - 8 Hour Exposure Limits
1 ppm TWA; 5 mg/m3 TWA
Canada - British Columbia - 15 Minute Exposure Limits
2 ppm STEL; 10 mg/m3 STEL
Canada - Ontario - OHSA - TWAEVs
1 ppm TWAEV; 8.6 mg/m3 TWAEV
Canada - Ontario - OHSA - STEVs
2 ppm STEV; 17 mg/m3 STEV
Canada - Quebec - Time-Weighted Average Exposure Values
1 ppm TWAEV; 8.6 mg/m3 TWAEV
Canada - Quebec - Short-term Exposure Values
2 ppm STEV; 17 mg/m3 STEV
United Kingdom - Occupational Exposure Standards - TWAs
1 ppm TWA; 5 mg/m3 TWA
United Kingdom - Occupational Exposure Standards - STELs
2 ppm STEL; 10 mg/m3 STEL
Israel - Time Weighted Averages
1 ppm TWA; 8.6 mg/m3 TWA
Israel - Short Term Exposure Limits
2 ppm STEL; 17 mg/m3 STEL
Israel - Action Levels
0.5 ppm AL; 4.3 mg/m3 AL
Mexico - Instruction No. 10 - TWAs
1 ppm TWA; 5 mg/m3 TWA
Mexico - Instruction No. 10 - STELs
2 ppm STEL; 10 mg/m3 STEL

STATE LISTS
California - Exposure Limits - PELs
1 ppm PEL; 5 mg/m3 PEL
California - Exposure Limits - STELs
2 ppm STEL; 10 mg/m3 STEL
California - Directors List of Hazardous Substances (8 CCR 339)
[present]
Florida Hazardous Substance List
[present]
Massachusetts Right To Know List
extraordinarily hazardous
Minnesota Hazardous Substance List
[present]
Pennsylvania Right to Know List
[present]

PROPOSED REGULATIONS
TSCA - ITC 31st Report Priority Testing List
designated to be tested

DIBUTYL PHOSPHITE   1809-19-4

HEALTH AND SAFETY LISTS
NFPA - Flash Points
flash point = 120 degrees F (49 degrees C)
NFPA - Hazard Identification Ratings
health-3; flammability-2; reactivity-0

INTERNATIONAL LISTS
Canada - WHMIS: Ingredient Disclosure
1% item 517 (1403)
STATE LISTS
Florida Hazardous Substance List
[present]

Massachusetts Right To Know List
[present]

Pennsylvania Right to Know List
[present]

DIBUTYL PHTHALATE   84-74-2
SEE ALSO:
F039-HAZARDOUS WASTES
CAT FOOD
ALKYL PHTHALATES

HEALTH AND SAFETY LISTS
ACGIH 1995 - Time Weighted Averages
5 mg/m3 TWA

U.S. DOT - Appendix B - Marine Pollutants
DOT regulated marine pollutant

U.S. DOT - Appendix A Table 1 - Hazardous Substances
final RQ = 10 pounds (4.54 kg)

NFPA - Flash Points
flash point = 315 degrees F (157 degrees C)

NFPA - Hazard Identification Ratings
health-0; flammability-1; reactivity-0

NIOSH - Selected LD50s and LC50s
Oral, rat: LD50 = 8000 mg/kg

NIOSH 1990 - Pocket Guide - RELs
5 mg/m3 TWA

NIOSH 1990 - Pocket Guide - IDLHs
9300 mg/m3 IDLH

NIOSH 1990 - Pocket Guide - Target organs
GI tract, respiratory system

NTP Chemical Status Reports - Testing Status and NTIS Number
Prechronic studies in progress; post peer review technical reports in progress

OSHA - Vacated PELs - Time Weighted Averages
5 mg/m3 TWA

OSHA - Final PELs - Time Weighted Averages
5 mg/m3 TWA

ENVIRONMENTAL LISTS
ATSDR Priority List
Rank (of 275): 030

CERCLA/SARA - Section 313 - Emission Reporting
form R reporting required for 1.0% de minimus concentration

CERCLA/SARA - Hazardous Substances and their Reportable Quantities
final RQ = 10 pounds (4.54 kg)

Clean Air Act (1990) - List of Hazardous Air Contaminants
[present]

Clean Water Act - Hazardous Substances
[present]

Clean Water Act - Priority Pollutants
[present]

List of Pesticide Product Inert Ingredients
[present]

RCRA - U Series Wastes
waste number U069

RCRA - Hazardous Constituents-Appendix VIII
waste number U069

RCRA - Basis for Listing - Appendix VII
Included in waste stream: F039

RCRA - Substances Banned From Land Disposal
[present]

RCRA - TSD Facilities Ground Water Monitoring
TM 8060 = 5 ug/L PQL;   TM 8270 = 10 ug/L PQL

RCRA - Universal Treatment Standards (LDR)
WW: 0.057 mg/l;   NWW: 28 mg/kg

TSCA - Code of Federal Regulations Citations
40 CFR 712.30(d);   40 CFR 716.120(c);   40 CFR 799.5000

TSCA - PAIR - Reporting List
Reporting Date: November 19, 1982

TSCA - Substances Subject to Testing Consent Orders
Test for: Environmental Effects

TSCA - Section 12(b) - Export Notification
export notification required - Section 4

INTERNATIONAL LISTS
Australian Exposure Standards - Time Weighted Averages
5 mg/m3 TWA

Canada - WHMIS: Ingredient Disclosure
1% item 518 (1416)

Canada - NPRI (National Pollutant Release Inventory)
[present]

Canada - CEPA - Priority Substances List
estimated time for completion of assessment reports: 4 years

Canada - Alberta - 8 Hour Occupational Exposure Limit
5 mg/m3 TWA

Canada - Alberta - 15 Minute Occupational Exposure Limit
10 mg/m3 STEL

Canada - British Columbia - 8 Hour Exposure Limits
5 mg/m3 TWA

Canada - British Columbia - 15 Minute Exposure Limits
10 mg/m3 STEL

Canada - Ontario - OHSA - TWAEVs
5 mg/m3 TWAEV

Canada - Quebec - Time-Weighted Average Exposure Values
5 mg/m3 TWAEV

United Kingdom - Occupational Exposure Standards - TWAs
5 mg/m3 TWA

United Kingdom - Occupational Exposure Standards - STELs
10 mg/m3 STEL

Israel - Time Weighted Averages
5 mg/m3 TWA

Israel - Action Levels
2.5 mg/m3 AL

Mexico - Instruction No. 10 - TWAs
5 mg/m3 TWA

Mexico - Instruction No. 10 - STELs
10 mg/m3 STEL

Mexico - Wastewater - Organic Toxic Pollutants and Heavy Metals
Listed under [Phthalate Esters]

STATE LISTS
California - Air Bill 2588 Appendix A-I
6/91

California - Exposure Limits - PELs
5 mg/m3 PEL

Florida Hazardous Substance List
[present]

Massachusetts Right To Know List
[present]

Minnesota Hazardous Substance List
[present]

Pennsylvania Right to Know List
environmental hazard

PROPOSED REGULATIONS
  Canada - Ontario - Proposed Occupational TWAEVs
  3 mg/m3 TWAEV
  Canada - Ontario - Proposed Occupational STEVs
  5 mg/m3 STEV

DI-N-BUTYL SEBACATE 109-43-3
  HEALTH AND SAFETY LISTS
    NFPA - Flash Points
    flash point = 353 degrees F (178 degrees C)
    NFPA - Hazard Identification Ratings
    health-1; flammability-1; reactivity-0
    NIOSH - Selected LD50s and LC50s
    Oral, rat: LD50 = 16 gm/kg

N.N-DIBUTYL STEARAMIDE 5831-88-9
  HEALTH AND SAFETY LISTS
    NFPA - Flash Points
    flash point = 420 degrees F (216 degrees C)
    NFPA - Hazard Identification Ratings
    health-0; flammability-1; reactivity-0

DI-N-BUTYL SUCCINATE 141-03-7
  HEALTH AND SAFETY LISTS
    NIOSH - Selected LD50s and LC50s
    Oral, rat: LD50 = 8000 mg/kg

N-DIBUTYL TARTRATE 87-92-3
  HEALTH AND SAFETY LISTS
    NFPA - Flash Points
    flash point = 195 degrees F (91 degrees C)
    NFPA - Hazard Identification Ratings
    health-0; flammability-2; reactivity-0

DIBUTYL THIOUREA 109-46-6
  HEALTH AND SAFETY LISTS
    NIOSH - Selected LD50s and LC50s
    Oral, rat: LD50 = 350 mg/kg
  ENVIRONMENTAL LISTS
    List of Pesticide Product Inert Ingredients
    [present]

DIBUTYLTIN BIS (ISOOCTYL MALEATE)-2-BUTENOIC ACID, 4,4'-[(DIBUTYLSTANNYLENE)BIS(OXY)]BIS[4-OXO-, DIISOOCTYL ESTER, (Z,Z)- 25168-21-2
  ENVIRONMENTAL LISTS
    TSCA - Code of Federal Regulations Citations
    40 CFR 712.30(g); 40 CFR 716.120(a)
    TSCA - PAIR - Reporting List
    Reporting Date: October 8, 1984
    TSCA - Health and Safety Reporting List
    Effective Date: January 3, 1983

DIBUTYLTIN BIS(ISOOCTYL MERCAPTOACETATE) 25168-24-5
  SEE ALSO:
    ALKYLTIN COMPOUNDS
    TIN
  HEALTH AND SAFETY LISTS
    NIOSH - Selected LD50s and LC50s
    Oral, rat: LD50 = 500 mg/kg
  ENVIRONMENTAL LISTS
    TSCA - Code of Federal Regulations Citations
    40 CFR 712.30(g); 40 CFR 716.120

TSCA - PAIR - Reporting List
Reporting Date: October 8, 1984
  INTERNATIONAL LISTS
    Canada - WHMIS: Ingredient Disclosure
    1% item 195 (300)

DIBUTYLTIN S,S'-BIS(ISOOCTYL MERCAPTOAC-ETATE) 26636-01-1
  SEE ALSO:
    ALKYLTIN COMPOUNDS
  ENVIRONMENTAL LISTS
    TSCA - Code of Federal Regulations Citations
    40 CFR 712.30(g); 40 CFR 716.120(c)
    TSCA - PAIR - Reporting List
    Reporting Date: October 8, 1984

DIBUTYLTIN BIS(LAURYL MERCAPTIDE) 1185-81-5
  ENVIRONMENTAL LISTS
    TSCA - Code of Federal Regulations Citations
    40 CFR 712.30(g); 40 CFR 716.120(a)
    TSCA - PAIR - Reporting List
    Reporting Date: October 8, 1984
    TSCA - Health and Safety Reporting List
    Effective Date: January 3, 1983

DI-N-BUTYLTIN BIS(METHYL MALEATE) 15546-11-9
  SEE ALSO:
    TIN
  HEALTH AND SAFETY LISTS
    NIOSH - Selected LD50s and LC50s
    Oral, rat: LD50 = 62 mg/kg
  INTERNATIONAL LISTS
    Canada - WHMIS: Ingredient Disclosure
    1% item 526 (1141)

DIBUTYLTIN DIACETATE 1067-33-0
  SEE ALSO:
    TIN
    TIN
    TIN ORGANIC COMPOUNDS
  HEALTH AND SAFETY LISTS
    NIOSH - Selected LD50s and LC50s
    Oral, rat: LD50 = 32 mg/kg
    NTP Chemical Status Reports - Testing Status and NTIS Number
    Technical reports printed (PB291567/AS)
    NTP Chemical Status Reports - Evidence of Carcinogenicity
    male rat-negative; female rat-inadequate; male mice-negative; female mice-negative
  INTERNATIONAL LISTS
    Canada - WHMIS: Ingredient Disclosure
    1% item 519 (618)
  STATE LISTS
    Massachusetts Right To Know List
    [present]

DIBUTYLTIN DICHLORIDE 683-18-1
  SEE ALSO:
    TIN
  INTERNATIONAL LISTS
    Canada - WHMIS: Ingredient Disclosure
    1% item 525 (682)

DI-N-BUTYLTIN DI-2-ETHYLHEXANOATE   2781-10-4
SEE ALSO:
    TIN

HEALTH AND SAFETY LISTS
    NIOSH - Selected LD50s and LC50s
    Oral, rat: LD50 = 200 mg/kg

INTERNATIONAL LISTS
    Canada - WHMIS: Ingredient Disclosure
    1% item 521 (705)

DIBUTYLTIN DILAURATE   77-58-7
SEE ALSO:
    TIN ORGANIC COMPOUNDS
    TIN
    TIN

HEALTH AND SAFETY LISTS
    NIOSH - Selected LD50s and LC50s
    Oral, rat: LD50 = 175 mg/kg

ENVIRONMENTAL LISTS
    TSCA - Code of Federal Regulations Citations
    40 CFR 712.30(g); 40 CFR 716.120(a)
    TSCA - PAIR - Reporting List
    Reporting Date: October 8, 1984
    TSCA - Health and Safety Reporting List
    Effective Date: January 3, 1983

INTERNATIONAL LISTS
    Canada - WHMIS: Ingredient Disclosure
    1% item 522 (723)

DI-N-BUTYLTIN DI(MONOBUTYL)MALEATE   15546-16-4
SEE ALSO:
    TIN

INTERNATIONAL LISTS
    Canada - WHMIS: Ingredient Disclosure
    1% item 523 (747)

DIBUTYLTIN DISTEARATE   5847-55-2
SEE ALSO:
    TIN

INTERNATIONAL LISTS
    Canada - WHMIS: Ingredient Disclosure
    1% item 524 (785)

DIBUTYLTIN MALEATE   78-04-6
SEE ALSO:
    TIN

INTERNATIONAL LISTS
    Canada - WHMIS: Ingredient Disclosure
    1% item 520 (1070)

DIBUTYLTIN OXIDE   818-08-6
SEE ALSO:
    TIN
    TIN
    TIN ORGANIC COMPOUNDS

HEALTH AND SAFETY LISTS
    NIOSH - Selected LD50s and LC50s
    Oral, rat: LD50 = 44900 ug/kg

INTERNATIONAL LISTS
    Canada - WHMIS: Ingredient Disclosure
    1% item 527 (1308)

DIBUTYLTIN SULFIDE   4253-22-9
SEE ALSO:
    TIN

INTERNATIONAL LISTS
    Canada - WHMIS: Ingredient Disclosure
    1% item 528 (791)

N,N-DIBUTYLTOLUENESULFONAMIDE   RR-00872-4
HEALTH AND SAFETY LISTS
    NFPA - Flash Points
    flash point = 330 degrees F (166 degrees C)
    NFPA - Hazard Identification Ratings
    health-0; flammability-1; reactivity-0

DIBUTYL XANTHOGEN DISULFIDE   105-77-1
HEALTH AND SAFETY LISTS
    NIOSH - Selected LD50s and LC50s
    Oral, mouse: LD50 = 2700 mg/kg

DI(C8-18)ALKYL DIMETHYL AMMONIUM CHLORIDE   73398-64-8
ENVIRONMENTAL LISTS
    List of Pesticide Product Inert Ingredients
    [present]

DI-C10-16-ALKYL DIMETHYL AMMONIUM CHLORIDE   68153-33-3
ENVIRONMENTAL LISTS
    List of Pesticide Product Inert Ingredients
    [present]

DICAMBA   1918-00-9
HEALTH AND SAFETY LISTS
    U.S. DOT - Appendix A Table 1 - Hazardous Substances
    final RQ = 1000 pounds (454 kg)
    NIOSH - Selected LD50s and LC50s
    Oral, rat: LD50 = 1039 mg/kg

ENVIRONMENTAL LISTS
    CERCLA/SARA - Section 313 - Emission Reporting
    form R reporting required
    CERCLA/SARA - Hazardous Substances and their Reportable Quantities
    final RQ = 1000 pounds (454 kg)
    Clean Water Act - Hazardous Substances
    [present]

INTERNATIONAL LISTS
    Canada - Drinking Water Quality - MACs
    0.12 mg/L MAC

STATE LISTS
    California - Directors List of Hazardous Substances (8 CCR 339)
    [present]
    Massachusetts Right To Know List
    [present]
    NJ Right to Know List (Total)
    sn 0634
    Pennsylvania Right to Know List
    environmental hazard

PROPOSED REGULATIONS
    Safe Drinking Water Act - Priority list
    [present]

DICARBOXYLIC ACID MONOESTER   RR-00258-4

ENVIRONMENTAL LISTS

TSCA - Chemicals with Significant New Use Rules
PMN number: P-83-255

DICETYL PEROXYDICARBONATE   26322-14-5

HEALTH AND SAFETY LISTS

U.S. DOT - Organic Peroxides Table
Organic peroxide UN3116

STATE LISTS

NJ Right to Know List (Total)
sn 0635

DICHLOBENIL   1194-65-6

HEALTH AND SAFETY LISTS

U.S. DOT - Appendix A Table 1 - Hazardous Substances
final RQ = 100 pounds (45.4 kg)

NIOSH - Selected LD50s and LC50s
Oral, rat: LD50 = 2710 mg/kg Skin, rabbit: LD50 = 1350 mg/kg

ENVIRONMENTAL LISTS

CERCLA/SARA - Hazardous Substances and their Reportable Quantities
final RQ = 100 pounds (45.4 kg)

Clean Water Act - Hazardous Substances
[present]

STATE LISTS

California - Directors List of Hazardous Substances (8 CCR 339)
[present]

Massachusetts Right To Know List
[present]

NJ Right to Know List (Total)
sn 0636

Pennsylvania Right to Know List
environmental hazard

DICHLOFENTHION   97-17-6

HEALTH AND SAFETY LISTS

U.S. DOT - Appendix B - Marine Pollutants
DOT regulated severe marine pollutant

DICHLONE   117-80-6

HEALTH AND SAFETY LISTS

U.S. DOT - Appendix A Table 1 - Hazardous Substances
final RQ = 1 pound (0.454 kg)

NIOSH - Selected LD50s and LC50s
Oral, rat: LD50 = 160 mg/kg Skin, rabbit: LD50 = 5000 mg/kg

ENVIRONMENTAL LISTS

CERCLA/SARA - Hazardous Substances and their Reportable Quantities
final RQ = 1 pound (0.454 kg)

Clean Water Act - Hazardous Substances
[present]

STATE LISTS

California - Directors List of Hazardous Substances (8 CCR 339)
[present]

Massachusetts Right To Know List
[present]

NJ Right to Know List (Total)
sn 0637

Pennsylvania Right to Know List
environmental hazard

N,N'-DICHLORAZODICARBONAMIDINE   RR-01446-0

HEALTH AND SAFETY LISTS

U.S. DOT - Hazard Classes
Forbidden from transport by the DOT

DICHLOROACETALDEHYDE   79-02-7

INTERNATIONAL LISTS

Canada - WHMIS: Ingredient Disclosure
1% item 529 (649)

DICHLOROACETIC ACID   79-43-6

HEALTH AND SAFETY LISTS

U.S. DOT - Substances From 49 CFR 172.101
regulated by DOT (UN1764)

U.S. DOT - Hazard Classes
DOT hazard class = 8

NIOSH - Selected LD50s and LC50s
Oral, rat: LD50 = 2820 mg/kg Skin, rabbit: LD50 = 510 mg/kg

ENVIRONMENTAL LISTS

List of Pesticide Product Inert Ingredients
[present]

STATE LISTS

NJ Right to Know List (Total)
sn 0638

NJ Special Hazardous Substances
(corrosive)

DICHLOROACETONITRILE   3018-12-0

HEALTH AND SAFETY LISTS

IARC - Group 3 (not classifiable)
[present]

PROPOSED REGULATIONS

Safe Drinking Water Act - Priority list
[present]

DICHLOROACETYL CHLORIDE   79-36-7

HEALTH AND SAFETY LISTS

U.S. DOT - Substances From 49 CFR 172.101
regulated by DOT (UN1765)

U.S. DOT - Hazard Classes
DOT hazard class = 8

NFPA - Flash Points
flash point = 151 degrees F (66 degrees C)

NFPA - Hazard Identification Ratings
health-3; flammability-2; reactivity-2 (avoid use of water)

NIOSH - Selected LD50s and LC50s
Oral, rat: LD50 = 2460 mg/kg Skin, rabbit: LD50 = 650 mg/kg

ENVIRONMENTAL LISTS

Clean Water Act - Hazardous Substances
[present]

INTERNATIONAL LISTS

Canada - WHMIS: Ingredient Disclosure
1% item 531 (492)

STATE LISTS

California - Directors List of Hazardous Substances (8 CCR 339)
[present]

Florida Hazardous Substance List
[present]

Massachusetts Right To Know List
[present]

NJ Right to Know List (Total)
sn 0639

NJ Special Hazardous Substances
(corrosive)
Pennsylvania Right to Know List
[present]

DICHLOROACETYLENE   7572-29-4

HEALTH AND SAFETY LISTS

ACGIH 1995 - Ceiling Limits
(C 0.1) ppm;   (C 0.39) mg/m3
U.S. DOT - Hazard Classes
Forbidden from transport by the DOT
IARC - Group 3 (not classifiable)
[present]
NIOSH - Selected LD50s and LC50s
Inhalation, mouse: LC50 = 19 ppm 6 hr
OSHA - Vacated PELs - Ceiling Limits
C 0.1 ppm;   C 0.4 mg/m3
OSHA - List of Highly Hazardous Chemicals
threshhold quantity = 250 pounds

INTERNATIONAL LISTS

Australian Exposure Standards - Time Weighted Averages
Peak Limitation: 0.1 ppm;   0.39 mg/m3
Australian Exposure Standards - Carcinogens
suspected carcinogen
Canada - WHMIS: Ingredient Disclosure
1% item 532 (651)
Canada - Alberta - Ceiling Occupational Exposure Limit
C 0.1 ppm;   C 0.39 mg/m3
Canada - British Columbia - Ceiling Exposure Limits
C 0.1 ppm;   C 0.4 mg/m3
Canada - Ontario - OHSA - CEVs
0.1 ppm CEV;   0.4 mg/m3 CEV
Canada - Quebec - Ceiling Limits
P 0.1 ppm;   P 0.39 mg/m3
United Kingdom - Occupational Exposure Standards - STELs
0.1 ppm STEL;   0.4 mg/m3 STEL
German (DFG) - Carcinogens
animal evidence of carcinogenicity
Israel - Ceiling Exposure Limits
C 0.1 ppm;   C 0.39 mg/m3
Mexico - Instruction No. 10 - Ceiling Limits
P 0.1 ppm;   P 0.4mg/m3

STATE LISTS

California - Exposure Limits - Ceilings
C 0.1 ppm;   C 0.4 mg/m3
California - Directors List of Hazardous Substances (8 CCR 339)
[present]
Florida Hazardous Substance List
[present]
Massachusetts Right To Know List
[present]
Minnesota Hazardous Substance List
[present]
Pennsylvania Right to Know List
[present]

PROPOSED REGULATIONS

ACGIH 1995 - Notice of Intended Changes
C 0.1 ppm;   C 0.39 mg/m3;   A3-animal carcinogen

3,4-DICHLOROANILINE   95-76-1

SEE ALSO:
ANILINE AND CHLORO-, BROMO-, AND/OR NITROANILINES

HEALTH AND SAFETY LISTS

NFPA - Flash Points
flash point = 331 degrees F (166 degrees C)
NFPA - Hazard Identification Ratings
health-3;   flammability-1;   reactivity-0
NIOSH - Selected LD50s and LC50s
Oral, rat: LD50 = 648 mg/kg Skin, cat: LD50 = 700 mg/kg

ENVIRONMENTAL LISTS

TSCA - Code of Federal Regulations Citations
40 CFR 712.30(d);   40 CFR 716.120(c);   40 CFR 799.5000
TSCA - PAIR - Reporting List
Reporting Date: November 19, 1982
TSCA - Substances Subject to Testing Consent Orders
Test for: Health Effects
TSCA - Section 12(b) - Export Notification
export notification required - Section 4

INTERNATIONAL LISTS

Canada - WHMIS: Ingredient Disclosure
1% item 534 (653)

STATE LISTS

Florida Hazardous Substance List
[present]
Massachusetts Right To Know List
[present]
Pennsylvania Right to Know List
[present]

DICHLOROANILINE   27134-27-6

HEALTH AND SAFETY LISTS

U.S. DOT - Substances From 49 CFR 172.101
regulated by DOT (UN1590)
U.S. DOT - Hazard Classes
DOT hazard class = 6.1

ENVIRONMENTAL LISTS

CAA - HON Rule - SOCMI Chemicals
compliance by Oct. 24, 1994
List of Pesticide Product Inert Ingredients
[present]

INTERNATIONAL LISTS

Canada - WHMIS: Ingredient Disclosure
1% item 533 (652)

STATE LISTS

NJ Right to Know List (Total)
sn 0641

DICHLOROANILINES   RR-01278-2

HEALTH AND SAFETY LISTS

U.S. DOT - Appendix B - Marine Pollutants
DOT regulated marine pollutant

DICHLOROBENZALKONIUM CHLORIDE   8023-53-8

HEALTH AND SAFETY LISTS

NIOSH - Selected LD50s and LC50s
Oral, rat: LD50 = 730 mg/kg

STATE LISTS

Massachusetts Right To Know List
[present]

O-DICHLOROBENZENE   95-50-1
SEE ALSO:
K042-HAZARDOUS WASTES
F002-HAZARDOUS WASTES
F039-HAZARDOUS WASTES

HEALTH AND SAFETY LISTS
ACGIH 1995 - Time Weighted Averages
25 ppm TWA; 150 mg/m3 TWA

ACGIH 1995 - Short Term Exposure Limits
50 ppm STEL; 301 mg/m3 STEL

AIHA - Odor Threshold Values
geometric mean air odor threshold = 0.70 ppm (detectable)

U.S. DOT - Substances From 49 CFR 172.101
regulated by DOT (UN1591)

U.S. DOT - Hazard Classes
DOT hazard class = 6.1

U.S. DOT - Appendix B - Marine Pollutants
DOT regulated marine pollutant

U.S. DOT - Appendix A Table 1 - Hazardous Substances
final RQ = 100 pounds (45.4 kg)

IARC - Group 3 (not classifiable)
[present]

NFPA - Flash Points
flash point = 151 degrees F (66 degrees C)

NFPA - Hazard Identification Ratings
health-2; flammability-2; reactivity-0

NIOSH - Selected LD50s and LC50s
Oral, rat: LD50 = 500 mg/kg

NIOSH 1990 - Pocket Guide - RELs
C 50 ppm; C 300 mg/m3

NIOSH 1990 - Pocket Guide - IDLHs
1000 ppm IDLH

NIOSH 1990 - Pocket Guide - Target organs
liver, kidneys, skin, eyes,

NTP Chemical Status Reports - Testing Status and NTIS Number
Technical reports printed (PB86144888/AS)

NTP Chemical Status Reports - Evidence of Carcinogenicity
male rat-negative; female rat-negative; male mice-negative; female mice-negative

OSHA - Vacated PELs - Ceiling Limits
C 50 ppm; C 300 mg/m3

OSHA - Final PELs - Ceiling Limits
C 50 ppm; C 300 mg/m3

ENVIRONMENTAL LISTS
ATSDR Priority List
Rank (of 275): 149

CERCLA/SARA - Section 313 - Emission Reporting
form R reporting required for 1.0% de minimus concentration

CERCLA/SARA - Hazardous Substances and their Reportable Quantities
final RQ = 100 pounds (45.4 kg)

CAA - HON Rule - SOCMI Chemicals
compliance by Oct. 24, 1995

Clean Water Act - Hazardous Substances
[present] (Listed under 'Dichlorobenzene')

Clean Water Act - Priority Pollutants
[present]

Clean Water Act - Toxic Pollutants
[present] (Listed under 'Dichlorobenzenes')

Safe Drinking Water Act - MCLs
MCL = 0.6 mg/L

Safe Drinking Water Act - MCLGs
MCLG = 0.6 mg/L

Safe Drinking Water Act - Monitoring
monitoring required

List of Pesticide Product Inert Ingredients
[present]

RCRA - U Series Wastes
waste number U070

RCRA - Hazardous Constituents-Appendix VIII
waste number U070

RCRA - Basis for Listing - Appendix VII
Included in waste streams: F002, F039, K042

RCRA - Substances Banned From Land Disposal
[present]

RCRA - TSD Facilities Ground Water Monitoring
TM 8010 = 2 ug/L PQL; TM 8020 = 5 ug/L PQL; TM 8120 = 10 ug/L PQL; TM 8270 = 10 ug/L PQL

RCRA - Universal Treatment Standards (LDR)
WW: 0.088 mg/l; NWW: 6.0 mg/kg

TSCA - Code of Federal Regulations Citations
40 CFR 712.30(d); 40 CFR 716.120(c); 40 CFR 799.1052

TSCA - PAIR - Reporting List
Reporting Date: November 19, 1982

TSCA - HDD/HDF - Precursors Required for Reporting
[present]

TSCA - Multichemical Test Rules - Waste Constituents
hydrolysis testing for Chemical Fate

TSCA - Chemical Test Rules
Testing required by: manufacturers; processors (40 CFR 799.1052) (Listed under 'Dichlorobenzenes')

TSCA - Section 12(b) - Export Notification
export notification required - Section 4

INTERNATIONAL LISTS
Australian Exposure Standards - Time Weighted Averages
Peak Limitation: 50 ppm; 301 mg/m3

Canada - WHMIS: Ingredient Disclosure
1% item 536 (655)

Canada - NPRI (National Pollutant Release Inventory)
[present]

Canada - CEPA - Priority Substances List
estimated time for completion of assessment reports: 4 years

Canada - Drinking Water Quality - MACs
0.2 mg/L MAC

Canada - Drinking Water Quality - AOs
<= 0.003 mg/L AO

Canada - Alberta - Ceiling Occupational Exposure Limit
C 50 ppm; C 300 mg/m3

Canada - British Columbia - Ceiling Exposure Limits
C 50 ppm; C 300 mg/m3

Canada - Ontario - OHSA - CEVs
50 ppm CEV; 300 mg/m3 CEV

Canada - Quebec - Ceiling Limits
P 50 ppm; P 301 mg/m3

Canada - Quebec - Skin Designations
absorbed through the skin

United Kingdom - Occupational Exposure Standards - STELs
50 ppm STEL; 300 mg/m3 STEL

German (DFG) - MAK Values
50 ppm MAK; 300 mg/m3 MAK

German (DFG) - Peak Limitations
2 x normal MAK (30 min. average value); don't exceed 4 times during shift

German (DFG) - Skin/Sensitizers
danger of cutaneous absorption

German (DFG) - Pregnancy
no risk to embryo/fetus is exposure limits adhered to

Israel - Ceiling Exposure Limits
  C 50 ppm; C 301 mg/m3

Mexico - Instruction No. 10 - TWAs
  50 ppm TWA; 300 mg/m3 TWA

Mexico - Wastewater - Organic Toxic Pollutants and Heavy Metals
  Listed under [Dichlorobenzenes]

STATE LISTS

California - Air Bill 2588 Appendix A-I
  6/91

California - Exposure Limits - Ceilings
  C 50 ppm; C 300 mg/m3

California - Directors List of Hazardous Substances (8 CCR 339)
  [present]

Florida Hazardous Substance List
  [present]

Massachusetts Right To Know List
  [present]

Minnesota Hazardous Substance List
  [present]

NJ Right to Know List (Total)
  sn 0642

Pennsylvania Right to Know List
  environmental hazard

P-DICHLOROBENZENE   106-46-7
SEE ALSO:
F039-HAZARDOUS WASTES
N-(2-PHENOXYETHYL) ANILINE
F002-HAZARDOUS WASTES
K150-HAZARDOUS WASTES
COCONUT OIL, POLYMER WITH ISOPHTHALIC ACID, TRIMELITTIC ANHY
K149-HAZARDOUS WASTES

HEALTH AND SAFETY LISTS

ACGIH 1995 - Time Weighted Averages
  10 ppm TWA; 60 mg/m3 TWA

ACGIH 1995 - Carcinogens
  A3-animal carcinogen

AIHA - Odor Threshold Values
  geometric mean air odor threshold = 0.12 ppm (detectable)

U.S. DOT - Substances From 49 CFR 172.101
  regulated by DOT (UN1592)

U.S. DOT - Hazard Classes
  DOT hazard class = 6.1

U.S. DOT - Appendix B - Marine Pollutants
  DOT regulated marine pollutant

U.S. DOT - Appendix A Table 1 - Hazardous Substances
  final RQ = 100 pounds (45.4 kg)

IARC - Group 2B (sufficient animal data)
  [present]

NFPA - Flash Points
  flash point = 150 degrees F (66 degrees C)

NFPA - Hazard Identification Ratings
  health-2; flammability-2; reactivity-0

NIOSH - Selected LD50s and LC50s
  Oral, rat: LD50 = 500 mg/kg

NIOSH 1990 - Pocket Guide - IDLHs
  1000 ppm IDLH (not considering carcinogenic effects)

NIOSH 1990 - Pocket Guide - Carcinogens
  occupational carcinogen

NIOSH 1990 - Pocket Guide - Target organs
  liver, respiratory system, eyes, kidneys, skin

NTP Chemical Status Reports - Testing Status and NTIS Number
  Technical reports printed (PB87208617/AS)

NTP Chemical Status Reports - Evidence of Carcinogenicity
  male rat-clear evidence; female rat-no evidence; male mice-clear evidence; female mice-clear evidence

NTP Seventh Report - Suspect Carcinogens
  suspect carcinogen

OSHA - Vacated PELs - Time Weighted Averages
  75 ppm TWA; 450 mg/m3 TWA

OSHA - Vacated PELs - Short Term Exposure Limits
  110 ppm STEL; 675 mg/m3 STEL

OSHA - Final PELs - Time Weighted Averages
  75 ppm TWA; 450 mg/m3 TWA

OSHA - Possible Select Carcinogens
  [present]

ENVIRONMENTAL LISTS

ATSDR Priority List
  Rank (of 275): 142

CERCLA/SARA - Section 313 - Emission Reporting
  form R reporting required for 0.1% de minimus concentration

CERCLA/SARA - Hazardous Substances and their Reportable Quantities
  final RQ = 100 pounds (45.4 kg)

Clean Air Act (1990) - List of Hazardous Air Contaminants
  [present]

CAA - HON Rule - SOCMI Chemicals
  compliance by Oct. 24, 1995

CAA - HON Rule - Organic HAPs
  [present]

Clean Water Act - Hazardous Substances
  [present] (Listed under 'Dichlorobenzene')

Clean Water Act - Priority Pollutants
  [present]

Clean Water Act - Toxic Pollutants
  [present] (Listed under 'Dichlorobenzenes')

Safe Drinking Water Act - MCLs
  MCL = 0.075 mg/L

Safe Drinking Water Act - MCLGs
  MCLG = 0.075 mg/L

RCRA - D Series - Maximum Concentration of Contaminants
  waste number D027; regulatory level = 7.5 mg/L

RCRA - D Series - Chronic Toxicity Reference Levels
  chronic toxicity reference level = 0.075 mg/L

RCRA - U Series Wastes
  waste number U072

RCRA - Hazardous Constituents-Appendix VIII
  waste number U072

RCRA - Basis for Listing - Appendix VII
  Included in waste streams: F039, K149, K150

RCRA - Substances Banned From Land Disposal
  [present]

RCRA - TSD Facilities Ground Water Monitoring
  TM 8010 = 2 ug/L PQL; TM 8020 = 5 ug/L PQL; TM 8120 = 15 ug/L PQL; TM 8270 = 10 ug/L PQL

RCRA - Universal Treatment Standards (LDR)
  WW: 0.090 mg/l; NWW: 6.0 mg/kg

TSCA - Code of Federal Regulations Citations
  40 CFR 712.30(d); 40 CFR 716.120(c); 40 CFR 799.1052

TSCA - PAIR - Reporting List
  Reporting Date: November 19, 1982

TSCA - HDD/HDF - Precursors Required for Reporting
  [present]

TSCA - Chemical Test Rules
  Testing required by manufacturers; processors (40 CFR 799.1052) (Listed under 'Dichlorobenzenes')

TSCA - Section 12(b) - Export Notification
export notification required - Section 4

INTERNATIONAL LISTS

Australian Exposure Standards - Time Weighted Averages
75 ppm TWA; 451 mg/m3 TWA

Australian Exposure Standards - Short Term Exposure Limits
110 ppm STEL; 661 mg/m3 STEL

Canada - WHMIS: Ingredient Disclosure
1% item 537 (656)

Canada - NPRI (National Pollutant Release Inventory)
[present]

Canada - CEPA - Priority Substances List
estimated time for completion of assessment reports: 4 years

Canada - Drinking Water Quality - MACs
0.005 mg/L MAC

Canada - Drinking Water Quality - AOs
<= 0.001 mg/L AO

Canada - Alberta - 8 Hour Occupational Exposure Limit
75 ppm TWA; 450 mg/m3 TWA

Canada - Alberta - 15 Minute Occupational Exposure Limit
110 ppm STEL; 660 mg/m3 STEL

Canada - British Columbia - 8 Hour Exposure Limits
75 ppm TWA; 450 mg/m3 TWA

Canada - British Columbia - 15 Minute Exposure Limits
110 ppm STEL; 675 mg/m3 STEL

Canada - Ontario - OHSA - TWAEVs
75 ppm TWAEV; 450 mg/m3 TWAEV

Canada - Ontario - OHSA - STEVs
110 ppm STEV; 660 mg/m3 STEV

Canada - Quebec - Time-Weighted Average Exposure Values
75 ppm TWAEV; 450 mg/m3 TWAEV

Canada - Quebec - Short-term Exposure Values
110 ppm STEV; 660 mg/m3 STEV

Canada - Quebec - Carcinogens
C3 carcinogen: effect detected in animals

United Kingdom - Occupational Exposure Standards - TWAs
75 ppm TWA; 450 mg/m3 TWA

United Kingdom - Occupational Exposure Standards - STELs
110 ppm STEL; 675 mg/m3 STEL

German (DFG) - MAK Values
50 ppm MAK; 300 mg/m3 MAK

German (DFG) - Peak Limitations
2 x normal MAK (30 min. average value); don't exceed 4 times during shift

German (DFG) - Pregnancy
no risk to embryo/fetus if exposure limits adhered to

Israel - Time Weighted Averages
75 ppm TWA; 451 mg/m3 TWA

Israel - Short Term Exposure Limits
110 ppm STEL; 661 mg/m3 STEL

Israel - Action Levels
37.5 ppm AL; 225.5 mg/m3 AL

Mexico - Instruction No. 10 - TWAs
75 ppm TWA; 450 mg/m3 TWA

Mexico - Instruction No. 10 - STELs
110 ppm STEL; 675 mg/m3 STEL

Mexico - Wastewater - Organic Toxic Pollutants and Heavy Metals
Listed under [Dichlorobenzenes]

STATE LISTS

California - Air Bill 2588 Appendix A-I
known or potential carcinogen

California - Prop. 65 - Cancer list
carcinogen - initial date 1/1/89

California - Prop. 65 - No Significant Risk Levels
no significant risk level = 20 ug/day

California - Exposure Limits - PELs
75 ppm PEL; 450 mg/m3 PEL

California - Exposure Limits - STELs
110 ppm STEL; 675 mg/m3 STEL

California - Exposure Limits - Ceilings
C 200 ppm

California - Directors List of Hazardous Substances (8 CCR 339)
[present]

Florida Hazardous Substance List
[present]

Massachusetts Right To Know List
[present]

Minnesota Hazardous Substance List
carcinogen

NJ Right to Know List (Total)
sn 0643

Pennsylvania Right to Know List
environmental hazard; special hazardous substance

Pennsylvania RTK - Special Hazardous Substances
[present]

1,3-DICHLOROBENZENE　　　　　　　　　　541-73-1
SEE ALSO:
F039-HAZARDOUS WASTES

HEALTH AND SAFETY LISTS

U.S. DOT - Appendix B - Marine Pollutants
DOT regulated marine pollutant

U.S. DOT - Appendix A Table 1 - Hazardous Substances
final RQ = 100 pounds (45.4 kg)

ENVIRONMENTAL LISTS

ATSDR Priority List
Rank (of 275): 195

CERCLA/SARA - Section 313 - Emission Reporting
form R reporting required for 1.0% de minimus concentration

CERCLA/SARA - Hazardous Substances and their Reportable Quantities
final RQ = 100 pounds (45.4 kg)

CAA - HON Rule - SOCMI Chemicals
compliance by Oct. 24, 1994

Clean Water Act - Priority Pollutants
[present]

Clean Water Act - Toxic Pollutants
[present] (Listed under 'Dichlorobenzenes')

Safe Drinking Water Act - Monitoring
monitoring required

RCRA - U Series Wastes
waste number U071

RCRA - Hazardous Constituents-Appendix VIII
waste number U071

RCRA - Basis for Listing - Appendix VII
Included in waste stream: F039

RCRA - Substances Banned From Land Disposal
[present]

RCRA - TSD Facilities Ground Water Monitoring
TM 8010 = 5 ug/L PQL; TM 8020 = 5 ug/L PQL; TM 8120 = 10 ug/L PQL; TM 8270 = 10 ug/L PQL

RCRA - Universal Treatment Standards (LDR)
WW: 0.036 mg/l; NWW: 6.0 mg/kg

TSCA - Code of Federal Regulations Citations
40 CFR 712.30(d); 40 CFR 716.120(c)

TSCA - PAIR - Reporting List
Reporting Date: November 19, 1982

INTERNATIONAL LISTS
  Canada - WHMIS: Ingredient Disclosure
    1% item 535 (654)
  Mexico - Wastewater - Organic Toxic Pollutants and Heavy Metals
    Listed under [Dichlorobenzenes]
STATE LISTS
  California - Air Bill 2588 Appendix A-I
    6/91
  California - Directors List of Hazardous Substances (8 CCR 339)
    [present]
  Massachusetts Right To Know List
    [present]
  NJ Right to Know List (Total)
    sn 2301
  Pennsylvania Right to Know List
    environmental hazard
PROPOSED REGULATIONS
  Safe Drinking Water Act - Priority list
    [present]

DICHLOROBENZENE (MIXED ISOMERS)   **25321-22-6**
SEE ALSO:
  K105-HAZARDOUS WASTES
  F025-HAZARDOUS WASTES
  F024-HAZARDOUS WASTES
  K085-HAZARDOUS WASTES
HEALTH AND SAFETY LISTS
  U.S. DOT - Appendix B - Marine Pollutants
    DOT regulated marine pollutant
  U.S. DOT - Appendix A Table 1 - Hazardous Substances
    final RQ = 100 pounds (45.4 kg)
ENVIRONMENTAL LISTS
  ATSDR Priority List
    Rank (of 275): 203
  CERCLA/SARA - Section 313 - Emission Reporting
    form R reporting required for 0.1% de minimus concentration
  CERCLA/SARA - Hazardous Substances and their Reportable Quantities
    final RQ = 100 pounds (45.4 kg)
  Clean Water Act - Hazardous Substances
    [present] (Listed under 'Dichlorobenzene')
  RCRA - Hazardous Constituents-Appendix VIII
    hazardous constituent - no waste number
  RCRA - Basis for Listing - Appendix VII
    Included in waste streams: F024, F025, K085, K105
INTERNATIONAL LISTS
  Mexico - Drinking Water - Ecological Criteria
    0.4 mg/l
STATE LISTS
  California - Air Bill 2588 Appendix A-I
    6/91
  Massachusetts Right To Know List
    [present]
  NJ Right to Know List (Total)
    sn 2321
  Pennsylvania Right to Know List
    environmental hazard

3,3'-DICHLOROBENZIDINE   **91-94-1**
HEALTH AND SAFETY LISTS
  ACGIH 1995 - Skin Designations
    skin - potential for cutaneous absorption

ACGIH 1995 - Carcinogens
  A2-suspected human carcinogen
U.S. DOT - Appendix A Table 1 - Hazardous Substances
  final RQ = 1 pound (0.454 kg)
IARC - Group 2B (sufficient animal data)
  [present]
NIOSH - Selected LD50s and LC50s
  Oral, rat: LD50 = 5250 mg/kg
NIOSH 1990 - Pocket Guide - Carcinogens
  occupational carcinogen
NIOSH 1990 - Pocket Guide - Target organs
  bladder, liver, lungs, skin, GI tract
NIOSH - Health Standards - Exposure Limits
  use 29 CFR 1910.1007
NIOSH - Health Standards - Health Effects and Precautions
  has produced tumors of the liver, bladder, and lungs in animals
NIOSH - Health Standards - Carcinogenic Chemicals
  potential human carcinogen
NTP Seventh Report - Suspect Carcinogens
  suspect carcinogen
OSHA - 29 CFR 1910 Specifically Regulated Chemicals
  Cancer suspect agent (see 29 CFR 1910.1007) (includes salts of 3,3'-Dichlorobenzidine)
OSHA - Select Carcinogens
  [present]
OSHA - Possible Select Carcinogens
  [present]
ENVIRONMENTAL LISTS
  ATSDR Priority List
    Rank (of 275): 058
  CERCLA/SARA - Section 313 - Emission Reporting
    form R reporting required for 0.1% de minimus concentration
  CERCLA/SARA - Hazardous Substances and their Reportable Quantities
    final RQ = 1 pound (0.454 kg)
  Clean Air Act (1990) - List of Hazardous Air Contaminants
    [present]
  CAA - HON Rule - SOCMI Chemicals
    compliance by Oct. 24, 1994
  CAA - HON Rule - Organic HAPs
    [present]
  Clean Water Act - Priority Pollutants
    [present]
  Clean Water Act - Toxic Pollutants
    [present]
  EPA - Carcinogen Hazard Ranking for RQ Adjustment
    Hazard ranking = Medium
  RCRA - U Series Wastes
    waste number U073
  RCRA - Hazardous Constituents-Appendix VIII
    waste number U073
  RCRA - Substances Banned From Land Disposal
    [present]
  RCRA - TSD Facilities Ground Water Monitoring
    TM 8270 = 20 ug/L PQL
INTERNATIONAL LISTS
  Australian Exposure Standards - Time Weighted Averages
    control to the lowest practical level
  Australian Exposure Standards - Skin Effects
    skin absorption
  Australian Exposure Standards - Carcinogens
    probable carcinogen
  Canada - WHMIS: Ingredient Disclosure
    0.1% item 538 (657)

Canada - CEPA - Priority Substances List
estimated time for completion of assessment reports: 5 years

Canada - Alberta - Skin Designation
can be absorbed through the intact skin

Canada - Alberta - Designated Substances
designated substance - requires code of practice

Canada - British Columbia - 8 Hour Exposure Limits
carcinogen with no permitted exposure or contact by any route

Canada - British Columbia - Skin Notations
skin - potential for skin absorption

Canada - British Columbia - Carcinogens
carcinogen with no permitted exposure or contact by any route

Canada - Quebec - Time-Weighted Average Exposure Values
substance of which the recirculation is prohibited

Canada - Quebec - Skin Designations
absorbed through the skin

Canada - Quebec - Carcinogens
C2 carcinogen: effect suspected in humans

German (DFG) - Skin/Sensitizers
danger of cutaneous absorption

German (DFG) - Carcinogens
animal evidence of carcinogenicity

Mexico - Wastewater - Organic Toxic Pollutants and Heavy Metals
Listed under [Dichlorobenzidines]

STATE LISTS

California - Air Bill 2588 Appendix A-I
known or potential carcinogen

California - Prop. 65 - Cancer list
carcinogen - initial date 10/1/87

California - Prop. 65 - No Significant Risk Levels
no significant risk level = 0.6 ug/day

California - Exposure Limits - Skin Notation
material may be absorbed through the skin, eyes or mucous membrane

California - Exposure Limits - Carcinogens
cancer-suspect agent (at a concentration >= 1.0%) (includes 3,3'-dichlorobenzidine salts)

California - Directors List of Hazardous Substances (8 CCR 339)
[present] (includes its salts)

Florida Hazardous Substance List
[present]

Massachusetts Right To Know List
carcinogen; extraordinarily hazardous

Minnesota Hazardous Substance List
skin (includes salts)

NJ Right to Know List (Total)
sn 0644

NJ Special Hazardous Substances
(carcinogen; mutagen)

Pennsylvania Right to Know List
environmental hazard; special hazardous substance

Pennsylvania RTK - Special Hazardous Substances
[present]

3,3'-DICHLOROBENZIDINE DIHYDROCHLORIDE 612-83-9

HEALTH AND SAFETY LISTS

NTP Seventh Report - Suspect Carcinogens
suspect carcinogen

OSHA - Possible Select Carcinogens
[present]

ENVIRONMENTAL LISTS

CERCLA/SARA - Section 313 - Emission Reporting
form R reporting required

3,3'-DICHLOROBENZIDINE SULFATE 64969-34-2

ENVIRONMENTAL LISTS

CERCLA/SARA - Section 313 - Emission Reporting
form R reporting required

5,6-DICHLORO-2-BENZOTHIAZOLAMINE 24072-75-1

HEALTH AND SAFETY LISTS

NTP Chemical Status Reports - Testing Status and NTIS Number
Prechronic studies for which toxicity technical reports were not prepared

3,4-DICHLOROBENZOTRIFLUORIDE 328-84-7

ENVIRONMENTAL LISTS

TSCA - Code of Federal Regulations Citations
40 CFR 712.30(i); 40 CFR 716.120(a); 40 CFR 799.5000

TSCA - PAIR - Reporting List
Reporting Date: July 8, 1985

TSCA - Health and Safety Reporting List
Effective Date: May 8, 1985; Sunset Date: November 9, 1993

TSCA - Substances Subject to Testing Consent Orders
Test for: Environmental Effects and Chemical Fate

TSCA - Section 12(b) - Export Notification
export notification required - Section 4

2,4-DICHLOROBENZOYL PEROXIDE 133-14-2

HEALTH AND SAFETY LISTS

U.S. DOT - Hazard Classes
Forbidden from transport by the DOT

STATE LISTS

NJ Right to Know List (Total)
sn 0645

DI-4-CHLOROBENZOYL PEROXIDE RR-01753-8

HEALTH AND SAFETY LISTS

U.S. DOT - Organic Peroxides Table
Organic peroxide UN3102; UN3106; exempt

DICHLOROBROMOETHANE 73506-91-9

ENVIRONMENTAL LISTS

ATSDR Priority List
Rank (of 275): 259

DICHLOROBROMOMETHANE 75-27-4
SEE ALSO:
F039-HAZARDOUS WASTES

HEALTH AND SAFETY LISTS

U.S. DOT - Appendix A Table 1 - Hazardous Substances
final RQ = 5000 pounds (2270 kg)

IARC - Group 2B (sufficient animal data)
[present]

NIOSH - Selected LD50s and LC50s
Oral, rat: LD50 = 916 mg/kg

NTP Chemical Status Reports - Testing Status and NTIS Number
Technical reports printed (PB88168687)

NTP Chemical Status Reports - Evidence of Carcinogenicity
male rat-clear evidence; female rat-clear evidence; male mice-clear evidence; female mice-clear evidence

NTP Seventh Report - Suspect Carcinogens
suspect carcinogen

OSHA - Possible Select Carcinogens
[present]

ENVIRONMENTAL LISTS

CERCLA/SARA - Section 313 - Emission Reporting
form R reporting required for 1.0% de minimus concentration

CERCLA/SARA - Hazardous Substances and their Reportable Quantities
final RQ = 5000 pounds (2270 kg)

Clean Water Act - Priority Pollutants
[present]

Safe Drinking Water Act - Monitoring
monitoring required

RCRA - Basis for Listing - Appendix VII
Included in waste stream: F039

RCRA - TSD Facilities Ground Water Monitoring
TM 8010 = 1 ug/L PQL;   TM 8240 = 5 ug/L PQL

RCRA - Universal Treatment Standards (LDR)
WW: 0.35 mg/l;   NWW: 15 mg/kg

TSCA - Code of Federal Regulations Citations
40 CFR 716.120(a)

TSCA - Health and Safety Reporting List
Effective Date: June 1, 1987

INTERNATIONAL LISTS

Canada - WHMIS: Ingredient Disclosure
1% item 219 (325)

Mexico - Wastewater - Organic Toxic Pollutants and Heavy Metals
Listed under [Halomethanes]

STATE LISTS

California - Air Bill 2588 Appendix A-II
known or potential carcinogen: 9/90

California - Prop. 65 - Cancer list
carcinogen - initial date 1/1/90

California - Prop. 65 - No Significant Risk Levels
no significant risk level = 5 ug/day

California - Directors List of Hazardous Substances (8 CCR 339)
[present]

Massachusetts Right To Know List
[present]

NJ Right to Know List (Total)
sn 2894

Pennsylvania Right to Know List
environmental hazard

PROPOSED REGULATIONS

Safe Drinking Water Act - Priority list
[present]

2,3-DICHLOROBUTADIENE-1,3   1653-19-6

HEALTH AND SAFETY LISTS

NFPA - Flash Points
flash point = 50 degrees F (10 degrees C)

NFPA - Hazard Identification Ratings
health-3;   flammability-3;   reactivity-2

STATE LISTS

Florida Hazardous Substance List
[present]

Massachusetts Right To Know List
[present]

Pennsylvania Right to Know List
[present]

1,4-DICHLOROBUTANE   110-56-5

HEALTH AND SAFETY LISTS

NFPA - Flash Points
flash point = 126 degrees F (52 degrees C)

NFPA - Hazard Identification Ratings
health-3;   flammability-2;   reactivity-0

STATE LISTS

Florida Hazardous Substance List
[present]

Massachusetts Right To Know List
[present]

Pennsylvania Right to Know List
[present]

1,2-DICHLOROBUTANE   616-21-

HEALTH AND SAFETY LISTS

NFPA - Hazard Identification Ratings
health-2;   flammability-2;   reactivity-0

ENVIRONMENTAL LISTS

TSCA - Code of Federal Regulations Citations
40 CFR 712.30(x), 40 CFR 716.120(d)

TSCA - PAIR - Reporting List
Reporting Date: November 27, 1991

TSCA - Health and Safety Reporting List
Effective Date: September 30, 1991

STATE LISTS

Florida Hazardous Substance List
[present]

Massachusetts Right To Know List
[present]

Pennsylvania Right to Know List
[present]

2,3-DICHLOROBUTANE   7581-97-7

HEALTH AND SAFETY LISTS

NFPA - Flash Points
flash point = 194 degrees F (90 degrees C)

NFPA - Hazard Identification Ratings
health-2;   flammability-2;   reactivity-0

STATE LISTS

Florida Hazardous Substance List
[present]

Massachusetts Right To Know List
[present]

Pennsylvania Right to Know List
[present]

TRANS-1,4-DICHLOROBUTENE   110-57-6

HEALTH AND SAFETY LISTS

IARC - Group 3 (not classifiable)
[present]

NIOSH - Selected LD50s and LC50s
Inhalation, rat: LC50 = 86 ppm 4 hr

ENVIRONMENTAL LISTS

CERCLA/SARA - Section 302 Extremely Hazardous Substances and TPQs
TPQ = 500 pounds

CERCLA/SARA - Section 313 - Emission Reporting
form R reporting required

RCRA - TSD Facilities Ground Water Monitoring
TM 8240 = 5 ug/L PQL

STATE LISTS

Florida Hazardous Substance List
effective March 13, 1992

Massachusetts Right To Know List
extraordinarily hazardous

NJ Right to Know List (Total)
sn 2829

Pennsylvania Right to Know List
environmental hazard

PROPOSED REGULATIONS

CERCLA/SARA - Proposed Hazardous Substance Additions
proposed RQ = 1 pound (.454 kg)

CERCLA/SARA - 1989 Proposed RQ Adjustments
proposed RQ = 100 pounds (45.4 kg)

3,4-DICHLORO-1-BUTENE   760-23-6

ENVIRONMENTAL LISTS

CAA - HON Rule - SOCMI Chemicals
compliance by Jan. 23, 1995

TSCA - Code of Federal Regulations Citations
40 CFR 712.30(x);   40 CFR 712.120(d)

TSCA - PAIR - Reporting List
Reporting Date: November 27, 1991

TSCA - Health and Safety Reporting List
Effective Date: September 30, 1991

INTERNATIONAL LISTS

Canada - WHMIS: Ingredient Disclosure
0.1% item 542 (661)

1,4-DICHLORO-2-BUTENE   764-41-0

HEALTH AND SAFETY LISTS

ACGIH 1995 - Time Weighted Averages
0.005 ppm TWA;   0.025 mg/m3 TWA

ACGIH 1995 - Skin Designations
skin - potential for cutaneous absorption

ACGIH 1995 - Carcinogens
A2-suspected human carcinogen

U.S. DOT - Appendix A Table 1 - Hazardous Substances
final RQ = 1 pound (0.454 kg)

NIOSH - Selected LD50s and LC50s
Inhalation, mouse: LC50 = 920 mg/m3 (8 hr) Oral, rat: LD50 = 89 mg/kg Skin, rabbit: LD50 = 620 mg/kg

ENVIRONMENTAL LISTS

CERCLA/SARA - Section 313 - Emission Reporting
form R reporting required

CERCLA/SARA - Hazardous Substances and their Reportable Quantities
final RQ = 1 pound (0.454 kg)

CAA - HON Rule - SOCMI Chemicals
compliance by Oct. 23, 1995

RCRA - U Series Wastes
waste number U074 (Ignitable waste;   Toxic waste)

RCRA - Hazardous Constituents-Appendix VIII
waste number U074

RCRA - Substances Banned From Land Disposal
[present]

INTERNATIONAL LISTS

Canada - WHMIS: Ingredient Disclosure
1% item 541 (660)

German (DFG) - Carcinogens
animal evidence of carcinogenicity

STATE LISTS

California - Air Bill 2588 Appendix A-II
known or potential carcinogen: 9/90

California - Prop. 65 - Cancer list
carcinogen - initial date 1/1/90

Massachusetts Right To Know List
[present]

NJ Right to Know List (Total)
sn 3070

Pennsylvania Right to Know List
environmental hazard

1,3-DICHLOROBUTENE-2   926-57-8

HEALTH AND SAFETY LISTS

NFPA - Flash Points
flash point = 80 degrees F (27 degrees C)

NFPA - Hazard Identification Ratings
health-2;   flammability-3;   reactivity-0

NIOSH - Selected LD50s and LC50s
Inhalation, rat: LC50 = 3930 mg/m3 4 hr

STATE LISTS

Florida Hazardous Substance List
[present]

Massachusetts Right To Know List
[present]

Pennsylvania Right to Know List
[present]

1,3-DICHLORO-2-BUTENE   7415-31-8

HEALTH AND SAFETY LISTS

NFPA - Flash Points
flash point = 80 degrees F (27 degrees C)

NFPA - Hazard Identification Ratings
health-3;   flammability-3;   reactivity-2

STATE LISTS

Florida Hazardous Substance List
[present]

Massachusetts Right To Know List
[present]

DICHLOROBUTENE   11069-19-5

HEALTH AND SAFETY LISTS

U.S. DOT - Substances From 49 CFR 172.101
regulated by DOT (NA2920)

U.S. DOT - Hazard Classes
DOT hazard class = 8

INTERNATIONAL LISTS

Canada - WHMIS: Ingredient Disclosure
1% item 540 (659)

STATE LISTS

NJ Right to Know List (Total)
sn 0647

NJ Special Hazardous Substances
(corrosive)

3,4-DICHLOROBUTENE-1   64037-54-3

HEALTH AND SAFETY LISTS

NFPA - Flash Points
flash point = 113 degrees F (45 degrees C)

NFPA - Hazard Identification Ratings
health-3;   flammability-2;   reactivity-1

STATE LISTS

Florida Hazardous Substance List
[present]

Massachusetts Right To Know List
[present]

Pennsylvania Right to Know List
[present]

3,3'-DICHLORO-4,4'-DIAMINODIPHENYL ETHER   28434-86-8

HEALTH AND SAFETY LISTS

IARC - Group 2B (sufficient animal data)
[present] (Overall evaluation based only on evidence of carcinogenicity in monograph (16, 1978) or in Supplement 4)

OSHA - Possible Select Carcinogens
[present]

STATE LISTS

California - Air Bill 2588 Appendix A-II
known or potential carcinogen: 9/89

California - Prop. 65 - Cancer list
carcinogen - initial date 1/1/88

California - Directors List of Hazardous Substances (8 CCR 339)
[present]

Florida Hazardous Substance List
[present]

Massachusetts Right To Know List
carcinogen; extraordinarily hazardous

Minnesota Hazardous Substance List
carcinogen

Pennsylvania Right to Know List
special hazardous substance

Pennsylvania RTK - Special Hazardous Substances
[present]

2,7-DICHLORODIBENZO-P-DIOXIN   33857-26-0

HEALTH AND SAFETY LISTS

NTP Chemical Status Reports - Testing Status and NTIS Number
Technical reports printed (PB290570/AS)

NTP Chemical Status Reports - Evidence of Carcinogenicity
male rat-negative; female rat-negative; male mice-equivocal; female mice-negative

STATE LISTS

Massachusetts Right To Know List
[present]

1,2-DICHLORO-1,1-DIFLUOROETHANE   1649-08-7

HEALTH AND SAFETY LISTS

NTP Chemical Status Reports - Testing Status and NTIS Number
Prechronic studies completed: in review for further evaluation

ENVIRONMENTAL LISTS

CERCLA/SARA - Section 313 - Emission Reporting
form R reporting required

Class 2 Ozone Depletors
ozone depletion weight reserved

TSCA - Code of Federal Regulations Citations
40 CFR 716.120(c)

TSCA - Health and Safety Reporting List
Effective Date: October 15, 1990

TSCA - Section 12(b) - Export Notification
export notification required - Section 5

DICHLORODIFLUOROETHYLENE   27156-03-2

STATE LISTS

NJ Right to Know List (Total)
sn 0648

DICHLORODIFLUOROMETHANE   75-71-8
SEE ALSO:
F039-HAZARDOUS WASTES

HEALTH AND SAFETY LISTS

ACGIH 1995 - Time Weighted Averages
1000 ppm TWA; 4950 mg/m3 TWA

U.S. DOT - Substances From 49 CFR 172.101
regulated by DOT (UN1028)

U.S. DOT - Hazard Classes
DOT hazard class = 2.2

U.S. DOT - Appendix A Table 1 - Hazardous Substances
final RQ = 5000 pounds (2270 kg)

NIOSH - Selected LD50s and LC50s
Inhalation, rat: LC50 = 80 pph 30 mn

NIOSH 1990 - Pocket Guide - RELs
1000 ppm TWA; 4950 mg/m3 TWA

NIOSH 1990 - Pocket Guide - IDLHs
50,000 ppm IDLH

NIOSH 1990 - Pocket Guide - Target organs
CVS, PNS

OSHA - Vacated PELs - Time Weighted Averages
1000 ppm TWA; 4950 mg/m3 TWA

OSHA - Final PELs - Time Weighted Averages
1000 ppm TWA; 4950 mg/m3 TWA

ENVIRONMENTAL LISTS

CERCLA/SARA - Section 313 - Emission Reporting
form R reporting required for 1.0% de minimus concentration

CERCLA/SARA - Hazardous Substances and their Reportable Quantities
final RQ = 5000 pounds (2270 kg)

Class 1 Ozone Depletors
ozone depletion potential = 1.0

CAA - HON Rule - SOCMI Chemicals
compliance by Oct. 24, 1994

Safe Drinking Water Act - Monitoring
monitoring required at discretion of the state

List of Pesticide Product Inert Ingredients
[present]

RCRA - U Series Wastes
waste number U075

RCRA - Hazardous Constituents-Appendix VIII
waste number U075

RCRA - Basis for Listing - Appendix VII
Included in waste stream: F039

RCRA - Substances Banned From Land Disposal
[present]

RCRA - TSD Facilities Ground Water Monitoring
TM 8010 = 10 ug/L PQL;   TM 8240 = 5 ug/L PQL

RCRA - Universal Treatment Standards (LDR)
WW: 0.23 mg/l; NWW: 7.2 mg/kg

INTERNATIONAL LISTS

Australian Exposure Standards - Time Weighted Averages
1000 ppm TWA; 4950 mg/m3 TWA

Canada - WHMIS: Ingredient Disclosure
1% item 543 (662)

Canada - Alberta - 8 Hour Occupational Exposure Limit
1000 ppm TWA; 4950 mg/m3 TWA

Canada - Alberta - 15 Minute Occupational Exposure Limit
1250 ppm STEL; 6190 mg/m3 STEL

Canada - British Columbia - 8 Hour Exposure Limits
1000 ppm TWA; 4950 mg/m3 TWA

Canada - British Columbia - 15 Minute Exposure Limits
1250 ppm STEL; 6200 mg/m3 STEL

Canada - Ontario - OHSA - TWAEVs
1000 ppm TWAEV; 4940 mg/m3 TWAEV

Canada - Quebec - Time-Weighted Average Exposure Values
1000 ppm TWAEV; 4950 mg/m3 TWAEV

United Kingdom - Occupational Exposure Standards - TWAs
1000 ppm TWA; 4950 mg/m3 TWA

United Kingdom - Occupational Exposure Standards - STELs
   1250 ppm STEL; 6200 mg/m3 STEL
German (DFG) - MAK Values
   1000 ppm MAK; 5000 mg/m3 MAK
German (DFG) - Peak Limitations
   2 x normal MAK (1 hour momentary value);   don't exceed 3 times per shift
German (DFG) - Pregnancy
   no risk to embryo/fetus if exposure limits adhered to
Israel - Time Weighted Averages
   1000 ppm TWA; 4950 mg/m3 TWA
Israel - Action Levels
   500 ppm AL; 2475 mg/m3 AL
Mexico - Wastewater - Organic Toxic Pollutants and Heavy Metals
   Listed under [Halomethanes]

STATE LISTS

California - Exposure Limits - PELs
   1000 ppm PEL; 4950 mg/m3 PEL
California - Exposure Limits - Ceilings
   C 6200 ppm
California - Directors List of Hazardous Substances (8 CCR 339)
   [present]
Florida Hazardous Substance List
   [present]
Massachusetts Right To Know List
   [present]
Minnesota Hazardous Substance List
   [present]
NJ Right to Know List (Total)
   sn 0649
Pennsylvania Right to Know List
   environmental hazard

PROPOSED REGULATIONS

Safe Drinking Water Act - Priority list
   [present]
Canada - Ontario - Proposed Occupational TWAEVs
   500 ppm TWAEV; 2500 mg/m3 TWAEV
Canada - Ontario - Proposed Occupational STEVs
   750 ppm STEV; 4000 mg/m3 STEV

DICHLORODIFLUOROPROPANE   134190-52-6

ENVIRONMENTAL LISTS

Class 2 Ozone Depletors
   ozone depletion weight reserved

1,3-DICHLORO-5,5-DIMETHYL HYDANTOIN   118-52-5

HEALTH AND SAFETY LISTS

ACGIH 1995 - Time Weighted Averages
   0.2 mg/m3 TWA
ACGIH 1995 - Short Term Exposure Limits
   0.4 mg/m3 STEL
NIOSH - Selected LD50s and LC50s
   Oral, rat: LD50 = 542 mg/kg
NIOSH 1990 - Pocket Guide - RELs
   0.2 mg/m3 TWA; 0.4 mg/m3 STEL
NIOSH 1990 - Pocket Guide - Target organs
   eyes, respiratory system
OSHA - Vacated PELs - Time Weighted Averages
   0.2 mg/m3 TWA
OSHA - Vacated PELs - Short Term Exposure Limits
   0.4 mg/m3 STEL
OSHA - Final PELs - Time Weighted Averages
   0.2 mg/m3 TWA

INTERNATIONAL LISTS

Australian Exposure Standards - Time Weighted Averages
   0.2 mg/m3 TWA
Australian Exposure Standards - Short Term Exposure Limits
   0.4 mg/m3 STEL
Canada - WHMIS: Ingredient Disclosure
   1% item 544 (663)
Canada - Alberta - 8 Hour Occupational Exposure Limit
   0.2 mg/m3 TWA
Canada - Alberta - 15 Minute Occupational Exposure Limit
   0.4 mg/m3 STEL
Canada - British Columbia - 8 Hour Exposure Limits
   0.2 mg/m3 TWA
Canada - British Columbia - 15 Minute Exposure Limits
   0.4 mg/m3 STEL
Canada - Ontario - OHSA - TWAEVs
   0.2 mg/m3 TWAEV
Canada - Ontario - OHSA - STEVs
   0.4 mg/m3 STEV
Canada - Quebec - Time-Weighted Average Exposure Values
   0.2 mg/m3 TWAEV
Canada - Quebec - Short-term Exposure Values
   0.4 mg/m3 STEV
United Kingdom - Occupational Exposure Standards - TWAs
   0.2 mg/m3 TWA
United Kingdom - Occupational Exposure Standards - STELs
   0.4 mg/m3 STEL
Israel - Time Weighted Averages
   0.2 mg/m3 TWA
Israel - Short Term Exposure Limits
   0.4 mg/m3 STEL
Israel - Action Levels
   0.1 mg/m3 AL
Mexico - Instruction No. 10 - TWAs
   0.2 mg/m3 TWA
Mexico - Instruction No. 10 - STELs
   0.4 mg/m3 STEL

STATE LISTS

California - Exposure Limits - PELs
   0.2 mg/m3 PEL
California - Exposure Limits - STELs
   0.4 mg/m3 STEL
California - Directors List of Hazardous Substances (8 CCR 339)
   [present]
Florida Hazardous Substance List
   [present]
Massachusetts Right To Know List
   [present]
Minnesota Hazardous Substance List
   [present]
Pennsylvania Right to Know List
   [present]

1,1-DICHLOROETHANE 75-34-3

SEE ALSO:
K030-HAZARDOUS WASTES
BIS(2,4-DIMETHYLBUTYL) MALEATE
F025-HAZARDOUS WASTES
F024-HAZARDOUS WASTES
F039-HAZARDOUS WASTES
K020-HAZARDOUS WASTES
K019-HAZARDOUS WASTES

HEALTH AND SAFETY LISTS

ACGIH 1995 - Time Weighted Averages
100 ppm TWA; 405 mg/m3 TWA

AIHA - Odor Threshold Values
no geometric mean air odor threshold

U.S. DOT - Substances From 49 CFR 172.101
regulated by DOT (UN2362)

U.S. DOT - Hazard Classes
DOT hazard class = 3

U.S. DOT - Appendix B - Marine Pollutants
DOT regulated marine pollutant

U.S. DOT - Appendix A Table 1 - Hazardous Substances
final RQ = 1000 pounds (454 kg)

NFPA - Flash Points
flash point = 2 degrees F (-17 degrees C)

NFPA - Hazard Identification Ratings
health-2; flammability-3; reactivity-0

NIOSH - Selected LD50s and LC50s
Oral, rat: LD50 = 725 mg/kg

NIOSH 1990 - Pocket Guide - RELs
100 ppm TWA; 400 mg/m3 TWA

NIOSH 1990 - Pocket Guide - IDLHs
4000 ppm IDLH

NIOSH 1990 - Pocket Guide - Target organs
skin, liver, kidneys

NIOSH - Health Standards - Exposure Limits
Handle with caution in the workplace

NIOSH - Health Standards - Health Effects and Precautions
Central nervous system effects, possible liver and/or kidney damage

NTP Chemical Status Reports - Testing Status and NTIS Number
Technical reports printed (PB283345/AS)

NTP Chemical Status Reports - Evidence of Carcinogenicity
male rat-negative; female rat-equivocal; male mice-negative; female mice-equivocal

OSHA - Vacated PELs - Time Weighted Averages
100 ppm TWA; 400 mg/m3 TWA

OSHA - Final PELs - Time Weighted Averages
100 ppm TWA; 400 mg/m3 TWA

ENVIRONMENTAL LISTS

ATSDR Priority List
Rank (of 275): 114

CERCLA/SARA - Section 313 - Emission Reporting
form R reporting required

CERCLA/SARA - Hazardous Substances and their Reportable Quantities
final RQ = 1000 pounds (454 kg)

Clean Air Act (1990) - List of Hazardous Air Contaminants
[present]

CAA - HON Rule - Organic HAPs
[present]

Clean Water Act - Priority Pollutants
[present]

Safe Drinking Water Act - Monitoring
monitoring required

EPA - Master Testing List
[present]

RCRA - U Series Wastes
waste number U076

RCRA - Hazardous Constituents-Appendix VIII
waste number U076

RCRA - Basis for Listing - Appendix VII
Included in waste streams: F024, F025, F039

RCRA - Substances Banned From Land Disposal
[present]

RCRA - TSD Facilities Ground Water Monitoring
TM 8010 = 1 ug/L PQL; TM 8240 = 5 ug/L PQL

RCRA - Universal Treatment Standards (LDR)
WW: 0.059 mg/l; NWW: 6.0 mg/kg

TSCA - Code of Federal Regulations Citations
40 CFR 716.120(a); 40 CFR 799.5055(c), (d)(2)

TSCA - Health and Safety Reporting List
Effective Date: June 1, 1987

TSCA - Multichemical Test Rules - Waste Constituents
hydrolysis testing for Chemical Fate

TSCA - Section 12(b) - Export Notification
export notification required - Section 4

INTERNATIONAL LISTS

Australian Exposure Standards - Time Weighted Averages
200 ppm TWA; 810 mg/m3 TWA

Australian Exposure Standards - Short Term Exposure Limits
250 ppm STEL; 1010 mg/m3 STEL

Canada - WHMIS: Ingredient Disclosure
1% item 545 (664)

Canada - Alberta - 8 Hour Occupational Exposure Limit
200 ppm TWA; 810 mg/m3 TWA

Canada - Alberta - 15 Minute Occupational Exposure Limit
250 ppm STEL; 1010 mg/m3 STEL

Canada - British Columbia - 8 Hour Exposure Limits
200 ppm TWA; 810 mg/m3 TWA

Canada - British Columbia - 15 Minute Exposure Limits
250 ppm STEL; 1010 mg/m3 STEL

Canada - Ontario - OHSA - TWAEVs
200 ppm TWAEV; 810 mg/m3 TWAEV

Canada - Ontario - OHSA - STEVs
250 ppm STEV; 1010 mg/m3 STEV

Canada - Quebec - Time-Weighted Average Exposure Values
100 ppm TWAEV; 400 mg/m3 TWAEV

United Kingdom - Occupational Exposure Standards - TWAs
200 ppm TWA; 810 mg/m3 TWA

United Kingdom - Occupational Exposure Standards - STELs
400 ppm STEL; 1620 mg/m3 STEL

German (DFG) - MAK Values
100 ppm MAK; 400 mg/m3 MAK

German (DFG) - Peak Limitations
2 x normal MAK (30 min. average value); don't exceed 4 times during shift

German (DFG) - Pregnancy
classification not yet possible

Israel - Time Weighted Averages
200 ppm TWA; 810 mg/m3 TWA

Israel - Short Term Exposure Limits
250 ppm STEL; 1010 mg/m3 STEL

Israel - Action Levels
100 ppm AL; 405 mg/m3 AL

Mexico - Instruction No. 10 - TWAs
200 ppm TWA; 810 mg/m3 TWA

Mexico - Instruction No. 10 - STELs
250 ppm STEL; 1010 mg/m3 STEL

Mexico - Wastewater - Organic Toxic Pollutants and Heavy Metals
Listed under [Chlorinated Ethanes]

STATE LISTS

California - Air Bill 2588 Appendix A-I
known or potential carcinogen: 9/90

California - Prop. 65 - Cancer list
carcinogen - initial date 1/1/90

California - Prop. 65 - No Significant Risk Levels
no significant risk level = 100 ug/day

California - Exposure Limits - PELs
100 ppm PEL;   400 mg/m3 PEL

California - Directors List of Hazardous Substances (8 CCR 339)
[present]

Florida Hazardous Substance List
[present]

Massachusetts Right To Know List
[present]

Minnesota Hazardous Substance List
carcinogen

NJ Right to Know List (Total)
sn 0651

Pennsylvania Right to Know List
environmental hazard

PROPOSED REGULATIONS

Safe Drinking Water Act - Priority list
[present]

TSCA - ITC 32nd Report Priority Testing List
designated for dermal absorption testing

DICHLOROETHER   RR-01279-3

HEALTH AND SAFETY LISTS

U.S. DOT - Appendix B - Marine Pollutants
DOT regulated marine pollutant

CIS-1,2-DICHLOROETHYLENE   156-59-2

HEALTH AND SAFETY LISTS

NTP Chemical Status Reports - Testing Status and NTIS Number
Short term toxicity studies scheduled for peer review

ENVIRONMENTAL LISTS

ATSDR Priority List
Rank (of 275): 265

Safe Drinking Water Act - MCLs
MCL = 0.07 mg/L

Safe Drinking Water Act - MCLGs
MCLG = 0.07 mg/L

Safe Drinking Water Act - Monitoring
monitoring required

INTERNATIONAL LISTS

German (DFG) - MAK Values
200 ppm MAK;   790 mg/m3 MAK (Listed under '1,2-Dichloroethylene sym')

German (DFG) - Peak Limitations
2 x normal MAK (30 min. average value);   don't exceed 4 times during shift (Listed under '1,2-Dichloroethylene sym')

STATE LISTS

Florida Hazardous Substance List
[present]

Massachusetts Right To Know List
[present]

Pennsylvania Right to Know List
[present]

1,2-TRANS-DICHLOROETHYLENE   156-60-5
SEE ALSO:
F024-HAZARDOUS WASTES
F025-HAZARDOUS WASTES

HEALTH AND SAFETY LISTS

NIOSH - Selected LD50s and LC50s
Oral, mouse: LD50 = 2122 mg/kg

NTP Chemical Status Reports - Testing Status and NTIS Number
Prechronic studies completed: in review for further evaluation;   short term toxicity studies scheduled for peer review

ENVIRONMENTAL LISTS

ATSDR Priority List
Rank (of 275): 127

Clean Water Act - Priority Pollutants
[present]

Safe Drinking Water Act - MCLs
MCL = 0.1 mg/L

Safe Drinking Water Act - MCLGs
MCLG = 0.1 mg/L

Safe Drinking Water Act - Monitoring
monitoring required

RCRA - U Series Wastes
waste number U079

RCRA - Hazardous Constituents-Appendix VIII
waste number U079

RCRA - Basis for Listing - Appendix VII
Included in waste streams: F024, F025, F039

RCRA - Substances Banned From Land Disposal
[present]

RCRA - TSD Facilities Ground Water Monitoring
TM 8010 = 1 ug/L PQL;   TM 8240 = 5 ug/L PQL

RCRA - Universal Treatment Standards (LDR)
WW: 0.054 mg/l;   NWW: 30 mg/kg

INTERNATIONAL LISTS

German (DFG) - MAK Values
200 ppm MAK;   790 mg/m3 MAK (Listed under '1,2-Dichloroethylene sym')

German (DFG) - Peak Limitations
2 x normal MAK (30 min. average value);   don't exceed 4 times during shift (Listed under '1,2-Dichloroethylene sym')

Mexico - Instruction No. 10 - TWAs
200 ppm TWA;   790 mg/m3 TWA

Mexico - Instruction No. 10 - STELs
250 ppm STEL;   1000 mg/m3 STEL

Mexico - Wastewater - Organic Toxic Pollutants and Heavy Metals
Listed under [Dichloroethylenes]

Mexico - Drinking Water - Ecological Criteria
0.0003 mg/l Substance presents persistence, bioaccumulations or risk of cancer, reduce human exposure to a minimum

STATE LISTS

California - Directors List of Hazardous Substances (8 CCR 339)
[present]

Florida Hazardous Substance List
[present]

Massachusetts Right To Know List
[present]

Pennsylvania Right to Know List
environmental hazard

1,2-DICHLOROETHYLENE   540-59-0

HEALTH AND SAFETY LISTS

ACGIH 1995 - Time Weighted Averages
200 ppm TWA;   793 mg/m3 TWA

NFPA - Flash Points
flash point = 36 degrees F (2 degrees C)

NFPA - Hazard Identification Ratings
health-2; flammability-3; reactivity-2

NIOSH - Selected LD50s and LC50s
Oral, rat: LD50 = 770 mg/kg

NIOSH 1990 - Pocket Guide - RELs
200 ppm TWA;   790 mg/m3 TWA

NIOSH 1990 - Pocket Guide - IDLHs
4000 ppm IDLH

NIOSH 1990 - Pocket Guide - Target organs
eyes, CNS, respiratory system

NTP Chemical Status Reports - Testing Status and NTIS Number
Short term toxicity studies scheduled for peer review

OSHA - Vacated PELs - Time Weighted Averages
200 ppm TWA;   790 mg/m3 TWA

OSHA - Final PELs - Time Weighted Averages
200 ppm TWA;   790 mg/m3 TWA

ENVIRONMENTAL LISTS

ATSDR Priority List
Rank (of 275): 204

CERCLA/SARA - Section 313 - Emission Reporting
form R reporting required for 1.0% de minimus concentration

CAA - HON Rule - SOCMI Chemicals
compliance Jan. 23, 1995

Clean Water Act - Toxic Pollutants
[present] (Listed under 'Dichloroethylenes')

TSCA - Health and Safety Reporting List
Effective Date: March 11, 1994;   Sunset Date: March 11, 2004

INTERNATIONAL LISTS

Australian Exposure Standards - Time Weighted Averages
200 ppm TWA;   793 mg/m3 TWA

Canada - WHMIS: Ingredient Disclosure
1% item 546 (665)

Canada - Alberta - 8 Hour Occupational Exposure Limit
200 ppm TWA;   795 mg/m3 TWA

Canada - Alberta - 15 Minute Occupational Exposure Limit
250 ppm STEL;   995 mg/m3 STEL

Canada - British Columbia - 8 Hour Exposure Limits
200 ppm TWA;   790 mg/m3 TWA

Canada - British Columbia - 15 Minute Exposure Limits
250 ppm STEL;   1000 mg/m3 STEL

Canada - Ontario - OHSA - TWAEVs
200 ppm TWAEV;   790 mg/m3 TWAEV

Canada - Ontario - OHSA - STEVs
250 ppm STEV;   990 mg/m3 STEV

Canada - Quebec - Time-Weighted Average Exposure Values
200 ppm TWAEV;   793 mg/m3 TWAEV

United Kingdom - Occupational Exposure Standards - TWAs
200 ppm TWA;   790 mg/m3 TWA

United Kingdom - Occupational Exposure Standards - STELs
250 ppm STEL;   1000 mg/m3 STEL

German (DFG) - MAK Values
200 ppm MAK;   790 mg/m3 MAK (Listed under '1,2-Dichloroethylene sym')

German (DFG) - Peak Limitations
2 x normal MAK (30 min. average value);   don't exceed 4 times during shift (Listed under '1,2-Dichloroethylene sym')

Israel - Time Weighted Averages
200 ppm TWA;   793 mg/m3 TWA

Israel - Action Levels
100 ppm AL;   396.5 mg/m3 AL

STATE LISTS

California - Air Bill 2588 Appendix A-II
6/91

California - Exposure Limits - PELs
200 ppm PEL;   790 mg/m3 PEL

California - Directors List of Hazardous Substances (8 CCR 339)
[present]

Florida Hazardous Substance List
[present]

Massachusetts Right To Know List
[present]

Minnesota Hazardous Substance List
[present]

NJ Right to Know List (Total)
sn 0653

NJ Special Hazardous Substances
(flammable - third degree;   reactive - second degree)

Pennsylvania Right to Know List
environmental hazard

PROPOSED REGULATIONS

TSCA - ITC 32nd Report Priority Testing List
designated for dermal absorption testing

DICHLOROETHYLENES   25323-30-2

HEALTH AND SAFETY LISTS

U.S. DOT - Substances From 49 CFR 172.101
regulated by DOT (UN1150)

U.S. DOT - Hazard Classes
DOT hazard class = 3

NIOSH - Health Standards - Carcinogenic Chemicals
potential human carcinogen

ENVIRONMENTAL LISTS

RCRA - Hazardous Constituents-Appendix VIII
hazardous constituent - no waste number

DICHLOROETHYL OXIDE   RR-01280-6

HEALTH AND SAFETY LISTS

U.S. DOT - Appendix B - Marine Pollutants
DOT regulated marine pollutant

DICHLOROETHYL SULFIDE   RR-01410-8

STATE LISTS

NJ Right to Know List (Total)
sn 2259

DICHLOROFLUOROMETHANE   75-43-4

HEALTH AND SAFETY LISTS

ACGIH 1995 - Time Weighted Averages
10 ppm TWA;   42 mg/m3 TWA

U.S. DOT - Substances From 49 CFR 172.101
regulated by DOT (UN1029)

U.S. DOT - Hazard Classes
DOT hazard class = 2.2

NIOSH - Selected LD50s and LC50s
Inhalation, rat: LC50 = 49900 ppm 4 hr

NIOSH 1990 - Pocket Guide - RELs
10 ppm TWA;   40 mg/m3 TWA

NIOSH 1990 - Pocket Guide - IDLHs
50,000 ppm IDLH

NIOSH 1990 - Pocket Guide - Target organs
CVS, respiratory system

OSHA - Vacated PELs - Time Weighted Averages
10 ppm TWA;   40 mg/m3 TWA

OSHA - Final PELs - Time Weighted Averages
1000 ppm TWA;   4200 mg/m3 TWA

ENVIRONMENTAL LISTS

CERCLA/SARA - Section 313 - Emission Reporting
form R reporting required

Class 2 Ozone Depletors
ozone depletion weight reserved

List of Pesticide Product Inert Ingredients
[present]

TSCA - Code of Federal Regulations Citations
40 CFR 712.30(w);   40 CFR 716.120(a)

TSCA - PAIR - Reporting List
Reporting Date: June 13, 1989

TSCA - Health and Safety Reporting List
Effective Date: April 13, 1989

INTERNATIONAL LISTS

Australian Exposure Standards - Time Weighted Averages
10 ppm TWA;   42 mg/m3 TWA

Canada - WHMIS: Ingredient Disclosure
1% item 551 (668)

Canada - Alberta - 8 Hour Occupational Exposure Limit
10 ppm TWA;   42 mg/m3 TWA

Canada - Alberta - 15 Minute Occupational Exposure Limit
20 ppm STEL;   84 mg/m3 STEL

Canada - British Columbia - 8 Hour Exposure Limits
1000 ppm TWA;   4200 mg/m3 TWA

Canada - Ontario - OHSA - TWAEVs
10 ppm TWAEV;   42 mg/m3 TWAEV

Canada - Quebec - Time-Weighted Average Exposure Values
10 ppm TWAEV;   42 mg/m3 TWAEV

United Kingdom - Occupational Exposure Standards - TWAs
10 ppm TWA;   40 mg/m3 TWA

German (DFG) - MAK Values
10 ppm MAK;   45 mg/m3 MAK

German (DFG) - Peak Limitations
2 x normal MAK (30 min. average value);   don't exceed 4 times during shift

Israel - Time Weighted Averages
10 ppm TWA;   42 mg/m3 TWA

Israel - Action Levels
5 ppm AL;   21 mg/m3 AL

Mexico - Instruction No. 10 - TWAs
500 ppm TWA;   2100 mg/m3 TWA

STATE LISTS

California - Exposure Limits - PELs
10 ppm PEL;   42 mg/m3 PEL

California - Directors List of Hazardous Substances (8 CCR 339)
[present]

Florida Hazardous Substance List
[present]

Massachusetts Right To Know List
[present]

Minnesota Hazardous Substance List
[present]

NJ Right to Know List (Total)
sn 3109

Pennsylvania Right to Know List
[present]

DICHLOROFLUOROPROPANE   134237-45-9

ENVIRONMENTAL LISTS

Class 2 Ozone Depletors
ozone depletion weight reserved

1,3-DICHLORO-2,4-HEXADIENE   RR-00562-9

HEALTH AND SAFETY LISTS

NFPA - Flash Points
flash point = 168 degrees F (76 degrees C)

NFPA - Hazard Identification Ratings
flammability-2;   reactivity-0

1,3-DICHLORO-1,1,2,2,3,3,-HEXAFLUOROPROPANE   662-01-1

ENVIRONMENTAL LISTS

Class 1 Ozone Depletors
ozone depletion potential = 1.0

1,6-DICHLOROHEXANE   2163-00-0

HEALTH AND SAFETY LISTS

U.S. DOT - Appendix B - Marine Pollutants
DOT regulated marine pollutant

DICHLOROISOCYANURIC ACID   2782-57-2

HEALTH AND SAFETY LISTS

U.S. DOT - Substances From 49 CFR 172.101
regulated by DOT (UN2465)

U.S. DOT - Hazard Classes
DOT hazard class = 5.1

NIOSH - Selected LD50s and LC50s
Oral, rat: LD50 = 1173 mg/kg

STATE LISTS

Florida Hazardous Substance List
[present]

Massachusetts Right To Know List
[present]

NJ Right to Know List (Total)
sn 0654

Pennsylvania Right to Know List
[present]

DICHLOROISOCYANURIC ACID, SODIUM SALT   2893-78-9

INTERNATIONAL LISTS

Canada - WHMIS: Ingredient Disclosure
1% item 549 (667)

STATE LISTS

Florida Hazardous Substance List
[present]

Massachusetts Right To Know List
[present]

NJ Right to Know List (Total)
sn 1694

Pennsylvania Right to Know List
[present]

1,3-DICHLORO ISOPROPYL PHOSPHATE   13674-87-8

HEALTH AND SAFETY LISTS

NIOSH - Selected LD50s and LC50s
Oral, rat: LD50 = 1850 mg/kg

ENVIRONMENTAL LISTS

TSCA - Code of Federal Regulations Citations
40 CFR 712.30(w);   40 CFR 716.120(a)

TSCA - PAIR - Reporting List
Reporting Date: February 14, 1989

TSCA - Health and Safety Reporting List
Effective Date: December 16, 1988

DICHLOROMETHOTREXATE   **528-74-5**

HEALTH AND SAFETY LISTS

NTP Chemical Status Reports - Testing Status and NTIS Number
Chronic studies exist for which technical reports were not prepared

1,2-DICHLOROMETHOXYETHANE   **41683-62-9**

INTERNATIONAL LISTS

German (DFG) - Skin/Sensitizers
danger of cutaneous absorption

German (DFG) - Carcinogens
suspected carcinogen

2,6-DICHLORO-4-NITROANILINE   **99-30-9**

SEE ALSO:
ANILINE AND CHLORO-, BROMO-, AND/OR NITROANILINES

HEALTH AND SAFETY LISTS

NIOSH - Selected LD50s and LC50s
Oral, mouse: LD50 = 1500 mg/kg

ENVIRONMENTAL LISTS

CERCLA/SARA - Section 313 - Emission Reporting
form R reporting required

TSCA - Code of Federal Regulations Citations
40 CFR 712.30(d);   40 CFR 716.120(c);   40 CFR 799.5000

TSCA - PAIR - Reporting List
Reporting Date: November 19, 1982

TSCA - HDD/HDF - Precursors Required for Reporting
[present]

TSCA - Substances Subject to Testing Consent Orders
Test for: Environmental Effects

TSCA - Section 12(b) - Export Notification
export notification required - Section 4

INTERNATIONAL LISTS

Canada - WHMIS: Ingredient Disclosure
1% item 552 (669)

STATE LISTS

California - Directors List of Hazardous Substances (8 CCR 339)
[present] (Listed under 'Dicloran')

2,5-DICHLORONITROBENZENE   **89-61-2**

HEALTH AND SAFETY LISTS

NIOSH - Selected LD50s and LC50s
Oral, rat: LD50 = 1210 mg/kg

ENVIRONMENTAL LISTS

EPA - Master Testing List
[present]

TSCA - HDD/HDF - Precursors Required for Reporting
[present]

INTERNATIONAL LISTS

Canada - WHMIS: Ingredient Disclosure
1% item 553 (670)

3,4-DICHLORONITROBENZENE   **99-54-7**

HEALTH AND SAFETY LISTS

NIOSH - Selected LD50s and LC50s
Oral, rat: LD50 = 643 mg/kg

ENVIRONMENTAL LISTS

TSCA - Code of Federal Regulations Citations
40 CFR 712.30(x);   40 CFR 716.120(d)

TSCA - PAIR - Reporting List
Reporting Date: November 27, 1991

TSCA - Health and Safety Reporting List
Effective Date: September 30, 1991

TSCA - HDD/HDF - Precursors Required for Reporting
[present]

INTERNATIONAL LISTS

Canada - WHMIS: Ingredient Disclosure
1% item 554 (671)

1,1-DICHLORO-1-NITROETHANE   **594-72-9**

HEALTH AND SAFETY LISTS

ACGIH 1995 - Time Weighted Averages
2 ppm TWA;   12 mg/m3 TWA

U.S. DOT - Substances From 49 CFR 172.101
regulated by DOT (UN2650)

U.S. DOT - Hazard Classes
DOT hazard class = 6.1

NFPA - Flash Points
flash point = 168 degrees F (76 degrees C)

NFPA - Hazard Identification Ratings
health-2;   flammability-2;   reactivity-3

NIOSH - Selected LD50s and LC50s
Oral, rat: LD50 = 410 mg/kg

NIOSH 1990 - Pocket Guide - RELs
2 ppm TWA;   10 mg/m3 TWA

NIOSH 1990 - Pocket Guide - IDLHs
150 ppm IDLH

NIOSH 1990 - Pocket Guide - Target organs
lungs

OSHA - Vacated PELs - Time Weighted Averages
2 ppm TWA;   10 mg/m3 TWA

OSHA - Final PELs - Ceiling Limits
C 10 ppm;   C 60 mg/m3

INTERNATIONAL LISTS

Australian Exposure Standards - Time Weighted Averages
2 ppm TWA;   12 mg/m3 TWA

Canada - WHMIS: Ingredient Disclosure
1% item 555 (672)

Canada - Alberta - 8 Hour Occupational Exposure Limit
2 ppm TWA;   12 mg/m3 TWA

Canada - Alberta - 15 Minute Occupational Exposure Limit
10 ppm STEL;   59 mg/m3 STEL

Canada - British Columbia - Ceiling Exposure Limits
C 10 ppm;   C 60 mg/m3

Canada - Ontario - OHSA - TWAEVs
2 ppm TWAEV;   12 mg/m3 TWAEV

Canada - Quebec - Time-Weighted Average Exposure Values
2 ppm TWAEV;   12 mg/m3 TWAEV

German (DFG) - MAK Values
10 ppm MAK;   60 mg/m3 MAK

Israel - Time Weighted Averages
2 ppm TWA;   12 mg/m3 TWA

Israel - Action Levels
1 ppm AL;   6 mg/m3 AL

Mexico - Instruction No. 10 - TWAs
2 ppm TWA;   10 mg/m3 TWA

Mexico - Instruction No. 10 - STELs
10 ppm STEL;   60 mg/m3 STEL

STATE LISTS

California - Exposure Limits - PELs
2 ppm PEL;   10 mg/m3 PEL

California - Directors List of Hazardous Substances (8 CCR 339)
[present]

Florida Hazardous Substance List
[present]

Massachusetts Right To Know List
[present]

Minnesota Hazardous Substance List
[present]

NJ Right to Know List (Total)
sn 0655

NJ Special Hazardous Substances
(reactive - third degree)

Pennsylvania Right to Know List
[present]

4-[4-[(2,6-DICHLORO-4-NITROPHENYL)AZO]   17741-62-7
PHENYL]THIOMORPHOLINE,1,1-DIOXIDE-

ENVIRONMENTAL LISTS

TSCA - Code of Federal Regulations Citations
40 CFR 712.30(x); 40 CFR 716.120(d)

TSCA - PAIR - Reporting List
Reporting Date: November 27, 1991

TSCA - Health and Safety Reporting List
Effective Date: September 30, 1991

DICHLOROPENTAFLUOROPROPANE   127564-92-5

ENVIRONMENTAL LISTS

CERCLA/SARA - Section 313 - Emission Reporting
form R reporting required

1,1-DICHLORO-1,2,2,3,3-PENTAFLUOROPROPANE   13474-88-9
(HCFC-225CC)

ENVIRONMENTAL LISTS

CERCLA/SARA - Section 313 - Emission Reporting
form R reporting required

1,1-DICHLORO-1,2,3,3,3-PENTAFLUOROPROPANE   111512-56-2
(HCFC-225EB)

ENVIRONMENTAL LISTS

CERCLA/SARA - Section 313 - Emission Reporting
form R reporting required

1,2-DICHLORO-1,1,2,3,3-PENTAFLUOROPROPANE   422-44-6
(HCFC-225BB)

ENVIRONMENTAL LISTS

CERCLA/SARA - Section 313 - Emission Reporting
form R reporting required

1,2-DICHLORO-1,1,3,3,3-PENTAFLUOROPROPANE   431-86-7
(HCFC-225DA)

ENVIRONMENTAL LISTS

CERCLA/SARA - Section 313 - Emission Reporting
form R reporting required

1,3-DICHLORO-1,1,2,3,3-PENTAFLUOROPROPANE   136013-79-1
(HCFC-225EA)

ENVIRONMENTAL LISTS

CERCLA/SARA - Section 313 - Emission Reporting
form R reporting required

2,2-DICHLORO-1,1,1,3,3-PENTAFLUOROPROPANE   128903-21-9
(HCFC-225AA)

ENVIRONMENTAL LISTS

CERCLA/SARA - Section 313 - Emission Reporting
form R reporting required

2,3-DICHLORO-1,1,1,2,3-PENTAFLUOROPROPANE   422-48-0
(HCFC-225BA)

ENVIRONMENTAL LISTS

CERCLA/SARA - Section 313 - Emission Reporting
form R reporting required

1,3-DICHLORO-1,1,2,2,3-PENTAFLUOROPROPANE   507-55-1

ENVIRONMENTAL LISTS

CERCLA/SARA - Section 313 - Emission Reporting
form R reporting required

Class 2 Ozone Depletors
ozone depletion weight reserved

3,3-DICHLORO-1,1,1,2,2-PENTAFLUOROPROPANE   422-56-0

ENVIRONMENTAL LISTS

CERCLA/SARA - Section 313 - Emission Reporting
form R reporting required

Class 2 Ozone Depletors
ozone depletion weight reserved

1,5-DICHLOROPENTANE   628-76-2

HEALTH AND SAFETY LISTS

NFPA - Flash Points
flash point > 80 degrees F (27 degrees C)

NFPA - Hazard Identification Ratings
health-2; flammability-3; reactivity-0

STATE LISTS

Florida Hazardous Substance List
[present]

Massachusetts Right To Know List
[present]

Pennsylvania Right to Know List
[present]

DICHLOROPENTANE   30586-10-8

HEALTH AND SAFETY LISTS

U.S. DOT - Substances From 49 CFR 172.101
regulated by DOT (UN1152)

U.S. DOT - Hazard Classes
DOT hazard class = 3

NFPA - Flash Points
flash point = 106 degrees F (41 degrees C)

NFPA - Hazard Identification Ratings
health-2; flammability-2; reactivity-0

STATE LISTS

NJ Right to Know List (Total)
sn 0656

Pennsylvania Right to Know List
[present]

DICHLOROPHEN   97-23-4

HEALTH AND SAFETY LISTS

NIOSH - Selected LD50s and LC50s
Oral, rat: LD50 = 1506 mg/kg

ENVIRONMENTAL LISTS

CERCLA/SARA - Section 313 - Emission Reporting
form R reporting required

List of Pesticide Product Inert Ingredients
[present]

TSCA - Code of Federal Regulations Citations
40 CFR 716.120(a)

TSCA - Health and Safety Reporting List
Effective Date: June 1, 1987

INTERNATIONAL LISTS

Canada - WHMIS: Ingredient Disclosure
1% item 556 (673)

2,6-DICHLOROPHENOL   **87-65-0**
SEE ALSO:
   K043-HAZARDOUS WASTES
   TETRACHLORODIBENZO-P-DIOXINS
   F039-HAZARDOUS WASTES

HEALTH AND SAFETY LISTS

   U.S. DOT - Appendix A Table 1 - Hazardous Substances
   final RQ = 100 pounds (45.4 kg)

   NIOSH - Selected LD50s and LC50s
   Oral, rat: LD50 = 2940 mg/kg

ENVIRONMENTAL LISTS

   CERCLA/SARA - Hazardous Substances and their Reportable Quantities
   final RQ = 100 pounds (45.4 kg)

   RCRA - U Series Wastes
   waste number U082

   RCRA - Hazardous Constituents-Appendix VIII
   waste number U082

   RCRA - Basis for Listing - Appendix VII
   Included in waste streams: F039, K043

   RCRA - Substances Banned From Land Disposal
   [present]

   RCRA - TSD Facilities Ground Water Monitoring
   TM 8270 = 10 ug/L PQL

   RCRA - Universal Treatment Standards (LDR)
   WW: 0.044 mg/l; NWW: 14 mg/kg

   TSCA - HDD/HDF - Chemicals Required for Testing
   [present]

   TSCA - Section 12(b) - Export Notification
   export notification required - Section 4

INTERNATIONAL LISTS

   Canada - WHMIS: Ingredient Disclosure
   1% item 557 (674)

STATE LISTS

   Massachusetts Right To Know List
   [present]

   Pennsylvania Right to Know List
   environmental hazard

3,4-DICHLOROPHENOL   **95-77-2**
SEE ALSO:
   TETRACHLORODIBENZO-P-DIOXINS

ENVIRONMENTAL LISTS

   TSCA - HDD/HDF - Chemicals Required for Testing
   [present]

   TSCA - Section 12(b) - Export Notification
   export notification required - Section 4

2,4-DICHLOROPHENOL   **120-83-2**
SEE ALSO:
   F039-HAZARDOUS WASTES
   TETRACHLORODIBENZO-P-DIOXINS
   K099-HAZARDOUS WASTES
   K043-HAZARDOUS WASTES

HEALTH AND SAFETY LISTS

   U.S. DOT - Appendix A Table 1 - Hazardous Substances
   final RQ = 100 pounds (45.4 kg)

   NFPA - Flash Points
   flash point = 237 degrees F (114 degrees C)

   NFPA - Hazard Identification Ratings
   flammability-1; reactivity-0

   NIOSH - Selected LD50s and LC50s
   Oral, rat: LD50 = 580 mg/kg Skin, mammal: LD50 = 790 mg/kg

NTP Chemical Status Reports - Testing Status and NTIS Number
Technical reports printed (PB90106170/AS)

NTP Chemical Status Reports - Evidence of Carcinogenicity
male rat-no evidence; female rat-no evidence; male mice-no evidence; female mice-no evidence

ENVIRONMENTAL LISTS

   ATSDR Priority List
   Rank (of 275): 210

   CERCLA/SARA - Section 313 - Emission Reporting
   form R reporting required for 1.0% de minimus concentration

   CERCLA/SARA - Hazardous Substances and their Reportable Quantities
   final RQ = 100 pounds (45.4 kg)

   CAA - HON Rule - SOCMI Chemicals
   compliance by April 24, 1995

   Clean Water Act - Priority Pollutants
   [present]

   Clean Water Act - Toxic Pollutants
   [present]

   RCRA - U Series Wastes
   waste number U081

   RCRA - Hazardous Constituents-Appendix VIII
   waste number U081

   RCRA - Basis for Listing - Appendix VII
   Included in waste streams: F039, K043, K099

   RCRA - Substances Banned From Land Disposal
   [present]

   RCRA - TSD Facilities Ground Water Monitoring
   TM 8040 = 5 ug/L PQL; TM 8270 = 10 ug/L PQL

   RCRA - Universal Treatment Standards (LDR)
   WW: 0.044 mg/l; NWW: 14 mg/kg

   TSCA - Health and Safety Reporting List
   Effective Date: September 30, 1991

   TSCA - HDD/HDF - Chemicals Required for Testing
   [present]

   TSCA - Section 12(b) - Export Notification
   export notification required - Section 4

INTERNATIONAL LISTS

   Canada - NPRI (National Pollutant Release Inventory)
   [present]

   Canada - Drinking Water Quality - MACs
   0.9 mg/L MAC

   Canada - Drinking Water Quality - AOs
   <= 0.0003 mg/L AO

   Mexico - Wastewater - Organic Toxic Pollutants and Heavy Metals
   Listed under [Chlorinated Phenols]

   Mexico - Drinking Water - Ecological Criteria
   0.03 mg/l

STATE LISTS

   California - Air Bill 2588 Appendix A-I
   known or potential carcinogen: 6/91

   California - Directors List of Hazardous Substances (8 CCR 339)
   [present]

   Massachusetts Right To Know List
   [present]

   NJ Right to Know List (Total)
   sn 2344

   Pennsylvania Right to Know List
   environmental hazard

2,3-DICHLOROPHENOL　　　　　　　　　**576-24-9**

ENVIRONMENTAL LISTS

TSCA - HDD/HDF - Chemicals Required for Testing
[present]

TSCA - Section 12(b) - Export Notification
export notification required - Section 4

2,5-DICHLOROPHENOL　　　　　　　　　**583-78-8**

ENVIRONMENTAL LISTS

TSCA - HDD/HDF - Chemicals Required for Testing
[present]

TSCA - Section 12(b) - Export Notification
export notification required - Section 4

DICHLOROPHENOLS　　　　　　　　　**RR-01281-7**

HEALTH AND SAFETY LISTS

U.S. DOT - Appendix B - Marine Pollutants
DOT regulated marine pollutant

2,4-DICHLOROPHENOXYACETIC ACID DI-ETHANOLAMINE SALT　　　　　**RR-01282-8**

HEALTH AND SAFETY LISTS

U.S. DOT - Appendix B - Marine Pollutants
DOT regulated marine pollutant

2,4-DICHLOROPHENOXYACETIC ACID DIMETHY-LAMINE SALT　　　　　**RR-01283-9**

HEALTH AND SAFETY LISTS

U.S. DOT - Appendix B - Marine Pollutants
DOT regulated marine pollutant

2,4-DICHLOROPHENOXYACETIC ACID TRIISO-PROPYLAMINE SALT　　　　**RR-01284-0**

HEALTH AND SAFETY LISTS

U.S. DOT - Appendix B - Marine Pollutants
DOT regulated marine pollutant

2,4-DICHLOROPHENOXYACETIC ACID ESTER　**6341-97-5**

STATE LISTS

NJ Right to Know List (Total)
sn 0673

DICHLOROPHENOXYACETIC ACID ESTER　**28165-71-1**

STATE LISTS

NJ Right to Know List (Total)
sn 0672

2,4-DICHLOROPHENOXYACETIC ACID SODIUM SALT　　　　　**2702-72-9**

HEALTH AND SAFETY LISTS

NIOSH - Selected LD50s and LC50s
Oral, rat: LD50 = 555 mg/kg

ENVIRONMENTAL LISTS

CERCLA/SARA - Section 313 - Emission Reporting
form R reporting required

2[2,4-(DICHLOROPHENOXY)]-PROPIONIC ACID　**120-36-5**

ENVIRONMENTAL LISTS

CERCLA/SARA - Section 313 - Emission Reporting
form R reporting required

TSCA - HDD/HDF - Chemicals Required for Testing
[present]

TSCA - Section 12(b) - Export Notification
export notification required - Section 4

STATE LISTS

California - Directors List of Hazardous Substances (8 CCR 339)
[present]

DICHLOROPHENYLARSINE　　　　　　　**696-28-6**
SEE ALSO:
ARSENIC

HEALTH AND SAFETY LISTS

U.S. DOT - Appendix A Table 1 - Hazardous Substances
final RQ = 1 pound (0.454 kg)

NIOSH - Selected LD50s and LC50s
Inhalation, mouse: LC50 = 3300 mg/m3 (10 mn) Skin, rat: LD50 = 16 mg/kg

ENVIRONMENTAL LISTS

CERCLA/SARA - Section 302 Extremely Hazardous Substances and TPQs
TPQ = 500 pounds

CERCLA/SARA - Hazardous Substances and their Reportable Quantities
final RQ = 1 pound (0.454 kg)

RCRA - P Series Wastes
waste number P036

RCRA - Hazardous Constituents-Appendix VIII
waste number P036

RCRA - Substances Banned From Land Disposal
[present]

TSCA - Code of Federal Regulations Citations
40 CFR 716.120(a)

TSCA - Health and Safety Reporting List
Effective Date: March 7, 1986

STATE LISTS

Florida Hazardous Substance List
effective March 13, 1992

Massachusetts Right To Know List
extraordinarily hazardous

NJ Right to Know List (Total)
sn 1494

Pennsylvania Right to Know List
environmental hazard

2,6-DICHLORO-P-PHENYLENEDIAMINE　　**609-20-1**

HEALTH AND SAFETY LISTS

IARC - Group 3 (not classifiable)
[present]

NTP Chemical Status Reports - Testing Status and NTIS Number
Technical reports printed (PB82184052)

NTP Chemical Status Reports - Evidence of Carcinogenicity
male rat-negative; female rat-negative; male mice-positive; female mice-positive

STATE LISTS

California - Directors List of Hazardous Substances (8 CCR 339)
[present]

2,4-DICHLORO-1-PHENYL ISOCYANATE　　**2612-57-9**

STATE LISTS

NJ Right to Know List (Total)
sn 0662

DICHLOROPHENYL ISOCYANATE, ALL ISOMERS　**RR-00584-5**

HEALTH AND SAFETY LISTS

U.S. DOT - Substances From 49 CFR 172.101
regulated by DOT (UN2250)

U.S. DOT - Hazard Classes
DOT hazard class = 6.1

INTERNATIONAL LISTS
Canada - WHMIS: Ingredient Disclosure
0.1% item 558 (1039)

1,4-DICHLORO-2-PHENYL ISOCYANATE   5392-82-5
SEE ALSO:
DICHLOROPHENYL ISOCYANATE, ALL ISOMERS

STATE LISTS
NJ Right to Know List (Total)
sn 0661

1,3-DICHLORO-5-PHENYL ISOCYANATE   34893-92-0
SEE ALSO:
DICHLOROPHENYL ISOCYANATE, ALL ISOMERS

ENVIRONMENTAL LISTS
TSCA - Code of Federal Regulations Citations
40 CFR 712.30(x);   40 CFR 716.120(d)
TSCA - PAIR - Reporting List
Reporting Date: December 27, 1990
TSCA - Health and Safety Reporting List
Effective Date: October 29, 1990;   Sunset Date: November 9, 1993

STATE LISTS
NJ Right to Know List (Total)
sn 0660

1,3-DICHLORO-2-PHENYL ISOCYANATE   39920-37-1
STATE LISTS
NJ Right to Know List (Total)
sn 0659

1,2-DICHLORO-3-PHENYL ISOCYANATE   41195-90-8
STATE LISTS
NJ Right to Know List (Total)
sn 0657

1,2-DICHLOROPROPANE   78-87-5
SEE ALSO:
F039-HAZARDOUS WASTES

HEALTH AND SAFETY LISTS
ACGIH 1995 - Time Weighted Averages
75 ppm TWA;   347 mg/m3 TWA
ACGIH 1995 - Short Term Exposure Limits
110 ppm STEL;   508 mg/m3 STEL
AIHA - Odor Threshold Values
geometric mean air odor threshold = 0.26 ppm (detectable);   0.52 ppm (recognizable)
U.S. DOT - Substances From 49 CFR 172.101
regulated by DOT (UN1279)
U.S. DOT - Hazard Classes
DOT hazard class = 3
U.S. DOT - Appendix B - Marine Pollutants
DOT regulated marine pollutant
U.S. DOT - Appendix A Table 1 - Hazardous Substances
final RQ = 1000 pounds (454 kg)
IARC - Group 3 (not classifiable)
[present]
NFPA - Flash Points
flash point = 60 degrees F (16 degrees C)
NFPA - Hazard Identification Ratings
health-2;   flammability-3;   reactivity-0
NIOSH - Selected LD50s and LC50s
Oral, rat: LD50 = 2196 mg/kg Skin, rabbit: LD50 = 8750 mg/kg
NIOSH 1990 - Pocket Guide - IDLHs
2000 ppm IDLH (not considering carcinogenic effects)

NIOSH 1990 - Pocket Guide - Carcinogens
occupational carcinogen
NIOSH 1990 - Pocket Guide - Target organs
skin, eyes, respiratory system, liver, kidneys
NTP Chemical Status Reports - Testing Status and NTIS Number
Technical reports printed (PB87114443/AS)
NTP Chemical Status Reports - Evidence of Carcinogenicity
male rat-no evidence;   female rat-equivocal evidence;   male mice-some evidence;   female mice-some evidence
OSHA - Vacated PELs - Time Weighted Averages
75 ppm TWA;   350 mg/m3 TWA
OSHA - Vacated PELs - Short Term Exposure Limits
110 ppm STEL;   510 mg/m3 STEL
OSHA - Final PELs - Time Weighted Averages
75 ppm TWA;   350 mg/m3 TWA

ENVIRONMENTAL LISTS
ATSDR Priority List
Rank (of 275): 219
CERCLA/SARA - Section 313 - Emission Reporting
form R reporting required for 1.0% de minimus concentration
CERCLA/SARA - Hazardous Substances and their Reportable Quantities
final RQ = 1000 pounds (454 kg)
Clean Air Act (1990) - List of Hazardous Air Contaminants
[present]
CAA - HON Rule - SOCMI Chemicals
compliance by July 24, 1995
CAA - HON Rule - Organic HAPs
[present]
Clean Water Act - Hazardous Substances
[present] (Listed under 'Dichloropropane')
Clean Water Act - Priority Pollutants
[present]
Safe Drinking Water Act - MCLs
MCL = 0.005 mg/L
Safe Drinking Water Act - MCLGs
MCLG = Zero
Safe Drinking Water Act - Monitoring
monitoring required
RCRA - U Series Wastes
waste number U083
RCRA - Hazardous Constituents-Appendix VIII
waste number U083
RCRA - Basis for Listing - Appendix VII
Included in waste stream: F039
RCRA - Substances Banned From Land Disposal
[present]
RCRA - TSD Facilities Ground Water Monitoring
TM 8010 = 0.5 ug/L PQL;   TM 8240 = 5 ug/L PQL
RCRA - Universal Treatment Standards (LDR)
WW: 0.85 mg/l;   NWW: 18 mg/kg
TSCA - Code of Federal Regulations Citations
40 CFR 712.30(d);   40 CFR 716.120(a);   40 CFR 799.1550
TSCA - PAIR - Reporting List
Reporting Date: November 19, 1982
TSCA - Health and Safety Reporting List
Effective Date: October 4, 1982
TSCA - Chemical Test Rules
Testing required by: manufacturers;   processors (40 CFR 799.1550)
TSCA - Section 12(b) - Export Notification
export notification required - Section 4

INTERNATIONAL LISTS
Australian Exposure Standards - Time Weighted Averages
75 ppm TWA;   347 mg/m3 TWA

Australian Exposure Standards - Short Term Exposure Limits
110 ppm STEL;   508 mg/m3 STEL

Canada - WHMIS: Ingredient Disclosure
1% item 1361 (683)

Canada - NPRI (National Pollutant Release Inventory)
[present]

Canada - Alberta - 8 Hour Occupational Exposure Limit
75 ppm TWA;   345 mg/m3 TWA

Canada - Alberta - 15 Minute Occupational Exposure Limit
110 ppm STEL;   510 mg/m3 STEL

Canada - British Columbia - 8 Hour Exposure Limits
75 ppm TWA;   350 mg/m3 TWA

Canada - British Columbia - 15 Minute Exposure Limits
110 ppm STEL;   510 mg/m3 STEL

Canada - Ontario - OHSA - TWAEVs
75 ppm TWAEV;   350 mg/m3 TWAEV

Canada - Ontario - OHSA - STEVs
110 ppm STEV;   510 mg/m3 STEV

Canada - Quebec - Time-Weighted Average Exposure Values
75 ppm TWAEV;   350 mg/m3 TWAEV

Canada - Quebec - Short-term Exposure Values
110 ppm STEV;   508 mg/m3 STEV

German (DFG) - MAK Values
75 ppm MAK;   350 mg/m3 MAK

German (DFG) - Peak Limitations
5 x normal MAK (30 min. average value);   don't exceed 2 times during shift

German (DFG) - Carcinogens
suspected carcinogen

Israel - Time Weighted Averages
75 ppm TWA;   347 mg/m3 TWA

Israel - Short Term Exposure Limits
110 ppm STEL;   508 mg/m3 STEL

Israel - Action Levels
37.5 ppm AL;   173.5 mg/m3 AL

Mexico - Instruction No. 10 - TWAs
75 ppm TWA;   350 mg/m3 TWA

Mexico - Instruction No. 10 - STELs
110 ppm STEL;   510 mg/m3 STEL

Mexico - Wastewater - Organic Toxic Pollutants and Heavy Metals
Listed under [Dichloropropanes and Dichloropropenes]

Mexico - Drinking Water - Ecological Criteria
None given.

STATE LISTS

California - Air Bill 2588 Appendix A-I
known or potential carcinogen: 9/90

California - Prop. 65 - Cancer list
carcinogen - initial date 1/1/90

California - Exposure Limits - PELs
75 ppm PEL;   350 mg/m3 PEL

California - Exposure Limits - STELs
110 ppm STEL;   510 mg/m3 STEL

California - Directors List of Hazardous Substances (8 CCR 339)
[present] (Listed under 'Dichloropropanes')

Florida Hazardous Substance List
[present]

Massachusetts Right To Know List
[present]

Minnesota Hazardous Substance List
[present]

NJ Right to Know List (Total)
sn 0664

NJ Special Hazardous Substances
(flammable - third degree)

Pennsylvania Right to Know List
environmental hazard

1,1-DICHLOROPROPANE   78-99-9

HEALTH AND SAFETY LISTS

U.S. DOT - Appendix A Table 1 - Hazardous Substances
final RQ = 1000 pounds (454 kg) (Listed under 'Dichloropropane')

NIOSH - Selected LD50s and LC50s
Oral, rat: LD50 = 6500 mg/kg Skin, rabbit: LD50 = 14 gm/kg

ENVIRONMENTAL LISTS

CERCLA/SARA - Hazardous Substances and their Reportable Quantities
final RQ = 1000 pounds (454 kg) (Listed under 'Dichloropropane')

Clean Water Act - Hazardous Substances
[present] (Listed under 'Dichloropropane')

TSCA - Code of Federal Regulations Citations
40 CFR 716.120(a)

TSCA - Health and Safety Reporting List
Effective Date: March 7, 1986

STATE LISTS

California - Directors List of Hazardous Substances (8 CCR 339)
[present] (Listed under 'Dichloropropanes')

Massachusetts Right To Know List
[present]

Pennsylvania Right to Know List
environmental hazard

1,3-DICHLOROPROPANE   142-28-9

HEALTH AND SAFETY LISTS

U.S. DOT - Appendix B - Marine Pollutants
DOT regulated marine pollutant

U.S. DOT - Appendix A Table 1 - Hazardous Substances
final RQ = 1000 pounds (454 kg) (Listed under 'Dichloropropane')

ENVIRONMENTAL LISTS

CERCLA/SARA - Hazardous Substances and their Reportable Quantities
final RQ = 1000 pounds (454 kg) (Listed under 'Dichloropropane')

Clean Water Act - Hazardous Substances
[present] (Listed under 'Dichloropropane')

Safe Drinking Water Act - Monitoring
monitoring required

TSCA - Code of Federal Regulations Citations
40 CFR 712.30(w);   40 CFR 716.120(a)

TSCA - PAIR - Reporting List
Reporting Date: June 13, 1989

TSCA - Health and Safety Reporting List
Effective Date: March 7, 1986

STATE LISTS

California - Directors List of Hazardous Substances (8 CCR 339)
[present] (Listed under 'Dichloropropanes')

Massachusetts Right To Know List
[present]

Pennsylvania Right to Know List
environmental hazard

PROPOSED REGULATIONS

Safe Drinking Water Act - Priority list
[present]

DICHLOROPROPANE   26638-19-7
SEE ALSO:
F024-HAZARDOUS WASTES
F025-HAZARDOUS WASTES

HEALTH AND SAFETY LISTS
U.S. DOT - Appendix A Table 1 - Hazardous Substances
final RQ = 1000 pounds (454 kg)

ENVIRONMENTAL LISTS
CERCLA/SARA - Hazardous Substances and their Reportable Quantities
final RQ = 1000 pounds (454 kg)

Clean Water Act - Hazardous Substances
[present]

Clean Water Act - Toxic Pollutants
[present]

RCRA - Hazardous Constituents-Appendix VIII
hazardous constituent - no waste number

RCRA - Basis for Listing - Appendix VII
Included in waste streams: F024, F025

STATE LISTS
California - Directors List of Hazardous Substances (8 CCR 339)
[present]

Massachusetts Right To Know List
[present]

Pennsylvania Right to Know List
environmental hazard

DICHLOROPROPANE-DICHLOROPROPENE (MIXTURE)   8003-19-8

HEALTH AND SAFETY LISTS
U.S. DOT - Appendix A Table 1 - Hazardous Substances
final RQ = 100 pounds (45.4 kg) (this RQ is subject to change when the study of potential carcinogenicity is completed)
NIOSH - Selected LD50s and LC50s
Inhalation, rat: LC50 = 1000 ppm 4 hr Oral, rat: LD50 = 140 mg/kg Skin, rat: LD50 = 2100 mg/kg

ENVIRONMENTAL LISTS
CERCLA/SARA - Hazardous Substances and their Reportable Quantities
final RQ = 100 pounds (45.4 kg) (this RQ is subject to change when the study of potential carcinogenicity is completed)
Clean Water Act - Hazardous Substances
[present]

STATE LISTS
California - Directors List of Hazardous Substances (8 CCR 339)
[present]

Massachusetts Right To Know List
[present]

NJ Right to Know List (Total)
sn 2983

Pennsylvania Right to Know List
environmental hazard

1,3-DICHLORO-2-PROPANOL   96-23-1

HEALTH AND SAFETY LISTS
U.S. DOT - Substances From 49 CFR 172.101
regulated by DOT (UN2750)
U.S. DOT - Hazard Classes
DOT hazard class = 6.1
NFPA - Flash Points
flash point = 165 degrees F (74 degrees C)
NFPA - Hazard Identification Ratings
health-2; flammability-2; reactivity-0

NIOSH - Selected LD50s and LC50s
Oral, rat: LD50 = 110 mg/kg Skin, rabbit: LD50 = 800 mg/kg

ENVIRONMENTAL LISTS
TSCA - Code of Federal Regulations Citations
40 CFR 799.5055(c), (d)(1), (e)(1)
TSCA - Multichemical Test Rules - Waste Constituents
soil adsorption testing for chemical fate;   subchronic toxicity testing for Health Effects
TSCA - Section 12(b) - Export Notification
export notification required - Section 4

INTERNATIONAL LISTS
Canada - WHMIS: Ingredient Disclosure
1% item 560 (676)
German (DFG) - Carcinogens
animal evidence of carcinogenicity

STATE LISTS
Florida Hazardous Substance List
[present]

Massachusetts Right To Know List
[present]

NJ Right to Know List (Total)
sn 0665

Pennsylvania Right to Know List
[present]

DICHLOROPROPANOLS   26545-73-3
SEE ALSO:
K017-HAZARDOUS WASTES

ENVIRONMENTAL LISTS
RCRA - Hazardous Constituents-Appendix VIII
hazardous constituent - no waste number
RCRA - Basis for Listing - Appendix VII
Included in waste stream: K017

2,3-DICHLOROPROPENE   78-88-6

HEALTH AND SAFETY LISTS
U.S. DOT - Appendix A Table 1 - Hazardous Substances
final RQ = 100 pounds (45.4 kg) (Listed under 'Dichloropropene')
NFPA - Flash Points
flash point = 59 degrees F (15 degrees C)
NFPA - Hazard Identification Ratings
health-3; flammability-3; reactivity-0
NIOSH - Selected LD50s and LC50s
Inhalation, mouse: LC50 = 3100 mg/m3 (2 hr) Oral, rat: LD50 = 320 mg/kg Skin, rabbit: LD50 = 1580 mg/kg
NTP Chemical Status Reports - Testing Status and NTIS Number
Project leader assigned/study in design;   prechronic studies for which toxicity technical reports were not prepared

ENVIRONMENTAL LISTS
CERCLA/SARA - Section 313 - Emission Reporting
form R reporting required for 1.0% de minimus concentration
CERCLA/SARA - Hazardous Substances and their Reportable Quantities
final RQ = 100 pounds (45.4 kg) (Listed under 'Dichloropropene')
Clean Water Act - Hazardous Substances
[present] (Listed under 'Dichloropropene')
List of Pesticide Product Inert Ingredients
[present]
TSCA - Code of Federal Regulations Citations
40 CFR 716.120(a)
TSCA - Health and Safety Reporting List
Effective Date: June 1, 1987

STATE LISTS

California - Air Bill 2588 Appendix A-II
6/91

California - Directors List of Hazardous Substances (8 CCR 339)
[present] (Listed under 'Dichloropropenes')

Massachusetts Right To Know List
[present]

NJ Right to Know List (Total)
sn 2929

Pennsylvania Right to Know List
environmental hazard

1,3-DICHLOROPROPENE 542-75-6

HEALTH AND SAFETY LISTS

ACGIH 1995 - Time Weighted Averages
1 ppm TWA; 4.5 mg/m3 TWA

ACGIH 1995 - Skin Designations
skin - potential for cutaneous absorption

U.S. DOT - Appendix A Table 1 - Hazardous Substances
final RQ = 100 pounds (45.4 kg)

IARC - Group 2B (sufficient animal data)
[present]

NFPA - Flash Points
flash point = 95 degrees F (35 degrees C)

NFPA - Hazard Identification Ratings
health-2; flammability-3; reactivity-0

NIOSH - Selected LD50s and LC50s
Inhalation, mouse: LC50 = 4650 mg/m3 (2 hr) Oral, rat: LD50 = 250 mg/kg Skin, rabbit: LD50 = 504 mg/kg

NTP Chemical Status Reports - Testing Status and NTIS Number
Technical reports printed (PB85230449/AS)

NTP Chemical Status Reports - Evidence of Carcinogenicity
male rat-clear evidence; female rat-some evidence; male mice-inadequate; female mice-clear evidence

NTP Seventh Report - Suspect Carcinogens
suspect carcinogen

OSHA - Vacated PELs - Time Weighted Averages
1 ppm TWA; 5 mg/m3 TWA

OSHA - Vacated PELs - Skin Designation
Prevent or reduce skin absorption

OSHA - Possible Select Carcinogens
[present]

ENVIRONMENTAL LISTS

CERCLA/SARA - Section 313 - Emission Reporting
form R reporting required for 0.1% de minimus concentration

CERCLA/SARA - Hazardous Substances and their Reportable Quantities
final RQ = 100 pounds (45.4 kg)

Clean Air Act (1990) - List of Hazardous Air Contaminants
[present]

CAA - HON Rule - SOCMI Chemicals
compliance by Jan. 23, 1995

CAA - HON Rule - Organic HAPs
[present]

Clean Water Act - Hazardous Substances
[present] (Listed under 'Dichloropropene')

Clean Water Act - Priority Pollutants
[present]

Safe Drinking Water Act - Monitoring
monitoring required

RCRA - U Series Wastes
waste number U084

RCRA - Hazardous Constituents-Appendix VIII
waste number U084

RCRA - Substances Banned From Land Disposal
[present]

TSCA - Code of Federal Regulations Citations
40 CFR 716.120(a)

TSCA - Health and Safety Reporting List
Effective Date: June 1, 1987

INTERNATIONAL LISTS

Australian Exposure Standards - Time Weighted Averages
1 ppm TWA; 4.5 mg/m3 TWA

Australian Exposure Standards - Skin Effects
skin absorption

Australian Exposure Standards - Carcinogens
suspected carcinogen

Canada - WHMIS: Ingredient Disclosure
1% item 561 (677)

Canada - Alberta - 8 Hour Occupational Exposure Limit
1 ppm TWA; 4.5 mg/m3 TWA

Canada - Alberta - 15 Minute Occupational Exposure Limit
10 ppm STEL; 45 mg/m3 STEL

Canada - Alberta - Skin Designation
can be absorbed through the intact skin

Canada - Ontario - OHSA - TWAEVs
1 ppm TWAEV; 5 mg/m3 TWAEV

Canada - Ontario - OHSA - Skin Notations
absorption through skin, eyes, or mucous membranes

Canada - Quebec - Time-Weighted Average Exposure Values
1 ppm TWAEV; 4.5 mg/m3 TWAEV

Canada - Quebec - Skin Designations
absorbed through the skin

Canada - Quebec - Carcinogens
C3 carcinogen: effect detected in animals

German (DFG) - Carcinogens
animal evidence of carcinogenicity (cis and trans isomers)

Israel - Time Weighted Averages
1 ppm TWA; 4.5 mg/m3 TWA

Israel - Action Levels
0.5 ppm AL; 2.25 mg/m3 AL

Mexico - Wastewater - Organic Toxic Pollutants and Heavy Metals
Listed under [Dichloropropanes and Dichloropropenes]

Mexico - Drinking Water - Ecological Criteria
0.09 mg/l

STATE LISTS

California - Air Bill 2588 Appendix A-I
known or potential carcinogen

California - Prop. 65 - Cancer list
carcinogen - initial date 1/1/89

California - Exposure Limits - PELs
1 ppm PEL; 5 mg/m3 PEL

California - Exposure Limits - Skin Notation
material may be absorbed through the skin, eyes or mucous membrane

California - Directors List of Hazardous Substances (8 CCR 339)
[present] (Listed under 'Dichloropropenes')

Florida Hazardous Substance List
[present]

Massachusetts Right To Know List
carcinogen; extraordinarily hazardous

Minnesota Hazardous Substance List
carcinogen

NJ Right to Know List (Total)
sn 0666

NJ Special Hazardous Substances
(flammable - third degree)

Pennsylvania Right to Know List
environmental hazard; special hazardous substance
Pennsylvania RTK - Special Hazardous Substances
[present]

PROPOSED REGULATIONS
Safe Drinking Water Act - Priority list
[present]

1,1-DICHLOROPROPENE   563-58-6
ENVIRONMENTAL LISTS
Safe Drinking Water Act - Monitoring
monitoring required
TSCA - Code of Federal Regulations Citations
40 CFR 716.120(a)
TSCA - Health and Safety Reporting List
Effective Date: March 7, 1986

STATE LISTS
Massachusetts Right To Know List
[present]

PROPOSED REGULATIONS
Safe Drinking Water Act - Priority list
[present]

CIS-1,3-DICHLOROPROPENE   10061-01-5
SEE ALSO:
F039-HAZARDOUS WASTES

ENVIRONMENTAL LISTS
ATSDR Priority List
Rank (of 275): 231
RCRA - Basis for Listing - Appendix VII
Included in waste stream: F039
RCRA - TSD Facilities Ground Water Monitoring
TM 8010 = 20 ug/L PQL; TM 8240 = 5 ug/L PQL
RCRA - Universal Treatment Standards (LDR)
WW: 0.036 mg/l; NWW: 18 mg/kg

STATE LISTS
Florida Hazardous Substance List
[present]
Massachusetts Right To Know List
carcinogen; extraordinarily hazardous

TRANS-1,3-DICHLOROPROPENE   10061-02-6
SEE ALSO:
F039-HAZARDOUS WASTES

ENVIRONMENTAL LISTS
ATSDR Priority List
Rank (of 275): 238
CERCLA/SARA - Section 313 - Emission Reporting
form R reporting required
RCRA - Basis for Listing - Appendix VII
Included in waste stream: F039
RCRA - TSD Facilities Ground Water Monitoring
TM 8010 = 5 ug/L PQL; TM 8240 = 5 ug/L PQL
RCRA - Universal Treatment Standards (LDR)
WW: 0.036 mg/l; NWW: 18 mg/kg

STATE LISTS
Florida Hazardous Substance List
[present]
Massachusetts Right To Know List
carcinogen; extraordinarily hazardous

DICHLOROPROPENE   26952-23-
SEE ALSO:
F025-HAZARDOUS WASTES
F024-HAZARDOUS WASTES

HEALTH AND SAFETY LISTS
U.S. DOT - Substances From 49 CFR 172.101
regulated by DOT (UN2047)
U.S. DOT - Hazard Classes
DOT hazard class = 3
U.S. DOT - Appendix A Table 1 - Hazardous Substances
final RQ = 100 pounds (45.4 kg)

ENVIRONMENTAL LISTS
CERCLA/SARA - Hazardous Substances and their Reportable Quantities
final RQ = 100 pounds (45.4 kg)
Clean Water Act - Hazardous Substances
[present]
Clean Water Act - Toxic Pollutants
[present]
RCRA - Hazardous Constituents-Appendix VIII
hazardous constituent - no waste number
RCRA - Basis for Listing - Appendix VII
Included in waste streams: F024, F025
TSCA - Code of Federal Regulations Citations
40 CFR 716.120(a)
TSCA - Health and Safety Reporting List
Effective Date: June 1, 1987

STATE LISTS
California - Directors List of Hazardous Substances (8 CCR 339)
[present]
Minnesota Hazardous Substance List
skin
Pennsylvania Right to Know List
environmental hazard

2,2-DICHLOROPROPIONIC ACID   75-99-0
HEALTH AND SAFETY LISTS
ACGIH 1995 - Time Weighted Averages
1 ppm TWA; 5.8 mg/m3 TWA
U.S. DOT - Appendix A Table 1 - Hazardous Substances
final RQ = 5000 pounds (2270 kg)
NIOSH - Selected LD50s and LC50s
Oral, rat: LD50 = 970 mg/kg
OSHA - Vacated PELs - Time Weighted Averages
1 ppm TWA; 6 mg/m3 TWA

ENVIRONMENTAL LISTS
CERCLA/SARA - Hazardous Substances and their Reportable Quantities
final RQ = 5000 pounds (2270 kg)
Clean Water Act - Hazardous Substances
[present]
Safe Drinking Water Act - MCLs
MCL = 0.2 mg/L
Safe Drinking Water Act - MCLGs
MCLG = 0.2 mg/L

INTERNATIONAL LISTS
Australian Exposure Standards - Time Weighted Averages
1 ppm TWA; 5.8 mg/m3 TWA
Canada - Alberta - 8 Hour Occupational Exposure Limit
1 ppm TWA; 6 mg/m3 TWA
Canada - Alberta - 15 Minute Occupational Exposure Limit
2 ppm STEL; 12 mg/m3 STEL

Canada - Ontario - OHSA - TWAEVs
1 ppm TWAEV; 6 mg/m3 TWAEV

Canada - Quebec - Time-Weighted Average Exposure Values
1 ppm TWAEV; 5.8 mg/m3 TWAEV

German (DFG) - MAK Values
1 ppm MAK; 6 mg/m3 MAK

Israel - Time Weighted Averages
1 ppm TWA; 5.8 mg/m3 TWA

Israel - Action Levels
0.5 ppm AL; 2.9 mg/m3 AL

STATE LISTS

California - Exposure Limits - PELs
1 ppm PEL; 6 mg/m3 PEL

California - Directors List of Hazardous Substances (8 CCR 339)
[present]

Florida Hazardous Substance List
[present]

Massachusetts Right To Know List
[present]

Minnesota Hazardous Substance List
[present]

NJ Right to Know List (Total)
sn 0668

NJ Special Hazardous Substances
(corrosive)

Pennsylvania Right to Know List
environmental hazard

2,2-DICHLOROPROPIONIC ACID, SODIUM SALT   127-20-8

INTERNATIONAL LISTS

German (DFG) - MAK Values
1 ppm MAK; 6 mg/m3 MAK (Listed under '2,2-Dichloropropionic acid')

3,5-DICHLOROSALICYLIC ACID   320-72-9

ENVIRONMENTAL LISTS

TSCA - HDD/HDF - Chemicals Required for Testing
[present]

TSCA - Section 12(b) - Export Notification
export notification required - Section 4

DICHLOROSILANE   4109-96-0

HEALTH AND SAFETY LISTS

U.S. DOT - Substances From 49 CFR 172.101
regulated by DOT (UN2189)

U.S. DOT - Hazard Classes
DOT hazard class = 2.3

U.S. DOT - Substances Which Are Poisonous by Inhalation
gaseous hazardous material poisonous by inhalation (UN2189)

NFPA - Flash Points
flash point = -35 degrees F (-37 degrees C)

NFPA - Hazard Identification Ratings
health-3; flammability-4; reactivity-2 (avoid water as an extinguishing agent)

OSHA - List of Highly Hazardous Chemicals
threshhold quantity = 2500 pounds

ENVIRONMENTAL LISTS

CAA - Flammable Substances for Accidental Release Prevention
threshold quantity = 10,000 lbs

INTERNATIONAL LISTS

Canada - WHMIS: Ingredient Disclosure
1% item 562 (678)

STATE LISTS

NJ Right to Know List (Total)
sn 0670

Pennsylvania Right to Know List
[present]

DICHLOROTETRAFLUOROETHANE   76-14-2

HEALTH AND SAFETY LISTS

ACGIH 1995 - Time Weighted Averages
1000 ppm TWA; 6990 mg/m3 TWA

NIOSH - Selected LD50s and LC50s
Inhalation, rat: LC50 = 72 pph 30 mn

NIOSH 1990 - Pocket Guide - RELs
1000 ppm TWA; 7000 mg/m3 TWA

NIOSH 1990 - Pocket Guide - IDLHs
50,000 ppm IDLH

NIOSH 1990 - Pocket Guide - Target organs
CVS, respiratory system

OSHA - Vacated PELs - Time Weighted Averages
1000 ppm TWA; 7000 mg/m3 TWA

OSHA - Final PELs - Time Weighted Averages
1000 ppm TWA; 7000 mg/m3 TWA

ENVIRONMENTAL LISTS

CERCLA/SARA - Section 313 - Emission Reporting
form R reporting required for 1.0% de minimus concentration

Class 1 Ozone Depletors
ozone depletion potential = 1.0

List of Pesticide Product Inert Ingredients
[present]

INTERNATIONAL LISTS

Australian Exposure Standards - Time Weighted Averages
1000 ppm TWA; 6990 mg/m3 TWA

Canada - WHMIS: Ingredient Disclosure
1% item 563 (679)

Canada - Alberta - 8 Hour Occupational Exposure Limit
1000 ppm TWA; 6990 mg/m3 TWA

Canada - Alberta - 15 Minute Occupational Exposure Limit
1250 ppm STEL; 8740 mg/m3 STEL

Canada - British Columbia - 8 Hour Exposure Limits
1000 ppm TWA; 7000 mg/m3 TWA

Canada - British Columbia - 15 Minute Exposure Limits
1250 ppm STEL; 8750 mg/m3 STEL

Canada - Ontario - OHSA - TWAEVs
1000 ppm TWAEV; 6985 mg/m3 TWAEV

Canada - Quebec - Time-Weighted Average Exposure Values
1000 ppm TWAEV; 6990 mg/m3 TWAEV

German (DFG) - MAK Values
1000 ppm MAK; 7000 mg/m3 MAK

German (DFG) - Peak Limitations
2 x normal MAK (1 hour momentary value); don't exceed 3 times per shift

Israel - Time Weighted Averages
1000 ppm TWA; 6990 mg/m3 TWA

Israel - Action Levels
500 ppm AL; 3495 mg/m3 AL

Mexico - Instruction No. 10 - TWAs
1000 ppm TWA; 7000 mg/m3 TWA

Mexico - Instruction No. 10 - STELs
1250 ppm STEL; 8760 mg/m3 STEL

STATE LISTS

California - Exposure Limits - PELs
1000 ppm PEL; 7000 mg/m3 PEL

California - Directors List of Hazardous Substances (8 CCR 339)
[present]

Florida Hazardous Substance List
[present]

Massachusetts Right To Know List
[present]

Minnesota Hazardous Substance List
[present]

NJ Right to Know List (Total)
sn 0671

DICHLOROTETRAFLUOROETHANE   1320-37-2

HEALTH AND SAFETY LISTS

U.S. DOT - Substances From 49 CFR 172.101
regulated by DOT (UN1958)

U.S. DOT - Hazard Classes
DOT hazard class = 2.2

ENVIRONMENTAL LISTS

CAA - HON Rule - SOCMI Chemicals
compliance by Oct. 23, 1995

INTERNATIONAL LISTS

United Kingdom - Occupational Exposure Standards - TWAs
1000 ppm TWA; 7000 mg/m3 TWA

United Kingdom - Occupational Exposure Standards - STELs
1250 ppm STEL; 8750 mg/m3 STEL

STATE LISTS

Pennsylvania Right to Know List
[present]

DICHLOROTETRAFLUOROPROPANE   127564-83-4

ENVIRONMENTAL LISTS

Class 2 Ozone Depletors
ozone depletion weight reserved

2,4-DICHLOROTOLUENE   95-73-8

ENVIRONMENTAL LISTS

EPA - Master Testing List
[present]

2,6-DICHLOROTOLUENE   118-69-4

ENVIRONMENTAL LISTS

EPA - Master Testing List
[present]

DICHLOROTRIAZINETRIONE AND ITS SALTS   RR-00255-1

STATE LISTS

NJ Right to Know List (Total)
sn 2322

1,1-DICHLORO-1,2,2-TRIFLUOROETHANE (HCFC-   812-04-4
123B)

ENVIRONMENTAL LISTS

CERCLA/SARA - Section 313 - Emission Reporting
form R reporting required

DICHLOROTRIFLUOROETHANE   34077-87-7

ENVIRONMENTAL LISTS

CERCLA/SARA - Section 313 - Emission Reporting
form R reporting required

DICHLORO-1,1,2-TRIFLUOROETHANE   90454-18-5

ENVIRONMENTAL LISTS

CERCLA/SARA - Section 313 - Emission Reporting
form R reporting required

1,2-DICHLORO-1,1,2-TRIFLUOROETHANE (HCFC-   354-23-
123A)

ENVIRONMENTAL LISTS

CERCLA/SARA - Section 313 - Emission Reporting
form R reporting required

DICHLOROTRIFLUOROPROPANE   134237-43-

ENVIRONMENTAL LISTS

Class 2 Ozone Depletors
ozone depletion weight reserved

3,5-DICHLORO-2,4,6-TRIFLUOROPYRIDINE   1737-93-

HEALTH AND SAFETY LISTS

U.S. DOT - Substances From 49 CFR 172.101
regulated by DOT (NA2810)

U.S. DOT - Hazard Classes
DOT hazard class = 6.1

U.S. DOT - Substances Which Are Poisonous by Inhalation
liquid hazardous material poisonous by inhalation

DICHLOROTRIPHENYLANTIMONY   594-31-0

SEE ALSO:
ANTIMONY

INTERNATIONAL LISTS

Canada - WHMIS: Ingredient Disclosure
1% item 564 (680)

DICHLOROVINYLCHLOROARSINE   RR-01411-9

HEALTH AND SAFETY LISTS

U.S. DOT - Hazard Classes
Forbidden from transport by the DOT

2,2'-DICHLORO-P-XYLENE   623-25-6

HEALTH AND SAFETY LISTS

NIOSH - Selected LD50s and LC50s
Oral, rat: LD50 = 1780 mg/kg

DICHLORVOS   62-73-7

HEALTH AND SAFETY LISTS

ACGIH 1995 - Time Weighted Averages
0.1 ppm TWA; 0.90 mg/m3 TWA

ACGIH 1995 - Skin Designations
skin - potential for cutaneous absorption

U.S. DOT - Appendix B - Marine Pollutants
DOT regulated severe marine pollutant

U.S. DOT - Appendix A Table 1 - Hazardous Substances
final RQ = 10 pounds (4.54 kg)

IARC - Group 2B (sufficient animal data)
[present]

NFPA - Flash Points
flash point = 350 degrees F (177 degrees C)

NFPA - Hazard Identification Ratings
health-3; flammability-1

NIOSH - Selected LD50s and LC50s
*Inhalation, rat: LC50 = 15 mg/m3 4 hr Oral, rat: LD50 = 25 mg/kg
Skin, rat: LD50 = 70400 ug/kg*

NIOSH 1990 - Pocket Guide - RELs
1 mg/m3 TWA

NIOSH 1990 - Pocket Guide - IDLHs
200 mg/m3 IDLH

NIOSH 1990 - Pocket Guide - Target organs
respiratory system, CVS, CNS, eyes, skin, blood cholinesterase

NIOSH 1990 - Pocket Guide - Skin list
Potential for dermal absorption

NTP Chemical Status Reports - Testing Status and NTIS Number
Technical reports printed (PB270937/AS) (PB90198508/AS)

NTP Chemical Status Reports - Evidence of Carcinogenicity
PB270937/AS: male rat-negative; female rat-negative; male mice-negative; female mice-negative; PB90198508/AS: male rat-some evidence; female rat-equivocal evidence; male mice-some evidence; female mice-clear evidence

OSHA - Vacated PELs - Time Weighted Averages
1 mg/m3 TWA

OSHA - Vacated PELs - Skin Designation
Prevent or reduce skin absorption

OSHA - Final PELs - Time Weighted Averages
1 mg/m3 TWA

OSHA - Final PELs - Skin Notations
prevent or reduce skin absorption

OSHA - Possible Select Carcinogens
[present]

ENVIRONMENTAL LISTS

ATSDR Priority List
Rank (of 275): 217

CERCLA/SARA - Section 302 Extremely Hazardous Substances and TPQs
TPQ = 1000 pounds

CERCLA/SARA - Section 313 - Emission Reporting
form R reporting required for 1.0% de minimus concentration

CERCLA/SARA - Hazardous Substances and their Reportable Quantities
final RQ = 10 pounds (4.54 kg)

Clean Air Act (1990) - List of Hazardous Air Contaminants
[present]

Clean Water Act - Hazardous Substances
[present]

INTERNATIONAL LISTS

Australian Exposure Standards - Time Weighted Averages
0.1 ppm TWA; 0.9 mg/m3 TWA

Australian Exposure Standards - Skin Effects
skin absorption

Canada - Alberta - 8 Hour Occupational Exposure Limit
0.1 ppm TWA; 0.9 mg/m3 TWA

Canada - Alberta - 15 Minute Occupational Exposure Limit
0.3 ppm STEL; 2.7 mg/m3 STEL

Canada - Alberta - Skin Designation
can be absorbed through the intact skin

Canada - British Columbia - 8 Hour Exposure Limits
0.1 ppm TWA; 1 mg/m3 TWA

Canada - British Columbia - 15 Minute Exposure Limits
0.3 ppm STEL; 3 mg/m3 STEL

Canada - British Columbia - Skin Notations
skin - potential for skin absorption

Canada - Ontario - OHSA - TWAEVs
0.1 ppm TWAEV; 0.9 mg/m3 TWAEV

Canada - Ontario - OHSA - Skin Notations
absorption through skin, eyes, or mucous membranes

Canada - Quebec - Time-Weighted Average Exposure Values
0.1 ppm TWAEV; 0.9 mg/m3 TWAEV

Canada - Quebec - Skin Designations
absorbed through the skin

United Kingdom - Occupational Exposure Standards - TWAs
0.1 ppm TWA; 1 mg/m3 TWA

United Kingdom - Occupational Exposure Standards - STELs
0.3 ppm STEL; 3 mg/m3 STEL

United Kingdom - Occupational Exposure Standards - Notes
can be absorbed through skin

German (DFG) - MAK Values
0.1 ppm MAK; 1 mg/m3 MAK

German (DFG) - Peak Limitations
10 x normal MAK (30 min average value); don't exceed during shift

German (DFG) - Skin/Sensitizers
danger of cutaneous absorption

German (DFG) - Pregnancy
no risk to embryo/fetus if exposure limits adhered to

Israel - Time Weighted Averages
0.1 ppm TWA; 0.90 mg/m3 TWA

Israel - Action Levels
0.05 ppm AL; 0.45 mg/m3 AL

Mexico - Instruction No. 10 - TWAs
0.16 ppm TWA; 1.5 mg/m3 TWA

Mexico - Instruction No. 10 - Skin designation
skin - potential for cutaneous absorption

STATE LISTS

California - Air Bill 2588 Appendix A-I
known or potential carcinogen: 9/89

California - Prop. 65 - Cancer list
carcinogen - initial date 1/1/89

California - Prop. 65 - No Significant Risk Levels
no significant risk level = 2 ug/day

California - Exposure Limits - PELs
0.1 ppm PEL; 1 mg/m3 PEL

California - Exposure Limits - Skin Notation
material may be absorbed through the skin, eyes or mucous membrane

California - Directors List of Hazardous Substances (8 CCR 339)
[present]

Florida Hazardous Substance List
[present]

Massachusetts Right To Know List
extraordinarily hazardous; neurotoxin

Minnesota Hazardous Substance List
skin

NJ Right to Know List (Total)
sn 0674

Pennsylvania Right to Know List
environmental hazard

DICHROMATES   RR-01620-6

INTERNATIONAL LISTS

Canada - Ontario - OHSA - TWAEVs
as Cr: 0.05 mg/m3 TWAEV

PROPOSED REGULATIONS

Canada - Ontario - Proposed Occupational TWAEVs
as Cr: 0.02 mg/m3 TWAEV

DICLORAN   102-30-7

STATE LISTS

California - Directors List of Hazardous Substances (8 CCR 339)
[present]

DI(COCO ALKYL) DIMETHYL AMMONIUM CHLORIDE   61789-77-3

ENVIRONMENTAL LISTS

List of Pesticide Product Inert Ingredients
[present]

DICOFOL   115-32-2

HEALTH AND SAFETY LISTS

U.S. DOT - Appendix A Table 1 - Hazardous Substances
final RQ = 10 pounds (4.54 kg)

IARC - Group 3 (not classifiable)
[present]

NIOSH - Selected LD50s and LC50s
Oral, rat: LD50 = 575 mg/kg Skin, rat: LD50 = 100 mg/kg

NTP Chemical Status Reports - Testing Status and NTIS Number
Technical reports printed (PB286206/AS)

NTP Chemical Status Reports - Evidence of Carcinogenicity
male rat-negative; female rat-negative; male mice-positive; female mice-negative

ENVIRONMENTAL LISTS

ATSDR Priority List
Rank (of 275): 107

CERCLA/SARA - Section 313 - Emission Reporting
form R reporting required for 1.0% de minimus concentration

CERCLA/SARA - Hazardous Substances and their Reportable Quantities
final RQ = 10 pounds (4.54 kg)

Clean Water Act - Hazardous Substances
[present]

STATE LISTS

California - Air Bill 2588 Appendix A-I
6/91

California - Directors List of Hazardous Substances (8 CCR 339)
[present]

Massachusetts Right To Know List
carcinogen; extraordinarily hazardous

NJ Right to Know List (Total)
sn 0675

Pennsylvania Right to Know List
environmental hazard

DICROTOPHOS   141-66-2

HEALTH AND SAFETY LISTS

ACGIH 1995 - Time Weighted Averages
0.25 mg/m3 TWA

ACGIH 1995 - Skin Designations
skin - potential for cutaneous absorption

U.S. DOT - Appendix B - Marine Pollutants
DOT regulated marine pollutant

NIOSH - Selected LD50s and LC50s
Inhalation, rat: LC50 = 90 mg/m3 4 hr Oral, rat: LD50 = 13 mg/kg Skin, rat: LD50 = 42 mg/kg

OSHA - Vacated PELs - Time Weighted Averages
0.25 mg/m3 TWA

OSHA - Vacated PELs - Skin Designation
Prevent or reduce skin absorption

ENVIRONMENTAL LISTS

CERCLA/SARA - Section 302 Extremely Hazardous Substances and TPQs
TPQ = 100 pounds

INTERNATIONAL LISTS

Australian Exposure Standards - Time Weighted Averages
0.25 mg/m3 TWA

Australian Exposure Standards - Skin Effects
skin absorption

Canada - Alberta - 8 Hour Occupational Exposure Limit
0.25 mg/m3 TWA

Canada - Alberta - 15 Minute Occupational Exposure Limit
0.75 mg/m3 STEL

Canada - Alberta - Skin Designation
can be absorbed through the intact skin

Canada - British Columbia - 8 Hour Exposure Limits
0.25 mg/m3 TWA

Canada - British Columbia - Skin Notations
skin - potential for skin absorption

Canada - Ontario - OHSA - TWAEVs
0.25 mg/m3 TWAEV

Canada - Ontario - OHSA - Skin Notations
absorption through skin, eyes, or mucous membranes

Canada - Quebec - Time-Weighted Average Exposure Values
0.25 mg/m3 TWAEV

Canada - Quebec - Skin Designations
absorbed through the skin

Israel - Time Weighted Averages
0.25 mg/m3 TWA

Israel - Action Levels
0.125 mg/m3 AL

Mexico - Instruction No. 10 - TWAs
0.25 mg/m3 TWA

Mexico - Instruction No. 10 - Skin designation
skin - potential for cutaneous absorption

STATE LISTS

California - Exposure Limits - PELs
0.25 mg/m3 PEL

California - Exposure Limits - Skin Notation
material may be absorbed through the skin, eyes or mucous membrane

California - Directors List of Hazardous Substances (8 CCR 339)
[present]

Florida Hazardous Substance List
[present]

Massachusetts Right To Know List
extraordinarily hazardous; neurotoxin

Minnesota Hazardous Substance List
skin

NJ Right to Know List (Total)
sn 0676

Pennsylvania Right to Know List
environmental hazard

PROPOSED REGULATIONS

CERCLA/SARA - Proposed Hazardous Substance Additions
proposed RQ = 1 pound (.454 kg)

CERCLA/SARA - 1989 Proposed RQ Adjustments
proposed RQ = 100 pounds (45.4 kg)

DICUMAROL   66-76-2

STATE LISTS

California - Prop. 65 - Developmental Toxicity
developmental toxicity - initial date 10/1/92

DICUMYL PEROXIDE   80-43-3

HEALTH AND SAFETY LISTS

U.S. DOT - Organic Peroxides Table
Organic peroxide UN3110; exempt at less than or equal to 42%

NIOSH - Selected LD50s and LC50s
Oral, rat: LD50 = 4100 mg/kg

ENVIRONMENTAL LISTS

EPA - Master Testing List
[present]

STATE LISTS

NJ Right to Know List (Total)
sn 0677

DICYANOETHYL DIETHYLENETRIAMINE 74849-88-0

ENVIRONMENTAL LISTS

List of Pesticide Product Inert Ingredients
[present]

DICYCLOHEXYLAMINE 101-83-7

HEALTH AND SAFETY LISTS

U.S. DOT - Substances From 49 CFR 172.101
regulated by DOT (UN2565)

U.S. DOT - Hazard Classes
DOT hazard class = 8

NFPA - Flash Points
flash point > 210 degrees F (99 degrees C)

NFPA - Hazard Identification Ratings
health-3; flammability-1; reactivity-0

NIOSH - Selected LD50s and LC50s
Oral, rat: LD50 = 373 mg/kg

INTERNATIONAL LISTS

Canada - WHMIS: Ingredient Disclosure
0.1% item 565 (690)

STATE LISTS

Florida Hazardous Substance List
[present]

Massachusetts Right To Know List
[present]

NJ Right to Know List (Total)
sn 0678

NJ Special Hazardous Substances
(corrosive)

Pennsylvania Right to Know List
[present]

DICYCLOHEXYLAMMONIUM NITRITE 3129-91-7

HEALTH AND SAFETY LISTS

U.S. DOT - Substances From 49 CFR 172.101
regulated by DOT (UN2687)

U.S. DOT - Hazard Classes
DOT hazard class = 4.1

NIOSH - Selected LD50s and LC50s
Oral, rat: LD50 = 284 mg/kg

DICYCLOHEXYLAMMONIUM NITRATE 3882-06-2

HEALTH AND SAFETY LISTS

NIOSH - Selected LD50s and LC50s
Oral, mouse: LD50 = 225 mg/kg

INTERNATIONAL LISTS

Canada - WHMIS: Ingredient Disclosure
1% item 566 (1204)

STATE LISTS

NJ Right to Know List (Total)
sn 0679

N,N-DICYCLOHEXYL-2-BENZOTHIAZOLE SULFENAMIDE 4979-32-2

HEALTH AND SAFETY LISTS

NIOSH - Selected LD50s and LC50s
Oral, rat: LD50 = 6420 mg/kg

ENVIRONMENTAL LISTS

EPA - Master Testing List
[present]

DICYCLOHEXYLCARBODIIMIDE 538-75-0

HEALTH AND SAFETY LISTS

NTP Chemical Status Reports - Testing Status and NTIS Number
Prechronic studies in progress; approved for toxicology/carcinogenesis study

DICYCLOHEXYL PEROXY-DICARBONATE 1561-49-5

HEALTH AND SAFETY LISTS

U.S. DOT - Organic Peroxides Table
Organic peroxide UN3112; UN3114

STATE LISTS

NJ Right to Know List (Total)
sn 0680

DICYCLOHEXYL PEROXY-DICARBONATE 1705-60-8

HEALTH AND SAFETY LISTS

U.S. DOT - Organic Peroxides Table
Organic peroxide UN3106

N,N'-DICYCLOHEXYLTHIOUREA 1212-29-9

HEALTH AND SAFETY LISTS

NTP Chemical Status Reports - Testing Status and NTIS Number
Technical reports printed (PB281539/AS)

NTP Chemical Status Reports - Evidence of Carcinogenicity
male rat-negative; female rat-negative; male mice-negative; female mice-negative

STATE LISTS

Massachusetts Right To Know List
[present]

DICYCLOPENTADIENE 77-73-6

HEALTH AND SAFETY LISTS

ACGIH 1995 - Time Weighted Averages
5 ppm TWA; 27 mg/m3 TWA

AIHA - Odor Threshold Values
geometric mean air odor threshold = 0.011 ppm (detectable); 0.020 ppm (recognizable)

U.S. DOT - Substances From 49 CFR 172.101
regulated by DOT (UN2048)

U.S. DOT - Hazard Classes
DOT hazard class = 3

NFPA - Flash Points
flash point = 90 degrees F (32 degrees C)

NFPA - Hazard Identification Ratings
health-1; flammability-3; reactivity-1

NIOSH - Selected LD50s and LC50s
Inhalation, rat: LC50 = 359 ppm 4 hr Oral, rat: LD50 = 353 mg/kg
Skin, rabbit: LD50 = 5080 mg/kg

NTP Chemical Status Reports - Testing Status and NTIS Number
Project leader assigned/study in design

OSHA - Vacated PELs - Time Weighted Averages
5 ppm TWA; 30 mg/m3 TWA

ENVIRONMENTAL LISTS

CERCLA/SARA - Section 313 - Emission Reporting
form R reporting required

List of Pesticide Product Inert Ingredients
[present]

INTERNATIONAL LISTS

Australian Exposure Standards - Time Weighted Averages
5 ppm TWA; 27 mg/m3 TWA

Canada - WHMIS: Ingredient Disclosure
1% item 567 (691)

Canada - Alberta - 8 Hour Occupational Exposure Limit
5 ppm TWA;   27 mg/m3 TWA

Canada - Alberta - 15 Minute Occupational Exposure Limit
10 ppm STEL;   54 mg/m3 STEL

Canada - British Columbia - 8 Hour Exposure Limits
5 ppm TWA;   30 mg/m3 TWA

Canada - Ontario - OHSA - TWAEVs
5 ppm TWAEV;   27 mg/m3 TWAEV

Canada - Quebec - Time-Weighted Average Exposure Values
5 ppm TWAEV;   27 mg/m3 TWAEV

United Kingdom - Occupational Exposure Standards - TWAs
5 ppm TWA;   30 mg/m3 TWA

German (DFG) - Peak Limitations
2 x normal MAK (5 min momentary value);   don't exceed 8 times during shift

Israel - Time Weighted Averages
5 ppm TWA;   27 mg/m3 TWA

Israel - Action Levels
2.5 ppm AL;   13.5 mg/m3 AL

Mexico - Instruction No. 10 - TWAs
5 ppm TWA;   30 mg/m3 TWA

STATE LISTS

California - Exposure Limits - PELs
5 ppm PEL;   30 mg/m3 PEL

California - Directors List of Hazardous Substances (8 CCR 339)
[present]

Florida Hazardous Substance List
[present]

Massachusetts Right To Know List
[present]

Minnesota Hazardous Substance List
[present]

NJ Right to Know List (Total)
sn 0681

NJ Special Hazardous Substances
(flammable - third degree)

Pennsylvania Right to Know List
[present]

DICYCLOPENTADIENE, POLYMER WITH (MIXED STYRENE AND ALPHA-METHYLSTYRENE), (MIXED INDENE AND METHYL INDENE), AND VINYL TOLUENE   RR-01065-1

ENVIRONMENTAL LISTS

List of Pesticide Product Inert Ingredients
[present]

DICYCLOPENTADIENYL IRON   102-54-5

HEALTH AND SAFETY LISTS

ACGIH 1995 - Time Weighted Averages
10 mg/m3 TWA

NIOSH - Selected LD50s and LC50s
Oral, rat: LD50 = 1320 mg/kg

NTP Chemical Status Reports - Testing Status and NTIS Number
Prechronic studies for which toxicity technical reports were not prepared

OSHA - Vacated PELs - Time Weighted Averages
total dust: 10 mg/m3 TWA;   respirable fraction: 5 mg/m3 TWA

OSHA - Final PELs - Time Weighted Averages
total dust: 15 mg/m3 TWA;   respirable fraction: 5 mg/m3 TWA

INTERNATIONAL LISTS

Australian Exposure Standards - Time Weighted Averages
10 mg/m3 TWA

Canada - WHMIS: Ingredient Disclosure
1% item 763 (880)

Canada - Alberta - 8 Hour Occupational Exposure Limit
10 mg/m3 TWA

Canada - Alberta - 15 Minute Occupational Exposure Limit
20 mg/m3 STEL

Canada - British Columbia - 8 Hour Exposure Limits
10 mg/m3 TWA

Canada - British Columbia - 15 Minute Exposure Limits
20 mg/m3 STEL

Canada - Ontario - OHSA - TWAEVs
10 mg/m3 TWAEV

Canada - Quebec - Time-Weighted Average Exposure Values
10 mg/m3 TWAEV

United Kingdom - Occupational Exposure Standards - TWAs
10 mg/m3 TWA

United Kingdom - Occupational Exposure Standards - STELs
20 mg/m3 STEL

Israel - Time Weighted Averages
10 mg/m3 TWA

Israel - Action Levels
5 mg/m3 AL

Mexico - Instruction No. 10 - TWAs
10 mg/m3 TWA

Mexico - Instruction No. 10 - STELs
20 mg/m3 STEL

STATE LISTS

California - Exposure Limits - PELs
total dust: 10 mg/m3 PEL;   respirable fraction: 5 mg/m3 PEL

California - Directors List of Hazardous Substances (8 CCR 339)
[present]

Florida Hazardous Substance List
[present]

Massachusetts Right To Know List
[present]

Minnesota Hazardous Substance List
[present]

Pennsylvania Right to Know List
[present]

DICYCLOHEXYL ADIPATE   849-99-0

HEALTH AND SAFETY LISTS

NIOSH - Selected LD50s and LC50s
Oral, rat: LD50 = 16 gm/kg

DIDECYLDIMETHYLAMMONIUM CHLORIDE   7173-51-5

HEALTH AND SAFETY LISTS

NIOSH - Selected LD50s and LC50s
Oral, rat: LD50 = 84 mg/kg

DIDECYL ETHER   2456-28-2

HEALTH AND SAFETY LISTS

NFPA - Hazard Identification Ratings
health-0;   flammability-1;   reactivity-0

2',3'-DIDEOXYCYTIDINE (AIDS INITIATIVE)   7481-89-2

HEALTH AND SAFETY LISTS

NTP Chemical Status Reports - Testing Status and NTIS Number
Prechronic studies completed: in review for further evaluation; prechronic studies exist for which toxicity technical reports were not prepared;   prechronic studies in progress

2',3'-DIDEOXYINOSINE (AIDS INITIATIVE)   69655-05-6

HEALTH AND SAFETY LISTS

NTP Chemical Status Reports - Testing Status and NTIS Number
Prechronic studies for which toxicity technical reports were not prepared

2,2-DI(4,4-DI-T-BUTYL-PEROXY CYCLOHEXYL)- RR-00256-2
PROPANE 42%
STATE LISTS
NJ Right to Know List (Total)
sn 2325

DI-2,4-DICHLOROBENZOYL PEROXIDE   RR-00257-3
HEALTH AND SAFETY LISTS
U.S. DOT - Organic Peroxides Table
Organic peroxide UN3102; UN3106
STATE LISTS
NJ Right to Know List (Total)
sn 2326; sn 2327; sn 2328

DIDYMIUM NITRATE   RR-00581-2
HEALTH AND SAFETY LISTS
U.S. DOT - Substances From 49 CFR 172.101
regulated by DOT (UN1465)
U.S. DOT - Hazard Classes
DOT hazard class = 5.1
STATE LISTS
NJ Right to Know List (Total)
sn 2330

DIELDRIN   60-57-1
SEE ALSO:
F039-HAZARDOUS WASTES
HEALTH AND SAFETY LISTS
ACGIH 1995 - Time Weighted Averages
0.25 mg/m3 TWA
ACGIH 1995 - Skin Designations
skin - potential for cutaneous absorption
U.S. DOT - Substances From 49 CFR 172.101
regulated by DOT (NA2761)
U.S. DOT - Hazard Classes
DOT hazard class = 6.1
U.S. DOT - Appendix B - Marine Pollutants
DOT regulated severe marine pollutant
U.S. DOT - Appendix A Table 1 - Hazardous Substances
final RQ = 1 pound (0.454 kg)
IARC - Group 3 (not classifiable)
[present]
NIOSH - Selected LD50s and LC50s
Inhalation, rat: LC50 = 43 mg/m3 4 hr Oral, rat: LD50 = 38300 ug/kg Skin, rat: LD50 = 10 mg/kg
NIOSH 1990 - Pocket Guide - RELs
0.25 mg/m3 TWA
NIOSH 1990 - Pocket Guide - IDLHs
450 mg/m3 IDLH (not considering carcinogenic effects)
NIOSH 1990 - Pocket Guide - Carcinogens
occupational carcinogen
NIOSH 1990 - Pocket Guide - Target organs
CNS, liver, skin, kidneys
NIOSH 1990 - Pocket Guide - Skin list
Potential for dermal absorption
NIOSH - Health Standards - Exposure Limits
reduce exposure to lowest reliably detectable concentration (Listed under 'Aldrin/dieldrin')
NIOSH - Health Standards - Health Effects and Precautions
has produced tumors of the lungs, liver, thyroid, and adrenal glands in animals (prevent skin contact) (Listed under 'Aldrin/dieldrin')

NIOSH - Health Standards - Carcinogenic Chemicals
potential human carcinogen (Listed under 'Aldrin/dieldrin')
NTP Chemical Status Reports - Testing Status and NTIS Number
Technical reports printed (PB275666/AS) (PB275676/AS)
NTP Chemical Status Reports - Evidence of Carcinogenicity
PB275676/AS: male rat-negative; female rat-negative; PB275666/AS: male rat-negative; female rat-negative; male mice-equivocal; female mice-negative
OSHA - Vacated PELs - Time Weighted Averages
0.025 mg/m3 TWA
OSHA - Vacated PELs - Skin Designation
Prevent or reduce skin absorption
OSHA - Final PELs - Time Weighted Averages
0.25 mg/m3 TWA
OSHA - Final PELs - Skin Notations
prevent or reduce skin absorption
ENVIRONMENTAL LISTS
ATSDR Priority List
Rank (of 275): 020
CERCLA/SARA - Hazardous Substances and their Reportable Quantities
final RQ = 1 pound (0.454 kg)
Clean Water Act - Hazardous Substances
[present]
Clean Water Act - Priority Pollutants
[present]
Clean Water Act - Toxic Pollutants
[present] (Listed under 'Aldrin/dieldrin')
EPA - Carcinogen Hazard Ranking for RQ Adjustment
Hazard ranking = High
RCRA - P Series Wastes
waste number P037
RCRA - Hazardous Constituents-Appendix VIII
waste number P037
RCRA - Basis for Listing - Appendix VII
Included in waste stream: F039
RCRA - Substances Banned From Land Disposal
[present]
RCRA - TSD Facilities Ground Water Monitoring
TM 8080 = 0.05 ug/L PQL; TM 8270 = 10 ug/L PQL
RCRA - Universal Treatment Standards (LDR)
WW: 0.017 mg/l; NWW: 0.13 mg/kg
INTERNATIONAL LISTS
Australian Exposure Standards - Time Weighted Averages
0.25 mg/m3 TWA
Australian Exposure Standards - Skin Effects
skin absorption
Canada - CEPA Schedule II Part II - Toxic Substances (Export)
[present]
Canada - Alberta - 8 Hour Occupational Exposure Limit
0.25 mg/m3 TWA
Canada - Alberta - 15 Minute Occupational Exposure Limit
0.75 mg/m3 STEL
Canada - Alberta - Skin Designation
can be absorbed through the intact skin
Canada - British Columbia - 8 Hour Exposure Limits
0.25 mg/m3 TWA
Canada - British Columbia - 15 Minute Exposure Limits
0.75 mg/m3 STEL
Canada - British Columbia - Skin Notations
skin - potential for skin absorption
Canada - Ontario - OHSA - TWAEVs
0.25 mg/m3 TWAEV

Canada - Ontario - OHSA - Skin Notations
absorption through skin, eyes, or mucous membranes

Canada - Quebec - Time-Weighted Average Exposure Values
0.25 mg/m3 TWAEV

Canada - Quebec - Skin Designations
absorbed through the skin

United Kingdom - Occupational Exposure Standards - TWAs
0.25 mg/m3 TWA

United Kingdom - Occupational Exposure Standards - STELs
0.75 mg/m3 STEL

United Kingdom - Occupational Exposure Standards - Notes
can be absorbed through skin

German (DFG) - MAK Values
total dust: 0.25 mg/m3 MAK

German (DFG) - Peak Limitations
10 x normal MAK (30 min average value); don't exceed during shift

German (DFG) - Skin/Sensitizers
danger of cutaneous absorption

Israel - Time Weighted Averages
0.25 mg/m3 TWA

Israel - Action Levels
0.125 mg/m3 AL

Mexico - Instruction No. 10 - TWAs
0.25 mg/m3 TWA

Mexico - Instruction No. 10 - STELs
0.75 mg/m3 STEL

Mexico - Instruction No. 10 - Skin designation
skin - potential for cutaneous absorption

Mexico - Wastewater - Organic Toxic Pollutants and Heavy Metals
Listed under [Pesticides and Metabolites]

Mexico - Drinking Water - Ecological Criteria
0.0000007 mg/l Substance presents persistence, bioaccumulations or risk of cancer, reduce human exposure to a minimum; This level has been extrapolated by using a mathematic model

STATE LISTS

California - Air Bill 2588 Appendix A-II
known or potential carcinogen: 9/89

California - Prop. 65 - Cancer list
carcinogen - initial date 7/1/88

California - Prop. 65 - No Significant Risk Levels
no significant risk level = 0.04 ug/day

California - Exposure Limits - PELs
0.25 mg/m3 PEL

California - Exposure Limits - Skin Notation
material may be absorbed through the skin, eyes or mucous membrane

California - Directors List of Hazardous Substances (8 CCR 339)
[present]

Florida Hazardous Substance List
[present]

Massachusetts Right To Know List
carcinogen; extraordinarily hazardous

Minnesota Hazardous Substance List
skin

NJ Right to Know List (Total)
sn 0683

NJ Special Hazardous Substances
(carcinogen; mutagen)

Pennsylvania Right to Know List
environmental hazard

DIENOESTROL   84-17-3

HEALTH AND SAFETY LISTS

IARC - Group Unspecified
[present] (Listed under 'Nonsteroidal oestrogens')

STATE LISTS

California - Air Bill 2588 Appendix A-II
known or potential carcinogen: 9/90

California - Prop. 65 - Cancer list
carcinogen - initial date 1/1/90

California - Directors List of Hazardous Substances (8 CCR 339)
[present]

Florida Hazardous Substance List
[present]

Massachusetts Right To Know List
carcinogen; extraordinarily hazardous

Minnesota Hazardous Substance List
carcinogen

NJ Right to Know List (Total)
sn 0684

NJ Special Hazardous Substances
(carcinogen)

Pennsylvania Right to Know List
special hazardous substance

Pennsylvania RTK - Special Hazardous Substances
[present]

DL-DIEPOXYBUTANE   298-18-

HEALTH AND SAFETY LISTS

NIOSH - Selected LD50s and LC50s
Inhalation, rat: LC50 = 56 ppm 4 hr Oral, rat: LD50 = 210 mg/kg Skin, rabbit: LD50 = 800 mg/kg

DIEPOXYBUTANE   1464-53-

HEALTH AND SAFETY LISTS

U.S. DOT - Appendix A Table 1 - Hazardous Substances
final RQ = 10 pounds (4.54 kg)

IARC - Group 2B (sufficient animal data)
[present] (Overall evaluation based only on evidence of carcinogenicity in monograph (11, 1976) or in Supplement 4)

NIOSH - Selected LD50s and LC50s
Inhalation, rat: LC50 = 90 ppm 4 hr Oral, rat: LD50 = 78 mg/kg Skin, rabbit: LD50 = 80 mg/kg

NTP Seventh Report - Suspect Carcinogens
suspect carcinogen

OSHA - Possible Select Carcinogens
[present]

ENVIRONMENTAL LISTS

CERCLA/SARA - Section 302 Extremely Hazardous Substances and TPQs
TPQ = 500 pounds

CERCLA/SARA - Section 313 - Emission Reporting
form R reporting required for 0.1% de minimus concentration

CERCLA/SARA - Hazardous Substances and their Reportable Quantities
final RQ = 10 pounds (4.54 kg)

EPA - Carcinogen Hazard Ranking for RQ Adjustment
Hazard ranking = Medium

RCRA - U Series Wastes
waste number U085 (Ignitable waste; Toxic waste)

RCRA - Hazardous Constituents-Appendix VIII
waste number U085

RCRA - Substances Banned From Land Disposal
[present]

TSCA - Code of Federal Regulations Citations
40 CFR 712.30(d); 40 CFR 716.120(a)

TSCA - PAIR - Reporting List
Reporting Date: November 19, 1982

TSCA - Health and Safety Reporting List
Effective Date: October 4, 1982

INTERNATIONAL LISTS

Canada - WHMIS: Ingredient Disclosure
0.1% item 568 (692)

STATE LISTS

California - Air Bill 2588 Appendix A-II
known or potential carcinogen

California - Prop. 65 - Cancer list
carcinogen - initial date 1/1/88

California - Directors List of Hazardous Substances (8 CCR 339)
[present] (exempt when part of a cured epoxy or rubber)

Florida Hazardous Substance List
[present]

Massachusetts Right To Know List
carcinogen;   extraordinarily hazardous

Minnesota Hazardous Substance List
carcinogen

NJ Right to Know List (Total)
sn 0685

NJ Special Hazardous Substances
(carcinogen)

Pennsylvania Right to Know List
environmental hazard;   special hazardous substance

Pennsylvania RTK - Special Hazardous Substances
[present]

DIESEL ENGINE EXHAUST   RR-00270-0

HEALTH AND SAFETY LISTS

IARC - Group 2A (limited human data)
[present] (Includes both whole and extracts of diesel engine exhaust)

NIOSH - Health Standards - Exposure Limits
reduce exposure to lowest feasible concentration

NIOSH - Health Standards - Health Effects and Precautions
Lung cancer, respiratory sytem effects, eye irritation

NIOSH - Health Standards - Carcinogenic Chemicals
potential human carcinogen

OSHA - Possible Select Carcinogens
[present]

INTERNATIONAL LISTS

German (DFG) - Carcinogens
animal evidence of carcinogenicity

STATE LISTS

California - Air Bill 2588 Appendix A-I
particulate matter, known or potential carcinogen: 9/90;   total organic gas, known or potential carcinogen: 9/90

California - Prop. 65 - Cancer list
carcinogen - initial date 10/1/90

DIESEL OIL (PETROLEUM)   68334-30-5

HEALTH AND SAFETY LISTS

U.S. DOT - Substances From 49 CFR 172.101
regulated by DOT (NA1993)

U.S. DOT - Hazard Classes
DOT hazard class = 3

IARC - Group 2B (sufficient animal data)
[present] (Listed under 'Diesel fuels')

NFPA - Flash Points
(flash point) 1-D: 100 degrees F (38 degrees C) or legal;   2-D: 125 degrees F (52 degrees C) or legal;   4-D: 130 degrees F (54 degrees C) or legal

NFPA - Hazard Identification Ratings
health-0;   flammability-2;   reactivity-0

NTP Chemical Status Reports - Evidence of Carcinogenicity
male mice-equivocal evidence;   female mice-equivocal evidence

OSHA - Possible Select Carcinogens
[present]

STATE LISTS

California - Air Bill 2588 Appendix A-I
known or potential carcinogen: 6/91

NJ Right to Know List (Total)
sn 2444

Pennsylvania Right to Know List
[present]

DIETHANOLAMIDE OF LINOLEIC ACID   56863-02-6

ENVIRONMENTAL LISTS

List of Pesticide Product Inert Ingredients
[present]

DIETHANOLAMINE   111-42-2

HEALTH AND SAFETY LISTS

ACGIH 1995 - Time Weighted Averages
0.46 ppm TWA;   2 mg/m3 TWA

ACGIH 1995 - Skin Designations
skin - potential for cutaneous absorption

NFPA - Flash Points
flash point = 342 degrees F (172 degrees C)

NFPA - Hazard Identification Ratings
health-1;   flammability-1;   reactivity-0

NIOSH - Selected LD50s and LC50s
Oral, rat: LD50 = 710 mg/kg Skin, rabbit: LD50 = 12200 mg/kg

NTP Chemical Status Reports - Testing Status and NTIS Number
Technical reports printed (PB93-133999);   two year studies: pathology quality assessment in progress

OSHA - Vacated PELs - Time Weighted Averages
3 ppm TWA;   15 mg/m3 TWA

ENVIRONMENTAL LISTS

CERCLA/SARA - Section 313 - Emission Reporting
form R reporting required for 1.0% de minimus concentration

CERCLA/SARA - Hazardous Substances and their Reportable Quantities
final RQ = 1 pound (.454 kg)

Clean Air Act (1990) - List of Hazardous Air Contaminants
[present]

CAA - HON Rule - SOCMI Chemicals
compliance by Oct. 24, 1994

CAA - HON Rule - Organic HAPs
[present]

EPA - Master Testing List
[present]

List of Pesticide Product Inert Ingredients
[present]

TSCA - Code of Federal Regulations Citations
40 CFR 712.30(w);   40 CFR 716.120(a)

TSCA - PAIR - Reporting List
Reporting Date: June 13, 1989

TSCA - Health and Safety Reporting List
Effective Date: April 13, 1989

INTERNATIONAL LISTS

Australian Exposure Standards - Time Weighted Averages
3 ppm TWA;   13 mg/m3 TWA

Canada - WHMIS: Ingredient Disclosure
1% item 569 (693)

Canada - NPRI (National Pollutant Release Inventory)
[present]

Canada - Alberta - 8 Hour Occupational Exposure Limit
3 ppm TWA; 13 mg/m3 TWA

Canada - Alberta - 15 Minute Occupational Exposure Limit
6 ppm STEL; 26 mg/m3 STEL

Canada - Ontario - OHSA - TWAEVs
3 ppm TWAEV; 13 mg/m3 TWAEV

Canada - Quebec - Time-Weighted Average Exposure Values
3 ppm TWAEV; 13 mg/m3 TWAEV

United Kingdom - Occupational Exposure Standards - TWAs
3 ppm TWA; 15 mg/m3 TWA

Israel - Time Weighted Averages
3 ppm TWA; 13 mg/m3 TWA

Israel - Action Levels
1.5 ppm AL; 6.5 mg/m3 AL

STATE LISTS

California - Air Bill 2588 Appendix A-I
6/91

California - Exposure Limits - PELs
3 ppm PEL; 15 mg/m3 PEL

California - Directors List of Hazardous Substances (8 CCR 339)
[present]

Florida Hazardous Substance List
[present]

Massachusetts Right To Know List
[present]

Minnesota Hazardous Substance List
[present]

NJ Right to Know List (Total)
sn 0686

Pennsylvania Right to Know List
environmental hazard

DIETHANOLAMINE ETHYLENEDIAMINETETRAACETATE   68133-37-9

ENVIRONMENTAL LISTS

List of Pesticide Product Inert Ingredients
[present]

DIETHANOLAMINE LAURATE   7487-79-8

ENVIRONMENTAL LISTS

List of Pesticide Product Inert Ingredients
[present]

DIETHANOLAMMONIUM DODECYL SULFATE   143-00-0

ENVIRONMENTAL LISTS

List of Pesticide Product Inert Ingredients
[present]

DIETHANOL NITROSAMINE DINITRATE   RR-01412-0

HEALTH AND SAFETY LISTS

U.S. DOT - Hazard Classes
Forbidden from transport by the DOT

DIETHATYL ETHYL   38727-55-8

ENVIRONMENTAL LISTS

CERCLA/SARA - Section 313 - Emission Reporting
form R reporting required

2,2-DIETHOXYACETOPHENONE   6175-45-7

HEALTH AND SAFETY LISTS

NIOSH - Selected LD50s and LC50s
Oral, rat: LD50 = 5660 mg/kg Skin, rabbit: LD50 = 11300 mg/kg

DIETHOXYMETHANE   462-95-3

HEALTH AND SAFETY LISTS

U.S. DOT - Substances From 49 CFR 172.101
regulated by DOT (UN2373)

U.S. DOT - Hazard Classes
DOT hazard class = 3

NIOSH - Selected LD50s and LC50s
Oral, rabbit: LD50 = 2604 mg/kg

STATE LISTS

NJ Right to Know List (Total)
sn 0687

2,5-DIETHOXY-4-MORPHOLINOBENZENEDIAZO-NIUM ZINC CHLORIDE   RR-01360-5

HEALTH AND SAFETY LISTS

U.S. DOT - Substances From 49 CFR 172.101
regulated by DOT (UN3036)

U.S. DOT - Hazard Classes
DOT hazard class = 4.1

3,3-DIETHOXYPROPENE   3054-95-3

HEALTH AND SAFETY LISTS

U.S. DOT - Substances From 49 CFR 172.101
regulated by DOT (UN2374)

U.S. DOT - Hazard Classes
DOT hazard class = 3

STATE LISTS

NJ Right to Know List (Total)
sn 0688

N,N-DIETHYLACETOACETAMIDE   2235-46-3

HEALTH AND SAFETY LISTS

NFPA - Flash Points
flash point = 250 degrees F (121 degrees C)

NFPA - Hazard Identification Ratings
health-0; flammability-1; reactivity-0

DIETHYL ACETOACETATE   1619-57-4

HEALTH AND SAFETY LISTS

NFPA - Flash Points
flash point = 170 degrees F (77 degrees C)

NFPA - Hazard Identification Ratings
health-2; flammability-2; reactivity-0

STATE LISTS

Florida Hazardous Substance List
[present]

Massachusetts Right To Know List
[present]

Pennsylvania Right to Know List
[present]

DIETHYL ADIPATE   141-28-6

INTERNATIONAL LISTS

Canada - WHMIS: Ingredient Disclosure
0.1% item 570 (163)

DIETHYLALUMINUM CHLORIDE   96-10-6

HEALTH AND SAFETY LISTS

NFPA - Flash Points
ignites spontaneously in air

NFPA - Hazard Identification Ratings
health-3; flammability-4; reactivity-3 (do not use water, foam or halogenated extinguishing agents)

OSHA - List of Highly Hazardous Chemicals
threshhold quantity = 5000 pounds

STATE LISTS

Florida Hazardous Substance List
[present]

Massachusetts Right To Know List
[present]

NJ Right to Know List (Total)
sn 0689

NJ Special Hazardous Substances
(flammable - third degree; reactive - third degree)

Pennsylvania Right to Know List
[present]

DIETHYLALUMINUM HYDRIDE   871-27-2

HEALTH AND SAFETY LISTS

NFPA - Flash Points
ignites spontaneously in air

NFPA - Hazard Identification Ratings
flammability-3; reactivity-3 (do not use water, foam or halogenated extinguishing agents)

STATE LISTS

Florida Hazardous Substance List
[present]

Massachusetts Right To Know List
[present]

Pennsylvania Right to Know List
[present]

DIETHYLAMINE   109-89-7

HEALTH AND SAFETY LISTS

ACGIH 1995 - Time Weighted Averages
5 ppm TWA; 15 mg/m3 TWA

ACGIH 1995 - Short Term Exposure Limits
15 ppm STEL; 45 mg/m3 STEL

ACGIH 1995 - Skin Designations
skin - potential for cutaneous absorption

ACGIH 1995 - Carcinogens
A4-not classifiable as a human carcinogen

AIHA - Odor Threshold Values
geometric mean air odor threshold = 0.053 ppm (detectable); 0.75 ppm (recognizable)

U.S. DOT - Substances From 49 CFR 172.101
regulated by DOT (UN1154)

U.S. DOT - Hazard Classes
DOT hazard class = 3

U.S. DOT - Appendix A Table 1 - Hazardous Substances
final RQ = 100 pounds (45.4 kg)

NFPA - Flash Points
flash point = -9 degrees F (-23 degrees C)

NFPA - Hazard Identification Ratings
health-3; flammability-3; reactivity-0

NIOSH - Selected LD50s and LC50s
Inhalation, rat: LC50 = 4000 ppm 4 hr Oral, rat: LD50 = 540 mg/kg Skin, rabbit: LD50 = 820 mg/kg

NIOSH 1990 - Pocket Guide - RELs
10 ppm TWA; 30 mg/m3 TWA; 25 ppm STEL; 75 mg/m3 STEL

NIOSH 1990 - Pocket Guide - IDLHs
2000 ppm IDLH

NIOSH 1990 - Pocket Guide - Target organs
skin, eyes, respiratory system

OSHA - Vacated PELs - Time Weighted Averages
10 ppm TWA; 30 mg/m3 TWA

OSHA - Vacated PELs - Short Term Exposure Limits
25 ppm STEL; 75 mg/m3 STEL

OSHA - Final PELs - Time Weighted Averages
25 ppm TWA; 75 mg/m3 TWA

ENVIRONMENTAL LISTS

CERCLA/SARA - Hazardous Substances and their Reportable Quantities
final RQ = 100 pounds (45.4 kg)

CAA - HON Rule - SOCMI Chemicals
compliance by July 24, 1995

Clean Water Act - Hazardous Substances
[present]

List of Pesticide Product Inert Ingredients
[present]

TSCA - Code of Federal Regulations Citations
40 CFR 716.120(a)

TSCA - Health and Safety Reporting List
Effective Date: June 1, 1987

INTERNATIONAL LISTS

Australian Exposure Standards - Time Weighted Averages
10 ppm TWA; 30 mg/m3 TWA

Australian Exposure Standards - Short Term Exposure Limits
25 ppm STEL; 75 mg/m3 STEL

Canada - WHMIS: Ingredient Disclosure
1% item 571 (694)

Canada - Alberta - 8 Hour Occupational Exposure Limit
10 ppm TWA; 30 mg/m3 TWA

Canada - Alberta - 15 Minute Occupational Exposure Limit
25 ppm STEL; 75 mg/m3 STEL

Canada - British Columbia - 8 Hour Exposure Limits
25 ppm TWA; 75 mg/m3 TWA

Canada - Ontario - OHSA - TWAEVs
10 ppm TWAEV; 30 mg/m3 TWAEV

Canada - Ontario - OHSA - STEVs
25 ppm STEV; 75 mg/m3 STEV

Canada - Quebec - Time-Weighted Average Exposure Values
10 ppm TWAEV; 30 mg/m3 TWAEV

Canada - Quebec - Short-term Exposure Values
25 ppm STEV; 75 mg/m3 STEV

United Kingdom - Occupational Exposure Standards - TWAs
10 ppm TWA; 30 mg/m3 TWA

United Kingdom - Occupational Exposure Standards - STELs
25 ppm STEL; 75 mg/m3 STEL

German (DFG) - MAK Values
10 ppm MAK; 30 mg/m3 MAK

German (DFG) - Peak Limitations
2 x normal MAK (10 min momentary value); don't exceed 4 times per shift

Israel - Time Weighted Averages
10 ppm TWA; 30 mg/m3 TWA

Israel - Short Term Exposure Limits
25 ppm STEL; 75 mg/m3 STEL

Israel - Action Levels
5 ppm AL; 15 mg/m3 AL

Mexico - Instruction No. 10 - TWAs
10 ppm TWA; 30 mg/m3 TWA

Mexico - Instruction No. 10 - STELs
25 ppm STEL; 75 mg/m3 STEL

STATE LISTS

California - Exposure Limits - PELs
10 ppm PEL; 30 mg/m3 PEL

California - Exposure Limits - STELs
25 ppm STEL; 75 mg/m3 STEL

California - Directors List of Hazardous Substances (8 CCR 339)
[present]

Florida Hazardous Substance List
[present]

Massachusetts Right To Know List
[present]

Minnesota Hazardous Substance List
[present]

NJ Right to Know List (Total)
sn 0690

NJ Special Hazardous Substances
(flammable - third degree)

Pennsylvania Right to Know List
environmental hazard

2-DIETHYLAMINOETHANOL   100-37-8

HEALTH AND SAFETY LISTS

ACGIH 1995 - Time Weighted Averages
2 ppm TWA; 9.6 mg/m3 TWA

ACGIH 1995 - Skin Designations
skin - potential for cutaneous absorption

AIHA - Odor Threshold Values
geometric mean air odor threshold = 0.011 ppm (detectable); 0.040 ppm (recognizable)

U.S. DOT - Substances From 49 CFR 172.101
regulated by DOT (UN2686)

U.S. DOT - Hazard Classes
DOT hazard class = 3

NFPA - Flash Points
flash point = 140 degrees F (60 degrees C)

NFPA - Hazard Identification Ratings
health-3; flammability-2; reactivity-0

NIOSH - Selected LD50s and LC50s
Inhalation, mouse: LC50 = 5000 mg/m3 (8 hr) Oral, rat: LD50 = 1300 mg/kg Skin, guinea pig: LD50 = 884 mg/kg

NIOSH 1990 - Pocket Guide - RELs
10 ppm TWA; 50 mg/m3 TWA

NIOSH 1990 - Pocket Guide - IDLHs
500 ppm IDLH

NIOSH 1990 - Pocket Guide - Target organs
skin, eyes, respiratory system

NIOSH 1990 - Pocket Guide - Skin list
Potential for dermal absorption

OSHA - Vacated PELs - Time Weighted Averages
10 ppm TWA; 50 mg/m3 TWA

OSHA - Vacated PELs - Skin Designation
Prevent or reduce skin absorption

OSHA - Final PELs - Time Weighted Averages
10 ppm TWA; 50 mg/m3 TWA

OSHA - Final PELs - Skin Notations
prevent or reduce skin absorption

ENVIRONMENTAL LISTS

List of Pesticide Product Inert Ingredients
[present]

INTERNATIONAL LISTS

Australian Exposure Standards - Time Weighted Averages
10 ppm TWA; 48 mg/m3 TWA

Australian Exposure Standards - Skin Effects
skin absorption

Canada - WHMIS: Ingredient Disclosure
1% item 572 (695)

Canada - Alberta - 8 Hour Occupational Exposure Limit
10 ppm TWA; 48 mg/m3 TWA

Canada - Alberta - 15 Minute Occupational Exposure Limit
20 ppm STEL; 96 mg/m3 STEL

Canada - Alberta - Skin Designation
can be absorbed through the intact skin

Canada - British Columbia - 8 Hour Exposure Limits
10 ppm TWA; 50 mg/m3 TWA

Canada - British Columbia - Skin Notations
skin - potential for skin absorption

Canada - Ontario - OHSA - TWAEVs
10 ppm TWAEV; 48 mg/m3 TWAEV

Canada - Ontario - OHSA - Skin Notations
absorption through skin, eyes, or mucous membranes

Canada - Quebec - Time-Weighted Average Exposure Values
10 ppm TWAEV; 48 mg/m3 TWAEV

Canada - Quebec - Skin Designations
absorbed through the skin

United Kingdom - Occupational Exposure Standards - TWAs
10 ppm TWA; 50 mg/m3 TWA

United Kingdom - Occupational Exposure Standards - Notes
can be absorbed through skin

German (DFG) - MAK Values
10 ppm MAK; 50 mg/m3 MAK

German (DFG) - Skin/Sensitizers
danger of cutaneous absorption

Israel - Time Weighted Averages
10 ppm TWA; 48 mg/m3 TWA

Israel - Action Levels
5 ppm AL; 24 mg/m3 AL

Mexico - Instruction No. 10 - TWAs
10 ppm TWA; 50 mg/m3 TWA

Mexico - Instruction No. 10 - Skin designation
skin - potential for cutaneous absorption

STATE LISTS

California - Exposure Limits - PELs
10 ppm PEL; 50 mg/m3 PEL

California - Exposure Limits - Skin Notation
material may be absorbed through the skin, eyes or mucous membrane

California - Directors List of Hazardous Substances (8 CCR 339)
[present]

Florida Hazardous Substance List
[present]

Massachusetts Right To Know List
[present]

Minnesota Hazardous Substance List
skin

NJ Right to Know List (Total)
sn 0691

Pennsylvania Right to Know List
[present]

DIETHYLAMINO ETHANOLAMINE   58145-14-5

ENVIRONMENTAL LISTS

List of Pesticide Product Inert Ingredients
[present]

2-(DIETHYLAMINO)ETHYL ACRYLATE   2426-54-2

HEALTH AND SAFETY LISTS

NFPA - Flash Points
flash point = 195 degrees F (91 degrees C)

NFPA - Hazard Identification Ratings
health-2; flammability-2; reactivity-1

ENVIRONMENTAL LISTS

TSCA - Code of Federal Regulations Citations
40 CFR 712.30(d)

TSCA - PAIR - Reporting List
Reporting Date: November 19, 1982

STATE LISTS

Florida Hazardous Substance List
[present]

Massachusetts Right To Know List
[present]

Pennsylvania Right to Know List
[present]

2-(N,N-DIETHYLAMINO)ETHYL METHACRYLATE   105-16-8

ENVIRONMENTAL LISTS

TSCA - Code of Federal Regulations Citations
40 CFR 712.30(d)

TSCA - PAIR - Reporting List
Reporting Date: November 19, 1982

INTERNATIONAL LISTS

Canada - WHMIS: Ingredient Disclosure
1% item 573 (1089)

2-(DIETHYLAMINO)ETHYL METHACRYLATE, POLYMER WITH DODECYL METHACRYLATE, STYRENE, HEXADEXYL METHACRYLATE AND TETRADECYL METHACRYLATE   64399-38-8

ENVIRONMENTAL LISTS

List of Pesticide Product Inert Ingredients
[present]

3-(DIETHYLAMINO)PROPYLAMINE   104-78-9

HEALTH AND SAFETY LISTS

U.S. DOT - Substances From 49 CFR 172.101
regulated by DOT (UN2684)

U.S. DOT - Hazard Classes
DOT hazard class = 8

NFPA - Flash Points
flash point = 138 degrees F (59 degrees C)

NFPA - Hazard Identification Ratings
health-2; flammability-2; reactivity-0

NIOSH - Selected LD50s and LC50s
Oral, rat: LD50 = 1410 mg/kg Skin, rabbit: LD50 = 750 mg/kg

INTERNATIONAL LISTS

Canada - WHMIS: Ingredient Disclosure
1% item 574 (696)

STATE LISTS

Florida Hazardous Substance List
[present]

Massachusetts Right To Know List
[present]

NJ Right to Know List (Total)
sn 0692

NJ Special Hazardous Substances
(corrosive)

Pennsylvania Right to Know List
[present]

N,N-DIETHYLANILINE   91-66-7

HEALTH AND SAFETY LISTS

U.S. DOT - Substances From 49 CFR 172.101
regulated by DOT (UN2432)

U.S. DOT - Hazard Classes
DOT hazard class = 6.1

NFPA - Flash Points
flash point = 185 degrees F (85 degrees C)

NFPA - Hazard Identification Ratings
health-3; flammability-2; reactivity-0

NIOSH - Selected LD50s and LC50s
Inhalation, rat: LC50 = 1920 mg/m3 4 hr Oral, rat: LD50 = 782 mg/kg

INTERNATIONAL LISTS

Canada - WHMIS: Ingredient Disclosure
1% item 575 (697)

STATE LISTS

Florida Hazardous Substance List
[present]

Massachusetts Right To Know List
[present]

NJ Right to Know List (Total)
sn 0693

Pennsylvania Right to Know List
[present]

2,6-DIETHYLANILINE   579-66-8

ENVIRONMENTAL LISTS

CAA - HON Rule - SOCMI Chemicals
compliance by Oct. 23, 1995

P-DIETHYL BENZENE   105-05-5

HEALTH AND SAFETY LISTS

NFPA - Flash Points
flash point = 132 degrees F (55 degrees C)

NFPA - Hazard Identification Ratings
health-2; flammability-2; reactivity-0

ENVIRONMENTAL LISTS

EPA - Master Testing List
[present]

STATE LISTS

Florida Hazardous Substance List
[present]

Massachusetts Right To Know List
[present]

Pennsylvania Right to Know List
[present]

O-DIETHYL BENZENE   135-01-3

HEALTH AND SAFETY LISTS

NFPA - Flash Points
flash point = 135 degrees F (57 degrees C)

NFPA - Hazard Identification Ratings
health-2; flammability-2; reactivity-0

STATE LISTS

Florida Hazardous Substance List
[present]

Massachusetts Right To Know List
[present]

Pennsylvania Right to Know List
[present]

M-DIETHYL BENZENE   141-93-5

HEALTH AND SAFETY LISTS

NFPA - Flash Points
flash point = 133 degrees F (56 degrees C)

NFPA - Hazard Identification Ratings
health-2; flammability-2; reactivity-0

STATE LISTS

Florida Hazardous Substance List
[present]

Massachusetts Right To Know List
[present]

Pennsylvania Right to Know List
[present]

DIETHYLBENZENE   25340-17-4

HEALTH AND SAFETY LISTS

U.S. DOT - Substances From 49 CFR 172.101
regulated by DOT (UN2049)

U.S. DOT - Hazard Classes
DOT hazard class = 3

INTERNATIONAL LISTS

Canada - WHMIS: Ingredient Disclosure
1% item 576 (698)

STATE LISTS

NJ Right to Know List (Total)
sn 0694

N,N-DIETHYL-1,3-BUTANEDIAMINE   32280-46-9

HEALTH AND SAFETY LISTS

NFPA - Flash Points
flash point = 115 degrees F (46 degrees C)

NFPA - Hazard Identification Ratings
health-2; flammability-2; reactivity-0

STATE LISTS

Florida Hazardous Substance List
[present]

Massachusetts Right To Know List
[present]

Pennsylvania Right to Know List
[present]

DI-2-ETHYLBUTYL PHTHALATE   RR-00832-2

HEALTH AND SAFETY LISTS

NFPA - Flash Points
flash point = 381 degrees F (194 degrees C)

NFPA - Hazard Identification Ratings
health-0; flammability-1; reactivity-0

DIETHYLCARBAMODITHIOIC ACID SODIUM SALT   20624-25-3
TRIHYDRATE

HEALTH AND SAFETY LISTS

NIOSH - Selected LD50s and LC50s
Oral, rat: LD50 = 1500 mg/kg

DIETHYLCARBAMAZINE CITRATE   1642-54-2

HEALTH AND SAFETY LISTS

NIOSH - Selected LD50s and LC50s
Inhalation, rat: LC50 = 309 mg/m3 4 hr Oral, rat: LD50 = 1400 mg/kg

STATE LISTS

Florida Hazardous Substance List
effective March 13, 1992

Massachusetts Right To Know List
extraordinarily hazardous

NJ Right to Know List (Total)
sn 2332

Pennsylvania Right to Know List
environmental hazard

PROPOSED REGULATIONS

CERCLA/SARA - Proposed Hazardous Substance Additions
proposed RQ = 1 pound (.454 kg)

CERCLA/SARA - 1989 Proposed RQ Adjustments
proposed RQ = 100 pounds (45.4 kg)

DIETHYLCARBAMOYL CHLORIDE   88-10-

HEALTH AND SAFETY LISTS

NFPA - Flash Points
flash point = 325 to 342 degrees F (163 to 172 degrees C)

NFPA - Hazard Identification Ratings
health-2; flammability-1; reactivity-2 (avoid use of water)

INTERNATIONAL LISTS

German (DFG) - Carcinogens
suspected carcinogen

STATE LISTS

Florida Hazardous Substance List
[present]

Massachusetts Right To Know List
[present]

Pennsylvania Right to Know List
[present]

DIETHYL CARBONATE   105-58-

HEALTH AND SAFETY LISTS

U.S. DOT - Substances From 49 CFR 172.101
regulated by DOT (UN2366)

U.S. DOT - Hazard Classes
DOT hazard class = 3

NFPA - Flash Points
flash point = 77 degrees F (25 degrees C)

NFPA - Hazard Identification Ratings
health-2; flammability-3; reactivity-1

STATE LISTS

Florida Hazardous Substance List
[present]

Massachusetts Right To Know List
[present]

NJ Right to Know List (Total)
sn 0697

NJ Special Hazardous Substances
(flammable - third degree)

Pennsylvania Right to Know List
[present]

DIETHYL CHLOROPHOSPHATE   814-49-3

HEALTH AND SAFETY LISTS

NIOSH - Selected LD50s and LC50s
Oral, rat: LD50 = 11 mg/kg Skin, rabbit: LD50 = 7900 ug/kg

ENVIRONMENTAL LISTS

CERCLA/SARA - Section 302 Extremely Hazardous Substances and TPQs
TPQ = 500 pounds

STATE LISTS

Florida Hazardous Substance List
effective March 13, 1992

Massachusetts Right To Know List
extraordinarily hazardous

NJ Right to Know List (Total)
sn 2333

Pennsylvania Right to Know List
environmental hazard

PROPOSED REGULATIONS

CERCLA/SARA - Proposed Hazardous Substance Additions
proposed RQ = 1 pound (.454 kg)

CERCLA/SARA - 1989 Proposed RQ Adjustments
proposed RQ = 100 pounds (45.4 kg)

DIETHYLCYCLOHEXANE   1331-43-7

HEALTH AND SAFETY LISTS

NFPA - Flash Points
flash point = 120 degrees F (49 degrees C)

NFPA - Hazard Identification Ratings
health-2; flammability-2; reactivity-0

STATE LISTS

Florida Hazardous Substance List
[present]

Massachusetts Right To Know List
[present]

Pennsylvania Right to Know List
[present]

DIETHYLDICHLOROSILANE   1719-53-5

HEALTH AND SAFETY LISTS

U.S. DOT - Substances From 49 CFR 172.101
regulated by DOT (UN1767)

U.S. DOT - Hazard Classes
DOT hazard class = 8

INTERNATIONAL LISTS

Canada - WHMIS: Ingredient Disclosure
1% item 578 (701)

STATE LISTS

Massachusetts Right To Know List
[present]

NJ Right to Know List (Total)
sn 0698

NJ Special Hazardous Substances
(corrosive)

DIETHYLDIISOCYANATOBENZENE   134190-37-7

ENVIRONMENTAL LISTS

CERCLA/SARA - Section 313 - Emission Reporting
form R reporting required; (Listed under 'Diisocyanates')

DIETHYLENEDIAMINE   RR-01162-1

HEALTH AND SAFETY LISTS

U.S. DOT - Substances From 49 CFR 172.101
regulated by DOT (UN2685)

U.S. DOT - Hazard Classes
DOT hazard class = 8

NFPA - Flash Points
flash point = 144 degrees F (62 degrees C)

DIETHYLENE GLYCOL   111-46-6

HEALTH AND SAFETY LISTS

AIHA - WEEL - Time Weighted Averages
total: 55 ppm TWA; aerosol only: 10 mg/m3 TWA

NFPA - Flash Points
flash point = 255 degrees F (124 degrees C)

NFPA - Hazard Identification Ratings
health-1; flammability-1; reactivity-0

NIOSH - Selected LD50s and LC50s
Oral, rat: LD50 = 12565 mg/kg Skin, rabbit: LD50 = 11890 mg/kg

ENVIRONMENTAL LISTS

CAA - HON Rule - SOCMI Chemicals
compliance by Oct. 24, 1994

EPA - Master Testing List
[present]

List of Pesticide Product Inert Ingredients
[present]

INTERNATIONAL LISTS

United Kingdom - Occupational Exposure Standards - TWAs
23 ppm TWA; 100 mg/m3 TWA

STATE LISTS

California - Air Bill 2588 Appendix A-I
9/90

Minnesota Hazardous Substance List
[present]

Pennsylvania Right to Know List
[present]

DIETHYLENE GLYCOL BIS (ALLYLCARBONATE)   142-22-3

HEALTH AND SAFETY LISTS

NFPA - Flash Points
flash point = 378 degrees F (192 degrees C)

NFPA - Hazard Identification Ratings
health-1; flammability-1; reactivity-0

DIETHYLENE GLYCOL BIS(3-AMINOPROPYL) ETHER   4246-51-9

HEALTH AND SAFETY LISTS

NIOSH - Selected LD50s and LC50s
Oral, rat: LD50 = 4290 mg/kg Skin, rabbit: LD50 = 2500 mg/kg

DIETHYLENE GLYCOL BIS (2-BUTOXYETHYL CARBONATE)   RR-00840-2

HEALTH AND SAFETY LISTS

NFPA - Flash Points
flash point = 379 degrees F (193 degrees C)

NFPA - Hazard Identification Ratings
health-1; flammability-1; reactivity-1

DIETHYLENE GLYCOL BIS (BUTYL CARBONATE)   RR-00841-3

HEALTH AND SAFETY LISTS

NFPA - Flash Points
flash point = 372 degrees F (189 degrees C)

NFPA - Hazard Identification Ratings
health-1; flammability-1; reactivity-1

DIETHYLENE GLYCOL BIS (PHENYLCARBONATE)   RR-00842-4

HEALTH AND SAFETY LISTS

NFPA - Flash Points
flash point = 460 degrees F (238 degrees C)

NFPA - Hazard Identification Ratings
health-0; flammability-1; reactivity-1

DIETHYLENE GLYCOL DIACETATE   628-68-2

HEALTH AND SAFETY LISTS

NFPA - Flash Points
flash point = 275 degrees F (135 degrees C)

NFPA - Hazard Identification Ratings
health-1; flammability-1; reactivity-0

DIETHYLENE GLYCOL DIBENZOATE   120-55-8

HEALTH AND SAFETY LISTS

NFPA - Flash Points
flash point = 450 degrees F (232 degrees C)

NFPA - Hazard Identification Ratings
health-0;   flammability-1;   reactivity-0

NIOSH - Selected LD50s and LC50s
Oral, rat: LD50 = 2830 mg/kg Skin, rabbit: LD50 = 20 gm/kg

DIETHYLENE GLYCOL DIBUTYL ETHER   112-73-2
SEE ALSO:
GLYCOL ETHERS

HEALTH AND SAFETY LISTS

NFPA - Flash Points
flash point = 245 degrees F (118 degrees C)

NFPA - Hazard Identification Ratings
health-1;   flammability-1;   reactivity-0

NIOSH - Selected LD50s and LC50s
Oral, rat: LD50 = 3900 mg/kg Skin, rabbit: LD50 = 4040 mg/kg

ENVIRONMENTAL LISTS

CAA - HON Rule - SOCMI Chemicals
compliance by Oct. 24, 1994

DIETHYLENE GLYCOL DIETHYL ETHER   112-36-7

HEALTH AND SAFETY LISTS

NFPA - Flash Points
flash point = 180 degrees F (82 degrees C)

NFPA - Hazard Identification Ratings
health-1;   flammability-2;   reactivity-0

NIOSH - Selected LD50s and LC50s
Oral, rat: LD50 = 4970 mg/kg

ENVIRONMENTAL LISTS

CAA - HON Rule - SOCMI Chemicals
compliance by Oct. 24, 1994

INTERNATIONAL LISTS

Canada - WHMIS: Ingredient Disclosure
1% item 579 (813)

DIETHYLENE GLYCOL DIETHYL LEVULINATE   RR-00843-5

HEALTH AND SAFETY LISTS

NFPA - Flash Points
flash point = 340 degrees F (171 degrees C)

NFPA - Hazard Identification Ratings
health-0;   flammability-1;   reactivity-0

DIETHYLENE GLYCOL DIMETHACRYLATE   2358-84-1

ENVIRONMENTAL LISTS

TSCA - Code of Federal Regulations Citations
40 CFR 712.30(d)

TSCA - PAIR - Reporting List
Reporting Date: November 19, 1982

INTERNATIONAL LISTS

Canada - WHMIS: Ingredient Disclosure
1% item 580 (725)

DIETHYLENE GLYCOL DIMETHYL ETHER   111-96-6
SEE ALSO:
GLYCOL ETHERS

HEALTH AND SAFETY LISTS

NFPA - Flash Points
flash point = 153 degrees F (67 degrees C)

NFPA - Hazard Identification Ratings
health-1;   flammability-2;   reactivity-1

ENVIRONMENTAL LISTS

CAA - HON Rule - SOCMI Chemicals
compliance by Oct. 24, 1994

TSCA - Code of Federal Regulations Citations
40 CFR 712.30(x);   40 CFR 716.120(d)

TSCA - PAIR - Reporting List
Reporting Date: November 27, 1991

TSCA - Health and Safety Reporting List
Effective Date: September 30, 1991

INTERNATIONAL LISTS

German (DFG) - MAK Values
5 ppm MAK;   27 mg/m3 MAK

German (DFG) - Peak Limitations
2 x normal MAK (30 min., average value) don't exceed 4 times during shift

German (DFG) - Skin/Sensitizers
danger of cutaneous absorption

German (DFG) - Pregnancy
risk to embryo/fetus probable

STATE LISTS

California - Air Bill 2588 Appendix A-I
9/90

California - Directors List of Hazardous Substances (8 CCR 339)
[present]

DIETHYLENEGLYCOL DINITRATE   693-21-0

HEALTH AND SAFETY LISTS

U.S. DOT - Substances From 49 CFR 172.101
regulated by DOT (UN0075)

U.S. DOT - Hazard Classes
Forbidden from transport by the DOT

NIOSH - Selected LD50s and LC50s
Oral, rat: LD50 = 777 mg/kg

STATE LISTS

NJ Right to Know List (Total)
sn 0699

DIETHYLENE GLYCOL DIPROPIONATE   RR-00844-6

HEALTH AND SAFETY LISTS

NFPA - Flash Points
flash point = 260 degrees F (127 degrees C)

NFPA - Hazard Identification Ratings
health-1;   flammability-1;   reactivity-0

DIETHYLENE GLYCOL ETHYL ETHER PHTHALATE   RR-00845-7

HEALTH AND SAFETY LISTS

NFPA - Flash Points
flash point = 406 degrees F (208 degrees C)

NFPA - Hazard Identification Ratings
health-0;   flammability-1;   reactivity-0

DIETHYLENE GLYCOL METHYL ETHER ACETATE   629-38-9

HEALTH AND SAFETY LISTS

NFPA - Flash Points
flash point = 180 degrees F (82 degrees C)

NFPA - Hazard Identification Ratings
health-0;   flammability-2;   reactivity-0

ENVIRONMENTAL LISTS

CAA - HON Rule - SOCMI Chemicals
compliance by Oct. 23, 1995

DIETHYLENE GLYCOL MONOBUTYL ETHER   112-34-5
SEE ALSO:
  GLYCOL ETHERS

HEALTH AND SAFETY LISTS
  NFPA - Flash Points
    *flash point = 172 degrees F (78 degrees C); *See list description*
  NFPA - Hazard Identification Ratings
    health-1; flammability-2; reactivity-0;
  NIOSH - Selected LD50s and LC50s
    Oral, rat: LD50 = 6560 mg/kg Skin, rabbit: LD50 = 4120 mg/kg

ENVIRONMENTAL LISTS
  CAA - HON Rule - SOCMI Chemicals
    compliance by Oct. 24, 1994
  List of Pesticide Product Inert Ingredients
    [present]
  TSCA - Code of Federal Regulations Citations
    40 CFR 799.1560
  TSCA - Chemical Test Rules
    Testing required by: manufacturers; importers; processors (40 CFR 799.1560)
  TSCA - Section 12(b) - Export Notification
    export notification required - Section 4

INTERNATIONAL LISTS
  Canada - WHMIS: Ingredient Disclosure
    1% item 581 (814)
  German (DFG) - MAK Values
    100 mg/m3 MAK
  German (DFG) - Peak Limitations
    2 x normal MAK (5 min momentary value); don't exceed 8 times during shift
  German (DFG) - Pregnancy
    no risk to embryo/fetus if exposure limits adhered to

STATE LISTS
  California - Air Bill 2588 Appendix A-I
    9/90

DIETHYLENE GLYCOL MONOETHYL ETHER   111-90-0
SEE ALSO:
  GLYCOL ETHERS

HEALTH AND SAFETY LISTS
  AIHA - WEEL - Time Weighted Averages
    25 ppm TWA; 140 mg/m3 TWA
  NFPA - Flash Points
    flash point = 196 degrees F (91 degrees C)
  NFPA - Hazard Identification Ratings
    health-1; flammability-1; reactivity-0
  NIOSH - Selected LD50s and LC50s
    Oral, rat: LD50 = 5500 mg/kg Skin, rat: LD50 = 6000 mg/kg

ENVIRONMENTAL LISTS
  CAA - HON Rule - SOCMI Chemicals
    compliance by Oct. 24, 1994
  List of Pesticide Product Inert Ingredients
    [present]
  TSCA - Code of Federal Regulations Citations
    40 CFR 712.30(w); 40 CFR 716.120(a)
  TSCA - PAIR - Reporting List
    Reporting Date: June 13, 1989
  TSCA - Health and Safety Reporting List
    Effective Date: April 13, 1989

INTERNATIONAL LISTS
  Canada - WHMIS: Ingredient Disclosure
    1% item 582 (815)

STATE LISTS
  California - Air Bill 2588 Appendix A-I
    9/90

DIETHYLENE GLYCOL MONOETHYL ETHER ACETATE   RR-01164-3

HEALTH AND SAFETY LISTS
  NFPA - Flash Points
    flash point = 225 degrees F (107 degrees C)
  NFPA - Hazard Identification Ratings
    health-1; flammability-1; reactivity-0

DIETHYLENE GLYCOL MONOHEXYL ETHER   112-59-4
SEE ALSO:
  GLYCOL ETHERS

HEALTH AND SAFETY LISTS
  NIOSH - Selected LD50s and LC50s
    Oral, rat: LD50 = 4920 mg/kg Skin, rabbit: LD50 = 1500 mg/kg

ENVIRONMENTAL LISTS
  CAA - HON Rule - SOCMI Chemicals
    compliance by Oct. 23, 1995

INTERNATIONAL LISTS
  Canada - WHMIS: Ingredient Disclosure
    1% item 583 (816)

DIETHYLENE GLYCOL MONOMETHYL ETHER   111-77-3
SEE ALSO:
  GLYCOL ETHERS

HEALTH AND SAFETY LISTS
  NFPA - Flash Points
    flash point = 205 degrees F (96 degrees C)
  NFPA - Hazard Identification Ratings
    health-2; flammability-2; reactivity-0
  NIOSH - Selected LD50s and LC50s
    Oral, rat: LD50 = 9210 mg/kg Skin, rabbit: LD50 = 650 mg/kg

ENVIRONMENTAL LISTS
  CAA - HON Rule - SOCMI Chemicals
    compliance by Oct. 24, 1994
  List of Pesticide Product Inert Ingredients
    [present]
  TSCA - Code of Federal Regulations Citations
    40 CFR 712.30(w); 40 CFR 716.120(a)
  TSCA - PAIR - Reporting List
    Reporting Date: June 13, 1989
  TSCA - Health and Safety Reporting List
    Effective Date: April 13, 1989

INTERNATIONAL LISTS
  Canada - WHMIS: Ingredient Disclosure
    1% item 584 (817)

STATE LISTS
  California - Air Bill 2588 Appendix A-I
    9/90
  Florida Hazardous Substance List
    [present]
  Massachusetts Right To Know List
    [present]
  Pennsylvania Right to Know List
    [present]

DIETHYLENE GLYCOL MONOMETHYL ETHER FORMAL   5405-88-9

HEALTH AND SAFETY LISTS

NFPA - Flash Points
flash point = 310 degrees F (154 degrees C)

NFPA - Hazard Identification Ratings
health-1; flammability-1; reactivity-0

DIETHYLENE GLYCOL MONOPHENYL ETHER   104-68-7

HEALTH AND SAFETY LISTS

NIOSH - Selected LD50s and LC50s
Oral, rat: LD50 = 2140 mg/kg Skin, rabbit: LD50 = 2120 mg/kg

INTERNATIONAL LISTS

Canada - WHMIS: Ingredient Disclosure
1% item 585 (818)

DIETHYLENE GLYCOL PHTHALATE   RR-00846-8

HEALTH AND SAFETY LISTS

NFPA - Flash Points
flash point = 343 degrees F (173 degrees C)

NFPA - Hazard Identification Ratings
health-0; flammability-1; reactivity-0

DIETHYLENE TRIAMINE   111-40-0

HEALTH AND SAFETY LISTS

ACGIH 1995 - Time Weighted Averages
1 ppm TWA; 4.2 mg/m3 TWA

ACGIH 1995 - Skin Designations
skin - potential for cutaneous absorption

U.S. DOT - Substances From 49 CFR 172.101
regulated by DOT (UN2079)

U.S. DOT - Hazard Classes
DOT hazard class = 8

NFPA - Flash Points
flash point = 208 degrees F (98 degrees C)

NFPA - Hazard Identification Ratings
health-3; flammability-1; reactivity-0

NIOSH - Selected LD50s and LC50s
Oral, rat: LD50 = 1080 mg/kg Skin, guinea pig: LD50 = 162 mg/kg

OSHA - Vacated PELs - Time Weighted Averages
1 ppm TWA; 4 mg/m3 TWA

ENVIRONMENTAL LISTS

EPA - Master Testing List
[present]

List of Pesticide Product Inert Ingredients
[present]

TSCA - Code of Federal Regulations Citations
40 CFR 712.30(d); 40 CFR 716.120(a); 40 CFR 799.1575

TSCA - PAIR - Reporting List
Reporting Date: November 19, 1982

TSCA - Health and Safety Reporting List
Effective Date: April 29, 1983

TSCA - Chemical Test Rules
Testing required by: manufacturers; processors (40 CFR 799.1575)

TSCA - Section 12(b) - Export Notification
export notification required - Section 4

INTERNATIONAL LISTS

Australian Exposure Standards - Time Weighted Averages
1 ppm TWA; 4.2 mg/m3 TWA

Australian Exposure Standards - Skin Effects
skin absorption

Canada - WHMIS: Ingredient Disclosure
0.1% item 586 (704)

Canada - Alberta - 8 Hour Occupational Exposure Limit
1 ppm TWA; 4 mg/m3 TWA

Canada - Alberta - 15 Minute Occupational Exposure Limit
3 ppm STEL; 13 mg/m3 STEL

Canada - Alberta - Skin Designation
can be absorbed through the intact skin

Canada - British Columbia - 8 Hour Exposure Limits
1 ppm TWA; 4 mg/m3 TWA

Canada - British Columbia - Skin Notations
skin - potential for skin absorption

Canada - Ontario - OHSA - TWAEVs
1 ppm TWAEV; 4 mg/m3 TWAEV

Canada - Ontario - OHSA - Skin Notations
absorption through skin, eyes, or mucous membranes

Canada - Quebec - Time-Weighted Average Exposure Values
1 ppm TWAEV; 4.2 mg/m3 TWAEV

Canada - Quebec - Skin Designations
absorbed through the skin

United Kingdom - Occupational Exposure Standards - TWAs
1 ppm TWA; 4 mg/m3 TWA

United Kingdom - Occupational Exposure Standards - Notes
can be absorbed through skin

Israel - Time Weighted Averages
1 ppm TWA; 4.2 mg/m3 TWA

Israel - Action Levels
0.5 ppm AL; 2.1 mg/m3 AL

Mexico - Instruction No. 10 - TWAs
1 ppm TWA; 4 mg/m3 TWA

STATE LISTS

California - Exposure Limits - PELs
1 ppm PEL; 4 mg/m3 PEL

California - Exposure Limits - Skin Notation
material may be absorbed through the skin, eyes or mucous membrane

California - Directors List of Hazardous Substances (8 CCR 339)
[present]

Florida Hazardous Substance List
[present]

Massachusetts Right To Know List
[present]

Minnesota Hazardous Substance List
skin

NJ Right to Know List (Total)
sn 0700

NJ Special Hazardous Substances
(corrosive)

Pennsylvania Right to Know List
[present]

DIETHYLENETRIAMINEPENTAACETIC ACID, DISODIUM IRON(III) SALT   19529-38-5

ENVIRONMENTAL LISTS

List of Pesticide Product Inert Ingredients
[present]

N,N-DIETHYLETHANAMINE PHOSPHATE   35365-94-7

ENVIRONMENTAL LISTS

List of Pesticide Product Inert Ingredients
[present]

DIETHYLETHYLENE DIAMINE   100-36-7

HEALTH AND SAFETY LISTS

NFPA - Flash Points
flash point = 115 degrees F (46 degrees C)

NFPA - Hazard Identification Ratings
health-3; flammability-2; reactivity-0
NIOSH - Selected LD50s and LC50s
Oral, rat: LD50 = 2830 mg/kg Skin, rabbit: LD50 = 820 mg/kg

STATE LISTS
Florida Hazardous Substance List
[present]
Massachusetts Right To Know List
[present]
NJ Right to Know List (Total)
sn 0702
Pennsylvania Right to Know List
[present]

DIETHYL GLYCOL   16484-86-9

HEALTH AND SAFETY LISTS
NFPA - Flash Points
flash point = 95 degrees F (35 degrees C)
NFPA - Hazard Identification Ratings
flammability-3; reactivity-0

DIETHYLGOLD BROMIDE   26645-10-3

HEALTH AND SAFETY LISTS
U.S. DOT - Hazard Classes
Forbidden from transport by the DOT

DI(2-ETHYLHEXYL)AMINE   106-20-7

HEALTH AND SAFETY LISTS
NFPA - Flash Points
flash point = 270 degrees F (132 degrees C)
NFPA - Hazard Identification Ratings
health-3; flammability-1; reactivity-0
NIOSH - Selected LD50s and LC50s
Oral, rat: LD50 = 1640 mg/kg Skin, rabbit: LD50 = 1190 mg/kg

INTERNATIONAL LISTS
Canada - WHMIS: Ingredient Disclosure
1% item 587 (706)

STATE LISTS
Florida Hazardous Substance List
[present]
Massachusetts Right To Know List
[present]
Pennsylvania Right to Know List
[present]

DI-2-ETHYL HEXYL AZELATE   103-24-2

HEALTH AND SAFETY LISTS
NFPA - Flash Points
flash point = 440 degrees F (227 degrees C)
NFPA - Hazard Identification Ratings
health-0; flammability-1; reactivity-0
NIOSH - Selected LD50s and LC50s
Oral, rat: LD50 = 8720 mg/kg Skin, rabbit: LD50 = 20 gm/kg

ENVIRONMENTAL LISTS
List of Pesticide Product Inert Ingredients
[present]

DI-2-ETHYLHEXYL ETHER   10143-60-9

ENVIRONMENTAL LISTS
List of Pesticide Product Inert Ingredients
[present]

DI(2-ETHYLHEXYL)FUMARATE   141-02-6

INTERNATIONAL LISTS
Canada - WHMIS: Ingredient Disclosure
1% item 588 (925)

DI(2-ETHYLHEXYL)PEROXY-DICARBONATE   16111-62-9

HEALTH AND SAFETY LISTS
U.S. DOT - Organic Peroxides Table
Organic peroxide UN3113; UN3115; UN3117
NIOSH - Selected LD50s and LC50s
Oral, rat: LD50 = 1020 mg/kg

STATE LISTS
NJ Right to Know List (Total)
sn 0703

DI(2-ETHYLHEXYL)PHOSPHORIC ACID   298-07-7

HEALTH AND SAFETY LISTS
NFPA - Flash Points
flash point = 385 degrees F (196 degrees C)

ENVIRONMENTAL LISTS
TSCA - Code of Federal Regulations Citations
40 CFR 712.30(x); 40 CFR 716.120(d)
TSCA - PAIR - Reporting List
Reporting Date: December 27, 1990
TSCA - Health and Safety Reporting List
Effective Date: October 29, 1990

INTERNATIONAL LISTS
Canada - WHMIS: Ingredient Disclosure
1% item 589 (81)

STATE LISTS
NJ Right to Know List (Total)
sn 2334

DI(2-ETHYLHEXYL)PHTHALATE   117-81-7
SEE ALSO:
CAT FOOD
ALKYL PHTHALATES
F039-HAZARDOUS WASTES

HEALTH AND SAFETY LISTS
ACGIH 1995 - Time Weighted Averages
5 mg/m3 TWA
ACGIH 1995 - Short Term Exposure Limits
10 mg/m3 STEL
U.S. DOT - Appendix A Table 1 - Hazardous Substances
final RQ = 100 pounds (45.4 kg)
IARC - Group 2B (sufficient animal data)
[present] (Overall evaluation based only on evidence of carcinogenicity in monograph (29, 1982) or in Supplement 4)
NIOSH - Selected LD50s and LC50s
Oral, rat: LD50 = 30600 mg/kg Skin, guinea pig: LD50 = 10 gm/kg
NIOSH 1990 - Pocket Guide - RELs
5 mg/m3 TWA; 10 mg/m3 STEL
NIOSH 1990 - Pocket Guide - Carcinogens
occupational carcinogen
NIOSH 1990 - Pocket Guide - Target organs
eyes, GI tract, upper respiratory system
NIOSH - Health Standards - Exposure Limits
reduce exposure to lowest feasible concentration
NIOSH - Health Standards - Health Effects and Precautions
has produced liver tumors in animals (replace with less toxic materials)
NIOSH - Health Standards - Carcinogenic Chemicals
potential human carcinogen

NTP Chemical Status Reports - Testing Status and NTIS Number
Technical reports printed (PB82184011)

NTP Chemical Status Reports - Evidence of Carcinogenicity
male rat-positive; female rat-positive; male mice-positive; female mice-positive

NTP Seventh Report - Suspect Carcinogens
suspect carcinogen

OSHA - Vacated PELs - Time Weighted Averages
5 mg/m3 TWA

OSHA - Vacated PELs - Short Term Exposure Limits
10 mg/m3 STEL

OSHA - Final PELs - Time Weighted Averages
5 mg/m3 TWA

OSHA - Possible Select Carcinogens
[present]

ENVIRONMENTAL LISTS

ATSDR Priority List
Rank (of 275): 063

CERCLA/SARA - Section 313 - Emission Reporting
form R reporting required for 0.1% de minimus concentration

CERCLA/SARA - Hazardous Substances and their Reportable Quantities
final RQ = 100 pounds (45.4 kg)

Clean Air Act (1990) - List of Hazardous Air Contaminants
[present]

Clean Water Act - Priority Pollutants
[present]

Safe Drinking Water Act - MCLs
MCL = 0.006 mg/L

Safe Drinking Water Act - MCLGs
MCLG = Zero

EPA - Carcinogen Hazard Ranking for RQ Adjustment
Hazard ranking = Low

List of Pesticide Product Inert Ingredients
[present]

RCRA - U Series Wastes
waste number U028

RCRA - Hazardous Constituents-Appendix VIII
waste number U028

RCRA - Basis for Listing - Appendix VII
Included in waste stream: F039

RCRA - Substances Banned From Land Disposal
[present]

RCRA - TSD Facilities Ground Water Monitoring
TM 8060 = 20 ug/L PQL; TM 8270 = 10 ug/L PQL

RCRA - Universal Treatment Standards (LDR)
WW: 0.28 mg/l; NWW: 28 mg/kg

TSCA - Code of Federal Regulations Citations
40 CFR 712.30(d); 40 CFR 716.120(c); 40 CFR 799.5000

TSCA - PAIR - Reporting List
Reporting Date: November 19, 1982

TSCA - Substances Subject to Testing Consent Orders
Test for: Chemical Fate

TSCA - Section 12(b) - Export Notification
export notification required - Section 4

INTERNATIONAL LISTS

Australian Exposure Standards - Time Weighted Averages
5 mg/m3 TWA

Australian Exposure Standards - Short Term Exposure Limits
10 mg/m3 STEL

Canada - WHMIS: Ingredient Disclosure
0.1% item 590 (1418)

Canada - NPRI (National Pollutant Release Inventory)
[present]

Canada - CEPA - Priority Substances List
estimated time for completion of assessment reports: 4 years

Canada - Alberta - 8 Hour Occupational Exposure Limit
5 mg/m3 TWA

Canada - Alberta - 15 Minute Occupational Exposure Limit
10 mg/m3 STEL

Canada - British Columbia - 8 Hour Exposure Limits
5 mg/m3 TWA

Canada - British Columbia - 15 Minute Exposure Limits
10 mg/m3 STEL

Canada - Ontario - OHSA - TWAEVs
3 mg/m3 TWAEV

Canada - Ontario - OHSA - STEVs
10 mg/m3 STEV

Canada - Quebec - Time-Weighted Average Exposure Values
5 mg/m3 TWAEV

Canada - Quebec - Short-term Exposure Values
10 ppm STEV

Canada - Quebec - Carcinogens
C3 carcinogen: effect detected in animals

United Kingdom - Occupational Exposure Standards - TWAs
5 mg/m3 TWA

United Kingdom - Occupational Exposure Standards - STELs
10 mg/m3 STEL

German (DFG) - MAK Values
total dust: 10 mg/m3 MAK

German (DFG) - Peak Limitations
10 x normal MAK (30 min average value); don't exceed during shift

German (DFG) - Pregnancy
no risk to embryo/fetus if exposure limits adhered to

Israel - Time Weighted Averages
5 mg/m3 TWA

Israel - Short Term Exposure Limits
10 mg/m3 STEL

Israel - Action Levels
2.5 mg/m3 AL

Mexico - Instruction No. 10 - TWAs
5 mg/m3 TWA

Mexico - Instruction No. 10 - STELs
10 mg/m3 STEL

Mexico - Wastewater - Organic Toxic Pollutants and Heavy Metals
Listed under [Phthalate Esters]

STATE LISTS

California - Air Bill 2588 Appendix A-I
known or potential carcinogen

California - Prop. 65 - Cancer list
carcinogen - initial date 1/1/88

California - Prop. 65 - No Significant Risk Levels
no significant risk level = 80 ug/day

California - Exposure Limits - PELs
5 mg/m3 PEL

California - Exposure Limits - STELs
10 mg/m3 STEL

Massachusetts Right To Know List
carcinogen; extraordinarily hazardous

Minnesota Hazardous Substance List
[present]

NJ Right to Know List (Total)
sn 0238

NJ Special Hazardous Substances
(carcinogen)

Pennsylvania Right to Know List
environmental hazard; special hazardous substance

Pennsylvania RTK - Special Hazardous Substances
  [present]

1,2-DIETHYLHYDRAZINE   1615-80-1

HEALTH AND SAFETY LISTS

U.S. DOT - Appendix A Table 1 - Hazardous Substances
  final RQ = 10 pounds (4.54 kg)
IARC - Group 2B (sufficient animal data)
  [present] (Overall evaluation based only on evidence of carcinogenicity in monograph (4, 1974) or in Supplement 4)
OSHA - Possible Select Carcinogens
  [present]

ENVIRONMENTAL LISTS

CERCLA/SARA - Hazardous Substances and their Reportable Quantities
  final RQ = 10 pounds (4.54 kg)
EPA - Carcinogen Hazard Ranking for RQ Adjustment
  Hazard ranking = Medium
RCRA - U Series Wastes
  waste number U086
RCRA - Hazardous Constituents-Appendix VIII
  waste number U086
RCRA - Substances Banned From Land Disposal
  [present]

INTERNATIONAL LISTS

Canada - WHMIS: Ingredient Disclosure
  0.1% item 591 (707)

STATE LISTS

California - Air Bill 2588 Appendix A-II
  known or potential carcinogen
California - Prop. 65 - Cancer list
  carcinogen - initial date 1/1/88
California - Directors List of Hazardous Substances (8 CCR 339)
  [present]
Florida Hazardous Substance List
  [present]
Massachusetts Right To Know List
  carcinogen; extraordinarily hazardous
Minnesota Hazardous Substance List
  carcinogen
NJ Right to Know List (Total)
  sn 1007
NJ Special Hazardous Substances
  (carcinogen)
Pennsylvania Right to Know List
  environmental hazard; special hazardous substance
Pennsylvania RTK - Special Hazardous Substances
  [present]

DIETHYL HYDROXYLAMINE   3710-84-7

ENVIRONMENTAL LISTS

List of Pesticide Product Inert Ingredients
  [present]

DIETHYL KETONE   96-22-0

HEALTH AND SAFETY LISTS

ACGIH 1995 - Time Weighted Averages
  200 ppm TWA; 705 mg/m3 TWA
AIHA - Odor Threshold Values
  geometric mean air odor threshold = 2.8 ppm (detectable); 14 ppm (recognizable)
U.S. DOT - Substances From 49 CFR 172.101
  regulated by DOT (UN1156)

U.S. DOT - Hazard Classes
  DOT hazard class = 3
NFPA - Flash Points
  flash point = 55 degrees F (13 degrees C)
NFPA - Hazard Identification Ratings
  health-1; flammability-3; reactivity-0
NIOSH - Selected LD50s and LC50s
  Oral, rat: LD50 = 2140 mg/kg Skin, rabbit: LD50 = 20 gm/kg
OSHA - Vacated PELs - Time Weighted Averages
  200 ppm TWA; 705 mg/m3 TWA

INTERNATIONAL LISTS

Australian Exposure Standards - Time Weighted Averages
  200 ppm TWA; 705 mg/m3 TWA
Canada - WHMIS: Ingredient Disclosure
  1% item 592 (700)
Canada - Alberta - 8 Hour Occupational Exposure Limit
  200 ppm TWA; 705 mg/m3 TWA
Canada - Alberta - 15 Minute Occupational Exposure Limit
  250 ppm STEL; 881 mg/m3 STEL
Canada - Ontario - OHSA - TWAEVs
  200 ppm TWAEV; 705 mg/m3 TWAEV
Canada - Quebec - Time-Weighted Average Exposure Values
  200 ppm TWAEV; 705 mg/m3 TWAEV
United Kingdom - Occupational Exposure Standards - TWAs
  200 ppm TWA; 700 mg/m3 TWA
United Kingdom - Occupational Exposure Standards - STELs
  250 ppm STEL; 875 mg/m3 STEL
Israel - Time Weighted Averages
  200 ppm TWA; 705 mg/m3 TWA
Israel - Action Levels
  100 ppm AL; 352.5 mg/m3 AL

STATE LISTS

California - Exposure Limits - PELs
  200 ppm PEL; 705 mg/m3 PEL
California - Directors List of Hazardous Substances (8 CCR 339)
  [present]
Florida Hazardous Substance List
  [present]
Massachusetts Right To Know List
  [present]
Minnesota Hazardous Substance List
  [present]
NJ Right to Know List (Total)
  sn 0704
NJ Special Hazardous Substances
  (flammable - third degree)
Pennsylvania Right to Know List
  [present]

N,N-DIETHYLLAURAMIDE   3352-87-2

HEALTH AND SAFETY LISTS

NFPA - Flash Points
  flash point > 150 degrees F (66 degrees C)
NFPA - Hazard Identification Ratings
  flammability-2; reactivity-0

DIETHYLMAGNESIUM   557-18-6

STATE LISTS

NJ Right to Know List (Total)
  sn 0705

DIETHYL MALEATE   141-05-9
HEALTH AND SAFETY LISTS
NFPA - Flash Points
flash point = 250 degrees F (121 degrees C)
NFPA - Hazard Identification Ratings
health-1; flammability-1; reactivity-0
NIOSH - Selected LD50s and LC50s
Oral, rat: LD50 = 3200 mg/kg Skin, rabbit: LD50 = 4000 mg/kg

DIETHYL MALONATE   105-53-3
HEALTH AND SAFETY LISTS
NFPA - Flash Points
flash point = 200 degrees F (93 degrees C)
NFPA - Hazard Identification Ratings
health-0; flammability-1; reactivity-0
NIOSH - Selected LD50s and LC50s
Oral, rat: LD50 = 15 gm/kg

DIETHYLMERCURY   627-44-1
SEE ALSO:
MERCURY
INTERNATIONAL LISTS
Canada - WHMIS: Ingredient Disclosure
0.1% item 593 (708)

O,O-DIETHYL S-METHYL DITHIOPHOSPHATE   3288-58-2
HEALTH AND SAFETY LISTS
U.S. DOT - Appendix A Table 1 - Hazardous Substances
final RQ = 5000 pounds (2270 kg)
ENVIRONMENTAL LISTS
CERCLA/SARA - Hazardous Substances and their Reportable Quantities
final RQ = 5000 pounds (2270 kg)
RCRA - U Series Wastes
waste number U087
RCRA - Hazardous Constituents-Appendix VIII
waste number U087
RCRA - Substances Banned From Land Disposal
[present]
TSCA - Code of Federal Regulations Citations
40 CFR 716.120(a)
TSCA - Health and Safety Reporting List
Effective Date: March 7, 1986
STATE LISTS
Massachusetts Right To Know List
[present]
Pennsylvania Right to Know List
environmental hazard

O,O-DIETHYL O-[4-METHYLSULFINYL)PHENYL]   115-91-3
PHOSPHOROTHIOATE (FENSULFOTHION)
STATE LISTS
California - Directors List of Hazardous Substances (8 CCR 339)
[present]

DIETHYL-P-NITROPHENYL PHOSPHATE   311-45-5
HEALTH AND SAFETY LISTS
U.S. DOT - Appendix B - Marine Pollutants
DOT regulated marine pollutant
U.S. DOT - Appendix A Table 1 - Hazardous Substances
final RQ = 100 pounds (45.4 kg)
NIOSH - Selected LD50s and LC50s
Oral, rat: LD50 = 1800 ug/kg Skin, rabbit: LD50 = 5 mg/kg

ENVIRONMENTAL LISTS
CERCLA/SARA - Hazardous Substances and their Reportable Quantities
final RQ = 100 pounds (45.4 kg)
RCRA - P Series Wastes
waste number P041
RCRA - Hazardous Constituents-Appendix VIII
waste number P041
RCRA - Substances Banned From Land Disposal
[present]
STATE LISTS
Massachusetts Right To Know List
[present]
NJ Right to Know List (Total)
sn 1457
Pennsylvania Right to Know List
environmental hazard

3,3-DIETHYLPENTANE   1067-20-5
HEALTH AND SAFETY LISTS
NFPA - Hazard Identification Ratings
health-0; flammability-3; reactivity-0
STATE LISTS
Florida Hazardous Substance List
[present]
Massachusetts Right To Know List
[present]
Pennsylvania Right to Know List
[present]

DIETHYL PEROXIDE   628-37-5
HEALTH AND SAFETY LISTS
NFPA - Hazard Identification Ratings
flammability-4; reactivity-4
STATE LISTS
Florida Hazardous Substance List
[present]
Massachusetts Right To Know List
[present]
Pennsylvania Right to Know List
[present]

DIETHYL PEROXYDICARBONATE   14666-78-5
HEALTH AND SAFETY LISTS
U.S. DOT - Hazard Classes
Forbidden from transport by the DOT
U.S. DOT - Organic Peroxides Table
Organic peroxide UN3115
STATE LISTS
NJ Right to Know List (Total)
sn 0706

DIETHYL PEROXYDICARBONATE   15042-77-0
HEALTH AND SAFETY LISTS
U.S. DOT - Organic Peroxides Table
Organic peroxide UN3102

DI(P-ETHYLPHENYL)DICHLOROETHANE   72-56-0
HEALTH AND SAFETY LISTS
NIOSH - Selected LD50s and LC50s
Oral, rat: LD50 = 6600 mg/kg
NTP Chemical Status Reports - Testing Status and NTIS Number
Technical reports printed (PB290582/AS)

NTP Chemical Status Reports - Evidence of Carcinogenicity
male rat-negative; female rat-negative; male mice-negative; female mice-equivocal
STATE LISTS
California - Directors List of Hazardous Substances (8 CCR 339)
[present]

Massachusetts Right To Know List
[present]

DIETHYL PHOSPHITE   762-04-9
ENVIRONMENTAL LISTS
List of Pesticide Product Inert Ingredients
[present]

N,N-DIETHYL-P-PHOSPHORIC ACID   93-05-0
STATE LISTS
Massachusetts Right To Know List
[present]

DIETHYL PHTHALATE   84-66-2
SEE ALSO:
  CAT FOOD
  F039-HAZARDOUS WASTES
  ALKYL PHTHALATES

HEALTH AND SAFETY LISTS
ACGIH 1995 - Time Weighted Averages
5 mg/m3 TWA

U.S. DOT - Appendix A Table 1 - Hazardous Substances
final RQ = 1000 pounds (454 kg)

NFPA - Flash Points
flash point = 322 degrees F (161 degrees C)

NFPA - Hazard Identification Ratings
health-0; flammability-1; reactivity-0

NIOSH - Selected LD50s and LC50s
Oral, rat: LD50 = 8600 mg/kg

NTP Chemical Status Reports - Testing Status and NTIS Number
Galley or camera copy technical reports in progress

OSHA - Vacated PELs - Time Weighted Averages
5 mg/m3 TWA

ENVIRONMENTAL LISTS
ATSDR Priority List
Rank (of 275): 243

CERCLA/SARA - Section 313 - Emission Reporting
form R reporting required for 1.0% de minimus concentration

CERCLA/SARA - Hazardous Substances and their Reportable Quantities
final RQ = 1000 pounds (454 kg)

Clean Water Act - Priority Pollutants
[present]

List of Pesticide Product Inert Ingredients
[present]

RCRA - U Series Wastes
waste number U088

RCRA - Hazardous Constituents-Appendix VIII
waste number U088

RCRA - Basis for Listing - Appendix VII
Included in waste stream: F039

RCRA - Substances Banned From Land Disposal
[present]

RCRA - TSD Facilities Ground Water Monitoring
TM 8060 = 5 ug/L PQL; TM 8270 = 10 ug/L PQL

RCRA - Universal Treatment Standards (LDR)
WW: 0.20 mg/l; NWW: 28 mg/kg

TSCA - Code of Federal Regulations Citations
40 CFR 712.30(d); 40 CFR 716.120(c)

INTERNATIONAL LISTS
Australian Exposure Standards - Time Weighted Averages
5 mg/m3 TWA

Canada - WHMIS: Ingredient Disclosure
0.1% item 594 (1417)

Canada - NPRI (National Pollutant Release Inventory)
[present]

Canada - Alberta - 8 Hour Occupational Exposure Limit
5 mg/m3 TWA

Canada - Alberta - 15 Minute Occupational Exposure Limit
10 mg/m3 STEL

Canada - British Columbia - 8 Hour Exposure Limits
5 mg/m3 TWA

Canada - British Columbia - 15 Minute Exposure Limits
10 mg/m3 STEL

Canada - Ontario - OHSA - TWAEVs
5 mg/m3 TWAEV

Canada - Quebec - Time-Weighted Average Exposure Values
5 mg/m3 TWAEV

United Kingdom - Occupational Exposure Standards - TWAs
5 mg/m3 TWA

United Kingdom - Occupational Exposure Standards - STELs
10 mg/m3 STEL

Israel - Time Weighted Averages
5 mg/m3 TWA

Israel - Action Levels
2.5 mg/m3 AL

Mexico - Instruction No. 10 - TWAs
5 mg/m3 TWA

Mexico - Instruction No. 10 - STELs
10 mg/m3 STEL

Mexico - Wastewater - Organic Toxic Pollutants and Heavy Metals
Listed under [Phthalate Esters]

Mexico - Drinking Water - Ecological Criteria
350.0 mg/l

STATE LISTS
California - Air Bill 2588 Appendix A-II
6/91

California - Exposure Limits - PELs
5 mg/m3 PEL

Florida Hazardous Substance List
[present]

Massachusetts Right To Know List
[present]

Minnesota Hazardous Substance List
[present]

NJ Right to Know List (Total)
sn 0707

Pennsylvania Right to Know List
environmental hazard

PROPOSED REGULATIONS
TSCA - ITC 32nd Report Priority Testing List
designated for dermal absorption testing

Canada - Ontario - Proposed Occupational TWAEVs
3 mg/m3 TWAEV

Canada - Ontario - Proposed Occupational STEVs
5 mg/m3 STEV

2,2-DIETHYL-1,3-PROPANEDIOL   115-76-4

HEALTH AND SAFETY LISTS

NFPA - Flash Points
flash point = 215 degrees F (102 degrees C)

NFPA - Hazard Identification Ratings
health-2; flammability-1; reactivity-0

NIOSH - Selected LD50s and LC50s
Oral, rat: LD50 = 850 mg/kg Skin, rabbit: LD50 = 4240 mg/kg

STATE LISTS

Florida Hazardous Substance List
[present]

Massachusetts Right To Know List
[present]

Pennsylvania Right to Know List
[present]

O,O-DIETHYL O-PYRAZINYL PHOSPHOROTHIATE   297-97-2

HEALTH AND SAFETY LISTS

U.S. DOT - Appendix A Table 1 - Hazardous Substances
final RQ = 100 pounds (45.4 kg)

NIOSH - Selected LD50s and LC50s
Oral, rat: LD50 = 3500 ug/kg Skin, rat: LD50 = 8 mg/kg

ENVIRONMENTAL LISTS

CERCLA/SARA - Section 302 Extremely Hazardous Substances and TPQs
TPQ = 500 pounds

CERCLA/SARA - Hazardous Substances and their Reportable Quantities
final RQ = 100 pounds (45.4 kg)

RCRA - P Series Wastes
waste number P040

RCRA - Hazardous Constituents-Appendix VIII
waste number P040

RCRA - Substances Banned From Land Disposal
[present]

RCRA - TSD Facilities Ground Water Monitoring
TM 8270 = 10 ug/L PQL

STATE LISTS

Florida Hazardous Substance List
effective March 13, 1992

Massachusetts Right To Know List
extraordinarily hazardous

Pennsylvania Right to Know List
environmental hazard

DIETHYL PYROCARBONATE   1609-47-8

HEALTH AND SAFETY LISTS

NIOSH - Selected LD50s and LC50s
Oral, rat: LD50 = 850 mg/kg

DIETHYL SEBACATE   110-40-7

HEALTH AND SAFETY LISTS

NIOSH - Selected LD50s and LC50s
Oral, rat: LD50 = 14470 mg/kg

N,N-DIETHYLSTEARAMIDE   RR-00873-1

HEALTH AND SAFETY LISTS

NFPA - Flash Points
flash point = 375 degrees F (191 degrees C)

NFPA - Hazard Identification Ratings
health-0; flammability-1; reactivity-0

DIETHYLSTILBESTROL   56-53-1

HEALTH AND SAFETY LISTS

U.S. DOT - Appendix A Table 1 - Hazardous Substances
final RQ = 1 pound (0.454 kg)

IARC - Group 1 (carcinogenic to humans)
[present] (Listed under 'nonsteroidal oestrogens')

NTP Seventh Report - Known Carcinogens
known carcinogen

OSHA - Select Carcinogens
[present]

ENVIRONMENTAL LISTS

CERCLA/SARA - Hazardous Substances and their Reportable Quantities
final RQ = 1 pound (0.454 kg)

EPA - Carcinogen Hazard Ranking for RQ Adjustment
Hazard ranking = High

RCRA - U Series Wastes
waste number U089

RCRA - Hazardous Constituents-Appendix VIII
waste number U089

RCRA - Substances Banned From Land Disposal
[present]

TSCA - Chemicals with Significant New Use Rules
[present]

TSCA - Section 12(b) - Export Notification
export notification required - Section 5

STATE LISTS

California - Air Bill 2588 Appendix A-II
known or potential carcinogen

California - Prop. 65 - Cancer list
carcinogen - initial date 2/27/87

California - Prop. 65 - Developmental Toxicity
developmental toxicity - initial date 7/1/87

California - Prop. 65 - No Significant Risk Levels
no significant risk level = 0.002 ug/day

California - Directors List of Hazardous Substances (8 CCR 339)
[present]

Florida Hazardous Substance List
[present]

Massachusetts Right To Know List
carcinogen; extraordinarily hazardous; teratogen

Minnesota Hazardous Substance List
carcinogen

NJ Right to Know List (Total)
sn 0709

NJ Special Hazardous Substances
(carcinogen; mutagen; teratogen)

Pennsylvania Right to Know List
environmental hazard; special hazardous substance

Pennsylvania RTK - Special Hazardous Substances
[present]

DIETHYL SUCCINATE   123-25-1

HEALTH AND SAFETY LISTS

NFPA - Flash Points
flash point = 195 degrees F (90 degrees C)

NFPA - Hazard Identification Ratings
health-1; flammability-1; reactivity-0

NIOSH - Selected LD50s and LC50s
Oral, rat: LD50 = 8530 mg/kg

DIETHYL SULFATE   64-67-5

HEALTH AND SAFETY LISTS

U.S. DOT - Substances From 49 CFR 172.101
regulated by DOT (UN1594)

U.S. DOT - Hazard Classes
DOT hazard class = 6.1

IARC - Group 2A (limited human data)
[present]

NFPA - Flash Points
flash point = 220 degrees F (104 degrees C)

NFPA - Hazard Identification Ratings
health-3; flammability-1; reactivity-1

NIOSH - Selected LD50s and LC50s
Oral, rat: LD50 = 880 mg/kg Skin, rabbit: LD50 = 600 mg/kg

NTP Seventh Report - Suspect Carcinogens
suspect carcinogen

OSHA - Possible Select Carcinogens
[present]

ENVIRONMENTAL LISTS

CERCLA/SARA - Section 313 - Emission Reporting
form R reporting required for 0.1% de minimus concentration

CERCLA/SARA - Hazardous Substances and their Reportable Quantities
final RQ = 1 pound (.454 kg)

Clean Air Act (1990) - List of Hazardous Air Contaminants
[present]

CAA - HON Rule - SOCMI Chemicals
compliance by Jan. 23, 1995

CAA - HON Rule - Organic HAPs
[present]

TSCA - Code of Federal Regulations Citations
40 CFR 712.30(d)

TSCA - PAIR - Reporting List
Reporting Date: November 19, 1982

INTERNATIONAL LISTS

Canada - WHMIS: Ingredient Disclosure
0.1% item 595 (1521)

Canada - NPRI (National Pollutant Release Inventory)
[present]

German (DFG) - Carcinogens
animal evidence of carcinogenicity

STATE LISTS

California - Air Bill 2588 Appendix A-I
known or potential carcinogen

California - Prop. 65 - Cancer list
carcinogen - initial date 1/1/88

California - Directors List of Hazardous Substances (8 CCR 339)
[present]

Florida Hazardous Substance List
[present]

Massachusetts Right To Know List
carcinogen; extraordinarily hazardous

Minnesota Hazardous Substance List
carcinogen

NJ Right to Know List (Total)
sn 0710

NJ Special Hazardous Substances
(carcinogen; mutagen)

Pennsylvania Right to Know List
environmental hazard; special hazardous substance

Pennsylvania RTK - Special Hazardous Substances
[present]

DIETHYL SULFIDE   352-93-2

HEALTH AND SAFETY LISTS

U.S. DOT - Substances From 49 CFR 172.101
regulated by DOT (UN2375)

U.S. DOT - Hazard Classes
DOT hazard class = 3

NIOSH - Selected LD50s and LC50s
Oral, rat: LD50 = 5930 mg/kg

INTERNATIONAL LISTS

Canada - WHMIS: Ingredient Disclosure
1% item 596 (1549)

STATE LISTS

NJ Right to Know List (Total)
sn 0711

DIETHYL TARTRATE   87-91-2

HEALTH AND SAFETY LISTS

NFPA - Flash Points
flash point = 200 degrees F (93 degrees C)

NFPA - Hazard Identification Ratings
health-0; flammability-1; reactivity-0

DIETHYL TEREPHTHALATE   636-09-9

HEALTH AND SAFETY LISTS

NFPA - Flash Points
flash point = 243 degrees F (117 degrees C)

NFPA - Hazard Identification Ratings
health-0; flammability-1; reactivity-0

DIETHYLTHIOPHOSPHORYL CHLORIDE   2524-04-1

HEALTH AND SAFETY LISTS

U.S. DOT - Substances From 49 CFR 172.101
regulated by DOT (UN2751)

U.S. DOT - Hazard Classes
DOT hazard class = 8

NIOSH - Selected LD50s and LC50s
Inhalation, rat: LC50 = 20 ppm 4 hr Oral, mouse: LD50 = 910 mg/kg

ENVIRONMENTAL LISTS

EPA - Master Testing List
[present]

INTERNATIONAL LISTS

Canada - WHMIS: Ingredient Disclosure
1% item 597 (493)

STATE LISTS

NJ Right to Know List (Total)
sn 0712

N,N'-DIETHYLTHIOUREA   105-55-5

HEALTH AND SAFETY LISTS

NIOSH - Selected LD50s and LC50s
Oral, rat: LD50 = 316 mg/kg

NTP Chemical Status Reports - Testing Status and NTIS Number
Technical reports printed (PB288626/AS)

NTP Chemical Status Reports - Evidence of Carcinogenicity
male rat-positive; female rat-positive; male mice-negative; female mice-negative

ENVIRONMENTAL LISTS

List of Pesticide Product Inert Ingredients
[present]

STATE LISTS

Massachusetts Right To Know List
carcinogen; extraordinarily hazardous

N,N-DIETHYL-M-TOLUAMIDE   134-62-3
HEALTH AND SAFETY LISTS
NIOSH - Selected LD50s and LC50s
Inhalation, rat: LC50 = 5950 mg/m3 8 hr Oral, rat: LD50 = 1950 mg/kg Skin, rat: LD50 = 5000 mg/kg

DIETHYLZINC   557-20-0
SEE ALSO:
ZINC
HEALTH AND SAFETY LISTS
U.S. DOT - Substances From 49 CFR 172.101
regulated by DOT (UN1366)

U.S. DOT - Hazard Classes
DOT hazard class = 4.2

NFPA - Flash Points
ignites spontaneously in air

NFPA - Hazard Identification Ratings
health-3; flammability-4; reactivity-3 (do not use water, foam or halogenated extinguishing agents)

OSHA - List of Highly Hazardous Chemicals
threshhold quantity = 10,000 pounds

STATE LISTS
Florida Hazardous Substance List
[present]

Massachusetts Right To Know List
[present]

NJ Right to Know List (Total)
sn 0713

NJ Special Hazardous Substances
(flammable - third degree; reactive - third degree)

Pennsylvania Right to Know List
environmental hazard

DIFLUBENZURON   35367-38-5
ENVIRONMENTAL LISTS
CERCLA/SARA - Section 313 - Emission Reporting
form R reporting required

STATE LISTS
Massachusetts Right To Know List
[present]

DIFLUORODIBROMOMETHANE   75-61-6
HEALTH AND SAFETY LISTS
ACGIH 1995 - Time Weighted Averages
100 ppm TWA; 858 mg/m3 TWA

U.S. DOT - Substances From 49 CFR 172.101
regulated by DOT (UN1941)

U.S. DOT - Hazard Classes
DOT hazard class = 9

NIOSH 1990 - Pocket Guide - RELs
100 ppm TWA; 860 mg/m3 TWA

NIOSH 1990 - Pocket Guide - IDLHs
2500 ppm IDLH

NIOSH 1990 - Pocket Guide - Target organs
respiratory system

OSHA - Vacated PELs - Time Weighted Averages
100 ppm TWA; 860 mg/m3 TWA

OSHA - Final PELs - Time Weighted Averages
100 ppm TWA; 860 mg/m3 TWA

INTERNATIONAL LISTS
Australian Exposure Standards - Time Weighted Averages
100 ppm TWA; 858 mg/m3 TWA

Canada - WHMIS: Ingredient Disclosure
1% item 598 (709)

Canada - Alberta - 8 Hour Occupational Exposure Limit
100 ppm TWA; 858 mg/m3 TWA

Canada - Alberta - 15 Minute Occupational Exposure Limit
150 ppm STEL; 1287 mg/m3 STEL

Canada - British Columbia - 8 Hour Exposure Limits
100 ppm TWA; 860 mg/m3 TWA

Canada - British Columbia - 15 Minute Exposure Limits
150 ppm STEL; 1290 mg/m3 STEL

Canada - Ontario - OHSA - TWAEVs
100 ppm TWAEV; 860 mg/m3 TWAEV

Canada - Quebec - Time-Weighted Average Exposure Values
100 ppm TWAEV; 858 mg/m3 TWAEV

United Kingdom - Occupational Exposure Standards - TWAs
100 ppm TWA; 860 mg/m3 TWA

United Kingdom - Occupational Exposure Standards - STELs
150 ppm STEL; 1290 mg/m3 STEL

German (DFG) - MAK Values
100 ppm MAK; 860 mg/m3 MAK

German (DFG) - Peak Limitations
2 x normal MAK (30 min. average value); don't exceed 4 times during shift

Israel - Time Weighted Averages
100 ppm TWA; 858 mg/m3 TWA

Israel - Action Levels
50 ppm AL; 429 mg/m3 AL

Mexico - Instruction No. 10 - TWAs
100 ppm TWA; 860 mg/m3 TWA

Mexico - Instruction No. 10 - STELs
150 ppm STEL; 1290 mg/m3 STEL

STATE LISTS
California - Exposure Limits - PELs
100 ppm PEL; 860 mg/m3 PEL

California - Directors List of Hazardous Substances (8 CCR 339)
[present]

Florida Hazardous Substance List
[present]

Massachusetts Right To Know List
[present]

Minnesota Hazardous Substance List
[present]

NJ Right to Know List (Total)
sn 0714

Pennsylvania Right to Know List
[present]

1,1-DIFLUOROETHANE   75-37-6
HEALTH AND SAFETY LISTS
AIHA - WEEL - Time Weighted Averages
1000 ppm TWA; 2700 mg/m3 TWA

U.S. DOT - Substances From 49 CFR 172.101
regulated by DOT (UN1030)

U.S. DOT - Hazard Classes
DOT hazard class = 2.1

NIOSH - Selected LD50s and LC50s
Inhalation, mouse: LC50 = 977 gm/m3 (2 hr)

ENVIRONMENTAL LISTS
CAA - Flammable Substances for Accidental Release Prevention
threshold quantity = 10,000 lbs

List of Pesticide Product Inert Ingredients
[present]

TSCA - Code of Federal Regulations Citations
40 CFR 712.30(w); 40 CFR 716.120(a)
TSCA - PAIR - Reporting List
Reporting Date: June 13, 1989
TSCA - Health and Safety Reporting List
Effective Date: April 13, 1989

STATE LISTS

Massachusetts Right To Know List
[present]
NJ Right to Know List (Total)
sn 0715

DIFLUOROPHOSPHORIC ACID   13779-41-4

HEALTH AND SAFETY LISTS

U.S. DOT - Substances From 49 CFR 172.101
regulated by DOT (UN1768)
U.S. DOT - Hazard Classes
DOT hazard class = 8

INTERNATIONAL LISTS

Canada - WHMIS: Ingredient Disclosure
1% item 599 (82)

STATE LISTS

NJ Right to Know List (Total)
sn 0716
NJ Special Hazardous Substances
(corrosive)

DIGITOXIN   71-63-6

HEALTH AND SAFETY LISTS

NIOSH - Selected LD50s and LC50s
Oral, rat: LD50 = 56 mg/kg

ENVIRONMENTAL LISTS

CERCLA/SARA - Section 302 Extremely Hazardous Substances and
TPQs
TPQ = 100/10,000 pounds

STATE LISTS

Florida Hazardous Substance List
effective March 13, 1992
Massachusetts Right To Know List
extraordinarily hazardous
NJ Right to Know List (Total)
sn 2336
Pennsylvania Right to Know List
environmental hazard

PROPOSED REGULATIONS

CERCLA/SARA - Proposed Hazardous Substance Additions
proposed RQ = 1 pound (.454 kg)
CERCLA/SARA - 1989 Proposed RQ Adjustments
proposed RQ = 100 pounds (45.4 kg)

DIGYLCERYL STEARATE   1323-83-7

ENVIRONMENTAL LISTS

List of Pesticide Product Inert Ingredients
[present]

DIGLYCIDYL ETHER (DGE)   2238-07-5
SEE ALSO:
GLYCIDOL (OXIRANEMETHANOL) AND ITS DERIVATIVES

HEALTH AND SAFETY LISTS

ACGIH 1995 - Time Weighted Averages
0.1 ppm TWA; 0.53 mg/m3 TWA

NIOSH - Selected LD50s and LC50s
Inhalation, mouse: LC50 = 30 ppm 4 hr Oral, rat: LD50 = 450 mg/kg
Skin, rabbit: LD50 = 1500 mg/kg
NIOSH 1990 - Pocket Guide - RELs
0.1 ppm TWA; 0.5 mg/m3 TWA
NIOSH 1990 - Pocket Guide - IDLHs
25 ppm IDLH (not considering carcinogenic effects)
NIOSH 1990 - Pocket Guide - Carcinogens
occupational carcinogen
NIOSH 1990 - Pocket Guide - Target organs
skin, eyes, respiratory system
NIOSH - Health Standards - Exposure Limits
C (15 min) 0.2 ppm; C (15 min) 1 mg/m3 (Listed under 'Glycidyl
ethers')
NIOSH - Health Standards - Health Effects and Precautions
has produced skin tumors in animals; skin and mucous membrane ef-
fects; sensitization potential; possible hematopoietic and reproductive
system effects (Medical monitoring required) (Listed under 'Glycidyl
ethers')
NIOSH - Health Standards - Carcinogenic Chemicals
potential human carcinogen (Listed under 'Glycidyl ethers')
OSHA - Vacated PELs - Time Weighted Averages
0.1 ppm TWA; 0.5 mg/m3 TWA
OSHA - Final PELs - Ceiling Limits
C 0.5 ppm; C 2.8 mg/m3

ENVIRONMENTAL LISTS

CERCLA/SARA - Section 302 Extremely Hazardous Substances and
TPQs
TPQ = 1000 pounds
EPA - Master Testing List
[present]
TSCA - Code of Federal Regulations Citations
40 CFR 712.30(d); 40 CFR 716.120(c)
TSCA - PAIR - Reporting List
Reporting Date: November 19, 1982

INTERNATIONAL LISTS

Australian Exposure Standards - Time Weighted Averages
0.1 ppm TWA; 0.53 mg/m3 TWA
Canada - WHMIS: Ingredient Disclosure
0.1% item 600 (819)
Canada - Alberta - 8 Hour Occupational Exposure Limit
0.1 ppm TWA; 0.5 mg/m3 TWA
Canada - Alberta - 15 Minute Occupational Exposure Limit
0.3 ppm STEL; 1.5 mg/m3 STEL
Canada - British Columbia - Ceiling Exposure Limits
C 0.5 ppm; C 3 mg/m3
Canada - Ontario - OHSA - TWAEVs
0.1 ppm TWAEV; 0.53 mg/m3 TWAEV
Canada - Quebec - Time-Weighted Average Exposure Values
0.1 ppm TWAEV; 0.53 mg/m3 TWAEV
United Kingdom - Occupational Exposure Standards - TWAs
0.1 ppm TWA; 0.6 mg/m3 TWA
German (DFG) - MAK Values
0.1 ppm MAK; 0.6 mg/m3 MAK
German (DFG) - Peak Limitations
2 x normal MAK (5 min momentary value); don't exceed 8 times
during shift
German (DFG) - Carcinogens
suspected carcinogen
Israel - Time Weighted Averages
0.1 ppm TWA; 0.53 mg/m3 TWA
Israel - Action Levels
0.05 ppm AL; 0.265 mg/m3 AL
Mexico - Instruction No. 10 - TWAs
0.1 ppm TWA; 0.5 mg/m3 TWA

STATE LISTS

California - Exposure Limits - PELs
0.1 ppm PEL; 0.5 mg/m3 PEL

California - Directors List of Hazardous Substances (8 CCR 339)
[present]

Florida Hazardous Substance List
[present]

Massachusetts Right To Know List
extraordinarily hazardous

Minnesota Hazardous Substance List
carcinogen

NJ Right to Know List (Total)
sn 0717

Pennsylvania Right to Know List
environmental hazard

PROPOSED REGULATIONS

CERCLA/SARA - Proposed Hazardous Substance Additions
proposed RQ = 1 pound (.454 kg)

CERCLA/SARA - 1989 Proposed RQ Adjustments
proposed RQ = 100 pounds (45.4 kg)

TSCA - Proposed Testing Rule for Glycidyl Ethers
member of Glycidyl subcategory V-A

DIGLYCIDYL ETHER OF DISUBSTITUTED CARBOPOLYCYCLE   RR-00935-8

ENVIRONMENTAL LISTS

TSCA - Chemicals with Significant New Use Rules
PMN number: P-88-837

DIGLYCIDYL RESORCINOL ETHER   101-90-6

HEALTH AND SAFETY LISTS

IARC - Group 2B (sufficient animal data)
[present] (Overall evaluation based only on evidence of carcinogenicity in monograph (36, 1985) or in Supplement 4)

NIOSH - Selected LD50s and LC50s
Oral, rat: LD50 = 2570 mg/kg

NTP Chemical Status Reports - Testing Status and NTIS Number
Technical reports printed (PB87146734/AS)

NTP Chemical Status Reports - Evidence of Carcinogenicity
male rat-positive; female rat-positive; male mice-positive; female mice-positive

NTP Seventh Report - Suspect Carcinogens
suspect carcinogen

OSHA - Possible Select Carcinogens
[present]

ENVIRONMENTAL LISTS

CERCLA/SARA - Section 313 - Emission Reporting
form R reporting required

EPA - Master Testing List
[present]

TSCA - Code of Federal Regulations Citations
40 CFR 712.30(d); 40 CFR 716.120(c)

TSCA - PAIR - Reporting List
Reporting Date: November 19, 1982

INTERNATIONAL LISTS

Canada - WHMIS: Ingredient Disclosure
1% item 1380 (834)

German (DFG) - Skin/Sensitizers
danger of sensitization

German (DFG) - Carcinogens
animal evidence of carcinogenicity

STATE LISTS

California - Air Bill 2588 Appendix A-II
known or potential carcinogen

California - Prop. 65 - Cancer list
carcinogen - initial date 7/1/89

California - Prop. 65 - No Significant Risk Levels
no significant risk level = 0.4 ug/day

California - Directors List of Hazardous Substances (8 CCR 339)
[present]

Florida Hazardous Substance List
effective March 13, 1992

Massachusetts Right To Know List
carcinogen; extraordinarily hazardous

Pennsylvania Right to Know List
special hazardous substance

Pennsylvania RTK - Special Hazardous Substances
[present]

PROPOSED REGULATIONS

TSCA - Proposed Testing Rule for Glycidyl Ethers
member of Glycidyl subcategory VI-C

DIGLYCOL CHLOROFORMATE   RR-00847-9

HEALTH AND SAFETY LISTS

NFPA - Flash Points
flash point = 295 degrees F (146 degrees C)

NFPA - Hazard Identification Ratings
health-0; flammability-1; reactivity-0

DIGLYCOL CHLOROHYDRIN   6288-89-7

HEALTH AND SAFETY LISTS

NFPA - Flash Points
flash point = 225 degrees F (107 degrees C)

NFPA - Hazard Identification Ratings
health-0; flammability-1; reactivity-0

DIGLYCOL DIACETATE   RR-00848-0

HEALTH AND SAFETY LISTS

NFPA - Flash Points
flash point = 255 degrees F (124 degrees C)

NFPA - Hazard Identification Ratings
health-0; flammability-1; reactivity-0

DIGLYCOL DILEVULINATE   RR-00849-1

HEALTH AND SAFETY LISTS

NFPA - Flash Points
flash point = 340 degrees F (171 degrees C)

NFPA - Hazard Identification Ratings
health-0; flammability-1

DIGLYCOLIC ACID   110-99-6

ENVIRONMENTAL LISTS

List of Pesticide Product Inert Ingredients
[present]

DIGLYCOL LAURATE   141-20-8

HEALTH AND SAFETY LISTS

NFPA - Flash Points
flash point = 290 degrees F (143 degrees C)

NFPA - Hazard Identification Ratings
health-0; flammability-1; reactivity-0

DIGLYCOL MONOSTEARATE   106-11-6
  ENVIRONMENTAL LISTS
    List of Pesticide Product Inert Ingredients
      [present]

DIGOXIN   20830-75-5
  HEALTH AND SAFETY LISTS
    NIOSH - Selected LD50s and LC50s
      Oral, mouse: LD50 = 17780 ug/kg
  ENVIRONMENTAL LISTS
    CERCLA/SARA - Section 302 Extremely Hazardous Substances and
    TPQs
      TPQ = 10/10,000 pounds
  STATE LISTS
    Florida Hazardous Substance List
      effective March 13, 1992
    Massachusetts Right To Know List
      extraordinarily hazardous
    NJ Right to Know List (Total)
      sn 2337
    Pennsylvania Right to Know List
      environmental hazard
  PROPOSED REGULATIONS
    CERCLA/SARA - Proposed Hazardous Substance Additions
      proposed RQ = 1 pound (.454 kg)
    CERCLA/SARA - 1989 Proposed RQ Adjustments
      proposed RQ = 10 pounds (4.54 kg)

DI(HEPTYL-, NONYL-) PHTHALATE (BRANCHED   111381-89-6
AND LINEAR ISOMERS)
  ENVIRONMENTAL LISTS
    TSCA - Code of Federal Regulations Citations
      40 CFR 799.5025

DI[HEPTYL, NONYL, UNDECYL] PHTHALATE   RR-00277-7
(D711P)
  ENVIRONMENTAL LISTS
    TSCA - Substances Subject to Testing Consent Orders
      *Test for: Environmental Effects (This chemical is tested as a mixture
      of the branched and linear isomers of diheptyl phthalate, dinonyl
      phthalate, di(heptyl,nonyl) phthalate, diundecyl phthalate, di(heptyl,
      undecyl) phthalate, and di(nonyl,undecyl phthalate))*

DIHEPTYL PHTHALATE   3648-21-3
  INTERNATIONAL LISTS
    Canada - WHMIS: Ingredient Disclosure
      0.1% item 601 (1419)

DI(HEPTYL-, UNDECYL-) PHTHALATE (BRANCHED   111381-90-9
AND LINEAR ISOMERS)
  ENVIRONMENTAL LISTS
    TSCA - Code of Federal Regulations Citations
      40 CFR 799.5025

DIHEXYLAMINE   143-16-8
  HEALTH AND SAFETY LISTS
    NFPA - Flash Points
      flash point = 220 degrees F (104 degrees C)
    NFPA - Hazard Identification Ratings
      health-2; flammability-1; reactivity-0
    NIOSH - Selected LD50s and LC50s
      Oral, rat: LD50 = 380 mg/kg Skin, rabbit: LD50 = 170 mg/kg

  STATE LISTS
    Florida Hazardous Substance List
      [present]
    Massachusetts Right To Know List
      [present]
    Pennsylvania Right to Know List
      [present]

DI-N-HEXYL AZELATE   109-31-9
  HEALTH AND SAFETY LISTS
    NIOSH - Selected LD50s and LC50s
      Oral, rat: LD50 = 16 gm/kg

DIHEXYL PHTHALATE (MIXED ISOMERS)   68515-50-4
  ENVIRONMENTAL LISTS
    TSCA - Code of Federal Regulations Citations
      40 CFR 799.5000
    TSCA - Substances Subject to Testing Consent Orders
      Test for: Environmental Effects
    TSCA - Section 12(b) - Export Notification
      export notification required - Section 4

3,4-DIHYDROCOUMARIN   119-84-6
  HEALTH AND SAFETY LISTS
    NTP Chemical Status Reports - Testing Status and NTIS Number
      *Prechronic studies for which toxicity technical reports were not pre-
      pared; Technical reports printed (no NTIS number given)*

DIHYDROGEN POTASSIUM PHOSPHATE   7778-77-0
  ENVIRONMENTAL LISTS
    List of Pesticide Product Inert Ingredients
      [present]

1,1-DIHYDROPERFLUOROPROPANOL   422-05-9
  HEALTH AND SAFETY LISTS
    NIOSH - Selected LD50s and LC50s
      *Inhalation, mouse: LC50 = 10 gm/m3 8 hr Oral, mouse: LD50 =
      1000 mg/kg*

2,2-DIHYDROPEROXY PROPANE   2614-76-8
  HEALTH AND SAFETY LISTS
    U.S. DOT - Organic Peroxides Table
      Organic peroxide UN3102
  STATE LISTS
    NJ Right to Know List (Total)
      sn 0719

DIHYDROPYRAN   110-87-2
  HEALTH AND SAFETY LISTS
    U.S. DOT - Substances From 49 CFR 172.101
      regulated by DOT (UN2376)
    U.S. DOT - Hazard Classes
      DOT hazard class = 3
    NFPA - Flash Points
      flash point = 0 degrees F (-18 degrees C)
    NFPA - Hazard Identification Ratings
      health-2; flammability-3; reactivity-0
  STATE LISTS
    Florida Hazardous Substance List
      [present]
    Massachusetts Right To Know List
      [present]
    NJ Right to Know List (Total)
      sn 0720

NJ Special Hazardous Substances
(flammable - third degree)
Pennsylvania Right to Know List
[present]

DIHYDRO-2H-PYRAN   25512-65-6

STATE LISTS

Pennsylvania Right to Know List
[present]

1,2-DIHYDRO-3,6-PYRIDAZINEDIONE   123-33-1

HEALTH AND SAFETY LISTS

U.S. DOT - Appendix A Table 1 - Hazardous Substances
final RQ = 5000 pounds (2270 kg)
IARC - Group 3 (not classifiable)
[present]

ENVIRONMENTAL LISTS

CERCLA/SARA - Hazardous Substances and their Reportable Quantities
final RQ = 5000 pounds (2270 kg)
CAA - HON Rule - SOCMI Chemicals
compliance by Oct. 24, 1994
RCRA - U Series Wastes
waste number U148
RCRA - Hazardous Constituents-Appendix VIII
waste number U148
RCRA - Substances Banned From Land Disposal
[present]
TSCA - Code of Federal Regulations Citations
40 CFR 799.5055(c),(d)(1),(2)
TSCA - Multichemical Test Rules - Waste Constituents
soil adsorption and hydrolysis testing for Chemical Fate
TSCA - Section 12(b) - Export Notification
export notification required - Section 4

STATE LISTS

Massachusetts Right To Know List
[present]
Pennsylvania Right to Know List
environmental hazard

DIHYDROSAFROLE   94-58-6

HEALTH AND SAFETY LISTS

U.S. DOT - Appendix A Table 1 - Hazardous Substances
final RQ = 10 pounds (4.54 kg)
IARC - Group 2B (sufficient animal data)
[present] (Overall evaluation based only on evidence of carcinogenicity in monograph (10, 1976) or in Supplement 4)
NIOSH - Selected LD50s and LC50s
Oral, rat: LD50 = 2260 mg/kg
OSHA - Possible Select Carcinogens
[present]

ENVIRONMENTAL LISTS

CERCLA/SARA - Section 313 - Emission Reporting
form R reporting required
CERCLA/SARA - Hazardous Substances and their Reportable Quantities
final RQ = 10 pounds (4.54 kg)
EPA - Carcinogen Hazard Ranking for RQ Adjustment
Hazard ranking = Medium
RCRA - U Series Wastes
waste number U090
RCRA - Hazardous Constituents-Appendix VIII
waste number U090

RCRA - Substances Banned From Land Disposal
[present]
TSCA - Code of Federal Regulations Citations
40 CFR 799.5055(c), (d)(2)
TSCA - Multichemical Test Rules - Waste Constituents
hydrolysis testing for Chemical Fate
TSCA - Section 12(b) - Export Notification
export notification required - Section 4

STATE LISTS

California - Air Bill 2588 Appendix A-II
known or potential carcinogen
California - Prop. 65 - Cancer list
carcinogen - initial date 1/1/88
California - Prop. 65 - No Significant Risk Levels
no significant risk level = 20 ug/day
California - Directors List of Hazardous Substances (8 CCR 339)
[present]
Florida Hazardous Substance List
[present]
Massachusetts Right To Know List
carcinogen; extraordinarily hazardous
Minnesota Hazardous Substance List
carcinogen
NJ Special Hazardous Substances
(carcinogen)
Pennsylvania Right to Know List
environmental hazard; special hazardous substance
Pennsylvania RTK - Special Hazardous Substances
[present]

DIHYDROSTREPTOMYCIN   128-46-1

STATE LISTS

Massachusetts Right To Know List
teratogen
NJ Right to Know List (Total)
sn 0721

1,2-DIHYDRO-2,2,4-TRIMETHYLQUINOLIN   147-47-7
(MONOMER)

HEALTH AND SAFETY LISTS

NIOSH - Selected LD50s and LC50s
Oral, rat: LD50 = 2000 mg/kg
NTP Chemical Status Reports - Testing Status and NTIS Number
Two year studies: pathology working in progress; two year studies scheduled for peer review; prechronic studies for which toxicity technical reports were not prepared

1,2-DIHYDRO-2,2,4-TRIMETHYLQUINOLINE   26780-96-1
POLYMER

HEALTH AND SAFETY LISTS

NTP Chemical Status Reports - Testing Status and NTIS Number
Prechronic studies for which toxicity technical reports were not prepared

1,4-DIHYDROXY-9,10-ANTHRACENEDIONE   81-64-1

INTERNATIONAL LISTS

Canada - WHMIS: Ingredient Disclosure
1% item 602 (712)

1,2-DIHYDROXYANTHRAQUINONE   72-48-0

HEALTH AND SAFETY LISTS

NIOSH - Selected LD50s and LC50s
Oral, bird: LD50 = 316 mg/kg

ENVIRONMENTAL LISTS
  CAA - HON Rule - SOCMI Chemicals
  compliance by Oct. 23, 1995
INTERNATIONAL LISTS
  Canada - WHMIS: Ingredient Disclosure
  1% item 603 (713)

1,8-DIHYDROXYANTHRAQUINONE   117-10-2

HEALTH AND SAFETY LISTS
  IARC - Group 2B (sufficient animal data)
  [present]
  OSHA - Possible Select Carcinogens
  [present]
INTERNATIONAL LISTS
  Canada - WHMIS: Ingredient Disclosure
  1% item 604 (714)
STATE LISTS
  California - Prop. 65 - Cancer list
  carcinogen - initial date 1/1/92
  California - Prop. 65 - No Significant Risk Levels
  no significant risk level = 9 ug/day
PROPOSED REGULATIONS
  NTP - Proposed Additions to Annual Report on Carcinogens
  proposed as a suspect carcinogen for the NTP 9th report

DIHYDROXYBENZOIC ACID (RESORCYLIC ACID)   27138-57-4

ENVIRONMENTAL LISTS
  CAA - HON Rule - SOCMI Chemicals
  compliance by Oct. 23, 1995

2,4-DIHYDROXYBENZOPHENONE   131-56-6

HEALTH AND SAFETY LISTS
  NIOSH - Selected LD50s and LC50s
  Oral, rat: LD50 = 7220 mg/kg
INTERNATIONAL LISTS
  Canada - WHMIS: Ingredient Disclosure
  1% item 605 (715)

DI(1-HYDROXYCYCLO-HEXYL)PEROXIDE   1758-61-8

HEALTH AND SAFETY LISTS
  U.S. DOT - Organic Peroxides Table
  Organic peroxide UN3106
INTERNATIONAL LISTS
  German (DFG) - Skin/Sensitizers
  very strong effects on skin
STATE LISTS
  NJ Right to Know List (Total)
  sn 0723

N-(2,4-DIHYDROXY-3,3-DIMETHYL-1-OXOBUTYL)-   137-08-6
.BETA.-ALANINE, CALCIUM SALT (2:1), (R)-
HEALTH AND SAFETY LISTS
  NIOSH - Selected LD50s and LC50s
  Oral, mouse: LD50 = 10 gm/kg

N,N-(2-DIHYDROXYETHYL)GLYCINE   150-25-4

ENVIRONMENTAL LISTS
  List of Pesticide Product Inert Ingredients
  [present]

2,3-DIHYDROXYPROPYL-3'-(HEXYLTHIO)   67859-56-7
PROPIONATE
ENVIRONMENTAL LISTS
  List of Pesticide Product Inert Ingredients
  [present]

1,8-DIHYDROXY-2,4,5,7-   517-92-0
TETRANITROANTHRAQUINONE
HEALTH AND SAFETY LISTS
  U.S. DOT - Hazard Classes
  Forbidden from transport by the DOT

DI-(1-HYDROXYTETRAZOLE)   RR-01403-9

HEALTH AND SAFETY LISTS
  U.S. DOT - Hazard Classes
  Forbidden from transport by the DOT

DIIDOACETYLENE   624-74-8

HEALTH AND SAFETY LISTS
  U.S. DOT - Hazard Classes
  Forbidden from transport by the DOT

DIIDOACETYLENE   RR-01413-1

STATE LISTS
  NJ Right to Know List (Total)
  sn 2385; sn 2386

1-(DIIODOMETHYL)SULFONYL-4-METHYL   20018-09-1
BENZENE
ENVIRONMENTAL LISTS
  TSCA - Code of Federal Regulations Citations
  40 CFR 712.30(x); 40 CFR 716.120(d)
  TSCA - PAIR - Reporting List
  Reporting Date: November 27, 1991
  TSCA - Health and Safety Reporting List
  Effective Date: September 30, 1991

DI-ISOBUTRYL PEROXIDE   3437-84-1

HEALTH AND SAFETY LISTS
  U.S. DOT - Organic Peroxides Table
  Organic peroxide UN3111;   UN3115
STATE LISTS
  NJ Right to Know List (Total)
  sn 0726

DIISOBUTYL ADIPATE   141-04-8

HEALTH AND SAFETY LISTS
  NIOSH - Selected LD50s and LC50s
  Oral, guinea pig: LD50 = 12300 mg/kg
ENVIRONMENTAL LISTS
  List of Pesticide Product Inert Ingredients
  [present]
INTERNATIONAL LISTS
  Canada - WHMIS: Ingredient Disclosure
  0.1% item 606 (165)

DIISOBUTYLALUMINUM HYDRIDE   1191-15-7

HEALTH AND SAFETY LISTS
  NFPA - Flash Points
  ignites spontaneously in air
  NFPA - Hazard Identification Ratings
  *flammability-3; reactivity-3 (do not use water, foam or halogenated
  extinguishing agents)*

STATE LISTS

Florida Hazardous Substance List
[present]

Massachusetts Right To Know List
[present]

Pennsylvania Right to Know List
[present]

DI-ISOBUTYLAMINE   110-96-3

HEALTH AND SAFETY LISTS

U.S. DOT - Substances From 49 CFR 172.101
regulated by DOT (UN2361)

U.S. DOT - Hazard Classes
DOT hazard class = 3

NFPA - Flash Points
flash point = 85 degrees F (29 degrees C)

NFPA - Hazard Identification Ratings
health-3; flammability-3; reactivity-0

NIOSH - Selected LD50s and LC50s
Oral, rat: LD50 = 258 mg/kg

STATE LISTS

Florida Hazardous Substance List
[present]

Massachusetts Right To Know List
[present]

NJ Right to Know List (Total)
sn 0724

NJ Special Hazardous Substances
(flammable - third degree)

Pennsylvania Right to Know List
[present]

DIISOBUTYL CARBITOL   108-82-7

HEALTH AND SAFETY LISTS

NFPA - Flash Points
flash point = 165 degrees F (74 degrees C)

NFPA - Hazard Identification Ratings
health-1; flammability-2; reactivity-0

DIISOBUTYLENE   25167-70-8

HEALTH AND SAFETY LISTS

AIHA - WEEL - Time Weighted Averages
600 ppm TWA; 2760 mg/m3 TWA

U.S. DOT - Substances From 49 CFR 172.101
regulated by DOT (UN2050)

U.S. DOT - Hazard Classes
DOT hazard class = 3

NFPA - Flash Points
flash point = 23 degrees F (-5 degrees C)

NFPA - Hazard Identification Ratings
health-1; flammability-3; reactivity-0

ENVIRONMENTAL LISTS

List of Pesticide Product Inert Ingredients
[present]

STATE LISTS

Florida Hazardous Substance List
[present]

Massachusetts Right To Know List
[present]

Minnesota Hazardous Substance List
[present]

NJ Right to Know List (Total)
sn 0725

NJ Special Hazardous Substances
(flammable - third degree)

Pennsylvania Right to Know List
[present]

DIISOBUTYL KETONE   108-83-8

HEALTH AND SAFETY LISTS

ACGIH 1995 - Time Weighted Averages
25 ppm TWA; 145 mg/m3 TWA

U.S. DOT - Substances From 49 CFR 172.101
regulated by DOT (UN1157)

U.S. DOT - Hazard Classes
DOT hazard class = 3

NFPA - Flash Points
flash point = 120 degrees F (49 degrees C)

NFPA - Hazard Identification Ratings
health-1; flammability-2; reactivity-0

NIOSH - Selected LD50s and LC50s
Oral, rat: LD50 = 5750 mg/kg Skin, rabbit: LD50 = 16 gm/kg

NIOSH 1990 - Pocket Guide - RELs
25 ppm TWA; 150 mg/m3 TWA

NIOSH 1990 - Pocket Guide - IDLHs
2000 ppm IDLH

NIOSH 1990 - Pocket Guide - Target organs
skin, eyes, respiratory system

NIOSH - Health Standards - Exposure Limits
25 ppm TWA; 140 mg/m3 TWA (Listed under 'Ketones')

NIOSH - Health Standards - Health Effects and Precautions
Irritation; liver, kidney, and nervous system effects (Urinalysis required) (Listed under 'Ketones')

OSHA - Vacated PELs - Time Weighted Averages
25 ppm TWA; 150 mg/m3 TWA

OSHA - Final PELs - Time Weighted Averages
50 ppm TWA; 290 mg/m3 TWA

ENVIRONMENTAL LISTS

EPA - Master Testing List
[present]

List of Pesticide Product Inert Ingredients
[present]

INTERNATIONAL LISTS

Australian Exposure Standards - Time Weighted Averages
25 ppm TWA; 145 mg/m3 TWA

Canada - WHMIS: Ingredient Disclosure
1% item 607 (716)

Canada - Alberta - 8 Hour Occupational Exposure Limit
25 ppm TWA; 145 mg/m3 TWA

Canada - Alberta - 15 Minute Occupational Exposure Limit
38 ppm STEL; 220 mg/m3 STEL

Canada - British Columbia - 8 Hour Exposure Limits
25 ppm TWA; 150 mg/m3 TWA

Canada - Ontario - OHSA - TWAEVs
25 ppm TWAEV; 145 mg/m3 TWAEV

Canada - Quebec - Time-Weighted Average Exposure Values
25 ppm TWAEV; 145 mg/m3 TWAEV

United Kingdom - Occupational Exposure Standards - TWAs
25 ppm TWA; 150 mg/m3 TWA

German (DFG) - MAK Values
50 ppm MAK; 290 mg/m3 MAK

Israel - Time Weighted Averages
25 ppm TWA; 145 mg/m3 TWA

Israel - Action Levels
12.5 ppm AL; 72.5 mg/m3 AL

Mexico - Instruction No. 10 - TWAs
25 ppm TWA;   150 mg/m3 TWA

STATE LISTS

California - Exposure Limits - PELs
25 ppm PEL;   150 mg/m3 PEL

California - Directors List of Hazardous Substances (8 CCR 339)
[present]

Florida Hazardous Substance List
[present]

Massachusetts Right To Know List
[present]

Minnesota Hazardous Substance List
[present]

NJ Right to Know List (Total)
sn 0760

Pennsylvania Right to Know List
[present]

DIISOBUTYLNAPHTHALENESULFONIC ACID,   27213-90-7
SODIUM SALT

ENVIRONMENTAL LISTS

List of Pesticide Product Inert Ingredients
[present]

DIISOCYANATES   RR-00547-0

STATE LISTS

Minnesota Hazardous Substance List
[present]

4,4'-DIISOCYANATODIPHENYL ETHER   4128-73-8

ENVIRONMENTAL LISTS

CERCLA/SARA - Section 313 - Emission Reporting
form R reporting required;   (Listed under 'Diisocyanates')

DIISODECYL ADIPATE   27178-16-1

HEALTH AND SAFETY LISTS

NFPA - Flash Points
flash point = 225 degrees F (107 degrees C)

NFPA - Hazard Identification Ratings
health-0;   flammability-1;   reactivity-0

DIISODECYL PHTHALATE   26761-40-0
SEE ALSO:
ALKYL PHTHALATES

HEALTH AND SAFETY LISTS

NFPA - Flash Points
flash point = 450 degrees F (232 degrees C)

NFPA - Hazard Identification Ratings
health-0;   flammability-1;   reactivity-0

NIOSH - Selected LD50s and LC50s
Oral, rat: LD50 = 64 gm/kg

ENVIRONMENTAL LISTS

List of Pesticide Product Inert Ingredients
[present]

TSCA - Code of Federal Regulations Citations
40 CFR 712.30(d);   40 CFR 716.120(c);   40 CFR 799.5000

TSCA - PAIR - Reporting List
Reporting Date: November 19, 1982

TSCA - Substances Subject to Testing Consent Orders
Test for: Chemical Fate

TSCA - Section 12(b) - Export Notification
export notification required - Section 4

INTERNATIONAL LISTS

United Kingdom - Occupational Exposure Standards - TWAs
5 mg/m3 TWA

DIISODECYL PHTHALATE (MIXED ISOMERS)   68515-49-1

ENVIRONMENTAL LISTS

TSCA - Code of Federal Regulations Citations
40 CFR 799.5000

TSCA - Substances Subject to Testing Consent Orders
Test for: Chemical Fate

TSCA - Section 12(b) - Export Notification
export notification required - Section 4

DIISONONYL PHTHALATE   28553-12-0
SEE ALSO:
ALKYL PHTHALATES

ENVIRONMENTAL LISTS

List of Pesticide Product Inert Ingredients
[present]

TSCA - Code of Federal Regulations Citations
40 CFR 716.120(c)

INTERNATIONAL LISTS

United Kingdom - Occupational Exposure Standards - TWAs
5 mg/m3 TWA

DIISOOCTYL ACID PHOSPHATE   27215-10-7

HEALTH AND SAFETY LISTS

U.S. DOT - Substances From 49 CFR 172.101
regulated by DOT (UN1902)

U.S. DOT - Hazard Classes
DOT hazard class = 8

ENVIRONMENTAL LISTS

TSCA - Code of Federal Regulations Citations
40 CFR 712.30(x);   40 CFR 716.120(d)

TSCA - PAIR - Reporting List
Reporting Date: December 27, 1990

TSCA - Health and Safety Reporting List
Effective Date: October 29, 1990;   Sunset Date: November 9, 1993

INTERNATIONAL LISTS

Canada - WHMIS: Ingredient Disclosure
1% item 608 (1393)

STATE LISTS

NJ Right to Know List (Total)
sn 0727

NJ Special Hazardous Substances
(corrosive)

DIISOOCTYL PHOSPHITE   3658-48-8

HEALTH AND SAFETY LISTS

NIOSH - Selected LD50s and LC50s
Oral, rat: LD50 = 11900 mg/kg Skin, rabbit: LD50 = 4500 mg/kg

DIISOOCTYL PHTHALATE   27554-26-3
SEE ALSO:
ALKYL PHTHALATES

HEALTH AND SAFETY LISTS

NFPA - Flash Points
flash point = 450 degrees F (232 degrees C)

NFPA - Hazard Identification Ratings
health-0;   flammability-1;   reactivity-0

NIOSH - Selected LD50s and LC50s
Oral, rat: LD50 = 22 gm/kg Skin, rabbit: LD50 = 13 gm/kg

ENVIRONMENTAL LISTS

TSCA - Code of Federal Regulations Citations
40 CFR 716.120(c)

INTERNATIONAL LISTS

United Kingdom - Occupational Exposure Standards - TWAs
5 mg/m3 TWA

DIISOPROPANOLAMINE   110-97-4

HEALTH AND SAFETY LISTS

NFPA - Flash Points
flash point = 260 degrees F (127 degrees C)

NFPA - Hazard Identification Ratings
health-2; flammability-1; reactivity-0

NIOSH - Selected LD50s and LC50s
Oral, rat: LD50 = 6720 mg/kg

ENVIRONMENTAL LISTS

List of Pesticide Product Inert Ingredients
[present]

INTERNATIONAL LISTS

Canada - WHMIS: Ingredient Disclosure
1% item 609 (720)

STATE LISTS

Florida Hazardous Substance List
[present]

Massachusetts Right To Know List
[present]

Pennsylvania Right to Know List
[present]

DIISOPROPYLAMINE   108-18-9

HEALTH AND SAFETY LISTS

ACGIH 1995 - Time Weighted Averages
5 ppm TWA; 21 mg/m3 TWA

ACGIH 1995 - Skin Designations
skin - potential for cutaneous absorption

AIHA - Odor Threshold Values
geometric mean air odor threshold = 0.13 ppm (detectable); 0.38 ppm (recognizable)

U.S. DOT - Substances From 49 CFR 172.101
regulated by DOT (UN1158)

U.S. DOT - Hazard Classes
DOT hazard class = 3

NFPA - Flash Points
flash point = 30 degrees F (-1 degrees C)

NFPA - Hazard Identification Ratings
health-3; flammability-3; reactivity-0

NIOSH - Selected LD50s and LC50s
Inhalation, rat: LC50 = 4800 mg/m3 2 hr Oral, rat: LD50 = 770 mg/kg

NIOSH 1990 - Pocket Guide - RELs
5 ppm TWA; 20 mg/m3 TWA

NIOSH 1990 - Pocket Guide - IDLHs
1000 ppm IDLH

NIOSH 1990 - Pocket Guide - Target organs
skin, eyes, respiratory system

NIOSH 1990 - Pocket Guide - Skin list
Potential for dermal absorption

OSHA - Vacated PELs - Time Weighted Averages
5 ppm TWA; 20 mg/m3 TWA

OSHA - Vacated PELs - Skin Designation
Prevent or reduce skin absorption

OSHA - Final PELs - Time Weighted Averages
5 ppm TWA; 20 mg/m3 TWA

OSHA - Final PELs - Skin Notations
prevent or reduce skin absorption

INTERNATIONAL LISTS

Australian Exposure Standards - Time Weighted Averages
5 ppm TWA; 21 mg/m3 TWA

Australian Exposure Standards - Skin Effects
skin absorption

Canada - WHMIS: Ingredient Disclosure
1% item 610 (721)

Canada - Alberta - 8 Hour Occupational Exposure Limit
5 ppm TWA; 21 mg/m3 TWA

Canada - Alberta - 15 Minute Occupational Exposure Limit
10 ppm STEL; 41 mg/m3 STEL

Canada - Alberta - Skin Designation
can be absorbed through the intact skin

Canada - British Columbia - 8 Hour Exposure Limits
5 ppm TWA; 20 mg/m3 TWA

Canada - British Columbia - Skin Notations
skin - potential for skin absorption

Canada - Ontario - OHSA - TWAEVs
5 ppm TWAEV; 20 mg/m3 TWAEV

Canada - Ontario - OHSA - Skin Notations
absorption through skin, eyes, or mucous membranes

Canada - Quebec - Time-Weighted Average Exposure Values
5 ppm TWAEV; 21 mg/m3 TWAEV

Canada - Quebec - Skin Designations
absorbed through the skin

United Kingdom - Occupational Exposure Standards - TWAs
5 ppm TWA; 20 mg/m3 TWA

United Kingdom - Occupational Exposure Standards - Notes
can be absorbed through skin

Israel - Time Weighted Averages
5 ppm TWA; 21 mg/m3 TWA

Israel - Action Levels
2.5 ppm AL; 10.5 mg/m3 AL

Mexico - Instruction No. 10 - TWAs
5 ppm TWA; 20 mg/m3 TWA

Mexico - Instruction No. 10 - Skin designation
skin - potential for cutaneous absorption

STATE LISTS

California - Exposure Limits - PELs
5 ppm PEL; 20 mg/m3 PEL

California - Exposure Limits - Skin Notation
material may be absorbed through the skin, eyes or mucous membrane

California - Directors List of Hazardous Substances (8 CCR 339)
[present]

Florida Hazardous Substance List
[present]

Massachusetts Right To Know List
[present]

Minnesota Hazardous Substance List
skin

NJ Right to Know List (Total)
sn 0728

NJ Special Hazardous Substances
(flammable - third degree)

Pennsylvania Right to Know List
[present]

1,3-DIISOPROPYLBENZENE   99-62-7

HEALTH AND SAFETY LISTS

NIOSH - Selected LD50s and LC50s
Oral, rat: LD50 = 7400 mg/kg

DIISOPROPYL BENZENE 25321-09-9

HEALTH AND SAFETY LISTS

U.S. DOT - Appendix B - Marine Pollutants
DOT regulated marine pollutant

NFPA - Flash Points
flash point = 170 degrees F (77 degrees C)

NFPA - Hazard Identification Ratings
health-0; flammability-2; reactivity-0

NIOSH - Selected LD50s and LC50s
Oral, rat: LD50 = 6500 mg/kg Skin, rabbit: LD50 = 16 gm/kg

DIISOPROPYLBENZENE HYDROPEROXIDE 26762-93-6

HEALTH AND SAFETY LISTS

U.S. DOT - Hazard Classes
Forbidden from transport by the DOT

STATE LISTS

NJ Right to Know List (Total)
sn 2338

DIISOPROPYL BIPHENYL 69009-90-1

ENVIRONMENTAL LISTS

List of Pesticide Product Inert Ingredients
[present]

TSCA - Code of Federal Regulations Citations
40 CFR 712.30(k); 40 CFR 716.120(a)

TSCA - PAIR - Reporting List
Reporting Date: August 27, 1984

TSCA - Health and Safety Reporting List
Effective Date: June 28, 1984

DIISOPROPYLCARBODIIMIDE 693-13-0

HEALTH AND SAFETY LISTS

NTP Chemical Status Reports - Testing Status and NTIS Number
Prechronic studies in progress; approved for toxicology/carcinogenesis study

N,N-DIISOPROPYL ETHANOLAMINE 96-80-0

HEALTH AND SAFETY LISTS

NFPA - Flash Points
flash point = 175 degrees F (79 degrees C)

NFPA - Hazard Identification Ratings
health-1; flammability-2; reactivity-0

NIOSH - Selected LD50s and LC50s
Oral, rat: LD50 = 1070 mg/kg Skin, rabbit: LD50 = 450 mg/kg

INTERNATIONAL LISTS

Canada - WHMIS: Ingredient Disclosure
1% item 611 (722)

STATE LISTS

NJ Special Hazardous Substances
(corrosive)

DIISOPROPYL MALEATE 10099-70-4

HEALTH AND SAFETY LISTS

NFPA - Flash Points
flash point = 220 degrees F (104 degrees C)

NFPA - Hazard Identification Ratings
health-1; flammability-1; reactivity-0

DIISOPROPYLPHENOLS 27923-56-4

ENVIRONMENTAL LISTS

List of Pesticide Product Inert Ingredients
[present]

DIISOPROPYL SULFATE 2973-10-6

STATE LISTS

California - Prop. 65 - Cancer list
carcinogen - initial date 4/1/93

DIISOPROPYL SULFATE RR-01553-2

HEALTH AND SAFETY LISTS

IARC - Group 2B (sufficient animal data)
[present]

OSHA - Possible Select Carcinogens
[present]

DI-ISOTRIDECYL PEROXYDICARBONATE RR-01516-7

STATE LISTS

NJ Right to Know List (Total)
sn 2339

DIISOTRIDECYL PEROXYDICARBONATE, TECHNICALLY PURE 82065-80-3

HEALTH AND SAFETY LISTS

U.S. DOT - Organic Peroxides Table
Organic peroxide UN3115

DIKETENE 674-82-8

HEALTH AND SAFETY LISTS

U.S. DOT - Substances From 49 CFR 172.101
regulated by DOT (UN2521)

U.S. DOT - Hazard Classes
DOT hazard class = 3

U.S. DOT - Substances Which Are Poisonous by Inhalation
liquid hazardous material poisonous by inhalation (inhibited form) (UN2521)

NFPA - Flash Points
flash point = 93 degrees F (34 degrees C)

NFPA - Hazard Identification Ratings
health-4; flammability-2; reactivity-2

NIOSH - Selected LD50s and LC50s
Oral, rat: LD50 = 560 mg/kg Skin, rabbit: LD50 = 2830 mg/kg

INTERNATIONAL LISTS

Canada - WHMIS: Ingredient Disclosure
1% item 612 (648)

STATE LISTS

Florida Hazardous Substance List
[present]

Massachusetts Right To Know List
[present]

NJ Right to Know List (Total)
sn 0732

NJ Special Hazardous Substances
(reactive - second degree)

Pennsylvania Right to Know List
[present]

DILAURYL BETA-THIODIPROPIONATE 123-28-4

ENVIRONMENTAL LISTS

List of Pesticide Product Inert Ingredients
[present]

INTERNATIONAL LISTS

Canada - WHMIS: Ingredient Disclosure
1% item 613 (1614)

DI-LINEAR 79 PHTHALATE RR-01192-7

INTERNATIONAL LISTS

United Kingdom - Occupational Exposure Standards - TWAs
5 mg/m3 TWA

DILINOLEIC ACID   6144-28-1

ENVIRONMENTAL LISTS

List of Pesticide Product Inert Ingredients
[present]

DIMEFOX   115-26-4

HEALTH AND SAFETY LISTS

NIOSH - Selected LD50s and LC50s
*Inhalation, rat: LC50 = 2 gm/m3 10 mn Oral, rat: LD50 = 1 mg/kg
Skin, rat: LD50 = 2 mg/kg*

ENVIRONMENTAL LISTS

CERCLA/SARA - Section 302 Extremely Hazardous Substances and
TPQs
TPQ = 500 pounds

STATE LISTS

Florida Hazardous Substance List
effective March 13, 1992

Massachusetts Right To Know List
extraordinarily hazardous

NJ Right to Know List (Total)
sn 2342

Pennsylvania Right to Know List
environmental hazard

PROPOSED REGULATIONS

CERCLA/SARA - Proposed Hazardous Substance Additions
proposed RQ = 1 pound (.454 kg)

CERCLA/SARA - 1989 Proposed RQ Adjustments
proposed RQ = 100 pounds (45.4 kg)

DIMER ACIDS, POLYMER WITH POLYALKYLENE   RR-00163-8
GLYCOL, BISPHENOL A-DIGLYCERYLETHER AND
ALKYLENEPOLYOLS POLYGLYCIDYLETHERS

ENVIRONMENTAL LISTS

TSCA - Chemicals with Significant New Use Rules
PMN numbers: P-86-628

2,5-DIMERCAPTO-1,3,4-THIADIAZOLE, ALKYL   RR-00909-6
POLYCARBOXYLATE

ENVIRONMENTAL LISTS

TSCA - Chemicals with Significant New Use Rules
PMN number: P-88-1460

BETA, BETA-DIMETHACRYLIC ACID   541-47-9

HEALTH AND SAFETY LISTS

NIOSH - Selected LD50s and LC50s
Oral, rat: LD50 = 3560 mg/kg

DIMETHIPIN [2,3,-DIHYDRO-5,6-DIMETHYL-1,4-   55290-64-7
DITHIIN-1,1,4,4-TETRAOXIDE]

ENVIRONMENTAL LISTS

CERCLA/SARA - Section 313 - Emission Reporting
form R reporting required

DIMETHISTERONE   79-64-1

HEALTH AND SAFETY LISTS

IARC - Group Unspecified
[present] (Listed under 'Progestins')

DIMETHISTERONE AND OESTROGENS   RR-00102-5

HEALTH AND SAFETY LISTS

IARC - Group Unspecified
[present] (Listed under 'Sequential oral contraceptives')

DIMETHOATE   60-51-5

HEALTH AND SAFETY LISTS

U.S. DOT - Appendix B - Marine Pollutants
DOT regulated severe marine pollutant

U.S. DOT - Appendix A Table 1 - Hazardous Substances
final RQ = 10 pounds (4.54 kg)

NIOSH - Selected LD50s and LC50s
Oral, rat: LD50 = 152 mg/kg Skin, rat: LD50 = 353 mg/kg

NTP Chemical Status Reports - Testing Status and NTIS Number
Technical reports printed (PB264367/AS)

NTP Chemical Status Reports - Evidence of Carcinogenicity
*male rat-negative; female rat-negative; male mice-negative; female
mice-negative*

ENVIRONMENTAL LISTS

CERCLA/SARA - Section 302 Extremely Hazardous Substances and
TPQs
TPQ = 500/10,000 pounds

CERCLA/SARA - Section 313 - Emission Reporting
form R reporting required

CERCLA/SARA - Hazardous Substances and their Reportable Quantities
final RQ = 10 pounds (4.54 kg)

RCRA - P Series Wastes
waste number P044

RCRA - Hazardous Constituents-Appendix VIII
waste number P044

RCRA - Substances Banned From Land Disposal
[present]

RCRA - TSD Facilities Ground Water Monitoring
TM 8270 = 10 ug/L PQL

INTERNATIONAL LISTS

Canada - Drinking Water Quality - IMACs
0.02 mg/L IMAC

STATE LISTS

Florida Hazardous Substance List
effective March 13, 1992

Massachusetts Right To Know List
extraordinarily hazardous

NJ Right to Know List (Total)
sn 0733

NJ Special Hazardous Substances
(mutagen)

Pennsylvania Right to Know List
environmental hazard

DIMETHOXANE   828-00-2

HEALTH AND SAFETY LISTS

IARC - Group 3 (not classifiable)
[present]

NIOSH - Selected LD50s and LC50s
Oral, rat: LD50 = 1930 mg/kg

NTP Chemical Status Reports - Testing Status and NTIS Number
Technical reports printed (PB90220096/AS)

NTP Chemical Status Reports - Evidence of Carcinogenicity
*male rat-no evidence; female rat-no evidence; male mice-equivocal
evidence; female mice-no evidence*

ENVIRONMENTAL LISTS

List of Pesticide Product Inert Ingredients
[present]

TSCA - Code of Federal Regulations Citations
40 CFR 716.120(a)

TSCA - Health and Safety Reporting List
Effective Date: June 1, 1987

STATE LISTS

California - Directors List of Hazardous Substances (8 CCR 339)
[present]

2,5-DIMETHOXYANILINE   102-56-7

HEALTH AND SAFETY LISTS

NFPA - Flash Points
flash point = 302 degrees F (150 degrees C)

NFPA - Hazard Identification Ratings
health-2; flammability-1; reactivity-0

STATE LISTS

Florida Hazardous Substance List
[present]

Massachusetts Right To Know List
[present]

Pennsylvania Right to Know List
[present]

2,4-DIMETHOXYANILINE HYDROCHLORIDE   54150-69-5

HEALTH AND SAFETY LISTS

NTP Chemical Status Reports - Testing Status and NTIS Number
Technical reports printed (PB288625/AS)

NTP Chemical Status Reports - Evidence of Carcinogenicity
male rat-negative; female rat-negative; male mice-negative; female mice-negative

3,3'-DIMETHOXYBENZIDINE   119-90-4

HEALTH AND SAFETY LISTS

U.S. DOT - Appendix A Table 1 - Hazardous Substances
final RQ = 100 pounds (45.4 kg)

IARC - Group 2B (sufficient animal data)
[present]

NFPA - Flash Points
flash point = 403 degrees F (206 degrees C)

NFPA - Hazard Identification Ratings
flammability-1; reactivity-0

NIOSH - Selected LD50s and LC50s
Oral, rat: LD50 = 1920 mg/kg

NTP Seventh Report - Suspect Carcinogens
suspect carcinogen

OSHA - Possible Select Carcinogens
[present]

ENVIRONMENTAL LISTS

CERCLA/SARA - Section 313 - Emission Reporting
form R reporting required for 0.1% de minimus concentration

CERCLA/SARA - Hazardous Substances and their Reportable Quantities
final RQ = 100 pounds (45.4 kg)

Clean Air Act (1990) - List of Hazardous Air Contaminants
[present]

EPA - Carcinogen Hazard Ranking for RQ Adjustment
Hazard ranking = Medium

RCRA - U Series Wastes
waste number U091

RCRA - Hazardous Constituents-Appendix VIII
waste number U091

RCRA - Substances Banned From Land Disposal
[present]

INTERNATIONAL LISTS

Canada - WHMIS: Ingredient Disclosure
0.1% item 614 (727)

German (DFG) - Carcinogens
animal evidence of carcinogenicity

STATE LISTS

California - Air Bill 2588 Appendix A-I
known or potential carcinogen

California - Prop. 65 - Cancer list
carcinogen - initial date 1/1/88

California - Directors List of Hazardous Substances (8 CCR 339)
[present]

Florida Hazardous Substance List
[present]

Massachusetts Right To Know List
carcinogen; extraordinarily hazardous

Minnesota Hazardous Substance List
carcinogen (includes o-dianisidine based dyes)

NJ Right to Know List (Total)
sn 0734

NJ Special Hazardous Substances
(carcinogen)

Pennsylvania Right to Know List
environmental hazard; special hazardous substance

Pennsylvania RTK - Special Hazardous Substances
[present]

3,3'-DIMETHOXYBENZIDINE HYDROCHLORIDE (O-DIANISIDINE HYDROCHLORIDE)   111984-09-9

ENVIRONMENTAL LISTS

CERCLA/SARA - Section 313 - Emission Reporting
form R reporting required

3,3'-DIMETHOXYBENZIDINE DIHYDROCHLORIDE   20325-40-0

HEALTH AND SAFETY LISTS

NTP Chemical Status Reports - Testing Status and NTIS Number
Technical reports printed (PB90241076)

NTP Chemical Status Reports - Evidence of Carcinogenicity
male rat-clear evidence; female rat-clear evidence

NTP Seventh Report - Suspect Carcinogens
suspect carcinogen

ENVIRONMENTAL LISTS

CERCLA/SARA - Section 313 - Emission Reporting
form R reporting required

STATE LISTS

California - Air Bill 2588 Appendix A-II
known or potential carcinogen: 6/91

California - Prop. 65 - Cancer list
carcinogen - initial date 10/1/90

3,3'-DIMETHOXYBENZIDINE-4,4'-DIISOCYANATE   91-93-0

HEALTH AND SAFETY LISTS

IARC - Group 3 (not classifiable)
[present]

NTP Chemical Status Reports - Testing Status and NTIS Number
Technical reports printed (PB290154/AS)

NTP Chemical Status Reports - Evidence of Carcinogenicity
male rat-positive; female rat-positive; male mice-negative; female mice-negative

ENVIRONMENTAL LISTS

CERCLA/SARA - Section 313 - Emission Reporting
form R reporting required; (Listed under 'Diisocyanates')

STATE LISTS
California - Directors List of Hazardous Substances (8 CCR 339)
[present]

Massachusetts Right To Know List
carcinogen; extraordinarily hazardous

2,5-DIMETHOXYCHLOROBENZENE   2100-42-7

HEALTH AND SAFETY LISTS
NFPA - Flash Points
flash point = 243 degrees F (117 degrees C)

NFPA - Hazard Identification Ratings
health-2; flammability-1; reactivity-0

STATE LISTS
Florida Hazardous Substance List
[present]

Massachusetts Right To Know List
[present]

Pennsylvania Right to Know List
[present]

1,1-DIMETHOXYETHANE   534-15-6

HEALTH AND SAFETY LISTS
U.S. DOT - Substances From 49 CFR 172.101
regulated by DOT (UN2377)

U.S. DOT - Hazard Classes
DOT hazard class = 3

NIOSH - Selected LD50s and LC50s
Inhalation, rat: LC50 = 3000 ppm 4 hr Oral, rat: LD50 = 6500 mg/kg Skin, rabbit: LD50 = 20 gm/kg

ENVIRONMENTAL LISTS
TSCA - Code of Federal Regulations Citations
40 CFR 712.30(d)

TSCA - PAIR - Reporting List
Reporting Date: November 19, 1982

STATE LISTS
NJ Right to Know List (Total)
sn 0735

1,1-DIMETHOXY-2-PROPENE   RR-01163-2

HEALTH AND SAFETY LISTS
U.S. DOT - Substances Which Are Poisonous by Inhalation
liquid hazardous material poisonous by inhalation

4,7-DIMETHOXY-5-(2-PROPENYL)-1,3-   523-80-8
BENZODIOXOLE

ENVIRONMENTAL LISTS
List of Pesticide Product Inert Ingredients
[present]

DIMETHYL-P-PHENYLENEDIAMINE   99-98-9

ENVIRONMENTAL LISTS
CERCLA/SARA - Section 302 Extremely Hazardous Substances and TPQs
TPQ = 10/10,000 pounds

STATE LISTS
Florida Hazardous Substance List
effective March 13, 1992

Massachusetts Right To Know List
extraordinarily hazardous

NJ Right to Know List (Total)
sn 2348

Pennsylvania Right to Know List
environmental hazard

PROPOSED REGULATIONS
CERCLA/SARA - Proposed Hazardous Substance Additions
proposed RQ = 1 pound (.454 kg)

CERCLA/SARA - 1989 Proposed RQ Adjustments
proposed RQ = 1 pound (.454 kg)

DIMETHYL ACETAMIDE   127-19-

HEALTH AND SAFETY LISTS
ACGIH 1995 - Time Weighted Averages
10 ppm TWA; 36 mg/m3 TWA

ACGIH 1995 - Skin Designations
skin - potential for cutaneous absorption

NFPA - Flash Points
flash point = 158 degrees F (70 degrees C)

NFPA - Hazard Identification Ratings
health-2; flammability-2; reactivity-0

NIOSH - Selected LD50s and LC50s
Inhalation, mouse: LC50 = 7200 mg/m3 (8 hr) Oral, rat: LD50 = 5000 mg/kg Skin, mouse: LD50 = 9600 mg/kg

NIOSH 1990 - Pocket Guide - RELs
10 ppm TWA; 35 mg/m3 TWA

NIOSH 1990 - Pocket Guide - IDLHs
400 ppm IDLH

NIOSH 1990 - Pocket Guide - Target organs
liver, skin

NIOSH 1990 - Pocket Guide - Skin list
Potential for dermal absorption

OSHA - Vacated PELs - Time Weighted Averages
10 ppm TWA; 35 mg/m3 TWA

OSHA - Vacated PELs - Skin Designation
Prevent or reduce skin absorption

OSHA - Final PELs - Time Weighted Averages
10 ppm TWA; 35 mg/m3 TWA

OSHA - Final PELs - Skin Notations
prevent or reduce skin absorption

ENVIRONMENTAL LISTS
EPA - Master Testing List
[present]

INTERNATIONAL LISTS
Australian Exposure Standards - Time Weighted Averages
10 ppm TWA; 36 mg/m3 TWA

Australian Exposure Standards - Skin Effects
skin absorption

Canada - WHMIS: Ingredient Disclosure
1% item 615 (728)

Canada - Alberta - 8 Hour Occupational Exposure Limit
10 ppm TWA; 36 mg/m3 TWA

Canada - Alberta - 15 Minute Occupational Exposure Limit
15 ppm STEL; 53 mg/m3 STEL

Canada - Alberta - Skin Designation
can be absorbed through the intact skin

Canada - British Columbia - 8 Hour Exposure Limits
10 ppm TWA; 35 mg/m3 TWA

Canada - British Columbia - 15 Minute Exposure Limits
15 ppm STEL; 50 mg/m3 STEL

Canada - British Columbia - Skin Notations
skin - potential for skin absorption

Canada - Ontario - OHSA - TWAEVs
10 ppm TWAEV; 36 mg/m3 TWAEV

Canada - Ontario - OHSA - Skin Notations
absorption through skin, eyes, or mucous membranes

Canada - Quebec - Time-Weighted Average Exposure Values
10 ppm TWAEV; 36 mg/m3 TWAEV

Canada - Quebec - Skin Designations
absorbed through the skin
United Kingdom - Occupational Exposure Standards - TWAs
10 ppm TWA; 35 mg/m3 TWA
United Kingdom - Occupational Exposure Standards - STELs
15 ppm STEL; 50 mg/m3 STEL
United Kingdom - Occupational Exposure Standards - Notes
can be absorbed through skin
German (DFG) - MAK Values
10 ppm MAK; 35 mg/m3 MAK
German (DFG) - Peak Limitations
2 x normal MAK (30 min. average value); don't exceed 4 times during shift
German (DFG) - Skin/Sensitizers
danger of cutaneous absorption
German (DFG) - Pregnancy
no risk to embryo/fetus if exposure limits adhered to
Israel - Time Weighted Averages
10 ppm TWA; 36 mg/m3 TWA
Israel - Action Levels
5 ppm AL; 18 mg/m3 AL
Mexico - Instruction No. 10 - TWAs
10 ppm TWA; 35 mg/m3 TWA
Mexico - Instruction No. 10 - STELs
15 ppm STEL; 50 mg/m3 STEL

STATE LISTS

California - Exposure Limits - PELs
10 ppm PEL; 35 mg/m3 PEL
California - Exposure Limits - Skin Notation
material may be absorbed through the skin, eyes or mucous membrane
California - Directors List of Hazardous Substances (8 CCR 339)
[present]
Florida Hazardous Substance List
[present]
Massachusetts Right To Know List
[present]
Minnesota Hazardous Substance List
skin
Pennsylvania Right to Know List
[present]

PROPOSED REGULATIONS

ACGIH 1995 - Proposed Biological Exposure Indices
N-methylacetamide in urine: 30 mg/g creatinine, end of shift at end of workweek

DIMETHYLAMINE   124-40-3

HEALTH AND SAFETY LISTS

ACGIH 1995 - Time Weighted Averages
5 ppm TWA; 9.2 mg/m3 TWA
ACGIH 1995 - Short Term Exposure Limits
15 ppm STEL; 27.6 mg/m3 STEL
AIHA - Odor Threshold Values
no geometric mean air odor threshold
U.S. DOT - Substances From 49 CFR 172.101
regulated by DOT (UN1160, UN1032)
U.S. DOT - Hazard Classes
DOT hazard class = 3
U.S. DOT - Appendix A Table 1 - Hazardous Substances
final RQ = 1000 pounds (454 kg)
NFPA - Flash Points
gas (no flash point given)
NFPA - Hazard Identification Ratings
health-3; flammability-4; reactivity-0

NIOSH - Selected LD50s and LC50s
Inhalation, rat: LC50 = 4540 ppm 6 hr Oral, rat: LD50 = 698 mg/kg
NIOSH 1990 - Pocket Guide - RELs
10 ppm TWA; 18 mg/m3 TWA
NIOSH 1990 - Pocket Guide - IDLHs
2000 ppm IDLH
NIOSH 1990 - Pocket Guide - Target organs
skin, eyes, respiratory system
OSHA - Vacated PELs - Time Weighted Averages
10 ppm TWA; 18 mg/m3 TWA
OSHA - Final PELs - Time Weighted Averages
10 ppm TWA; 18 mg/m3 TWA
OSHA - List of Highly Hazardous Chemicals
anhydrous: threshhold quantity = 2500 pounds

ENVIRONMENTAL LISTS

CERCLA/SARA - Section 313 - Emission Reporting
form R reporting required
CERCLA/SARA - Hazardous Substances and their Reportable Quantities
final RQ = 1000 pounds (454 kg)
CAA - Flammable Substances for Accidental Release Prevention
threshold quantity = 10,000 lbs
CAA - HON Rule - SOCMI Chemicals
compliance by July 24, 1995
Clean Water Act - Hazardous Substances
[present]
List of Pesticide Product Inert Ingredients
[present]
RCRA - U Series Wastes
waste number U092 (Ignitable waste)
RCRA - Hazardous Constituents-Appendix VIII
waste number U092 (Ignitable waste)
RCRA - Substances Banned From Land Disposal
[present]

INTERNATIONAL LISTS

Australian Exposure Standards - Time Weighted Averages
10 ppm TWA; 18 mg/m3 TWA
Canada - WHMIS: Ingredient Disclosure
1% item 616 (729)
Canada - Alberta - 8 Hour Occupational Exposure Limit
10 ppm TWA; 18 mg/m3 TWA
Canada - Alberta - 15 Minute Occupational Exposure Limit
20 ppm STEL; 36 mg/m3 STEL
Canada - British Columbia - 8 Hour Exposure Limits
10 ppm TWA; 18 mg/m3 TWA
Canada - Ontario - OHSA - TWAEVs
10 ppm TWAEV; 18 mg/m3 TWAEV
Canada - Quebec - Time-Weighted Average Exposure Values
10 ppm TWAEV; 18 mg/m3 TWAEV
United Kingdom - Occupational Exposure Standards - TWAs
10 ppm TWA; 18 mg/m3 TWA
German (DFG) - MAK Values
2 ppm MAK; 4 mg/m3 MAK
German (DFG) - Peak Limitations
2 x normal MAK (10 min momentary value); don't exceed 4 times per shift
Israel - Time Weighted Averages
10 ppm TWA; 18 mg/m3 TWA
Israel - Action Levels
5 ppm AL; 9 mg/m3 AL
Mexico - Instruction No. 10 - TWAs
10 ppm TWA; 18 mg/m3 TWA

STATE LISTS

California - Exposure Limits - PELs
10 ppm PEL;   18 mg/m3 PEL

California - Directors List of Hazardous Substances (8 CCR 339)
[present]

Florida Hazardous Substance List
[present]

Massachusetts Right To Know List
[present]

Minnesota Hazardous Substance List
[present]

NJ Right to Know List (Total)
sn 0737

NJ Special Hazardous Substances
(flammable - fourth degree)

Pennsylvania Right to Know List
environmental hazard

DIMETHYLAMINE DICAMBA   2300-66-5

ENVIRONMENTAL LISTS

CERCLA/SARA - Section 313 - Emission Reporting
form R reporting required

DIMETHYLAMINE (2,4-DICHLOROPHENOXY) ACETATE   2008-39-1

HEALTH AND SAFETY LISTS

NIOSH - Selected LD50s and LC50s
Oral, rat: LD50 = 300 mg/kg Skin, rabbit: LD50 = 2115 mg/kg

DIMETHYLAMINE DODECYLBENZENESULFONATE   37452-11-2

ENVIRONMENTAL LISTS

List of Pesticide Product Inert Ingredients
[present]

DIMETHYLAMINE ETHYLENEDIAMINETETRAACETATE   73455-30-8

ENVIRONMENTAL LISTS

List of Pesticide Product Inert Ingredients
[present]

DIMETHYLAMINE HYDROCHLORIDE   506-59-2

HEALTH AND SAFETY LISTS

NIOSH - Selected LD50s and LC50s
Oral, rat: LD50 = 1070 mg/kg

DIMETHYLAMINOACETONITRILE   926-64-7

HEALTH AND SAFETY LISTS

U.S. DOT - Substances From 49 CFR 172.101
regulated by DOT (UN2378)

U.S. DOT - Hazard Classes
DOT hazard class = 3

NIOSH - Selected LD50s and LC50s
Oral, rat: LD50 = 50 mg/kg Skin, rabbit: LD50 = 170 mg/kg

INTERNATIONAL LISTS

Canada - WHMIS: Ingredient Disclosure
1% item 617 (730)

STATE LISTS

NJ Right to Know List (Total)
sn 0738

4-DIMETHYLAMINOAZOBENZENE   60-11-7

SEE ALSO:
F039-HAZARDOUS WASTES

HEALTH AND SAFETY LISTS

U.S. DOT - Appendix A Table 1 - Hazardous Substances
final RQ = 10 pounds (4.54 kg)

IARC - Group 2B (sufficient animal data)
[present] (Overall evaluation based only on evidence of carcinogenicity in monograph (8, 1975) or in Supplement 4)

NIOSH - Selected LD50s and LC50s
Oral, rat: LD50 = 200 mg/kg

NIOSH 1990 - Pocket Guide - Carcinogens
occupational carcinogen

NIOSH 1990 - Pocket Guide - Target organs
liver, skin, bladder

NIOSH - Health Standards - Exposure Limits
use 29 CFR 1910.1015

NIOSH - Health Standards - Health Effects and Precautions
has produced tumors of the liver and bladder in animals

NIOSH - Health Standards - Carcinogenic Chemicals
potential human carcinogen

NTP Seventh Report - Suspect Carcinogens
suspect carcinogen

OSHA - 29 CFR 1910 Specifically Regulated Chemicals
Cancer suspect agent (see 29 CFR 1910.1015)

OSHA - Select Carcinogens
[present]

OSHA - Possible Select Carcinogens
[present]

ENVIRONMENTAL LISTS

CERCLA/SARA - Section 313 - Emission Reporting
form R reporting required for 0.1% de minimus concentration

CERCLA/SARA - Hazardous Substances and their Reportable Quantities
final RQ = 10 pounds (4.54 kg)

Clean Air Act (1990) - List of Hazardous Air Contaminants
[present]

EPA - Carcinogen Hazard Ranking for RQ Adjustment
Hazard ranking = Medium

RCRA - U Series Wastes
waste number U093

RCRA - Hazardous Constituents-Appendix VIII
waste number U093

RCRA - Basis for Listing - Appendix VII
Included in waste stream: F039

RCRA - Substances Banned From Land Disposal
[present]

RCRA - TSD Facilities Ground Water Monitoring
TM 8270 = 10 ug/L PQL

RCRA - Universal Treatment Standards (LDR)
WW: 0.13 mg/l; NWW: Not applicable

INTERNATIONAL LISTS

Canada - WHMIS: Ingredient Disclosure
0.1% item 618 (731)

STATE LISTS

California - Air Bill 2588 Appendix A-I
[present]

California - Prop. 65 - Cancer list
carcinogen - initial date 1/1/88

California - Prop. 65 - No Significant Risk Levels
no significant risk level = 0.2 ug/day

California - Exposure Limits - Carcinogens
cancer-suspect agent (at a concentration >= 1.0%)

California - Directors List of Hazardous Substances (8 CCR 339)
  [present]
Florida Hazardous Substance List
  [present]
Massachusetts Right To Know List
  carcinogen; extraordinarily hazardous
Minnesota Hazardous Substance List
  carcinogen
NJ Right to Know List (Total)
  sn 0739
NJ Special Hazardous Substances
  (carcinogen; mutagen; teratogen)
Pennsylvania Right to Know List
  environmental hazard; special hazardous substance
Pennsylvania RTK - Special Hazardous Substances
  [present]

PARA-DIMETHYLAMINOAZOBENZENEDIAZO   RR-01532-7
SODIUM SULPHONATE
  HEALTH AND SAFETY LISTS
    IARC - Group 3 (not classifiable)
      [present]

4-(DIMETHYLAMINO)BENZOIC ACID, 2-ETHYL-   21245-02-3
HEXYL ESTER
  ENVIRONMENTAL LISTS
    List of Pesticide Product Inert Ingredients
      [present]

4-DIMETHYLAMINO-6-(2-DIMETHY-   RR-00765-8
LAMINOETHOXY)TOLUENE-2-DIAZONIUMZINC
CHLORIDE
  HEALTH AND SAFETY LISTS
    U.S. DOT - Substances From 49 CFR 172.101
      regulated by DOT (UN3039)
    U.S. DOT - Hazard Classes
      DOT hazard class = 4.1

3-DIMETHYLAMINO-N,N-   17268-47-2
DIMETHYLPROPIONAMIDE
  HEALTH AND SAFETY LISTS
    NIOSH - Selected LD50s and LC50s
      Oral, rat: LD50 = 3080 mg/kg Skin, rabbit: LD50 = 790 mg/kg
  INTERNATIONAL LISTS
    Canada - WHMIS: Ingredient Disclosure
      1% item 619 (732)

2-(DIMETHYLAMINO) ETHANOL   108-01-0
  HEALTH AND SAFETY LISTS
    U.S. DOT - Substances From 49 CFR 172.101
      regulated by DOT (UN2051)
    U.S. DOT - Hazard Classes
      DOT hazard class = 3
    NFPA - Flash Points
      flash point = 105 degrees F (41 degrees C)
    NFPA - Hazard Identification Ratings
      health-2; flammability-2; reactivity-0
    NIOSH - Selected LD50s and LC50s
      Inhalation, mouse: LC50 = 3250 mg/m3 (8 hr) Oral, rat: LD50 = 2 gm/kg Skin, rabbit: LD50 = 1370 mg/kg
  ENVIRONMENTAL LISTS
    CAA - HON Rule - SOCMI Chemicals
      compliance by Oct. 24, 1994
    EPA - Master Testing List
      [present]

STATE LISTS
  Florida Hazardous Substance List
    [present]
  Massachusetts Right To Know List
    [present]
  Pennsylvania Right to Know List
    [present]

DIMETHYLAMINOETHOXYETHANOL   1704-62-7
  HEALTH AND SAFETY LISTS
    NIOSH - Selected LD50s and LC50s
      Oral, rat: LD50 = 2460 mg/kg Skin, rabbit: LD50 = 1410 mg/kg

N,N-DIMETHYLAMINOETHYL METHACRYLATE   2867-47-2
  HEALTH AND SAFETY LISTS
    U.S. DOT - Substances From 49 CFR 172.101
      regulated by DOT (UN2522)
    U.S. DOT - Hazard Classes
      DOT hazard class = 6.1
    NFPA - Flash Points
      flash point = 165 degrees F (74 degrees C)
    NFPA - Hazard Identification Ratings
      health-2; flammability-2; reactivity-0
    NIOSH - Selected LD50s and LC50s
      Inhalation, rat: LC50 = 620 mg/m3 4 hr Oral, rat: LD50 = 1751 mg/kg
  ENVIRONMENTAL LISTS
    TSCA - Code of Federal Regulations Citations
      40 CFR 712.30(d)
    TSCA - PAIR - Reporting List
      Reporting Date: November 19, 1982
  INTERNATIONAL LISTS
    Canada - WHMIS: Ingredient Disclosure
      1% item 620 (1090)
  STATE LISTS
    Florida Hazardous Substance List
      [present]
    Massachusetts Right To Know List
      [present]
    NJ Right to Know List (Total)
      sn 0740
    Pennsylvania Right to Know List
      [present]

N-[2-(DIMETHYLAMINO)ETHYL]-N,N'.N'-   3030-47-5
TRIMETHYL-1,2-ETHANEDIAMINE
  HEALTH AND SAFETY LISTS
    NIOSH - Selected LD50s and LC50s
      Oral, rat: LD50 = 1630 mg/kg Skin, rabbit: LD50 = 280 mg/kg

TRANS-2-((DIMETHYLAMINO)METHYLIMINO)-5-(2-   55738-54-0
(5-NITRO-2-FURYL)VINYL)-1,3,4-OXADIAZOLE
  HEALTH AND SAFETY LISTS
    IARC - Group 2B (sufficient animal data)
      [present] (Overall evaluation based only on evidence of carcinogenicity in monograph (7, 1974) or in Supplement 4)
    OSHA - Possible Select Carcinogens
      [present]
  STATE LISTS
    California - Air Bill 2588 Appendix A-II
      known or potential carcinogen
    California - Prop. 65 - Cancer list
      carcinogen - initial date 1/1/88

California - Prop. 65 - No Significant Risk Levels
no significant risk level = 2 ug/day

California - Directors List of Hazardous Substances (8 CCR 339)
[present]

Florida Hazardous Substance List
[present]

Massachusetts Right To Know List
carcinogen; extraordinarily hazardous

Minnesota Hazardous Substance List
carcinogen

Pennsylvania Right to Know List
special hazardous substance

Pennsylvania RTK - Special Hazardous Substances
[present]

DIMETHYLAMINOMETHYLPHENOL   25338-55-0

ENVIRONMENTAL LISTS

List of Pesticide Product Inert Ingredients
[present]

BETA-DIMETHYLAMINOPROPIONITRILE   1738-25-6

HEALTH AND SAFETY LISTS

NFPA - Flash Points
flash point = 149 degrees F (65 degrees C)

NFPA - Hazard Identification Ratings
flammability-2; reactivity-1

NIOSH - Selected LD50s and LC50s
Oral, rat: LD50 = 2600 mg/kg Skin, rabbit: LD50 = 1410 mg/kg

3-(DIMETHYLAMINO)-PROPYLAMINE   109-55-7

HEALTH AND SAFETY LISTS

NFPA - Flash Points
flash point = 100 degrees F (38 degrees C)

NFPA - Hazard Identification Ratings
health-3; flammability-2; reactivity-0

NIOSH - Selected LD50s and LC50s
Oral, rat: LD50 = 1870 mg/kg

ENVIRONMENTAL LISTS

EPA - Master Testing List
[present]

STATE LISTS

Florida Hazardous Substance List
[present]

Massachusetts Right To Know List
[present]

Pennsylvania Right to Know List
[present]

[[[3-(DIMETHYLAMINO)PROPYL]IMINO]   71113-21-8
BIS(METHYLENE)]BISPHOSPHONIC ACID,
MONOHYDROCHLORIDE

ENVIRONMENTAL LISTS

List of Pesticide Product Inert Ingredients
[present]

4,4'-DIMETHYLANGELICIN PLUS ULTRAVIOLET A   22975-76-4
RADIATION

HEALTH AND SAFETY LISTS

IARC - Group 3 (not classifiable)
[present]

4,5'-DIMETHYLANGELICIN PLUS ULTRAVIOLET A   RR-01533-8
RADIATION

HEALTH AND SAFETY LISTS

IARC - Group 3 (not classifiable)
[present]

DIMETHYLANILINE   121-6

HEALTH AND SAFETY LISTS

ACGIH 1995 - Time Weighted Averages
5 ppm TWA; 25 mg/m3 TWA

ACGIH 1995 - Short Term Exposure Limits
10 ppm STEL; 50 mg/m3 STEL

ACGIH 1995 - Skin Designations
skin - potential for cutaneous absorption

AIHA - Odor Threshold Values
no geometric mean air odor threshold

U.S. DOT - Substances From 49 CFR 172.101
regulated by DOT (UN2253)

U.S. DOT - Hazard Classes
DOT hazard class = 6.1

IARC - Group 3 (not classifiable)
[present]

NFPA - Flash Points
flash point = 145 degrees F (63 degrees C);   C.P.: 165 degrees F degrees C)

NFPA - Hazard Identification Ratings
health-3; flammability-2; reactivity-0

NIOSH - Selected LD50s and LC50s
Oral, rat: LD50 = 1410 mg/kg Skin, rabbit: LD50 = 1770 mg/kg

NIOSH 1990 - Pocket Guide - RELs
5 ppm TWA; 25 mg/m3 TWA; 10 ppm STEL; 50 mg/m3 STE.

NIOSH 1990 - Pocket Guide - IDLHs
100 ppm IDLH

NIOSH 1990 - Pocket Guide - Target organs
blood, kidneys, liver, CVS

NIOSH 1990 - Pocket Guide - Skin list
Potential for dermal absorption

NTP Chemical Status Reports - Testing Status and NTIS Number
Technical reports printed (PB90227240/AS)

NTP Chemical Status Reports - Evidence of Carcinogenicity
male rat-some evidence; female rat-no evidence; male rat-no e dence; female rat-equivocal evidence

OSHA - Vacated PELs - Time Weighted Averages
5 ppm TWA; 25 mg/m3 TWA

OSHA - Vacated PELs - Short Term Exposure Limits
10 ppm STEL; 50 mg/m3 STEL

OSHA - Vacated PELs - Skin Designation
Prevent or reduce skin absorption

OSHA - Final PELs - Time Weighted Averages
5 ppm TWA; 25 mg/m3 TWA

OSHA - Final PELs - Skin Notations
prevent or reduce skin absorption

ENVIRONMENTAL LISTS

CERCLA/SARA - Section 313 - Emission Reporting
form R reporting required for 1.0% de minimus concentration

CERCLA/SARA - Hazardous Substances and their Reportable Qua
tities
final RQ = 1 pound (.454 kg)

Clean Air Act (1990) - List of Hazardous Air Contaminants
[present]

CAA - HON Rule - SOCMI Chemicals
compliance by April 24, 1995

CAA - HON Rule - Organic HAPs
[present]

EPA - Master Testing List
[present]

INTERNATIONAL LISTS

Australian Exposure Standards - Time Weighted Averages
5 ppm TWA; 25 mg/m3 TWA

Australian Exposure Standards - Short Term Exposure Limits
10 ppm STEL; 50 mg/m3 STEL

Australian Exposure Standards - Skin Effects
skin absorption

Canada - WHMIS: Ingredient Disclosure
1% item 621 (733)

Canada - NPRI (National Pollutant Release Inventory)
[present]

Canada - Alberta - 8 Hour Occupational Exposure Limit
5 ppm TWA; 25 mg/m3 TWA

Canada - Alberta - 15 Minute Occupational Exposure Limit
10 ppm STEL; 50 mg/m3 STEL

Canada - Alberta - Skin Designation
can be absorbed through the intact skin

Canada - British Columbia - 8 Hour Exposure Limits
5 ppm TWA; 25 mg/m3 TWA

Canada - British Columbia - 15 Minute Exposure Limits
10 ppm STEL; 50 mg/m3 STEL

Canada - British Columbia - Skin Notations
skin - potential for skin absorption

Canada - Ontario - OHSA - TWAEVs
5 ppm TWAEV; 25 mg/m3 TWAEV

Canada - Ontario - OHSA - STEVs
10 ppm STEV; 50 mg/m3 STEV

Canada - Ontario - OHSA - Skin Notations
absorption through skin, eyes, or mucous membranes

Canada - Quebec - Time-Weighted Average Exposure Values
5 ppm TWAEV; 25 mg/m3 TWAEV

Canada - Quebec - Short-term Exposure Values
10 ppm STEV; 50 mg/m3 STEV

Canada - Quebec - Skin Designations
absorbed through the skin

United Kingdom - Occupational Exposure Standards - TWAs
5 ppm TWA; 25 mg/m3 TWA

United Kingdom - Occupational Exposure Standards - STELs
10 ppm STEL; 50 mg/m3 STEL

United Kingdom - Occupational Exposure Standards - Notes
can be absorbed through skin

German (DFG) - MAK Values
5 ppm MAK; 25 mg/m3 MAK

German (DFG) - Peak Limitations
2 x normal MAK (30 min. average value); don't exceed 4 times during shift

German (DFG) - Skin/Sensitizers
danger of cutaneous absorption

German (DFG) - Carcinogens
suspected carcinogen

Israel - Time Weighted Averages
5 ppm TWA; 25 mg/m3 TWA

Israel - Short Term Exposure Limits
10 ppm STEL; 50 mg/m3 STEL

Israel - Action Levels
2.5 ppm AL; 12.5 mg/m3 AL

Mexico - Instruction No. 10 - TWAs
5 ppm TWA; 25 mg/m3 TWA

Mexico - Instruction No. 10 - STELs
10 ppm STEL; 50 mg/m3 STEL

Mexico - Instruction No. 10 - Skin designation
skin - potential for cutaneous absorption

STATE LISTS

California - Air Bill 2588 Appendix A-I
6/91

California - Exposure Limits - PELs
5 ppm PEL; 25 mg/m3 PEL

California - Exposure Limits - STELs
10 ppm STEL; 50 mg/m3 STEL

California - Exposure Limits - Skin Notation
material may be absorbed through the skin, eyes or mucous membrane

California - Directors List of Hazardous Substances (8 CCR 339)
[present]

Florida Hazardous Substance List
[present]

Massachusetts Right To Know List
[present]

Minnesota Hazardous Substance List
skin

NJ Right to Know List (Total)
sn 0741

Pennsylvania Right to Know List
environmental hazard

PROPOSED REGULATIONS

TSCA - ITC 33rd Report Priority Testing List
designated for testing

3,5-DIMETHYLANILINE   108-69-0

INTERNATIONAL LISTS

Canada - CEPA - Priority Substances List
estimated time for completion of assessment reports: 5 years

DIMETHYL ANTHRANILATE   85-91-6

HEALTH AND SAFETY LISTS

NFPA - Flash Points
flash point = 195 degrees F (91 degrees C)

NFPA - Hazard Identification Ratings
health-1; flammability-2; reactivity-0

7,12-DIMETHYLBENZ(A)ANTHRACENE   57-97-6

HEALTH AND SAFETY LISTS

U.S. DOT - Appendix A Table 1 - Hazardous Substances
final RQ = 1 pound (0.454 kg)

NIOSH - Selected LD50s and LC50s
Oral, rat: LD50 = 327 mg/kg

ENVIRONMENTAL LISTS

CERCLA/SARA - Section 313 - Emission Reporting
form R reporting required; (Listed under 'Polycyclic aromatic compounds')

CERCLA/SARA - Hazardous Substances and their Reportable Quantities
final RQ = 1 pound (0.454 kg)

EPA - Carcinogen Hazard Ranking for RQ Adjustment
Hazard ranking = High

RCRA - U Series Wastes
waste number U094

RCRA - Hazardous Constituents-Appendix VIII
waste number U094

RCRA - Substances Banned From Land Disposal
[present]

RCRA - TSD Facilities Ground Water Monitoring
TM 8270 = 10 ug/L PQL

INTERNATIONAL LISTS

Canada - WHMIS: Ingredient Disclosure
0.1% item 622 (734)

STATE LISTS

California - Air Bill 2588 Appendix A-I
known or potential carcinogen: 9/90

California - Prop. 65 - Cancer list
carcinogen - initial date 1/1/90

California - Prop. 65 - No Significant Risk Levels
no significant risk level = 0.003 ug/day

Florida Hazardous Substance List
[present]

Massachusetts Right To Know List
carcinogen;   extraordinarily hazardous

NJ Special Hazardous Substances
(carcinogen; mutagen; teratogen)

Pennsylvania Right to Know List
environmental hazard;   special hazardous substance

Pennsylvania RTK - Special Hazardous Substances
[present]

3,3'-DIMETHYLBENZIDINE   119-93-7

HEALTH AND SAFETY LISTS

ACGIH 1995 - Skin Designations
skin - potential for cutaneous absorption

ACGIH 1995 - Carcinogens
A2-suspected human carcinogen

U.S. DOT - Appendix A Table 1 - Hazardous Substances
final RQ = 1 pound (0.454 kg)

IARC - Group 2B (sufficient animal data)
[present] (Overall evaluation based only on evidence of carcinogenicity in monograph (1, 1972) or in Supplement 4)

NIOSH - Selected LD50s and LC50s
Oral, rat: LD50 = 404 mg/kg

NIOSH - Health Standards - Exposure Limits
C (60 min) 20 ug/m3

NIOSH - Health Standards - Health Effects and Precautions
Bladder cancer; nasal irritation (Urine testing required; prevent skin contact)

NIOSH - Health Standards - Carcinogenic Chemicals
potential human carcinogen

NTP Seventh Report - Suspect Carcinogens
suspect carcinogen

OSHA - Possible Select Carcinogens
[present]

ENVIRONMENTAL LISTS

CERCLA/SARA - Section 313 - Emission Reporting
form R reporting required for 0.1% de minimus concentration

CERCLA/SARA - Hazardous Substances and their Reportable Quantities
final RQ = 1 pound (0.454 kg)

Clean Air Act (1990) - List of Hazardous Air Contaminants
[present]

CAA - HON Rule - SOCMI Chemicals
compliance by Jan. 23, 1995

CAA - HON Rule - Organic HAPs
[present]

EPA - Carcinogen Hazard Ranking for RQ Adjustment
Hazard ranking = Medium

RCRA - U Series Wastes
waste number U095

RCRA - Hazardous Constituents-Appendix VIII
waste number U095

RCRA - Substances Banned From Land Disposal
[present]

RCRA - TSD Facilities Ground Water Monitoring
TM 8270 = 10 ug/L PQL

INTERNATIONAL LISTS

Australian Exposure Standards - Time Weighted Averages
control to the lowest practical level

Australian Exposure Standards - Skin Effects
skin absorption

Australian Exposure Standards - Carcinogens
probable carcinogen

Canada - WHMIS: Ingredient Disclosure
0.1% item 1577 (1621)

Canada - Alberta - Designated Substances
designated substance - requires code of practice

Canada - Quebec - Time-Weighted Average Exposure Values
substance of which the recirculation is prohibited

Canada - Quebec - Skin Designations
absorbed through the skin

Canada - Quebec - Carcinogens
C2 carcinogen: effect suspected in humans

German (DFG) - Carcinogens
animal evidence of carcinogenicity

STATE LISTS

California - Air Bill 2588 Appendix A-I
known or potential carcinogen

California - Prop. 65 - Cancer list
carcinogen - initial date 1/1/88

California - Directors List of Hazardous Substances (8 CCR 339)
[present]

Florida Hazardous Substance List
[present]

Massachusetts Right To Know List
carcinogen;   extraordinarily hazardous

Minnesota Hazardous Substance List
carcinogen

NJ Right to Know List (Total)
sn 0742

NJ Special Hazardous Substances
(carcinogen; mutagen)

Pennsylvania Right to Know List
environmental hazard;   special hazardous substance

Pennsylvania RTK - Special Hazardous Substances
[present]

3,3'-DIMETHYLBENZIDINE DIHYDROFLUORIDE (O-   41766-75-0
TOLUIDINE DIHYDROFLUORIDE)

ENVIRONMENTAL LISTS

CERCLA/SARA - Section 313 - Emission Reporting
form R reporting required

3,3'-DIMETHYLBENZIDINE DIHYDROCHLORIDE   612-82-8

HEALTH AND SAFETY LISTS

NTP Chemical Status Reports - Testing Status and NTIS Number
Technical reports printed (PB92103779)

ENVIRONMENTAL LISTS

CERCLA/SARA - Section 313 - Emission Reporting
form R reporting required

STATE LISTS

California - Prop. 65 - Cancer list
carcinogen - initial date 4/1/92

DI(2-METHYLBENZOYL)PEROXIDE   3034-79-5

HEALTH AND SAFETY LISTS

U.S. DOT - Organic Peroxides Table
Organic peroxide UN3112

STATE LISTS
NJ Right to Know List (Total)
sn 0743

2,2-DIMETHYLBENZYL ALCOHOL   617-94-7

HEALTH AND SAFETY LISTS
NIOSH - Selected LD50s and LC50s
Oral, rat: LD50 = 1300 mg/kg Skin, rabbit: LD50 = 4300 mg/kg

DIMETHYLBENZYLCARBINYL ACETATE   151-05-3

HEALTH AND SAFETY LISTS
NFPA - Flash Points
flash point = 205 degrees F (96 degrees C)
NFPA - Hazard Identification Ratings
health-1; flammability-1; reactivity-0

DIMETHYL BENZYL HYDROGENATED TALLOW AMMONIUM CHLORIDE   61789-72-8

ENVIRONMENTAL LISTS
List of Pesticide Product Inert Ingredients
[present]

DIMETHYLBENZYLOCTADECYLAMMONIUM CHLORIDE   122-19-0

HEALTH AND SAFETY LISTS
NIOSH - Selected LD50s and LC50s
Oral, rat: LD50 = 1250 mg/kg
ENVIRONMENTAL LISTS
List of Pesticide Product Inert Ingredients
[present]

2,3-DIMETHYLBUTANE   79-29-8

HEALTH AND SAFETY LISTS
U.S. DOT - Substances From 49 CFR 172.101
regulated by DOT (UN2457)
U.S. DOT - Hazard Classes
DOT hazard class = 3
NFPA - Flash Points
flash point = -20 degrees F (-29 degrees C)
NFPA - Hazard Identification Ratings
health-1; flammability-3; reactivity-0
INTERNATIONAL LISTS
Canada - WHMIS: Ingredient Disclosure
1% item 623 (735)
STATE LISTS
Florida Hazardous Substance List
[present]
Massachusetts Right To Know List
[present]
NJ Right to Know List (Total)
sn 0744
Pennsylvania Right to Know List
[present]

2,3-DIMETHYL-1-BUTENE   563-78-0

HEALTH AND SAFETY LISTS
NFPA - Flash Points
flash point < -4 degrees F (-20 degrees C)
NFPA - Hazard Identification Ratings
health-0; flammability-3; reactivity-0
STATE LISTS
Florida Hazardous Substance List
[present]

Massachusetts Right To Know List
[present]
Pennsylvania Right to Know List
[present]

2,3-DIMETHYL-2-BUTENE   563-79-1

HEALTH AND SAFETY LISTS
NFPA - Flash Points
flash point < -4 degrees F (-20 degrees C)
NFPA - Hazard Identification Ratings
health-0; flammability-3; reactivity-0
STATE LISTS
Florida Hazardous Substance List
[present]
Massachusetts Right To Know List
[present]
Pennsylvania Right to Know List
[present]

1,3-DIMETHYLBUTYLAMINE   108-09-8

HEALTH AND SAFETY LISTS
U.S. DOT - Substances From 49 CFR 172.101
regulated by DOT (UN2379)
U.S. DOT - Hazard Classes
DOT hazard class = 3
NFPA - Flash Points
flash point = 55 degrees F (13 degrees C)
NFPA - Hazard Identification Ratings
health-2; flammability-3; reactivity-0
STATE LISTS
Florida Hazardous Substance List
[present]
Massachusetts Right To Know List
[present]
NJ Right to Know List (Total)
sn 0745
NJ Special Hazardous Substances
(flammable - third degree)
Pennsylvania Right to Know List
[present]

1,3-DIMETHYLBUTYL GLYCIDYL ETHER   68134-06-5

ENVIRONMENTAL LISTS
EPA - Master Testing List
[present]
PROPOSED REGULATIONS
TSCA - Proposed Testing Rule for Glycidyl Ethers
member of Glycidyl subcategory I-A

2,4-DIMETHYL-6-TERT-BUTYLPHENOL   1879-09-0

HEALTH AND SAFETY LISTS
NIOSH - Selected LD50s and LC50s
Oral, mouse: LD50 = 530 mg/kg
ENVIRONMENTAL LISTS
EPA - Master Testing List
[present]

N,N-DIMETHYLCAPRAMIDE   14433-76-2

ENVIRONMENTAL LISTS
List of Pesticide Product Inert Ingredients
[present]

N,N-DIMETHYLCAPRYLAMIDE   **1118-92-9**

ENVIRONMENTAL LISTS

List of Pesticide Product Inert Ingredients
[present]

DIMETHYLCARBAMOYL CHLORIDE   **79-44-7**

HEALTH AND SAFETY LISTS

ACGIH 1995 - Carcinogens
A2-suspected human carcinogen

U.S. DOT - Substances From 49 CFR 172.101
regulated by DOT (UN2262)

U.S. DOT - Hazard Classes
DOT hazard class = 8

U.S. DOT - Appendix A Table 1 - Hazardous Substances
final RQ = 1 pound (0.454 kg)

IARC - Group 2A (limited human data)
[present] (Other relevant data, as given in Supplement 7, influenced the making of the overall evaluation)

NIOSH - Selected LD50s and LC50s
Inhalation, rat: LC50 = 180 ppm 6 hr Oral, rat: LD50 = 1000 mg/kg

NTP Seventh Report - Suspect Carcinogens
suspect carcinogen

OSHA - Possible Select Carcinogens
[present]

ENVIRONMENTAL LISTS

CERCLA/SARA - Section 313 - Emission Reporting
form R reporting required for 0.1% de minimus concentration

CERCLA/SARA - Hazardous Substances and their Reportable Quantities
final RQ = 1 pound (0.454 kg)

Clean Air Act (1990) - List of Hazardous Air Contaminants
[present]

EPA - Carcinogen Hazard Ranking for RQ Adjustment
Hazard ranking = High

RCRA - U Series Wastes
waste number U097

RCRA - Hazardous Constituents-Appendix VIII
waste number U097

RCRA - Substances Banned From Land Disposal
[present]

INTERNATIONAL LISTS

Australian Exposure Standards - Time Weighted Averages
control to the lowest practical level

Australian Exposure Standards - Carcinogens
probable carcinogen

Canada - WHMIS: Ingredient Disclosure
0.1% item 624 (494)

Canada - Alberta - Designated Substances
designated substance - requires code of practice

Canada - British Columbia - 8 Hour Exposure Limits
carcinogen with no established permitted concentration

Canada - British Columbia - Carcinogens
carcinogen with no established permitted concentration

Canada - Quebec - Time-Weighted Average Exposure Values
substance of which the recirculation is prohibited

Canada - Quebec - Carcinogens
C2 carcinogen: effect suspected in humans

German (DFG) - Carcinogens
animal evidence of carcinogenicity

STATE LISTS

California - Air Bill 2588 Appendix A-I
known or potential carcinogen

California - Prop. 65 - Cancer list
carcinogen - initial date 1/1/88

California - Prop. 65 - No Significant Risk Levels
no significant risk level = 0.05 ug/day

California - Directors List of Hazardous Substances (8 CCR 339)
[present]

Florida Hazardous Substance List
[present]

Massachusetts Right To Know List
carcinogen; extraordinarily hazardous

Minnesota Hazardous Substance List
carcinogen

NJ Right to Know List (Total)
sn 0746

NJ Special Hazardous Substances
(carcinogen; mutagen)

Pennsylvania Right to Know List
environmental hazard; special hazardous substance

Pennsylvania RTK - Special Hazardous Substances
[present]

DIMETHYL CARBONATE   **616-38-**

HEALTH AND SAFETY LISTS

U.S. DOT - Substances From 49 CFR 172.101
regulated by DOT (UN1161)

U.S. DOT - Hazard Classes
DOT hazard class = 3

NFPA - Flash Points
flash point = 66 degrees F (19 degrees C)

NFPA - Hazard Identification Ratings
health-3; flammability-3; reactivity-0

NIOSH - Selected LD50s and LC50s
Oral, rat: LD50 = 13 gm/kg

STATE LISTS

Florida Hazardous Substance List
[present]

Massachusetts Right To Know List
[present]

NJ Right to Know List (Total)
sn 0747

NJ Special Hazardous Substances
(flammable - third degree)

Pennsylvania Right to Know List
[present]

N,N-DIMETHYL (COCONUT OIL ALKYL) AMINE OXIDE   **61788-90-**

ENVIRONMENTAL LISTS

List of Pesticide Product Inert Ingredients
[present]

DIMETHYLCYANAMIDE   **1467-79-**

HEALTH AND SAFETY LISTS

NFPA - Flash Points
flash point = 160 degrees F (71 degrees C)

NFPA - Hazard Identification Ratings
health-4; flammability-2; reactivity-1

STATE LISTS

Florida Hazardous Substance List
[present]

Massachusetts Right To Know List
[present]

Pennsylvania Right to Know List
[present]

1,2-DIMETHYLCYCLOHEXANE   583-57-3

HEALTH AND SAFETY LISTS

NFPA - Hazard Identification Ratings
health-0; reactivity-0

1,4-DIMETHYLCYCLOHEXANE   589-90-2

HEALTH AND SAFETY LISTS

NFPA - Flash Points
flash point = 52 degrees F (11 degrees C)

NFPA - Hazard Identification Ratings
health-1; flammability-3; reactivity-0

STATE LISTS

Florida Hazardous Substance List
[present]

Massachusetts Right To Know List
[present]

NJ Right to Know List (Total)
sn 0748

NJ Special Hazardous Substances
(flammable - third degree)

Pennsylvania Right to Know List
[present]

1,3-DIMETHYL CYCLOHEXANE   591-21-9

HEALTH AND SAFETY LISTS

NFPA - Flash Points
flash point around 50 degrees F (10 degrees C)

NFPA - Hazard Identification Ratings
health-0; flammability-3; reactivity-0

STATE LISTS

Florida Hazardous Substance List
[present]

Massachusetts Right To Know List
[present]

Pennsylvania Right to Know List
[present]

1,4-DIMETHYLCYCLOHEXANE-CIS   624-29-3

HEALTH AND SAFETY LISTS

NFPA - Flash Points
flash point = 61 degrees F (16 degrees C)

NFPA - Hazard Identification Ratings
health-0; flammability-3; reactivity-0

STATE LISTS

Florida Hazardous Substance List
[present]

Massachusetts Right To Know List
[present]

Pennsylvania Right to Know List
[present]

1,4-DIMETHYLCYCLOHEXANE-TRANS   2207-04-7

HEALTH AND SAFETY LISTS

NFPA - Flash Points
flash point = 51 degrees F (11 degrees C)

NFPA - Hazard Identification Ratings
health-0; flammability-3; reactivity-0

STATE LISTS

Florida Hazardous Substance List
[present]

Massachusetts Right To Know List
[present]

Pennsylvania Right to Know List
[present]

DIMETHYLCYCLOHEXANES   RR-01348-9

HEALTH AND SAFETY LISTS

U.S. DOT - Substances From 49 CFR 172.101
regulated by DOT (UN2263)

U.S. DOT - Hazard Classes
DOT hazard class = 3

N,N-DIMETHYLCYCLOHEXYLAMINE   98-94-2

HEALTH AND SAFETY LISTS

U.S. DOT - Substances From 49 CFR 172.101
regulated by DOT (UN2264)

U.S. DOT - Hazard Classes
DOT hazard class = 8

NIOSH - Selected LD50s and LC50s
Inhalation, rat: LC50 = 1889 mg/m3 2 hr Oral, rat: LD50 = 348 mg/kg

ENVIRONMENTAL LISTS

List of Pesticide Product Inert Ingredients
[present]

INTERNATIONAL LISTS

Canada - WHMIS: Ingredient Disclosure
1% item 625 (736)

STATE LISTS

NJ Right to Know List (Total)
sn 0749

NJ Special Hazardous Substances
(corrosive)

DIMETHYL DECALIN   RR-00850-4

HEALTH AND SAFETY LISTS

NFPA - Flash Points
flash point = 184 degrees F (84 degrees C)

NFPA - Hazard Identification Ratings
health-0; flammability-2; reactivity-0

N,N-DIMETHYLDECYLAMINE OXIDE   2605-79-0

ENVIRONMENTAL LISTS

List of Pesticide Product Inert Ingredients
[present]

**2,5-DIMETHYL-2,5-DI(2-ETHYLHEXANOYLPEROXY)
-HEXANE**   13052-09-0

STATE LISTS

NJ Right to Know List (Total)
sn 3112

**DIMETHYL DIALKYL AMMONIUM CHLORIDE
POWDER**   RR-01066-2

ENVIRONMENTAL LISTS

List of Pesticide Product Inert Ingredients
[present]

2,5-DIMETHYL-2,5-DI-(BENZOYLPEROXY)HEXANE   2618-77-1

HEALTH AND SAFETY LISTS

U.S. DOT - Organic Peroxides Table
Organic peroxide UN3102; UN3104; UN3105; UN3106

STATE LISTS

NJ Right to Know List (Total)
sn 0750

2,5-DIMETHYL-2,5-DI(TERT-BUTYLPEROXY) HEXANE   78-63-7

STATE LISTS

NJ Right to Know List (Total)
sn 0757

2,5-DIMETHYL-2,5-DI(TERT-BUTYLPEROXY) HEXYNE-3   1068-27-5

HEALTH AND SAFETY LISTS

U.S. DOT - Organic Peroxides Table
Organic peroxide UN3103;   UN3106

STATE LISTS

NJ Right to Know List (Total)
sn 2346;   technically pure: sn 2345

DIMETHYL DICHLORACETAL   97-97-2

HEALTH AND SAFETY LISTS

NFPA - Flash Points
flash point = 111 degrees F (44 degrees C)

NFPA - Hazard Identification Ratings
health-2; flammability-2; reactivity-0

STATE LISTS

Florida Hazardous Substance List
[present]

Massachusetts Right To Know List
[present]

Pennsylvania Right to Know List
[present]

N,N-DIMETHYL DICHLOROACETAMIDE   56343-50-1

ENVIRONMENTAL LISTS

List of Pesticide Product Inert Ingredients
[present]

DIMETHYLDICHLOROSILANE   75-78-5

HEALTH AND SAFETY LISTS

U.S. DOT - Substances From 49 CFR 172.101
regulated by DOT (UN1162)

U.S. DOT - Hazard Classes
DOT hazard class = 3

NFPA - Flash Points
flash point < 70 degrees F (21 degrees C)

NFPA - Hazard Identification Ratings
health-3; flammability-3; reactivity-1 (decomposes in water)

NIOSH - Selected LD50s and LC50s
Inhalation, rat: LC50 = 930 ppm 4 hr

OSHA - List of Highly Hazardous Chemicals
threshhold quantity = 1000 pounds

ENVIRONMENTAL LISTS

CERCLA/SARA - Section 302 Extremely Hazardous Substances and TPQs
TPQ = 500 pounds

CERCLA/SARA - Section 313 - Emission Reporting
form R reporting required

CAA -Toxic Substances for Accidental Release Prevention
threshold quantity = 5,000 lbs

EPA - Master Testing List
[present]

INTERNATIONAL LISTS

Canada - WHMIS: Ingredient Disclosure
1% item 626 (737)

STATE LISTS

Florida Hazardous Substance List
[present]

Massachusetts Right To Know List
extraordinarily hazardous

NJ Right to Know List (Total)
sn 0752

NJ Special Hazardous Substances
(corrosive; flammable - third degree)

Pennsylvania Right to Know List
environmental hazard

PROPOSED REGULATIONS

CERCLA/SARA - Proposed Hazardous Substance Additions
proposed RQ = 1 pound (.454 kg)

CERCLA/SARA - 1989 Proposed RQ Adjustments
proposed RQ = 100 pounds (45.4 kg)

DIMETHYLDIETHOXYSILANE   78-62-6

HEALTH AND SAFETY LISTS

U.S. DOT - Substances From 49 CFR 172.101
regulated by DOT (UN2380)

U.S. DOT - Hazard Classes
DOT hazard class = 3

NIOSH - Selected LD50s and LC50s
Oral, rat: LD50 = 9280 mg/kg

INTERNATIONAL LISTS

Canada - WHMIS: Ingredient Disclosure
1% item 627 (738)

STATE LISTS

NJ Right to Know List (Total)
sn 0753

2,5-DIMETHYL-2,5-DI-(2-ETHYLHEXANOYLPEROXY) HEXANE   RR-01754-9

HEALTH AND SAFETY LISTS

U.S. DOT - Organic Peroxides Table
Organic peroxide UN3115

2,5-DIMETHYL-2,5-DIHYDRO-PEROXYHEXANE   3025-88-5

HEALTH AND SAFETY LISTS

U.S. DOT - Hazard Classes
Forbidden from transport by the DOT

U.S. DOT - Organic Peroxides Table
Organic peroxide UN3104

STATE LISTS

NJ Right to Know List (Total)
sn 0754

1,3-DIMETHYL-5,5-DIMETHYLHYDANTOIN   15414-89-8

ENVIRONMENTAL LISTS

List of Pesticide Product Inert Ingredients
[present]

DIMETHYLDIOCTADECYLAMMONIUM BENTONITE   73138-28-0

ENVIRONMENTAL LISTS

List of Pesticide Product Inert Ingredients
[present]

DIMETHYLDIOXANE   25136-55-4

HEALTH AND SAFETY LISTS

U.S. DOT - Substances From 49 CFR 172.101
regulated by DOT (UN2707)

U.S. DOT - Hazard Classes
DOT hazard class = 3

NFPA - Flash Points
flash point = 75 degrees F (24 degrees C)

NFPA - Hazard Identification Ratings
health-2; flammability-3; reactivity-0

NIOSH - Selected LD50s and LC50s
Oral, rat: LD50 = 3000 mg/kg

STATE LISTS

Florida Hazardous Substance List
[present]

Massachusetts Right To Know List
[present]

NJ Right to Know List (Total)
sn 0755

NJ Special Hazardous Substances
(flammable - third degree)

Pennsylvania Right to Know List
[present]

1,3-DIMETHYL-1,3-DIPHENYLCYCLOBUTANE RR-00785-2

HEALTH AND SAFETY LISTS

NFPA - Flash Points
flash point = 289 degrees F (143 degrees C)

NFPA - Hazard Identification Ratings
health-0; flammability-1; reactivity-0

DIMETHYLDISULFIDE 624-92-0

HEALTH AND SAFETY LISTS

U.S. DOT - Substances From 49 CFR 172.101
regulated by DOT (UN2381)

U.S. DOT - Hazard Classes
DOT hazard class = 3

NIOSH - Selected LD50s and LC50s
Inhalation, rat: LC50 = 15850 ug/m3 (2 hr)

ENVIRONMENTAL LISTS

TSCA - Code of Federal Regulations Citations
40 CFR 704.225(a)

TSCA - CAIR - Reporting List
reporting required by: manufacturer; importer

STATE LISTS

Massachusetts Right To Know List
extraordinarily hazardous

NJ Right to Know List (Total)
sn 0756

NJ Special Hazardous Substances
(flammable - fourth degree)

Pennsylvania Right to Know List
environmental hazard

PROPOSED REGULATIONS

CERCLA/SARA - Proposed Hazardous Substance Additions
proposed RQ = 1 pound (.454 kg)

CERCLA/SARA - 1989 Proposed RQ Adjustments
proposed RQ = 10 pounds (4.54 kg)

DIMETHYL ETHER 115-10-6

HEALTH AND SAFETY LISTS

AIHA - WEEL - Time Weighted Averages
500 ppm TWA; 942 mg/m3 TWA

U.S. DOT - Substances From 49 CFR 172.101
regulated by DOT (UN1033)

U.S. DOT - Hazard Classes
DOT hazard class = 2.1

NFPA - Flash Points
gas (no flash point given)

NFPA - Hazard Identification Ratings
health-1; flammability-4; reactivity-1

NIOSH - Selected LD50s and LC50s
Inhalation, rat: LC50 = 308 gm/m3 8 hr

ENVIRONMENTAL LISTS

CAA - Flammable Substances for Accidental Release Prevention
threshold quantity = 10,000 lbs

CAA - HON Rule - SOCMI Chemicals
compliance by July 24, 1995

List of Pesticide Product Inert Ingredients
[present]

INTERNATIONAL LISTS

German (DFG) - MAK Values
1000 ppm MAK; 1910 mg/m3 MAK

German (DFG) - Peak Limitations
2 x normal MAK (1 hour momentary value); don't exceed 3 times per shift

German (DFG) - Pregnancy
classification not yet possible

STATE LISTS

Florida Hazardous Substance List
[present]

Massachusetts Right To Know List
[present]

Minnesota Hazardous Substance List
[present]

NJ Right to Know List (Total)
sn 0758

NJ Special Hazardous Substances
(flammable - fourth degree)

Pennsylvania Right to Know List
[present]

DIMETHYLETHOXYSILOXANE 14857-34-2

PROPOSED REGULATIONS

ACGIH 1995 - Notice of Intended Changes
0.5 ppm TWA; 2.1 mg/m3 TWA; 1.5 ppm STEL; 6.4 mg/m3 STEL

N,N-DIMETHYL ETHYLAMINE 598-56-1

INTERNATIONAL LISTS

United Kingdom - Occupational Exposure Standards - TWAs
10 ppm TWA; 30 mg/m3 TWA

United Kingdom - Occupational Exposure Standards - STELs
15 ppm STEL; 45 mg/m3 STEL

German (DFG) - MAK Values
25 ppm MAK; 75 mg/m3 MAK

German (DFG) - Peak Limitations
2 x normal MAK (10 min momentary value); don't exceed 4 times per shift

2,4-DIMETHYL-3-ETHYL PENTANE 1068-87-7

HEALTH AND SAFETY LISTS

NFPA - Flash Points
flash point = 734 degrees F (390 degrees C)

NFPA - Hazard Identification Ratings
health-0; flammability-3; reactivity-0

STATE LISTS

Florida Hazardous Substance List
[present]

Massachusetts Right To Know List
[present]

Pennsylvania Right to Know List
[present]

DIMETHYLFORMAMIDE   68-12-2

HEALTH AND SAFETY LISTS

ACGIH 1995 - Time Weighted Averages
10 ppm TWA; 30 mg/m3 TWA

ACGIH 1995 - Skin Designations
skin - potential for cutaneous absorption

ACGIH 1995 - Biological Exposure Indices
N-Methylformamide in urine: (40 mg/g creatinine), end of shift

AIHA - Odor Threshold Values
no geometric mean air odor threshold

U.S. DOT - Substances From 49 CFR 172.101
regulated by DOT (UN2265)

U.S. DOT - Hazard Classes
DOT hazard class = 3

IARC - Group 2B (sufficient animal data)
[present]

NFPA - Flash Points
flash point = 136 degrees F (58 degrees C)

NFPA - Hazard Identification Ratings
health-1; flammability-2; reactivity-0

NIOSH - Selected LD50s and LC50s
Inhalation, mouse: LC50 = 9400 mg/m3 (2 hr) Oral, rat: LD50 = 2800 mg/kg Skin, rat: LD50 = 5 gm/kg

NIOSH 1990 - Pocket Guide - RELs
10 ppm TWA; 30 mg/m3 TWA

NIOSH 1990 - Pocket Guide - IDLHs
3500 ppm IDLH

NIOSH 1990 - Pocket Guide - Target organs
liver, kidneys, CVS, skin

NIOSH 1990 - Pocket Guide - Skin list
Potential for dermal absorption

NTP Chemical Status Reports - Testing Status and NTIS Number
Technical reports printed (PB93-131936)

OSHA - Vacated PELs - Time Weighted Averages
10 ppm TWA; 30 mg/m3 TWA

OSHA - Vacated PELs - Skin Designation
Prevent or reduce skin absorption

OSHA - Final PELs - Time Weighted Averages
10 ppm TWA; 30 mg/m3 TWA

OSHA - Final PELs - Skin Notations
prevent or reduce skin absorption

OSHA - Possible Select Carcinogens
[present]

ENVIRONMENTAL LISTS

CERCLA/SARA - Section 313 - Emission Reporting
form R reporting required

CERCLA/SARA - Hazardous Substances and their Reportable Quantities
final RQ = 1 pound (.454 kg)

Clean Air Act (1990) - List of Hazardous Air Contaminants
[present]

CAA - HON Rule - SOCMI Chemicals
compliance by Jan. 23, 1995

CAA - HON Rule - Organic HAPs
[present]

List of Pesticide Product Inert Ingredients
[present]

TSCA - Code of Federal Regulations Citations
40 CFR 712.30(w); 40 CFR 716.120(a)

TSCA - PAIR - Reporting List
Reporting Date: June 13, 1989

TSCA - Health and Safety Reporting List
Effective Date: April 13, 1989

INTERNATIONAL LISTS

Australian Exposure Standards - Time Weighted Averages
10 ppm TWA; 30 mg/m3 TWA

Australian Exposure Standards - Skin Effects
skin absorption

Canada - WHMIS: Ingredient Disclosure
1% item 628 (741)

Canada - Alberta - 8 Hour Occupational Exposure Limit
10 ppm TWA; 30 mg/m3 TWA

Canada - Alberta - 15 Minute Occupational Exposure Limit
20 ppm STEL; 60 mg/m3 STEL

Canada - Alberta - Skin Designation
can be absorbed through the intact skin

Canada - British Columbia - 8 Hour Exposure Limits
10 ppm TWA; 30 mg/m3 TWA

Canada - British Columbia - 15 Minute Exposure Limits
20 ppm STEL; 60 mg/m3 STEL

Canada - British Columbia - Skin Notations
skin - potential for skin absorption

Canada - Ontario - OHSA - TWAEVs
10 ppm TWAEV; 30 mg/m3 TWAEV

Canada - Ontario - OHSA - Skin Notations
absorption through skin, eyes, or mucous membranes

Canada - Quebec - Time-Weighted Average Exposure Values
10 ppm TWAEV; 30 mg/m3 TWAEV

Canada - Quebec - Skin Designations
absorbed through the skin

Canada - Quebec - Carcinogens
C2 carcinogen: effect suspected in humans

United Kingdom - Occupational Exposure Standards - TWAs
10 ppm TWA; 30 mg/m3 TWA

United Kingdom - Occupational Exposure Standards - STELs
20 ppm STEL; 60 mg/m3 STEL

United Kingdom - Occupational Exposure Standards - Notes
can be absorbed through skin

German (DFG) - MAK Values
10 ppm MAK; 30 mg/m3 MAK

German (DFG) - Peak Limitations
2 x normal MAK (30 min. average value); don't exceed 4 times during shift

German (DFG) - Skin/Sensitizers
danger of cutaneous absorption

German (DFG) - Pregnancy
risk to embryo/fetus probable

Israel - Time Weighted Averages
10 ppm TWA; 30 mg/m3 TWA

Israel - Action Levels
5 ppm AL; 15 mg/m3 AL

Mexico - Instruction No. 10 - TWAs
10 ppm TWA; 30 mg/m3 TWA

Mexico - Instruction No. 10 - STELs
20 ppm STEL; 60 mg/m3 STEL

STATE LISTS

California - Air Bill 2588 Appendix A-I
known or potential carcinogen: 9/90

California - Exposure Limits - PELs
10 ppm PEL; 30 mg/m3 PEL

California - Exposure Limits - Skin Notation
material may be absorbed through the skin, eyes or mucous membranes

California - Directors List of Hazardous Substances (8 CCR 339)
[present]

Florida Hazardous Substance List
[present]

Massachusetts Right To Know List
[present]

Minnesota Hazardous Substance List
[present]

NJ Right to Know List (Total)
sn 0759

Pennsylvania Right to Know List
[present]

PROPOSED REGULATIONS

ACGIH 1995 - Proposed Biological Exposure Indices
N-methylformamide in urine: 20 mg/g creatinine, end of shift

2,5-DIMETHYLFURAN   625-86-5

HEALTH AND SAFETY LISTS

NFPA - Flash Points
flash point = 45 degrees F (7 degrees C)

NFPA - Hazard Identification Ratings
health-2; flammability-3; reactivity-0

STATE LISTS

Florida Hazardous Substance List
[present]

Massachusetts Right To Know List
[present]

Pennsylvania Right to Know List
[present]

DIMETHYL GLYCOL PHTHALATE   RR-00851-5

HEALTH AND SAFETY LISTS

NFPA - Flash Points
flash point = 369 degrees F (187 degrees C)

NFPA - Hazard Identification Ratings
health-0; flammability-1; reactivity-0

3,3-DIMETHYLHEPTANE   4032-86-4

HEALTH AND SAFETY LISTS

NFPA - Hazard Identification Ratings
health-0; flammability-3; reactivity-0

STATE LISTS

Florida Hazardous Substance List
[present]

Massachusetts Right To Know List
[present]

Pennsylvania Right to Know List
[present]

N,N-DIMETHYL-1-HEXADECYLAMINE   112-69-6

ENVIRONMENTAL LISTS

List of Pesticide Product Inert Ingredients
[present]

2,3-DIMETHYLHEXANE   584-94-1

HEALTH AND SAFETY LISTS

NFPA - Flash Points
flash point = 45 degrees F (7 degrees C)

NFPA - Hazard Identification Ratings
health-0; flammability-3; reactivity-0

STATE LISTS

Florida Hazardous Substance List
[present]

Massachusetts Right To Know List
[present]

Pennsylvania Right to Know List
[present]

2,4-DIMETHYLHEXANE   589-43-5

HEALTH AND SAFETY LISTS

NFPA - Flash Points
flash point = 50 degrees F (10 degrees C)

NFPA - Hazard Identification Ratings
health-0; flammability-3; reactivity-0

STATE LISTS

Florida Hazardous Substance List
[present]

Massachusetts Right To Know List
[present]

Pennsylvania Right to Know List
[present]

5,5-DIMETHYLHYDANTOIN   77-71-4

HEALTH AND SAFETY LISTS

NIOSH - Selected LD50s and LC50s
Oral, rat: LD50 = 7800 mg/kg

ENVIRONMENTAL LISTS

List of Pesticide Product Inert Ingredients
[present]

1,1-DIMETHYLHYDRAZINE   57-14-7

SEE ALSO:
K110-HAZARDOUS WASTES
K107-HAZARDOUS WASTES
K109-HAZARDOUS WASTES
K108-HAZARDOUS WASTES
CYANIDE COMPOUNDS

HEALTH AND SAFETY LISTS

ACGIH 1995 - Time Weighted Averages
(0.5) ppm TWA; (1.2) mg/m3 TWA

ACGIH 1995 - Skin Designations
skin - potential for cutaneous absorption

ACGIH 1995 - Carcinogens
(A2)-suspected human carcinogen

AIHA - Odor Threshold Values
geometric mean air odor threshold = 9.2 ppm (detectable)

U.S. DOT - Substances From 49 CFR 172.101
regulated by DOT (UN2382, UN1163)

U.S. DOT - Hazard Classes
symmetrical: DOT hazard class = 3; unsymmetrical: DOT hazard class = 6.1

U.S. DOT - Substances Which Are Poisonous by Inhalation
liquid hazardous material poisonous by inhalation (UN2382, UN1163)

U.S. DOT - Appendix A Table 1 - Hazardous Substances
final RQ = 10 pounds (4.54 kg)

IARC - Group 2B (sufficient animal data)
[present] (Overall evaluation based only on evidence of carcinogenicity in monograph (4, 1974) or in Supplement 4)

NFPA - Flash Points
flash point = 5 degrees F (-15 degrees C)

NFPA - Hazard Identification Ratings
health-4; flammability-3; reactivity-1

NIOSH - Selected LD50s and LC50s
Inhalation, rat: LC50 = 252 ppm 4 hr Oral, rat: LD50 = 122 mg/kg Skin, guinea pig: LD50 = 1329 mg/kg

NIOSH 1990 - Pocket Guide - RELs
C 0.006 ppm (2 hr); C 0.15 mg/m3 (2 hr)

NIOSH 1990 - Pocket Guide - IDLHs
50 ppm IDLH (not considering carcinogenic effects)

NIOSH 1990 - Pocket Guide - Carcinogens
occupational carcinogen

NIOSH 1990 - Pocket Guide - Target organs
CNS, liver, GI tract, blood, eyes, skin, respiratory system

NTP Seventh Report - Suspect Carcinogens
suspect carcinogen

OSHA - Vacated PELs - Time Weighted Averages
0.5 ppm TWA; 1 mg/m3 TWA

OSHA - Vacated PELs - Skin Designation
Prevent or reduce skin absorption

OSHA - Final PELs - Time Weighted Averages
0.5 ppm TWA; 1 mg/m3 TWA

OSHA - Final PELs - Skin Notations
prevent or reduce skin absorption

OSHA - Possible Select Carcinogens
[present]

OSHA - List of Highly Hazardous Chemicals
threshhold quantity = 1000 pounds

ENVIRONMENTAL LISTS

CERCLA/SARA - Section 302 Extremely Hazardous Substances and TPQs
TPQ = 1000 pounds

CERCLA/SARA - Section 313 - Emission Reporting
form R reporting required for 0.1% de minimus concentration

CERCLA/SARA - Hazardous Substances and their Reportable Quantities
final RQ = 10 pounds (4.54 kg)

Clean Air Act (1990) - List of Hazardous Air Contaminants
[present]

CAA -Toxic Substances for Accidental Release Prevention
threshold quantity = 15,000 lbs

CAA - HON Rule - SOCMI Chemicals
compliance by Jan. 23, 1995

CAA - HON Rule - Organic HAPs
[present]

EPA - Carcinogen Hazard Ranking for RQ Adjustment
Hazard ranking = Medium

RCRA - U Series Wastes
waste number U098

RCRA - Hazardous Constituents-Appendix VIII
waste number U098

RCRA - Basis for Listing - Appendix VII
Included in waste streams: K107, K108, K109, K110

RCRA - Substances Banned From Land Disposal
[present]

INTERNATIONAL LISTS

Australian Exposure Standards - Time Weighted Averages
0.5 ppm TWA; 1.2 mg/m3 TWA

Australian Exposure Standards - Skin Effects
skin absorption

Australian Exposure Standards - Carcinogens
probable carcinogen

Canada - WHMIS: Ingredient Disclosure
0.1% item 629 (742)

Canada - Alberta - 8 Hour Occupational Exposure Limit
0.5 ppm TWA; 1.3 mg/m3 TWA

Canada - Alberta - 15 Minute Occupational Exposure Limit
1 ppm STEL; 2.6 mg/m3 STEL

Canada - Alberta - Skin Designation
can be absorbed through the intact skin

Canada - Alberta - Designated Substances
designated substance - requires code of practice

Canada - British Columbia - 8 Hour Exposure Limits
0.5 ppm TWA; 1 mg/m3 TWA

Canada - British Columbia - 15 Minute Exposure Limits
1 ppm STEL; 2 mg/m3 STEL

Canada - British Columbia - Skin Notations
skin - potential for skin absorption

Canada - Ontario - OHSA - TWAEVs
0.5 ppm TWAEV; 1.2 mg/m3 TWAEV

Canada - Ontario - OHSA - Skin Notations
absorption through skin, eyes, or mucous membranes

Canada - Quebec - Time-Weighted Average Exposure Values
0.5 ppm TWAEV; 1.2 mg/m3 TWAEV (substance of which the recirculation is prohibited)

Canada - Quebec - Skin Designations
absorbed through the skin

Canada - Quebec - Carcinogens
C2 carcinogen: effect suspected in humans

German (DFG) - Skin/Sensitizers
danger of cutaneous absorption; danger of sensitization (skin or respiratory)

German (DFG) - Carcinogens
animal evidence of carcinogenicity

Israel - Time Weighted Averages
(0.5) ppm TWA; (1.2) mg/m3 TWA

Israel - Action Levels
0.25 ppm AL; 0.6 mg/m3 AL

Mexico - Instruction No. 10 - TWAs
0.5 ppm TWA; 1 mg/m3 TWA

Mexico - Instruction No. 10 - STELs
1 ppm STEL; 2 mg/m3 STEL

Mexico - Instruction No. 10 - Skin designation
skin - potential for cutaneous absorption

Mexico - Instruction No. 10 - Carcinogens
potential carcinogen in humans - limited epidemiological evidence

STATE LISTS

California - Air Bill 2588 Appendix A-I
known or potential carcinogen

California - Prop. 65 - Cancer list
carcinogen - initial date 10/1/89

California - Exposure Limits - PELs
0.5 ppm PEL; 1 mg/m3 PEL

California - Exposure Limits - Skin Notation
material may be absorbed through the skin, eyes or mucous membrane

California - Directors List of Hazardous Substances (8 CCR 339)
[present]

Florida Hazardous Substance List
[present]

Massachusetts Right To Know List
carcinogen; extraordinarily hazardous

Minnesota Hazardous Substance List
carcinogen; skin

NJ Right to Know List (Total)
sn 0761

NJ Special Hazardous Substances
(carcinogen; corrosive; flammable - third degree; mutagen)

Pennsylvania Right to Know List
environmental hazard; special hazardous substance

Pennsylvania RTK - Special Hazardous Substances
[present]

PROPOSED REGULATIONS

ACGIH 1995 - Notice of Intended Changes
(skin) 0.01 ppm TWA; 0.025 mg/m3 TWA; A3-animal carcinogen

1,2-DIMETHYLHYDRAZINE 540-73-8

HEALTH AND SAFETY LISTS

U.S. DOT - Substances From 49 CFR 172.101
regulated by DOT (UN2382)

U.S. DOT - Hazard Classes
DOT hazard class = 3

U.S. DOT - Appendix A Table 1 - Hazardous Substances
final RQ = 1 pound (0.454 kg)

IARC - Group 2B (sufficient animal data)
[present] (Overall evaluation based only on evidence of carcinogenicity in monograph (4, 1974) or in Supplement 4)

NIOSH - Selected LD50s and LC50s
Oral, rat: LD50 = 100 mg/kg

OSHA - Possible Select Carcinogens
[present]

ENVIRONMENTAL LISTS

CERCLA/SARA - Hazardous Substances and their Reportable Quantities
final RQ = 1 pound (0.454 kg)

EPA - Carcinogen Hazard Ranking for RQ Adjustment
Hazard ranking = High

RCRA - U Series Wastes
waste number U099

RCRA - Hazardous Constituents-Appendix VIII
waste number U099

RCRA - Substances Banned From Land Disposal
[present]

INTERNATIONAL LISTS

Canada - WHMIS: Ingredient Disclosure
1% item 630 (743)

German (DFG) - Skin/Sensitizers
danger of cutaneous absorption; danger of sensitization (skin or respiratory)

German (DFG) - Carcinogens
animal evidence of carcinogenicity

STATE LISTS

California - Air Bill 2588 Appendix A-II
known or potential carcinogen

California - Prop. 65 - Cancer list
carcinogen - initial date 1/1/88

California - Prop. 65 - No Significant Risk Levels
no sginificant risk level = 0.001 ug/day

California - Directors List of Hazardous Substances (8 CCR 339)
[present]

Florida Hazardous Substance List
[present]

Massachusetts Right To Know List
carcinogen; extraordinarily hazardous

Minnesota Hazardous Substance List
carcinogen

NJ Right to Know List (Total)
sn 1008

NJ Special Hazardous Substances
(carcinogen; mutagen)

Pennsylvania Right to Know List
environmental hazard; special hazardous substance

Pennsylvania RTK - Special Hazardous Substances
[present]

SYM-DIMETHYLHYDRAZINE DIHYDROCHLORIDE 306-37-6

HEALTH AND SAFETY LISTS

NIOSH - Selected LD50s and LC50s
Oral, rat: LD50 = 100 mg/kg

DIMETHYL HYDROGEN PHOSPHITE 868-85-9

HEALTH AND SAFETY LISTS

IARC - Group 3 (not classifiable)
[present]

NIOSH - Selected LD50s and LC50s
Oral, rat: LD50 = 3050 mg/kg Skin, rabbit: LD50 = 2400 mg/kg

NTP Chemical Status Reports - Testing Status and NTIS Number
Technical reports printed (PB86144805/AS)

NTP Chemical Status Reports - Evidence of Carcinogenicity
male rat-clear evidence; female rat-equivocal evidence; male mice-no evidence; female mice-no evidence

INTERNATIONAL LISTS

German (DFG) - Carcinogens
suspected carcinogen

DIMETHYL HYDROGEN PHOSPHITE 920-37-6

INTERNATIONAL LISTS

German (DFG) - Carcinogens
suspected carcinogen

DIMETHYLHYDROPOLYSILOXANE 68037-59-2

ENVIRONMENTAL LISTS

TSCA - PAIR - Reporting List
Effective Date: October 12, 1993; Reporting Date: February 28, 1994

TSCA - Health and Safety Reporting List
Effective Date: October 12, 1993; Sunset Date: October 12, 2003

N,N-DIMETHYLISOOCTADECANAMINE, N-OXIDE 125972-19-2

ENVIRONMENTAL LISTS

List of Pesticide Product Inert Ingredients
[present]

DIMETHYL ISOPHTHALATE 1459-93-4

HEALTH AND SAFETY LISTS

NFPA - Flash Points
flash point = 280 degrees F (138 degrees C)

NFPA - Hazard Identification Ratings
health-0; flammability-1; reactivity-0

NIOSH - Selected LD50s and LC50s
Oral, rat: LD50 = 4390 mg/kg

N,N-DIMETHYLISOPROPANOLAMINE 108-16-7

HEALTH AND SAFETY LISTS

NFPA - Flash Points
flash point = 95 degrees F (35 degrees C)

NFPA - Hazard Identification Ratings
health-2; flammability-3; reactivity-0

NIOSH - Selected LD50s and LC50s
Oral, rat: LD50 = 1890 mg/kg

STATE LISTS

Florida Hazardous Substance List
[present]

Massachusetts Right To Know List
[present]

Pennsylvania Right to Know List
[present]

DIMETHYLMAGNESIUM 2999-74-8

STATE LISTS

NJ Right to Know List (Total)
sn 0762

DIMETHYL MALEATE   **624-48-6**

HEALTH AND SAFETY LISTS

NFPA - Flash Points
flash point = 235 degrees F (113 degrees C)

NFPA - Hazard Identification Ratings
health-1; flammability-1; reactivity-0

DIMETHYL MALONATE   **108-59-8**

HEALTH AND SAFETY LISTS

NIOSH - Selected LD50s and LC50s
Oral, rat: LD50 = 5331 mg/kg

DIMETHYLMERCURY   **593-74-8**
SEE ALSO:
MERCURY

INTERNATIONAL LISTS

Canada - WHMIS: Ingredient Disclosure
0.1% item 631 (744)

STATE LISTS

California - Air Bill 2588 Appendix A-I
[present]

NJ Right to Know List (Total)
sn 0763

NJ Special Hazardous Substances
(teratogen)

DIMETHYL METHYLPHOSPHONATE   **756-79-6**

HEALTH AND SAFETY LISTS

NTP Chemical Status Reports - Testing Status and NTIS Number
Technical reports printed (PB88168695)

NTP Chemical Status Reports - Evidence of Carcinogenicity
male rat-some evidence; female rat-no evidence; male mice-inadequate; female mice-no evidence

2,2-DIMETHYL-3-(2-METHYL-1-PROPENYL)CYCLO-   **28434-00-6**
PROPANECARBOXYLIC ACID, 2-METHYL-4-OXO-3-
(2-PROPENYL)-2-CYCLOPENTEN-1-YL ESTER

HEALTH AND SAFETY LISTS

NIOSH - Selected LD50s and LC50s
Inhalation, rat: LC50 = 1600 mg/m3 3 hr Oral, rat: LD50 = 430 mg/kg

DIMETHYLMETHYL 3,3,3-TRIFLUOROPROPYL   **115361-68-7**
SILOXANE

ENVIRONMENTAL LISTS

TSCA - PAIR - Reporting List
Effective Date: October 12, 1993; Reporting Date: February 28, 1994

TSCA - Health and Safety Reporting List
Effective Date: October 12, 1993; Sunset Date: October 12, 2003

DIMETHYLMETHYLVINYLSILOXANE   **67762-94-1**

ENVIRONMENTAL LISTS

TSCA - PAIR - Reporting List
Effective Date: October 12, 1993; Reporting Date: February 28, 1994

2,6-DIMETHYLMORPHOLINE   **141-91-3**

ENVIRONMENTAL LISTS

TSCA - PAIR - Reporting List
Effective Date: October 12, 1993; Reporting Date: February 28, 1994

DIMETHYL MORPHOLINOPHOSPHORAMIDATE   **597-25-1**

ENVIRONMENTAL LISTS

TSCA - PAIR - Reporting List
Effective Date: October 12, 1993; Reporting Date: February 2̇ 1994

DIMETHYLNAPHTHALENESULFONIC ACID,   **27178-87-**
SODIUM SALT

ENVIRONMENTAL LISTS

TSCA - PAIR - Reporting List
Effective Date: October 12, 1993; Reporting Date: February 2̇ 1994

2,8-DIMETHYL-5-NONANONE   **2050-99-**

ENVIRONMENTAL LISTS

TSCA - PAIR - Reporting List
Effective Date: October 12, 1993; Reporting Date: February 2̇ 1994

N,N-DIMETHYLOCTADECYLAMINE OXIDE   **2571-88-**

ENVIRONMENTAL LISTS

TSCA - PAIR - Reporting List
Effective Date: October 12, 1993; Reporting Date: February 2̇ 1994

3,4-DIMETHYLOCTANE   **15869-92-**

ENVIRONMENTAL LISTS

TSCA - PAIR - Reporting List
Effective Date: October 12, 1993; Reporting Date: February 2̇ 1994

2,3-DIMETHYLOCTANE   **RR-00531-**

ENVIRONMENTAL LISTS

TSCA - PAIR - Reporting List
Effective Date: October 12, 1993; Reporting Date: February 28 1994

TSCA - Health and Safety Reporting List
Effective Date: October 12, 1993; Sunset Date: October 12, 2003

N,N-DIMETHYLOCTYLAMINE   **7378-99-6**

HEALTH AND SAFETY LISTS

NFPA - Flash Points
flash point = 112 degrees F (44 degrees C)

NFPA - Hazard Identification Ratings
health-2; flammability-2; reactivity-0

STATE LISTS

Florida Hazardous Substance List
[present]

Massachusetts Right To Know List
[present]

Pennsylvania Right to Know List
[present]

N,N-DIMETHYLOLEAMIDE   **2664-42-8**

HEALTH AND SAFETY LISTS

NTP Chemical Status Reports - Testing Status and NTIS Number
Technical reports printed (PB86186491/AS)

NTP Chemical Status Reports - Evidence of Carcinogenicity
male rat-some evidence; female rat-some evidence; male mice-no evidence; female mice-no evidence

N,N-DIMETHYLOLEYLAMINE OXIDE   **14351-50-9**

ENVIRONMENTAL LISTS

List of Pesticide Product Inert Ingredients
[present]

DIMETHYLOLUREA   **140-95-4**

ENVIRONMENTAL LISTS

List of Pesticide Product Inert Ingredients
[present]

DIMETHYLOL UREA - FORMALDEHYDE -   62073-57-8
MONOMETHYLOL UREA POLYMER
ENVIRONMENTAL LISTS
List of Pesticide Product Inert Ingredients
[present]

4,4-DIMETHYLOXAZOLIDINE   51200-87-4
HEALTH AND SAFETY LISTS
NFPA - Flash Points
flash point < 131 degrees F (55 degrees C)
NFPA - Hazard Identification Ratings
health-0; flammability-2; reactivity-0

2,3-DIMETHYLPENTALDEHYDE   RR-00784-1
HEALTH AND SAFETY LISTS
NFPA - Flash Points
flash point < 131 degrees F (55 degrees C)
NFPA - Hazard Identification Ratings
health-0; flammability-2; reactivity-0

2,4-DIMETHYLPENTANE   108-08-7
ENVIRONMENTAL LISTS
List of Pesticide Product Inert Ingredients
[present]

2,3-DIMETHYLPENTANE   565-59-3
ENVIRONMENTAL LISTS
List of Pesticide Product Inert Ingredients
[present]

2,4-DIMETHYL-3-PENTANOL   600-36-2
ENVIRONMENTAL LISTS
List of Pesticide Product Inert Ingredients
[present]

1,4-DIMETHYLPHENANTHRENE   22349-59-3
HEALTH AND SAFETY LISTS
NIOSH - Selected LD50s and LC50s
Oral, rat: LD50 = 3400 mg/kg
ENVIRONMENTAL LISTS
List of Pesticide Product Inert Ingredients
[present]

3,4-DIMETHYLPHENOL   95-65-8
ENVIRONMENTAL LISTS
List of Pesticide Product Inert Ingredients
[present]

2,4-DIMETHYLPHENOL   105-67-9
HEALTH AND SAFETY LISTS
NIOSH - Selected LD50s and LC50s
Inhalation, rat: LC50 = 11700 mg/m3 (8 hr) Oral, rat: LD50 = 950 mg/kg Skin, rabbit: LD50 = 1400 mg/kg

2,6-DIMETHYLPHENOL   576-26-1
HEALTH AND SAFETY LISTS
NFPA - Flash Points
flash point = 94 degrees F (34 degrees C)
NFPA - Hazard Identification Ratings
health-2; flammability-3; reactivity-0

DIMETHYLPHENOLS   RR-01285-1
HEALTH AND SAFETY LISTS
NFPA - Flash Points
flash point = 10 degrees F (-12 degrees C)
NFPA - Hazard Identification Ratings
health-0; flammability-3; reactivity-0
STATE LISTS
Florida Hazardous Substance List
[present]
Massachusetts Right To Know List
[present]
Pennsylvania Right to Know List
[present]

ALPHA,ALPHA-DIMETHYLPHENETHYLAMINE   122-09-8
HEALTH AND SAFETY LISTS
NFPA - Flash Points
flash point < 20 degrees F (-7 degrees C)
NFPA - Hazard Identification Ratings
health-0; flammability-3; reactivity-0
STATE LISTS
Florida Hazardous Substance List
[present]
Massachusetts Right To Know List
[present]
Pennsylvania Right to Know List
[present]

3,3-DIMETHYL-1-PHENYLTRIAZENE   7227-91-0
HEALTH AND SAFETY LISTS
NFPA - Flash Points
flash point = 120 degrees F (49 degrees C)
NFPA - Hazard Identification Ratings
health-0; flammability-2; reactivity-0

DIMETHYL PHOSPHOROCHLORIDOTHIOATE   2524-03-0
HEALTH AND SAFETY LISTS
IARC - Group 3 (not classifiable)
[present]

DIMETHYL P-PHTHALATE   120-61-6
HEALTH AND SAFETY LISTS
NIOSH - Selected LD50s and LC50s
Oral, rat: LD50 = 727 mg/kg
ENVIRONMENTAL LISTS
TSCA - Code of Federal Regulations Citations
40 CFR 712.30(x); 40 CFR 716.120(d)
TSCA - PAIR - Reporting List
Reporting Date: November 27, 1991
TSCA - Health and Safety Reporting List
Effective Date: September 30, 1991

DIMETHYL PHTHALATE   131-11-3
SEE ALSO:
K001-HAZARDOUS WASTES
F039-HAZARDOUS WASTES
HEALTH AND SAFETY LISTS
U.S. DOT - Appendix A Table 1 - Hazardous Substances
final RQ = 100 pounds (45.4 kg)
NIOSH - Selected LD50s and LC50s
Oral, rat: LD50 = 3200 mg/kg Skin, rat: LD50 = 1040 mg/kg

ENVIRONMENTAL LISTS

ATSDR Priority List
Rank (of 275): 152

CERCLA/SARA - Section 313 - Emission Reporting
form R reporting required for 1.0% de minimus concentration

CERCLA/SARA - Hazardous Substances and their Reportable Quantities
final RQ = 100 pounds (45.4 kg)

Clean Water Act - Priority Pollutants
[present]

Clean Water Act - Toxic Pollutants
[present]

RCRA - U Series Wastes
waste number U101

RCRA - Hazardous Constituents-Appendix VIII
waste number U101

RCRA - Basis for Listing - Appendix VII
Included in waste streams: F039, K001

RCRA - Substances Banned From Land Disposal
[present]

RCRA - TSD Facilities Ground Water Monitoring
TM 8040 = 5 ug/L PQL; TM 8270 = 10 ug/L PQL

RCRA - Universal Treatment Standards (LDR)
WW: 0.036 mg/l; NWW: 14 mg/kg

INTERNATIONAL LISTS

Mexico - Wastewater - Organic Toxic Pollutants and Heavy Metals
Listed under [Organic Toxic Pollutants]

Mexico - Drinking Water - Ecological Criteria
0.4 mg/l

STATE LISTS

California - Air Bill 2588 Appendix A-II
6/91

California - Directors List of Hazardous Substances (8 CCR 339)
[present]

Massachusetts Right To Know List
[present]

NJ Right to Know List (Total)
sn 0764

Pennsylvania Right to Know List
environmental hazard

DIMETHYLPIPERAZINE, CIS-   106-55-8

HEALTH AND SAFETY LISTS

NIOSH - Selected LD50s and LC50s
Oral, rat: LD50 = 296 mg/kg Skin, mouse: LD50 = 920 mg/kg

ENVIRONMENTAL LISTS

CERCLA/SARA - Section 313 - Emission Reporting
form R reporting required

EPA - Master Testing List
[present]

PROPOSED REGULATIONS

TSCA - ITC 33rd Report Priority Testing List
designated for testing

DIMETHYL PIPERAZINE-CIS   6284-84-0

HEALTH AND SAFETY LISTS

U.S. DOT - Appendix B - Marine Pollutants
DOT regulated marine pollutant

DIMETHYLPOLYSILOXANES   68037-74-1

HEALTH AND SAFETY LISTS

U.S. DOT - Appendix A Table 1 - Hazardous Substances
final RQ = 5000 pounds (2270 kg)

ENVIRONMENTAL LISTS

CERCLA/SARA - Hazardous Substances and their Reportable Quantities
final RQ = 5000 pounds (2270 kg)

RCRA - P Series Wastes
waste number P046

RCRA - Hazardous Constituents-Appendix VIII
waste number P046

RCRA - Substances Banned From Land Disposal
[present]

RCRA - TSD Facilities Ground Water Monitoring
TM 8270 = 10 ug/L PQL

TSCA - Code of Federal Regulations Citations
40 CFR 716.120(a)

TSCA - Health and Safety Reporting List
Effective Date: March 7, 1986

STATE LISTS

Massachusetts Right To Know List
[present]

NJ Right to Know List (Total)
sn 3016

Pennsylvania Right to Know List
environmental hazard

2,2-DIMETHYLPROPANE   463-82-

HEALTH AND SAFETY LISTS

NIOSH - Selected LD50s and LC50s
Oral, rat: LD50 = 310 mg/kg

2,2-DIMETHYLPROPANOIC ACID   75-98-

HEALTH AND SAFETY LISTS

U.S. DOT - Substances From 49 CFR 172.101
regulated by DOT (UN2267)

U.S. DOT - Hazard Classes
DOT hazard class = 8

NIOSH - Selected LD50s and LC50s
Inhalation, rat: LC50 = 340 mg/m3 4 hr Oral, mouse: LD50 = 180 mg/kg

ENVIRONMENTAL LISTS

CERCLA/SARA - Section 302 Extremely Hazardous Substances and TPQs
TPQ = 500 pounds

CERCLA/SARA - Section 313 - Emission Reporting
form R reporting required

EPA - Master Testing List
[present]

INTERNATIONAL LISTS

Canada - WHMIS: Ingredient Disclosure
1% item 632 (1408)

STATE LISTS

Florida Hazardous Substance List
effective March 13, 1992

Massachusetts Right To Know List
extraordinarily hazardous

NJ Right to Know List (Total)
sn 0770

NJ Special Hazardous Substances
(corrosive)

Pennsylvania Right to Know List
environmental hazard

PROPOSED REGULATIONS

CERCLA/SARA - Proposed Hazardous Substance Additions
proposed RQ = 1 pound (.454 kg)

CERCLA/SARA - 1989 Proposed RQ Adjustments
proposed RQ = 100 pounds (45.4 kg)

DIMETHYLPROPYLAMINE   926-63-6

HEALTH AND SAFETY LISTS

AIHA - WEEL - Time Weighted Averages
total: 10 mg/m3 TWA; respirable: 5 mg/m3 TWA

NFPA - Flash Points
flash point = 308 degrees F (153 degrees C)

NFPA - Hazard Identification Ratings
health-1; flammability-1; reactivity-0

NIOSH - Selected LD50s and LC50s
Oral, rat: LD50 = 14400 mg/kg

NTP Chemical Status Reports - Testing Status and NTIS Number
Technical reports printed (PB299903/AS)

NTP Chemical Status Reports - Evidence of Carcinogenicity
male rat-negative; female rat-negative; male mice-equivocal; female mice-negative

ENVIRONMENTAL LISTS

CAA - HON Rule - SOCMI Chemicals
compliance by Jan. 23, 1995

EPA - Master Testing List
[present]

INTERNATIONAL LISTS

Canada - WHMIS: Ingredient Disclosure
1% item 634 (1422)

STATE LISTS

Minnesota Hazardous Substance List
[present]

1,1-DIMETHYLPROPYL PEROXYESTER   RR-00169-4
SEE ALSO:
ALKYL PHTHALATES
F039-HAZARDOUS WASTES
CAT FOOD

HEALTH AND SAFETY LISTS

ACGIH 1995 - Time Weighted Averages
5 mg/m3 TWA

U.S. DOT - Appendix A Table 1 - Hazardous Substances
final RQ = 5000 pounds (2270 kg)

NFPA - Flash Points
flash point = 295 degrees F (146 degrees C)

NFPA - Hazard Identification Ratings
health-0; flammability-1; reactivity-0

NIOSH - Selected LD50s and LC50s
Oral, rat: LD50 = 6800 mg/kg

NIOSH 1990 - Pocket Guide - RELs
5 mg/m3 TWA

NIOSH 1990 - Pocket Guide - IDLHs
9300 mg/m3 IDLH

NIOSH 1990 - Pocket Guide - Target organs
GI tract, respiratory system

NTP Chemical Status Reports - Testing Status and NTIS Number
Galley or camera copy technical reports in progress

OSHA - Vacated PELs - Time Weighted Averages
5 mg/m3 TWA

OSHA - Final PELs - Time Weighted Averages
5 mg/m3 TWA

ENVIRONMENTAL LISTS

CERCLA/SARA - Section 313 - Emission Reporting
form R reporting required for 1.0% de minimus concentration

CERCLA/SARA - Hazardous Substances and their Reportable Quantities
final RQ = 5000 pounds (2270 kg)

Clean Air Act (1990) - List of Hazardous Air Contaminants
[present]

CAA - HON Rule - Organic HAPs
[present]

Clean Water Act - Priority Pollutants
[present]

List of Pesticide Product Inert Ingredients
[present]

RCRA - U Series Wastes
waste number U102

RCRA - Hazardous Constituents-Appendix VIII
waste number U102

RCRA - Basis for Listing - Appendix VII
Included in waste stream: F039

RCRA - Substances Banned From Land Disposal
[present]

RCRA - TSD Facilities Ground Water Monitoring
TM 8060 = 5 ug/L PQL; TM 8270 = 10 ug/L PQL

RCRA - Universal Treatment Standards (LDR)
WW: 0.047 mg/l; NWW: 28 mg/kg

TSCA - Code of Federal Regulations Citations
40 CFR 712.30(d); 40 CFR 716.120(c); 40 CFR 799.5000

TSCA - PAIR - Reporting List
Reporting Date: November 19, 1982

TSCA - Substances Subject to Testing Consent Orders
Test for: Environmental Effects

TSCA - Section 12(b) - Export Notification
export notification required - Section 4

INTERNATIONAL LISTS

Australian Exposure Standards - Time Weighted Averages
5 mg/m3 TWA

Canada - WHMIS: Ingredient Disclosure
1% item 633 (1420)

Canada - NPRI (National Pollutant Release Inventory)
[present]

Canada - Alberta - 8 Hour Occupational Exposure Limit
5 mg/m3 TWA

Canada - Alberta - 15 Minute Occupational Exposure Limit
10 mg/m3 STEL

Canada - British Columbia - 8 Hour Exposure Limits
5 mg/m3 TWA

Canada - British Columbia - 15 Minute Exposure Limits
10 mg/m3 STEL

Canada - Ontario - OHSA - TWAEVs
5 mg/m3 TWAEV

Canada - Quebec - Time-Weighted Average Exposure Values
5 mg/m3 TWAEV

United Kingdom - Occupational Exposure Standards - TWAs
5 mg/m3 TWA

United Kingdom - Occupational Exposure Standards - STELs
10 mg/m3 STEL

Israel - Time Weighted Averages
5 mg/m3 TWA

Israel - Action Levels
2.5 mg/m3 AL

Mexico - Instruction No. 10 - TWAs
5 mg/m3 TWA

Mexico - Instruction No. 10 - STELs
10 mg/m3 STEL

Mexico - Wastewater - Organic Toxic Pollutants and Heavy Metals
Listed under [Phthalate Esters]

Mexico - Drinking Water - Ecological Criteria
313.0 mg/l

STATE LISTS

California - Air Bill 2588 Appendix A-I
6/91

California - Exposure Limits - PELs
5 mg/m3 PEL

Florida Hazardous Substance List
[present]

Massachusetts Right To Know List
[present]

Minnesota Hazardous Substance List
[present]

NJ Right to Know List (Total)
sn 0765

Pennsylvania Right to Know List
environmental hazard

PROPOSED REGULATIONS

Canada - Ontario - Proposed Occupational TWAEVs
3 mg/m3 TWAEV

Canada - Ontario - Proposed Occupational STEVs
5 mg/m3 STEV

2,5-DIMETHYLPYRAZINE   123-32-0

HEALTH AND SAFETY LISTS

NFPA - Flash Points
flash point = 155 degrees F (68 degrees C)

NFPA - Hazard Identification Ratings
health-2; flammability-2; reactivity-0

STATE LISTS

Pennsylvania Right to Know List
[present]

DIMETHYL SEBACATE   106-79-6

STATE LISTS

Massachusetts Right To Know List
[present]

DIMETHYL SILICONE   9006-65-9

ENVIRONMENTAL LISTS

TSCA - PAIR - Reporting List
Effective Date: June 14, 1993

TSCA - Health and Safety Reporting List
Effective Date: October 12, 1993; Sunset Date: October 12, 2003

DIMETHYL SILICONE POLYMER WITH SILICA   67762-90-7

HEALTH AND SAFETY LISTS

U.S. DOT - Substances From 49 CFR 172.101
regulated by DOT (UN2044)

U.S. DOT - Hazard Classes
DOT hazard class = 2.1

NFPA - Flash Points
gas (no flash point given)

NFPA - Hazard Identification Ratings
health-0; flammability-4; reactivity-0

ENVIRONMENTAL LISTS

CAA - Flammable Substances for Accidental Release Prevention
threshold quantity = 10,000 lbs

INTERNATIONAL LISTS

German (DFG) - MAK Values
1000 ppm MAK; 2950 mg/m3 MAK (Listed under 'Pentane')

German (DFG) - Peak Limitations
2 x normal MAK (1 hour momentary value); don't exceed 3 time
per shift (Listed under 'Pentane')

STATE LISTS

Florida Hazardous Substance List
[present]

Massachusetts Right To Know List
[present]

NJ Right to Know List (Total)
sn 0766

NJ Special Hazardous Substances
(flammable - fourth degree)

Pennsylvania Right to Know List
[present]

DIMETHYL-3-SUBSTITUTED HETEROMONOCYCLE   RR-01571

HEALTH AND SAFETY LISTS

NIOSH - Selected LD50s and LC50s
Oral, rat: LD50 = 900 mg/kg Skin, rat: LD50 = 1900 mg/kg

ENVIRONMENTAL LISTS

EPA - Master Testing List
[present]

DIMETHYL-3-SUBSTITUTED HETEROMONOCYCLE   RR-01572-
AMINE

HEALTH AND SAFETY LISTS

U.S. DOT - Substances From 49 CFR 172.101
regulated by DOT (UN2266)

U.S. DOT - Hazard Classes
DOT hazard class = 3

INTERNATIONAL LISTS

Canada - WHMIS: Ingredient Disclosure
1% item 635 (745)

STATE LISTS

NJ Right to Know List (Total)
sn 0767

NJ Special Hazardous Substances
(corrosive)

DIMETHYLSULFAMOYL CHLORIDE   13360-57-1

ENVIRONMENTAL LISTS

TSCA - Chemicals with Significant New Use Rules
PMN number: P-85-680

DIMETHYL SULFATE   77-78-1

HEALTH AND SAFETY LISTS

NFPA - Flash Points
flash point = 147 degrees F (64 degrees C)

NFPA - Hazard Identification Ratings
flammability-2; reactivity-0

DIMETHYL SULFIDE   75-18-3

HEALTH AND SAFETY LISTS

NFPA - Flash Points
flash point = 293 degrees F (145 degrees C)

NFPA - Hazard Identification Ratings
health-0; flammability-1; reactivity-0

DIMETHYL SULFOLANE   1003-78-7

ENVIRONMENTAL LISTS

TSCA - PAIR - Reporting List
Effective Date: June 14, 1993

DIMETHYLSULFONE   **67-71-0**

ENVIRONMENTAL LISTS

List of Pesticide Product Inert Ingredients
[present]

TSCA - PAIR - Reporting List
Effective Date: October 12, 1993;   Reporting Date: February 28, 1994

TSCA - Health and Safety Reporting List
Effective Date: October 12, 1993; Sunset Date: October 12, 2003

DIMETHYL SULFOXIDE   **67-68-5**

ENVIRONMENTAL LISTS

TSCA - Chemicals with Significant New Use Rules
PMN number: P-91-1322

N,N-DIMETHYL TALL-OIL FATTY AMIDES   **68308-74-7**

ENVIRONMENTAL LISTS

TSCA - Chemicals with Significant New Use Rules
PMN number: P-91-1323

N,N-DIMETHYLTETRADECYLAMINE OXIDE   **3332-27-2**

INTERNATIONAL LISTS

German (DFG) - Carcinogens
animal evidence of carcinogenicity

DIMETHYL THIOPHOSPHORYL CHLORIDE   **993-12-4**
SEE ALSO:
AMYL OXALATE
K131-HAZARDOUS WASTES

HEALTH AND SAFETY LISTS

ACGIH 1995 - Time Weighted Averages
0.1 ppm TWA; 0.52 mg/m3 TWA

ACGIH 1995 - Skin Designations
skin - potential for cutaneous absorption

ACGIH 1995 - Carcinogens
A2-suspected human carcinogen

U.S. DOT - Substances From 49 CFR 172.101
regulated by DOT (UN1595)

U.S. DOT - Hazard Classes
DOT hazard class = 6.1

U.S. DOT - Substances Which Are Poisonous by Inhalation
liquid hazardous material poisonous by inhalation (UN1595)

U.S. DOT - Appendix A Table 1 - Hazardous Substances
final RQ = 100 pounds (45.4 kg)

IARC - Group 2A (limited human data)
[present] (Other relevant data, as given in Supplement 7, influenced the making of the overall evaluation)

NFPA - Flash Points
flash point = 182 degrees F (83 degrees C)

NFPA - Hazard Identification Ratings
health-4; flammability-2; reactivity-0

NIOSH - Selected LD50s and LC50s
Inhalation, rat: LC50 = 45 mg/m3 4 hr Oral, rat: LD50 = 205 mg/kg

NIOSH 1990 - Pocket Guide - RELs
0.1 ppm TWA; 0.5 mg/m3 TWA

NIOSH 1990 - Pocket Guide - IDLHs
10 ppm IDLH (not considering carcinogenic effects)

NIOSH 1990 - Pocket Guide - Carcinogens
occupational carcinogen

NIOSH 1990 - Pocket Guide - Target organs
eyes, respiratory system, liver, kidneys, CNS, skin

NIOSH 1990 - Pocket Guide - Skin list
Potential for dermal absorption

NTP Seventh Report - Suspect Carcinogens
suspect carcinogen

OSHA - Vacated PELs - Time Weighted Averages
0.1 ppm TWA; 0.5 mg/m3 TWA

OSHA - Vacated PELs - Skin Designation
Prevent or reduce skin absorption

OSHA - Final PELs - Time Weighted Averages
1 ppm TWA; 5 mg/m3 TWA

OSHA - Final PELs - Skin Notations
prevent or reduce skin absorption

OSHA - Possible Select Carcinogens
[present]

ENVIRONMENTAL LISTS

CERCLA/SARA - Section 302 Extremely Hazardous Substances and TPQs
TPQ = 500 pounds

CERCLA/SARA - Section 313 - Emission Reporting
form R reporting required for 0.1% de minimus concentration

CERCLA/SARA - Hazardous Substances and their Reportable Quantities
final RQ = 100 pounds (45.4 kg)

Clean Air Act (1990) - List of Hazardous Air Contaminants
[present]

CAA - HON Rule - SOCMI Chemicals
compliance by Oct. 24, 1994

CAA - HON Rule - Organic HAPs
[present]

EPA - Carcinogen Hazard Ranking for RQ Adjustment
Hazard ranking = Medium

RCRA - U Series Wastes
waste number U103

RCRA - Hazardous Constituents-Appendix VIII
waste number U103

RCRA - Basis for Listing - Appendix VII
Included in waste stream: K131

RCRA - Substances Banned From Land Disposal
[present]

TSCA - Code of Federal Regulations Citations
40 CFR 712.30(d)

TSCA - PAIR - Reporting List
Reporting Date: November 19, 1982

TSCA - Health and Safety Reporting List
Effective Date: March 11, 1994; Sunset Date: March 11, 2004

INTERNATIONAL LISTS

Australian Exposure Standards - Time Weighted Averages
0.1 ppm TWA; 0.52 mg/m3 TWA

Australian Exposure Standards - Skin Effects
skin absorption

Australian Exposure Standards - Carcinogens
probable carcinogen

Canada - WHMIS: Ingredient Disclosure
0.1% item 636 (1522)

Canada - NPRI (National Pollutant Release Inventory)
[present]

Canada - Alberta - 8 Hour Occupational Exposure Limit
0.1 ppm TWA; 0.52 mg/m3 TWA

Canada - Alberta - 15 Minute Occupational Exposure Limit
0.3 ppm STEL; 1.6 mg/m3 STEL

Canada - Alberta - Skin Designation
can be absorbed through the intact skin

Canada - Alberta - Designated Substances
designated substance - requires code of practice

Canada - British Columbia - Ceiling Exposure Limits
C 0.1 ppm;   C 0.5 mg/m3

Canada - British Columbia - Skin Notations
skin - potential for skin absorption

Canada - British Columbia - Carcinogens
carcinogen - 0.1 ppm TWA;   0.5 mg/m3 TWA

Canada - Ontario - OHSA - TWAEVs
0.1 ppm TWAEV;   0.5 mg/m3 TWAEV

Canada - Ontario - OHSA - Skin Notations
absorption through skin, eyes, or mucous membranes

Canada - Quebec - Time-Weighted Average Exposure Values
0.1 ppm TWAEV;   0.52 mg/m3 TWAEV

Canada - Quebec - Skin Designations
absorbed through the skin

Canada - Quebec - Carcinogens
C2 carcinogen: effect suspected in humans

German (DFG) - Skin/Sensitizers
danger of cutaneous absorption

German (DFG) - Carcinogens
animal evidence of carcinogenicity

Israel - Time Weighted Averages
0.1 ppm TWA;   0.52 mg/m3 TWA

Israel - Action Levels
0.05 ppm AL;   0.26 mg/m3 AL

Mexico - Instruction No. 10 - TWAs
1 ppm TWA;   5 mg/m3 TWA

Mexico - Instruction No. 10 - Skin designation
skin - potential for cutaneous absorption

Mexico - Instruction No. 10 - Carcinogens
potential carcinogen in humans - limited epidemiological evidence

STATE LISTS

California - Air Bill 2588 Appendix A-I
known or potential carcinogen

California - Prop. 65 - Cancer list
carcinogen - initial date 1/1/88

California - Exposure Limits - PELs
0.1 ppm PEL;   0.5 mg/m3 PEL

California - Exposure Limits - Skin Notation
material may be absorbed through the skin, eyes or mucous membrane

California - Directors List of Hazardous Substances (8 CCR 339)
[present]

Florida Hazardous Substance List
[present]

Massachusetts Right To Know List
carcinogen;   extraordinarily hazardous

Minnesota Hazardous Substance List
carcinogen;   skin

NJ Right to Know List (Total)
sn 0768

NJ Special Hazardous Substances
(carcinogen;   corrosive;   mutagen)

Pennsylvania Right to Know List
environmental hazard;   special hazardous substance

Pennsylvania RTK - Special Hazardous Substances
[present]

PROPOSED REGULATIONS

TSCA - ITC 32nd Report Priority Testing List
designated for dermal absorption testing

2,5-DIMETHYL-2,5-DI-(3,5,5-TRI-METHYLHEX-ANOYLPEROXY) HEXANE   RR-01755

HEALTH AND SAFETY LISTS

U.S. DOT - Substances From 49 CFR 172.101
regulated by DOT (UN1164)

U.S. DOT - Hazard Classes
DOT hazard class = 3

NFPA - Flash Points
flash point < 0 degrees F (-18 degrees C)

NFPA - Hazard Identification Ratings
health-1;   flammability-4;   reactivity-0

NIOSH - Selected LD50s and LC50s
Inhalation, rat: LC50 = 40250 ppm 8 hr Oral, rat: LD50 = 53 mg/kg

INTERNATIONAL LISTS

Canada - Alberta - 8 Hour Occupational Exposure Limit
50 ppm TWA;   135 mg/m3 TWA

Canada - Alberta - 15 Minute Occupational Exposure Limit
75 ppm STEL;   202 mg/m3 STEL

Canada - Alberta - Skin Designation
can be absorbed through the intact skin

STATE LISTS

Florida Hazardous Substance List
[present]

Massachusetts Right To Know List
extraordinarily hazardous

NJ Right to Know List (Total)
sn 0769

NJ Special Hazardous Substances
(reactive - fourth degree)

Pennsylvania Right to Know List
environmental hazard

PROPOSED REGULATIONS

CERCLA/SARA - Proposed Hazardous Substance Additions
proposed RQ = 1 pound (.454 kg)

CERCLA/SARA - 1989 Proposed RQ Adjustments
proposed RQ = 100 pounds (45.4 kg)

DIMETHYLVINYLCHLORIDE (DMVC)   513-37-1

HEALTH AND SAFETY LISTS

NIOSH - Selected LD50s and LC50s
Oral, mouse: LD50 = 140 mg/kg

DIMETHYLZINC   544-97-8

ENVIRONMENTAL LISTS

TSCA - Code of Federal Regulations Citations
40 CFR 712.30(x);   40 CFR 716.120(d)

TSCA - PAIR - Reporting List
Reporting Date: November 27, 1991

TSCA - Health and Safety Reporting List
Effective Date: September 30, 1991

DIMETILAN   644-64-4

HEALTH AND SAFETY LISTS

NFPA - Flash Points
flash point = 203 degrees F (95 degrees C)

NFPA - Hazard Identification Ratings
health-1;   flammability-1;   reactivity-0

NIOSH - Selected LD50s and LC50s
Oral, rat: LD50 = 14500 mg/kg

ENVIRONMENTAL LISTS

List of Pesticide Product Inert Ingredients
[present]

INTERNATIONAL LISTS

Canada - WHMIS: Ingredient Disclosure
1% item 637 (746)

DIMETRIDAZOLE   RR-01239-5

ENVIRONMENTAL LISTS

List of Pesticide Product Inert Ingredients
[present]

DIMYRISTYL PEROXYDI-CARBONATE   53220-22-7

ENVIRONMENTAL LISTS

List of Pesticide Product Inert Ingredients
[present]

DI-(1-NAPHTHOYL) PEROXIDE   29903-04-6

INTERNATIONAL LISTS

Canada - WHMIS: Ingredient Disclosure
1% item 638 (495)

N,N'-DI-BETA-NAPHTHYL-P-PHENYLENE-DIAMINE   93-46-9

HEALTH AND SAFETY LISTS

U.S. DOT - Organic Peroxides Table
Organic peroxide UN3105

DINITOLMIDE   148-01-6

HEALTH AND SAFETY LISTS

NIOSH - Selected LD50s and LC50s
Inhalation, mouse: LC50 = 181 gm/m3 (8 hr)

NTP Chemical Status Reports - Testing Status and NTIS Number
Technical reports printed (PB87115184)

NTP Chemical Status Reports - Evidence of Carcinogenicity
male rat-clear evidence; female rat-clear evidence; male mice-clear evidence; female mice-clear evidence

NTP Seventh Report - Suspect Carcinogens
suspect carcinogen

OSHA - Possible Select Carcinogens
[present]

STATE LISTS

California - Air Bill 2588 Appendix A-II
known or potential carcinogen: 9/89

California - Prop. 65 - Cancer list
carcinogen - initial date 7/1/89

California - Prop. 65 - No Significant Risk Levels
no significant risk level = 20 ug/day

Florida Hazardous Substance List
effective March 13, 1992

2,4-DINITROANILINE   97-02-9
SEE ALSO:
ZINC

HEALTH AND SAFETY LISTS

U.S. DOT - Substances From 49 CFR 172.101
regulated by DOT (UN1370)

U.S. DOT - Hazard Classes
DOT hazard class = 4.2

STATE LISTS

NJ Right to Know List (Total)
sn 0771

DINITROANILINE   26471-56-7

HEALTH AND SAFETY LISTS

NIOSH - Selected LD50s and LC50s
Oral, rat: LD50 = 25 mg/kg Skin, rat: LD50 = 600 mg/kg

ENVIRONMENTAL LISTS

CERCLA/SARA - Section 302 Extremely Hazardous Substances and TPQs
TPQ = 500/10,000 pounds

STATE LISTS

Florida Hazardous Substance List
effective March 13, 1992

Massachusetts Right To Know List
extraordinarily hazardous

NJ Right to Know List (Total)
sn 2349

Pennsylvania Right to Know List
environmental hazard

PROPOSED REGULATIONS

CERCLA/SARA - Proposed Hazardous Substance Additions
proposed RQ = 1 pound (.454 kg)

CERCLA/SARA - 1989 Proposed RQ Adjustments
proposed RQ = 100 pounds (45.4 kg)

M-DINITROBENZENE   99-65-0

ENVIRONMENTAL LISTS

TSCA - Chemicals with Significant New Use Rules
PMN number: P-90-1308

P-DINITROBENZENE   100-25-4

HEALTH AND SAFETY LISTS

U.S. DOT - Organic Peroxides Table
Organic peroxide UN3116; UN3119

STATE LISTS

NJ Right to Know List (Total)
sn 0772

O-DINITROBENZENE   528-29-0

HEALTH AND SAFETY LISTS

U.S. DOT - Hazard Classes
Forbidden from transport by the DOT

DINITROBENZENE (MIXED)   25154-54-5

INTERNATIONAL LISTS

Canada - WHMIS: Ingredient Disclosure
1% item 639 (748)

DINITROCHLOROBENZENE   25567-67-3

HEALTH AND SAFETY LISTS

ACGIH 1995 - Time Weighted Averages
5 mg/m3 TWA

NIOSH - Selected LD50s and LC50s
Oral, rat: LD50 = 600 mg/kg

OSHA - Vacated PELs - Time Weighted Averages
5 mg/m3 TWA

INTERNATIONAL LISTS

Australian Exposure Standards - Time Weighted Averages
5 mg/m3 TWA

Canada - Alberta - 8 Hour Occupational Exposure Limit
5 mg/m3 TWA

Canada - Alberta - 15 Minute Occupational Exposure Limit
10 mg/m3 STEL

Canada - British Columbia - 8 Hour Exposure Limits
5 mg/m3 TWA

Canada - British Columbia - 15 Minute Exposure Limits
10 mg/m3 STEL

Canada - Ontario - OHSA - TWAEVs
5 mg/m3 TWAEV

Canada - Ontario - OHSA - STEVs
10 mg/m3 STEV

Canada - Quebec - Time-Weighted Average Exposure Values
5 mg/m3 TWAEV

Israel - Time Weighted Averages
5 mg/m3 TWA

Israel - Action Levels
2.5 mg/m3 AL

Mexico - Instruction No. 10 - TWAs
5 mg/m3 TWA

STATE LISTS

California - Exposure Limits - PELs
5 mg/m3 PEL

California - Directors List of Hazardous Substances (8 CCR 339)
[present]

Florida Hazardous Substance List
[present]

Massachusetts Right To Know List
[present]

Minnesota Hazardous Substance List
[present]

Pennsylvania Right to Know List
[present]

4,6-DINITRO-O-CRESOL   534-52-1
SEE ALSO:
ANILINE AND CHLORO-, BROMO-, AND/OR NITROANILINES

HEALTH AND SAFETY LISTS

NFPA - Flash Points
flash point = 435 degrees F (224 degrees C)

NFPA - Hazard Identification Ratings
health-3; flammability-1; reactivity-3

NIOSH - Selected LD50s and LC50s
Oral, rat: LD50 = 418 mg/kg

OSHA - List of Highly Hazardous Chemicals
threshhold quantity = 5000 pounds

ENVIRONMENTAL LISTS

TSCA - Code of Federal Regulations Citations
40 CFR 712.30(d); 40 CFR 716.120(c); 40 CFR 799.5000

TSCA - PAIR - Reporting List
Reporting Date: November 19, 1982

TSCA - Substances Subject to Testing Consent Orders
Test for: Health Effects

TSCA - Section 12(b) - Export Notification
export notification required - Section 4

INTERNATIONAL LISTS

Canada - WHMIS: Ingredient Disclosure
1% item 641 (753)

STATE LISTS

Florida Hazardous Substance List
[present]

Massachusetts Right To Know List
[present]

Pennsylvania Right to Know List
[present]

4,6-DINITRO-O-CYCLOHEXYLPHENOL   131-89-5

HEALTH AND SAFETY LISTS

U.S. DOT - Substances From 49 CFR 172.101
regulated by DOT (UN1596)

U.S. DOT - Hazard Classes
DOT hazard class = 6.1

INTERNATIONAL LISTS

Canada - WHMIS: Ingredient Disclosure
1% item 640 (752)

STATE LISTS

NJ Right to Know List (Total)
sn 0776

NJ Special Hazardous Substances
(reactive - third degree)

DINITRO-7,8-DIMETHYLGLYCOLURIL   RR-01414-
SEE ALSO:
K025-HAZARDOUS WASTES

HEALTH AND SAFETY LISTS

ACGIH 1995 - Time Weighted Averages
0.15 ppm TWA; 1 mg/m3 TWA

ACGIH 1995 - Skin Designations
skin - potential for cutaneous absorption

U.S. DOT - Appendix A Table 1 - Hazardous Substances
final RQ = 100 pounds (45.4 kg) (Listed under 'Dinitrobenzene (mixed)')

NIOSH - Selected LD50s and LC50s
Oral, rat: LD50 = 83 mg/kg

NIOSH 1990 - Pocket Guide - RELs
1 mg/m3 TWA (Listed under 'Dinitrobenzene (all isomers)')

NIOSH 1990 - Pocket Guide - IDLHs
200 mg/m3 IDLH (Listed under 'Dinitrobenzene (all isomers)')

NIOSH 1990 - Pocket Guide - Target organs
blood, liver, CVS, eyes, CNS (Listed under 'Dinitrobenzene (all isomers)')

NIOSH 1990 - Pocket Guide - Skin list
Potential for dermal absorption (Listed under 'Dinitrobenzene (all isomers)')

OSHA - Vacated PELs - Time Weighted Averages
1 mg/m3 TWA (Listed under 'Dinitrobenzene (all isomers)')

OSHA - Vacated PELs - Skin Designation
Prevent or reduce skin absorption (Listed under 'Dinitrobenzene (all isomers)')

ENVIRONMENTAL LISTS

ATSDR Priority List
Rank (of 275): 275

CERCLA/SARA - Section 313 - Emission Reporting
form R reporting required for 1.0% de minimus concentration

CERCLA/SARA - Hazardous Substances and their Reportable Quantities
final RQ = 100 pounds (45.4 kg) (Listed under 'Dinitrobenzene (mixed)')

Clean Water Act - Hazardous Substances
[present] (Listed under 'Dinitrobenzene (mixed)')

RCRA - Basis for Listing - Appendix VII
Included in waste stream: K025

RCRA - TSD Facilities Ground Water Monitoring
TM 8270 = 10 ug/L PQL

TSCA - Health and Safety Reporting List
Effective Date: September 30, 1991

INTERNATIONAL LISTS

Australian Exposure Standards - Time Weighted Averages
0.15 ppm TWA; 1 mg/m3 TWA

Australian Exposure Standards - Skin Effects
skin absorption

Canada - WHMIS: Ingredient Disclosure
1% item 642 (754)

Israel - Time Weighted Averages
0.15 ppm TWA; 1 mg/m3 TWA

Israel - Action Levels
0.075 ppm AL; 0.5 mg/m3 AL

STATE LISTS

California - Air Bill 2588 Appendix A-II
6/91

California - Prop. 65 - Reproductive - Male
male reproductive toxicity - initial date 7/1/90

California - Exposure Limits - PELs
0.15 ppm PEL; 1 mg/m3 PEL (Listed under 'Dinitrobenzene (all isomers)')

California - Exposure Limits - Skin Notation
material may be absorbed through the skin, eyes or mucous membrane (Listed under 'Dinitrobenzene (all isomers)')

California - Directors List of Hazardous Substances (8 CCR 339)
[present] (Listed under 'Dinitrobenzene, all isomers')

Florida Hazardous Substance List
[present]

Massachusetts Right To Know List
[present]

Minnesota Hazardous Substance List
skin

NJ Right to Know List (Total)
sn 3017

Pennsylvania Right to Know List
environmental hazard

PROPOSED REGULATIONS

TSCA - ITC 32nd Report Priority Testing List
designated for dermal absorption testing

TSCA - ITC 33rd Report Priority Testing List
recommended for testing

TSCA - ITC 34th Report Priority Testing List
recommended for testing

1,3-DINITRO-5,5-DIMETHYL HYDANTOIN   RR-01367-2
SEE ALSO:
F039-HAZARDOUS WASTES

HEALTH AND SAFETY LISTS

ACGIH 1995 - Time Weighted Averages
0.15 ppm TWA; 1 mg/m3 TWA

ACGIH 1995 - Skin Designations
skin - potential for cutaneous absorption

U.S. DOT - Appendix A Table 1 - Hazardous Substances
final RQ = 100 pounds (45.4 kg) (Listed under 'Dinitrobenzene (mixed)')

NIOSH 1990 - Pocket Guide - RELs
1 mg/m3 TWA (Listed under 'Dinitrobenzene (all isomers)')

NIOSH 1990 - Pocket Guide - IDLHs
200 mg/m3 IDLH (Listed under 'Dinitrobenzene (all isomers)')

NIOSH 1990 - Pocket Guide - Target organs
blood, liver, CVS, eyes, CNS (Listed under 'Dinitrobenzene (all isomers)')

NIOSH 1990 - Pocket Guide - Skin list
Potential for dermal absorption (Listed under 'Dinitrobenzene (all isomers)')

OSHA - Vacated PELs - Time Weighted Averages
1 mg/m3 TWA (Listed under 'Dinitrobenzene (all isomers)')

OSHA - Vacated PELs - Skin Designation
Prevent or reduce skin absorption (Listed under 'Dinitrobenzene (all isomers)')

ENVIRONMENTAL LISTS

CERCLA/SARA - Section 313 - Emission Reporting
form R reporting required for 1.0% de minimus concentration

CERCLA/SARA - Hazardous Substances and their Reportable Quantities
final RQ = 100 pounds (45.4 kg) (Listed under 'Dinitrobenzene (mixed)')

Clean Water Act - Hazardous Substances
[present] (Listed under 'Dinitrobenzene (mixed)')

RCRA - Basis for Listing - Appendix VII
Included in waste stream: F039

RCRA - Universal Treatment Standards (LDR)
WW: 0.32 mg/l; NWW: 2.3 mg/kg

TSCA - PAIR - Reporting List
Effective Date: January 26, 1994; Reporting Date: March 28, 1994

TSCA - Health and Safety Reporting List
Effective Date: January 26, 1994; Sunset Date: January 26, 2004

INTERNATIONAL LISTS

Australian Exposure Standards - Time Weighted Averages
0.15 ppm TWA; 1 mg/m3 TWA

Australian Exposure Standards - Skin Effects
skin absorption

Canada - WHMIS: Ingredient Disclosure
1% item 644 (756)

Israel - Time Weighted Averages
0.15 ppm TWA; 1 mg/m3 TWA

Israel - Action Levels
0.075 ppm AL; 0.5 mg/m3 AL

STATE LISTS

California - Air Bill 2588 Appendix A-II
6/91

California - Prop. 65 - Reproductive - Male
male reproductive toxicity - initial date 7/1/90

California - Exposure Limits - PELs
0.15 ppm PEL; 1 mg/m3 PEL (Listed under 'Dinitrobenzene (all isomers)')

California - Exposure Limits - Skin Notation
material may be absorbed through the skin, eyes or mucous membrane (Listed under 'Dinitrobenzene (all isomers)')

California - Directors List of Hazardous Substances (8 CCR 339)
[present] (Listed under 'Dinitrobenzene, all isomers')

Florida Hazardous Substance List
[present]

Massachusetts Right To Know List
[present]

Minnesota Hazardous Substance List
skin

NJ Right to Know List (Total)
sn 3019

Pennsylvania Right to Know List
environmental hazard

PROPOSED REGULATIONS

TSCA - ITC 31st Report Priority Testing List
designated to be tested

1,3-DINITRO-4,5-DINITROSOBENZENE   RR-01366-1

HEALTH AND SAFETY LISTS

ACGIH 1995 - Time Weighted Averages
0.15 ppm TWA; 1 mg/m3 TWA

ACGIH 1995 - Skin Designations
skin - potential for cutaneous absorption

U.S. DOT - Appendix A Table 1 - Hazardous Substances
final RQ = 100 pounds (45.4 kg) (Listed under 'Dinitrobenzene (mixed)')

NFPA - Flash Points
flash point = 302 degrees F (150 degrees C)

NFPA - Hazard Identification Ratings
health-3; flammability-1; reactivity-4

NIOSH 1990 - Pocket Guide - RELs
1 mg/m3 TWA (Listed under 'Dinitrobenzene (all isomers)')

NIOSH 1990 - Pocket Guide - IDLHs
200 mg/m3 IDLH (Listed under 'Dinitrobenzene (all isomers)')

NIOSH 1990 - Pocket Guide - Target organs
blood, liver, CVS, CNS, eyes (Listed under 'Dinitrobenzene (all isomers)')

NIOSH 1990 - Pocket Guide - Skin list
Potential for dermal absorption (Listed under 'Dinitrobenzene (all isomers)')

OSHA - Vacated PELs - Time Weighted Averages
1 mg/m3 TWA (Listed under 'Dinitrobenzene (all isomers)')

OSHA - Vacated PELs - Skin Designation
Prevent or reduce skin absorption (Listed under 'Dinitrobenzene (all isomers)')

OSHA - Final PELs - Time Weighted Averages
1 mg/m3 TWA (Listed under 'Dinitrobenzene (all isomers)')

OSHA - Final PELs - Skin Notations
prevent or reduce skin absorption

ENVIRONMENTAL LISTS

CERCLA/SARA - Section 313 - Emission Reporting
form R reporting required for 1.0% de minimus concentration

CERCLA/SARA - Hazardous Substances and their Reportable Quantities
final RQ = 100 pounds (45.4 kg) (Listed under 'Dinitrobenzene (mixed)')

Clean Water Act - Hazardous Substances
[present] (Listed under 'Dinitrobenzene (mixed)')

INTERNATIONAL LISTS

Australian Exposure Standards - Time Weighted Averages
0.15 ppm TWA;   1 mg/m3 TWA

Australian Exposure Standards - Skin Effects
skin absorption

Canada - WHMIS: Ingredient Disclosure
1% item 643 (755)

Israel - Time Weighted Averages
0.15 ppm TWA;   1 mg/m3 TWA

Israel - Action Levels
0.075 ppm AL;   0.5 mg/m3 AL

STATE LISTS

California - Air Bill 2588 Appendix A-II
6/91

California - Prop. 65 - Reproductive - Male
male reproductive toxicity - initial date 7/1/90

California - Exposure Limits - PELs
0.15 ppm PEL;   1 mg/m3 PEL (Listed under 'Dinitrobenzene (all isomers)')

California - Exposure Limits - Skin Notation
material may be absorbed through the skin, eyes or mucous membrane (Listed under 'Dinitrobenzene (all isomers)')

California - Directors List of Hazardous Substances (8 CCR 339)
[present] (Listed under 'Dinitrobenzene, all isomers')

Florida Hazardous Substance List
[present]

Massachusetts Right To Know List
[present]

Minnesota Hazardous Substance List
skin

NJ Right to Know List (Total)
sn 3018

Pennsylvania Right to Know List
environmental hazard

PROPOSED REGULATIONS

TSCA - ITC 32nd Report Priority Testing List
designated for dermal absorption testing

1,1-DINITROETHANE

HEALTH AND SAFETY LISTS

U.S. DOT - Substances From 49 CFR 172.101
regulated by DOT (UN1597)

U.S. DOT - Hazard Classes
DOT hazard class = 6.1

U.S. DOT - Appendix A Table 1 - Hazardous Substances
final RQ = 100 pounds (45.4 kg)

NIOSH 1990 - Pocket Guide - RELs
1 mg/m3 TWA

NIOSH 1990 - Pocket Guide - IDLHs
200 mg/m3 IDLH

NIOSH 1990 - Pocket Guide - Target organs
blood, liver, CVS, eyes, CNS

NIOSH 1990 - Pocket Guide - Skin list
Potential for dermal absorption

OSHA - Vacated PELs - Time Weighted Averages
1 mg/m3 TWA

OSHA - Vacated PELs - Skin Designation
Prevent or reduce skin absorption

ENVIRONMENTAL LISTS

CERCLA/SARA - Hazardous Substances and their Reportable Quantities
final RQ = 100 pounds (45.4 kg)

CAA - HON Rule - SOCMI Chemicals
compliance by Oct. 24, 1994

Clean Water Act - Hazardous Substances
[present]

RCRA - Hazardous Constituents-Appendix VIII
hazardous constituent - no waste number

INTERNATIONAL LISTS

Canada - Alberta - 8 Hour Occupational Exposure Limit
0.15 ppm TWA;   1 mg/m3 TWA

Canada - Alberta - 15 Minute Occupational Exposure Limit
0.5 ppm STEL;   3.4 mg/m3 STEL

Canada - Alberta - Skin Designation
can be absorbed through the intact skin

Canada - British Columbia - 8 Hour Exposure Limits
0.15 ppm TWA;   1 mg/m3 TWA

Canada - British Columbia - 15 Minute Exposure Limits
0.5 ppm STEL;   3 mg/m3 STEL

Canada - British Columbia - Skin Notations
skin - potential for skin absorption

Canada - Ontario - OHSA - TWAEVs
0.15 ppm TWAEV;   1.0 mg/m3 TWAEV

Canada - Ontario - OHSA - Skin Notations
absorption through skin, eyes, or mucous membranes

Canada - Quebec - Time-Weighted Average Exposure Values
0.15 ppm TWAEV;   1 mg/m3 TWAEV

Canada - Quebec - Skin Designations
absorbed through the skin

United Kingdom - Occupational Exposure Standards - TWAs
0.15 ppm TWA;   1 mg/m3 TWA

United Kingdom - Occupational Exposure Standards - STELs
0.5 ppm STEL;   3 mg/m3 STEL

United Kingdom - Occupational Exposure Standards - Notes
can be absorbed through skin

German (DFG) - Skin/Sensitizers
danger of cutaneous absorption

German (DFG) - Carcinogens
suspected carcinogen

Mexico - Instruction No. 10 - TWAs
0.15 ppm TWA

Mexico - Instruction No. 10 - STELs
0.5 ppm STEL; 3 mg/m3 STEL

Mexico - Instruction No. 10 - Skin designation
skin - potential for cutaneous absorption

STATE LISTS

California - Air Bill 2588 Appendix A-II
9/90

California - Directors List of Hazardous Substances (8 CCR 339)
[present]

Massachusetts Right To Know List
[present]

NJ Right to Know List (Total)
sn 0777

Pennsylvania Right to Know List
environmental hazard

1,2-DINITROETHANE   7570-26-5

HEALTH AND SAFETY LISTS

U.S. DOT - Appendix B - Marine Pollutants
DOT regulated marine pollutant

NFPA - Flash Points
flash point = 382 degrees F (194 degrees C)

NFPA - Hazard Identification Ratings
health-3; flammability-1; reactivity-4

STATE LISTS

Florida Hazardous Substance List
[present]

Massachusetts Right To Know List
[present]

NJ Right to Know List (Total)
sn 0778

NJ Special Hazardous Substances
(reactive - fourth degree)

Pennsylvania Right to Know List
[present]

3,9-DINITROFLUORANTHENE   22506-53-2
SEE ALSO:
F039-HAZARDOUS WASTES

HEALTH AND SAFETY LISTS

ACGIH 1995 - Time Weighted Averages
0.2 mg/m3 TWA

ACGIH 1995 - Skin Designations
skin - potential for cutaneous absorption

U.S. DOT - Substances From 49 CFR 172.101
regulated by DOT (UN1598)

U.S. DOT - Hazard Classes
DOT hazard class = 6.1

U.S. DOT - Appendix B - Marine Pollutants
DOT regulated marine pollutant

U.S. DOT - Appendix A Table 1 - Hazardous Substances
final RQ = 10 pounds (4.54 kg)

NIOSH - Selected LD50s and LC50s
Oral, rat: LD50 = 10 mg/kg Skin, rat: LD50 = 200 mg/kg

NIOSH 1990 - Pocket Guide - RELs
0.2 mg/m3 TWA

NIOSH 1990 - Pocket Guide - IDLHs
5 mg/m3 IDLH

NIOSH 1990 - Pocket Guide - Target organs
CVS, eyes, endocrine system

NIOSH 1990 - Pocket Guide - Skin list
Potential for dermal absorption

NIOSH - Health Standards - Exposure Limits
0.2 mg/m3 TWA

NIOSH - Health Standards - Health Effects and Precautions
Central nervous system and metabolic effects (Blood and urine monitoring required; prevent skin and eye contact; possible delayed effects)

OSHA - Vacated PELs - Time Weighted Averages
0.2 mg/m3 TWA

OSHA - Vacated PELs - Skin Designation
Prevent or reduce skin absorption

OSHA - Final PELs - Time Weighted Averages
0.2 mg/m3 TWA

OSHA - Final PELs - Skin Notations
prevent or reduce skin absorption

ENVIRONMENTAL LISTS

ATSDR Priority List
Rank (of 275): 097

CERCLA/SARA - Section 302 Extremely Hazardous Substances and TPQs
TPQ = 10/10,000 pounds

CERCLA/SARA - Section 313 - Emission Reporting
form R reporting required for 1.0% de minimus concentration

CERCLA/SARA - Hazardous Substances and their Reportable Quantities
final RQ = 10 pounds (4.54 kg)

Clean Air Act (1990) - List of Hazardous Air Contaminants
[present]

Clean Water Act - Priority Pollutants
[present]

RCRA - P Series Wastes
waste number P047

RCRA - Hazardous Constituents-Appendix VIII
waste number P047

RCRA - Basis for Listing - Appendix VII
Included in waste stream: F039

RCRA - Substances Banned From Land Disposal
[present]

RCRA - TSD Facilities Ground Water Monitoring
TM 8040 = 150 ug/L PQL; TM 8270 = 50 ug/L PQL

RCRA - Universal Treatment Standards (LDR)
WW: 0.28 mg/l; NWW: 160 mg/kg

INTERNATIONAL LISTS

Australian Exposure Standards - Time Weighted Averages
0.2 mg/m3 TWA

Australian Exposure Standards - Skin Effects
skin absorption

Canada - WHMIS: Ingredient Disclosure
1% item 645 (757)

Canada - NPRI (National Pollutant Release Inventory)
[present]

Canada - Alberta - 8 Hour Occupational Exposure Limit
0.2 mg/m3 TWA

Canada - Alberta - 15 Minute Occupational Exposure Limit
0.6 mg/m3 STEL

Canada - Alberta - Skin Designation
can be absorbed through the intact skin

Canada - British Columbia - 8 Hour Exposure Limits
0.2 mg/m3 TWA

Canada - British Columbia - 15 Minute Exposure Limits
0.6 mg/m3 STEL

Canada - British Columbia - Skin Notations
skin - potential for skin absorption

Canada - Ontario - OHSA - TWAEVs
0.2 mg/m3 TWAEV

Canada - Ontario - OHSA - Skin Notations
absorption through skin, eyes, or mucous membranes
Canada - Quebec - Time-Weighted Average Exposure Values
0.2 mg/m3 TWAEV
Canada - Quebec - Skin Designations
absorbed through the skin
United Kingdom - Occupational Exposure Standards - TWAs
0.2 mg/m3 TWA
United Kingdom - Occupational Exposure Standards - STELs
0.6 mg/m3 STEL
United Kingdom - Occupational Exposure Standards - Notes
can be absorbed through skin
German (DFG) - MAK Values
total dust: 0.2 mg/m3 MAK
German (DFG) - Peak Limitations
2 x normal MAK (30 min. average value); don't exceed 4 times during shift
German (DFG) - Skin/Sensitizers
danger of cutaneous absorption
Israel - Time Weighted Averages
0.2 mg/m3 TWA
Israel - Action Levels
0.1 mg/m3 AL
Mexico - Instruction No. 10 - TWAs
0.2 mg/m3 TWA
Mexico - Instruction No. 10 - STELs
0.6 mg/m3 STEL
Mexico - Instruction No. 10 - Skin designation
skin - potential for cutaneous absorption
Mexico - Wastewater - Organic Toxic Pollutants and Heavy Metals
Listed under [Nitrophenols]
Mexico - Drinking Water - Ecological Criteria
0.01 mg/l This level has been extrapolated by using a mathematic model
STATE LISTS
California - Air Bill 2588 Appendix A-I
6/91
California - Exposure Limits - PELs
0.2 mg/m3 PEL
California - Exposure Limits - Skin Notation
material may be absorbed through the skin, eyes or mucous membrane
California - Directors List of Hazardous Substances (8 CCR 339)
[present]
Florida Hazardous Substance List
[present]
Massachusetts Right To Know List
extraordinarily hazardous
Minnesota Hazardous Substance List
skin
NJ Right to Know List (Total)
sn 0779
Pennsylvania Right to Know List
environmental hazard

3,7-DINITROFLUORANTHENE   105735-71-5
HEALTH AND SAFETY LISTS
U.S. DOT - Appendix A Table 1 - Hazardous Substances
final RQ = 100 pounds (45.4 kg)
NIOSH - Selected LD50s and LC50s
Oral, rat: LD50 = 65 mg/kg
ENVIRONMENTAL LISTS
CERCLA/SARA - Hazardous Substances and their Reportable Quantities
final RQ = 100 pounds (45.4 kg)

RCRA - P Series Wastes
waste number P034
RCRA - Hazardous Constituents-Appendix VIII
waste number P034
RCRA - Substances Banned From Land Disposal
[present]
STATE LISTS
Massachusetts Right To Know List
[present]
NJ Right to Know List (Total)
sn 0774
Pennsylvania Right to Know List
environmental hazard

DINITROGLYCOLURIL   RR-01333-2
HEALTH AND SAFETY LISTS
U.S. DOT - Hazard Classes
Forbidden from transport by the DOT

DINITROMETHANE   625-76-3
HEALTH AND SAFETY LISTS
U.S. DOT - Hazard Classes
Forbidden from transport by the DOT

DINITRONAPHTHALENES   27478-34-8
HEALTH AND SAFETY LISTS
U.S. DOT - Hazard Classes
Forbidden from transport by the DOT

DINITROPHENATES ALKALI METALS, DRY OR WETTED   RR-00262-0
HEALTH AND SAFETY LISTS
U.S. DOT - Hazard Classes
Forbidden from transport by the DOT

2,4-DINITROPHENOL   51-28-5
HEALTH AND SAFETY LISTS
U.S. DOT - Hazard Classes
Forbidden from transport by the DOT

2,5-DINITROPHENOL   329-71-5
HEALTH AND SAFETY LISTS
IARC - Group 3 (not classifiable)
[present]

2,6-DINITROPHENOL   573-56-8
HEALTH AND SAFETY LISTS
IARC - Group 3 (not classifiable)
[present]

DINITROPHENOL   25550-58-7
HEALTH AND SAFETY LISTS
U.S. DOT - Substances From 49 CFR 172.101
regulated by DOT (UN0489)
U.S. DOT - Hazard Classes
DOT hazard class = 1.1D

DINITROPHENOLATES   RR-00263-1
HEALTH AND SAFETY LISTS
U.S. DOT - Hazard Classes
Forbidden from transport by the DOT

2,4-DINITROPHENYLHYDRAZINE   119-26-6

INTERNATIONAL LISTS

German (DFG) - Carcinogens
suspected carcinogen

DINITROPROPYLENE GLYCOL   RR-01415-3

STATE LISTS

NJ Right to Know List (Total)
sn 2350

1,6-DINITROPYRENE   42397-64-8

SEE ALSO:
K001-HAZARDOUS WASTES
F039-HAZARDOUS WASTES

HEALTH AND SAFETY LISTS

U.S. DOT - Appendix A Table 1 - Hazardous Substances
final RQ = 10 pounds (4.54 kg)

NIOSH - Selected LD50s and LC50s
Oral, rat: LD50 = 30 mg/kg

ENVIRONMENTAL LISTS

ATSDR Priority List
Rank (of 275): 080

CERCLA/SARA - Section 313 - Emission Reporting
form R reporting required for 1.0% de minimus concentration

CERCLA/SARA - Hazardous Substances and their Reportable Quantities
final RQ = 10 pounds (4.54 kg)

Clean Air Act (1990) - List of Hazardous Air Contaminants
[present]

CAA - HON Rule - SOCMI Chemicals
compliance by April 24, 1995

CAA - HON Rule - Organic HAPs
[present]

Clean Water Act - Hazardous Substances
[present] (Listed under 'Dinitrophenol')

Clean Water Act - Priority Pollutants
[present]

RCRA - P Series Wastes
waste number P048

RCRA - Hazardous Constituents-Appendix VIII
waste number P048

RCRA - Basis for Listing - Appendix VII
Included in waste streams: F039, K001

RCRA - Substances Banned From Land Disposal
[present]

RCRA - TSD Facilities Ground Water Monitoring
TM 8040 = 150 ug/L PQL; TM 8270 = 50 ug/L PQL

RCRA - Universal Treatment Standards (LDR)
WW: 0.12 mg/l; NWW: 160 mg/kg

TSCA - Code of Federal Regulations Citations
40 CFR 712.30(x); 40 CFR 716.120(d)

TSCA - PAIR - Reporting List
Reporting Date: November 27, 1991

TSCA - Health and Safety Reporting List
Effective Date: September 30, 1991

INTERNATIONAL LISTS

Canada - WHMIS: Ingredient Disclosure
0.1% item 647 (761)

Mexico - Wastewater - Organic Toxic Pollutants and Heavy Metals
Listed under [Nitrophenols]

Mexico - Drinking Water - Ecological Criteria
0.07 mg/l

STATE LISTS

California - Air Bill 2588 Appendix A-I
6/91

California - Directors List of Hazardous Substances (8 CCR 339)
[present] (Listed under 'Dinitrophenols')

Massachusetts Right To Know List
[present]

NJ Right to Know List (Total)
sn 2950

Pennsylvania Right to Know List
environmental hazard

PROPOSED REGULATIONS

Safe Drinking Water Act - Priority list
[present]

1,8-DINITROPYRENE   42397-65-9

HEALTH AND SAFETY LISTS

U.S. DOT - Appendix A Table 1 - Hazardous Substances
final RQ = 10 pounds (4.54 kg) (Listed under 'Dinitrophenol')

ENVIRONMENTAL LISTS

CERCLA/SARA - Hazardous Substances and their Reportable Quantities
final RQ = 10 pounds (4.54 kg) (Listed under 'Dinitrophenol')

Clean Water Act - Hazardous Substances
[present] (Listed under 'Dinitrophenol')

STATE LISTS

California - Directors List of Hazardous Substances (8 CCR 339)
[present] (Listed under 'Dinitrophenols')

Massachusetts Right To Know List
[present]

NJ Right to Know List (Total)
sn 2951

Pennsylvania Right to Know List
environmental hazard

1,3-DINITROPYRENE   75321-20-9

HEALTH AND SAFETY LISTS

U.S. DOT - Appendix A Table 1 - Hazardous Substances
final RQ = 10 pounds (4.54 kg) (Listed under 'Dinitrophenol')

ENVIRONMENTAL LISTS

CERCLA/SARA - Hazardous Substances and their Reportable Quantities
final RQ = 10 pounds (4.54 kg) (Listed under 'Dinitrophenol')

Clean Water Act - Hazardous Substances
[present] (Listed under 'Dinitrophenol')

STATE LISTS

California - Directors List of Hazardous Substances (8 CCR 339)
[present] (Listed under 'Dinitrophenols')

Massachusetts Right To Know List
[present]

NJ Right to Know List (Total)
sn 2953

Pennsylvania Right to Know List
environmental hazard

2,4-DINITRORESORCINOL   519-44-8

HEALTH AND SAFETY LISTS

U.S. DOT - Substances From 49 CFR 172.101
regulated by DOT (UN1320, UN0076, UN1599)

U.S. DOT - Hazard Classes
DOT hazard class = 4.1

U.S. DOT - Appendix B - Marine Pollutants
DOT regulated marine pollutant

U.S. DOT - Appendix A Table 1 - Hazardous Substances
final RQ = 10 pounds (4.54 kg)

ENVIRONMENTAL LISTS

CERCLA/SARA - Hazardous Substances and their Reportable Quantities
final RQ = 10 pounds (4.54 kg)

INTERNATIONAL LISTS

Canada - WHMIS: Ingredient Disclosure
1% item 646 (760)

STATE LISTS

Massachusetts Right To Know List
[present]

NJ Right to Know List (Total)
sn 0780

Pennsylvania Right to Know List
environmental hazard

4,6-DINITRORESORCINOL   616-74-0

HEALTH AND SAFETY LISTS

U.S. DOT - Substances From 49 CFR 172.101
regulated by DOT (UN0077, UN1321)

U.S. DOT - Hazard Classes
with less than 15% water: DOT hazard class = 1.3C; with 15% or more water: DOT hazard class = 4.1

U.S. DOT - Appendix B - Marine Pollutants
DOT regulated marine pollutant

STATE LISTS

NJ Right to Know List (Total)
sn 2351

DINITRORESORCINOL   35860-51-6

HEALTH AND SAFETY LISTS

NIOSH - Selected LD50s and LC50s
Oral, rat: LD50 = 654 mg/kg

INTERNATIONAL LISTS

Canada - WHMIS: Ingredient Disclosure
1% item 648 (762)

2,4-DINITRORESORCINOL   RR-01385-4

HEALTH AND SAFETY LISTS

U.S. DOT - Hazard Classes
Forbidden from transport by the DOT

4,6-DINITRORESORCINOL   RR-01387-6

HEALTH AND SAFETY LISTS

IARC - Group 2B (sufficient animal data)
[present]

OSHA - Possible Select Carcinogens
[present]

STATE LISTS

California - Air Bill 2588 Appendix A-I
known or potential carcinogen: 6/91

California - Prop. 65 - Cancer list
carcinogen - initial date 10/1/90

California - Directors List of Hazardous Substances (8 CCR 339)
[present]

PROPOSED REGULATIONS

NTP - Proposed Additions to Annual Report on Carcinogens
proposed as a suspect carcinogen for NTP 8th report

DINITROSOBENZENE   25550-55–

HEALTH AND SAFETY LISTS

IARC - Group 2B (sufficient animal data)
[present]

OSHA - Possible Select Carcinogens
[present]

STATE LISTS

California - Air Bill 2588 Appendix A-I
known or potential carcinogen: 6/91

California - Prop. 65 - Cancer list
carcinogen - initial date 10/1/90

California - Directors List of Hazardous Substances (8 CCR 339)
[present]

PROPOSED REGULATIONS

NTP - Proposed Additions to Annual Report on Carcinogens
proposed as a suspect carcinogen for NTP 8th report

DINITROSOBENZYLAMIDINE AND SALTS   RR-01416-4

HEALTH AND SAFETY LISTS

IARC - Group 3 (not classifiable)
[present]

DINITROSOPENTAMETHYLENETETRAMINE   101-25-7

HEALTH AND SAFETY LISTS

U.S. DOT - Hazard Classes
Forbidden from transport by the DOT

1,4-DINITROSOPIPERAZINE   140-79-4

HEALTH AND SAFETY LISTS

U.S. DOT - Hazard Classes
Forbidden from transport by the DOT

2,2-DINITROSTILBENE   RR-01373-0

HEALTH AND SAFETY LISTS

U.S. DOT - Substances From 49 CFR 172.101
regulated by DOT (UN1322)

U.S. DOT - Hazard Classes
DOT hazard class = 4.1

STATE LISTS

NJ Right to Know List (Total)
sn 0781

1,4-DINITRO-1,1,4,4-TETRAMETHYLOLBUTANETETRANITRATE   RR-01368-3

STATE LISTS

NJ Right to Know List (Total)
sn 2217

2,4-DINITROTOLUENE   121-14-2

STATE LISTS

NJ Right to Know List (Total)
sn 0350

2,4-DINITROTOLUENE   121-20-0

HEALTH AND SAFETY LISTS

U.S. DOT - Substances From 49 CFR 172.101
regulated by DOT (UN0406)

U.S. DOT - Hazard Classes
DOT hazard class = 1.3C

STATE LISTS

NJ Right to Know List (Total)
sn 0782

2,6-DINITROTOLUENE   606-20-2

HEALTH AND SAFETY LISTS
    U.S. DOT - Hazard Classes
        Forbidden from transport by the DOT

3,4-DINITROTOLUENE   610-39-9

HEALTH AND SAFETY LISTS
    U.S. DOT - Substances From 49 CFR 172.101
        regulated by DOT (UN2972)
    U.S. DOT - Hazard Classes
        DOT hazard class = 4.1
    IARC - Group 3 (not classifiable)
        [present]
    NIOSH - Selected LD50s and LC50s
        Oral, rat: LD50 = 940 mg/kg

ENVIRONMENTAL LISTS
    List of Pesticide Product Inert Ingredients
        [present]

INTERNATIONAL LISTS
    Canada - WHMIS: Ingredient Disclosure
        1% item 649 (763)

DINITROTOLUENE (MIXED ISOMERS)   25321-14-6

INTERNATIONAL LISTS
    Canada - WHMIS: Ingredient Disclosure
        0.1% item 650 (764)

2,4-DINITRO-1,3,5-TRIMETHYLBENZENE   608-50-4

HEALTH AND SAFETY LISTS
    U.S. DOT - Hazard Classes
        Forbidden from transport by the DOT

DI-(BETA-NITROXYETHYL) AMMONIUM NITRATE   RR-01404-0

HEALTH AND SAFETY LISTS
    U.S. DOT - Hazard Classes
        Forbidden from transport by the DOT

A,A'-DI-(NITROXY) METHYLETHER   33453-96-2
    SEE ALSO:
        K111-HAZARDOUS WASTES
        F039-HAZARDOUS WASTES
        K025-HAZARDOUS WASTES

HEALTH AND SAFETY LISTS
    U.S. DOT - Appendix A Table 1 - Hazardous Substances
        final RQ = 10 pounds (4.54 kg)
    NFPA - Flash Points
        flash point = 404 degrees F (207 degrees C)
    NFPA - Hazard Identification Ratings
        health-3; flammability-1; reactivity-3
    NIOSH - Selected LD50s and LC50s
        Oral, rat: LD50 = 268 mg/kg
    NTP Chemical Status Reports - Testing Status and NTIS Number
        Technical reports printed (PB280990/AS)
    NTP Chemical Status Reports - Evidence of Carcinogenicity
        male rat-positive; female rat-positive; male mice-negative; female mice-negative

ENVIRONMENTAL LISTS
    ATSDR Priority List
        Rank (of 275): 103
    CERCLA/SARA - Section 313 - Emission Reporting
        form R reporting required for 1.0% de minimus concentration

CERCLA/SARA - Hazardous Substances and their Reportable Quantities
    final RQ = 10 pounds (4.54 kg)
Clean Air Act (1990) - List of Hazardous Air Contaminants
    [present]
CAA - HON Rule - SOCMI Chemicals
    compliance by April 24, 1995
CAA - HON Rule - Organic HAPs
    [present]
Clean Water Act - Hazardous Substances
    [present] (Listed under 'Dinitrotoluene')
Clean Water Act - Priority Pollutants
    [present]
EPA - Carcinogen Hazard Ranking for RQ Adjustment
    Hazard ranking = Medium
EPA - Master Testing List
    [present]
RCRA - D Series - Maximum Concentration of Contaminants
    waste number D030; regulatory level = 0.13 mg/L
RCRA - D Series - Chronic Toxicity Reference Levels
    chronic toxicity reference level = 0.0005 mg/L
RCRA - U Series Wastes
    waste number U105
RCRA - Hazardous Constituents-Appendix VIII
    waste number U105
RCRA - Basis for Listing - Appendix VII
    Included in waste streams: F039, K025, K111
RCRA - Substances Banned From Land Disposal
    [present]
RCRA - TSD Facilities Ground Water Monitoring
    TM 8090 = 0.2 ug/L PQL; TM 8270 = 10 ug/L PQL
RCRA - Universal Treatment Standards (LDR)
    WW: 0.32 mg/l; NWW: 140 mg/kg
TSCA - Health and Safety Reporting List
    Effective Date: March 11, 1994; Sunset Date: March 11, 2004

INTERNATIONAL LISTS
    Australian Exposure Standards - Time Weighted Averages
        1.5 mg/m3 TWA
    Australian Exposure Standards - Skin Effects
        skin absorption
    Canada - WHMIS: Ingredient Disclosure
        1% item 652 (766)
    Canada - NPRI (National Pollutant Release Inventory)
        [present]
    Canada - Alberta - 8 Hour Occupational Exposure Limit
        1.5 mg/m3 TWA
    Canada - Alberta - 15 Minute Occupational Exposure Limit
        5 mg/m3 STEL
    Canada - Alberta - Skin Designation
        can be absorbed through the intact skin
    Mexico - Instruction No. 10 - TWAs
        5 mg/m3 TWA
    Mexico - Instruction No. 10 - STELs
        5 mg/m3 STEL
    Mexico - Instruction No. 10 - Skin designation
        skin - potential for cutaneous absorption
    Mexico - Wastewater - Organic Toxic Pollutants and Heavy Metals
        Listed under [Dinitrotoluenes]
    Mexico - Drinking Water - Ecological Criteria
        0.001 mg/l Substance presents persistence, bioaccumulations or risk of cancer, reduce human exposure to a minimum; This level has been extrapolated by using a mathematic model

STATE LISTS

California - Air Bill 2588 Appendix A-I
known or potential carcinogen: 9/89

California - Prop. 65 - Cancer list
carcinogen - initial date 7/1/88

California - Prop. 65 - No Significant Risk Levels
no significant risk level - 2 ug/day

California - Exposure Limits - PELs
1.5 mg/m3 PEL

California - Exposure Limits - Skin Notation
material may be absorbed through the skin, eyes or mucous membrane

California - Directors List of Hazardous Substances (8 CCR 339)
[present]

Florida Hazardous Substance List
[present]

Massachusetts Right To Know List
carcinogen; extraordinarily hazardous

Minnesota Hazardous Substance List
carcinogen; skin

NJ Right to Know List (Total)
sn 0783

NJ Special Hazardous Substances
(carcinogen; mutagen; reactive - third degree)

Pennsylvania Right to Know List
environmental hazard

PROPOSED REGULATIONS

Safe Drinking Water Act - Priority list
[present]

TSCA - ITC 32nd Report Priority Testing List
designated for dermal absorption testing

1,9-DINITROXY PENTAMETHYLENE-2,4,6,8-   RR-01370-7
TETRAMINE

STATE LISTS

Massachusetts Right To Know List
[present]

DINOBUTON   973-21-7
SEE ALSO:
F039-HAZARDOUS WASTES

HEALTH AND SAFETY LISTS

U.S. DOT - Appendix A Table 1 - Hazardous Substances
final RQ - 100 pounds (45.4 kg)

NIOSH - Selected LD50s and LC50s
Oral, rat: LD50 - 177 mg/kg

ENVIRONMENTAL LISTS

ATSDR Priority List
Rank (of 275): 224

CERCLA/SARA - Section 313 - Emission Reporting
form R reporting required for 1.0% de minimus concentration

CERCLA/SARA - Hazardous Substances and their Reportable Quantities
final RQ - 100 pounds (45.4 kg)

Clean Water Act - Hazardous Substances
[present] (Listed under 'Dinitrotoluene')

Clean Water Act - Priority Pollutants
[present]

EPA - Carcinogen Hazard Ranking for RQ Adjustment
Hazard ranking - Low

RCRA - U Series Wastes
waste number U106

RCRA - Hazardous Constituents-Appendix VIII
waste number U106

RCRA - Basis for Listing - Appendix VII
Included in waste stream: F039

RCRA - Substances Banned From Land Disposal
[present]

RCRA - TSD Facilities Ground Water Monitoring
TM 8090 - 0.1 ug/L PQL; TM 8270 - 10 ug/L PQL

RCRA - Universal Treatment Standards (LDR)
WW: 0.55 mg/l; NWW: 28 mg/kg

INTERNATIONAL LISTS

Canada - NPRI (National Pollutant Release Inventory)
[present]

Mexico - Wastewater - Organic Toxic Pollutants and Heavy Metals
Listed under [Dinitrotoluenes]

STATE LISTS

California - Air Bill 2588 Appendix A-I
6/91

California - Directors List of Hazardous Substances (8 CCR 339)
[present]

Massachusetts Right To Know List
[present]

NJ Right to Know List (Total)
sn 0784

Pennsylvania Right to Know List
environmental hazard

PROPOSED REGULATIONS

Safe Drinking Water Act - Priority list
[present]

DINOCAP   39300-45-3

HEALTH AND SAFETY LISTS

U.S. DOT - Appendix A Table 1 - Hazardous Substances
final RQ - 10 pounds (4.54 kg) (Listed under 'Dinitrotoluene')

NIOSH - Selected LD50s and LC50s
Oral, rat: LD50 - 177 mg/kg

ENVIRONMENTAL LISTS

CERCLA/SARA - Hazardous Substances and their Reportable Quantities
final RQ - 10 pounds (4.54 kg) (Listed under 'Dinitrotoluene')

Clean Water Act - Hazardous Substances
[present] (Listed under 'Dinitrotoluene')

STATE LISTS

California - Directors List of Hazardous Substances (8 CCR 339)
[present] (Listed under 'Dinitrotoluenes, all isomers')

Massachusetts Right To Know List
[present]

Pennsylvania Right to Know List
environmental hazard

DI-N-NONANOYL PEROXIDE   762-13-0

HEALTH AND SAFETY LISTS

ACGIH 1995 - Time Weighted Averages
0.15 mg/m3 TWA

ACGIH 1995 - Skin Designations
skin - potential for cutaneous absorption

ACGIH 1995 - Carcinogens
A2-suspected human carcinogen

U.S. DOT - Substances From 49 CFR 172.101
regulated by DOT (UN2038, UN1600)

U.S. DOT - Hazard Classes
DOT hazard class - 6.1

U.S. DOT - Appendix A Table 1 - Hazardous Substances
final RQ - 10 pounds (4.54 kg)

NIOSH 1990 - Pocket Guide - RELs
1.5 mg/m3 TWA

NIOSH 1990 - Pocket Guide - IDLHs
200 mg/m3 IDLH (not considering carcinogenic effects)

NIOSH 1990 - Pocket Guide - Carcinogens
occupational carcinogen

NIOSH 1990 - Pocket Guide - Target organs
blood, liver, CVS

NIOSH 1990 - Pocket Guide - Skin list
Potential for dermal absorption

NIOSH - Health Standards - Exposure Limits
reduce exposure to lowest feasible concentration

NIOSH - Health Standards - Health Effects and Precautions
has produced tumors of the liver, skin, and kidneys in animals; reproductive effects (Prevent skin contact)

NIOSH - Health Standards - Carcinogenic Chemicals
potential human carcinogen

OSHA - Vacated PELs - Time Weighted Averages
1.5 mg/m3 TWA

OSHA - Vacated PELs - Skin Designation
Prevent or reduce skin absorption

OSHA - Final PELs - Time Weighted Averages
1.5 mg/m3 TWA

OSHA - Final PELs - Skin Notations
prevent or reduce skin absorption

ENVIRONMENTAL LISTS

ATSDR Priority List
Rank (of 275): 261

CERCLA/SARA - Section 313 - Emission Reporting
form R reporting required for 1.0% de minimus concentration

CERCLA/SARA - Hazardous Substances and their Reportable Quantities
final RQ = 10 pounds (4.54 kg)

Clean Water Act - Hazardous Substances
[present] (Listed under 'Dinitrotoluene')

Clean Water Act - Toxic Pollutants
[present]

EPA - Carcinogen Hazard Ranking for RQ Adjustment
Hazard ranking = Medium

INTERNATIONAL LISTS

Canada - WHMIS: Ingredient Disclosure
1% item 651 (765)

Canada - NPRI (National Pollutant Release Inventory)
[present]

Canada - British Columbia - 8 Hour Exposure Limits
1.5 mg/m3 TWA

Canada - British Columbia - 15 Minute Exposure Limits
5 mg/m3 STEL

Canada - British Columbia - Skin Notations
skin - potential for skin absorption

Canada - Ontario - OHSA - TWAEVs
1.5 mg/m3 TWAEV

Canada - Ontario - OHSA - Skin Notations
absorption through skin, eyes, or mucous membranes

Canada - Quebec - Time-Weighted Average Exposure Values
1.5 mg/m3 TWAEV

Canada - Quebec - Skin Designations
absorbed through the skin

German (DFG) - Skin/Sensitizers
danger of cutaneous absorption

German (DFG) - Carcinogens
animal evidence of carcinogenicity

Israel - Time Weighted Averages
1.5 mg/m3 TWA

Israel - Action Levels
0.75 mg/m3 AL

STATE LISTS

California - Air Bill 2588 Appendix A-I
6/91

California - Directors List of Hazardous Substances (8 CCR 339)
[present]

Massachusetts Right To Know List
[present]

NJ Right to Know List (Total)
sn 2985

Pennsylvania Right to Know List
environmental hazard

DINONYLPHENOL, ETHOXYLATED, PHOSPHATED   39464-64-7

HEALTH AND SAFETY LISTS

U.S. DOT - Hazard Classes
Forbidden from transport by the DOT

O,P-DINONYLPHENOL, ETHOXYLATED, PHOSPHATED, AMMONIUM SALT   RR-01067-3

HEALTH AND SAFETY LISTS

U.S. DOT - Hazard Classes
Forbidden from transport by the DOT

O,P-DINONYLPHENOL, ETHOXYLATED, PHOSPHATED, CALCIUM SALT   RR-01068-4

HEALTH AND SAFETY LISTS

U.S. DOT - Hazard Classes
Forbidden from transport by the DOT

O,P-DINONYLPHENOL, ETHOXYLATED, PHOSPHATED, POTASSIUM SALT   72067-21-1

HEALTH AND SAFETY LISTS

U.S. DOT - Hazard Classes
Forbidden from transport by the DOT

O,P-DINONYLPHENOL, ETHOXYLATED, PHOSPHATED, SODIUM SALT   70903-62-7

HEALTH AND SAFETY LISTS

U.S. DOT - Appendix B - Marine Pollutants
DOT regulated marine pollutant

O,P-DINONYLPHENOL, ETHOXYLATED, PHOSPHATED, ZINC SALT   RR-01069-5

ENVIRONMENTAL LISTS

CERCLA/SARA - Section 313 - Emission Reporting
form R reporting required

STATE LISTS

California - Air Bill 2588 Appendix A-II
9/90

California - Prop. 65 - Developmental Toxicity
developmental toxicity - initial date 4/1/90

DI(NONYL-, UNDECYL-) PHTHALATE (BRANCHED AND LINEAR ISOMERS)   111381-91-0

HEALTH AND SAFETY LISTS

U.S. DOT - Organic Peroxides Table
Organic peroxide UN3116

STATE LISTS

NJ Right to Know List (Total)
sn 0785

DINOTERB   1420-07-1
  ENVIRONMENTAL LISTS
    List of Pesticide Product Inert Ingredients
      [present]

DIOCO ALKYLAMINE   61789-76-2
  ENVIRONMENTAL LISTS
    List of Pesticide Product Inert Ingredients
      [present]

DI-N-OCTANOYL PEROXIDE   762-16-3
  ENVIRONMENTAL LISTS
    List of Pesticide Product Inert Ingredients
      [present]

P,P'-DIOCTYLDIPHENYLAMINE   101-67-7
  ENVIRONMENTAL LISTS
    List of Pesticide Product Inert Ingredients
      [present]

DIOCTYL ETHER   629-82-3
  ENVIRONMENTAL LISTS
    List of Pesticide Product Inert Ingredients
      [present]

DIOCTYL MALEATE   2915-53-9
  ENVIRONMENTAL LISTS
    List of Pesticide Product Inert Ingredients
      [present]

DI-N-OCTYL PHTHALATE   117-84-0
  ENVIRONMENTAL LISTS
    TSCA - Code of Federal Regulations Citations
      40 CFR 799.5025

DIOCTYL SODIUM SULFOSUCCINATE   577-11-7
  HEALTH AND SAFETY LISTS
    NIOSH - Selected LD50s and LC50s
      Oral, rat: LD50 = 62 mg/kg Skin, guinea pig: LD50 = 150 mg/kg
  ENVIRONMENTAL LISTS
    CERCLA/SARA - Section 302 Extremely Hazardous Substances and TPQs
      TPQ = 500/10,000 pounds
  STATE LISTS
    Florida Hazardous Substance List
      effective March 13, 1992
    Massachusetts Right To Know List
      extraordinarily hazardous
    NJ Right to Know List (Total)
      sn 2355
    Pennsylvania Right to Know List
      environmental hazard
  PROPOSED REGULATIONS
    CERCLA/SARA - Proposed Hazardous Substance Additions
      proposed RQ = 1 pound (.454 kg)
    CERCLA/SARA - 1989 Proposed RQ Adjustments
      proposed RQ = 100 pounds (45.4 kg)

DIOCTYL SODIUM SULFOSUCCINATE   1639-66-3
  ENVIRONMENTAL LISTS
    List of Pesticide Product Inert Ingredients
      [present]

DIOCTYL SODIUM SULFOSUCCINATE   20727-33-7
  HEALTH AND SAFETY LISTS
    U.S. DOT - Organic Peroxides Table
      Organic peroxide UN3114
  STATE LISTS
    NJ Right to Know List (Total)
      sn 0786

DIOCTYL TEREPHTHALATE   4654-26-6
  HEALTH AND SAFETY LISTS
    NIOSH - Selected LD50s and LC50s
      Oral, rat: LD50 = 8000 mg/kg

DIOCTYL TIN   26401-97-8
  HEALTH AND SAFETY LISTS
    NFPA - Flash Points
      flash point > 212 degrees F (100 degrees C)
    NFPA - Hazard Identification Ratings
      health-0; flammability-1; reactivity-0

DIOCTYLTIN MALEATE   16091-18-2
  ENVIRONMENTAL LISTS
    List of Pesticide Product Inert Ingredients
      [present]

DIOCTYLTIN BIS(2-ETHYLHEXYL THIOGLYCO-   15571-58-1
LATE)
  SEE ALSO:
    F039-HAZARDOUS WASTES
    ALKYL PHTHALATES
  HEALTH AND SAFETY LISTS
    U.S. DOT - Appendix A Table 1 - Hazardous Substances
      final RQ = 5000 pounds (2270 kg)
    NFPA - Flash Points
      flash point = 420 degrees F (215 degrees C)
    NFPA - Hazard Identification Ratings
      health-0; flammability-1; reactivity-0
    NIOSH - Selected LD50s and LC50s
      Oral, mouse: LD50 = 6513 mg/kg
  ENVIRONMENTAL LISTS
    ATSDR Priority List
      Rank (of 275): 249
    CERCLA/SARA - Hazardous Substances and their Reportable Quantities
      final RQ = 5000 pounds (2270 kg)
    Clean Water Act - Priority Pollutants
      [present]
    List of Pesticide Product Inert Ingredients
      [present]
    RCRA - U Series Wastes
      waste number U107
    RCRA - Hazardous Constituents-Appendix VIII
      waste number U107
    RCRA - Basis for Listing - Appendix VII
      Included in waste stream: F039
    RCRA - Substances Banned From Land Disposal
      [present]
    RCRA - TSD Facilities Ground Water Monitoring
      TM 8060 = 30 ug/L PQL; TM 8270 = 10 ug/L PQL
    RCRA - Universal Treatment Standards (LDR)
      WW: 0.017 mg/l; NWW: 28 mg/kg
    TSCA - Code of Federal Regulations Citations
      40 CFR 712.30(d); 40 CFR 716.120(c)

TSCA - PAIR - Reporting List
Reporting Date: November 19, 1982

INTERNATIONAL LISTS

Canada - WHMIS: Ingredient Disclosure
1% item 653 (1423)

Canada - NPRI (National Pollutant Release Inventory)
[present]

Canada - CEPA - Priority Substances List
estimated time for completion of assessment reports: 4 years

Mexico - Wastewater - Organic Toxic Pollutants and Heavy Metals
Listed under [Phthalate Esters]

STATE LISTS

California - Air Bill 2588 Appendix A-II
6/91

California - Directors List of Hazardous Substances (8 CCR 339)
[present]

Massachusetts Right To Know List
[present]

NJ Right to Know List (Total)
sn 0787

Pennsylvania Right to Know List
environmental hazard

DIOCTYLTIN BIS(ISOOCTYL MALEATE)   33568-99-9

HEALTH AND SAFETY LISTS

NIOSH - Selected LD50s and LC50s
Oral, rat: LD50 = 1900 mg/kg

ENVIRONMENTAL LISTS

List of Pesticide Product Inert Ingredients
[present]

INTERNATIONAL LISTS

Canada - WHMIS: Ingredient Disclosure
1% item 654 (1542)

DIOCTYLTIN DICHLORIDE   3542-36-7

ENVIRONMENTAL LISTS

List of Pesticide Product Inert Ingredients
[present]

DIOCTYLTIN OXIDE   870-08-6

ENVIRONMENTAL LISTS

List of Pesticide Product Inert Ingredients
[present]

DIOXACARB   6988-21-2

ENVIRONMENTAL LISTS

List of Pesticide Product Inert Ingredients
[present]

1,4-DIOXANE   123-91-1
SEE ALSO:
TIN

HEALTH AND SAFETY LISTS

NIOSH - Selected LD50s and LC50s
Oral, rat: LD50 = 1277 mg/kg

INTERNATIONAL LISTS

German (DFG) - MAK Values
total dust: 0.1 mg/m3 MAK (Listed under 'Tin compounds, organic')

German (DFG) - Peak Limitations
2 x normal MAK (30 min. average value); don't exceed 4 times during shift (Listed under 'Tin compounds, organic')

German (DFG) - Skin/Sensitizers
danger of cutaneous absorption (Listed under 'Tin compounds, organic')

1,5-DIOXASPIRO [5.5] UNDECANE-3,3-DICAR-   110843-98-6
BOXYLIC ACID BIS (1,2,2,6,6-PENTAMETHYL-4-
PIPERIDINYL) ESTER
SEE ALSO:
TIN

HEALTH AND SAFETY LISTS

NIOSH - Selected LD50s and LC50s
Oral, rat: LD50 = 4500 mg/kg

INTERNATIONAL LISTS

German (DFG) - MAK Values
total dust: 0.1 mg/m3 MAK (Listed under 'Tin compounds, organic')

German (DFG) - Peak Limitations
2 x normal MAK (30 min. average value); don't exceed 4 times during shift (Listed under 'Tin compounds, organic')

German (DFG) - Skin/Sensitizers
danger of cutaneous absorption (Listed under 'Tin compounds, organic')

DIOXATHION   78-34-2

INTERNATIONAL LISTS

German (DFG) - MAK Values
total dust: 0.1 mg/m3 MAK (Listed under 'Tin compounds, organic')

German (DFG) - Peak Limitations
2 x normal MAK (30 min. average value); don't exceed 4 times during shift (Listed under 'Tin compounds, organic')

German (DFG) - Skin/Sensitizers
danger of cutaneous absorption (Listed under 'Tin compounds, organic')

DIOXINS, POLYHALOGENATED DIBENZO-P-   RR-01476-6

INTERNATIONAL LISTS

German (DFG) - MAK Values
total dust: 0.1 mg/m3 MAK (Listed under 'Tin compounds, organic')

German (DFG) - Peak Limitations
2 x normal MAK (30 min. average value); don't exceed 4 times during shift (Listed under 'Tin compounds, organic')

German (DFG) - Skin/Sensitizers
danger of cutaneous absorption (Listed under 'Tin compounds, organic')

DIOXOLANE   100-79-8

INTERNATIONAL LISTS

German (DFG) - MAK Values
total dust: 0.1 mg/m3 MAK (Listed under 'Tin compounds, organic')

German (DFG) - Peak Limitations
2 x normal MAK (30 min. average value); don't exceed 4 times during shift (Listed under 'Tin compounds, organic')

German (DFG) - Skin/Sensitizers
danger of cutaneous absorption (Listed under 'Tin compounds, organic')

1,3-DIOXOLANE   646-06-0

INTERNATIONAL LISTS

German (DFG) - MAK Values
total dust: 0.1 mg/m3 MAK (Listed under 'Tin compounds, organic')

German (DFG) - Peak Limitations
2 x normal MAK (30 min. average value); don't exceed 4 times during shift (Listed under 'Tin compounds, organic')

German (DFG) - Skin/Sensitizers
danger of cutaneous absorption (Listed under 'Tin compounds, organic')

DIPENTENE   **138-86-3**

HEALTH AND SAFETY LISTS

U.S. DOT - Appendix B - Marine Pollutants
DOT regulated marine pollutant

DIPEROXYAZELAIC ACID   **1941-79-3**
SEE ALSO:
F039-HAZARDOUS WASTES

HEALTH AND SAFETY LISTS

ACGIH 1995 - Time Weighted Averages
25 ppm TWA; 90 mg/m3 TWA

ACGIH 1995 - Skin Designations
skin - potential for cutaneous absorption

AIHA - Odor Threshold Values
geometric mean air odor threshold = 12 ppm (detectable); 22 ppm
(recognizable)

U.S. DOT - Substances From 49 CFR 172.101
regulated by DOT (UN1165)

U.S. DOT - Hazard Classes
DOT hazard class = 3

U.S. DOT - Appendix A Table 1 - Hazardous Substances
final RQ = 100 pounds (45.4 kg)

IARC - Group 2B (sufficient animal data)
[present]

NFPA - Flash Points
flash point = 54 degrees F (12 degrees C)

NFPA - Hazard Identification Ratings
health-2; flammability-3; reactivity-1

NIOSH - Selected LD50s and LC50s
Inhalation, rat: LC50 = 46 gm/m3 2 hr Oral, rat: LD50 = 4200
mg/kg Skin, rabbit: LD50 = 7600 mg/kg

NIOSH 1990 - Pocket Guide - RELs
C 1 ppm (30 min); C 3.6 mg/m3 (30 min)

NIOSH 1990 - Pocket Guide - IDLHs
2000 ppm IDLH (not considering carcinogenic effects)

NIOSH 1990 - Pocket Guide - Carcinogens
occupational carcinogen

NIOSH 1990 - Pocket Guide - Target organs
liver, kidneys, skin, eyes

NIOSH - Health Standards - Exposure Limits
C (30 min) 1 ppm; C (30 min) 3.6 mg/m3

NIOSH - Health Standards - Health Effects and Precautions
has produced tumors of liver, lungs, and nasal cavity in animals;
effects on liver and kidney (Blood and urine monitoring required;
prevent skin contact)

NIOSH - Health Standards - Carcinogenic Chemicals
potential human carcinogen

NTP Chemical Status Reports - Testing Status and NTIS Number
Technical reports printed (PB285711/AS)

NTP Chemical Status Reports - Evidence of Carcinogenicity
male rat-positive; female rat-positive; male mice-positive; female
mice-positive

NTP Seventh Report - Suspect Carcinogens
suspect carcinogen

OSHA - Vacated PELs - Time Weighted Averages
25 ppm TWA; 90 mg/m3 TWA

OSHA - Vacated PELs - Skin Designation
Prevent or reduce skin absorption

OSHA - Final PELs - Time Weighted Averages
100 ppm TWA; 360 mg/m3

OSHA - Final PELs - Skin Notations
prevent or reduce skin absorption

OSHA - Possible Select Carcinogens
[present]

ENVIRONMENTAL LISTS

ATSDR Priority List
Rank (of 275): 245

CERCLA/SARA - Section 313 - Emission Reporting
form R reporting required for 0.1% de minimus concentration

CERCLA/SARA - Hazardous Substances and their Reportable Quantities
final RQ = 100 pounds (45.4 kg)

Clean Air Act (1990) - List of Hazardous Air Contaminants
[present]

CAA - HON Rule - SOCMI Chemicals
compliance by April 24, 1995

CAA - HON Rule - Organic HAPs
[present]

EPA - Carcinogen Hazard Ranking for RQ Adjustment
Hazard ranking = Low

RCRA - U Series Wastes
waste number U108

RCRA - Hazardous Constituents-Appendix VIII
waste number U108

RCRA - Basis for Listing - Appendix VII
Included in waste stream: F039

RCRA - Substances Banned From Land Disposal
[present]

RCRA - TSD Facilities Ground Water Monitoring
TM 8015 = 150 ug/L PQL

RCRA - Universal Treatment Standards (LDR)
WW: Not applicable; NWW: 170 mg/kg

INTERNATIONAL LISTS

Australian Exposure Standards - Time Weighted Averages
25 ppm TWA; 90 mg/m3 TWA

Australian Exposure Standards - Skin Effects
skin absorption

Canada - WHMIS: Ingredient Disclosure
0.1% item 655 (767)

Canada - NPRI (National Pollutant Release Inventory)
[present]

Canada - Alberta - 8 Hour Occupational Exposure Limit
25 ppm TWA; 90 mg/m3 TWA

Canada - Alberta - 15 Minute Occupational Exposure Limit
100 ppm STEL; 360 mg/m3 STEL

Canada - Alberta - Skin Designation
can be absorbed through the intact skin

Canada - British Columbia - 8 Hour Exposure Limits
50 ppm TWA; 180 mg/m3 TWA

Canada - British Columbia - Skin Notations
skin - potential for skin absorption

Canada - Ontario - OHSA - TWAEVs
25 ppm TWAEV; 90 mg/m3 TWAEV

Canada - Ontario - OHSA - Skin Notations
absorption through skin, eyes, or mucous membranes

Canada - Quebec - Time-Weighted Average Exposure Values
25 ppm TWAEV; 90 mg/m3 TWAEV

Canada - Quebec - Skin Designations
absorbed through the skin

Canada - Quebec - Carcinogens
C3 carcinogen: effect detected in animals

United Kingdom - Occupational Exposure Standards - TWAs
25 ppm TWA; 90 mg/m3 TWA

United Kingdom - Occupational Exposure Standards - STELs
100 ppm STEL; 360 mg/m3 STEL

United Kingdom - Occupational Exposure Standards - Notes
can be absorbed through skin

German (DFG) - MAK Values
50 ppm MAK; 180 mg/m3 MAK

German (DFG) - Peak Limitations
2 x normal MAK (30 min. average value); don't exceed 4 times during shift

German (DFG) - Skin/Sensitizers
danger of cutaneous absorption

German (DFG) - Carcinogens
suspected carcinogen

German (DFG) - Pregnancy
classification not yet possible

Israel - Time Weighted Averages
25 ppm TWA; 90 mg/m3 TWA

Israel - Action Levels
12.5 ppm AL; 45 mg/m3 AL

Mexico - Instruction No. 10 - TWAs
25 ppm TWA; 50 mg/m3 TWA

Mexico - Instruction No. 10 - STELs
100 ppm STEL; 360 mg/m3 STEL

Mexico - Instruction No. 10 - Skin designation
skin - potential for cutaneous absorption

STATE LISTS

California - Air Bill 2588 Appendix A-I
known or potential carcinogen

California - Prop. 65 - Cancer list
carcinogen - initial date 1/1/88

California - Prop. 65 - No Significant Risk Levels
no significant risk level = 30 ug/day

California - Exposure Limits - PELs
25 ppm PEL; 90 mg/m3 PEL

California - Exposure Limits - Skin Notation
material may be absorbed through the skin, eyes or mucous membrane

California - Directors List of Hazardous Substances (8 CCR 339)
[present]

Florida Hazardous Substance List
[present]

Massachusetts Right To Know List
carcinogen; extraordinarily hazardous

Minnesota Hazardous Substance List
carcinogen; skin

NJ Right to Know List (Total)
sn 0789

NJ Special Hazardous Substances
(carcinogen; flammable - third degree)

Pennsylvania Right to Know List
environmental hazard; special hazardous substance

Pennsylvania RTK - Special Hazardous Substances
[present]

PROPOSED REGULATIONS

Canada - Ontario - Proposed Occupational TWAEVs
5 ppm TWAEV; 18 mg/m3 TWAEV

DIPEROXY DODECANE DIACID   RR-01756-1

ENVIRONMENTAL LISTS

TSCA - Section 12(b) - Export Notification
P-91-1361; export notification required - Section 5

DIPEROXY KETAL   RR-01655-7

HEALTH AND SAFETY LISTS

ACGIH 1995 - Time Weighted Averages
0.2 mg/m3 TWA

ACGIH 1995 - Skin Designations
skin - potential for cutaneous absorption

U.S. DOT - Appendix B - Marine Pollutants
DOT regulated marine pollutant

NIOSH - Selected LD50s and LC50s
Inhalation, rat: LC50 = 1398 mg/m3 1 hr Oral, rat: LD50 = 20 mg/kg Skin, rat: LD50 = 63 mg/kg

NTP Chemical Status Reports - Testing Status and NTIS Number
Technical reports printed (PB286185/AS)

NTP Chemical Status Reports - Evidence of Carcinogenicity
male rat-negative; female rat-negative; male mice-negative; female mice-negative

OSHA - Vacated PELs - Time Weighted Averages
0.2 mg/m3 TWA

OSHA - Vacated PELs - Skin Designation
Prevent or reduce skin absorption

ENVIRONMENTAL LISTS

CERCLA/SARA - Section 302 Extremely Hazardous Substances and TPQs
TPQ = 500 pounds

INTERNATIONAL LISTS

Australian Exposure Standards - Time Weighted Averages
0.2 mg/m3 TWA

Australian Exposure Standards - Skin Effects
skin absorption

Canada - Alberta - 8 Hour Occupational Exposure Limit
0.2 mg/m3 TWA

Canada - Alberta - 15 Minute Occupational Exposure Limit
0.6 mg/m3 STEL

Canada - Alberta - Skin Designation
can be absorbed through the intact skin

Canada - British Columbia - 8 Hour Exposure Limits
0.2 mg/m3 TWA

Canada - British Columbia - Skin Notations
skin - potential for skin absorption

Canada - Ontario - OHSA - TWAEVs
0.2 mg/m3 TWAEV

Canada - Ontario - OHSA - Skin Notations
absorption through skin, eyes, or mucous membranes

Canada - Quebec - Time-Weighted Average Exposure Values
0.2 mg/m3 TWAEV

Canada - Quebec - Skin Designations
absorbed through the skin

United Kingdom - Occupational Exposure Standards - TWAs
0.2 mg/m3 TWA

United Kingdom - Occupational Exposure Standards - Notes
can be absorbed through skin

Israel - Time Weighted Averages
0.2 mg/m3 TWA

Israel - Action Levels
0.1 mg/m3 AL

Mexico - Instruction No. 10 - TWAs
0.2 mg/m3 TWA

Mexico - Instruction No. 10 - Skin designation
skin - potential for cutaneous absorption

STATE LISTS

California - Exposure Limits - PELs
0.2 mg/m3 PEL

California - Exposure Limits - Skin Notation
material may be absorbed through the skin, eyes or mucous membrane

California - Directors List of Hazardous Substances (8 CCR 339)
[present]

Florida Hazardous Substance List
[present]

Massachusetts Right To Know List
extraordinarily hazardous; neurotoxin

Minnesota Hazardous Substance List
skin

NJ Right to Know List (Total)
sn 0790

Pennsylvania Right to Know List
environmental hazard

PROPOSED REGULATIONS

CERCLA/SARA - Proposed Hazardous Substance Additions
proposed RQ = 1 pound (.454 kg)

CERCLA/SARA - 1989 Proposed RQ Adjustments
proposed RQ = 100 pounds (45.4 kg)

DIPHACINONE   82-66-6

ENVIRONMENTAL LISTS

EPA - Master Testing List
[present]

STATE LISTS

California - Air Bill 2588 Appendix A-I
total, with or without isomers reported: known or potential carcinogen

DIPHENAMIDE   957-51-7

HEALTH AND SAFETY LISTS

NFPA - Flash Points
flash point = 35 degrees F (2 degrees C)

NFPA - Hazard Identification Ratings
health-2; flammability-3; reactivity-2

STATE LISTS

Florida Hazardous Substance List
[present]

DIPHENHYDRAMINE HYDROCHLORIDE   147-24-0

HEALTH AND SAFETY LISTS

U.S. DOT - Substances From 49 CFR 172.101
regulated by DOT (UN1166)

U.S. DOT - Hazard Classes
DOT hazard class = 3

NIOSH - Selected LD50s and LC50s
Inhalation, rat: LC50 = 20650 mg/m3 (4 hr) Oral, rat: LD50 = 3000 mg/kg Skin, rabbit: LD50 = 8480 mg/kg

ENVIRONMENTAL LISTS

CAA - HON Rule - SOCMI Chemicals
compliance by Oct. 24, 1994

TSCA - Code of Federal Regulations Citations
40 CFR 712.30(g); 40 CFR 716.120(a)

TSCA - PAIR - Reporting List
Reporting Date: October 8, 1984

TSCA - Health and Safety Reporting List
Effective Date: January 3, 1983

INTERNATIONAL LISTS

Canada - WHMIS: Ingredient Disclosure
1% item 656 (768)

STATE LISTS

Massachusetts Right To Know List
[present]

NJ Right to Know List (Total)
sn 0791

NJ Special Hazardous Substances
(flammable - third degree; reactive - second degree)

Pennsylvania Right to Know List
[present]

1,4-DIPHENOXYBENZENE   3061-36-

HEALTH AND SAFETY LISTS

U.S. DOT - Substances From 49 CFR 172.101
regulated by DOT (UN2052)

U.S. DOT - Hazard Classes
DOT hazard class = 3

NFPA - Flash Points
flash point = 113 degrees F (45 degrees C)

NFPA - Hazard Identification Ratings
health-0; flammability-2; reactivity-0

ENVIRONMENTAL LISTS

List of Pesticide Product Inert Ingredients
[present]

INTERNATIONAL LISTS

Canada - WHMIS: Ingredient Disclosure
1% item 657 (777)

STATE LISTS

NJ Right to Know List (Total)
sn 0792

DI-(2 PHENOXYETHYL) PEROXYDICARBONATE   RR-01757-2

HEALTH AND SAFETY LISTS

U.S. DOT - Organic Peroxides Table
Organic peroxide UN3116

STATE LISTS

NJ Right to Know List (Total)
sn 0793

DIPHENYLACETIC ACID   117-34-0

HEALTH AND SAFETY LISTS

U.S. DOT - Organic Peroxides Table
Organic peroxide UN3116;   exempt

DIPHENYLAMINE   122-39-4

ENVIRONMENTAL LISTS

TSCA - Chemicals with Significant New Use Rules
PMN number: P-92-1394

DIPHENYLAMINECHLORO-ARSINE   578-94-9

HEALTH AND SAFETY LISTS

U.S. DOT - Appendix B - Marine Pollutants
DOT regulated marine pollutant

NIOSH - Selected LD50s and LC50s
Oral, rat: LD50 = 1500 ug/kg

ENVIRONMENTAL LISTS

CERCLA/SARA - Section 302 Extremely Hazardous Substances and TPQs
TPQ = 10/10,000 pounds

STATE LISTS

Florida Hazardous Substance List
effective March 13, 1992

Massachusetts Right To Know List
extraordinarily hazardous

Pennsylvania Right to Know List
environmental hazard

PROPOSED REGULATIONS

CERCLA/SARA - Proposed Hazardous Substance Additions
proposed RQ = 1 pound (.454 kg)

CERCLA/SARA - 1989 Proposed RQ Adjustments
proposed RQ = 10 pounds (4.54 kg)

1,1-DIPHENYLBUTANE   RR-00120-7

ENVIRONMENTAL LISTS

CERCLA/SARA - Section 313 - Emission Reporting
form R reporting required

DIPHENYLCHLOROARSINE   712-48-1

HEALTH AND SAFETY LISTS

NTP Chemical Status Reports - Testing Status and NTIS Number
Technical reports printed (PB90219437/AS)

NTP Chemical Status Reports - Evidence of Carcinogenicity
male rat-equivocal evidence; female rat-equivocal evidence; male mice-no evidence; female mice-no evidence

DIPHENYL DECYL PHOSPHITE   3287-06-7

ENVIRONMENTAL LISTS

TSCA - PAIR - Reporting List
Reporting date: June 10, 1993

TSCA - Health and Safety Reporting List
Effective Date: April 12, 1993

2,4'-DIPHENYLDIAMINE   492-17-1

HEALTH AND SAFETY LISTS

U.S. DOT - Organic Peroxides Table
Organic peroxide UN3102; UN3106

DIPHENYL DICHLOROSILANE   80-10-4

HEALTH AND SAFETY LISTS

NIOSH - Selected LD50s and LC50s
Oral, rat: LD50 = 5540 mg/kg

DIPHENYLDODECYL PHOSPHITE   RR-00853-7
SEE ALSO:
K083-HAZARDOUS WASTES
K104-HAZARDOUS WASTES

HEALTH AND SAFETY LISTS

ACGIH 1995 - Time Weighted Averages
10 mg/m3 TWA

AIHA - Odor Threshold Values
no geometric mean air odor threshold

NFPA - Flash Points
flash point = 307 degrees F (153 degrees C)

NFPA - Hazard Identification Ratings
health-3; flammability-1; reactivity-0

NIOSH - Selected LD50s and LC50s
Oral, guinea pig: LD50 = 300 mg/kg

OSHA - Vacated PELs - Time Weighted Averages
10 mg/m3 TWA

ENVIRONMENTAL LISTS

CERCLA/SARA - Section 313 - Emission Reporting
form R reporting required

CAA - HON Rule - SOCMI Chemicals
compliance by April 24, 1995

RCRA - Hazardous Constituents-Appendix VIII
hazardous constituent - no waste number

RCRA - Basis for Listing - Appendix VII
Included in waste streams: K083, K104

RCRA - TSD Facilities Ground Water Monitoring
TM 8270 = 10 ug/L PQL

RCRA - Universal Treatment Standards (LDR)
WW: 0.92 mg/l; NNW: 13 mg/kg (difficult to distinguish from diphenylnitrosamine)

TSCA - Health and Safety Reporting List
Effective Date: March 11, 1994; Sunset Date: March 11, 2004

INTERNATIONAL LISTS

Australian Exposure Standards - Time Weighted Averages
10 mg/m3 TWA

Canada - WHMIS: Ingredient Disclosure
0.1% item 659 (778)

Canada - Alberta - 8 Hour Occupational Exposure Limit
10 mg/m3 TWA

Canada - Alberta - 15 Minute Occupational Exposure Limit
20 mg/m3 STEL

Canada - British Columbia - 8 Hour Exposure Limits
10 mg/m3 TWA

Canada - British Columbia - 15 Minute Exposure Limits
20 mg/m3 STEL

Canada - Ontario - OHSA - TWAEVs
10 mg/m3 TWAEV

Canada - Quebec - Time-Weighted Average Exposure Values
10 mg/m3 TWAEV

United Kingdom - Occupational Exposure Standards - TWAs
10 mg/m3 TWA

United Kingdom - Occupational Exposure Standards - STELs
20 mg/m3 STEL

Israel - Time Weighted Averages
10 mg/m3 TWA

Israel - Action Levels
5 mg/m3 AL

Mexico - Instruction No. 10 - TWAs
10 mg/m3 TWA

Mexico - Instruction No. 10 - STELs
20 mg/m3 STEL

STATE LISTS

California - Exposure Limits - PELs
10 mg/m3 PEL

California - Directors List of Hazardous Substances (8 CCR 339)
[present]

Florida Hazardous Substance List
[present]

Massachusetts Right To Know List
[present]

Minnesota Hazardous Substance List
[present]

Pennsylvania Right to Know List
environmental hazard

PROPOSED REGULATIONS

TSCA - ITC 32nd Report Priority Testing List
designated for dermal absorption testing

1,2-DIPHENYLETHANE (SYM)   103-29-7
SEE ALSO:
ARSENIC

HEALTH AND SAFETY LISTS

U.S. DOT - Substances From 49 CFR 172.101
regulated by DOT (UN1698)

U.S. DOT - Hazard Classes
DOT hazard class = 6.1

U.S. DOT - Appendix B - Marine Pollutants
DOT regulated severe marine pollutant

STATE LISTS

NJ Right to Know List (Total)
sn 0797

1,1-DIPHENYLETHANE (UNS)   38888-98-1

HEALTH AND SAFETY LISTS

NFPA - Flash Points
flash point > 212 degrees F (100 degrees C)

NFPA - Hazard Identification Ratings
health-0; flammability-1; reactivity-0

DIPHENYL ETHER/BIPHENYL MIXTURE   RR-00209-5

SEE ALSO:
ARSENIC

HEALTH AND SAFETY LISTS

U.S. DOT - Substances From 49 CFR 172.101
regulated by DOT (UN1699)

U.S. DOT - Hazard Classes
DOT hazard class = 6.1

U.S. DOT - Appendix B - Marine Pollutants
DOT regulated severe marine pollutant

STATE LISTS

NJ Right to Know List (Total)
sn 0798

DIPHENYL ETHER/BIPHENYL MIXTURE   RR-01748-1

ENVIRONMENTAL LISTS

List of Pesticide Product Inert Ingredients
[present]

1,3-DIPHENYLGUANIDINE   102-06-7

HEALTH AND SAFETY LISTS

IARC - Group 3 (not classifiable)
[present]

DIPHENYLHYDANTOIN, SODIUM SALT   630-93-3

HEALTH AND SAFETY LISTS

U.S. DOT - Substances From 49 CFR 172.101
regulated by DOT (UN1769)

U.S. DOT - Hazard Classes
DOT hazard class = 8

NFPA - Flash Points
flash point = 288 degrees F (142 degrees C)

NFPA - Hazard Identification Ratings
health-3; flammability-1; reactivity-0

INTERNATIONAL LISTS

Canada - WHMIS: Ingredient Disclosure
1% item 660 (779)

STATE LISTS

Florida Hazardous Substance List
[present]

Massachusetts Right To Know List
[present]

NJ Right to Know List (Total)
sn 0799

NJ Special Hazardous Substances
(corrosive)

Pennsylvania Right to Know List
[present]

DIPHENYLHYDRAZINE   55299-18-8

HEALTH AND SAFETY LISTS

NFPA - Flash Points
flash point = 425 degrees F (218 degrees C)

NFPA - Hazard Identification Ratings
health-0; flammability-1; reactivity-0

DIPHENYL ISODECYL PHOSPHITE   26544-23-

HEALTH AND SAFETY LISTS

NFPA - Flash Points
flash point = 264 degrees F (129 degrees C)

NFPA - Hazard Identification Ratings
health-0; flammability-1; reactivity-0

DIPHENYLMERCURY   587-85-

HEALTH AND SAFETY LISTS

NFPA - Flash Points
flash point > 212 degrees F (100 degrees C)

NFPA - Hazard Identification Ratings
health-0; flammability-1; reactivity-0

DIPHENYLMETHANE   101-81-

INTERNATIONAL LISTS

German (DFG) - MAK Values
vapor: 1 ppm MAK; 7 mg/m3 MAK

2,4'-DIPHENYLMETHANE DIISOCYANATE   5873-54-

INTERNATIONAL LISTS

German (DFG) - MAK Values
fine dust: 4 mg/m3 MAK

DIPHENYLMETHANE DIISOCYANATE (MDI)   RR-01224-8
MODIFIED

HEALTH AND SAFETY LISTS

NIOSH - Selected LD50s and LC50s
Oral, rat: LD50 = 507 mg/kg

NTP Chemical Status Reports - Testing Status and NTIS Number
Post peer review technical reports in progress

DIPHENYL METHYL BROMIDE   776-74-9

STATE LISTS

California - Air Bill 2588 Appendix A-I
known or potential carcinogen

California - Prop. 65 - Cancer list
carcinogen - initial date 1/1/88

California - Directors List of Hazardous Substances (8 CCR 339)
[present]

Pennsylvania Right to Know List
special hazardous substance

Pennsylvania RTK - Special Hazardous Substances
[present]

DIPHENYLNITROSAMINE   86-30-6

STATE LISTS

California - Directors List of Hazardous Substances (8 CCR 339)
[present]

DIPHENYL OXIDE AND BIPHENYL PHENYL ETHER   RR-01286-2
MIXTURES

ENVIRONMENTAL LISTS

List of Pesticide Product Inert Ingredients
[present]

DIPHENYLOXIDE-4,4'-DISULFOHYDRAZIDE   80-51-3

SEE ALSO:
MERCURY

INTERNATIONAL LISTS

Canada - WHMIS: Ingredient Disclosure
1% item 662 (781)

1,1-DIPHENYLPENTANE   RR-00774-9

HEALTH AND SAFETY LISTS

NFPA - Flash Points
flash point = 266 degrees F (130 degrees C)

NFPA - Hazard Identification Ratings
health-1; flammability-1; reactivity-0

ENVIRONMENTAL LISTS

CAA - HON Rule - SOCMI Chemicals
compliance by Oct. 24, 1994

N,N'-DIPHENYL-P-PHENYLENEDIAMINE   74-31-7

ENVIRONMENTAL LISTS

TSCA - Code of Federal Regulations Citations
40 CFR 712.30(x); 40 CFR 716.120(a)

TSCA - PAIR - Reporting List
Reporting Date: December 27, 1990

TSCA - Health and Safety Reporting List
Effective Date: June 1, 1987; Sunset Date: November 9, 1993

DIPHENYL PHOSPHITE   4712-55-4

ENVIRONMENTAL LISTS

TSCA - Chemicals with Significant New Use Rules
PMN number: P-92-294

DIPHENYL PHTHALATE   84-62-8

HEALTH AND SAFETY LISTS

U.S. DOT - Substances From 49 CFR 172.101
regulated by DOT (UN1770)

U.S. DOT - Hazard Classes
DOT hazard class = 8

INTERNATIONAL LISTS

Canada - WHMIS: Ingredient Disclosure
1% item 664 (335)

STATE LISTS

NJ Right to Know List (Total)
sn 0801

NJ Special Hazardous Substances
(corrosive)

1,1-DIPHENYLPROPANE   RR-00627-9

HEALTH AND SAFETY LISTS

U.S. DOT - Appendix A Table 1 - Hazardous Substances
final RQ = 100 pounds (45.4 kg)

IARC - Group 3 (not classifiable)
[present]

NIOSH - Selected LD50s and LC50s
Oral, rat: LD50 = 1650 mg/kg

NTP Chemical Status Reports - Testing Status and NTIS Number
Technical reports printed (PB298275/AS)

NTP Chemical Status Reports - Evidence of Carcinogenicity
male rat-positive; female rat-positive; male mice-negative; female mice-negative

ENVIRONMENTAL LISTS

ATSDR Priority List
Rank (of 275): 162

CERCLA/SARA - Section 313 - Emission Reporting
form R reporting required for 1.0% de minimus concentration

CERCLA/SARA - Hazardous Substances and their Reportable Quantities
final RQ = 100 pounds (45.4 kg)

Clean Water Act - Priority Pollutants
[present]

RCRA - TSD Facilities Ground Water Monitoring
TM 8270 = 10 ug/L PQL

RCRA - Universal Treatment Standards (LDR)
WW: 0.92 mg/l; NWW: 13 mg/kg (difficult to distinguish from diphenylamine)

INTERNATIONAL LISTS

Canada - WHMIS: Ingredient Disclosure
0.1% item 1193 (1254)

Canada - NPRI (National Pollutant Release Inventory)
[present]

Mexico - Wastewater - Organic Toxic Pollutants and Heavy Metals
Listed under [Nitrosamines]

Mexico - Drinking Water - Ecological Criteria
0.05 mg/l Substance presents persistence, bioaccumulations or risk of cancer, reduce human exposure to a minimum; This level has been extrapolated by using a mathematic model

STATE LISTS

California - Air Bill 2588 Appendix A-II
known or potential carcinogen: 9/89

California - Prop. 65 - Cancer list
carcinogen - initial date 4/1/88

California - Prop. 65 - No Significant Risk Levels
no significant risk level = 80 ug/day

California - Directors List of Hazardous Substances (8 CCR 339)
[present]

Massachusetts Right To Know List
carcinogen; extraordinarily hazardous

NJ Right to Know List (Total)
sn 1408

Pennsylvania Right to Know List
environmental hazard

DIPHENYLSULFONE   127-63-9

HEALTH AND SAFETY LISTS

U.S. DOT - Appendix B - Marine Pollutants
DOT regulated marine pollutant

DIPHENYL-2,4,6-TRIMETHYLBENZOYL PHOSPHINE OXIDE   75980-60-8

HEALTH AND SAFETY LISTS

U.S. DOT - Substances From 49 CFR 172.101
regulated by DOT (UN2951)

U.S. DOT - Hazard Classes
DOT hazard class = 4.1

ENVIRONMENTAL LISTS

TSCA - Code of Federal Regulations Citations
40 CFR 712.30(x); 40 CFR 716.120(d)

TSCA - PAIR - Reporting List
Reporting Date: November 27, 1991

TSCA - Health and Safety Reporting List
Effective Date: September 30, 1991

DIPHENYL (O-XENYL) PHOSPHATE   RR-00852-6

HEALTH AND SAFETY LISTS

NFPA - Flash Points
flash point > 212 degrees F (100 degrees C)

NFPA - Hazard Identification Ratings
health-0; flammability-1; reactivity-0

DIPHOSPHORIC ACID, DISODIUM SALT   7758-16-9

HEALTH AND SAFETY LISTS

NIOSH - Selected LD50s and LC50s
Oral, rat: LD50 = 2370 mg/kg

INTERNATIONAL LISTS
Canada - WHMIS: Ingredient Disclosure
0.1% item 665 (782)

DIPICRYL SULFIDE   2217-06-3

ENVIRONMENTAL LISTS
List of Pesticide Product Inert Ingredients
[present]

DIPOTASSIUM ENDOTHALL [7-OXABICYCLO(2.2.1)   2164-07-0
HEPTANE-2,3-DICARBOXYLIC ACID, DIPOTASSIUM
SALT]

HEALTH AND SAFETY LISTS
NFPA - Flash Points
flash point = 435 degrees F (224 degrees C)
NFPA - Hazard Identification Ratings
health-0;   flammability-1;   reactivity-0
NIOSH - Selected LD50s and LC50s
Oral, rat: LD50 = 8 gm/kg

DIPOTASSIUM HEXABROMOPLATINATE   16920-93-7

HEALTH AND SAFETY LISTS
NFPA - Flash Points
flash point > 212 degrees F (100 degrees C)
NFPA - Hazard Identification Ratings
health-0;   flammability-1;   reactivity-0

DIPOTASSIUM HEXACHLOROPALLADATE   16919-73-6

HEALTH AND SAFETY LISTS
NIOSH - Selected LD50s and LC50s
Oral, rat: LD50 = 1390 mg/kg
ENVIRONMENTAL LISTS
TSCA - Code of Federal Regulations Citations
40 CFR 712.30(x);   40 CFR 716.120(d)
TSCA - PAIR - Reporting List
Reporting Date: November 27, 1991
TSCA - Health and Safety Reporting List
Effective Date: September 30, 1991

DIPOTASSIUM PHOSPHATE   7758-11-4

ENVIRONMENTAL LISTS
TSCA - Section 12(b) - Export Notification
P-87-586;   export notification required - proposed Section 5

DIPOTASSIUM 12-SULFATO-9-OCTADECENOATE   53404-44-7

HEALTH AND SAFETY LISTS
NFPA - Flash Points
flash point = 437 degrees F (225 degrees C)
NFPA - Hazard Identification Ratings
health-0;   flammability-1;   reactivity-0

DIPOTASSIUM TETRABROMOPALLADATE   13826-93-2

HEALTH AND SAFETY LISTS
NIOSH - Selected LD50s and LC50s
Oral, mouse: LD50 = 2650 mg/kg
ENVIRONMENTAL LISTS
List of Pesticide Product Inert Ingredients
[present]

DIPROPYLALUMINUM HYDRIDE   2036-15-9

HEALTH AND SAFETY LISTS
U.S. DOT - Substances From 49 CFR 172.101
regulated by DOT (UN2852, UN0401)

U.S. DOT - Hazard Classes
DOT hazard class = 1.1D
STATE LISTS
NJ Right to Know List (Total)
sn 0802

DIPROPYLAMINE   142-84-7

ENVIRONMENTAL LISTS
CERCLA/SARA - Section 313 - Emission Reporting
form R reporting required

4-DIPROPYLAMINOBENZENEDIAZONIUM ZINC   RR-00766-9
CHLORIDE

INTERNATIONAL LISTS
Canada - WHMIS: Ingredient Disclosure
0.1% item 666 (943)

DIPROPYLENE GLYCOL   25265-71-8

INTERNATIONAL LISTS
Canada - WHMIS: Ingredient Disclosure
1% item 667 (950)

DIPROPYLENE GLYCOL DIBENZOATE   94-51-9

ENVIRONMENTAL LISTS
List of Pesticide Product Inert Ingredients
[present]

DIPROPYLENE GLYCOL METHYL ETHER   12002-25-4

ENVIRONMENTAL LISTS
List of Pesticide Product Inert Ingredients
[present]

DIPROPYLENE GLYCOL METHYL ETHER   20324-32-7

INTERNATIONAL LISTS
Canada - WHMIS: Ingredient Disclosure
1% item 668 (1568)

DIPROPYLENE GLYCOL MONOETHYL ETHER   30025-38-8

HEALTH AND SAFETY LISTS
NFPA - Flash Points
ignites spontaneously in air
NFPA - Hazard Identification Ratings
flammability-3;   reactivity-3 (do not use water, foam or halogenated extinguishing agents)
STATE LISTS
Florida Hazardous Substance List
[present]
Massachusetts Right To Know List
[present]
Pennsylvania Right to Know List
[present]

DIPROPYLENE GLYCOL MONOMETHYL ETHER   34590-94-8

HEALTH AND SAFETY LISTS
U.S. DOT - Substances From 49 CFR 172.101
regulated by DOT (UN2383)
U.S. DOT - Hazard Classes
DOT hazard class = 3
U.S. DOT - Appendix A Table 1 - Hazardous Substances
final RQ = 5000 pounds (2270 kg)
NFPA - Flash Points
flash point = 63 degrees F (17 degrees C)
NFPA - Hazard Identification Ratings
health-3;   flammability-3;   reactivity-0

NIOSH - Selected LD50s and LC50s
Inhalation, rat: LC50 = 4400 mg/m3 4 hr Oral, rat: LD50 = 930 mg/kg Skin, rabbit: LD50 = 1250 mg/kg

ENVIRONMENTAL LISTS

CERCLA/SARA - Hazardous Substances and their Reportable Quantities
final RQ = 5000 pounds (2270 kg)

RCRA - U Series Wastes
waste number U110 (Ignitable waste)

RCRA - Hazardous Constituents-Appendix VIII
waste number U110 (Ignitable waste)

RCRA - Substances Banned From Land Disposal
[present]

TSCA - Code of Federal Regulations Citations
40 CFR 716.120(a)

TSCA - Health and Safety Reporting List
Effective Date: March 7, 1986

STATE LISTS

Florida Hazardous Substance List
[present]

Massachusetts Right To Know List
[present]

NJ Right to Know List (Total)
sn 0803

NJ Special Hazardous Substances
(flammable - third degree)

Pennsylvania Right to Know List
environmental hazard

DIPROPYLENETRIAMINE AMIDE OF TALL-OIL FATTY ACID REACTED WITH TALL-OIL FATTY ACID-POLYETHYLENE GLYCOL ESTER AND MALEIC ANHYDRIDE   RR-01070-8

HEALTH AND SAFETY LISTS

U.S. DOT - Substances From 49 CFR 172.101
regulated by DOT (UN3034)

U.S. DOT - Hazard Classes
DOT hazard class = 4.1

DIPROPYL ETHER   111-43-3

HEALTH AND SAFETY LISTS

NFPA - Flash Points
flash point = 250 degrees F (121 degrees C)

NFPA - Hazard Identification Ratings
health-0; flammability-1; reactivity-0

NTP Chemical Status Reports - Testing Status and NTIS Number
Approved for toxicology/carcinogenesis study; short term toxicity studies scheduled for peer review

ENVIRONMENTAL LISTS

List of Pesticide Product Inert Ingredients
[present]

STATE LISTS

California - Air Bill 2588 Appendix A-I
9/90

Pennsylvania Right to Know List
[present]

DIPROPYL ISOCINCHOMERONATE   136-45-8

HEALTH AND SAFETY LISTS

NIOSH - Selected LD50s and LC50s
Oral, rat: LD50 = 9800 mg/kg

DIPROPYL KETONE   123-19-3

HEALTH AND SAFETY LISTS

NFPA - Flash Points
flash point = 186 degrees F (86 degrees C)

NFPA - Hazard Identification Ratings
health-0; flammability-2; reactivity-0

OSHA - Final PELs - Time Weighted Averages
100 ppm TWA; 600 mg/m3 TWA

OSHA - Final PELs - Skin Notations
prevent or reduce skin absorption

DI-N-PROPYL PEROXYDICARBONATE   16066-38-9

ENVIRONMENTAL LISTS

TSCA - PAIR - Reporting List
Effective Date: January 26, 1994; Reporting Date: March 28, 1994

TSCA - Health and Safety Reporting List
Effective Date: January 26, 1994; Sunset Date: January 26, 2004

PROPOSED REGULATIONS

TSCA - ITC 31st Report Priority Testing List
recommended for testing

DIQUAT   2764-72-9

ENVIRONMENTAL LISTS

List of Pesticide Product Inert Ingredients
[present]

DIQUAT DIBROMIDE   85-00-7

HEALTH AND SAFETY LISTS

ACGIH 1995 - Time Weighted Averages
100 ppm TWA; 606 mg/m3 TWA

ACGIH 1995 - Short Term Exposure Limits
150 ppm STEL; 909 mg/m3 STEL

ACGIH 1995 - Skin Designations
skin - potential for cutaneous absorption

NIOSH - Selected LD50s and LC50s
Oral, rat: LD50 = 5135 mg/kg Skin, rat: LD50 = 9500 mg/kg

NIOSH 1990 - Pocket Guide - RELs
100 ppm TWA; 600 mg/m3 TWA; 150 ppm STEL; 900 mg/m3 STEL

NIOSH 1990 - Pocket Guide - Target organs
eyes, respiratory system

NIOSH 1990 - Pocket Guide - Skin list
Potential for dermal absorption

OSHA - Vacated PELs - Time Weighted Averages
100 ppm TWA; 600 mg/m3 TWA

OSHA - Vacated PELs - Short Term Exposure Limits
150 ppm STEL; 900 mg/m3 STEL

OSHA - Vacated PELs - Skin Designation
Prevent or reduce skin absorption

ENVIRONMENTAL LISTS

List of Pesticide Product Inert Ingredients
[present]

TSCA - Code of Federal Regulations Citations
40 CFR 712.30(w); 40 CFR 716.120(a)

TSCA - PAIR - Reporting List
Reporting Date: June 13, 1989

TSCA - Health and Safety Reporting List
Effective Date: April 13, 1989

INTERNATIONAL LISTS

Australian Exposure Standards - Time Weighted Averages
100 ppm TWA; 606 mg/m3 TWA

Australian Exposure Standards - Short Term Exposure Limits
150 ppm STEL; 909 mg/m3 STEL

Australian Exposure Standards - Skin Effects
skin absorption

Canada - WHMIS: Ingredient Disclosure
1% item 669 (821)

Canada - Alberta - 8 Hour Occupational Exposure Limit
100 ppm TWA; 606 mg/m3 TWA

Canada - Alberta - 15 Minute Occupational Exposure Limit
150 ppm STEL; 909 mg/m3 STEL

Canada - British Columbia - 8 Hour Exposure Limits
100 ppm TWA; 600 mg/m3 TWA

Canada - British Columbia - 15 Minute Exposure Limits
150 ppm STEL; 900 mg/m3 STEL

Canada - British Columbia - Skin Notations
skin - potential for skin absorption

Canada - Ontario - OHSA - TWAEVs
100 ppm TWAEV; 605 mg/m3 TWAEV

Canada - Ontario - OHSA - STEVs
150 ppm STEV; 910 mg/m3 STEV

German (DFG) - MAK Values
50 ppm MAK; 300 mg/m3 MAK

German (DFG) - Peak Limitations
2 x normal MAK (5 min momentary value); don't exceed 8 times during shift

Israel - Time Weighted Averages
100 ppm TWA; 606 mg/m3 TWA

Israel - Short Term Exposure Limits
150 ppm STEL; 909 mg/m3 STEL

Israel - Action Levels
50 ppm AL; 303 mg/m3 AL

Mexico - Instruction No. 10 - TWAs
100 ppm TWA; 600 mg/m3 TWA

Mexico - Instruction No. 10 - STELs
150 ppm STEL; 800 mg/m3 STEL

STATE LISTS

California - Air Bill 2588 Appendix A-I
9/90

California - Exposure Limits - PELs
100 ppm PEL; 600 mg/m3 PEL

California - Exposure Limits - STELs
150 ppm STEL; 900 mg/m3 STEL

California - Exposure Limits - Skin Notation
material may be absorbed through the skin, eyes or mucous membrane

California - Directors List of Hazardous Substances (8 CCR 339)
[present]

Florida Hazardous Substance List
[present]

Massachusetts Right To Know List
[present]

Minnesota Hazardous Substance List
[present]

Pennsylvania Right to Know List
[present]

DIRECT BLUE 53   314-13-6

ENVIRONMENTAL LISTS

List of Pesticide Product Inert Ingredients
[present]

N,N'-DISALICYLIDENE-1,2-PROPANEDIAMINE   94-91-7

HEALTH AND SAFETY LISTS

U.S. DOT - Substances From 49 CFR 172.101
regulated by DOT (UN2384)

U.S. DOT - Hazard Classes
DOT hazard class = 3

NFPA - Flash Points
flash point = 70 degrees F (21 degrees C)

NFPA - Hazard Identification Ratings
flammability-3; reactivity-0

STATE LISTS

Florida Hazardous Substance List
[present]

Massachusetts Right To Know List
[present]

NJ Right to Know List (Total)
sn 0805

NJ Special Hazardous Substances
(flammable - third degree)

Pennsylvania Right to Know List
[present]

DISILOXANE, 1,1,3,3-TETRAMETHYL-1,3-BIS[3-   126-80-
OXIRANYLMETHOXY)PROPYL]-

ENVIRONMENTAL LISTS

CERCLA/SARA - Section 313 - Emission Reporting
form R reporting required

DISINFECTANTS, N.O.S.   RR-00264-

HEALTH AND SAFETY LISTS

ACGIH 1995 - Time Weighted Averages
50 ppm TWA; 233 mg/m3 TWA

U.S. DOT - Substances From 49 CFR 172.101
regulated by DOT (UN2710)

U.S. DOT - Hazard Classes
DOT hazard class = 3

NFPA - Flash Points
flash point = 120 degrees F (49 degrees C)

NFPA - Hazard Identification Ratings
health-2; flammability-2; reactivity-0

NIOSH - Selected LD50s and LC50s
Oral, rat: LD50 = 3730 mg/kg Skin, rabbit: LD50 = 5660 mg/kg

OSHA - Vacated PELs - Time Weighted Averages
50 ppm TWA; 235 mg/m3 TWA

INTERNATIONAL LISTS

Australian Exposure Standards - Time Weighted Averages
50 ppm TWA; 233 mg/m3 TWA

Canada - WHMIS: Ingredient Disclosure
1% item 670 (783)

Canada - Alberta - 8 Hour Occupational Exposure Limit
50 ppm TWA; 234 mg/m3 TWA

Canada - Alberta - 15 Minute Occupational Exposure Limit
75 ppm STEL; 351 mg/m3 STEL

Canada - Ontario - OHSA - TWAEVs
50 ppm TWAEV; 233 mg/m3 TWAEV

Israel - Time Weighted Averages
50 ppm TWA; 233 mg/m3 TWA

Israel - Action Levels
25 ppm AL; 116.5 mg/m3 AL

STATE LISTS

California - Exposure Limits - PELs
50 ppm PEL; 235 mg/m3 PEL

California - Directors List of Hazardous Substances (8 CCR 339)
[present]

Florida Hazardous Substance List
[present]

Massachusetts Right To Know List
[present]

Minnesota Hazardous Substance List
[present]

NJ Right to Know List (Total)
sn 0806

Pennsylvania Right to Know List
[present]

DISODIUM CALCIUM EDTA   62-33-9

HEALTH AND SAFETY LISTS

U.S. DOT - Organic Peroxides Table
Organic peroxide UN3113

NIOSH - Selected LD50s and LC50s
Oral, rat: LD50 = 3400 mg/kg Skin, rabbit: LD50 = 3500 mg/kg

STATE LISTS

NJ Right to Know List (Total)
sn 0788

DISODIUM CYANODITHIOIMIDOCARBONATE   138-93-2

HEALTH AND SAFETY LISTS

ACGIH 1995 - Time Weighted Averages
total dust/particulate: 0.5 mg/m3 TWA; respirable fraction: 0.1 mg/m3 TWA

ACGIH 1995 - Skin Designations
skin - potential for cutaneous absorption

U.S. DOT - Appendix A Table 1 - Hazardous Substances
final RQ = 1000 pounds (454 kg)

ENVIRONMENTAL LISTS

CERCLA/SARA - Hazardous Substances and their Reportable Quantities
final RQ = 1000 pounds (454 kg)

Clean Water Act - Hazardous Substances
[present] (Listed under 'Diquat')

INTERNATIONAL LISTS

Canada - Drinking Water Quality - MACs
0.07 mg/L MAC

Canada - British Columbia - 8 Hour Exposure Limits
0.5 mg/m3 TWA

Canada - British Columbia - 15 Minute Exposure Limits
1 mg/m3 STEL

Canada - Quebec - Time-Weighted Average Exposure Values
0.5 mg/m3 TWA

STATE LISTS

California - Directors List of Hazardous Substances (8 CCR 339)
[present]

Massachusetts Right To Know List
[present]

NJ Right to Know List (Total)
sn 0807

Pennsylvania Right to Know List
environmental hazard

DISODIUM DODECYLDIPHENYL ETHER   28519-02-0
DISULFONATE

SEE ALSO:
DIQUAT

HEALTH AND SAFETY LISTS

U.S. DOT - Appendix A Table 1 - Hazardous Substances
final RQ = 1000 pounds (454 kg) (Listed under 'Diquat')

NIOSH - Selected LD50s and LC50s
Oral, rat: LD50 = 120 mg/kg Skin, rat: LD50 = 433 mg/kg

OSHA - Vacated PELs - Time Weighted Averages
0.5 mg/m3 TWA

ENVIRONMENTAL LISTS

CERCLA/SARA - Hazardous Substances and their Reportable Quantities
final RQ = 1000 pounds (454 kg) (Listed under 'Diquat')

Clean Water Act - Hazardous Substances
[present]

Safe Drinking Water Act - MCLs
MCL = 0.02 mg/L

Safe Drinking Water Act - MCLGs
MCLG = 0.02 mg/L

INTERNATIONAL LISTS

Australian Exposure Standards - Time Weighted Averages
0.5 mg/m3 TWA

Canada - Alberta - 8 Hour Occupational Exposure Limit
0.5 mg/m3 TWA

Canada - Alberta - 15 Minute Occupational Exposure Limit
1 mg/m3 STEL

Canada - Ontario - OHSA - TWAEVs
0.5 mg/m3 TWAEV

United Kingdom - Occupational Exposure Standards - TWAs
0.5 mg/m3 TWA

United Kingdom - Occupational Exposure Standards - STELs
1 mg/m3 STEL

Israel - Time Weighted Averages
0.5 mg/m3 TWA

Israel - Action Levels
0.25 mg/m3 AL

Mexico - Instruction No. 10 - TWAs
0.5 mg/m3 TWA

Mexico - Instruction No. 10 - STELs
1 mg/m3 STEL

STATE LISTS

California - Exposure Limits - PELs
0.5 mg/m3 PEL

California - Directors List of Hazardous Substances (8 CCR 339)
[present] (Listed under 'Diquat')

Florida Hazardous Substance List
[present]

Massachusetts Right To Know List
[present]

Minnesota Hazardous Substance List
[present]

NJ Right to Know List (Total)
sn 0808

Pennsylvania Right to Know List
environmental hazard

DISODIUM DODECYLIMIDAZOLINIUM   RR-01071-9
DICARBOXYLATE

HEALTH AND SAFETY LISTS

IARC - Group 3 (not classifiable)
[present]

INTERNATIONAL LISTS

Canada - WHMIS: Ingredient Disclosure
1% item 672 (313)

STATE LISTS

California - Directors List of Hazardous Substances (8 CCR 339)
[present]

DISODIUM 4-DODECYL-2,4'-OXYDIBENZENESULFONATE   7575-62-4

HEALTH AND SAFETY LISTS
NIOSH - Selected LD50s and LC50s
Oral, rat: LD50 = 4560 mg/kg

ENVIRONMENTAL LISTS
List of Pesticide Product Inert Ingredients
[present]

DISODIUM AR-(DODECYLPHENOXY)BENZENE DISULFONATE   12068-17-6
SEE ALSO:
GLYCIDOL (OXIRANEMETHANOL) AND ITS DERIVATIVES

ENVIRONMENTAL LISTS
EPA - Master Testing List
[present]
TSCA - Code of Federal Regulations Citations
40 CFR 716.120(c)

PROPOSED REGULATIONS
TSCA - Proposed Testing Rule for Glycidyl Ethers
member of Glycidyl subcategory III-C

DISODIUM EDTA   139-33-3

HEALTH AND SAFETY LISTS
U.S. DOT - Substances From 49 CFR 172.101
regulated by DOT (UN1601, UN1903, UN3142)
U.S. DOT - Hazard Classes
DOT hazard class = 6.1; corrosive: DOT hazard class = 8

STATE LISTS
NJ Right to Know List (Total)
corrosive, liquid: sn 2361; poisonous, liquid or solid: sn 2362

DISODIUM [(ETHYLENEDINITRILO)TETRAACETATO] ZINC   14025-21-9

HEALTH AND SAFETY LISTS
NIOSH - Selected LD50s and LC50s
Oral, rabbit: LD50 = 7000 mg/kg

ENVIRONMENTAL LISTS
List of Pesticide Product Inert Ingredients
[present]

INTERNATIONAL LISTS
Canada - WHMIS: Ingredient Disclosure
0.1% item 676 (798)

DISODIUM HYDROGEN ARSENATE   7778-43-0

ENVIRONMENTAL LISTS
CERCLA/SARA - Section 313 - Emission Reporting
form R reporting required

DISODIUM HYDROGEN ARSENATE   10048-95-0

ENVIRONMENTAL LISTS
List of Pesticide Product Inert Ingredients
[present]

DISODIUM IRON(II) ETHYLENEDIAMINETETRAACETATE   14729-89-6

ENVIRONMENTAL LISTS
List of Pesticide Product Inert Ingredients
[present]

DISODIUM MANGANESE ETHYLENEDIAMINETETRAACETATE   15375-84-5

ENVIRONMENTAL LISTS

List of Pesticide Product Inert Ingredients
[present]

DISODIUM ORTHOPHOSPHATE HEPTAHYDRATE   7782-85-6

ENVIRONMENTAL LISTS
List of Pesticide Product Inert Ingredients
[present]

DISODIUM SALT OF LAURYL ALCOHOL POLYETHYLENE GLYCOL ETHER SULFOSUCCINATE   39354-45-5

HEALTH AND SAFETY LISTS
NIOSH - Selected LD50s and LC50s
Oral, rat: LD50 = 2000 mg/kg

ENVIRONMENTAL LISTS
List of Pesticide Product Inert Ingredients
[present]

DISPERSANT GAS, N.O.S.   RR-00267-5

ENVIRONMENTAL LISTS
List of Pesticide Product Inert Ingredients
[present]

DISTEARYLDIMETHYLAMMONIUM CHLORIDE   107-64-2
SEE ALSO:
ARSENIC
INORGANIC ARSENIC

INTERNATIONAL LISTS
Canada - WHMIS: Ingredient Disclosure
1% item 677 (259)

STATE LISTS
Florida Hazardous Substance List
[present]
Massachusetts Right To Know List
[present]

DISTEARYL PEROXYDI-CARBONATE   52326-66-6

HEALTH AND SAFETY LISTS
NTP Seventh Report - Known Carcinogens
known carcinogen (Listed under 'Arsenic and certain arsenic compounds')
OSHA - Select Carcinogens
[present]

DISTILLATE (LIGHT) FUELS   RR-01550-9

ENVIRONMENTAL LISTS
List of Pesticide Product Inert Ingredients
[present]

DISTILLATES (PETROLEUM), C(3-6), POLYMERS WITH STYRENE AND MIXED TERPENES   RR-00939-2

ENVIRONMENTAL LISTS
List of Pesticide Product Inert Ingredients
[present]

DISTILLATES (PETROLEUM), CATALYTIC REFORMER FRACTIONATORRESIDUE, INTERMEDIATE BOILING   68477-30-5

HEALTH AND SAFETY LISTS
NIOSH - Selected LD50s and LC50s
Oral, rat: LD50 = 12930 mg/kg

DISTILLATES (PETROLEUM), SOLVENT-DEWAXED LIGHT NAPHTHENIC   64742-64-9

ENVIRONMENTAL LISTS

List of Pesticide Product Inert Ingredients
[present]

DISTILLATES (PETROLEUM), SOLVENT REFINED LIGHT NAPHTHENIC   64741-97-5

STATE LISTS

NJ Right to Know List (Total)
sn 2364; flammable: sn 2363

DISTILLATES (PETROLEUM), STRAIGHT-RUN MIDDLE   64741-44-2

ENVIRONMENTAL LISTS

EPA - Master Testing List
[present]

List of Pesticide Product Inert Ingredients
[present]

DISUBSTITUTED ALKYL TRIAZINES   RR-00196-7

HEALTH AND SAFETY LISTS

U.S. DOT - Organic Peroxides Table
Organic peroxide UN3106

STATE LISTS

NJ Right to Know List (Total)
sn 0810

DISUBSTITUTED DIAMINO ANISOLE   RR-00265-3

HEALTH AND SAFETY LISTS

IARC - Group 3 (not classifiable)
[present]

DISUBSTITUTED DIPHENYLSULFONE   RR-01675-1

ENVIRONMENTAL LISTS

TSCA - Chemicals with Significant New Use Rules
PMN number: P-89-676

DISUBSTITUTED NITROBENZENE   RR-00921-2

ENVIRONMENTAL LISTS

List of Pesticide Product Inert Ingredients
[present]

DISUBSTITUTED PHENOXAZINE, CHLOROMETA-LATE SALT   RR-01206-6

STATE LISTS

Massachusetts Right To Know List
carcinogen; extraordinarily hazardous

FATTY ACID POLYAMINE CONDENSATE, PH-SOPHORIC ACID ESTER SALT   RR-01207-7

STATE LISTS

Massachusetts Right To Know List
carcinogen; extraordinarily hazardous

DISUBSTITUTED PHENYLAZO TRISUBSTITUTED NAPHTHALENE   RR-01213-5

HEALTH AND SAFETY LISTS

U.S. DOT - Substances From 49 CFR 172.101
regulated by DOT (UN1202)

U.S. DOT - Hazard Classes
DOT hazard class = 3

NFPA - Flash Points
flash point = 150+ degrees F (66+ degrees C)

NFPA - Hazard Identification Ratings
health-0; flammability-2; reactivity-0

STATE LISTS

NJ Right to Know List (Total)
sn 2452

DISUCCINIC ACID PEROXIDE, 72% IN WATER   RR-00268-6

ENVIRONMENTAL LISTS

TSCA - Chemicals with Significant New Use Rules
PMN numbers: P-85-932; P-85-933

DISULFIRAM   97-77-8

ENVIRONMENTAL LISTS

TSCA - Chemicals with Significant New Use Rules
PMN number: P-83-822

DISULFONIC ACID ROSIN AMINE SALT OF A BEN-ZIDINE DERIVATIVE   RR-00906-3

ENVIRONMENTAL LISTS

TSCA - Chemicals with Significant New Use Rules
PMN number: P-92-1119

DISULFOTON   298-04-4

ENVIRONMENTAL LISTS

TSCA - Chemicals with Significant New Use Rules
PMN number: P-84-860

DISULFUROUS ACID, DISODIUM SALT   7757-74-6

ENVIRONMENTAL LISTS

TSCA - Chemicals with Significant New Use Rules
PMN number: P-90-2

DITALLOW DIMETHYL AMMONIUM CHLORIDE   68783-78-8

ENVIRONMENTAL LISTS

TSCA - Chemicals with Significant New Use Rules
PMN numbers: P-90-1984; P-90-1985

DITHIAZANINE IODIDE   514-73-8

ENVIRONMENTAL LISTS

TSCA - Chemicals with Significant New Use Rules
PMN number: P-91-328

DITHIAZANINE IODIDE   521-10-8

STATE LISTS

NJ Right to Know List (Total)
sn 2365

2,5-DITHIOBIUREA   142-46-1

HEALTH AND SAFETY LISTS

ACGIH 1995 - Time Weighted Averages
2 mg/m3 TWA

IARC - Group 3 (not classifiable)
[present]

NIOSH - Selected LD50s and LC50s
Oral, rat: LD50 = 500 mg/kg

NTP Chemical Status Reports - Testing Status and NTIS Number
Technical reports printed (PB298514/AS)

NTP Chemical Status Reports - Evidence of Carcinogenicity
male rat-negative; female rat-negative; male mice-negative; female mice-negative

OSHA - Vacated PELs - Time Weighted Averages
2 mg/m3 TWA

ENVIRONMENTAL LISTS

TSCA - PAIR - Reporting List
Effective Date: January 26, 1994;   Reporting Date: March 28, 1994

TSCA - Health and Safety Reporting List
Effective Date: January 26, 1994;   Sunset Date: January 26, 2004

INTERNATIONAL LISTS

Australian Exposure Standards - Time Weighted Averages
2 mg/m3 TWA

Canada - WHMIS: Ingredient Disclosure
1% item 1520 (793)

Canada - Alberta - 8 Hour Occupational Exposure Limit
2 mg/m3 TWA

Canada - Alberta - 15 Minute Occupational Exposure Limit
5 mg/m3 STEL

Canada - British Columbia - 8 Hour Exposure Limits
2 mg/m3 TWA

Canada - British Columbia - 15 Minute Exposure Limits
5 mg/m3 STEL

Canada - Ontario - OHSA - TWAEVs
2 mg/m3 TWAEV

Canada - Quebec - Time-Weighted Average Exposure Values
2 mg/m3 TWAEV

German (DFG) - MAK Values
total dust: 2 mg/m3 MAK

German (DFG) - Peak Limitations
10 x normal MAK (30 min average value);   don't exceed during shift

Israel - Time Weighted Averages
2 mg/m3 TWA

Israel - Action Levels
1 mg/m3 AL

Mexico - Instruction No. 10 - TWAs
2 mg/m3 TWA

Mexico - Instruction No. 10 - STELs
3 mg/m3 STEL

STATE LISTS

California - Exposure Limits - PELs
2 mg/m3 PEL

California - Directors List of Hazardous Substances (8 CCR 339)
[present]

Florida Hazardous Substance List
[present]

Massachusetts Right To Know List
[present]

Minnesota Hazardous Substance List
[present]

NJ Right to Know List (Total)
sn 0811

Pennsylvania Right to Know List
[present]

PROPOSED REGULATIONS

TSCA - ITC 31st Report Priority Testing List
designated to be tested

DITHIOBIURET   541-53-7

ENVIRONMENTAL LISTS

TSCA - Chemicals with Significant New Use Rules
PMN number: P-87-1337

DITHIOCARBAMATE PESTICIDES, N.O.S.   RR-00269-
SEE ALSO:
F039-HAZARDOUS WASTES

HEALTH AND SAFETY LISTS

ACGIH 1995 - Time Weighted Averages
0.1 mg/m3 TWA

ACGIH 1995 - Skin Designations
skin - potential for cutaneous absorption

U.S. DOT - Appendix B - Marine Pollutants
DOT regulated marine pollutant

U.S. DOT - Appendix A Table 1 - Hazardous Substances
final RQ = 1 pound (0.454 kg)

NIOSH - Selected LD50s and LC50s
Inhalation, rat: LC50 = 200 mg/m3 8 hr Oral, rat: LD50 = 2 mg/k
Skin, rat: LD50 = 6 mg/kg

OSHA - Vacated PELs - Time Weighted Averages
0.1 mg/m3 TWA

OSHA - Vacated PELs - Skin Designation
Prevent or reduce skin absorption

ENVIRONMENTAL LISTS

ATSDR Priority List
Rank (of 275): 036

CERCLA/SARA - Section 302 Extremely Hazardous Substances and TPQs
TPQ = 500 pounds

CERCLA/SARA - Hazardous Substances and their Reportable Quantities
final RQ = 1 pound (0.454 kg)

Clean Water Act - Hazardous Substances
[present]

RCRA - P Series Wastes
waste number P039

RCRA - Hazardous Constituents-Appendix VIII
waste number P039

RCRA - Basis for Listing - Appendix VII
Included in waste stream: F039

RCRA - Substances Banned From Land Disposal
[present]

RCRA - TSD Facilities Ground Water Monitoring
TM 8140 = 2 ug/L PQL;   TM 8270 = 10 ug/L PQL

RCRA - Universal Treatment Standards (LDR)
WW: 0.017 mg/l;   NWW: 6.2 mg/kg

INTERNATIONAL LISTS

Australian Exposure Standards - Time Weighted Averages
0.1 mg/m3 TWA

Canada - Alberta - 8 Hour Occupational Exposure Limit
0.1 mg/m3 TWA

Canada - Alberta - 15 Minute Occupational Exposure Limit
0.3 mg/m3 STEL

Canada - British Columbia - 8 Hour Exposure Limits
0.1 mg/m3 TWA

Canada - British Columbia - 15 Minute Exposure Limits
0.3 mg/m3 STEL

Canada - British Columbia - Skin Notations
skin - potential for skin absorption

Canada - Ontario - OHSA - TWAEVs
0.1 mg/m3 TWAEV

Canada - Quebec - Time-Weighted Average Exposure Values
0.1 mg/m3 TWAEV

United Kingdom - Occupational Exposure Standards - TWAs
0.1 mg/m3 TWA

United Kingdom - Occupational Exposure Standards - STELs
0.3 mg/m3 STEL

Israel - Time Weighted Averages
0.1 mg/m3 TWA
Israel - Action Levels
0.05 mg/m3 AL
Mexico - Instruction No. 10 - TWAs
0.1 mg/m3 TWA
Mexico - Instruction No. 10 - STELs
0.3 mg/m3 STEL

STATE LISTS

California - Exposure Limits - PELs
0.1 mg/m3 PEL
California - Exposure Limits - Skin Notation
material may be absorbed through the skin, eyes or mucous membrane
California - Directors List of Hazardous Substances (8 CCR 339)
[present]
Florida Hazardous Substance List
[present]
Massachusetts Right To Know List
extraordinarily hazardous; neurotoxin
Minnesota Hazardous Substance List
[present]
NJ Right to Know List (Total)
sn 0812
Pennsylvania Right to Know List
environmental hazard

4,4'-DITHIODIMORPHOLINE   103-34-4

STATE LISTS

Pennsylvania Right to Know List
[present]

DITHRANOL   480-22-8

ENVIRONMENTAL LISTS

List of Pesticide Product Inert Ingredients
[present]

1,3-DI-O-TOLYLGUANIDINE   97-39-2

HEALTH AND SAFETY LISTS

NIOSH - Selected LD50s and LC50s
Oral, mouse: LD50 = 20 mg/kg

ENVIRONMENTAL LISTS

CERCLA/SARA - Section 302 Extremely Hazardous Substances and TPQs
TPQ = 500/10,000 pounds

STATE LISTS

Florida Hazardous Substance List
effective March 13, 1992
Massachusetts Right To Know List
extraordinarily hazardous
Pennsylvania Right to Know List
environmental hazard

PROPOSED REGULATIONS

CERCLA/SARA - Proposed Hazardous Substance Additions
proposed RQ = 1 pound (.454 kg)
CERCLA/SARA - 1989 Proposed RQ Adjustments
proposed RQ = 100 pounds (45.4 kg)

DITRIDECYL PHTHALATE (MIXED ISOMERS)   68515-47-9

STATE LISTS

Massachusetts Right To Know List
teratogen

DI(3,5,5-TRIMETHYL-1,2-DIOXOLANYL-3) PEROX-   RR-00273-3
IDE, PASTE

HEALTH AND SAFETY LISTS

NTP Chemical Status Reports - Testing Status and NTIS Number
Technical reports printed (PB291534/AS)
NTP Chemical Status Reports - Evidence of Carcinogenicity
male rat-negative; female rat-negative; male mice-negative; female mice-equivocal

STATE LISTS

Massachusetts Right To Know List
[present]

DI(3,5,5-TRIMETHYLHEXANOYL) PEROXIDE   3851-87-4

HEALTH AND SAFETY LISTS

U.S. DOT - Appendix A Table 1 - Hazardous Substances
final RQ = 100 pounds (45.4 kg)
NIOSH - Selected LD50s and LC50s
Oral, rat: LD50 = 5 mg/kg

ENVIRONMENTAL LISTS

CERCLA/SARA - Section 302 Extremely Hazardous Substances and TPQs
TPQ = 100/10,000 pounds
CERCLA/SARA - Section 313 - Emission Reporting
form R reporting required
CERCLA/SARA - Hazardous Substances and their Reportable Quantities
final RQ = 100 pounds (45.4 kg)
RCRA - P Series Wastes
waste number P049
RCRA - Hazardous Constituents-Appendix VIII
waste number P049
RCRA - Substances Banned From Land Disposal
[present]

STATE LISTS

Florida Hazardous Substance List
effective March 13, 1992
Massachusetts Right To Know List
extraordinarily hazardous
NJ Right to Know List (Total)
sn 2368
Pennsylvania Right to Know List
environmental hazard

DI(3,5,5-TRIMETHYLHEXANOYL) PEROXIDE   4511-39-1

HEALTH AND SAFETY LISTS

U.S. DOT - Substances From 49 CFR 172.101
regulated by DOT (UN2771, UN2772, UN3005, UN3006)
U.S. DOT - Hazard Classes
toxic or toxic, flammable: DOT hazard class = 6.1; flammable, toxic: DOT hazard class = 3

STATE LISTS

NJ Right to Know List (Total)
sn 2369; sn 2370; sn 2371; sn 2372

DIURON   330-54-1

HEALTH AND SAFETY LISTS

NIOSH - Selected LD50s and LC50s
Oral, mouse: LD50= 1660 mg/kg

DIVINYL ACETYLENE   821-08-9

HEALTH AND SAFETY LISTS

IARC - Group 3 (not classifiable)
[present]

DIVINYL BENZENE   108-57-6

HEALTH AND SAFETY LISTS

NIOSH - Selected LD50s and LC50s
Oral, rat: LD50 = 500 mg/kg

ENVIRONMENTAL LISTS

CAA - HON Rule - SOCMI Chemicals
compliance by April 24, 1995

DIVINYL BENZENE   1321-74-0

ENVIRONMENTAL LISTS

TSCA - Code of Federal Regulations Citations
40 CFR 799.5000

TSCA - Substances Subject to Testing Consent Orders
Test for: Chemical Fate

TSCA - Section 12(b) - Export Notification
export notification required - Section 4

DIVINYL ETHER   109-93-3

HEALTH AND SAFETY LISTS

U.S. DOT - Organic Peroxides Table
Organic peroxide UN3116

STATE LISTS

NJ Right to Know List (Total)
sn 2373

DI(O-XENYL) PHENYL PHOSPHATE   RR-00831-1

HEALTH AND SAFETY LISTS

U.S. DOT - Organic Peroxides Table
Organic peroxide UN3115

STATE LISTS

NJ Right to Know List (Total)
sn 2374

DOCOSAMETHYLCYLCLOUNDECASILOXANE   18766-38-6

HEALTH AND SAFETY LISTS

U.S. DOT - Organic Peroxides Table
Organic peroxide UN3105

DOCOSAMETHYLDECASILOXANE   556-70-7

HEALTH AND SAFETY LISTS

ACGIH 1995 - Time Weighted Averages
10 mg/m3 TWA

U.S. DOT - Appendix A Table 1 - Hazardous Substances
final RQ = 100 pounds (45.4 kg)

NIOSH - Selected LD50s and LC50s
Oral, rat: LD50 = 1017 mg/kg

OSHA - Vacated PELs - Time Weighted Averages
10 mg/m3 TWA

ENVIRONMENTAL LISTS

CERCLA/SARA - Section 313 - Emission Reporting
form R reporting required

CERCLA/SARA - Hazardous Substances and their Reportable Quantities
final RQ = 100 pounds (45.4 kg)

Clean Water Act - Hazardous Substances
[present]

INTERNATIONAL LISTS

Australian Exposure Standards - Time Weighted Averages
10 mg/m3 TWA

Canada - Drinking Water Quality - MACs
0.15 mg/L MAC

Canada - Alberta - 8 Hour Occupational Exposure Limit
10 mg/m3 TWA

Canada - Alberta - 15 Minute Occupational Exposure Limit
20 mg/m3 STEL

Canada - British Columbia - 8 Hour Exposure Limits
10 mg/m3 TWA

Canada - Ontario - OHSA - TWAEVs
10 mg/m3 TWAEV

Canada - Quebec - Time-Weighted Average Exposure Values
10 mg/m3 TWAEV

United Kingdom - Occupational Exposure Standards - TWAs
10 mg/m3 TWA

Israel - Time Weighted Averages
10 mg/m3 TWA

Israel - Action Levels
5 mg/m3 AL

Mexico - Instruction No. 10 - TWAs
10 mg/m3 TWA

STATE LISTS

California - Exposure Limits - PELs
10 mg/m3 PEL

California - Directors List of Hazardous Substances (8 CCR 339)
[present]

Florida Hazardous Substance List
[present]

Massachusetts Right To Know List
[present]

Minnesota Hazardous Substance List
[present]

NJ Right to Know List (Total)
sn 0819

Pennsylvania Right to Know List
environmental hazard

DODECAMETHYLCYCLOHEXASILOXANE   540-97-6

HEALTH AND SAFETY LISTS

NFPA - Flash Points
flash point < -4 degrees F (-20 degrees C)

NFPA - Hazard Identification Ratings
flammability-3; reactivity-3

STATE LISTS

Florida Hazardous Substance List
[present]

Massachusetts Right To Know List
[present]

Pennsylvania Right to Know List
[present]

DODECAMETHYLPENTASILOXANE   141-63-9

HEALTH AND SAFETY LISTS

NFPA - Flash Points
flash point = 169 degrees F (76 degrees C)

NFPA - Hazard Identification Ratings
health-1; flammability-2; reactivity-2

NIOSH - Selected LD50s and LC50s
Oral, rat: LD50 = 4640 mg/kg

INTERNATIONAL LISTS

Canada - WHMIS: Ingredient Disclosure
1% item 678 (794)

Canada - Alberta - 8 Hour Occupational Exposure Limit
10 ppm TWA; 53 mg/m3 TWA

Canada - Alberta - 15 Minute Occupational Exposure Limit
20 ppm STEL; 106 mg/m3 STEL

Canada - Ontario - OHSA - TWAEVs
10 ppm TWAEV; 53 mg/m3 TWAEV

United Kingdom - Occupational Exposure Standards - TWAs
10 ppm TWA; 50 mg/m3 TWA

STATE LISTS

California - Directors List of Hazardous Substances (8 CCR 339)
[present]

Florida Hazardous Substance List
[present]

Massachusetts Right To Know List
[present]

NJ Special Hazardous Substances
(reactive - second degree)

Pennsylvania Right to Know List
[present]

DODECANE   112-40-3

HEALTH AND SAFETY LISTS

ACGIH 1995 - Time Weighted Averages
10 ppm TWA; 53 mg/m3 TWA

NTP Chemical Status Reports - Testing Status and NTIS Number
Assigned to laboratory for toxicology/carcinogenesis study

OSHA - Vacated PELs - Time Weighted Averages
10 ppm TWA; 50 mg/m3 TWA

INTERNATIONAL LISTS

Australian Exposure Standards - Time Weighted Averages
10 ppm TWA; 53 mg/m3 TWA

Canada - Quebec - Time-Weighted Average Exposure Values
10 ppm TWAEV; 53 mg/m3 TWAEV

Israel - Time Weighted Averages
10 ppm TWA; 53 mg/m3 TWA

Israel - Action Levels
5 ppm AL; 26.5 mg/m3 AL

STATE LISTS

California - Exposure Limits - PELs
10 ppm PEL; 50 mg/m3 PEL

Massachusetts Right To Know List
[present]

Minnesota Hazardous Substance List
[present]

Pennsylvania Right to Know List
[present]

DODECANEDIOIC ACID   693-23-2

HEALTH AND SAFETY LISTS

U.S. DOT - Substances From 49 CFR 172.101
regulated by DOT (UN1167)

U.S. DOT - Hazard Classes
DOT hazard class = 3

NFPA - Flash Points
flash point < -22 degrees F (-30 degrees C)

NFPA - Hazard Identification Ratings
health-2; flammability-4; reactivity-2

STATE LISTS

Florida Hazardous Substance List
[present]

Massachusetts Right To Know List
[present]

NJ Right to Know List (Total)
sn 0821

NJ Special Hazardous Substances
(flammable - third degree; reactive - second degree)

Pennsylvania Right to Know List
[present]

1-DODECANETHIOL   112-55-0

HEALTH AND SAFETY LISTS

NFPA - Flash Points
flash point = 482 degrees F (250 degrees C)

NFPA - Hazard Identification Ratings
health-0; flammability-1; reactivity-0

DODECANOIC ACID, 2,3-DIHYDROXYPROPYL ESTER   142-18-7

ENVIRONMENTAL LISTS

TSCA - PAIR - Reporting List
Effective Date: October 12, 1993; Reporting Date: February 28, 1994

TSCA - Health and Safety Reporting List
Effective Date: October 12, 1993; Sunset Date: October 12, 2003

DODECANOIC ACID, METHYL ESTER   111-82-0

ENVIRONMENTAL LISTS

TSCA - PAIR - Reporting List
Effective Date: October 12, 1993; Reporting Date: February 28, 1994

TSCA - Health and Safety Reporting List
Effective Date: October 12, 1993; Sunset Date: October 12, 2003

DODECANOIC ACID, PENTYL ESTER   5350-03-8

ENVIRONMENTAL LISTS

TSCA - PAIR - Reporting List
Effective Date: October 12, 1993; Reporting Date: February 28, 1994

TSCA - Health and Safety Reporting List
Effective Date: October 12, 1993; Sunset Date: October 12, 2003

1-DODECENE   112-41-4

ENVIRONMENTAL LISTS

TSCA - PAIR - Reporting List
Effective Date: October 12, 1993; Reporting Date: February 28, 1994

TSCA - Health and Safety Reporting List
Effective Date: October 12, 1993; Sunset Date: October 12, 2003

DODECENYLSUCCINIC ACID, MONOTRIDECYL ESTER   85081-53-4

HEALTH AND SAFETY LISTS

NFPA - Flash Points
flash point = 165 degrees F (74 degrees C)

NFPA - Hazard Identification Ratings
health-0; flammability-2; reactivity-0

DODECENYLSUCCINIC ANHYDRIDE   19780-11-1

ENVIRONMENTAL LISTS

CAA - HON Rule - SOCMI Chemicals
compliance by Oct. 24, 1994

EPA - Master Testing List
[present]

DODECYL PHENOL (BRANCHED)   121158-58-5

HEALTH AND SAFETY LISTS

NFPA - Flash Points
flash point = 262 degrees F (128 degrees C)

NFPA - Hazard Identification Ratings
health-2; flammability-1; reactivity-0

NIOSH - Health Standards - Exposure Limits
C (15 min) 0.5 ppm; C (15 min) 4.1 mg/m3 (Listed under 'Thiols')

NIOSH - Health Standards - Health Effects and Precautions
*Irritation; eye, skin, blood, and nervous system effects (Blood and
urine monitoring required; prevent skin contact) (Listed under 'Thiols')*

ENVIRONMENTAL LISTS

List of Pesticide Product Inert Ingredients
[present]

STATE LISTS

Pennsylvania Right to Know List
[present]

DODECYL ALCOHOL, ETHOXYLATED AND   26183-44-8
SULFATED

HEALTH AND SAFETY LISTS

NIOSH - Selected LD50s and LC50s
Oral, rat: LD50 = 53 gm/kg

DODECYLAMINE   124-22-1

ENVIRONMENTAL LISTS

List of Pesticide Product Inert Ingredients
[present]

DODECYL AND HIGHER ALIPHATIC KETONES   70955-37-2

STATE LISTS

Pennsylvania Right to Know List
[present]

DODECYLANILINE   28675-17-4

ENVIRONMENTAL LISTS

EPA - Master Testing List
[present]

DODECYL BENZENE (CRUDE)   123-01-3

ENVIRONMENTAL LISTS

List of Pesticide Product Inert Ingredients
[present]

DODECYL BENZENESULFONATE   1886-81-3

ENVIRONMENTAL LISTS

List of Pesticide Product Inert Ingredients
[present]

DODECYLBENZENESULFONIC ACID   27176-87-0

ENVIRONMENTAL LISTS

CAA - HON Rule - SOCMI Chemicals
compliance by Oct. 23, 1995

DODECYLBENZENESULFONIC ACID, N-(2-   68084-55-9
AMINOETHYL)ETHANOLAMINE SALT

ENVIRONMENTAL LISTS

List of Pesticide Product Inert Ingredients
[present]

DODECYLBENZENESULFONIC ACID, AMMONIUM   1331-61-9
SALT

HEALTH AND SAFETY LISTS

NIOSH - Selected LD50s and LC50s
Oral, rat: LD50 = 1020 mg/kg

DODECYLBENZENESULFONIC ACID, BUTYLAMINE   12068-09-6
SALT

ENVIRONMENTAL LISTS

List of Pesticide Product Inert Ingredients
[present]

DODECYLBENZENESULFONIC ACID, DI-   26545-53-
ETHANOLAMINE SALT

ENVIRONMENTAL LISTS

CAA - HON Rule - SOCMI Chemicals
compliance by Oct. 23, 1995

DODECYLBENZENESULFONIC ACID, DIISOPROPY-   29061-61-
LAMINE SALT

HEALTH AND SAFETY LISTS

NFPA - Flash Points
flash point = 285 degrees F (141 degrees C)

NFPA - Hazard Identification Ratings
health-1; flammability-1; reactivity-0

ENVIRONMENTAL LISTS

CAA - HON Rule - SOCMI Chemicals
compliance by Oct. 23, 1995

EPA - Master Testing List
[present]

DODECYLBENZENESULFONIC ACID, N,N-   60816-39-
DIMETHYL-1,3-PROPANEDIAMINESALT

HEALTH AND SAFETY LISTS

NIOSH - Selected LD50s and LC50s
Oral, rat: LD50 = 650 mg/kg

DODECYLBENZENESULFONIC ACID, ETHYLENEDI-   67952-66-3
AMINE SALT

HEALTH AND SAFETY LISTS

U.S. DOT - Substances From 49 CFR 172.101
regulated by DOT (NA2584)

U.S. DOT - Hazard Classes
DOT hazard class = 8

U.S. DOT - Appendix A Table 1 - Hazardous Substances
final RQ = 1000 pounds (454 kg)

ENVIRONMENTAL LISTS

CERCLA/SARA - Hazardous Substances and their Reportable Quantities
final RQ = 1000 pounds (454 kg)

Clean Water Act - Hazardous Substances
[present]

List of Pesticide Product Inert Ingredients
[present]

INTERNATIONAL LISTS

Canada - WHMIS: Ingredient Disclosure
1% item 679 (83)

STATE LISTS

California - Directors List of Hazardous Substances (8 CCR 339)
[present]

Massachusetts Right To Know List
[present]

NJ Right to Know List (Total)
sn 0822

NJ Special Hazardous Substances
(corrosive)

Pennsylvania Right to Know List
environmental hazard

DODECYLBENZENESULFONIC ACID, ISOPROPY-   26264-05-1
LAMINE SALT

ENVIRONMENTAL LISTS

List of Pesticide Product Inert Ingredients
[present]

DODECYLBENZENESULFONIC ACID, MORPHO-   12068-08-5
LINE SALT
  ENVIRONMENTAL LISTS
    List of Pesticide Product Inert Ingredients
    [present]

DODECYLBENZENESULFONIC ACID, 1,3-PROPANE-   60816-37-7
DIAMINE SALT
  ENVIRONMENTAL LISTS
    List of Pesticide Product Inert Ingredients
    [present]

DODECYLBENZENESULFONIC ACID, 1,1,2,3-TE-   RR-01073-1
TRAMETHYLBUTYLAMINE SALT
  ENVIRONMENTAL LISTS
    List of Pesticide Product Inert Ingredients
    [present]

DODECYLBENZENESULFONIC ACID, STRONTIUM   12068-15-4
SALT
  ENVIRONMENTAL LISTS
    List of Pesticide Product Inert Ingredients
    [present]

DODECYLBENZENESULFONIC ACID, TRIETHY-   29061-63-0
LAMINE SALT
  ENVIRONMENTAL LISTS
    List of Pesticide Product Inert Ingredients
    [present]

DODECYLBENZENESULFONIC ACID, ZINC SALT   12068-16-5
  ENVIRONMENTAL LISTS
    List of Pesticide Product Inert Ingredients
    [present]

DODECYLBETAINE   683-10-3
  ENVIRONMENTAL LISTS
    List of Pesticide Product Inert Ingredients
    [present]

N-DODECYLCAPROLACTAM   59227-89-3
  ENVIRONMENTAL LISTS
    List of Pesticide Product Inert Ingredients
    [present]

DODECYLDI(AMINOETHYL)GLYCINE   6843-97-6
  ENVIRONMENTAL LISTS
    List of Pesticide Product Inert Ingredients
    [present]

DODECYL DIMETHYL BENZYL AMMONIUM   RR-01072-0
NAPHTHENATE
  ENVIRONMENTAL LISTS
    List of Pesticide Product Inert Ingredients
    [present]

DODECYLGUANIDINE MONOACETATE   2439-10-3
  ENVIRONMENTAL LISTS
    List of Pesticide Product Inert Ingredients
    [present]

3,3'-(DODECYLIMINO)DIPROPIONIC ACID, DIS-   3655-00-3
ODIUM SALT
  ENVIRONMENTAL LISTS

    List of Pesticide Product Inert Ingredients
    [present]

TERT-DODECYL MERCAPTAN   25103-58-6
  ENVIRONMENTAL LISTS
    List of Pesticide Product Inert Ingredients
    [present]

DODECYL 2-METHYLACRYLATE POLYMER   25719-52-2
  ENVIRONMENTAL LISTS
    List of Pesticide Product Inert Ingredients
    [present]

4-DODECYLOXY-2-HYDROXYBENZOPHENONE   2985-59-3
  ENVIRONMENTAL LISTS
    List of Pesticide Product Inert Ingredients
    [present]

DODECYLOXYPOLY(ETHYLENEOXY) ETHYL SUL-   9004-82-4
FATE, SODIUM SALT
  ENVIRONMENTAL LISTS
    List of Pesticide Product Inert Ingredients
    [present]

DODECYLPHENOL   27193-86-8
  ENVIRONMENTAL LISTS
    List of Pesticide Product Inert Ingredients
    [present]

DODECYLPHENOXYBENZENE DISULFONIC ACID   67993-50-4
  HEALTH AND SAFETY LISTS
    NIOSH - Selected LD50s and LC50s
    Oral, rat: LD50 = 566 mg/kg Skin, rabbit: LD50 = 1500 mg/kg
  ENVIRONMENTAL LISTS
    CERCLA/SARA - Section 313 - Emission Reporting
    form R reporting required

ALPHA-(DODECYLPHENYL)-OMEGA-HYDROXY-   9014-92-0
POLY(OXY-1,2-ETHANEDIYL)
  ENVIRONMENTAL LISTS
    List of Pesticide Product Inert Ingredients
    [present]

N-DODECYLSARCOSINE, SODIUM SALT   7631-98-3
  HEALTH AND SAFETY LISTS
    NFPA - Flash Points
    flash point = 205 degrees F (96 degrees C)
    NFPA - Hazard Identification Ratings
    health-2; flammability-1; reactivity-0
  STATE LISTS
    Florida Hazardous Substance List
    [present]
    Massachusetts Right To Know List
    [present]
    Pennsylvania Right to Know List
    [present]

DODECYL SULFATE, MAGNESIUM SALT   3097-08-3
  ENVIRONMENTAL LISTS
    List of Pesticide Product Inert Ingredients
    [present]

DODECYL SULFURIC ACID, POTASSIUM SALT   4706-78-9

HEALTH AND SAFETY LISTS

NFPA - Flash Points
flash point = 498 degrees F (254 degrees C)

NFPA - Hazard Identification Ratings
flammability-1;   reactivity-0

N-DODECYL-N-TETRADECYL BETA-ALANINE   3614-12-8

ENVIRONMENTAL LISTS

List of Pesticide Product Inert Ingredients
[present]

DODECYL TRICHLOROSILANE   4484-72-4

HEALTH AND SAFETY LISTS

U.S. DOT - Appendix B - Marine Pollutants
DOT regulated severe marine pollutant

NFPA - Flash Points
flash point = 325 degrees F (163 degrees C)

NFPA - Hazard Identification Ratings
health-0;   flammability-1;   reactivity-0

ENVIRONMENTAL LISTS

CAA - HON Rule - SOCMI Chemicals
compliance by April 24, 1995

EPA - Master Testing List
[present]

List of Pesticide Product Inert Ingredients
[present]

PROPOSED REGULATIONS

TSCA - Proposed Substances for Developmental/Reproductive Testing
proposed testing for: Developmental Toxicity - oral

DOLOMITE [CAMG(CO3)2]   16389-88-1

ENVIRONMENTAL LISTS

List of Pesticide Product Inert Ingredients
[present]

DOTRIACONTAMETHYLPENTADECASILOXANE   2471-11-6

ENVIRONMENTAL LISTS

List of Pesticide Product Inert Ingredients
[present]

INTERNATIONAL LISTS

Canada - WHMIS: Ingredient Disclosure
1% item 680 (796)

DOUGLAS FIR BARK   RR-01074-2

ENVIRONMENTAL LISTS

List of Pesticide Product Inert Ingredients
[present]

DOXYCYCLINE   564-25-0

ENVIRONMENTAL LISTS

List of Pesticide Product Inert Ingredients
[present]

DOXYCYCLINE CALCIUM   94088-85-4

ENVIRONMENTAL LISTS

List of Pesticide Product Inert Ingredients
[present]

DOXYCYCLINE HYCLATE   24390-14-5

ENVIRONMENTAL LISTS

List of Pesticide Product Inert Ingredients
[present]

DOXYCYCLINE MONOHYDRATE   17086-28-

HEALTH AND SAFETY LISTS

U.S. DOT - Substances From 49 CFR 172.101
regulated by DOT (UN1771)

U.S. DOT - Hazard Classes
DOT hazard class = 8

INTERNATIONAL LISTS

Canada - WHMIS: Ingredient Disclosure
1% item 681 (797)

STATE LISTS

NJ Right to Know List (Total)
sn 0823

NJ Special Hazardous Substances
(corrosive)

DOXYLAMINE   469-21-

ENVIRONMENTAL LISTS

List of Pesticide Product Inert Ingredients
[present]

DRAZOXOLON   5707-69-

ENVIRONMENTAL LISTS

TSCA - PAIR - Reporting List
Effective Date: June 14, 1993

TSCA - Health and Safety Reporting List
Effective Date: October 12, 1993;   Sunset Date: October 12, 2003

DRIED BLOOD   68911-49-

ENVIRONMENTAL LISTS

List of Pesticide Product Inert Ingredients
[present]

DULCIN   150-69-

STATE LISTS

California - Air Bill 2588 Appendix A-II
9/90

California - Prop. 65 - Developmental Toxicity
developmental toxicity (internal use) - initial date 7/1/90

DYPNONE   1322-90-

STATE LISTS

California - Prop. 65 - Developmental Toxicity
internal use: developmental toxicity - initial date 1/1/92

DYSPROSIUM 155   14982-00-

STATE LISTS

California - Prop. 65 - Developmental Toxicity
developmental toxicity (internal use) - initial date 10/1/91

DYSPROSIUM 157   14981-97-6

STATE LISTS

California - Prop. 65 - Developmental Toxicity
developmental toxicity (internal use) - initial date 10/1/91

DYSPROSIUM 159   14280-34-3

HEALTH AND SAFETY LISTS

NTP Chemical Status Reports - Testing Status and NTIS Number
Technical reports printed (no NTIS number given)

DYSPROSIUM 165   13967-64-

HEALTH AND SAFETY LISTS

U.S. DOT - Appendix B - Marine Pollutants
DOT regulated marine pollutant

DYSPROSIUM 166   15840-01-4

ENVIRONMENTAL LISTS
  List of Pesticide Product Inert Ingredients
  [present]

ECDYSTERONE   5289-74-7

HEALTH AND SAFETY LISTS
  IARC - Group 3 (not classifiable)
  [present]

EDIFENPHOS   17109-49-8

HEALTH AND SAFETY LISTS
  NFPA - Flash Points
  flash point = 350 degrees F (177 degrees C)
  NFPA - Hazard Identification Ratings
  health-1; flammability-1; reactivity-0

EICOSAMETHYLCYCLODECASILOXANE   18772-36-6

HEALTH AND SAFETY LISTS
  U.S. DOT - Appendix A Table 2 - Radionuclides
  final RQ = 100 curies (3.7E 12 Bq)

ENVIRONMENTAL LISTS
  CERCLA/SARA List of Radionuclides (Appendix B) and Their Reportable Quantities
  final RQ = 100 curies (3.7E 12 Bq)

EICOSAMETHYLNONASILOXANE   2652-13-3

HEALTH AND SAFETY LISTS
  U.S. DOT - Appendix A Table 2 - Radionuclides
  final RQ = 100 curies (3.7E 12 Bq)

ENVIRONMENTAL LISTS
  CERCLA/SARA List of Radionuclides (Appendix B) and Their Reportable Quantities
  final RQ = 100 curies (3.7E 12 Bq)

EICOSANE   112-95-8

HEALTH AND SAFETY LISTS
  U.S. DOT - Appendix A Table 2 - Radionuclides
  final RQ = 100 curies (3.7E 12 Bq)

ENVIRONMENTAL LISTS
  CERCLA/SARA List of Radionuclides (Appendix B) and Their Reportable Quantities
  final RQ = 100 curies (3.7E 12 Bq)

N-EICOSANOYL-N-METHYLTAURINE, SODIUM SALT   26885-07-4

HEALTH AND SAFETY LISTS
  U.S. DOT - Appendix A Table 2 - Radionuclides
  final RQ = 1000 curies (3.7E 13 Bq)

ENVIRONMENTAL LISTS
  CERCLA/SARA List of Radionuclides (Appendix B) and Their Reportable Quantities
  final RQ = 1000 curies (3.7E 13 Bq)

EICOSYLOXYPOLY(ETHYLENEOXY) ETHANOL   26636-39-5

HEALTH AND SAFETY LISTS
  U.S. DOT - Appendix A Table 2 - Radionuclides
  final RQ = 10 curies (3.7E 11 Bq)

ENVIRONMENTAL LISTS
  CERCLA/SARA List of Radionuclides (Appendix B) and Their Reportable Quantities
  final RQ = 10 curies (3.7E 11 Bq)

EINSTEINIUM 250   26150-38-9

HEALTH AND SAFETY LISTS
  NTP Chemical Status Reports - Testing Status and NTIS Number
  Project leader assigned/study in design

EINSTEINIUM 251   26250-43-1

HEALTH AND SAFETY LISTS
  U.S. DOT - Appendix B - Marine Pollutants
  DOT regulated marine pollutant

EINSTEINIUM 253   15840-02-5

ENVIRONMENTAL LISTS
  TSCA - PAIR - Reporting List
  Effective Date: October 12, 1993; Reporting Date: February 28, 1994
  TSCA - Health and Safety Reporting List
  Effective Date: October 12, 1993; Sunset Date: October 12, 2003

EINSTEINIUM 254   15840-03-6

ENVIRONMENTAL LISTS
  TSCA - PAIR - Reporting List
  Effective Date: October 12, 1993; Reporting Date: February 28, 1994
  TSCA - Health and Safety Reporting List
  Effective Date: October 12, 1993; Sunset Date: October 12, 2003

EINSTEINIUM 254M   RR-00464-8

HEALTH AND SAFETY LISTS
  NFPA - Flash Points
  flash point > 212 degrees F (100 degrees C)
  NFPA - Hazard Identification Ratings
  flammability-1; reactivity-0

ELMIRON (SODIUM PENTOSANPOLYSULFATE)   37319-17-8

ENVIRONMENTAL LISTS
  List of Pesticide Product Inert Ingredients
  [present]

EMERY   112-62-9

ENVIRONMENTAL LISTS
  List of Pesticide Product Inert Ingredients
  [present]

EMERY   12415-34-8

HEALTH AND SAFETY LISTS
  U.S. DOT - Appendix A Table 2 - Radionuclides
  final RQ = 10 curies (3.7E 11 Bq)

ENVIRONMENTAL LISTS
  CERCLA/SARA List of Radionuclides (Appendix B) and Their Reportable Quantities
  final RQ = 10 curies (3.7E 11 Bq)

EMERY   57407-26-8

HEALTH AND SAFETY LISTS
  U.S. DOT - Appendix A Table 2 - Radionuclides
  final RQ = 1000 curies (3.7E 13 Bq)

ENVIRONMENTAL LISTS
  CERCLA/SARA List of Radionuclides (Appendix B) and Their Reportable Quantities
  final RQ = 1000 curies (3.7E 13 Bq)

EMETINE DIHYDROCHLORIDE   316-42-7

HEALTH AND SAFETY LISTS

U.S. DOT - Appendix A Table 2 - Radionuclides
final RQ = 10 curies (3.7E 11 Bq)

ENVIRONMENTAL LISTS

CERCLA/SARA List of Radionuclides (Appendix B) and Their Reportable Quantities
final RQ = 10 curies (3.7E 11 Bq)

EMODIN   518-82-1

HEALTH AND SAFETY LISTS

U.S. DOT - Appendix A Table 2 - Radionuclides
final RQ = 0.1 curies (3.7E 9 Bq)

ENVIRONMENTAL LISTS

CERCLA/SARA List of Radionuclides (Appendix B) and Their Reportable Quantities
final RQ = 0.1 curies (3.7E 9 Bq)

ENDOSULFAN   115-29-7

HEALTH AND SAFETY LISTS

U.S. DOT - Appendix A Table 2 - Radionuclides
final RQ = 1 curie (3.7E 10 Bq)

ENVIRONMENTAL LISTS

CERCLA/SARA List of Radionuclides (Appendix B) and Their Reportable Quantities
final RQ = 1 curie (3.7E 10 Bq)

ALPHA-ENDOSULFAN   959-98-8

HEALTH AND SAFETY LISTS

NTP Chemical Status Reports - Testing Status and NTIS Number
Approved for toxicology/carcinogenesis study

BETA-ENDOSULFAN   33213-65-9

ENVIRONMENTAL LISTS

List of Pesticide Product Inert Ingredients
[present]

INTERNATIONAL LISTS

Australian Exposure Standards - Time Weighted Averages
dust: 10 mg/m3 TWA

Canada - Alberta - 8 Hour Occupational Exposure Limit
respirable mass: 5 mg/m3 TWA; total mass: 10 mg/m3 TWA

Canada - British Columbia - 8 Hour Exposure Limits
nuisance dust: 10 mg/m3 TWA

Canada - British Columbia - 15 Minute Exposure Limits
20 mg/m3 STEL

Canada - Quebec - Time-Weighted Average Exposure Values
10 mg/m3 TWAEV

United Kingdom - Occupational Exposure Standards - TWAs
total inhalable dust: 10 mg/m3 TWA; respirable dust: 5 mg/m3 TWA

Mexico - Instruction No. 10 - TWAs
10 mg/m3 TWA; (nuisance particulate)

Mexico - Instruction No. 10 - STELs
20 mg/m3 STEL

STATE LISTS

Minnesota Hazardous Substance List
[present] (includes inert or nuisance dust)

ENDOSULFAN METABOLITES   RR-00564-1

HEALTH AND SAFETY LISTS

OSHA - Vacated PELs - Time Weighted Averages
total dust: 10 mg/m3 TWA; respirable fraction: 5 mg/m3 TWA

OSHA - Final PELs - Time Weighted Averages
total dust: 15 mg/m3 TWA; respirable fraction: 5 mg/m3 TWA

STATE LISTS

Pennsylvania Right to Know List
[present]

ENDOSULFAN SULFATE   1031-07-8

INTERNATIONAL LISTS

Canada - Ontario - OHSA - TWAEVs
total dust: 10 mg/m3 TWAEV (listed as nuisance particulate)

ENDOTHALL   145-73-3

HEALTH AND SAFETY LISTS

NIOSH - Selected LD50s and LC50s
Oral, rat: LD50 = 12 ug/kg

NTP Chemical Status Reports - Testing Status and NTIS Number
Technical reports printed (PB278891/AS)

NTP Chemical Status Reports - Evidence of Carcinogenicity
male rat-inadequate; female rat-inadequate; male mice-inadequate; female mice-inadequate

ENVIRONMENTAL LISTS

CERCLA/SARA - Section 302 Extremely Hazardous Substances and TPQs
TPQ = 1/10,000 pounds

STATE LISTS

Florida Hazardous Substance List
effective March 13, 1992

Massachusetts Right To Know List
extraordinarily hazardous

NJ Right to Know List (Total)
sn 2387

Pennsylvania Right to Know List
environmental hazard

PROPOSED REGULATIONS

CERCLA/SARA - Proposed Hazardous Substance Additions
proposed RQ = 1 pound (.454 kg)

CERCLA/SARA - 1989 Proposed RQ Adjustments
proposed RQ = 1 pounds (.454 kg)

ENDOTHION   2778-04-3

HEALTH AND SAFETY LISTS

NTP Chemical Status Reports - Testing Status and NTIS Number
Two year studies in progress

ENDRIN   72-20-8

HEALTH AND SAFETY LISTS

ACGIH 1995 - Time Weighted Averages
0.1 mg/m3 TWA

ACGIH 1995 - Skin Designations
skin - potential for cutaneous absorption

U.S. DOT - Appendix B - Marine Pollutants
DOT regulated severe marine pollutant

U.S. DOT - Appendix A Table 1 - Hazardous Substances
final RQ = 1 pound (0.454 kg)

NIOSH - Selected LD50s and LC50s
Inhalation, rat: LC50 = 80 mg/m3 4 hr Oral, rat: LD50 = 18 mg/kg Skin, rat: LD50 = 74 mg/kg

NTP Chemical Status Reports - Testing Status and NTIS Number
Technical reports printed (PB281731/AS)

NTP Chemical Status Reports - Evidence of Carcinogenicity
male rat-inadequate; female rat-negative; male mice-inadequate; female mice-negative

OSHA - Vacated PELs - Time Weighted Averages
0.1 mg/m3 TWA

OSHA - Vacated PELs - Skin Designation
Prevent or reduce skin absorption

ENVIRONMENTAL LISTS

ATSDR Priority List
Rank (of 275): 059

CERCLA/SARA - Section 302 Extremely Hazardous Substances and TPQs
TPQ = 10/10,000 pounds

CERCLA/SARA - Hazardous Substances and their Reportable Quantities
final RQ = 1 pound (0.454 kg)

Clean Water Act - Hazardous Substances
[present]

Clean Water Act - Toxic Pollutants
[present]

RCRA - P Series Wastes
waste number P050

RCRA - Hazardous Constituents-Appendix VIII
waste number P050

RCRA - Substances Banned From Land Disposal
[present]

INTERNATIONAL LISTS

Australian Exposure Standards - Time Weighted Averages
0.1 mg/m3 TWA

Australian Exposure Standards - Skin Effects
skin absorption

Canada - Alberta - 8 Hour Occupational Exposure Limit
0.1 mg/m3 TWA

Canada - Alberta - 15 Minute Occupational Exposure Limit
0.3 mg/m3 STEL

Canada - Alberta - Skin Designation
can be absorbed through the intact skin

Canada - British Columbia - 8 Hour Exposure Limits
0.1 mg/m3 TWA

Canada - British Columbia - 15 Minute Exposure Limits
0.3 mg/m3 STEL

Canada - British Columbia - Skin Notations
skin - potential for skin absorption

Canada - Ontario - OHSA - TWAEVs
0.1 mg/m3 TWAEV

Canada - Ontario - OHSA - Skin Notations
absorption through skin, eyes, or mucous membranes

Canada - Quebec - Time-Weighted Average Exposure Values
0.1 mg/m3 TWAEV

Canada - Quebec - Skin Designations
absorbed through the skin

United Kingdom - Occupational Exposure Standards - TWAs
0.1 mg/m3 TWA

United Kingdom - Occupational Exposure Standards - STELs
0.3 mg/m3 STEL

United Kingdom - Occupational Exposure Standards - Notes
can be absorbed through skin

Israel - Time Weighted Averages
0.1 mg/m3 TWA

Israel - Action Levels
0.05 mg/m3 AL

Mexico - Instruction No. 10 - TWAs
0.1 mg/m3 TWA

Mexico - Instruction No. 10 - STELs
0.3 mg/m3 STEL

Mexico - Instruction No. 10 - Skin designation
skin - potential for cutaneous absorption

Mexico - Drinking Water - Ecological Criteria
(Alpha and Beta): 0.07 mg/l Substance presents persistence, bioaccumulations or risk of cancer, reduce human exposure to a minimum

STATE LISTS

California - Exposure Limits - PELs
0.1 mg/m3 PEL

California - Exposure Limits - Skin Notation
material may be absorbed through the skin, eyes or mucous membrane

California - Directors List of Hazardous Substances (8 CCR 339)
[present]

Florida Hazardous Substance List
[present]

Massachusetts Right To Know List
extraordinarily hazardous

Minnesota Hazardous Substance List
skin

NJ Right to Know List (Total)
sn 0824;   metabolites: sn 2987

Pennsylvania Right to Know List
environmental hazard

ENDRIN ALDEHYDE   7421-93-4
SEE ALSO:
F039-HAZARDOUS WASTES

HEALTH AND SAFETY LISTS

U.S. DOT - Appendix A Table 1 - Hazardous Substances
final RQ = 1 pound (0.454 kg)

ENVIRONMENTAL LISTS

ATSDR Priority List
Rank (of 275): 033

CERCLA/SARA - Hazardous Substances and their Reportable Quantities
final RQ = 1 pound (0.454 kg)

Clean Water Act - Priority Pollutants
[present]

RCRA - Basis for Listing - Appendix VII
Included in waste stream: F039

RCRA - TSD Facilities Ground Water Monitoring
TM 8080 = 0.1 ug/L PQL; TM 8250 = 10 ug/L PQL

RCRA - Universal Treatment Standards (LDR)
WW: 0.023 mg/l; NWW: 0.066 mg/kg

INTERNATIONAL LISTS

Mexico - Wastewater - Organic Toxic Pollutants and Heavy Metals
Listed under [Endosulfan and Metabolites]

STATE LISTS

California - Directors List of Hazardous Substances (8 CCR 339)
[present]

Massachusetts Right To Know List
[present]

NJ Right to Know List (Total)
sn 3014

Pennsylvania Right to Know List
environmental hazard

ENDRIN KETONE   53494-70-5
SEE ALSO:
F039-HAZARDOUS WASTES

HEALTH AND SAFETY LISTS

U.S. DOT - Appendix A Table 1 - Hazardous Substances
final RQ = 1 pound (0.454 kg)

ENVIRONMENTAL LISTS

ATSDR Priority List
Rank (of 275): 041

CERCLA/SARA - Hazardous Substances and their Reportable Quantities
final RQ = 1 pound (0.454 kg)

Clean Water Act - Priority Pollutants
[present]

RCRA - Basis for Listing - Appendix VII
Included in waste stream: F039

RCRA - TSD Facilities Ground Water Monitoring
TM 8080 = 0.05 ug/L PQL

RCRA - Universal Treatment Standards (LDR)
WW: 0.029 mg/l; NWW: 0.13 mg/kg

INTERNATIONAL LISTS

Mexico - Wastewater - Organic Toxic Pollutants and Heavy Metals
Listed under [Endosulfan and Metabolites]

STATE LISTS

California - Directors List of Hazardous Substances (8 CCR 339)
[present]

Massachusetts Right To Know List
[present]

NJ Right to Know List (Total)
sn 3015

Pennsylvania Right to Know List
environmental hazard

ENDRIN METABOLITES   **RR-00565-2**

STATE LISTS

Pennsylvania Right to Know List
environmental hazard

ENFLURANE   **13838-16-9**
SEE ALSO:
F039-HAZARDOUS WASTES

HEALTH AND SAFETY LISTS

U.S. DOT - Appendix A Table 1 - Hazardous Substances
final RQ = 1 pound (0.454 kg)

ENVIRONMENTAL LISTS

ATSDR Priority List
Rank (of 275): 040

CERCLA/SARA - Hazardous Substances and their Reportable Quantities
final RQ = 1 pound (0.454 kg)

Clean Water Act - Priority Pollutants
[present]

RCRA - Basis for Listing - Appendix VII
Included in waste stream: F039

RCRA - TSD Facilities Ground Water Monitoring
TM 8080 = 0.5 ug/L PQL; TM 8270 = 10 ug/L PQL

RCRA - Universal Treatment Standards (LDR)
WW: 0.029 mg/l; NWW: 0.13 mg/kg

INTERNATIONAL LISTS

Mexico - Wastewater - Organic Toxic Pollutants and Heavy Metals
Listed under [Endosulfan and Metabolites]

STATE LISTS

California - Directors List of Hazardous Substances (8 CCR 339)
[present]

Massachusetts Right To Know List
[present]

NJ Right to Know List (Total)
sn 2988

Pennsylvania Right to Know List
environmental hazard

EOSIN   **15086-94-9**

HEALTH AND SAFETY LISTS

U.S. DOT - Appendix A Table 1 - Hazardous Substances
final RQ = 1000 pounds (454 kg)

ENVIRONMENTAL LISTS

CERCLA/SARA - Hazardous Substances and their Reportable Quantities
final RQ = 1000 pounds (454 kg)

Safe Drinking Water Act - MCLs
MCL = 0.1 mg/L

Safe Drinking Water Act - MCLGs
MCLG = 0.1 mg/L

RCRA - P Series Wastes
waste number P088

RCRA - Hazardous Constituents-Appendix VIII
waste number P088

RCRA - Substances Banned From Land Disposal
[present]

TSCA - Code of Federal Regulations Citations
40 CFR 712.30(d)

TSCA - PAIR - Reporting List
Reporting Date: November 19, 1982

STATE LISTS

Massachusetts Right To Know List
[present]

Pennsylvania Right to Know List
environmental hazard

EPHEDRINE   **299-42-3**

HEALTH AND SAFETY LISTS

NIOSH - Selected LD50s and LC50s
Oral, rat: LD50 = 23 mg/kg Skin, rat: LD50 = 130 mg/kg

ENVIRONMENTAL LISTS

CERCLA/SARA - Section 302 Extremely Hazardous Substances and TPQs
TPQ = 500/10,000 pounds

STATE LISTS

Florida Hazardous Substance List
effective March 13, 1992

Massachusetts Right To Know List
extraordinarily hazardous

NJ Right to Know List (Total)
sn 2389

Pennsylvania Right to Know List
environmental hazard

PROPOSED REGULATIONS

CERCLA/SARA - Proposed Hazardous Substance Additions
proposed RQ = 1 pound (.454 kg)

CERCLA/SARA - 1989 Proposed RQ Adjustments
proposed RQ = 100 pounds (45.4 kg)

EPHEDRINE SALTS, OPTICAL ISOMERS, AND   **RR-01773-2**
SALTS OF OPTICAL ISOMERS
SEE ALSO:
F039-HAZARDOUS WASTES

HEALTH AND SAFETY LISTS

ACGIH 1995 - Time Weighted Averages
0.1 mg/m3 TWA

ACGIH 1995 - Skin Designations
skin - potential for cutaneous absorption

U.S. DOT - Appendix B - Marine Pollutants
DOT regulated severe marine pollutant

U.S. DOT - Appendix A Table 1 - Hazardous Substances
final RQ = 1 pound (0.454 kg)

IARC - Group 3 (not classifiable)
[present]

NIOSH - Selected LD50s and LC50s
Oral, rat: LD50 = 3 mg/kg Skin, rat: LD50 = 12 mg/kg

NIOSH 1990 - Pocket Guide - RELs
0.1 mg/m3 TWA

NIOSH 1990 - Pocket Guide - IDLHs
2000 mg/m3 IDLH

NIOSH 1990 - Pocket Guide - Target organs
CNS, liver

NIOSH 1990 - Pocket Guide - Skin list
Potential for dermal absorption

NTP Chemical Status Reports - Testing Status and NTIS Number
Technical reports printed (PB288461/AS)

NTP Chemical Status Reports - Evidence of Carcinogenicity
male rat-negative; female rat-negative; male mice-negative; female mice-negative

OSHA - Vacated PELs - Time Weighted Averages
0.1 mg/m3 TWA

OSHA - Vacated PELs - Skin Designation
Prevent or reduce skin absorption

OSHA - Final PELs - Time Weighted Averages
0.1 mg/m3 TWA

OSHA - Final PELs - Skin Notations
prevent or reduce skin absorption

ENVIRONMENTAL LISTS

ATSDR Priority List
Rank (of 275): 045

CERCLA/SARA - Section 302 Extremely Hazardous Substances and TPQs
TPQ = 500/10,000 pounds

CERCLA/SARA - Hazardous Substances and their Reportable Quantities
final RQ = 1 pound (0.454 kg)

Clean Water Act - Hazardous Substances
[present]

Clean Water Act - Priority Pollutants
[present]

Clean Water Act - Toxic Pollutants
[present]

Safe Drinking Water Act - MCLs
MCL = 0.0002 mg/L

Safe Drinking Water Act - MCLGs
MCLG = 0.002 mg/L

RCRA - D Series - Maximum Concentration of Contaminants
waste number D012;   regulatory level = 0.02 mg/L

RCRA - D Series - Chronic Toxicity Reference Levels
chronic toxicity reference level = 0.0002 mg/L

RCRA - P Series Wastes
waste number P051

RCRA - Hazardous Constituents-Appendix VIII
waste number P051

RCRA - Basis for Listing - Appendix VII
Included in waste stream: F039

RCRA - Substances Banned From Land Disposal
[present]

RCRA - TSD Facilities Ground Water Monitoring
TM 8080 = 0.1 ug/L PQL;   TM 8250 = 10 ug/L PQL

RCRA - Universal Treatment Standards (LDR)
WW: 0.0028 mg/l; NWW: 0.13 mg/kg

TSCA - Code of Federal Regulations Citations
40 CFR 704.102;   40 CFR 799.5055 (c),(d)(2)

TSCA - Multichemical Test Rules - Waste Constituents
hydrolysis testing for Chemical Fate

TSCA - Section 12(b) - Export Notification
export notification required - Section 4

INTERNATIONAL LISTS

Australian Exposure Standards - Time Weighted Averages
0.1 mg/m3 TWA

Australian Exposure Standards - Skin Effects
skin absorption

Canada - CEPA Schedule II Part II - Toxic Substances (Export)
[present]

Canada - Alberta - 8 Hour Occupational Exposure Limit
0.1 mg/m3 TWA

Canada - Alberta - 15 Minute Occupational Exposure Limit
0.3 mg/m3 STEL

Canada - Alberta - Skin Designation
can be absorbed through the intact skin

Canada - British Columbia - 8 Hour Exposure Limits
0.1 mg/m3 TWA

Canada - British Columbia - 15 Minute Exposure Limits
0.3 mg/m3 STEL

Canada - British Columbia - Skin Notations
skin - potential for skin absorption

Canada - Ontario - OHSA - TWAEVs
0.1 mg/m3 TWAEV

Canada - Ontario - OHSA - Skin Notations
absorption through skin, eyes, or mucous membranes

Canada - Quebec - Time-Weighted Average Exposure Values
0.1 mg/m3 TWAEV

Canada - Quebec - Skin Designations
absorbed through the skin

United Kingdom - Occupational Exposure Standards - TWAs
0.1 mg/m3 TWA

United Kingdom - Occupational Exposure Standards - STELs
0.3 mg/m3 STEL

United Kingdom - Occupational Exposure Standards - Notes
can be absorbed through skin

German (DFG) - MAK Values
total dust: 0.1 mg/m3 MAK

German (DFG) - Peak Limitations
10 x normal MAK (30 min average value);   don't exceed during shift

German (DFG) - Skin/Sensitizers
danger of cutaneous absorption

Israel - Time Weighted Averages
0.1 mg/m3 TWA

Israel - Action Levels
0.05 mg/m3 AL

Mexico - Instruction No. 10 - TWAs
0.1 mg/m3 TWA

Mexico - Instruction No. 10 - STELs
0.3 mg/m3 STEL

Mexico - Instruction No. 10 - Skin designation
skin - potential for cutaneous absorption

Mexico - Wastewater - Organic Toxic Pollutants and Heavy Metals
Listed under [Endrin and Metabolites]

Mexico - Drinking Water - Ecological Criteria
0.001 mg/l

STATE LISTS

California - Exposure Limits - PELs
0.1 mg/m3 PEL

California - Exposure Limits - Skin Notation
material may be absorbed through the skin, eyes or mucous membrane

California - Directors List of Hazardous Substances (8 CCR 339)
[present]

Florida Hazardous Substance List
[present]

Massachusetts Right To Know List
extraordinarily hazardous

Minnesota Hazardous Substance List
skin

NJ Right to Know List (Total)
sn 0825; metabolites: sn 2991

Pennsylvania Right to Know List
environmental hazard

EPHEDRINE SULFATE   134-72-5
SEE ALSO:
F039-HAZARDOUS WASTES

HEALTH AND SAFETY LISTS
U.S. DOT - Appendix A Table 1 - Hazardous Substances
final RQ = 1 pound (0.454 kg)

ENVIRONMENTAL LISTS
ATSDR Priority List
Rank (of 275): 145

CERCLA/SARA - Hazardous Substances and their Reportable Quantities
final RQ = 1 pound (0.454 kg)

Clean Water Act - Priority Pollutants
[present]

RCRA - Basis for Listing - Appendix VII
Included in waste stream: F039

RCRA - TSD Facilities Ground Water Monitoring
TM 8080 = 0.2 ug/L PQL;   TM 8270 = 10 ug/L PQL

RCRA - Universal Treatment Standards (LDR)
WW: 0.025 mg/l;   NWW: 0.13 mg/kg

INTERNATIONAL LISTS
Mexico - Wastewater - Organic Toxic Pollutants and Heavy Metals
Listed under [Endrin and Metabolites]

STATE LISTS
California - Directors List of Hazardous Substances (8 CCR 339)
[present]

Massachusetts Right To Know List
[present]

NJ Right to Know List (Total)
sn 2990

Pennsylvania Right to Know List
environmental hazard

EPIBROMOHYDRIN   3132-64-7
ENVIRONMENTAL LISTS
ATSDR Priority List
Rank (of 275): 066

EPICHLOROHYDRIN   106-89-8
STATE LISTS
Pennsylvania Right to Know List
environmental hazard

EPICHLOROHYDRIN, POLYMER WITH   26658-42-4
TETRAETHYLENEPENTAMINE
HEALTH AND SAFETY LISTS
ACGIH 1995 - Time Weighted Averages
75 ppm TWA;   566 mg/m3 TWA

NIOSH - Selected LD50s and LC50s
Inhalation, rat: LC50 = 14000 ppm 3 hr Oral, rat: LD50 = 5450 mg/kg

NIOSH - Health Standards - Exposure Limits
C (1 hr) 2 ppm (Listed under 'Waste anasthetic gases and vapors')

NIOSH - Health Standards - Health Effects and Precautions
Reproductive system effects and audio-visual performance decremen (Advise workers of potential effects) (Listed under 'Waste anasthet gases and vapors')

INTERNATIONAL LISTS
Australian Exposure Standards - Time Weighted Averages
0.5 ppm TWA;   3.8 mg/m3 TWA

Canada - Alberta - 8 Hour Occupational Exposure Limit
10 ppm TWA;   76 mg/m3 TWA

Canada - Alberta - 15 Minute Occupational Exposure Limit
15 ppm STEL;   113 mg/m3 STEL

Canada - British Columbia - 8 Hour Exposure Limits
2 ppm TWA

Canada - Ontario - OHSA - TWAEVs
2 ppm TWAEV;   16 mg/m3 TWAEV

Canada - Quebec - Time-Weighted Average Exposure Values
75 ppm TWAEV;   566 mg/m3 TWAEV

German (DFG) - MAK Values
20 ppm MAK;   150 mg/m3 MAK

German (DFG) - Peak Limitations
2 x normal MAK (30 min., average value) don't exceed 4 times durin shift

German (DFG) - Pregnancy
no risk to embryo/fetus if exposure limits adhered to

Israel - Time Weighted Averages
75 ppm TWA;   566 mg/m3 TWA

Israel - Action Levels
37.5 ppm AL;   284 mg/m3 AL

STATE LISTS
California - Exposure Limits - PELs
2 ppm PEL;   15 mg/m3 PEL

Massachusetts Right To Know List
[present]

Minnesota Hazardous Substance List
[present]

EPINEPHRINE HYDROCHLORIDE   55-31-2
HEALTH AND SAFETY LISTS
IARC - Group 3 (not classifiable)
[present]

EPN   2104-64-5
HEALTH AND SAFETY LISTS
FDA - Controlled Substances Act - Precursor chemicals
Threshold by base weight = 1 kilogram

EPOXIDIZED COPOLYMER OF PHENOL AND SUB-   RR-01255-5
STITUTED PHENOL
HEALTH AND SAFETY LISTS
FDA - Controlled Substances Act - Precursor chemicals
Threshold by base weight = 1 kilogram

EPOXIDIZED LINSEED OIL   8016-11-3
HEALTH AND SAFETY LISTS
NTP Chemical Status Reports - Testing Status and NTIS Number
Technical reports printed (PB86247285/AS)

NTP Chemical Status Reports - Evidence of Carcinogenicity
male rat-no evidence;   female rat-no evidence;   male mice-no evidence;   female mice-no evidence

EPOXIDIZED POLYBUTENE   RR-00907-4

SEE ALSO:
  HALOGENATED ALKYL EPOXIDES

HEALTH AND SAFETY LISTS

U.S. DOT - Substances From 49 CFR 172.101
  regulated by DOT (UN2558)

U.S. DOT - Hazard Classes
  DOT hazard class = 6.1

ENVIRONMENTAL LISTS

TSCA - Code of Federal Regulations Citations
  40 CFR 712.30(d); 40 CFR 716.120(c); 40 CFR 721.950

TSCA - PAIR - Reporting List
  Reporting Date: November 19, 1982

TSCA - Chemicals with Significant New Use Rules
  [present]

TSCA - Section 12(b) - Export Notification
  export notification required - Section 5

INTERNATIONAL LISTS

Canada - WHMIS: Ingredient Disclosure
  1% item 682 (800)

STATE LISTS

NJ Right to Know List (Total)
  sn 0827

EPOXIDIZED SOYBEAN OIL   8013-07-8

SEE ALSO:
  K017-HAZARDOUS WASTES

HEALTH AND SAFETY LISTS

ACGIH 1995 - Time Weighted Averages
  (2) ppm TWA; (7.6) mg/m3 TWA

ACGIH 1995 - Skin Designations
  skin - potential for cutaneous absorption

AIHA - Odor Threshold Values
  no geometric mean air odor threshold

U.S. DOT - Substances From 49 CFR 172.101
  regulated by DOT (UN2023)

U.S. DOT - Hazard Classes
  DOT hazard class = 6.1

U.S. DOT - Appendix A Table 1 - Hazardous Substances
  final RQ = 100 pounds (45.4 kg)

IARC - Group 2A (limited human data)
  [present] (Other relevant data, as given in Supplement 7, influenced the making of the overall evaluation)

NFPA - Flash Points
  flash point = 88 degrees F (31 degrees C)

NFPA - Hazard Identification Ratings
  health-3; flammability-3; reactivity-2

NIOSH - Selected LD50s and LC50s
  Inhalation, rat: LC50 = 250 ppm 8 hr Oral, rat: LD50 = 90 mg/kg Skin, rabbit: LD50 = 515 mg/kg

NIOSH 1990 - Pocket Guide - RELs
  Reduce exposure to lowest feasible concentration

NIOSH 1990 - Pocket Guide - IDLHs
  250 ppm IDLH (not considering carcinogenic effects)

NIOSH 1990 - Pocket Guide - Carcinogens
  occupational carcinogen

NIOSH 1990 - Pocket Guide - Target organs
  respiratory system, skin, kidneys

NIOSH - Health Standards - Exposure Limits
  minimize occupational exposure

NIOSH - Health Standards - Health Effects and Precautions
  Respiratory cancer; mutagenesis; reproductive, skin, kidney, liver, and respiratory effects (Prevent skin contact)

NIOSH - Health Standards - Carcinogenic Chemicals
  potential human carcinogen

NTP Seventh Report - Suspect Carcinogens
  suspect carcinogen

OSHA - Vacated PELs - Time Weighted Averages
  2 ppm TWA; 8 mg/m3 TWA

OSHA - Vacated PELs - Skin Designation
  Prevent or reduce skin absorption

OSHA - Final PELs - Time Weighted Averages
  5 ppm TWA; 19 mg/m3 TWA

OSHA - Final PELs - Skin Notations
  prevent or reduce skin absorption

OSHA - Possible Select Carcinogens
  [present]

ENVIRONMENTAL LISTS

CERCLA/SARA - Section 302 Extremely Hazardous Substances and TPQs
  TPQ = 1000 pounds

CERCLA/SARA - Section 313 - Emission Reporting
  form R reporting required for 0.1% de minimus concentration

CERCLA/SARA - Hazardous Substances and their Reportable Quantities
  final RQ = 100 pounds (45.4 kg)

Clean Air Act (1990) - List of Hazardous Air Contaminants
  [present]

CAA -Toxic Substances for Accidental Release Prevention
  threshold quantity = 20,000 lbs

CAA - HON Rule - SOCMI Chemicals
  compliance by Oct. 24, 1994

CAA - HON Rule - Organic HAPs
  [present]

Clean Water Act - Hazardous Substances
  [present]

Safe Drinking Water Act - MCLs
  MCL = treatment technique (polymer addition practices)

Safe Drinking Water Act - MCLGs
  MCLG = Zero

EPA - Carcinogen Hazard Ranking for RQ Adjustment
  Hazard ranking = Low

RCRA - U Series Wastes
  waste number U041

RCRA - Hazardous Constituents-Appendix VIII
  waste number U041

RCRA - Basis for Listing - Appendix VII
  Included in waste stream: K017

RCRA - Substances Banned From Land Disposal
  [present]

TSCA - Code of Federal Regulations Citations
  40 CFR 712.30(d); 40 CFR 716.120(a)

TSCA - PAIR - Reporting List
  Reporting Date: November 19, 1982

TSCA - Health and Safety Reporting List
  Effective Date: October 4, 1982

INTERNATIONAL LISTS

Australian Exposure Standards - Time Weighted Averages
  2 ppm TWA; 7.6 mg/m3 TWA

Australian Exposure Standards - Skin Effects
  skin absorption

Australian Exposure Standards - Carcinogens
  probable carcinogen

Canada - WHMIS: Ingredient Disclosure
  0.1% item 683 (801)

Canada - NPRI (National Pollutant Release Inventory)
[present]

Canada - Alberta - 8 Hour Occupational Exposure Limit
2 ppm TWA;   7.6 mg/m3 TWA

Canada - Alberta - 15 Minute Occupational Exposure Limit
5 ppm STEL;   19 mg/m3 STEL

Canada - Alberta - Skin Designation
can be absorbed through the intact skin

Canada - British Columbia - 8 Hour Exposure Limits
5 ppm TWA;   20 mg/m3 TWA

Canada - British Columbia - 15 Minute Exposure Limits
10 ppm STEL;   40 mg/m3 STEL

Canada - British Columbia - Skin Notations
skin - potential for skin absorption

Canada - British Columbia - Carcinogens
carcinogen - 5 ppm TWA;   20 mg/m3 TWA

Canada - Ontario - OHSA - TWAEVs
2 ppm TWAEV;   7.6 mg/m3 TWAEV

Canada - Ontario - OHSA - Skin Notations
absorption through skin, eyes, or mucous membranes

Canada - Quebec - Time-Weighted Average Exposure Values
2 ppm TWAEV;   7.6 mg/m3 TWAEV (substance of which the recirculation is prohibited)

Canada - Quebec - Skin Designations
absorbed through the skin

Canada - Quebec - Carcinogens
C2 carcinogen: effect suspected in humans

German (DFG) - Skin/Sensitizers
danger of cutaneous absorption

German (DFG) - Carcinogens
animal evidence of carcinogenicity

Israel - Time Weighted Averages
(2) ppm TWA;   (7.6) mg/m3 TWA

Israel - Action Levels
1 ppm AL;   3.8 mg/m3 AL

Mexico - Instruction No. 10 - TWAs
2 ppm TWA;   10 mg/m3 TWA

Mexico - Instruction No. 10 - STELs
5 ppm STEL;   20 mg/m3 STEL

STATE LISTS

California - Air Bill 2588 Appendix A-I
known or potential carcinogen

California - Prop. 65 - Cancer list
carcinogen - initial date 10/1/87

California - Prop. 65 - No Significant Risk Levels
no significant risk level = 9 ug/day

California - Exposure Limits - PELs
2 ppm PEL;   7.6 mg/m3 PEL

California - Exposure Limits - Skin Notation
material may be absorbed through the skin, eyes or mucous membrane

California - Directors List of Hazardous Substances (8 CCR 339)
[present] (exempt when part of a cured epoxy or rubber)

Florida Hazardous Substance List
[present]

Massachusetts Right To Know List
carcinogen;   extraordinarily hazardous

Minnesota Hazardous Substance List
carcinogen;   skin

NJ Right to Know List (Total)
sn 0828

NJ Special Hazardous Substances
(carcinogen;   mutagen;   reactive - second degree)

Pennsylvania Right to Know List
environmental hazard;   special hazardous substance

Pennsylvania RTK - Special Hazardous Substances
[present]

PROPOSED REGULATIONS

ACGIH 1995 - Notice of Intended Changes
(skin) 0.1 ppm TWA;   0.38 mg/m3 TWA;   A2-suspected human carcinogen

EPOXIDIZED TALL-OIL FATTY ACIDS WITH TALL-OIL ROSIN   RR-01075-

ENVIRONMENTAL LISTS

List of Pesticide Product Inert Ingredients
[present]

3,4-EPOXYCYCLOHEXANECARBOXYLIC ACID (3,4-EPOXYCYCLOHEXYLMETHYL) ESTER   2386-87-

HEALTH AND SAFETY LISTS

NTP Chemical Status Reports - Testing Status and NTIS Number
Technical reports printed (PB91142323/AS)

NTP Chemical Status Reports - Evidence of Carcinogenicity
male rat-inadequate;   female rat-inadequate;   male mice-inadequate female mice-inadequate

EPOXY ETHYLOXY PROPANE   4016-11-9

HEALTH AND SAFETY LISTS

ACGIH 1995 - Time Weighted Averages
0.1 mg/m3 TWA

ACGIH 1995 - Skin Designations
skin - potential for cutaneous absorption

U.S. DOT - Appendix B - Marine Pollutants
DOT regulated severe marine pollutant

NIOSH - Selected LD50s and LC50s
Oral, rat: LD50 = 7 mg/kg Skin, rat: LD50 = 25 mg/kg

NIOSH 1990 - Pocket Guide - RELs
0.5 mg/m3 TWA

NIOSH 1990 - Pocket Guide - IDLHs
50 mg/m3 IDLH

NIOSH 1990 - Pocket Guide - Target organs
respiratory system, CVS, CNS, eyes, skin, blood cholinesterase

NIOSH 1990 - Pocket Guide - Skin list
Potential for dermal absorption

OSHA - Vacated PELs - Time Weighted Averages
0.5 mg/m3 TWA

OSHA - Vacated PELs - Skin Designation
Prevent or reduce skin absorption

OSHA - Final PELs - Time Weighted Averages
0.5 mg/m3 TWA

OSHA - Final PELs - Skin Notations
prevent or reduce skin absorption

ENVIRONMENTAL LISTS

CERCLA/SARA - Section 302 Extremely Hazardous Substances and TPQs
TPQ = 100/10,000 pounds

INTERNATIONAL LISTS

Australian Exposure Standards - Time Weighted Averages
0.5 mg/m3 TWA

Australian Exposure Standards - Skin Effects
skin absorption

Canada - Alberta - 8 Hour Occupational Exposure Limit
0.5 mg/m3 TWA

Canada - Alberta - 15 Minute Occupational Exposure Limit
2 mg/m3 STEL

Canada - Alberta - Skin Designation
can be absorbed through the intact skin

Canada - British Columbia - 8 Hour Exposure Limits
0.5 mg/m3 TWA

Canada - British Columbia - 15 Minute Exposure Limits
2 mg/m3 STEL

Canada - British Columbia - Skin Notations
skin - potential for skin absorption

Canada - Ontario - OHSA - TWAEVs
0.5 mg/m3 TWAEV

Canada - Ontario - OHSA - Skin Notations
absorption through skin, eyes, or mucous membranes

Canada - Quebec - Time-Weighted Average Exposure Values
0.5 mg/m3 TWAEV

Canada - Quebec - Skin Designations
absorbed through the skin

German (DFG) - MAK Values
total dust: 0.5 mg/m3 MAK

German (DFG) - Peak Limitations
10 x normal MAK (30 min average value);   don't exceed during shift

German (DFG) - Skin/Sensitizers
danger of cutaneous absorption

Israel - Time Weighted Averages
0.5 mg/m3 TWA

Israel - Action Levels
0.25 mg/m3 AL

Mexico - Instruction No. 10 - TWAs
0.5 mg/m3 TWA

Mexico - Instruction No. 10 - STELs
2 mg/m3 STEL

STATE LISTS

California - Exposure Limits - PELs
0.5 mg/m3 PEL

California - Exposure Limits - Skin Notation
material may be absorbed through the skin, eyes or mucous membrane

California - Directors List of Hazardous Substances (8 CCR 339)
[present]

Florida Hazardous Substance List
[present]

Massachusetts Right To Know List
extraordinarily hazardous

Minnesota Hazardous Substance List
skin

NJ Right to Know List (Total)
sn 0829

Pennsylvania Right to Know List
environmental hazard

PROPOSED REGULATIONS

CERCLA/SARA - Proposed Hazardous Substance Additions
proposed RQ = 1 pound (.454 kg)

CERCLA/SARA - 1989 Proposed RQ Adjustments
proposed RQ = 100 pounds (45.4 kg)

3,4-EPOXY-6-METHYLCYCLOHEXYLMETHYL-3,4-   141-37-7
EPOXY-6-METHYLCYCLOHEXANE CARBOXYLATE

ENVIRONMENTAL LISTS

TSCA - Chemicals with Significant New Use Rules
PMN number: P-91-598

EPOXY RESIN   RR-01000-4

ENVIRONMENTAL LISTS

List of Pesticide Product Inert Ingredients
[present]

CIS-9,10-EPOXYSTEARIC ACID   2443-39-2

ENVIRONMENTAL LISTS

TSCA - Chemicals with Significant New Use Rules
PMN number: P-88-217

ERBIUM 161   14967-67-0

HEALTH AND SAFETY LISTS

NIOSH - Selected LD50s and LC50s
Oral, rat: LD50 = 40 gm/kg

ENVIRONMENTAL LISTS

List of Pesticide Product Inert Ingredients
[present]

ERBIUM 165   14041-43-1

ENVIRONMENTAL LISTS

List of Pesticide Product Inert Ingredients
[present]

ERBIUM 169   15840-13-8

HEALTH AND SAFETY LISTS

NIOSH - Selected LD50s and LC50s
Oral, rat: LD50 = 4490 mg/kg

ERBIUM 171   14391-45-8
SEE ALSO:
GLYCIDOL (OXIRANEMETHANOL) AND ITS DERIVATIVES

HEALTH AND SAFETY LISTS

U.S. DOT - Substances From 49 CFR 172.101
regulated by DOT (UN2752)

U.S. DOT - Hazard Classes
DOT hazard class = 3

ENVIRONMENTAL LISTS

EPA - Master Testing List
[present]

TSCA - Code of Federal Regulations Citations
40 CFR 712.30(d);   40 CFR 716.120(c)

TSCA - PAIR - Reporting List
Reporting Date: November 19, 1982

STATE LISTS

NJ Right to Know List (Total)
sn 0830

PROPOSED REGULATIONS

TSCA - Proposed Testing Rule for Glycidyl Ethers
member of Glycidyl subcategory I-A

ERBIUM 172   15840-14-9

HEALTH AND SAFETY LISTS

IARC - Group 3 (not classifiable)
[present]

STATE LISTS

California - Directors List of Hazardous Substances (8 CCR 339)
[present]

ERGOCALCIFEROL   50-14-6

ENVIRONMENTAL LISTS

TSCA - Chemicals with Significant New Use Rules
PMN number: P-84-1167

STATE LISTS

California - Air Bill 2588 Appendix A-I
9/89

ERGONOVINE   60-79-7

HEALTH AND SAFETY LISTS

IARC - Group 3 (not classifiable)
[present]

ERGONOVINE SALTS   **RR-01774-3**
  HEALTH AND SAFETY LISTS
    U.S. DOT - Appendix A Table 2 - Radionuclides
    final RQ = 100 curies (3.7E 12 Bq)
  ENVIRONMENTAL LISTS
    CERCLA/SARA List of Radionuclides (Appendix B) and Their Reportable Quantities
    final RQ = 100 curies (3.7E 12 Bq)

ERGOTAMINE   **113-15-5**
  HEALTH AND SAFETY LISTS
    U.S. DOT - Appendix A Table 2 - Radionuclides
    final RQ = 1000 curies (3.7E 13 Bq)
  ENVIRONMENTAL LISTS
    CERCLA/SARA List of Radionuclides (Appendix B) and Their Reportable Quantities
    final RQ = 1000 curies (3.7E 13 Bq)

ERGOTAMINE SALTS   **RR-01775-4**
  HEALTH AND SAFETY LISTS
    U.S. DOT - Appendix A Table 2 - Radionuclides
    final RQ = 100 curies (3.7E 12 Bq)
  ENVIRONMENTAL LISTS
    CERCLA/SARA List of Radionuclides (Appendix B) and Their Reportable Quantities
    final RQ = 100 curies (3.7E 12 Bq)

ERGOTAMINE TARTRATE   **379-79-3**
  HEALTH AND SAFETY LISTS
    U.S. DOT - Appendix A Table 2 - Radionuclides
    final RQ = 100 curies (3.7E 12 Bq)
  ENVIRONMENTAL LISTS
    CERCLA/SARA List of Radionuclides (Appendix B) and Their Reportable Quantities
    final RQ = 100 curies (3.7E 12 Bq)

ERIONITE   **12510-42-8**
  HEALTH AND SAFETY LISTS
    U.S. DOT - Appendix A Table 2 - Radionuclides
    final RQ = 10 curies (3.7E 11 Bq)
  ENVIRONMENTAL LISTS
    CERCLA/SARA List of Radionuclides (Appendix B) and Their Reportable Quantities
    final RQ = 10 curies (3.7E 11 Bq)

ERIONITE   **66733-21-9**
  HEALTH AND SAFETY LISTS
    NFPA - Flash Points
    flash point = 55 degrees F (13 degrees C)
    NFPA - Hazard Identification Ratings
    health-2; flammability-3; reactivity-0
  ENVIRONMENTAL LISTS
    CERCLA/SARA - Section 302 Extremely Hazardous Substances and TPQs
    TPQ = 1000/10,000 pounds
  STATE LISTS
    Massachusetts Right To Know List
    extraordinarily hazardous
    NJ Right to Know List (Total)
    sn 2391
    Pennsylvania Right to Know List
    environmental hazard

PROPOSED REGULATIONS
  CERCLA/SARA - Proposed Hazardous Substance Additions
  proposed RQ = 1 pound (.454 kg)
  CERCLA/SARA - 1989 Proposed RQ Adjustments
  proposed RQ = 100 pounds (45.4 kg)

ERYTHORBIC ACID   **89-65-**
  HEALTH AND SAFETY LISTS
    FDA - Controlled Substances Act - Precursor chemicals
    Threshold by base weight = 10 grams

ERYTHROMYCIN   **114-07-**
  HEALTH AND SAFETY LISTS
    FDA - Controlled Substances Act - Precursor chemicals
    Threshold by base weight = 10 grams

ERYTHROMYCIN STEARATE   **643-22-**
  HEALTH AND SAFETY LISTS
    FDA - Controlled Substances Act - Precursor chemicals
    Threshold by base weight = 20 grams

ESSENTIAL OILS   **8022-96-6**
  HEALTH AND SAFETY LISTS
    FDA - Controlled Substances Act - Precursor chemicals
    Threshold by base weight = 20 grams

ESTRADIOL-17B   **50-28-2**
  ENVIRONMENTAL LISTS
    CERCLA/SARA - Section 302 Extremely Hazardous Substances and TPQs
    TPQ = 500/10,000 pounds
  STATE LISTS
    California - Air Bill 2588 Appendix A-II
    9/90
    California - Prop. 65 - Developmental Toxicity
    developmental toxicity - initial date 4/1/90
    Florida Hazardous Substance List
    effective March 13, 1992
    Massachusetts Right To Know List
    extraordinarily hazardous
    NJ Right to Know List (Total)
    sn 2392
    Pennsylvania Right to Know List
    environmental hazard
  PROPOSED REGULATIONS
    CERCLA/SARA - Proposed Hazardous Substance Additions
    proposed RQ = 1 pound (.454 kg)
    CERCLA/SARA - 1989 Proposed RQ Adjustments
    proposed RQ = 100 pounds (45.4 kg)

ESTRADIOL MUSTARD   **22966-79-6**
  HEALTH AND SAFETY LISTS
    NTP Seventh Report - Known Carcinogens
    known carcinogen
    OSHA - Select Carcinogens
    [present]
  ENVIRONMENTAL LISTS
    TSCA - Chemicals with Significant New Use Rules
    [present]
    TSCA - Section 12(b) - Export Notification
    export notification required - Section 5

INTERNATIONAL LISTS
  German (DFG) - Carcinogens
  proven carcinogen

STATE LISTS
  California - Air Bill 2588 Appendix A-I
  known or potential carcinogen
  California - Prop. 65 - Cancer list
  carcinogen - initial date 10/1/88
  California - Directors List of Hazardous Substances (8 CCR 339)
  [present]

STROGENS, NON-STEROIDAL   RR-01742-5

HEALTH AND SAFETY LISTS
  IARC - Group 1 (carcinogenic to humans)
  [present]
  OSHA - Select Carcinogens
  [present]

ENVIRONMENTAL LISTS
  TSCA - Chemicals with Significant New Use Rules
  [present]

INTERNATIONAL LISTS
  Canada - Quebec - Time-Weighted Average Exposure Values
  PROHIBITED USE
  Canada - Quebec - Carcinogens
  C1 carcinogen: effect detected in humans

STATE LISTS
  Massachusetts Right To Know List
  carcinogen; extraordinarily hazardous
  Minnesota Hazardous Substance List
  carcinogen

STROGENS, STEROIDAL   RR-01743-6

ENVIRONMENTAL LISTS
  List of Pesticide Product Inert Ingredients
  [present]

STRONE   53-16-7

HEALTH AND SAFETY LISTS
  AIHA - WEEL - Time Weighted Averages
  3 mg/m3 TWA

STATE LISTS
  Minnesota Hazardous Substance List
  [present]

THANAMINE, 2,2-DIMETHOXY-N-METHYL-   122-07-6

HEALTH AND SAFETY LISTS
  NTP Chemical Status Reports - Testing Status and NTIS Number
  Technical reports printed (PB89178537/AS)
  NTP Chemical Status Reports - Evidence of Carcinogenicity
  male rat-no evidence; female rat-no evidence; male mice-no evidence; female mice-no evidence

ETHANAMINIUM, 2-AMINO-N-(2-AMINOETHYL)-N-   68153-35-5
2-HYDROXYETHYL)-N-METHYL-, N,N'-DITALLOW
ACYL DERIVATIVES, METHYL SULFATES (SALTS)

ENVIRONMENTAL LISTS
  List of Pesticide Product Inert Ingredients
  [present]

ETHANAMINIUM, 2-[[2-CYANO-3-[4-(DIETHY-   64992-16-1
LAMINO)PHENYL]-1-OXO-2-PROPENYL]OXY]-N,N,
N-TRIMETHYL-, CHLORIDE

HEALTH AND SAFETY LISTS
  NTP Seventh Report - Suspect Carcinogens
  suspect carcinogen (Listed under 'Estrogens (not conjugated)')
  OSHA - Possible Select Carcinogens
  [present]

STATE LISTS
  California - Air Bill 2588 Appendix A-II
  known or potential carcinogen
  California - Prop. 65 - Cancer list
  carcinogen - initial date 1/1/88
  California - Prop. 65 - No Significant Risk Levels
  no significant risk level = 0.02 ug/day
  California - Directors List of Hazardous Substances (8 CCR 339)
  [present]
  Florida Hazardous Substance List
  [present]
  Massachusetts Right To Know List
  carcinogen; extraordinarily hazardous
  Minnesota Hazardous Substance List
  carcinogen
  NJ Right to Know List (Total)
  sn 0832
  NJ Special Hazardous Substances
  (carcinogen)
  Pennsylvania Right to Know List
  special hazardous substance
  Pennsylvania RTK - Special Hazardous Substances
  [present]

ETHANE   74-84-0

HEALTH AND SAFETY LISTS
  IARC - Group 3 (not classifiable)
  [present]
  NTP Chemical Status Reports - Testing Status and NTIS Number
  Technical reports printed (PB285787/AS)
  NTP Chemical Status Reports - Evidence of Carcinogenicity
  male rat-negative; female rat-negative; male mice-positive; female mice-positive

STATE LISTS
  California - Directors List of Hazardous Substances (8 CCR 339)
  [present]
  Massachusetts Right To Know List
  carcinogen; extraordinarily hazardous

ETHANE, 1,2-BIS(2-CHLOROETHOXY)-   112-26-5

STATE LISTS
  California - Air Bill 2588 Appendix A-II
  known or potential carcinogen

ETHANE, 2-CHLORO-1,1,1,2-TETRAFLUORO-   2837-89-0

STATE LISTS
  California - Air Bill 2588 Appendix A-II
  known or potential carcinogen

ETHANE DICHLORIDE   1300-21-6

HEALTH AND SAFETY LISTS
  IARC - Group Unspecified
  [present] (Listed under 'Steroidal oestrogens')
  NTP Seventh Report - Suspect Carcinogens
  suspect carcinogen (Listed under 'Estrogens (not conjugated)')

OSHA - Possible Select Carcinogens
[present]

STATE LISTS

California - Air Bill 2588 Appendix A-II
known or potential carcinogen

California - Prop. 65 - Cancer list
carcinogen - initial date 1/1/88

Florida Hazardous Substance List
[present]

Massachusetts Right To Know List
carcinogen; extraordinarily hazardous

Minnesota Hazardous Substance List
carcinogen

NJ Right to Know List (Total)
sn 0833

NJ Special Hazardous Substances
(carcinogen)

Pennsylvania Right to Know List
special hazardous substance

Pennsylvania RTK - Special Hazardous Substances
[present]

ETHANE, 1,1-DICHLORO-1-FLUORO-   1717-00-6

ENVIRONMENTAL LISTS

TSCA - Code of Federal Regulations Citations
40 CFR 712.30(d)

TSCA - PAIR - Reporting List
Reporting Date: November 19, 1982

ETHANE, 2,2-DICHLORO-1,1,1-TRIFLUORO-   306-83-2

ENVIRONMENTAL LISTS

TSCA - Code of Federal Regulations Citations
40 CFR 712.30(w); 40 CFR 716.120(a)

TSCA - PAIR - Reporting List
Reporting Date: August 18, 1988

TSCA - Health and Safety Reporting List
Effective Date: June 20, 1988; Sunset Date: November 9, 1993

ETHANEDIMIDIC ACID   RR-01242-0

ENVIRONMENTAL LISTS

TSCA - PAIR - Reporting List
Effective Date: January 26, 1994; Reporting Date: March 28, 1994

TSCA - Health and Safety Reporting List
Effective Date: January 26, 1994; Sunset Date: January 26, 2004

ETHANEDIOIC ACID, DIHYDRATE   6153-56-6

HEALTH AND SAFETY LISTS

ACGIH 1995 - Time Weighted Averages
simple asphyxiant

AIHA - Odor Threshold Values
no geometric mean air odor threshold

U.S. DOT - Substances From 49 CFR 172.101
regulated by DOT (UN1035, UN1961)

U.S. DOT - Hazard Classes
DOT hazard class = 2.1

NFPA - Flash Points
gas (no flash point given)

NFPA - Hazard Identification Ratings
health-1; flammability-4; reactivity-0

ENVIRONMENTAL LISTS

CAA - Flammable Substances for Accidental Release Prevention
threshold quantity = 10,000 lbs

List of Pesticide Product Inert Ingredients
[present]

INTERNATIONAL LISTS

Australian Exposure Standards - Time Weighted Averages
Asphyxiant at < 18% oxygen by volume; explosion hazard

Canada - British Columbia - 8 Hour Exposure Limits
asphyxiant substance

Canada - Ontario - OHSA - TWAEVs
simple asphyxiant

Canada - Quebec - Time-Weighted Average Exposure Values
simple asphyxiant

Israel - Time Weighted Averages
Asphyxiant

Mexico - Instruction No. 10 - TWAs
simple asphyxiant

STATE LISTS

California - Exposure Limits - PELs
asphyxiant (limit depends on level of oxygen)

Massachusetts Right To Know List
[present]

Minnesota Hazardous Substance List
[present]

NJ Right to Know List (Total)
sn 0834

NJ Special Hazardous Substances
(flammable - fourth degree)

Pennsylvania Right to Know List
[present]

1,2-ETHANEDIOL BIS(4-METHYLBENZENESUL-   6315-52-2
FONATE)

HEALTH AND SAFETY LISTS

NFPA - Flash Points
flash point = 250 degrees F (121 degrees C)

NFPA - Hazard Identification Ratings
health-2; flammability-1; reactivity-0

NIOSH - Selected LD50s and LC50s
Oral, rat: LD50 = 250 mg/kg Skin, rabbit: LD50 = 1410 mg/kg

STATE LISTS

Florida Hazardous Substance List
[present]

Massachusetts Right To Know List
[present]

Pennsylvania Right to Know List
[present]

1,2-ETHANEDIOL, DIACETATE   111-55-7

HEALTH AND SAFETY LISTS

AIHA - WEEL - Time Weighted Averages
1000 ppm TWA

ENVIRONMENTAL LISTS

CERCLA/SARA - Section 313 - Emission Reporting
form R reporting required

Class 2 Ozone Depletors
ozone depletion weight = 0.02

TSCA - Code of Federal Regulations Citations
40 CFR 721.1006

TSCA - Health and Safety Reporting List
Effective Date: October 15, 1990; Sunset Date: November 9, 1993

TSCA - Chemicals with Significant New Use Rules
PMN number: P-88-1763

TSCA - Section 12(b) - Export Notification
P-88-1763; export notification required - Section 5

1,1-ETHANEDIOL, 2,2,2-TRICHLORO-   302-17-0

  HEALTH AND SAFETY LISTS

    NIOSH - Selected LD50s and LC50s
    Oral, rat: LD50 = 1120 mg/kg Skin, rabbit: LD50 = 3890 mg/kg

  ENVIRONMENTAL LISTS

    ATSDR Priority List
    Rank (of 275): 220

ETHANEDITHIOAMIDE   79-40-3

  HEALTH AND SAFETY LISTS

    AIHA - WEEL - Time Weighted Averages
    500 ppm TWA

  ENVIRONMENTAL LISTS

    CERCLA/SARA - Section 313 - Emission Reporting
    form R reporting required
    Class 2 Ozone Depletors
    ozone depletion weight = 0.12
    TSCA - Code of Federal Regulations Citations
    40 CFR 721.1007
    TSCA - Health and Safety Reporting List
    Effective Date: October 15, 1990; Sunset Date: November 9, 1993
    TSCA - Chemicals with Significant New Use Rules
    PMN numbers: P-88-1303; P-88-2177; P-90-212
    TSCA - Section 12(b) - Export Notification
    P-88-1303, P-88-2177, P-90-212; export notification required - Section 5; proposed Section 5

1,2-ETHANEDIYLBISCARBAMODITHIOIC ACID   111-54-6

  ENVIRONMENTAL LISTS

    CERCLA/SARA - Section 313 - Emission Reporting
    form R reporting required
    Class 2 Ozone Depletors
    ozone depletion weight = 0.02
    TSCA - Code of Federal Regulations Citations
    40 CFR 716.120
    TSCA - Health and Safety Reporting List
    Effective Date: October 15, 1990; Sunset Date: November 9, 1993

  INTERNATIONAL LISTS

    German (DFG) - Carcinogens
    suspected carcinogen

N,N'-1,2-ETHANEDIYLBIS-, POLYMER WITH 2,4, 6-TRICHLORO-1,3,5-TRIAZINE, REACTION PRODUCTS WITH N-BUTYL-2,2,6,6-TETRAMETHYL-4-PIPERIDINAMINE   RR-00918-7

  ENVIRONMENTAL LISTS

    TSCA - Chemicals with Significant New Use Rules
    PMN numbers: P-90-1472; P-90-1473

1,2-ETHANEDIYL TETRAKIS(2-CHLORO-1-METHYLETHYLENE) PHOSPHATE   34621-99-3

  STATE LISTS

    Pennsylvania Right to Know List
    [present]

ETHANE, 1,1'-SELENOBIS-   627-53-2

  ENVIRONMENTAL LISTS

    TSCA - Chemicals with Significant New Use Rules
    PMN number: P-93-1193
    TSCA - Section 12(b) - Export Notification
    P-93-1193; export notification required - Section 5

ETHANESULFONIC ACID, SODIUM SALT   3039-83-6

  HEALTH AND SAFETY LISTS

    NFPA - Flash Points
    flash point = 191 degrees F (88 degrees C)
    NFPA - Hazard Identification Ratings
    health-1; flammability-1; reactivity-0
    NIOSH - Selected LD50s and LC50s
    Oral, rat: LD50 = 6850 mg/kg Skin, rabbit: LD50 = 8480 mg/kg

  ENVIRONMENTAL LISTS

    CAA - HON Rule - SOCMI Chemicals
    compliance by Oct. 24, 1995

  STATE LISTS

    Pennsylvania Right to Know List
    [present]

ETHANESULFONIC ACID, 2-[2-[2-[4-(1,1,3,3-TE-TRAMETHYLBUTYL)PHENOXY]ETHOXY]ETHOXY]-, SODIUM SALT   2917-94-4

  HEALTH AND SAFETY LISTS

    NIOSH - Selected LD50s and LC50s
    Oral, rat: LD50 = 480 mg/kg
    NTP Chemical Status Reports - Testing Status and NTIS Number
    Prechronic studies in progress

ETHANESULFONYL CHLORIDE, 2-CHLORO-   1622-32-8

  HEALTH AND SAFETY LISTS

    NIOSH - Selected LD50s and LC50s
    Oral, mouse: LD50 = 350 mg/kg

ETHANE, 1,1,1-TRIFLUORO   420-46-2

  HEALTH AND SAFETY LISTS

    U.S. DOT - Appendix A Table 1 - Hazardous Substances
    final RQ = 5000 pounds (2270 kg)

  ENVIRONMENTAL LISTS

    CERCLA/SARA - Section 313 - Emission Reporting
    form R reporting required
    CERCLA/SARA - Hazardous Substances and their Reportable Quantities
    final RQ = 5000 pounds (2270 kg)
    RCRA - U Series Wastes
    waste number U114
    RCRA - Hazardous Constituents-Appendix VIII
    waste number U114
    RCRA - Substances Banned From Land Disposal
    [present]

  STATE LISTS

    Massachusetts Right To Know List
    [present]
    Pennsylvania Right to Know List
    environmental hazard

ETHANOL,2-[1-[[2-[2-[[(4-METHYLPHENYL) SULFONYL]OXY]ETHOXY]ETHOXY] METHYL]-2-(PROPENYLOXY)ETHOXY]-,4-METHYLBENZENESULFONATE   124029-00-1

  ENVIRONMENTAL LISTS

    TSCA - Chemicals with Significant New Use Rules
    PMN number: P-89-632

ETHANOLAMINE   141-43-5

  ENVIRONMENTAL LISTS

    TSCA - PAIR - Reporting List
    Effective Date: June 14, 1993

TSCA - Health and Safety Reporting List
Effective Date: June 14, 1993

ETHANOL AMINE DINITRATE   RR-01417-5

HEALTH AND SAFETY LISTS

NFPA - Hazard Identification Ratings
health-2; reactivity-0

STATE LISTS

Florida Hazardous Substance List
[present]

Massachusetts Right To Know List
[present]

Pennsylvania Right to Know List
[present]

ETHANOL, 2-(2-BUTOXYETHOXY)-, PHOSPHATE (3:1)   7332-46-9

ENVIRONMENTAL LISTS

EPA - Master Testing List
[present]

ETHANOL, 2,2'-[2-BUTYNE-1,4-DIYLBIS(OXY)]BIS-   1606-85-5

ENVIRONMENTAL LISTS

List of Pesticide Product Inert Ingredients
[present]

ETHANOL, 2-CHLORO-, ACETATE   542-58-5

HEALTH AND SAFETY LISTS

NIOSH - Selected LD50s and LC50s
Inhalation, rat: LC50 = 420 mg/m3 4 hr Oral, rat: LD50 = 240 mg/kg

ENVIRONMENTAL LISTS

CERCLA/SARA - Section 302 Extremely Hazardous Substances and TPQs
TPQ = 500 pounds

STATE LISTS

Florida Hazardous Substance List
effective March 13, 1992

Massachusetts Right To Know List
extraordinarily hazardous

Pennsylvania Right to Know List
environmental hazard

PROPOSED REGULATIONS

CERCLA/SARA - Proposed Hazardous Substance Additions
proposed RQ = 1 pound (.454 kg)

CERCLA/SARA - 1989 Proposed RQ Adjustments
proposed RQ = 100 pounds (45.4 kg)

ETHANOL, 2-CHLORO-, PHOSPHATE (3:1)   115-96-8

ENVIRONMENTAL LISTS

TSCA - Chemicals with Significant New Use Rules
PMN number: P-92-341

TSCA - Section 12(b) - Export Notification
P-92-341; export notification required - Section 5

ETHANOL, 2-CHLORO-, PHOSPHITE (3:1)   140-08-9

ENVIRONMENTAL LISTS

TSCA - Chemicals with Significant New Use Rules
PMN number: P-93-1204

TSCA - Section 12(b) - Export Notification
P-93-1204; export notification required - Section 5

ETHANOL, 2-(CYCLOHEXYLAMINO)-   2842-38-

HEALTH AND SAFETY LISTS

ACGIH 1995 - Time Weighted Averages
3 ppm TWA; 7.5 mg/m3 TWA

ACGIH 1995 - Short Term Exposure Limits
6 ppm STEL; 15 mg/m3 STEL

U.S. DOT - Substances From 49 CFR 172.101
regulated by DOT (UN2491)

U.S. DOT - Hazard Classes
DOT hazard class = 8

NFPA - Flash Points
flash point = 186 degrees F (86 degrees C)

NFPA - Hazard Identification Ratings
health-3; flammability-2; reactivity-0

NIOSH - Selected LD50s and LC50s
Oral, rat: LD50 = 2050 mg/kg Skin, rat: LD50 = 1500 mg/kg

NIOSH 1990 - Pocket Guide - RELs
3 ppm TWA; 8 mg/m3 TWA; 6 ppm STEL; 15 mg/m3 STEL

NIOSH 1990 - Pocket Guide - IDLHs
1000 ppm IDLH

NIOSH 1990 - Pocket Guide - Target organs
skin, eyes, respiratory system

OSHA - Vacated PELs - Time Weighted Averages
3 ppm TWA; 8 mg/m3 TWA

OSHA - Vacated PELs - Short Term Exposure Limits
6 ppm STEL; 15 mg/m3 STEL

OSHA - Final PELs - Time Weighted Averages
3 ppm TWA; 6 mg/m3 TWA

ENVIRONMENTAL LISTS

CAA - HON Rule - SOCMI Chemicals
compliance by Oct. 24, 1994

List of Pesticide Product Inert Ingredients
[present]

INTERNATIONAL LISTS

Australian Exposure Standards - Time Weighted Averages
3 ppm TWA; 7.5 mg/m3 TWA

Australian Exposure Standards - Short Term Exposure Limits
6 ppm STEL; 15 mg/m3 STEL

Canada - WHMIS: Ingredient Disclosure
1% item 1096 (1170)

Canada - Alberta - 8 Hour Occupational Exposure Limit
3 ppm TWA; 7.5 mg/m3 TWA

Canada - Alberta - 15 Minute Occupational Exposure Limit
6 ppm STEL; 15 mg/m3 STEL

Canada - British Columbia - 8 Hour Exposure Limits
3 ppm TWA; 8 mg/m3 TWA

Canada - British Columbia - 15 Minute Exposure Limits
6 ppm STEL; 15 mg/m3 STEL

Canada - Ontario - OHSA - TWAEVs
3 ppm TWAEV; 7.5 mg/m3 TWAEV

Canada - Ontario - OHSA - STEVs
6 ppm STEV; 15 mg/m3 STEV

Canada - Quebec - Time-Weighted Average Exposure Values
3 ppm TWAEV; 7.5 mg/m3 TWAEV

Canada - Quebec - Short-term Exposure Values
6 ppm STEV; 15 mg/m3 STEV

United Kingdom - Occupational Exposure Standards - TWAs
3 ppm TWA; 8 mg/m3 TWA

United Kingdom - Occupational Exposure Standards - STELs
6 ppm STEL; 15 mg/m3 STEL

German (DFG) - MAK Values
3 ppm MAK; 8 mg/m3 MAK

German (DFG) - Peak Limitations
5 x normal MAK (30 min. average value); don't exceed 2 times during shift

Israel - Time Weighted Averages
3 ppm TWA; 7.5 mg/m3 TWA

Israel - Short Term Exposure Limits
6 ppm STEL; 15 mg/m3 STEL

Israel - Action Levels
1.5 ppm AL; 3.75 mg/m3 AL

Mexico - Instruction No. 10 - TWAs
3 ppm TWA; 8 mg/m3 TWA

Mexico - Instruction No. 10 - STELs
6 ppm STEL; 15 mg/m3 STEL

STATE LISTS

California - Exposure Limits - PELs
3 ppm PEL; 8 mg/m3 PEL

California - Exposure Limits - STELs
6 ppm STEL; 15 mg/m3 STEL

California - Directors List of Hazardous Substances (8 CCR 339)
[present]

Florida Hazardous Substance List
[present]

Massachusetts Right To Know List
[present]

Minnesota Hazardous Substance List
[present]

NJ Right to Know List (Total)
sn 0835

NJ Special Hazardous Substances
(corrosive)

Pennsylvania Right to Know List
[present]

ETHANOL, 2-(2,4-DIAMINOPHENOXY)-,   66422-95-5
DIHYDROCHLORIDE

HEALTH AND SAFETY LISTS

U.S. DOT - Hazard Classes
Forbidden from transport by the DOT

ETHANOL, 1,2-DIBROMO-, ACETATE   24442-57-7

ENVIRONMENTAL LISTS

TSCA - Code of Federal Regulations Citations
40 CFR 712.30(x); 40 CFR 716.120(d)

TSCA - PAIR - Reporting List
Reporting Date: December 27, 1990

TSCA - Health and Safety Reporting List
Effective Date: October 29, 1990; Sunset Date: November 9, 1993

ETHANOL, 1,2-DICHLORO-, ACETATE   10140-87-1

ENVIRONMENTAL LISTS

List of Pesticide Product Inert Ingredients
[present]

ETHANOL, 2-(ETHYLPHENYLAMINO)-   92-50-2

HEALTH AND SAFETY LISTS

NFPA - Flash Points
flash point = 151 degrees F (66 degrees C)

NFPA - Hazard Identification Ratings
health-2; flammability-2; reactivity-0

STATE LISTS

Florida Hazardous Substance List
[present]

Massachusetts Right To Know List
[present]

Pennsylvania Right to Know List
[present]

ETHANOL, 2,2'-(HEXYLAMINO)BIS-,   6752-33-6

HEALTH AND SAFETY LISTS

IARC - Group 3 (not classifiable)
[present]

NIOSH - Selected LD50s and LC50s
Oral, rat; LD50 = 1230 mg/kg

NTP Chemical Status Reports - Testing Status and NTIS Number
Technical reports printed (PB92105147/AS)

NTP Chemical Status Reports - Evidence of Carcinogenicity
male rat-clear evidence; female rat-clear evidence; male mice-equivocal evidence; female mice-equivocal evidence

ENVIRONMENTAL LISTS

EPA - Master Testing List
[present]

TSCA - Code of Federal Regulations Citations
40 CFR 704.225(a); 40 CFR 712.30(w); 40 CFR 716.120(a)

TSCA - CAIR - Reporting List
reporting required by: manufacturer; importer

TSCA - PAIR - Reporting List
Reporting Date: February 14, 1989

TSCA - Health and Safety Reporting List
Effective Date: December 16, 1988; Sunset Date: November 9, 1993

STATE LISTS

California - Prop. 65 - Cancer list
carcinogen - initial date 4/1/92

ETHANOL, 2-(METHYLPHENYLAMINO)-   93-90-3

ENVIRONMENTAL LISTS

TSCA - Code of Federal Regulations Citations
40 CFR 712.30(d)

TSCA - PAIR - Reporting List
Reporting Date: November 19, 1982

TSCA - Health and Safety Reporting List
Effective Date: April 29, 1983

PROPOSED REGULATIONS

TSCA - ITC 33rd Report Priority Testing List
recommended with intent-to-designate

TSCA - ITC 34th Report Priority Testing List
recommended with intent-to-designate

ETHANOL, 2-[2-(2-METHYLPROPOXY)ETHOXY]-   18912-80-6

HEALTH AND SAFETY LISTS

NFPA - Flash Points
flash point = 249 degrees F (121 degrees C)

NFPA - Hazard Identification Ratings
health-3; flammability-1; reactivity-0

STATE LISTS

Florida Hazardous Substance List
[present]

Massachusetts Right To Know List
[present]

Pennsylvania Right to Know List
[present]

ETHANOL,2,2'-[OXYBIS(2,1-ETHANEDIYLOXY)]BIS-,   **19249-03-7**
BIS(4-METHYLBENZENESULFONATE)
SEE ALSO:
    ETHANOL, 2-(2,4-DIAMINOPHENOXY)-, DIHYDROCHLORIDE
ENVIRONMENTAL LISTS
    TSCA - Code of Federal Regulations Citations
    40 CFR 712.30(f);   40 CFR 716.120
    TSCA - PAIR - Reporting List
    Reporting Date: August 17, 1983

ETHANOL,2,2'-[OXYBIS(2,1-ETHANEDIYLOXY)]BIS,   **37860-51-8**
BIS(4-METHYLBENZENESULFONATE))
ENVIRONMENTAL LISTS
    TSCA - Code of Federal Regulations Citations
    40 CFR 712.30(d)
    TSCA - PAIR - Reporting List
    Reporting Date: November 19, 1982

ETHANOL,2,2'-[[1-[(2-PROPENYLOXY)METHYL]-1,   **114719-15-2**
2-ETHANEDIYL]BIS(OXY)]BIS-,BIS(4-METHYLBEN-
ZENESULFONATE)
ENVIRONMENTAL LISTS
    CERCLA/SARA - Section 302 Extremely Hazardous Substances and
    TPQs
    TPQ = 1000 pounds
STATE LISTS
    Florida Hazardous Substance List
    effective March 13, 1992
    Massachusetts Right To Know List
    extraordinarily hazardous
    NJ Right to Know List (Total)
    sn 2394
    Pennsylvania Right to Know List
    environmental hazard
PROPOSED REGULATIONS
    CERCLA/SARA - Proposed Hazardous Substance Additions
    proposed RQ = 1 pound (.454 kg)
    CERCLA/SARA - 1989 Proposed RQ Adjustments
    proposed RQ = 100 pounds (45.4 kg)

ETHANOL, 2,2'-THIOBIS-   **111-48-8**
HEALTH AND SAFETY LISTS
    NFPA - Flash Points
    flash point = 270 degrees F (132 degrees C)
    NFPA - Hazard Identification Ratings
    health-2; flammability-1; reactivity-0
STATE LISTS
    Florida Hazardous Substance List
    [present]
    Massachusetts Right To Know List
    [present]
    Pennsylvania Right to Know List
    [present]

ETHENE, HOMOPOLYMER, OXIDIZED   **68441-17-8**
ENVIRONMENTAL LISTS
    TSCA - Chemicals with Significant New Use Rules
    PMN number: P-91-1346
    TSCA - Section 12(b) - Export Notification
    P-93-1193; export notification required - Section 5

ETHENE, (2-METHOXYETHOXY)-   **1663-35-**
HEALTH AND SAFETY LISTS
    NFPA - Flash Points
    flash point = 280 degrees F (138 degrees C)
    NFPA - Hazard Identification Ratings
    health-2; flammability-1; reactivity-0
STATE LISTS
    Florida Hazardous Substance List
    [present]
    Massachusetts Right To Know List
    [present]
    Pennsylvania Right to Know List
    [present]

ETHENE, PENTAFLUORO-   **354-33-6**
HEALTH AND SAFETY LISTS
    NFPA - Flash Points
    flash point = 222 degrees F (106 degrees C)
    NFPA - Hazard Identification Ratings
    health-1; flammability-1; reactivity-0
    NIOSH - Selected LD50s and LC50s
    Oral, rat: LD50 = 4920 mg/kg Skin, rabbit: LD50 = 3560 mg/kg

ETHENE, TRIFLUORO-   **359-11-5**
ENVIRONMENTAL LISTS
    TSCA - Chemicals with Significant New Use Rules
    PMN number: P-93-1195
    TSCA - Section 12(b) - Export Notification
    P-93-1195; export notification required - Section 5

ETHENOL   **557-75-5**
ENVIRONMENTAL LISTS
    TSCA - Chemicals with Significant New Use Rules
    PMN number: P-93-1196

ETHINYLESTRADIOL   **57-63-6**
ENVIRONMENTAL LISTS
    TSCA - Chemicals with Significant New Use Rules
    PMN number: P-93-1203
    TSCA - Section 12(b) - Export Notification
    P-93-1203; export notification required - Section 5

ETHION   **563-12-2**
HEALTH AND SAFETY LISTS
    NFPA - Flash Points
    flash point = 320 degrees F (160 degrees C)
    NFPA - Hazard Identification Ratings
    health-2; flammability-1; reactivity-0
    NIOSH - Selected LD50s and LC50s
    Oral, rat: LD50 = 6610 mg/kg
STATE LISTS
    Florida Hazardous Substance List
    [present]
    Massachusetts Right To Know List
    [present]
    Pennsylvania Right to Know List
    [present]

ETHIONAMIDE   **536-33-4**
ENVIRONMENTAL LISTS
    List of Pesticide Product Inert Ingredients
    [present]

ETHOPROPHOS   13194-48-4

HEALTH AND SAFETY LISTS

NFPA - Flash Points
flash point = 64 degrees F (18 degrees C)

NFPA - Hazard Identification Ratings
health-0; flammability-3; reactivity-0

STATE LISTS

Florida Hazardous Substance List
[present]

Massachusetts Right To Know List
[present]

Pennsylvania Right to Know List
[present]

ETHOXYBENZOTHIAZOLE DISULFIDE   RR-01241-9

ENVIRONMENTAL LISTS

TSCA - Code of Federal Regulations Citations
40 CFR 716.120

TSCA - Health and Safety Reporting List
Effective Date: October 15, 1990; Sunset Date: November 9, 1993

TSCA - Chemicals with Significant New Use Rules
PMN number: P-91-1392

TSCA - Section 12(b) - Export Notification
export notification required - Section 5

2-ETHOXY-3,4-DIHYDRO-2H-PYRAN   103-75-3
SEE ALSO:
FLUOROALKENES

ENVIRONMENTAL LISTS

TSCA - Code of Federal Regulations Citations
40 CFR 712.30(f); 40 CFR 716.120(c)

TSCA - PAIR - Reporting List
Reporting Date: August 17, 1983

2-ETHOXYETHANOL   110-80-5

HEALTH AND SAFETY LISTS

NIOSH - Selected LD50s and LC50s
Oral, rat: LD50 = 64 mg/kg

2-ETHOXYETHYL ACETATE   111-15-9

HEALTH AND SAFETY LISTS

IARC - Group Unspecified
[present] (Listed under 'Steroidal oestrogens')

NIOSH - Selected LD50s and LC50s
Oral, rat: LD50 = 2952 mg/kg

NTP Seventh Report - Suspect Carcinogens
suspect carcinogen (Listed under 'Estrogens (not conjugated)')

OSHA - Possible Select Carcinogens
[present]

STATE LISTS

California - Air Bill 2588 Appendix A-II
known or potential carcinogen

California - Prop. 65 - Cancer list
carcinogen - initial date 1/1/88

California - Prop. 65 - Developmental Toxicity
developmental toxicity (when mixed with Norethisterone) - initial date 4/1/90

California - Directors List of Hazardous Substances (8 CCR 339)
[present]

Florida Hazardous Substance List
[present]

Massachusetts Right To Know List
carcinogen; extraordinarily hazardous

Minnesota Hazardous Substance List
carcinogen

NJ Right to Know List (Total)
sn 0836

NJ Special Hazardous Substances
(carcinogen; mutagen)

Pennsylvania Right to Know List
special hazardous substance

Pennsylvania RTK - Special Hazardous Substances
[present]

ETHOXYLATED ABIETYLAMINE   51344-60-6

HEALTH AND SAFETY LISTS

ACGIH 1995 - Time Weighted Averages
0.4 mg/m3 TWA

ACGIH 1995 - Skin Designations
skin - potential for cutaneous absorption

U.S. DOT - Appendix B - Marine Pollutants
DOT regulated severe marine pollutant

U.S. DOT - Appendix A Table 1 - Hazardous Substances
final RQ = 10 pounds (4.54 kg)

NIOSH - Selected LD50s and LC50s
Oral, rat: LD50 = 13 mg/kg Skin, rat: LD50 = 62 mg/kg

OSHA - Vacated PELs - Time Weighted Averages
0.4 mg/m3 TWA

OSHA - Vacated PELs - Skin Designation
Prevent or reduce skin absorption

ENVIRONMENTAL LISTS

ATSDR Priority List
Rank (of 275): 194

CERCLA/SARA - Section 302 Extremely Hazardous Substances and TPQs
TPQ = 1000 pounds

CERCLA/SARA - Hazardous Substances and their Reportable Quantities
final RQ = 10 pounds (4.54 kg)

Clean Water Act - Hazardous Substances
[present]

INTERNATIONAL LISTS

Australian Exposure Standards - Time Weighted Averages
0.4 mg/m3 TWA

Australian Exposure Standards - Skin Effects
skin absorption

Canada - Alberta - 8 Hour Occupational Exposure Limit
0.4 mg/m3 TWA

Canada - Alberta - 15 Minute Occupational Exposure Limit
1.2 mg/m3 STEL

Canada - Alberta - Skin Designation
can be absorbed through the intact skin

Canada - British Columbia - 8 Hour Exposure Limits
0.4 mg/m3 TWA

Canada - British Columbia - Skin Notations
skin - potential for skin absorption

Canada - Ontario - OHSA - TWAEVs
0.4 mg/m3 TWAEV

Canada - Ontario - OHSA - Skin Notations
absorption through skin, eyes, or mucous membranes

Canada - Quebec - Time-Weighted Average Exposure Values
0.4 mg/m3 TWAEV

Canada - Quebec - Skin Designations
absorbed through the skin

Israel - Time Weighted Averages
0.4 mg/m3 TWA

Israel - Action Levels
0.2 mg/m3 AL

Mexico - Instruction No. 10 - TWAs
0.4 mg/m3 TWA

Mexico - Instruction No. 10 - Skin designation
skin - potential for cutaneous absorption

STATE LISTS

California - Exposure Limits - PELs
0.4 mg/m3 PEL

California - Exposure Limits - Skin Notation
material may be absorbed through the skin, eyes or mucous membrane

California - Directors List of Hazardous Substances (8 CCR 339)
[present]

Florida Hazardous Substance List
[present]

Massachusetts Right To Know List
extraordinarily hazardous; neurotoxin

Minnesota Hazardous Substance List
skin

NJ Right to Know List (Total)
sn 0837

Pennsylvania Right to Know List
environmental hazard

ETHOXYLATED CASTOR OIL   61791-12-6

HEALTH AND SAFETY LISTS

IARC - Group 3 (not classifiable)
[present]

NIOSH - Selected LD50s and LC50s
Oral, rat: LD50 = 1320 mg/kg

NTP Chemical Status Reports - Testing Status and NTIS Number
Technical reports printed (PB285193/AS)

NTP Chemical Status Reports - Evidence of Carcinogenicity
male rat-negative; female rat-negative; male mice-negative; female mice-negative

STATE LISTS

California - Directors List of Hazardous Substances (8 CCR 339)
[present]

NJ Right to Know List (Total)
sn 0838

NJ Special Hazardous Substances
(teratogen)

ETHOXYLATED COCONUT OIL ALKYL AMINE   61791-14-8

HEALTH AND SAFETY LISTS

U.S. DOT - Appendix B - Marine Pollutants
DOT regulated marine pollutant

NIOSH - Selected LD50s and LC50s
Oral, rat: LD50 = 34 mg/kg Skin, rat: LD50 = 60 mg/kg

ENVIRONMENTAL LISTS

CERCLA/SARA - Section 302 Extremely Hazardous Substances and TPQs
TPQ = 1000 pounds

CERCLA/SARA - Section 313 - Emission Reporting
form R reporting required

STATE LISTS

California - Directors List of Hazardous Substances (8 CCR 339)
[present]

Florida Hazardous Substance List
effective March 13, 1992

Massachusetts Right To Know List
extraordinarily hazardous; neurotoxin

NJ Right to Know List (Total)
sn 2395

Pennsylvania Right to Know List
environmental hazard

PROPOSED REGULATIONS

CERCLA/SARA - Proposed Hazardous Substance Additions
proposed RQ = 1 pound (.454 kg)

CERCLA/SARA - 1989 Proposed RQ Adjustments
proposed RQ = 1000 pounds (454 kg)

ETHOXYLATED HYDROABIETHYL ALCOHOL   61524-98-

ENVIRONMENTAL LISTS

TSCA - Chemicals with Significant New Use Rules
PMN number: P-90-1384

ETHOXYLATED LINEAR C12-15-SEC-ALCOHOL   68511-39-
SULFATE

HEALTH AND SAFETY LISTS

NFPA - Flash Points
flash point = 111 degrees F (44 degrees C)

NFPA - Hazard Identification Ratings
health-2; flammability-2; reactivity-1

INTERNATIONAL LISTS

Canada - WHMIS: Ingredient Disclosure
1% item 685 (846)

STATE LISTS

Massachusetts Right To Know List
[present]

Pennsylvania Right to Know List
[present]

ETHOXYLATED METHYL GLUCOSIDE   68239-42-9
SEE ALSO:
GLYCOL ETHERS
F005-HAZARDOUS WASTES

HEALTH AND SAFETY LISTS

ACGIH 1995 - Time Weighted Averages
5 ppm TWA; 18 mg/m3 TWA

ACGIH 1995 - Skin Designations
skin - potential for cutaneous absorption

ACGIH 1995 - Biological Exposure Indices
2-Ethoxyacetic acid in urine: 100 mg/g creatinine, end of shift at end of workweek

AIHA - Odor Threshold Values
geometric mean air odor threshold = 2.7 ppm (detectable); 6.5 ppm (recognizable)

U.S. DOT - Substances From 49 CFR 172.101
regulated by DOT (UN1171)

U.S. DOT - Hazard Classes
DOT hazard class = 3

U.S. DOT - Appendix A Table 1 - Hazardous Substances
final RQ = 1000 pounds (454 kg)

NFPA - Flash Points
flash point = 110 degrees F (43 degrees C)

NFPA - Hazard Identification Ratings
health-2; flammability-2; reactivity-0

NIOSH - Selected LD50s and LC50s
Inhalation, rat: LC50 = 2000 ppm 7 hr Oral, rat: LD50 = 3000 mg/kg Skin, rat: LD50 = 3900 mg/kg

NIOSH 1990 - Pocket Guide - RELs
Reduce exposure to lowest feasible concentration

NIOSH 1990 - Pocket Guide - IDLHs
6000 ppm IDLH (not applicable because of the NIOSH REL)

NIOSH 1990 - Pocket Guide - Target organs
in animals: lungs, eyes, blood, kidneys, liver

NIOSH - Health Standards - Exposure Limits
reduce exposure to lowest feasible concentration (Listed under 'Glycol ethers')

NIOSH - Health Standards - Health Effects and Precautions
Male and female reproductive effects; teratogenicity (Prevent skin contact) (Listed under 'Glycol ethers')

NTP Chemical Status Reports - Testing Status and NTIS Number
Technical reports printed (PB94-118106)

OSHA - Vacated PELs - Time Weighted Averages
200 ppm TWA; 740 mg/m3 TWA

OSHA - Vacated PELs - Skin Designation
Prevent or reduce skin absorption

OSHA - Final PELs - Time Weighted Averages
200 ppm TWA; 740 mg/m3 TWA

OSHA - Final PELs - Skin Notations
prevent or reduce skin absorption

ENVIRONMENTAL LISTS

CERCLA/SARA - Section 313 - Emission Reporting
form R reporting required for 1.0% de minimus concentration

CERCLA/SARA - Hazardous Substances and their Reportable Quantities
final RQ = 1000 pounds (454 kg)

CAA - HON Rule - SOCMI Chemicals
compliance by Oct. 24, 1994

EPA - Master Testing List
[present]

RCRA - U Series Wastes
waste number U359

RCRA - Hazardous Constituents-Appendix VIII
waste number U359

RCRA - Basis for Listing - Appendix VII
Included in waste stream: F005

RCRA - Substances Banned From Land Disposal
[present]

TSCA - Multichemical Test Rules - Neurotoxicity
administrative stay for neurotoxicity tests effective June 27, 1994

INTERNATIONAL LISTS

Australian Exposure Standards - Time Weighted Averages
5 ppm TWA; 18 mg/m3 TWA

Australian Exposure Standards - Skin Effects
skin absorption

Canada - WHMIS: Ingredient Disclosure
0.1% item 686 (847)

Canada - NPRI (National Pollutant Release Inventory)
[present]

Canada - Alberta - 8 Hour Occupational Exposure Limit
5 ppm TWA; 18 mg/m3 TWA

Canada - Alberta - 15 Minute Occupational Exposure Limit
10 ppm STEL; 37 mg/m3 STEL

Canada - Alberta - Skin Designation
can be absorbed through the intact skin

Canada - British Columbia - 8 Hour Exposure Limits
100 ppm TWA; 370 mg/m3 TWA

Canada - British Columbia - 15 Minute Exposure Limits
150 ppm STEL; 560 mg/m3 STEL

Canada - British Columbia - Skin Notations
skin - potential for skin absorption

Canada - Ontario - OHSA - TWAEVs
5 ppm TWAEV; 18 mg/m3 TWAEV

Canada - Ontario - OHSA - Skin Notations
absorption through skin, eyes, or mucous membranes

Canada - Quebec - Time-Weighted Average Exposure Values
5 ppm TWAEV; 18 mg/m3 TWAEV

Canada - Quebec - Skin Designations
absorbed through the skin

United Kingdom - Maximum Exposure Limits - TWAs
10 ppm TWA; 37 mg/m3 TWA

United Kingdom - Maximum Exposure Limits - Notes
can be absorbed through skin

German (DFG) - MAK Values
5 ppm MAK; 19 mg/m3 MAK

German (DFG) - Peak Limitations
2 x normal MAK (30 min. average value); don't exceed 4 times during shift

German (DFG) - Skin/Sensitizers
danger of cutaneous absorption

German (DFG) - Pregnancy
risk to embryo/fetus probable

Israel - Time Weighted Averages
5 ppm TWA; 18 mg/m3 TWA

Israel - Action Levels
2.5 ppm AL; 9 mg/m3 AL

Mexico - Instruction No. 10 - TWAs
50 ppm TWA; 185 mg/m3 TWA

Mexico - Instruction No. 10 - STELs
100 ppm STEL; 370 mg/m3 STEL

Mexico - Instruction No. 10 - Skin designation
skin - potential for cutaneous absorption

STATE LISTS

California - Prop. 65 - Developmental Toxicity
developmental toxicity - initial date 1/1/89

California - Prop. 65 - Reproductive - Male
male reproductive toxicity - initial date 1/1/89

California - Exposure Limits - PELs
5 ppm PEL; 18 mg/m3 PEL

California - Exposure Limits - Skin Notation
material may be absorbed through the skin, eyes or mucous membrane

California - Directors List of Hazardous Substances (8 CCR 339)
[present]

Florida Hazardous Substance List
[present]

Massachusetts Right To Know List
teratogen

Minnesota Hazardous Substance List
skin

NJ Right to Know List (Total)
sn 0839

Pennsylvania Right to Know List
environmental hazard

PROPOSED REGULATIONS

Proposed Amendments to OSHA Regulated Chemicals
proposed PEL: 0.5 ppm TWA; 2.5 ppm Excursion Limit; 0.25 ppm TWA Action Level

ETHOXYLATED P-NONYL PHENOL   26027-38-3
SEE ALSO:
GLYCOL ETHERS

HEALTH AND SAFETY LISTS

ACGIH 1995 - Time Weighted Averages
5 ppm TWA; 27 mg/m3 TWA

ACGIH 1995 - Skin Designations
skin - potential for cutaneous absorption

ACGIH 1995 - Biological Exposure Indices
2-Ethoxyacetic acid in urine: 100 mg/g creatinine, end of shift at end of workweek

AIHA - Odor Threshold Values
geometric mean air odor threshold = 0.060 ppm (detectable); 0.13 ppm (recognizable)

U.S. DOT - Substances From 49 CFR 172.101
regulated by DOT (UN1172)

U.S. DOT - Hazard Classes
DOT hazard class = 3

NFPA - Flash Points
*flash point = 124 degrees F (52 degrees C); *See list description*

NFPA - Hazard Identification Ratings
health-1; flammability-2

NIOSH - Selected LD50s and LC50s
Oral, rat: LD50 = 2900 mg/kg Skin, rabbit: LD50 = 10500 mg/kg

NIOSH 1990 - Pocket Guide - RELs
Reduce exposure to lowest feasible concentration

NIOSH 1990 - Pocket Guide - IDLHs
2500 ppm IDLH (not applicable because of the NIOSH REL)

NIOSH 1990 - Pocket Guide - Target organs
eyes, GI tract, respiratory system

OSHA - Vacated PELs - Time Weighted Averages
100 ppm TWA; 540 mg/m3 TWA

OSHA - Vacated PELs - Skin Designation
Prevent or reduce skin absorption

OSHA - Final PELs - Time Weighted Averages
100 ppm TWA; 540 mg/m3 TWA

OSHA - Final PELs - Skin Notations
prevent or reduce skin absorption

ENVIRONMENTAL LISTS

CAA - HON Rule - SOCMI Chemicals
compliance by Oct. 24, 1994

List of Pesticide Product Inert Ingredients
[present]

INTERNATIONAL LISTS

Australian Exposure Standards - Time Weighted Averages
5 ppm TWA; 27 mg/m3 TWA

Australian Exposure Standards - Skin Effects
skin absorption

Canada - WHMIS: Ingredient Disclosure
0.1% item 687 (16)

Canada - NPRI (National Pollutant Release Inventory)
[present]

Canada - Alberta - 8 Hour Occupational Exposure Limit
5 ppm TWA; 27 mg/m3 TWA

Canada - Alberta - 15 Minute Occupational Exposure Limit
10 ppm STEL; 54 mg/m3 STEL

Canada - Alberta - Skin Designation
can be absorbed through the intact skin

Canada - British Columbia - 8 Hour Exposure Limits
100 ppm TWA; 540 mg/m3 TWA

Canada - British Columbia - 15 Minute Exposure Limits
150 ppm STEL; 810 mg/m3 STEL

Canada - British Columbia - Skin Notations
skin - potential for skin absorption

Canada - Ontario - OHSA - TWAEVs
5 ppm TWAEV; 27 mg/m3 TWAEV

Canada - Ontario - OHSA - Skin Notations
absorption through skin, eyes, or mucous membranes

Canada - Quebec - Time-Weighted Average Exposure Values
5 ppm TWAEV; 27 mg/m3 TWAEV

Canada - Quebec - Skin Designations
absorbed through the skin

United Kingdom - Maximum Exposure Limits - TWAs
10 ppm TWA; 54 mg/m3 TWA

United Kingdom - Maximum Exposure Limits - Notes
can be absorbed through skin

German (DFG) - MAK Values
5 ppm MAK; 27 mg/m3 MAK

German (DFG) - Peak Limitations
2 x normal MAK (30 min. average value); don't exceed 4 times durin shift

German (DFG) - Skin/Sensitizers
danger of cutaneous absorption

German (DFG) - Pregnancy
risk to embryo/fetus probable

Israel - Time Weighted Averages
5 ppm TWA; 27 mg/m3 TWA

Israel - Action Levels
2.5 ppm AL; 13.5 mg/m3 AL

Mexico - Instruction No. 10 - TWAs
50 ppm TWA; 270 mg/m3 TWA

Mexico - Instruction No. 10 - STELs
100 ppm STEL; 540 mg/m3 STEL

Mexico - Instruction No. 10 - Skin designation
skin - potential for cutaneous absorption

STATE LISTS

California - Air Bill 2588 Appendix A-I
9/90

California - Prop. 65 - Developmental Toxicity
developmental toxicity - initial date 1/1/93

California - Prop. 65 - Reproductive - Male
male reproductive toxicity - initial date 1/1/93

California - Exposure Limits - PELs
5 ppm PEL; 27 mg/m3 PEL

California - Exposure Limits - Skin Notation
material may be absorbed through the skin, eyes or mucous membrane

California - Directors List of Hazardous Substances (8 CCR 339)
[present]

Florida Hazardous Substance List
[present]

Massachusetts Right To Know List
[present]

Minnesota Hazardous Substance List
skin

NJ Right to Know List (Total)
sn 0840

Pennsylvania Right to Know List
[present]

PROPOSED REGULATIONS

Proposed Amendments to OSHA Regulated Chemicals
proposed PEL: 0.5 ppm TWA; 2.5 ppm Excursion Limit; 0.25 ppm TWA Action Level

ETHOXYLATED OCTYLPHENOL   9036-19-5

ENVIRONMENTAL LISTS

List of Pesticide Product Inert Ingredients
[present]

ETHOXYLATED SORBITAN POLYSORBATE   RR-01076-4

ENVIRONMENTAL LISTS

List of Pesticide Product Inert Ingredients
[present]

ETHOXYLATED SORBITOL HEXAESTER OF TALL-   61790-90-7
OIL ACIDS

ENVIRONMENTAL LISTS

List of Pesticide Product Inert Ingredients
[present]

ETHOXYLATED SORBITOL PENTAESTER OF TALL- OIL ACIDS   RR-01077-5

ENVIRONMENTAL LISTS

     List of Pesticide Product Inert Ingredients
        [present]

ETHOXYLATED SUBSTITUTED NAPHTHOL   RR-01665-9

ENVIRONMENTAL LISTS

     List of Pesticide Product Inert Ingredients
        [present]

ETHOXYLATED N-(TALLOW ALKYL)TRIMETHY- LENE DIAMINES   61790-85-0

ENVIRONMENTAL LISTS

     List of Pesticide Product Inert Ingredients
        [present]

ETHOXYLATED TALLOW AMINE HYDROCHLORIDE   68132-78-5

ENVIRONMENTAL LISTS

     List of Pesticide Product Inert Ingredients
        [present]

3-ETHOXYPROPIONALDEHYDE   2806-85-1

HEALTH AND SAFETY LISTS

     NIOSH - Selected LD50s and LC50s
        Oral, rat: LD50 = 4190 mg/kg

ENVIRONMENTAL LISTS

     List of Pesticide Product Inert Ingredients
        [present]

INTERNATIONAL LISTS

     Canada - WHMIS: Ingredient Disclosure
        1% item 1326 (832)

3-ETHOXYPROPIONIC ACID   1331-11-9

ENVIRONMENTAL LISTS

     List of Pesticide Product Inert Ingredients
        [present]

ETHOXYQUIN   91-53-2

ENVIRONMENTAL LISTS

     List of Pesticide Product Inert Ingredients
        [present]

ETHYL ABIETATE   631-71-0

ENVIRONMENTAL LISTS

     List of Pesticide Product Inert Ingredients
        [present]

N-ETHYLACETAMIDE   625-50-3

ENVIRONMENTAL LISTS

     TSCA - Chemicals with Significant New Use Rules
        PMN number: P-88-2484

N-ETHYL ACETANILIDE   529-65-7

ENVIRONMENTAL LISTS

     List of Pesticide Product Inert Ingredients
        [present]

ETHYL ACETATE   141-78-6

ENVIRONMENTAL LISTS

     List of Pesticide Product Inert Ingredients
        [present]

ETHYL ACETOACETATE   141-97-9

HEALTH AND SAFETY LISTS

     NFPA - Flash Points
        flash point = 100 degrees F (38 degrees C)

     NFPA - Hazard Identification Ratings
        health-2; flammability-2; reactivity-0

STATE LISTS

     Florida Hazardous Substance List
        [present]

     Massachusetts Right To Know List
        [present]

     Pennsylvania Right to Know List
        [present]

ETHYL ACETYLENE   107-00-6

HEALTH AND SAFETY LISTS

     NFPA - Flash Points
        flash point = 225 degrees F (107 degrees C)

     NFPA - Hazard Identification Ratings
        health-2; flammability-1; reactivity-0

STATE LISTS

     Florida Hazardous Substance List
        [present]

     Massachusetts Right To Know List
        [present]

ETHYL ACETYL GLYCOLATE   RR-00855-9

HEALTH AND SAFETY LISTS

     NIOSH - Selected LD50s and LC50s
        Oral, rat: LD50 = 800 mg/kg

     NTP Chemical Status Reports - Testing Status and NTIS Number
        Prechronic studies completed: chemicals in review for further evaluation

ETHYLACROLEIN   RR-01165-4

HEALTH AND SAFETY LISTS

     NFPA - Flash Points
        flash point = 352 degrees F (178 degrees C)

     NFPA - Hazard Identification Ratings
        health-0; flammability-1; reactivity-0

ETHYL ACRYLATE   140-88-5

HEALTH AND SAFETY LISTS

     NFPA - Flash Points
        flash point = 230 degrees F (110 degrees C)

     NFPA - Hazard Identification Ratings
        health-1; flammability-1; reactivity-0

ETHYL ACRYLATE AND 2-ETHYLHEXYL ACRY- LATE COPOLYMER   26376-86-3

HEALTH AND SAFETY LISTS

     NFPA - Flash Points
        flash point = 126 degrees F (52 degrees C)

     NFPA - Hazard Identification Ratings
        health-0; flammability-2; reactivity-0

ETHYL ACRYLATE-METHACRYLIC ACID-STYRENE POLYMER   25035-68-1

SEE ALSO:

     F003-HAZARDOUS WASTES
     F039-HAZARDOUS WASTES

HEALTH AND SAFETY LISTS

     ACGIH 1995 - Time Weighted Averages
        400 ppm TWA; 1440 mg/m3 TWA

AIHA - Odor Threshold Values
geometric mean air odor threshold = 18 ppm (detectable);   32 ppm (recognizable)

U.S. DOT - Substances From 49 CFR 172.101
regulated by DOT (UN1173)

U.S. DOT - Hazard Classes
DOT hazard class = 3

U.S. DOT - Appendix A Table 1 - Hazardous Substances
final RQ = 5000 pounds (2270 kg)

NFPA - Flash Points
flash point = 24 degrees F (-4 degrees C)

NFPA - Hazard Identification Ratings
health-1; flammability-3; reactivity-0

NIOSH - Selected LD50s and LC50s
Inhalation, rat: LC50 = 1600 ppm 8 hr Oral, rat: LD50 = 5620 mg/kg

NIOSH 1990 - Pocket Guide - RELs
400 ppm TWA; 1400 mg/m3 TWA

NIOSH 1990 - Pocket Guide - IDLHs
10,000 ppm IDLH

NIOSH 1990 - Pocket Guide - Target organs
eyes, skin, respiratory system

OSHA - Vacated PELs - Time Weighted Averages
400 ppm TWA; 1400 mg/m3 TWA

OSHA - Final PELs - Time Weighted Averages
400 ppm TWA; 1400 mg/m3 TWA

ENVIRONMENTAL LISTS

CERCLA/SARA - Hazardous Substances and their Reportable Quantities
final RQ = 5000 pounds (2270 kg)

EPA - Master Testing List
[present]

List of Pesticide Product Inert Ingredients
[present]

RCRA - U Series Wastes
waste number U112 (Ignitable waste)

RCRA - Hazardous Constituents-Appendix VIII
waste number U112 (Ignitable waste)

RCRA - Basis for Listing - Appendix VII
Included in waste stream: F039

RCRA - Substances Banned From Land Disposal
[present]

RCRA - Universal Treatment Standards (LDR)
WW: 0.34 mg/l; NWW: 33 mg/kg

TSCA - Multichemical Test Rules - Neurotoxicity
administrative stay for neurotoxicity tests effective June 27, 1994

INTERNATIONAL LISTS

Australian Exposure Standards - Time Weighted Averages
400 ppm TWA; 1440 mg/m3 TWA

Australian Exposure Standards - Under Review
exposure limits under review

Canada - WHMIS: Ingredient Disclosure
1% item 689 (17)

Canada - Alberta - 8 Hour Occupational Exposure Limit
400 ppm TWA; 1441 mg/m3 TWA

Canada - Alberta - 15 Minute Occupational Exposure Limit
500 ppm STEL; 1801 mg/m3 STEL

Canada - British Columbia - 8 Hour Exposure Limits
400 ppm TWA; 1400 mg/m3 TWA

Canada - Ontario - OHSA - TWAEVs
400 ppm TWAEV; 1440 mg/m3 TWAEV

Canada - Quebec - Time-Weighted Average Exposure Values
400 ppm TWAEV; 1440 mg/m3 TWAEV

United Kingdom - Occupational Exposure Standards - TWAs
400 ppm TWA; 1400 mg/m3 TWA

German (DFG) - MAK Values
400 ppm MAK; 1400 mg/m3 MAK

German (DFG) - Peak Limitations
2 x normal MAK (5 min momentary value);   don't exceed 8 time during shift

Israel - Time Weighted Averages
400 ppm TWA; 1440 mg/m3 TWA

Israel - Action Levels
200 ppm AL; 720 mg/m3 AL

Mexico - Instruction No. 10 - TWAs
400 ppm TWA; 1400 mg/m3 TWA

STATE LISTS

California - Exposure Limits - PELs
400 ppm PEL; 1400 mg/m3 PEL

California - Directors List of Hazardous Substances (8 CCR 339)
[present]

Florida Hazardous Substance List
[present]

Massachusetts Right To Know List
[present]

Minnesota Hazardous Substance List
[present]

NJ Right to Know List (Total)
sn 0841

NJ Special Hazardous Substances
(flammable - third degree)

Pennsylvania Right to Know List
environmental hazard

PROPOSED REGULATIONS

TSCA - ITC 33rd Report Priority Testing List
designated for testing

ETHYL ACRYLATE-METHYL METHACRYLATE POLYMER   9010-88-2

HEALTH AND SAFETY LISTS

NFPA - Flash Points
flash point = 135 degrees F (57 degrees C)

NFPA - Hazard Identification Ratings
health-2; flammability-2; reactivity-0

NIOSH - Selected LD50s and LC50s
Oral, rat: LD50 = 3980 mg/kg

STATE LISTS

Florida Hazardous Substance List
[present]

Massachusetts Right To Know List
[present]

Pennsylvania Right to Know List
[present]

ETHYL ACRYLATE POLYMER   9003-32-1

HEALTH AND SAFETY LISTS

U.S. DOT - Substances From 49 CFR 172.101
regulated by DOT (UN2452)

U.S. DOT - Hazard Classes
DOT hazard class = 2.1

ENVIRONMENTAL LISTS

CAA - Flammable Substances for Accidental Release Prevention
threshold quantity = 10,000 lbs

STATE LISTS

NJ Right to Know List (Total)
sn 0842

ETHYL ALCOHOL   64-17-5

HEALTH AND SAFETY LISTS

NFPA - Flash Points
flash point = 180 degrees F (82 degrees C)

NFPA - Hazard Identification Ratings
health-0; flammability-2; reactivity-0

ETHYL ALCOHOL AND WATER   RR-00802-6

HEALTH AND SAFETY LISTS

U.S. DOT - Substances Which Are Poisonous by Inhalation
liquid hazardous material poisonous by inhalation

ETHYLALUMINUM DICHLORIDE   563-43-9

HEALTH AND SAFETY LISTS

ACGIH 1995 - Time Weighted Averages
5 ppm TWA; 20 mg/m3 TWA

ACGIH 1995 - Short Term Exposure Limits
15 ppm STEL; 61 mg/m3 STEL

ACGIH 1995 - Carcinogens
A2-suspected human carcinogen

AIHA - Odor Threshold Values
geometric mean air odor threshold = 0.00024 ppm (detectable); 0.00037 ppm (recognizable)

U.S. DOT - Substances From 49 CFR 172.101
regulated by DOT (UN1917)

U.S. DOT - Hazard Classes
DOT hazard class = 3

U.S. DOT - Appendix B - Marine Pollutants
inhibited form: DOT regulated marine pollutant

U.S. DOT - Appendix A Table 1 - Hazardous Substances
final RQ = 1000 pounds (454 kg)

IARC - Group 2B (sufficient animal data)
[present] (Overall evaluation based only on evidence of carcinogenicity in monograph (39, 1986) or in Supplement 4)

NFPA - Flash Points
flash point = 50 degrees F (10 degrees C)

NFPA - Hazard Identification Ratings
health-2; flammability-3; reactivity-2

NIOSH - Selected LD50s and LC50s
Inhalation, rat: LC50 = 2180 ppm 4 hr Oral, rat: LD50 = 800 mg/kg Skin, rabbit: LD50 = 1834 mg/kg

NIOSH 1990 - Pocket Guide - IDLHs
2000 ppm IDLH (not considering carcinogenic effects)

NIOSH 1990 - Pocket Guide - Carcinogens
occupational carcinogen

NIOSH 1990 - Pocket Guide - Target organs
eyes, skin, respiratory system

NTP Chemical Status Reports - Testing Status and NTIS Number
Technical reports printed (PB87204061/AS)

NTP Chemical Status Reports - Evidence of Carcinogenicity
male rat-positive; female rat-positive; male mice-positive; female mice-positive

NTP Seventh Report - Suspect Carcinogens
suspect carcinogen

OSHA - Vacated PELs - Time Weighted Averages
5 ppm TWA; 20 mg/m3 TWA

OSHA - Vacated PELs - Short Term Exposure Limits
25 ppm STEL; 100 mg/m3 STEL

OSHA - Vacated PELs - Skin Designation
Prevent or reduce skin absorption

OSHA - Final PELs - Time Weighted Averages
25 ppm TWA; 100 mg/m3 TWA

OSHA - Final PELs - Skin Notations
prevent or reduce skin absorption

OSHA - Possible Select Carcinogens
[present]

ENVIRONMENTAL LISTS

CERCLA/SARA - Section 313 - Emission Reporting
form R reporting required for 0.1% de minimus concentration

CERCLA/SARA - Hazardous Substances and their Reportable Quantities
final RQ = 1000 pounds (454 kg)

Clean Air Act (1990) - List of Hazardous Air Contaminants
[present]

CAA - HON Rule - SOCMI Chemicals
compliance by Jan. 23, 1995

CAA - HON Rule - Organic HAPs
[present]

RCRA - U Series Wastes
waste number U113 (Ignitable waste)

RCRA - Hazardous Constituents-Appendix VIII
waste number U113 (Ignitable waste)

RCRA - Substances Banned From Land Disposal
[present]

TSCA - Code of Federal Regulations Citations
40 CFR 712.30(d); 40 CFR 716.120(a)

TSCA - PAIR - Reporting List
Reporting Date: November 19, 1982

TSCA - Health and Safety Reporting List
Effective Date: April 13, 1989

INTERNATIONAL LISTS

Australian Exposure Standards - Time Weighted Averages
Peak Limitation: 5 ppm; 20 mg/m3

Australian Exposure Standards - Skin Effects
sensitiser

Australian Exposure Standards - Under Review
exposure limits under review

Canada - WHMIS: Ingredient Disclosure
1% item 690 (156)

Canada - NPRI (National Pollutant Release Inventory)
[present]

Canada - Alberta - 8 Hour Occupational Exposure Limit
5 ppm TWA; 20 mg/m3 TWA

Canada - Alberta - 15 Minute Occupational Exposure Limit
25 ppm STEL; 102 mg/m3 STEL

Canada - Alberta - Skin Designation
can be absorbed through the intact skin

Canada - British Columbia - 8 Hour Exposure Limits
25 ppm TWA; 100 mg/m3 TWA

Canada - British Columbia - Skin Notations
skin - potential for skin absorption

Canada - Ontario - OHSA - TWAEVs
5 ppm TWAEV; 20 mg/m3 TWAEV

Canada - Ontario - OHSA - Skin Notations
absorption through skin, eyes, or mucous membranes

Canada - Quebec - Time-Weighted Average Exposure Values
5 ppm TWAEV; 20 mg/m3 TWAEV

Canada - Quebec - Short-term Exposure Values
15 ppm STEV; 61 mg/m3 STEV

Canada - Quebec - Carcinogens
C3 carcinogen: effect detected in animals

United Kingdom - Occupational Exposure Standards - TWAs
5 ppm TWA; 20 mg/m3 TWA

United Kingdom - Occupational Exposure Standards - STELs
15 ppm STEL; 60 mg/m3 STEL

United Kingdom - Occupational Exposure Standards - Notes
can be absorbed through skin

German (DFG) - MAK Values
5 ppm MAK;   20 mg/m3 MAK

German (DFG) - Peak Limitations
2 x normal MAK (5 min momentary value);   don't exceed 8 times during shift

German (DFG) - Skin/Sensitizers
danger of sensitization (skin or respiratory)

German (DFG) - Pregnancy
classification not yet possible

Israel - Time Weighted Averages
5 ppm TWA;   20 mg/m3 TWA

Israel - Short Term Exposure Limits
15 ppm STEL;   61 mg/m3 STEL

Israel - Action Levels
2.5 ppm AL;   10 mg/m3 AL

Mexico - Instruction No. 10 - TWAs
5 ppm TWA;   20 mg/m3 TWA

Mexico - Instruction No. 10 - STELs
25 ppm STEL;   100 mg/m3 STEL

Mexico - Instruction No. 10 - Skin designation
skin - potential for cutaneous absorption

STATE LISTS

California - Air Bill 2588 Appendix A-I
known or potential carcinogen

California - Prop. 65 - Cancer list
carcinogen - initial date 7/1/89

California - Exposure Limits - PELs
5 ppm PEL;   20 mg/m3 PEL

California - Exposure Limits - STELs
25 ppm STEL;   100 mg/m3 STEL

California - Exposure Limits - Skin Notation
material may be absorbed through the skin, eyes or mucous membrane

California - Directors List of Hazardous Substances (8 CCR 339)
[present]

Florida Hazardous Substance List
[present]

Massachusetts Right To Know List
carcinogen;   extraordinarily hazardous

Minnesota Hazardous Substance List
carcinogen;   skin

NJ Right to Know List (Total)
sn 0843

NJ Special Hazardous Substances
(flammable - third degree;   reactive - second degree)

Pennsylvania Right to Know List
environmental hazard;   special hazardous substance

Pennsylvania RTK - Special Hazardous Substances
[present]

ETHYL ALUMINUM SESQUICHLORIDE   **12075-68-2**

ENVIRONMENTAL LISTS

List of Pesticide Product Inert Ingredients
[present]

ETHYLAMINE   **75-04-7**

ENVIRONMENTAL LISTS

List of Pesticide Product Inert Ingredients
[present]

ETHYLAMINE, DISTILLATION PRODUCTS   **79171-09-8**

ENVIRONMENTAL LISTS

List of Pesticide Product Inert Ingredients
[present]

ETHYLAMINE, DISTILLATION RESIDUES   **79771-09-**

ENVIRONMENTAL LISTS

List of Pesticide Product Inert Ingredients
[present]

ETHYLAMINE SALTS   **RR-01783-4**

HEALTH AND SAFETY LISTS

ACGIH 1995 - Time Weighted Averages
1000 ppm TWA;   1880 mg/m3 TWA

AIHA - Odor Threshold Values
geometric mean air odor threshold = 180 ppm (detectable);   100 ppm (recognizable)

U.S. DOT - Substances From 49 CFR 172.101
regulated by DOT (UN1170, NA1986)

U.S. DOT - Hazard Classes
DOT hazard class = 3

NFPA - Flash Points
flash point = 55 degrees F (13 degrees C) (flash point rises when mixed with increasingly greater percentages of water)

NFPA - Hazard Identification Ratings
health-0;   flammability-3;   reactivity-0

NIOSH - Selected LD50s and LC50s
Inhalation, rat: LC50 = 20000 ppm 10 hr Oral, rat: LD50 = 7060 mg/kg

OSHA - Vacated PELs - Time Weighted Averages
1000 ppm TWA;   1900 mg/m3 TWA

OSHA - Final PELs - Time Weighted Averages
1000 ppm TWA;   1900 mg/m3 TWA

ENVIRONMENTAL LISTS

List of Pesticide Product Inert Ingredients
[present]

INTERNATIONAL LISTS

Australian Exposure Standards - Time Weighted Averages
1000 ppm TWA;   1880 mg/m3 TWA

Canada - WHMIS: Ingredient Disclosure
0.1% item 684 (805)

Canada - Alberta - 8 Hour Occupational Exposure Limit
1000 ppm TWA;   1884 mg/m3 TWA

Canada - Alberta - 15 Minute Occupational Exposure Limit
1250 ppm STEL;   2355 mg/m3 STEL

Canada - British Columbia - 8 Hour Exposure Limits
1000 ppm TWA;   1900 mg/m3 TWA

Canada - Ontario - OHSA - TWAEVs
1000 ppm TWAEV;   1900 mg/m3 TWAEV

Canada - Quebec - Time-Weighted Average Exposure Values
1000 ppm TWAEV;   1880 mg/m3 TWAEV

United Kingdom - Occupational Exposure Standards - TWAs
1000 ppm TWA;   1900 mg/m3 TWA

German (DFG) - MAK Values
1000 ppm MAK;   1900 mg/m3 MAK

German (DFG) - Peak Limitations
2 x normal MAK (1 hour momentary value);   don't exceed 3 times per shift

German (DFG) - Pregnancy
no risk to embryo/fetus if exposure limits adhered

Israel - Time Weighted Averages
1000 ppm TWA;   1880 mg/m3 TWA

Israel - Action Levels
500 ppm AL;   940 mg/m3 AL

Mexico - Instruction No. 10 - TWAs
1000 ppm TWA;   1900 mg/m3 TWA

STATE LISTS

California - Prop. 65 - Developmental Toxicity
developmental toxicity (when in alcoholic beverages) - initial date 10/1/87

California - Exposure Limits - PELs
1000 ppm PEL; 1900 mg/m3 PEL

California - Directors List of Hazardous Substances (8 CCR 339)
[present] (refers to solutions greater than or equal to 25% which are not beverage alcohols)

Florida Hazardous Substance List
[present]

Massachusetts Right To Know List
teratogen

Minnesota Hazardous Substance List
[present]

NJ Right to Know List (Total)
sn 0844

NJ Special Hazardous Substances
(flammable - third degree)

Pennsylvania Right to Know List
[present]

ETHYL P-AMINOBENZOATE
94-09-7

STATE LISTS

Pennsylvania Right to Know List
[present]

2-ETHYLAMINOETHANOL
110-73-6

HEALTH AND SAFETY LISTS

NFPA - Flash Points
fumes vigorously in air; may ignite spontaneously

NFPA - Hazard Identification Ratings
health-3; flammability-3; reactivity-3 (do not use water, foam or halogenated extinguishing agents)

STATE LISTS

Florida Hazardous Substance List
[present]

Massachusetts Right To Know List
[present]

NJ Right to Know List (Total)
sn 0845

NJ Special Hazardous Substances
(flammable - third degree; reactive - third degree)

Pennsylvania Right to Know List
[present]

ETHYL AMYL KETONE
106-68-3

HEALTH AND SAFETY LISTS

NFPA - Flash Points
flash point = -4 degrees F (-20 degrees C) ignites spontaneously in air

NFPA - Hazard Identification Ratings
flammability-3; reactivity-3 (do not use water, foam or halogenated extinguishing agents)

STATE LISTS

Florida Hazardous Substance List
[present]

Massachusetts Right To Know List
[present]

NJ Right to Know List (Total)
sn 0846

NJ Special Hazardous Substances
(flammable - third degree; reactive - third degree)

Pennsylvania Right to Know List
[present]

ETHYL AMYL KETONE
541-85-5

HEALTH AND SAFETY LISTS

ACGIH 1995 - Time Weighted Averages
5 ppm TWA; 9.2 mg/m3 TWA

ACGIH 1995 - Short Term Exposure Limits
15 ppm STEL; 27.6 mg/m3 STEL

ACGIH 1995 - Skin Designations
skin - potential for cutaneous absorption

AIHA - Odor Threshold Values
geometric mean air odor threshold = 0.27 ppm (detectable); 1.7 ppm (recognizable)

U.S. DOT - Substances From 49 CFR 172.101
regulated by DOT (UN2270, UN1036)

U.S. DOT - Hazard Classes
DOT hazard class = 2.1

U.S. DOT - Appendix A Table 1 - Hazardous Substances
final RQ = 100 pounds (45.4 kg)

FDA - Controlled Substances Act - Precursor chemicals
Threshold by base weight = 1 kilogram

NFPA - Flash Points
70% aqueous solution: flash point < 0 degrees F (-18 degrees C)

NFPA - Hazard Identification Ratings
health-3; flammability-4; reactivity-0

NIOSH - Selected LD50s and LC50s
Inhalation, mammal: LC50 = 2300 mg/m3 (8 hr) Oral, rat: 400 mg/kg Skin, rabbit: LD50 = 390 mg/kg

NIOSH 1990 - Pocket Guide - RELs
10 ppm TWA; 18 mg/m3 TWA

NIOSH 1990 - Pocket Guide - IDLHs
4000 ppm IDLH

NIOSH 1990 - Pocket Guide - Target organs
eyes, skin, respiratory system

OSHA - Vacated PELs - Time Weighted Averages
10 ppm TWA; 18 mg/m3 TWA

OSHA - Final PELs - Time Weighted Averages
10 ppm TWA; 18 mg/m3 TWA

OSHA - List of Highly Hazardous Chemicals
threshhold quantity = 7500 pounds

ENVIRONMENTAL LISTS

CERCLA/SARA - Hazardous Substances and their Reportable Quantities
final RQ = 100 pounds (45.4 kg)

CAA - Flammable Substances for Accidental Release Prevention
threshold quantity = 10,000 lbs

CAA - HON Rule - SOCMI Chemicals
compliance by Oct. 23, 1995

Clean Water Act - Hazardous Substances
[present]

TSCA - Code of Federal Regulations Citations
40 CFR 716.120(a)

TSCA - Health and Safety Reporting List
Effective Date: June 1, 1987

INTERNATIONAL LISTS

Australian Exposure Standards - Time Weighted Averages
10 ppm TWA; 18 mg/m3 TWA

Canada - WHMIS: Ingredient Disclosure
1% item 691 (848)

Canada - Alberta - 8 Hour Occupational Exposure Limit
10 ppm TWA; 18 mg/m3 TWA

Canada - Alberta - 15 Minute Occupational Exposure Limit
20 ppm STEL; 37 mg/m3 STEL

Canada - British Columbia - 8 Hour Exposure Limits
10 ppm TWA; 18 mg/m3 TWA

Canada - Ontario - OHSA - TWAEVs
5 ppm TWAEV;   9 mg/m3 TWAEV

Canada - Quebec - Time-Weighted Average Exposure Values
10 ppm TWAEV;   18 mg/m3 TWAEV

United Kingdom - Occupational Exposure Standards - TWAs
10 ppm TWA;   18 mg/m3 TWA

United Kingdom - Occupational Exposure Standards - STELs
10 ppm STEL;   18 mg/m3 STEL

German (DFG) - MAK Values
10 ppm MAK;   18 mg/m3 MAK

German (DFG) - Peak Limitations
2 x normal MAK (10 min momentary value);   don't exceed 4 times per shift

Israel - Time Weighted Averages
10 ppm TWA;   18 mg/m3 TWA

Israel - Action Levels
5 ppm AL;   9 mg/m3 AL

Mexico - Instruction No. 10 - TWAs
10 ppm TWA;   18 mg/m3 TWA

STATE LISTS

California - Exposure Limits - PELs
10 ppm PEL;   18 mg/m3 PEL

California - Directors List of Hazardous Substances (8 CCR 339)
[present]

Florida Hazardous Substance List
[present]

Massachusetts Right To Know List
[present]

Minnesota Hazardous Substance List
[present]

NJ Right to Know List (Total)
sn 0847

NJ Special Hazardous Substances
(flammable - fourth degree)

Pennsylvania Right to Know List
environmental hazard

N-ETHYLANILINE   103-69-5

ENVIRONMENTAL LISTS

TSCA - Chemicals with Significant New Use Rules
[present]

2-ETHYLANILINE   578-54-1

ENVIRONMENTAL LISTS

TSCA - Code of Federal Regulations Citations
40 CFR 721.1250

TSCA - Section 12(b) - Export Notification
P-80-290;   export notification required - Section 5

2-ETHYLANTHRAQUINONE   84-51-5

HEALTH AND SAFETY LISTS

FDA - Controlled Substances Act - Precursor chemicals
Threshold by base weight = 1 kilogram

ETHYLBENZENE   100-41-4

HEALTH AND SAFETY LISTS

NIOSH - Selected LD50s and LC50s
Oral, bird: LD50 = 56 mg/kg

ENVIRONMENTAL LISTS

List of Pesticide Product Inert Ingredients
[present]

ETHYL BENZOYLACETATE   94-02-

HEALTH AND SAFETY LISTS

NFPA - Flash Points
flash point = 160 degrees F (71 degrees C)

NFPA - Hazard Identification Ratings
health-1;   flammability-2;   reactivity-0

NIOSH - Selected LD50s and LC50s
Oral, rat: LD50 = 1000 mg/kg Skin, rabbit: LD50 = 360 mg/kg

ENVIRONMENTAL LISTS

List of Pesticide Product Inert Ingredients
[present]

INTERNATIONAL LISTS

Canada - WHMIS: Ingredient Disclosure
1% item 692 (849)

ETHYLBENZYLANILINE   92-59-

HEALTH AND SAFETY LISTS

AIHA - Odor Threshold Values
no geometric mean air odor threshold

U.S. DOT - Substances From 49 CFR 172.101
regulated by DOT (UN2271)

U.S. DOT - Hazard Classes
DOT hazard class = 3

ENVIRONMENTAL LISTS

List of Pesticide Product Inert Ingredients
[present]

INTERNATIONAL LISTS

Canada - WHMIS: Ingredient Disclosure
1% item 694 (851)

Canada - Alberta - 8 Hour Occupational Exposure Limit
25 ppm TWA;   131 mg/m3 TWA

Canada - Alberta - 15 Minute Occupational Exposure Limit
38 ppm STEL;   196 mg/m3 STEL

Canada - Quebec - Time-Weighted Average Exposure Values
25 ppm TWAEV;   131 mg/m3 TWAEV

Mexico - Instruction No. 10 - TWAs
25 ppm TWA;   130 mg/m3 TWA

STATE LISTS

Florida Hazardous Substance List
[present]

Pennsylvania Right to Know List
[present]

ETHYLBENZYLTOLUIDINE   119-94-8

HEALTH AND SAFETY LISTS

ACGIH 1995 - Time Weighted Averages
25 ppm TWA;   131 mg/m3 TWA

NIOSH - Selected LD50s and LC50s
Oral, rat: LD50 = 3500 mg/kg

NIOSH 1990 - Pocket Guide - RELs
25 ppm TWA;   130 mg/m3 TWA

NIOSH 1990 - Pocket Guide - IDLHs
3000 ppm IDLH

NIOSH 1990 - Pocket Guide - Target organs
eyes, skin, CNS, respiratory system

OSHA - Vacated PELs - Time Weighted Averages
25 ppm TWA;   130 mg/m3 TWA

OSHA - Final PELs - Time Weighted Averages
25 ppm TWA;   130 mg/m3 TWA

INTERNATIONAL LISTS

Australian Exposure Standards - Time Weighted Averages
25 ppm TWA;   131 mg/m3 TWA

Canada - WHMIS: Ingredient Disclosure
1% item 693 (850)
Canada - British Columbia - 8 Hour Exposure Limits
25 ppm TWA; 130 mg/m3 TWA
Canada - Ontario - OHSA - TWAEVs
25 ppm TWAEV; 130 mg/m3 TWAEV
United Kingdom - Occupational Exposure Standards - TWAs
25 ppm TWA; 130 mg/m3 TWA
Israel - Time Weighted Averages
25 ppm TWA; 131 mg/m3 TWA
Israel - Action Levels
12.5 ppm AL; 65.5 mg/m3 AL

STATE LISTS

California - Exposure Limits - PELs
25 ppm PEL; 130 mg/m3 PEL
California - Directors List of Hazardous Substances (8 CCR 339)
[present]
Florida Hazardous Substance List
[present]
Massachusetts Right To Know List
[present]
Minnesota Hazardous Substance List
[present]
NJ Right to Know List (Total)
sn 0848
Pennsylvania Right to Know List
[present]

N-ETHYLBENZYLTOLUIDINE, ALL ISOMERS   RR-00623-5

HEALTH AND SAFETY LISTS

U.S. DOT - Substances From 49 CFR 172.101
regulated by DOT (UN2272)
U.S. DOT - Hazard Classes
DOT hazard class = 6.1
NFPA - Flash Points
flash point = 185 degrees F (85 degrees C)
NFPA - Hazard Identification Ratings
health-3; flammability-2; reactivity-0
NIOSH - Selected LD50s and LC50s
Inhalation, rat: LC50 = 1480 mg/m3 4 hr Oral, rat: LD50 = 334 mg/kg Skin, rat: LD50 = 4700 mg/kg

ENVIRONMENTAL LISTS

CAA - HON Rule - SOCMI Chemicals
compliance by April 24, 1995

INTERNATIONAL LISTS

Canada - WHMIS: Ingredient Disclosure
1% item 696 (853)

STATE LISTS

Florida Hazardous Substance List
[present]
Massachusetts Right To Know List
[present]
NJ Right to Know List (Total)
sn 0850
Pennsylvania Right to Know List
[present]

ETHYLBIS(2-CHLOROETHYL)AMINE   538-07-8

HEALTH AND SAFETY LISTS

U.S. DOT - Substances From 49 CFR 172.101
regulated by DOT (UN2273)
U.S. DOT - Hazard Classes
DOT hazard class = 6.1

NIOSH - Selected LD50s and LC50s
Oral, rat: LD50 = 1260 mg/kg

ENVIRONMENTAL LISTS

CAA - HON Rule - SOCMI Chemicals
compliance by April 24, 1995

INTERNATIONAL LISTS

Canada - WHMIS: Ingredient Disclosure
1% item 695 (852)

STATE LISTS

NJ Right to Know List (Total)
sn 0849

ETHYL BIXIN   6895-43-8

ENVIRONMENTAL LISTS

TSCA - Code of Federal Regulations Citations
40 CFR 712.30(x); 40 CFR 716.120(d)
TSCA - PAIR - Reporting List
Reporting Date: November 27, 1991
TSCA - Health and Safety Reporting List
Effective Date: September 30, 1991

ETHYL BORATE   34099-73-5
SEE ALSO:
F003-HAZARDOUS WASTES
F039-HAZARDOUS WASTES

HEALTH AND SAFETY LISTS

ACGIH 1995 - Time Weighted Averages
100 ppm TWA; 434 mg/m3 TWA
ACGIH 1995 - Short Term Exposure Limits
125 ppm STEL; 543 mg/m3 STEL
ACGIH 1995 - Biological Exposure Indices
Mandelic acid in urine: 1.5 g/g creatinine, end of shift at end of workweek (Ns); Ethyl benzene in end-exhaled air (Sq)
AIHA - Odor Threshold Values
no geometric mean air odor threshold
U.S. DOT - Substances From 49 CFR 172.101
regulated by DOT (UN1175)
U.S. DOT - Hazard Classes
DOT hazard class = 3
U.S. DOT - Appendix A Table 1 - Hazardous Substances
final RQ = 1000 pounds (454 kg)
NFPA - Flash Points
flash point = 70 degrees F (21 degrees C)
NFPA - Hazard Identification Ratings
health-2; flammability-3; reactivity-0
NIOSH - Selected LD50s and LC50s
Oral, rat: LD50 = 3500 mg/kg Skin, rabbit: LD50 = 17800 mg/kg
NIOSH 1990 - Pocket Guide - RELs
100 ppm TWA; 435 mg/m3 TWA; 125 ppm STEL; 545 mg/m3 STEL
NIOSH 1990 - Pocket Guide - IDLHs
2000 ppm IDLH
NIOSH 1990 - Pocket Guide - Target organs
eyes, upper respiratory system, skin, CNS
NTP Chemical Status Reports - Testing Status and NTIS Number
Technical reports printed (PB93-149722); two year studies: pathology quality assessment in progress
OSHA - Vacated PELs - Time Weighted Averages
100 ppm TWA; 435 mg/m3 TWA
OSHA - Vacated PELs - Short Term Exposure Limits
125 ppm STEL; 545 mg/m3 STEL
OSHA - Final PELs - Time Weighted Averages
100 ppm TWA; 435 mg/m3 TWA

ENVIRONMENTAL LISTS

ATSDR Priority List
Rank (of 275): 073

CERCLA/SARA - Section 313 - Emission Reporting
form R reporting required for 1.0% de minimus concentration

CERCLA/SARA - Hazardous Substances and their Reportable Quantities
final RQ = 1000 pounds (454 kg)

Clean Air Act (1990) - List of Hazardous Air Contaminants
[present]

CAA - HON Rule - SOCMI Chemicals
compliance by Oct. 24, 1994

CAA - HON Rule - Organic HAPs
[present]

Clean Water Act - Hazardous Substances
[present]

Clean Water Act - Priority Pollutants
[present]

Clean Water Act - Toxic Pollutants
[present]

Safe Drinking Water Act - MCLs
MCL = 0.7 mg/L

Safe Drinking Water Act - Monitoring
monitoring required

List of Pesticide Product Inert Ingredients
[present]

RCRA - Basis for Listing - Appendix VII
Included in waste stream: F039

RCRA - TSD Facilities Ground Water Monitoring
TM 8020 = 2 ug/L PQL;   TM 8240 = 5 ug/L PQL

RCRA - Universal Treatment Standards (LDR)
WW: 0.057 mg/l;   NWW: 10 mg/kg

TSCA - Code of Federal Regulations Citations
40 CFR 712.30(d),(u);   40 CFR 716.120(a)

TSCA - PAIR - Reporting List
Reporting Dates: November 19, 1982;   August 18, 1987

TSCA - Health and Safety Reporting List
Effective Date: June 19, 1987

INTERNATIONAL LISTS

Australian Exposure Standards - Time Weighted Averages
100 ppm TWA;   434 mg/m3 TWA

Australian Exposure Standards - Short Term Exposure Limits
125 ppm STEL;   543 mg/m3 STEL

Australian Exposure Standards - Under Review
exposure limits under review

Canada - WHMIS: Ingredient Disclosure
0.1% item 697 (854)

Canada - NPRI (National Pollutant Release Inventory)
[present]

Canada - Drinking Water Quality - AOs
<= 0.0024 mg/L AO

Canada - Alberta - 8 Hour Occupational Exposure Limit
100 ppm TWA;   434 mg/m3 TWA

Canada - Alberta - 15 Minute Occupational Exposure Limit
125 ppm STEL;   542 mg/m3 STEL

Canada - British Columbia - 8 Hour Exposure Limits
100 ppm TWA;   435 mg/m3 TWA

Canada - British Columbia - 15 Minute Exposure Limits
125 ppm STEL;   545 mg/m3 STEL

Canada - Ontario - OHSA - TWAEVs
100 ppm TWAEV;   435 mg/m3 TWAEV

Canada - Ontario - OHSA - STEVs
125 ppm STEV;   540 mg/m3 STEV

Canada - Quebec - Time-Weighted Average Exposure Values
100 ppm TWAEV;   434 mg/m3 TWAEV

Canada - Quebec - Short-term Exposure Values
125 ppm STEV;   543 mg/m3 STEV

United Kingdom - Occupational Exposure Standards - TWAs
100 ppm TWA;   435 mg/m3 TWA

United Kingdom - Occupational Exposure Standards - STELs
125 ppm STEL;   545 mg/m3 STEL

German (DFG) - MAK Values
100 ppm MAK;   440 mg/m3 MAK

German (DFG) - Peak Limitations
2 x normal MAK (5 min momentary value);   don't exceed 8 time. during shift

German (DFG) - Skin/Sensitizers
danger of cutaneous absorption

German (DFG) - Pregnancy
classification not yet possible

Israel - Time Weighted Averages
100 ppm TWA;   434 mg/m3 TWA

Israel - Short Term Exposure Limits
125 ppm STEL;   543 mg/m3 STEL

Israel - Action Levels
50 ppm AL;   217 mg/m3 AL

Mexico - Instruction No. 10 - TWAs
100 ppm TWA;   435 mg/m3 TWA

Mexico - Instruction No. 10 - STELs
125 ppm STEL;   545 mg/m3 STEL

Mexico - Wastewater - Organic Toxic Pollutants and Heavy Metals
Listed under [Organic Toxic Pollutants]

Mexico - Drinking Water - Ecological Criteria
1.4 mg/l

STATE LISTS

California - Air Bill 2588 Appendix A-I
6/91

California - Exposure Limits - PELs
100 ppm PEL;   435 mg/m3 PEL

California - Exposure Limits - STELs
125 ppm STEL;   545 mg/m3 STEL

California - Directors List of Hazardous Substances (8 CCR 339)
[present]

Florida Hazardous Substance List
[present]

Massachusetts Right To Know List
[present]

Minnesota Hazardous Substance List
[present]

NJ Right to Know List (Total)
sn 0851

NJ Special Hazardous Substances
(flammable - third degree)

Pennsylvania Right to Know List
environmental hazard

PROPOSED REGULATIONS

Canada - Ontario - Proposed Occupational TWAEVs
50 ppm TWAEV;   200 mg/m3 TWAEV

Canada - Ontario - Proposed Occupational STEVs
100 ppm STEV;   450 mg/m3 STEV

ETHYL BROMIDE   74-96-4

HEALTH AND SAFETY LISTS

NFPA - Flash Points
flash point = 285 degrees F (141 degrees C)

NFPA - Hazard Identification Ratings
health-0; flammability-1; reactivity-0

NIOSH - Selected LD50s and LC50s
Oral, mouse: LD50 = 6800 mg/kg

ETHYL BROMOACETATE   105-36-2

HEALTH AND SAFETY LISTS

U.S. DOT - Substances From 49 CFR 172.101
regulated by DOT (UN2274)

U.S. DOT - Hazard Classes
DOT hazard class = 6.1

NFPA - Flash Points
flash point = 284 degrees F (140 degrees C)

NFPA - Hazard Identification Ratings
health-2; flammability-1; reactivity-0

INTERNATIONAL LISTS

Canada - WHMIS: Ingredient Disclosure
1% item 698 (855)

STATE LISTS

Florida Hazardous Substance List
[present]

Massachusetts Right To Know List
[present]

NJ Right to Know List (Total)
sn 0852

Pennsylvania Right to Know List
[present]

2-ETHYL-1-BUTENE   760-21-4

STATE LISTS

NJ Right to Know List (Total)
sn 0853

ETHYL BUTOXY ETHANOL   4468-93-3

HEALTH AND SAFETY LISTS

U.S. DOT - Substances From 49 CFR 172.101
regulated by DOT (UN2753)

U.S. DOT - Hazard Classes
DOT hazard class = 6.1

INTERNATIONAL LISTS

Canada - WHMIS: Ingredient Disclosure
1% item 699 (856)

3-(2-ETHYLBUTOXY) PROPIONIC ACID   10213-74-8

HEALTH AND SAFETY LISTS

NIOSH - Selected LD50s and LC50s
*Inhalation, rat: LC50 = 750 mg/m3 10 mn Oral, rat: LD50 = 2500
ug/kg Skin, rat: LD50 = 17 mg/kg*

ENVIRONMENTAL LISTS

CERCLA/SARA - Section 302 Extremely Hazardous Substances and
TPQs
TPQ = 500 pounds

STATE LISTS

Florida Hazardous Substance List
effective March 13, 1992

Massachusetts Right To Know List
extraordinarily hazardous

NJ Right to Know List (Total)
sn 2396

Pennsylvania Right to Know List
environmental hazard

PROPOSED REGULATIONS

CERCLA/SARA - Proposed Hazardous Substance Additions
proposed RQ = 1 pound (.454 kg)

CERCLA/SARA - 1989 Proposed RQ Adjustments
proposed RQ = 100 pounds (45.4 kg)

2-ETHYLBUTYL ACRYLATE   3953-10-4

ENVIRONMENTAL LISTS

List of Pesticide Product Inert Ingredients
[present]

ETHYLBUTYLAMINE   13360-63-9

HEALTH AND SAFETY LISTS

NFPA - Flash Points
flash point = 52 degrees F (11 degrees C)

NFPA - Hazard Identification Ratings
health-2; flammability-3; reactivity-0

STATE LISTS

Florida Hazardous Substance List
[present]

Massachusetts Right To Know List
[present]

NJ Right to Know List (Total)
sn 0854

NJ Special Hazardous Substances
(flammable - third degree)

ETHYL BUTYL CARBONATE   30714-78-4

HEALTH AND SAFETY LISTS

ACGIH 1995 - Time Weighted Averages
5 ppm TWA; 22 mg/m3 TWA

ACGIH 1995 - Skin Designations
skin - potential for cutaneous absorption

ACGIH 1995 - Carcinogens
A2-suspected human carcinogen

U.S. DOT - Substances From 49 CFR 172.101
regulated by DOT (UN1891)

U.S. DOT - Hazard Classes
DOT hazard class = 6.1

IARC - Group 3 (not classifiable)
[present]

NFPA - Flash Points
no flash point

NFPA - Hazard Identification Ratings
health-2; flammability-1; reactivity-0

NIOSH - Selected LD50s and LC50s
Inhalation, mouse: LC50 = 16230 ppm (1 hr) Oral, rat: 1350 mg/kg

NIOSH 1990 - Pocket Guide - RELs
See Appendix D

NIOSH 1990 - Pocket Guide - IDLHs
3500 ppm IDLH

NIOSH 1990 - Pocket Guide - Target organs
skin, liver, kidneys, respiratory system, CVS, CNS

NTP Chemical Status Reports - Testing Status and NTIS Number
Technical reports printed (PB90219445/AS)

NTP Chemical Status Reports - Evidence of Carcinogenicity
*male rat-some evidence; female rat-equivocal evidence; male mice-
equivocal evidence; female mice-clear evidence*

OSHA - Vacated PELs - Time Weighted Averages
200 ppm TWA; 890 mg/m3 TWA

OSHA - Vacated PELs - Short Term Exposure Limits
250 ppm STEL; 1110 mg/m3 STEL

OSHA - Final PELs - Time Weighted Averages
200 ppm TWA; 890 mg/m3 TWA

ENVIRONMENTAL LISTS

TSCA - Health and Safety Reporting List
Effective Date: March 11, 1994; Sunset Date: March 11, 2004

INTERNATIONAL LISTS

Australian Exposure Standards - Time Weighted Averages
200 ppm TWA; 891 mg/m3 TWA

Australian Exposure Standards - Short Term Exposure Limits
250 ppm STEL; 1110 mg/m3 STEL

Canada - WHMIS: Ingredient Disclosure
1% item 700 (336)

Canada - Alberta - 8 Hour Occupational Exposure Limit
200 ppm TWA; 891 mg/m3 TWA

Canada - Alberta - 15 Minute Occupational Exposure Limit
250 ppm STEL; 1114 mg/m3 STEL

Canada - British Columbia - 8 Hour Exposure Limits
200 ppm TWA; 890 mg/m3 TWA

Canada - British Columbia - 15 Minute Exposure Limits
250 ppm STEL; 1110 mg/m3 STEL

Canada - Ontario - OHSA - TWAEVs
200 ppm TWAEV; 890 mg/m3 TWAEV

Canada - Ontario - OHSA - STEVs
250 ppm STEV; 1110 mg/m3 STEV

Canada - Quebec - Time-Weighted Average Exposure Values
200 ppm TWAEV; 891 mg/m3 TWAEV

Canada - Quebec - Short-term Exposure Values
250 ppm STEV; 1110 mg/m3 STEV

United Kingdom - Occupational Exposure Standards - TWAs
200 ppm TWA; 890 mg/m3 TWA

United Kingdom - Occupational Exposure Standards - STELs
250 ppm STEL; 1110 mg/m3 STEL

German (DFG) - Carcinogens
animal evidence of carcinogenicity

Israel - Time Weighted Averages
(200) ppm TWA; (891) mg/m3 TWA

Israel - Short Term Exposure Limits
(250) ppm STEL; (1110) mg/m3 STEL

Israel - Action Levels
100 ppm AL; 445.5 mg/m3 AL

Mexico - Instruction No. 10 - TWAs
200 ppm TWA; 890 mg/m3 TWA

Mexico - Instruction No. 10 - STELs
250 ppm STEL; 1110 mg/m3 STEL

Mexico - Drinking Water - Ecological Criteria
0.002 mg/l

STATE LISTS

California - Exposure Limits - PELs
200 ppm PEL; 890 mg/m3 PEL

California - Exposure Limits - STELs
250 ppm STEL; 1110 mg/m3 STEL

California - Directors List of Hazardous Substances (8 CCR 339)
[present]

Florida Hazardous Substance List
[present]

Massachusetts Right To Know List
[present]

Minnesota Hazardous Substance List
[present]

NJ Right to Know List (Total)
sn 0855

Pennsylvania Right to Know List
[present]

PROPOSED REGULATIONS

TSCA - ITC 32nd Report Priority Testing List
designated for dermal absorption testing

ETHYL BUTYL ETHER   628-81-

HEALTH AND SAFETY LISTS

U.S. DOT - Substances From 49 CFR 172.101
regulated by DOT (UN1603)

U.S. DOT - Hazard Classes
DOT hazard class = 6.1

NFPA - Flash Points
flash point = 118 degrees F (48 degrees C)

NFPA - Hazard Identification Ratings
flammability-2; reactivity-0

INTERNATIONAL LISTS

Canada - WHMIS: Ingredient Disclosure
1% item 701 (322)

STATE LISTS

NJ Right to Know List (Total)
sn 0856

ETHYL TERT-BUTYL ETHER   637-92-

HEALTH AND SAFETY LISTS

NFPA - Flash Points
flash point < -4 degrees F (-20 degrees C)

NFPA - Hazard Identification Ratings
health-0; flammability-3; reactivity-0

STATE LISTS

Florida Hazardous Substance List
[present]

Massachusetts Right To Know List
[present]

Pennsylvania Right to Know List
[present]

ETHYL BUTYL KETONE   106-35-4

HEALTH AND SAFETY LISTS

NFPA - Flash Points
flash point = 180 degrees F (82 degrees C)

NFPA - Hazard Identification Ratings
health-0; flammability-2; reactivity-0

NIOSH - Selected LD50s and LC50s
Oral, rat: LD50 = 1910 mg/kg Skin, rabbit: LD50 = 320 mg/kg

2-ETHYL-2-BUTYL-1,3-PROPANEDIOL   115-84-4

HEALTH AND SAFETY LISTS

NFPA - Flash Points
flash point = 280 degrees F (138 degrees C)

NFPA - Hazard Identification Ratings
health-2; flammability-1; reactivity-0

STATE LISTS

Florida Hazardous Substance List
[present]

Massachusetts Right To Know List
[present]

Pennsylvania Right to Know List
[present]

2-ETHYLBUTYRALDEHYDE   97-96-1

HEALTH AND SAFETY LISTS

NFPA - Flash Points
flash point = 125 degrees F (52 degrees C)

NFPA - Hazard Identification Ratings
health-2; flammability-2; reactivity-0

ENVIRONMENTAL LISTS

TSCA - Code of Federal Regulations Citations
40 CFR 712.30(d)

TSCA - PAIR - Reporting List
Reporting Date: November 19, 1982

STATE LISTS

Florida Hazardous Substance List
[present]

Massachusetts Right To Know List
[present]

Pennsylvania Right to Know List
[present]

ETHYL BUTYRATE   105-54-4

HEALTH AND SAFETY LISTS

NFPA - Flash Points
flash point = 64 degrees F (18 degrees C)

NFPA - Hazard Identification Ratings
health-3; flammability-3; reactivity-0

STATE LISTS

Florida Hazardous Substance List
[present]

Massachusetts Right To Know List
[present]

Pennsylvania Right to Know List
[present]

2-ETHYLBUTYRIC ACID   88-09-5

HEALTH AND SAFETY LISTS

NFPA - Flash Points
flash point = 122 degrees F (50 degrees C)

NFPA - Hazard Identification Ratings
health-2; flammability-2; reactivity-1

STATE LISTS

Florida Hazardous Substance List
[present]

Massachusetts Right To Know List
[present]

Pennsylvania Right to Know List
[present]

ETHYL CAPRYLATE   106-32-1

HEALTH AND SAFETY LISTS

U.S. DOT - Substances From 49 CFR 172.101
regulated by DOT (UN1179)

U.S. DOT - Hazard Classes
DOT hazard class = 3

NFPA - Flash Points
flash point = 40 degrees F (4 degrees C)

NFPA - Hazard Identification Ratings
health-2; flammability-3; reactivity-0

NIOSH - Selected LD50s and LC50s
Oral, rat: LD50 = 1870 mg/kg

STATE LISTS

Florida Hazardous Substance List
[present]

Massachusetts Right To Know List
[present]

NJ Right to Know List (Total)
sn 0859

NJ Special Hazardous Substances
(flammable - third degree)

Pennsylvania Right to Know List
[present]

ETHYLCELLULOSE   9004-57-3

PROPOSED REGULATIONS

TSCA - ITC 34th Report Priority Testing List
recommended for health effects testing

ETHYL CENTRALITE   85-98-3

HEALTH AND SAFETY LISTS

ACGIH 1995 - Time Weighted Averages
50 ppm TWA; 234 mg/m3 TWA

NFPA - Flash Points
flash point = 115 degrees F (46 degrees C)

NFPA - Hazard Identification Ratings
health-1; flammability-2; reactivity-0

NIOSH - Selected LD50s and LC50s
Oral, rat: LD50 = 2760 mg/kg

NIOSH 1990 - Pocket Guide - RELs
50 ppm TWA; 230 mg/m3 TWA

NIOSH 1990 - Pocket Guide - IDLHs
3000 ppm IDLH

NIOSH 1990 - Pocket Guide - Target organs
eyes, skin, respiratory system

OSHA - Vacated PELs - Time Weighted Averages
50 ppm TWA; 230 mg/m3 TWA

OSHA - Final PELs - Time Weighted Averages
50 ppm TWA; 230 mg/m3 TWA

INTERNATIONAL LISTS

Australian Exposure Standards - Time Weighted Averages
50 ppm TWA; 234 mg/m3 TWA

Canada - WHMIS: Ingredient Disclosure
1% item 702 (857)

Canada - Alberta - 8 Hour Occupational Exposure Limit
50 ppm TWA; 234 mg/m3 TWA

Canada - Alberta - 15 Minute Occupational Exposure Limit
75 ppm STEL; 350 mg/m3 STEL

Canada - British Columbia - 8 Hour Exposure Limits
50 ppm TWA; 230 mg/m3 TWA

Canada - British Columbia - 15 Minute Exposure Limits
75 ppm STEL; 345 mg/m3 STEL

Canada - Ontario - OHSA - TWAEVs
50 ppm TWAEV; 230 mg/m3 TWAEV

Canada - Quebec - Time-Weighted Average Exposure Values
50 ppm TWAEV; 234 mg/m3 TWAEV

United Kingdom - Occupational Exposure Standards - TWAs
50 ppm TWA; 230 mg/m3 TWA

United Kingdom - Occupational Exposure Standards - STELs
75 ppm STEL; 345 mg/m3 STEL

Israel - Time Weighted Averages
50 ppm TWA; 234 mg/m3 TWA

Israel - Action Levels
25 ppm AL; 117 mg/m3 AL

Mexico - Instruction No. 10 - TWAs
50 ppm TWA; 230 mg/m3 TWA

Mexico - Instruction No. 10 - STELs
75 ppm STEL; 345 mg/m3 STEL

STATE LISTS

California - Exposure Limits - PELs
50 ppm PEL; 230 mg/m3 PEL

California - Directors List of Hazardous Substances (8 CCR 339)
[present]
Florida Hazardous Substance List
[present]
Massachusetts Right To Know List
[present]
Minnesota Hazardous Substance List
[present]
NJ Right to Know List (Total)
sn 0860
Pennsylvania Right to Know List
[present]

ETHYL CHLORIDE   75-00-3

HEALTH AND SAFETY LISTS
NFPA - Flash Points
flash point = 280 degrees F (138 degrees C)
NFPA - Hazard Identification Ratings
health-2; flammability-1; reactivity-0
STATE LISTS
Florida Hazardous Substance List
[present]
Massachusetts Right To Know List
[present]
Pennsylvania Right to Know List
[present]

ETHYL CHLOROACETATE   105-39-5

HEALTH AND SAFETY LISTS
U.S. DOT - Substances From 49 CFR 172.101
regulated by DOT (UN1178)
U.S. DOT - Hazard Classes
DOT hazard class = 3
NFPA - Flash Points
flash point = 70 degrees F (21 degrees C)
NFPA - Hazard Identification Ratings
health-2; flammability-3; reactivity-1
NIOSH - Selected LD50s and LC50s
Oral, rat: LD50 = 3980 mg/kg
STATE LISTS
Florida Hazardous Substance List
[present]
Massachusetts Right To Know List
[present]
NJ Right to Know List (Total)
sn 0861
NJ Special Hazardous Substances
(flammable - third degree)
Pennsylvania Right to Know List
[present]

ETHYL CHLOROFORMATE   541-41-3

HEALTH AND SAFETY LISTS
U.S. DOT - Substances From 49 CFR 172.101
regulated by DOT (UN1180)
U.S. DOT - Hazard Classes
DOT hazard class = 3
NFPA - Flash Points
flash point = 75 degrees F (24 degrees C)
NFPA - Hazard Identification Ratings
health-0; flammability-3; reactivity-0
NIOSH - Selected LD50s and LC50s
Oral, rat: LD50 = 13 gm/kg

INTERNATIONAL LISTS
Canada - WHMIS: Ingredient Disclosure
1% item 703 (373)
STATE LISTS
Florida Hazardous Substance List
[present]
Massachusetts Right To Know List
[present]
NJ Right to Know List (Total)
sn 0862
NJ Special Hazardous Substances
(flammable - third degree)
Pennsylvania Right to Know List
[present]

ETHYL-2-CHLOROPROPIONATE   535-13-7

HEALTH AND SAFETY LISTS
NFPA - Flash Points
flash point = 210 degrees F (99 degrees C)
NFPA - Hazard Identification Ratings
health-2; flammability-1; reactivity-0
INTERNATIONAL LISTS
Canada - WHMIS: Ingredient Disclosure
1% item 704 (84)
STATE LISTS
Florida Hazardous Substance List
[present]
Massachusetts Right To Know List
[present]
Pennsylvania Right to Know List
[present]

ETHYL CHLOROTHIOFORMATE   2812-73-9

HEALTH AND SAFETY LISTS
NFPA - Flash Points
flash point = 175 degrees F (79 degrees C)
NFPA - Hazard Identification Ratings
health-2; flammability-2; reactivity-0
STATE LISTS
Florida Hazardous Substance List
[present]
Massachusetts Right To Know List
[present]
Pennsylvania Right to Know List
[present]

S-ETHYL CHLOROTHIOFORMATE   2941-64-2

ENVIRONMENTAL LISTS
CAA - HON Rule - SOCMI Chemicals
compliance by Oct. 23, 1995
List of Pesticide Product Inert Ingredients
[present]

ETHYL CINNAMATE   103-36-6

HEALTH AND SAFETY LISTS
NFPA - Flash Points
flash point = 302 degrees F (150 degrees C)
NFPA - Hazard Identification Ratings
health-1; flammability-1; reactivity-0

THYL CROTONATE   **623-70-1**
SEE ALSO:
   F039-HAZARDOUS WASTES
HEALTH AND SAFETY LISTS
   ACGIH 1995 - Time Weighted Averages
      (1000) ppm TWA;   (2640) mg/m3 TWA
   U.S. DOT - Substances From 49 CFR 172.101
      regulated by DOT (UN1037)
   U.S. DOT - Hazard Classes
      DOT hazard class = 2.1
   U.S. DOT - Appendix A Table 1 - Hazardous Substances
      final RQ = 100 pounds (45.4 kg)
   IARC - Group 3 (not classifiable)
      [present]
   NFPA - Flash Points
      flash point = -58 degrees F (-50 degrees C)
   NFPA - Hazard Identification Ratings
      health-1; flammability-4; reactivity-0
   NIOSH - Selected LD50s and LC50s
      Inhalation, rat: LC50 = 160 gm/m3 2 hr
   NIOSH 1990 - Pocket Guide - RELs
      Handle with caution in the workplace; See Appendix D
   NIOSH 1990 - Pocket Guide - IDLHs
      20,000 ppm IDLH
   NIOSH 1990 - Pocket Guide - Target organs
      liver, kidneys, respiratory system, CVS
   NIOSH - Health Standards - Exposure Limits
      Handle with caution in the workplace
   NIOSH - Health Standards - Health Effects and Precautions
      Central nervous system effects, possible liver and/or kidney effects
   NTP Chemical Status Reports - Testing Status and NTIS Number
      Technical reports printed (PB90225053/AS)
   NTP Chemical Status Reports - Evidence of Carcinogenicity
      male rat-equivocal evidence; female rat-equivocal evidence; male mice-inadequate; female mice-clear evidence
   OSHA - Vacated PELs - Time Weighted Averages
      1000 ppm TWA; 2600 mg/m3 TWA
   OSHA - Final PELs - Time Weighted Averages
      1000 ppm TWA; 2600 mg/m3 TWA
ENVIRONMENTAL LISTS
   ATSDR Priority List
      Rank (of 275): 147
   CERCLA/SARA - Section 313 - Emission Reporting
      form R reporting required for 1.0% de minimus concentration
   CERCLA/SARA - Hazardous Substances and their Reportable Quantities
      final RQ = 100 pounds (45.4 kg)
   Clean Air Act (1990) - List of Hazardous Air Contaminants
      [present]
   CAA - Flammable Substances for Accidental Release Prevention
      threshold quantity = 10,000 lbs
   CAA - HON Rule - SOCMI Chemicals
      compliance by July 24, 1995
   CAA - HON Rule - Organic HAPs
      [present]
   Clean Water Act - Priority Pollutants
      [present]
   Safe Drinking Water Act - Monitoring
      monitoring required
   EPA - Master Testing List
      [present]
   List of Pesticide Product Inert Ingredients
      [present]

   RCRA - Basis for Listing - Appendix VII
      Included in waste stream: F039
   RCRA - TSD Facilities Ground Water Monitoring
      TM 8010 = 5 ug/L PQL;   TM 8240 = 10 ug/L PQL
   RCRA - Universal Treatment Standards (LDR)
      WW: 0.27 mg/l;   NWW: 6.0 mg/kg
   TSCA - Code of Federal Regulations Citations
      40 CFR 716.120(a)
   TSCA - PAIR - Reporting List
      Reporting Date: June 13, 1989
   TSCA - Health and Safety Reporting List
      Effective Date: June 1, 1987
   TSCA - Section 12(b) - Export Notification
      export notification required - Section 4
INTERNATIONAL LISTS
   Australian Exposure Standards - Time Weighted Averages
      1000 ppm TWA; 2640 mg/m3 TWA
   Canada - WHMIS: Ingredient Disclosure
      1% item 705 (497)
   Canada - NPRI (National Pollutant Release Inventory)
      [present]
   Canada - Alberta - 8 Hour Occupational Exposure Limit
      1000 ppm TWA; 2639 mg/m3 TWA
   Canada - Alberta - 15 Minute Occupational Exposure Limit
      1250 ppm STEL; 3299 mg/m3 STEL
   Canada - British Columbia - 8 Hour Exposure Limits
      1000 ppm TWA; 2600 mg/m3 TWA
   Canada - British Columbia - 15 Minute Exposure Limits
      1250 ppm STEL; 3250 mg/m3 STEL
   Canada - Ontario - OHSA - TWAEVs
      1000 ppm TWAEV; 2635 mg/m3 TWAEV
   Canada - Quebec - Time-Weighted Average Exposure Values
      1000 ppm TWAEV; 2640 mg/m3 TWAEV
   United Kingdom - Occupational Exposure Standards - TWAs
      1000 ppm TWA; 2600 mg/m3 TWA
   United Kingdom - Occupational Exposure Standards - STELs
      1250 ppm STEL; 3250 mg/m3 STEL
   German (DFG) - Carcinogens
      suspected carcinogen
   Israel - Time Weighted Averages
      1000 ppm TWA; 2640 mg/m3 TWA
   Israel - Action Levels
      500 ppm AL; 1320 mg/m3 AL
   Mexico - Instruction No. 10 - TWAs
      1000 ppm TWA; 2600 mg/m3 TWA
   Mexico - Instruction No. 10 - STELs
      1250 ppm STEL; 3250 mg/m3 STEL
   Mexico - Wastewater - Organic Toxic Pollutants and Heavy Metals
      Listed under [Chlorinated Ethanes]
STATE LISTS
   California - Air Bill 2588 Appendix A-I
      [present]
   California - Prop. 65 - Cancer list
      carcinogen - initial date 7/1/90
   California - Exposure Limits - PELs
      1000 ppm PEL; 2600 mg/m3 PEL
   California - Directors List of Hazardous Substances (8 CCR 339)
      [present]
   Florida Hazardous Substance List
      [present]
   Massachusetts Right To Know List
      [present]

Minnesota Hazardous Substance List
[present]
NJ Right to Know List (Total)
sn 0863
NJ Special Hazardous Substances
(flammable - fourth degree)
Pennsylvania Right to Know List
environmental hazard

PROPOSED REGULATIONS
ACGIH 1995 - Notice of Intended Changes
(skin) 100 ppm TWA;   264 mg/m3 TWA;   A3-animal carcinogen
Safe Drinking Water Act - Priority list
[present]

ETHYL CROTONATE   10544-63-5

HEALTH AND SAFETY LISTS
U.S. DOT - Substances From 49 CFR 172.101
regulated by DOT (UN1181)
U.S. DOT - Hazard Classes
DOT hazard class = 6.1
NFPA - Flash Points
flash point = 147 degrees F (64 degrees C)
NFPA - Hazard Identification Ratings
flammability-3;   reactivity-0
NIOSH - Selected LD50s and LC50s
Skin, rabbit: LD50 = 230 mg/kg

ENVIRONMENTAL LISTS
CAA - HON Rule - SOCMI Chemicals
compliance by Jan. 23, 1995

INTERNATIONAL LISTS
Canada - WHMIS: Ingredient Disclosure
1% item 706 (406)

STATE LISTS
Florida Hazardous Substance List
[present]
Massachusetts Right To Know List
[present]
NJ Right to Know List (Total)
sn 0864
NJ Special Hazardous Substances
(flammable - third degree)
Pennsylvania Right to Know List
[present]

ETHYL CYANOACETATE   105-56-6

HEALTH AND SAFETY LISTS
U.S. DOT - Substances From 49 CFR 172.101
regulated by DOT (UN1182)
U.S. DOT - Hazard Classes
DOT hazard class = 6.1
U.S. DOT - Substances Which Are Poisonous by Inhalation
liquid hazardous material poisonous by inhalation (UN1182)
NFPA - Flash Points
flash point = 61 degrees F (16 degrees C)
NFPA - Hazard Identification Ratings
health-4;   flammability-3;   reactivity-1
NIOSH - Selected LD50s and LC50s
Inhalation, rat: LC50 = 145 ppm 1 hr Oral, rat: LD50 = 270 mg/kg
Skin, rabbit: LD50 = 7120 mg/kg

ENVIRONMENTAL LISTS
CERCLA/SARA - Section 313 - Emission Reporting
form R reporting required for 1.0% de minimus concentration

INTERNATIONAL LISTS
Canada - WHMIS: Ingredient Disclosure
1% item 707 (436)
Canada - NPRI (National Pollutant Release Inventory)
[present]
United Kingdom - Occupational Exposure Standards - TWAs
1 ppm TWA;   4.4 mg/m3 TWA

STATE LISTS
California - Air Bill 2588 Appendix A-II
6/91
Florida Hazardous Substance List
[present]
Massachusetts Right To Know List
[present]
NJ Right to Know List (Total)
sn 0865
NJ Special Hazardous Substances
(corrosive;   flammable - third degree)
Pennsylvania Right to Know List
environmental hazard

ETHYL CYANOACRYLATE   7085-85-(

HEALTH AND SAFETY LISTS
U.S. DOT - Substances From 49 CFR 172.101
regulated by DOT (UN2935)
U.S. DOT - Hazard Classes
DOT hazard class = 3

STATE LISTS
NJ Right to Know List (Total)
sn 0866

ETHYL CYCLOBUTANE   4806-61-5

HEALTH AND SAFETY LISTS
U.S. DOT - Substances From 49 CFR 172.101
regulated by DOT (UN2826)
U.S. DOT - Hazard Classes
DOT hazard class = 8
U.S. DOT - Substances Which Are Poisonous by Inhalation
liquid hazardous material poisonous by inhalation (UN2826)
U.S. DOT - Appendix B - Marine Pollutants
DOT regulated marine pollutant

INTERNATIONAL LISTS
Canada - WHMIS: Ingredient Disclosure
1% item 708 (460)

STATE LISTS
NJ Right to Know List (Total)
sn 0867
NJ Special Hazardous Substances
(corrosive)

ETHYL CYCLOHEXANE   1678-91-7

HEALTH AND SAFETY LISTS
U.S. DOT - Substances From 49 CFR 172.101
regulated by DOT (UN2826)
U.S. DOT - Hazard Classes
DOT hazard class = 8

N-ETHYLCYCLOHEXYLAMINE   5459-93-8

HEALTH AND SAFETY LISTS
NIOSH - Selected LD50s and LC50s
Oral, rat: LD50 = 4000 mg/kg

ETHYL CYCLOPENTANE   1640-89-7

HEALTH AND SAFETY LISTS

NFPA - Flash Points
flash point = 36 degrees F (2 degrees C)

NFPA - Hazard Identification Ratings
health-2; flammability-3; reactivity-0

INTERNATIONAL LISTS

Canada - WHMIS: Ingredient Disclosure
1% item 709 (576)

STATE LISTS

Florida Hazardous Substance List
[present]

Massachusetts Right To Know List
[present]

NJ Right to Know List (Total)
sn 0869

NJ Special Hazardous Substances
(flammable - third degree)

ETHYL DECANOATE   110-38-3

HEALTH AND SAFETY LISTS

U.S. DOT - Substances From 49 CFR 172.101
regulated by DOT (UN1862)

U.S. DOT - Hazard Classes
DOT hazard class = 3

NIOSH - Selected LD50s and LC50s
Oral, rat: LD50 = 3 gm/kg

STATE LISTS

Pennsylvania Right to Know List
[present]

ETHYL-3,3-DI(TERT-BUTYL-PEROXY)BUTYRATE   RR-00287-9

HEALTH AND SAFETY LISTS

U.S. DOT - Substances From 49 CFR 172.101
regulated by DOT (UN2666)

U.S. DOT - Hazard Classes
DOT hazard class = 6.1

NFPA - Flash Points
flash point = 230 degrees F (110 degrees C)

NFPA - Hazard Identification Ratings
health-2; flammability-1; reactivity-0

ENVIRONMENTAL LISTS

CAA - HON Rule - SOCMI Chemicals
compliance by Oct. 23, 1995

INTERNATIONAL LISTS

Canada - WHMIS: Ingredient Disclosure
1% item 711 (584)

STATE LISTS

Florida Hazardous Substance List
[present]

Massachusetts Right To Know List
[present]

NJ Right to Know List (Total)
sn 0870

Pennsylvania Right to Know List
[present]

ETHYL DICHLOROARSINE   598-14-1

HEALTH AND SAFETY LISTS

NTP Chemical Status Reports - Testing Status and NTIS Number
Project leader assigned/study in design

ENVIRONMENTAL LISTS

TSCA - PAIR - Reporting List
Effective Date: January 26, 1994; Reporting Date: March 28, 1994

TSCA - Health and Safety Reporting List
Effective Date: January 26, 1994; Sunset Date: January 26, 2004

ETHYL DICHLOROSILANE   1789-58-8

HEALTH AND SAFETY LISTS

NFPA - Flash Points
flash point < 4 degrees F (-16 degrees C)

NFPA - Hazard Identification Ratings
health-1; flammability-3; reactivity-0

STATE LISTS

Florida Hazardous Substance List
[present]

Massachusetts Right To Know List
[present]

Pennsylvania Right to Know List
[present]

N-ETHYLDIETHANOLAMINE   139-87-7

HEALTH AND SAFETY LISTS

NFPA - Flash Points
flash point = 95 degrees F (35 degrees C)

NFPA - Hazard Identification Ratings
health-1; flammability-3; reactivity-0

STATE LISTS

Florida Hazardous Substance List
[present]

Massachusetts Right To Know List
[present]

Pennsylvania Right to Know List
[present]

ETHYL 3,3-DIMETHACRYLATE   638-10-8

HEALTH AND SAFETY LISTS

NFPA - Flash Points
flash point = 86 degrees F (30 degrees C)

NFPA - Hazard Identification Ratings
health-3; flammability-3; reactivity-0

STATE LISTS

Florida Hazardous Substance List
[present]

Massachusetts Right To Know List
[present]

Pennsylvania Right to Know List
[present]

N-ETHYL-N,N-DIMETHYL-1-DODECAMINIUM ETHYL SULFATE   3006-13-1

HEALTH AND SAFETY LISTS

NFPA - Flash Points
flash point < 70 degrees F (21 degrees C)

NFPA - Hazard Identification Ratings
health-1; flammability-3; reactivity-0

STATE LISTS

Florida Hazardous Substance List
[present]

Massachusetts Right To Know List
[present]

Pennsylvania Right to Know List
[present]

ETHYL DIPROPYLTHIOCARBAMATE [EPTC]   759-94-4
HEALTH AND SAFETY LISTS
NFPA - Flash Points
flash point > 212 degrees F (100 degrees C)

NFPA - Hazard Identification Ratings
health-0; flammability-1; reactivity-0

ETHYLENE   74-85-1
STATE LISTS
NJ Right to Know List (Total)
sn 2397; sn 2398; technically pure: sn 2399

ETHYLENE GLYCOL MONOCTYL ETHER   10020-43-6
SEE ALSO:
ARSENIC

HEALTH AND SAFETY LISTS
U.S. DOT - Substances From 49 CFR 172.101
regulated by DOT (UN1892)

U.S. DOT - Hazard Classes
DOT hazard class = 6.1

U.S. DOT - Substances Which Are Poisonous by Inhalation
liquid hazardous material poisonous by inhalation (UN1892)

U.S. DOT - Appendix B - Marine Pollutants
DOT regulated marine pollutant

NIOSH - Selected LD50s and LC50s
Inhalation, mouse: LC50 = 1555 mg/m3 (10 mn)

STATE LISTS
NJ Right to Know List (Total)
sn 0871

ETHYLENE BIS(5,6-DIBROMONORBORNANE-2,3-DICARBOXIMIDE)   52907-07-0
HEALTH AND SAFETY LISTS
U.S. DOT - Substances From 49 CFR 172.101
regulated by DOT (UN1183)

U.S. DOT - Hazard Classes
DOT hazard class = 4.3

NFPA - Flash Points
flash point = 30 degrees F (-1 degrees C)

NFPA - Hazard Identification Ratings
health-3; flammability-3; reactivity-0

STATE LISTS
Florida Hazardous Substance List
[present]

Massachusetts Right To Know List
[present]

NJ Right to Know List (Total)
sn 0872

NJ Special Hazardous Substances
(corrosive; flammable - third degree)

Pennsylvania Right to Know List
[present]

ETHYLENE BIS DITHIOCARBAMATE   142-59-6
HEALTH AND SAFETY LISTS
NFPA - Flash Points
flash point = 280 degrees F (138 degrees C)

NFPA - Hazard Identification Ratings
health-2; flammability-1; reactivity-0

ENVIRONMENTAL LISTS
List of Pesticide Product Inert Ingredients
[present]

STATE LISTS
Florida Hazardous Substance List
[present]

Massachusetts Right To Know List
[present]

Pennsylvania Right to Know List
[present]

N,N'-ETHYLENEBIS(12-HYDROXYOCTADE-CANAMIDE)   123-26-2
HEALTH AND SAFETY LISTS
NIOSH - Selected LD50s and LC50s
Oral, rat: LD50 = 11600 mg/kg

ETHYLENE BIS(OXYETHYLENE) DIACETATE   111-21-7
ENVIRONMENTAL LISTS
List of Pesticide Product Inert Ingredients
[present]

ETHYLENE BIS(PENTABROMOPHENOXIDE)   61262-53-1
ENVIRONMENTAL LISTS
CERCLA/SARA - Section 313 - Emission Reporting
form R reporting required

N,N'-ETHYLENEBIS(STEARAMIDE)   110-30-5
HEALTH AND SAFETY LISTS
ACGIH 1995 - Time Weighted Averages
simple asphyxiant

AIHA - Odor Threshold Values
geometric mean air odor threshold = 270 ppm (detectable); 420 ppm (recognizable)

U.S. DOT - Substances From 49 CFR 172.101
regulated by DOT (UN1962, UN1038)

U.S. DOT - Hazard Classes
DOT hazard class = 2.1

IARC - Group 3 (not classifiable)
[present]

NFPA - Flash Points
gas (no flash point given)

NFPA - Hazard Identification Ratings
health-1; flammability-4; reactivity-2

NIOSH - Selected LD50s and LC50s
Inhalation, mouse: LC50 = 95 pph 8 hr

ENVIRONMENTAL LISTS
CERCLA/SARA - Section 313 - Emission Reporting
form R reporting required for 1.0% de minimus concentration

CAA - Flammable Substances for Accidental Release Prevention
threshold quantity = 10,000 lbs

EPA - Master Testing List
[present]

INTERNATIONAL LISTS
Australian Exposure Standards - Time Weighted Averages
Asphyxiant at < 18% oxygen by volume; explosion hazard

Canada - NPRI (National Pollutant Release Inventory)
[present]

Canada - British Columbia - 8 Hour Exposure Limits
asphyxiant substance

Canada - Ontario - OHSA - TWAEVs
simple asphyxiant

Canada - Quebec - Time-Weighted Average Exposure Values
simple asphyxiant

German (DFG) - Carcinogens
suspected carcinogen

Israel - Time Weighted Averages
Asphyxiant

Mexico - Instruction No. 10 - TWAs
simple asphyxiant

STATE LISTS

California - Air Bill 2588 Appendix A-I
6/91

California - Exposure Limits - PELs
asphyxiant (limit depends on level of oxygen)

Florida Hazardous Substance List
[present]

Massachusetts Right To Know List
[present]

Minnesota Hazardous Substance List
[present]

NJ Right to Know List (Total)
sn 0873

NJ Special Hazardous Substances
(flammable - fourth degree; reactive - second degree)

Pennsylvania Right to Know List
environmental hazard

ETHYLENE BIS(TETRABROMOPHTHALIMIDE)   32588-76-4

ENVIRONMENTAL LISTS

CAA - HON Rule - SOCMI Chemicals
compliance by Oct. 23, 1995

ETHYLENE CARBONATE   96-49-1

ENVIRONMENTAL LISTS

TSCA - PAIR - Reporting List
Effective Date: January 26, 1994 Reporting Date: March 28, 1994

TSCA - Health and Safety Reporting List
Effective Date: January 26, 1994; Sunset Date: January 26, 2004

ETHYLENE CHLOROHYDRIN   107-07-3

HEALTH AND SAFETY LISTS

U.S. DOT - Appendix B - Marine Pollutants
DOT regulated marine pollutant

NIOSH - Selected LD50s and LC50s
Oral, rat: LD50 = 395 mg/kg

ENVIRONMENTAL LISTS

CERCLA/SARA - Section 313 - Emission Reporting
form R reporting required

STATE LISTS

Florida Hazardous Substance List
[present]

Massachusetts Right To Know List
carcinogen; extraordinarily hazardous

ETHYLENE CYANOHYDRIN   109-78-4

ENVIRONMENTAL LISTS

List of Pesticide Product Inert Ingredients
[present]

ETHYLENEDIAMINE   107-15-3

HEALTH AND SAFETY LISTS

NFPA - Flash Points
flash point = 345 degrees F (174 degrees C)

NFPA - Hazard Identification Ratings
health-0; flammability-1; reactivity-0

NIOSH - Selected LD50s and LC50s
Oral, rat: LD50 = 22600 mg/kg Skin, rabbit: LD50 = 8 gm/kg

ENVIRONMENTAL LISTS

TSCA - Code of Federal Regulations Citations
40 CFR 712.30(j); 40 CFR 716.120(a)

TSCA - PAIR - Reporting List
Reporting Date: March 13, 1984

TSCA - Health and Safety Reporting List
Effective Date: January 13, 1984

ETHYLENE DIAMINE DIPERCHLORATE   15718-71-5

ENVIRONMENTAL LISTS

TSCA - Code of Federal Regulations Citations
40 CFR 712.30(w); 40 CFR 716.120(a)

TSCA - PAIR - Reporting List
Reporting Date: March 12, 1990

TSCA - Health and Safety Reporting List
Effective Date: January 11, 1990

ETHYLENEDIAMINE TETRAACETIC ACID (EDTA)   60-00-4

ENVIRONMENTAL LISTS

EPA - Master Testing List
[present]

List of Pesticide Product Inert Ingredients
[present]

ETHYLENEDIAMINETETRAACETIC ACID, POTAS-SIUM SALT   7379-27-3

ENVIRONMENTAL LISTS

TSCA - Code of Federal Regulations Citations
40 CFR 712.30(w); 40 CFR 716.120(a)

TSCA - PAIR - Reporting List
Reporting Date: March 12, 1990

TSCA - Health and Safety Reporting List
Effective Date: January 11, 1990

ETHYLENEDIAMINETETRAACETIC ACID, TRI-ETHYLAMINE SALT   60816-63-9

HEALTH AND SAFETY LISTS

NFPA - Flash Points
flash point = 290 degrees F (143 degrees C)

NFPA - Hazard Identification Ratings
health-2; flammability-1; reactivity-1

NIOSH - Selected LD50s and LC50s
Oral, rat: LD50 = 10 gm/kg

ENVIRONMENTAL LISTS

CAA - HON Rule - SOCMI Chemicals
compliance by Oct. 24, 1994

STATE LISTS

Florida Hazardous Substance List
[present]

Massachusetts Right To Know List
[present]

Pennsylvania Right to Know List
[present]

ETHYLENEDIAMINETETRAACETIC ACID, TRIPOTASSIUM SALT   17572-97-3

HEALTH AND SAFETY LISTS

ACGIH 1995 - Ceiling Limits
C 1 ppm; C 3.3 mg/m3

ACGIH 1995 - Skin Designations
skin - potential for cutaneous absorption

U.S. DOT - Substances From 49 CFR 172.101
regulated by DOT (UN1135)

U.S. DOT - Hazard Classes
DOT hazard class = 6.1

U.S. DOT - Substances Which Are Poisonous by Inhalation
liquid hazardous material poisonous by inhalation (UN1135)

NFPA - Flash Points
flash point = 140 degrees F (60 degrees C)

NFPA - Hazard Identification Ratings
health-4; flammability-2; reactivity-0

NIOSH - Selected LD50s and LC50s
Inhalation, rat: LC50 = 290 mg/m3 8 hr Oral, rat: LD50 = 71 mg/kg Skin, rat: LD50 = 84 mg/kg

NIOSH 1990 - Pocket Guide - RELs
C 1 ppm; C 3 mg/m3

NIOSH 1990 - Pocket Guide - IDLHs
10 ppm IDLH

NIOSH 1990 - Pocket Guide - Target organs
respiratory system, liver, kidneys, skin, CNS, CVS

NIOSH 1990 - Pocket Guide - Skin list
Potential for dermal absorption

NTP Chemical Status Reports - Testing Status and NTIS Number
Technical reports printed (PB86145513/AS)

NTP Chemical Status Reports - Evidence of Carcinogenicity
male rat-no evidence; female rat-no evidence; male mice-no evidence; female mice-no evidence

OSHA - Vacated PELs - Ceiling Limits
C 1 ppm; C 3 mg/m3

OSHA - Vacated PELs - Skin Designation
Prevent or reduce skin absorption

OSHA - Final PELs - Time Weighted Averages
5 ppm TWA; 16 mg/m3 TWA

OSHA - Final PELs - Skin Notations
prevent or reduce skin absorption

ENVIRONMENTAL LISTS

CERCLA/SARA - Section 302 Extremely Hazardous Substances and TPQs
TPQ = 500 pounds

INTERNATIONAL LISTS

Australian Exposure Standards - Time Weighted Averages
Peak Limitation: 1 ppm; 3.3 mg/m3

Australian Exposure Standards - Skin Effects
skin absorption

Canada - WHMIS: Ingredient Disclosure
1% item 712 (402)

Canada - Alberta - Ceiling Occupational Exposure Limit
C 1 ppm; C 3.3 mg/m3

Canada - Alberta - Skin Designation
can be absorbed through the intact skin

Canada - British Columbia - 8 Hour Exposure Limits
1 ppm TWA; 3 mg/m3 TWA

Canada - British Columbia - Ceiling Exposure Limits
C 1 ppm; C 3 mg/m3

Canada - British Columbia - Skin Notations
skin - potential for skin absorption

Canada - Ontario - OHSA - CEVs
1 ppm CEV; 3.3 mg/m3 CEV

Canada - Ontario - OHSA - Skin Notations
absorption through skin, eyes, or mucous membranes

Canada - Quebec - Ceiling Limits
P 1 ppm; P 3.3 mg/m3

Canada - Quebec - Skin Designations
absorbed through the skin

German (DFG) - MAK Values
1 ppm MAK; 3 mg/m3 MAK

German (DFG) - Peak Limitations
5 x normal MAK (30 min. average value); don't exceed 2 times during shift

German (DFG) - Skin/Sensitizers
danger of cutaneous absorption

German (DFG) - Pregnancy
no risk to embryo/fetus if exposure limits adhered to

Israel - Ceiling Exposure Limits
C 1 ppm; C 3.3 mg/m3

Mexico - Instruction No. 10 - TWAs
1 ppm TWA; 3 mg/m3 TWA

Mexico - Instruction No. 10 - Skin designation
skin - potential for cutaneous absorption

STATE LISTS

California - Exposure Limits - Ceilings
C 1 ppm; C 3 mg/m3

California - Exposure Limits - Skin Notation
material may be absorbed through the skin, eyes or mucous membranes

California - Directors List of Hazardous Substances (8 CCR 339)
[present]

Florida Hazardous Substance List
[present]

Massachusetts Right To Know List
extraordinarily hazardous

Minnesota Hazardous Substance List
skin

NJ Right to Know List (Total)
sn 0874

NJ Special Hazardous Substances
(mutagen)

Pennsylvania Right to Know List
environmental hazard

PROPOSED REGULATIONS

CERCLA/SARA - Proposed Hazardous Substance Additions
proposed RQ = 1 pound (.454 kg)

CERCLA/SARA - 1989 Proposed RQ Adjustments
proposed RQ = 100 pounds (45.4 kg)

ETHYLENEDIAMINETETRA(METHYLENEPHOS-PHONIC) ACID   1429-50-1

HEALTH AND SAFETY LISTS

NFPA - Flash Points
flash point = 265 degrees F (129 degrees C)

NFPA - Hazard Identification Ratings
health-1; flammability-1; reactivity-2

NIOSH - Selected LD50s and LC50s
Oral, rat: LD50 = 10 gm/kg Skin, rabbit: LD50 = 5000 mg/kg

STATE LISTS

Florida Hazardous Substance List
[present]

Massachusetts Right To Know List
[present]

Pennsylvania Right to Know List
[present]

ETHYLENE DIBROMIDE (1,2-DIBROMOETHANE)   106-93-4

HEALTH AND SAFETY LISTS

ACGIH 1995 - Time Weighted Averages
10 ppm TWA; 25 mg/m3 TWA

ACGIH 1995 - Skin Designations
skin - potential for cutaneous absorption

U.S. DOT - Substances From 49 CFR 172.101
regulated by DOT (UN1604)

U.S. DOT - Hazard Classes
DOT hazard class = 8

U.S. DOT - Appendix A Table 1 - Hazardous Substances
final RQ = 5000 pounds (2270 kg)

NFPA - Flash Points
flash point = 104 degrees F (40 degrees C); anhydrous 76%: flash point = 150 degrees F (66 degrees C)

NFPA - Hazard Identification Ratings
health-3; flammability-2; reactivity-0

NIOSH - Selected LD50s and LC50s
Inhalation, mouse: LC50 = 300 mg/m3 (8 hr) Oral, rat: LD50 = 500 mg/kg Skin, rabbit: LD50 = 730 mg/kg

NIOSH 1990 - Pocket Guide - RELs
10 ppm TWA; 25 mg/m3 TWA

NIOSH 1990 - Pocket Guide - IDLHs
2000 ppm IDLH

NIOSH 1990 - Pocket Guide - Target organs
respiratory system, liver, kidneys, skin

OSHA - Vacated PELs - Time Weighted Averages
10 ppm TWA; 25 mg/m3 TWA

OSHA - Final PELs - Time Weighted Averages
10 ppm TWA; 25 mg/m3 TWA

ENVIRONMENTAL LISTS

CERCLA/SARA - Section 302 Extremely Hazardous Substances and TPQs
TPQ = 10,000 pounds

CERCLA/SARA - Hazardous Substances and their Reportable Quantities
final RQ = 5000 pounds (2270 kg)

CAA - Toxic Substances for Accidental Release Prevention
threshold quantity = 20,000 lbs

CAA - HON Rule - SOCMI Chemicals
compliance by Jan. 23, 1995

Clean Water Act - Hazardous Substances
[present]

List of Pesticide Product Inert Ingredients
[present]

INTERNATIONAL LISTS

Australian Exposure Standards - Time Weighted Averages
10 ppm TWA; 25 mg/m3 TWA

Australian Exposure Standards - Skin Effects
sensitiser

Canada - WHMIS: Ingredient Disclosure
0.1% item 713 (859)

Canada - Alberta - 8 Hour Occupational Exposure Limit
10 ppm TWA; 26 mg/m3 TWA

Canada - Alberta - 15 Minute Occupational Exposure Limit
20 ppm STEL; 51 mg/m3 STEL

Canada - British Columbia - 8 Hour Exposure Limits
10 ppm TWA; 25 mg/m3 TWA

Canada - Ontario - OHSA - TWAEVs
10 ppm TWAEV; 25 mg/m3 TWAEV

Canada - Quebec - Time-Weighted Average Exposure Values
10 ppm TWAEV; 25 mg/m3 TWAEV

United Kingdom - Occupational Exposure Standards - TWAs
10 ppm TWA; 25 mg/m3 TWA

German (DFG) - MAK Values
10 ppm MAK; 25 mg/m3 MAK

German (DFG) - Peak Limitations
2 x normal MAK (30 min. average value); don't exceed 4 times during shift

German (DFG) - Pregnancy
classification not yet possible

Israel - Time Weighted Averages
10 ppm TWA; 25 mg/m3 TWA

Israel - Action Levels
5 ppm AL; 12.5 mg/m3 AL

Mexico - Instruction No. 10 - TWAs
10 ppm TWA; 25 mg/m3 TWA

Mexico - Instruction No. 10 - STELs
3 mg/m3 STEL

STATE LISTS

California - Exposure Limits - PELs
10 ppm PEL; 25 mg/m3 PEL

California - Directors List of Hazardous Substances (8 CCR 339)
[present]

Florida Hazardous Substance List
[present]

Massachusetts Right To Know List
extraordinarily hazardous

Minnesota Hazardous Substance List
[present]

NJ Right to Know List (Total)
sn 0875

NJ Special Hazardous Substances
(corrosive)

Pennsylvania Right to Know List
environmental hazard

ETHYLENE DIBROMIDE AND METHYL BROMIDE MIXTURES, LIQUID   RR-01288-4

HEALTH AND SAFETY LISTS

U.S. DOT - Hazard Classes
Forbidden from transport by the DOT

ETHYLENE(5,6-DIBROMONORBORNANE-2,3-DICARBOXIMIDE)   41291-34-3

HEALTH AND SAFETY LISTS

U.S. DOT - Appendix A Table 1 - Hazardous Substances
final RQ = 5000 pounds (2270 kg)

ENVIRONMENTAL LISTS

CERCLA/SARA - Hazardous Substances and their Reportable Quantities
final RQ = 5000 pounds (2270 kg)

CAA - HON Rule - SOCMI Chemicals
compliance by Oct. 23, 1995

Clean Water Act - Hazardous Substances
[present]

List of Pesticide Product Inert Ingredients
[present]

STATE LISTS

California - Directors List of Hazardous Substances (8 CCR 339)
[present]

Massachusetts Right To Know List
[present]

NJ Right to Know List (Total)
sn 0876

Pennsylvania Right to Know List
environmental hazard

ETHYLENE DICHLORIDE   107-06-2

ENVIRONMENTAL LISTS

List of Pesticide Product Inert Ingredients
[present]

ETHYLENE FLUOROHYDRIN   371-62-0

ENVIRONMENTAL LISTS

List of Pesticide Product Inert Ingredients
[present]

ETHYLENE GLYCOL   107-21-1

ENVIRONMENTAL LISTS

List of Pesticide Product Inert Ingredients
[present]

ETHYLENE GLYCOL ACETATE   542-59-6

ENVIRONMENTAL LISTS

TSCA - Code of Federal Regulations Citations
40 CFR 704.95

ETHYLENE GLYCOL BIS(SEMIFORMAL)   3586-55-8
SEE ALSO:
K136-HAZARDOUS WASTES
K118-HAZARDOUS WASTES
F039-HAZARDOUS WASTES
K117-HAZARDOUS WASTES

HEALTH AND SAFETY LISTS

ACGIH 1995 - Skin Designations
skin - potential for cutaneous absorption

ACGIH 1995 - Carcinogens
A2-suspected human carcinogen

U.S. DOT - Substances From 49 CFR 172.101
regulated by DOT (UN1605)

U.S. DOT - Hazard Classes
DOT hazard class = 6.1

U.S. DOT - Substances Which Are Poisonous by Inhalation
liquid hazardous material poisonous by inhalation (UN1605)

U.S. DOT - Appendix B - Marine Pollutants
DOT regulated marine pollutant

U.S. DOT - Appendix A Table 1 - Hazardous Substances
final RQ = 1 pound (0.454 kg)

IARC - Group 2A (limited human data)
[present] (Other relevant data, as given in Supplement 7, influenced the making of the overall evaluation)

NIOSH - Selected LD50s and LC50s
Oral, rat: LD50 = 108 mg/kg Skin, rat: LD50 = 300 mg/kg

NIOSH 1990 - Pocket Guide - RELs
0.045 ppm TWA; C 0.13 ppm (15 min)

NIOSH 1990 - Pocket Guide - IDLHs
400 ppm IDLH (not considering carcinogenic effects)

NIOSH 1990 - Pocket Guide - Carcinogens
occupational carcinogen

NIOSH 1990 - Pocket Guide - Target organs
respiratory system, liver, kidneys, skin, eyes

NIOSH - Health Standards - Exposure Limits
0.045 ppm TWA (8 hr); 0.38 mg/m3 TWA (8 hr); C (15 min) 0.13 ppm; C (15 min) 1 mg/m3

NIOSH - Health Standards - Health Effects and Precautions
mutagenesis; damage to skin, eyes, heart, liver, spleen, and reproductive, respiratory, and central nervous system (Warn workers of hazards; prevent skin contact)

NIOSH - Health Standards - Carcinogenic Chemicals
potential human carcinogen

NTP Chemical Status Reports - Testing Status and NTIS Number
Technical reports printed (PB288428/AS) (PB82181710)

NTP Chemical Status Reports - Evidence of Carcinogenicity
PB288428/AS: male rat-positive; female rat-positive; male mice-positive; female mice-positive; PB82181710: male rat-positive; female rat-positive; male mice-positive; female mice-positive

NTP Seventh Report - Suspect Carcinogens
suspect carcinogen

OSHA - Vacated PELs - Time Weighted Averages
20 ppm TWA

OSHA - Vacated PELs - Short Term Exposure Limits
50 ppm STEL (5 minutes)

OSHA - Vacated PELs - Ceiling Limits
C 100 ppm

OSHA - Final PELs - Time Weighted Averages
20 ppm TWA; C 30 ppm

OSHA - Final PELs - Ceiling Limits
C 30 ppm

OSHA - Possible Select Carcinogens
[present]

ENVIRONMENTAL LISTS

ATSDR Priority List
Rank (of 275): 046

CERCLA/SARA - Section 313 - Emission Reporting
form R reporting required for 0.1% de minimus concentration

CERCLA/SARA - Hazardous Substances and their Reportable Quantities
final RQ = 1 pound (0.454 kg)

Clean Air Act (1990) - List of Hazardous Air Contaminants
[present]

CAA - HON Rule - SOCMI Chemicals
compliance by Oct. 24, 1994

CAA - HON Rule - Organic HAPs
[present]

Clean Water Act - Hazardous Substances
[present]

Safe Drinking Water Act - MCLs
MCL = 0.00005 mg/L

Safe Drinking Water Act - MCLGs
MCLG = Zero

Safe Drinking Water Act - Monitoring
monitoring required

EPA - Carcinogen Hazard Ranking for RQ Adjustment
Hazard ranking = High

RCRA - U Series Wastes
waste number U067

RCRA - Hazardous Constituents-Appendix VIII
waste number U067

RCRA - Basis for Listing - Appendix VII
Included in waste streams: F039, K117, K118, K136

RCRA - Substances Banned From Land Disposal
[present]

RCRA - TSD Facilities Ground Water Monitoring
TM 8010 = 10 ug/L PQL; TM 8240 = 5 ug/L PQL

RCRA - Universal Treatment Standards (LDR)
WW: 0.028 mg/l; NWW: 15 mg/kg

INTERNATIONAL LISTS

Australian Exposure Standards - Time Weighted Averages
control to the lowest practical level

Australian Exposure Standards - Skin Effects
skin absorption

Australian Exposure Standards - Carcinogens
probable carcinogen

Canada - WHMIS: Ingredient Disclosure
0.1% item 714 (337)

Canada - CEPA Schedule II Part II - Toxic Substances (Export)
[present]

Canada - Alberta - 8 Hour Occupational Exposure Limit
5 ppm TWA; 38 mg/m3 TWA

Canada - Alberta - 15 Minute Occupational Exposure Limit
10 ppm STEL; 77 mg/m3 STEL

Canada - Alberta - Skin Designation
can be absorbed through the intact skin

Canada - Alberta - Designated Substances
designated substance - requires code of practice

Canada - British Columbia - 8 Hour Exposure Limits
20 ppm TWA; 155 mg/m3 TWA

Canada - British Columbia - 15 Minute Exposure Limits
30 ppm STEL; 230 mg/m3 STEL

Canada - British Columbia - Skin Notations
skin - potential for skin absorption

Canada - British Columbia - Carcinogens
carcinogen - 20 ppm TWA; 155 mg/m3 TWA

Canada - Quebec - Time-Weighted Average Exposure Values
20 ppm TWAEV; 155 mg/m3 TWAEV (substance of which the recirculation is prohibited)

Canada - Quebec - Skin Designations
absorbed through the skin

Canada - Quebec - Carcinogens
C2 carcinogen: effect suspected in humans

United Kingdom - Maximum Exposure Limits - TWAs
0.5 ppm TWA; 4 mg/m3 TWA

United Kingdom - Maximum Exposure Limits - Notes
can be absorbed through skin

German (DFG) - Skin/Sensitizers
danger of cutaneous absorption

German (DFG) - Carcinogens
animal evidence of carcinogenicity

Israel - Time Weighted Averages
1 ppm TWA

Mexico - Instruction No. 10 - Skin designation
skin - potential for cutaneous absorption

Mexico - Instruction No. 10 - Carcinogens
potential carcinogen in humans - limited epidemiological evidence

STATE LISTS

California - Air Bill 2588 Appendix A-I
known or potential carcinogen

California - Prop. 65 - Cancer list
carcinogen - initial date 7/1/87

California - Prop. 65 - No Significant Risk Levels
ingestion: no significant risk level = 0.2 ug/day; inhalation: no significant risk level = 3 ug/day

California - Exposure Limits - Ceilings
C 0.13 ppm; C 1 mg/m3

California - Exposure Limits - Skin Notation
material may be absorbed through the skin, eyes or mucous membrane

California - Directors List of Hazardous Substances (8 CCR 339)
[present] (refers to any mixture containing greater than 0.1% EDB)

Florida Hazardous Substance List
[present]

Massachusetts Right To Know List
carcinogen; extraordinarily hazardous

Minnesota Hazardous Substance List
carcinogen; skin

NJ Special Hazardous Substances
(carcinogen; mutagen)

Pennsylvania Right to Know List
environmental hazard; special hazardous substance

Pennsylvania RTK - Special Hazardous Substances
[present]

ETHYLENE GLYCOL DIBUTYL ETHER   112-48-1

HEALTH AND SAFETY LISTS

U.S. DOT - Appendix B - Marine Pollutants
DOT regulated marine pollutant

ETHYLENE GLYCOL DIETHYL ETHER   629-14-1

ENVIRONMENTAL LISTS

TSCA - Code of Federal Regulations Citations
40 CFR 712.30(w); 40 CFR 716.120(a)

TSCA - PAIR - Reporting List
Reporting Date: March 12, 1990

TSCA - Health and Safety Reporting List
Effective Date: January 11, 1990

ETHYLENE GLYCOL DIFORMATE   629-15-2

SEE ALSO:
K096-HAZARDOUS WASTES
F039-HAZARDOUS WASTES
F024-HAZARDOUS WASTES
BIS(2,4-DIMETHYLBUTYL) MALEATE
K018-HAZARDOUS WASTES
F025-HAZARDOUS WASTES
O-TOLYL P-TOLUENE SULFONATE
K029-HAZARDOUS WASTES
CHLORINATED ETHANES

HEALTH AND SAFETY LISTS

ACGIH 1995 - Time Weighted Averages
10 ppm TWA; 40 mg/m3 TWA

AIHA - Odor Threshold Values
geometric mean air odor threshold = 26 ppm (detectable); 87 ppm (recognizable)

U.S. DOT - Substances From 49 CFR 172.101
regulated by DOT (UN1184)

U.S. DOT - Hazard Classes
DOT hazard class = 3

U.S. DOT - Appendix B - Marine Pollutants
DOT regulated marine pollutant

U.S. DOT - Appendix A Table 1 - Hazardous Substances
final RQ = 100 pounds (45.4 kg)

IARC - Group 2B (sufficient animal data)
[present] (Overall evaluation based only on evidence of carcinogenicity in monograph (20, 1979) or in Supplement 4)

NFPA - Flash Points
flash point = 56 degrees F (13 degrees C)

NFPA - Hazard Identification Ratings
health-2; flammability-3; reactivity-0

NIOSH - Selected LD50s and LC50s
Inhalation, rat: LC50 = 1000 ppm 7 hr Oral, rat: LD50 = 670 mg/kg Skin, rabbit: LD50 = 3890 mg/kg

NIOSH 1990 - Pocket Guide - RELs
1 ppm TWA; 4 mg/m3 TWA; 2 ppm STEL; 8 mg/m3 STEL

NIOSH 1990 - Pocket Guide - IDLHs
1000 ppm IDLH (not considering carcinogenic effects)

NIOSH 1990 - Pocket Guide - Carcinogens
occupational carcinogen

NIOSH 1990 - Pocket Guide - Target organs
kidneys, liver, eyes, skin, CNS

NIOSH - Health Standards - Exposure Limits
1 ppm TWA; 4 mg/m3 TWA; C (15 min) 2 ppm; C (15 min) 8 mg/m3

NIOSH - Health Standards - Health Effects and Precautions
nervous system, respiratory, cardiovascular, and liver effects (Nursing infants of exposed mothers are at risk)

NIOSH - Health Standards - Carcinogenic Chemicals
potential human carcinogen

NTP Chemical Status Reports - Testing Status and NTIS Number
Technical reports printed (PB91185363) (PB285968/AS)

NTP Chemical Status Reports - Evidence of Carcinogenicity
PB285968/AS: male rat-positive; female rat-positive; male mice-positive; female mice-positive

NTP Seventh Report - Suspect Carcinogens
suspect carcinogen

OSHA - Vacated PELs - Time Weighted Averages
1 ppm TWA; 4 mg/m3 TWA

OSHA - Vacated PELs - Short Term Exposure Limits
2 ppm STEL; 8 mg/m3 STEL

OSHA - Final PELs - Time Weighted Averages
50 ppm TWA; C 100 ppm

OSHA - Final PELs - Ceiling Limits
C 100 ppm

OSHA - Possible Select Carcinogens
[present]

ENVIRONMENTAL LISTS

ATSDR Priority List
Rank (of 275): 071

CERCLA/SARA - Section 313 - Emission Reporting
form R reporting required for 0.1% de minimus concentration

CERCLA/SARA - Hazardous Substances and their Reportable Quantities
final RQ = 100 pounds (45.4 kg)

Clean Air Act (1990) - List of Hazardous Air Contaminants
[present]

CAA - HON Rule - SOCMI Chemicals
compliance by Oct. 24, 1994

CAA - HON Rule - Organic HAPs
[present]

Clean Water Act - Hazardous Substances
[present]

Clean Water Act - Priority Pollutants
[present]

Safe Drinking Water Act - MCLs
MCL = 0.005 mg/L

Safe Drinking Water Act - MCLGs
MCLG = Zero

EPA - Carcinogen Hazard Ranking for RQ Adjustment
Hazard ranking = Low

RCRA - D Series - Maximum Concentration of Contaminants
waste number D028; regulatory level = 0.5 mg/L

RCRA - D Series - Chronic Toxicity Reference Levels
chronic toxicity reference level = 0.005 mg/L

RCRA - U Series Wastes
waste number U077

RCRA - Hazardous Constituents-Appendix VIII
waste number U077

RCRA - Basis for Listing - Appendix VII
Included in waste streams: F024, F025, F039, K018, K019, K020, K029, K030, K096

RCRA - Substances Banned From Land Disposal
[present]

RCRA - TSD Facilities Ground Water Monitoring
TM 8010 = 0.5 ug/L PQL; TM 8240 = 5 ug/L PQL

RCRA - Universal Treatment Standards (LDR)
WW: 0.21 mg/l; NWW: 6.0 mg/kg

TSCA - Code of Federal Regulations Citations
40 CFR 716.120(a)

TSCA - Health and Safety Reporting List
Effective Date: June 1, 1987

INTERNATIONAL LISTS

Australian Exposure Standards - Time Weighted Averages
10 ppm TWA; 40 mg/m3 TWA

Australian Exposure Standards - Under Review
exposure limits under review

Canada - WHMIS: Ingredient Disclosure
1% item 715 (498)

Canada - NPRI (National Pollutant Release Inventory)
[present]

Canada - CEPA Schedule II Part II - Toxic Substances (Export)
[present]

Canada - CEPA - Priority Substances List
estimated time for completion of assessment reports: 4 years

Canada - Drinking Water Quality - IMACs
0.005 mg/L IMAC

Canada - Alberta - 8 Hour Occupational Exposure Limit
10 ppm TWA; 40 mg/m3 TWA

Canada - Alberta - 15 Minute Occupational Exposure Limit
15 ppm STEL; 60 mg/m3 STEL

Canada - British Columbia - 8 Hour Exposure Limits
50 ppm TWA; 200 mg/m3 TWA

Canada - British Columbia - 15 Minute Exposure Limits
75 ppm STEL; 300 mg/m3 STEL

Canada - Ontario - OHSA - TWAEVs
10 ppm TWAEV; 40 mg/m3 TWAEV

Canada - Quebec - Time-Weighted Average Exposure Values
1 ppm TWAEV; 4 mg/m3 TWAEV

Canada - Quebec - Short-term Exposure Values
2 ppm STEV; 8 mg/m3 STEV

Canada - Quebec - Carcinogens
C2 carcinogen: effect suspected in humans

German (DFG) - Carcinogens
animal evidence of carcinogenicity

Israel - Time Weighted Averages
10 ppm TWA; 40 mg/m3 TWA

Israel - Action Levels
5 ppm AL; 20 mg/m3 AL

Mexico - Wastewater - Organic Toxic Pollutants and Heavy Metals
Listed under [Chlorinated Ethanes]

Mexico - Drinking Water - Ecological Criteria
0.005 mg/l Substance presents persistence, bioaccumulations or risk of cancer, reduce human exposure to a minimum

STATE LISTS

California - Air Bill 2588 Appendix A-I
known or potential carcinogen

California - Prop. 65 - Cancer list
carcinogen - initial date 10/1/87

California - Prop. 65 - No Significant Risk Levels
no significant risk level = 10 ug/day

California - Exposure Limits - PELs
1 ppm PEL; 4 mg/m3 PEL

California - Exposure Limits - STELs
2 ppm STEL; 8 mg/m3 STEL

California - Exposure Limits - Ceilings
C 200 ppm

California - Directors List of Hazardous Substances (8 CCR 339)
[present]

Florida Hazardous Substance List
[present]

Massachusetts Right To Know List
carcinogen; extraordinarily hazardous

Minnesota Hazardous Substance List
carcinogen

NJ Right to Know List (Total)
sn 0652

NJ Special Hazardous Substances
(carcinogen; flammable - third degree; mutagen)

Pennsylvania Right to Know List
environmental hazard; special hazardous substance

Pennsylvania RTK - Special Hazardous Substances
[present]

PROPOSED REGULATIONS

Canada - Ontario - Proposed Occupational TWAEVs
1 ppm TWAEV; 4 mg/m3 TWAEV

Canada - Ontario - Proposed Occupational STEVs
5 ppm STEV; 20 mg/m3 STEV

ETHYLENE GLYCOL DIMETHYL ETHER   110-71-4

HEALTH AND SAFETY LISTS

NIOSH - Selected LD50s and LC50s
Inhalation, rat: LC50 = 200 mg/m3 10 mn

OSHA - List of Highly Hazardous Chemicals
threshhold quantity = 100 pounds

ENVIRONMENTAL LISTS

CERCLA/SARA - Section 302 Extremely Hazardous Substances and TPQs
TPQ = 10 pounds

STATE LISTS

Florida Hazardous Substance List
effective March 13, 1992

Massachusetts Right To Know List
extraordinarily hazardous

NJ Right to Know List (Total)
sn 2400

Pennsylvania Right to Know List
environmental hazard

PROPOSED REGULATIONS

CERCLA/SARA - Proposed Hazardous Substance Additions
proposed RQ = 1 pound (.454 kg)

CERCLA/SARA - 1989 Proposed RQ Adjustments
proposed RQ = 10 pounds (4.54 kg)

ETHYLENE GLYCOL DINITRATE   628-96-6

HEALTH AND SAFETY LISTS

ACGIH 1995 - Ceiling Limits
vapor and mist: (C 50) ppm; (C 127) mg/m3

NFPA - Flash Points
flash point = 232 degrees F (111 degrees C)

NFPA - Hazard Identification Ratings
health-1; flammability-1; reactivity-0

NIOSH - Selected LD50s and LC50s
Oral, rat: LD50 = 4700 mg/kg Skin, rabbit: LD50 = 9530 mg/kg

NTP Chemical Status Reports - Testing Status and NTIS Number
Technical reports printed (no NTIS number given)

NTP Chemical Status Reports - Evidence of Carcinogenicity
male mice: no evidence; female mice: no evidence

OSHA - Vacated PELs - Ceiling Limits
C 50 ppm; C 125 mg/m3

ENVIRONMENTAL LISTS

CERCLA/SARA - Section 313 - Emission Reporting
form R reporting required for 1.0% de minimus concentration

CERCLA/SARA - Hazardous Substances and their Reportable Quantities
final RQ = 1 pound (.454 kg)

Clean Air Act (1990) - List of Hazardous Air Contaminants
[present]

CAA - HON Rule - SOCMI Chemicals
compliance by Oct. 24, 1995

CAA - HON Rule - Organic HAPs
[present]

EPA - Master Testing List
[present]

List of Pesticide Product Inert Ingredients
[present]

INTERNATIONAL LISTS

Australian Exposure Standards - Time Weighted Averages
vapour: 60 mg/m3 TWA

Australian Exposure Standards - Short Term Exposure Limits
vapour: 120 mg/m3 STEL

Canada - WHMIS: Ingredient Disclosure
1% item 716 (860)

Canada - NPRI (National Pollutant Release Inventory)
[present]

Canada - Alberta - Ceiling Occupational Exposure Limit
C 50 ppm; C 127 mg/m3

Canada - British Columbia - 8 Hour Exposure Limits
particulate: 10 mg/m3 TWA; vapour: 100 ppm TWA; 250 mg/m3 TWA

Canada - British Columbia - 15 Minute Exposure Limits
particulate: 20 mg/m3 STEL; vapour: 125 ppm STEL; 325 mg/m3 STEL

Canada - Ontario - OHSA - CEVs
50 ppm CEV; 127 mg/m3 CEV

Canada - Quebec - Ceiling Limits
P 50 ppm; P 127 mg/m3

United Kingdom - Occupational Exposure Standards - TWAs
particulate: 10 mg/m3 TWA; vapour: 60 mg/m3 TWA

United Kingdom - Occupational Exposure Standards - STELs
vapour: 125 mg/m3 STEL

German (DFG) - Peak Limitations
2 x normal MAK (5 min momentary value); don't exceed 8 times during shift

German (DFG) - Skin/Sensitizers
danger of cutaneous absorption

German (DFG) - Pregnancy
no risk to embryo/fetus is exposure limits are adhered to

Israel - Ceiling Exposure Limits
C 50 ppm; C 127 mg/m3

Mexico - Instruction No. 10 - TWAs
particulate: 10 mg/m3 TWA vapor: 50 ppm TWA; 125 mg/m3 TWA

Mexico - Instruction No. 10 - STELs
as particulate: 20 mg/m3 STEL

STATE LISTS

California - Air Bill 2588 Appendix A-I
6/91

California - Exposure Limits - Ceilings
vapor: C 50 ppm; C 125 mg/m3

California - Directors List of Hazardous Substances (8 CCR 339)
[present] (exempt when vapors or particulates are formed due to work practices or procedures)

Florida Hazardous Substance List
[present]

Massachusetts Right To Know List
[present]

Minnesota Hazardous Substance List
[present] (particulate and vapor)

NJ Right to Know List (Total)
sn 0878

Pennsylvania Right to Know List
environmental hazard

PROPOSED REGULATIONS

ACGIH 1995 - Notice of Intended Changes
as an aerosol: C 39.4 ppm; C 100 mg/m3: A4- not classifiable as a human carcinogen

ETHYLENE GLYCOL DISTEARATE   627-83-8

HEALTH AND SAFETY LISTS

NFPA - Flash Points
flash point = 215 degrees F (102 degrees C)

NFPA - Hazard Identification Ratings
health-0; flammability-1; reactivity-0

NIOSH - Selected LD50s and LC50s
Oral, rat: LD50 = 8250 mg/kg

ENVIRONMENTAL LISTS

CAA - HON Rule - SOCMI Chemicals
compliance by Oct. 23, 1995

ETHYLENE GLYCOL ETHER OF PINENE   53404-49-2

ENVIRONMENTAL LISTS

List of Pesticide Product Inert Ingredients
[present]

ETHYLENE GLYCOL ETHYLBUTYL ETHER   RR-00856-0

HEALTH AND SAFETY LISTS

NFPA - Flash Points
flash point = 185 degrees F (85 degrees C)

NFPA - Hazard Identification Ratings
health-1; flammability-2; reactivity-0

ENVIRONMENTAL LISTS

CAA - HON Rule - SOCMI Chemicals
compliance by Oct. 23, 1995

ETHYLENE GLYCOL ETHYLHEXYL ETHER   RR-00857-1

HEALTH AND SAFETY LISTS

U.S. DOT - Substances From 49 CFR 172.101
regulated by DOT (UN1153)

U.S. DOT - Hazard Classes
DOT hazard class = 3

NFPA - Flash Points
flash point = 95 degrees F (35 degrees C)

NFPA - Hazard Identification Ratings
health-1; flammability-3; reactivity-0

NIOSH - Selected LD50s and LC50s
Oral, rat: LD50 = 4390 mg/kg

ENVIRONMENTAL LISTS

CAA - HON Rule - SOCMI Chemicals
compliance by Oct. 24, 1994

STATE LISTS

California - Air Bill 2588 Appendix A-I
9/90

Florida Hazardous Substance List
[present]

Massachusetts Right To Know List
[present]

NJ Right to Know List (Total)
sn 0879

NJ Special Hazardous Substances
(flammable - third degree)

Pennsylvania Right to Know List
[present]

ETHYLENE GLYCOL MONOACRYLATE   818-61-1

HEALTH AND SAFETY LISTS

NFPA - Flash Points
flash point = 200 degrees F (93 degrees C)

NFPA - Hazard Identification Ratings
health-1; flammability-2; reactivity-0 (decomposes in water)

INTERNATIONAL LISTS

Canada - WHMIS: Ingredient Disclosure
1% item 717 (711)

ETHYLENE GLYCOL MONOBENZYL ETHER   622-08-2

HEALTH AND SAFETY LISTS

U.S. DOT - Substances From 49 CFR 172.101
regulated by DOT (UN2252)

U.S. DOT - Hazard Classes
DOT hazard class = 3

NFPA - Flash Points
flash point = 29 degrees F (-2 degrees C)

NFPA - Hazard Identification Ratings
health-2; flammability-2; reactivity-0

ENVIRONMENTAL LISTS

CAA - HON Rule - SOCMI Chemicals
compliance by Oct. 24, 1994

STATE LISTS

California - Air Bill 2588 Appendix A-I
9/90

Florida Hazardous Substance List
[present]

Massachusetts Right To Know List
[present]

Pennsylvania Right to Know List
[present]

ETHYLENE GLYCOL, MONOBUTYL ETHER   112-07-2
ACETATE

HEALTH AND SAFETY LISTS

ACGIH 1995 - Time Weighted Averages
0.05 ppm TWA;   0.31 mg/m3 TWA

ACGIH 1995 - Skin Designations
skin - potential for cutaneous absorption

U.S. DOT - Hazard Classes
Forbidden from transport by the DOT

NIOSH - Selected LD50s and LC50s
Oral, rat: LD50 = 616 mg/kg

NIOSH 1990 - Pocket Guide - RELs
0.1 mg/m3 STEL

NIOSH 1990 - Pocket Guide - IDLHs
500 mg/m3 IDLH

NIOSH 1990 - Pocket Guide - Target organs
CVS, blood, skin

NIOSH 1990 - Pocket Guide - Skin list
Potential for dermal absorption

NIOSH - Health Standards - Exposure Limits
C (20 min) 0.1 mg/m3 (recommended for substance alone or with Nitroglycerin) (Listed under 'Nitroglycerin')

NIOSH - Health Standards - Health Effects and Precautions
Circulatory system effects (Prevent skin contact) (Listed under 'Nitro-glycerin')

OSHA - Vacated PELs - Short Term Exposure Limits
0.1 mg/m3 STEL (not in effect as a result of reconsideration for the industrial sector of civilian manufacture)

OSHA - Vacated PELs - Skin Designation
Prevent or reduce skin absorption

OSHA - Final PELs - Ceiling Limits
C 0.2 ppm;   C 1 mg/m3

OSHA - Final PELs - Skin Notations
prevent or reduce skin absorption

INTERNATIONAL LISTS

Australian Exposure Standards - Time Weighted Averages
0.05 ppm TWA;   0.31 mg/m3 TWA

Australian Exposure Standards - Skin Effects
skin absorption

Canada - WHMIS: Ingredient Disclosure
1% item 718 (749)

Canada - Alberta - 8 Hour Occupational Exposure Limit
0.02 ppm TWA; 0.12 mg/m3 TWA

Canada - Alberta - 15 Minute Occupational Exposure Limit
0.05 ppm STEL; 0.30 mg/m3 STEL

Canada - Alberta - Skin Designation
can be absorbed through the intact skin

Canada - British Columbia - Ceiling Exposure Limits
C 0.2 ppm; C 2 mg/m3

Canada - British Columbia - Skin Notations
skin - potential for skin absorption

Canada - Ontario - OHSA - TWAEVs
0.05 ppm TWAEV; 0.31 mg/m3 TWAEV

Canada - Ontario - OHSA - Skin Notations
absorption through skin, eyes, or mucous membranes

Canada - Quebec - Ceiling Limits
P 0.2 ppm; P 1.24 mg/m3

Canada - Quebec - Skin Designations
absorbed through the skin

United Kingdom - Occupational Exposure Standards - TWAs
0.2 ppm TWA; 1.2 mg/m3 TWA

United Kingdom - Occupational Exposure Standards - STELs
0.2 ppm STEL; 1.2 mg/m3 STEL

United Kingdom - Occupational Exposure Standards - Notes
can be absorbed through skin

German (DFG) - MAK Values
0.05 ppm MAK; 0.3 mg/m3 MAK (when skin contact does not occur)

German (DFG) - Peak Limitations
2 x normal MAK (30 min. average value); don't exceed 4 times during shift

German (DFG) - Skin/Sensitizers
danger of cutaneous absorption

Israel - Time Weighted Averages
0.05 ppm TWA; 0.31 mg/m3 TWA

Israel - Action Levels
0.025 ppm AL; 0.155 mg/m3 AL

Mexico - Instruction No. 10 - TWAs
0.05 ppm TWA; 0.3 mg/m3 TWA

Mexico - Instruction No. 10 - STELs
0.1 ppm STEL; 0.6 mg/m3 STEL

Mexico - Instruction No. 10 - Skin designation
skin - potential for cutaneous absorption

STATE LISTS

California - Exposure Limits - PELs
when mixed with nitroglycerin: 0.05 ppm PEL

California - Exposure Limits - STELs
0.1 mg/m3 STEL

California - Exposure Limits - Skin Notation
material may be absorbed through the skin, eyes or mucous membrane

California - Directors List of Hazardous Substances (8 CCR 339)
[present]

Florida Hazardous Substance List
[present]

Massachusetts Right To Know List
[present]

Minnesota Hazardous Substance List
skin

Pennsylvania Right to Know List
[present]

PROPOSED REGULATIONS

Canada - Ontario - Proposed Occupational TWAEVs
0.03 ppm TWAEV; 0.20 mg/m3 TWAEV

Canada - Ontario - Proposed Occupational STEVs
0.1 ppm STEV; 0.6 mg/m3 STEV

ETHYLENE GLYCOL MONOETHYL ETHER ACRYLATE   106-74-1

ENVIRONMENTAL LISTS

List of Pesticide Product Inert Ingredients
[present]

ETHYLENE GLYCOL MONOHEXYL ETHER   112-25-4

ENVIRONMENTAL LISTS

List of Pesticide Product Inert Ingredients
[present]

ETHYLENE GLYCOL MONOISOBUTYL ETHER   4439-24-1

HEALTH AND SAFETY LISTS

NFPA - Flash Points
flash point = 180 degrees F (85 degrees C)

NFPA - Hazard Identification Ratings
health-1; flammability-2; reactivity-0

ETHYLENE GLYCOL MONOMETHYL ETHER ACETAL   RR-00858-2

HEALTH AND SAFETY LISTS

NFPA - Flash Points
flash point = 230 degrees F (110 degrees C)

NFPA - Hazard Identification Ratings
health-0; flammability-1; reactivity-0

ETHYLENE GLYCOL MONOMETHYL ETHER FORMAL   RR-00859-3

HEALTH AND SAFETY LISTS

NFPA - Flash Points
flash point = 214 degrees F (101 degrees C); *See list description

NFPA - Hazard Identification Ratings
health-2; flammability-1; reactivity-2

NIOSH - Selected LD50s and LC50s
Oral, rat: LD50 = 650 mg/kg Skin, rabbit: LD50 = 1010 mg/kg

ENVIRONMENTAL LISTS

TSCA - Code of Federal Regulations Citations
40 CFR 712.30(d)

TSCA - PAIR - Reporting List
Reporting Date: November 19, 1982

INTERNATIONAL LISTS

Canada - WHMIS: Ingredient Disclosure
1% item 857 (159)

STATE LISTS

Florida Hazardous Substance List
[present]

Massachusetts Right To Know List
[present]

Pennsylvania Right to Know List
[present]

ETHYLENE GLYCOL MONOPHENYL ETHER   122-99-6
SEE ALSO:
GLYCOL ETHERS

HEALTH AND SAFETY LISTS

NFPA - Flash Points
flash point = 265 degrees F (129 degrees C)

NFPA - Hazard Identification Ratings
health-2; flammability-1; reactivity-0; *See list description

INTERNATIONAL LISTS
  Canada - WHMIS: Ingredient Disclosure
  1% item 720 (823)

STATE LISTS
  Florida Hazardous Substance List
  [present]

  Massachusetts Right To Know List
  [present]

  Pennsylvania Right to Know List
  [present]

ETHYLENE GLYCOL MONOPROPYL ETHER   2807-30-9
SEE ALSO:
  GLYCOL ETHERS

HEALTH AND SAFETY LISTS
  NFPA - Flash Points
  flash point = 160 degrees F (71 degrees C)

  NFPA - Hazard Identification Ratings
  health-1; flammability-2; reactivity-0

  NIOSH - Selected LD50s and LC50s
  Oral, rat: LD50 = 2400 mg/kg Skin, rabbit: LD50 = 1500 mg/kg

ENVIRONMENTAL LISTS
  CAA - HON Rule - SOCMI Chemicals
  compliance by Oct. 24, 1994

INTERNATIONAL LISTS
  German (DFG) - MAK Values
  20 ppm MAK; 135 mg/m3 MAK

  German (DFG) - Peak Limitations
  2 x normal MAK (30 min. average value); don't exceed 4 times during shift

  German (DFG) - Skin/Sensitizers
  danger of cutaneous absorption

  German (DFG) - Pregnancy
  no risk to embryo/fetus if exposure limits adhered to

ETHYLENE GLYCOL POLYMER WITH FUMARIC   68152-55-6
ACID & ROSIN
SEE ALSO:
  GLYCOL ETHERS

ENVIRONMENTAL LISTS
  TSCA - Code of Federal Regulations Citations
  40 CFR 712.30(d)

  TSCA - PAIR - Reporting List
  Reporting Date: November 19, 1982

INTERNATIONAL LISTS
  Canada - WHMIS: Ingredient Disclosure
  1% item 722 (155)

ETHYLENEIMINE   151-56-4
SEE ALSO:
  GLYCOL ETHERS

HEALTH AND SAFETY LISTS
  NIOSH - Selected LD50s and LC50s
  Oral, rat: LD50 = 1480 mg/kg Skin, rabbit: LD50 = 890 mg/kg

ENVIRONMENTAL LISTS
  CAA - HON Rule - SOCMI Chemicals
  compliance by Oct. 23, 1995

ETHYLENE OXIDE   75-21-8

HEALTH AND SAFETY LISTS
  NFPA - Flash Points
  flash point = 136 degrees F (58 degrees F)

  NFPA - Hazard Identification Ratings
  health-2; flammability-2

STATE LISTS
  Florida Hazardous Substance List
  [present]

  Massachusetts Right To Know List
  [present]

  Pennsylvania Right to Know List
  [present]

ETHYLENE OXIDE ADDUCT OF FATTY ACID ES-   RR-01214-
TER WITH PENTAERYTHRITOL
HEALTH AND SAFETY LISTS
  NFPA - Flash Points
  flash point = 200 degrees F (93 degrees C)

  NFPA - Hazard Identification Ratings
  health-1; flammability-2

ETHYLENE OXIDE-NONYLPHENOL POLYMER   9016-45-
HEALTH AND SAFETY LISTS
  NFPA - Flash Points
  flash point = 155 degrees F (68 degrees C)

  NFPA - Hazard Identification Ratings
  health-1; flammability-2

ETHYLENE OXIDE-PROPYLENE OXIDE COPOLY-   9038-95-
MER MONOBUTYL ETHER
SEE ALSO:
  GLYCOL ETHERS

HEALTH AND SAFETY LISTS
  NFPA - Flash Points
  *flash point = 260 degrees F (127 degrees C); *See list description*

  NFPA - Hazard Identification Ratings
  health-0; flammability-1; reactivity-0

  NIOSH - Selected LD50s and LC50s
  Oral, rat: LD50 = 1260 mg/kg Skin, rabbit: LD50 = 5000 mg/kg

ENVIRONMENTAL LISTS
  CAA - HON Rule - SOCMI Chemicals
  compliance by Oct. 24, 1994

  TSCA - Code of Federal Regulations Citations
  40 CFR 712.30(h); 40 CFR 716.120(a)

  TSCA - PAIR - Reporting List
  Reporting Date: September 20, 1983

  TSCA - Health and Safety Reporting List
  Effective Date: July 1, 1983

INTERNATIONAL LISTS
  Canada - WHMIS: Ingredient Disclosure
  1% item 723 (825)

ETHYLENE OXIDE-PROPYLENE OXIDE COPOLY-   11111-34-5
MER ETHYLENEDIAMINE ETHER
SEE ALSO:
  GLYCOL ETHERS

HEALTH AND SAFETY LISTS
  NIOSH - Selected LD50s and LC50s
  Inhalation, mouse: LC50 = 1530 ppm 7 hr Oral, rat: LD50 = 3089 mg/kg Skin, guinea pig: LD50 = 1 gm/kg

ENVIRONMENTAL LISTS
  CAA - HON Rule - SOCMI Chemicals
  compliance by Oct. 24, 1994

  List of Pesticide Product Inert Ingredients
  [present]

STATE LISTS
  California - Air Bill 2588 Appendix A-I
  9/90

ETHYLENE-PROPYLENE POLYMER   9010-79-1

ENVIRONMENTAL LISTS

List of Pesticide Product Inert Ingredients
[present]

ETHYLENE SULFIDE   420-12-2

HEALTH AND SAFETY LISTS

ACGIH 1995 - Time Weighted Averages
0.5 ppm TWA; 0.88 mg/m3 TWA

ACGIH 1995 - Skin Designations
skin - potential for cutaneous absorption

AIHA - Odor Threshold Values
no geometric mean air odor threshold

U.S. DOT - Substances From 49 CFR 172.101
regulated by DOT (UN1185)

U.S. DOT - Hazard Classes
DOT hazard class = 6.1

U.S. DOT - Substances Which Are Poisonous by Inhalation
liquid hazardous material poisonous by inhalation (inhibited form) (UN1185)

U.S. DOT - Appendix A Table 1 - Hazardous Substances
final RQ = 1 pound (0.454 kg)

IARC - Group 3 (not classifiable)
[present]

NFPA - Flash Points
flash point = 12 degrees F (-11 degrees C)

NFPA - Hazard Identification Ratings
health-4; flammability-3; reactivity-3

NIOSH - Selected LD50s and LC50s
Inhalation, rat: LC50 = 100 mg/m3 2 hr Oral, rat: LD50 = 15 mg/kg Skin, guinea pig: LD50 = 14 mg/kg

NIOSH 1990 - Pocket Guide - IDLHs
100 ppm IDLH (not considering carcinogenic effects)

NIOSH 1990 - Pocket Guide - Carcinogens
occupational carcinogen

NIOSH 1990 - Pocket Guide - Target organs
eyes, lungs, skin, liver, kidneys

NIOSH - Health Standards - Exposure Limits
use 29 CFR 1910.1012

NIOSH - Health Standards - Health Effects and Precautions
has produced tumors of the liver and lung in animals (Stringent workplace controls and medical monitoring required)

NIOSH - Health Standards - Carcinogenic Chemicals
potential human carcinogen

OSHA - 29 CFR 1910 Specifically Regulated Chemicals
Cancer suspect agent (see 29 CFR 1910.1012)

OSHA - Final PELs - Time Weighted Averages
0.5 ppm TWA; 1 mg/m3 TWA

OSHA - Final PELs - Skin Notations
prevent or reduce skin absorption

OSHA - Select Carcinogens
[present]

OSHA - List of Highly Hazardous Chemicals
threshhold quantity = 1000 pounds

ENVIRONMENTAL LISTS

CERCLA/SARA - Section 302 Extremely Hazardous Substances and TPQs
TPQ = 500 pounds

CERCLA/SARA - Section 313 - Emission Reporting
form R reporting required for 0.1% de minimus concentration

CERCLA/SARA - Hazardous Substances and their Reportable Quantities
final RQ = 1 pound (0.454 kg)

Clean Air Act (1990) - List of Hazardous Air Contaminants
[present]

CAA -Toxic Substances for Accidental Release Prevention
threshold quantity = 10,000 lbs

CAA - HON Rule - SOCMI Chemicals
compliance by Jan. 23, 1995

EPA - Carcinogen Hazard Ranking for RQ Adjustment
Hazard ranking = High

RCRA - P Series Wastes
waste number P054

RCRA - Hazardous Constituents-Appendix VIII
waste number P054

RCRA - Substances Banned From Land Disposal
[present]

INTERNATIONAL LISTS

Australian Exposure Standards - Time Weighted Averages
0.5 ppm TWA; 0.88 mg/m3 TWA

Australian Exposure Standards - Skin Effects
skin absorption

Australian Exposure Standards - Carcinogens
suspected carcinogen

Canada - WHMIS: Ingredient Disclosure
1% item 724 (861)

Canada - Alberta - 8 Hour Occupational Exposure Limit
0.5 ppm TWA; 0.90 mg/m3 TWA

Canada - Alberta - 15 Minute Occupational Exposure Limit
1.5 ppm STEL; 2.7 mg/m3 STEL

Canada - Alberta - Skin Designation
can be absorbed through the intact skin

Canada - British Columbia - 8 Hour Exposure Limits
0.5 ppm TWA; 1 mg/m3 TWA

Canada - British Columbia - Skin Notations
skin - potential for skin absorption

Canada - Ontario - OHSA - TWAEVs
0.5 ppm TWAEV; 0.9 mg/m3 TWAEV

Canada - Ontario - OHSA - Skin Notations
absorption through skin, eyes, or mucous membranes

Canada - Quebec - Time-Weighted Average Exposure Values
0.5 ppm TWAEV; 0.88 mg/m3 TWAEV

Canada - Quebec - Skin Designations
absorbed through the skin

German (DFG) - Skin/Sensitizers
danger of cutaneous absorption

German (DFG) - Carcinogens
animal evidence of carcinogenicity

Israel - Time Weighted Averages
0.5 ppm TWA; 0.88 mg/m3 TWA

Israel - Action Levels
0.25 ppm AL; 0.44 mg/m3 AL

Mexico - Instruction No. 10 - TWAs
0.5 ppm TWA; 1 mg/m3 TWA

Mexico - Instruction No. 10 - Skin designation
skin - potential for cutaneous absorption

STATE LISTS

California - Air Bill 2588 Appendix A-I
6/91

California - Prop. 65 - Cancer list
carcinogen - initial date 1/1/88

California - Prop. 65 - No Significant Risk Levels
no significant risk level = 0.01 ug/day

California - Exposure Limits - PELs
0.5 ppm PEL; 1 mg/m3 PEL

California - Exposure Limits - Skin Notation
material may be absorbed through the skin, eyes or mucous membrane

California - Exposure Limits - Carcinogens
cancer-suspect agent (at a concentration >= 1.0%)

California - Directors List of Hazardous Substances (8 CCR 339)
[present]

Florida Hazardous Substance List
[present]

Massachusetts Right To Know List
carcinogen; extraordinarily hazardous

Minnesota Hazardous Substance List
carcinogen; skin

NJ Right to Know List (Total)
sn 0881

NJ Special Hazardous Substances
(carcinogen; flammable - third degree; mutagen; reactive - third degree)

Pennsylvania Right to Know List
environmental hazard

ETHYLENE THIOUREA   96-45-7
SEE ALSO:
F039-HAZARDOUS WASTES

HEALTH AND SAFETY LISTS

ACGIH 1995 - Time Weighted Averages
1 ppm TWA; 1.8 mg/m3 TWA

ACGIH 1995 - Carcinogens
A2-suspected human carcinogen

AIHA - Odor Threshold Values
geometric mean air odor threshold = 420 ppm (detectable); 490 ppm (recognizable)

U.S. DOT - Substances From 49 CFR 172.101
regulated by DOT (UN1040)

U.S. DOT - Hazard Classes
DOT hazard class = 2.3

U.S. DOT - Substances Which Are Poisonous by Inhalation
gaseous hazardous material poisonous by inhalation (UN1040)

U.S. DOT - Appendix A Table 1 - Hazardous Substances
final RQ = 10 pounds (4.54 kg)

IARC - Group 2A (limited human data)
[present]

NFPA - Flash Points
flash point = -20 degrees F (-29 degrees C)

NFPA - Hazard Identification Ratings
health-3; flammability-4; reactivity-3 (vapors explode)

NIOSH - Selected LD50s and LC50s
Inhalation, rat: LC50 = 800 ppm 4 hr Oral, rat: LD50 = 72 mg/kg

NIOSH 1990 - Pocket Guide - RELs
< 0.1 ppm TWA; 0.18 mg/m3 TWA; C 5 ppm (10 min/day); C 9 mg/m3 (10 min/day)

NIOSH 1990 - Pocket Guide - IDLHs
800 ppm IDLH (not considering carcinogenic effects)

NIOSH 1990 - Pocket Guide - Carcinogens
occupational carcinogen

NIOSH 1990 - Pocket Guide - Target organs
eyes, blood, liver, CNS, kidneys, respiratory system

NIOSH - Health Standards - Exposure Limits
< 0.1 ppm TWA (8 hr); < 0.18 mg/m3 TWA (8 hr); C (10 min/day) 5 ppm; C (10 min/day) 9 mg/m3

NIOSH - Health Standards - Health Effects and Precautions
Peritoneal cancer, leukemia, mutagenesis, reproductive effects (Blood monitoring and medical counseling recommended)

NIOSH - Health Standards - Carcinogenic Chemicals
potential human carcinogen

NTP Chemical Status Reports - Testing Status and NTIS Number
Technical reports printed (PB88169859)

NTP Chemical Status Reports - Evidence of Carcinogenicity
male mice-clear evidence; female mice-clear evidence

NTP Seventh Report - Suspect Carcinogens
suspect carcinogen

OSHA - 29 CFR 1910 Specifically Regulated Chemicals
1 ppm TWA PEL; 0.5 ppm TWA action level; 5 ppm excursion limit (15 min); Cancer hazard and reproductive hazard (see 29 CFR 1910.1047)

OSHA - Select Carcinogens
[present]

OSHA - Possible Select Carcinogens
[present]

OSHA - List of Highly Hazardous Chemicals
threshhold quantity = 5000 pounds

ENVIRONMENTAL LISTS

CERCLA/SARA - Section 302 Extremely Hazardous Substances and TPQs
TPQ = 1000 pounds

CERCLA/SARA - Section 313 - Emission Reporting
form R reporting required for 0.1% de minimus concentration

CERCLA/SARA - Hazardous Substances and their Reportable Quantities
final RQ = 10 pounds (4.54 kg)

Clean Air Act (1990) - List of Hazardous Air Contaminants
[present]

CAA -Toxic Substances for Accidental Release Prevention
threshold quantity = 10,000 lbs

CAA - HON Rule - SOCMI Chemicals
compliance by Oct. 24, 1994

CAA - HON Rule - Organic HAPs
[present]

EPA - Carcinogen Hazard Ranking for RQ Adjustment
Hazard ranking = Medium

RCRA - U Series Wastes
waste number U115 (Ignitable waste; Toxic waste)

RCRA - Hazardous Constituents-Appendix VIII
waste number U115

RCRA - Basis for Listing - Appendix VII
Included in waste stream: F039

RCRA - Substances Banned From Land Disposal
[present]

RCRA - Universal Treatment Standards (LDR)
WW: 0.12 mg/l; NWW: Not applicable

TSCA - Code of Federal Regulations Citations
40 CFR 712.30(d); 40 CFR 716.120(a)

TSCA - PAIR - Reporting List
Reporting Date: November 19, 1982

TSCA - Health and Safety Reporting List
Effective Date: October 4, 1982

INTERNATIONAL LISTS

Australian Exposure Standards - Time Weighted Averages
1 ppm TWA; 1.8 mg/m3 TWA

Australian Exposure Standards - Carcinogens
probable carcinogen

Canada - WHMIS: Ingredient Disclosure
0.1% item 725 (1310)

Canada - NPRI (National Pollutant Release Inventory)
[present]

Canada - Alberta - 8 Hour Occupational Exposure Limit
1 ppm TWA; 1.8 mg/m3 TWA

Canada - Alberta - 15 Minute Occupational Exposure Limit
5 ppm STEL; 9 mg/m3 STEL

Canada - Alberta - Designated Substances
  designated substance - requires code of practice
Canada - British Columbia - 8 Hour Exposure Limits
  0.1 ppm
Canada - British Columbia - 15 Minute Exposure Limits
  1 ppm STEL
Canada - British Columbia - Skin Notations
  skin - potential for skin absorption
Canada - British Columbia - Carcinogens
  carcinogen - 0.1 ppm TWA;   1 ppm STEL
Canada - Ontario - OHSA - TWAEVs
  1 ppm TWAEV;   1.8 mg/m3 TWAEV (designated substance regulation)
Canada - Ontario - OHSA - STEVs
  10 ppm STEV;   18 mg/m3 STEV (designated substance regulations)
Canada - Ontario - OHSA - Designated Substances
  1 ppm TWAEV;   1.8 mg/m3 TWAEV;   See Ontario Reg. 841 for full information.
Canada - Quebec - Time-Weighted Average Exposure Values
  1 ppm TWAEV;   1.8 mg/m3 TWAEV (substance of which the recirculation is prohibited)
Canada - Quebec - Carcinogens
  C2 carcinogen: effect suspected in humans
United Kingdom - Maximum Exposure Limits - TWAs
  5 ppm TWA;   10 mg/m3 TWA
German (DFG) - Skin/Sensitizers
  danger of cutaneous absorption
German (DFG) - Carcinogens
  animal evidence of carcinogenicity
Israel - Time Weighted Averages
  1 ppm TWA;   1.8 mg/m3 TWA
Israel - Ceiling Exposure Limits
  C 1 ppm
Israel - Action Levels
  0.5 ppm AL
Mexico - Instruction No. 10 - TWAs
  1 ppm TWA;   2 mg/m3 TWA
Mexico - Instruction No. 10 - Carcinogens
  potential carcinogen in humans - limited epidemiological evidence

STATE LISTS

California - Air Bill 2588 Appendix A-I
  known or potential carcinogen
California - Prop. 65 - Cancer list
  carcinogen - initial date 7/1/87
California - Prop. 65 - Reproductive - Female
  female reproductive toxicity - initial date 2/27/87
California - Prop. 65 - No Significant Risk Levels
  no significant risk level = 2 ug/day;   NOEL = 20 ug/day
California - Exposure Limits - PELs
  1 ppm PEL;   2 mg/m3 PEL;   avoid inhalation, ingestion and eye or skin contact;   see also section 5220 for respiratory specifications and chronic effects
California - Exposure Limits - STELs
  5 ppm STEL
California - Directors List of Hazardous Substances (8 CCR 339)
  [present] (exempt when part of a cured epoxy or rubber)
Florida Hazardous Substance List
  [present]
Massachusetts Right To Know List
  carcinogen;   extraordinarily hazardous
Minnesota Hazardous Substance List
  carcinogen
NJ Right to Know List (Total)
  sn 0882

NJ Special Hazardous Substances
  (carcinogen;   mutagen;   reactive - third degree)
Pennsylvania Right to Know List
  environmental hazard;   special hazardous substance
Pennsylvania RTK - Special Hazardous Substances
  [present]

PROPOSED REGULATIONS

Canada - Ontario - Proposed Occupational STEVs
  5 ppm STEV;   9 mg/m3 STEV

ETHYLENE-VINYL ACETATE POLYMER   24937-78-8

ENVIRONMENTAL LISTS

TSCA - Chemicals with Significant New Use Rules
  PMN number: P-91-442

N-ETHYLEPHEDRINE   RR-01789-0

HEALTH AND SAFETY LISTS

NIOSH - Selected LD50s and LC50s
  Oral, rat: LD50 = 1310 mg/kg Skin, rabbit: LD50 = 2000 mg/kg

ENVIRONMENTAL LISTS

List of Pesticide Product Inert Ingredients
  [present]

N-ETHYLEPHEDRINE SALTS, OPTICAL ISOMERS,   RR-01786-7
AND SALTS OF OPTICAL ISOMERS

HEALTH AND SAFETY LISTS

NIOSH - Selected LD50s and LC50s
  Oral, rat: LD50 = 12 gm/kg Skin, rabbit: LD50 = 14 gm/kg

ENVIRONMENTAL LISTS

List of Pesticide Product Inert Ingredients
  [present]

ETHYL ETHER   60-29-7

ENVIRONMENTAL LISTS

List of Pesticide Product Inert Ingredients
  [present]

ETHYL 3-ETHOXYPROPANOATE   763-69-9

ENVIRONMENTAL LISTS

List of Pesticide Product Inert Ingredients
  [present]

ETHYL FLUID   RR-01287-3

HEALTH AND SAFETY LISTS

IARC - Group 3 (not classifiable)
  [present]
NIOSH - Selected LD50s and LC50s
  Inhalation, rat: LC50 = 690 ppm 6 hr Oral, rat: LD50 = 178 mg/kg

STATE LISTS

California - Directors List of Hazardous Substances (8 CCR 339)
  [present]

ETHYL FLUORIDE   353-36-6
SEE ALSO:
  K125-HAZARDOUS WASTES
  K123-HAZARDOUS WASTES
  K124-HAZARDOUS WASTES
  K126-HAZARDOUS WASTES

HEALTH AND SAFETY LISTS

U.S. DOT - Appendix A Table 1 - Hazardous Substances
  final RQ = 10 pounds (4.54 kg)
IARC - Group 2B (sufficient animal data)
  [present]

NIOSH - Selected LD50s and LC50s
Oral, rat: LD50 - 265 mg/kg

NIOSH - Health Standards - Exposure Limits
use in encapsulated form in industry; minimize worker exposure

NIOSH - Health Standards - Health Effects and Precautions
has produced tumors of the liver, thyroid, and lymphatic system in animals; potential human teratogen (Inform workers of hazards; give special attention to thyroid function tests)

NIOSH - Health Standards - Carcinogenic Chemicals
potential human carcinogen

NTP Chemical Status Reports - Testing Status and NTIS Number
Technical reports printed (PB92191618)

NTP Chemical Status Reports - Evidence of Carcinogenicity
male rat: clear evidence; female rat: clear evidence; male mice: clear evidence; female mice: clear evidence

NTP Seventh Report - Suspect Carcinogens
suspect carcinogen

OSHA - Possible Select Carcinogens
[present]

ENVIRONMENTAL LISTS

CERCLA/SARA - Section 313 - Emission Reporting
form R reporting required for 0.1% de minimus concentration

CERCLA/SARA - Hazardous Substances and their Reportable Quantities
final RQ - 10 pounds (4.54 kg)

Clean Air Act (1990) - List of Hazardous Air Contaminants
[present]

EPA - Carcinogen Hazard Ranking for RQ Adjustment
Hazard ranking - Medium

RCRA - U Series Wastes
waste number U116

RCRA - Hazardous Constituents-Appendix VIII
waste number U116

RCRA - Basis for Listing - Appendix VII
Included in waste streams: K123, K124, K125, K126

RCRA - Substances Banned From Land Disposal
[present]

INTERNATIONAL LISTS

Canada - WHMIS: Ingredient Disclosure
0.1% item 726 (862)

Canada - NPRI (National Pollutant Release Inventory)
[present]

STATE LISTS

California - Air Bill 2588 Appendix A-I
known or potential carcinogen

California - Prop. 65 - Cancer list
carcinogen - initial date 1/1/88

California - Prop. 65 - Developmental Toxicity
developmental toxicity - initial date 1/1/93

California - Prop. 65 - No Significant Risk Levels
no significant risk level - 20 ug/day

California - Directors List of Hazardous Substances (8 CCR 339)
[present]

Florida Hazardous Substance List
[present]

Massachusetts Right To Know List
carcinogen; extraordinarily hazardous

Minnesota Hazardous Substance List
carcinogen

NJ Right to Know List (Total)
sn 0883

NJ Special Hazardous Substances
(carcinogen; teratogen)

Pennsylvania Right to Know List
environmental hazard; special hazardous substance

Pennsylvania RTK - Special Hazardous Substances
[present]

PROPOSED REGULATIONS

Safe Drinking Water Act - Priority list
[present]

ETHYL FORMATE   109-94-

ENVIRONMENTAL LISTS

List of Pesticide Product Inert Ingredients
[present]

ETHYLHEXADECYLDIMETHYL-AMMONIUM   124-03-
BROMIDE

HEALTH AND SAFETY LISTS

FDA - Controlled Substances Act - Precursor chemicals
Threshold by base weight - 1 kilogram

4-ETHYL-4-HEXADECYLMORPHOLINIUM, ETHYL   78-21-
SULFATE

HEALTH AND SAFETY LISTS

FDA - Controlled Substances Act - Precursor chemicals
Threshold by base weight - 1 kilogram

2-ETHYLHEXALDEHYDE   123-05-
SEE ALSO:
F039-HAZARDOUS WASTES
F003-HAZARDOUS WASTES

HEALTH AND SAFETY LISTS

ACGIH 1995 - Time Weighted Averages
400 ppm TWA; 1210 mg/m3 TWA

ACGIH 1995 - Short Term Exposure Limits
500 ppm STEL; 1520 mg/m3 STEL

U.S. DOT - Substances From 49 CFR 172.101
regulated by DOT (UN1155)

U.S. DOT - Hazard Classes
DOT hazard class - 3

U.S. DOT - Appendix A Table 1 - Hazardous Substances
final RQ - 100 pounds (45.4 kg)

FDA - Controlled Substances Act - Essential chemicals
Import/Export threshold volume - 500 gallons; weight - 1,364 kilograms; Domestic threshold volume - 50 gallons; weight - 135.8 kilograms

NFPA - Flash Points
flash point - -49 degrees F (-45 degrees C)

NFPA - Hazard Identification Ratings
health-1; flammability-4; reactivity-1

NIOSH - Selected LD50s and LC50s
Inhalation, rat: LC50 - 73000 ppm 2 hr Oral, rat: LD50 - 1215 mg/kg

NIOSH 1990 - Pocket Guide - RELs
See Appendix D

NIOSH 1990 - Pocket Guide - IDLHs
19,000 ppm IDLH(lower explosive level)

NIOSH 1990 - Pocket Guide - Target organs
CNS, skin, respiratory system, eyes

OSHA - Vacated PELs - Time Weighted Averages
400 ppm TWA; 1200 mg/m3 TWA

OSHA - Vacated PELs - Short Term Exposure Limits
500 ppm STEL; 1500 mg/m3 STEL

OSHA - Final PELs - Time Weighted Averages
400 ppm TWA; 1200 mg/m3 TWA

ENVIRONMENTAL LISTS

ATSDR Priority List
Rank (of 275): 225

CERCLA/SARA - Hazardous Substances and their Reportable Quantities
final RQ = 100 pounds (45.4 kg)

CAA - Flammable Substances for Accidental Release Prevention
threshold quantity = 10,000 lbs

EPA - Master Testing List
[present]

RCRA - U Series Wastes
waste number U117 (Ignitable waste)

RCRA - Hazardous Constituents-Appendix VIII
waste number U117 (Ignitable waste)

RCRA - Basis for Listing - Appendix VII
Included in waste stream: F039

RCRA - Substances Banned From Land Disposal
[present]

RCRA - Universal Treatment Standards (LDR)
WW: 0.12 mg/l; NWW: 160 mg/kg

TSCA - PAIR - Reporting List
Effective Date: January 26, 1994; Reporting Date: March 28, 1994

TSCA - Health and Safety Reporting List
Effective Date: January 26, 1994; Sunset Date: January 26, 2004

TSCA - Multichemical Test Rules - Neurotoxicity
administrative stay for neurotoxicity tests effective June 27, 1994

INTERNATIONAL LISTS

Australian Exposure Standards - Time Weighted Averages
400 ppm TWA; 1210 mg/m3 TWA

Australian Exposure Standards - Short Term Exposure Limits
500 ppm STEL; 1520 mg/m3 STEL

Canada - WHMIS: Ingredient Disclosure
1% item 727 (863)

Canada - Alberta - 8 Hour Occupational Exposure Limit
400 ppm TWA; 1213 mg/m3 TWA

Canada - Alberta - 15 Minute Occupational Exposure Limit
500 ppm STEL; 1516 mg/m3 STEL

Canada - British Columbia - 8 Hour Exposure Limits
400 ppm TWA; 1200 mg/m3 TWA

Canada - British Columbia - 15 Minute Exposure Limits
500 ppm STEL; 1500 mg/m3 STEL

Canada - Ontario - OHSA - TWAEVs
400 ppm TWAEV; 1210 mg/m3 TWAEV

Canada - Ontario - OHSA - STEVs
500 ppm STEV; 1515 mg/m3 STEV

Canada - Quebec - Time-Weighted Average Exposure Values
400 ppm TWAEV; 1210 mg/m3 TWAEV

Canada - Quebec - Short-term Exposure Values
500 ppm STEV; 1520 mg/m3 STEV

United Kingdom - Occupational Exposure Standards - TWAs
400 ppm TWA; 1200 mg/m3 TWA

United Kingdom - Occupational Exposure Standards - STELs
500 ppm STEL; 1500 mg/m3 STEL

German (DFG) - MAK Values
400 ppm MAK; 1200 mg/m3 MAK

German (DFG) - Peak Limitations
2 x normal MAK (30 min. average value); don't exceed 4 times during shift

German (DFG) - Pregnancy
classification not yet possible

Israel - Time Weighted Averages
400 ppm TWA; 1210 mg/m3 TWA

Israel - Short Term Exposure Limits
500 ppm STEL; 1520 mg/m3 STEL

Israel - Action Levels
200 ppm AL; 605 mg/m3 AL

Mexico - Instruction No. 10 - TWAs
400 ppm TWA; 1200 mg/m3 TWA

Mexico - Instruction No. 10 - STELs
500 ppm STEL; 1500 mg/m3 STEL

STATE LISTS

California - Exposure Limits - PELs
400 ppm PEL; 1200 mg/m3 PEL

California - Exposure Limits - STELs
500 ppm STEL; 1500 mg/m3 STEL

California - Directors List of Hazardous Substances (8 CCR 339)
[present]

Florida Hazardous Substance List
[present]

Massachusetts Right To Know List
[present]

Minnesota Hazardous Substance List
[present]

NJ Right to Know List (Total)
sn 0701

NJ Special Hazardous Substances
(flammable - fourth degree)

Pennsylvania Right to Know List
environmental hazard

PROPOSED REGULATIONS

TSCA - ITC 31st Report Priority Testing List
designated to be tested

2-ETHYL-1,3-HEXANEDIOL　　　　　　　　　　94-96-2

HEALTH AND SAFETY LISTS

NIOSH - Selected LD50s and LC50s
Oral, rat: LD50 = 5000 mg/kg Skin, rabbit: LD50 = 10 gm/kg

ENVIRONMENTAL LISTS

List of Pesticide Product Inert Ingredients
[present]

ETHYL HEXANOATE　　　　　　　　　　　　123-66-0

HEALTH AND SAFETY LISTS

U.S. DOT - Appendix B - Marine Pollutants
DOT regulated marine pollutant

STATE LISTS

NJ Right to Know List (Total)
sn 2568

2-ETHYLHEXANOIC ACID　　　　　　　　　149-57-5

HEALTH AND SAFETY LISTS

U.S. DOT - Substances From 49 CFR 172.101
regulated by DOT (UN2453)

U.S. DOT - Hazard Classes
DOT hazard class = 2.1

NFPA - Hazard Identification Ratings
flammability-4; reactivity-0

STATE LISTS

Florida Hazardous Substance List
[present]

Massachusetts Right To Know List
[present]

NJ Right to Know List (Total)
sn 0884

NJ Special Hazardous Substances
(flammable - fourth degree)

Pennsylvania Right to Know List
[present]

2-ETHYLHEXANOIC ACID, MANGANESE SALT   15956-58-8

HEALTH AND SAFETY LISTS

ACGIH 1995 - Time Weighted Averages
100 ppm TWA; 303 mg/m3 TWA

AIHA - Odor Threshold Values
no geometric mean air odor threshold

U.S. DOT - Substances From 49 CFR 172.101
regulated by DOT (UN1190)

U.S. DOT - Hazard Classes
DOT hazard class = 3

NFPA - Flash Points
flash point = -4 degrees F (-20 degrees C)

NFPA - Hazard Identification Ratings
health-2; flammability-3; reactivity-0

NIOSH - Selected LD50s and LC50s
Oral, rat: LD50 = 1850 mg/kg Skin, rabbit: LD50 = 20 gm/kg

NIOSH 1990 - Pocket Guide - RELs
100 ppm TWA; 300 mg/m3 TWA

NIOSH 1990 - Pocket Guide - IDLHs
8000 ppm IDLH

NIOSH 1990 - Pocket Guide - Target organs
eyes, respiratory system

OSHA - Vacated PELs - Time Weighted Averages
100 ppm TWA; 300 mg/m3 TWA

OSHA - Final PELs - Time Weighted Averages
100 ppm TWA; 300 mg/m3 TWA

INTERNATIONAL LISTS

Australian Exposure Standards - Time Weighted Averages
100 ppm TWA; 303 mg/m3 TWA

Canada - WHMIS: Ingredient Disclosure
1% item 728 (922)

Canada - Alberta - 8 Hour Occupational Exposure Limit
100 ppm TWA; 303 mg/m3 TWA

Canada - Alberta - 15 Minute Occupational Exposure Limit
150 ppm STEL; 454 mg/m3 STEL

Canada - British Columbia - 8 Hour Exposure Limits
100 ppm TWA; 300 mg/m3 TWA

Canada - British Columbia - 15 Minute Exposure Limits
150 ppm STEL; 450 mg/m3 STEL

Canada - Ontario - OHSA - TWAEVs
100 ppm TWAEV; 300 mg/m3 TWAEV

Canada - Quebec - Time-Weighted Average Exposure Values
100 ppm TWAEV; 303 mg/m3 TWAEV

United Kingdom - Occupational Exposure Standards - TWAs
100 ppm TWA; 300 mg/m3 TWA

United Kingdom - Occupational Exposure Standards - STELs
150 ppm STEL; 450 mg/m3 STEL

German (DFG) - MAK Values
100 ppm MAK; 300 mg/m3 MAK

German (DFG) - Peak Limitations
2 x normal MAK (5 min momentary value); don't exceed 8 times during shift

German (DFG) - Pregnancy
classification not yet possible

Israel - Time Weighted Averages
100 ppm TWA; 303 mg/m3 TWA

Israel - Action Levels
50 ppm AL; 151.5 mg/m3 AL

Mexico - Instruction No. 10 - TWAs
100 ppm TWA; 300 mg/m3 TWA

Mexico - Instruction No. 10 - STELs
150 ppm STEL; 450 mg/m3 STEL

STATE LISTS

California - Exposure Limits - PELs
100 ppm PEL; 300 mg/m3 PEL

California - Directors List of Hazardous Substances (8 CCR 339)
[present]

Florida Hazardous Substance List
[present]

Massachusetts Right To Know List
[present]

Minnesota Hazardous Substance List
[present]

NJ Right to Know List (Total)
sn 0885

NJ Special Hazardous Substances
(flammable - third degree)

Pennsylvania Right to Know List
[present]

2-ETHYLHEXANOIC ACID, NICKEL SALT   7580-31-6

HEALTH AND SAFETY LISTS

NIOSH - Selected LD50s and LC50s
Oral, rat: LD50 = 500 mg/kg

2-ETHYLHEXANOL   104-76-7

ENVIRONMENTAL LISTS

List of Pesticide Product Inert Ingredients
[present]

2-ETHYLHEXYL ACETATE   103-09-3

HEALTH AND SAFETY LISTS

NFPA - Flash Points
flash point = 112 degrees F (44 degrees C)

NFPA - Hazard Identification Ratings
health-2; flammability-2; reactivity-1

NIOSH - Selected LD50s and LC50s
Oral, rat: LD50 = 3730 mg/kg Skin, rabbit: LD50 = 5040 mg/kg

ENVIRONMENTAL LISTS

TSCA - Code of Federal Regulations Citations
40 CFR 712.30(x); 40 CFR 716.120(d)

TSCA - PAIR - Reporting List
Reporting Date: November 27, 1991

TSCA - Health and Safety Reporting List
Effective Date: September 30, 1991

INTERNATIONAL LISTS

Canada - WHMIS: Ingredient Disclosure
1% item 729 (864)

STATE LISTS

Florida Hazardous Substance List
[present]

Massachusetts Right To Know List
[present]

NJ Right to Know List (Total)
sn 0886

Pennsylvania Right to Know List
[present]

2-ETHYLHEXYL ACRYLATE   103-11-7

HEALTH AND SAFETY LISTS

NFPA - Flash Points
*flash point = 230 degrees F (110 degrees C); *See list description*

NFPA - Hazard Identification Ratings
health-1; flammability-1; reactivity-0
NIOSH - Selected LD50s and LC50s
Oral, rat: LD50 = 1400 mg/kg Skin, guinea pig: LD50 = 9422 mg/kg

INTERNATIONAL LISTS
Canada - WHMIS: Ingredient Disclosure
1% item 730 (865)

2-ETHYLHEXYL ACRYLATE, POLYMER WITH VINYL TOLUENE   60381-61-5

HEALTH AND SAFETY LISTS
NFPA - Flash Points
flash point = 120 degrees F (49 degrees C)
NFPA - Hazard Identification Ratings
health-2; flammability-2; reactivity-0

INTERNATIONAL LISTS
Canada - WHMIS: Ingredient Disclosure
1% item 731 (967)

STATE LISTS
Florida Hazardous Substance List
[present]
Massachusetts Right To Know List
[present]
NJ Right to Know List (Total)
sn 0858
Pennsylvania Right to Know List
[present]

2-ETHYLHEXYLAMINE   104-75-6

HEALTH AND SAFETY LISTS
NFPA - Flash Points
flash point = 245 degrees F (118 degrees C)
NFPA - Hazard Identification Ratings
health-1; flammability-1; reactivity-0
NIOSH - Selected LD50s and LC50s
Oral, rat: LD50 = 3000 mg/kg Skin, rabbit: LD50 = 1260 mg/kg

ENVIRONMENTAL LISTS
EPA - Master Testing List
[present]
List of Pesticide Product Inert Ingredients
[present]
TSCA - Code of Federal Regulations Citations
40 CFR 712.30(k); 40 CFR 716.120(a); 40 CFR 799.1650
TSCA - PAIR - Reporting List
Reporting Date: August 27, 1984
TSCA - Health and Safety Reporting List
Effective Date: June 28, 1984
TSCA - Chemical Test Rules
Testing required by: manufacturers; processors (40 CFR 799.1650)
TSCA - Section 12(b) - Export Notification
export notification required - Section 4

INTERNATIONAL LISTS
Canada - WHMIS: Ingredient Disclosure
1% item 732 (85)

2-ETHYLHEXYL CHLORIDE   123-04-6

SEE ALSO:
MANGANESE
MANGANESE COMPOUNDS, N.O.S.

ENVIRONMENTAL LISTS
List of Pesticide Product Inert Ingredients
[present]

2-ETHYLHEXYL CHLOROFORMATE   24468-13-1

ENVIRONMENTAL LISTS
List of Pesticide Product Inert Ingredients
[present]

N-2-(ETHYLHEXYL)-CYCLOHEXYLAMINE   5432-61-1

HEALTH AND SAFETY LISTS
NFPA - Flash Points
flash point = 164 degrees F (73 degrees C)
NFPA - Hazard Identification Ratings
health-2; flammability-2; reactivity-0
NIOSH - Selected LD50s and LC50s
Oral, rat: LD50 = 2049 mg/kg Skin, rabbit: LD50 = 1970 mg/kg

ENVIRONMENTAL LISTS
EPA - Master Testing List
[present]
List of Pesticide Product Inert Ingredients
[present]
TSCA - Code of Federal Regulations Citations
40 CFR 716.120(a); 40 CFR 799.1645
TSCA - Health and Safety Reporting List
Effective Date: June 1, 1987
TSCA - Chemical Test Rules
Testing required by: manufacturers; processors (40 CFR 799.1645)
TSCA - Section 12(b) - Export Notification
export notification required - Section 4

INTERNATIONAL LISTS
Canada - WHMIS: Ingredient Disclosure
1% item 734 (173)

STATE LISTS
Florida Hazardous Substance List
[present]
Massachusetts Right To Know List
[present]
Pennsylvania Right to Know List
[present]

PROPOSED REGULATIONS
TSCA - Proposed Substances for Developmental/Reproductive Testing
proposed testing for: Developmental Toxicity - oral

2-ETHYLHEXYL ETHER   5756-43-4

HEALTH AND SAFETY LISTS
NFPA - Flash Points
flash point = 160 degrees F (71 degrees C)
NFPA - Hazard Identification Ratings
health-2; flammability-2; reactivity-0
NIOSH - Selected LD50s and LC50s
Oral, rat: LD50 = 3000 mg/kg

STATE LISTS
Florida Hazardous Substance List
[present]
Massachusetts Right To Know List
[present]
Pennsylvania Right to Know List
[present]

2-ETHYLHEXYL METHACRYLATE   688-84-6

HEALTH AND SAFETY LISTS
NFPA - Flash Points
flash point = 180 degrees F (82 degrees C)
NFPA - Hazard Identification Ratings
health-2; flammability-2; reactivity-2

NIOSH - Selected LD50s and LC50s
Oral, rat: LD50 = 5660 mg/kg Skin, rabbit: LD50 = 8480 mg/kg

ENVIRONMENTAL LISTS

CAA - HON Rule - SOCMI Chemicals
compliance by Jan. 23, 1995

List of Pesticide Product Inert Ingredients
[present]

TSCA - Code of Federal Regulations Citations
40 CFR 712.30(d)

TSCA - PAIR - Reporting List
Reporting Date: November 19, 1982

INTERNATIONAL LISTS

Canada - WHMIS: Ingredient Disclosure
1% item 733 (157)

STATE LISTS

Florida Hazardous Substance List
[present]

Massachusetts Right To Know List
[present]

Pennsylvania Right to Know List
[present]

2-ETHYLHEXYL-P-METHOXYCINNAMATE   5466-77-3

ENVIRONMENTAL LISTS

List of Pesticide Product Inert Ingredients
[present]

2-(2-ETHYLHEXYLOXY)ETHANOL   1559-35-9

HEALTH AND SAFETY LISTS

U.S. DOT - Substances From 49 CFR 172.101
regulated by DOT (UN2276)

U.S. DOT - Hazard Classes
DOT hazard class = 8

NFPA - Flash Points
flash point = 140 degrees F (60 degrees C)

NFPA - Hazard Identification Ratings
health-2; flammability-2; reactivity-0

NIOSH - Selected LD50s and LC50s
Oral, rat: LD50 = 450 mg/kg Skin, rabbit: LD50 = 600 mg/kg

INTERNATIONAL LISTS

Canada - WHMIS: Ingredient Disclosure
1% item 735 (866)

STATE LISTS

Florida Hazardous Substance List
[present]

Massachusetts Right To Know List
[present]

NJ Right to Know List (Total)
sn 0888

Pennsylvania Right to Know List
[present]

2-ETHYLHEXYL SALICYLATE   118-60-5

HEALTH AND SAFETY LISTS

NFPA - Flash Points
flash point = 140 degrees F (60 degrees C)

NFPA - Hazard Identification Ratings
health-2; flammability-2; reactivity-0

STATE LISTS

Florida Hazardous Substance List
[present]

Massachusetts Right To Know List
[present]

Pennsylvania Right to Know List
[present]

2-ETHYLHEXYL SODIUM SULFATE   126-92-1

HEALTH AND SAFETY LISTS

U.S. DOT - Substances From 49 CFR 172.101
regulated by DOT (UN2748)

U.S. DOT - Hazard Classes
DOT hazard class = 6.1

INTERNATIONAL LISTS

Canada - WHMIS: Ingredient Disclosure
1% item 736 (437)

United Kingdom - Occupational Exposure Standards - TWAs
1 ppm TWA; 7.9 mg/m3 TWA

STATE LISTS

NJ Right to Know List (Total)
sn 0889

NJ Special Hazardous Substances
(corrosive)

2-ETHYLHEXYL SULFATE   72214-01-8

HEALTH AND SAFETY LISTS

NFPA - Flash Points
flash point = 265 degrees F (129 degrees C)

NFPA - Hazard Identification Ratings
health-2; flammability-1; reactivity-0

STATE LISTS

Florida Hazardous Substance List
[present]

Massachusetts Right To Know List
[present]

Pennsylvania Right to Know List
[present]

ETHYL HYDROPEROXIDE   3031-74-1

HEALTH AND SAFETY LISTS

NFPA - Flash Points
flash point = 235 degrees F (113 degrees C)

NFPA - Hazard Identification Ratings
health-1; flammability-1; reactivity-0

ETHYL P-HYDROXYBENZOATE   120-47-8

ENVIRONMENTAL LISTS

TSCA - Code of Federal Regulations Citations
40 CFR 712.30(d)

TSCA - PAIR - Reporting List
Reporting Date: November 19, 1982

INTERNATIONAL LISTS

Canada - WHMIS: Ingredient Disclosure
1% item 737 (1092)

ETHYL HYDROXYETHYL CELLULOSE   9004-58-4

ENVIRONMENTAL LISTS

List of Pesticide Product Inert Ingredients
[present]

ETHYLIDENE NORBORNENE   16219-75-3

HEALTH AND SAFETY LISTS

NIOSH - Selected LD50s and LC50s
Oral, rat: LD50 = 3080 mg/kg Skin, rabbit: LD50 = 2120 mg/kg

INTERNATIONAL LISTS
Canada - WHMIS: Ingredient Disclosure
1% item 738 (867)

ETHYL IODIDE 75-03-6
ENVIRONMENTAL LISTS
List of Pesticide Product Inert Ingredients
[present]

ETHYL ISOBUTYRATE 97-62-1
HEALTH AND SAFETY LISTS
NIOSH - Selected LD50s and LC50s
Oral, rat: LD50 = 4 gm/kg
ENVIRONMENTAL LISTS
List of Pesticide Product Inert Ingredients
[present]

ETHYL ISOCYANATE 109-90-0
ENVIRONMENTAL LISTS
List of Pesticide Product Inert Ingredients
[present]

ETHYL LACTATE 97-64-3
HEALTH AND SAFETY LISTS
U.S. DOT - Hazard Classes
Forbidden from transport by the DOT

ETHYL MERCAPTAN 75-08-1
HEALTH AND SAFETY LISTS
NIOSH - Selected LD50s and LC50s
Oral, mouse: LD50 = 3 gm/kg
ENVIRONMENTAL LISTS
List of Pesticide Product Inert Ingredients
[present]

ETHYL MERCAPTOACETATE 623-51-8
ENVIRONMENTAL LISTS
List of Pesticide Product Inert Ingredients
[present]

ETHYLMERCURIC CHLORIDE 107-27-7
HEALTH AND SAFETY LISTS
ACGIH 1995 - Ceiling Limits
C 5 ppm; C 25 mg/m3
NIOSH - Selected LD50s and LC50s
*Inhalation, rat: LC50 = 4000 ppm 4 hr Oral, rat: LD50 = 2527
mg/kg Skin, rabbit: LD50 = 8189 mg/kg*
OSHA - Vacated PELs - Ceiling Limits
C 5 ppm; C 25 mg/m3 (Enforcement indefinitely stayed)
INTERNATIONAL LISTS
Australian Exposure Standards - Time Weighted Averages
Peak Limitation: 5 ppm; 25 mg/m3
Canada - WHMIS: Ingredient Disclosure
1% item 739 (868)
Canada - Alberta - Ceiling Occupational Exposure Limit
C 5 ppm; C 25 mg/m3
Canada - British Columbia - Ceiling Exposure Limits
C 5 ppm; C 25 mg/m3
Canada - Ontario - OHSA - CEVs
5 ppm CEV; 25 mg/m3 CEV
Canada - Quebec - Ceiling Limits
P 5 ppm; P 25 mg/m3

Israel - Ceiling Exposure Limits
C 5 ppm; C 25 mg/m3
Mexico - Instruction No. 10 - TWAs
5 ppm TWA; 25 mg/m3 TWA
STATE LISTS
California - Exposure Limits - Ceilings
C 5 ppm; C 25 mg/m3
California - Directors List of Hazardous Substances (8 CCR 339)
[present]
Florida Hazardous Substance List
[present]
Massachusetts Right To Know List
[present]
Minnesota Hazardous Substance List
[present]
Pennsylvania Right to Know List
[present]

ETHYLMERCURIC PHOSPHATE 2235-25-8
HEALTH AND SAFETY LISTS
NIOSH - Selected LD50s and LC50s
Inhalation, rat: LC50 = 65000 mg/m3 (30 mn)

N-(ETHYLMERCURIC)-P-TOLUENESULPHONANNILIDE 517-16-8
HEALTH AND SAFETY LISTS
U.S. DOT - Substances From 49 CFR 172.101
regulated by DOT (UN2385)
U.S. DOT - Hazard Classes
DOT hazard class = 3
NFPA - Flash Points
flash point < 70 degrees F (21 degrees C)
NFPA - Hazard Identification Ratings
health-0; flammability-3; reactivity-0
INTERNATIONAL LISTS
Canada - WHMIS: Ingredient Disclosure
1% item 740 (1034)
STATE LISTS
Florida Hazardous Substance List
[present]
Massachusetts Right To Know List
[present]
NJ Right to Know List (Total)
sn 0891
NJ Special Hazardous Substances
(flammable - third degree)
Pennsylvania Right to Know List
[present]

ETHYL METHACRYLATE 97-63-2
HEALTH AND SAFETY LISTS
U.S. DOT - Substances From 49 CFR 172.101
regulated by DOT (UN2481)
U.S. DOT - Hazard Classes
DOT hazard class = 3
U.S. DOT - Substances Which Are Poisonous by Inhalation
liquid hazardous material poisonous by inhalation (UN2481)
ENVIRONMENTAL LISTS
TSCA - Code of Federal Regulations Citations
40 CFR 712.30(x); 40 CFR 716.120(d)
TSCA - PAIR - Reporting List
Reporting Date: December 27, 1990

TSCA - Health and Safety Reporting List
Effective Date: October 29, 1990

INTERNATIONAL LISTS

Canada - WHMIS: Ingredient Disclosure
0.1% item 741 (1040)

STATE LISTS

NJ Right to Know List (Total)
sn 0892

ETHYL METHANESULFONATE   62-50-0

HEALTH AND SAFETY LISTS

U.S. DOT - Substances From 49 CFR 172.101
regulated by DOT (UN1192)

U.S. DOT - Hazard Classes
DOT hazard class = 3

NFPA - Flash Points
flash point = 115 degrees F (46 degrees C);   technical grade: flash point = 131 degrees F (55 degrees C)

NFPA - Hazard Identification Ratings
health-2; flammability-2; reactivity-0

NIOSH - Selected LD50s and LC50s
Oral, mouse: LD50 = 2500 mg/kg

ENVIRONMENTAL LISTS

List of Pesticide Product Inert Ingredients
[present]

STATE LISTS

Florida Hazardous Substance List
[present]

Massachusetts Right To Know List
[present]

NJ Right to Know List (Total)
sn 0893

Pennsylvania Right to Know List
[present]

7-ETHYL-2-METHYL-4-HENDECANOL   RR-00622-4

HEALTH AND SAFETY LISTS

ACGIH 1995 - Time Weighted Averages
0.5 ppm TWA; 1.3 mg/m3 TWA

AIHA - Odor Threshold Values
geometric mean air odor threshold = 0.00035 ppm (detectable); 0.00040 ppm (recognizable)

U.S. DOT - Substances From 49 CFR 172.101
regulated by DOT (UN2363)

U.S. DOT - Hazard Classes
DOT hazard class = 3

NFPA - Flash Points
flash point < 0 degrees F (-18 degrees C)

NFPA - Hazard Identification Ratings
health-2; flammability-4; reactivity-0

NIOSH - Selected LD50s and LC50s
Inhalation, rat: LC50 = 4420 ppm 4 hr Oral, rat: LD50 = 682 mg/kg

NIOSH 1990 - Pocket Guide - RELs
C 0.5 ppm (15 min); C 1.3 mg/m3 (15 min)

NIOSH 1990 - Pocket Guide - IDLHs
2500 ppm IDLH

NIOSH 1990 - Pocket Guide - Target organs
respiratory system; in animals: liver, kidneys

NIOSH - Health Standards - Exposure Limits
C (15 min) 0.5 ppm; C (15 min) 1.3 mg/m3 (Listed under 'Thiols')

NIOSH - Health Standards - Health Effects and Precautions
Irritation; eye, skin, blood, and nervous system effects (Blood and urine monitoring required; prevent skin contact) (Listed under 'Thiols')

OSHA - Vacated PELs - Time Weighted Averages
0.5 ppm TWA; 1 mg/m3 TWA

OSHA - Final PELs - Ceiling Limits
C 10 ppm; C 25 mg/m3

ENVIRONMENTAL LISTS

CAA - Flammable Substances for Accidental Release Prevention
threshold quantity = 10,000 lbs

INTERNATIONAL LISTS

Australian Exposure Standards - Time Weighted Averages
0.5 ppm TWA; 1.3 mg/m3 TWA

Canada - WHMIS: Ingredient Disclosure
1% item 742 (870)

Canada - Alberta - 8 Hour Occupational Exposure Limit
0.5 ppm TWA; 1.3 mg/m3 TWA

Canada - Alberta - 15 Minute Occupational Exposure Limit
2 ppm STEL; 5.1 mg/m3 STEL

Canada - British Columbia - 8 Hour Exposure Limits
air contaminant: 3 ppm TWA; 7.6 mg/m3 TWA

Canada - British Columbia - Ceiling Exposure Limits
C 3 ppm; C 7.6 mg/m3

Canada - Ontario - OHSA - TWAEVs
0.5 ppm TWAEV; 1.3 mg/m3 TWAEV

Canada - Quebec - Time-Weighted Average Exposure Values
0.5 ppm TWAEV; 1.3 mg/m3 TWAEV

United Kingdom - Occupational Exposure Standards - TWAs
0.5 ppm TWA; 1 mg/m3 TWA

United Kingdom - Occupational Exposure Standards - STELs
2 ppm STEL; 3 mg/m3 STEL

German (DFG) - MAK Values
0.5 ppm MAK; 1 mg/m3 MAK

German (DFG) - Peak Limitations
2 x normal MAK (10 min momentary value); don't exceed 4 times per shift

Israel - Time Weighted Averages
0.5 ppm TWA; 1.3 mg/m3 TWA

Israel - Action Levels
0.25 ppm AL; 0.65 mg/m3 AL

Mexico - Instruction No. 10 - TWAs
0.95 ppm TWA; 2 mg/m3 TWA

Mexico - Instruction No. 10 - STELs
2 ppm STEL; 3 mg/m3 STEL

STATE LISTS

California - Exposure Limits - PELs
0.5 ppm PEL; 1 mg/m3 PEL

California - Directors List of Hazardous Substances (8 CCR 339)
[present]

Florida Hazardous Substance List
[present]

Massachusetts Right To Know List
[present]

Minnesota Hazardous Substance List
[present]

NJ Right to Know List (Total)
sn 0894

NJ Special Hazardous Substances
(flammable - fourth degree)

Pennsylvania Right to Know List
[present]

2-ETHYL-4-METHYL-1-PENTANOL   106-67-2

HEALTH AND SAFETY LISTS

NIOSH - Selected LD50s and LC50s
Oral, rat: LD50 = 178 mg/kg

-[N-ETHYL-4-[[6-(METHYLSULFONYL)-2-BENZOTH-   16588-67-3
AZOLYL]AZO]-M-TOLUIDINO]-PROPIONITRILE
SEE ALSO:
   MERCURY

HEALTH AND SAFETY LISTS
   NIOSH - Selected LD50s and LC50s
     Oral, rat: LD50 = 40 mg/kg Skin, rat: LD50 = 200 mg/kg

STATE LISTS
   Florida Hazardous Substance List
     [present]
   NJ Right to Know List (Total)
     sn 0895
   NJ Special Hazardous Substances
     (teratogen)

N-ETHYLMORPHOLINE   100-74-3
SEE ALSO:
   MERCURY

STATE LISTS
   Massachusetts Right To Know List
     [present]
   NJ Right to Know List (Total)
     sn 0896

1-ETHYLNAPHTHALENE   1127-76-0
SEE ALSO:
   MERCURY

HEALTH AND SAFETY LISTS
   NIOSH - Selected LD50s and LC50s
     Oral, rat: LD50 = 100 mg/kg

STATE LISTS
   NJ Right to Know List (Total)
     sn 1337

ETHYL NITRATE   625-58-1
SEE ALSO:
   F039-HAZARDOUS WASTES

HEALTH AND SAFETY LISTS
   U.S. DOT - Substances From 49 CFR 172.101
     regulated by DOT (UN2277)
   U.S. DOT - Hazard Classes
     DOT hazard class = 3
   U.S. DOT - Appendix A Table 1 - Hazardous Substances
     final RQ = 1000 pounds (454 kg)
   NFPA - Flash Points
     flash point = 68 degrees F (20 degrees C)
   NFPA - Hazard Identification Ratings
     health-2; flammability-3; reactivity-0
   NIOSH - Selected LD50s and LC50s
     Inhalation, rat: LC50 = 8300 ppm 4 hr Oral, rat: LD50 = 14800 mg/kg

ENVIRONMENTAL LISTS
   CERCLA/SARA - Hazardous Substances and their Reportable Quantities
     final RQ = 1000 pounds (454 kg)
   List of Pesticide Product Inert Ingredients
     [present]
   RCRA - U Series Wastes
     waste number U118
   RCRA - Hazardous Constituents-Appendix VIII
     waste number U118
   RCRA - Basis for Listing - Appendix VII
     Included in waste stream: F039

RCRA - Substances Banned From Land Disposal
     [present]
RCRA - TSD Facilities Ground Water Monitoring
     TM 8015 = 10 ug/L PQL; TM 8240 = 5 ug/L PQL; TM 8270 = 10 ug/L PQL
RCRA - Universal Treatment Standards (LDR)
     WW: 0.14 mg/l; NWW: 160 mg/kg
TSCA - Code of Federal Regulations Citations
     40 CFR 712.30(d); 40 CFR 799.5055(c), (d)(2)
TSCA - PAIR - Reporting List
     Reporting Date: November 19, 1982
TSCA - Multichemical Test Rules - Waste Constituents
     hydrolysis testing for Chemical Fate
TSCA - Section 12(b) - Export Notification
     export notification required - Section 4

INTERNATIONAL LISTS
   Canada - WHMIS: Ingredient Disclosure
     0.1% item 995 (1091)
   Canada - Alberta - 8 Hour Occupational Exposure Limit
     100 ppm TWA; 392 mg/m3 TWA
   Canada - Alberta - 15 Minute Occupational Exposure Limit
     125 ppm STEL; 490 mg/m3 STEL

STATE LISTS
   Florida Hazardous Substance List
     [present]
   Massachusetts Right To Know List
     [present]
   NJ Right to Know List (Total)
     sn 0897
   NJ Special Hazardous Substances
     (flammable - third degree)
   Pennsylvania Right to Know List
     environmental hazard

ETHYL NITRITE   109-95-5
HEALTH AND SAFETY LISTS
   U.S. DOT - Appendix A Table 1 - Hazardous Substances
     final RQ = 1 pound (0.454 kg)
   IARC - Group 2B (sufficient animal data)
     [present] (Overall evaluation based only on evidence of carcinogenicity in monograph (7, 1974) or in Supplement 4)
   NTP Seventh Report - Suspect Carcinogens
     suspect carcinogen
   OSHA - Possible Select Carcinogens
     [present]

ENVIRONMENTAL LISTS
   CERCLA/SARA - Hazardous Substances and their Reportable Quantities
     final RQ = 1 pound (0.454 kg)
   EPA - Carcinogen Hazard Ranking for RQ Adjustment
     Hazard ranking = High
   RCRA - U Series Wastes
     waste number U119
   RCRA - Hazardous Constituents-Appendix VIII
     waste number U119
   RCRA - Substances Banned From Land Disposal
     [present]
   RCRA - TSD Facilities Ground Water Monitoring
     TM 8270 = 10 ug/L PQL
   TSCA - Chemicals with Significant New Use Rules
     [present]
   TSCA - Section 12(b) - Export Notification
     export notification required - Section 5

INTERNATIONAL LISTS

Canada - WHMIS: Ingredient Disclosure
1% item 743 (1101)

STATE LISTS

California - Air Bill 2588 Appendix A-II
known or potential carcinogen

California - Prop. 65 - Cancer list
carcinogen - initial date 1/1/88

California - Directors List of Hazardous Substances (8 CCR 339)
[present]

Florida Hazardous Substance List
[present]

Massachusetts Right To Know List
carcinogen; extraordinarily hazardous

Minnesota Hazardous Substance List
carcinogen

NJ Special Hazardous Substances
(carcinogen; mutagen)

Pennsylvania Right to Know List
environmental hazard; special hazardous substance

Pennsylvania RTK - Special Hazardous Substances
[present]

4,4'-(2-ETHYL-2-NITROTRIMETHYLENE)-   1854-23-5
DIMORPHOLINE

HEALTH AND SAFETY LISTS

NFPA - Flash Points
flash point = 285 degrees F (141 degrees C)

NFPA - Hazard Identification Ratings
health-0; flammability-1; reactivity-0

ETHYL NONANOATE   123-29-5

HEALTH AND SAFETY LISTS

NFPA - Flash Points
flash point = 158 degrees F (70 degrees C)

NFPA - Hazard Identification Ratings
health-1; flammability-2

NIOSH - Selected LD50s and LC50s
Oral, rat: LD50 = 4290 mg/kg

3-ETHYLOCTANE   5881-17-4

ENVIRONMENTAL LISTS

TSCA - Code of Federal Regulations Citations
40 CFR 712.30(x); 40 CFR 716.120(d)

TSCA - PAIR - Reporting List
Reporting Date: November 27, 1991

TSCA - Health and Safety Reporting List
Effective Date: September 30, 1991

4-ETHYLOCTANE   15869-86-0

HEALTH AND SAFETY LISTS

ACGIH 1995 - Time Weighted Averages
5 ppm TWA; 24 mg/m3 TWA

ACGIH 1995 - Skin Designations
skin - potential for cutaneous absorption

AIHA - Odor Threshold Values
geometric mean air odor threshold = 0.085 ppm (detectable); 0.25 ppm (recognizable)

NFPA - Flash Points
flash point = 90 degrees F (32 degrees C)

NFPA - Hazard Identification Ratings
health-2; flammability-3; reactivity-0

NIOSH - Selected LD50s and LC50s
Inhalation, mouse: LC50 = 18000 mg/m3 (2 hr) Oral, rat: LD50 1780 mg/kg

NIOSH 1990 - Pocket Guide - RELs
5 ppm TWA; 23 mg/m3 TWA

NIOSH 1990 - Pocket Guide - IDLHs
2000 ppm IDLH

NIOSH 1990 - Pocket Guide - Target organs
eyes, skin, respiratory system

NIOSH 1990 - Pocket Guide - Skin list
Potential for dermal absorption

OSHA - Vacated PELs - Time Weighted Averages
5 ppm TWA; 23 mg/m3 TWA

OSHA - Vacated PELs - Skin Designation
Prevent or reduce skin absorption

OSHA - Final PELs - Time Weighted Averages
20 ppm TWA; 94 mg/m3 TWA

OSHA - Final PELs - Skin Notations
prevent or reduce skin absorption

INTERNATIONAL LISTS

Australian Exposure Standards - Time Weighted Averages
5 ppm TWA; 24 mg/m3 TWA

Australian Exposure Standards - Skin Effects
skin absorption

Canada - WHMIS: Ingredient Disclosure
1% item 745 (872)

Canada - Alberta - 8 Hour Occupational Exposure Limit
5 ppm TWA; 23 mg/m3 TWA

Canada - Alberta - 15 Minute Occupational Exposure Limit
20 ppm STEL; 94 mg/m3 STEL

Canada - Alberta - Skin Designation
can be absorbed through the intact skin

Canada - British Columbia - 8 Hour Exposure Limits
20 ppm TWA; 94 mg/m3 TWA

Canada - British Columbia - Skin Notations
skin - potential for skin absorption

Canada - Ontario - OHSA - TWAEVs
5 ppm TWAEV; 23 mg/m3 TWAEV

Canada - Ontario - OHSA - Skin Notations
absorption through skin, eyes, or mucous membranes

Canada - Quebec - Time-Weighted Average Exposure Values
5 ppm TWAEV; 24 mg/m3 TWAEV

Canada - Quebec - Skin Designations
absorbed through the skin

United Kingdom - Occupational Exposure Standards - TWAs
5 ppm TWA; 23 mg/m3 TWA

United Kingdom - Occupational Exposure Standards - STELs
20 ppm STEL; 95 mg/m3 STEL

United Kingdom - Occupational Exposure Standards - Notes
can be absorbed through skin

Israel - Time Weighted Averages
5 ppm TWA; 24 mg/m3 TWA

Israel - Action Levels
2.5 ppm AL; 12 mg/m3 AL

Mexico - Instruction No. 10 - TWAs
20 ppm TWA; 95 mg/m3 TWA

Mexico - Instruction No. 10 - Skin designation
skin - potential for cutaneous absorption

STATE LISTS

California - Exposure Limits - PELs
5 ppm PEL; 23 mg/m3 PEL

California - Exposure Limits - Skin Notation
material may be absorbed through the skin, eyes or mucous membrane

California - Directors List of Hazardous Substances (8 CCR 339)
[present]

Florida Hazardous Substance List
[present]

Massachusetts Right To Know List
[present]

Minnesota Hazardous Substance List
skin

NJ Right to Know List (Total)
sn 1338

Pennsylvania Right to Know List
[present]

ETHYL ORTHOFORMATE   122-51-0

HEALTH AND SAFETY LISTS

NFPA - Hazard Identification Ratings
health-0; flammability-1; reactivity-0

ETHYL OXALATE   95-92-1

HEALTH AND SAFETY LISTS

NFPA - Flash Points
flash point = 50 degrees F (10 degrees C)

NFPA - Hazard Identification Ratings
health-2; flammability-3; reactivity-4

STATE LISTS

Florida Hazardous Substance List
[present]

Massachusetts Right To Know List
[present]

NJ Right to Know List (Total)
sn 0898

NJ Special Hazardous Substances
(flammable - third degree; reactive - fourth degree)

Pennsylvania Right to Know List
[present]

ETHYL PERCHLORATE   22750-93-2

HEALTH AND SAFETY LISTS

U.S. DOT - Substances From 49 CFR 172.101
regulated by DOT (UN1194)

U.S. DOT - Hazard Classes
DOT hazard class = 3

NFPA - Flash Points
flash point = -31 degrees F (-35 degrees C)

NFPA - Hazard Identification Ratings
health-3; flammability-4; reactivity-4

OSHA - List of Highly Hazardous Chemicals
threshhold quantity = 5000 pounds

ENVIRONMENTAL LISTS

CAA - Flammable Substances for Accidental Release Prevention
threshold quantity = 10,000 lbs

STATE LISTS

Florida Hazardous Substance List
[present]

Massachusetts Right To Know List
[present]

NJ Right to Know List (Total)
sn 0899

NJ Special Hazardous Substances
(flammable - fourth degree; reactive - fourth degree)

Pennsylvania Right to Know List
[present]

O-ETHYLPHENOL   90-00-6

INTERNATIONAL LISTS

German (DFG) - MAK Values
0.5 ppm MAK; 0.6 mg/m3 MAK

ETHYL PHENYL KETONE   93-55-0

INTERNATIONAL LISTS

Canada - WHMIS: Ingredient Disclosure
1% item 746 (1275)

ETHYL PHENYLACETATE   101-97-3

HEALTH AND SAFETY LISTS

NFPA - Hazard Identification Ratings
health-0; flammability-2; reactivity-0

ETHYL PHENYL DICHLOROSILANE   1125-27-5

HEALTH AND SAFETY LISTS

NFPA - Hazard Identification Ratings
health-0; flammability-2; reactivity-0

ETHYLPHOSPHONOTHIOIC-DICHLORIDE   993-43-1

HEALTH AND SAFETY LISTS

U.S. DOT - Substances From 49 CFR 172.101
regulated by DOT (UN2524)

U.S. DOT - Hazard Classes
DOT hazard class = 3

NFPA - Flash Points
flash point = 86 degrees F (30 degrees C)

NFPA - Hazard Identification Ratings
health-0; flammability-3; reactivity-0

NIOSH - Selected LD50s and LC50s
Oral, rat: LD50 = 2920 mg/kg Skin, rabbit: LD50 = 20 gm/kg

ENVIRONMENTAL LISTS

List of Pesticide Product Inert Ingredients
[present]

INTERNATIONAL LISTS

Canada - WHMIS: Ingredient Disclosure
1% item 747 (1289)

STATE LISTS

Florida Hazardous Substance List
[present]

Massachusetts Right To Know List
[present]

NJ Right to Know List (Total)
sn 0900

Pennsylvania Right to Know List
[present]

ETHYL PHOSPHONOUS DICHLORIDE   1498-40-4

HEALTH AND SAFETY LISTS

U.S. DOT - Substances From 49 CFR 172.101
regulated by DOT (UN2525)

U.S. DOT - Hazard Classes
DOT hazard class = 6.1

NFPA - Flash Points
flash point = 168 degrees F (76 degrees C)

NFPA - Hazard Identification Ratings
health-0; flammability-2; reactivity-0

NIOSH - Selected LD50s and LC50s
Oral, rat: LD50 = 400 mg/kg

INTERNATIONAL LISTS
Canada - WHMIS: Ingredient Disclosure
1% item 748 (1293)
STATE LISTS
NJ Right to Know List (Total)
sn 0901

ETHYL PHOSPHORODICHLORIDATE   1498-51-7
HEALTH AND SAFETY LISTS
U.S. DOT - Hazard Classes
Forbidden from transport by the DOT

5-ETHYL-2-PICOLINE   RR-01319-4
HEALTH AND SAFETY LISTS
NIOSH - Selected LD50s and LC50s
Oral, mouse: LD50 = 600 mg/kg

1-ETHYL PIPERDINE   766-09-6
HEALTH AND SAFETY LISTS
NFPA - Flash Points
flash point = 210 degrees F (99 degrees C)
NFPA - Hazard Identification Ratings
flammability-1;   reactivity-0

ETHYL PROPENYL ETHER   RR-01171-2
HEALTH AND SAFETY LISTS
NFPA - Flash Points
flash point = 210 degrees F (99 degrees C)
NFPA - Hazard Identification Ratings
health-0;   flammability-1
NIOSH - Selected LD50s and LC50s
Oral, rat: LD50 = 3300 mg/kg

ETHYL PROPIONATE   105-37-3
HEALTH AND SAFETY LISTS
U.S. DOT - Substances From 49 CFR 172.101
regulated by DOT (UN2435)
U.S. DOT - Hazard Classes
DOT hazard class = 8
INTERNATIONAL LISTS
Canada - WHMIS: Ingredient Disclosure
1% item 749 (873)
STATE LISTS
NJ Right to Know List (Total)
sn 0902
NJ Special Hazardous Substances
(corrosive)

2-ETHYL-3-PROPYLACROLEIN   645-62-5
HEALTH AND SAFETY LISTS
U.S. DOT - Substances From 49 CFR 172.101
regulated by DOT (NA2927)
U.S. DOT - Hazard Classes
DOT hazard class = 6.1
U.S. DOT - Substances Which Are Poisonous by Inhalation
liquid hazardous material poisonous by inhalation (anhydrous form)
(NA2927)
STATE LISTS
NJ Right to Know List (Total)
sn 0903

2-ETHYL-3-PROPYLACRYLIC ACID   5309-52-
HEALTH AND SAFETY LISTS
U.S. DOT - Substances From 49 CFR 172.101
regulated by DOT (NA2845)
U.S. DOT - Hazard Classes
DOT hazard class = 6.1
U.S. DOT - Substances Which Are Poisonous by Inhalation
liquid hazardous material poisonous by inhalation (anhydrous form
(NA2845)
INTERNATIONAL LISTS
Canada - WHMIS: Ingredient Disclosure
1% item 750 (499)
STATE LISTS
NJ Right to Know List (Total)
sn 0904
NJ Special Hazardous Substances
(corrosive)

ETHYL PROPYL ETHER   628-32-0
HEALTH AND SAFETY LISTS
U.S. DOT - Substances From 49 CFR 172.101
regulated by DOT (NA2927)
U.S. DOT - Hazard Classes
DOT hazard class = 6.1
U.S. DOT - Substances Which Are Poisonous by Inhalation
liquid hazardous material poisonous by inhalation (NA2927)
ENVIRONMENTAL LISTS
TSCA - Code of Federal Regulations Citations
40 CFR 712.30(x);   40 CFR 716.120(d)
TSCA - PAIR - Reporting List
Reporting Date: December 27, 1990
TSCA - Health and Safety Reporting List
Effective Date: October 29, 1990;   Sunset Date: November 9, 1993
INTERNATIONAL LISTS
Canada - WHMIS: Ingredient Disclosure
1% item 751 (1409)
STATE LISTS
NJ Right to Know List (Total)
sn 0905
NJ Special Hazardous Substances
(corrosive)

N-ETHYLPSEUDOEPHEDRINE   RR-01790-3
HEALTH AND SAFETY LISTS
U.S. DOT - Appendix B - Marine Pollutants
DOT regulated marine pollutant

N-ETHYLPSEUDOEPHEDRINE SALTS, OPTICAL   RR-01788-9
ISOMERS, AND SALTS OFOPTICAL ISOMERS
HEALTH AND SAFETY LISTS
U.S. DOT - Substances From 49 CFR 172.101
regulated by DOT (UN2386)
U.S. DOT - Hazard Classes
DOT hazard class = 3
STATE LISTS
NJ Right to Know List (Total)
sn 0906

ETHYL SALICYLATE   118-61-6
HEALTH AND SAFETY LISTS
NFPA - Flash Points
flash point > 19 degrees F (> -7 degrees C)

NFPA - Hazard Identification Ratings
health-2; flammability-3; reactivity-1

ETHYL SELENAC   5456-28-0

HEALTH AND SAFETY LISTS

U.S. DOT - Substances From 49 CFR 172.101
regulated by DOT (UN1195)

U.S. DOT - Hazard Classes
DOT hazard class = 3

NFPA - Flash Points
flash point = 54 degrees F (12 degrees C)

NFPA - Hazard Identification Ratings
flammability-3; reactivity-0

NIOSH - Selected LD50s and LC50s
Oral, rabbit: LD50 = 3500 mg/kg

INTERNATIONAL LISTS

Canada - WHMIS: Ingredient Disclosure
1% item 752 (1448)

STATE LISTS

Florida Hazardous Substance List
[present]

Massachusetts Right To Know List
[present]

NJ Right to Know List (Total)
sn 0907

NJ Special Hazardous Substances
(flammable - third degree)

Pennsylvania Right to Know List
[present]

ETHYL SILICATE   78-10-4

HEALTH AND SAFETY LISTS

U.S. DOT - Appendix B - Marine Pollutants
DOT regulated marine pollutant

NFPA - Flash Points
flash point = 155 degrees F (68 degrees C)

NFPA - Hazard Identification Ratings
health-2; flammability-2; reactivity-1

NIOSH - Selected LD50s and LC50s
Oral, rat: LD50 = 3000 mg/kg

STATE LISTS

Florida Hazardous Substance List
[present]

Massachusetts Right To Know List
[present]

Pennsylvania Right to Know List
[present]

ETHYLSULPHURIC ACID   540-82-9

HEALTH AND SAFETY LISTS

NFPA - Flash Points
flash point = 330 degrees F (166 degrees C)

NFPA - Hazard Identification Ratings
health-2; flammability-1; reactivity-1

STATE LISTS

Florida Hazardous Substance List
[present]

Massachusetts Right To Know List
[present]

Pennsylvania Right to Know List
[present]

ETHYL TELLURAC   20941-65-5

HEALTH AND SAFETY LISTS

U.S. DOT - Substances From 49 CFR 172.101
regulated by DOT (UN2615)

U.S. DOT - Hazard Classes
DOT hazard class = 3

NFPA - Flash Points
flash point < -4 degrees F (-20 degrees C)

NFPA - Hazard Identification Ratings
health-1; flammability-3; reactivity-0

STATE LISTS

Florida Hazardous Substance List
[present]

Massachusetts Right To Know List
[present]

NJ Right to Know List (Total)
sn 0908

NJ Special Hazardous Substances
(flammable - third degree)

Pennsylvania Right to Know List
[present]

ETHYL THIOCYANATE   542-90-5

HEALTH AND SAFETY LISTS

FDA - Controlled Substances Act - Precursor chemicals
Threshold by base weight = 1 kilogram

2-(ETHYLTHIO)ETHANOL   110-77-0

HEALTH AND SAFETY LISTS

FDA - Controlled Substances Act - Precursor chemicals
Threshold by base weight = 1 kilogram

M-ETHYLTOLUENE   620-14-4

HEALTH AND SAFETY LISTS

NIOSH - Selected LD50s and LC50s
Oral, rat: LD50 = 1320 mg/kg

P-ETHYLTOLUENE   622-96-8

HEALTH AND SAFETY LISTS

IARC - Group 3 (not classifiable)
[present]

ETHYL P-TOLUENE SULFONAMIDE   80-39-7

HEALTH AND SAFETY LISTS

ACGIH 1995 - Time Weighted Averages
10 ppm TWA; 85 mg/m3 TWA

AIHA - Odor Threshold Values
geometric mean air odor threshold = 3.6 ppm (detectable); 5.0 ppm (recognizable)

U.S. DOT - Substances From 49 CFR 172.101
regulated by DOT (UN1292)

U.S. DOT - Hazard Classes
DOT hazard class = 3

NFPA - Flash Points
flash point = 125 degrees F (52 degrees C)

NFPA - Hazard Identification Ratings
health-2; flammability-2; reactivity-0

NIOSH - Selected LD50s and LC50s
Oral, rat: LD50 = 6270 mg/kg Skin, rabbit: LD50 = 5878 mg/kg

NIOSH 1990 - Pocket Guide - RELs
10 ppm TWA; 85 mg/m3 TWA

NIOSH 1990 - Pocket Guide - IDLHs
1000 ppm IDLH

NIOSH 1990 - Pocket Guide - Target organs
respiratory system, liver, kidneys, blood, skin
OSHA - Vacated PELs - Time Weighted Averages
10 ppm TWA; 85 mg/m3 TWA
OSHA - Final PELs - Time Weighted Averages
100 ppm TWA; 850 mg/m3 TWA

INTERNATIONAL LISTS

Australian Exposure Standards - Time Weighted Averages
10 ppm TWA; 85 mg/m3 TWA
Canada - WHMIS: Ingredient Disclosure
1% item 753 (1485)
Canada - Alberta - 8 Hour Occupational Exposure Limit
10 ppm TWA; 85 mg/m3 TWA
Canada - Alberta - 15 Minute Occupational Exposure Limit
30 ppm STEL; 256 mg/m3 STEL
Canada - British Columbia - 8 Hour Exposure Limits
100 ppm TWA; 850 mg/m3 TWA
Canada - Ontario - OHSA - TWAEVs
10 ppm TWAEV; 85 mg/m3 TWAEV
Canada - Quebec - Time-Weighted Average Exposure Values
10 ppm TWAEV; 85 mg/m3 TWAEV
United Kingdom - Occupational Exposure Standards - TWAs
10 ppm TWA; 85 mg/m3 TWA
United Kingdom - Occupational Exposure Standards - STELs
30 ppm STEL; 255 mg/m3 STEL
German (DFG) - MAK Values
20 ppm MAK; 170 mg/m3 MAK
German (DFG) - Peak Limitations
2 x normal MAK (5 min momentary value); don't exceed 8 times during shift
Israel - Time Weighted Averages
10 ppm TWA; 85 mg/m3 TWA
Israel - Action Levels
5 ppm AL; 42.5 mg/m3 AL
Mexico - Instruction No. 10 - TWAs
10 ppm TWA; 85 mg/m3 TWA
Mexico - Instruction No. 10 - STELs
30 ppm STEL; 255 mg/m3 STEL

STATE LISTS

California - Exposure Limits - PELs
10 ppm PEL; 85 mg/m3 PEL
California - Directors List of Hazardous Substances (8 CCR 339)
[present]
Florida Hazardous Substance List
[present]
Massachusetts Right To Know List
[present]
Minnesota Hazardous Substance List
[present]
NJ Right to Know List (Total)
sn 0909
Pennsylvania Right to Know List
[present]

N-ETHYLTOLUENESULFONAMIDE   8047-99-2

HEALTH AND SAFETY LISTS

U.S. DOT - Substances From 49 CFR 172.101
regulated by DOT (UN2571)
U.S. DOT - Hazard Classes
DOT hazard class = 8

INTERNATIONAL LISTS

Canada - WHMIS: Ingredient Disclosure
1% item 754 (86)

STATE LISTS

NJ Right to Know List (Total)
sn 0910
NJ Special Hazardous Substances
(corrosive)

ETHYL P-TOLUENE SULFONATE   80-40-

HEALTH AND SAFETY LISTS

IARC - Group 3 (not classifiable)
[present]
NTP Chemical Status Reports - Testing Status and NTIS Number
Technical reports printed (PB298513/AS)
NTP Chemical Status Reports - Evidence of Carcinogenicity
male rat-equivocal; female rat-negative; male mice-equivocal; fe male mice-equivocal

STATE LISTS

Massachusetts Right To Know List
carcinogen; extraordinarily hazardous

N-ETHYL-O-TOLUIDINE   94-68-

ENVIRONMENTAL LISTS

CERCLA/SARA - Section 302 Extremely Hazardous Substances an TPQs
TPQ = 10,000 pounds

STATE LISTS

Florida Hazardous Substance List
effective March 13, 1992
Massachusetts Right To Know List
extraordinarily hazardous
NJ Right to Know List (Total)
sn 2402
Pennsylvania Right to Know List
environmental hazard

PROPOSED REGULATIONS

CERCLA/SARA - Proposed Hazardous Substance Additions
proposed RQ = 1 pound (.454 kg)
CERCLA/SARA - 1989 Proposed RQ Adjustments
proposed RQ = 1000 pounds (454 kg)

N-ETHYL-M-TOLUIDINE   102-27-2

INTERNATIONAL LISTS

Canada - WHMIS: Ingredient Disclosure
1% item 755 (875)

N-ETHYL-P-TOLUIDINE   622-57-1
SEE ALSO:
AROMATIC C9 FRACTION FROM PETROLEUM REFINING

HEALTH AND SAFETY LISTS

NFPA - Hazard Identification Ratings
flammability-2; reactivity-0

ENVIRONMENTAL LISTS

TSCA - Code of Federal Regulations Citations
40 CFR 716.120(a),(b); 40 CFR 716.1203
TSCA - Health and Safety Reporting List
Effective Date: April 29, 1983

N-ETHYLTOLUIDINES   RR-01356-9
SEE ALSO:
AROMATIC C9 FRACTION FROM PETROLEUM REFINING

HEALTH AND SAFETY LISTS

NFPA - Hazard Identification Ratings
flammability-2; reactivity-0

ENVIRONMENTAL LISTS
 TSCA - Code of Federal Regulations Citations
  40 CFR 716.120(a),(b);   40 CFR 716.1203
 TSCA - Health and Safety Reporting List
  Effective Date: April 29, 1983

ETHYL VINYL KETONE   1629-58-9
 HEALTH AND SAFETY LISTS
  NFPA - Flash Points
   flash point = 260 degrees F (127 degrees C)
  NFPA - Hazard Identification Ratings
   flammability-1; reactivity-0
 ENVIRONMENTAL LISTS
  List of Pesticide Product Inert Ingredients
   [present]

ETHYNE, ETHOXY-   927-80-0
 ENVIRONMENTAL LISTS
  List of Pesticide Product Inert Ingredients
   [present]

ETHYNODIOL DIACETATE   297-76-7
 HEALTH AND SAFETY LISTS
  NFPA - Flash Points
   flash point = 316 degrees F (158 degrees C)
  NFPA - Hazard Identification Ratings
   flammability-1; reactivity-0

ETHYNODIOL DIACETATE AND OESTROGENS   RR-00100-3
 INTERNATIONAL LISTS
  Canada - WHMIS: Ingredient Disclosure
   1% item 757 (877)

ETIDRONATE DISODIUM   7414-83-7
 INTERNATIONAL LISTS
  Canada - WHMIS: Ingredient Disclosure
   1% item 756 (876)

ETOPOSIDE   33419-42-0
 INTERNATIONAL LISTS
  Canada - WHMIS: Ingredient Disclosure
   1% item 758 (878)
 STATE LISTS
  NJ Right to Know List (Total)
   sn 0911

ETRETINATE   54350-48-0
 HEALTH AND SAFETY LISTS
  U.S. DOT - Substances From 49 CFR 172.101
   regulated by DOT (UN2754)
  U.S. DOT - Hazard Classes
   DOT hazard class = 6.1

EUGENOL   97-53-0
 HEALTH AND SAFETY LISTS
  NTP Chemical Status Reports - Testing Status and NTIS Number
   Project leader assigned/study in design

EUROPIUM 145   14981-86-3
 HEALTH AND SAFETY LISTS
  NFPA - Flash Points
   flash point < 20 degrees F (-7 degrees C)

NFPA - Hazard Identification Ratings
 health-2; flammability-3; reactivity-1
STATE LISTS
 Florida Hazardous Substance List
  [present]
 Massachusetts Right To Know List
  [present]
 Pennsylvania Right to Know List
  [present]

EUROPIUM 146   14907-88-1
 HEALTH AND SAFETY LISTS
  IARC - Group Unspecified
   [present] (Listed under 'Progestins')

EUROPIUM 147   14191-78-7
 HEALTH AND SAFETY LISTS
  IARC - Group Unspecified
   [present] (Listed under 'Combined oral contraceptives')

EUROPIUM 148   15840-15-0
 HEALTH AND SAFETY LISTS
  NIOSH - Selected LD50s and LC50s
   Oral, rat: LD50 = 1340 mg/kg

EUROPIUM 149   14907-89-2
 STATE LISTS
  California - Air Bill 2588 Appendix A-II
   9/90
  California - Prop. 65 - Developmental Toxicity
   developmental toxicity - initial date 7/1/90

EUROPIUM 150   15840-16-1
 STATE LISTS
  California - Air Bill 2588 Appendix A-II
   [present]
  California - Prop. 65 - Developmental Toxicity
   developmental toxicity - initial date 7/1/87

EUROPIUM 152   14683-23-9
 HEALTH AND SAFETY LISTS
  IARC - Group 3 (not classifiable)
   [present]
  NTP Chemical Status Reports - Testing Status and NTIS Number
   Technical reports printed (PB84186402)
  NTP Chemical Status Reports - Evidence of Carcinogenicity
   male rat-negative; female rat-negative; male mice-equivocal; female mice-equivocal
 STATE LISTS
  California - Directors List of Hazardous Substances (8 CCR 339)
   [present]

EUROPIUM 152M   RR-00463-7
 HEALTH AND SAFETY LISTS
  U.S. DOT - Appendix A Table 2 - Radionuclides
   final RQ = 10 curies (3.7E 11 Bq)
 ENVIRONMENTAL LISTS
  CERCLA/SARA List of Radionuclides (Appendix B) and Their Reportable Quantities
   final RQ = 10 curies (3.7E 11 Bq)

EUROPIUM 154   15585-10-1
HEALTH AND SAFETY LISTS
U.S. DOT - Appendix A Table 2 - Radionuclides
final RQ = 10 curies (3.7E 11 Bq)
ENVIRONMENTAL LISTS
CERCLA/SARA List of Radionuclides (Appendix B) and Their Reportable Quantities
final RQ = 10 curies (3.7E 11 Bq)

EUROPIUM 155   14391-16-3
HEALTH AND SAFETY LISTS
U.S. DOT - Appendix A Table 2 - Radionuclides
final RQ = 10 curies (3.7E 11 Bq)
ENVIRONMENTAL LISTS
CERCLA/SARA List of Radionuclides (Appendix B) and Their Reportable Quantities
final RQ = 10 curies (3.7E 11 Bq)

EUROPIUM 156   14280-35-4
HEALTH AND SAFETY LISTS
U.S. DOT - Appendix A Table 2 - Radionuclides
final RQ = 10 curies (3.7E 11 Bq)
ENVIRONMENTAL LISTS
CERCLA/SARA List of Radionuclides (Appendix B) and Their Reportable Quantities
final RQ = 10 curies (3.7E 11 Bq)

EUROPIUM 157   14280-36-5
HEALTH AND SAFETY LISTS
U.S. DOT - Appendix A Table 2 - Radionuclides
final RQ = 100 curies (3.7E 12 Bq)
ENVIRONMENTAL LISTS
CERCLA/SARA List of Radionuclides (Appendix B) and Their Reportable Quantities
final RQ = 100 curies (3.7E 12 Bq)

EUROPIUM 158   14041-40-8
HEALTH AND SAFETY LISTS
U.S. DOT - Appendix A Table 2 - Radionuclides
12.6 hour half-life: Final RQ = 1000 curies (3.7E 13 Bq);   34.2 year half-life: Final RQ = 10 curies (3.7E 11 Bq)
ENVIRONMENTAL LISTS
CERCLA/SARA List of Radionuclides (Appendix B) and Their Reportable Quantities
12.6 hour half-life: Final RQ = 1000 curies (3.7E 13 Bq);   34.2 year half-life: Final RQ = 10 curies (3.7E 11 Bq)

EUROPIUM CHLORIDE   10025-76-0
HEALTH AND SAFETY LISTS
U.S. DOT - Appendix A Table 2 - Radionuclides
final RQ = 10 curies (3.7E 11 Bq)
ENVIRONMENTAL LISTS
CERCLA/SARA List of Radionuclides (Appendix B) and Their Reportable Quantities
final RQ = 10 curies (3.7E 11 Bq)

EXT D & C RED NO. 3   6252-76-2
HEALTH AND SAFETY LISTS
U.S. DOT - Appendix A Table 2 - Radionuclides
final RQ = 100 curies (3.7E 12 Bq)

ENVIRONMENTAL LISTS
CERCLA/SARA List of Radionuclides (Appendix B) and Their Reportable Quantities
final RQ = 100 curies (3.7E 12 Bq)

EXTRACTS (PETROLEUM), MIDDLE DISTILLATE   64742-06-
SOLVENT
HEALTH AND SAFETY LISTS
U.S. DOT - Appendix A Table 2 - Radionuclides
final RQ = 10 curies (3.7E 11 Bq)
ENVIRONMENTAL LISTS
CERCLA/SARA List of Radionuclides (Appendix B) and Their Reportable Quantities
final RQ = 10 curies (3.7E 11 Bq)

F001-HAZARDOUS WASTES   RR-00635-
HEALTH AND SAFETY LISTS
U.S. DOT - Appendix A Table 2 - Radionuclides
final RQ = 10 curies (3.7E 11 Bq)
ENVIRONMENTAL LISTS
CERCLA/SARA List of Radionuclides (Appendix B) and Their Reportable Quantities
final RQ = 10 curies (3.7E 11 Bq)

F002-HAZARDOUS WASTES   RR-00636-0
HEALTH AND SAFETY LISTS
U.S. DOT - Appendix A Table 2 - Radionuclides
final RQ = 10 curies (3.7E 11 Bq)
ENVIRONMENTAL LISTS
CERCLA/SARA List of Radionuclides (Appendix B) and Their Reportable Quantities
final RQ = 10 curies (3.7E 11 Bq)

F003-HAZARDOUS WASTES   RR-00637-1
HEALTH AND SAFETY LISTS
U.S. DOT - Appendix A Table 2 - Radionuclides
final RQ = 10 curies (3.7E 11 Bq)
ENVIRONMENTAL LISTS
CERCLA/SARA List of Radionuclides (Appendix B) and Their Reportable Quantities
final RQ = 10 curies (3.7E 11 Bq)

F004-HAZARDOUS WASTES   RR-00638-2
HEALTH AND SAFETY LISTS
U.S. DOT - Appendix A Table 2 - Radionuclides
final RQ = 1000 curies (3.7E 13 Bq)
ENVIRONMENTAL LISTS
CERCLA/SARA List of Radionuclides (Appendix B) and Their Reportable Quantities
final RQ = 1000 curies (3.7E 13 Bq)

F005-HAZARDOUS WASTES   RR-00639-3
HEALTH AND SAFETY LISTS
NIOSH - Selected LD50s and LC50s
Oral, mouse: LD50 = 3527 mg/kg

F006-HAZARDOUS WASTES   RR-00640-6
ENVIRONMENTAL LISTS
List of Pesticide Product Inert Ingredients
[present]

F007-HAZARDOUS WASTES   RR-00641-7

HEALTH AND SAFETY LISTS

NFPA - Flash Points
flash point = 275 degrees F (135 degrees C)

NFPA - Hazard Identification Ratings
health-0; flammability-2; reactivity-0

F008-HAZARDOUS WASTES   RR-00642-8

HEALTH AND SAFETY LISTS

U.S. DOT - Appendix A Table 1 - Hazardous Substances
final RQ = 1 pound (0.454 kg) (the RQ is subject to change when the assessment of potential carcinogenicity is completed)

ENVIRONMENTAL LISTS

CERCLA/SARA - Hazardous Substances and their Reportable Quantities
final RQ = 1 pound (0.454 kg) (the RQ is subject to change when the assessment of potential carcinogenicity is completed)

RCRA - F Series Wastes
Toxic waste

RCRA - Substances Banned From Land Disposal
[present]

F009-HAZARDOUS WASTES   RR-00643-9

HEALTH AND SAFETY LISTS

U.S. DOT - Appendix A Table 1 - Hazardous Substances
final RQ = 10 pounds (4.54 kg)

ENVIRONMENTAL LISTS

CERCLA/SARA - Hazardous Substances and their Reportable Quantities
final RQ = 10 pounds (4.54 kg)

RCRA - F Series Wastes
Toxic waste

RCRA - Substances Banned From Land Disposal
[present]

F010-HAZARDOUS WASTES   RR-00644-0

HEALTH AND SAFETY LISTS

U.S. DOT - Appendix A Table 1 - Hazardous Substances
final RQ = 100 pounds (45.4 kg)

ENVIRONMENTAL LISTS

CERCLA/SARA - Hazardous Substances and their Reportable Quantities
final RQ = 100 pounds (45.4 kg)

RCRA - F Series Wastes
Ignitable waste

RCRA - Substances Banned From Land Disposal
[present]

F011-HAZARDOUS WASTES   RR-00645-1

HEALTH AND SAFETY LISTS

U.S. DOT - Appendix A Table 1 - Hazardous Substances
final RQ = 1000 pounds (454 kg)

ENVIRONMENTAL LISTS

CERCLA/SARA - Hazardous Substances and their Reportable Quantities
final RQ = 1000 pounds (454 kg)

RCRA - F Series Wastes
Toxic waste

RCRA - Substances Banned From Land Disposal
[present]

F012-HAZARDOUS WASTES   RR-00646-2

HEALTH AND SAFETY LISTS

U.S. DOT - Appendix A Table 1 - Hazardous Substances
final RQ = 100 pounds (45.4 kg)

ENVIRONMENTAL LISTS

CERCLA/SARA - Hazardous Substances and their Reportable Quantities
final RQ = 100 pounds (45.4 kg)

RCRA - F Series Wastes
Ignitable waste; Toxic waste

RCRA - Substances Banned From Land Disposal
[present]

F019-HAZARDOUS WASTES   RR-00647-3

HEALTH AND SAFETY LISTS

U.S. DOT - Appendix A Table 1 - Hazardous Substances
final RQ = 10 pounds (4.54 kg)

ENVIRONMENTAL LISTS

CERCLA/SARA - Hazardous Substances and their Reportable Quantities
final RQ = 10 pounds (4.54 kg)

RCRA - F Series Wastes
Toxic Waste

RCRA - Substances Banned From Land Disposal
[present]

F020-HAZARDOUS WASTES   RR-00648-4

HEALTH AND SAFETY LISTS

U.S. DOT - Appendix A Table 1 - Hazardous Substances
final RQ = 10 pounds (4.54 kg)

ENVIRONMENTAL LISTS

CERCLA/SARA - Hazardous Substances and their Reportable Quantities
final RQ = 10 pounds (4.54 kg)

RCRA - F Series Wastes
Reactive waste; Toxic waste

RCRA - Substances Banned From Land Disposal
[present] (except if it is underground injected)

F021-HAZARDOUS WASTES   RR-00649-5

HEALTH AND SAFETY LISTS

U.S. DOT - Appendix A Table 1 - Hazardous Substances
final RQ = 10 pounds (4.54 kg)

ENVIRONMENTAL LISTS

CERCLA/SARA - Hazardous Substances and their Reportable Quantities
final RQ = 10 pounds (4.54 kg)

RCRA - F Series Wastes
Reactive waste; Toxic waste

RCRA - Substances Banned From Land Disposal
[present]

F022-HAZARDOUS WASTES   RR-00650-8

HEALTH AND SAFETY LISTS

U.S. DOT - Appendix A Table 1 - Hazardous Substances
final RQ = 10 pounds (4.54 kg)

ENVIRONMENTAL LISTS

CERCLA/SARA - Hazardous Substances and their Reportable Quantities
final RQ = 10 pounds (4.54 kg)

RCRA - F Series Wastes
Reactive waste; Toxic waste

RCRA - Substances Banned From Land Disposal
[present]

F023-HAZARDOUS WASTES   RR-00651-9

HEALTH AND SAFETY LISTS

U.S. DOT - Appendix A Table 1 - Hazardous Substances
final RQ = 10 pounds (4.54 kg)

ENVIRONMENTAL LISTS

CERCLA/SARA - Hazardous Substances and their Reportable Quantities
final RQ = 10 pounds (4.54 kg)

RCRA - F Series Wastes
Reactive waste; Toxic waste

RCRA - Substances Banned From Land Disposal
[present]

F024-HAZARDOUS WASTES   RR-00652-0

HEALTH AND SAFETY LISTS

U.S. DOT - Appendix A Table 1 - Hazardous Substances
final RQ = 10 pounds (4.54 kg)

ENVIRONMENTAL LISTS

CERCLA/SARA - Hazardous Substances and their Reportable Quantities
final RQ = 10 pounds (4.54 kg)

RCRA - F Series Wastes
Reactive waste; Toxic waste

RCRA - Substances Banned From Land Disposal
[present]

F025-HAZARDOUS WASTES   RR-00755-6

HEALTH AND SAFETY LISTS

U.S. DOT - Appendix A Table 1 - Hazardous Substances
final RQ = 10 pounds (4.54 kg)

ENVIRONMENTAL LISTS

CERCLA/SARA - Hazardous Substances and their Reportable Quantities
final RQ = 10 pounds (4.54 kg)

RCRA - F Series Wastes
Toxic waste

RCRA - Substances Banned From Land Disposal
[present]

F026-HAZARDOUS WASTES   RR-00653-1

HEALTH AND SAFETY LISTS

U.S. DOT - Appendix A Table 1 - Hazardous Substances
final RQ = 10 pounds (4.54 kg)

ENVIRONMENTAL LISTS

CERCLA/SARA - Hazardous Substances and their Reportable Quantities
final RQ = 10 pounds (4.54 kg)

RCRA - F Series Wastes
Toxic waste

RCRA - Substances Banned From Land Disposal
[present]

F027-HAZARDOUS WASTES   RR-00654-2

HEALTH AND SAFETY LISTS

U.S. DOT - Appendix A Table 1 - Hazardous Substances
final RQ = 1 pound (0.454 kg)

ENVIRONMENTAL LISTS

CERCLA/SARA - Hazardous Substances and their Reportable Quantities
final RQ = 1 pound (0.454 kg)

RCRA - F Series Wastes
Acute hazardous waste

RCRA - Substances Banned From Land Disposal
[present]

F028-HAZARDOUS WASTES   RR-00655-

HEALTH AND SAFETY LISTS

U.S. DOT - Appendix A Table 1 - Hazardous Substances
final RQ = 1 pound (0.454 kg)

ENVIRONMENTAL LISTS

CERCLA/SARA - Hazardous Substances and their Reportable Quantities
final RQ = 1 pound (0.454 kg)

RCRA - F Series Wastes
Acute hazardous waste

RCRA - Substances Banned From Land Disposal
[present]

F032-HAZARDOUS WASTES   RR-00106-9

HEALTH AND SAFETY LISTS

U.S. DOT - Appendix A Table 1 - Hazardous Substances
final RQ = 1 pound (0.454 kg)

ENVIRONMENTAL LISTS

CERCLA/SARA - Hazardous Substances and their Reportable Quantities
final RQ = 1 pound (0.454 kg)

RCRA - F Series Wastes
Acute hazardous waste

RCRA - Substances Banned From Land Disposal
[present]

F034-HAZARDOUS WASTES   RR-00105-8

HEALTH AND SAFETY LISTS

U.S. DOT - Appendix A Table 1 - Hazardous Substances
final RQ = 1 pound (0.454 kg)

ENVIRONMENTAL LISTS

CERCLA/SARA - Hazardous Substances and their Reportable Quantities
final RQ = 1 pound (0.454 kg)

RCRA - F Series Wastes
Acute hazardous waste

RCRA - Substances Banned From Land Disposal
[present]

F035-HAZARDOUS WASTES   RR-00104-7

HEALTH AND SAFETY LISTS

U.S. DOT - Appendix A Table 1 - Hazardous Substances
final RQ = 1 pound (0.454 kg)

ENVIRONMENTAL LISTS

CERCLA/SARA - Hazardous Substances and their Reportable Quantities
final RQ = 1 pound (0.454 kg)

RCRA - F Series Wastes
Toxic waste

RCRA - Substances Banned From Land Disposal
[present]

F037-HAZARDOUS WASTES   RR-00292-6

ENVIRONMENTAL LISTS

RCRA - F Series Wastes
Toxic waste

F038-HAZARDOUS WASTES   RR-00293-7

HEALTH AND SAFETY LISTS

U.S. DOT - Appendix A Table 1 - Hazardous Substances
final RQ = 1 pound (0.454 kg)

ENVIRONMENTAL LISTS

CERCLA/SARA - Hazardous Substances and their Reportable Quantities
final RQ = 1 pound (0.454 kg)

RCRA - F Series Wastes
Acute hazardous waste

RCRA - Substances Banned From Land Disposal
[present]

F039-HAZARDOUS WASTES   RR-00291-5

HEALTH AND SAFETY LISTS

U.S. DOT - Appendix A Table 1 - Hazardous Substances
final RQ = 1 pound (0.454 kg)

ENVIRONMENTAL LISTS

CERCLA/SARA - Hazardous Substances and their Reportable Quantities
final RQ = 1 pound (0.454 kg)

RCRA - F Series Wastes
Acute hazardous waste

RCRA - Substances Banned From Land Disposal
[present]

FAMPHUR   52-85-7

HEALTH AND SAFETY LISTS

U.S. DOT - Appendix A Table 1 - Hazardous Substances
final RQ = 1 pound (0.454 kg)

ENVIRONMENTAL LISTS

CERCLA/SARA - Hazardous Substances and their Reportable Quantities
final RQ = 1 pound (0.454 kg)

RCRA - F Series Wastes
Toxic waste

RCRA - Substances Banned From Land Disposal
[present]

FAST GREEN FCF   2353-45-9

HEALTH AND SAFETY LISTS

U.S. DOT - Appendix A Table 1 - Hazardous Substances
final RQ = 1 pound (0.454 kg)

ENVIRONMENTAL LISTS

CERCLA/SARA - Hazardous Substances and their Reportable Quantities
final RQ = 1 pound (0.454 kg)

RCRA - F Series Wastes
Toxic waste

FATTY ACID AMINE CONDENSATE, POLYCAR-   RR-01225-9
BOXYLIC ACID SALTS

HEALTH AND SAFETY LISTS

U.S. DOT - Appendix A Table 1 - Hazardous Substances
final RQ = 1 pound (0.454 kg)

ENVIRONMENTAL LISTS

CERCLA/SARA - Hazardous Substances and their Reportable Quantities
final RQ = 1 pound (0.454 kg)

RCRA - F Series Wastes
Toxic waste

FATTY ACID, AMINE SALT   RR-00914-3

HEALTH AND SAFETY LISTS

U.S. DOT - Appendix A Table 1 - Hazardous Substances
final RQ = 1 pound (0.454 kg)

ENVIRONMENTAL LISTS

CERCLA/SARA - Hazardous Substances and their Reportable Quantities
final RQ = 1 pound (0.454 kg)

RCRA - F Series Wastes
Toxic waste

FATTY ACID, ESTER WITH STYRENATED PHENOL,   RR-01236-2
ETHYLENE OXIDE ADDUCT

HEALTH AND SAFETY LISTS

U.S. DOT - Appendix A Table 1 - Hazardous Substances
Statuatory RQ = 1 pound (the agency may adjust the statuatory RQ for this hazardous substance in a future rulemaking; until then the statuatory RQ applies)

ENVIRONMENTAL LISTS

CERCLA/SARA - Hazardous Substances and their Reportable Quantities
Statuatory RQ = 1 pound (the agency may adjust the statuatory RQ for this hazardous substance in a future rulemaking; until then the statuatory RQ applies)

RCRA - F Series Wastes
Toxic waste

FATTY ACIDS, C6-19-BRANCHED, MANGANESE   68551-42-8
SALTS

HEALTH AND SAFETY LISTS

U.S. DOT - Appendix A Table 1 - Hazardous Substances
Statuatory RQ = 1 pound (the agency may adjust the statuatory RQ for this hazardous substance in a future rulemaking; until then the statuatory RQ applies)

ENVIRONMENTAL LISTS

CERCLA/SARA - Hazardous Substances and their Reportable Quantities
Statuatory RQ = 1 pound (the agency may adjust the statuatory RQ for this hazardous substance in a future rulemaking; until then the statuatory RQ applies)

RCRA - F Series Wastes
Toxic waste

FATTY ACIDS, C8-18 AND C18-UNSATURATED   67701-05-7

ENVIRONMENTAL LISTS

RCRA - F Series Wastes
Toxic Waste

RCRA - Substances Banned From Land Disposal
[present]

FATTY ACIDS, C18-UNSATURATED, DIMERS   61788-89-4
SEE ALSO:
F039-HAZARDOUS WASTES

HEALTH AND SAFETY LISTS

U.S. DOT - Appendix A Table 1 - Hazardous Substances
final RQ = 1000 pounds (454 kg)

ENVIRONMENTAL LISTS

CERCLA/SARA - Section 313 - Emission Reporting
form R reporting required

CERCLA/SARA - Hazardous Substances and their Reportable Quantities
final RQ = 1000 pounds (454 kg)

RCRA - P Series Wastes
waste number P097

RCRA - Hazardous Constituents-Appendix VIII
waste number P097
RCRA - Basis for Listing - Appendix VII
Included in waste stream: F039
RCRA - Substances Banned From Land Disposal
[present]
RCRA - TSD Facilities Ground Water Monitoring
TM 8270 = 10 ug/L PQL
RCRA - Universal Treatment Standards (LDR)
WW: 0.017 mg/l; NWW: 15 mg/kg
STATE LISTS
Massachusetts Right To Know List
[present]
Pennsylvania Right to Know List
environmental hazard

FATTY ACIDS, COCO, HYDROGENATED   68938-15-8
HEALTH AND SAFETY LISTS
IARC - Group 3 (not classifiable)
[present]
ENVIRONMENTAL LISTS
List of Pesticide Product Inert Ingredients
[present]
STATE LISTS
California - Directors List of Hazardous Substances (8 CCR 339)
[present]

FATTY ACIDS, COCO, METHYL ESTERS   61788-59-8
ENVIRONMENTAL LISTS
TSCA - Chemicals with Significant New Use Rules
PMN number: P-92-445

FATTY ACIDS, COCO, MONOESTERS WITH SORBITAN   68154-36-9
ENVIRONMENTAL LISTS
TSCA - Chemicals with Significant New Use Rules
PMN number: P-88-1889

FATTY ACIDS, COCO, REACTION PRODUCTS WITH 2-[(2-AMINOETHYL)AMINO] ETHANOL, ALKYLATION PRODUCTS WITH METHYL ACRYLATE, SODIUM SALTS   RR-01078-6
ENVIRONMENTAL LISTS
TSCA - Chemicals with Significant New Use Rules
PMN number: P-90-364

FATTY ACIDS, COCO, 2-SULFOETHYL ESTERS, SODIUM SALTS   61789-32-0
ENVIRONMENTAL LISTS
List of Pesticide Product Inert Ingredients
[present]

FATTY ACIDS, DEHYDRATED CASTOR-OIL, POLYMERS WITH MALEIC ANHYDRIDE, TRIETHANOLAMINE SALTS   RR-01079-7
ENVIRONMENTAL LISTS
List of Pesticide Product Inert Ingredients
[present]

FATTY ACIDS, TALL OIL, EPOXIDIZED, 2-ETHYL-HEXYL ESTERS   61789-01-3
ENVIRONMENTAL LISTS
List of Pesticide Product Inert Ingredients
[present]

FATTY ACIDS, TALL-OIL, ESTERS WITH ETHYLENE GLYCOL   68187-85-9
ENVIRONMENTAL LISTS
List of Pesticide Product Inert Ingredients
[present]

FATTY ACIDS, TALL-OIL, ESTERS WITH PENTAERYTHRITOL   68188-27-2
ENVIRONMENTAL LISTS
List of Pesticide Product Inert Ingredients
[present]

FATTY ACIDS, TALL-OIL, MIXED ESTERS WITH GLYCEROL AND POLYETHYLENE GLYCOL   68650-09-9
ENVIRONMENTAL LISTS
List of Pesticide Product Inert Ingredients
[present]

FATTY ACIDS, TALLOW, SODIUM SALTS   8052-48-0
ENVIRONMENTAL LISTS
List of Pesticide Product Inert Ingredients
[present]

FATTY AMIDE   RR-01208-8
ENVIRONMENTAL LISTS
List of Pesticide Product Inert Ingredients
[present]

FD&C BLUE NO.2   6408-78-2
ENVIRONMENTAL LISTS
List of Pesticide Product Inert Ingredients
[present]

FD&C RED 40   25956-17-6
ENVIRONMENTAL LISTS
List of Pesticide Product Inert Ingredients
[present]

FD&C YELLOW 5   1934-21-0
ENVIRONMENTAL LISTS
List of Pesticide Product Inert Ingredients
[present]

FD & C YELLOW NO. 6   2783-94-0
ENVIRONMENTAL LISTS
List of Pesticide Product Inert Ingredients
[present]

FD&C BLUE NO. 1 (ALUMINUM SALT)   15792-67-3
ENVIRONMENTAL LISTS
List of Pesticide Product Inert Ingredients
[present]

FELDSPAR   68476-25-5
ENVIRONMENTAL LISTS
List of Pesticide Product Inert Ingredients
[present]

FENAMINOSULF (LESAN)   140-56-7
ENVIRONMENTAL LISTS
TSCA - Chemicals with Significant New Use Rules
PMN number: P-91-87

FENAMINPHOS   22224-92-6
ENVIRONMENTAL LISTS
  List of Pesticide Product Inert Ingredients
  [present]

FENARIMOL [.ALPHA.-(2-CHLOROPHENYL)   60168-88-9
.ALPHA.-4-CHLOROPHENYL)-5-PYRIM-
IDINEMETHANOL]
ENVIRONMENTAL LISTS
  List of Pesticide Product Inert Ingredients
  [present]

FENCHONE   1195-79-5
HEALTH AND SAFETY LISTS
  NIOSH - Selected LD50s and LC50s
  Oral, mouse: LD50 = 12750 mg/kg
ENVIRONMENTAL LISTS
  List of Pesticide Product Inert Ingredients
  [present]

ALPHA-FENCHYL ALCOHOL   512-13-0
HEALTH AND SAFETY LISTS
  IARC - Group 3 (not classifiable)
  [present]
  NTP Chemical Status Reports - Testing Status and NTIS Number
  Technical reports printed (PB82117433)
  NTP Chemical Status Reports - Evidence of Carcinogenicity
  male rat-negative; female rat-negative; male mice-negative; female
  mice-negative
ENVIRONMENTAL LISTS
  List of Pesticide Product Inert Ingredients
  [present]

FENITROTHION   122-14-5
ENVIRONMENTAL LISTS
  List of Pesticide Product Inert Ingredients
  [present]

FENOXAPROP ETHYL [2-(4-((6-CHLORO-2-BENZOX-   66441-23-4
AZOLYLEN)OXY)PHENOXY) PROPANOIC ACID,
ETHYL ESTER]
ENVIRONMENTAL LISTS
  List of Pesticide Product Inert Ingredients
  [present]

FENOXYCARB [2-(4-PHENOXYPHENOXY)ETHYL]   72490-01-8
CARBAMIC ACID ETHYL ESTER
HEALTH AND SAFETY LISTS
  NIOSH - Selected LD50s and LC50s
  Oral, rat: LD50 = 60 mg/kg
  NTP Chemical Status Reports - Testing Status and NTIS Number
  Technical reports printed (PB287443/AS)
  NTP Chemical Status Reports - Evidence of Carcinogenicity
  male rat-negative; female rat-negative; male mice-negative; female
  mice-negative
STATE LISTS
  NJ Right to Know List (Total)
  sn 0913

FENPROPATHRIN   39515-41-8
HEALTH AND SAFETY LISTS
  ACGIH 1995 - Time Weighted Averages
  0.1 mg/m3 TWA

ACGIH 1995 - Skin Designations
  skin - potential for cutaneous absorption
U.S. DOT - Appendix B - Marine Pollutants
  DOT regulated marine pollutant
NIOSH - Selected LD50s and LC50s
  Inhalation, rat: LC50 = 91 mg/m3 4 hr Oral, rat: LD50 = 8 mg/kg
  Skin, rat: LD50 = 500 mg/kg
OSHA - Vacated PELs - Time Weighted Averages
  0.1 mg/m3 TWA
OSHA - Vacated PELs - Skin Designation
  Prevent or reduce skin absorption
ENVIRONMENTAL LISTS
  CERCLA/SARA - Section 302 Extremely Hazardous Substances and
  TPQs
  TPQ = 10/10,000 pounds
INTERNATIONAL LISTS
  Australian Exposure Standards - Time Weighted Averages
  0.1 mg/m3 TWA
  Australian Exposure Standards - Skin Effects
  skin absorption
  Canada - Alberta - 8 Hour Occupational Exposure Limit
  0.1 mg/m3 TWA
  Canada - Alberta - 15 Minute Occupational Exposure Limit
  0.3 mg/m3 STEL
  Canada - Alberta - Skin Designation
  can be absorbed through the intact skin
  Canada - Alberta - Designated Substances
  designated substance - requires code of practice
  Canada - Ontario - OHSA - TWAEVs
  0.1 mg/m3 TWAEV
  Canada - Ontario - OHSA - Skin Notations
  absorption through skin, eyes, or mucous membranes
  Canada - Quebec - Time-Weighted Average Exposure Values
  0.1 mg/m3 TWAEV
  Canada - Quebec - Skin Designations
  absorbed through the skin
  Israel - Time Weighted Averages
  0.1 mg/m3 TWA
  Israel - Action Levels
  0.05 mg/m3 AL
STATE LISTS
  California - Exposure Limits - PELs
  0.1 mg/m3 PEL
  California - Exposure Limits - Skin Notation
  material may be absorbed through the skin, eyes or mucous membrane
  California - Directors List of Hazardous Substances (8 CCR 339)
  [present]
  Florida Hazardous Substance List
  effective March 13, 1992
  Massachusetts Right To Know List
  carcinogen; extraordinarily hazardous
  Minnesota Hazardous Substance List
  skin
  NJ Right to Know List (Total)
  sn 0914
  Pennsylvania Right to Know List
  environmental hazard
PROPOSED REGULATIONS
  CERCLA/SARA - Proposed Hazardous Substance Additions
  proposed RQ = 1 pound (.454 kg)
  CERCLA/SARA - 1989 Proposed RQ Adjustments
  proposed RQ = 10 pounds (4.54 kg)

FENSULFOTHION   **115-90-2**

ENVIRONMENTAL LISTS

CERCLA/SARA - Section 313 - Emission Reporting
form R reporting required

FENTHION   **55-38-9**

HEALTH AND SAFETY LISTS

NIOSH - Selected LD50s and LC50s
Oral, rat: LD50 = 4400 mg/kg

FENURON-TCA   **4482-55-7**

INTERNATIONAL LISTS

Canada - WHMIS: Ingredient Disclosure
1% item 759 (174)

FENVALERATE   **51630-58-1**

HEALTH AND SAFETY LISTS

U.S. DOT - Appendix B - Marine Pollutants
DOT regulated severe marine pollutant

NIOSH - Selected LD50s and LC50s
*Inhalation, rat: LC50 = 378 mg/m3 4 hr Oral, rat: LD50 = 242
mg/kg Skin, rat: LD50 = 750 mg/kg*

STATE LISTS

Florida Hazardous Substance List
effective March 13, 1992

Massachusetts Right To Know List
extraordinarily hazardous

NJ Right to Know List (Total)
sn 2410

Pennsylvania Right to Know List
environmental hazard

PROPOSED REGULATIONS

CERCLA/SARA - Proposed Hazardous Substance Additions
proposed RQ = 1 pound (.454 kg)

CERCLA/SARA - 1989 Proposed RQ Adjustments
proposed RQ = 100 pounds (45.4 kg)

FERBAM   **14484-64-1**

ENVIRONMENTAL LISTS

CERCLA/SARA - Section 313 - Emission Reporting
form R reporting required

FERMIUM 252   **15756-90-8**

ENVIRONMENTAL LISTS

CERCLA/SARA - Section 313 - Emission Reporting
form R reporting required

FERMIUM 253   **18396-20-8**

HEALTH AND SAFETY LISTS

U.S. DOT - Appendix B - Marine Pollutants
DOT regulated severe marine pollutant

ENVIRONMENTAL LISTS

CERCLA/SARA - Section 313 - Emission Reporting
form R reporting required

FERMIUM 254   **15750-23-9**

HEALTH AND SAFETY LISTS

ACGIH 1995 - Time Weighted Averages
0.1 mg/m3 TWA

U.S. DOT - Appendix B - Marine Pollutants
DOT regulated marine pollutant

NIOSH - Selected LD50s and LC50s
Oral, rat: LD50 = 2 mg/kg Skin, rat: LD50 = 3 mg/kg

OSHA - Vacated PELs - Time Weighted Averages
0.1 mg/m3 TWA

ENVIRONMENTAL LISTS

CERCLA/SARA - Section 302 Extremely Hazardous Substances an
TPQs
TPQ = 500 pounds

INTERNATIONAL LISTS

Australian Exposure Standards - Time Weighted Averages
0.1 mg/m3 TWA

Canada - Alberta - 8 Hour Occupational Exposure Limit
0.1 mg/m3 TWA

Canada - Alberta - 15 Minute Occupational Exposure Limit
0.3 mg/m3 STEL

Canada - British Columbia - 8 Hour Exposure Limits
0.1 mg/m3 TWA

Canada - Ontario - OHSA - TWAEVs
0.1 mg/m3 TWAEV

Canada - Quebec - Time-Weighted Average Exposure Values
0.1 mg/m3 TWAEV

Israel - Time Weighted Averages
0.1 mg/m3 TWA

Israel - Action Levels
0.05 mg/m3 AL

Mexico - Instruction No. 10 - TWAs
0.1 mg/m3 TWA

STATE LISTS

California - Exposure Limits - PELs
0.1 mg/m3 PEL

California - Directors List of Hazardous Substances (8 CCR 339)
[present]

Florida Hazardous Substance List
[present]

Massachusetts Right To Know List
extraordinarily hazardous; neurotoxin

Minnesota Hazardous Substance List
[present]

NJ Right to Know List (Total)
sn 0915

Pennsylvania Right to Know List
environmental hazard

PROPOSED REGULATIONS

CERCLA/SARA - Proposed Hazardous Substance Additions
proposed RQ = 1 pound (.454 kg)

CERCLA/SARA - 1989 Proposed RQ Adjustments
proposed RQ = 100 pounds (45.4 kg)

FERMIUM 255   **15750-24-0**

HEALTH AND SAFETY LISTS

ACGIH 1995 - Time Weighted Averages
0.2 mg/m3 TWA

ACGIH 1995 - Skin Designations
skin - potential for cutaneous absorption

U.S. DOT - Appendix B - Marine Pollutants
DOT regulated severe marine pollutant

NIOSH - Selected LD50s and LC50s
Oral, rat: LD50 = 180 mg/kg Skin, rat: LD50 = 330 mg/kg

NTP Chemical Status Reports - Testing Status and NTIS Number
Technical reports printed (PB293832/AS)

NTP Chemical Status Reports - Evidence of Carcinogenicity
*male rat-negative; female rat-negative; male mice-equivocal; female
mice-negative*

OSHA - Vacated PELs - Time Weighted Averages
0.2 mg/m3 TWA

OSHA - Vacated PELs - Skin Designation
Prevent or reduce skin absorption

ENVIRONMENTAL LISTS
CERCLA/SARA - Section 313 - Emission Reporting
form R reporting required

INTERNATIONAL LISTS
Australian Exposure Standards - Time Weighted Averages
0.2 mg/m3 TWA

Australian Exposure Standards - Skin Effects
skin absorption

Canada - Alberta - 8 Hour Occupational Exposure Limit
0.1 mg/m3 TWA

Canada - Alberta - 15 Minute Occupational Exposure Limit
0.3 mg/m3 STEL

Canada - Alberta - Skin Designation
can be absorbed through the intact skin

Canada - Alberta - Designated Substances
designated substance - requires code of practice

Canada - Ontario - OHSA - TWAEVs
0.1 mg/m3 TWAEV

Canada - Ontario - OHSA - Skin Notations
absorption through skin, eyes, or mucous membranes

Canada - Quebec - Time-Weighted Average Exposure Values
0.2 mg/m3 TWAEV

Canada - Quebec - Skin Designations
absorbed through the skin

German (DFG) - MAK Values
total dust: 0.2 mg/m3 MAK

German (DFG) - Peak Limitations
10 x normal MAK (30 min average value); don't exceed during shift

German (DFG) - Skin/Sensitizers
danger of cutaneous absorption

Israel - Time Weighted Averages
0.2 mg/m3 TWA

Israel - Action Levels
0.1 mg/m3 AL

STATE LISTS
California - Exposure Limits - PELs
0.2 mg/m3 PEL

California - Exposure Limits - Skin Notation
material may be absorbed through the skin, eyes or mucous membrane

California - Directors List of Hazardous Substances (8 CCR 339)
[present]

Florida Hazardous Substance List
[present]

Massachusetts Right To Know List
neurotoxin

Minnesota Hazardous Substance List
[present]

NJ Right to Know List (Total)
sn 0916

Pennsylvania Right to Know List
[present]

FERMIUM 257   15750-26-2

STATE LISTS
California - Directors List of Hazardous Substances (8 CCR 339)
[present]

FERRIC AMMONIUM CITRATE   1185-57-5

HEALTH AND SAFETY LISTS
IARC - Group 3 (not classifiable)
[present]

NIOSH - Selected LD50s and LC50s
Oral, rat: LD50 = 451 mg/kg Skin, rabbit: LD50 = 2500 mg/kg

ENVIRONMENTAL LISTS
CERCLA/SARA - Section 313 - Emission Reporting
form R reporting required

STATE LISTS
Massachusetts Right To Know List
[present]

FERRIC AMMONIUM OXALATE   14221-47-7

HEALTH AND SAFETY LISTS
ACGIH 1995 - Time Weighted Averages
10 mg/m3 TWA

IARC - Group 3 (not classifiable)
[present]

NIOSH - Selected LD50s and LC50s
Oral, rat: LD50 = 4000 mg/kg

NIOSH 1990 - Pocket Guide - RELs
10 mg/m3 TWA

NIOSH 1990 - Pocket Guide - Target organs
skin, GI tract, respiratory system

OSHA - Vacated PELs - Time Weighted Averages
total dust: 10 mg/m3 TWA

OSHA - Final PELs - Time Weighted Averages
total dust: 15 mg/m3 TWA

ENVIRONMENTAL LISTS
CERCLA/SARA - Section 313 - Emission Reporting
form R reporting required

INTERNATIONAL LISTS
Australian Exposure Standards - Time Weighted Averages
10 mg/m3 TWA

Canada - Alberta - 8 Hour Occupational Exposure Limit
10 mg/m3 TWA

Canada - Alberta - 15 Minute Occupational Exposure Limit
20 mg/m3 STEL

Canada - British Columbia - 8 Hour Exposure Limits
10 mg/m3 TWA

Canada - British Columbia - 15 Minute Exposure Limits
20 mg/m3 STEL

Canada - Ontario - OHSA - TWAEVs
10 mg/m3 TWAEV

Canada - Quebec - Time-Weighted Average Exposure Values
10 mg/m3 TWAEV

United Kingdom - Occupational Exposure Standards - TWAs
10 mg/m3 TWA

United Kingdom - Occupational Exposure Standards - STELs
20 mg/m3 STEL

German (DFG) - MAK Values
total dust: 15 mg/m3 MAK

Israel - Time Weighted Averages
10 mg/m3 TWA

Israel - Action Levels
5 mg/m3 AL

Mexico - Instruction No. 10 - TWAs
10 mg/m3 TWA

Mexico - Instruction No. 10 - STELs
20 mg/m3 STEL

STATE LISTS
California - Exposure Limits - PELs
10 mg/m3 PEL

California - Directors List of Hazardous Substances (8 CCR 339)
[present]

Florida Hazardous Substance List
[present]

Massachusetts Right To Know List
[present]

Minnesota Hazardous Substance List
[present]

NJ Right to Know List (Total)
sn 0917

Pennsylvania Right to Know List
[present]

FERRIC AMMONIUM OXALATE, UNSPECIFIED HYDRATE   55488-87-4

HEALTH AND SAFETY LISTS

U.S. DOT - Appendix A Table 2 - Radionuclides
final RQ = 10 curies (3.7E 11 Bq)

ENVIRONMENTAL LISTS

CERCLA/SARA List of Radionuclides (Appendix B) and Their Reportable Quantities
final RQ = 10 curies (3.7E 11 Bq)

FERRIC AMMONIUM SULFATE   10138-04-2

HEALTH AND SAFETY LISTS

U.S. DOT - Appendix A Table 2 - Radionuclides
final RQ = 10 curies (3.7E 11 Bq)

ENVIRONMENTAL LISTS

CERCLA/SARA List of Radionuclides (Appendix B) and Their Reportable Quantities
final RQ = 10 curies (3.7E 11 Bq)

FERRIC ARSENATE   10102-49-5

HEALTH AND SAFETY LISTS

U.S. DOT - Appendix A Table 2 - Radionuclides
final RQ = 10 curies (3.7E 11 Bq)

ENVIRONMENTAL LISTS

CERCLA/SARA List of Radionuclides (Appendix B) and Their Reportable Quantities
final RQ = 10 curies (3.7E 11 Bq)

FERRIC ARSENITE   63989-69-5

HEALTH AND SAFETY LISTS

U.S. DOT - Appendix A Table 2 - Radionuclides
final RQ = 10 curies (3.7E 11 Bq)

ENVIRONMENTAL LISTS

CERCLA/SARA List of Radionuclides (Appendix B) and Their Reportable Quantities
final RQ = 10 curies (3.7E 11 Bq)

FERRIC CHLORIDE   7705-08-0

HEALTH AND SAFETY LISTS

U.S. DOT - Appendix A Table 2 - Radionuclides
final RQ = 100 curies (3.7E 12 Bq)

ENVIRONMENTAL LISTS

CERCLA/SARA List of Radionuclides (Appendix B) and Their Reportable Quantities
final RQ = 100 curies (3.7E 12 Bq)

FERRIC EDTA   17099-81-9

HEALTH AND SAFETY LISTS

U.S. DOT - Appendix A Table 1 - Hazardous Substances
final RQ = 1000 pounds (454 kg)

ENVIRONMENTAL LISTS

CERCLA/SARA - Hazardous Substances and their Reportable Quantities
final RQ = 1000 pounds (454 kg)

Clean Water Act - Hazardous Substances
[present]

STATE LISTS

Massachusetts Right To Know List
[present]

NJ Right to Know List (Total)
sn 0918

Pennsylvania Right to Know List
environmental hazard

FERRIC FERROCYANIDE   14038-43-8

SEE ALSO:
OXALIC ACID, AMMONIUM IRON (3+) SALT (3:3:1)

INTERNATIONAL LISTS

Canada - WHMIS: Ingredient Disclosure
1% item 760 (1292)

STATE LISTS

NJ Right to Know List (Total)
sn 0919

FERRIC FLUORIDE   7783-50-8

SEE ALSO:
OXALIC ACID, AMMONIUM IRON (3+) SALT (3:3:1)

HEALTH AND SAFETY LISTS

U.S. DOT - Appendix A Table 1 - Hazardous Substances
final RQ = 1000 pounds (454 kg) (Listed under 'Ferric ammonium oxalate')

ENVIRONMENTAL LISTS

CERCLA/SARA - Hazardous Substances and their Reportable Quantities
final RQ = 1000 pounds (454 kg) (Listed under 'Ferric ammonium oxalate')

Clean Water Act - Hazardous Substances
[present] (Listed under 'Ferric ammonium oxalate')

STATE LISTS

Massachusetts Right To Know List
[present]

Pennsylvania Right to Know List
environmental hazard

FERRIC NITRATE.9H20   7782-61-8

ENVIRONMENTAL LISTS

List of Pesticide Product Inert Ingredients
[present]

FERRIC NITRATE   10421-48-4

SEE ALSO:
ARSENIC

HEALTH AND SAFETY LISTS

U.S. DOT - Substances From 49 CFR 172.101
regulated by DOT (UN1606)

U.S. DOT - Hazard Classes
DOT hazard class = 6.1

U.S. DOT - Appendix B - Marine Pollutants
DOT regulated marine pollutant

STATE LISTS

NJ Right to Know List (Total)
sn 0920

FERRIC SULFATE   10028-22-5
  SEE ALSO:
    ARSENIC
  HEALTH AND SAFETY LISTS
    U.S. DOT - Substances From 49 CFR 172.101
      regulated by DOT (UN1607)
    U.S. DOT - Hazard Classes
      DOT hazard class = 6.1
    U.S. DOT - Appendix B - Marine Pollutants
      DOT regulated marine pollutant
  STATE LISTS
    NJ Right to Know List (Total)
      sn 0921

FERROCERIUM   69523-06-4
  SEE ALSO:
    IRON
  HEALTH AND SAFETY LISTS
    U.S. DOT - Substances From 49 CFR 172.101
      regulated by DOT (UN2582, UN1773)
    U.S. DOT - Hazard Classes
      DOT hazard class = 8
    U.S. DOT - Appendix A Table 1 - Hazardous Substances
      final RQ = 1000 pounds (454 kg)
    NIOSH - Selected LD50s and LC50s
      Oral, rat: LD50 = 1872 mg/kg
  ENVIRONMENTAL LISTS
    CERCLA/SARA - Hazardous Substances and their Reportable Quantities
      final RQ = 1000 pounds (454 kg)
    Clean Water Act - Hazardous Substances
      [present]
    List of Pesticide Product Inert Ingredients
      [present]
  STATE LISTS
    Massachusetts Right To Know List
      [present]
    NJ Right to Know List (Total)
      sn 1034
    NJ Special Hazardous Substances
      (C0)
    Pennsylvania Right to Know List
      environmental hazard

FERROMANGANESE   12604-53-4
  ENVIRONMENTAL LISTS
    List of Pesticide Product Inert Ingredients
      [present]
  INTERNATIONAL LISTS
    Canada - WHMIS: Ingredient Disclosure
      1% item 761 (799)

FERROSILICON   8049-17-0
  ENVIRONMENTAL LISTS
    List of Pesticide Product Inert Ingredients
      [present]

FERROUS AMMONIUM SULFATE.6H20   7783-85-9
  HEALTH AND SAFETY LISTS
    U.S. DOT - Appendix A Table 1 - Hazardous Substances
      final RQ = 100 pounds (45.4 kg)

  ENVIRONMENTAL LISTS
    CERCLA/SARA - Hazardous Substances and their Reportable Quantities
      final RQ = 100 pounds (45.4 kg)
    Clean Water Act - Hazardous Substances
      [present]
  STATE LISTS
    Massachusetts Right To Know List
      [present]
    NJ Right to Know List (Total)
      sn 0923
    Pennsylvania Right to Know List
      environmental hazard

FERROUS AMMONIUM SULFATE   10045-89-3
  HEALTH AND SAFETY LISTS
    NIOSH - Selected LD50s and LC50s
      Oral, rat: LD50 = 3250 mg/kg

FERROUS ARSENATE   10102-50-8
  SEE ALSO:
    IRON
  HEALTH AND SAFETY LISTS
    U.S. DOT - Substances From 49 CFR 172.101
      regulated by DOT (UN1466)
    U.S. DOT - Hazard Classes
      DOT hazard class = 5.1
    U.S. DOT - Appendix A Table 1 - Hazardous Substances
      final RQ = 1000 pounds (454 kg)
  ENVIRONMENTAL LISTS
    CERCLA/SARA - Hazardous Substances and their Reportable Quantities
      final RQ = 1000 pounds (454 kg)
    Clean Water Act - Hazardous Substances
      [present]
  STATE LISTS
    Massachusetts Right To Know List
      [present]
    NJ Right to Know List (Total)
      sn 0924
    Pennsylvania Right to Know List
      environmental hazard

FERROUS CHLORIDE   7758-94-3
  SEE ALSO:
    IRON
  HEALTH AND SAFETY LISTS
    U.S. DOT - Appendix A Table 1 - Hazardous Substances
      final RQ = 1000 pounds (454 kg)
  ENVIRONMENTAL LISTS
    CERCLA/SARA - Hazardous Substances and their Reportable Quantities
      final RQ = 1000 pounds (454 kg)
    Clean Water Act - Hazardous Substances
      [present]
    List of Pesticide Product Inert Ingredients
      [present]
  STATE LISTS
    Massachusetts Right To Know List
      [present]
    NJ Right to Know List (Total)
      sn 0925

Pennsylvania Right to Know List
environmental hazard

FERROUS IRON (II) CHLORIDE.4H2O   13478-10-9

HEALTH AND SAFETY LISTS

U.S. DOT - Substances From 49 CFR 172.101
regulated by DOT (UN1323)

U.S. DOT - Hazard Classes
DOT hazard class = 4.1

STATE LISTS

NJ Right to Know List (Total)
sn 0926

FERROUS LACTATE   5905-52-2

ENVIRONMENTAL LISTS

TSCA - Code of Federal Regulations Citations
40 CFR 712.30(w)

TSCA - PAIR - Reporting List
Reporting Date: July 13, 1988

FERROUS OXIDE   1345-25-1

HEALTH AND SAFETY LISTS

U.S. DOT - Substances From 49 CFR 172.101
regulated by DOT (UN1408)

U.S. DOT - Hazard Classes
DOT hazard class = 4.3

INTERNATIONAL LISTS

Canada - WHMIS: Ingredient Disclosure
1% item 764 (881)

STATE LISTS

NJ Right to Know List (Total)
sn 0927

FERROUS SULFATE   7720-78-7

HEALTH AND SAFETY LISTS

NIOSH - Selected LD50s and LC50s
Oral, rat: LD50 = 3250 mg/kg

FERROUS SULFATE HEPTAHYDRATE   7782-63-0

HEALTH AND SAFETY LISTS

U.S. DOT - Appendix A Table 1 - Hazardous Substances
final RQ = 1000 pounds (454 kg)

ENVIRONMENTAL LISTS

CERCLA/SARA - Hazardous Substances and their Reportable Quantities
final RQ = 1000 pounds (454 kg)

Clean Water Act - Hazardous Substances
[present]

List of Pesticide Product Inert Ingredients
[present]

STATE LISTS

Massachusetts Right To Know List
[present]

NJ Right to Know List (Total)
sn 0928

Pennsylvania Right to Know List
environmental hazard

FERROVANADIUM   12604-58-

SEE ALSO:
ARSENIC

HEALTH AND SAFETY LISTS

U.S. DOT - Substances From 49 CFR 172.101
regulated by DOT (UN1608)

U.S. DOT - Hazard Classes
DOT hazard class = 6.1

U.S. DOT - Appendix B - Marine Pollutants
DOT regulated marine pollutant

STATE LISTS

NJ Right to Know List (Total)
sn 0929

FERTILIZER   RR-01080-

HEALTH AND SAFETY LISTS

U.S. DOT - Substances From 49 CFR 172.101
regulated by DOT (NA1759, NA1760)

U.S. DOT - Hazard Classes
DOT hazard class = 8

U.S. DOT - Appendix A Table 1 - Hazardous Substances
final RQ = 100 pounds (45.4 kg)

ENVIRONMENTAL LISTS

CERCLA/SARA - Hazardous Substances and their Reportable Quantities
final RQ = 100 pounds (45.4 kg)

Clean Water Act - Hazardous Substances
[present]

INTERNATIONAL LISTS

Canada - WHMIS: Ingredient Disclosure
1% item 765 (542)

STATE LISTS

Massachusetts Right To Know List
[present]

NJ Right to Know List (Total)
sn 0930

NJ Special Hazardous Substances
(corrosive)

Pennsylvania Right to Know List
environmental hazard

FERTILIZER AMMONIATING SOLUTION   RR-00297-1

HEALTH AND SAFETY LISTS

NIOSH - Selected LD50s and LC50s
Oral, rat: LD50 = 984 mg/kg

FIBER, ANIMAL OR VEGETABLE   RR-00298-2

ENVIRONMENTAL LISTS

List of Pesticide Product Inert Ingredients
[present]

FIBROUS GLASS, MICROFIBRES   RR-01808-6

SEE ALSO:
IRON

ENVIRONMENTAL LISTS

List of Pesticide Product Inert Ingredients
[present]

INTERNATIONAL LISTS

German (DFG) - MAK Values
fine dust: 6 mg/m3 MAK (Listed under 'Iron oxide')

FIBROUS GLASS   RR-00002-2
SEE ALSO:
  IRON
HEALTH AND SAFETY LISTS
U.S. DOT - Appendix A Table 1 - Hazardous Substances
  final RQ = 1000 pounds (454 kg)
NIOSH - Selected LD50s and LC50s
  Oral, rat: LD50 = 319 mg/kg
ENVIRONMENTAL LISTS
CERCLA/SARA - Hazardous Substances and their Reportable Quantities
  final RQ = 1000 pounds (454 kg)
Clean Water Act - Hazardous Substances
  [present]
List of Pesticide Product Inert Ingredients
  [present]
INTERNATIONAL LISTS
Canada - WHMIS: Ingredient Disclosure
  1% item 766 (1536)
STATE LISTS
Massachusetts Right To Know List
  [present]
NJ Right to Know List (Total)
  sn 0931
Pennsylvania Right to Know List
  environmental hazard

FINE DUST CONTAINING GREATER THAN 1% QUARTZ   RR-00089-5
HEALTH AND SAFETY LISTS
U.S. DOT - Appendix A Table 1 - Hazardous Substances
  final RQ = 1000 pounds (454 kg) (Listed under 'Ferrous sulfate')
NIOSH - Selected LD50s and LC50s
  Oral, mouse: LD50 = 1520 mg/kg
ENVIRONMENTAL LISTS
CERCLA/SARA - Hazardous Substances and their Reportable Quantities
  final RQ = 1000 pounds (454 kg) (Listed under 'Ferrous sulfate')
Clean Water Act - Hazardous Substances
  [present] (Listed under 'Ferrous sulfate')
List of Pesticide Product Inert Ingredients
  [present]
STATE LISTS
Massachusetts Right To Know List
  [present]
Pennsylvania Right to Know List
  environmental hazard

FIREMASTER 680   37853-59-1
HEALTH AND SAFETY LISTS
ACGIH 1995 - Time Weighted Averages
  1 mg/m3 TWA
ACGIH 1995 - Short Term Exposure Limits
  3 mg/m3 STEL
NIOSH 1990 - Pocket Guide - RELs
  1 mg/m3 TWA; 3 mg/m3 STEL
NIOSH 1990 - Pocket Guide - Target organs
  eyes, respiratory system
OSHA - Vacated PELs - Time Weighted Averages
  1 mg/m3 TWA
OSHA - Vacated PELs - Short Term Exposure Limits
  3 mg/m3 STEL

OSHA - Final PELs - Time Weighted Averages
  1 mg/m3 TWA
INTERNATIONAL LISTS
Australian Exposure Standards - Time Weighted Averages
  1 mg/m3 TWA
Australian Exposure Standards - Short Term Exposure Limits
  3 mg/m3 STEL
Canada - WHMIS: Ingredient Disclosure
  1% item 767 (882)
Canada - Alberta - 8 Hour Occupational Exposure Limit
  1 mg/m3 TWA
Canada - Alberta - 15 Minute Occupational Exposure Limit
  3 mg/m3 STEL
Canada - British Columbia - 8 Hour Exposure Limits
  1 mg/m3 TWA
Canada - British Columbia - 15 Minute Exposure Limits
  3.0 mg/m3 STEL
Canada - Ontario - OHSA - TWAEVs
  1 mg/m3 TWAEV
Canada - Ontario - OHSA - STEVs
  3 mg/m3 STEV
Canada - Quebec - Time-Weighted Average Exposure Values
  1 mg/m3 TWAEV
Canada - Quebec - Short-term Exposure Values
  3 mg/m3 STEV
German (DFG) - MAK Values
  total dust: 1 mg/m3 MAK
Israel - Time Weighted Averages
  1 mg/m3 TWA
Israel - Short Term Exposure Limits
  3 mg/m3 STEL
Israel - Action Levels
  0.5 mg/m3 AL
Mexico - Instruction No. 10 - TWAs
  1 mg/m3 TWA
Mexico - Instruction No. 10 - STELs
  3 mg/m3 STEL
STATE LISTS
California - Exposure Limits - PELs
  dust: 1 mg/m3 PEL
California - Exposure Limits - STELs
  dust: 3 mg/m3 STEL
California - Directors List of Hazardous Substances (8 CCR 339)
  [present]
Florida Hazardous Substance List
  [present]
Massachusetts Right To Know List
  [present]
Minnesota Hazardous Substance List
  [present] includes
Pennsylvania Right to Know List
  [present]

FIREMASTER BP-6   59536-65-1
ENVIRONMENTAL LISTS
List of Pesticide Product Inert Ingredients
  [present]

FIREMASTER FF-1   67774-32-7
HEALTH AND SAFETY LISTS
U.S. DOT - Substances From 49 CFR 172.101
  regulated by DOT (UN1043)

U.S. DOT - Hazard Classes
DOT hazard class = 2.2
STATE LISTS
NJ Right to Know List (Total)
sn 2877

FISH MEAL   **RR-01081-1**
HEALTH AND SAFETY LISTS
U.S. DOT - Substances From 49 CFR 172.101
regulated by DOT (UN1373)
U.S. DOT - Hazard Classes
DOT hazard class = 4.2

FISH OIL   **8016-13-5**
INTERNATIONAL LISTS
Canada - Quebec - Time-Weighted Average Exposure Values
1 fibre/cm3 TWAEV

FLAMMABLE GAS, N.O.S.   **RR-00306-5**
HEALTH AND SAFETY LISTS
ACGIH 1995 - Time Weighted Averages
10 mg/m3 TWA
NIOSH - Health Standards - Exposure Limits
3 million fibers/m3 TWA (fibers < 3.5 um in diameter and > 10 um
long); 5 mg/m3 TWA (total fibrous glass)
NIOSH - Health Standards - Health Effects and Precautions
Eye, skin, and respiratory effects
INTERNATIONAL LISTS
Canada - WHMIS: Ingredient Disclosure
1% item 768 (884)
Canada - Alberta - 8 Hour Occupational Exposure Limit
1 f/cm3 TWA
Canada - British Columbia - 8 Hour Exposure Limits
nuisance dust: 10 mg/m3 TWA
Canada - Ontario - OHSA - TWAEVs
10 mg/m3 TWAEV
Canada - Quebec - Time-Weighted Average Exposure Values
total dust: 10 mg/m3 TWAEV; respirable fraction: 5 mg/m3 TWAEV
German (DFG) - Carcinogens
animal evidence of carcinogenicity
Israel - Time Weighted Averages
10 mg/m3 TWA
Israel - Action Levels
5 mg/m3 AL
Mexico - Instruction No. 10 - TWAs
10 mg/m3 TWA
STATE LISTS
California - Directors List of Hazardous Substances (8 CCR 339)
[present] (only known hazardous effect is as a mechanical irritant)
Massachusetts Right To Know List
[present] exempt when encapsulated or if particulates are not present
and cannot be substantially generated through use of the product.
Pennsylvania Right to Know List
[present]

FLAMMABLE LIQUIDS, N.O.S.   **RR-00309-8**
INTERNATIONAL LISTS
German (DFG) - MAK Values
fine dust: 4 mg/m3 MAK (Listed under 'Quartz')

FLAMMABLE SOLIDS, N.O.S.   **RR-00312-3**
ENVIRONMENTAL LISTS
EPA - Master Testing List
[present]

TSCA - Code of Federal Regulations Citations
40 CFR 712.30(d),(w); 40 CFR 716.120(a); 40 CFR 766.35
TSCA - PAIR - Reporting List
Reporting Dates: November 19, 1982; March 12, 1990
TSCA - Health and Safety Reporting List
Effective Date: January 11, 1990
TSCA - HDD/HDF - Chemicals Required for Testing
[present]
TSCA - Section 12(b) - Export Notification
export notification required - Section 4

FLOUR   **RR-01082-2**
SEE ALSO:
POLYBROMINATED BIPHENYLS (PBB)(GENERIC)
STATE LISTS
Florida Hazardous Substance List
[present]
Massachusetts Right To Know List
[present]
NJ Special Hazardous Substances
(carcinogen)
Pennsylvania Right to Know List
special hazardous substance
Pennsylvania RTK - Special Hazardous Substances
[present]

FLUAZIFOP-BUTYL [2-[4-[[5-(TRIFLUOROMETHYL)   **69806-50-4**
-2-PYRIDINYL]OXY]-PHENOXY]PROPANOIC ACID,
BUTYL ESTER
SEE ALSO:
POLYBROMINATED BIPHENYLS (PBB)(GENERIC)
HEALTH AND SAFETY LISTS
NTP Chemical Status Reports - Testing Status and NTIS Number
Technical reports printed (PB83240473)
NTP Chemical Status Reports - Evidence of Carcinogenicity
male rat-positive; female rat-positive; male mice-positive; female
mice-positive
NTP Seventh Report - Suspect Carcinogens
suspect carcinogen (Listed under 'Polybrominated biphenyls')
ENVIRONMENTAL LISTS
ATSDR Priority List
Rank (of 275): 128
INTERNATIONAL LISTS
Canada - WHMIS: Ingredient Disclosure
0.1% item 1322 (1439)
STATE LISTS
NJ Special Hazardous Substances
(carcinogen)

FLUCYTHRINATE   **70124-77-5**
HEALTH AND SAFETY LISTS
U.S. DOT - Substances From 49 CFR 172.101
regulated by DOT (UN1374, UN2216)
U.S. DOT - Hazard Classes
stabilized: DOT hazard class = 9; unstabilized: DOT hazard class =
4.2
ENVIRONMENTAL LISTS
List of Pesticide Product Inert Ingredients
[present]

FLUE DUST, POISONOUS   **67711-90-4**
HEALTH AND SAFETY LISTS
NFPA - Flash Points
flash point = 420 degrees F (216 degrees C)

NFPA - Hazard Identification Ratings
health-0; flammability-1; reactivity-0

ENVIRONMENTAL LISTS
List of Pesticide Product Inert Ingredients
[present]

FLUE GASSES, FERROUS METAL, BLAST FURNACE · 65996-68-1

STATE LISTS
NJ Right to Know List (Total)
sn 2422; sn 2250

FLUENETIL · 4301-50-2

HEALTH AND SAFETY LISTS
U.S. DOT - Substances From 49 CFR 172.101
regulated by DOT (UN1992, UN1993, UN2924)
U.S. DOT - Hazard Classes
DOT hazard class = 3

STATE LISTS
NJ Right to Know List (Total)
sn 2423; sn 2425; sn 2426; sn 2427

FLUOBORATE COMPOUNDS, N.O.S. · RR-00598-1

HEALTH AND SAFETY LISTS
U.S. DOT - Substances From 49 CFR 172.101
regulated by DOT (UN1325, UN2925, UN2926)
U.S. DOT - Hazard Classes
DOT hazard class = 4.1

STATE LISTS
NJ Right to Know List (Total)
sn 2428; sn 2429; sn 2430

FLUOMETURON · 2164-17-2

ENVIRONMENTAL LISTS
List of Pesticide Product Inert Ingredients
[present]

FLUORANTHENE · 206-44-0

ENVIRONMENTAL LISTS
CERCLA/SARA - Section 313 - Emission Reporting
form R reporting required

FLUORAPATITE · 1306-05-4

STATE LISTS
Massachusetts Right To Know List
neurotoxin

FLUORENE · 86-73-7

STATE LISTS
NJ Right to Know List (Total)
sn 2432

FLUORENE SUBSTITUTED AROMATIC AMINE · RR-01247-5

HEALTH AND SAFETY LISTS
NFPA - Hazard Identification Ratings
health-2; flammability-4; reactivity-0

STATE LISTS
Massachusetts Right To Know List
[present]
Pennsylvania Right to Know List
[present]

4,4'-(9H-FLUOREN-9-YLIDENE)BIS · RR-00934-7

HEALTH AND SAFETY LISTS
NIOSH - Selected LD50s and LC50s
Oral, rat: LD50 = 6 mg/kg

ENVIRONMENTAL LISTS
CERCLA/SARA - Section 302 Extremely Hazardous Substances and TPQs
TPQ = 100/10,000 pounds

STATE LISTS
Florida Hazardous Substance List
effective March 13, 1992
Massachusetts Right To Know List
extraordinarily hazardous
NJ Right to Know List (Total)
sn 2433
Pennsylvania Right to Know List
environmental hazard

PROPOSED REGULATIONS
CERCLA/SARA - Proposed Hazardous Substance Additions
proposed RQ = 1 pound (.454 kg)
CERCLA/SARA - 1989 Proposed RQ Adjustments
proposed RQ = 100 pounds (45.4 kg)

FLUORESCEIN · 2321-07-5

INTERNATIONAL LISTS
Canada - WHMIS: Ingredient Disclosure
1% item 769 (885)

FLUORESCEIN MERCURIC ACETATE · 3570-80-7

HEALTH AND SAFETY LISTS
IARC - Group 3 (not classifiable)
[present]
NIOSH - Selected LD50s and LC50s
Oral, rat: LD50 = 6416 mg/kg
NTP Chemical Status Reports - Testing Status and NTIS Number
Technical reports printed (PB80217904)
NTP Chemical Status Reports - Evidence of Carcinogenicity
male rat-negative; female rat-negative; male mice-equivocal; female mice-negative

ENVIRONMENTAL LISTS
CERCLA/SARA - Section 313 - Emission Reporting
form R reporting required for 1.0% de minimus concentration

STATE LISTS
California - Air Bill 2588 Appendix A-II
6/91
Massachusetts Right To Know List
[present]
NJ Right to Know List (Total)
sn 0935
Pennsylvania Right to Know List
environmental hazard

FLUORESCEIN, 2',4',5',7'-TETRAIODO, DISODIUM SALT · 16423-68-0

SEE ALSO:
F039-HAZARDOUS WASTES
K035-HAZARDOUS WASTES
K001-HAZARDOUS WASTES

HEALTH AND SAFETY LISTS
U.S. DOT - Appendix A Table 1 - Hazardous Substances
final RQ = 100 pounds (45.4 kg)
IARC - Group 3 (not classifiable)
[present]

NIOSH - Selected LD50s and LC50s
Oral, rat: LD50 = 2000 mg/kg Skin, rabbit: LD50 = 3180 mg/kg

ENVIRONMENTAL LISTS

ATSDR Priority List
Rank (of 275): 100

CERCLA/SARA - Hazardous Substances and their Reportable Quantities
final RQ = 100 pounds (45.4 kg)

CAA - HON Rule - SOCMI Chemicals
compliance by Oct. 23, 1995

Clean Water Act - Priority Pollutants
[present]

Clean Water Act - Toxic Pollutants
[present]

RCRA - U Series Wastes
waste number U120

RCRA - Hazardous Constituents-Appendix VIII
waste number U120

RCRA - Basis for Listing - Appendix VII
Included in waste streams: F039, K001, K035

RCRA - Substances Banned From Land Disposal
[present] (wastewaters)

RCRA - TSD Facilities Ground Water Monitoring
TM 8100 = 200 ug/L PQL; TM 8270 = 10 ug/L PQL

RCRA - Universal Treatment Standards (LDR)
WW: 0.068 mg/l; NWW: 3.4 mg/kg

INTERNATIONAL LISTS

Canada - WHMIS: Ingredient Disclosure
1% item 770 (892)

Mexico - Wastewater - Organic Toxic Pollutants and Heavy Metals
Listed under [Organic Toxic Pollutants]

Mexico - Drinking Water - Ecological Criteria
0.04 mg/l

STATE LISTS

California - Directors List of Hazardous Substances (8 CCR 339)
[present]

Massachusetts Right To Know List
[present]

Pennsylvania Right to Know List
environmental hazard

FLUORIDES   16984-48-8

INTERNATIONAL LISTS

Canada - WHMIS: Ingredient Disclosure
1% item 771 (893)

FLUORIDES, INORGANIC   RR-00599-2
SEE ALSO:
F039-HAZARDOUS WASTES

HEALTH AND SAFETY LISTS

U.S. DOT - Appendix A Table 1 - Hazardous Substances
final RQ = 5000 pounds (2270 kg)

IARC - Group 3 (not classifiable)
[present]

ENVIRONMENTAL LISTS

ATSDR Priority List
Rank (of 275): 237

CERCLA/SARA - Hazardous Substances and their Reportable Quantities
final RQ = 5000 pounds (2270 kg)

Clean Water Act - Priority Pollutants
[present]

RCRA - Basis for Listing - Appendix VII
Included in waste stream: F039

RCRA - TSD Facilities Ground Water Monitoring
TM 8100 = 200 ug/L PQL; TM 8270 = 10 ug/L PQL

RCRA - Universal Treatment Standards (LDR)
WW: 0.059 mg/l; NWW: 3.4 mg/kg

INTERNATIONAL LISTS

Mexico - Wastewater - Organic Toxic Pollutants and Heavy Metals
Listed under [Aromatic Hydrocarbons]

STATE LISTS

Massachusetts Right To Know List
[present]

Pennsylvania Right to Know List
environmental hazard

FLUORINE   7782-41-

ENVIRONMENTAL LISTS

TSCA - Chemicals with Significant New Use Rules
PMN number: P-91-43

FLUORINE 18   13981-56-

ENVIRONMENTAL LISTS

TSCA - Chemicals with Significant New Use Rules
PMN number: P-88-831

FLUOROACETAMIDE   640-19-7

ENVIRONMENTAL LISTS

List of Pesticide Product Inert Ingredients
[present]

FLUOROACETIC ACID   144-49-0
SEE ALSO:
MERCURY

INTERNATIONAL LISTS

Canada - WHMIS: Ingredient Disclosure
1% item 772 (37)

FLUOROACETYL CHLORIDE   359-06-8

ENVIRONMENTAL LISTS

List of Pesticide Product Inert Ingredients
[present]

FLUOROALKENES   RR-00276-6
SEE ALSO:
F039-HAZARDOUS WASTES

HEALTH AND SAFETY LISTS

ACGIH 1995 - Time Weighted Averages
as F: 2.5 mg/m3 TWA

ACGIH 1995 - Biological Exposure Indices
Fluorides in urine: 3 mg/g creatinine, prior to shift (B, Ns), 10 mg/g creatinine, end of shift (B, Ns)

NIOSH 1990 - Pocket Guide - RELs
as F: 2.5 mg/m3 TWA

NIOSH 1990 - Pocket Guide - IDLHs
as F: 500 mg/m3 IDLH

NIOSH 1990 - Pocket Guide - Target organs
eyes, respiratory system, CNS, skeleton, kidneys, skin

OSHA - Vacated PELs - Time Weighted Averages
as F: 2.5 mg/m3 TWA

OSHA - Final PELs - Time Weighted Averages
as F: 2.5 mg/m3 TWA

ENVIRONMENTAL LISTS

ATSDR Priority List
Rank (of 275): 240

Safe Drinking Water Act - MCLs
MCL = 4.0 mg/L

Safe Drinking Water Act - MCLGs
MCLG = 4.0 mg/L

Safe Drinking Water Act - SMCLs
SMCL = 2.0 mg/L

RCRA - Basis for Listing - Appendix VII
Included in waste stream: F039

RCRA - Universal Treatment Standards (LDR)
WW: 35 mg/l; NWW: Not applicable

INTERNATIONAL LISTS

Australian Exposure Standards - Time Weighted Averages
as F: 2.5 mg/m3 TWA

Canada - CEPA Schedule III Part II - Restricted Substances (Ocean Dumping)
[present]

Canada - Drinking Water Quality - MACs
1.5 mg/L MAC

Canada - Alberta - 8 Hour Occupational Exposure Limit
as F: 2.5 mg/m3 TWA

Canada - Alberta - 15 Minute Occupational Exposure Limit
as F: 5 mg/m3 STEL

Canada - British Columbia - 8 Hour Exposure Limits
as F: 2.5 mg/m3 TWA

Canada - Ontario - OHSA - TWAEVs
as fluoride: 2.5 mg/m3 TWAEV

Canada - Quebec - Time-Weighted Average Exposure Values
as F: 2.5 mg/m3 TWAEV

United Kingdom - Occupational Exposure Standards - TWAs
as F: 2.5 mg/m3 TWA

German (DFG) - MAK Values
total dust, as fluorine: 2.5 mg/m3 MAK

German (DFG) - Peak Limitations
2 x normal MAK (30 min. average value); don't exceed 4 times during shift; with hydrogen fluoride: 2 x normal MAK (5 min. momentary value); don't exceed 8 times during shift

Israel - Time Weighted Averages
as F: 2.5 mg/m3 TWA

Israel - Action Levels
as F: 1.25 mg/m3 AL

Mexico - Instruction No. 10 - TWAs
2.5 mg/m3 TWA

Mexico - Drinking Water - Ecological Criteria
1.5 mg/l

STATE LISTS

California - Air Bill 2588 Appendix A-I
9/89

California - Exposure Limits - PELs
as F: 2.5 mg/m3 PEL

California - Directors List of Hazardous Substances (8 CCR 339)
[present]

Minnesota Hazardous Substance List
[present] as F

Pennsylvania Right to Know List
[present]

2-FLUOROANILINE   348-54-9

HEALTH AND SAFETY LISTS

IARC - Group 3 (not classifiable)
[present] (when used in drinking water)

NIOSH - Health Standards - Exposure Limits
2.5 mg F/m3 TWA

NIOSH - Health Standards - Health Effects and Precautions
Kidney and bone effects (Urine monitoring required)

INTERNATIONAL LISTS

Canada - WHMIS: Ingredient Disclosure
1% item 773 (915)

Canada - CEPA - Priority Substances List
estimated time for completion of assessment reports: 4 years

STATE LISTS

California - Directors List of Hazardous Substances (8 CCR 339)
[present]

Minnesota Hazardous Substance List
[present]

4-FLUOROANILINE   371-40-4

HEALTH AND SAFETY LISTS

ACGIH 1995 - Time Weighted Averages
1 ppm TWA; 1.6 mg/m3 TWA

ACGIH 1995 - Short Term Exposure Limits
2 ppm STEL; 3.1 mg/m3 STEL

AIHA - Odor Threshold Values
no geometric mean air odor threshold

U.S. DOT - Substances From 49 CFR 172.101
regulated by DOT (UN1045)

U.S. DOT - Hazard Classes
DOT hazard class = 2.3

U.S. DOT - Substances Which Are Poisonous by Inhalation
gaseous hazardous material poisonous by inhalation (compressed or refrigerated liquid) (UN1045)

U.S. DOT - Appendix A Table 1 - Hazardous Substances
final RQ = 10 pounds (4.54 kg)

NIOSH - Selected LD50s and LC50s
Inhalation, rat: LC50 = 185 ppm 1 hr

NIOSH 1990 - Pocket Guide - RELs
0.1 ppm TWA; 0.2 mg/m3 TWA

NIOSH 1990 - Pocket Guide - IDLHs
25 ppm IDLH

NIOSH 1990 - Pocket Guide - Target organs
eyes, respiratory system, skin; in animals: liver, kidneys

OSHA - Vacated PELs - Time Weighted Averages
0.1 ppm TWA; 0.2 mg/m3 TWA

OSHA - Final PELs - Time Weighted Averages
0.1 ppm TWA; 0.2 mg/m3 TWA

OSHA - List of Highly Hazardous Chemicals
threshhold quantity = 1000 pounds

ENVIRONMENTAL LISTS

ATSDR Priority List
Rank (of 275): 211

CERCLA/SARA - Section 302 Extremely Hazardous Substances and TPQs
TPQ = 500 pounds

CERCLA/SARA - Section 313 - Emission Reporting
form R reporting required

CERCLA/SARA - Hazardous Substances and their Reportable Quantities
final RQ = 10 pounds (4.54 kg)

CAA -Toxic Substances for Accidental Release Prevention
threshold quantity = 1,000 lbs

RCRA - P Series Wastes
waste number P056

RCRA - Hazardous Constituents-Appendix VIII
waste number P056

RCRA - Substances Banned From Land Disposal
[present]

INTERNATIONAL LISTS

Australian Exposure Standards - Time Weighted Averages
1 ppm TWA; 1.6 mg/m3 TWA

Australian Exposure Standards - Short Term Exposure Limits
2 ppm STEL; 3.1 mg/m3 STEL

Australian Exposure Standards - Under Review
exposure limits under review

Canada - WHMIS: Ingredient Disclosure
1% item 774 (891)

Canada - Alberta - 8 Hour Occupational Exposure Limit
1 ppm TWA; 1.6 mg/m3 TWA

Canada - Alberta - 15 Minute Occupational Exposure Limit
2 ppm STEL; 3.1 mg/m3 STEL

Canada - British Columbia - 8 Hour Exposure Limits
1 ppm TWA; 2 mg/m3 TWA

Canada - British Columbia - 15 Minute Exposure Limits
2 ppm STEL; 4 mg/m3 STEL

Canada - Ontario - OHSA - TWAEVs
1 ppm TWAEV; 1.6 mg/m3 TWAEV

Canada - Ontario - OHSA - STEVs
2 ppm STEV; 3.1 mg/m3 STEV

Canada - Quebec - Time-Weighted Average Exposure Values
0.1 ppm TWAEV; 0.2 mg/m3 TWAEV

United Kingdom - Occupational Exposure Standards - STELs
1 ppm STEL; 1.5 mg/m3 STEL

German (DFG) - MAK Values
0.1 ppm MAK; 0.2 mg/m3 MAK

German (DFG) - Peak Limitations
*2 x normal MAK (5 min momentary value);   don't exceed 8 times
during shift*

Israel - Time Weighted Averages
1 ppm TWA; 1.6 mg/m3 TWA

Israel - Short Term Exposure Limits
2 ppm STEL; 3.1 mg/m3 STEL

Israel - Action Levels
0.5 ppm AL; 0.8 mg/m3 AL

Mexico - Instruction No. 10 - TWAs
1 ppm TWA; 2 mg/m3 TWA

Mexico - Instruction No. 10 - STELs
2 ppm STEL; 4 mg/m3 STEL

STATE LISTS

California - Exposure Limits - PELs
0.1 ppm PEL; 0.2 mg/m3 PEL

California - Directors List of Hazardous Substances (8 CCR 339)
[present]

Florida Hazardous Substance List
[present]

Massachusetts Right To Know List
extraordinarily hazardous

Minnesota Hazardous Substance List
[present]

NJ Right to Know List (Total)
sn 0937

NJ Special Hazardous Substances
(reactive - third degree)

Pennsylvania Right to Know List
environmental hazard

FLUOROANILINES   RR-01357-0

HEALTH AND SAFETY LISTS

U.S. DOT - Appendix A Table 2 - Radionuclides
final RQ = 1000 curies (3.7E 13 Bq)

ENVIRONMENTAL LISTS

CERCLA/SARA List of Radionuclides (Appendix B) and Their Reportable Quantities
final RQ = 1000 curies (3.7E 13 Bq)

FLUOROBENZENE   462-06-

HEALTH AND SAFETY LISTS

U.S. DOT - Appendix A Table 1 - Hazardous Substances
final RQ = 100 pounds (45.4 kg)

NIOSH - Selected LD50s and LC50s
Oral, rat: LD50 = 5750 ug/kg Skin, rat: LD50 = 80 mg/kg

ENVIRONMENTAL LISTS

CERCLA/SARA - Section 302 Extremely Hazardous Substances and TPQs
TPQ = 100/10,000 pounds

CERCLA/SARA - Hazardous Substances and their Reportable Quantities
final RQ = 100 pounds (45.4 kg)

RCRA - P Series Wastes
waste number P057

RCRA - Hazardous Constituents-Appendix VIII
waste number P057

RCRA - Substances Banned From Land Disposal
[present]

TSCA - Code of Federal Regulations Citations
40 CFR 799.5055(c),(e)(1)

TSCA - Multichemical Test Rules - Waste Constituents
subchronic toxicity testing for Health Effects

TSCA - Section 12(b) - Export Notification
export notification required - Section 4

STATE LISTS

California - Directors List of Hazardous Substances (8 CCR 339)
[present]

Florida Hazardous Substance List
effective March 13, 1992

Massachusetts Right To Know List
extraordinarily hazardous

NJ Right to Know List (Total)
sn 2434

Pennsylvania Right to Know List
environmental hazard

FLUOROBORIC ACID   16872-11-0

HEALTH AND SAFETY LISTS

U.S. DOT - Substances From 49 CFR 172.101
regulated by DOT (UN2642)

U.S. DOT - Hazard Classes
DOT hazard class = 6.1

NIOSH - Selected LD50s and LC50s
Oral, rat: LD50 = 4680 ug/kg

ENVIRONMENTAL LISTS

CERCLA/SARA - Section 302 Extremely Hazardous Substances and TPQs
TPQ = 10/10,000 pounds

STATE LISTS

Florida Hazardous Substance List
effective March 13, 1992

Massachusetts Right To Know List
extraordinarily hazardous

NJ Right to Know List (Total)
sn 0938

Pennsylvania Right to Know List
environmental hazard

PROPOSED REGULATIONS
  CERCLA/SARA - Proposed Hazardous Substance Additions
    proposed RQ = 1 pound (.454 kg)
  CERCLA/SARA - 1989 Proposed RQ Adjustments
    proposed RQ = 10 pounds (4.54 kg)

1-FLUORO-2-BROMOBENZENE   1072-85-1

ENVIRONMENTAL LISTS
  CERCLA/SARA - Section 302 Extremely Hazardous Substances and TPQs
    TPQ = 10 pounds

STATE LISTS
  Florida Hazardous Substance List
    effective March 13, 1992
  Massachusetts Right To Know List
    extraordinarily hazardous
  NJ Right to Know List (Total)
    sn 2435
  Pennsylvania Right to Know List
    environmental hazard

PROPOSED REGULATIONS
  CERCLA/SARA - Proposed Hazardous Substance Additions
    proposed RQ = 1 pound (.454 kg)
  CERCLA/SARA - 1989 Proposed RQ Adjustments
    proposed RQ = 10 pounds (4.54 kg)

1-FLUORO-3-BROMOBENZENE   1073-06-9

ENVIRONMENTAL LISTS
  TSCA - Health and Safety Reporting List
    Effective Date: April 29, 1991 (This category is defined as fluoroalkenes of the general formula: $C_nH_{2n}F_x$ where 'n' equals 2 to 3 and 'x' equals 1 to 6)

FLUOROSILICATES, N.O.S.   RR-00314-5

INTERNATIONAL LISTS
  Canada - WHMIS: Ingredient Disclosure
    1% item 775 (894)

STATE LISTS
  NJ Right to Know List (Total)
    sn 2436

FLUOROSILICIC ACID   16961-83-4

INTERNATIONAL LISTS
  Canada - WHMIS: Ingredient Disclosure
    1% item 776 (895)

STATE LISTS
  NJ Right to Know List (Total)
    sn 2437

FLUOROSULFONIC ACID   7789-21-1

HEALTH AND SAFETY LISTS
  U.S. DOT - Substances From 49 CFR 172.101
    regulated by DOT (UN2941)
  U.S. DOT - Hazard Classes
    DOT hazard class = 6.1

O-FLUOROTOLUENE   95-52-3

HEALTH AND SAFETY LISTS
  U.S. DOT - Substances From 49 CFR 172.101
    regulated by DOT (UN2387)
  U.S. DOT - Hazard Classes
    DOT hazard class = 3

NFPA - Flash Points
    flash point = 5 degrees F (-15 degrees C)
NFPA - Hazard Identification Ratings
    flammability-3; reactivity-0
NIOSH - Selected LD50s and LC50s
    Inhalation, rat: LC50 = 26908 mg/m3 (8 hr) Oral, rat: LD50 = 4399 mg/kg

STATE LISTS
  Florida Hazardous Substance List
    [present]
  Massachusetts Right To Know List
    [present]
  NJ Right to Know List (Total)
    sn 0939
  NJ Special Hazardous Substances
    (flammable - third degree)
  Pennsylvania Right to Know List
    [present]

FLUOROTOLUENE   25496-08-6

HEALTH AND SAFETY LISTS
  U.S. DOT - Substances From 49 CFR 172.101
    regulated by DOT (UN1775)
  U.S. DOT - Hazard Classes
    DOT hazard class = 8

STATE LISTS
  NJ Special Hazardous Substances
    (corrosive)

FLUOROURACIL   51-21-8

INTERNATIONAL LISTS
  Canada - WHMIS: Ingredient Disclosure
    1% item 777 (896)

FLUORPHLOGOPITE (MG3K[ALF2O(SIO3)3])   12003-38-2

INTERNATIONAL LISTS
  Canada - WHMIS: Ingredient Disclosure
    1% item 778 (897)

FLUOXYMESTERONE   76-43-7

HEALTH AND SAFETY LISTS
  U.S. DOT - Substances From 49 CFR 172.101
    regulated by DOT (UN2856)
  U.S. DOT - Hazard Classes
    DOT hazard class = 6.1

STATE LISTS
  NJ Right to Know List (Total)
    sn 2438

FLURAZEPAM   17617-23-1

HEALTH AND SAFETY LISTS
  U.S. DOT - Substances From 49 CFR 172.101
    regulated by DOT (UN1778)
  U.S. DOT - Hazard Classes
    DOT hazard class = 8

INTERNATIONAL LISTS
  Canada - WHMIS: Ingredient Disclosure
    1% item 779 (87)

STATE LISTS
  Massachusetts Right To Know List
    [present]
  NJ Right to Know List (Total)
    sn 1665

FLURAZEPAM HYDROCHLORIDE   1172-18-5

HEALTH AND SAFETY LISTS

U.S. DOT - Substances From 49 CFR 172.101
regulated by DOT (UN1777)

U.S. DOT - Hazard Classes
DOT hazard class = 8

INTERNATIONAL LISTS

Canada - WHMIS: Ingredient Disclosure
1% item 780 (88)

STATE LISTS

NJ Right to Know List (Total)
sn 0941

NJ Special Hazardous Substances
(corrosive)

FLUTAMIDE   13311-84-7

HEALTH AND SAFETY LISTS

NIOSH - Selected LD50s and LC50s
Oral, bird: LD50 = 100 mg/kg

FLUVALINATE [N-[2-CHLORO-4-(TRIFLUO-   69409-94-5
ROMETHYL)PHENYL]-DL-VALINE(+)-CYANO (3-
PHENOXYPHENYL)METHYL ESTER

HEALTH AND SAFETY LISTS

U.S. DOT - Substances From 49 CFR 172.101
regulated by DOT (UN2388)

U.S. DOT - Hazard Classes
DOT hazard class = 3

STATE LISTS

NJ Right to Know List (Total)
sn 0942

FOAMING AGENTS   RR-00064-6

HEALTH AND SAFETY LISTS

IARC - Group 3 (not classifiable)
[present]

NIOSH - Selected LD50s and LC50s
Oral, rat: LD50 = 230 mg/kg

ENVIRONMENTAL LISTS

CERCLA/SARA - Section 302 Extremely Hazardous Substances and
TPQs
TPQ = 500/10,000 pounds

CERCLA/SARA - Section 313 - Emission Reporting
form R reporting required

STATE LISTS

California - Air Bill 2588 Appendix A-II
9/89

California - Prop. 65 - Developmental Toxicity
developmental toxicity - initial date 1/1/89

Florida Hazardous Substance List
effective March 13, 1992

Massachusetts Right To Know List
extraordinarily hazardous; teratogen

NJ Right to Know List (Total)
sn 1966

NJ Special Hazardous Substances
(mutagen; teratogen)

Pennsylvania Right to Know List
environmental hazard

PROPOSED REGULATIONS

CERCLA/SARA - Proposed Hazardous Substance Additions
proposed RQ = 1 pound (.454 kg)

CERCLA/SARA - 1989 Proposed RQ Adjustments
proposed RQ = 100 pounds (45.4 kg)

FOLPET   133-07-

ENVIRONMENTAL LISTS

List of Pesticide Product Inert Ingredients
[present]

FOMESAFEN   72178-02-

STATE LISTS

California - Air Bill 2588 Appendix A-II
9/90

California - Prop. 65 - Developmental Toxicity
developmental toxicity - initial date 4/1/90

FONOFOS   944-22-

HEALTH AND SAFETY LISTS

NIOSH - Selected LD50s and LC50s
Oral, rat: LD50 = 796 mg/kg

STATE LISTS

NJ Right to Know List (Total)
sn 0944

NJ Special Hazardous Substances
(teratogen)

FORMALDEHYDE   50-00-(

STATE LISTS

California - Prop. 65 - Developmental Toxicity
developmental toxicity - initial date 10/1/92

FORMALDEHYDE, CONDENSATED POLYOXYETHY-   RR-01235-
LENE FATTY ACID, ESTERWITH STYRENATED
PHENOL, ETHYLENE OXIDE ADDUCT

STATE LISTS

California - Air Bill 2588 Appendix A-II
9/90

California - Prop. 65 - Developmental Toxicity
developmental toxicity - initial date 7/1/90

FORMALDEHYDE CYANOHYDRIN   107-16-4

ENVIRONMENTAL LISTS

CERCLA/SARA - Section 313 - Emission Reporting
form R reporting required

FORMALDEHYDE-NAPHTHALENESULFONIC ACID   9084-06-4
POLYMER SODIUM SALT

ENVIRONMENTAL LISTS

Safe Drinking Water Act - SMCLs
SMCL = 0.5 mg/L

FORMALDEHYDE, POLYMER WITH   RR-00995-0
(CHLOROMETHYL)OXIRANE, 4,4'-(1-METHYL
ETHYLIDENE) BIS(2,6-DIBROMOPHENOL) AND
PHENOL, 2-METHYL-2-PROPENOATE

HEALTH AND SAFETY LISTS

NIOSH - Selected LD50s and LC50s
Oral, rat: LD50 = 7540 mg/kg

ENVIRONMENTAL LISTS

CERCLA/SARA - Section 313 - Emission Reporting
form R reporting required

STATE LISTS

California - Air Bill 2588 Appendix A-II
known or potential carcinogen: 9/89

California - Prop. 65 - Cancer list
carcinogen - initial date 1/1/89

California - Prop. 65 - No Significant Risk Levels
no significant risk level = 200 ug/day

California - Directors List of Hazardous Substances (8 CCR 339)
[present]

FORMALDEHYDE, POLYMER WITH NONYLPHE-NOL AND OXIRANE   55845-06-2

ENVIRONMENTAL LISTS

CERCLA/SARA - Section 313 - Emission Reporting
form R reporting required

PROPOSED REGULATIONS

Safe Drinking Water Act - Priority list
[present]

FORMAMIDE   75-12-7

HEALTH AND SAFETY LISTS

ACGIH 1995 - Time Weighted Averages
0.1 mg/m3 TWA

ACGIH 1995 - Skin Designations
skin - potential for cutaneous absorption

U.S. DOT - Appendix B - Marine Pollutants
DOT regulated severe marine pollutant

NIOSH - Selected LD50s and LC50s
Oral, rat: LD50 = 3 mg/kg Skin, rat: LD50 = 147 mg/kg

OSHA - Vacated PELs - Time Weighted Averages
0.1 mg/m3 TWA

OSHA - Vacated PELs - Skin Designation
Prevent or reduce skin absorption

ENVIRONMENTAL LISTS

CERCLA/SARA - Section 302 Extremely Hazardous Substances and TPQs
TPQ = 500 pounds

INTERNATIONAL LISTS

Australian Exposure Standards - Time Weighted Averages
0.1 mg/m3 TWA

Australian Exposure Standards - Skin Effects
skin absorption

Canada - Alberta - 8 Hour Occupational Exposure Limit
0.1 mg/m3 TWA

Canada - Alberta - 15 Minute Occupational Exposure Limit
0.3 mg/m3 STEL

Canada - Alberta - Skin Designation
can be absorbed through the intact skin

Canada - British Columbia - 8 Hour Exposure Limits
0.1 mg/m3 TWA

Canada - Ontario - OHSA - TWAEVs
0.1 mg/m3 TWAEV

Canada - Ontario - OHSA - Skin Notations
absorption through skin, eyes, or mucous membranes

Canada - Quebec - Time-Weighted Average Exposure Values
0.1 mg/m3 TWAEV

Canada - Quebec - Skin Designations
absorbed through the skin

Israel - Time Weighted Averages
0.1 mg/m3 TWA

Israel - Action Levels
0.05 mg/m3 AL

Mexico - Instruction No. 10 - TWAs
0.1 mg/m3 TWA

STATE LISTS

California - Exposure Limits - PELs
0.1 mg/m3 PEL

California - Exposure Limits - Skin Notation
material may be absorbed through the skin, eyes or mucous membrane

California - Directors List of Hazardous Substances (8 CCR 339)
[present]

Florida Hazardous Substance List
[present]

Massachusetts Right To Know List
extraordinarily hazardous; neurotoxin

Minnesota Hazardous Substance List
skin

NJ Right to Know List (Total)
sn 0945

Pennsylvania Right to Know List
environmental hazard

PROPOSED REGULATIONS

CERCLA/SARA - Proposed Hazardous Substance Additions
proposed RQ = 1 pound (.454 kg)

CERCLA/SARA - 1989 Proposed RQ Adjustments
proposed RQ = 100 pounds (45.4 kg)

FORMETANATE   23422-53-9

SEE ALSO:
K040-HAZARDOUS WASTES
K038-HAZARDOUS WASTES
K009-HAZARDOUS WASTES
K010-HAZARDOUS WASTES

HEALTH AND SAFETY LISTS

ACGIH 1995 - Ceiling Limits
C 0.3 ppm; C 0.37 mg/m3

ACGIH 1995 - Carcinogens
A2-suspected human carcinogen

AIHA - Odor Threshold Values
no geometric mean air odor threshold

U.S. DOT - Substances From 49 CFR 172.101
regulated by DOT (UN1198, UN2209)

U.S. DOT - Hazard Classes
DOT hazard class = 9

U.S. DOT - Appendix A Table 1 - Hazardous Substances
final RQ = 100 pounds (45.4 kg)

IARC - Group 2A (limited human data)
[present]

NFPA - Flash Points
gas: no flash point given; 37% methanol-free: flash point = 185 degrees F (85 degrees C); 37% to 15% methanol: flash point = 122 degress F (50 degrees C)

NFPA - Hazard Identification Ratings
gas: health-3; flammability-4; reactivity-0; 37% methanol-free: health-3; flammability-2; reactivity-0; 37% to 15% methanol: health-3; flammability-2; reactivity-0

NIOSH - Selected LD50s and LC50s
Inhalation, rat: LC50 = 590 mg/m3 8 hr Oral, rat: LD50 = 800 mg/kg Skin, rabbit: LD50 = 270 mg/kg

NIOSH 1990 - Pocket Guide - RELs
0.016 ppm TWA; C 0.1 ppm (15 min)

NIOSH 1990 - Pocket Guide - IDLHs
30 ppm IDLH (not considering carcinogenic effects)

NIOSH 1990 - Pocket Guide - Carcinogens
occupational carcinogen

NIOSH 1990 - Pocket Guide - Target organs
eyes, skin, respiratory system

NIOSH - Health Standards - Exposure Limits
0.016 ppm TWA (8 hr); C (15 min) 0.1 ppm (this limit represents the lowest reliably quantifiable concentration)

NIOSH - Health Standards - Health Effects and Precautions
Nasal cancer (Implement medical monitoring; protect skin)

NIOSH - Health Standards - Carcinogenic Chemicals
potential human carcinogen

NTP Chemical Status Reports - Testing Status and NTIS Number
Prechronic studies for which toxicity technical reports were not prepared

NTP Seventh Report - Suspect Carcinogens
suspect carcinogen

OSHA - 29 CFR 1910 Specifically Regulated Chemicals
0.75 ppm TWA PEL; 2 ppm STEL; 0.5 ppm TWA action level; Irritant and potential cancer hazard (29 CFR 1910.1048)

OSHA - Vacated PELs - Time Weighted Averages
3 ppm TWA (unless specified in 1910.1048)

OSHA - Vacated PELs - Short Term Exposure Limits
10 ppm STEL (30 min) (unless specified in 1910.1048)

OSHA - Vacated PELs - Ceiling Limits
C 5 ppm (unless specified in 1910.1048)

OSHA - Select Carcinogens
[present]

OSHA - Possible Select Carcinogens
[present]

OSHA - List of Highly Hazardous Chemicals
threshhold quantity = 1000 pounds

ENVIRONMENTAL LISTS

ATSDR Priority List
Rank (of 275): 221

CERCLA/SARA - Section 302 Extremely Hazardous Substances and TPQs
TPQ = 500 pounds

CERCLA/SARA - Section 313 - Emission Reporting
form R reporting required for 0.1% de minimus concentration

CERCLA/SARA - Hazardous Substances and their Reportable Quantities
final RQ = 100 pounds (45.4 kg)

Clean Air Act (1990) - List of Hazardous Air Contaminants
[present]

CAA -Toxic Substances for Accidental Release Prevention
threshold quantity = 15,000 lbs

CAA - HON Rule - SOCMI Chemicals
compliance by Oct. 24, 1994

CAA - HON Rule - Organic HAPs
[present]

Clean Water Act - Hazardous Substances
[present]

EPA - Carcinogen Hazard Ranking for RQ Adjustment
Hazard ranking = Medium

EPA - Master Testing List
[present]

List of Pesticide Product Inert Ingredients
[present]

RCRA - U Series Wastes
waste number U122

RCRA - Hazardous Constituents-Appendix VIII
waste number U122

RCRA - Basis for Listing - Appendix VII
Included in waste streams: K009, K010, K038, K040

RCRA - Substances Banned From Land Disposal
[present]

INTERNATIONAL LISTS

Australian Exposure Standards - Time Weighted Averages
1 ppm TWA; 1.2 mg/m3 TWA

Australian Exposure Standards - Short Term Exposure Limits
2 ppm STEL; 2.5 mg/m3 STEL

Australian Exposure Standards - Skin Effects
sensitiser

Australian Exposure Standards - Carcinogens
probable carcinogen

Canada - WHMIS: Ingredient Disclosure
0.1% item 781 (918)

Canada - NPRI (National Pollutant Release Inventory)
[present]

Canada - Alberta - Ceiling Occupational Exposure Limit
C 2 ppm; C 2.4 mg/m3

Canada - British Columbia - Ceiling Exposure Limits
C 2 ppm; C 3 mg/m3

Canada - Ontario - OHSA - TWAEVs
1 ppm TWAEV; 1.5 mg/m3 TWAEV

Canada - Ontario - OHSA - STEVs
2 ppm STEV; 3 mg/m3 STEV

Canada - Quebec - Ceiling Limits
P 2 ppm; P 3 mg/m3

Canada - Quebec - Carcinogens
C2 carcinogen: effect suspected in humans

United Kingdom - Maximum Exposure Limits - TWAs
2 ppm TWA; 2.5 mg/m3 TWA

United Kingdom - Maximum Exposure Limits - STELs
2 ppm STEL; 2.5 mg/m3 STEL

German (DFG) - MAK Values
0.5 ppm MAK; 0.6 mg/m3 MAK

German (DFG) - Peak Limitations
2 x normal MAK (5 min momentary value); don't exceed 8 times during shift

German (DFG) - Skin/Sensitizers
danger of sensitization (skin or respiratory)

German (DFG) - Carcinogens
suspected carcinogen

German (DFG) - Pregnancy
no risk to embryo/fetus if exposure limits are adhered to

Israel - Short Term Exposure Limits
(2) ppm STEL; (2.5) mg/m3 STEL

Israel - Ceiling Exposure Limits
C 1 ppm

Israel - Action Levels
0.5 ppm AL

Mexico - Instruction No. 10 - Ceiling Limits
P 2 ppm; P 3 mg/m3

STATE LISTS

California - Air Bill 2588 Appendix A-I
known or potential carcinogen

California - Prop. 65 - Cancer list
carcinogen - initial date 1/1/88

California - Prop. 65 - No Significant Risk Levels
no significant risk level = 40 ug/day

California - Exposure Limits - PELs
1 ppm PEL; avoid inhalation and skin contact; see also section 5217 for chronic effects

California - Exposure Limits - STELs
2 ppm STEL

California - Directors List of Hazardous Substances (8 CCR 339)
[present]

Florida Hazardous Substance List
[present]

Massachusetts Right To Know List
carcinogen; extraordinarily hazardous

Minnesota Hazardous Substance List
carcinogen

NJ Right to Know List (Total)
sn 0946

NJ Special Hazardous Substances
(carcinogen; mutagen)

Pennsylvania Right to Know List
environmental hazard; special hazardous substance

Pennsylvania RTK - Special Hazardous Substances
[present]

PROPOSED REGULATIONS

Canada - Ontario - Proposed Occupational TWAEVs
0.5 ppm TWAEV; 0.6 mg/m3 TWAEV

FORMIC ACID   64-18-6

ENVIRONMENTAL LISTS

TSCA - Chemicals with Significant New Use Rules
PMN number: P-90-360

FORMOTHION   2540-82-1

HEALTH AND SAFETY LISTS

NIOSH - Selected LD50s and LC50s
Oral, rat: LD50 = 16 mg/kg Skin, rabbit: LD50 = 5 mg/kg

NIOSH - Health Standards - Exposure Limits
C (15 min) 2 ppm; C (15 min) 5 mg/m3 (Listed under 'Nitriles')

NIOSH - Health Standards - Health Effects and Precautions
Hepatic, renal, respiratory, cardiovascular, gastrointestinal, and nervous system effects (Periodic chest X-ray and pulmonary function testing required; prevent skin and eye contact; make first-aid kits and personnel available during use) (Listed under 'Nitriles')

ENVIRONMENTAL LISTS

CERCLA/SARA - Section 302 Extremely Hazardous Substances and TPQs
TPQ = 1000 pounds

STATE LISTS

Florida Hazardous Substance List
effective March 13, 1992

Massachusetts Right To Know List
extraordinarily hazardous

Minnesota Hazardous Substance List
[present]

NJ Right to Know List (Total)
sn 0962

Pennsylvania Right to Know List
environmental hazard

PROPOSED REGULATIONS

CERCLA/SARA - Proposed Hazardous Substance Additions
proposed RQ = 1 pound (.454 kg)

CERCLA/SARA - 1989 Proposed RQ Adjustments
proposed RQ = 1000 pounds (454 kg)

FORMPARANATE   17702-57-7

HEALTH AND SAFETY LISTS

NIOSH - Selected LD50s and LC50s
Oral, rat: LD50 = 3800 mg/kg

ENVIRONMENTAL LISTS

List of Pesticide Product Inert Ingredients
[present]

2-(2-FORMYLHYDRAZINO)-4-(5-NITRO-2-FURYL) THIAZOLE   3570-75-0

ENVIRONMENTAL LISTS

TSCA - Chemicals with Significant New Use Rules
PMN number: P-90-667

FOSTHIETAN   21548-32-3

ENVIRONMENTAL LISTS

List of Pesticide Product Inert Ingredients
[present]

FRANCIUM 222   36840-25-2

HEALTH AND SAFETY LISTS

ACGIH 1995 - Time Weighted Averages
10 ppm TWA; 18 mg/m3 TWA

ACGIH 1995 - Skin Designations
skin - potential for cutaneous absorption

NFPA - Flash Points
flash point = 310 degrees F (154 degrees C)

NFPA - Hazard Identification Ratings
health-2; flammability-1

NIOSH - Selected LD50s and LC50s
Oral, rat: LD50 = 6000 mg/kg

NTP Chemical Status Reports - Testing Status and NTIS Number
Project leader assigned/study in design

OSHA - Vacated PELs - Time Weighted Averages
20 ppm TWA; 30 mg/m3 TWA

OSHA - Vacated PELs - Short Term Exposure Limits
30 ppm STEL; 45 mg/m3 STEL

ENVIRONMENTAL LISTS

CAA - HON Rule - SOCMI Chemicals
compliance by Jan. 23, 1995

TSCA - Code of Federal Regulations Citations
40 CFR 712.30(d); 40 CFR 716.120(a)

TSCA - PAIR - Reporting List
Reporting Date: November 19, 1982

TSCA - Health and Safety Reporting List
Effective Date: April 29, 1983

INTERNATIONAL LISTS

Australian Exposure Standards - Time Weighted Averages
10 ppm TWA; 18 mg/m3 TWA

Australian Exposure Standards - Skin Effects
skin absorption

Canada - WHMIS: Ingredient Disclosure
0.1% item 782 (919)

Canada - Alberta - 8 Hour Occupational Exposure Limit
20 ppm TWA; 37 mg/m3 TWA

Canada - Alberta - 15 Minute Occupational Exposure Limit
30 ppm STEL; 55 mg/m3 STEL

Canada - British Columbia - 8 Hour Exposure Limits
20 ppm TWA; 30 mg/m3 TWA

Canada - British Columbia - 15 Minute Exposure Limits
30 ppm STEL; 45 mg/m3 STEL

Canada - Ontario - OHSA - TWAEVs
10 ppm TWAEV; 15 mg/m3 TWAEV

Canada - Ontario - OHSA - Skin Notations
absorption through skin, eyes, or mucous membranes

Canada - Quebec - Time-Weighted Average Exposure Values
10 ppm TWAEV; 18 mg/m3 TWAEV

Canada - Quebec - Skin Designations
absorbed through the skin

United Kingdom - Occupational Exposure Standards - TWAs
20 ppm TWA; 30 mg/m3 TWA

United Kingdom - Occupational Exposure Standards - STELs
30 ppm STEL;   45 mg/m3 STEL

Israel - Time Weighted Averages
10 ppm TWA;   18 mg/m3 TWA

Israel - Action Levels
5 ppm AL;   9 mg/m3 AL

Mexico - Instruction No. 10 - TWAs
20 ppm TWA;   30 mg/m3 TWA

Mexico - Instruction No. 10 - STELs
30 ppm STEL;   45 mg/m3 STEL

STATE LISTS

California - Exposure Limits - PELs
10 ppm PEL;   18 mg/m3 PEL

California - Exposure Limits - Skin Notation
material may be absorbed through the skin, eyes or mucous membrane

California - Directors List of Hazardous Substances (8 CCR 339)
[present]

Florida Hazardous Substance List
[present]

Massachusetts Right To Know List
[present]

Minnesota Hazardous Substance List
skin

NJ Right to Know List (Total)
sn 0947

Pennsylvania Right to Know List
[present]

FRANCIUM 223   15756-98-6

HEALTH AND SAFETY LISTS

U.S. DOT - Appendix B - Marine Pollutants
DOT regulated marine pollutant

NIOSH - Selected LD50s and LC50s
Oral, mouse: LD50 = 18 mg/kg Skin, rabbit: LD50 = 10200 mg/kg

ENVIRONMENTAL LISTS

CERCLA/SARA - Section 302 Extremely Hazardous Substances and TPQs
TPQ = 500/10,000 pounds

STATE LISTS

Massachusetts Right To Know List
extraordinarily hazardous

NJ Right to Know List (Total)
sn 2862

Pennsylvania Right to Know List
environmental hazard

PROPOSED REGULATIONS

CERCLA/SARA - Proposed Hazardous Substance Additions
proposed RQ = 1 pound (.454 kg)

CERCLA/SARA - 1989 Proposed RQ Adjustments
proposed RQ = 100 pounds (45.4 kg)

FRIANITE   RR-01083-3
SEE ALSO:
K010-HAZARDOUS WASTES
K009-HAZARDOUS WASTES

HEALTH AND SAFETY LISTS

ACGIH 1995 - Time Weighted Averages
5 ppm TWA;   9.4 mg/m3 TWA

ACGIH 1995 - Short Term Exposure Limits
10 ppm STEL;   19 mg/m3 STEL

AIHA - Odor Threshold Values
no geometric mean air odor threshold

U.S. DOT - Substances From 49 CFR 172.101
regulated by DOT (UN1779)

U.S. DOT - Hazard Classes
DOT hazard class = 8

U.S. DOT - Appendix A Table 1 - Hazardous Substances
final RQ = 5000 pounds (2270 kg)

NFPA - Flash Points
flash point = 156 degrees F (69 degrees C);   90% solution: flash poi = 122 degrees F (50 degrees C)

NFPA - Hazard Identification Ratings
health-3;   flammability-2;   reactivity-0

NIOSH - Selected LD50s and LC50s
Inhalation, rat: LC50 = 15 gm/m3 15 mn Oral, rat: LD50 = 110 mg/kg

NIOSH 1990 - Pocket Guide - RELs
5 ppm TWA;   9 mg/m3 TWA

NIOSH 1990 - Pocket Guide - IDLHs
30 ppm IDLH

NIOSH 1990 - Pocket Guide - Target organs
respiratory system, skin, kidneys, liver, eyes

NTP Chemical Status Reports - Testing Status and NTIS Number
Technical reports printed (PB93-149730)

OSHA - Vacated PELs - Time Weighted Averages
5 ppm TWA;   9 mg/m3 TWA

OSHA - Final PELs - Time Weighted Averages
5 ppm TWA;   9 mg/m3 TWA

ENVIRONMENTAL LISTS

CERCLA/SARA - Section 313 - Emission Reporting
form R reporting required

CERCLA/SARA - Hazardous Substances and their Reportable Quantities
final RQ = 5000 pounds (2270 kg)

CAA - HON Rule - SOCMI Chemicals
compliance by Jan. 23, 1995

Clean Water Act - Hazardous Substances
[present]

List of Pesticide Product Inert Ingredients
[present]

RCRA - U Series Wastes
waste number U123 (Corrosive waste;   Toxic waste)

RCRA - Hazardous Constituents-Appendix VIII
waste number U123

RCRA - Basis for Listing - Appendix VII
Included in waste streams: K009, K010

RCRA - Substances Banned From Land Disposal
[present]

INTERNATIONAL LISTS

Australian Exposure Standards - Time Weighted Averages
5 ppm TWA;   9.4 mg/m3 TWA

Australian Exposure Standards - Short Term Exposure Limits
10 ppm STEL;   19 mg/m3 STEL

Canada - WHMIS: Ingredient Disclosure
1% item 783 (89)

Canada - Alberta - 8 Hour Occupational Exposure Limit
5 ppm TWA;   9 mg/m3 TWA

Canada - Alberta - 15 Minute Occupational Exposure Limit
10 ppm STEL;   18 mg/m3 STEL

Canada - British Columbia - 8 Hour Exposure Limits
5 ppm TWA;   9 mg/m3 TWA

Canada - Ontario - OHSA - TWAEVs
5 ppm TWAEV;   9.4 mg/m3 TWAEV

Canada - Quebec - Time-Weighted Average Exposure Values
5 ppm TWAEV;   9.4 mg/m3 TWAEV

Canada - Quebec - Short-term Exposure Values
10 ppm STEV; 19 mg/m3 STEV

United Kingdom - Occupational Exposure Standards - TWAs
5 ppm TWA; 9 mg/m3 TWA

German (DFG) - MAK Values
5 ppm MAK; 9 mg/m3 MAK

German (DFG) - Peak Limitations
2 x normal MAK (5 min momentary value); don't exceed 8 times during shift

Israel - Time Weighted Averages
(5) ppm TWA; (9.4) mg/m3 TWA

Israel - Action Levels
2.5 ppm AL; 4.7 mg/m3 AL

Mexico - Instruction No. 10 - TWAs
5 ppm TWA; 9 mg/m3 TWA

STATE LISTS

California - Exposure Limits - PELs
5 ppm PEL; 9 mg/m3 PEL

California - Directors List of Hazardous Substances (8 CCR 339)
[present]

Florida Hazardous Substance List
[present]

Massachusetts Right To Know List
[present]

Minnesota Hazardous Substance List
[present]

NJ Right to Know List (Total)
sn 0948

NJ Special Hazardous Substances
(corrosive)

Pennsylvania Right to Know List
environmental hazard

PROPOSED REGULATIONS

Canada - Ontario - Proposed Occupational TWAEVs
3 ppm TWAEV; 5 mg/m3 TWAEV

Canada - Ontario - Proposed Occupational STEVs
5 ppm STEV; 9 mg/m3 STEV

FUBERIDAZOLE   3878-19-1

HEALTH AND SAFETY LISTS

NIOSH - Selected LD50s and LC50s
Inhalation, mouse: LC50 = 27 mg/m3 Oral, rat: LD50 = 250 mg/kg Skin, rat: LD50 = 353 mg/kg

ENVIRONMENTAL LISTS

CERCLA/SARA - Section 302 Extremely Hazardous Substances and TPQs
TPQ = 100 pounds

STATE LISTS

Florida Hazardous Substance List
effective March 13, 1992

Massachusetts Right To Know List
extraordinarily hazardous

NJ Right to Know List (Total)
sn 2439

Pennsylvania Right to Know List
environmental hazard

PROPOSED REGULATIONS

CERCLA/SARA - Proposed Hazardous Substance Additions
proposed RQ = 1 pound (.454 kg)

CERCLA/SARA - 1989 Proposed RQ Adjustments
proposed RQ = 10 pounds (4.54 kg)

FUEL, AVIATION, TURBINE ENGINE   RR-00315-6

HEALTH AND SAFETY LISTS

NIOSH - Selected LD50s and LC50s
Oral, rat: LD50 = 7200 ug/kg

ENVIRONMENTAL LISTS

CERCLA/SARA - Section 302 Extremely Hazardous Substances and TPQs
TPQ = 100/10,000 pounds

STATE LISTS

Florida Hazardous Substance List
effective March 13, 1992

Massachusetts Right To Know List
extraordinarily hazardous

Pennsylvania Right to Know List
environmental hazard

PROPOSED REGULATIONS

CERCLA/SARA - Proposed Hazardous Substance Additions
proposed RQ = 1 pound (.454 kg)

CERCLA/SARA - 1989 Proposed RQ Adjustments
proposed RQ = 100 pounds (45.4 kg)

FUEL CONTAINING TOXIC SUBSTANCES THAT   RR-01602-4
ARE DANGEROUS GOODS

HEALTH AND SAFETY LISTS

IARC - Group 2B (sufficient animal data)
[present] (Overall evaluation based only on evidence of carcinogenicity in monograph (7, 1974) or in Supplement 4)

OSHA - Possible Select Carcinogens
[present]

STATE LISTS

California - Air Bill 2588 Appendix A-II
known or potential carcinogen

California - Prop. 65 - Cancer list
carcinogen - initial date 1/1/88

California - Prop. 65 - No Significant Risk Levels
no significant risk level = 0.3 ug/day

California - Directors List of Hazardous Substances (8 CCR 339)
[present]

Florida Hazardous Substance List
[present]

Massachusetts Right To Know List
carcinogen; extraordinarily hazardous

Minnesota Hazardous Substance List
carcinogen

Pennsylvania Right to Know List
special hazardous substance

Pennsylvania RTK - Special Hazardous Substances
[present]

FUEL GASES   68476-26-6

HEALTH AND SAFETY LISTS

NIOSH - Selected LD50s and LC50s
Oral, rat: LD50 = 4700 ug/kg Skin, rabbit: LD50 = 27400 ug/kg

ENVIRONMENTAL LISTS

CERCLA/SARA - Section 302 Extremely Hazardous Substances and TPQs
TPQ = 500 pounds

STATE LISTS

Florida Hazardous Substance List
effective March 13, 1992

Massachusetts Right To Know List
extraordinarily hazardous

NJ Right to Know List (Total)
sn 2441

Pennsylvania Right to Know List
environmental hazard

PROPOSED REGULATIONS

CERCLA/SARA - Proposed Hazardous Substance Additions
proposed RQ = 1 pound (.454 kg)

CERCLA/SARA - 1989 Proposed RQ Adjustments
proposed RQ = 100 pounds (45.4 kg)

FUEL GASES, WATER GAS   8021-92-9

HEALTH AND SAFETY LISTS

U.S. DOT - Appendix A Table 2 - Radionuclides
final RQ = 100 curies (3.7E 12 Bq)

ENVIRONMENTAL LISTS

CERCLA/SARA List of Radionuclides (Appendix B) and Their Reportable Quantities
final RQ = 100 curies (3.7E 12 Bq)

FUEL OIL   68476-33-5

HEALTH AND SAFETY LISTS

U.S. DOT - Appendix A Table 2 - Radionuclides
final RQ = 100 curies (3.7E 12 Bq)

ENVIRONMENTAL LISTS

CERCLA/SARA List of Radionuclides (Appendix B) and Their Reportable Quantities
final RQ = 100 curies (3.7E 12 Bq)

FUEL OIL NO. 2   68476-30-2

ENVIRONMENTAL LISTS

List of Pesticide Product Inert Ingredients
[present]

FUEL OIL NO. 4   68476-31-3

HEALTH AND SAFETY LISTS

NIOSH - Selected LD50s and LC50s
Inhalation, rat: LC50 = 330 mg/m3 4 hr Oral, rat: LD50 = 1100 mg/kg Skin, rat: LD50 = 500 mg/kg

ENVIRONMENTAL LISTS

CERCLA/SARA - Section 302 Extremely Hazardous Substances and TPQs
TPQ = 100/10,000 pounds

STATE LISTS

Massachusetts Right To Know List
extraordinarily hazardous

NJ Right to Know List (Total)
sn 2442

Pennsylvania Right to Know List
environmental hazard

PROPOSED REGULATIONS

CERCLA/SARA - Proposed Hazardous Substance Additions
proposed RQ = 1 pound (.454 kg)

CERCLA/SARA - 1989 Proposed RQ Adjustments
proposed RQ = 100 pounds (45.4 kg)

FUEL OIL NO. 5   RR-00860-6

HEALTH AND SAFETY LISTS

U.S. DOT - Substances From 49 CFR 172.101
regulated by DOT (UN1863)

U.S. DOT - Hazard Classes
DOT hazard class = 3

STATE LISTS

NJ Right to Know List (Total)
sn 2443

FUEL OIL NO. 6   68553-00-

INTERNATIONAL LISTS

Canada - CEPA Schedule I - Toxic Substances
prohibition of importation and exportation

FUEL, PYROPHORIC, N.O.S.   RR-00317-8

HEALTH AND SAFETY LISTS

U.S. DOT - Substances From 49 CFR 172.101
regulated by DOT (NA1993)

U.S. DOT - Hazard Classes
DOT hazard class = 3

NFPA - Hazard Identification Ratings
health-2; flammability-4; reactivity-0

STATE LISTS

NJ Right to Know List (Total)
sn 2595

Pennsylvania Right to Know List
[present]

FULLERS EARTH   8031-18-3

STATE LISTS

Pennsylvania Right to Know List
[present]

FULMINATING GOLD   RR-01418-6

STATE LISTS

NJ Right to Know List (Total)
sn 2444

Pennsylvania Right to Know List
[present]

FULMINATING PLATINUM   RR-01419-7

HEALTH AND SAFETY LISTS

NFPA - Flash Points
flash point = 126 to 204 degrees F (52 to 96 degrees C)

NFPA - Hazard Identification Ratings
health-0; flammability-2; reactivity-0

FULMINATING SILVER   5610-59-3

HEALTH AND SAFETY LISTS

NFPA - Flash Points
flash point = 142 to 240 degrees F (61 to 116 degrees C)

NFPA - Hazard Identification Ratings
health-0; flammability-2; reactivity-0

FUMARIC ACID   110-17-8

HEALTH AND SAFETY LISTS

NFPA - Flash Points
light: flash point = 156 to 336 degrees F (69 to 169 degrees C); heavy: flash point = 160 to 250 degrees F (71 to 121 degrees C)

NFPA - Hazard Identification Ratings
health-0; flammability-2; reactivity-0

FUMARIC ACID, DIETHYL ESTER   623-91-6

HEALTH AND SAFETY LISTS

NFPA - Flash Points
flash point = 150 to 270 degrees F (66 to 132 degrees C)

NFPA - Hazard Identification Ratings
health-0; flammability-2; reactivity-0

ENVIRONMENTAL LISTS

List of Pesticide Product Inert Ingredients
[present]

FUMARONITRILE 764-42-1

STATE LISTS

NJ Right to Know List (Total)
sn 2445

FUMARYL CHLORIDE 627-63-4

ENVIRONMENTAL LISTS

List of Pesticide Product Inert Ingredients
[present]

INTERNATIONAL LISTS

German (DFG) - Carcinogens
animal evidence of carcinogenicity

FUMONISIN B1 116355-83-0

HEALTH AND SAFETY LISTS

U.S. DOT - Hazard Classes
Forbidden from transport by the DOT

FUMONISIN B2 116355-84-1

HEALTH AND SAFETY LISTS

U.S. DOT - Hazard Classes
Forbidden from transport by the DOT

FUNGICIDES, N.O.S. RR-00318-9
SEE ALSO:
SILVER

HEALTH AND SAFETY LISTS

U.S. DOT - Hazard Classes
Forbidden from transport by the DOT

2,5-FURADIONE POLYMER WITH ETHENE 9006-26-2

HEALTH AND SAFETY LISTS

U.S. DOT - Appendix A Table 1 - Hazardous Substances
final RQ = 5000 pounds (2270 kg)
NIOSH - Selected LD50s and LC50s
Oral, rat: LD50 = 10700 mg/kg Skin, rabbit: LD50 = 20 gm/kg

ENVIRONMENTAL LISTS

CERCLA/SARA - Hazardous Substances and their Reportable Quantities
final RQ = 5000 pounds (2270 kg)
CAA - HON Rule - SOCMI Chemicals
compliance by Oct. 24, 1994
Clean Water Act - Hazardous Substances
[present]
List of Pesticide Product Inert Ingredients
[present]

INTERNATIONAL LISTS

Canada - WHMIS: Ingredient Disclosure
1% item 784 (90)

STATE LISTS

California - Directors List of Hazardous Substances (8 CCR 339)
[present] (exempt except when present as a dust)
Massachusetts Right To Know List
[present]
NJ Right to Know List (Total)
sn 0949
Pennsylvania Right to Know List
environmental hazard

FURAN 110-00-9

HEALTH AND SAFETY LISTS

NFPA - Flash Points
flash point = 220 degrees F (104 degrees C)
NFPA - Hazard Identification Ratings
health-1; flammability-1; reactivity-0
NIOSH - Selected LD50s and LC50s
Oral, rat: LD50 = 1780 mg/kg

ENVIRONMENTAL LISTS

EPA - Master Testing List
[present]

FURANS, POLYHALOGENATED DIBENZO- RR-01477-7

HEALTH AND SAFETY LISTS

NIOSH - Selected LD50s and LC50s
Oral, rat: LD50 = 132 mg/kg

FURAZOLIDONE 67-45-8

HEALTH AND SAFETY LISTS

U.S. DOT - Substances From 49 CFR 172.101
regulated by DOT (UN1780)
U.S. DOT - Hazard Classes
DOT hazard class = 8
NIOSH - Selected LD50s and LC50s
Oral, rat: LD50 = 810 mg/kg Skin, rabbit: LD50 = 1410 mg/kg

INTERNATIONAL LISTS

Canada - WHMIS: Ingredient Disclosure
1% item 785 (500)

STATE LISTS

NJ Right to Know List (Total)
sn 0951
NJ Special Hazardous Substances
(corrosive)

FURFURAL 98-01-1

HEALTH AND SAFETY LISTS

IARC - Group Unspecified
[present] (Listed under 'Toxins derived from Fusarium moniliforme')
NTP Chemical Status Reports - Testing Status and NTIS Number
Prechronic studies in progress; prechronic studies for which toxicity technical reports were not prepared

FURFURYL ACETATE 623-17-6

HEALTH AND SAFETY LISTS

IARC - Group Unspecified
[present] (Listed under 'Toxins derived from Fusarium moniliforme')

FURFURYL ALCOHOL 98-00-0

STATE LISTS

NJ Right to Know List (Total)
sn 2446; sn 2447

FURFURYLAMINE 617-89-0

ENVIRONMENTAL LISTS

List of Pesticide Product Inert Ingredients
[present]

FURMECYCLOX 60568-05-0

HEALTH AND SAFETY LISTS

AIHA - WEEL - Time Weighted Averages
exposure should be minimized to the fullest extent possible
U.S. DOT - Substances From 49 CFR 172.101
regulated by DOT (UN2389)

U.S. DOT - Hazard Classes
DOT hazard class = 3

U.S. DOT - Appendix A Table 1 - Hazardous Substances
final RQ = 100 pounds (45.4 kg)

NFPA - Flash Points
flash point < 32 degrees F (0 degrees C)

NFPA - Hazard Identification Ratings
health-1; flammability-4; reactivity-1

NIOSH - Selected LD50s and LC50s
Inhalation, mouse: LC50 = 120 mg/m3 (1 hr)

NTP Chemical Status Reports - Testing Status and NTIS Number
Technical reports printed (no NTIS number given)

NTP Chemical Status Reports - Evidence of Carcinogenicity
male rat: clear evidence; female rat: clear evidence; male mice: clear evidence; female mice: clear evidence

OSHA - List of Highly Hazardous Chemicals
threshhold quantity = 500 pounds

ENVIRONMENTAL LISTS

CERCLA/SARA - Section 302 Extremely Hazardous Substances and TPQs
TPQ = 500 pounds

CERCLA/SARA - Hazardous Substances and their Reportable Quantities
final RQ = 100 pounds (45.4 kg)

CAA -Toxic Substances for Accidental Release Prevention
threshold quantity = 5,000 lbs

RCRA - U Series Wastes
waste number U124 (Ignitable waste)

RCRA - Hazardous Constituents-Appendix VIII
waste number U124 (Ignitable waste)

RCRA - Substances Banned From Land Disposal
[present]

STATE LISTS

California - Prop. 65 - Cancer list
carcinogen - initial date 10/1/93

Florida Hazardous Substance List
[present]

Massachusetts Right To Know List
extraordinarily hazardous

NJ Right to Know List (Total)
sn 0952

NJ Special Hazardous Substances
(flammable - fourth degree)

Pennsylvania Right to Know List
environmental hazard

PROPOSED REGULATIONS

NTP - Proposed Additions to Annual Report on Carcinogens
proposed as a suspect carcinogen for NTP 8th report

2-FUROIC ACID 26447-28-9

ENVIRONMENTAL LISTS

EPA - Master Testing List
[present]

FUROSEMIDE 54-31-9

HEALTH AND SAFETY LISTS

IARC - Group 3 (not classifiable)
[present]

STATE LISTS

California - Air Bill 2588 Appendix A-II
known or potential carcinogen: 9/90

California - Prop. 65 - Cancer list
carcinogen - initial date 1/1/90

FURYLFURAMIDE (AF-2) 3688-53-

HEALTH AND SAFETY LISTS

ACGIH 1995 - Time Weighted Averages
2 ppm TWA; 7.9 mg/m3 TWA

ACGIH 1995 - Skin Designations
skin - potential for cutaneous absorption

ACGIH 1995 - Biological Exposure Indices
Total furoic acid in urine: 200 mg/g creatinine, end of shift (B, Ns)

AIHA - Odor Threshold Values
no geometric mean air odor threshold

U.S. DOT - Substances From 49 CFR 172.101
regulated by DOT (UN1199)

U.S. DOT - Hazard Classes
DOT hazard class = 3

U.S. DOT - Appendix A Table 1 - Hazardous Substances
final RQ = 5000 pounds (2270 kg)

NFPA - Flash Points
flash point = 140 degrees F (60 degrees C)

NFPA - Hazard Identification Ratings
health-3; flammability-2; reactivity-0

NIOSH - Selected LD50s and LC50s
Inhalation, rat: LC50 - 153 ppm 4 hr Oral, rat: LD50 - 65 mg/kg

NIOSH 1990 - Pocket Guide - RELs
See Appendix D

NIOSH 1990 - Pocket Guide - IDLHs
250 ppm IDLH

NIOSH 1990 - Pocket Guide - Target organs
eyes, skin, respiratory system

NTP Chemical Status Reports - Testing Status and NTIS Number
Technical reports printed (PB91108662)

NTP Chemical Status Reports - Evidence of Carcinogenicity
male rat-some evidence; female rat-no evidence; male mice-clear evidence; female mice-some evidence

OSHA - Vacated PELs - Time Weighted Averages
2 ppm TWA; 8 mg/m3 TWA

OSHA - Vacated PELs - Skin Designation
Prevent or reduce skin absorption

OSHA - Final PELs - Time Weighted Averages
5 ppm TWA; 20 mg/m3 TWA

OSHA - Final PELs - Skin Notations
prevent or reduce skin absorption

ENVIRONMENTAL LISTS

CERCLA/SARA - Hazardous Substances and their Reportable Quantities
final RQ = 5000 pounds (2270 kg)

Clean Water Act - Hazardous Substances
[present]

RCRA - U Series Wastes
waste number U125 (Ignitable waste)

RCRA - Hazardous Constituents-Appendix VIII
waste number U125 (Ignitable waste)

RCRA - Substances Banned From Land Disposal
[present]

TSCA - Code of Federal Regulations Citations
40 CFR 712.30(x); 40 CFR 716.120(a)

TSCA - PAIR - Reporting List
Reporting Date: November 27, 1991

TSCA - Health and Safety Reporting List
Effective Date: June 1, 1987

INTERNATIONAL LISTS

Australian Exposure Standards - Time Weighted Averages
2 ppm TWA; 7.9 mg/m3 TWA

Australian Exposure Standards - Skin Effects
skin absorption
Canada - WHMIS: Ingredient Disclosure
1% item 786 (926)
Canada - Alberta - 8 Hour Occupational Exposure Limit
2 ppm TWA; 8 mg/m3 TWA
Canada - Alberta - 15 Minute Occupational Exposure Limit
10 ppm STEL; 39 mg/m3 STEL
Canada - Alberta - Skin Designation
can be absorbed through the intact skin
Canada - British Columbia - 8 Hour Exposure Limits
5 ppm TWA; 20 mg/m3 TWA
Canada - British Columbia - 15 Minute Exposure Limits
15 ppm STEL; 60 mg/m3 STEL
Canada - British Columbia - Skin Notations
skin - potential for skin absorption
Canada - Ontario - OHSA - TWAEVs
2 ppm TWAEV; 8 mg/m3 TWAEV
Canada - Ontario - OHSA - Skin Notations
absorption through skin, eyes, or mucous membranes
Canada - Quebec - Time-Weighted Average Exposure Values
2 ppm TWAEV; 7.9 mg/m3 TWAEV
Canada - Quebec - Skin Designations
absorbed through the skin
United Kingdom - Occupational Exposure Standards - TWAs
2 ppm TWA; 8 mg/m3 TWA
United Kingdom - Occupational Exposure Standards - STELs
10 ppm STEL; 40 mg/m3 STEL
United Kingdom - Occupational Exposure Standards - Notes
can be absorbed through skin
German (DFG) - Skin/Sensitizers
danger of cutaneous absorption
German (DFG) - Carcinogens
suspected carcinogen
Israel - Time Weighted Averages
2 ppm TWA; 7.9 mg/m3 TWA
Israel - Action Levels
1 ppm AL; 3.95 mg/m3 AL
Mexico - Instruction No. 10 - TWAs
2 ppm TWA; 8 mg/m3 TWA
Mexico - Instruction No. 10 - STELs
10 ppm STEL; 40 mg/m3 STEL
Mexico - Instruction No. 10 - Skin designation
skin - potential for cutaneous absorption

STATE LISTS
California - Exposure Limits - PELs
2 ppm PEL; 8 mg/m3 PEL
California - Exposure Limits - Skin Notation
material may be absorbed through the skin, eyes or mucous membrane
California - Directors List of Hazardous Substances (8 CCR 339)
[present]
Florida Hazardous Substance List
[present]
Massachusetts Right To Know List
[present]
Minnesota Hazardous Substance List
skin
NJ Right to Know List (Total)
sn 0953
Pennsylvania Right to Know List
environmental hazard

FUSARENON-X   23255-69-8
HEALTH AND SAFETY LISTS
NFPA - Flash Points
flash point = 185 degrees F (85 degrees C)
NFPA - Hazard Identification Ratings
health-1; flammability-2; reactivity-1

FUSARIN C   79748-81-5
HEALTH AND SAFETY LISTS
ACGIH 1995 - Time Weighted Averages
10 ppm TWA; 40 mg/m3 TWA
ACGIH 1995 - Short Term Exposure Limits
15 ppm STEL; 60 mg/m3 STEL
ACGIH 1995 - Skin Designations
skin - potential for cutaneous absorption
AIHA - Odor Threshold Values
geometric mean air odor threshold = 8.0 ppm (detectable)
U.S. DOT - Substances From 49 CFR 172.101
regulated by DOT (UN2874)
U.S. DOT - Hazard Classes
DOT hazard class = 6.1
NFPA - Flash Points
flash point = 167 degrees F (75 degrees C)
NFPA - Hazard Identification Ratings
health-1; flammability-2; reactivity-1
NIOSH - Selected LD50s and LC50s
Inhalation, rat: LC50 = 233 ppm 4 hr Oral, rat: LD50 = 88300 ug/kg
Skin, rabbit: LD50 = 400 mg/kg
NIOSH 1990 - Pocket Guide - RELs
10 ppm TWA; 40 mg/m3 TWA; 15 ppm STEL; 60 mg/m3 STEL
NIOSH 1990 - Pocket Guide - IDLHs
250 ppm IDLH
NIOSH 1990 - Pocket Guide - Target organs
respiratory system
NIOSH 1990 - Pocket Guide - Skin list
Potential for dermal absorption
NIOSH - Health Standards - Exposure Limits
50 ppm TWA; 200 mg/m3 TWA
NIOSH - Health Standards - Health Effects and Precautions
Respiratory effects
NTP Chemical Status Reports - Testing Status and NTIS Number
Prechronic studies for which toxicity technical reports were not prepared; Two year studies: laboratory study report in preparation
OSHA - Vacated PELs - Time Weighted Averages
10 ppm TWA; 40 mg/m3 TWA
OSHA - Vacated PELs - Short Term Exposure Limits
15 ppm STEL; 60 mg/m3 STEL
OSHA - Vacated PELs - Skin Designation
Prevent or reduce skin absorption
OSHA - Final PELs - Time Weighted Averages
50 ppm TWA; 200 mg/m3 TWA

INTERNATIONAL LISTS
Australian Exposure Standards - Time Weighted Averages
10 ppm TWA; 40 mg/m3 TWA
Australian Exposure Standards - Short Term Exposure Limits
15 ppm STEL; 60 mg/m3 STEL
Australian Exposure Standards - Skin Effects
skin absorption
Canada - WHMIS: Ingredient Disclosure
1% item 787 (175)
Canada - Alberta - 8 Hour Occupational Exposure Limit
10 ppm TWA; 40 mg/m3 TWA

Canada - Alberta - 15 Minute Occupational Exposure Limit
15 ppm STEL; 60 mg/m3 STEL

Canada - Alberta - Skin Designation
can be absorbed through the intact skin

Canada - British Columbia - 8 Hour Exposure Limits
5 ppm TWA; 20 mg/m3 TWA

Canada - British Columbia - 15 Minute Exposure Limits
10 ppm STEL; 40 mg/m3 STEL

Canada - British Columbia - Skin Notations
skin - potential for skin absorption

Canada - Ontario - OHSA - TWAEVs
10 ppm TWAEV; 40 mg/m3 TWAEV

Canada - Ontario - OHSA - STEVs
15 ppm STEV; 60 mg/m3 STEV

Canada - Ontario - OHSA - Skin Notations
absorption through skin, eyes, or mucous membranes

Canada - Quebec - Time-Weighted Average Exposure Values
10 ppm TWAEV; 40 mg/m3 TWAEV

Canada - Quebec - Short-term Exposure Values
15 ppm STEV; 60 mg/m3 STEV

Canada - Quebec - Skin Designations
absorbed through the skin

United Kingdom - Occupational Exposure Standards - TWAs
5 ppm TWA; 20 mg/m3 TWA

United Kingdom - Occupational Exposure Standards - STELs
15 ppm STEL; 60 mg/m3 STEL

United Kingdom - Occupational Exposure Standards - Notes
can be absorbed through skin

German (DFG) - MAK Values
10 ppm MAK; 40 mg/m3 MAK

Israel - Time Weighted Averages
10 ppm TWA; 40 mg/m3 TWA

Israel - Short Term Exposure Limits
15 ppm STEL; 60 mg/m3 STEL

Israel - Action Levels
5 ppm AL; 20 mg/m3 AL

Mexico - Instruction No. 10 - TWAs
10 ppm TWA; 40 mg/m3 TWA

Mexico - Instruction No. 10 - STELs
15 ppm STEL; 60 mg/m3 STEL

Mexico - Instruction No. 10 - Skin designation
skin - potential for cutaneous absorption

STATE LISTS

California - Exposure Limits - PELs
10 ppm PEL; 40 mg/m3 PEL

California - Exposure Limits - STELs
15 ppm STEL; 60 mg/m3 STEL

California - Exposure Limits - Skin Notation
material may be absorbed through the skin, eyes or mucous membrane

California - Directors List of Hazardous Substances (8 CCR 339)
[present]

Florida Hazardous Substance List
[present]

Massachusetts Right To Know List
[present]

Minnesota Hazardous Substance List
skin

NJ Right to Know List (Total)
sn 0954

Pennsylvania Right to Know List
[present]

FUSED SILICA   7699-41-

HEALTH AND SAFETY LISTS

U.S. DOT - Substances From 49 CFR 172.101
regulated by DOT (UN2526)

U.S. DOT - Hazard Classes
DOT hazard class = 3

NFPA - Flash Points
flash point = 99 degrees F (37 degrees C)

NFPA - Hazard Identification Ratings
flammability-3; reactivity-0

STATE LISTS

Florida Hazardous Substance List
[present]

Massachusetts Right To Know List
[present]

NJ Right to Know List (Total)
sn 0955

NJ Special Hazardous Substances
(flammable - third degree)

Pennsylvania Right to Know List
[present]

FUSEL OIL   8013-75-

STATE LISTS

California - Air Bill 2588 Appendix A-II
known or potential carcinogen: 9/90

California - Prop. 65 - Cancer list
carcinogen - initial date 1/1/90

California - Prop. 65 - No Significant Risk Levels
no significant risk level = 20 ug/day

GADOLINIUM 145   23315-89-1

HEALTH AND SAFETY LISTS

NIOSH - Selected LD50s and LC50s
Oral, mouse: LD50 = 1000 mg/kg

GADOLINIUM 146   14952-32-0

HEALTH AND SAFETY LISTS

IARC - Group 3 (not classifiable)
[present]

NTP Chemical Status Reports - Testing Status and NTIS Number
Technical reports printed (PB90106162/AS)

NTP Chemical Status Reports - Evidence of Carcinogenicity
male rat-equivocal evidence; female rat-no evidence; male mice-no evidence; female mice-some evidence

GADOLINIUM 147   14952-31-9

HEALTH AND SAFETY LISTS

IARC - Group 2B (sufficient animal data)
[present] (Overall evaluation based only on evidence of carcinogenicity in monograph (31, 1983) or in Supplement 4)

OSHA - Possible Select Carcinogens
[present]

INTERNATIONAL LISTS

Canada - WHMIS: Ingredient Disclosure
0.1% item 788 (927)

STATE LISTS

California - Air Bill 2588 Appendix A-II
known or potential carcinogen

California - Prop. 65 - Cancer list
carcinogen - initial date 7/1/87

California - Prop. 65 - No Significant Risk Levels
no significant risk level = 3 ug/day

California - Directors List of Hazardous Substances (8 CCR 339)
[present]
Massachusetts Right To Know List
carcinogen; extraordinarily hazardous
Minnesota Hazardous Substance List
carcinogen
Pennsylvania Right to Know List
special hazardous substance
Pennsylvania RTK - Special Hazardous Substances
[present]

GADOLINIUM 148   14119-21-2

HEALTH AND SAFETY LISTS
IARC - Group Unspecified
[present] (Listed under 'Toxins derived from Fusarium graminearum, F. culmorum and F. crookwellense')

GADOLINIUM 149   14937-16-7

HEALTH AND SAFETY LISTS
IARC - Group Unspecified
[present] (Listed under 'Toxins derived from Fusarium moniliforme')

GADOLINIUM 151   14937-17-8

INTERNATIONAL LISTS
Canada - Alberta - 8 Hour Occupational Exposure Limit
respirable mass: 0.1 mg/m3; total mass: 0.3 mg/m3 (See additional requirements in Part 5)
German (DFG) - MAK Values
fine dust: 0.3 mg/m3 MAK
German (DFG) - Pregnancy
no risk to embryo/fetus if exposure limits adhered to

GADOLINIUM 152   14867-54-0

HEALTH AND SAFETY LISTS
U.S. DOT - Substances From 49 CFR 172.101
regulated by DOT (UN1201)
U.S. DOT - Hazard Classes
DOT hazard class = 3

STATE LISTS
NJ Right to Know List (Total)
sn 3114

GADOLINIUM 153   14276-65-4

HEALTH AND SAFETY LISTS
U.S. DOT - Appendix A Table 2 - Radionuclides
final RQ = 100 curies (3.7E 12 Bq)

ENVIRONMENTAL LISTS
CERCLA/SARA List of Radionuclides (Appendix B) and Their Reportable Quantities
final RQ = 100 curies (3.7E 12 Bq)

GADOLINIUM 159   14041-42-0

HEALTH AND SAFETY LISTS
U.S. DOT - Appendix A Table 2 - Radionuclides
final RQ = 10 curies (3.7E 11 Bq)

ENVIRONMENTAL LISTS
CERCLA/SARA List of Radionuclides (Appendix B) and Their Reportable Quantities
final RQ = 10 curies (3.7E 11 Bq)

GADOLINIUM NITRATE   10168-81-7

HEALTH AND SAFETY LISTS
U.S. DOT - Appendix A Table 2 - Radionuclides
final RQ = 10 curies (3.7E 11 Bq)

ENVIRONMENTAL LISTS
CERCLA/SARA List of Radionuclides (Appendix B) and Their Reportable Quantities
final RQ = 10 curies (3.7E 11 Bq)

GADOLINIUM TRICHLORIDE   10138-52-0

HEALTH AND SAFETY LISTS
U.S. DOT - Appendix A Table 2 - Radionuclides
final RQ = 0.001 curies (3.7E 7 Bq)

ENVIRONMENTAL LISTS
CERCLA/SARA List of Radionuclides (Appendix B) and Their Reportable Quantities
final RQ = 0.001 curies (3.7E 7 Bq)

GALACTSAN TRINITRATE   RR-01420-0

HEALTH AND SAFETY LISTS
U.S. DOT - Appendix A Table 2 - Radionuclides
final RQ = 100 curies (3.7E 12 Bq)

ENVIRONMENTAL LISTS
CERCLA/SARA List of Radionuclides (Appendix B) and Their Reportable Quantities
final RQ = 100 curies (3.7E 12 Bq)

GALLIC ACID   149-91-7

HEALTH AND SAFETY LISTS
U.S. DOT - Appendix A Table 2 - Radionuclides
final RQ = 100 curies (3.7E 12 Bq)

ENVIRONMENTAL LISTS
CERCLA/SARA List of Radionuclides (Appendix B) and Their Reportable Quantities
final RQ = 100 curies (3.7E 12 Bq)

GALLIUM   7440-55-3

HEALTH AND SAFETY LISTS
U.S. DOT - Appendix A Table 2 - Radionuclides
final RQ = 0.001 curies (3.7E 7 Bq)

ENVIRONMENTAL LISTS
CERCLA/SARA List of Radionuclides (Appendix B) and Their Reportable Quantities
final RQ = 0.001 curies (3.7E 7 Bq)

GALLIUM 65   16922-44-4

HEALTH AND SAFETY LISTS
U.S. DOT - Appendix A Table 2 - Radionuclides
final RQ = 10 curies (3.7E 11 Bq)

ENVIRONMENTAL LISTS
CERCLA/SARA List of Radionuclides (Appendix B) and Their Reportable Quantities
final RQ = 10 curies (3.7E 11 Bq)

GALLIUM 66   14119-08-5

HEALTH AND SAFETY LISTS
U.S. DOT - Appendix A Table 2 - Radionuclides
final RQ = 1000 curies (3.7E 13 Bq)

ENVIRONMENTAL LISTS
CERCLA/SARA List of Radionuclides (Appendix B) and Their Reportable Quantities
final RQ = 1000 curies (3.7E 13 Bq)

GALLIUM 67   14119-09-6

HEALTH AND SAFETY LISTS
NIOSH - Selected LD50s and LC50s
Oral, rat: LD50 = 3805 mg/kg

GALLIUM 68   15757-14-9

INTERNATIONAL LISTS

Canada - WHMIS: Ingredient Disclosure
1% item 789 (1655)

GALLIUM 70   14391-74-3

HEALTH AND SAFETY LISTS

U.S. DOT - Hazard Classes
Forbidden from transport by the DOT

GALLIUM 72   13982-22-4

HEALTH AND SAFETY LISTS

NIOSH - Selected LD50s and LC50s
Oral, rabbit: LD50 = 5 gm/kg

GALLIUM 73   15034-51-2

HEALTH AND SAFETY LISTS

U.S. DOT - Substances From 49 CFR 172.101
regulated by DOT (UN2803)

U.S. DOT - Hazard Classes
DOT hazard class = 8

INTERNATIONAL LISTS

Canada - WHMIS: Ingredient Disclosure
1% item 791 (931)

STATE LISTS

NJ Right to Know List (Total)
sn 0956

NJ Special Hazardous Substances
(corrosive)

GALLIUM ARSENIDE   1303-00-0

HEALTH AND SAFETY LISTS

U.S. DOT - Appendix A Table 2 - Radionuclides
final RQ = 1000 curies (3.7E 13 Bq)

ENVIRONMENTAL LISTS

CERCLA/SARA List of Radionuclides (Appendix B) and Their Reportable Quantities
final RQ = 1000 curies (3.7E 13 Bq)

GALLIUM(III) NITRATE   13494-90-1

HEALTH AND SAFETY LISTS

U.S. DOT - Appendix A Table 2 - Radionuclides
final RQ = 10 curies (3.7E 11 Bq)

ENVIRONMENTAL LISTS

CERCLA/SARA List of Radionuclides (Appendix B) and Their Reportable Quantities
final RQ = 10 curies (3.7E 11 Bq)

GALLIUM OXIDE   12024-21-4

HEALTH AND SAFETY LISTS

U.S. DOT - Appendix A Table 2 - Radionuclides
final RQ = 100 curies (3.7E 12 Bq)

ENVIRONMENTAL LISTS

CERCLA/SARA List of Radionuclides (Appendix B) and Their Reportable Quantities
final RQ = 100 curies (3.7E 12 Bq)

GALLIUM PHOSPHIDE   12063-98-8

HEALTH AND SAFETY LISTS

U.S. DOT - Appendix A Table 2 - Radionuclides
final RQ = 1000 curies (3.7E 13 Bq)

ENVIRONMENTAL LISTS

CERCLA/SARA List of Radionuclides (Appendix B) and Their Reportable Quantities
final RQ = 1000 curies (3.7E 13 Bq)

GALLIUM TRICHLORIDE   13450-90-

HEALTH AND SAFETY LISTS

U.S. DOT - Appendix A Table 2 - Radionuclides
final RQ = 1000 curies (3.7E 13 Bq)

ENVIRONMENTAL LISTS

CERCLA/SARA List of Radionuclides (Appendix B) and Their Reportable Quantities
final RQ = 1000 curies (3.7E 13 Bq)

GASOHOL   RR-00322-

HEALTH AND SAFETY LISTS

U.S. DOT - Appendix A Table 2 - Radionuclides
final RQ = 10 curies (3.7E 11 Bq)

ENVIRONMENTAL LISTS

CERCLA/SARA List of Radionuclides (Appendix B) and Their Reportable Quantities
final RQ = 10 curies (3.7E 11 Bq)

GASOLINE   8006-61-

HEALTH AND SAFETY LISTS

U.S. DOT - Appendix A Table 2 - Radionuclides
final RQ = 100 curies (3.7E 12 Bq)

ENVIRONMENTAL LISTS

CERCLA/SARA List of Radionuclides (Appendix B) and Their Reportable Quantities
final RQ = 100 curies (3.7E 12 Bq)

GASOLINE (CASINGHEAD)   RR-00861-7
SEE ALSO:
ARSENIC
GALLIUM

HEALTH AND SAFETY LISTS

NIOSH - Health Standards - Exposure Limits
C (15 min) 2 ug As/m3

NIOSH - Health Standards - Health Effects and Precautions
may dissociate in the body, releasing Gallium and Inorgannic arsenic

NIOSH - Health Standards - Carcinogenic Chemicals
potential human carcinogen

NTP Chemical Status Reports - Testing Status and NTIS Number
Two year studies in progress; prechronic studies for which toxicity technical reports were not prepared

GASOLINE ENGINE EXHAUST (CONDEN-   RR-00266-4
SATES/EXTRACTS)
SEE ALSO:
GALLIUM

INTERNATIONAL LISTS

Canada - WHMIS: Ingredient Disclosure
1% item 792 (1205)

GASOLINE ENGINE EXHAUST   RR-01738-9
SEE ALSO:
GALLIUM

HEALTH AND SAFETY LISTS

NIOSH - Selected LD50s and LC50s
Oral, mouse: LD50 = 10 gm/kg

NTP Chemical Status Reports - Testing Status and NTIS Number
Prechronic studies for which toxicity technical reports were not prepared

GASOLINE, MOTOR FUEL   86290-81-5
SEE ALSO:
GALLIUM

HEALTH AND SAFETY LISTS
NIOSH - Selected LD50s and LC50s
Oral, mouse: LD50 = 8 gm/kg

GASOLINE, NATURAL   68425-31-0
SEE ALSO:
GALLIUM

ENVIRONMENTAL LISTS
CERCLA/SARA - Section 302 Extremely Hazardous Substances and TPQs
TPQ = 500/10,000 pounds

INTERNATIONAL LISTS
Canada - WHMIS: Ingredient Disclosure
1% item 790 (501)

STATE LISTS
Florida Hazardous Substance List
effective March 13, 1992
Massachusetts Right To Know List
extraordinarily hazardous
Pennsylvania Right to Know List
environmental hazard

PROPOSED REGULATIONS
CERCLA/SARA - Proposed Hazardous Substance Additions
proposed RQ = 1 pound (.454 kg)
CERCLA/SARA - 1989 Proposed RQ Adjustments
proposed RQ = 100 pounds (45.4 kg)

GAS, PRODUCER   8006-20-0
HEALTH AND SAFETY LISTS
U.S. DOT - Substances From 49 CFR 172.101
regulated by DOT (NA1203)
U.S. DOT - Hazard Classes
DOT hazard class = 3

STATE LISTS
NJ Right to Know List (Total)
sn 2451

GELATIN   9000-70-8
HEALTH AND SAFETY LISTS
ACGIH 1995 - Time Weighted Averages
300 ppm TWA; 890 mg/m3 TWA
ACGIH 1995 - Short Term Exposure Limits
500 ppm STEL; 1480 mg/m3 STEL
U.S. DOT - Substances From 49 CFR 172.101
regulated by DOT (UN1203)
U.S. DOT - Hazard Classes
DOT hazard class = 3
U.S. DOT - Appendix B - Marine Pollutants
DOT regulated marine pollutant
IARC - Group 2B (sufficient animal data)
[present]
NFPA - Flash Points
flash point = -45 degrees F (-43 degrees C); 100 octane: flash point = -36 degrees F (-38 degrees C); aviation grade 100-130: flash point = -50 degrees F (-46 degrees C); aviation grade 115-145: flash point = -50 degrees F (-46 degrees C)
NFPA - Hazard Identification Ratings
health-1; flammability-3; reactivity-0
OSHA - Vacated PELs - Time Weighted Averages
300 ppm TWA; 900 mg/m3 TWA

OSHA - Vacated PELs - Short Term Exposure Limits
500 ppm STEL; 1500 mg/m3 STEL
OSHA - Possible Select Carcinogens
[present]

INTERNATIONAL LISTS
Canada - WHMIS: Ingredient Disclosure
1% item 793 (802)
Canada - Alberta - 8 Hour Occupational Exposure Limit
300 ppm TWA; 900 mg/m3 TWA
Canada - Alberta - 15 Minute Occupational Exposure Limit
500 ppm STEL; 1500 mg/m3 STEL
Canada - British Columbia - 8 Hour Exposure Limits
(air contaminant): 500 ppm TWA; 625 mg/m3 TWA
Canada - Ontario - OHSA - TWAEVs
900 mg/m3 TWAEV (listed as an agent of variable composition)
Canada - Ontario - OHSA - STEVs
1500 mg/m3 STEV (listed as 'Agents of variable composition')
Canada - Quebec - Time-Weighted Average Exposure Values
300 ppm TWAEV; 890 mg/m3 (See standards applicable to its main consituents, in particular benzene, aromatic hydrocarbons and additives).
Canada - Quebec - Short-term Exposure Values
500 ppm STEV; 1480 mg/m3 STEV
Canada - Quebec - Carcinogens
C3 carcinogen: effect detected in animals
Israel - Time Weighted Averages
300 ppm TWA; 890 mg/m3 TWA
Israel - Short Term Exposure Limits
500 ppm STEL; 1480 mg/m3 STEL
Israel - Action Levels
150 ppm AL; 445 mg/m3 AL

STATE LISTS
California - Air Bill 2588 Appendix A-I
vapors: known or potential carcinogen
California - Exposure Limits - PELs
300 ppm PEL; 900 mg/m3 PEL
California - Exposure Limits - STELs
500 ppm STEL; 1500 mg/m3 STEL
California - Directors List of Hazardous Substances (8 CCR 339)
[present] (exempt when used as fuel)
Florida Hazardous Substance List
[present]
Massachusetts Right To Know List
[present]
Minnesota Hazardous Substance List
[present]
NJ Right to Know List (Total)
sn 0957
NJ Special Hazardous Substances
(flammable - third degree)

GERANIOL   106-24-1
HEALTH AND SAFETY LISTS
NFPA - Flash Points
flash point = 0 degrees F (-18 degrees C) (or less)
NFPA - Hazard Identification Ratings
health-1; flammability-4; reactivity-0

STATE LISTS
NJ Right to Know List (Total)
sn 2220

GERANYL ACETATE   105-87-3

HEALTH AND SAFETY LISTS

IARC - Group 2B (sufficient animal data)
[present]

OSHA - Possible Select Carcinogens
[present]

STATE LISTS

California - Air Bill 2588 Appendix A-I
known or potential carcinogen: 6/91

California - Prop. 65 - Cancer list
carcinogen - initial date 10/1/90

GERANYL BUTYRATE   106-29-6

STATE LISTS

California - Air Bill 2588 Appendix A-I
particulate matter: known or potential carcinogen: 9/90;   total organic gas:

GERANYL FORMATE   105-86-2

HEALTH AND SAFETY LISTS

U.S. DOT - Appendix B - Marine Pollutants
DOT regulated marine pollutant

INTERNATIONAL LISTS

Australian Exposure Standards - Time Weighted Averages
900 mg/m3 TWA

Australian Exposure Standards - Under Review
exposure limits under review

STATE LISTS

NJ Right to Know List (Total)
sn 2569

Pennsylvania Right to Know List
[present]

GERANYL PROPIONATE   105-90-8

HEALTH AND SAFETY LISTS

U.S. DOT - Substances From 49 CFR 172.101
regulated by DOT (UN1257)

U.S. DOT - Hazard Classes
DOT hazard class = 3

STATE LISTS

Pennsylvania Right to Know List
[present]

GERMANIUM   7440-56-4

HEALTH AND SAFETY LISTS

NFPA - Hazard Identification Ratings
health-2; flammability-4; reactivity-0

STATE LISTS

Florida Hazardous Substance List
[present]

Massachusetts Right To Know List
[present]

Pennsylvania Right to Know List
[present]

GERMANIUM 66   15756-84-0

ENVIRONMENTAL LISTS

List of Pesticide Product Inert Ingredients
[present]

GERMANIUM 67   15756-76-

HEALTH AND SAFETY LISTS

NFPA - Flash Points
flash point > 212 degrees F (100 degrees C)

NFPA - Hazard Identification Ratings
health-0; flammability-1; reactivity-0

NIOSH - Selected LD50s and LC50s
Oral, rat: LD50 = 3600 mg/kg

GERMANIUM 68   15756-77-

HEALTH AND SAFETY LISTS

NFPA - Flash Points
flash point > 212 degrees F (100 degrees C)

NFPA - Hazard Identification Ratings
health-0; flammability-1; reactivity-0

NTP Chemical Status Reports - Testing Status and NTIS Number
Technical reports printed (PB88174313/AS)

NTP Chemical Status Reports - Evidence of Carcinogenicity
male rat-negative; female rat-negative; male mice-negative; female mice-negative

GERMANIUM 69   15034-49-8

HEALTH AND SAFETY LISTS

NFPA - Flash Points
flash point > 212 degrees F (100 degrees C)

NFPA - Hazard Identification Ratings
health-0; flammability-1; reactivity-0

GERMANIUM 71   14374-81-3

HEALTH AND SAFETY LISTS

NFPA - Flash Points
flash point = 185 degrees F (85 degrees C)

NFPA - Hazard Identification Ratings
health-0; flammability-2; reactivity-0

GERMANIUM 75   14687-40-2

HEALTH AND SAFETY LISTS

NFPA - Flash Points
flash point > 212 degrees F (100 degrees C)

NFPA - Hazard Identification Ratings
health-0; flammability-1; reactivity-0

GERMANIUM 77   14687-59-3

INTERNATIONAL LISTS

Mexico - Wastewater - Organic Toxic Pollutants and Heavy Metals
Listed under [Heavy Metals]

GERMANIUM 78   15756-83-9

HEALTH AND SAFETY LISTS

U.S. DOT - Appendix A Table 2 - Radionuclides
final RQ = 100 curies (3.7E 12 Bq)

ENVIRONMENTAL LISTS

CERCLA/SARA List of Radionuclides (Appendix B) and Their Reportable Quantities
final RQ = 100 curies (3.7E 12 Bq)

GERMANIUM OXIDE   1310-53-8

HEALTH AND SAFETY LISTS

U.S. DOT - Appendix A Table 2 - Radionuclides
final RQ = 1000 curies (3.7E 13 Bq)

ENVIRONMENTAL LISTS
  CERCLA/SARA List of Radionuclides (Appendix B) and Their Reportable Quantities
  final RQ = 1000 curies (3.7E 13 Bq)

GERMANIUM TETRACHLORIDE   10038-98-9
  HEALTH AND SAFETY LISTS
    U.S. DOT - Appendix A Table 2 - Radionuclides
    final RQ = 10 curies (3.7E 11 Bq)

  ENVIRONMENTAL LISTS
    CERCLA/SARA List of Radionuclides (Appendix B) and Their Reportable Quantities
    final RQ = 10 curies (3.7E 11 Bq)

GERMANIUM TETRAHYDRIDE   7782-65-2
  HEALTH AND SAFETY LISTS
    U.S. DOT - Appendix A Table 2 - Radionuclides
    final RQ = 10 curies (3.7E 11 Bq)

  ENVIRONMENTAL LISTS
    CERCLA/SARA List of Radionuclides (Appendix B) and Their Reportable Quantities
    final RQ = 10 curies (3.7E 11 Bq)

GIBBERELLIC ACID   77-06-5
  HEALTH AND SAFETY LISTS
    U.S. DOT - Appendix A Table 2 - Radionuclides
    final RQ = 1000 curies (3.7E 13 Bq)

  ENVIRONMENTAL LISTS
    CERCLA/SARA List of Radionuclides (Appendix B) and Their Reportable Quantities
    final RQ = 1000 curies (3.7E 13 Bq)

GILSONITE   12002-43-6
  HEALTH AND SAFETY LISTS
    U.S. DOT - Appendix A Table 2 - Radionuclides
    final RQ = 1000 curies (3.7E 13 Bq)

  ENVIRONMENTAL LISTS
    CERCLA/SARA List of Radionuclides (Appendix B) and Their Reportable Quantities
    final RQ = 1000 curies (3.7E 13 Bq)

GLASS FILAMENTS   RR-01545-2
  HEALTH AND SAFETY LISTS
    U.S. DOT - Appendix A Table 2 - Radionuclides
    final RQ = 10 curies (3.7E 11 Bq)

  ENVIRONMENTAL LISTS
    CERCLA/SARA List of Radionuclides (Appendix B) and Their Reportable Quantities
    final RQ = 10 curies (3.7E 11 Bq)

GLASS, OXIDE   65997-17-3
  HEALTH AND SAFETY LISTS
    U.S. DOT - Appendix A Table 2 - Radionuclides
    final RQ = 1000 curies (3.7E 13 Bq)

  ENVIRONMENTAL LISTS
    CERCLA/SARA List of Radionuclides (Appendix B) and Their Reportable Quantities
    final RQ = 1000 curies (3.7E 13 Bq)

GLU-P-1   67730-11-4
  HEALTH AND SAFETY LISTS
    NIOSH - Selected LD50s and LC50s
    Oral, rat: LD50 = 1250 mg/kg

GLU-P-2   67730-10-3
  HEALTH AND SAFETY LISTS
    U.S. DOT - Substances Which Are Poisonous by Inhalation
    liquid hazardous material poisonous by inhalation

D-GLUCITOL   50-70-4
  HEALTH AND SAFETY LISTS
    ACGIH 1995 - Time Weighted Averages
    0.2 ppm TWA; 0.63 mg/m3 TWA
    U.S. DOT - Substances From 49 CFR 172.101
    regulated by DOT (UN2192)
    U.S. DOT - Hazard Classes
    DOT hazard class = 2.3
    U.S. DOT - Substances Which Are Poisonous by Inhalation
    gaseous hazardous material poisonous by inhalation (UN2192)
    OSHA - Vacated PELs - Time Weighted Averages
    0.2 ppm TWA; 0.6 mg/m3 TWA

  INTERNATIONAL LISTS
    Australian Exposure Standards - Time Weighted Averages
    0.2 ppm TWA; 0.63 mg/m3 TWA
    Canada - WHMIS: Ingredient Disclosure
    1% item 794 (1595)
    Canada - Alberta - 8 Hour Occupational Exposure Limit
    0.2 ppm TWA; 0.63 mg/m3 TWA
    Canada - Alberta - 15 Minute Occupational Exposure Limit
    0.6 ppm STEL; 1.9 mg/m3 STEL
    Canada - British Columbia - 8 Hour Exposure Limits
    0.2 ppm TWA; 0.6 mg/m3 TWA
    Canada - British Columbia - 15 Minute Exposure Limits
    0.6 ppm STEL; 1.8 mg/m3 STEL
    Canada - Ontario - OHSA - TWAEVs
    0.2 ppm TWAEV; 0.63 mg/m3 TWAEV
    Canada - Quebec - Time-Weighted Average Exposure Values
    0.2 ppm TWAEV; 0.63 mg/m3 TWAEV
    United Kingdom - Occupational Exposure Standards - TWAs
    0.2 ppm TWA; 0.6 mg/m3 TWA
    United Kingdom - Occupational Exposure Standards - STELs
    0.6 ppm STEL; 1.8 mg/m3 STEL
    Israel - Time Weighted Averages
    0.2 ppm TWA; 0.63 mg/m3 TWA
    Israel - Action Levels
    0.1 ppm AL; 0.315 mg/m3 AL
    Mexico - Instruction No. 10 - TWAs
    0.2 ppm TWA; 0.6 mg/m3 TWA
    Mexico - Instruction No. 10 - STELs
    0.6 ppm STEL; 1.8 mg/m3 STEL

  STATE LISTS
    California - Exposure Limits - PELs
    0.2 ppm PEL; 0.6 mg/m3 PEL
    California - Directors List of Hazardous Substances (8 CCR 339)
    [present]
    Florida Hazardous Substance List
    [present]
    Massachusetts Right To Know List
    [present]
    Minnesota Hazardous Substance List
    [present]
    NJ Right to Know List (Total)
    sn 0958
    Pennsylvania Right to Know List
    [present]

ALPHA,DELTA-GLUCOHEPTONIC ACID, SODIUM   10094-62-9
SALT, DIHYDRATE
  ENVIRONMENTAL LISTS
    List of Pesticide Product Inert Ingredients
    [present]

GLUCONIC ACID   526-95-4
  ENVIRONMENTAL LISTS
    List of Pesticide Product Inert Ingredients
    [present]

GLUCONIC ACID, MONOPOTASSIUM SALT   299-27-4
  HEALTH AND SAFETY LISTS
    IARC - Group 3 (not classifiable)
    [present]

D-GLUCONIC ACID, POTASSIUM SALT   35087-77-5
  STATE LISTS
    Minnesota Hazardous Substance List
    [present]

DELTA-GLUCONOLACTONE   90-80-2
  HEALTH AND SAFETY LISTS
    IARC - Group 2B (sufficient animal data)
    *[present] (Overall evaluation based only on evidence of carcinogenicity
    in monograph (40, 1986) or in Supplement 4)*
    OSHA - Possible Select Carcinogens
    [present]
  STATE LISTS
    California - Air Bill 2588 Appendix A-II
    known or potential carcinogen
    California - Prop. 65 - Cancer list
    carcinogen - initial date 1/1/90
    California - Prop. 65 - No Significant Risk Levels
    no significant risk level = 0.1 ug/day
    California - Directors List of Hazardous Substances (8 CCR 339)
    [present]
    Massachusetts Right To Know List
    carcinogen; extraordinarily hazardous
    Minnesota Hazardous Substance List
    carcinogen

BETA-D-GLUCOPYRANOSIDE, 1-O-OCTYL   29836-26-8
  HEALTH AND SAFETY LISTS
    IARC - Group 2B (sufficient animal data)
    *[present] (Overall evaluation based only on evidence of carcinogenicity
    in monograph (40, 1986) or in Supplement 4)*
    OSHA - Possible Select Carcinogens
    [present]
  STATE LISTS
    California - Air Bill 2588 Appendix A-II
    known or potential carcinogen
    California - Prop. 65 - Cancer list
    carcinogen - initial date 1/1/90
    California - Prop. 65 - No Significant Risk Levels
    no significant risk level = 0.5 ug/day
    California - Directors List of Hazardous Substances (8 CCR 339)
    [present]
    Massachusetts Right To Know List
    carcinogen; extraordinarily hazardous
    Minnesota Hazardous Substance List
    carcinogen

GLUCOSE   50-99-
  HEALTH AND SAFETY LISTS
    NIOSH - Selected LD50s and LC50s
    Oral, rat: LD50 = 15900 mg/kg
  ENVIRONMENTAL LISTS
    List of Pesticide Product Inert Ingredients
    [present]

GLUCOSE PENTAPROPIONATE   RR-00862-
  ENVIRONMENTAL LISTS
    List of Pesticide Product Inert Ingredients
    [present]

GLUE (AS DEPOLYMERIZED ANIMAL COLLAGEN)   68476-37-
  ENVIRONMENTAL LISTS
    List of Pesticide Product Inert Ingredients
    [present]

ALPHA-GLUTAMIC ACID   56-86-
  HEALTH AND SAFETY LISTS
    NIOSH - Selected LD50s and LC50s
    Oral, rat: LD50 = 10380 mg/kg
  ENVIRONMENTAL LISTS
    List of Pesticide Product Inert Ingredients
    [present]

GLUTAMIC ACID, SODIUM SALT   142-47-2
  ENVIRONMENTAL LISTS
    List of Pesticide Product Inert Ingredients
    [present]

GLUTARALDEHYDE   111-30-8
  ENVIRONMENTAL LISTS
    List of Pesticide Product Inert Ingredients
    [present]

GLUTARIC ACID   110-94-1
  ENVIRONMENTAL LISTS
    List of Pesticide Product Inert Ingredients
    [present]

GLUTARIC ANHYDRIDE   108-55-4
  HEALTH AND SAFETY LISTS
    NIOSH - Selected LD50s and LC50s
    Oral, rat: LD50 = 25800 mg/kg
  ENVIRONMENTAL LISTS
    List of Pesticide Product Inert Ingredients
    [present]

GLUTATHIONE   70-18-8
  HEALTH AND SAFETY LISTS
    NFPA - Flash Points
    flash point = 509 degrees F (265 degrees C)
    NFPA - Hazard Identification Ratings
    health-1; flammability-1; reactivity-0

GLYCERALDEHYDE   367-47-5
  ENVIRONMENTAL LISTS
    List of Pesticide Product Inert Ingredients
    [present]

GLYCERIN   56-81-5
  ENVIRONMENTAL LISTS

List of Pesticide Product Inert Ingredients
[present]

GLYCEROL 1,3-DIGLYCIDYL ETHER   3568-29-4

ENVIRONMENTAL LISTS

List of Pesticide Product Inert Ingredients
[present]

GLYCEROL-1,3-DINITRATE   623-87-0

HEALTH AND SAFETY LISTS

ACGIH 1995 - Ceiling Limits
C 0.2 ppm; C 0.82 mg/m3

NIOSH - Selected LD50s and LC50s
Inhalation, rat: LC50 = 5000 ppm 4 hr Oral, rat: LD50 = 134 mg/kg
Skin, rabbit: LD50 = 2560 mg/kg

NTP Chemical Status Reports - Testing Status and NTIS Number
Two year studies in progress; technical reports printed (no NTIS number given)

OSHA - Vacated PELs - Ceiling Limits
C 0.2 ppm; C 0.8 mg/m3

ENVIRONMENTAL LISTS

CAA - HON Rule - SOCMI Chemicals
compliance by July 24, 1995

List of Pesticide Product Inert Ingredients
[present]

TSCA - Code of Federal Regulations Citations
40 CFR 712.30(x); 40 CFR 716.120(d)

TSCA - PAIR - Reporting List
Reporting Date: November 27, 1991

TSCA - Health and Safety Reporting List
Effective Date: September 30, 1991

INTERNATIONAL LISTS

Australian Exposure Standards - Time Weighted Averages
Peak Limitation: 0.2 ppm; 0.82 mg/m3

Australian Exposure Standards - Skin Effects
sensitiser

Australian Exposure Standards - Under Review
exposure limits under review

Canada - WHMIS: Ingredient Disclosure
1% item 795 (932)

Canada - Alberta - Ceiling Occupational Exposure Limit
C 0.2 ppm; C 0.82 mg/m3

Canada - British Columbia - Ceiling Exposure Limits
C 0.25 mg/m3

Canada - Ontario - OHSA - CEVs
0.2 ppm CEV; 0.8 mg/m3 CEV

Canada - Quebec - Ceiling Limits
P 0.2 ppm; P 0.82 mg/m3

United Kingdom - Occupational Exposure Standards - STELs
0.2 ppm STEL; 0.7 mg/m3 STEL

German (DFG) - MAK Values
0.1 ppm MAK; 0.4 mg/m3 MAK

German (DFG) - Peak Limitations
2 x normal MAK (5 min momentary value); don't exceed 8 times during shift

German (DFG) - Skin/Sensitizers
danger of sensitization (skin or respiratory)

German (DFG) - Pregnancy
no risk to embryo/fetus if exposure limits adhered to

Israel - Ceiling Exposure Limits
C 0.2 ppm; C 0.82 mg/m3

Mexico - Instruction No. 10 - Ceiling Limits
P 0.2 ppm; P 0.7 mg/m3

STATE LISTS

California - Air Bill 2588 Appendix A-I
[present]

California - Exposure Limits - Ceilings
C 0.2 ppm; C 0.82 mg/m3

California - Directors List of Hazardous Substances (8 CCR 339)
[present]

Florida Hazardous Substance List
[present]

Massachusetts Right To Know List
[present]

Minnesota Hazardous Substance List
[present]

Pennsylvania Right to Know List
[present]

PROPOSED REGULATIONS

Canada - Ontario - Proposed Occupational CEVs
0.25 mg/m3 CEV

GLYCEROL GLUCONATE TRINITRATE   RR-01421-1

HEALTH AND SAFETY LISTS

NIOSH - Selected LD50s and LC50s
Oral, mouse: LD50 = 6000 mg/kg

GLYCEROL LACTATE TRINITRATE   RR-01422-2

HEALTH AND SAFETY LISTS

NIOSH - Selected LD50s and LC50s
Oral, rat: LD50 = 4460 mg/kg Skin, rabbit: LD50 = 1780 mg/kg

GLYCEROL MONOOLEATE   25496-72-4

HEALTH AND SAFETY LISTS

NIOSH - Selected LD50s and LC50s
Oral, mouse: LD50 = 5 gm/kg

GLYCEROL TRIACETATE   102-76-1

ENVIRONMENTAL LISTS

CAA - HON Rule - SOCMI Chemicals
compliance by Oct. 23, 1995

GLYCERYL P-AMINOBENZOATE   136-44-7

HEALTH AND SAFETY LISTS

ACGIH 1995 - Time Weighted Averages
10 mg/m3 TWA (total dust/particulate)

NFPA - Flash Points
flash point = 390 degrees F (199 degrees C)

NFPA - Hazard Identification Ratings
health-1; flammability-1; reactivity-0

NIOSH - Selected LD50s and LC50s
Oral, rat: LD50 = 12600 mg/kg

OSHA - Vacated PELs - Time Weighted Averages
total dust: 10 mg/m3 TWA; respirable fraction: 5 mg/m3 TWA

OSHA - Final PELs - Time Weighted Averages
total dust: 15 mg/m3 TWA; respirable fraction: 5 mg/m3 TWA

ENVIRONMENTAL LISTS

CAA - HON Rule - SOCMI Chemicals
compliance by Jan. 23, 1995

List of Pesticide Product Inert Ingredients
[present]

INTERNATIONAL LISTS

Australian Exposure Standards - Time Weighted Averages
10 mg/m3 TWA

Canada - Alberta - 8 Hour Occupational Exposure Limit
10 mg/m3 TWA

Canada - Alberta - 15 Minute Occupational Exposure Limit
   20 mg/m3 STEL
Canada - British Columbia - 8 Hour Exposure Limits
   nuisance dust, mist, and fume: 10 mg/m3 TWA
Canada - Ontario - OHSA - TWAEVs
   total dust: 10 mg/m3 TWAEV (listed as nuisance particulates)
Canada - Quebec - Time-Weighted Average Exposure Values
   10 mg/m3 TWAEV
United Kingdom - Occupational Exposure Standards - TWAs
   mist: 10 mg/m3 TWA
Israel - Time Weighted Averages
   10 mg/m3 TWA (total dust/particulate)
Israel - Action Levels
   5 mg/m3 AL (total dust/particulate)
Mexico - Instruction No. 10 - TWAs
   10 mg/m3 TWA;   (nuisance particulate)

STATE LISTS
Minnesota Hazardous Substance List
   [present] (includes inert or nuisance dust)
Pennsylvania Right to Know List
   [present]

GLYCERYL DIOLEATE   25637-84-7

ENVIRONMENTAL LISTS
EPA - Master Testing List
   [present]

PROPOSED REGULATIONS
TSCA - Proposed Testing Rule for Glycidyl Ethers
   member of Glycidyl subcategory V-A

GLYCERYL MONOACETATE   26446-35-5

HEALTH AND SAFETY LISTS
U.S. DOT - Hazard Classes
   Forbidden from transport by the DOT

GLYCERYL MONORICINOLEATE   1323-38-2

HEALTH AND SAFETY LISTS
U.S. DOT - Hazard Classes
   Forbidden from transport by the DOT

GLYCERYL MONOSTEARATE   31566-31-1

HEALTH AND SAFETY LISTS
U.S. DOT - Hazard Classes
   Forbidden from transport by the DOT

GLYCERYL MONOTHIOGLYCOLATE   30618-84-9

ENVIRONMENTAL LISTS
List of Pesticide Product Inert Ingredients
   [present]

GLYCERYL STEARATE   123-94-4

HEALTH AND SAFETY LISTS
NFPA - Flash Points
   flash point = 280 degrees F (138 degrees C)
NFPA - Hazard Identification Ratings
   health-1; flammability-1; reactivity-0
NIOSH - Selected LD50s and LC50s
   Oral, rat: LD50 = 3000 mg/kg

ENVIRONMENTAL LISTS
List of Pesticide Product Inert Ingredients
   [present]

GLYCERYL TRIBUTYRATE   60-01-5

ENVIRONMENTAL LISTS
List of Pesticide Product Inert Ingredients
   [present]

GLYCERYL TRIPROPIONATE   139-45-

ENVIRONMENTAL LISTS
List of Pesticide Product Inert Ingredients
   [present]

GLYCERYL TRIS(12-HYDROXYSTEARATE)   139-44-6

ENVIRONMENTAL LISTS
List of Pesticide Product Inert Ingredients
   [present]

GLYCIDOL   556-52-5

ENVIRONMENTAL LISTS
List of Pesticide Product Inert Ingredients
   [present]

GLYCIDOL (OXIRANEMETHANOL) AND ITS   RR-00275-5
DERIVATIVES

ENVIRONMENTAL LISTS
List of Pesticide Product Inert Ingredients
   [present]

GLYCIDOXYPROPYLTRIMETHOXYSILANE   2530-83-8

INTERNATIONAL LISTS
German (DFG) - Skin/Sensitizers
   danger of sensitization

GLYCIDYL ACRYLATE   106-90-1

STATE LISTS
California - Exposure Limits - PELs
   10 mg/m3 PEL

GLYCIDYLALDEHYDE   765-34-4

HEALTH AND SAFETY LISTS
NFPA - Flash Points
   flash point = 356 degrees F (180 degrees C)
NFPA - Hazard Identification Ratings
   health-0; flammability-1; reactivity-0

GLYCIDYL ETHERS   RR-00551-6

HEALTH AND SAFETY LISTS
NFPA - Flash Points
   flash point = 332 degrees F (167 degrees C)
NFPA - Hazard Identification Ratings
   health-0; flammability-1; reactivity-0

GLYCIDYL METHACRYLATE   106-91-2

ENVIRONMENTAL LISTS
List of Pesticide Product Inert Ingredients
   [present]

GLYCIDYL OLEATE   5431-33-4

HEALTH AND SAFETY LISTS
ACGIH 1995 - Time Weighted Averages
   25 ppm TWA; 76 mg/m3 TWA
NIOSH - Selected LD50s and LC50s
   Inhalation, rat: LC50 = 580 ppm 8 hr Oral, rat: LD50 = 420 mg/kg
   Skin, rabbit: LD50 = 1980 mg/kg
NIOSH 1990 - Pocket Guide - RELs
   25 ppm TWA; 75 mg/m3 TWA

NIOSH 1990 - Pocket Guide - IDLHs
500 ppm IDLH

NIOSH 1990 - Pocket Guide - Target organs
eyes, skin, CNS, respiratory system

NTP Chemical Status Reports - Testing Status and NTIS Number
Technical reports printed (PB90259094)

NTP Chemical Status Reports - Evidence of Carcinogenicity
male rat-clear evidence; female rat-clear evidence; male mice-clear evidence; female mice-clear evidence

NTP Seventh Report - Suspect Carcinogens
suspect carcinogen

OSHA - Vacated PELs - Time Weighted Averages
25 ppm TWA; 75 mg/m3 TWA

OSHA - Final PELs - Time Weighted Averages
50 ppm TWA; 150 mg/m3 TWA

ENVIRONMENTAL LISTS

EPA - Master Testing List
[present]

TSCA - Code of Federal Regulations Citations
40 CFR 712.30(d);　40 CFR 716.120(c)

TSCA - PAIR - Reporting List
Reporting Date: November 19, 1982

INTERNATIONAL LISTS

Australian Exposure Standards - Time Weighted Averages
25 ppm TWA; 76 mg/m3 TWA

Canada - WHMIS: Ingredient Disclosure
1% item 797 (934)

Canada - Alberta - 8 Hour Occupational Exposure Limit
25 ppm TWA; 76 mg/m3 TWA

Canada - Alberta - 15 Minute Occupational Exposure Limit
100 ppm STEL; 303 mg/m3 STEL

Canada - British Columbia - 8 Hour Exposure Limits
50 ppm TWA; 150 mg/m3 TWA

Canada - British Columbia - 15 Minute Exposure Limits
75 ppm STEL; 225 mg/m3 STEL

Canada - Ontario - OHSA - TWAEVs
25 ppm TWAEV; 76 mg/m3 TWAEV

Canada - Quebec - Time-Weighted Average Exposure Values
25 ppm TWAEV; 76 mg/m3 TWAEV

German (DFG) - MAK Values
50 ppm MAK; 150 mg/m3 MAK

German (DFG) - Peak Limitations
2 x normal MAK (5 min momentary value);　don't exceed 8 times during shift

Israel - Time Weighted Averages
25 ppm TWA; 76 mg/m3 TWA

Israel - Action Levels
12.5 ppm AL; 38 mg/m3 AL

Mexico - Instruction No. 10 - TWAs
25 ppm TWA; 75 mg/m3 TWA

Mexico - Instruction No. 10 - STELs
100 ppm STEL; 300 mg/m3 STEL

STATE LISTS

California - Air Bill 2588 Appendix A-II
known or potential carcinogen: 9/90

California - Prop. 65 - Cancer list
carcinogen - initial date 7/1/90

California - Exposure Limits - PELs
25 ppm PEL; 75 mg/m3 PEL

California - Directors List of Hazardous Substances (8 CCR 339)
[present] (exempt when part of a cured epoxy or rubber)

Florida Hazardous Substance List
[present]

Massachusetts Right To Know List
[present]

Minnesota Hazardous Substance List
[present]

NJ Special Hazardous Substances
(mutagen)

Pennsylvania Right to Know List
environmental hazard

PROPOSED REGULATIONS

TSCA - Proposed Testing Rule for Glycidyl Ethers
subject to neurotoxicity, reproductive and fertility effects and mutagenicity testing

GLYCIDYL STEARATE　　　　　　　　　7460-84-6

ENVIRONMENTAL LISTS

TSCA - Health and Safety Reporting List
Effective Date: October 4, 1982

GLYCIDYL TRIMETHYL AMMONIUM CHLORIDE　　3033-77-0

SEE ALSO:
GLYCIDOL (OXIRANEMETHANOL) AND ITS DERIVATIVES

HEALTH AND SAFETY LISTS

NIOSH - Selected LD50s and LC50s
Oral, rat: LD50 = 23 gm/kg Skin, rabbit: LD50 = 3970 mg/kg

ENVIRONMENTAL LISTS

EPA - Master Testing List
[present]

TSCA - Code of Federal Regulations Citations
40 CFR 712.30(d);　40 CFR 716.120(c)

PROPOSED REGULATIONS

TSCA - Proposed Testing Rule for Glycidyl Ethers
subject to neurotoxicity, reproductive and fertility effects, screening subcategory and mutagenicity testing (results apply to all members of Glycidyl subcategory III-A)

GLYCINE　　　　　　　　　　　　　　56-40-6

SEE ALSO:
GLYCIDOL (OXIRANEMETHANOL) AND ITS DERIVATIVES

HEALTH AND SAFETY LISTS

NFPA - Flash Points
flash point = 141 degrees F (61 degrees C)

NFPA - Hazard Identification Ratings
health-0; flammability-2; reactivity-0

ENVIRONMENTAL LISTS

EPA - Master Testing List
[present]

TSCA - Code of Federal Regulations Citations
40 CFR 712.30(d);　40 CFR 716.120(c)

TSCA - PAIR - Reporting List
Reporting Date: November 19, 1982

INTERNATIONAL LISTS

Canada - WHMIS: Ingredient Disclosure
1% item 798 (158)

PROPOSED REGULATIONS

TSCA - Proposed Testing Rule for Glycidyl Ethers
subject to screening subcategory and mutagenicity testing (results apply to all members of Glycidyl subcategory VII-B)

GLYCINE OF TALL-OIL FATTY ACIDS　　RR-01084-4

HEALTH AND SAFETY LISTS

U.S. DOT - Substances From 49 CFR 172.101
regulated by DOT (UN2622)

U.S. DOT - Hazard Classes
DOT hazard class = 3

U.S. DOT - Appendix A Table 1 - Hazardous Substances
final RQ = 10 pounds (4.54 kg)

IARC - Group 2B (sufficient animal data)
[present] (Overall evaluation based only on evidence of carcinogenicity in monograph (11, 1976) or in Supplement 4)

NIOSH - Selected LD50s and LC50s
Skin, rabbit: LD50 = 249 mg/kg

OSHA - Possible Select Carcinogens
[present]

ENVIRONMENTAL LISTS

CERCLA/SARA - Hazardous Substances and their Reportable Quantities
final RQ = 10 pounds (4.54 kg)

EPA - Carcinogen Hazard Ranking for RQ Adjustment
Hazard ranking = Medium

RCRA - U Series Wastes
waste number U126

RCRA - Hazardous Constituents-Appendix VIII
waste number U126

RCRA - Substances Banned From Land Disposal
[present]

INTERNATIONAL LISTS

Canada - WHMIS: Ingredient Disclosure
1% item 796 (933)

STATE LISTS

California - Air Bill 2588 Appendix A-II
known or potential carcinogen

California - Prop. 65 - Cancer list
carcinogen - initial date 1/1/88

California - Directors List of Hazardous Substances (8 CCR 339)
[present]

Florida Hazardous Substance List
[present]

Massachusetts Right To Know List
carcinogen; extraordinarily hazardous

Minnesota Hazardous Substance List
carcinogen

NJ Right to Know List (Total)
sn 0961

NJ Special Hazardous Substances
(carcinogen)

Pennsylvania Right to Know List
environmental hazard

GLYCINE, N-(PHOSPHONOMETHYL)- 1071-83-6

STATE LISTS

Minnesota Hazardous Substance List
carcinogen

GLYCOL DIMERCAPTOACETATE 123-81-9

SEE ALSO:
GLYCIDOL (OXIRANEMETHANOL) AND ITS DERIVATIVES

HEALTH AND SAFETY LISTS

NIOSH - Selected LD50s and LC50s
Oral, rat: LD50 = 597 mg/kg Skin, rabbit: LD50 = 469 mg/kg

ENVIRONMENTAL LISTS

EPA - Master Testing List
[present]

TSCA - Code of Federal Regulations Citations
40 CFR 712.30(d); 40 CFR 716.120(c)

TSCA - PAIR - Reporting List
Reporting Date: November 19, 1982

PROPOSED REGULATIONS

TSCA - Proposed Testing Rule for Glycidyl Ethers
subject to subchronic toxicity and screening subcategory testing (resul apply to all members of Glycidyl subcategory VII-B)

GLYCOL ETHERS RR-00067-

HEALTH AND SAFETY LISTS

IARC - Group 3 (not classifiable)
[present]

GLYCOLIC ACID, SODIUM SALT 2836-32-

HEALTH AND SAFETY LISTS

IARC - Group 3 (not classifiable)
[present]

GLYCOL MONOBENZOATE RR-00956-

INTERNATIONAL LISTS

German (DFG) - Skin/Sensitizers
danger of cutaneous absorption; danger of sensitization (skin o respiratory)

German (DFG) - Carcinogens
animal evidence of carcinogenicity

GLYCOL, POLYETHYLENE, 3-SULFO-2-HYDROX-YPROPYL-P-(1,1,3,3-TETRAMETHYLBUTYL) PHENYL ETHER, SODIUM SALT RR-00958-

HEALTH AND SAFETY LISTS

NIOSH - Selected LD50s and LC50s
Oral, rat: LD50 = 7930 mg/kg

ENVIRONMENTAL LISTS

CAA - HON Rule - SOCMI Chemicals
compliance by Jan. 23, 1995

GLYCOL STEARATE 111-60-4

ENVIRONMENTAL LISTS

List of Pesticide Product Inert Ingredients
[present]

GLYCOURIL 496-46-8

HEALTH AND SAFETY LISTS

NIOSH - Selected LD50s and LC50s
Oral, rat: LD50 = 470 mg/kg

NTP Chemical Status Reports - Testing Status and NTIS Number
Technical reports printed (no NTIS number given)

ENVIRONMENTAL LISTS

Safe Drinking Water Act - MCLs
MCL = 0.7 mg/L

Safe Drinking Water Act - MCLGs
MCLG = 0.7 mg/L

INTERNATIONAL LISTS

Canada - Drinking Water Quality - IMACs
0.28 mg/L IMAC

GLYOXAL 107-22-2

HEALTH AND SAFETY LISTS

NFPA - Flash Points
flash point = 396 degrees F (202 degrees C)

NFPA - Hazard Identification Ratings
health-2; flammability-1; reactivity-0

STATE LISTS
  Florida Hazardous Substance List
  [present]
  Massachusetts Right To Know List
  [present]
  Pennsylvania Right to Know List
  [present]

GOLD 193   13982-20-2

HEALTH AND SAFETY LISTS
  U.S. DOT - Appendix A Table 1 - Hazardous Substances
  final RQ = 1 pound (.454 kg)
  NIOSH - Health Standards - Exposure Limits
  reduce exposure to lowest feasible concentration
  NIOSH - Health Standards - Health Effects and Precautions
  male and female reproductive effects; teratogenicity (Prevent skin contact)

ENVIRONMENTAL LISTS
  CERCLA/SARA - Section 313 - Emission Reporting
  form R reporting required for 1.0% de minimus concentration (applies to R-(OCH2CH2)n-OR' ethers, where n = 1,2, or 3'; R=alkyl C7 or less or R = phenyl or alkyl subst. phenyl; R' = H or alkyl C7 or less, or OR' consisting of carboxylic acid ester, sulfate, phosphate, nitrate, or sulfonate)
  CERCLA/SARA - Hazardous Substances and their Reportable Quantities
  final RQ = 1 pound (.454 kg)
  Clean Air Act (1990) - List of Hazardous Air Contaminants
  [present] (includes mono- and di- ethers of Ethylene glycol, and Triethylene glycol as specified by referenced formula)
  CAA - HON Rule - Organic HAPs
  Includes mono- and di- ethers of ethylene glycol, diethylene glycol, and triethylene glycol R-(OCH2CH2)n-OR' where: n=1,2, or 3; R = alkyl or aryl groups; and R'=R, H, or groups which, when removed, yield glycol ethers with the structure: R-(OCH2CH2)n-OH. Polymers are excluded from the glycol cat.

STATE LISTS
  California - Air Bill 2588 Appendix A-I
  [present]
  NJ Right to Know List (Total)
  sn 3138
  Pennsylvania Right to Know List
  environmental hazard

GOLD 194   15756-89-5

ENVIRONMENTAL LISTS
  List of Pesticide Product Inert Ingredients
  [present]

GOLD 195   14320-93-5

ENVIRONMENTAL LISTS
  TSCA - Chemicals with Significant New Use Rules
  PMN number: P-90-1357

GOLD 198   10043-49-9

ENVIRONMENTAL LISTS
  TSCA - Chemicals with Significant New Use Rules
  PMN number: P-90-1565

GOLD 198M   RR-00462-6

ENVIRONMENTAL LISTS
  List of Pesticide Product Inert Ingredients
  [present]

GOLD 199   14391-11-8

ENVIRONMENTAL LISTS
  List of Pesticide Product Inert Ingredients
  [present]

GOLD 200   20091-45-6

HEALTH AND SAFETY LISTS
  NIOSH - Selected LD50s and LC50s
  Oral, rat: LD50 = 1100 mg/kg Skin, guinea pig: LD50 = 6600 mg/kg
  NTP Chemical Status Reports - Testing Status and NTIS Number
  Short term toxicity studies scheduled for peer review

ENVIRONMENTAL LISTS
  CAA - HON Rule - SOCMI Chemicals
  compliance by Jan. 23, 1995
  EPA - Master Testing List
  [present]
  List of Pesticide Product Inert Ingredients
  [present]
  TSCA - Code of Federal Regulations Citations
  40 CFR 712.30(x); 40 CFR 716.120(d)
  TSCA - PAIR - Reporting List
  Reporting Date: November 27, 1991
  TSCA - Health and Safety Reporting List
  Effective Date: September 30, 1991

INTERNATIONAL LISTS
  Canada - WHMIS: Ingredient Disclosure
  1% item 799 (935)

GOLD 200M   RR-00461-5

HEALTH AND SAFETY LISTS
  U.S. DOT - Appendix A Table 2 - Radionuclides
  final RQ = 100 curies (3.7E 12 Bq)

ENVIRONMENTAL LISTS
  CERCLA/SARA List of Radionuclides (Appendix B) and Their Reportable Quantities
  final RQ = 100 curies (3.7E 12 Bq)

GOLD 201   23238-59-7

HEALTH AND SAFETY LISTS
  U.S. DOT - Appendix A Table 2 - Radionuclides
  final RQ = 10 curies (3.7E 11 Bq)

ENVIRONMENTAL LISTS
  CERCLA/SARA List of Radionuclides (Appendix B) and Their Reportable Quantities
  final RQ = 10 curies (3.7E 11 Bq)

GOLD SODIUM THIOSULFATE   10233-88-2

HEALTH AND SAFETY LISTS
  U.S. DOT - Appendix A Table 2 - Radionuclides
  final RQ = 100 curies (3.7E 12 Bq)

ENVIRONMENTAL LISTS
  CERCLA/SARA List of Radionuclides (Appendix B) and Their Reportable Quantities
  final RQ = 100 curies (3.7E 12 Bq)

GOLD TRICHLORIDE   13453-07-1

HEALTH AND SAFETY LISTS
  U.S. DOT - Appendix A Table 2 - Radionuclides
  final RQ = 100 curies (3.7E 12 Bq)

ENVIRONMENTAL LISTS

CERCLA/SARA List of Radionuclides (Appendix B) and Their Reportable Quantities
final RQ = 100 curies (3.7E 12 Bq)

GRAIN DUST   RR-00014-6

HEALTH AND SAFETY LISTS

U.S. DOT - Appendix A Table 2 - Radionuclides
final RQ = 10 curies (3.7E 11 Bq)

ENVIRONMENTAL LISTS

CERCLA/SARA List of Radionuclides (Appendix B) and Their Reportable Quantities
final RQ = 10 curies (3.7E 11 Bq)

GRAPES (POMACE)   RR-01085-5

HEALTH AND SAFETY LISTS

U.S. DOT - Appendix A Table 2 - Radionuclides
final RQ = 100 curies (3.7E 12 Bq)

ENVIRONMENTAL LISTS

CERCLA/SARA List of Radionuclides (Appendix B) and Their Reportable Quantities
final RQ = 100 curies (3.7E 12 Bq)

GRAPHITE   7782-42-5

HEALTH AND SAFETY LISTS

U.S. DOT - Appendix A Table 2 - Radionuclides
final RQ = 1000 curies (3.7E 13 Bq)

ENVIRONMENTAL LISTS

CERCLA/SARA List of Radionuclides (Appendix B) and Their Reportable Quantities
final RQ = 1000 curies (3.7E 13 Bq)

GRAPHITE, SYNTHETIC   RR-00012-4

HEALTH AND SAFETY LISTS

U.S. DOT - Appendix A Table 2 - Radionuclides
final RQ = 10 curies (3.7E 11 Bq)

ENVIRONMENTAL LISTS

CERCLA/SARA List of Radionuclides (Appendix B) and Their Reportable Quantities
final RQ = 10 curies (3.7E 11 Bq)

GRAVEL   RR-01086-6

HEALTH AND SAFETY LISTS

U.S. DOT - Appendix A Table 2 - Radionuclides
final RQ = 1000 curies (3.7E 13 Bq)

ENVIRONMENTAL LISTS

CERCLA/SARA List of Radionuclides (Appendix B) and Their Reportable Quantities
final RQ = 1000 curies (3.7E 13 Bq)

GRISEOFULVIN   126-07-8
SEE ALSO:
GOLD

INTERNATIONAL LISTS

Canada - WHMIS: Ingredient Disclosure
1% item 800 (1618)

GROUTS   RR-01768-5
SEE ALSO:
GOLD

INTERNATIONAL LISTS

Canada - WHMIS: Ingredient Disclosure
1% item 801 (1657)

GUAIAC GUM   9000-29-

HEALTH AND SAFETY LISTS

ACGIH 1995 - Time Weighted Averages
4 mg/m3 TWA (total dust/particulate)

OSHA - Vacated PELs - Time Weighted Averages
10 mg/m3 TWA

OSHA - Final PELs - Time Weighted Averages
10 mg/m3 TWA

INTERNATIONAL LISTS

Australian Exposure Standards - Time Weighted Averages
4 mg/m3 TWA

Canada - Alberta - 8 Hour Occupational Exposure Limit
total mass: 4 mg/m3 TWA

Canada - Ontario - OHSA - TWAEVs
4 mg/m3 TWAEV

Canada - Quebec - Time-Weighted Average Exposure Values
4 ppm TWAEV

United Kingdom - Maximum Exposure Limits - TWAs
10 mg/m3 TWA

Israel - Time Weighted Averages
4 mg/m3 TWA (total dust/particulate)

Israel - Action Levels
2 mg/m3 AL (total dust/particulate)

STATE LISTS

California - Exposure Limits - PELs
4 mg/m3 PEL

Minnesota Hazardous Substance List
[present] (Oat, Wheat, and Barley)

Pennsylvania Right to Know List
[present]

GUAIACOL   90-05-1

ENVIRONMENTAL LISTS

List of Pesticide Product Inert Ingredients
[present]

M-GUAIACOL   150-19-6

HEALTH AND SAFETY LISTS

ACGIH 1995 - Time Weighted Averages
all forms except grahpite fibers, respirable dust: 2 mg/m3 TWA

NIOSH 1990 - Pocket Guide - RELs
respirable dust: 2.5 mg/m3 TWA

OSHA - Vacated PELs - Time Weighted Averages
respirable dust: 2.5 mg/m3 TWA

OSHA - Final PELs - Time Weighted Averages
see Table Z-3

ENVIRONMENTAL LISTS

List of Pesticide Product Inert Ingredients
[present]

INTERNATIONAL LISTS

Australian Exposure Standards - Time Weighted Averages
10 mg/m3 TWA

Canada - Alberta - 8 Hour Occupational Exposure Limit
respirable mass: 2.5 mg/m3 TWA; total mass: 5 mg/m3 TWA

Canada - British Columbia - 8 Hour Exposure Limits
nuisance dust: 10 mg/m3 TWA

Canada - Ontario - OHSA - TWAEVs
total dust: 5 mg/m3 TWAEV; respirable dust: 2.5 mg/m3 TWAEV (listed as mineral dust)

Canada - Quebec - Time-Weighted Average Exposure Values
natural: 2.5 mg/m3 TWAEV; synthetic: 10 mg/m3 TWAEV

United Kingdom - Occupational Exposure Standards - TWAs
total inhalable dust: 10 mg/m3 TWA; respirable dust: 5 mg/m3 TWA

German (DFG) - MAK Values
fine dust (with < 1% quartz): 6 mg/m3 MAK

German (DFG) - Pregnancy
no risk to embryo/fetus if exposure limits adhered to

Israel - Time Weighted Averages
(2.5) mg/m3 TWA (this TLV is for the respirable fraction of Graphite dust)

Israel - Action Levels
respirable dust: 1.25 mg/m3 AL

STATE LISTS

California - Exposure Limits - PELs
natural respirable dust: 2.5 mg/m3 PEL; synthetic total dust: 10 mg/m3 PEL; respirable fraction: 5 mg/m3 PEL

California - Directors List of Hazardous Substances (8 CCR 339)
[present] (exempt when inhalable dusts or particulates are generated by use)

Florida Hazardous Substance List
[present]

Massachusetts Right To Know List
[present] Exempt when encapsulated or if particulates are not present and cannot be substantially generated through use of the product.

Minnesota Hazardous Substance List
[present]

Pennsylvania Right to Know List
[present]

GUANAZOLE   1455-77-2

HEALTH AND SAFETY LISTS

OSHA - Vacated PELs - Time Weighted Averages
total dust: 10 mg/m3 TWA; respirable fraction: 5 mg/m3 TWA

OSHA - Final PELs - Time Weighted Averages
total dust: 15 mg/m3 TWA; respirable fraction: 5 mg/m3 TWA

INTERNATIONAL LISTS

Canada - Alberta - 8 Hour Occupational Exposure Limit
respirable mass: 5 mg/m3 TWA; total mass: 10 mg/m3 TWA

Canada - Ontario - OHSA - TWAEVs
total dust: 5 mg/m3 TWAEV (listed as nuisance particulates)

Israel - Time Weighted Averages
(10) mg/m3 TWA (The value is for total dust containing no asbestos and < 1% crystalline silica)

Israel - Action Levels
5 mg/m3 AL

Mexico - Instruction No. 10 - TWAs
10 mg/m3 TWA; (nuisance particulate)

STATE LISTS

Pennsylvania Right to Know List
[present]

GUANIDINE, CYANO   461-58-5

ENVIRONMENTAL LISTS

List of Pesticide Product Inert Ingredients
[present]

GUANIDINE HYDROCHLORIDE   50-01-1

HEALTH AND SAFETY LISTS

IARC - Group 2B (sufficient animal data)
[present] (Degree of evidence in animals revised on the basis of data that appeared after the most recent monograph and/or on the basis of present criteria)

OSHA - Possible Select Carcinogens
[present]

STATE LISTS

California - Air Bill 2588 Appendix A-I
known or potential carcinogen

California - Prop. 65 - Cancer list
carcinogen - initial date 1/1/90

California - Directors List of Hazardous Substances (8 CCR 339)
[present]

Minnesota Hazardous Substance List
carcinogen

GUANIDINE MONONITRATE   506-93-4

ENVIRONMENTAL LISTS

TSCA - Section 12(b) - Export Notification
export notification required - Section 6 In this context, grouts are defined as those chemical grouts that contain either Acrylamide or N-Methylolacrylamide (40 CFR 764.125).

GUANIDINE NITRATE (VAN)   52470-25-4

HEALTH AND SAFETY LISTS

NIOSH - Selected LD50s and LC50s
Oral, guinea pig: LD50 = 1120 mg/kg

GUANYL NITROSOAMINOGUANYLIDENE HYDRAZINE   RR-01327-4

HEALTH AND SAFETY LISTS

NIOSH - Selected LD50s and LC50s
Inhalation, mouse: LC50 = 7570 mg/m3 (8 hr) Oral, rat: LD50 = 725 mg/kg Skin, rabbit: LD50 = 4600 mg/kg

INTERNATIONAL LISTS

Canada - WHMIS: Ingredient Disclosure
1% item 802 (929)

GUANYL NITROSOAMINOGUANYLIDENE HYDRAZINE   RR-01424-4

INTERNATIONAL LISTS

Canada - WHMIS: Ingredient Disclosure
1% item 803 (930)

GUANYL NITROSOAMINOGUANYLTETRAZENE   RR-01328-5

HEALTH AND SAFETY LISTS

NTP Chemical Status Reports - Testing Status and NTIS Number
Chronic studies exist for which technical reports were not prepared

GUAR GUM   9000-30-0

ENVIRONMENTAL LISTS

List of Pesticide Product Inert Ingredients
[present]

L-GULITOL   6706-59-8

HEALTH AND SAFETY LISTS

NIOSH - Selected LD50s and LC50s
Oral, rat: LD50 = 475 mg/kg

GUM ARABIC   9000-01-5

ENVIRONMENTAL LISTS

List of Pesticide Product Inert Ingredients
[present]

STATE LISTS

NJ Right to Know List (Total)
sn 0964

GUM GHATTI   9000-28-6
   HEALTH AND SAFETY LISTS
      U.S. DOT - Substances From 49 CFR 172.101
         regulated by DOT (UN1467)
      U.S. DOT - Hazard Classes
         DOT hazard class = 5.1

GUM THUS   8050-07-5
   HEALTH AND SAFETY LISTS
      U.S. DOT - Substances From 49 CFR 172.101
         regulated by DOT (UN0113)
      U.S. DOT - Hazard Classes
         DOT hazard class = 1.1A

GUM TRAGACANTH   9000-65-1
   HEALTH AND SAFETY LISTS
      U.S. DOT - Hazard Classes
         Forbidden from transport by the DOT

GUTTA-PERCHA SOLUTION   RR-00767-0
   HEALTH AND SAFETY LISTS
      U.S. DOT - Substances From 49 CFR 172.101
         regulated by DOT (UN0114)
      U.S. DOT - Hazard Classes
         DOT hazard class = 1.1A

GYPSUM (CA(SO4).2H2O)   13397-24-5
   HEALTH AND SAFETY LISTS
      NIOSH - Selected LD50s and LC50s
         Oral, rat: LD50 = 7060 mg/kg
      NTP Chemical Status Reports - Testing Status and NTIS Number
         Technical reports printed (PB82202813)
      NTP Chemical Status Reports - Evidence of Carcinogenicity
         *male rat-negative; female rat-negative; male mice-negative; female
         mice-negative*
   ENVIRONMENTAL LISTS
      List of Pesticide Product Inert Ingredients
         [present]

GYROMITRIN   16568-02-8
   ENVIRONMENTAL LISTS
      List of Pesticide Product Inert Ingredients
         [present]

HAFNIUM   7440-58-6
   HEALTH AND SAFETY LISTS
      NIOSH - Selected LD50s and LC50s
         Oral, rabbit: LD50 = 8000 mg/kg
      NTP Chemical Status Reports - Testing Status and NTIS Number
         Technical reports printed (PB82229584)
      NTP Chemical Status Reports - Evidence of Carcinogenicity
         *male rat-negative; female rat-negative; male mice-negative; female
         mice-negative*
   ENVIRONMENTAL LISTS
      List of Pesticide Product Inert Ingredients
         [present]

HAFNIUM 170   14922-51-1
   HEALTH AND SAFETY LISTS
      NIOSH - Selected LD50s and LC50s
         Oral, rat: LD50 = 17 gm/kg

   ENVIRONMENTAL LISTS
      List of Pesticide Product Inert Ingredients
         [present]

HAFNIUM 172   14093-11-
   ENVIRONMENTAL LISTS
      List of Pesticide Product Inert Ingredients
         [present]

HAFNIUM 173   15757-23-
   HEALTH AND SAFETY LISTS
      NIOSH - Selected LD50s and LC50s
         Oral, rat: LD50 = 16400 mg/kg
   ENVIRONMENTAL LISTS
      List of Pesticide Product Inert Ingredients
         [present]
   INTERNATIONAL LISTS
      Canada - WHMIS: Ingredient Disclosure
         0.1% item 804 (936)

HAFNIUM 175   15750-13-
   STATE LISTS
      NJ Right to Know List (Total)
         sn 2457

HAFNIUM 177M   RR-00460-4
   HEALTH AND SAFETY LISTS
      OSHA - Vacated PELs - Time Weighted Averages
         total dust: 15 mg/m3 TWA; respirable fraction: 5 mg/m3 TWA
   INTERNATIONAL LISTS
      Canada - Quebec - Time-Weighted Average Exposure Values
         total dust: 10 mg/m3 TWAEV; respirable dust: 5 mg/m3
   STATE LISTS
      Pennsylvania Right to Know List
         [present]

HAFNIUM 178M   RR-00458-0
   HEALTH AND SAFETY LISTS
      IARC - Group 3 (not classifiable)
         [present]
   STATE LISTS
      California - Air Bill 2588 Appendix A-II
         known or potential carcinogen
      California - Prop. 65 - Cancer list
         carcinogen - initial date 1/1/88
      California - Prop. 65 - No Significant Risk Levels
         no significant risk level = 0.07 ug/day
      California - Directors List of Hazardous Substances (8 CCR 339)
         [present]
      Massachusetts Right To Know List
         carcinogen; extraordinarily hazardous
      Minnesota Hazardous Substance List
         carcinogen
      Pennsylvania Right to Know List
         special hazardous substance
      Pennsylvania RTK - Special Hazardous Substances
         [present]

HAFNIUM 179M   RR-00451-3
   HEALTH AND SAFETY LISTS
      ACGIH 1995 - Time Weighted Averages
         0.5 mg/m3 TWA

U.S. DOT - Substances From 49 CFR 172.101
regulated by DOT (UN2545, UN1326)

U.S. DOT - Hazard Classes
DOT hazard class = 4.2

NIOSH 1990 - Pocket Guide - RELs
as Hf: 0.5 mg/m3 TWA

NIOSH 1990 - Pocket Guide - Target organs
eyes, skin, mucous membrane

OSHA - Vacated PELs - Time Weighted Averages
0.5 mg/m3 TWA

OSHA - Final PELs - Time Weighted Averages
0.5 mg/m3 TWA

INTERNATIONAL LISTS

Australian Exposure Standards - Time Weighted Averages
0.5 mg/m3 TWA

Canada - WHMIS: Ingredient Disclosure
1% item 805 (938)

Canada - Alberta - 8 Hour Occupational Exposure Limit
0.5 mg/m3 TWA

Canada - Alberta - 15 Minute Occupational Exposure Limit
1.5 mg/m3 STEL

Canada - British Columbia - 8 Hour Exposure Limits
0.5 mg/m3 TWA

Canada - British Columbia - 15 Minute Exposure Limits
1.5 mg/m3 STEL

Canada - Ontario - OHSA - TWAEVs
0.5 mg/m3 TWAEV

Canada - Quebec - Time-Weighted Average Exposure Values
0.5 mg/m3 TWAEV

United Kingdom - Occupational Exposure Standards - TWAs
0.5 mg/m3 TWA

United Kingdom - Occupational Exposure Standards - STELs
1.5 mg/m3 STEL

German (DFG) - MAK Values
total dust: 0.5 mg/m3 MAK

German (DFG) - Peak Limitations
10 x normal MAK (30 min average value); don't exceed during shift

Israel - Time Weighted Averages
0.5 mg/m3 TWA

Israel - Action Levels
0.25 mg/m3 AL

Mexico - Instruction No. 10 - TWAs
0.5 mg/m3 TWA

Mexico - Instruction No. 10 - STELs
15 mg/m3 STEL

STATE LISTS

California - Exposure Limits - PELs
0.5 mg/m3 PEL

California - Directors List of Hazardous Substances (8 CCR 339)
[present]

Florida Hazardous Substance List
[present]

Massachusetts Right To Know List
[present]

Minnesota Hazardous Substance List
[present]

NJ Right to Know List (Total)
sn 0967

Pennsylvania Right to Know List
[present]

HAFNIUM 180M   RR-00449-9

HEALTH AND SAFETY LISTS

U.S. DOT - Appendix A Table 2 - Radionuclides
final RQ = 100 curies (3.7E 12 Bq)

ENVIRONMENTAL LISTS

CERCLA/SARA List of Radionuclides (Appendix B) and Their Reportable Quantities
final RQ = 100 curies (3.7E 12 Bq)

HAFNIUM 181   14900-21-1

HEALTH AND SAFETY LISTS

U.S. DOT - Appendix A Table 2 - Radionuclides
final RQ = 1 curie (3.7E 10 Bq)

ENVIRONMENTAL LISTS

CERCLA/SARA List of Radionuclides (Appendix B) and Their Reportable Quantities
final RQ = 1 curie (3.7E 10 Bq)

HAFNIUM 182   29492-85-1

HEALTH AND SAFETY LISTS

U.S. DOT - Appendix A Table 2 - Radionuclides
final RQ = 100 curies (3.7E 12 Bq)

ENVIRONMENTAL LISTS

CERCLA/SARA List of Radionuclides (Appendix B) and Their Reportable Quantities
final RQ = 100 curies (3.7E 12 Bq)

HAFNIUM 182M   RR-00448-8

HEALTH AND SAFETY LISTS

U.S. DOT - Appendix A Table 2 - Radionuclides
final RQ = 100 curies (3.7E 12 Bq)

ENVIRONMENTAL LISTS

CERCLA/SARA List of Radionuclides (Appendix B) and Their Reportable Quantities
final RQ = 100 curies (3.7E 12 Bq)

HAFNIUM 183   15832-40-3

HEALTH AND SAFETY LISTS

U.S. DOT - Appendix A Table 2 - Radionuclides
final RQ = 1000 curies (3.7E 13 Bq)

ENVIRONMENTAL LISTS

CERCLA/SARA List of Radionuclides (Appendix B) and Their Reportable Quantities
final RQ = 1000 curies (3.7E 13 Bq)

HAFNIUM 184   29687-28-3

HEALTH AND SAFETY LISTS

U.S. DOT - Appendix A Table 2 - Radionuclides
final RQ = 0.1 curies (3.7E 9 Bq)

ENVIRONMENTAL LISTS

CERCLA/SARA List of Radionuclides (Appendix B) and Their Reportable Quantities
final RQ = 0.1 curies (3.7E 9 Bq)

HAIR COLORANTS, PERSONAL USE OF   RR-01701-6

HEALTH AND SAFETY LISTS

U.S. DOT - Appendix A Table 2 - Radionuclides
final RQ = 100 curies (3.7E 12 Bq)

ENVIRONMENTAL LISTS

CERCLA/SARA List of Radionuclides (Appendix B) and Their Reportable Quantities
final RQ = 100 curies (3.7E 12 Bq)

HAIRDRESSER OR BARBER, OCCUPATIONAL EXPOSURES AS   RR-01702-7

HEALTH AND SAFETY LISTS

U.S. DOT - Appendix A Table 2 - Radionuclides
final RQ = 100 curies (3.7E 12 Bq)

ENVIRONMENTAL LISTS

CERCLA/SARA List of Radionuclides (Appendix B) and Their Reportable Quantities
final RQ = 100 curies (3.7E 12 Bq)

HALAZEPAM   23092-17-3

HEALTH AND SAFETY LISTS

U.S. DOT - Appendix A Table 2 - Radionuclides
final RQ = 10 curies (3.7E 11 Bq)

ENVIRONMENTAL LISTS

CERCLA/SARA List of Radionuclides (Appendix B) and Their Reportable Quantities
final RQ = 10 curies (3.7E 11 Bq)

HALLOYSITE   12298-43-0

HEALTH AND SAFETY LISTS

U.S. DOT - Appendix A Table 2 - Radionuclides
final RQ = 0.1 curies (3.7E 9 Bq)

ENVIRONMENTAL LISTS

CERCLA/SARA List of Radionuclides (Appendix B) and Their Reportable Quantities
final RQ = 0.1 curies (3.7E 9 Bq)

HALOALKYL EPOXIDE   428-25-1

HEALTH AND SAFETY LISTS

U.S. DOT - Appendix A Table 2 - Radionuclides
final RQ = 100 curies (3.7E 12 Bq)

ENVIRONMENTAL LISTS

CERCLA/SARA List of Radionuclides (Appendix B) and Their Reportable Quantities
final RQ = 100 curies (3.7E 12 Bq)

HALOALKYL SUBSTITUTED CYCLIC ETHERS   RR-00171-8

HEALTH AND SAFETY LISTS

U.S. DOT - Appendix A Table 2 - Radionuclides
final RQ = 100 curies (3.7E 12 Bq)

ENVIRONMENTAL LISTS

CERCLA/SARA List of Radionuclides (Appendix B) and Their Reportable Quantities
final RQ = 100 curies (3.7E 12 Bq)

HALOETHERS   RR-00325-8

HEALTH AND SAFETY LISTS

U.S. DOT - Appendix A Table 2 - Radionuclides
final RQ = 100 curies (3.7E 12 Bq)

ENVIRONMENTAL LISTS

CERCLA/SARA List of Radionuclides (Appendix B) and Their Reportable Quantities
final RQ = 100 curies (3.7E 12 Bq)

HALOGENATED ACRYLONITRILE   RR-00972-3

HEALTH AND SAFETY LISTS

IARC - Group 3 (not classifiable)
[present]

HALOGENATED ALKYL EPOXIDES   RR-00274-4

HEALTH AND SAFETY LISTS

IARC - Group 2A (limited human data)
[present]

OSHA - Possible Select Carcinogens
[present]

HALOGENATED BIPHENYL GLYCIDYL ETHERS   RR-01246-

STATE LISTS

California - Air Bill 2588 Appendix A-II
9/90

California - Prop. 65 - Developmental Toxicity
developmental toxicity - initial date 7/1/90

NJ Right to Know List (Total)
sn 0968

NJ Special Hazardous Substances
(teratogen)

HALOGENATED ETHANES   RR-01636-

INTERNATIONAL LISTS

German (DFG) - Carcinogens
suspected carcinogen

HALOGENATED ETHANES CLASS STUDY   RR-01634-2

ENVIRONMENTAL LISTS

TSCA - Code of Federal Regulations Citations
40 CFR 716.120(c)

HALOGENATED PHENYL ALKANE   RR-01667-1

ENVIRONMENTAL LISTS

TSCA - Chemicals with Significant New Use Rules
PMN numbers: P-85-367; P-85-368; P-85-369

HALOGENATED PHOSPHATE ESTERS   RR-01641-1

ENVIRONMENTAL LISTS

Clean Water Act - Toxic Pollutants
[present]

STATE LISTS

California - Directors List of Hazardous Substances (8 CCR 339)
[present]

Pennsylvania Right to Know List
environmental hazard

HALOMETHANES   RR-00327-0

ENVIRONMENTAL LISTS

TSCA - Chemicals with Significant New Use Rules
PMN number: P-90-299

HALONITROBENZOIC ACID, SUBSTITUTED   RR-00929-0

ENVIRONMENTAL LISTS

TSCA - Health and Safety Reporting List
Effective Date: October 4, 1991 (halogenated noncyclic aliphatic hydrocarbons with one or more epoxy functional groups)

HALOPHENYL SULFONAMIDE SALT   RR-01176-7

ENVIRONMENTAL LISTS

TSCA - Chemicals with Significant New Use Rules
PMN numbers: P-90-1844; P-90-1845; P-90-1846

HALOTHANE   151-67-7

HEALTH AND SAFETY LISTS

NTP Chemical Status Reports - Testing Status and NTIS Number
Project leader assigned/study in design

HAY   RR-00328-1

HEALTH AND SAFETY LISTS

NTP Chemical Status Reports - Testing Status and NTIS Number
Project leader assigned/study in design

IC BLUE 1   2784-94-3

ENVIRONMENTAL LISTS

TSCA - Chemicals with Significant New Use Rules
PMN number: P-89-867

IC BLUE 2   33229-34-4

ENVIRONMENTAL LISTS

TSCA - Chemicals with Significant New Use Rules
PMN number: P-86-1662

IC RED NO. 3   2871-01-4

ENVIRONMENTAL LISTS

Clean Water Act - Toxic Pollutants
[present]

RCRA - Hazardous Constituents-Appendix VIII
hazardous constituent - no waste number

INTERNATIONAL LISTS

Mexico - Drinking Water - Ecological Criteria
0.002 mg/l Substance presents persistence, bioaccumulations or risk of cancer, reduce human exposure to a minimum; This level has been extrapolated by using a mathematic model

STATE LISTS

California - Directors List of Hazardous Substances (8 CCR 339)
[present]

Pennsylvania Right to Know List
environmental hazard

IC YELLOW 4   59820-43-8

ENVIRONMENTAL LISTS

TSCA - Chemicals with Significant New Use Rules
PMN number: P-86-1098

HEAVY AROMATIC BOTTOMS   64741-67-9

ENVIRONMENTAL LISTS

TSCA - Chemicals with Significant New Use Rules
PMN number: P-90-1730

HEAVY AROMATIC DISTILLATE (PETROLEUM)   67891-79-6

HEALTH AND SAFETY LISTS

ACGIH 1995 - Time Weighted Averages
50 ppm TWA; 404 mg/m3 TWA

NIOSH - Selected LD50s and LC50s
Inhalation, mouse: LC50 = 22000 ppm (10 mn) Oral, rat: LD50 = 5680 mg/kg

NIOSH - Health Standards - Exposure Limits
C (1 hr) 2 ppm (Listed under 'Waste anasthetic gases and vapors')

NIOSH - Health Standards - Health Effects and Precautions
Reproductive system effects and audio-visual performance decrements (Advise workers of potential effects) (Listed under 'Waste anasthetic gases')

INTERNATIONAL LISTS

Australian Exposure Standards - Time Weighted Averages
0.5 ppm TWA; 4.1 mg/m3 TWA

Canada - Alberta - 8 Hour Occupational Exposure Limit
10 ppm TWA; 80 mg/m3 TWA

Canada - Alberta - 15 Minute Occupational Exposure Limit
15 ppm STEL; 120 mg/m3 STEL

Canada - British Columbia - 8 Hour Exposure Limits
2 ppm TWA

Canada - Ontario - OHSA - TWAEVs
2 ppm TWAEV; 16 mg/m3 TWAEV

Canada - Quebec - Time-Weighted Average Exposure Values
50 ppm TWAEV; 404 mg/m3 TWAEV

German (DFG) - MAK Values
5 ppm MAK; 40 mg/m3 MAK

German (DFG) - Peak Limitations
2 x normal MAK (30 min. average value); don't exceed 4 times during shift

German (DFG) - Pregnancy
risk to embryo/fetus probable

Israel - Time Weighted Averages
50 ppm TWA; 404 mg/m3 TWA

Israel - Action Levels
2.5 ppm AL;

STATE LISTS

California - Exposure Limits - PELs
2 ppm PEL; 16 mg/m3 PEL

Massachusetts Right To Know List
[present]

Minnesota Hazardous Substance List
[present]

NJ Right to Know List (Total)
sn 0969

HEAVY NAPHTHENIC DISTILLATE SOLVENT EXTRACT   64742-11-6

STATE LISTS

NJ Right to Know List (Total)
sn 2459

HECTORITE   12173-47-6

HEALTH AND SAFETY LISTS

IARC - Group 2B (sufficient animal data)
[present]

NTP Chemical Status Reports - Testing Status and NTIS Number
Technical reports printed (PB86114683/AS)

NTP Chemical Status Reports - Evidence of Carcinogenicity
male rat-equivocal evidence; female rat-some evidence; male mice-clear evidence; female mice-clear evidence

OSHA - Possible Select Carcinogens
[present]

STATE LISTS

California - Air Bill 2588 Appendix A-II
known or potential carcinogen: 9/89

California - Prop. 65 - Cancer list
carcinogen - initial date 7/1/89

California - Prop. 65 - No Significant Risk Levels
no significant risk level = 10 ug/day

California - Directors List of Hazardous Substances (8 CCR 339)
[present]

HELIUM   7440-59-7

HEALTH AND SAFETY LISTS

IARC - Group 3 (not classifiable)
[present]

NTP Chemical Status Reports - Testing Status and NTIS Number
Technical reports printed (PB86108339/AS)

NTP Chemical Status Reports - Evidence of Carcinogenicity
male rat-no evidence; female rat-no evidence; male mice-no evidence; female mice-no evidence

HEMATITE   1317-60-8

HEALTH AND SAFETY LISTS

IARC - Group 3 (not classifiable)
[present]

NTP Chemical Status Reports - Testing Status and NTIS Number
Technical reports printed (PB86188075/AS)

NTP Chemical Status Reports - Evidence of Carcinogenicity
male rat-no evidence; female rat-no evidence; male mice-equivocal evidence; female mice-inadequate

HEPTACHLOR   76-44-8

HEALTH AND SAFETY LISTS

IARC - Group 3 (not classifiable)
[present]

NTP Chemical Status Reports - Testing Status and NTIS Number
Technical reports printed (PB93-123883)

NTP Chemical Status Reports - Evidence of Carcinogenicity
male rat: equivocal evidence; female rat: no evidence; male mice: no evidence; female mice: no evidence

HEPTACHLOR EPOXIDE   1024-57-3

ENVIRONMENTAL LISTS

List of Pesticide Product Inert Ingredients
[present]

HEPTACHLOR METABOLITES   RR-00566-3

ENVIRONMENTAL LISTS

List of Pesticide Product Inert Ingredients
[present]

HEPTACHLOROBENZO-P-DIOXIN   37871-00-4

STATE LISTS

Massachusetts Right To Know List
carcinogen; extraordinarily hazardous

1,2,3,4,6,7,8-HEPTACHLORODIBENZO-P-DIOXIN   35822-46-9

ENVIRONMENTAL LISTS

List of Pesticide Product Inert Ingredients
[present]

HEPTACHLORODIBENZO-P-DIOXINS   RR-01174-5

HEALTH AND SAFETY LISTS

ACGIH 1995 - Time Weighted Averages
simple asphyxiant

U.S. DOT - Substances From 49 CFR 172.101
regulated by DOT (UN1963, UN1046)

U.S. DOT - Hazard Classes
DOT hazard class = 2.2

INTERNATIONAL LISTS

Australian Exposure Standards - Time Weighted Averages
Asphyxiant at < 18% oxygen by volume

Canada - British Columbia - 8 Hour Exposure Limits
asphyxiant substance

Canada - Ontario - OHSA - TWAEVs
simple asphyxiant

Canada - Quebec - Time-Weighted Average Exposure Values
simple asphyxiant

Israel - Time Weighted Averages
Asphyxiant

Mexico - Instruction No. 10 - TWAs
simple asphyxiant

STATE LISTS

California - Exposure Limits - PELs
asphyxiant (limit depends on level of oxygen)

Florida Hazardous Substance List
[present]

Massachusetts Right To Know List
[present]

Minnesota Hazardous Substance List
[present]

NJ Right to Know List (Total)
sn 0972

Pennsylvania Right to Know List
[present]

1,2,3,4,7,8,9-HEPTACHLORO DIBENZOFURAN   38998-75-

HEALTH AND SAFETY LISTS

IARC - Group 3 (not classifiable)
[present]

1,2,3,4,6,7,8-HEPTACHLORODIBENZOFURAN   67562-39-

SEE ALSO:
F039-HAZARDOUS WASTES
K097-HAZARDOUS WASTES
PINE PITCH

HEALTH AND SAFETY LISTS

ACGIH 1995 - Time Weighted Averages
0.05 mg/m3 TWA

ACGIH 1995 - Skin Designations
skin - potential for cutaneous absorption

ACGIH 1995 - Carcinogens
A3-animal carcinogen

U.S. DOT - Appendix B - Marine Pollutants
DOT regulated severe marine pollutant

U.S. DOT - Appendix A Table 1 - Hazardous Substances
final RQ = 1 pound (0.454 kg)

IARC - Group 2B (sufficient animal data)
[present]

NIOSH - Selected LD50s and LC50s
Oral, rat: LD50 = 40 mg/kg Skin, rat: LD50 = 119 mg/kg

NIOSH 1990 - Pocket Guide - RELs
0.5 mg/m3 TWA

NIOSH 1990 - Pocket Guide - IDLHs
700 mg/m3 IDLH (not considering carcinogenic effects)

NIOSH 1990 - Pocket Guide - Carcinogens
occupational carcinogen

NIOSH 1990 - Pocket Guide - Target organs
in animals: CNS, liver

NIOSH 1990 - Pocket Guide - Skin list
Potential for dermal absorption

NTP Chemical Status Reports - Testing Status and NTIS Number
Technical reports printed (PB271967/AS)

NTP Chemical Status Reports - Evidence of Carcinogenicity
male rat-negative; female rat-equivocal; male mice-positive; female mice-positive

OSHA - Vacated PELs - Time Weighted Averages
0.5 mg/m3 TWA

OSHA - Vacated PELs - Skin Designation
Prevent or reduce skin absorption

OSHA - Final PELs - Time Weighted Averages
0.5 mg/m3 TWA

OSHA - Final PELs - Skin Notations
prevent or reduce skin absorption

OSHA - Possible Select Carcinogens
[present]

ENVIRONMENTAL LISTS

ATSDR Priority List
Rank (of 275): 029

CERCLA/SARA - Section 313 - Emission Reporting
form R reporting required for 1.0% de minimus concentration

CERCLA/SARA - Hazardous Substances and their Reportable Quantities
final RQ = 1 pound (0.454 kg)

Clean Air Act (1990) - List of Hazardous Air Contaminants
[present]

Clean Water Act - Hazardous Substances
[present]

Clean Water Act - Priority Pollutants
[present]

Clean Water Act - Toxic Pollutants
[present]

Safe Drinking Water Act - MCLs
MCL = 0.0004 mg/L

Safe Drinking Water Act - MCLGs
MCLG = Zero

EPA - Carcinogen Hazard Ranking for RQ Adjustment
Hazard ranking = High

RCRA - D Series - Maximum Concentration of Contaminants
waste number D031; regulatory level = 0.008 mg/L

RCRA - D Series - Chronic Toxicity Reference Levels
chronic toxicity reference level = 0.00008 mg/L

RCRA - P Series Wastes
waste number P059

RCRA - Hazardous Constituents-Appendix VIII
waste number P059

RCRA - Basis for Listing - Appendix VII
Included in waste streams: F039, K097

RCRA - Substances Banned From Land Disposal
[present]

RCRA - TSD Facilities Ground Water Monitoring
TM 8080 = 0.05 ug/L PQL; TM 8270 = 10 ug/L PQL

RCRA - Universal Treatment Standards (LDR)
WW: 0.0012 mg/l; NWW: 0.066 mg/kg

INTERNATIONAL LISTS

Australian Exposure Standards - Time Weighted Averages
0.5 mg/m3 TWA

Australian Exposure Standards - Skin Effects
skin absorption

Australian Exposure Standards - Under Review
exposure limits under review

Canada - Drinking Water Quality - MACs
0.003 mg/L MAC

Canada - Alberta - 8 Hour Occupational Exposure Limit
0.5 mg/m3 TWA

Canada - Alberta - 15 Minute Occupational Exposure Limit
2 mg/m3 STEL

Canada - Alberta - Skin Designation
can be absorbed through the intact skin

Canada - British Columbia - 8 Hour Exposure Limits
0.5 mg/m3 TWA

Canada - British Columbia - 15 Minute Exposure Limits
2 mg/m3 STEL

Canada - British Columbia - Skin Notations
skin - potential for skin absorption

Canada - Ontario - OHSA - TWAEVs
0.5 mg/m3 TWAEV

Canada - Ontario - OHSA - Skin Notations
absorption through skin, eyes, or mucous membranes

Canada - Quebec - Time-Weighted Average Exposure Values
0.5 mg/m3 TWAEV

Canada - Quebec - Skin Designations
absorbed through the skin

German (DFG) - MAK Values
total dust: 0.5 mg/m3 MAK

German (DFG) - Peak Limitations
10 x normal MAK (30 min average value); don't exceed during shift

German (DFG) - Skin/Sensitizers
danger of cutaneous absorption

German (DFG) - Carcinogens
suspected carcinogen

Israel - Time Weighted Averages
(0.5) mg/m3 TWA

Israel - Action Levels
0.25 mg/m3 AL

Mexico - Instruction No. 10 - TWAs
0.5 mg/m3 TWA

Mexico - Instruction No. 10 - STELs
2 mg/m3 STEL

Mexico - Instruction No. 10 - Skin designation
skin - potential for cutaneous absorption

Mexico - Wastewater - Organic Toxic Pollutants and Heavy Metals
Listed under [Heptachlor and Metabolites]

STATE LISTS

California - Air Bill 2588 Appendix A-I
known or potential carcinogen: 9/89

California - Prop. 65 - Cancer list
carcinogen - initial date 7/1/88

California - Prop. 65 - No Significant Risk Levels
no significant risk level = 0.2 ug/day

California - Exposure Limits - PELs
0.5 mg/m3 PEL

California - Exposure Limits - Skin Notation
material may be absorbed through the skin, eyes or mucous membrane

California - Directors List of Hazardous Substances (8 CCR 339)
[present]

Florida Hazardous Substance List
[present]

Massachusetts Right To Know List
carcinogen; extraordinarily hazardous

Minnesota Hazardous Substance List
skin

NJ Right to Know List (Total)
sn 0974; metabolites: sn 2996

NJ Special Hazardous Substances
(carcinogen)

Pennsylvania Right to Know List
environmental hazard

HEPTACHLORODIBENZOFURANS   RR-01173-4
SEE ALSO:
F039-HAZARDOUS WASTES

HEALTH AND SAFETY LISTS

ACGIH 1995 - Time Weighted Averages
0.05 mg/m3 TWA

ACGIH 1995 - Skin Designations
skin - potential for cutaneous absorption

ACGIH 1995 - Carcinogens
A3-animal carcinogen

U.S. DOT - Appendix A Table 1 - Hazardous Substances
final RQ = 1 pound (0.454 kg)

NIOSH - Selected LD50s and LC50s
Oral, rat: LD50 = 47 mg/kg

ENVIRONMENTAL LISTS

ATSDR Priority List
Rank (of 275): 044

CERCLA/SARA - Hazardous Substances and their Reportable Quantities
final RQ = 1 pound (0.454 kg)

Clean Water Act - Priority Pollutants
[present]
Safe Drinking Water Act - MCLs
MCL = 0.0002 mg/L
Safe Drinking Water Act - MCLGs
MCLG = Zero
EPA - Carcinogen Hazard Ranking for RQ Adjustment
Hazard ranking = High
RCRA - Hazardous Constituents-Appendix VIII
hazardous constituent - no waste number
RCRA - Basis for Listing - Appendix VII
Included in waste stream: F039
RCRA - TSD Facilities Ground Water Monitoring
TM 8080 = 1 ug/L PQL;   TM 8270 = 10 ug/L PQL
RCRA - Universal Treatment Standards (LDR)
WW: 0.016 mg/l;   NWW: 0.066 mg/kg
INTERNATIONAL LISTS
Mexico - Wastewater - Organic Toxic Pollutants and Heavy Metals
Listed under [Heptachlor and Metabolites]
STATE LISTS
California - Air Bill 2588 Appendix A-II
known or potential carcinogen: 9/89
California - Prop. 65 - Cancer list
carcinogen - initial date 7/1/88
California - Prop. 65 - No Significant Risk Levels
no significant risk level = 0.08 ug/day
California - Directors List of Hazardous Substances (8 CCR 339)
[present]
Massachusetts Right To Know List
[present]
NJ Right to Know List (Total)
sn 2997
Pennsylvania Right to Know List
environmental hazard

HEPTACHLOROFLUOROPROPANE   135401-87-5
STATE LISTS
Pennsylvania Right to Know List
environmental hazard

3-(((HEPTADECAFLUOROOCTYL)SULFONYL)   1652-63-7
AMINO)-N,N,N-TRIMETHYL-1-PROPANAMINIUM
IODIDE
ENVIRONMENTAL LISTS
ATSDR Priority List
Rank (of 275): 106

HEPTADECANOL   52783-44-5
STATE LISTS
California - Air Bill 2588 Appendix A-I
known or potential carcinogen

2-(8-HEPTADECENYL)-4,5-DIHYDRO-1H-IMIDA-   62449-33-6
ZOLE-1-ETHANOL, MONOHYDROCHLORIDE, (Z)-

ENVIRONMENTAL LISTS
RCRA - Hazardous Constituents-Appendix VIII
hazardous constituent - no waste number

2-(8-HEPTADECENYL)-4-METHYL-2-OXAZOLINE-4-   14408-42-5
METHANOL
ENVIRONMENTAL LISTS
ATSDR Priority List
Rank (of 275): 111

2-HEPTADECYL-2-IMIDAZOLINE   105-28-2
ENVIRONMENTAL LISTS
ATSDR Priority List
Rank (of 275): 256
STATE LISTS
California - Air Bill 2588 Appendix A-I
known or potential carcinogen

2-HEPTADECYL-1-METHYL-1-(2-STEAROYLAMIDO)   13470-50-3
ETHYL-2-IMIDAZOLINIUM METHYL SULFATE
ENVIRONMENTAL LISTS
RCRA - Hazardous Constituents-Appendix VIII
hazardous constituent - no waste number

HEPTADECYL SULFATE, SODIUM SALT   5910-79-2
ENVIRONMENTAL LISTS
Class 1 Ozone Depletors
ozone depletion potential = 1.0

HEPTANAL   111-71-7
ENVIRONMENTAL LISTS
List of Pesticide Product Inert Ingredients
[present]

HEPTANAL, 2-(PHENYLMETHYLENE)-   122-40-7
HEALTH AND SAFETY LISTS
NFPA - Flash Points
flash point = 310 degrees F (154 degrees C)
NFPA - Hazard Identification Ratings
health-0; flammability-1; reactivity-0

1-HEPTANAMINE   111-68-2
ENVIRONMENTAL LISTS
List of Pesticide Product Inert Ingredients
[present]

HEPTANE (N-)   142-82-5
ENVIRONMENTAL LISTS
List of Pesticide Product Inert Ingredients
[present]

1-HEPTANETHIOL   1639-09-4
HEALTH AND SAFETY LISTS
NIOSH - Selected LD50s and LC50s
Oral, rat: LD50 = 3170 mg/kg

HEPTANOIC ACID   111-14-8
ENVIRONMENTAL LISTS
List of Pesticide Product Inert Ingredients
[present]

2-HEPTANOL   543-49-7
ENVIRONMENTAL LISTS
List of Pesticide Product Inert Ingredients
[present]

3-HEPTANOL   589-82-2
HEALTH AND SAFETY LISTS
U.S. DOT - Substances From 49 CFR 172.101
regulated by DOT (UN3056)
U.S. DOT - Hazard Classes
DOT hazard class = 3
NIOSH - Selected LD50s and LC50s
Oral, rat: LD50 = 14 gm/kg

ENVIRONMENTAL LISTS

TSCA - Code of Federal Regulations Citations
40 CFR 712.30(x); 40 CR 716.120(d)

TSCA - PAIR - Reporting List
Reporting Date: November 27, 1991

TSCA - Health and Safety Reporting List
Effective Date: September 30, 1991

3,6,9,12,15,18,21-HEPTAOXATETRATRIAOCTANOIC   104503-68-6
ACID, SODIUM SALT

ENVIRONMENTAL LISTS

TSCA - Code of Federal Regulations Citations
40 CFR 712.30(x); 40 CFR 716.120(d)

TSCA - PAIR - Reporting List
Reporting Date: November 27, 1991

TSCA - Health and Safety Reporting List
Effective Date: September 30, 1991

5-HEPTENAL, 2,6-DIMETHYL-   106-72-9

HEALTH AND SAFETY LISTS

NFPA - Flash Points
flash point = 130 degrees F (54 degrees C)

NFPA - Hazard Identification Ratings
health-2; flammability-2; reactivity-0

STATE LISTS

Florida Hazardous Substance List
[present]

Massachusetts Right To Know List
[present]

Pennsylvania Right to Know List
[present]

1-HEPTENE   592-76-7

HEALTH AND SAFETY LISTS

ACGIH 1995 - Time Weighted Averages
400 ppm TWA; 1640 mg/m3 TWA

ACGIH 1995 - Short Term Exposure Limits
500 ppm STEL; 2050 mg/m3 STEL

AIHA - Odor Threshold Values
geometric mean air odor threshold = 230 ppm (detectable); 330 ppm (recognizable)

U.S. DOT - Substances From 49 CFR 172.101
regulated by DOT (UN1206)

U.S. DOT - Hazard Classes
DOT hazard class = 3

NFPA - Flash Points
flash point = 25 degrees F (-4 degrees C)

NFPA - Hazard Identification Ratings
health-1; flammability-3; reactivity-0

NIOSH 1990 - Pocket Guide - RELs
85 ppm TWA; 350 mg/m3 TWA; C 440 ppm (15 min); C 1800 mg/m3 (15 min)

NIOSH 1990 - Pocket Guide - IDLHs
5000 ppm IDLH

NIOSH 1990 - Pocket Guide - Target organs
skin, PNS, respiratory system

NIOSH - Health Standards - Exposure Limits
85 ppm TWA; 350 mg/m3 TWA; C (15 min) 440 ppm ; C (15 min) 1800 mg/m3 (Listed under 'Alkanes (C5-C8)')

NIOSH - Health Standards - Health Effects and Precautions
Skin and nervous system effects (Listed under 'Alkanes (C5-C8)')

OSHA - Vacated PELs - Time Weighted Averages
400 ppm TWA; 1600 mg/m3 TWA

OSHA - Vacated PELs - Short Term Exposure Limits
500 ppm STEL; 2000 mg/m3 STEL

OSHA - Final PELs - Time Weighted Averages
500 ppm TWA; 2000 mg/m3 TWA

ENVIRONMENTAL LISTS

TSCA - PAIR - Reporting List
Effective Date: January 26, 1994; Reporting Date: March 28, 1994

TSCA - Health and Safety Reporting List
Effective Date: January 26, 1994; Sunset Date: January 26, 2004

INTERNATIONAL LISTS

Australian Exposure Standards - Time Weighted Averages
400 ppm TWA; 1640 mg/m3 TWA

Australian Exposure Standards - Short Term Exposure Limits
500 ppm STEL; 2050 mg/m3 STEL

Canada - WHMIS: Ingredient Disclosure
1% item 806 (940)

Canada - Alberta - 8 Hour Occupational Exposure Limit
400 ppm TWA; 1640 mg/m3 TWA

Canada - Alberta - 15 Minute Occupational Exposure Limit
500 ppm STEL; 2049 mg/m3 STEL

Canada - British Columbia - 8 Hour Exposure Limits
400 ppm TWA; 1600 mg/m3 TWA

Canada - British Columbia - 15 Minute Exposure Limits
500 ppm STEL; 2000 mg/m3 STEL

Canada - Ontario - OHSA - TWAEVs
400 ppm TWAEV; 1635 mg/m3 TWAEV

Canada - Ontario - OHSA - STEVs
500 ppm STEV; 2045 mg/m3 STEV

Canada - Quebec - Time-Weighted Average Exposure Values
400 ppm TWAEV; 1640 mg/m3 TWAEV

Canada - Quebec - Short-term Exposure Values
500 ppm STEV; 2050 mg/m3 STEV

United Kingdom - Occupational Exposure Standards - TWAs
400 ppm TWA; 1600 mg/m3 TWA

United Kingdom - Occupational Exposure Standards - STELs
500 ppm STEL; 2000 mg/m3 STEL

German (DFG) - MAK Values
500 ppm MAK; 2000 mg/m3 MAK (all isomers included)

German (DFG) - Peak Limitations
2 x normal MAK (30 min. average value); don't exceed 4 times during shift

Israel - Time Weighted Averages
400 ppm TWA; 1640 mg/m3 TWA

Israel - Short Term Exposure Limits
500 ppm STEL; 2050 mg/m3 STEL

Israel - Action Levels
200 ppm AL; 820 mg/m3 AL

Mexico - Instruction No. 10 - TWAs
400 ppm TWA; 1600 mg/m3 TWA

Mexico - Instruction No. 10 - STELs
500 ppm STEL; 2000 mg/m3 STEL

Mexico - Instruction No. 10 - Skin designation
skin - potential for cutaneous absorption

STATE LISTS

California - Exposure Limits - PELs
400 ppm PEL; 1600 mg/m3 PEL

California - Exposure Limits - STELs
500 ppm STEL; 2000 mg/m3 STEL

California - Directors List of Hazardous Substances (8 CCR 339)
[present]

Florida Hazardous Substance List
[present]

Massachusetts Right To Know List
[present]

Minnesota Hazardous Substance List
[present]

NJ Right to Know List (Total)
sn 1339

NJ Special Hazardous Substances
(flammable - third degree)

Pennsylvania Right to Know List
[present]

PROPOSED REGULATIONS

TSCA - ITC 31st Report Priority Testing List
designated to be tested

Canada - Ontario - Proposed Occupational TWAEVs
200 ppm TWAEV;   800 mg/m3 TWAEV

Canada - Ontario - Proposed Occupational STEVs
300 ppm STEV;   1200 mg/m3 STEV

HEPTENE   25339-56-4

HEALTH AND SAFETY LISTS

NIOSH - Health Standards - Exposure Limits
C (15 min) 0.5 ppm;   C (15 min) 2.7 mg/m3 (Listed under 'Thiols')

NIOSH - Health Standards - Health Effects and Precautions
Irritation;   eye, skin, blood, and nervous system effects (Blood and urine monitoring required;   prevent skin contact) (Listed under 'Thiols')

HEPTENE   81624-04-6

HEALTH AND SAFETY LISTS

NIOSH - Selected LD50s and LC50s
Oral, rat: LD50 = 7000 mg/kg

3-HEPTENE (MIXED CIS- AND TRANS- ISOMERS)   RR-00519-6

HEALTH AND SAFETY LISTS

NFPA - Flash Points
flash point = 160 degrees F (71 degrees C)

NFPA - Hazard Identification Ratings
health-0;   flammability-2;   reactivity-0

INTERNATIONAL LISTS

Canada - WHMIS: Ingredient Disclosure
1% item 807 (941)

HEPTENOPHOS   23560-59-0

HEALTH AND SAFETY LISTS

NFPA - Flash Points
flash point = 140 degrees F (60 degrees C)

NFPA - Hazard Identification Ratings
health-0;   flammability-2;   reactivity-0

INTERNATIONAL LISTS

Canada - WHMIS: Ingredient Disclosure
1% item 808 (942)

N-HEPTYL ALCOHOL   111-70-6

ENVIRONMENTAL LISTS

TSCA - Code of Federal Regulations Citations
40 CFR 721.1137

TSCA - Chemicals with Significant New Use Rules
PMN number: P-90-489

TSCA - Section 12(b) - Export Notification
P-90-489;   export notification required - Section 5

HEPTYLENE-2-TRANS   14686-13-6

ENVIRONMENTAL LISTS

TSCA - Code of Federal Regulations Citations
40 CFR 712.30(x);   40 CFR 716.120(d)

TSCA - PAIR - Reporting List
Reporting Date: November 27, 1991

TSCA - Health and Safety Reporting List
Effective Date: September 30, 1991

HEROIN   561-27-

STATE LISTS

Pennsylvania Right to Know List
[present]

HETEROCYCLIC ALDEHYDE IMINE   RR-00959-

HEALTH AND SAFETY LISTS

U.S. DOT - Substances From 49 CFR 172.101
regulated by DOT (UN2278)

U.S. DOT - Hazard Classes
DOT hazard class = 3

NFPA - Flash Points
flash point < 32 degrees F (0 degrees C)

NFPA - Hazard Identification Ratings
health-0;   flammability-3;   reactivity-0

STATE LISTS

Florida Hazardous Substance List
[present]

Massachusetts Right To Know List
[present]

Pennsylvania Right to Know List
[present]

HEXAAMMONIUM MOLYBDATE   12027-67-7

STATE LISTS

NJ Right to Know List (Total)
sn 0976

NJ Special Hazardous Substances
(flammable - third degree)

HEXABROMOBIPHENYL   36355-01-8

HEALTH AND SAFETY LISTS

NFPA - Flash Points
flash point = 21 degrees F (-6 degrees C)

NFPA - Hazard Identification Ratings
health-0;   flammability-3;   reactivity-0

STATE LISTS

Pennsylvania Right to Know List
[present]

2,2',4,4',5,5'-HEXABROMO-1,1'-BIPHENYL   59080-40-9

HEALTH AND SAFETY LISTS

U.S. DOT - Appendix B - Marine Pollutants
DOT regulated marine pollutant

HEXABROMOCYCLODODECANE   3194-55-6

HEALTH AND SAFETY LISTS

NIOSH - Selected LD50s and LC50s
Inhalation, mouse: LC50 - 6600 mg/k3 (8 hr) Oral, rat: LD50 = 500 mg/kg Skin, rabbit: LD50 = 2 gm/kg

HEXACHLOROACETONE   116-16-5

HEALTH AND SAFETY LISTS

NFPA - Flash Points
flash point < 32 degrees F (0 degrees C)

NFPA - Hazard Identification Ratings
health-0;   flammability-3;   reactivity-0

STATE LISTS

Florida Hazardous Substance List
[present]

Massachusetts Right To Know List
[present]

Pennsylvania Right to Know List
[present]

HEXACHLOROBENZENE   118-74-1

STATE LISTS

Massachusetts Right To Know List
teratogen

HEXACHLOROBUTADIENE   87-68-3

ENVIRONMENTAL LISTS

TSCA - Chemicals with Significant New Use Rules
PMN number: P-90-1624

HEXACHLOROCYCLOHEXANE (MIXED ISOMERS)   608-73-1
SEE ALSO:
MOLYBDENUM

INTERNATIONAL LISTS

Canada - WHMIS: Ingredient Disclosure
1% item 809 (1164)

HEXACHLOROCYCLOPENTADIENE   77-47-4

HEALTH AND SAFETY LISTS

IARC - Group 2B (sufficient animal data)
[present]

NTP Seventh Report - Suspect Carcinogens
suspect carcinogen

OSHA - Possible Select Carcinogens
[present]

ENVIRONMENTAL LISTS

TSCA - Code of Federal Regulations Citations
40 CFR 721.600

TSCA - Chemicals with Significant New Use Rules
[present]

TSCA - Section 12(b) - Export Notification
export notification required - Section 5

STATE LISTS

Massachusetts Right To Know List
carcinogen; extraordinarily hazardous

Minnesota Hazardous Substance List
carcinogen

1,2,3,7,8,9-HEXACHLORODIBENZO-P-DIOXIN   19408-74-3

STATE LISTS

Massachusetts Right To Know List
carcinogen; extraordinarily hazardous

HEXACHLORODIBENZODIOXIN   34465-46-8

ENVIRONMENTAL LISTS

EPA - Master Testing List
[present]

TSCA - Code of Federal Regulations Citations
40 CFR 712.30(d),(w);   40 CFR 716.120(a)

TSCA - PAIR - Reporting List
Reporting Dates: November 19, 1982;   March 12, 1990

TSCA - Health and Safety Reporting List
Effective Date: January 11, 1990

1,2,3,4,7,8-HEXACHLORODIBENZO-P-DIOXIN   39227-28-6

HEALTH AND SAFETY LISTS

U.S. DOT - Substances From 49 CFR 172.101
regulated by DOT (UN2662)

U.S. DOT - Hazard Classes
DOT hazard class = 6.1

NIOSH - Selected LD50s and LC50s
Inhalation, rat: LC50 = 360 ppm 6 hr Oral, rat: LD50 = 1290 mg/kg
Skin, rat: LD50 = 2980 mg/kg

INTERNATIONAL LISTS

Canada - WHMIS: Ingredient Disclosure
1% item 810 (944)

STATE LISTS

NJ Right to Know List (Total)
sn 0977

1,2,3,6,7,8-HEXACHLORODIBENZO-P-DIOXIN   57653-85-7
SEE ALSO:
F022-HAZARDOUS WASTES
F039-HAZARDOUS WASTES
K085-HAZARDOUS WASTES
K018-HAZARDOUS WASTES
K042-HAZARDOUS WASTES
F025-HAZARDOUS WASTES
K030-HAZARDOUS WASTES
K016-HAZARDOUS WASTES
F024-HAZARDOUS WASTES
K149-HAZARDOUS WASTES
K150-HAZARDOUS WASTES
K151-HAZARDOUS WASTES
F026-HAZARDOUS WASTES

HEALTH AND SAFETY LISTS

ACGIH 1995 - Time Weighted Averages
0.025 mg/m3 TWA

ACGIH 1995 - Skin Designations
skin - potential for cutaneous absorption

ACGIH 1995 - Carcinogens
A3-animal carcinogen

U.S. DOT - Substances From 49 CFR 172.101
regulated by DOT (UN2729)

U.S. DOT - Hazard Classes
DOT hazard class = 6.1

U.S. DOT - Appendix A Table 1 - Hazardous Substances
final RQ = 10 pounds (4.54 kg)

IARC - Group 2B (sufficient animal data)
[present]

NIOSH - Selected LD50s and LC50s
Inhalation, rat: LC50 = 3600 mg/m3 8 hr Oral, rat: LD50 = 10000 mg/kg

NTP Seventh Report - Suspect Carcinogens
suspect carcinogen

OSHA - Possible Select Carcinogens
[present]

ENVIRONMENTAL LISTS

ATSDR Priority List
Rank (of 275): 092

CERCLA/SARA - Section 313 - Emission Reporting
form R reporting required for 0.1% de minimus concentration

CERCLA/SARA - Hazardous Substances and their Reportable Quantities
final RQ = 10 pounds (4.54 kg)

Clean Air Act (1990) - List of Hazardous Air Contaminants
[present]

CAA - HON Rule - SOCMI Chemicals
compliance by Jan. 23, 1995

CAA - HON Rule - Organic HAPs
[present]

Clean Water Act - Priority Pollutants
[present]

Safe Drinking Water Act - MCLs
MCL = 0.001 mg/L

Safe Drinking Water Act - MCLGs
MCLG = Zero

EPA - Carcinogen Hazard Ranking for RQ Adjustment
Hazard ranking = Medium

RCRA - D Series - Maximum Concentration of Contaminants
waste number D032;   regulatory level = 0.13 mg/L

RCRA - D Series - Chronic Toxicity Reference Levels
chronic toxicity reference level = 0.0002 mg/L

RCRA - U Series Wastes
waste number U127

RCRA - Hazardous Constituents-Appendix VIII
waste number U127

RCRA - Basis for Listing - Appendix VII
Included in waste streams: F024, F025, F039, K016, K018, K030, K042, K085, K149, K150, K151

RCRA - Substances Banned From Land Disposal
[present]

RCRA - TSD Facilities Ground Water Monitoring
TM 8120 = 0.5 ug/L PQL;   TM 8270 = 10 ug/L PQL

RCRA - Universal Treatment Standards (LDR)
WW: 0.055 mg/l;   NWW: 10 mg/kg

INTERNATIONAL LISTS

Canada - WHMIS: Ingredient Disclosure
1% item 811 (945)

Canada - CEPA - Priority Substances List
estimated time for completion of assessment reports: 3 years

Mexico - Wastewater - Organic Toxic Pollutants and Heavy Metals
Listed under [Chlorinated Benzenes]

Mexico - Drinking Water - Ecological Criteria
0.00001 mg/l This level has been extrapolated by using a mathematic model

STATE LISTS

California - Air Bill 2588 Appendix A-I
known or potential carcinogen

California - Prop. 65 - Cancer list
carcinogen - initial date 10/1/87

California - Prop. 65 - Developmental Toxicity
developmental toxicity - initial date 1/1/89

California - Prop. 65 - No Significant Risk Levels
no significant risk level = 0.4 ug/day

Florida Hazardous Substance List
[present]

Massachusetts Right To Know List
carcinogen;   extraordinarily hazardous;   teratogen

Minnesota Hazardous Substance List
carcinogen

NJ Right to Know List (Total)
sn 0978

NJ Special Hazardous Substances
(carcinogen)

Pennsylvania Right to Know List
environmental hazard;   special hazardous substance

Pennsylvania RTK - Special Hazardous Substances
[present]

HEXACHLORODIBENZO-P-DIOXINS   RR-00509-4
SEE ALSO:
F024-HAZARDOUS WASTES
F039-HAZARDOUS WASTES
F025-HAZARDOUS WASTES
K030-HAZARDOUS WASTES
K018-HAZARDOUS WASTES
K016-HAZARDOUS WASTES
PROPYLCYCLOHEXANE
BIS(2,4-DIMETHYLBUTYL) MALEATE

HEALTH AND SAFETY LISTS

ACGIH 1995 - Time Weighted Averages
0.02 ppm TWA;   0.21 mg/m3 TWA

ACGIH 1995 - Skin Designations
skin - potential for cutaneous absorption

ACGIH 1995 - Carcinogens
A2-suspected human carcinogen

U.S. DOT - Substances From 49 CFR 172.101
regulated by DOT (UN2279)

U.S. DOT - Hazard Classes
DOT hazard class = 6.1

U.S. DOT - Appendix B - Marine Pollutants
DOT regulated severe marine pollutant

U.S. DOT - Appendix A Table 1 - Hazardous Substances
final RQ = 10 pounds (4.54 kg)

IARC - Group 3 (not classifiable)
[present]

NFPA - Hazard Identification Ratings
health-2;   flammability-1;   reactivity-1

NIOSH - Selected LD50s and LC50s
Oral, rat: LD50 = 90 mg/kg Skin, rabbit: LD50 = 1211 mg/kg

NTP Chemical Status Reports - Testing Status and NTIS Number
Technical reports printed (PB91185884)

OSHA - Vacated PELs - Time Weighted Averages
0.02 ppm TWA;   0.24 mg/m3 TWA

ENVIRONMENTAL LISTS

ATSDR Priority List
Rank (of 275): 014

CERCLA/SARA - Section 313 - Emission Reporting
form R reporting required for 1.0% de minimus concentration

CERCLA/SARA - Hazardous Substances and their Reportable Quantities
final RQ = 10 pounds (4.54 kg)

Clean Air Act (1990) - List of Hazardous Air Contaminants
[present]

CAA - HON Rule - SOCMI Chemicals
compliance by Jan. 23, 1995

CAA - HON Rule - Organic HAPs
[present]

Clean Water Act - Priority Pollutants
[present]

Clean Water Act - Toxic Pollutants
[present]

Safe Drinking Water Act - Monitoring
monitoring required at discretion of the state

RCRA - D Series - Maximum Concentration of Contaminants
waste number D033;   regulatory level = 0.5 mg/L

RCRA - D Series - Chronic Toxicity Reference Levels
chronic toxicity reference level = 0.005 mg/L

RCRA - U Series Wastes
waste number U128

RCRA - Hazardous Constituents-Appendix VIII
waste number U128

RCRA - Basis for Listing - Appendix VII
Included in waste streams: F024, F025, F039, K016, K018, K030

RCRA - Substances Banned From Land Disposal
[present]

RCRA - TSD Facilities Ground Water Monitoring
TM 8120 = 5 ug/L PQL; TM 8270 = 10 ug/L PQL

RCRA - Universal Treatment Standards (LDR)
WW: 0.055 mg/l; NWW: 5.6 mg/kg

TSCA - Code of Federal Regulations Citations
40 CFR 712.30(d); 40 CFR 716.120(a)

TSCA - PAIR - Reporting List
Reporting Date: November 19, 1982

TSCA - Health and Safety Reporting List
Effective Date: October 4, 1982

INTERNATIONAL LISTS

Australian Exposure Standards - Time Weighted Averages
0.02 ppm TWA; 0.21 mg/m3 TWA

Australian Exposure Standards - Skin Effects
skin absorption

Australian Exposure Standards - Carcinogens
suspected carcinogen

Canada - WHMIS: Ingredient Disclosure
0.1% item 812 (946)

Canada - Alberta - 8 Hour Occupational Exposure Limit
0.02 ppm TWA; 0.21 mg/m3 TWA

Canada - Alberta - 15 Minute Occupational Exposure Limit
0.06 ppm STEL; 0.64 mg/m3 STEL

Canada - Alberta - Designated Substances
designated substance - requires code of practice

Canada - Ontario - OHSA - TWAEVs
0.02 ppm TWAEV; 0.21 mg/m3 TWAEV

Canada - Ontario - OHSA - Skin Notations
absorption through skin, eyes, or mucous membranes

Canada - Quebec - Time-Weighted Average Exposure Values
0.02 ppm TWAEV; 0.21 mg/m3 TWAEV

Canada - Quebec - Skin Designations
absorbed through the skin

Canada - Quebec - Carcinogens
C2 carcinogen: effect suspected in humans

German (DFG) - Skin/Sensitizers
danger of cutaneous absorption

German (DFG) - Carcinogens
suspected carcinogen

Israel - Time Weighted Averages
0.02 ppm TWA; 0.21 mg/m3 TWA

Israel - Action Levels
0.01 ppm AL; 0.105 mg/m3 AL

Mexico - Wastewater - Organic Toxic Pollutants and Heavy Metals
Listed under [Organic Toxic Pollutants]

Mexico - Drinking Water - Ecological Criteria
0.004 mg/l Substance presents persistence, bioaccumulations or risk of cancer, reduce human exposure to a minimum; This level has been extrapolated by using a mathematic model

STATE LISTS

California - Air Bill 2588 Appendix A-I
6/91

California - Exposure Limits - PELs
0.02 ppm PEL; 0.24 mg/m3 PEL

California - Exposure Limits - Skin Notation
material may be absorbed through the skin, eyes or mucous membrane

California - Directors List of Hazardous Substances (8 CCR 339)
[present]

Florida Hazardous Substance List
[present]

Massachusetts Right To Know List
carcinogen; extraordinarily hazardous

Minnesota Hazardous Substance List
carcinogen

NJ Right to Know List (Total)
sn 0979

NJ Special Hazardous Substances
(carcinogen)

Pennsylvania Right to Know List
environmental hazard

PROPOSED REGULATIONS

Safe Drinking Water Act - Priority list
[present]

1,2,3,4,7,8,9-HEXACHLORODIBENZOFURAN   55673-89-7
SEE ALSO:
    F024-HAZARDOUS WASTES

HEALTH AND SAFETY LISTS

IARC - Group Unspecified
[present] (Listed under 'Hexachlorocyclohexanes (HCH)')

NIOSH - Selected LD50s and LC50s
Oral, rat: LD50 = 100 mg/kg Skin, rat: LD50 = 900 mg/kg

NTP Seventh Report - Suspect Carcinogens
technical grade: suspect carcinogen (Listed under 'Lindane and other hexachlorocyclohexane isomers')

OSHA - Possible Select Carcinogens
[present]

ENVIRONMENTAL LISTS

ATSDR Priority List
Rank (of 275): 134

Clean Water Act - Toxic Pollutants
[present]

RCRA - Basis for Listing - Appendix VII
Included in waste stream: F024

INTERNATIONAL LISTS

German (DFG) - MAK Values
total dust: 0.5 mg/m3 MAK (technical mixture of alpha and gamma isomers)

German (DFG) - Skin/Sensitizers
danger of cutaneous absorption (mixture of alpha and beta isomers)

Mexico - Drinking Water - Ecological Criteria
None given Substance presents persistence, bioaccumulations or risk of cancer, reduce human exposure to a minimum

STATE LISTS

California - Air Bill 2588 Appendix A-I
known or potential carcinogen

California - Prop. 65 - Cancer list
carcinogen (includes technical grade) - initial date 10/1/89

California - Prop. 65 - No Significant Risk Levels
technical grade: no significant risk level = 0.2 ug/day

California - Directors List of Hazardous Substances (8 CCR 339)
[present]

Florida Hazardous Substance List
[present]

Massachusetts Right To Know List
carcinogen; extraordinarily hazardous

Minnesota Hazardous Substance List
carcinogen (includes Lindane)

Pennsylvania Right to Know List
special hazardous substance

Pennsylvania RTK - Special Hazardous Substances
[present]

HEXACHLORODIBENZOFURAN   55684-94-1

SEE ALSO:
K033-HAZARDOUS WASTES
K034-HAZARDOUS WASTES
F025-HAZARDOUS WASTES
BIS(2,4-DIMETHYLBUTYL) MALEATE
F024-HAZARDOUS WASTES
F039-HAZARDOUS WASTES
K032-HAZARDOUS WASTES

HEALTH AND SAFETY LISTS

ACGIH 1995 - Time Weighted Averages
0.01 ppm TWA; 0.11 mg/m3 TWA

U.S. DOT - Substances From 49 CFR 172.101
regulated by DOT (UN2646)

U.S. DOT - Hazard Classes
DOT hazard class = 6.1

U.S. DOT - Substances Which Are Poisonous by Inhalation
liquid hazardous material poisonous by inhalation (UN2646)

U.S. DOT - Appendix A Table 1 - Hazardous Substances
final RQ = 10 pounds (4.54 kg)

NIOSH - Selected LD50s and LC50s
*Inhalation, rat: LC50 = 1600 ppb 4 hr Oral, rat: LD50 = 113 mg/kg
Skin, rabbit: LD50 = 430 mg/kg*

NTP Chemical Status Reports - Testing Status and NTIS Number
Technical reports printed (no NTIS number given)

OSHA - Vacated PELs - Time Weighted Averages
0.01 ppm TWA; 0.1 mg/m3 TWA

ENVIRONMENTAL LISTS

ATSDR Priority List
Rank (of 275): 126

CERCLA/SARA - Section 302 Extremely Hazardous Substances and TPQs
TPQ = 100 pounds

CERCLA/SARA - Section 313 - Emission Reporting
form R reporting required for 1.0% de minimus concentration

CERCLA/SARA - Hazardous Substances and their Reportable Quantities
final RQ = 10 pounds (4.54 kg)

Clean Air Act (1990) - List of Hazardous Air Contaminants
[present]

Clean Water Act - Hazardous Substances
[present]

Clean Water Act - Priority Pollutants
[present]

Clean Water Act - Toxic Pollutants
[present]

Safe Drinking Water Act - MCLs
MCL = 0.05 mg/L

Safe Drinking Water Act - MCLGs
MCLG = 0.05 mg/L

RCRA - U Series Wastes
waste number U130

RCRA - Hazardous Constituents-Appendix VIII
waste number U130

RCRA - Basis for Listing - Appendix VII
Included in waste streams: F024, F025, F039, K032, K033, K034

RCRA - Substances Banned From Land Disposal
[present]

RCRA - TSD Facilities Ground Water Monitoring
TM 8120 = 5 ug/L PQL; TM 8270 = 10 ug/L PQL

RCRA - Universal Treatment Standards (LDR)
WW: 0.057 mg/l; NWW: 2.4 mg/kg

TSCA - Code of Federal Regulations Citations
40 CFR 712.30(d); 40 CFR 716.120(a)

TSCA - PAIR - Reporting List
Reporting Date: November 19, 1982

TSCA - Health and Safety Reporting List
Effective Date: October 4, 1982

INTERNATIONAL LISTS

Australian Exposure Standards - Time Weighted Averages
0.01 ppm TWA; 0.11 mg/m3 TWA

Canada - WHMIS: Ingredient Disclosure
1% item 813 (947)

Canada - NPRI (National Pollutant Release Inventory)
[present]

Canada - Alberta - 8 Hour Occupational Exposure Limit
0.01 ppm TWA; 0.11 mg/m3 TWA

Canada - Alberta - 15 Minute Occupational Exposure Limit
0.03 ppm STEL; 0.34 mg/m3 STEL

Canada - British Columbia - 8 Hour Exposure Limits
0.01 ppm TWA; 0.1 mg/m3 TWA

Canada - British Columbia - 15 Minute Exposure Limits
0.03 ppm STEL; 0.3 mg/m3 STEL

Canada - Ontario - OHSA - TWAEVs
0.01 ppm TWAEV; 0.11 mg/m3 TWAEV

Canada - Quebec - Time-Weighted Average Exposure Values
0.01 ppm TWAEV; 0.11 mg/m3 TWAEV

Israel - Time Weighted Averages
0.01 ppm TWA; 0.11 mg/m3 TWA

Israel - Action Levels
0.005 ppm AL; 0.05 mg/m3 AL

Mexico - Instruction No. 10 - TWAs
0.01 ppm TWA; 0.1 mg/m3 TWA

Mexico - Instruction No. 10 - STELs
0.03 ppm STEL; 0.3 mg/m3 STEL

Mexico - Wastewater - Organic Toxic Pollutants and Heavy Metals
Listed under [Organic Toxic Pollutants]

Mexico - Drinking Water - Ecological Criteria
0.001 mg/l

STATE LISTS

California - Air Bill 2588 Appendix A-I
[present]

California - Exposure Limits - PELs
0.01 ppm PEL; 0.11 mg/m3 PEL

California - Directors List of Hazardous Substances (8 CCR 339)
[present]

Florida Hazardous Substance List
[present]

Massachusetts Right To Know List
extraordinarily hazardous

Minnesota Hazardous Substance List
[present]

NJ Right to Know List (Total)
sn 0980

NJ Special Hazardous Substances
(corrosive)

Pennsylvania Right to Know List
environmental hazard

1,2,3,6,7,8-HEXACHLORO DIBENZOFURAN   57117-44-9

STATE LISTS

California - Air Bill 2588 Appendix A-I
known or potential carcinogen

1,2,3,4,7,8-HEXACHLORO DIBENZOFURAN 70648-26-9
ENVIRONMENTAL LISTS
ATSDR Priority List
Rank (of 275): 130
STATE LISTS
California - Prop. 65 - Cancer list
carcinogen - initial date 4/1/88
California - Prop. 65 - No Significant Risk Levels
no significant risk level = 0.0002 ug/day
Massachusetts Right To Know List
carcinogen; extraordinarily hazardous

1,2,3,7,8,9-HEXACHLORO DIBENZOFURAN 72918-21-9
STATE LISTS
California - Air Bill 2588 Appendix A-I
known or potential carcinogen

2,3,4,6,7,8-HEXACHLORO DIBENZOFURANS 60851-34-5
HEALTH AND SAFETY LISTS
NTP Chemical Status Reports - Testing Status and NTIS Number
Technical reports printed (PB80124844) (PB80124836)
NTP Chemical Status Reports - Evidence of Carcinogenicity
PB80124836: male mice-negative; female mice-negative; PB80124844: male rat-equivocal; female rat-positive; male mice-positive; female mice-positive
STATE LISTS
California - Air Bill 2588 Appendix A-I
known or potential carcinogen

HEXACHLORODIBENZOFURANS RR-00505-0
SEE ALSO:
F039-HAZARDOUS WASTES
ENVIRONMENTAL LISTS
RCRA - Hazardous Constituents-Appendix VIII
hazardous constituent - no waste number
RCRA - Basis for Listing - Appendix VII
Included in waste streams: F021, F022, F026, F027, F028, F039
RCRA - Universal Treatment Standards (LDR)
WW: 0.000063 mg/l; NWW: 0.001 mg/kg

HEXACHLORODIFLUOROPROPANE 134452-44-1
STATE LISTS
California - Air Bill 2588 Appendix A-I
known or potential carcinogen

HEXACHLOROETHANE 67-72-1
ENVIRONMENTAL LISTS
ATSDR Priority List
Rank (of 275): 133

HEXACHLOROFLUOROPROPANE 134237-35-7
STATE LISTS
California - Air Bill 2588 Appendix A-I
known or potential carcinogen

HEXACHLORONAPHTHALENE 1335-87-1
STATE LISTS
California - Air Bill 2588 Appendix A-I
known or potential carcinogen

1,2,3,4,7,7-HEXACHLORONORBORNADIENE 3389-71-7
STATE LISTS

California - Air Bill 2588 Appendix A-I
known or potential carcinogen

HEXACHLOROPHENE 70-30-4
STATE LISTS
California - Air Bill 2588 Appendix A-I
known or potential carcinogen

HEXACHLOROPROPENE 1888-71-7
SEE ALSO:
F039-HAZARDOUS WASTES
ENVIRONMENTAL LISTS
RCRA - Hazardous Constituents-Appendix VIII
hazardous constituent - no waste number
RCRA - Basis for Listing - Appendix VII
Included in waste streams: F021, F022, F026, F027, F028, F039
RCRA - Universal Treatment Standards (LDR)
WW: 0.000063 mg/l; NWW: 0.001 mg/kg

HEXACOSAMETHYLCYCLOTRIDECASILOXANE 23732-94-7
ENVIRONMENTAL LISTS
Class 1 Ozone Depletors
ozone depletion potential = 1.0

HEXACOSAMETHYLDODECASILOXANE 2471-08-1
SEE ALSO:
F025-HAZARDOUS WASTES
K073-HAZARDOUS WASTES
F039-HAZARDOUS WASTES
CHLORINATED ETHANES
K030-HAZARDOUS WASTES
BIS(2,4-DIMETHYLBUTYL) MALEATE
SEC-BUTYLCYCLOHEXANE
F024-HAZARDOUS WASTES
K016-HAZARDOUS WASTES
HEALTH AND SAFETY LISTS
ACGIH 1995 - Time Weighted Averages
1 ppm TWA; 9.7 mg/m3 TWA
ACGIH 1995 - Skin Designations
skin - potential for cutaneous absorption
ACGIH 1995 - Carcinogens
A2-suspected human carcinogen
U.S. DOT - Appendix A Table 1 - Hazardous Substances
final RQ = 100 pounds (45.4 kg)
IARC - Group 3 (not classifiable)
[present]
NIOSH - Selected LD50s and LC50s
Oral, rat: LD50 = 4460 mg/kg
NIOSH 1990 - Pocket Guide - RELs
1 ppm TWA; 10 mg/m3 TWA
NIOSH 1990 - Pocket Guide - IDLHs
300 ppm IDLH (not considering carcinogenic effects)
NIOSH 1990 - Pocket Guide - Carcinogens
occupational carcinogen
NIOSH 1990 - Pocket Guide - Target organs
eyes
NIOSH 1990 - Pocket Guide - Skin list
Potential for dermal absorption
NIOSH - Health Standards - Exposure Limits
reduce exposure to lowest feasible concentration
NIOSH - Health Standards - Health Effects and Precautions
has produced liver tumors in animals (Prevent skin contact)
NIOSH - Health Standards - Carcinogenic Chemicals
potential human carcinogen

NTP Chemical Status Reports - Testing Status and NTIS Number
Technical reports printed (PB282668/AS) (PB90170895/AS); Prechronic studies completed: in review for further evaluation

NTP Chemical Status Reports - Evidence of Carcinogenicity
PB282668/AS: male rat-negative; female rat-negative; male mice-positive; female mice-positive; PB90170895/AS: male rat-clear evidence; female rat-no evidence

NTP Seventh Report - Suspect Carcinogens
suspect carcinogen

OSHA - Vacated PELs - Time Weighted Averages
1 ppm TWA; 10 mg/m3 TWA

OSHA - Vacated PELs - Skin Designation
Prevent or reduce skin absorption

OSHA - Final PELs - Time Weighted Averages
1 ppm TWA; 10 mg/m3 TWA

OSHA - Final PELs - Skin Notations
prevent or reduce skin absorption

ENVIRONMENTAL LISTS

ATSDR Priority List
Rank (of 275): 181

CERCLA/SARA - Section 313 - Emission Reporting
form R reporting required for 1.0% de minimus concentration

CERCLA/SARA - Hazardous Substances and their Reportable Quantities
final RQ = 100 pounds (45.4 kg)

Clean Air Act (1990) - List of Hazardous Air Contaminants
[present]

CAA - HON Rule - SOCMI Chemicals
compliance by Jan. 23, 1995

CAA - HON Rule - Organic HAPs
[present]

Clean Water Act - Priority Pollutants
[present]

RCRA - D Series - Maximum Concentration of Contaminants
waste number D034; regulatory level = 3.0 mg/L

RCRA - D Series - Chronic Toxicity Reference Levels
chronic toxicity reference level = 0.03 mg/L

RCRA - U Series Wastes
waste number U131

RCRA - Hazardous Constituents-Appendix VIII
waste number U131

RCRA - Basis for Listing - Appendix VII
Included in waste streams: F024, F025, F039, K016, K030, K073

RCRA - Substances Banned From Land Disposal
[present]

RCRA - TSD Facilities Ground Water Monitoring
TM 8120 = 0.5 ug/L PQL; TM 8270 = 1 ug/L PQL

RCRA - Universal Treatment Standards (LDR)
WW: 0.055 mg/l; NWW: 30 mg/kg

TSCA - Code of Federal Regulations Citations
40 CFR 712.30(f); 40 CFR 716.120(a)

TSCA - PAIR - Reporting List
Reporting Date: November 19, 1982

TSCA - Health and Safety Reporting List
Effective Date: April 29, 1983

INTERNATIONAL LISTS

Australian Exposure Standards - Time Weighted Averages
1 ppm TWA; 9.7 mg/m3 TWA

Canada - WHMIS: Ingredient Disclosure
1% item 814 (948)

Canada - NPRI (National Pollutant Release Inventory)
[present]

Canada - Alberta - 8 Hour Occupational Exposure Limit
10 ppm TWA; 97 mg/m3 TWA

Canada - Alberta - 15 Minute Occupational Exposure Limit
15 ppm STEL; 145 mg/m3 STEL

Canada - British Columbia - 8 Hour Exposure Limits
1 ppm TWA; 10 mg/m3 TWA

Canada - British Columbia - 15 Minute Exposure Limits
3 ppm STEL; 30 mg/m3 STEL

Canada - British Columbia - Skin Notations
skin - potential for skin absorption

Canada - Ontario - OHSA - TWAEVs
10 ppm TWAEV; 97 mg/m3 TWAEV

Canada - Quebec - Time-Weighted Average Exposure Values
1 ppm TWAEV; 9.7 mg/m3 TWAEV

Canada - Quebec - Skin Designations
absorbed through the skin

United Kingdom - Occupational Exposure Standards - TWAs
vapour: 5 ppm TWA; 50 mg/m3 TWA; total inhalable dust: 1(mg/m3 TWA; respirable dust: 5 mg/m3 TWA

German (DFG) - MAK Values
1 ppm MAK; 10 mg/m3 MAK

Israel - Time Weighted Averages
(1) ppm TWA; (9.7) mg/m3 TWA

Israel - Action Levels
0.5 ppm AL; 4.85 mg/m3 AL

Mexico - Instruction No. 10 - TWAs
10 ppm TWA; 100 mg/m3 TWA

Mexico - Instruction No. 10 - Skin designation
skin - potential for cutaneous absorption

Mexico - Wastewater - Organic Toxic Pollutants and Heavy Metals
Listed under [Chlorinated Ethanes]

Mexico - Drinking Water - Ecological Criteria
0.02 mg/l Substance presents persistence, bioaccumulations or risk of cancer, reduce human exposure to a minimum; This level has been extrapolated by using a mathematic model

STATE LISTS

California - Air Bill 2588 Appendix A-I
known or potential carcinogen: 9/90

California - Prop. 65 - Cancer list
carcinogen - initial date 7/1/90

California - Prop. 65 - No Significant Risk Levels
no significant risk level = 20 ug/day

California - Exposure Limits - PELs
1 ppm PEL; 10 mg/m3 PEL

California - Exposure Limits - Skin Notation
material may be absorbed through the skin, eyes or mucous membrane

California - Directors List of Hazardous Substances (8 CCR 339)
[present]

Florida Hazardous Substance List
[present]

Massachusetts Right To Know List
carcinogen; extraordinarily hazardous

Minnesota Hazardous Substance List
carcinogen

NJ Right to Know List (Total)
sn 0981

NJ Special Hazardous Substances
(carcinogen)

Pennsylvania Right to Know List
environmental hazard

PROPOSED REGULATIONS

Safe Drinking Water Act - Priority list
[present]

HEXADECAMETHYLCYCLOOCTASILOXANE   556-68-3

ENVIRONMENTAL LISTS

  Class 2 Ozone Depletors
    ozone depletion weight reserved

HEXADECAMETHYLHEPTASILOXANE   541-01-5

HEALTH AND SAFETY LISTS

  ACGIH 1995 - Time Weighted Averages
    0.2 mg/m3 TWA

  ACGIH 1995 - Skin Designations
    skin - potential for cutaneous absorption

  NIOSH 1990 - Pocket Guide - RELs
    0.2 mg/m3 TWA

  NIOSH 1990 - Pocket Guide - IDLHs
    2 mg/m3 IDLH

  NIOSH 1990 - Pocket Guide - Target organs
    liver, skin

  NIOSH 1990 - Pocket Guide - Skin list
    Potential for dermal absorption

  OSHA - Vacated PELs - Time Weighted Averages
    0.2 mg/m3 TWA

  OSHA - Vacated PELs - Skin Designation
    Prevent or reduce skin absorption

  OSHA - Final PELs - Time Weighted Averages
    0.2 mg/m3 TWA

  OSHA - Final PELs - Skin Notations
    prevent or reduce skin absorption

ENVIRONMENTAL LISTS

  CERCLA/SARA - Section 313 - Emission Reporting
    form R reporting required for 1.0% de minimus concentration

  TSCA - Code of Federal Regulations Citations
    40 CFR 704.83; 40 CFR 712.30(d); 40 CFR 716.120(a)

  TSCA - PAIR - Reporting List
    Reporting Date: November 19, 1982

  TSCA - Health and Safety Reporting List
    Effective Date: October 4, 1982

INTERNATIONAL LISTS

  Australian Exposure Standards - Time Weighted Averages
    0.2 mg/m3 TWA

  Australian Exposure Standards - Skin Effects
    skin absorption

  Canada - WHMIS: Ingredient Disclosure
    1% item 815 (949)

  Canada - Alberta - 8 Hour Occupational Exposure Limit
    0.02 mg/m3 TWA

  Canada - Alberta - 15 Minute Occupational Exposure Limit
    0.06 mg/m3 STEL

  Canada - Alberta - Skin Designation
    can be absorbed through the intact skin

  Canada - British Columbia - 8 Hour Exposure Limits
    0.2 mg/m3 TWA

  Canada - British Columbia - 15 Minute Exposure Limits
    0.6 mg/m3 STEL

  Canada - British Columbia - Skin Notations
    skin - potential for skin absorption

  Canada - Ontario - OHSA - TWAEVs
    0.2 mg/m3 TWAEV

  Canada - Ontario - OHSA - Skin Notations
    absorption through skin, eyes, or mucous membranes

  Canada - Quebec - Time-Weighted Average Exposure Values
    0.2 mg/m3 TWAEV

  Canada - Quebec - Skin Designations
    absorbed through the skin

  Israel - Time Weighted Averages
    0.2 mg/m3 TWA

  Israel - Action Levels
    0.1 mg/m3 AL

  Mexico - Instruction No. 10 - TWAs
    0.2 mg/m3 TWA

  Mexico - Instruction No. 10 - Skin designation
    skin - potential for cutaneous absorption

STATE LISTS

  California - Air Bill 2588 Appendix A-II
    6/91

  California - Exposure Limits - PELs
    0.2 mg/m3 PEL

  California - Exposure Limits - Skin Notation
    material may be absorbed through the skin, eyes or mucous membrane

  California - Directors List of Hazardous Substances (8 CCR 339)
    [present]

  Florida Hazardous Substance List
    [present]

  Massachusetts Right To Know List
    [present]

  Minnesota Hazardous Substance List
    skin

  NJ Right to Know List (Total)
    sn 0982

  Pennsylvania Right to Know List
    environmental hazard

HEXADECANE   544-76-3

ENVIRONMENTAL LISTS

  TSCA - Code of Federal Regulations Citations
    40 CFR 704.102; 40 CFR 712.30(j); 40 CFR 716.120(a); 40 CFR 721.1150

  TSCA - PAIR - Reporting List
    Reporting Date: March 13, 1983

  TSCA - Health and Safety Reporting List
    Effective Date: January 13, 1984

  TSCA - Section 12(b) - Export Notification
    export notification required - Section 5

1-HEXADECANETHIOL   2917-26-2

HEALTH AND SAFETY LISTS

  U.S. DOT - Substances From 49 CFR 172.101
    regulated by DOT (UN2875)

  U.S. DOT - Hazard Classes
    DOT hazard class = 6.1

  U.S. DOT - Appendix A Table 1 - Hazardous Substances
    final RQ = 100 pounds (45.4 kg)

  IARC - Group 3 (not classifiable)
    [present]

  NIOSH - Selected LD50s and LC50s
    Inhalation, rat: LC50 = 340 mg/m3 8 hr Oral, rat: LD50 = 57 mg/kg Skin, rat: LD50 = 1840 mg/kg

  NTP Chemical Status Reports - Testing Status and NTIS Number
    Technical reports printed (PB279525/AS)

  NTP Chemical Status Reports - Evidence of Carcinogenicity
    male rat-negative; female rat-negative

ENVIRONMENTAL LISTS

  CERCLA/SARA - Section 313 - Emission Reporting
    form R reporting required

  CERCLA/SARA - Hazardous Substances and their Reportable Quantities
    final RQ = 100 pounds (45.4 kg)

RCRA - U Series Wastes
waste number U132

RCRA - Hazardous Constituents-Appendix VIII
waste number U132

RCRA - Substances Banned From Land Disposal
[present]

RCRA - TSD Facilities Ground Water Monitoring
TM 8270 = 10 ug/L PQL

INTERNATIONAL LISTS

Canada - WHMIS: Ingredient Disclosure
0.1% item 816 (951)

STATE LISTS

California - Directors List of Hazardous Substances (8 CCR 339)
[present]

Massachusetts Right To Know List
[present]

NJ Right to Know List (Total)
sn 0983

Pennsylvania Right to Know List
environmental hazard

TERT-HEXADECANETHIOL   25360-09-2
SEE ALSO:
F039-HAZARDOUS WASTES

HEALTH AND SAFETY LISTS

U.S. DOT - Appendix A Table 1 - Hazardous Substances
final RQ = 1000 pounds (454 kg)

ENVIRONMENTAL LISTS

CERCLA/SARA - Hazardous Substances and their Reportable Quantities
final RQ = 1000 pounds (454 kg)

RCRA - U Series Wastes
waste number U243

RCRA - Hazardous Constituents-Appendix VIII
waste number U243

RCRA - Basis for Listing - Appendix VII
Included in waste stream: F039

RCRA - Substances Banned From Land Disposal
[present]

RCRA - TSD Facilities Ground Water Monitoring
TM 8270 = 10 ug/L PQL

RCRA - Universal Treatment Standards (LDR)
WW: 0.035 mg/l; NWW: 30 mg/kg

TSCA - Code of Federal Regulations Citations
40 CFR 716.120(a)

TSCA - Health and Safety Reporting List
Effective Date: March 7, 1986

TSCA - Chemicals with Significant New Use Rules
[present]

TSCA - Section 12(b) - Export Notification
export notification required - Section 5

STATE LISTS

Massachusetts Right To Know List
[present]

Pennsylvania Right to Know List
environmental hazard

HEXADECYLENE-1   629-73-2

ENVIRONMENTAL LISTS

TSCA - PAIR - Reporting List
Effective Date: October 12, 1993;   Reporting Date: February 28, 1994

TSCA - Health and Safety Reporting List
Effective Date: October 12, 1993;   Sunset Date: October 12, 2003

HEXADECYL SULFATE, SODIUM SALT   1120-01-

ENVIRONMENTAL LISTS

TSCA - PAIR - Reporting List
Effective Date: October 12, 1993;   Reporting Date: February 28 1994

TSCA - Health and Safety Reporting List
Effective Date: October 12, 1993;   Sunset Date: October 12, 2003

HEXADECYLTRICHLOROSILANE   5894-60-

ENVIRONMENTAL LISTS

TSCA - PAIR - Reporting List
Effective Date: October 12, 1993;   Reporting Date: February 28 1994

TSCA - Health and Safety Reporting List
Effective Date: October 12, 1993;   Sunset Date: October 12, 2003

HEXADECYLTRIMETHYLAMMONIUM BROMIDE   57-09-

ENVIRONMENTAL LISTS

TSCA - PAIR - Reporting List
Effective Date: October 12, 1993;   Reporting Date: February 28 1994

TSCA - Health and Safety Reporting List
Effective Date: October 12, 1993;   Sunset Date: October 12, 2003

2,4-HEXADIENAL   80466-34-8

HEALTH AND SAFETY LISTS

NFPA - Flash Points
flash point > 212 degrees F (100 degrees C)

NFPA - Hazard Identification Ratings
health-0; flammability-1; reactivity-0

HEXADIENE   592-42-7

HEALTH AND SAFETY LISTS

NIOSH - Health Standards - Exposure Limits
C (15 min) 0.5 ppm; C (15 min) 5.3 mg/m3 (Listed under 'Thiols')

NIOSH - Health Standards - Health Effects and Precautions
Irritation; eye, skin, blood, and nervous system effects (Blood and urine monitoring required; prevent skin contact) (Listed under 'Thiols')

1,4-HEXADIENE   592-45-0

HEALTH AND SAFETY LISTS

NFPA - Flash Points
flash point = 265 degrees F (129 degrees C)

NFPA - Hazard Identification Ratings
health-0; flammability-1; reactivity-0

HEXADIENE   42296-74-2

HEALTH AND SAFETY LISTS

NFPA - Flash Points
flash point > 212 degrees F (100 degrees C)

NFPA - Hazard Identification Ratings
health-0; flammability-1; reactivity-0

HEXADIENES   RR-01352-5

ENVIRONMENTAL LISTS

List of Pesticide Product Inert Ingredients
[present]

HEXAETHYL TETRAPHOSPHATE   757-58-4

HEALTH AND SAFETY LISTS

U.S. DOT - Substances From 49 CFR 172.101
regulated by DOT (UN1781)

U.S. DOT - Hazard Classes
DOT hazard class = 8

NFPA - Flash Points
flash point = 295 degrees F (146 degrees C)
NFPA - Hazard Identification Ratings
health-3; flammability-1; reactivity-0
INTERNATIONAL LISTS
Canada - WHMIS: Ingredient Disclosure
1% item 817 (953)
STATE LISTS
Florida Hazardous Substance List
[present]
Massachusetts Right To Know List
[present]
Pennsylvania Right to Know List
[present]

HEXAFLUOROACETONE   684-16-2
HEALTH AND SAFETY LISTS
NIOSH - Selected LD50s and LC50s
Oral, rat: LD50 = 410 mg/kg

HEXAFLUOROACETONE HYDRATE   10543-95-0
HEALTH AND SAFETY LISTS
NFPA - Flash Points
flash point = 154 degrees F (68 degrees C)
NFPA - Hazard Identification Ratings
health-2; flammability-2; reactivity-0
STATE LISTS
Pennsylvania Right to Know List
[present]

HEXAFLUOROBENZENE   392-56-3
HEALTH AND SAFETY LISTS
AIHA - WEEL - Time Weighted Averages
100 ppm TWA; 337 mg/m3 TWA
STATE LISTS
Minnesota Hazardous Substance List
[present]

HEXAFLUOROETHANE   76-16-4
HEALTH AND SAFETY LISTS
NFPA - Flash Points
flash point = -6 degrees F (-21 degrees C)
NFPA - Hazard Identification Ratings
health-0; flammability-3; reactivity-0
ENVIRONMENTAL LISTS
CAA - HON Rule - SOCMI Chemicals
compliance by Jan. 23, 1995
STATE LISTS
Florida Hazardous Substance List
[present]
Massachusetts Right To Know List
[present]
Pennsylvania Right to Know List
[present]

HEXAFLUOROPHOSPHORIC ACID   16940-81-1
STATE LISTS
NJ Right to Know List (Total)
sn 0985
NJ Special Hazardous Substances
(flammable - third degree)

1,1,1,3,3,3-HEXAFLUORO-2-PROPANOL   920-66-1
HEALTH AND SAFETY LISTS
U.S. DOT - Substances From 49 CFR 172.101
regulated by DOT (UN2458)
U.S. DOT - Hazard Classes
DOT hazard class = 3

HEXAFLUOROPROPYLENE   116-15-4
HEALTH AND SAFETY LISTS
U.S. DOT - Substances From 49 CFR 172.101
regulated by DOT (UN1611)
U.S. DOT - Hazard Classes
DOT hazard class = 6.1
U.S. DOT - Substances Which Are Poisonous by Inhalation
gaseous hazardous material poisonous by inhalation (when mixed with compressed gas) (UN1612)
U.S. DOT - Appendix B - Marine Pollutants
DOT regulated marine pollutant
U.S. DOT - Appendix A Table 1 - Hazardous Substances
final RQ = 100 pounds (45.4 kg)
NIOSH - Selected LD50s and LC50s
Oral, rat: LD50 = 7 mg/kg
ENVIRONMENTAL LISTS
CERCLA/SARA - Hazardous Substances and their Reportable Quantities
final RQ = 100 pounds (45.4 kg)
RCRA - P Series Wastes
waste number P062
RCRA - Hazardous Constituents-Appendix VIII
waste number P062
RCRA - Substances Banned From Land Disposal
[present]
TSCA - Code of Federal Regulations Citations
40 CFR 716.120(a)
TSCA - Health and Safety Reporting List
Effective Date: March 7, 1986
STATE LISTS
Massachusetts Right To Know List
[present]
NJ Right to Know List (Total)
sn 0986
Pennsylvania Right to Know List
environmental hazard

HEXAFLUOROPROPYLENE OXIDE   428-59-1
HEALTH AND SAFETY LISTS
ACGIH 1995 - Time Weighted Averages
0.1 ppm TWA; 0.68 mg/m3 TWA
ACGIH 1995 - Skin Designations
skin - potential for cutaneous absorption
U.S. DOT - Substances From 49 CFR 172.101
regulated by DOT (UN2420, UN2552)
U.S. DOT - Hazard Classes
DOT hazard class = 2.3
U.S. DOT - Substances Which Are Poisonous by Inhalation
gaseous hazardous material poisonous by inhalation (UN2420)
NIOSH - Selected LD50s and LC50s
Inhalation, rat: LC50 = 275 ppm 3 hr
OSHA - Vacated PELs - Time Weighted Averages
0.1 ppm TWA; 0.7 mg/m3 TWA (Enforcement indefinitely stayed)
OSHA - Vacated PELs - Skin Designation
Prevent or reduce skin absorption

OSHA - List of Highly Hazardous Chemicals
threshhold quantity = 5000 pounds

INTERNATIONAL LISTS

Australian Exposure Standards - Time Weighted Averages
0.1 ppm TWA;   0.68 mg/m3 TWA

Australian Exposure Standards - Skin Effects
skin absorption

Canada - WHMIS: Ingredient Disclosure
1% item 818 (954)

Canada - Alberta - 8 Hour Occupational Exposure Limit
0.1 ppm TWA;   0.68 mg/m3 TWA

Canada - Alberta - 15 Minute Occupational Exposure Limit
0.3 ppm STEL;   2 mg/m3 STEL

Canada - British Columbia - 8 Hour Exposure Limits
0.1 ppm TWA;   0.7 mg/m3 TWA

Canada - British Columbia - 15 Minute Exposure Limits
0.3 ppm STEL;   2 mg/m3 STEL

Canada - Ontario - OHSA - TWAEVs
0.1 ppm TWAEV;   0.7 mg/m3 TWAEV

Canada - Ontario - OHSA - Skin Notations
absorption through skin, eyes, or mucous membranes

Canada - Quebec - Time-Weighted Average Exposure Values
0.1 ppm TWAEV;   0.68 mg/m3 TWAEV

Canada - Quebec - Skin Designations
absorbed through the skin

Israel - Time Weighted Averages
0.1 ppm TWA;   0.68 mg/m3 TWA

Israel - Action Levels
0.05 ppm AL;   0.34 mg/m3 AL

STATE LISTS

California - Exposure Limits - PELs
0.1 ppm PEL;   0.7 mg/m3 PEL

California - Exposure Limits - Skin Notation
material may be absorbed through the skin, eyes or mucous membrane

California - Directors List of Hazardous Substances (8 CCR 339)
[present]

Florida Hazardous Substance List
[present]

Massachusetts Right To Know List
[present]

Minnesota Hazardous Substance List
skin

NJ Right to Know List (Total)
sn 0987

Pennsylvania Right to Know List
[present]

HEXAHYDRO-1,3,5-TRIETHYL-S-TRIAZINE   7779-27-3

INTERNATIONAL LISTS

Canada - WHMIS: Ingredient Disclosure
1% item 819 (955)

STATE LISTS

NJ Right to Know List (Total)
sn 0988

HEXAKIS(METHOXYMETHYL)MELAMINE   3089-11-0

HEALTH AND SAFETY LISTS

NIOSH - Selected LD50s and LC50s
Inhalation, mouse: LC50 = 95 gm/m3 2 hr

HEXALDEHYDE   66-25-1

HEALTH AND SAFETY LISTS

U.S. DOT - Substances From 49 CFR 172.101
regulated by DOT (UN2193)

U.S. DOT - Hazard Classes
DOT hazard class = 2.2

STATE LISTS

NJ Right to Know List (Total)
sn 0989

HEXAMETHYLDISILIZANE   999-97-

HEALTH AND SAFETY LISTS

U.S. DOT - Substances From 49 CFR 172.101
regulated by DOT (UN1782)

U.S. DOT - Hazard Classes
DOT hazard class = 8

INTERNATIONAL LISTS

Canada - WHMIS: Ingredient Disclosure
1% item 820 (91)

STATE LISTS

NJ Right to Know List (Total)
sn 0990

NJ Special Hazardous Substances
(corrosive)

HEXAMETHYLDISILOXANE   107-46-0

HEALTH AND SAFETY LISTS

NIOSH - Selected LD50s and LC50s
Oral, mouse: LD50 = 600 mg/kg

HEXAMETHYLENEDIAMINE   124-09-4
SEE ALSO:
FLUOROALKENES

HEALTH AND SAFETY LISTS

U.S. DOT - Substances From 49 CFR 172.101
regulated by DOT (UN1858)

U.S. DOT - Hazard Classes
DOT hazard class = 2.2

NIOSH - Selected LD50s and LC50s
Inhalation, rat: LC50 = 11200 mg/m3 (4 hr)

ENVIRONMENTAL LISTS

EPA - Master Testing List
[present]

TSCA - Code of Federal Regulations Citations
40 CFR 712.30(d);   40 CFR 716.120(c);   40 CFR 799.1700(a)(1)

TSCA - PAIR - Reporting List
Reporting Date: November 19, 1982

TSCA - Chemical Test Rules
Testing required by: manufacturers (40 CFR 799.1700) (Listed under 'Fluoroalkenes')

TSCA - Section 12(b) - Export Notification
export notification required - Section 4

STATE LISTS

NJ Right to Know List (Total)
sn 0991

HEXAMETHYLENEDIAMINE, N,N'-DIBUTYL-   4835-11-4

HEALTH AND SAFETY LISTS

U.S. DOT - Substances From 49 CFR 172.101
regulated by DOT (NA1956)

U.S. DOT - Hazard Classes
DOT hazard class = 2.2

ENVIRONMENTAL LISTS

TSCA - Code of Federal Regulations Citations
40 CFR 704.102; 40 CFR 712.30(d); 40 CFR 716.120(a); 40 CFR 721.1175

TSCA - PAIR - Reporting List
Reporting Date: November 19, 1982

TSCA - Health and Safety Reporting List
Effective Date: October 4, 1982

TSCA - Chemicals with Significant New Use Rules
[present]

TSCA - Section 12(b) - Export Notification
export notification required - Section 5

STATE LISTS

NJ Right to Know List (Total)
sn 0992

HEXAMETHYLENE DIISOCYANATE   822-06-0

HEALTH AND SAFETY LISTS

NIOSH - Selected LD50s and LC50s
Oral, rat: LD50 = 316 mg/kg

HEXAMETHYLENE DIISOCYANATE BIURET   4035-89-6

HEALTH AND SAFETY LISTS

NIOSH - Selected LD50s and LC50s
Oral, rat: LD50 = 3080 mg/kg

ENVIRONMENTAL LISTS

TSCA - Code of Federal Regulations Citations
40 CFR 712.30(x); 40 CFR 716.120(d)

TSCA - PAIR - Reporting List
Reporting Date: November 27, 1991

TSCA - Health and Safety Reporting List
Effective Date: September 30, 1991

INTERNATIONAL LISTS

Canada - WHMIS: Ingredient Disclosure
1% item 821 (958)

HEXAMETHYLENE DIISOCYANATE HOMOPOLYMER   28182-81-2

HEALTH AND SAFETY LISTS

U.S. DOT - Substances From 49 CFR 172.101
regulated by DOT (UN1207)

U.S. DOT - Hazard Classes
DOT hazard class = 3

NFPA - Flash Points
flash point = 90 degrees F (32 degrees C)

NFPA - Hazard Identification Ratings
health-2; flammability-3; reactivity-1

NIOSH - Selected LD50s and LC50s
Oral, rat: LD50 = 4890 mg/kg

STATE LISTS

Florida Hazardous Substance List
[present]

Massachusetts Right To Know List
[present]

NJ Right to Know List (Total)
sn 0993

NJ Special Hazardous Substances
(flammable - third degree)

Pennsylvania Right to Know List
[present]

HEXAMETHYLENEIMINE   111-49-9

HEALTH AND SAFETY LISTS

NTP Chemical Status Reports - Testing Status and NTIS Number
Project leader assigned/study in design

ENVIRONMENTAL LISTS

TSCA - PAIR - Reporting List
Effective Date: October 12, 1993; Reporting Date: February 28, 1994

TSCA - Health and Safety Reporting List
Effective Date: October 12, 1993; Sunset Date: October 12, 2003

HEXAMETHYLENETETRAMINE HYDROGEN CHLORIDE   24360-05-2

ENVIRONMENTAL LISTS

TSCA - PAIR - Reporting List
Effective Date: June 14, 1993

TSCA - Health and Safety Reporting List
Effective Date: October 12, 1993; Sunset Date: October 12, 2003

HEXAMETHYLENE TRIPEROXIDE DIAMINE   283-66-9

HEALTH AND SAFETY LISTS

ACGIH 1995 - Time Weighted Averages
0.5 ppm TWA; 2.3 mg/m3 TWA

AIHA - WEEL - Time Weighted Averages
5 mg/m3 TWA

U.S. DOT - Substances From 49 CFR 172.101
regulated by DOT (UN2280, UN1783)

U.S. DOT - Hazard Classes
DOT hazard class = 8

NIOSH - Selected LD50s and LC50s
Oral, rat: LD50 = 750 mg/kg Skin, rabbit: LD50 = 1110 mg/kg

ENVIRONMENTAL LISTS

EPA - Master Testing List
[present]

INTERNATIONAL LISTS

Canada - WHMIS: Ingredient Disclosure
1% item 823 (959)

STATE LISTS

Massachusetts Right To Know List
[present]

NJ Right to Know List (Total)
sn 0994

NJ Special Hazardous Substances
(corrosive)

HEXAMETHYLOL BENZENE HEXANITRATE   105554-30-1

HEALTH AND SAFETY LISTS

NIOSH - Selected LD50s and LC50s
Inhalation, rat: LC50 = 220 mg/m3 4 hr

ENVIRONMENTAL LISTS

CERCLA/SARA - Section 302 Extremely Hazardous Substances and TPQs
TPQ = 500 pounds

TSCA - Code of Federal Regulations Citations
40 CFR 712.30(d)

TSCA - PAIR - Reporting List
Reporting Date: November 19, 1982

STATE LISTS

Florida Hazardous Substance List
effective March 13, 1992

Massachusetts Right To Know List
extraordinarily hazardous

NJ Right to Know List (Total)
sn 2462

Pennsylvania Right to Know List
environmental hazard

PROPOSED REGULATIONS

CERCLA/SARA - Proposed Hazardous Substance Additions
proposed RQ = 1 pound (.454 kg)

CERCLA/SARA - 1989 Proposed RQ Adjustments
proposed RQ = 100 pounds (45.4 kg)

HEXAMETHYLOLMELAMINE   531-18-0

HEALTH AND SAFETY LISTS

ACGIH 1995 - Time Weighted Averages
0.005 ppm TWA;   0.034 mg/m3 TWA

U.S. DOT - Substances From 49 CFR 172.101
regulated by DOT (UN2281)

U.S. DOT - Hazard Classes
DOT hazard class = 6.1

NIOSH - Selected LD50s and LC50s
Oral, rat: LD50 = 738 mg/kg Skin, rabbit: LD50 = 593 mg/kg

NIOSH - Health Standards - Exposure Limits
5 ppb TWA;   35 ug/m3 TWA;   C (10 min) 20 ppb;   C (10 min) 140 ug/m3 (Listed under 'Diisocyanates')

NIOSH - Health Standards - Health Effects and Precautions
Respiratory effects and sensitization, pulmonary irritation (Periodic chest X-ray and pulmonary function testing required) (Listed under 'Diisocyanates')

ENVIRONMENTAL LISTS

CERCLA/SARA - Section 313 - Emission Reporting
form R reporting required;   (Listed under 'Diisocyanates')

CERCLA/SARA - Hazardous Substances and their Reportable Quantities
final RQ = 1 pound (.454 kg)

Clean Air Act (1990) - List of Hazardous Air Contaminants
[present]

EPA - Master Testing List
[present]

List of Pesticide Product Inert Ingredients
[present]

TSCA - Code of Federal Regulations Citations
40 CFR 716.120(a);   40 CFR 712.30(w)

TSCA - PAIR - Reporting List
Reporting Date: August 18, 1988

TSCA - Health and Safety Reporting List
Effective Date: June 1, 1987

INTERNATIONAL LISTS

Canada - WHMIS: Ingredient Disclosure
0.1% item 824 (960)

Canada - Alberta - 8 Hour Occupational Exposure Limit
0.005 ppm TWA;   0.034 mg/m3 TWA

Canada - Alberta - Ceiling Occupational Exposure Limit
C 0.02 ppm;   C 0.138 mg/m3

Canada - Ontario - OHSA - TWAEVs
0.005 ppm TWAEV;   0.2 micromoles/m3 TWAEV (designated substance regulation)

Canada - Ontario - OHSA - CEVs
0.02 ppm CEV;   0.8 micromoles/m3 CEV (designated substance regulation)

Canada - Ontario - OHSA - Designated Substances
0.005 ppm TWAEV;   0.2 micromoles/m3 TWAEV;   See Ontario Reg. 842 for full information.

Canada - Quebec - Time-Weighted Average Exposure Values
0.005 ppm TWAEV;   0.034 mg/m3 TWAEV

German (DFG) - MAK Values
0.01 ppm MAK;   0.07 mg/m3 MAK

German (DFG) - Peak Limitations
2 x normal MAK (5 min momentary value);   don't exceed 8 time during shift

German (DFG) - Skin/Sensitizers
danger of sensitization (skin or respiratory)

Israel - Time Weighted Averages
0.005 ppm TWA;   0.034 mg/m3 TWA

Israel - Action Levels
0.0025 ppm AL;   0.017 mg/m3 AL

STATE LISTS

California - Air Bill 2588 Appendix A-I
6/91

California - Exposure Limits - PELs
0.005 ppm PEL;   0.034 mg/m3 PEL

California - Directors List of Hazardous Substances (8 CCR 339)
[present]

Massachusetts Right To Know List
[present]

Minnesota Hazardous Substance List
[present]

NJ Right to Know List (Total)
sn 0995

HEXAMETHYL PHOSPHORAMIDE   680-31-9

ENVIRONMENTAL LISTS

TSCA - Code of Federal Regulations Citations
40 CFR 712.30(x);   40 CFR 716.120(d)

TSCA - PAIR - Reporting List
Reporting Date: December 27, 1990

TSCA - Health and Safety Reporting List
Effective Date: October 29, 1990;   Sunset Date: November 9, 1993

3,3,6,6,9,9-HEXAMETHYL-1,2,4,5-   22397-33-7
TETRAOXOCYCLONONANE

ENVIRONMENTAL LISTS

List of Pesticide Product Inert Ingredients
[present]

HEXANAL, 3,5,5-TRIMETHYL-   5435-64-3

HEALTH AND SAFETY LISTS

U.S. DOT - Substances From 49 CFR 172.101
regulated by DOT (UN2493)

U.S. DOT - Hazard Classes
DOT hazard class = 3

NIOSH - Selected LD50s and LC50s
Inhalation, mouse: LC50 = 10800 mg/m3 (2 hr) Oral, rat: LD50 = 410 mg/kg

INTERNATIONAL LISTS

Canada - WHMIS: Ingredient Disclosure
1% item 825 (961)

STATE LISTS

Massachusetts Right To Know List
[present]

NJ Right to Know List (Total)
sn 2463

HEXANE   110-54-3

ENVIRONMENTAL LISTS

List of Pesticide Product Inert Ingredients
[present]

HEXANE (OTHER ISOMERS)   RR-01809-7

HEALTH AND SAFETY LISTS

U.S. DOT - Hazard Classes
Forbidden from transport by the DOT

HEXANE, CHLORO-   25495-90-3

HEALTH AND SAFETY LISTS

U.S. DOT - Hazard Classes
Forbidden from transport by the DOT

1,6-HEXANEDIAMINE, DIHYDROCHLORIDE   6055-52-3

INTERNATIONAL LISTS

Canada - WHMIS: Ingredient Disclosure
1% item 826 (962)

HEXANE, 1,6-DIISOCYANATO-2,4,4-TRIMETHYL-   15646-96-5

HEALTH AND SAFETY LISTS

ACGIH 1995 - Skin Designations
skin - potential for cutaneous absorption

ACGIH 1995 - Carcinogens
A2-suspected human carcinogen

IARC - Group 2B (sufficient animal data)
[present] (Overall evaluation based only on evidence of carcinogenicity in monograph [15,1977] or in Supplement 4)

NIOSH - Selected LD50s and LC50s
Oral, rat: LD50 = 2525 mg/kg Skin, guinea pig: LD50 = 1175 mg/kg

NTP Seventh Report - Suspect Carcinogens
suspect carcinogen

OSHA - Possible Select Carcinogens
[present]

ENVIRONMENTAL LISTS

CERCLA/SARA - Section 313 - Emission Reporting
form R reporting required for 0.1% de minimus concentration

CERCLA/SARA - Hazardous Substances and their Reportable Quantities
final RQ = 1 pound (.454 kg)

Clean Air Act (1990) - List of Hazardous Air Contaminants
[present]

TSCA - Code of Federal Regulations Citations
40 CFR 721.1200

TSCA - Chemicals with Significant New Use Rules
[present]

TSCA - Section 12(b) - Export Notification
export notification required - Section 5

INTERNATIONAL LISTS

Australian Exposure Standards - Time Weighted Averages
control to the lowest practical level

Australian Exposure Standards - Skin Effects
skin absorption

Australian Exposure Standards - Carcinogens
probable carcinogen

Canada - WHMIS: Ingredient Disclosure
0.1% item 822 (963)

Canada - Alberta - Designated Substances
designated substance - requires code of practice

Canada - British Columbia - 8 Hour Exposure Limits
carcinogen with no established permitted concentration

Canada - British Columbia - Skin Notations
skin - potential for skin absorption

Canada - British Columbia - Carcinogens
carcinogen with no established permitted concentration

Canada - Quebec - Time-Weighted Average Exposure Values
substance of which the recirculation is prohibited

Canada - Quebec - Skin Designations
absorbed through the skin

Canada - Quebec - Carcinogens
C2 carcinogen: effect suspected in humans

German (DFG) - Carcinogens
animal evidence of carcinogenicity

STATE LISTS

California - Air Bill 2588 Appendix A-I
known or potential carcinogen

California - Prop. 65 - Cancer list
carcinogen - initial date 1/1/88

California - Prop. 65 - Reproductive - Male
male reproductive toxicity - initial date 10/01/94

California - Directors List of Hazardous Substances (8 CCR 339)
[present]

Florida Hazardous Substance List
[present]

Massachusetts Right To Know List
carcinogen; extraordinarily hazardous

Minnesota Hazardous Substance List
carcinogen; skin

NJ Right to Know List (Total)
sn 0973

Pennsylvania Right to Know List
environmental hazard; special hazardous substance

Pennsylvania RTK - Special Hazardous Substances
[present]

HEXANE, 1,6-DIISOCYANATO-2,2,4-TRIMETHYL-   16938-22-0

HEALTH AND SAFETY LISTS

U.S. DOT - Organic Peroxides Table
Organic peroxide UN3102; UN3105; UN3106

STATE LISTS

NJ Right to Know List (Total)
sn 2465; sn 2466; technically pure: sn 2464

HEXANEDINITRILE, HYDROGENATED, HIGH-BOILING FRACTION, PHOSPHONOMETHYLATED   68955-64-6

HEALTH AND SAFETY LISTS

NIOSH - Selected LD50s and LC50s
Oral, rat: LD50 = 3240 mg/kg

ENVIRONMENTAL LISTS

TSCA - Code of Federal Regulations Citations
40 CFR 712.30(x); 40 CFR 716.120(d)

TSCA - PAIR - Reporting List
Reporting Date: November 27, 1991

TSCA - Health and Safety Reporting List
Effective Date: September 30, 1991

HEXANEDIOIC ACID, DIETHENYL ESTER   RR-01644-4

HEALTH AND SAFETY LISTS

ACGIH 1995 - Time Weighted Averages
50 ppm TWA; 176 mg/m3 TWA

ACGIH 1995 - Biological Exposure Indices
2,5-Hexanedione in urine: 5 mg/g creatinine, end of shift (Ns); N-Hexane in exhaled air: (Sq)

AIHA - Odor Threshold Values
no geometric mean air odor threshold

NFPA - Flash Points
flash point = -7 degrees F (-22 degrees C)

NFPA - Hazard Identification Ratings
health-1; flammability-3; reactivity-0

NIOSH - Selected LD50s and LC50s
Oral, rat: LD50 = 28710 mg/kg

NIOSH 1990 - Pocket Guide - RELs
50 ppm TWA;   180 mg/m3 TWA

NIOSH 1990 - Pocket Guide - IDLHs
5000 ppm IDLH

NIOSH 1990 - Pocket Guide - Target organs
skin, eyes, respiratory system

NIOSH - Health Standards - Exposure Limits
100 ppm TWA;   350 mg/m3 TWA;   C (15 min) 510 ppm;   C (15 min) 1800 mg/m3 (Listed under 'Alkanes (C5-C8)')

NIOSH - Health Standards - Health Effects and Precautions
Skin and nervous system effects (Listed under 'Alkanes (C5-C8)')

NTP Chemical Status Reports - Testing Status and NTIS Number
Technical reports printed (PB91185322)

OSHA - Vacated PELs - Time Weighted Averages
50 ppm TWA;   180 mg/m3 TWA

OSHA - Final PELs - Time Weighted Averages
500 ppm TWA;   1800 mg/m3 TWA

ENVIRONMENTAL LISTS

ATSDR Priority List
Rank (of 275): 182

CERCLA/SARA - Section 313 - Emission Reporting
form R reporting required

CERCLA/SARA - Hazardous Substances and their Reportable Quantities
final RQ = 1 pound (.454 kg)

Clean Air Act (1990) - List of Hazardous Air Contaminants
[present]

CAA - HON Rule - SOCMI Chemicals
compliance by Oct. 23, 1995

CAA - HON Rule - Organic HAPs
[present]

TSCA - Code of Federal Regulations Citations
40 CFR 799.2155

INTERNATIONAL LISTS

Australian Exposure Standards - Time Weighted Averages
50 ppm TWA;   180 mg/m3 TWA

Canada - WHMIS: Ingredient Disclosure
1% item 828 (965)

Canada - Alberta - 8 Hour Occupational Exposure Limit
as n-hexane: 50 ppm TWA;   176 mg/m3 TWA

Canada - Alberta - 15 Minute Occupational Exposure Limit
75 ppm STEL;   264 mg/m3 STEL

Canada - British Columbia - 8 Hour Exposure Limits
100 ppm TWA;   360 mg/m3 TWA

Canada - British Columbia - 15 Minute Exposure Limits
125 ppm STEL;   450 mg/m3 STEL

Canada - Ontario - OHSA - TWAEVs
50 ppm TWAEV;   176 mg/m3 TWAEV

Canada - Quebec - Time-Weighted Average Exposure Values
50 ppm TWAEV;   176 mg/m3 TWAEV

United Kingdom - Occupational Exposure Standards - TWAs
20 ppm TWA;   70 mg/m3 TWA

German (DFG) - MAK Values
50 ppm MAK;   180 mg/m3 MAK

German (DFG) - Peak Limitations
2 x normal MAK (30 min. average value);   don't exceed 4 times during shift

German (DFG) - Pregnancy
no risk to embryo/fetus if exposure limits are adhered to

Israel - Time Weighted Averages
50 ppm TWA;   176 mg/m3 TWA

Israel - Action Levels
25 ppm AL;   88 mg/m3 AL

Mexico - Instruction No. 10 - TWAs
100 ppm TWA;   300 mg/m3 TWA

STATE LISTS

California - Air Bill 2588 Appendix A-I
6/91

California - Exposure Limits - PELs
50 ppm PEL;   180 mg/m3 PEL

Florida Hazardous Substance List
[present]

Massachusetts Right To Know List
[present]

Minnesota Hazardous Substance List
[present]

NJ Right to Know List (Total)
sn 1340

NJ Special Hazardous Substances
(flammable - third degree)

Pennsylvania Right to Know List
[present]

PROPOSED REGULATIONS

Canada - Ontario - Proposed Occupational TWAEVs
25 ppm TWAEV;   90 mg/m3 TWAEV

HEXANEDIOIC ACID, POLYMER WITH 1,2-ETHANEDIOL AND 1,7-DIISOCYANATO-2,2,4(OR 2,4,4)-TRIMETHYLHEXANE, 2-HYDROXYETHYL-ACRYLATED-BLOCKED   RR-00960-9

INTERNATIONAL LISTS

Canada - Quebec - Time-Weighted Average Exposure Values
500 ppm TWAEV;   1760 mg/m3 TWAEV

Canada - Quebec - Short-term Exposure Values
1000 ppm STEV;   3500 mg/m3 STEV

1,6-HEXANEDIOL   629-11-8

STATE LISTS

Pennsylvania Right to Know List
[present]

2,5-HEXANEDIOL   2935-44-6

HEALTH AND SAFETY LISTS

NTP Chemical Status Reports - Testing Status and NTIS Number
Technical reports printed (no NTIS number)

1,6-HEXANEDIOL DIACRYLATE   13048-33-4

ENVIRONMENTAL LISTS

CERCLA/SARA - Section 313 - Emission Reporting
form R reporting required;   (Listed under 'Diisocyanates')

TSCA - Code of Federal Regulations Citations
40 CFR 712.30(x);   40 CFR 716.120(a)

TSCA - PAIR - Reporting List
Reporting Date: December 27, 1990

TSCA - Health and Safety Reporting List
Effective Date: June 1, 1987;   Sunset Date: November 9, 1993

2,5-HEXANEDIONE   110-13-4

ENVIRONMENTAL LISTS

CERCLA/SARA - Section 313 - Emission Reporting
form R reporting required;   (Listed under 'Diisocyanates')

TSCA - Code of Federal Regulations Citations
40 CFR 712.30(x);   40 CFR 716.120(a)

TSCA - PAIR - Reporting List
Reporting Date: December 27, 1990

TSCA - Health and Safety Reporting List
Effective Date: June 1, 1987;   Sunset Date: November 9, 1993

HEXANE ISOMERS   RR-00003-3

ENVIRONMENTAL LISTS

List of Pesticide Product Inert Ingredients
[present]

1-HEXANETHIOL   111-31-9

ENVIRONMENTAL LISTS

TSCA - Chemicals with Significant New Use Rules
PMN number: P-90-1564

HEXANE, 2,2,5-TRIMETHYL-   3522-94-9

ENVIRONMENTAL LISTS

TSCA - Chemicals with Significant New Use Rules
PMN number: P-90-1636

1,2,6-HEXANETRIOL   106-69-4

HEALTH AND SAFETY LISTS

NIOSH - Selected LD50s and LC50s
Oral, rat: LD50 = 3730 mg/kg

ENVIRONMENTAL LISTS

EPA - Master Testing List
[present]

HEXANITROAZOXY BENZENE   RR-01425-5

HEALTH AND SAFETY LISTS

NFPA - Flash Points
flash point = 230 degrees F (110 degrees C)

NFPA - Hazard Identification Ratings
health-2; flammability-1; reactivity-0

NIOSH - Selected LD50s and LC50s
Oral, rat: LD50 = 5000 mg/kg

STATE LISTS

Florida Hazardous Substance List
[present]

Massachusetts Right To Know List
[present]

Pennsylvania Right to Know List
[present]

2,2',4,4',6,6'-HEXANITRO-3,3'-   RR-01371-8
DIHYDROXYAZOBENZENE

HEALTH AND SAFETY LISTS

AIHA - WEEL - Time Weighted Averages
1 mg/m3 TWA

ENVIRONMENTAL LISTS

TSCA - Code of Federal Regulations Citations
40 CFR 712.30(d)

TSCA - PAIR - Reporting List
Reporting Date: November 19, 1982

STATE LISTS

Minnesota Hazardous Substance List
[present]

HEXANITRODIPHENYLAMINE   35860-31-2

HEALTH AND SAFETY LISTS

NFPA - Flash Points
flash point = 174 degrees F (79 degrees C)

NFPA - Hazard Identification Ratings
health-1; flammability-1; reactivity-0

NIOSH - Selected LD50s and LC50s
Oral, rat: LD50 = 2700 mg/kg Skin, guinea pig: LD50 = 6422 mg/kg

2,3',4,4',6,6'-HEXANITRODIPHENYLETHER   RR-01374-1

HEALTH AND SAFETY LISTS

ACGIH 1995 - Time Weighted Averages
500 ppm TWA; 1760 mg/m3 TWA

ACGIH 1995 - Short Term Exposure Limits
1000 ppm STEL; 3500 mg/m3 STEL

U.S. DOT - Substances From 49 CFR 172.101
regulated by DOT (UN1208)

U.S. DOT - Hazard Classes
DOT hazard class = 3

OSHA - Vacated PELs - Time Weighted Averages
500 ppm TWA; 1800 mg/m3 TWA

OSHA - Vacated PELs - Short Term Exposure Limits
1000 ppm STEL; 3600 mg/m3 STEL

INTERNATIONAL LISTS

Australian Exposure Standards - Time Weighted Averages
500 ppm TWA; 1760 mg/m3 TWA

Australian Exposure Standards - Short Term Exposure Limits
1000 ppm STEL; 3500 mg/m3 STEL

Canada - WHMIS: Ingredient Disclosure
1% item 827 (964)

Canada - Alberta - 8 Hour Occupational Exposure Limit
500 ppm TWA; 1760 mg/m3 TWA

Canada - Alberta - 15 Minute Occupational Exposure Limit
625 ppm STEL; 2203 mg/m3 STEL

Canada - Ontario - OHSA - TWAEVs
500 ppm TWAEV; 1760 mg/m3 TWAEV

Canada - Ontario - OHSA - STEVs
1000 ppm STEV; 3520 mg/m3 STEV

United Kingdom - Occupational Exposure Standards - TWAs
500 ppm TWA; 1800 mg/m3 TWA (does not include n-Hexane)

United Kingdom - Occupational Exposure Standards - STELs
1000 ppm STEL; 3600 mg/m3 STEL (does not include n-hexane)

Israel - Time Weighted Averages
500 ppm TWA; 1760 mg/m3 TWA

Israel - Short Term Exposure Limits
1000 ppm STEL; 3500 mg/m3 STEL

Israel - Action Levels
250 ppm AL; 880 mg/m3 AL

Mexico - Instruction No. 10 - TWAs
500 ppm TWA; 1800 mg/m3 TWA

Mexico - Instruction No. 10 - STELs
1000 ppm STEL; 3600 mg/m3 STEL

STATE LISTS

California - Exposure Limits - PELs
500 ppm PEL; 1800 mg/m3 PEL (does not include n-Hexane)

California - Exposure Limits - STELs
1000 ppm STEL; 3600 mg/m3 STEL

California - Directors List of Hazardous Substances (8 CCR 339)
[present]

N,N'-(HEXANITRODIPHENYL) ETHYLENE   RR-01445-9
DINITRAMINE

HEALTH AND SAFETY LISTS

NIOSH - Selected LD50s and LC50s
Inhalation, rat: LC50 = 1080 ppm 4 hr Oral, rat: LD50 = 1254 mg/kg

NIOSH - Health Standards - Exposure Limits
C (15 min) 0.5 ppm; C (15 min) 2.4 mg/m3 (Listed under 'Thiols')

NIOSH - Health Standards - Health Effects and Precautions
Irritation; eye, skin, blood, and nervous system effects (Blood and urine monitoring required; prevent skin contact) (Listed under 'Thiols')

STATE LISTS
  Minnesota Hazardous Substance List
  [present]

HEXANITRODIPHENYL UREA   RR-01426-6
  HEALTH AND SAFETY LISTS
    NFPA - Flash Points
    flash point = 55 degrees F (13 degrees C)
    NFPA - Hazard Identification Ratings
    health-2; flammability-3; reactivity-0
  STATE LISTS
    Florida Hazardous Substance List
    [present]
    Massachusetts Right To Know List
    [present]
    Pennsylvania Right to Know List
    [present]

HEXANITROETHANE   918-37-6
  HEALTH AND SAFETY LISTS
    NFPA - Flash Points
    flash point = 375 degrees F (191 degrees C)
    NFPA - Hazard Identification Ratings
    health-1; flammability-1; reactivity-0
    NIOSH - Selected LD50s and LC50s
    Oral, rat: LD50 = 15500 mg/kg
  ENVIRONMENTAL LISTS
    CAA - HON Rule - SOCMI Chemicals
    compliance by July 24, 1995

HEXANITROOXANILIDE   29135-62-4
  HEALTH AND SAFETY LISTS
    U.S. DOT - Hazard Classes
    Forbidden from transport by the DOT

HEXANITROSTILBINE   20062-22-0
  HEALTH AND SAFETY LISTS
    U.S. DOT - Hazard Classes
    Forbidden from transport by the DOT

HEXANOIC ACID, 2-ETHYL-, ETHENYL ESTER   RR-01648-8
  HEALTH AND SAFETY LISTS
    U.S. DOT - Substances From 49 CFR 172.101
    regulated by DOT (UN0079)
    U.S. DOT - Hazard Classes
    Forbidden from transport by the DOT
  INTERNATIONAL LISTS
    Canada - WHMIS: Ingredient Disclosure
    1% item 829 (966)
  STATE LISTS
    NJ Right to Know List (Total)
    sn 0999

HEXANOIC ACID, 2-ETHYL-, DIESTER WITH TE-   18268-70-7
TRAETHYLENE GLYCOL
  HEALTH AND SAFETY LISTS
    U.S. DOT - Hazard Classes
    Forbidden from transport by the DOT

HEXANOIC ACID, 2-ETHYL-, ETHENYL ESTER   94-04-2
  HEALTH AND SAFETY LISTS
    U.S. DOT - Hazard Classes
    Forbidden from transport by the DOT

N-HEXANOL   111-27-
  HEALTH AND SAFETY LISTS
    U.S. DOT - Hazard Classes
    Forbidden from transport by the DOT

HEXANOL   25917-35-
  HEALTH AND SAFETY LISTS
    U.S. DOT - Hazard Classes
    Forbidden from transport by the DOT

1-HEXANOL, 3,5,5-TRIMETHYL-   3452-97-
  HEALTH AND SAFETY LISTS
    U.S. DOT - Hazard Classes
    Forbidden from transport by the DOT

3-HEXANONE   589-38-
  HEALTH AND SAFETY LISTS
    U.S. DOT - Substances From 49 CFR 172.101
    regulated by DOT (UN0392)
    U.S. DOT - Hazard Classes
    DOT hazard class = 1.1D
  STATE LISTS
    NJ Right to Know List (Total)
    sn 1000

2-HEXANONE   591-78-6
  ENVIRONMENTAL LISTS
    TSCA - Chemicals with Significant New Use Rules
    PMN number: P-91-826

1,4,7,10,13,16-HEXAOXACYCLOOCTADECANE,2-[(2-   84812-04-4
PROPENYLOXY)METHYL]
  HEALTH AND SAFETY LISTS
    NIOSH - Selected LD50s and LC50s
    Oral, rat: LD50 = 18 gm/kg

HEXATRIACONTAMETHYLCYCLOOCTADECASILOXANE 13523-12-8
  HEALTH AND SAFETY LISTS
    NFPA - Flash Points
    flash point = 165 degrees F (74 degrees C)
    NFPA - Hazard Identification Ratings
    health-2; flammability-2; reactivity-2
  STATE LISTS
    Florida Hazardous Substance List
    [present]
    Massachusetts Right To Know List
    [present]
    Pennsylvania Right to Know List
    [present]

HEXATRIACONTAMETHYLHEPTADECASILOXANE   18844-04-7
  HEALTH AND SAFETY LISTS
    NFPA - Flash Points
    flash point = 145 degrees F (63 degrees C)
    NFPA - Hazard Identification Ratings
    health-1; flammability-2; reactivity-0
    NIOSH - Selected LD50s and LC50s
    Oral, rat: LD50 = 720 mg/kg Skin, rabbit: LD50 = 3100 mg/kg
  ENVIRONMENTAL LISTS
    List of Pesticide Product Inert Ingredients
    [present]

INTERNATIONAL LISTS

Canada - WHMIS: Ingredient Disclosure
1% item 832 (176)

STATE LISTS

NJ Right to Know List (Total)
sn 1001

Pennsylvania Right to Know List
[present]

HEXAVALENT CHROMIUM CHEMICALS   RR-01769-6

HEALTH AND SAFETY LISTS

U.S. DOT - Substances From 49 CFR 172.101
regulated by DOT (UN2282)

U.S. DOT - Hazard Classes
DOT hazard class = 3

HEXAZINONE   51235-04-2

HEALTH AND SAFETY LISTS

NFPA - Flash Points
flash point = 200 degrees F (93 degrees C)

NFPA - Hazard Identification Ratings
health-2; flammability-2; reactivity-0

STATE LISTS

Florida Hazardous Substance List
[present]

Massachusetts Right To Know List
[present]

Pennsylvania Right to Know List
[present]

2-HEXENAL   505-57-7

HEALTH AND SAFETY LISTS

NFPA - Flash Points
flash point = 95 degrees F (35 degrees C)

NFPA - Hazard Identification Ratings
health-1; flammability-3; reactivity-0

STATE LISTS

Florida Hazardous Substance List
[present]

Massachusetts Right To Know List
[present]

Pennsylvania Right to Know List
[present]

HEXENAL, 2-ETHYL-   26266-68-2

HEALTH AND SAFETY LISTS

ACGIH 1995 - Time Weighted Averages
5 ppm TWA; 20 mg/m3 TWA

ACGIH 1995 - Skin Designations
skin - potential for cutaneous absorption

AIHA - Odor Threshold Values
no geometric mean air odor threshold

NFPA - Flash Points
flash point = 77 degrees F (25 degrees C)

NFPA - Hazard Identification Ratings
health-2; flammability-3; reactivity-0

NIOSH - Selected LD50s and LC50s
Inhalation, rat: LC50 = 8000 ppm 4 hr Oral, rat: LD50 = 2590 mg/kg Skin, rabbit: LD50 = 4800 mg/kg

NIOSH 1990 - Pocket Guide - RELs
1 ppm TWA; 4 mg/m3 TWA

NIOSH 1990 - Pocket Guide - IDLHs
5000 ppm IDLH

NIOSH 1990 - Pocket Guide - Target organs
CNS, skin, respiratory system

NIOSH - Health Standards - Exposure Limits
1 ppm TWA; 4 mg/m3 TWA (Listed under 'Ketones')

NIOSH - Health Standards - Health Effects and Precautions
Irritation; liver, kidney, and nervous system effects (Urinalysis required; warn exposed workers about nervous system effects) (Listed under 'Ketones')

OSHA - Vacated PELs - Time Weighted Averages
5 ppm TWA; 20 mg/m3 TWA

OSHA - Final PELs - Time Weighted Averages
100 ppm TWA; 410 mg/m3 TWA

ENVIRONMENTAL LISTS

ATSDR Priority List
Rank (of 275): 068

RCRA - TSD Facilities Ground Water Monitoring
TM 8240 = 50 ug/L PQL

TSCA - Code of Federal Regulations Citations
40 CFR 721.1300

TSCA - Chemicals with Significant New Use Rules
[present]

TSCA - Section 12(b) - Export Notification
export notification required - Section 5

INTERNATIONAL LISTS

Australian Exposure Standards - Time Weighted Averages
5 ppm TWA; 20 mg/m3 TWA

Australian Exposure Standards - Under Review
exposure limits under review

Canada - WHMIS: Ingredient Disclosure
1% item 831 (968)

Canada - Alberta - 8 Hour Occupational Exposure Limit
5 ppm TWA; 20 mg/m3 TWA

Canada - Alberta - 15 Minute Occupational Exposure Limit
10 ppm STEL; 40 mg/m3 STEL

Canada - Alberta - Skin Designation
can be absorbed through the intact skin

Canada - British Columbia - 8 Hour Exposure Limits
25 ppm TWA; 100 mg/m3 TWA

Canada - British Columbia - 15 Minute Exposure Limits
40 ppm STEL; 165 mg/m3 STEL

Canada - Ontario - OHSA - TWAEVs
1 ppm TWAEV; 4 mg/m3 TWAEV

Canada - Quebec - Time-Weighted Average Exposure Values
5 ppm TWAEV; 20 mg/m3 TWAEV

Canada - Quebec - Skin Designations
absorbed through the skin

United Kingdom - Occupational Exposure Standards - TWAs
5 ppm TWA; 20 mg/m3 TWA

United Kingdom - Occupational Exposure Standards - Notes
can be absorbed through skin

German (DFG) - MAK Values
5 ppm MAK; 21 mg/m3 MAK

German (DFG) - Peak Limitations
2 x normal MAK (30 min. average value); don't exceed 4 times during shift

Israel - Time Weighted Averages
5 ppm TWA; 20 mg/m3 TWA

Israel - Action Levels
2.5 ppm AL; 10 mg/m3 AL

STATE LISTS

California - Exposure Limits - PELs
5 ppm PEL; 20 mg/m3 PEL

California - Exposure Limits - Skin Notation
material may be absorbed through the skin, eyes or mucous membrane

California - Directors List of Hazardous Substances (8 CCR 339)
[present]

Florida Hazardous Substance List
[present]

Massachusetts Right To Know List
[present]

Minnesota Hazardous Substance List
[present]

NJ Right to Know List (Total)
sn 1280

Pennsylvania Right to Know List
[present]

1-HEXENE   592-41-6

ENVIRONMENTAL LISTS

TSCA - Chemicals with Significant New Use Rules
PMN number: P-93-1208

TSCA - Section 12(b) - Export Notification
P-93-1208; export notification required - Section 5

2-HEXENE-CIS   7688-21-3

ENVIRONMENTAL LISTS

TSCA - PAIR - Reporting List
Effective Date: October 12, 1993; Reporting Date: February 28, 1994

TSCA - Health and Safety Reporting List
Effective Date: October 12, 1993; Sunset Date: October 12, 2003

2-HEXENE (MIXED CIS & TRANS ISOMERS)   592-43-8

ENVIRONMENTAL LISTS

TSCA - PAIR - Reporting List
Effective Date: October 12, 1993; Reporting Date: February 28, 1994

TSCA - Health and Safety Reporting List
Effective Date: October 12, 1993; Sunset Date: October 12, 2003

3-HEXEN-1-OL, (Z)-   928-96-1

ENVIRONMENTAL LISTS

TSCA - Section 12(b) - Export Notification
export notification required - Section 6 Hexavalent chromium is usually in the form of sodium dichromate, but also includes any combination of chemical substances containing hexavalent chromium including hexavalent chromium-based water treatment chemicals.

HEXOESTROL   84-16-2

ENVIRONMENTAL LISTS

CERCLA/SARA - Section 313 - Emission Reporting
form R reporting required

HEXOLITE   RR-01329-6

ENVIRONMENTAL LISTS

TSCA - Code of Federal Regulations Citations
40 CFR 712.30(x); 40 CFR 716.120(d)

TSCA - PAIR - Reporting List
Reporting Date: November 27, 1991

TSCA - Health and Safety Reporting List
Effective Date: September 30, 1991

SEC-HEXYL ACETATE   108-84-9

HEALTH AND SAFETY LISTS

U.S. DOT - Appendix B - Marine Pollutants
DOT regulated marine pollutant

ENVIRONMENTAL LISTS

TSCA - Code of Federal Regulations Citations
40 CFR 712.30(x); 40 CFR 716.120(d)

TSCA - PAIR - Reporting List
Reporting Date: November 27, 1991

TSCA - Health and Safety Reporting List
Effective Date: September 30, 1991

N-HEXYL ACETATE   142-92-

HEALTH AND SAFETY LISTS

U.S. DOT - Substances From 49 CFR 172.101
regulated by DOT (UN2370)

U.S. DOT - Hazard Classes
DOT hazard class = 3

NFPA - Flash Points
flash point < 20 degrees F (-7 degrees C)

NFPA - Hazard Identification Ratings
health-1; flammability-3; reactivity-0

ENVIRONMENTAL LISTS

EPA - Master Testing List
[present]

STATE LISTS

Florida Hazardous Substance List
[present]

Massachusetts Right To Know List
[present]

NJ Right to Know List (Total)
sn 1002

NJ Special Hazardous Substances
(flammable - third degree)

Pennsylvania Right to Know List
[present]

SEC-HEXYL ALCOHOL   97-95-

HEALTH AND SAFETY LISTS

NFPA - Flash Points
flash point < -4 degrees F (-20 degrees C)

NFPA - Hazard Identification Ratings
health-0; flammability-3; reactivity-0

STATE LISTS

Florida Hazardous Substance List
[present]

Massachusetts Right To Know List
[present]

N-HEXYLAMINE   111-26-

HEALTH AND SAFETY LISTS

NFPA - Flash Points
flash point < 20 degrees F (-7 degrees C)

NFPA - Hazard Identification Ratings
health-1; flammability-3; reactivity-0

STATE LISTS

Florida Hazardous Substance List
[present]

Massachusetts Right To Know List
[present]

Pennsylvania Right to Know List
[present]

HEXYLENE GLYCOL   107-41-

HEALTH AND SAFETY LISTS

NFPA - Flash Points
flash point = 130 degrees F (54 degrees C)

NFPA - Hazard Identification Ratings
health-1; flammability-2; reactivity-0

STATE LISTS

Pennsylvania Right to Know List
[present]

N-HEXYL ETHER   112-58-3

HEALTH AND SAFETY LISTS

IARC - Group Unspecified
[present] (Listed under 'Nonsteroidal oestrogens')

HEXYLMETHACRYLATE   142-09-6

HEALTH AND SAFETY LISTS

U.S. DOT - Substances From 49 CFR 172.101
regulated by DOT (UN0118)

U.S. DOT - Hazard Classes
DOT hazard class = 1.1D

HEXYL NEOPENTANOATE   5434-57-1

HEALTH AND SAFETY LISTS

ACGIH 1995 - Time Weighted Averages
50 ppm TWA; 295 mg/m3 TWA

AIHA - Odor Threshold Values
no geometric mean air odor threshold

U.S. DOT - Substances From 49 CFR 172.101
regulated by DOT (UN1233)

U.S. DOT - Hazard Classes
DOT hazard class = 3

NFPA - Flash Points
flash point = 113 degrees F (45 degrees C)

NFPA - Hazard Identification Ratings
health-1; flammability-2; reactivity-0

NIOSH - Selected LD50s and LC50s
Oral, rat: LD50 = 6160 mg/kg Skin, rabbit: LD50 = 20 gm/kg

NIOSH 1990 - Pocket Guide - RELs
50 ppm TWA; 300 mg/m3 TWA

NIOSH 1990 - Pocket Guide - IDLHs
4000 ppm IDLH

NIOSH 1990 - Pocket Guide - Target organs
CNS, eyes

OSHA - Vacated PELs - Time Weighted Averages
50 ppm TWA; 300 mg/m3 TWA

OSHA - Final PELs - Time Weighted Averages
50 ppm TWA; 300 mg/m3 TWA

INTERNATIONAL LISTS

Australian Exposure Standards - Time Weighted Averages
50 ppm TWA; 295 mg/m3 TWA

Canada - WHMIS: Ingredient Disclosure
1% item 835 (19)

Canada - Alberta - 8 Hour Occupational Exposure Limit
50 ppm TWA; 295 mg/m3 TWA

Canada - Alberta - 15 Minute Occupational Exposure Limit
75 ppm STEL; 440 mg/m3 STEL

Canada - British Columbia - 8 Hour Exposure Limits
50 ppm TWA; 300 mg/m3 TWA

Canada - Quebec - Time-Weighted Average Exposure Values
50 ppm TWAEV; 295 mg/m3 TWAEV

United Kingdom - Occupational Exposure Standards - TWAs
50 ppm TWA; 300 mg/m3 TWA

United Kingdom - Occupational Exposure Standards - STELs
100 ppm STEL; 600 mg/m3 STEL

German (DFG) - MAK Values
50 ppm MAK; 300 mg/m3 MAK

German (DFG) - Peak Limitations
2 x normal MAK (5 min momentary value); don't exceed 8 times during shift

Israel - Time Weighted Averages
50 ppm TWA; 295 mg/m3 TWA

Israel - Action Levels
25 ppm AL; 147.5 mg/m3 AL

Mexico - Instruction No. 10 - TWAs
50 ppm TWA; 300 mg/m3 TWA

STATE LISTS

California - Exposure Limits - PELs
50 ppm PEL; 300 mg/m3 PEL

California - Directors List of Hazardous Substances (8 CCR 339)
[present]

Massachusetts Right To Know List
[present]

Minnesota Hazardous Substance List
[present]

NJ Right to Know List (Total)
sn 1227

Pennsylvania Right to Know List
[present]

4-HEXYLRESORCINOL   136-77-6

HEALTH AND SAFETY LISTS

NFPA - Flash Points
flash point = 113 degrees F (45 degrees C)

NFPA - Hazard Identification Ratings
health-1; flammability-2; reactivity-0

NIOSH - Selected LD50s and LC50s
Oral, rat: LD50 = 42 gm/kg

INTERNATIONAL LISTS

Canada - WHMIS: Ingredient Disclosure
1% item 834 (18)

Canada - Ontario - OHSA - TWAEVs
50 ppm TWAEV; 294 mg/m3 TWAEV

STATE LISTS

Florida Hazardous Substance List
[present]

HEXYLTRICHLOROSILANE   928-65-4

HEALTH AND SAFETY LISTS

U.S. DOT - Substances From 49 CFR 172.101
regulated by DOT (UN2275)

U.S. DOT - Hazard Classes
DOT hazard class = 3

NFPA - Flash Points
flash point = 135 degrees F (57 degrees C)

NFPA - Hazard Identification Ratings
health-1; flammability-2; reactivity-0

NIOSH - Selected LD50s and LC50s
Oral, rat: LD50 = 1850 mg/kg Skin, rabbit: LD50 = 1260 mg/kg

INTERNATIONAL LISTS

Canada - WHMIS: Ingredient Disclosure
1% item 833 (177)

STATE LISTS

NJ Right to Know List (Total)
sn 0857

Pennsylvania Right to Know List
[present]

1-HEXYN-3-OL   105-31-7

HEALTH AND SAFETY LISTS

NFPA - Flash Points
flash point = 85 degrees F (29 degrees C)

NFPA - Hazard Identification Ratings
health-2; flammability-3; reactivity-0
NIOSH - Selected LD50s and LC50s
Oral, rat: LD50 = 670 mg/kg Skin, rabbit: LD50 = 420 mg/kg

INTERNATIONAL LISTS

Canada - WHMIS: Ingredient Disclosure
1% item 836 (969)

STATE LISTS

Florida Hazardous Substance List
[present]

Massachusetts Right To Know List
[present]

Pennsylvania Right to Know List
[present]

1-HEXYN-3-OL, 3,5-DIMETHYL-   107-54-0

HEALTH AND SAFETY LISTS

ACGIH 1995 - Ceiling Limits
C 25 ppm; C 121 mg/m3
NFPA - Flash Points
flash point = 205 degrees F (96 degrees C);   *See list description
NFPA - Hazard Identification Ratings
health-0; flammability-1; reactivity-0
NIOSH - Selected LD50s and LC50s
Oral, rat: LD50 = 3700 mg/kg Skin, rabbit: LD50 = 8560 mg/kg
OSHA - Vacated PELs - Ceiling Limits
C 25 ppm; C 125 mg/m3

ENVIRONMENTAL LISTS

List of Pesticide Product Inert Ingredients
[present]

INTERNATIONAL LISTS

Australian Exposure Standards - Time Weighted Averages
Peak Limitation: 25 ppm; 121 mg/m3
Canada - WHMIS: Ingredient Disclosure
1% item 837 (970)
Canada - Alberta - Ceiling Occupational Exposure Limit
C 25 ppm; C 120 mg/m3
Canada - British Columbia - Ceiling Exposure Limits
C 25 ppm; C 125 mg/m3
Canada - Ontario - OHSA - CEVs
25 ppm CEV; 120 mg/m3 CEV
Canada - Quebec - Ceiling Limits
P 25 ppm; P 121 mg/m3
United Kingdom - Occupational Exposure Standards - TWAs
25 ppm TWA; 125 mg/m3 TWA
United Kingdom - Occupational Exposure Standards - STELs
25 ppm STEL; 125 mg/m3 STEL
Israel - Ceiling Exposure Limits
C 25 ppm; C 121 mg/m3
Mexico - Instruction No. 10 - TWAs
25 ppm TWA; 125 mg/m3 TWA

STATE LISTS

California - Exposure Limits - Ceilings
C 25 ppm; C 125 mg/m3
California - Directors List of Hazardous Substances (8 CCR 339)
[present]
Florida Hazardous Substance List
[present]
Massachusetts Right To Know List
[present]
Minnesota Hazardous Substance List
[present]

Pennsylvania Right to Know List
[present]

HOELON   51338-27-

HEALTH AND SAFETY LISTS

NFPA - Flash Points
flash point = 170 degrees f (77 degrees C)
NFPA - Hazard Identification Ratings
health-2; flammability-2; reactivity-0
NIOSH - Selected LD50s and LC50s
Oral, rat: LD50 = 30900 mg/kg Skin, rabbit: LD50 = 6900 mg/kg

INTERNATIONAL LISTS

Canada - WHMIS: Ingredient Disclosure
1% item 838 (843)

STATE LISTS

Florida Hazardous Substance List
[present]

Massachusetts Right To Know List
[present]

Pennsylvania Right to Know List
[present]

HOLMIUM 155   15125-75-4

HEALTH AND SAFETY LISTS

NFPA - Flash Points
flash point = 180 degrees F (82 degrees C)
NFPA - Hazard Identification Ratings
health-0; flammability-2; reactivity-0

ENVIRONMENTAL LISTS

TSCA - Code of Federal Regulations Citations
40 CFR 712.30(d)
TSCA - PAIR - Reporting List
Reporting Date: November 19, 1982

INTERNATIONAL LISTS

Canada - WHMIS: Ingredient Disclosure
1% item 839 (1093)

HOLMIUM 157   15832-34-5

ENVIRONMENTAL LISTS

List of Pesticide Product Inert Ingredients
[present]

HOLMIUM 159   15750-02-4

HEALTH AND SAFETY LISTS

NIOSH - Selected LD50s and LC50s
Oral, rat: LD50 = 550 mg/kg
NTP Chemical Status Reports - Testing Status and NTIS Number
Technical reports printed (PB89128607/AS)
NTP Chemical Status Reports - Evidence of Carcinogenicity
male rat-no evidence; female rat-no evidence; male mice-equivocal evidence; female mice-no evidence

HOLMIUM 161   14391-20-9

HEALTH AND SAFETY LISTS

U.S. DOT - Substances From 49 CFR 172.101
regulated by DOT (UN1784)
U.S. DOT - Hazard Classes
DOT hazard class = 8

INTERNATIONAL LISTS

Canada - WHMIS: Ingredient Disclosure
1% item 840 (971)

STATE LISTS
  NJ Right to Know List (Total)
  sn 1004
  NJ Special Hazardous Substances
  (corrosive)

HOLMIUM 162   15700-49-9
HEALTH AND SAFETY LISTS
  NIOSH - Selected LD50s and LC50s
  Oral, mouse: LD50 = 210 mg/kg

HOLMIUM 162M   RR-00447-7
HEALTH AND SAFETY LISTS
  NFPA - Flash Points
  flash point = 135 degrees F (57 degrees C)
  NFPA - Hazard Identification Ratings
  health-0; flammability-2; reactivity-0
ENVIRONMENTAL LISTS
  List of Pesticide Product Inert Ingredients
  [present]

HOLMIUM 164   15749-97-0
ENVIRONMENTAL LISTS
  CERCLA/SARA - Section 313 - Emission Reporting
  form R reporting required
INTERNATIONAL LISTS
  Canada - Drinking Water Quality - MACs
  0.009 mg/L MAC
STATE LISTS
  Massachusetts Right To Know List
  [present]

HOLMIUM 164M   RR-00446-6
HEALTH AND SAFETY LISTS
  U.S. DOT - Appendix A Table 2 - Radionuclides
  final RQ = 1000 curies (3.7E 13 Bq)
ENVIRONMENTAL LISTS
  CERCLA/SARA List of Radionuclides (Appendix B) and Their Reportable Quantities
  final RQ = 1000 curies (3.7E 13 Bq)

HOLMIUM 166M   RR-00444-4
HEALTH AND SAFETY LISTS
  U.S. DOT - Appendix A Table 2 - Radionuclides
  final RQ = 1000 curies (3.7E 13 Bq)
ENVIRONMENTAL LISTS
  CERCLA/SARA List of Radionuclides (Appendix B) and Their Reportable Quantities
  final RQ = 1000 curies (3.7E 13 Bq)

HOLMIUM 167   15750-04-6
HEALTH AND SAFETY LISTS
  U.S. DOT - Appendix A Table 2 - Radionuclides
  final RQ = 1000 curies (3.7E 13 Bq)
ENVIRONMENTAL LISTS
  CERCLA/SARA List of Radionuclides (Appendix B) and Their Reportable Quantities
  final RQ = 1000 curies (3.7E 13 Bq)

HONEY   8028-66-8
HEALTH AND SAFETY LISTS
  U.S. DOT - Appendix A Table 2 - Radionuclides
  final RQ = 1000 curies (3.7E 13 Bq)

ENVIRONMENTAL LISTS
  CERCLA/SARA List of Radionuclides (Appendix B) and Their Reportable Quantities
  final RQ = 1000 curies (3.7E 13 Bq)

HYCANTHONE MESYLATE   23255-93-8
HEALTH AND SAFETY LISTS
  U.S. DOT - Appendix A Table 2 - Radionuclides
  final RQ = 1000 curies (3.7E 13 Bq)
ENVIRONMENTAL LISTS
  CERCLA/SARA List of Radionuclides (Appendix B) and Their Reportable Quantities
  final RQ = 1000 curies (3.7E 13 Bq)

HYDRALAZINE   86-54-4
HEALTH AND SAFETY LISTS
  U.S. DOT - Appendix A Table 2 - Radionuclides
  final RQ = 1000 curies (3.7E 13 Bq)
ENVIRONMENTAL LISTS
  CERCLA/SARA List of Radionuclides (Appendix B) and Their Reportable Quantities
  final RQ = 1000 curies (3.7E 13 Bq)

HYDRAMETHYLNON [TETRAHYDRO-5,5-DIMETHYL-2(1H)-PYRIMIDINONE[3-[4-(TRIFLUO-ROMETHYL)PHENYL]-1-[2-[4-(TRIFLUOROMETHYL)PHENYL]ETHENYL]-2-PROPENYLIDENE]HYDRA-ZONE]   67485-29-4
HEALTH AND SAFETY LISTS
  U.S. DOT - Appendix A Table 2 - Radionuclides
  final RQ = 1000 curies (3.7E 13 Bq)
ENVIRONMENTAL LISTS
  CERCLA/SARA List of Radionuclides (Appendix B) and Their Reportable Quantities
  final RQ = 1000 curies (3.7E 13 Bq)

HYDRATED SILICA   10279-57-9
HEALTH AND SAFETY LISTS
  U.S. DOT - Appendix A Table 2 - Radionuclides
  final RQ = 1000 curies (3.7E 13 Bq)
ENVIRONMENTAL LISTS
  CERCLA/SARA List of Radionuclides (Appendix B) and Their Reportable Quantities
  final RQ = 1000 curies (3.7E 13 Bq)

HYDRAZINE   302-01-2
HEALTH AND SAFETY LISTS
  U.S. DOT - Appendix A Table 2 - Radionuclides
  final RQ = 1 curie (3.7E 10 Bq)
ENVIRONMENTAL LISTS
  CERCLA/SARA List of Radionuclides (Appendix B) and Their Reportable Quantities
  final RQ = 1 curie (3.7E 10 Bq)

HYDRAZINE AZIDE   14546-44-2
HEALTH AND SAFETY LISTS
  U.S. DOT - Appendix A Table 2 - Radionuclides
  final RQ = 100 curies (3.7E 12 Bq)
ENVIRONMENTAL LISTS
  CERCLA/SARA List of Radionuclides (Appendix B) and Their Reportable Quantities
  final RQ = 100 curies (3.7E 12 Bq)

HYDRAZINECARBOXAMIDE   57-56-7
ENVIRONMENTAL LISTS
    List of Pesticide Product Inert Ingredients
    [present]

HYDRAZINECARBOXAMIDE, N,N'-1,6-HEX-AENDIYLBIS [2,2-DIMETHYL-   RR-00965-4
HEALTH AND SAFETY LISTS
    IARC - Group 3 (not classifiable)
    [present]

HYDRAZINE CHLORATE   134282-27-2
HEALTH AND SAFETY LISTS
    IARC - Group 3 (not classifiable)
    [present]
STATE LISTS
    California - Directors List of Hazardous Substances (8 CCR 339)
    [present]

HYDRAZINE CHLORATE   RR-01427-7
ENVIRONMENTAL LISTS
    CERCLA/SARA - Section 313 - Emission Reporting
    form R reporting required

HYDRAZINE DICARBONIC ACID DIAZIDE   67880-17-5
ENVIRONMENTAL LISTS
    List of Pesticide Product Inert Ingredients
    [present]

HYDRAZINE, 1,1-DIPHENYL   530-50-7
HEALTH AND SAFETY LISTS
    ACGIH 1995 - Time Weighted Averages
    (0.1) ppm TWA; (0.13) mg/m3 TWA
    ACGIH 1995 - Skin Designations
    skin - potential for cutaneous absorption
    ACGIH 1995 - Carcinogens
    (A2)-suspected human carcinogen
    AIHA - Odor Threshold Values
    geometric mean air odor threshold = 3.7 ppm (detectable)
    U.S. DOT - Substances From 49 CFR 172.101
    regulated by DOT (UN2029, UN2030)
    U.S. DOT - Hazard Classes
    DOT hazard class = 3
    U.S. DOT - Appendix A Table 1 - Hazardous Substances
    final RQ = 1 pound (0.454 kg)
    IARC - Group 2B (sufficient animal data)
    [present]
    NFPA - Flash Points
    anhydrous: flash point = 100 degrees F (38 degrees C)
    NFPA - Hazard Identification Ratings
    health-3; flammability-3; reactivity-3 (vapors explosive)
    NIOSH - Selected LD50s and LC50s
    Inhalation, rat: LC50 = 570 ppm 4 hr Oral, rat: LD50 = 60 mg/kg
    Skin, guinea pig: LD50 = 190 mg/kg
    NIOSH 1990 - Pocket Guide - RELs
    C 0.03 ppm (2 hr); C 0.04 mg/m3 (2 hr)
    NIOSH 1990 - Pocket Guide - IDLHs
    80 ppm IDLH (not considering carcinogenic effects)
    NIOSH 1990 - Pocket Guide - Carcinogens
    occupational carcinogen
    NIOSH 1990 - Pocket Guide - Target organs
    CNS, skin, eyes, respiratory system
    NIOSH - Health Standards - Exposure Limits
    C (2 hr) 0.03 ppm; C (2 hr) 0.04 mg/m3 (Listed under 'Hydrazines')

NIOSH - Health Standards - Health Effects and Precautions
has produced tumors of the lung, liver, blood vessels, and intestines i animals; blood, liver, and skin effects (Blood and urine monitorin and periodic chest X-ray required; bowel examination for worker over 40) (Listed under 'Hydrazines')
NIOSH - Health Standards - Carcinogenic Chemicals
potential human carcinogen (Listed under 'Hydrazines')
NTP Seventh Report - Suspect Carcinogens
suspect carcinogen
OSHA - Vacated PELs - Time Weighted Averages
0.1 ppm TWA; 0.1 mg/m3 TWA
OSHA - Vacated PELs - Skin Designation
Prevent or reduce skin absorption
OSHA - Final PELs - Time Weighted Averages
1 ppm TWA; 1.3 mg/m3 TWA
OSHA - Final PELs - Skin Notations
prevent or reduce skin absorption
OSHA - Possible Select Carcinogens
[present]
ENVIRONMENTAL LISTS
    ATSDR Priority List
    Rank (of 275): 165
    CERCLA/SARA - Section 302 Extremely Hazardous Substances and TPQs
    TPQ = 1000 pounds
    CERCLA/SARA - Section 313 - Emission Reporting
    form R reporting required for 0.1% de minimus concentration
    CERCLA/SARA - Hazardous Substances and their Reportable Quantities
    final RQ = 1 pound (0.454 kg)
    Clean Air Act (1990) - List of Hazardous Air Contaminants
    [present]
    CAA -Toxic Substances for Accidental Release Prevention
    threshold quantity = 15,000 lbs
    EPA - Carcinogen Hazard Ranking for RQ Adjustment
    Hazard ranking = High
    RCRA - U Series Wastes
    waste number U133 (Reactive waste; Toxic waste)
    RCRA - Hazardous Constituents-Appendix VIII
    waste number U133
    RCRA - Substances Banned From Land Disposal
    [present]
INTERNATIONAL LISTS
    Australian Exposure Standards - Time Weighted Averages
    0.1 ppm TWA; 0.13 mg/m3 TWA
    Australian Exposure Standards - Skin Effects
    skin absorption
    Australian Exposure Standards - Carcinogens
    probable carcinogen
    Canada - WHMIS: Ingredient Disclosure
    0.1% item 841 (978)
    Canada - NPRI (National Pollutant Release Inventory)
    [present]
    Canada - Alberta - 8 Hour Occupational Exposure Limit
    0.1 ppm TWA; 0.13 mg/m3 TWA
    Canada - Alberta - 15 Minute Occupational Exposure Limit
    0.3 ppm STEL; 0.39 mg/m3 STEL
    Canada - Alberta - Skin Designation
    can be absorbed through the intact skin
    Canada - Alberta - Designated Substances
    designated substance - requires code of practice
    Canada - British Columbia - 8 Hour Exposure Limits
    0.1 ppm TWA; 0.1 mg/m3 TWA

Canada - British Columbia - Skin Notations
  skin - potential for skin absorption
Canada - British Columbia - Carcinogens
  carcinogen - 0.1 ppm TWA; 0.1 mg/m3 TWA
Canada - Ontario - OHSA - TWAEVs
  0.1 ppm TWAEV; 0.13 mg/m3 TWAEV
Canada - Ontario - OHSA - Skin Notations
  absorption through skin, eyes, or mucous membranes
Canada - Quebec - Time-Weighted Average Exposure Values
  0.1 ppm TWAEV; 0.13 mg/m3 TWAEV (substance of which the recirculation is prohibited)
Canada - Quebec - Skin Designations
  absorbed through the skin
Canada - Quebec - Carcinogens
  C2 carcinogen: effect suspected in humans
German (DFG) - Skin/Sensitizers
  danger of cutaneous absorption; danger of sensitization (skin or respiratory)
German (DFG) - Carcinogens
  animal evidence of carcinogenicity
Israel - Time Weighted Averages
  (0.1) ppm TWA; (0.13) mg/m3 TWA
Israel - Action Levels
  0.05 ppm AL; 0.065 mg/m3 AL
Mexico - Instruction No. 10 - TWAs
  0.1 ppm TWA; 0.1 mg/m3 TWA
Mexico - Instruction No. 10 - Skin designation
  skin - potential for cutaneous absorption
Mexico - Instruction No. 10 - Carcinogens
  potential carcinogen in humans - limited epidemiological evidence

STATE LISTS

California - Air Bill 2588 Appendix A-I
  known or potential carcinogen
California - Prop. 65 - Cancer list
  carcinogen - initial date 1/1/88
California - Prop. 65 - No Significant Risk Levels
  no significant risk level = 0.04 ug/day
California - Exposure Limits - PELs
  0.1 ppm PEL; 0.1 mg/m3 PEL
California - Exposure Limits - Skin Notation
  material may be absorbed through the skin, eyes or mucous membrane
California - Directors List of Hazardous Substances (8 CCR 339)
  [present]
Florida Hazardous Substance List
  [present]
Massachusetts Right To Know List
  carcinogen; extraordinarily hazardous
Minnesota Hazardous Substance List
  carcinogen; skin
NJ Right to Know List (Total)
  sn 1006
NJ Special Hazardous Substances
  (carcinogen; corrosive; flammable - third degree; mutagen; reactive - second deg)
Pennsylvania Right to Know List
  environmental hazard; special hazardous substance
Pennsylvania RTK - Special Hazardous Substances
  [present]

PROPOSED REGULATIONS

ACGIH 1995 - Notice of Intended Changes
  (skin) 0.01 ppm TWA; 0.013 mg/m3 TWA; A3-animal carcinogen

HYDRAZINE, [4-[1-METHYLBUTOXY]PHENYL], MONOHYDROCHLORIDE   124993-63-1

HEALTH AND SAFETY LISTS
  U.S. DOT - Hazard Classes
    Forbidden from transport by the DOT

HYDRAZINE MONOHYDRATE   7803-57-8

HEALTH AND SAFETY LISTS
  NIOSH - Selected LD50s and LC50s
    Oral, mouse: LD50 = 176 mg/kg

ENVIRONMENTAL LISTS
  TSCA - Code of Federal Regulations Citations
    40 CFR 704.225(a)
  TSCA - CAIR - Reporting List
    reporting required by: manufacturer, distributor, importer, processor

HYDRAZINE PERCHLORATE   27978-54-7

ENVIRONMENTAL LISTS
  TSCA - Chemicals with Significant New Use Rules
    PMN number: P-87-1192

HYDRAZINE SELENATE   73506-32-8

HEALTH AND SAFETY LISTS
  U.S. DOT - Hazard Classes
    Forbidden from transport by the DOT

HYDRAZINE SULFATE   10034-93-2

STATE LISTS
  NJ Right to Know List (Total)
    sn 2390

HYDRAZOBENZENE   122-66-7

HEALTH AND SAFETY LISTS
  U.S. DOT - Hazard Classes
    Forbidden from transport by the DOT

HYDRAZOIC ACID   7782-79-8

ENVIRONMENTAL LISTS
  TSCA - Code of Federal Regulations Citations
    40 CFR 716.120(a)
  TSCA - Health and Safety Reporting List
    Effective Date: June 1, 1987

HYDRIDES, METAL, N.O.S.   RR-00333-8

ENVIRONMENTAL LISTS
  TSCA - Chemicals with Significant New Use Rules
    PMN number: P-90-558
  TSCA - Section 12(b) - Export Notification
    P-90-558; export notification required - Section 5

HYDRINDANE   496-10-6

HEALTH AND SAFETY LISTS
  NIOSH - Selected LD50s and LC50s
    Oral, rat: LD50 = 129 mg/kg

STATE LISTS
  NJ Right to Know List (Total)
    sn 3115

HYDRIODIC ACID   10034-85-2

HEALTH AND SAFETY LISTS
  U.S. DOT - Hazard Classes
    Forbidden from transport by the DOT

HYDROBROMOFLUOROCARBONS   **RR-01708-3**

HEALTH AND SAFETY LISTS

U.S. DOT - Hazard Classes
Forbidden from transport by the DOT

HYDROCARBON GASES, N.O.S.   **RR-00334-9**

HEALTH AND SAFETY LISTS

NIOSH - Selected LD50s and LC50s
Oral, rat: LD50 = 601 mg/kg

NTP Seventh Report - Suspect Carcinogens
suspect carcinogen

OSHA - Possible Select Carcinogens
[present]

ENVIRONMENTAL LISTS

CERCLA/SARA - Section 313 - Emission Reporting
form R reporting required for 0.1% de minimus concentration

STATE LISTS

California - Air Bill 2588 Appendix A-II
known or potential carcinogen

California - Prop. 65 - Cancer list
carcinogen - initial date 1/1/88

California - Prop. 65 - No Significant Risk Levels
no significant risk level = 0.2 ug/day

Florida Hazardous Substance List
[present]

Massachusetts Right To Know List
carcinogen; extraordinarily hazardous

Minnesota Hazardous Substance List
carcinogen

NJ Right to Know List (Total)
sn 2360

Pennsylvania Right to Know List
environmental hazard; special hazardous substance

Pennsylvania RTK - Special Hazardous Substances
[present]

HYDROCARBON OILS, WHITE MINERAL OIL   **8020-83-5**
SEE ALSO:
F039-HAZARDOUS WASTES

HEALTH AND SAFETY LISTS

U.S. DOT - Appendix A Table 1 - Hazardous Substances
final RQ = 10 pounds (4.54 kg)

NIOSH - Selected LD50s and LC50s
Oral, rat: LD50 = 301 mg/kg

NTP Chemical Status Reports - Testing Status and NTIS Number
Technical reports printed (PB285791/AS)

NTP Chemical Status Reports - Evidence of Carcinogenicity
male rat-positive; female rat-positive; male mice-negative; female mice-positive

NTP Seventh Report - Suspect Carcinogens
suspect carcinogen

OSHA - Possible Select Carcinogens
[present]

ENVIRONMENTAL LISTS

ATSDR Priority List
Rank (of 275): 154

CERCLA/SARA - Section 313 - Emission Reporting
form R reporting required for 0.1% de minimus concentration

CERCLA/SARA - Hazardous Substances and their Reportable Quantities
final RQ = 10 pounds (4.54 kg)

Clean Air Act (1990) - List of Hazardous Air Contaminants
[present]

CAA - HON Rule - Organic HAPs
[present]

Clean Water Act - Priority Pollutants
[present]

Clean Water Act - Toxic Pollutants
[present]

EPA - Carcinogen Hazard Ranking for RQ Adjustment
Hazard ranking = Medium

RCRA - U Series Wastes
waste number U109

RCRA - Hazardous Constituents-Appendix VIII
waste number U109

RCRA - Substances Banned From Land Disposal
[present]

RCRA - Universal Treatment Standards (LDR)
WW: 0.087 mg/l; NWW: Not applicable

TSCA - Code of Federal Regulations Citations
40 CFR 716.120(a)

TSCA - Health and Safety Reporting List
Effective Date: June 1, 1987

INTERNATIONAL LISTS

Canada - WHMIS: Ingredient Disclosure
0.1% item 842 (979)

German (DFG) - Carcinogens
animal evidence of carcinogenicity

Mexico - Wastewater - Organic Toxic Pollutants and Heavy Metals
Listed under [Organic Toxic Pollutants]

Mexico - Drinking Water - Ecological Criteria
0.0004 mg/l This level has been extrapolated by using a mathematic model

STATE LISTS

California - Air Bill 2588 Appendix A-I
known or potential carcinogen

California - Prop. 65 - Cancer list
carcinogen - initial date 1/1/88

California - Prop. 65 - No Significant Risk Levels
no significant risk level = 0.8 ug/day

Florida Hazardous Substance List
[present]

Massachusetts Right To Know List
carcinogen; extraordinarily hazardous

Minnesota Hazardous Substance List
carcinogen

NJ Right to Know List (Total)
sn 0800

NJ Special Hazardous Substances
(carcinogen)

Pennsylvania Right to Know List
environmental hazard; special hazardous substance

Pennsylvania RTK - Special Hazardous Substances
[present]

PROPOSED REGULATIONS

Safe Drinking Water Act - Priority list
[present]

HYDROCARBON PROPELLANT   **68476-40-4**

INTERNATIONAL LISTS

Canada - WHMIS: Ingredient Disclosure
1% item 843 (93)

United Kingdom - Occupational Exposure Standards - STELs
as vapour: 0.1 ppm STEL

German (DFG) - MAK Values
0.1 ppm MAK; 0.27 mg/m3 MAK

German (DFG) - Peak Limitations
2 x normal MAK (5 min momentary value); don't exceed 8 times during shift

HYDROCARBON WAXES (PETROLEUM), CLAY-TREATED MICROCRYSTALLINE   64742-42-3

HEALTH AND SAFETY LISTS

U.S. DOT - Substances From 49 CFR 172.101
regulated by DOT (UN1409)

U.S. DOT - Hazard Classes
DOT hazard class = 4.3

STATE LISTS

NJ Right to Know List (Total)
sn 2467

HYDROCHLOROTHIAZIDE   58-93-5

HEALTH AND SAFETY LISTS

NFPA - Hazard Identification Ratings
reactivity-0

HYDROCYANIC ACID (PRUSSIC)   RR-01428-8

SEE ALSO:
HALOGEN ACIDS

HEALTH AND SAFETY LISTS

U.S. DOT - Substances From 49 CFR 172.101
regulated by DOT (UN1787, UN2197)

U.S. DOT - Hazard Classes
anhydrous: DOT hazard class = 2.2; solution: DOT hazard class = 8

U.S. DOT - Substances Which Are Poisonous by Inhalation
gaseous hazardous material poisonous by inhalation (anhydrous form) (UN2197)

FDA - Controlled Substances Act - Precursor chemicals
Threshold by base weight = 1.7 kilograms

INTERNATIONAL LISTS

Canada - WHMIS: Ingredient Disclosure
1% item 848 (1025)

STATE LISTS

Florida Hazardous Substance List
[present]

Massachusetts Right To Know List
[present]

NJ Right to Know List (Total)
sn 1009

NJ Special Hazardous Substances
(corrosive)

Pennsylvania Right to Know List
[present]

HYDROGEN   1333-74-0

ENVIRONMENTAL LISTS

Class 1 Ozone Depletors
[present]

HYDROGENATED ARYLATED POLYDECENE   RR-00957-4

HEALTH AND SAFETY LISTS

U.S. DOT - Substances From 49 CFR 172.101
regulated by DOT (UN1964, UN1965)

U.S. DOT - Hazard Classes
DOT hazard class = 2.1

STATE LISTS

NJ Right to Know List (Total)
sn 2468; sn 2469

HYDROGENATED CASTOR OIL   8001-78-3

STATE LISTS

Pennsylvania Right to Know List
special hazardous substance

Pennsylvania RTK - Special Hazardous Substances
[present]

HYDROGENATED COTTONSEED OIL   68334-00-9

ENVIRONMENTAL LISTS

List of Pesticide Product Inert Ingredients
[present]

HYDROGENATED METHYL ABIETE   30968-45-7

ENVIRONMENTAL LISTS

List of Pesticide Product Inert Ingredients
[present]

HYDROGENATED SOYBEAN OIL   8016-70-4

HEALTH AND SAFETY LISTS

IARC - Group 3 (not classifiable)
[present]

NTP Chemical Status Reports - Testing Status and NTIS Number
Technical reports printed (PB90110156/AS)

NTP Chemical Status Reports - Evidence of Carcinogenicity
male rat-no evidence; female rat-no evidence; male mice-equivocal evidence; female mice-no evidence

HYDROGENATED TALLOW ALKYL AMINE ACETATE   61790-59-8

HEALTH AND SAFETY LISTS

U.S. DOT - Hazard Classes
Forbidden from transport by the DOT

HYDROGENATED TERPHENYLS   61788-32-7

HEALTH AND SAFETY LISTS

ACGIH 1995 - Time Weighted Averages
simple asphyxiant

U.S. DOT - Substances From 49 CFR 172.101
regulated by DOT (UN1966, UN1047, UN2600)

U.S. DOT - Hazard Classes
DOT hazard class = 2.1

NFPA - Flash Points
gas (no flash point given)

NFPA - Hazard Identification Ratings
health-0; flammability-4; reactivity-0

ENVIRONMENTAL LISTS

CAA - Flammable Substances for Accidental Release Prevention
threshold quantity = 10,000 lbs

INTERNATIONAL LISTS

Australian Exposure Standards - Time Weighted Averages
Asphyxiant at < 18% oxygen by volume; explosion hazard

Canada - British Columbia - 8 Hour Exposure Limits
asphyxiant substance

Canada - Ontario - OHSA - TWAEVs
simple asphyxiant

Canada - Quebec - Time-Weighted Average Exposure Values
simple asphyxiant

Israel - Time Weighted Averages
Asphyxiant

Mexico - Instruction No. 10 - TWAs
simple asphyxiant

STATE LISTS

California - Exposure Limits - PELs
asphyxiant (limit depends on level of oxygen)

California - Directors List of Hazardous Substances (8 CCR 339)
[present]

Florida Hazardous Substance List
[present]

Massachusetts Right To Know List
[present]

Minnesota Hazardous Substance List
[present]

NJ Right to Know List (Total)
sn 1010

NJ Special Hazardous Substances
(flammable - fourth degree)

Pennsylvania Right to Know List
[present]

HYDROGEN BROMIDE   10035-10-6

ENVIRONMENTAL LISTS

TSCA - Chemicals with Significant New Use Rules
PMN number: P-90-1454

HYDROGEN CHLORIDE   7647-01-0

HEALTH AND SAFETY LISTS

NFPA - Hazard Identification Ratings
health-0; flammability-1; reactivity-0

ENVIRONMENTAL LISTS

List of Pesticide Product Inert Ingredients
[present]

HYDROGEN CYANIDE   74-90-8

ENVIRONMENTAL LISTS

List of Pesticide Product Inert Ingredients
[present]

HYDROGEN FLUORIDE   7664-39-3

ENVIRONMENTAL LISTS

List of Pesticide Product Inert Ingredients
[present]

HYDROGEN PEROXIDE   7722-84-1

ENVIRONMENTAL LISTS

List of Pesticide Product Inert Ingredients
[present]

HYDROGEN SELENIDE   7783-07-5

ENVIRONMENTAL LISTS

List of Pesticide Product Inert Ingredients
[present]

HYDROGEN SULFIDE   7783-06-4

HEALTH AND SAFETY LISTS

ACGIH 1995 - Time Weighted Averages
0.5 ppm TWA; 4.9 mg/m3 TWA

OSHA - Vacated PELs - Time Weighted Averages
0.5 ppm TWA; 5 mg/m3 TWA

INTERNATIONAL LISTS

Australian Exposure Standards - Time Weighted Averages
0.5 ppm TWA; 4.9 mg/m3 TWA

Canada - Alberta - 8 Hour Occupational Exposure Limit
0.5 ppm TWA; 5 mg/m3 TWA

Canada - Alberta - 15 Minute Occupational Exposure Limit
1.5 ppm STEL; 15 mg/m3 STEL

Canada - British Columbia - 8 Hour Exposure Limits
0.5 ppm TWA; 5 mg/m3 TWA

Canada - Ontario - OHSA - TWAEVs
5 mg/m3 TWAEV (listed as agents of variable composition) (As sum of components assayed by chromatographic procedure with referenc to the bulk sample)

Canada - Quebec - Time-Weighted Average Exposure Values
0.5 ppm TWAEV; 4.9 mg/m3 TWAEV

Israel - Time Weighted Averages
0.5 ppm TWA; 4.9 mg/m3 TWA

Israel - Action Levels
0.25 ppm AL; 2.45 mg/m3 AL

Mexico - Instruction No. 10 - TWAs
0.5 ppm TWA; 5 mg/m3 TWA

STATE LISTS

California - Exposure Limits - PELs
0.5 ppm PEL; 5 mg/m3 PEL

California - Directors List of Hazardous Substances (8 CCR 339)
[present]

Massachusetts Right To Know List
[present]

Minnesota Hazardous Substance List
[present]

Pennsylvania Right to Know List
[present]

HYDROLYZED POLYACRYLONITRILE   25214-69-1
SEE ALSO:
HALOGEN ACIDS

HEALTH AND SAFETY LISTS

ACGIH 1995 - Ceiling Limits
C 3 ppm; C 9.9 mg/m3

U.S. DOT - Substances From 49 CFR 172.101
regulated by DOT (UN1048, UN1788)

U.S. DOT - Hazard Classes
anhydrous: DOT hazard class = 2.3; solution: DOT hazard class = 8

U.S. DOT - Substances Which Are Poisonous by Inhalation
gaseous hazardous material poisonous by inhalation (anhydrous form) (UN1048)

NIOSH - Selected LD50s and LC50s
Inhalation, rat: LC50 = 2858 ppm 1 hr

NIOSH 1990 - Pocket Guide - RELs
C 3 ppm; C 10 mg/m3

NIOSH 1990 - Pocket Guide - IDLHs
50 ppm IDLH

NIOSH 1990 - Pocket Guide - Target organs
eyes, skin, respiratory system

OSHA - Vacated PELs - Ceiling Limits
C 3 ppm; C 10 mg/m3

OSHA - Final PELs - Time Weighted Averages
3 ppm TWA; 10 mg/m3 TWA

OSHA - List of Highly Hazardous Chemicals
threshhold quantity = 5000 pounds

INTERNATIONAL LISTS

Australian Exposure Standards - Time Weighted Averages
Peak Limitation: 3 ppm; 9.9 mg/m3

Canada - WHMIS: Ingredient Disclosure
1% item 844 (338)

Canada - Alberta - Ceiling Occupational Exposure Limit
C 3 ppm; C 9.9 mg/m3

Canada - British Columbia - 8 Hour Exposure Limits
3 ppm TWA; 10 mg/m3 TWA

Canada - Ontario - OHSA - CEVs
3 ppm CEV; 10 mg/m3 CEV

Canada - Quebec - Ceiling Limits
P 3 ppm; P 9.9 mg/m3

United Kingdom - Occupational Exposure Standards - STELs
3 ppm STEL; 10 mg/m3 STEL

German (DFG) - MAK Values
5 ppm MAK; 17 mg/m3 MAK

German (DFG) - Peak Limitations
2 x normal MAK (5 min momentary value); don't exceed 8 times during shift

Israel - Ceiling Exposure Limits
C 3 ppm; C 9.9 mg/m3

Mexico - Instruction No. 10 - TWAs
3 ppm TWA; 10 mg/m3 TWA

STATE LISTS

California - Exposure Limits - Ceilings
C 3 ppm; C 10 mg/m3

California - Directors List of Hazardous Substances (8 CCR 339)
[present]

Florida Hazardous Substance List
[present]

Massachusetts Right To Know List
[present]

Minnesota Hazardous Substance List
[present]

NJ Right to Know List (Total)
sn 1011

NJ Special Hazardous Substances
(corrosive)

Pennsylvania Right to Know List
[present]

HYDROLYZED PROTEIN   9015-54-7

SEE ALSO:
HALOGEN ACIDS

HEALTH AND SAFETY LISTS

ACGIH 1995 - Ceiling Limits
C 5 ppm; C 7.5 mg/m3

AIHA - Odor Threshold Values
no geometric mean air odor threshold

U.S. DOT - Substances From 49 CFR 172.101
regulated by DOT (UN1789, UN1050, UN2186)

U.S. DOT - Hazard Classes
DOT hazard class = 8

U.S. DOT - Substances Which Are Poisonous by Inhalation
gaseous hazardous material poisonous by inhalation (anhydrous or refrigerated liquid) (UN1050, UN2186)

U.S. DOT - Appendix A Table 1 - Hazardous Substances
final RQ = 5000 pounds (2270 kg)

IARC - Group 3 (not classifiable)
[present]

NIOSH - Selected LD50s and LC50s
Inhalation, rat: LC50 = 3124 ppm 1 hr Oral, rabbit: LD50 = 900 mg/kg

NIOSH 1990 - Pocket Guide - RELs
C 5 ppm; C 7 mg/m3

NIOSH 1990 - Pocket Guide - IDLHs
100 ppm IDLH

NIOSH 1990 - Pocket Guide - Target organs
skin, eyes, respiratory system

OSHA - Vacated PELs - Ceiling Limits
C 5 ppm; C 7 mg/m3

OSHA - Final PELs - Ceiling Limits
C 5 ppm; C 7 mg/m3

OSHA - List of Highly Hazardous Chemicals
anhydrous: threshhold quantity = 5000 pounds

ENVIRONMENTAL LISTS

CERCLA/SARA - Section 302 Extremely Hazardous Substances and TPQs
TPQ = 500 pounds

CERCLA/SARA - Section 313 - Emission Reporting
form R reporting required for 1.0% de minimus concentration

CERCLA/SARA - Hazardous Substances and their Reportable Quantities
final RQ = 5000 pounds (2270 kg)

Clean Air Act (1990) - List of Hazardous Air Contaminants
[present]

CAA -Toxic Substances for Accidental Release Prevention
conc. 30% or greater: threshold quantity = 15,000 lbs; anhydrous: threshold quantity = 5,000 lbs

Clean Water Act - Hazardous Substances
[present]

List of Pesticide Product Inert Ingredients
[present]

INTERNATIONAL LISTS

Australian Exposure Standards - Time Weighted Averages
Peak Limitation: 5 ppm; 7.5 mg/m3

Canada - WHMIS: Ingredient Disclosure
1% item 845 (502)

Canada - NPRI (National Pollutant Release Inventory)
[present]

Canada - Alberta - Ceiling Occupational Exposure Limit
C 5 ppm; C 7.5 mg/m3

Canada - British Columbia - Ceiling Exposure Limits
C 5 ppm; C 7 mg/m3

Canada - Ontario - OHSA - CEVs
5 ppm CEV; 7.4 mg/m3 CEV

Canada - Quebec - Ceiling Limits
P 5 ppm; P 7.5 mg/m3

United Kingdom - Occupational Exposure Standards - STELs
5 ppm STEL; 7 mg/m3 STEL

German (DFG) - MAK Values
5 ppm MAK; 7 mg/m3 MAK

German (DFG) - Peak Limitations
2 x normal MAK (5 min momentary value); don't exceed 8 times during shift

German (DFG) - Pregnancy
no risk to embryo/fetus if exposure limits adhered to

Israel - Ceiling Exposure Limits
C 5 ppm; C 7.5 mg/m3

Mexico - Instruction No. 10 - TWAs
5 ppm TWA; 7 mg/m3 TWA

STATE LISTS

California - Air Bill 2588 Appendix A-I
[present]

California - Exposure Limits - Ceilings
C 5 ppm; C 7 mg/m3

California - Directors List of Hazardous Substances (8 CCR 339)
[present]

Florida Hazardous Substance List
[present]

Massachusetts Right To Know List
extraordinarily hazardous

Minnesota Hazardous Substance List
[present]

NJ Right to Know List (Total)
sn 1012;   gas only: sn 2909

NJ Special Hazardous Substances
(corrosive)

Pennsylvania Right to Know List
environmental hazard

HYDROQUINONE   123-31-9
SEE ALSO:
K013-HAZARDOUS WASTES
K011-HAZARDOUS WASTES

HEALTH AND SAFETY LISTS

ACGIH 1995 - Ceiling Limits
C 4.7 ppm;   C 5 mg/m3

ACGIH 1995 - Skin Designations
skin - potential for cutaneous absorption

U.S. DOT - Substances From 49 CFR 172.101
regulated by DOT (UN1613, UN1051, UN1614, NA1613)

U.S. DOT - Hazard Classes
DOT hazard class = 6.1

U.S. DOT - Substances Which Are Poisonous by Inhalation
liquid hazardous material poisonous by inhalation (anhydrous, stabilized form or aqueous solutions) (UN1051, UN1613)

U.S. DOT - Appendix B - Marine Pollutants
DOT regulated marine pollutant

U.S. DOT - Appendix A Table 1 - Hazardous Substances
final RQ = 10 pounds (4.54 kg)

NFPA - Flash Points
96%: flash point = 0 degrees F (-18 degrees C)

NFPA - Hazard Identification Ratings
96%: health-4;   flammability-4;   reactivity-2 (vapors extremely toxic)

NIOSH - Selected LD50s and LC50s
Inhalation, rat: LC50 = 484 ppm 5 mn Oral, mouse: LD50 = 3700 ug/kg

NIOSH 1990 - Pocket Guide - RELs
4.7 ppm STEL;   5 mg/m3 STEL

NIOSH 1990 - Pocket Guide - IDLHs
50 ppm IDLH

NIOSH 1990 - Pocket Guide - Target organs
CNS, CVS, liver, kidneys,

NIOSH 1990 - Pocket Guide - Skin list
Potential for dermal absorption

NIOSH - Health Standards - Exposure Limits
C (10 min) 4.7 ppm CN;   C (10 min) 5 mg CN/m3 (Listed under 'Hydrogen cyanide and cyanide salts')

NIOSH - Health Standards - Health Effects and Precautions
Thyroid, blood, and respiratory system effects (Prevent skin and eye contact;   make first-aid kits and personnel available during use) (Listed under 'Hydrogen cyanide and cyanide salts')

OSHA - Vacated PELs - Short Term Exposure Limits
4.7 ppm STEL;   5 mg/m3 STEL

OSHA - Vacated PELs - Skin Designation
Prevent or reduce skin absorption

OSHA - Final PELs - Time Weighted Averages
10 ppm TWA;   11 mg/m3 TWA

OSHA - Final PELs - Skin Notations
prevent or reduce skin absorption

OSHA - List of Highly Hazardous Chemicals
anhydrous: threshhold quantity = 1000 pounds

ENVIRONMENTAL LISTS

CERCLA/SARA - Section 302 Extremely Hazardous Substances and TPQs
TPQ = 100 pounds

CERCLA/SARA - Section 313 - Emission Reporting
form R reporting required for 1.0% de minimus concentration

CERCLA/SARA - Hazardous Substances and their Reportable Quantities
final RQ = 10 pounds (4.54 kg)

CAA -Toxic Substances for Accidental Release Prevention
threshold quantity = 2,500 lbs

Clean Water Act - Hazardous Substances
[present]

RCRA - P Series Wastes
waste number P063

RCRA - Hazardous Constituents-Appendix VIII
waste number P063

RCRA - Basis for Listing - Appendix VII
Included in waste streams: K011, K013

RCRA - Substances Banned From Land Disposal
[present]

TSCA - Code of Federal Regulations Citations
40 CFR 712.30(w)

TSCA - PAIR - Reporting List
Reporting Date: July 13, 1988

INTERNATIONAL LISTS

Australian Exposure Standards - Time Weighted Averages
Peak Limitation: 10 ppm;   11 mg/m3

Australian Exposure Standards - Skin Effects
skin absorption

Canada - WHMIS: Ingredient Disclosure
1% item 846 (593)

Canada - NPRI (National Pollutant Release Inventory)
[present]

Canada - Alberta - Ceiling Occupational Exposure Limit
C 10 ppm;   C 11 mg/m3

Canada - Alberta - Skin Designation
can be absorbed through the intact skin

Canada - British Columbia - 8 Hour Exposure Limits
10 ppm TWA;   11 mg/m3 TWA

Canada - British Columbia - 15 Minute Exposure Limits
15 ppm STEL;   16 mg/m3 STEL

Canada - British Columbia - Skin Notations
skin - potential for skin absorption

Canada - Ontario - OHSA - CEVs
10 ppm CEV;   11 mg/m3 CEV

Canada - Ontario - OHSA - Skin Notations
absorption through skin, eyes, or mucous membranes

Canada - Quebec - Ceiling Limits
P 10 ppm;   P 11 mg/m3

Canada - Quebec - Skin Designations
absorbed through the skin

United Kingdom - Maximum Exposure Limits - STELs
10 ppm STEL;   10 mg/m3 STEL

United Kingdom - Maximum Exposure Limits - Notes
can be absorbed through skin

German (DFG) - MAK Values
10 ppm MAK;   11 mg/m3 MAK

German (DFG) - Peak Limitations
2 x normal MAK (30 min. average value);   don't exceed 4 times during shift

German (DFG) - Skin/Sensitizers
danger of cutaneous absorption

Israel - Ceiling Exposure Limits
C 10 ppm;   C 11 mg/m3

Mexico - Instruction No. 10 - Ceiling Limits
P 10 ppm;   P 10 mg/m3

Mexico - Instruction No. 10 - Skin designation
skin - potential for cutaneous absorption

STATE LISTS

California - Air Bill 2588 Appendix A-I
[present]

California - Exposure Limits - STELs
4.7 ppm STEL; 5 mg/m3 STEL

California - Exposure Limits - Skin Notation
material may be absorbed through the skin, eyes or mucous membrane

California - Directors List of Hazardous Substances (8 CCR 339)
[present]

Florida Hazardous Substance List
[present]

Massachusetts Right To Know List
extraordinarily hazardous

Minnesota Hazardous Substance List
skin

NJ Right to Know List (Total)
sn 1013

NJ Special Hazardous Substances
(flammable - fourth degree; reactive - second degree)

Pennsylvania Right to Know List
environmental hazard

PROPOSED REGULATIONS

Canada - Ontario - Proposed Occupational CEVs
5 ppm CEV; 5 mg/m3 CEV

HYDROQUINONE DI-(BETA-HYDROXYETHYL)   104-38-1
ETHER
SEE ALSO:
HALOGEN ACIDS

HEALTH AND SAFETY LISTS

ACGIH 1995 - Ceiling Limits
as F: C 3 ppm; C 2.6 mg/m3

AIHA - Odor Threshold Values
no geometric mean air odor threshold

U.S. DOT - Substances From 49 CFR 172.101
regulated by DOT (UN1052, UN1790)

U.S. DOT - Hazard Classes
DOT hazard class = 8

U.S. DOT - Substances Which Are Poisonous by Inhalation
liquid hazardous material poisonous by inhalation (anhydrous form) (UN1052)

U.S. DOT - Appendix A Table 1 - Hazardous Substances
final RQ = 100 pounds (45.4 kg)

NIOSH - Selected LD50s and LC50s
Inhalation, rat: LC50 = 1276 ppm 1 hr

NIOSH 1990 - Pocket Guide - RELs
as F: 3 ppm TWA; 2.5 mg/m3 TWA; C 6 ppm (15 min); C 5 mg/m3 (15 min)

NIOSH 1990 - Pocket Guide - IDLHs
as F: 30 ppm IDLH

NIOSH 1990 - Pocket Guide - Target organs
eyes, skin, respiratory system

NIOSH - Health Standards - Exposure Limits
3 ppm TWA; 2.5 mg F/m3 TWA; C (15 min) 6 ppm; C (15 min) 5.0 mg F/m3

NIOSH - Health Standards - Health Effects and Precautions
Skin, eye, and airway irritation; bone effects (Periodic pelvic X-ray to detect changes in the osseous system (for males only) and urine testing required)

OSHA - Vacated PELs - Time Weighted Averages
as F: 3 ppm TWA

OSHA - Vacated PELs - Short Term Exposure Limits
as F: 6 ppm STEL

OSHA - Final PELs - Time Weighted Averages
3 ppm TWA

OSHA - List of Highly Hazardous Chemicals
anhydrous: threshhold quantity = 1000 pounds

ENVIRONMENTAL LISTS

CERCLA/SARA - Section 302 Extremely Hazardous Substances and TPQs
TPQ = 100 pounds

CERCLA/SARA - Section 313 - Emission Reporting
form R reporting required for 1.0% de minimus concentration

CERCLA/SARA - Hazardous Substances and their Reportable Quantities
final RQ = 100 pounds (45.4 kg)

Clean Air Act (1990) - List of Hazardous Air Contaminants
[present]

CAA -Toxic Substances for Accidental Release Prevention
conc. 50% or greater: threshold quantity = 1,000 lbs

Clean Water Act - Hazardous Substances
[present]

RCRA - U Series Wastes
waste number U134 (Corrosive waste; Toxic waste)

RCRA - Hazardous Constituents-Appendix VIII
waste number U134

RCRA - Substances Banned From Land Disposal
[present]

TSCA - Code of Federal Regulations Citations
40 CFR 712.30(w)

TSCA - PAIR - Reporting List
Reporting Date: July 13, 1988

INTERNATIONAL LISTS

Australian Exposure Standards - Time Weighted Averages
as F: Peak Limitation: 3 ppm; 2.6 mg/m3

Canada - WHMIS: Ingredient Disclosure
1% item 847 (906)

Canada - NPRI (National Pollutant Release Inventory)
[present]

Canada - Alberta - Ceiling Occupational Exposure Limit
C 3 ppm; C 2.3 mg/m3

Canada - British Columbia - 8 Hour Exposure Limits
3 ppm TWA; 2 mg/m3 TWA

Canada - Ontario - OHSA - CEVs
as F-: 3 ppm CEV; 2.5 mg/m3 CEV

Canada - Quebec - Ceiling Limits
P 3 ppm; P 2.6 mg/m3

United Kingdom - Occupational Exposure Standards - STELs
as F: 3 ppm STEL; 2.5 mg/m3 STEL

German (DFG) - MAK Values
3 ppm MAK; 2 mg/m3 MAK

German (DFG) - Peak Limitations
2 x normal MAK (5 min momentary value); don't exceed 8 times during shift

Israel - Ceiling Exposure Limits
as F: C 3 ppm; C 2.6 mg/m3

Mexico - Instruction No. 10 - TWAs
3 ppm TWA; 2.5 mg/m3 TWA

Mexico - Instruction No. 10 - STELs
6 ppm STEL; 5 mg/m3 STEL

STATE LISTS

California - Air Bill 2588 Appendix A-I
[present]

California - Exposure Limits - PELs
as F: 3 ppm PEL; 2.5 mg/m3 PEL

California - Exposure Limits - STELs
  as F: 6 ppm STEL

California - Directors List of Hazardous Substances (8 CCR 339)
  [present]

Florida Hazardous Substance List
  [present]

Massachusetts Right To Know List
  extraordinarily hazardous

Minnesota Hazardous Substance List
  [present] as F

NJ Right to Know List (Total)
  sn 1014

NJ Special Hazardous Substances
  (corrosive)

Pennsylvania Right to Know List
  environmental hazard

PROPOSED REGULATIONS

Canada - Ontario - Proposed Occupational CEVs
  2 ppm CEV;   1.7 mg/m3 CEV

HYDROQUINONE, DIMETHYL ETHER   150-78-7

HEALTH AND SAFETY LISTS

ACGIH 1995 - Time Weighted Averages
  1 ppm TWA;   1.4 mg/m3 TWA

U.S. DOT - Substances From 49 CFR 172.101
  regulated by DOT (UN2984, UN2015, UN2014)

U.S. DOT - Hazard Classes
  DOT hazard class = 5.1

IARC - Group 3 (not classifiable)
  [present]

NIOSH - Selected LD50s and LC50s
  Oral, mouse: LD50 = 2 gm/kg Skin, rat: LD50 = 4060 mg/kg

NIOSH 1990 - Pocket Guide - RELs
  1 ppm TWA;   1.4 mg/m3 TWA

NIOSH 1990 - Pocket Guide - IDLHs
  75 ppm IDLH

NIOSH 1990 - Pocket Guide - Target organs
  eyes, skin, respiratory system

OSHA - Vacated PELs - Time Weighted Averages
  1 ppm TWA;   1.4 mg/m3 TWA

OSHA - Final PELs - Time Weighted Averages
  1 ppm TWA;   1.4 mg/m3 TWA

OSHA - List of Highly Hazardous Chemicals
  52% by weight or more: threshhold quantity = 7500 pounds

ENVIRONMENTAL LISTS

CERCLA/SARA - Section 302 Extremely Hazardous Substances and TPQs
  concentration > 52%: TPQ = 1000 pounds

List of Pesticide Product Inert Ingredients
  [present]

INTERNATIONAL LISTS

Australian Exposure Standards - Time Weighted Averages
  1 ppm TWA;   1.4 mg/m3 TWA

Canada - WHMIS: Ingredient Disclosure
  1% item 849 (1365)

Canada - Alberta - 8 Hour Occupational Exposure Limit
  1 ppm TWA;   1.4 mg/m3 TWA

Canada - Alberta - 15 Minute Occupational Exposure Limit
  2 ppm STEL;   2.8 mg/m3 STEL

Canada - British Columbia - 8 Hour Exposure Limits
  1 ppm TWA;   1.5 mg/m3 TWA

Canada - British Columbia - 15 Minute Exposure Limits
  2 ppm STEL;   3 mg/m3 STEL

Canada - Ontario - OHSA - TWAEVs
  1 ppm TWAEV;   1.4 mg/m3 TWAEV

Canada - Quebec - Time-Weighted Average Exposure Values
  1 ppm TWAEV;   1.4 mg/m3 TWAEV

United Kingdom - Occupational Exposure Standards - TWAs
  1 ppm TWA;   1.5 mg/m3 TWA

United Kingdom - Occupational Exposure Standards - STELs
  2 ppm STEL;   3 mg/m3 STEL

German (DFG) - MAK Values
  1 ppm MAK;   1.4 mg/m3 MAK

German (DFG) - Peak Limitations
  2 x normal MAK (5 min momentary value);   don't exceed 8 time during shift

Israel - Time Weighted Averages
  1 ppm TWA;   1.4 mg/m3 TWA

Israel - Action Levels
  0.5 ppm AL;   0.7 mg/m3 AL

Mexico - Instruction No. 10 - TWAs
  1 ppm TWA;   3.5 mg/m3 TWA

Mexico - Instruction No. 10 - STELs
  2 ppm STEL;   3 mg/m3 STEL

STATE LISTS

California - Exposure Limits - PELs
  as H2O2: 1 ppm PEL;   1.4 mg/m3 PEL

California - Directors List of Hazardous Substances (8 CCR 339)
  [present]

Florida Hazardous Substance List
  [present]

Massachusetts Right To Know List
  extraordinarily hazardous

Minnesota Hazardous Substance List
  [present] (includes Hydrogen Peroxide (90%))

NJ Right to Know List (Total)
  sn 1015

NJ Special Hazardous Substances
  (corrosive;   mutagen;   reactive - third degree)

Pennsylvania Right to Know List
  environmental hazard

PROPOSED REGULATIONS

CERCLA/SARA - Proposed Hazardous Substance Additions
  proposed RQ = 1 pound (.454 kg)

CERCLA/SARA - 1989 Proposed RQ Adjustments
  proposed RQ = 1000 pounds (454 kg)

HYDROXYACETIC ACID   79-14-1
SEE ALSO:
SELENIUM

HEALTH AND SAFETY LISTS

ACGIH 1995 - Time Weighted Averages
  as Se: 0.05 ppm TWA;   0.16 mg/m3 TWA

AIHA - Odor Threshold Values
  no geometric mean air odor threshold

U.S. DOT - Substances From 49 CFR 172.101
  regulated by DOT (UN2202)

U.S. DOT - Hazard Classes
  DOT hazard class = 2.3

U.S. DOT - Substances Which Are Poisonous by Inhalation
  gaseous hazardous material poisonous by inhalation (anhydrous form) (UN2202)

NIOSH - Selected LD50s and LC50s
  Inhalation, guinea pig: LC50 = 300 ppb (8 hr)

NIOSH 1990 - Pocket Guide - RELs
  as Se: 0.05 ppm TWA;   0.2 mg/m3 TWA

NIOSH 1990 - Pocket Guide - IDLHs
as Se: 2 ppm IDLH

NIOSH 1990 - Pocket Guide - Target organs
eyes, respiratory system

OSHA - Vacated PELs - Time Weighted Averages
as Se: 0.05 ppm TWA; 0.2 mg/m3 TWA

OSHA - Final PELs - Time Weighted Averages
as Se: 0.05 ppm TWA; 0.2 mg/m3 TWA

OSHA - List of Highly Hazardous Chemicals
threshhold quantity = 150 pounds

ENVIRONMENTAL LISTS

CERCLA/SARA - Section 302 Extremely Hazardous Substances and TPQs
TPQ = 10 pounds

CAA -Toxic Substances for Accidental Release Prevention
threshold quantity = 500 lbs

INTERNATIONAL LISTS

Australian Exposure Standards - Time Weighted Averages
as Se: 0.05 ppm TWA; 0.16 mg/m3 TWA

Canada - WHMIS: Ingredient Disclosure
0.1% item 850 (1480)

Canada - Alberta - 8 Hour Occupational Exposure Limit
0.05 ppm TWA; 0.16 mg/m3 TWA

Canada - Alberta - 15 Minute Occupational Exposure Limit
as Se: 0.15 ppm STEL; 0.48 mg/m3 STEL

Canada - British Columbia - 8 Hour Exposure Limits
0.05 ppm TWA; 0.2 mg/m3 TWA

Canada - Ontario - OHSA - TWAEVs
as Se: 0.05 ppm TWAEV; 0.16 mg/m3 TWAEV

Canada - Quebec - Time-Weighted Average Exposure Values
0.05 ppm TWAEV; 0.16 mg/m3 TWAEV

United Kingdom - Occupational Exposure Standards - TWAs
as Se: 0.05 ppm TWA; 0.2 mg/m3 TWA

German (DFG) - MAK Values
0.05 ppm MAK; 0.2 mg/m3 MAK

German (DFG) - Peak Limitations
2 x normal MAK (30 min. average value); don't exceed 4 times during shift

Israel - Time Weighted Averages
as Se: 0.05 ppm TWA; 0.16 mg/m3 TWA

Israel - Action Levels
as Se: 0.025 ppm AL; 0.08 mg/m3 AL

Mexico - Instruction No. 10 - TWAs
0.05 ppm TWA; 0.2 mg/m3 TWA

STATE LISTS

California - Exposure Limits - PELs
as Se: 0.05 ppm PEL; 0.2 mg/m3 PEL

Florida Hazardous Substance List
[present]

Massachusetts Right To Know List
extraordinarily hazardous

Minnesota Hazardous Substance List
[present] as Se

NJ Right to Know List (Total)
sn 1016

Pennsylvania Right to Know List
environmental hazard

PROPOSED REGULATIONS

CERCLA/SARA - Proposed Hazardous Substance Additions
proposed RQ = 1 pound (.454 kg)

CERCLA/SARA - 1989 Proposed RQ Adjustments
proposed RQ = 1 pounds (.454 kg)

P-HYDROXYACETOPHENONE   99-93-4

HEALTH AND SAFETY LISTS

ACGIH 1995 - Time Weighted Averages
10 ppm TWA; 14 mg/m3 TWA

ACGIH 1995 - Short Term Exposure Limits
15 ppm STEL; 21 mg/m3 STEL

AIHA - Odor Threshold Values
geometric mean air odor threshold = 0.0094 ppm (detectable); 0.0045 ppm (recognizable)

U.S. DOT - Substances From 49 CFR 172.101
regulated by DOT (UN1053)

U.S. DOT - Hazard Classes
DOT hazard class = 2.3

U.S. DOT - Substances Which Are Poisonous by Inhalation
gaseous hazardous material poisonous by inhalation (anhydrous form) (UN1053)

U.S. DOT - Appendix A Table 1 - Hazardous Substances
final RQ = 100 pounds (45.4 kg)

NFPA - Flash Points
gas (no flash point given)

NFPA - Hazard Identification Ratings
health-4; flammability-4; reactivity-0

NIOSH - Selected LD50s and LC50s
Inhalation, rat: LC50 = 444 ppm 8 hr

NIOSH 1990 - Pocket Guide - RELs
C 10 ppm (10 min); C 15 mg/m3 (10 min)

NIOSH 1990 - Pocket Guide - IDLHs
300 ppm IDLH

NIOSH 1990 - Pocket Guide - Target organs
eyes, respiratory system

NIOSH - Health Standards - Exposure Limits
C (10 min) 10 ppm; C (10 min) 15 mg/m3

NIOSH - Health Standards - Health Effects and Precautions
Irritation; severe acute effects involving nervous and respiratory systems (Continuous monitoring required; evacuation required if exposure equals or exceeds 70 mg/m3 (47 ppm))

OSHA - Vacated PELs - Time Weighted Averages
10 ppm TWA; 14 mg/m3 TWA

OSHA - Vacated PELs - Short Term Exposure Limits
15 ppm STEL; 21 mg/m3 STEL

OSHA - Final PELs - Ceiling Limits
C 20 ppm

OSHA - List of Highly Hazardous Chemicals
threshhold quantity = 1500 pounds

ENVIRONMENTAL LISTS

ATSDR Priority List
Rank (of 275): 183

CERCLA/SARA - Section 302 Extremely Hazardous Substances and TPQs
TPQ = 500 pounds

CERCLA/SARA - Section 313 - Emission Reporting
form R reporting required

CERCLA/SARA - Hazardous Substances and their Reportable Quantities
final RQ = 100 pounds (45.4 kg)

Clean Air Act (1990) - List of Hazardous Air Contaminants
[present]

CAA -Toxic Substances for Accidental Release Prevention
threshold quantity = 10,000 lbs

Clean Water Act - Hazardous Substances
[present]

RCRA - U Series Wastes
waste number U135

RCRA - Hazardous Constituents-Appendix VIII
waste number U135

RCRA - Substances Banned From Land Disposal
[present]

INTERNATIONAL LISTS

Australian Exposure Standards - Time Weighted Averages
10 ppm TWA;   14 mg/m3 TWA

Australian Exposure Standards - Short Term Exposure Limits
15 ppm STEL;   21 mg/m3 STEL

Canada - WHMIS: Ingredient Disclosure
1% item 851 (1550)

Canada - Alberta - 8 Hour Occupational Exposure Limit
10 ppm TWA;   14 mg/m3 TWA

Canada - Alberta - 15 Minute Occupational Exposure Limit
15 ppm STEL;   21 mg/m3 STEL

Canada - Alberta - Ceiling Occupational Exposure Limit
C 20 ppm;   C 28 mg/m3

Canada - Alberta - Designated Substances
designated substance - requires code of practice

Canada - British Columbia - 8 Hour Exposure Limits
10 ppm TWA;   15 mg/m3 TWA

Canada - British Columbia - 15 Minute Exposure Limits
15 ppm STEL;   27 mg/m3 STEL

Canada - Ontario - OHSA - TWAEVs
10 ppm TWAEV;   14 mg/m3 TWAEV

Canada - Ontario - OHSA - STEVs
15 ppm STEV;   21 mg/m3 STEV

Canada - Quebec - Time-Weighted Average Exposure Values
10 ppm TWAEV;   14 mg/m3 TWAEV

Canada - Quebec - Short-term Exposure Values
15 ppm STEV;   21 mg/m3 STEV

United Kingdom - Occupational Exposure Standards - TWAs
10 ppm TWA;   14 mg/m3 TWA

United Kingdom - Occupational Exposure Standards - STELs
15 ppm STEL;   21 mg/m3 STEL

German (DFG) - MAK Values
10 ppm MAK;   15 mg/m3 MAK

German (DFG) - Peak Limitations
2 x normal MAK (10 min momentary value);   don't exceed 4 times per shift

Israel - Time Weighted Averages
10 ppm TWA;   14 mg/m3 TWA

Israel - Short Term Exposure Limits
15 ppm STEL;   21 mg/m3 STEL

Israel - Action Levels
5 ppm AL;   7 mg/m3 AL

Mexico - Instruction No. 10 - TWAs
10 ppm TWA;   14 mg/m3 TWA

Mexico - Instruction No. 10 - STELs
15 ppm STEL;   21 mg/m3 STEL

STATE LISTS

California - Air Bill 2588 Appendix A-I
[present]

California - Exposure Limits - PELs
10 ppm PEL;   14 mg/m3 PEL

California - Exposure Limits - STELs
15 ppm STEL;   21 mg/m3 STEL

California - Exposure Limits - Ceilings
C 50 ppm

California - Directors List of Hazardous Substances (8 CCR 339)
[present]

Florida Hazardous Substance List
[present]

Massachusetts Right To Know List
extraordinarily hazardous

Minnesota Hazardous Substance List
[present]

NJ Right to Know List (Total)
sn 1017

NJ Special Hazardous Substances
(flammable - fourth degree)

Pennsylvania Right to Know List
environmental hazard

HYDROXYADIPALDEHYDE   141-31-

ENVIRONMENTAL LISTS

List of Pesticide Product Inert Ingredients
[present]

HYDROXYALKYL METHACRYLATE, ALKYL ESTER   RR-00223-

ENVIRONMENTAL LISTS

List of Pesticide Product Inert Ingredients
[present]

4-HYDROXYAZOBENZENE   1689-82-3

HEALTH AND SAFETY LISTS

ACGIH 1995 - Time Weighted Averages
2 mg/m3 TWA

U.S. DOT - Substances From 49 CFR 172.101
regulated by DOT (UN2662)

U.S. DOT - Hazard Classes
DOT hazard class = 6.1

IARC - Group 3 (not classifiable)
[present]

NFPA - Flash Points
flash point = 329 degrees F (165 degrees C)

NFPA - Hazard Identification Ratings
*flammability-1;   reactivity-0;   *See list description*

NIOSH - Selected LD50s and LC50s
Oral, rat: LD50 = 320 mg/kg Skin, mammal: LD50 = 5970 mg/kg

NIOSH 1990 - Pocket Guide - RELs
C 2 mg/m3 (15 min)

NIOSH 1990 - Pocket Guide - Target organs
eyes, skin, CNS, respiratory system

NIOSH - Health Standards - Exposure Limits
C (15 min) 0.44 ppm;   C (15 min) 2 mg/m3

NIOSH - Health Standards - Health Effects and Precautions
Eye and skin effects .

NTP Chemical Status Reports - Testing Status and NTIS Number
Technical reports printed (PB90240839)

NTP Chemical Status Reports - Evidence of Carcinogenicity
male rat-some evidence;   female rat-some evidence;   male mice-no evidence;   female mice-some evidence

OSHA - Vacated PELs - Time Weighted Averages
2 mg/m3 TWA

OSHA - Final PELs - Time Weighted Averages
2 mg/m3 TWA

ENVIRONMENTAL LISTS

CERCLA/SARA - Section 302 Extremely Hazardous Substances and TPQs
TPQ = 500/10,000 pounds

CERCLA/SARA - Section 313 - Emission Reporting
form R reporting required for 1.0% de minimus eoncentration

CERCLA/SARA - Hazardous Substances and their Reportable Quantities
final RQ = 1 pound (.454 kg)